# Lecture Notes in Computer Science 9985

Commenced Publication in 1973
Founding and Former Series Editors:
Gerhard Goos, Juris Hartmanis, and Jan van Leeuwen

## Editorial Board

David Hutchison
*Lancaster University, Lancaster, UK*
Takeo Kanade
*Carnegie Mellon University, Pittsburgh, PA, USA*
Josef Kittler
*University of Surrey, Guildford, UK*
Jon M. Kleinberg
*Cornell University, Ithaca, NY, USA*
Friedemann Mattern
*ETH Zurich, Zurich, Switzerland*
John C. Mitchell
*Stanford University, Stanford, CA, USA*
Moni Naor
*Weizmann Institute of Science, Rehovot, Israel*
C. Pandu Rangan
*Indian Institute of Technology, Madras, India*
Bernhard Steffen
*TU Dortmund University, Dortmund, Germany*
Demetri Terzopoulos
*University of California, Los Angeles, CA, USA*
Doug Tygar
*University of California, Berkeley, CA, USA*
Gerhard Weikum
*Max Planck Institute for Informatics, Saarbrücken, Germany*

Martin Hirt · Adam Smith (Eds.)

# Theory of Cryptography

14th International Conference, TCC 2016-B
Beijing, China, October 31 – November 3, 2016
Proceedings, Part I

 Springer

*Editors*
Martin Hirt
Department of Computer Science
ETH Zurich
Zurich
Switzerland

Adam Smith
Pennsylvania State University
University Park, PA
USA

ISSN 0302-9743        ISSN 1611-3349   (electronic)
Lecture Notes in Computer Science
ISBN 978-3-662-53640-7        ISBN 978-3-662-53641-4   (eBook)
DOI 10.1007/978-3-662-53641-4

Library of Congress Control Number: 2016954934

LNCS Sublibrary: SL4 – Security and Cryptology

Printed on acid-free paper

This Springer imprint is published by Springer Nature
The registered company is Springer-Verlag GmbH Germany
The registered company address is: Heidelberger Platz 3, 14197 Berlin, Germany

# Preface

The 14th Theory of Cryptography Conference (TCC 2016-B) was held October 31 to November 3, 2016, at the Beijing Friendship Hotel in Beijing, China. It was sponsored by the International Association for Cryptographic Research (IACR) and organized in cooperation with State Key Laboratory of Information Security at the Institute of Information Engineering of the Chinese Academy of Sciences. The general chair was Dongdai Lin, and the honorary chair was Andrew Chi-Chih Yao.

The conference received 113 submissions, of which the Program Committee (PC) selected 45 for presentation (with three pairs of papers sharing a single presentation slot per pair). Of these, there were four whose authors were all students at the time of submission. The committee selected "Simulating Auxiliary Inputs, Revisited" by Maciej Skórski for the Best Student Paper award. Each submission was reviewed by at least three PC members, often more. The 25 PC members, all top researchers in our field, were helped by 154 external reviewers, who were consulted when appropriate. These proceedings consist of the revised version of the 45 accepted papers. The revisions were not reviewed, and the authors bear full responsibility for the content of their papers.

As in previous years, we used Shai Halevi's excellent Web review software, and are extremely grateful to him for writing it and for providing fast and reliable technical support whenever we had any questions. Based on the experience from the last two years, we used the interaction feature supported by the review software, where PC members may directly and anonymously interact with authors. The feature allowed the PC to ask specific technical questions that arose during the review process, for example, about suspected bugs. Authors were prompt and extremely helpful in their replies. We hope that it will continue to be used in the future.

This was the third year where TCC presented the Test of Time Award to an outstanding paper that was published at TCC at least eight years ago, making a significant contribution to the theory of cryptography, preferably with influence also in other areas of cryptography, theory, and beyond. The Test of Time Award Committee consisted of Tal Rabin (chair), Yuval Ishai, Daniele Micciancio, and Jesper Nielsen. They selected "Indifferentiability, Impossibility Results on Reductions, and Applications to the Random Oracle Methodology" by Ueli Maurer, Renato Renner, and Clemens Holenstein—which appeared in TCC 2004, the first edition of the conference—for introducing indifferentiability, a security notion that had "significant impact on both the theory of cryptography and the design of practical cryptosystems." Sadly, Clemens Holenstein passed away in 2012. He is survived by his wife and two sons. Maurer and Renner accepted the award on his behalf. The authors delivered a talk in a special session at TCC 2016-B. An invited paper by them, which was not reviewed, is included in these proceedings.

The conference featured two other invited talks, by Allison Bishop and Srini Devadas. In addition to regular papers and invited events, there was a rump session featuring short talks by attendees.

We are greatly indebted to many people who were involved in making TCC 2016-B a success. First of all, our sincere thanks to the most important contributors: all the authors who submitted papers to the conference. There were many more good submissions than we had space to accept. We would like to thank the PC members for their hard work, dedication, and diligence in reviewing the papers, verifying their correctness, and discussing their merits in depth. We are also thankful to the external reviewers for their volunteered hard work in reviewing papers and providing valuable expert feedback in response to specific queries. For running the conference itself, we are very grateful to Dongdai and the rest of the local Organizing Committee. Finally, we are grateful to the TCC Steering Committee, and especially Shai Halevi, for guidance and advice, as well as to the entire thriving and vibrant theoretical cryptography community. TCC exists for and because of that community, and we are proud to be a part of it.

November 2016

Martin Hirt
Adam Smith

# TCC 2016-B

## Theory of Cryptography Conference

Beijing, China
October 31 – November 3, 2016

Sponsored by the International Association for Cryptologic Research and organized in cooperation with the State Key Laboratory of Information Security, Institute of Information Engineering, Chinese Academy of Sciences.

## General Chair

Dongdai Lin            Chinese Academy of Sciences, China

## Honorary Chair

Andrew Chi-Chih Yao     Tsinghua University, China

## Program Committee

| | |
|---|---|
| Masayuki Abe | NTT, Japan |
| Divesh Aggarwal | NUS, Singapore |
| Andrej Bogdanov | Chinese University of Hong Kong, Hong Kong |
| Elette Boyle | IDC Herzliya, Israel |
| Anne Broadbent | University of Ottawa, Canada |
| Chris Brzuska | TU Hamburg, Germany |
| David Cash | Rutgers University, USA |
| Alessandro Chiesa | University of California, Berkeley, USA |
| Kai-Min Chung | Academia Sinica, Taiwan |
| Nico Döttling | University of California, Berkeley, USA |
| Sergey Gorbunov | University of Waterloo, Canada |
| Martin Hirt (Co-chair) | ETH Zurich, Switzerland |
| Abhishek Jain | Johns Hopkins University, USA |
| Huijia Lin | University of California, Santa Barbara, USA |
| Hemanta K. Maji | Purdue University, USA |
| Adam O'Neill | Georgetown University, USA |
| Rafael Pass | Cornell University, USA |
| Krzysztof Pietrzak | IST Austria, Austria |
| Manoj Prabhakaran | IIT Bombay, India |
| Renato Renner | ETH Zurich, Switzerland |
| Alon Rosen | IDC Herzliya, Israel |
| abhi shelat | Northeastern University, USA |
| Adam Smith (Co-chair) | Pennsylvania State University, USA |

Wei-Kai Lin
Helger Lipmaa
Feng-Hao Liu
Vadim Lyubashevsky
Mohammad Mahmoody
Giulio Malavolta
Alex J. Malozemoff
Daniel Masny
Takahiro Matsuda
Christian Matt
Patrick McCorry
Or Meir
Peihan Miao
Eric Miles
Pratyush Mishra
Ameer Mohammed
Payman Mohassel
Tal Moran
Kirill Morozov
Pratyay Mukherjee
Hai H. Nguyen
Ryo Nishimaki
Maciej Obremski
Miyako Ohkubo
Jiaxin Pan
Omkant Pandey
Omer Paneth
Valerio Pastro

Christopher Peikert
Oxana Poburinnaya
Bertram Poettering
Antigoni Polychroniadou
Christopher Portmann
Srini Raghuraman
Samuel Ranellucci
Vanishree Rao
Mariana Raykova
Joseph Renes
Leonid Reyzin
Silas Richelson
Mike Rosulek
Guy Rothblum
Ron Rothblum
Sajin Sasy
Alessandra Scafuro
Dominique Schröder
Karn Seth
Vladimir Shpilrain
Mark Simkin
Nigel Smart
Pratik Soni
Bing Sun
David Sutter
Björn Tackmann
Stefano Tessaro
Justin Thaler

Aishwarya
  Thiruvengadam
Junnichi Tomida
Rotem Tsabary
Margarita Vald
Prashant Vasudevan
Daniele Venturi
Damien Vergnaud
Jorge L. Villar
Dhinakaran
  Vinayagamurthy
Madars Virza
Ivan Visconti
Hoeteck Wee
Eyal Widder
David Wu
Keita Xagawa
Sophia Yakoubov
Takashi Yamakawa
Avishay Yanay
Arkady Yerukhimovich
Eylon Yogev
Mohammad Zaheri
Mark Zhandry
Hong-Sheng Zhou
Juba Ziani

# Contents – Part I

**TCC Test-of-Time Award**

From Indifferentiability to Constructive Cryptography (and Back) . . . . . . . . . 3
*Ueli Maurer and Renato Renner*

**Foundations**

Fast Pseudorandom Functions Based on Expander Graphs . . . . . . . . . . . . . . 27
*Benny Applebaum and Pavel Raykov*

3-Message Zero Knowledge Against Human Ignorance . . . . . . . . . . . . . . . . 57
*Nir Bitansky, Zvika Brakerski, Yael Kalai, Omer Paneth,*
*and Vinod Vaikuntanathan*

The GGM Function Family Is a Weakly One-Way Family of Functions . . . . 84
*Aloni Cohen and Saleet Klein*

On the (In)Security of SNARKs in the Presence of Oracles . . . . . . . . . . . . 108
*Dario Fiore and Anca Nitulescu*

Leakage Resilient One-Way Functions: The Auxiliary-Input Setting . . . . . . . 139
*Ilan Komargodski*

Simulating Auxiliary Inputs, Revisited . . . . . . . . . . . . . . . . . . . . . . . . . . 159
*Maciej Skórski*

**Unconditional Security**

Pseudoentropy: Lower-Bounds for Chain Rules and Transformations . . . . . . . 183
*Krzysztof Pietrzak and Maciej Skórski*

Oblivious Transfer from Any Non-trivial Elastic Noisy Channel via Secret
Key Agreement . . . . . . . . . . . . . . . . . . . . . . . . . . . . . . . . . . . . . . . . . . 204
*Ignacio Cascudo, Ivan Damgård, Felipe Lacerda,*
*and Samuel Ranellucci*

Simultaneous Secrecy and Reliability Amplification for a General Channel
Model . . . . . . . . . . . . . . . . . . . . . . . . . . . . . . . . . . . . . . . . . . . . . . . . . 235
*Russell Impagliazzo, Ragesh Jaiswal, Valentine Kabanets,*
*Bruce M. Kapron, Valerie King, and Stefano Tessaro*

Proof of Space from Stacked Expanders. . . . . . . . . . . . . . . . . . . . . . . . .     262
    Ling Ren and Srinivas Devadas

Perfectly Secure Message Transmission in Two Rounds. . . . . . . . . . . . . .     286
    Gabriele Spini and Gilles Zémor

## Foundations of Multi-Party Protocols

Almost-Optimally Fair Multiparty Coin-Tossing with Nearly
Three-Quarters Malicious. . . . . . . . . . . . . . . . . . . . . . . . . . . . . . . . . . .     307
    Bar Alon and Eran Omri

Binary AMD Circuits from Secure Multiparty Computation . . . . . . . . . . . .     336
    Daniel Genkin, Yuval Ishai, and Mor Weiss

Composable Security in the Tamper-Proof Hardware Model Under Minimal
Complexity . . . . . . . . . . . . . . . . . . . . . . . . . . . . . . . . . . . . . . . . . . . . .     367
    Carmit Hazay, Antigoni Polychroniadou,
    and Muthuramakrishnan Venkitasubramaniam

Composable Adaptive Secure Protocols Without Setup Under Polytime
Assumptions. . . . . . . . . . . . . . . . . . . . . . . . . . . . . . . . . . . . . . . . . . . .     400
    Carmit Hazay and Muthuramakrishnan Venkitasubramaniam

Adaptive Security of Yao's Garbled Circuits . . . . . . . . . . . . . . . . . . . . . .     433
    Zahra Jafargholi and Daniel Wichs

## Round Complexity and Efficiency of Multi-party Computation

Efficient Secure Multiparty Computation with Identifiable Abort. . . . . . . . .     461
    Carsten Baum, Emmanuela Orsini, and Peter Scholl

Secure Multiparty RAM Computation in Constant Rounds. . . . . . . . . . . . .     491
    Sanjam Garg, Divya Gupta, Peihan Miao, and Omkant Pandey

Constant-Round Maliciously Secure Two-Party Computation in the RAM
Model . . . . . . . . . . . . . . . . . . . . . . . . . . . . . . . . . . . . . . . . . . . . . . . . .     521
    Carmit Hazay and Avishay Yanai

More Efficient Constant-Round Multi-party Computation from BMR
and SHE . . . . . . . . . . . . . . . . . . . . . . . . . . . . . . . . . . . . . . . . . . . . . . .     554
    Yehuda Lindell, Nigel P. Smart, and Eduardo Soria-Vazquez

Cross and Clean: Amortized Garbled Circuits with Constant Overhead . . . . .     582
    Jesper Buus Nielsen and Claudio Orlandi

## Differential Privacy

Separating Computational and Statistical Differential Privacy
in the Client-Server Model . . . . . . . . . . . . . . . . . . . . . . . . . . . . . . . . . . . .    607
  Mark Bun, Yi-Hsiu Chen, and Salil Vadhan

Concentrated Differential Privacy: Simplifications, Extensions,
and Lower Bounds . . . . . . . . . . . . . . . . . . . . . . . . . . . . . . . . . . . . . . . . .    635
  Mark Bun and Thomas Steinke

Strong Hardness of Privacy from Weak Traitor Tracing . . . . . . . . . . . . . . .    659
  Lucas Kowalczyk, Tal Malkin, Jonathan Ullman, and Mark Zhandry

**Author Index** . . . . . . . . . . . . . . . . . . . . . . . . . . . . . . . . . . . . . . . . . . . .    691

# Contents – Part II

**Delegation and IP**

Delegating RAM Computations with Adaptive Soundness and Privacy . . . . . 3
*Prabhanjan Ananth, Yu-Chi Chen, Kai-Min Chung, Huijia Lin,*
*and Wei-Kai Lin*

Interactive Oracle Proofs . . . . . . . . . . . . . . . . . . . . . . . . . . . . 31
*Eli Ben-Sasson, Alessandro Chiesa, and Nicholas Spooner*

Adaptive Succinct Garbled RAM or: How to Delegate Your Database . . . . . . 61
*Ran Canetti, Yilei Chen, Justin Holmgren, and Mariana Raykova*

Delegating RAM Computations . . . . . . . . . . . . . . . . . . . . . . . 91
*Yael Kalai and Omer Paneth*

**Public-Key Encryption**

Standard Security Does Not Imply Indistinguishability Under Selective
Opening. . . . . . . . . . . . . . . . . . . . . . . . . . . . . . . . . . . . . 121
*Dennis Hofheinz, Vanishree Rao, and Daniel Wichs*

Public-Key Encryption with Simulation-Based Selective-Opening Security
and Compact Ciphertexts . . . . . . . . . . . . . . . . . . . . . . . . . . . 146
*Dennis Hofheinz, Tibor Jager, and Andy Rupp*

Towards Non-Black-Box Separations of Public Key Encryption and One
Way Function. . . . . . . . . . . . . . . . . . . . . . . . . . . . . . . . . 169
*Dana Dachman-Soled*

Post-Quantum Security of the Fujisaki-Okamoto and OAEP Transforms . . . . 192
*Ehsan Ebrahimi Targhi and Dominique Unruh*

Multi-key FHE from LWE, Revisited . . . . . . . . . . . . . . . . . . . . . 217
*Chris Peikert and Sina Shiehian*

**Obfuscation and Multilinear Maps**

Secure Obfuscation in a Weak Multilinear Map Model . . . . . . . . . . . . . 241
*Sanjam Garg, Eric Miles, Pratyay Mukherjee, Amit Sahai,*
*Akshayaram Srinivasan, and Mark Zhandry*

Virtual Grey-Boxes Beyond Obfuscation: A Statistical Security Notion
for Cryptographic Agents. . . . . . . . . . . . . . . . . . . . . . . . . . . . . . .    269
    *Shashank Agrawal, Manoj Prabhakaran, and Ching-Hua Yu*

## Attribute-Based Encryption

Deniable Attribute Based Encryption for Branching Programs from LWE . . .    299
    *Daniel Apon, Xiong Fan, and Feng-Hao Liu*

Targeted Homomorphic Attribute-Based Encryption . . . . . . . . . . . . . . . .    330
    *Zvika Brakerski, David Cash, Rotem Tsabary, and Hoeteck Wee*

Semi-adaptive Security and Bundling Functionalities Made Generic
and Easy . . . . . . . . . . . . . . . . . . . . . . . . . . . . . . . . . . . . . . . . . .    361
    *Rishab Goyal, Venkata Koppula, and Brent Waters*

## Functional Encryption

From Cryptomania to Obfustopia Through Secret-Key Functional
Encryption . . . . . . . . . . . . . . . . . . . . . . . . . . . . . . . . . . . . . . . . . .    391
    *Nir Bitansky, Ryo Nishimaki, Alain Passelègue, and Daniel Wichs*

Single-Key to Multi-Key Functional Encryption with Polynomial Loss . . . . .    419
    *Sanjam Garg and Akshayaram Srinivasan*

Compactness vs Collusion Resistance in Functional Encryption . . . . . . . . . .    443
    *Baiyu Li and Daniele Micciancio*

## Secret Sharing

Threshold Secret Sharing Requires a Linear Size Alphabet. . . . . . . . . . . . .    471
    *Andrej Bogdanov, Siyao Guo, and Ilan Komargodski*

How to Share a Secret, Infinitely . . . . . . . . . . . . . . . . . . . . . . . . . . . .    485
    *Ilan Komargodski, Moni Naor, and Eylon Yogev*

## New Models

Designing Proof of Human-Work Puzzles for Cryptocurrency and Beyond. . .    517
    *Jeremiah Blocki and Hong-Sheng Zhou*

Access Control Encryption: Enforcing Information Flow
with Cryptography . . . . . . . . . . . . . . . . . . . . . . . . . . . . . . . . . . . . .    547
    *Ivan Damgård, Helene Haagh, and Claudio Orlandi*

**Author Index** . . . . . . . . . . . . . . . . . . . . . . . . . . . . . . . . . . . . . . . .    577

# TCC Test-of-Time Award

# From Indifferentiability to Constructive Cryptography (and Back)

Ueli Maurer[1]([⊠]) and Renato Renner[2]

[1] Department of Computer Science, ETH Zurich, Zurich, Switzerland
maurer@inf.ethz.ch
[2] Department of Physics, ETH Zurich, Zurich, Switzerland
renner@phys.ethz.ch

**Abstract.** The concept of indifferentiability of systems, a generalized form of indistinguishability, was proposed in 2004 to provide a simplified and generalized explanation of impossibility results like the non-instantiability of random oracles by hash functions due to Canetti, Goldreich, and Halevi (STOC 1998). But indifferentiability is actually a constructive notion, leading to possibility results. For example, Coron et al. (Crypto 2005) argued that the soundness of the construction $C(f)$ of a hash function from a compression function $f$ can be demonstrated by proving that $C(R)$ is indifferentiable from a random oracle if $R$ is an ideal random compression function.

The purpose of this short paper is to describe how the indifferentiability notion was a precursor to the theory of constructive cryptography and thereby to provide a simplified and generalized treatment of indifferentiability as a special type of constructive statement.

## 1  Introduction

An important abstraction in cryptography, introduced by Bellare *et al.* [4], is the so-called random oracle model (ROM). A random oracle is an idealized resource or system available to all involved parties, with parameters $m$ and $n$, which behaves as if it contained a uniformly chosen function table $F : \{0,1\}^m \to \{0,1\}^n$ and, for every query $x \in \{0,1\}^m$ from any party, provides the function value $F(x)$ to that party. Other parties do not see the query $x$ nor the reply $F(x)$. A random oracle can also be defined for the countably infinite domain $\{0,1\}^*$ of all finite-length input strings, the resource usually meant in cryptography by the term "random oracle".

The idea behind the ROM is a natural decomposition idea often arising in cryptographic reasoning. On one hand one tries to construct, at least approximately, a random oracle from weaker resources (e.g. a shared random string), and on the other hand one uses the idealized resource of a random oracle to design secure protocols. The rationale is that if a well-designed hash function can be assumed to behave like a random oracle, then a cryptographic protocol proved secure in the ROM remains secure when the random oracle is replaced

© International Association for Cryptologic Research 2016
M. Hirt and A. Smith (Eds.): TCC 2016-B, Part I, LNCS 9985, pp. 3–24, 2016.
DOI: 10.1007/978-3-662-53641-4_1

by a hash function, thus composing two steps of reasoning. Analogous reasoning is, for example, applied if one proves a scheme secure assuming it has access to a uniformly random value (e.g., a shared secret key), and then argues that the random value can be replaced by a pseudo-random value without compromising security.

Two questions arise.

1. What exactly do we mean by composition of steps in the above reasoning and how can we make it mathematically sound? It turns out, as discussed in this paper, that the random oracle example requires a different and more sophisticated reasoning compared to the pseudo-randomness example.
2. Can a random oracle be constructed from a weaker resource, especially one that can realistically be assumed to be available in a given application context?

An important paper by Canetti *et al.* [6] showed that the random oracle model is not instantiable by any hash function. The approach taken in that paper was to devise a provably secure signature scheme $S$, which internally makes use of a secure signature scheme $S'$ and has access to a random oracle, such that $S$ is insecure if the random oracle is replaced by any hash function, even one devised in the future and in full knowledge of the random oracle. Intuitively, the reason for this impossibility is that the program code $p$ for a hash function can not contain more entropy than the length of $p$ and that therefore, if one accesses the random oracle for a number of arguments yielding more entropy than the length of $p$, then one can distinguish a black-box containing the random oracle from one containing the hash function.

This result raises some natural questions which were the starting point for the research leading to the paper [18] on indifferentiability.

1. How can this simple entropy argument be made precise, in view of the quite involved original proof of [6], and how can it be generalized?
2. What is a meaningful definition of the possibility (rather than impossibility) of such a construction, and which concrete constructions are indeed possible?
3. How can the construction notion be generalized to capture other cryptographic settings like encryption or message authentication?
4. How can one design complex cryptographic protocols such that their security proof follows simply from composition and the (generally simple) security proofs of the individual construction steps?

The answer to the second question turned out to be useful for the design of hash functions from a compression function (e.g. see [1,2,7,11,12]).

The third question asks for an understanding of the application of a cryptographic scheme like a symmetric or public-key encryption scheme, a message authentication scheme, or a digital signature scheme, as a construction of a resource from other resources. The question then is which resources one should consider and how cryptographic schemes can be understood as such constructions. Cryptographic resources provide a guarantee to honest parties in view of potentially dishonest parties behaving arbitrarily. Such arbitrary or unspecified behavior is often called "malicious". For example, a secure communication channel guarantees to the honest parties (the sender and the receiver) that an adversary can learn at most the length of the message. Note that, in the sense of a specification discussed later, it is not guaranteed that the adversary learns the message length, only that she does not learn more. For example, symmetric encryption can be understood as constructing a secure channel from an authenticated channel and a shared secret key, and message authentication can be understood as constructing an authenticated channel from an insecure channel and a shared secret key [16,17,19,20]. Similarly, public-key encryption can be understood as constructing a confidential channel from an insecure channel and an authenticated channel in the other direction [8].

The above approach to cryptography was proposed in [17], motivated by earlier approaches to achieving composition in cryptography, most notably Canetti's UC framework [5] and the reactive simulatability framework of Backes, Pfitzmann, and Waidner [3].

The outline of the paper is as follows. In Sect. 2, the general construction paradigm and composability is discussed. In Sect. 3, we introduce the type of resources relevant in cryptography. In Sect. 4, the cryptographic construction notion is introduced and a few simple construction statements are proved. In Sect. 5, a few impossibility results are proved which imply considerably strengthened versions of the impossibility of constructing a random oracle. In Sect. 6, the positive construction result of Coron et al. [9] is discussed in view of the new treatment appearing in this paper. In Sect. 7, it is mentioned that the construction notion of this paper directly leads to construction statements involving several parties, some of which are honest and some of which are dishonest. In Sect. 8, the relation of this paper to the original indifferentiability paper [18] is explained.

**A Word About Terminology.** The title of the original paper [17] proposing constructive cryptography was "Abstract cryptography". Two main aspects of that paper were (1) the proposal to use top-down abstraction in the spirit of algebra in cryptography (and more generally in computer science), and (2) to use the construction paradigm (see Sect. 2) in cryptography. Therefore, depending on which aspect is stressed, both "abstract cryptography" and "constructive cryptography" have been used in the literature to refer to this theory. The term constructive cryptography, which was first used in [16], seems more natural and captures the goal of the theory better, and we propose to use it from now on to avoid confusion.

## 2    The Construction Paradigm

### 2.1    Specifications and Constructions

In almost every engineering discipline one considers, explicitly or implicitly, the concept of a *specification* of an object or *resource*. Examples include the specification of a mechanical part (e.g. by lower and upper bounds on its dimensions, its weight, and material parameters) and the specification of a software module $M$ (e.g. by defining the functions that $M$ computes and possibly some accuracy guarantees and/or some timing guarantees).

A key task in such a discipline is to *construct*, from an object or resource satisfying a certain specification $\mathcal{R}$, an object or resource satisfying another (better or more valuable) specification $\mathcal{S}$. Such a construction is achieved by means of a *constructor* or recipe, say $\gamma$. One can then write

$$\mathcal{R} \xrightarrow{\gamma} \mathcal{S}.$$

For example, the designer of a software module $N$ making use of the module $M$ will provide a specification $\mathcal{S}$ which is guaranteed (and proved) to be satisfied by $N$, provided the underlying module $M$ satisfies specification $\mathcal{R}$.

As another example, in communication theory and information theory, a binary symmetric channel (BSC) is a well-known resource specification characterized by a maximal probability $p$ of flipping the transmitted bits (where the errors for all bits are independent). A good error-correcting code with $2^k$ codewords of length $n$ can be understood as constructing, from an $n$-bit BSC with parameter $p$, an error-free $k$-bit communication channel. More precisely, one only achieves a specification of a channel which is $\epsilon$-close to an error-free $k$-bit channel, for a small $\epsilon$ and a certain measure of closeness, i.e., for a metric on the set of channels, namely the worst-case (over messages) decoding error probability.

Typically one considers a certain set $\Gamma$ of constructors, possibly restricted in terms of efficiency or implementation cost. One is then interested in constructibility and also in non-constructibility statements, where $\mathcal{S}$ is not constructible from $\mathcal{R}$, denoted $\mathcal{R} \not\to \mathcal{S}$, if there exists no constructor $\gamma$ for which $\mathcal{R} \xrightarrow{\gamma} \mathcal{S}$:

$$\mathcal{R} \not\to \mathcal{S} :\Longleftrightarrow \neg \exists\, \gamma \in \Gamma : \mathcal{R} \xrightarrow{\gamma} \mathcal{S}.$$

One often wants to use several resources in a construction, i.e., one wants to consider a tuple of resources, for example a tuple of three resources satisfying specifications $\mathcal{R}_1$, $\mathcal{R}_2$, and $\mathcal{R}_3$, as a single resource. We denote such a combined resource specification as $[\mathcal{R}_1, \mathcal{R}_2, \mathcal{R}_3]$.

## 2.2   Composition

If we assume that constructors can be composed, where the constructor resulting from applying $\gamma$ and then $\gamma'$ is denoted as $\gamma' \circ \gamma$, then a very desirable and natural property is that the corresponding construction statements can be composed. Formally, this means that

$$\mathcal{R} \xrightarrow{\gamma} \mathcal{S} \wedge \mathcal{S} \xrightarrow{\gamma'} \mathcal{T} \Longrightarrow \mathcal{R} \xrightarrow{\gamma' \circ \gamma} \mathcal{T}.$$

For example, any construction requiring an error-free channel and resulting in a yet more useful resource should also be (approximately) correct if, instead of the error-free channel, the channel constructed by an error-correcting code from an error-prone channel is used. Whether or not this is indeed the case requires a formalization and a proof.

Another useful property of the construction notion is *context-insensitivity*: For any $\mathcal{U}$ and $\mathcal{V}$,

$$\mathcal{R} \xrightarrow{\gamma} \mathcal{S} \implies [\mathcal{U}_1, \ldots, \mathcal{U}_k, \mathcal{R}, \mathcal{V}_1, \ldots, \mathcal{V}_\ell] \xrightarrow{\gamma} [\mathcal{U}_1, \ldots, \mathcal{U}_k, \mathcal{S}, \mathcal{V}_1, \ldots, \mathcal{V}_\ell]$$

for any $\mathcal{R}$, $\mathcal{S}$, and $\mathcal{U}_1, \ldots, \mathcal{U}_k, \mathcal{V}_1, \ldots, \mathcal{V}_\ell$. The understanding here is that $\gamma$ "knows" which resource it needs to access.[1]

We point out that these properties may or may not be satisfied by a construction notion under consideration, and when investigating a concrete such notion one needs to prove that they are satisfied.

## 2.3 Sets as Specifications

The notion of a specification is abstract, but often a specification is understood as the subset of a universe $\Phi$ of objects, namely those that satisfy the specification. For example the specification of a BSC corresponds to the set of all channels where the bit-flipping probability of each bit is upper bounded by $p$ but otherwise arbitrary (and the flipping events are independent). As another example, a software specification may require only an approximative computation of certain results, and a concrete element of the specification is given by a fixed function that is within the accuracy bounds.

If a pseudo-metric $d$ on $\Phi$ is defined, a particular type of specification by sets are $\epsilon$-balls around a given object $R$, denoted

$$R^\epsilon = \{R' \mid R' \approx_\epsilon R\},$$

where we write $R' \approx_\epsilon R$ for $d(R, R') \leq \epsilon$. More generally,

$$\mathcal{R}^\epsilon = \{R' \mid \exists R \in \mathcal{R} : R' \approx_\epsilon R\} = \bigcup_{R \in \mathcal{R}} R^\epsilon,$$

A construction statement $\mathcal{R} \xrightarrow{\gamma} \mathcal{S}$ becomes stronger the larger the specification $\mathcal{R}$ (i.e., the less needs to be assumed about the given resource), and, analogously, the statement becomes stronger the smaller the specification $\mathcal{S}$, i.e., the more specific the guarantee about the constructed resource is. In other words, we have

$$\mathcal{R} \xrightarrow{\gamma} \mathcal{S} \implies \mathcal{R}' \xrightarrow{\gamma} \mathcal{S}'$$

if $\mathcal{R}' \subseteq \mathcal{R}$ and $\mathcal{S} \subseteq \mathcal{S}'$.

---

[1] Formally, the constructor $\gamma$ on the right side might involve some scheme for addressing the resource specified by $\mathcal{R}$ among all resources, and in this case it would have to be an adequately modified version of $\gamma$ on the left side (i.e., in $\mathcal{R} \xrightarrow{\gamma} \mathcal{S}$).

The situation is dual for impossibility results, which are a focus of [6, 18] and of this paper. Namely,

$$\mathcal{R} \not\longrightarrow \mathcal{S} \Longrightarrow \mathcal{R}' \not\longrightarrow \mathcal{S}'$$

if $\mathcal{R} \subseteq \mathcal{R}'$ and $\mathcal{S}' \subseteq \mathcal{S}$. In other words, the smaller $\mathcal{R}$ or the larger $\mathcal{S}$, the stronger is the impossibility statement. We will pay attention to trying to obtain strong possibility and impossibility results.

## 3   Cryptographic Resource Systems and Their Use

In this section we discuss the specific type of resource appearing in cryptographic statements.

### 3.1   Systems, Interfaces, Parties

Cryptographic resources can be modelled as systems with several interfaces. One can think of each interface as allowing one party to connect to the system and access the functionality provided by it, but this view is not strict. It is also possible that interfaces capture a more fine-grained capability and that several interfaces are assigned to the same party. Conversely, one could also consider several parties as accessing (sub-interfaces of) the same interface.

In a cryptographic context, one considers so-called "honest" and "dishonest" parties, where often all the dishonest parties are modeled as a single party, called "the adversary" or Eve.

For the purpose of this paper, it suffices to consider resources with two interfaces, where all honest parties (sometimes summarized as Alice) access the resource through the left interface and Eve accesses it from the right side.

More technically, in this paper we consider a specific type of system, namely discrete resource systems that can (possibly) take an input at any interface and provide an output at the same interface. Then a system can take another input at some interface and produce an output at that interface, etc. For this paper, we will not need a formalization of such discrete systems, but we refer to [15, 22]. The metric on the set of discrete systems is naturally defined via the optimal distinguishing advantage of a certain class of distinguishers.

### 3.2   Example Resource Systems

An example of such a resource is a uniform random function (URF) $\{0,1\}^m \to \{0,1\}^n$, accessible to all involved parties, which can be specified by considering a uniformly chosen function table $F : \{0,1\}^m \to \{0,1\}^n$ that can be accessed by giving as input a value $x$ and receiving as output the value $F(x)$.

When considering the above URF resource in a cryptographic context, even when restricted to a single honest party and a single adversary, the above specification is not adequate as it is on one hand too specific (it guarantees that the adversary can access the resource, while one does not want to give such a guarantee), and it is on the other hand not sufficiently specific in that one would want to additionally specify lower and upper bounds on the number of allowed queries (see later), as well as what is guaranteed to be hidden from the adversary. There are a number of such specifications which are natural, and we list a few of them below.

1. Alice can access the URF and Eve has no access to it.
2. Alice can access the URF and Eve has no access to it, but she potentially sees whenever Alice makes a query.
3. As before, but Eve can potentially also learn the values queried and obtained by Alice.
4. Alice and Eve can both access the URF and Eve obtains no other information (e.g. about Alice's access).
5. As before, but Eve can potentially also learn the values queried and obtained by Alice.

The fourth example is what is often called a (fixed input-length) random oracle which is accessible to all parties, whether honest or not, here restricted to a single honest party. One can also consider such a random oracle resource with arbitrary input-length, i.e., which for each input in $\{0,1\}^*$ returns a random value in $\{0,1\}^n$. An important question is from which resources a random oracle can or cannot be constructed. The impossibility result of [6] can be interpreted as the statement that a random oracle cannot be constructed from a fixed bit-string (the hash program) which can be probabilistically chosen.

## 3.3   Converters

A party can use a resource $R \in \Phi$ by applying to it a so-called *converter*[2] $\alpha$ which is, for example, a (state-full) protocol engine. A converter can be thought of as a system, with an inside and an outside interface, which is attached to the resource system. Application of a converter at interface $i$ transforms a resource $R$ into another resource which we denote by $\alpha^i R$, with the same set of interfaces as $R$.

More formally, we consider a set $\Sigma$ of objects, called converters. A converter $\alpha$, when applied as an interface $i$ of a resource, induces a function[3] $\Phi \to \Phi$ : $R \mapsto \alpha^i R$. Moreover, $\Sigma$ is equipped with a composition operation $\circ$ satisfying

$$(\beta \circ \alpha)^i R = \beta^i(\alpha^i R).$$

The set $\Sigma$ also contains a special element, the *identity converter* $id \in \Sigma$, which induces the identity function $\Phi \to \Phi$ (for any interface $i$) and simply stands for using the resource "as is". It satisfies

$$id \circ \alpha = \alpha \circ id = \alpha.$$

---

[2] The term "converter" is used because its application at an interface converts the interface into an interface with a different behavior.

[3] In general, one could consider partial function where the application of a converter at an interface need not always be defined. For the purpose of this paper there is no need to consider partial functions.

The set $\Sigma$ is closed under composition, i.e., $\Sigma \circ \Sigma = \Sigma$, where equality holds because $id \in \Sigma$.

For two-interface resources as used in this paper, if one (i.e., Alice) applies a converter $\alpha$ at the left interface of a resource $R$, the resulting resource is denoted as

$$\alpha R.$$

Similarly, if one (i.e., Eve) applies a converter $\beta$ at the right interface of a resource $R$, the resulting resource is denoted as

$$R\beta.$$

A key property we require, and which is typically satisfied, is that application of converters at the left and the right interface commute, i.e.,

$$(\alpha R)\beta = \alpha(R\beta),$$

which justifies to write $\alpha R\beta$ for the resulting resource.

A *resource specification* is simply a subset of $\mathcal{R} \subseteq \Phi$ containing those resources satisfying the specification. When no confusion can arise, we will also use the term resource for a resource specification. An element of $R \in \Phi$ can be understood as a singleton specification, i.e., as $\{R\}$.

Applying a converter $\alpha$ to a resource specification $\mathcal{R}$ is naturally defined as

$$\alpha\mathcal{R} = \{\alpha R \mid R \in \mathcal{R}\},$$

and analogously for $\mathcal{R}\beta$ and $\alpha\mathcal{R}\beta$.

## 3.4   Some Relevant Resource Specification Relaxations

The purpose of this section is to introduce a few generic types of relaxations of a resource specification $\mathcal{R}$ and to state some simple facts. We have already discussed $\epsilon$-balls $\mathcal{R}^\epsilon$.

The understanding is that a dishonest party can do something arbitrary, i.e., apply an arbitrary converter. For a specification $\mathcal{R}$, the specification capturing that it is unknown what happens at the right interface is

$$\mathcal{R}^* := \mathcal{R}\Sigma = \{R\beta \mid R \in \mathcal{R}, \beta \in \Sigma\},$$

where the symbol $*$ stands for an arbitrary converter. One can prove that

$$\mathcal{R} \subseteq \mathcal{R}^* = (\mathcal{R}^*)^*. \tag{1}$$

One can consider a special converter $\dashv$ which blocks the right interface, i.e., the resource $R \dashv$ only has a left interface. More technically speaking, for a resource $R \dashv$, a distinguisher sees only the left interface and has no access to the right interface. A resource $R$ is *right-outbound* if no converter attached to the right interface can have an effect at the left interface, i.e., if

$$R^* \dashv = R \dashv.$$

This means that no signalling from the right to the left interface of $R$ is possible. In this paper we do not need the dual left-outbound property.

For a given resource specification $\mathcal{R}$ one can consider the set, denoted $\mathcal{R}[\![$, of right-outbound resources $S$ compatible with (a resource in) $\mathcal{R}$ (only) at the left interface:

$$\mathcal{R}[\![ := \{S \mid S \text{ is right-outbound and } S\dashv \in \mathcal{R}\dashv\} = \{S \mid S^*\dashv = S\dashv \in \mathcal{R}\dashv\}.$$

For example, if $\mathcal{R}$ denotes the specification of a random oracle (which hides Alice's queries from Eve), then $\mathcal{R}[\![$ includes all resources that leak partial or all information about Alice's queries to Eve. An impossibility result stating that $\mathcal{R}[\![$ is not constructible is therefore a significantly stronger statement than that a standard random oracle is not constructible. One can prove that

$$\mathcal{R} \subseteq \mathcal{R}[\![ = (\mathcal{R}[\!])[\![. \tag{2}$$

## 3.5 Modeling Aspects: Resources vs. Converters

The implementation of a converter requires computational resources such as computing power, memory, and randomness. On one hand, how many resources an implementation requires seems relevant, and it appears generally better if a converter can be more efficiently implemented. On the other hand, one often makes statements that involve a quantification over all converters (e.g. all simulators), and such a quantification only makes sense if, by definition, the actual choice is irrelevant.[4]

In almost every scientific consideration, one intentionally ignores certain aspects as irrelevant and focuses on the particular ones considered relevant in the given context. What is relevant or irrelevant is generally a conscious choice. For example, in a computer science (or more specifically a cryptographic) context, one may or may not care to model the exact computational power available to a party. In particular, one may use an asymptotic model and only require that the number of computational steps is polynomially bounded in a security parameter.

The general guiding principle in constructive cryptography is that everything that is considered relevant for the analysis one wants to perform is modeled as part of the resource. In contrast, the choice of a converter is, by definition, irrelevant with regard to the entailed cost or complexity. If, for instance, computing power, memory, or randomness needed for a cryptographic construction is considered to matter, then it has to be explicitly modeled as part of the resource. To illustrate this point, we explain a few possible such explicit choices. Each can be thought of as a particular security model (e.g. computational or information-theoretic).

1. The term *information-theoretic security* is usually used when computation (at least by the adversary) is irrelevant. In such a case the converter set includes all systems, regardless of the computational complexity of implementing them.

---

[4] For a logical predicate $P$, the purpose of a statement of the form $\exists x \, P(x)$ is precisely to ignore *which* $x$ makes $P(x)$ true.

2. Even for information-theoretic security one may be interested in making nevertheless the memory requirements explicit (see [10]). In this case, memory is modeled as part of the resource and the converters are all systems that can compute arbitrary functions (regardless of the complexity) but cannot keep state between invocations.[5] Ristenpart et al. [23] pointed out an apparent problem with the indifferentiability notion of [18], but it was shown in [10] that this problem was only an artefact of the fact that Turing machines come, by definition, with an arbitrary amount of memory (the tape) and that therefore this model is not adequate in a setting, as that considered in [23], where memory is indeed a relevant resource.

3. If computing power is considered relevant, then one can consider converters that perform no computation by themselves but only connect systems and possibly input constants (for example a program). Any computational resource can be modeled as a (parallel) resource. Such a resource can either be a specific system with a certain behavior (e.g. a system encrypting messages), without reference to an implementation on a certain computational model. Alternatively, it could be a computer resource $C$ in some computational model, with an upper bound on the available computing power (for example called complexity), and which can run an arbitrary program up to that complexity bound. In this case, the converter inputs a program to $C$, and we consider it irrelevant (from a resource viewpoint) which program is used. Possibly, the specification of $C$ could involve an upper bound on the length of the program. In such a view, converters only route information, without performing computation.

4. If, for some notion of efficiency, efficient computing power is considered irrelevant, then one can consider $\Sigma$ to be the set of efficiently implementable converter systems. Typically in cryptography, efficient is defined as some form of polynomial-time notion, which of course, and unfortunately, requires now the objects to be defined asymptotically in some way. A main reason for using polynomial-time is that this notion, if properly defined, is closed.[6] We point out that polynomial-time is a specific choice that has its merits but for many statements need not be fixed.

Clearly, one could consider different converter sets for honest parties and for dishonest parties. For example, it would be natural to consider a notion of efficiency and a different, larger notion of feasibility, where the converters of honest parties must be efficiently implementable and the converters of dishonest parties must only be feasibly implementable. It does not really seem well-justified to use the same polynomial-time notion for both, except by tradition and possibly by the set of results one can prove for this choice.

---

[5] In this model, the memory required for a function computation is assumed to be free. Of course, one could also model this memory as a resource.

[6] More formally, converters $\alpha$ and $\beta$ from this particular set $\Sigma$ can be composed to a new converter, say $\alpha \circ \beta$, and this composition is closed in the sense that the function $\Phi \to \Phi$ induced by $\alpha \circ \beta$ is contained in the class of functions induced by converters in $\Sigma$.

# 4 Cryptographic Constructions for a Fixed Adversary Interface

## 4.1 Definition of Constructions and Some Lemmas

If a resource satisfying specification $\mathcal{R}$ is available, Alice can apply a converter $\pi$ to it, resulting in specification $\pi\mathcal{R}$. Often one wants to think about $\pi\mathcal{R}$ in a simpler way, namely in terms of a specification $\mathcal{S}$ such that $\pi\mathcal{R} \subseteq \mathcal{S}$. The guarantee given to Alice by the specification $\mathcal{S}$ is generally weaker than the specification $\pi\mathcal{R}$, but, in the usual sense of abstraction, this loss of information is accepted because $\mathcal{S}$ is a simpler (to use and work with) specification.

We can then say that a desired resource (specification) $\mathcal{S}$ is *constructed from* an assumed resource (specification) $\mathcal{R}$ by application of the converter $\pi \in \Sigma$ (which is the constructor). This is written as $\mathcal{R} \xrightarrow{\pi} \mathcal{S}$.

**Definition 1.** $\mathcal{R} \xrightarrow{\pi} \mathcal{S} :\Longleftrightarrow \pi\mathcal{R} \subseteq \mathcal{S}$.

**Lemma 1.** *This construction notion is composable:*

$$\mathcal{R} \xrightarrow{\pi} \mathcal{S} \wedge \mathcal{S} \xrightarrow{\pi'} \mathcal{T} \Longrightarrow \mathcal{R} \xrightarrow{\pi' \circ \pi} \mathcal{T}.$$

*Proof.* From the first condition $\pi\mathcal{R} \subseteq \mathcal{S}$ it follows that $\pi'\pi\mathcal{R} \subseteq \pi'\mathcal{S}$. Combining this with the second condition, $\pi'\mathcal{S} \subseteq \mathcal{T}$, we obtain $\pi'\pi\mathcal{R} \subseteq \mathcal{T}$, which was to be proved.  □

The following lemmas assert that the three specification relaxations discussed in Sect. 3.4 are compatible with the construction notion.

**Definition 2.** *A metric $d$ on $\Phi$ is called* non-expanding *if $d(\alpha R, \alpha S) \leq d(R, S)$ for all $\alpha$ and $d(R\beta, S\beta) \leq d(R, S)$ for all $\beta$.*

**Lemma 2.** *If the metric on $\Phi$ is non-expanding, then, for any $\epsilon > 0$,*

$$\mathcal{R} \xrightarrow{\pi} \mathcal{S} \Longrightarrow \mathcal{R}^\epsilon \xrightarrow{\pi} \mathcal{S}^\epsilon.$$

*Proof.* We need to show that if $R' \in \mathcal{R}^\epsilon$, i.e., $R' \approx_\epsilon R$ for some $R \in \mathcal{R}$, then $\pi R' \in \mathcal{S}^\epsilon$, i.e., $\pi R' \approx_\epsilon S$ for some $S \in \mathcal{S}$. The condition $\mathcal{R} \xrightarrow{\pi} \mathcal{S}$ guarantees that $\pi R = S$ for some $S \in \mathcal{S}$. For the same $S$ we have $\pi R' \approx_\epsilon S$ since $\pi R' \approx_\epsilon \pi R = S$ (due to the non-expanding property). This completes the proof.  □

The following lemmas are stated without proofs.

**Lemma 3.** $\mathcal{R} \xrightarrow{\pi} \mathcal{S} \Longrightarrow \mathcal{R}^* \xrightarrow{\pi} \mathcal{S}^*$.

**Lemma 4.** $\mathcal{R} \xrightarrow{\pi} \mathcal{S} \Longrightarrow \mathcal{R}[\![ \xrightarrow{\pi} \mathcal{S}[\![$.

## 4.2   Proving Constructions by Simulators

A line of reasoning often arising in cryptography, including [18], can be captured by the following system equation (see also [17]):

$$\pi R \approx_\epsilon S\sigma, \tag{3}$$

where the converter $\sigma$ is usually called a *simulator* (see discussion in Sect. 4.2). The usefulness of finding a simulator $\sigma$ satisfying the equation is that it implies a construction statement:

**Lemma 5.** *If the metric is non-expanding, then*

$$\exists \sigma \in \Sigma : \ \pi R \approx_\epsilon S\sigma \implies R \xrightarrow{\ \pi\ } (S^*)^\epsilon.$$

*Proof.* Since $\sigma \in \Sigma$ we have $S\sigma \in S\Sigma = S^*$. Hence $\pi R \approx_\epsilon S\sigma$ implies that $\pi R \subseteq (S\sigma)^\epsilon \subseteq (S^*)^\epsilon$, which is the definition of $R \xrightarrow{\ \pi\ } (S^*)^\epsilon$.          □

In the literature, the converter $\sigma$ in Eq. (3) is usually called a simulator. It is sometimes described as translating what an adversary could do in the real world (the left side of the equation), say $\beta$, into what she needs to do in the ideal world (the right side of the equation) to achieve the same (or something close to) what she would achieve in the real world, namely $\beta \circ \sigma$. Note that $\pi R\beta \approx_\epsilon S\sigma\beta$ due to the non-expanding property of the pseudo-metric.

We point out, however, that in contrast to most of the existing literature, the actual statement of interest (see Lemma 5) to us is not Eq. (3) itself, but the construction statement it implies. In particular, the simulator does not appear in the definition of a construction, and there can be interesting construction statements proved in different ways than by use of Lemma 5.

In view of Lemma 5, the notion of indifferentiability [18] can be understood as follows: $T$ is indifferentiable from $S$, within $\epsilon$, if $T \subseteq (S^*)^\epsilon$, where this is proved by demonstrating a simulator $\sigma$ such that $T \approx_\epsilon S\sigma$. If $T = \pi R$, this corresponds to the construction statement $R \xrightarrow{\ \pi\ } (S^*)^\epsilon$.

## 4.3   Computational Considerations

Often in cryptography, $\Sigma$ is the set of polynomial-time implementable converter systems. If the metric on $\Phi$ is chosen as the two-valued computational indistinguishability metric, then a polynomial-time converter can be absorbed into a poly-time distinguisher without leaving the distinguisher class, i.e., the metric is non-expanding.

In a concrete-security consideration, the efficiency loss of a reduction and therefore the concrete implementation complexity of $\sigma$ matters. In other words, a statement of the form (3) becomes more useful for a more efficient $\sigma$. This, however, does at first not seem to be compatible with the idea that converters in $\Sigma$ are considered free (of cost). Either a converter is free, or it is not. Let us explain how this contradiction is resolved in our approach.

More specifically, suppose we use model 3 described in Sect. 3.5, where $\Sigma$ are the converters that perform no computation. Suppose furthermore that one has shown that equality $\pi R = S\beta$ holds for some system $\beta$ that requires some computation, i.e., $\beta \notin \Sigma$. Then we can give the equation the following meaning. Let $\bar{\beta}$ be a system corresponding to the resource that behaves like $\beta$, with inside and outside interface both available to Eve (only at the right interface). Then one can rephrase the equation $\pi R = S\beta$ as

$$\pi R = [S, \bar{\beta}]\, \sigma,$$

where $\sigma$ is the trivial converter that simply connects $\bar{\beta}$ to $S$, i.e., such that

$$[S, \bar{\beta}]\, \sigma = S\beta.$$

In other words, any equation of the type $\pi R = S\beta$ can be turned into a construction statement of the form

$$R \xrightarrow{\pi} \left([S, \bar{\beta}]\right)^*$$

which makes the computational resource required for the "simulation" explicit.

# 5   Impossibility of Constructing a Random Oracle

As an example for an impossibility result, we show that a random oracle cannot be constructed, even if a source of public randomness is available. To state this more precisely, we use the following specifications.

- $\mathrm{PR}^k$ is *public randomness* of size $k$. The resource chooses $Z$ uniformly at random from the set $\{0,1\}^k$ of $k$-bit strings.[7] Any party can read $Z$.[8]
- $\mathrm{RO}^{m \to n}_{[q,q']}$ is a *random oracle* with input size $m$ and output size $n$. The resource chooses $F$ uniformly at random from the set of all functions from $\{0,1\}^m$ to $\{0,1\}^n$. Any party can submit queries $x \in \{0,1\}^m$ which are answered by $F(x)$. At least $q$ and at most $q'$ queries by any party are allowed.

As before, we assume that the set of resources is equipped with a (non-expanding) distance measure, $d$, defined as the maximum advantage of any distinguisher from a class $\mathcal{D}$.[9] The results derived below will be valid for any reasonable distinguisher class $\mathcal{D}$. The only requirement is that the execution of basic algorithms giving inputs and receiving outputs and performing equality checks, such as $D_1$ and $D_2$ below, are within the class $\mathcal{D}$.

We start with a basic impossibility result, which asserts that public randomness cannot be expanded.

---

[7] To keep the presentation simple we assume that the probability distribution of $Z$ is uniform; a generalization to arbitrary probability distributions is straightforward. This includes the case where $\mathrm{PR}^k$ is a fixed hash function program of length $k$.

[8] One may impose the additional restriction that the string $Z$ can only be read bit-wise, but this is not relevant for the considerations here.

[9] That is, $d(R, S) = \sup_{D \in \mathcal{D}} \Delta^D(R, S)$, where $\Delta^D(R, S)$ is the absolute value of the difference between the probability that $D$ returns 0 when connected to $R$ and the probability that it returns 0 when connected to $S$.

**Lemma 6.** *Let $k \in \mathbb{N}$ and $\epsilon < \frac{1}{4}$. Then*

$$\mathrm{PR}^k \;\not\longrightarrow\; \mathrm{PR}^{k+1} [\![ \, ^{\epsilon}.$$

*Proof.* As explained, we regard $\mathrm{PR}^k$ as a specification of a system with two interfaces (left and right), which model the access to the resource by the honest and the dishonest parties, respectively. It suffices to consider two honest parties, which we label by $A$ and $A'$, as well as one dishonest party, labelled by $E$. We recall that in this two-interface case, any constructor corresponds to a converter $\pi$ for the left interface, which can be understood as a pair of converters $\pi_A$ and $\pi_{A'}$ for the two honest parties.

We need to prove that

$$d(\pi\mathrm{PR}^k, \mathcal{R}) \geq \frac{1}{4}$$

for any converter $\pi$ and for any right-outbound resource $\mathcal{R}$ with the property $\mathcal{R} \dashv \subseteq \mathrm{PR}^{k+1} \dashv$. Because $d$ is non-expanding, it suffices to show that

$$d(\pi\mathrm{PR}^k\pi', \mathcal{R}\pi') \geq \frac{1}{4} \tag{4}$$

for some converter $\pi'$. We take $\pi'$ to be $\pi_{A'}$. More precisely, $\pi'$ answers a query by $E$ in the same way as $\pi$ would answer a query by $A'$. We then consider a distinguisher $D_1$ that executes the following simple algorithm and show that it can tell apart $\pi\mathrm{PR}^k\pi'$ and $\mathcal{R}\pi'$ with advantage at least $\frac{1}{4}$.

---

**Distinguisher $D_1$**

read the $(k+1)$-bit strings $Z_A$ and $Z_{A'}$ from the left interface;
read the $(k+1)$-bit string $Z_E$ from the right interface;
**if** $Z_A \neq Z_{A'}$ **then**
  | **return** 0; **halt** ;
**else if** $Z_A \neq Z_E$ **then**
  | **return** 1; **halt** ;
**return** 0

---

Suppose first that $D_1$ is connected to $\pi\mathrm{PR}^k\pi'$. It only returns 1 if $Z_A = Z_{A'} \neq Z_E$. By the definition of $\pi'$, the strings $Z_{A'}$ and $Z_E$ are generated by identical (possibly probabilistic, but independent) procedures. It follows that the probability of the event $Z_A = Z_{A'} \neq Z_E$ is upper bounded by

$$\Pr[Z_A = Z_{A'}] \Pr[Z_E \neq Z_{A'}] = \Pr[Z_A = Z_{A'}](1 - \Pr[Z_A = Z_{A'}]) \leq \frac{1}{4}$$

(since $\frac{1}{4}$ is the maximum of the function $x \mapsto x(1-x)$ for $0 \leq x \leq 1$). Hence $D_1$ returns 0 with probability at least $\frac{3}{4}$.

Conversely, in the case where $D_1$ is connected to $\mathcal{R}\pi'$, $Z_A = Z_{A'}$ holds by definition of $\mathcal{R}$, and $Z_A$ is a uniformly random $(k+1)$-bit string, whereas

$Z_E$ is a $(k+1)$-bit string computed by $\pi'$. Since $\pi'$ behaves by definition like $\pi_{A'}$ and thus takes as input only a $k$-bit string, $Z_E$ depends on a string $W$ of length at most $k$. $D_1$ only returns 0 if $Z_A = Z_E$. The probability of this event is upper bounded by the min-entropy of $Z_A$ conditioned on $W$, i.e., $\Pr[Z_A = Z_E] \leq 2^{-H_{\min}(Z_A|W)}$ (cf. Appendix). By (11), the chain rule for the min-entropy, we have $H_{\min}(Z_A|W) \geq H_{\min}(Z_A) - k = 1$, where we used that $W$ consists of at most $k$ bits. We conclude that $\Pr[Z_A = Z_E] \leq \frac{1}{2}$. Hence, when connected to $\mathcal{R}\pi'$, $D_1$ returns 0 with probability at most $\frac{1}{2}$. Combining this with the above shows that the distinguishing advantage is at least $\frac{1}{4}$, which implies (4). □

Lemma 6 states that public randomness cannot be expanded by a single bit, even if one would tolerate that Eve may learn something about what happens at the honest parties' interface (which is captured by "$[\![$"). This also suggests that one cannot construct a more powerful public randomness resource that allows to extract more than $k$ bits:

**Corollary 1.** *Let $k \in \mathbb{N}$ and $\epsilon < \frac{1}{4}$. Then*

$$\mathrm{PR}^k \;\not\longrightarrow\; \mathrm{RO}^{m \to 1}_{[q,\infty]} [\![\,^\epsilon$$

*unless $m < \log_2(k+1)$ or $q \leq k$.*

*Proof.* Suppose that

$$\mathrm{PR}^k \;\overset{\pi}{\longrightarrow}\; \mathrm{RO}^{m \to 1}_{[q,\infty]} [\![\,^\epsilon \tag{5}$$

holds for some constructor $\pi$. Let furthermore $\pi'$ be a constructor that simply outputs the first $\min(q, 2^m)$ entries of the function table of the random oracle, and thus achieves

$$\mathrm{RO}^{m \to 1}_{[q,\infty]} \;\overset{\pi'}{\longrightarrow}\; \mathrm{PR}^{\min(q,2^m)} [\![\,.$$

Using Lemma 4 as well as (2), this yields

$$\mathrm{RO}^{m \to 1}_{[q,\infty]} [\![ \;\overset{\pi'}{\longrightarrow}\; \mathrm{PR}^{\min(q,2^m)} [\![$$

and hence, using Lemma 2, also

$$\mathrm{RO}^{m \to 1}_{[q,\infty]} [\![\,^\epsilon \;\overset{\pi'}{\longrightarrow}\; \mathrm{PR}^{\min(q,2^m)} [\![\,^\epsilon. \tag{6}$$

By Lemma 1, the composition of constructions (5) and (6) gives

$$\mathrm{PR}^k \;\overset{\pi' \circ \pi}{\longrightarrow}\; \mathrm{PR}^{\min(q,2^m)} [\![\,^\epsilon.$$

Lemma 6 now implies that $\min(q, 2^m) < k+1$. □

We now proceed to a substantially stronger impossibility claim. Note that Corollary 1 only applies to cases where the total entropy that the honest parties can draw from the random oracle is strictly larger than the number $k$ of public random bits that are available. Theorem 1 below shows that this is not necessary for the impossibility result to hold. It asserts that even a weak random oracle that answers only a small number of queries (say, $q = 1024$), and thus only provides a small amount of entropy to the honest parties, cannot be constructed. In addition, the impossibility claim remains valid if one tolerates that the constructed random oracle leaks arbitrary information, e.g., about what happens at the honest parties' interface, to the adversary.

For simplicity, we restrict the statement to oracles with output size 1 (but it obviously implies a corresponding impossibility result for random oracles with larger output size).

**Theorem 1.** *For any* $k, m, q \in \mathbb{N}$ *and* $\epsilon \leq \frac{1}{2}$

$$\mathrm{PR}^k \;\not\longrightarrow\; \mathrm{RO}_{[q,\infty]}^{m \to 1} [\![\;^\epsilon$$

*unless* $m < \min(1 + \log_2 k, 10)$ *or* $q < 2^{10}$.

*Proof.* Set without loss of generality $q = 2^{10}$ and assume that $m \geq 1 + \log_2 k$ and $m \geq 10$. The proof proceeds analogously to that of Lemma 6, i.e., we show that

$$d(\pi \mathrm{PR}^k \pi', \mathcal{R}\pi') > \frac{1}{2}, \tag{7}$$

where $\mathcal{R}$ is a right-outbound resource such that $\mathcal{R} \dashv = \mathrm{RO}_{[q,\infty]}^{m \to 1} \dashv$, and where $\pi'$ is again a converter that reproduces the behavior of $\pi$ for one party. To establish this inequality we consider a distinguisher $D_2$ defined by the following simple algorithm and show that it can tell apart $\pi \mathrm{PR}^k \pi'$ and $\mathcal{R}\pi'$ with advantage strictly larger than $\frac{1}{2}$.

---

**Distinguisher $D_2$**

choose $q$ different values $X_1, \ldots, X_q$ at random from the set $\{0,1\}^m$ ;
**for** $j \in \{1, \ldots, q\}$ **do**
  | $A$ and $A'$ submit query $X_j$ and record the answers $Z_{A,j}$ and $Z_{A',j}$;
  | **if** $Z_{A,j} \neq Z_{A',j}$ **then return** 0; **halt** ;
**end**
**for** $j \in \{1, \ldots, q\}$ **do**
  | $E$ submits query $X_j$ and records the answer $Z_{E,j}$;
  | **if** $Z_{A,j} \neq Z_{E,j}$ **then return** 1; **halt** ;
**end**
**return** 0

---

We first treat the case where $D_2$ is connected to $\pi \mathrm{PR}^k \pi'$. $D_2$ only returns 1 if, for some $j \in \{1, \ldots, q\}$, $Z_{A,j} \neq Z_{E,j}$. Following the same reasoning as in the proof of Lemma 6, we can infer that the probability of this event is upper bounded by $\frac{1}{4}$. Hence, when connected to $\pi \mathrm{PR}^k \pi'$, $D_2$ returns 0 with probability at least $\frac{3}{4}$.

Conversely, if $D_2$ is connected to $\mathcal{R}\pi'$, the answers $Z_{A,j}$ and $Z_{A',j}$ received by the honest parties upon any query $X_j$ will agree by definition of $\mathcal{R}$. The distinguisher thus returns 0 only if they also coincide with the answers $Z_{E,j}$ received by a dishonest party $E$. This latter event only occurs if the tuple of answers $Z = (Z_{A,1}, \ldots, Z_{A,q})$ to all queries $X_1, \ldots, X_q$ is reproduced by the output of the converter $\pi'$. Since $\pi'$ carries out the same computation as $\pi$ for one party, this output depends on a string $W$ of length at most $k$. Because $Z$ can be regarded as a subset of $q$ bits chosen at random from $2^m \geq 2k$ uniform bits, Corollary 2 (see Appendix) asserts that $H_{\min}(Z|X_1 \cdots X_q W) > 2$. This implies that the success probability of any strategy for guessing $Z$ from $W$ is strictly smaller than $\frac{1}{4}$. Hence, if connected to $\mathcal{R}\pi'$, $D_2$ returns 0 with probability strictly smaller than $\frac{1}{4}$. Combining this with the above shows that $D_2$ has distinguishing advantage strictly larger than $\frac{1}{2}$, which establishes (7).     □

## 6   Construction Results

Coron *et al.* [9] showed that a random oracle with arbitrary input length and fixed output length $n$ can be constructed from a compression function with fixed input length and output length $n$. The latter is itself modelled as a random oracle. The following theorem is a variation of this result.[10]

**Theorem 2.** *For any* $n, \kappa, \ell, q, q' \in \mathbb{N}$ *and* $\epsilon = 2^{-n+1}q'^2$ *there is* $\pi$ *such that*

$$\mathrm{RO}_{[\ell q, q']}^{n+\kappa+\lceil \log_2 \ell \rceil \to n} \xrightarrow{\ \pi\ } \left( (\mathrm{RO}_{[q,q']}^{n+\ell\kappa \to n})^* \right)^{\ell\epsilon}. \tag{8}$$

We are going to provide a proof of Theorem 2 based on the following result.

**Lemma 7.** *For any* $n, a, q, q' \in \mathbb{N}$ *and* $\epsilon = 2^{-n+1}q'^2$

$$\left[ \mathrm{RO}_{[q,\infty]}^{a \to n}, \mathrm{RO}_{[q,q']}^{n+\kappa \to n} \right] \xrightarrow{\ \pi\ } \left( (\mathrm{RO}_{[q,q']}^{a+\kappa \to n})^* \right)^{\epsilon},$$

*where* $\pi$ *is the constructor which answers queries* $(x, y) \in \{0,1\}^a \times \{0,1\}^\kappa$ *with* $F^{n+\kappa \to n}(F^{a \to n}(x), y)$, *where* $F^{a \to n}$ *and* $F^{n+\kappa \to n}$ *are the functions defined by the two random oracles.*

*Proof.* As shown in [9]

$$d\left( \pi \left[ \mathrm{RO}_{[q,\infty]}^{a \to n}, \mathrm{RO}_{[q,q']}^{n+\kappa \to n} \right], \mathrm{RO}_{[q,q']}^{a+\kappa \to n} \sigma \right) \leq \epsilon$$

holds for a simulator $\sigma$ defined by the following algorithm.

---

[10] The result in [9] corresponding to Theorem 2 is weaker in that the error $\epsilon$ is multiplied with $\ell^2$ rather than $\ell$.

---

**Simulator $\sigma$**

**if** *query* $x \in \{0,1\}^a$ *to* $F^{a \to n}$ **then**
|    **return** *random* $v \in \{0,1\}^n$;
**else if** *query* $(v', y) \in \{0,1\}^n \times \{0,1\}^\kappa$ *to* $F^{n+\kappa \to n}$ **then**
|    **if** $v'$ *equals output of* $F^{a \to n}$ *for some previously queried* $x'$ **then**
|      **return** *answer of the resource to query* $(x', y)$
|    **else**
|      **return** *random* $z \in \{0,1\}^n$

---

The claim of the lemma then follows from Lemma 5.       □

*Proof. (of Theorem 2).* The construction that gives rise to (8) can be regarded as the concatenation of several more basic constructions. The first, $\pi_0$, a simple domain splitting step, constructs $\ell$ independent random oracles with identical domain from a single random oracle, whose input domain consists of $\lceil \log_2 \ell \rceil$ additional bits, i.e.,

$$\mathrm{RO}_{[\ell q, q']}^{n+\kappa+\lceil \log_2 \ell \rceil \to n} \xrightarrow{\pi_0} \underbrace{[\mathrm{RO}_{[q,q']}^{n+\kappa \to n}, \cdots, \mathrm{RO}_{[q,q']}^{n+\kappa \to n}]}_{\ell \text{ times}}{}^{*}. \tag{9}$$

This is achieved by converters which simply answer any query $x$ to the $j$th constructed random oracle by submitting the concatenation of $x$ and a binary encoding of $j$ to the given random oracle and then forwarding its answer.

For the next step, we invoke Lemma 7 with $a = n + j\kappa$, for $j \in \{1, \ldots, \ell-1\}$. This lemma, together with Lemmas 2 and 3, the fact that $(\mathcal{R}^\epsilon)^* \subseteq (\mathcal{R}^*)^\epsilon$, and (1), implies that there exists a constructor $\pi_j$ such that

$$\left[ \left( (\mathrm{RO}_{[q,\infty]}^{n+j\kappa \to n})^* \right)^{(j-1)\epsilon}, \mathrm{RO}_{[q,q']}^{n+\kappa \to n} \right] \xrightarrow{\pi_j} \left( (\mathrm{RO}_{[q,q']}^{n+(j+1)\kappa \to n})^* \right)^{j\epsilon}.$$

Recursive application of this construction gives

$$\left[ (\mathrm{RO}_{[q,\infty]}^{n+\kappa \to n})^*, \underbrace{\mathrm{RO}_{[q,q']}^{n+\kappa \to n}, \cdots, \mathrm{RO}_{[q,q']}^{n+\kappa \to n}}_{\ell-1 \text{ times}} \right] \xrightarrow{\pi_{\ell-1} \circ \cdots \circ \pi_1} \left( (\mathrm{RO}_{[q,q']}^{n+\ell\kappa \to n})^* \right)^{(\ell-1)\epsilon}.$$

Using $\mathrm{RO}_{[q,q']}^{n+\kappa \to n} \subseteq \mathrm{RO}_{[q,\infty]}^{n+\kappa \to n} \subseteq (\mathrm{RO}_{[q,\infty]}^{n+\kappa \to n})^*$ we can substitute the first term in the above construction statement to obtain

$$\left[ \underbrace{\mathrm{RO}_{[q,q']}^{n+\kappa \to n}, \cdots, \mathrm{RO}_{[q,q']}^{n+\kappa \to n}}_{\ell \text{ times}} \right] \xrightarrow{\pi_{\ell-1} \circ \cdots \circ \pi_1} \left( (\mathrm{RO}_{[q,q']}^{n+\ell\kappa \to n})^* \right)^{(\ell-1)\epsilon}.$$

Similarly to the above, this implies that

$$\left[ \underbrace{\mathrm{RO}_{[q,q']}^{n+\kappa \to n}, \cdots, \mathrm{RO}_{[q,q']}^{n+\kappa \to n}}_{\ell \text{ times}} \right]^* \xrightarrow{\pi_{\ell-1} \circ \cdots \circ \pi_1} \left( (\mathrm{RO}_{[q,q']}^{n+\ell\kappa \to n})^* \right)^{(\ell-1)\epsilon}. \tag{10}$$

Theorem 2 now follows by composing the constructions (9) and (10).       □

# 7 Generalization to Many Parties

We briefly sketch how the construction notion described in Sect. 4 directly leads to a construction notion for resources with several honest parties and an adversary, simply by considering the left interface as consisting of a sub-interface for each honest party and by considering the special type of converter (for the combined interface) as corresponding to a list of converters, one for each sub-interface. A typical case is the so-called Alice-Bob-Eve setting as discussed in [16,17] with two honest parties Alice and Bob. This model allows to capture many core cryptographic constructions, including the construction of a shared secret key, of an authenticated channel, and of a secure channel.

One can also capture a setting where various parties could be dishonest. Usually the terminology used is that a central adversary corrupts some of the parties. In other words, any party can possibly be honest or dishonest. A *protocol* is a tuple of converters, one for each potentially honest party, where the idea is that an honest party is guaranteed to apply the designated converter (i.e., to "follow the protocol"). One can then make a collection of construction statements, for each set of dishonest parties that needs to be considered, where for each such statements the honest parties' interfaces can be thought of as being grouped at the left side and the dishonest parties' interfaces are grouped at the right side.

# 8 Conclusions

The goal of this paper was to cover the essential aspects of the original indifferentiability paper [18], but in a more general and more adequate manner, leading to a general construction notion. The paper [18] contained basic ideas of constructive cryptography [17], but this is perhaps not apparent since [18] was mostly written in the tradition of the cryptography literature at the time: The objects considered were usually asymptotic in a security parameter, and the usual polynomial-time efficiency notion and the usual negligibility notion were used. It should be clear from [17] and this paper that fixing such a particular model is unnecessary. Moreover, indifferentiability was presented in [18] as a generalized form of indistinguishability, appearing as an intermediate step needed to define constructions (actually called reductions in [18]).

In view of the general construction notion presented in this paper, the indifferentiability notion corresponds to a specific construction type, for the special type $S^*$ of resource specifications, where, moreover, $S$ is right-outbound. Then $T$ is indifferentiable from $S$, within $\epsilon$, if $T \subseteq (S^*)^\epsilon$, where this is proved by demonstrating a simulator $\sigma$ (not called simulator in [18]) such that $T \approx_\epsilon S\sigma$. If $T = \pi R$, this corresponds to the construction statement $R \xrightarrow{\pi} (S^*)^\epsilon$. Demonstrating a simulator and applying Lemma 5 is only one of possibly several ways of proving construction statements, and simulators should therefore probably only appear in proofs, not in definitions.

**Acknowledgments.** We would like to thank the TCC Test-of-Time award committee for selecting our paper for the award of this instantiation of TCC. Very sadly, our coauthor Clemens Holenstein passed away in 2012 and could neither receive the award nor contribute to this paper. Discussions with many people have contributed immensely to shaping our described viewpoint of cryptography. Of particular help were discussions with Joël Alwen, Christian Badertscher, Ran Canetti, Sandro Coretti, Grégory Demay, Yevgeniy Dodis, Peter Gaži, Martin Hirt, Dennis Hofheinz, Daniel Jost, Christian Matt, Christopher Portmann, Phil Rogaway, Gregor Seiler, Björn Tackmann, Stefano Tessaro, Daniel Tschudi, Daniele Venturi, Stefan Wolf, and Vassilis Zikas.

## Appendix: Min-entropy sampling

The min-entropy of a random variable $X$ conditioned on another random variable $Y$, $H_{\min}(X|Y)$, is defined as (see, e.g., [14])

$$H_{\min}(X|Y) = -\log_2 \max_f \Pr[X = f(Y)],$$

where the maximum ranges over all functions $f$ from the alphabet $\mathcal{Y}$ of $Y$ to the alphabet $\mathcal{X}$ of $X$. Note that the expression in the logarithm on the right hand side can be interpreted as the maximum probability of correctly guessing $X$ from $Y$. The min-entropy has several natural properties analogous to the Shannon entropy. Among them is a chain rule, which implies

$$H_{\min}(X|Y) \geq H_{\min}(X) - \log_2 |\mathcal{Y}|. \tag{11}$$

The min-entropy of a sample chosen at random from a min-entropy source has been studied in [13,21,24]. Roughly speaking, one can show that the min-entropy of the sample is proportional to the sample size and the min-entropy of the source. We use a version of this statement due to Wullschleger, which provides explicit bounds [25].[11]

**Proposition 1.** *Let* $X \in \{0,1\}^n$ *and* $Z$ *be random variables and let* $T$ *be a uniformly chosen subset of* $\{1,\dots,n\}$ *of size* $|T|$. *Then*

$$\frac{H_{\min}(X_T|TZ)}{|T|} \geq f\left(\frac{H_{\min}(X|Z)}{n}\right) - \frac{5}{|T|},$$

*where* $f : [0,1] \to [0,1]$ *is a monotonically strictly increasing function such that* $f(1/2) > 1/144$.

**Corollary 2.** *Let* $X \in \{0,1\}^n$ *be uniformly distributed, let* $Z \in \{0,1\}^k$ *be an arbitrary random variable on* $k \leq n/2$ *bits, and let* $T$ *be a uniformly chosen subset of* $\{1,\dots,n\}$ *of size* $|T|$. *Then*

$$H_{\min}(X_T|TZ) > \frac{|T|}{144} - 5.$$

*Proof.* It follows from the chain rule (11) that conditioning on $k$ bits cannot decrease the min-entropy by more than $k$ bits, i.e.,

$$H_{\min}(X|Z) \geq H_{\min}(X) - k = n - k \geq n/2.$$

The claim then follows from Proposition 1. □

---

[11] Proposition 1 is a corollary of Theorem 1 of [25].

# References

1. Andreeva, E., Mennink, B., Preneel, B.: On the indifferentiability of the Grøstl hash function. In: Garay, J.A., Prisco, R. (eds.) SCN 2010. LNCS, vol. 6280, pp. 88–105. Springer, Heidelberg (2010). doi:10.1007/978-3-642-15317-4_7
2. Bertoni, G., Daemen, J., Peeters, M., Assche, G.: On the indifferentiability of the sponge construction. In: Smart, N. (ed.) EUROCRYPT 2008. LNCS, vol. 4965, pp. 181–197. Springer, Heidelberg (2008). doi:10.1007/978-3-540-78967-3_11
3. Backes, M., Pfitzmann, B., Waidner, M.: A general composition theorem for secure reactive systems. In: Naor, M. (ed.) TCC 2004. LNCS, vol. 2951, pp. 336–354. Springer, Heidelberg (2004). doi:10.1007/978-3-540-24638-1_19
4. Bellare, M., Rogaway, P.: Random oracles are practical: a paradigm for designing efficient protocols. In: ACM Conference on Computer and Communications Security, pp. 62–73 (1993)
5. Canetti, R., Universally composable security: a new paradigm for cryptographic protocols. In: Proceedings of the 42nd IEEE Annual Symposium on Foundations of Computer Science, FOCS 2001, pp. 136–145. IEEE Computer Society Press, October 2001. Full version, http://eprint.iacr.org/2000/067
6. Canetti, R., Goldreich, O., Halevi, S.: The random oracle methodology, revisited. In: Proceedings of the 30th Annual ACM Symposium on Theory of Computing, STOC 1998, pp. 209–218. ACM (1998)
7. Chang, D., Nandi, M.: Improved indifferentiability security analysis of chopMD hash function. In: Nyberg, K. (ed.) FSE 2008. LNCS, vol. 5086, pp. 429–443. Springer, Heidelberg (2008). doi:10.1007/978-3-540-71039-4_27
8. Coretti, S., Maurer, U., Tackmann, B.: Constructing confidential channels from authenticated channels—public-key encryption revisited. In: Sako, K., Sarkar, P. (eds.) ASIACRYPT 2013. LNCS, vol. 8269, pp. 134–153. Springer, Heidelberg (2013). doi:10.1007/978-3-642-42033-7_8
9. Coron, J.-S., Dodis, Y., Malinaud, C., Puniya, P.: Merkle-Damgård revisited: how to construct a hash function. In: Shoup, V. (ed.) CRYPTO 2005. LNCS, vol. 3621, pp. 430–448. Springer, Heidelberg (2005). doi:10.1007/11535218_26
10. Demay, G., Gaži, P., Hirt, M., Maurer, U.: Resource-restricted indifferentiability. In: Johansson, T., Nguyen, P.Q. (eds.) EUROCRYPT 2013. LNCS, vol. 7881, pp. 664–683. Springer, Heidelberg (2013). doi:10.1007/978-3-642-38348-9_39
11. Dodis, Y., Reyzin, L., Rivest, R.L., Shen, E.: Indifferentiability of permutation-based compression functions and tree-based modes of operation, with applications to MD6. In: Dunkelman, O. (ed.) FSE 2009. LNCS, vol. 5665, pp. 104–121. Springer, Heidelberg (2009). doi:10.1007/978-3-642-03317-9_7
12. Dodis, Y., Ristenpart, T., Steinberger, J., Tessaro, S.: To hash or not to hash again? (In)Differentiability results for $H^2$ and HMAC. In: Safavi-Naini, R., Canetti, R. (eds.) CRYPTO 2012. LNCS, vol. 7417, pp. 348–366. Springer, Heidelberg (2012)
13. König, R., Renner, R.: Sampling of min-entropy relative to quantum knowledge. IEEE Trans. Inf. Theor. **57**, 4760–4787 (2011)
14. König, R., Renner, R., Schaffner, C.: The operational meaning of min- and max-entropy. IEEE Trans. Inf. Theor. **55**, 4337–4347 (2009)
15. Maurer, U.: Indistinguishability of random systems. In: Knudsen, L.R. (ed.) EUROCRYPT 2002. LNCS, vol. 2332, pp. 110–132. Springer, Heidelberg (2002). doi:10.1007/3-540-46035-7_8
16. Maurer, U.: Constructive cryptography - a new paradigm for security definitions and proofs. In: Moedersheim, S., Palamidessi, C. (eds.) TOSCA 2011. LNCS, vol. 6993, pp. 33–56. Springer, Heidelberg (2011)

17. Maurer, U., Renner, R.: Abstract cryptography. In: Chazelle, B. (ed.) The Second Symposium on Innovations in Computer Science, ICS 2011, pp. 1–21. Tsinghua University Press, January 2011

18. Maurer, U., Renner, R., Holenstein, C.: Indifferentiability, impossibility results on reductions, and applications to the random Oracle methodology. In: Naor, M. (ed.) TCC 2004. LNCS, vol. 2951, pp. 21–39. Springer, Heidelberg (2004). doi:10.1007/978-3-540-24638-1_2

19. Maurer, U., Rüedlinger, A., Tackmann, B.: Confidentiality and integrity: a constructive perspective. In: Cramer, R. (ed.) TCC 2012. LNCS, vol. 7194, pp. 209–229. Springer, Heidelberg (2012). doi:10.1007/978-3-642-28914-9_12

20. Maurer, U., Tackmann, B.: On the soundness of authenticate-then-encrypt: formalizing the malleability of symmetric encryption. In: Proceedings of the 17th ACM Conference on Computer and Communication Security (ACM-CCS), pp. 505–515. ACM, October 2010

21. Nisan, N., Zuckerman, D.: Randomness is linear in space. J. Comput. Syst. Sci. **52**, 43–52 (1996)

22. Portmann, C., Matt, C., Maurer, U., Renner, R., Tackmann, B., Boxes, C.: Quantum information-processing systems closed under composition. eprint, arXiv:1512.02240 (2016)

23. Ristenpart, T., Shacham, H., Shrimpton, T.: Careful with composition: limitations of the indifferentiability framework. In: Paterson, K.G. (ed.) EUROCRYPT 2011. LNCS, vol. 6632, pp. 487–506. Springer, Heidelberg (2011). doi:10.1007/978-3-642-20465-4_27

24. Vadhan, S.P.: On constructing locally computable extractors and cryptosystems in the bounded storage model. In: Boneh, D. (ed.) CRYPTO 2003. LNCS, vol. 2729, pp. 61–77. Springer, Heidelberg (2003). doi:10.1007/978-3-540-45146-4_4

25. Wullschleger, J.: Bitwise quantum min-entropy sampling and new lower bounds for random access codes. In: Bacon, D., Martin-Delgado, M., Roetteler, M. (eds.) TQC 2011. LNCS, vol. 6745, pp. 164–173. Springer, Heidelberg (2014). doi:10.1007/978-3-642-54429-3_11

# Foundations

# Fast Pseudorandom Functions
# Based on Expander Graphs

Benny Applebaum[(✉)] and Pavel Raykov

School of Electrical Engineering, Tel-Aviv University, Tel Aviv, Israel
{bennyap,pavelraykov}@post.tau.ac.il

**Abstract.** We present direct constructions of pseudorandom function (PRF) families based on Goldreich's one-way function. Roughly speaking, we assume that non-trivial local mappings $f : \{0,1\}^n \rightarrow \{0,1\}^m$ whose input-output dependencies graph form an expander are hard to invert. We show that this one-wayness assumption yields PRFs with relatively low complexity. This includes weak PRFs which can be computed in linear time of $O(n)$ on a RAM machine with $O(\log n)$ word size, or by a depth-3 circuit with unbounded fan-in AND and OR gates (AC0 circuit), and standard PRFs that can be computed by a quasilinear size circuit or by a constant-depth circuit with unbounded fan-in AND, OR and Majority gates (TC0).

Our proofs are based on a new search-to-decision reduction for expander-based functions. This extends a previous reduction of the first author (STOC 2012) which was applicable for the special case of *random* local functions. Additionally, we present a new family of highly efficient hash functions whose output on exponentially many inputs jointly forms (with high probability) a good expander graph. These hash functions are based on the techniques of Miles and Viola (Crypto 2012). Although some of our reductions provide only relatively weak security guarantees, we believe that they yield novel approach for constructing PRFs, and therefore enrich the study of pseudorandomness.

## 1 Introduction

A pseudorandom function (PRF) is a family of efficiently computable functions with the property that the input-output behavior of a random instance of the family is "computationally indistinguishable" from that of a truly random function. Abstractly, such functions provide a "direct access" to an exponentially long pseudorandom string. Since their discovery by Goldreich, Goldwasser and

A full version of this paper is available in [AR16]. Research supported by the European Union's Horizon 2020 Programme (ERC-StG-2014-2020) under grant agreement no. 639813 ERC-CLC, ISF grant 1155/11, the Blavatnik Interdisciplinary Cyber Research Center and by the Check Point Institute for Information Security. This work was done in part while the first author was visiting the Simons Institute for the Theory of Computing, supported by the Simons Foundation and by the DIMACS/Simons Collaboration in Cryptography through NSF grant CNS-1523467.

© International Association for Cryptologic Research 2016
M. Hirt and A. Smith (Eds.): TCC 2016-B, Part I, LNCS 9985, pp. 27–56, 2016.
DOI: 10.1007/978-3-662-53641-4_2

Micali [GGM86], PRFs have played a central role in cryptography and complexity theory. Correspondingly, the question of minimizing the complexity of PRFs has attracted a considerable amount of attention.

Indeed, apart of being a fundamental object, fast PRFs are strongly motivated by a wide range of applications. Being the core component of symmetric cryptography, highly-efficient PRFs directly imply highly-efficient implementations of Private-Key cryptosystems, Message-Authentication Codes, and Identification Schemes. Fast pseudorandom objects (PRFs and PRGs) can be also used to speed-up several expensive Cryptomania-type applications. For example, secure computation protocols, functional encryption schemes, and program obfuscators that efficiently support a PRF functionality can be bootstrapped with relatively minor cost to general functionalities (cf., [DI05, IKOS08, GVW12, App14]). Interestingly, for these applications parallel-complexity (e.g., circuit depth) seems to be the main relevant complexity measure (affecting round complexity or the number of multilinear levels), while time (e.g., circuit size) is secondary. Another somewhat different motivation comes from the theory of computational complexity. PRFs with low-complexity shed light on the power of low-complexity functions, and partially explain our inability to analyze them. For example, the existence of PRFs in a complexity class $\mathcal{C}$ can be used to show that this class is not PAC-learnable [PW88, Val84] and that certain "natural proof" techniques will fail to prove circuit lower-bounds for functions in $\mathcal{C}$ [RR97]. Last, but not least, identifying the simplest construction of PRFs may provide valuable insights regarding the nature of computational intractability and the way it is achieved by a sequence of cheap and basic operations. This "magic" of hardness which arises from highly-efficient computation can be viewed as the essence of modern cryptography.

Being relatively complicated objects, a considerable research effort has been made to put PRFs on more solid ground at the form of simpler one-wayness assumptions (cf. [GGM86, HILL99, NR95, NR97, NRR00, LW09, BMR10, BPR12]). Annoyingly, the existence of a security reduction seem to incur a cost in efficiency. Indeed, existing theoretical constructions (either based on general primitives or on concrete intractability assumptions) are relatively slow compared to "practical constructions" whose security is based on first-order cryptanalytic principles rather than on a security reduction. As a concrete example, theoretical constructions of PRFs $F_k : \{0,1\}^n \to \{0,1\}^n$ have super-linear (or even quadratic) circuit size. In contrast, Miles and Viola [MV12] presented a candidate PRF which can be computed by a quasilinear circuit of size $\tilde{O}(n)$. (The notation $\tilde{O}(n)$ subsumes polylogarithmic factors.) Similarly, Akavia et al. [ABG+14] proposed a candidate for a weak PRF[1] which can be computed by a constant-depth circuit with unbounded fan-in AND, OR and XOR gates, whereas it is unknown how to construct such a weak PRF based on one-wayness assumption.

Our goal in this paper is to narrow the gap between provably-secure constructions and highly-efficient candidates. We present several constructions of

---

[1] A weak PRF is a relaxation of a PRF which is indistinguishable from a random function for an adversary whose queries are chosen uniformly at random.

pseudorandom functions with low-complexity, and show that their security can be reduced to variants of Goldreich's one-way function. Before introducing our constructions, let us present Goldreich's one-way function. (For more details see the survey [App15].)

## 1.1  Goldreich's One-Way Function

Let $n$ be an input length parameter, $m \geq n$ be an output length parameter and $d \ll n$ be a locality parameter. For a $d$-local predicate $P : \{0,1\}^d \rightarrow \{0,1\}$ and a sequence $G = (S_1, \ldots, S_m)$ of $d$-tuples over the set $[n] := \{1, \ldots, n\}$, we let $f_{G,P} : \{0,1\}^n \rightarrow \{0,1\}^m$ denote the mapping

$$z \mapsto (P(z[S_1]), \ldots, P(z[S_m])),$$

i.e., the $i$-th output bit is computed by applying the predicate $P$ to the input bits which are indexed by the $i$-th tuple $S_i$. Goldreich [Gol00] conjectured that for $m = n$ and possibly small value of $d$ (e.g., logarithmic or even constant), the function $f_{G,P}$ is one-way as long as the set system $(S_1, \ldots, S_m)$ is "highly expanding" and the predicate $P$ is sufficiently "non-degenerate". We elaborate on these two requirements.

*Expansion.* To formalize the expansion property let us think of $G = (S_1, \ldots, S_m)$ as a $d$-uniform hypergraph with $m$ hyperedges (which correspond to the outputs) over $n$ nodes (which correspond to the inputs). The expansion property essentially requires that every not-too-large subset of hyperedges is almost pair-wise disjoint. Formally, for a threshold $r$, the union of every set of $\ell \leq r$ hyperedges $S_{i_1}, \ldots, S_{i_\ell}$ should contain at least $(1 - \beta)d\ell$ nodes, i.e., $|\bigcup_{j=1}^{\ell} S_{i_j}| \geq (1 - \beta)d\ell$, where $\beta$ is some constant smaller than $\frac{1}{2}$ (e.g., 0.1).

*Secure Predicates.* A noticeable amount of research was devoted to studying the properties of "secure" predicates accumulating in several algebraic criteria (cf., [Ale03, MST03, ABW10, BQ12, ABR12, OW14, FPV15]). It is known for example, that in order to support an output length of $m = n^c$ the predicate $P$ must have *resiliency* of $k = \Omega(c)$, i.e., $P$ should be uncorrelated with any GF(2)-linear combination of at most $k$ of its inputs. Additionally, the predicate $P$ must have algebraic degree (as a GF(2) polynomial) of at least $c$. Moreover, $P$ must have high *rational degree* in the following sense: any polynomial $Q$ whose roots cover the roots of $P$ or its complement must have algebraic degree of $\Omega(c)$ [AL15]. An example for such a predicate (suggested in [AL15]) is the $d$-ary XOR-MAJ$_d$ predicate which partitions its input $w = (w_1, \ldots, w_d)$ into two parts $w_L = (w_1, \ldots, w_{\lfloor d/2 \rfloor})$ and $w_R = (w_{\lfloor d/2 \rfloor + 1}, \ldots, w_d)$, computes the XOR of the left part and the majority of the right part, and XOR's the results together.[2] This predicate achieves resiliency of $d/2$ and rational degree of $d/4$ and therefore seems to achieve security for $m = n^{\Omega(d)}$ outputs.

---

[2] In fact, it seems better to allocate a larger fraction of the inputs to the Majority part. See [AL15].

*Security.* Intuitively, large expansion (together with high resiliency) provide security against local algorithms that employ some form of divide-and-conquer approach. Due to the expansion of the input-output hypergraph, any small subset of the outputs gives very little information on the global solution $x$. High rational degree provides security against more global approaches which rely on different forms of linearization and algebraic attacks. These intuitions were formalized and proved for several classes of algorithms in previous works (cf. [AHI05, ABW10, CEMT14, ABR12, BR13, OW14]). Following these works, we make the following strong version of Goldreich's conjecture:

**Assumption 1 (Expander-based OWFs (Informal)).** *For some universal constant $\alpha \in (0,1)$ and every d-uniform hypergraph $G$ with $n$ nodes and $m < n^{\alpha d}$ hyperedges which is expanding for sets of size $r = n^{\Omega(1)}$, the function $f_{G,\text{XOR-MAJ}_d}$ cannot be inverted in polynomial time.*[3]

This assumption is consistent with known attacks. In fact, hardness results (against limited families of attacks) suggest that inversion is hard even for adversaries of complexity $\exp(r)$ where $r$ is the expansion threshold. We refer to this variant as the strong EOWF assumption. We further mention that although previous works mainly focused on the case where the locality $d$ is constant or logarithmic in $n$ (which is going to be our main setting here as well), it seems reasonable to conjecture that the assumption holds even for larger values of $d$ (e.g., $d = n^\delta$ for constant $\delta \in (0,1)$). Finally, we note that the expansion requirement implicitly puts restrictions on the values of $n, m$ and $d$. Roughly speaking, an expansion of $r = n^{1-\beta}$ requires $\Theta(1/\beta^2) \leq d \leq n^{\Theta(\beta)}$ and restricts $m$ to be at most $n^{\Theta(d\beta^2)}$.

## 1.2 Results and Techniques

We present several constructions of expander-based PRFs.

**Weak PRF.** Let $P$ be some $d$-ary predicate (e.g., XOR-MAJ$_d$). In our first construction $F_1$, we think of the input $x \in \{0,1\}^n$ as specifying a hypergraph $G_x$ and let the output $y$ be the value of $f_{G_x,P}$ applied to the collection key $k \in \{0,1\}^n$. Namely, we think of the data $x$ as specifying a computation that should be applied to $k$. The hypergraph $G_x$ is defined in the natural way: Partition $x$ to $(d \log n)$-size substrings, and view each substring as a $d$-tuple of elements in $[n]$ where each element is given in its binary representation. An adversary that makes $q$ queries $x_1, \ldots, x_q$ essentially sees the value of $f_{G,P}(k)$ where $G = \bigcup G_{x_i}$. When

---

[3] In Sect. 2 we provide a more general assumption which allows the hypergraph to be non-uniform, and is parameterized by an expansion parameter, by a predicate family $\mathcal{P}$ and by a concrete bound on the security of the function in terms of the (circuit) size of the adversary and its success probability. The above assumption is given here in a simplified form for ease of presentation.

the adversary is allowed to choose the queries, the outcome cannot be pseudo-random (think of the case where $G_{x_1}$ and $G_{x_2}$ share the same hyperedge). However, when the queries $x_1, \ldots, x_q$ are chosen at random (as in the setting of a weak PRF), the resulting hypergraph $G$ is a random hypergraph which is likely to be expanding. At this point, we can employ a search-to-decision reduction from [App13], which shows that for random hypergraphs $G$, one-wayness implies pseudorandomness. It follows that, for a proper choice of parameters (e.g., $d = \Omega(\log n)$), our assumption implies that the function $F_1$ is a weak PRF.[4]

This construction can be instantiated with different locality parameters $d$, ranging from $O(\log n)$ to $n^\delta$. In the logarithmic regime, this gives rise to a construction $F_1 : \{0,1\}^n \to \{0,1\}^{n/\log^2 n}$ which is computable in linear time of $O(n)$ on a RAM machine with $O(\log n)$ word size. Additionally, this function can be computed, for any fixed key $k$, by a depth-3 circuit with unbounded fan-in AND and OR gates (i.e., an $\mathbf{AC^0}$ circuit).[5] To the best of our knowledge this is the first construction of a weak PRF that achieves such efficiency guarantees.

*Concrete Security and Application to Learning.* The (strong) EOWF assumption implies that $F_1$ resists almost-exponential size adversaries (computable by circuits of size $t = \exp(n^{1-\beta})$ for any $\beta > 0$ as long as they make only $q = n^{O(d)}$ queries to the function. Hence, logarithmic locality provides only security against a quasi-polynomial number of queries (e.g., $\exp(\text{polylog}(n))$. Similarly, the distinguishing advantage of the adversary is only quasi-polynomial $\varepsilon = \exp(-\text{polylog}(n))$. While this setting of parameters may seem too weak for many cryptographic applications, it provides a useful theoretical insight. The classical learning algorithm of Linial, Mansour and Nisan [LMN93] shows that any $\mathbf{AC^0}$-computable weak PRF can be broken either with quasipolynomial distinguishing advantage or by making quasipolynomial number of queries. (In the computational learning terminology, $\mathbf{AC^0}$ functions are PAC-learnable under the uniform distribution using a quasipolynomial number of samples and time, or weakly learnable in polynomial-time with advantage $1/\text{polylog}(n)$ over $\frac{1}{2}$.) The LMN algorithm relies on the Fourier spectrum of $\mathbf{AC^0}$ functions, and the possibility of improving it to a polynomial-time algorithm is considered to be an important open problem in learning theory. Our construction suggests that this is impossible even for depth-3 circuits, and so the Fourier-based algorithm of [LMN93] is essentially optimal. To the best of our knowledge, this is the first hardness result for learning depth-3 $\mathbf{AC^0}$ circuits over the uniform distribution. Previous hardness results either apply to $\mathbf{AC^0}$ circuits of depth $d$ for *large* (unspecified) constant depth $d$ [Kha93], to depth-3 *arithmetic* circuits [KS09], or to depth-2 $\mathbf{AC^0}$ circuits but over a *non-uniform* distribution [ABW10, DLS14].

---

[4] Formally, Assumption 1 implies that for a random hypergraph $G$, the function $f_{G,P}$ is one-way (since such a hypergraph is likely to be expanding). Then, we can apply the result of [App13].

[5] When analyzing parallel-complexity it is common to restrict the attention for the case where the key is fixed, cf. [NR95, NR97, NRR00, LW09, BMR10, MV12, ABG+14].

**Reducing the Distinguishing Advantage.** Our second construction attempts to strengthen the distinguishing advantage $\varepsilon$ of $F_1$. In $F_1$ the hypergraph $G = \bigcup G_{x_i}$ fails to be expanding with quasipolynomial probability, and in this case pseudorandomness may be easily violated. As a concrete example note that, with probability $\Omega(n^{-d})$, the hypergraph $G$ contains a pair of identical hyperedges $S_i = S_j$, and so the corresponding outputs will be identical, and distinguishing (with constant advantage) becomes trivial.

Following [CEMT14], we observe that, although expansion is violated with quasipolynomial small probability, not all is lost, and, except for a tiny (almost exponentially small) probability, the hypergraph $G$ is *almost expanding* in the sense that after removing a small (say sub-linear) amount of hyperedges the remaining hypergraph is expanding. We use this combinatorial structure to argue that $f_{G,P}(k)$ can be partitioned into two functions $f_1$ and $f_2$, where the input-output hypergraph $G_1$ of $f_1$ is highly expanding and the function $f_2$ depends only on a relatively small (sub-linear) number of inputs. As a result we can show that, for such an almost-expander $G$, the distribution $f_{G,P}(U_n)$ is pseudorandom except for small number of "bad outputs".[6] In fact, the number of "bad outputs" is small enough to argue that each block of $f_{G,P}(U_n)$ (corresponding to the $i$-th query) has a large amount of "pseudoentropy". Hence, we can get a pseudorandom output (even for almost expanding hypergraphs) by adding a postprocessing stage in which a randomness extractor is applied to the output of $F_1$ (i.e., extraction is performed separately per each block of $f_{G,P}(U_n)$).

Formally, our second construction $F_2$ is keyed by a pair of $n$-bit strings $(k, s)$, and for a given input $x$, we output the value $\text{Ext}_s(f_{G_x,P}(k))$ where Ext is a strong seeded randomness extractor. Since there are linear-time computable extractors [IKOS08], the construction can be still implemented by a linear-time RAM machine. Moreover, since the extractor can be computed by a linear function (and therefore by a single layer of unbounded fan-in parity gates), the function $F_2$ can be computed by a constant-depth circuit with unbounded fan-in AND, OR and XOR gates (or even in $\mathbf{MOD}_2 \circ \mathbf{AC^0}$). We prove that the distinguishing advantage of the construction is almost exponentially-small. We do not know whether $F_2$ provides security against larger (say subexponential) number of queries, and leave it as an open question.

**Handling Non-random Inputs.** Our next goal is to move from the weak PRF setting in which the function is evaluated only over random inputs, to the standard setting where the queries can be chosen by the adversary.[7] It is natural to try to achieve this goal by introducing a preprocessing mapping $M$ that maps an input $x$ to a hypergraph $M(x)$ with the property that every set

---

[6] Technically, this requires an extension of our assumption to the case of non-uniform hypergraphs, and the ability to analyze the function with respect to new predicates (obtained by restricting some of the inputs of the original predicate).

[7] We do not use general transformations from weak PRFs to standard PRFs (e.g., [NR97]) since they make a linear number of calls to the underlying weak PRF and therefore incur at least a quadratic overhead in the size of the resulting circuit.

of $q$ queries $x_1 \ldots, x_q$ form together a hypergraph $G = \bigcup_i M(x_i)$ with good expansion properties. This approach faces two challenges. First, it is not clear at all how to implement the mapping $M$ (let alone in a very efficient way). Second, we can no longer rely on the standard search-to-decision reduction from [App13] since it applies only to randomly chosen hypergraphs (as opposed to arbitrary expanders).

*Search-to-Decision Reduction for Expander-Based Functions.* We solve the second challenge, by proving a new search-to-decision reduction that applies directly to expander hypergraphs. Namely, we show that if $f_{G,P}$ is one-way for every expander hypergraph $G$ (as conjectured by in Assumption 1) then it is also pseudorandom for every expander hypergraph. Technically, the original reduction of [App13] shows that if an adversary $A$ can distinguish $f_{G,P}(U_n)$ from a truly random string, then there exists an adversary $B$ that inverts $f_{H,P}(U_n)$ where $G$ and $H$ are random hypergraphs (with polynomially related parameters). This reduction strongly exploits the ability of $A$ to attack many different hypergraphs $G$. Roughly speaking, every attack on a hypergraph $G_i$ is translated into a small piece of information on the input $x$ (i.e., a noisy estimation on some bit $x_i$), and by accumulating the information gathered from different $G_i$'s the input $x$ is fully recovered.[8]

In contrast, in the new search-to-decision theorem we are given a distinguisher $A_G$ which succeeds only over some *fixed* expanding hypergraph $G$. First, we observe that one can slightly modify $G$ and define, for every index $i \in [n]$, a hypergraph $G_i$ such that given $y = f_{G_i}(x)$ the attacker $A_G$ can be used to obtain an estimation for the $i$-th bit of $x$. (This is already implicit in [App13].) One may therefore try to argue that the function $f_{\bigcup_i G_i, P}(x) = (f_{G_1}(x), \ldots, f_{G_n}(x))$ can be inverted by calling $A_G$ for each block separately. This is problematic for two reasons: (1) inversion may fail miserably since the calls to $A_G$ are all over statistically-dependent inputs (the same $x$ is being used); and (2) the resulting hypergraph $H = \bigcup_i G_i$ is non-expanding (due to the use of almost identical copies of the same hypergraph $G$), and so inversion over $H$ does not contradict the theorem.

Fortunately, both problems can be solved by randomizing each of the $G_i$'s (essentially by permuting the names of the inputs). By concatenating the randomized $G_i$'s, we get a probability distribution $\mathcal{D}(G)$ over hypergraphs which satisfies the following two properties: (1) a random hypergraph $H \overset{R}{\leftarrow} \mathcal{D}(G)$ is typically a good expander; and (2) Inverting $f_{H,P}$ for a random $H \overset{R}{\leftarrow} \mathcal{D}(G)$ reduces to inverting $f_{G,P}$. Since we work in a non-uniform model of adversaries (circuits), this suffices to prove the theorem. (See Sect. 3 for details.)

*Mapping Inputs to Expanders.* Going back to the first challenge, we still need to provide a mapping $M(x)$ which, when accumulated over different inputs, results

---

[8] An analogous use of public randomness appears in the seminal Goldreich-Levin theorem [GL89] which can be viewed as search-to-decision reduction for the keyed function $f_k(x) = (g(x), \langle x, k \rangle)$.

in a highly expanding hypergraph. Note that although $M$ operates on $n$-bit inputs, it should satisfy a global property that applies to collection of super-polynomial (or even exponential) number of inputs. Unfortunately, we do not know how to obtain such a mapping deterministically with a low computational cost. Instead, we show how to provide a family of mappings $M_\sigma$ with the property that for every fixed sequence of inputs $x_1, \ldots, x_q$ and for a random $\sigma$, the hypergraph $G = \bigcup_i M_\sigma(x_i)$ is highly expanding with all but exponentially small probability. The key idea is to note that in order to guarantee expansion for $r$-size sets, it suffices to make sure that each set of $r$ hyperedges of $G$ is (almost) uniformly distributed. This means that $M_\sigma$ should satisfy the following form of pseudorandomness: For a random $\sigma$, every subset of $R = rd\log(n)$ bits of the random variable $(M_\sigma(x))_{x \in \{0,1\}^n}$ should be statistically-close to uniform. This setting is somewhat non-standard: Efficiency is measured with respect to a single invocation of $M_\sigma$ (i.e., the complexity of generating a block of $m$ hyperedges), but pseudorandomness should hold for any set of $r$ hyperedges ($R$ bits) across different invocations.

We construct such a mapping $M_\sigma$ by tweaking a construction of Miles and Viola [MV12]. We view $\sigma \in \{0,1\}^{2n}$ as a pair of $GF(2^n)$ elements $\sigma_1, \sigma_2$, and map an input $x \in GF(2^n)$ to the $GF(2^n)$-element $(x + \sigma_1)^{-1} \cdot \sigma_2$. (The statistical analysis of $M_\sigma$ appears in Sect. 4.3.) The resulting function $F_3$ is keyed by $(k, \sigma, s)$ and for an input $x$ it outputs the value $\mathrm{Ext}_s(f_{M_\sigma(x),P}(k))$ where $M_\sigma(x)$ is parsed as a $d$-uniform hypergraph with $m = n/(d\log n)$ hyperedges and $d$ is treated as a parameter. Due to the high efficiency of $M$ (which consists of a single multiplication and a single inversion over $GF(2^n)$), the function $F_3$ can be computed by a quasilinear circuit $\tilde{O}(n)$ or by a constant-depth circuit with unbounded fan-in AND, OR, and Majority gates (i.e., $\mathbf{TC^0}$ circuit), for any choice of the locality parameter $d$.

The use of keyed mapping, allows us to prove security against a non-adaptive adversary whose $i$-th query is independent of the answers for the previous queries. We do not know whether the construction remains secure for adaptive adversaries, however, using the non-adaptive to adaptive transformation of [BH15], we can turn our function into a standard PRF without increasing the asymptotic cost of the construction (in terms of size and depth). We mention that the parallel complexity (i.e., $\mathbf{TC^0}$) seems essentially optimal for PRF and it matches the complexity of the best known PRF constructions based on number-theoretic or lattice assumptions [NR95, NR97, NRR00, BPR12].

*Concrete Security.* Recall that the locality parameter $d$ can vary from logarithmic to $n^\delta$ for some $\delta \in (0, 1)$. To get an expansion for sets of size $n^{1-\beta}$ (and therefore security against $\exp(n^{1-\beta})$-size circuits), we must restrict the number of queries $q$ to be smaller than $n^{d\beta^2}$. In addition, the locality $d$ should satisfy $4/\beta^2 < d < n^{\beta/4}$. Hence, polynomial locality $d = n^\delta$ allows to support sub-exponential number of queries while providing security against sub-exponential size circuits with respect to sub-exponential distinguishing advantage. Note that polynomial locality has also some effect on efficiency: The number of output bits per invocations decreases to $\tilde{O}(n/d)$ and so the computational cost per output bit is $\tilde{O}(d) = \tilde{O}(n^\delta)$. On the other extreme, a logarithmic value of $d$ achieves an almost-optimal complexity per bit (i.e., $\tilde{O}(1)$), and provides security against circuits of almost-exponential size ($\exp(n^{1-\beta})$ for every $\beta > 0$) which make a quasipolynomial number of queries.

*Security Beyond Expansion.* We do not know whether our analysis is tight. To the best of our knowledge, $F_3$ with logarithmic locality may achieve security even in the presence of sub-exponentially many queries. We remark that our analysis is somewhat pessimistic since it essentially assumes that the seed $s$ of the extractor and the seed $\sigma$ of the preprocessing mapping are both given to the adversary. Indeed, in this case the adversary sees the underlying hypergraph and, after sufficiently many queries, it can exploit its non-expanding properties. In contrast, when $s$ and, more importantly, $\sigma$ are not given, the adversary does not get a direct access to the hypergraph. One may assume that as long as $M$ somewhat hides the hypergraph $G$, lack of expansion cannot be used to break the system. The question of identifying the right (and minimal) notion of hiding remains open for future research.[9]

## 1.3   Related Candidate PRFs

It is instructive to compare the structure of our constructions to three somewhat related candidates for PRFs.

*The BFKL Candidate Weak-PRF* [BFKL93] Blum et al. conjectured that the function

$$f_{A,B} : x \mapsto \left( \bigoplus_{i \in A} x_i \right) \oplus \left( \mathrm{MAJ}_{j \in B}(x_j) \right),$$

is a weak PRF[10], where the key $(A, B)$ is a random pair of logarithmic size sets $A, B \subseteq [n]$. That is, the function $f_{A,B}$ takes an $n$-bit vector $x$, computes the parity of the bits of $x$ which are indexed by $A$ and the majority of the bits which are indexed by $B$, and outputs the XOR of the two results. This candidate is essentially dual to our first suggestion. Here the sets $A$ and $B$ are used as a secret key and the XOR-MAJ predicate is applied to a public random $x$ (the input to the weak PRF). In contrast, we use $x$ as a key (and keep it private) and let the input specify the graph structure. Observe that, unlike our construction, the key of Blum et al. can be described by a string of length $n^{O(\log n)}$ and so it can be broken in quasi-polynomial time and polynomially many samples. In contrast, we conjecture that, in the presence of polynomially many samples, our constructions resist attacks of sub-exponential (or even "almost" exponential) complexity of $\exp(n^{1-\beta})$.

---

[9] It is not hard to show that if $M$ by itself is a PRF then security holds for $F_3$. The hope is to get somewhat weaker form of hiding, ideally, one which can be satisfied by some concrete and highly-efficient mapping $M$ such as the one proposed here.

[10] In the terminology of learning theory this means that a random function from the family is hard to weakly-predict over the uniform distribution.

*Goldreich's Suggestion [Gol00].* In the paper which introduced the expander-based one-way functions (leading to Assumption 1), Goldreich suggested to construct a pseudorandom function by iterating the basic (length-preserving) OWF $f_{G,P} : \{0,1\}^n \to \{0,1\}^n$ a logarithmic number of times and letting the (secret) key specify the sequence of randomly chosen predicates. This construction yields a candidate PRF of circuit complexity $O(n \log n)$ and logarithmic depth. Analyzing the security of this candidate was left as an interesting open question.

*A Suggestion by Gowers [Gow96].* Gowers conjectured that, for sufficiently large polynomial $m(n)$, a random $m(n)$-depth Boolean circuit is a PRF. More accurately, each level of the circuit contains $n$ wires and a single gate $P : \{0,1\}^3 \to \{0,1\}^3$. For each level $\ell$ we select three random indices $(i, j, k) \in [n]$ and use the corresponding wires in the $\ell$-th layer as the incoming wires to the $\ell$-th gate, the output values of the gate are connected to the wires $(i, j, k)$ located at the next level. (All other wires simply copy the previous values to the next layer). When the gate $P$ computes a permutation (over three bits) the resulting circuits computes a permutation over $n$-bits. Letting the key consists of the description of the circuit (i.e., the wiring of the gates), yields a candidate pseudorandom permutation. Moreover, Gowers proved that the resulting collection is $\ell$-wise independent after $m = \text{poly}(n, \ell)$ levels. (The polynomial dependency in $n$ and $\ell$ was improved by [HMMR05, BH08].)

Unlike the constructions presented in this paper, it is currently unknown how to base any of the above candidates on a one-wayness assumption. Interestingly, all the above candidates (as well as the candidates of Miles and Viola [MV12] and Akavia et al. [ABG+14]) can be naturally viewed as letting the key $k$ specify a "simple" function $F_k$ which is then applied to the (public) input $x$. In contrast, in our construction every public input $x$ specifies a simple function $f_x$ that is applied to the key $k$. This approach is conceptually similar to the structure of the classical GGM construction [GGM86] which uses the input $x$ to specify a circuit (whose building blocks are length-doubling pseudorandom generators) that is applied to the key.

## 1.4    Conclusion

We presented several elementary constructions of pseudorandom functions. All our constructions follow a similar template: The input $x$ is mapped to a hypergraph $G_x$, which represents a simple (essentially single-layered) circuit $f_{G_x,P}$, the resulting circuit is applied to the key $k$, and the output is fed through some randomness extractor. We believe that this structure provides a new methodology for constructing pseudorandom functions which deserves to be further studied.

Following Goldreich, we conjecture that as long as the input-output relations is expanding the computation is hard to invert. We further show that such one-wayness leads to pseudorandomness by extending the techniques of [App13]. We believe that understanding this assumption, or more generally, relating the combinatorial structure of circuits to their cryptographic properties is a key question, which may eventually lead to faster and highly secure PRFs. Our proofs, which fall short of providing optimal security (in some cases they are very far from that), should be viewed as a first step in this direction.

Finally, we believe that the tools developed here (e.g., pseudoranodmness over imperfect expanders, the expander-based search-to-decision reduction, and the expander-generating hash function $M$) will turn out to be useful for future works in the field.

## 1.5 Organization

We begin with some standard preliminaries along with a basic hypergraph notation in Sect. 2. In Sect. 3 we give the new search-to-decision reduction that applies to arbitrary expander hypergraphs. The PRF constructions are described in Sect. 4.

## 2 Preliminaries

*General Preliminaries.* We let $[n]$ denote the set $\{1, \ldots, n\}$. For a string $x \in \{0,1\}^n$ and $i \in [n]$, we let $x[i]$ denote the $i$  bit of $x$. For a tuple $S = (i_1, \ldots, i_d)$, we let $x[S] = x[i_1, \ldots, i_d]$ denote the restriction of $x$ to indices in $S$, i.e., the string $x[i_1] \ldots x[i_d]$. For strings $x_1, \ldots, x_q$ we write $(x_i)_{i=1}^q$ to denote the concatenation of the strings $x_1 || \cdots || x_q$. We write $\log_d n$ to denote the logarithm of $n$ base $d$, if $d = 2$ we omit writing it explicitly. A function $\varepsilon(\cdot)$ is said to be negligible if $\varepsilon(n) < n^{-c}$ for any constant $c > 0$ and sufficiently large $n$. We will sometimes use $\mathrm{neg}(\cdot)$ to denote an unspecified negligible function. For a function $t(\cdot)$, we write $t = \tilde{O}(n)$, if $t = O(n \log^k(n))$ for some $k \in \mathbb{N}$.

*Probabilistic Notation.* For a probability distribution or random variable $X$ (resp., set), we write $x \xleftarrow{R} X$ to denote the operation of sampling a random $x$ according to $X$ (resp., sampled uniformly from $X$). We let $U_n$ (resp., $U_S$) denote a random variable uniformly distributed over $\{0,1\}^n$ (resp., over the set $S$). We write $\mathrm{supp}(X)$ to denote the support of the random variable $X$, i.e., $\mathrm{supp}(X) = \{x \mid \Pr[X = x] > 0\}$. The statistical distance between two probability distributions $X$ and $Y$, denoted $\Delta(X; Y)$, is defined as the maximum, over all functions $A$, of the distinguishing advantage $\Delta_A(X, Y) := |\Pr[A(X) = 1] - \Pr[A(Y) = 1]|$. We say that $X$ is $\varepsilon$-*statistically indistinguishable* from $Y$ if $\Delta(X; Y) \leq \varepsilon$ and write $X \overset{s}{\equiv}_\varepsilon Y$. The random variable $X$ is $(t, \varepsilon)$-*computationally indistinguishable* from $Y$ if for every circuit $A$ of size $t$, the distinguishing advantage $\Delta_A(X, Y)$ is at most $\varepsilon$, and we write $X \overset{c}{\equiv}_{t,\varepsilon} Y$.

*Cryptographic Primitives.* A random variable $X$ over $n$-bit strings is called $(t, \varepsilon)$-*pseudorandom* if $X \overset{c}{\equiv}_{t,\varepsilon} U_n$. A function $f : \{0,1\}^n \to \{0,1\}^m$ is $(t, \varepsilon)$ one-way if for every $t$-size adversary $A$ it holds that $\Pr_x[A(f(x)) \in f^{-1}(f(x))] < \varepsilon$.

**Definition 1 (PRF).** *A keyed function* $f : \mathcal{K} \times \mathcal{X} \to \mathcal{Y}$ *is called* $(q, t, \varepsilon)$-*pseudorandom if for any t-size circuit* $D^{(\cdot)}$ *aided with q oracle gates, the distinguishing advantage*

$$\left| \Pr_{k \xleftarrow{R} \mathcal{K}} [D^{f_k} = 1] - \Pr_{h \xleftarrow{R} \mathcal{H}} [D^h = 1] \right| \leq \varepsilon,$$

*where* $\mathcal{H}$ *is a set of all functions mapping inputs from* $\mathcal{X}$ *to* $\mathcal{Y}$. *An adversary is called* non-adaptive *if it generates all the queries at the beginning independently of the received responses from the oracle gates.*

A $(q, t, \varepsilon)$-PRF family is a sequence of keyed functions $\mathcal{F} = \{f_n : \mathcal{K}_n \times \mathcal{X}_n \to \mathcal{Y}_n\}$ equipped with an efficient key sampling algorithm and an efficient evaluation algorithm where each $f_n$ is $(q(n), t(n), \varepsilon(n))$-pseudorandom. We say that $\mathcal{F}$ is a $(q, t, \varepsilon)$ non-adaptive PRF (resp., weak PRF) if the above holds for non-adaptive adversaries (resp., for adversaries such that each of their queries is chosen independently and uniformly from $\mathcal{X}_n$).

*Low-Bias Generators.* We employ the following notions of low-bias and bitwise-independence generators. As in the case of PRFs, we view a two-argument function $f(k, x)$ as a keyed function whose first argument $k$ serves as a key. We emphasize this distinction by writing $f_k(x)$ for $f(k, x)$.

**Definition 2.** *Let* $g : \{0, 1\}^{\kappa} \times \{0, 1\}^m \to \{0, 1\}^n$ *be a keyed function. For* $x \in \{0, 1\}^m$, *let* $Y(x)$ *denote the random variable* $g_k(x)$ *induced by* $k \xleftarrow{R} \{0, 1\}^{\kappa}$, *and let* $\mathbf{Y}$ *denote the random variable* $(Y(x))_{x \in \{0,1\}^n}$ *where the same random key is used for all* $x$'s. *We say that* $g$ *is:*

- $(t, \varepsilon)$-*bitwise independent if every t-bit subset of* $\mathbf{Y}$ *is* $\varepsilon$-*close to uniform (in statistical distance), i.e., for every* $\ell \leq t$ *distinct indices* $i_1, \dots, i_\ell$ *we have that*

$$\Delta(U_\ell; (\mathbf{Y}[i_j])_{j=1}^\ell) \leq \varepsilon.$$

- $(t, \varepsilon)$-*biased over* GF(2) *if for every* $\ell \leq t$ *distinct indices* $\{i_1, \dots, i_\ell\}$, *we have that*

$$\left| \Pr \left[ \sum_{j=1}^\ell \mathbf{Y}[i_j] = 1 \right] - \frac{1}{2} \right| \leq \varepsilon,$$

*where the sum is computed over* GF(2).

- $(t, \varepsilon)$-*linear-fooling over* GF($2^n$) *if for every* $t$ *outputs* $Y(x_1), \dots, Y(x_t)$ *(parsed as elements of* GF($2^n$)*) of distinct* $x_1, \dots, x_t$, *every* $t$ *constants* $b_1, \dots, b_t$ *from* GF($2^n$) *(that are not all equal to zero), we have that*

$$\Delta \left( \sum_{i=1}^t b_i Y(x_i) \; ; \; U_{\mathrm{GF}(2^n)} \right) \leq \varepsilon.$$

*Sources and Extractors.* The *min-entropy* of a random variable $X$ is defined to be $\min_{x \in \text{supp}(X)} \log \frac{1}{\Pr[X=x]}$ and is denoted by $H_\infty(X)$. A keyed function $E : \mathcal{S} \times \mathcal{X} \to \mathcal{Y}$ is a *strong $(k, \varepsilon)$-extractor* if for every distribution $X$ over $\mathcal{X}$ with $H_\infty(X) \geq k$, it holds that $\Delta((s, \text{Ext}_s(x)) \ ; \ (s, U(\mathcal{Y}))) \leq \varepsilon$, where $s \overset{R}{\leftarrow} \mathcal{S}$, $x \overset{R}{\leftarrow} X$ and $\Delta(\cdot; \cdot)$ stands for statistical distance.

We consider the following notion of random sources that can be viewed as a convex combination of the traditional bit-fixing sources [CGH+85].

**Definition 3 (Generalized Bit-Fixing Source).** *A distribution $X$ over $\{0,1\}^n$ is a generalized $k$-bit-fixing source if there exist $k$ distinct indices $S$ such that $X[S]$ is distributed like $U_k$ and $X[[n]\backslash S]$ is independent from $X[S]$.*

We use the following simple lemma (whose proof is deferred to the full version [AR16]).

**Lemma 1.** *Let $\text{Ext}$ be a strong $(m - r, \delta)$-extractor for $m$-bit sources. Let $Z = Z_1 || \cdots || Z_q$ be a generalized $(qm-r)$-bit-fixing source, where each $|Z_i| = m$. Then for a uniformly chosen seed $s$, the random variable $(s, \text{Ext}_s(Z_1), \ldots, \text{Ext}_s(Z_q))$ is $(q \cdot \delta)$-statistically indistinguishable from uniform.*

*Hypergraphs.* An $(n, m)$-hypergraph $G$ is a hypergraph over vertices $[n]$ with $m$ hyperedges $(S_1, \ldots, S_m)$ where each hyperedge is viewed as a tuple $(i_1, \ldots, i_k)$, i.e., it is ordered and may contain duplications. It is sometimes convenient to think of a hypergraph $G$ as a bipartite graph, where the $n$ vertices represent the lower layer of the graph, the hyperedges represent the upper layer of the graph such that each hyperedge $S = (i_1, \ldots, i_k)$ is connected to the vertices $i_1, \ldots, i_k$. We say that $G$ is $d$-uniform (denoted by $(n, m, d)$-hypergraph) if all the hyperedges are of the same cardinality $d$. $G$ is *almost $d$-uniform* (denoted by $[n, m, d]$-hypergraph) if $d/2 < |S_i| \leq d$ for all $i \in [m]$. We let $\mathcal{G}_{n,m,d}$ denote the probability distribution over $(n, m, d)$-hypergraphs in which each of the $m$ hyperedges is chosen independently and uniformly at random from $[n]^d$. We say that a distribution over $(n, m, d)$-hypergraphs is $(k, \varepsilon)$-*random* if any $k$ hyperedges are $\varepsilon$-close (in statistical distance) to the uniform distribution $\mathcal{G}_{n,k,d}$. A distribution over hypergraphs is $(r, d, \varepsilon)$-random if any $s \leq r$ hyperedges $S_1, \ldots, S_s$ contain at least $sd$ entries that are $\varepsilon$-close to uniform.

For a set of hyperedges $T = \{S_1, \ldots, S_k\}$ we write $\Gamma(T)$ to denote the union of tuples $S_1, \ldots, S_k$ (where the union of tuples is naturally defined to be the set of all indices occuring in $S_1, \ldots, S_k$). Let $G\backslash T$ denote the hypergraph obtained from $G$ by removing hyperedges $T$ and updating the remaining hyperedges by deleting from them vertices that belong to $\Gamma(T)$. A hypergraph $G$ is an $(r, c)$-expander if for any set $I$ of hyperedges of size at most $r$ we have $\Gamma(I) \geq c|I|$. We refer to $r$ as "the expansion threshold" and to $c$ as "the expansion factor". A hypergraph $G$ is an $r_{\text{bad}}$-*imperfect $(r, c)$-expander* if there exists a subset of $G$'s hyperedges $I_{\text{bad}}$ of size $|I_{\text{bad}}| \leq r_{\text{bad}}$ such that $G\backslash I_{\text{bad}}$ is an $(r, c)$-expander.

It is well known that a random hypergraph is likely to be highly expanding. The following lemma (whose proof is deferred to the full version [AR16]) generalizes this fact to the case of $(r, d, \varepsilon)$-random hypergraphs and to the case of imperfect expansion. (Note that the failure probability drops down exponentially with the size of the imperfectness parameter $t$.)

**Lemma 2.** *Let $\beta$ be a constant in $(0,1)$ and $d \in \mathbb{N}$ such that $4/\beta^2 \leq d \leq n^{\beta/4}$. Let $r = n^{1-\beta}$ and $m \leq n^{d\beta^2/4}$. Let $t = t(n)$ be a non-negative function such that $t \leq r$. Then, a $(r+t, d, 2^{-\Omega(n)})$-random $(n,m)$-hypergraph $G$ is $t$-imperfect $(r, (1-\beta)d)$-expander except with probability $n^{-(t+1)d\beta^2/10}$.*

The *union* of an $(n, m_1)$-hypergraph $G = (S_1, \ldots, S_{m_1})$ and $(n, m_2)$-hypergraph $H = (R_1, \ldots, R_{m_2})$ is the $(n, m_1 + m_2)$-hypergraph $J = G \cup H$ whose hyperedges are $(S_1, \ldots, S_{m_1}, R_1, \ldots, R_{m_2})$. Since union is an associative operation, the union of $q$ hypergraphs $G_1 \cup \cdots \cup G_q$ is defined unambiguously.

## 2.1   Expander-Based Functions

For an $(n, m)$-hypergraph $G = (S_1, \ldots, S_m)$, a sequence of $m$ predicates $P = (P_1, \ldots, P_m)$ where $P_i : \{0,1\}^{|S_i|} \to \{0,1\}$, we let $f_{G,P} : \{0,1\}^n \to \{0,1\}^m$ denote the function that takes an input $x \in \{0,1\}^n$ and maps it to the $m$-bit string $(P_1(x[S_1]), \ldots, P_m(x[S_m]))$. (If all predicates are identical we simply write $f_{G,P}$.) In its most abstract form, our assumption is parameterized by an expansion parameter $\beta$ (that quantifies the "expansion loss"), and by a (possibly infinite) predicate family $\mathcal{P}$. Formally, the Expander-based OWF assumption (EOWF) and Expander-based PRG assumption (EPRG) are defined as follows.

**Definition 4 (EOWF and EPRG).** *The $\mathsf{EOWF}(\mathcal{P}, m, \beta, t, \varepsilon)$ assumption asserts that for every $[n, m, d]$-hypergraph $G = (S_1, \ldots, S_m)$ that is $(n^{1-\beta}, (1-\beta)d)$-expanding, and every sequence of predicates[11] $P = (P_i)_{i \in [m]}$ taken from $\mathcal{P}$, the function $f_{G,P}$ is $(t, \varepsilon)$ one-way. The $\mathsf{EPRG}(\mathcal{P}, m, \beta, t, \varepsilon)$ is defined similarly except that $f_{G,P}(U_n)$ is $(t, \varepsilon)$ pseudorandom.*

A considerable amount of research was devoted to studying the properties of "secure" predicates. (See [App15] and references therein.) These results suggest that for some predicates of logarithmic arity $d = \Theta(\log n)$, and some constant $\beta < \frac{1}{2}$, the $\mathsf{EOWF}(\mathcal{P}, m, \beta, t, \varepsilon)$ assumption holds for every polynomial $m, t$ and every inverse polynomial $\varepsilon$. We adopt this setting as our main intractability assumption and abbreviate this assumption by $\mathsf{EOWF}(\mathcal{P})$. Similarly, we let $\mathsf{EPRG}(\mathcal{P})$ denote the analogous assumption for pseudorandomness. In fact, known results suggest that for a proper family of predicates $\mathcal{P}$, every $d = d(n)$ and every $\beta < \frac{1}{2}$, the assumption holds against adversaries whose size $t$ and success probability $\varepsilon$ are exponential in the expansion threshold, i.e., $t = \exp(\Omega(n^{1-\beta}))$ and $\varepsilon = 1/t$, as long as the output length satisfies $m < n^{o(d)}$ or even $m < n^{\alpha d}$ for some constant $\alpha$. We refer to this variant of the assumption as the *strong* $\mathsf{EOWF}(\mathcal{P})$ and *strong* $\mathsf{EPRG}(\mathcal{P})$.

---

[11] Here and through the paper, we implicitly assume that for all $i \in [m]$ the arity of the $i$-th predicate $P_i$ matches the cardinality of the $i$-th hyperedge $S_i$ of $G$.

*Concrete Instantiation.* A candidate for such a secure predicate (that is suggested in [AL15]) is the $d$-ary XOR-MAJ$_d$ predicate which partitions its input $w = (w_1, \ldots, w_d)$ into two parts $w_L = (w_1, \ldots, w_{\lfloor d/2 \rfloor})$ and $w_R = (w_{\lfloor d/2 \rfloor+1}, \ldots, w_d)$, computes the XOR of the left part and the majority of the right part, and XOR's the results together. This predicate satisfies several useful properties such as high resiliency, high algebraic degree and high rational degree (see Sect. 1.1). In fact, these properties hold for the more general case of XOR-Threshold predicates defined by:

$$\text{XOR-TH}_{d,\alpha,\tau}(w_1, \ldots, w_d) = \left( \sum_{j=1}^{\lfloor \alpha d \rfloor} w_j > \tau \lfloor \alpha d \rfloor \right) \oplus \left( \bigoplus_{i=\lfloor \alpha d \rfloor+1}^{d} w_i \right),$$

where the first term evaluates to one if $w_1 + \cdots + w_{\lfloor \alpha d \rfloor} > \tau$ and to zero otherwise. We define[12] $\text{XOR-TH}_d = \{\text{XOR-TH}_{d,\alpha,\tau} : \forall \alpha, \tau \in (1/3, 2/3)\}$ and let $\text{XOR-TH} = \bigcup_{d \in \mathbb{N}} \text{XOR-TH}_d$. We conjecture that strong EOWF holds for this family predicates.

## 3    From One-Wayness to Pseudorandomness

In this section, we show that EPRG reduces to EOWF as long as the predicate family $\mathcal{P}$ is *sensitive*. The latter condition means that every $d$-ary predicate $P \in \mathcal{P}$ can be written as $P(w) = w_i \oplus P'(w)$ where $i$ is some input variable and $P'$ does not depend on $w_i$. (Namely, the predicate is fully sensitive to one of its coordinates.)

**Theorem 1.** *Let $\beta$ be a constant in $(0,1)$; and $d = d(n)$, $m = m(n)$ and $\varepsilon = \varepsilon(n)$ be such that:*

$$\frac{4}{\beta} \le d(1 - \beta) \le n^{\beta/4} \quad and \quad \frac{4nm^3 \ln n}{\varepsilon^2} \le n^{(\beta^2/4)(1-\beta)d},$$

*and $\mathcal{P}$ be a sensitive predicate family. Then, the EPRG$(\mathcal{P}, m, \beta, t, \varepsilon)$ assumption follows from the EOWF$(\mathcal{P}, m', \beta', t', \varepsilon')$ assumption where $m' = m \cdot O(n \ln nm^2/\varepsilon^2)$, $\beta' = 3\beta$, $t' = t \cdot O(n \ln nm^2/\varepsilon^2)$ and $\varepsilon' = \Omega(\varepsilon/(mn))$.*

Note that once $d(n)$ is logarithmic in $n$, the conditions in the theorem are satisfied for every polynomial $m = \text{poly}(n)$, every inverse polynomial $\varepsilon(n)$, and every constant $\beta$. We conclude the following corollary.

**Corollary 1.** *For every sensitive family of predicates $\mathcal{P}$, if EOWF$(\mathcal{P})$ holds then so does EPRG$(\mathcal{P})$. In particular, this holds for the special case of $\mathcal{P} = \text{XOR-TH}$.*

Note that if we plug in larger (super logarithmic) values of $d$ in Theorem 1, we can support larger (super-polynomial) values of $m$ and smaller values of $\varepsilon$ (at the expense of decreasing $\beta$ to some concrete constant).

---

[12] The constants $(1/3, 2/3)$ in the definition are somewhat arbitrary and it seems that any constants bounded away from 0 and 1 will do.

### 3.1   Proof of Theorem 1

Assume, towards a contradiction, that there exists a $t$-size adversary that breaks the pseudorandomness of $f_{G,P}$ with advantage $\varepsilon$ for some $[n, m, d]$-hypergraph $G$ which is $(n^{1-\beta}, (1-\beta)d)$-expanding and some sequence of sensitive predicates $P = (P_1, \ldots, P_m) \in \mathcal{P}^m$. Then, due to Yao's theorem [Yao82], there exists an adversary $A_G$ of similar complexity that predicts some bit of $f_{G,P}$ with advantage $\varepsilon_p = \varepsilon/m$. To simplify notation, we assume that $A_G$ predicts the last bit[13] of $f_{G,P}$. That is,

$$\Pr_{x \xleftarrow{R} \{0,1\}^n, y = f_{G,P}(x)} [A_G(y[1, \ldots, m-1]) = y[m]] - \frac{1}{2} \geq \varepsilon_p. \tag{1}$$

We will prove the following lemma.

**Lemma 3.** *Let $\kappa = 4\ln n / \varepsilon_p^2$, $m' = \kappa \cdot m \cdot n$ and $P' = P^{\kappa n} = (P_1, \ldots, P_m)^{\kappa n}$. There exists a distribution $\mathcal{D}$ over $(n, m', d)$-hypergraphs such that:*

1. *A hypergraph $H$ sampled from $\mathcal{D}$ is $(n^{1-3\beta}, (1-3\beta)d)$-expanding with probability $1 - 1/(n \ln n)$.*
2. *There exists an adversary $B$ of size $t' = O(\kappa \cdot n \cdot t)$ and a set of inputs $\text{Good} \subseteq \{0,1\}^n$ which contains at least $\varepsilon_p/2$-fraction of all $n$-bit strings, such that for every string $x \in \text{Good}$,*

$$\Pr_{H \xleftarrow{R} \mathcal{D}} [B(H, f_{H,P'}(x)) = x] \geq 1/(2n).$$

We show that Theorem 1 follows from Lemma 3. Call $H$ good if

$$\Pr_{x \xleftarrow{R} \{0,1\}^n} [B(H, f_{H,P'}(x)) = x | x \in \text{Good}] \geq 1/(3n).$$

By a Markov argument, a random $H \xleftarrow{R} \mathcal{D}$ is likely to be good with probability $\Omega(1/n)$. Combing this with the first item, it follows, by a union bound, that there exists a good $H$ which is also $(n^{1-3\beta}, (1-3\beta)d)$-expanding. By hardwiring $H$ to $B$, we get an adversary $B_H$ which inverts $f_{H,P'}$ with probability of at least

$$\Pr_{x \xleftarrow{R} \{0,1\}^n} [x \in \text{Good}] \cdot \Pr_{x \xleftarrow{R} \{0,1\}^n} [B_H(f_{H,P'}(x)) = x | x \in \text{Good}] \geq \Omega(\varepsilon_p/n) = \Omega(\varepsilon/(mn)),$$

contradicting the $\text{EOWF}(\mathcal{P}, m', 3\beta, t', \varepsilon/(mn))$ assumption. We move on to prove Lemma 3.

*Proof (Proof of Lemma 3).* Before describing the distribution $\mathcal{D}$, we need some additional notation. For a permutation $\pi : [n] \rightarrow [n]$ and a tuple $S = (i_1, \ldots, i_d) \subseteq [n]^d$, let $\pi(S)$ denote the tuple $(\pi(i_1), \ldots, \pi(i_d))$. For an $[n, m, d]$-hypergraph $G$ with the hyperedges $(S_1, \ldots, S_m)$, let $\pi(G)$ denote a $[n, m, d]$-hypergraph with the hyperedges $(\pi(S_1), \ldots, \pi(S_m))$. For a string $x \in \{0,1\}^n$, let $\pi(x)$ denote the bit-string whose coordinates are permuted under $\pi$. We define the distribution $\mathcal{D}$ based on the hypergraph $G$ via the following procedure: (Fig. 1)

1. Take $[n, m, d]$-hypergraph $G$ as an input. Let $\ell^* \in [n]$ denote the first index of the last hyperedge of $G$.
2. Sample a random index $\tau \xleftarrow{R} [n]$. For each $j \in [n]$, let $\pi_1^j, \ldots, \pi_\kappa^j$ be $\kappa = 4 \ln n / \varepsilon_p^2$ random permutations over $[n]$ subject to $\pi_i^j(\ell^*) = \tau$.
3. For each $j \in [n]$ and $i \in [\kappa]$, let $G_i^j$ be the hypergraph $\pi_i^j(G)$ modified such that the first entry of its last hyperedge is set to $j$.
4. The output of $\mathcal{D}$ is the hypergraph $H = \bigcup_{j \in [n], i \in [\kappa]} G_i^j$.

**Fig. 1.** The distribution $\mathcal{D}$

We start by proving the first item of Lemma 3. Consider the distribution $\mathcal{D}'$ resulting from generating $\kappa \cdot n$ uniform and independent permutations $\phi_i^j$ ($j \in [n], i \in [\kappa]$), and outputting the hypergraph $H' = \cup_{i,j} H'_{i,j}$ where $H'_{i,j} = \phi_i^j(G)$. Observe that $\mathcal{D}$ can be viewed as a two step process in which: (1) $H'$ is sampled from $\mathcal{D}'$; and (2) We modify at most two nodes in every hyperedge of $H'$ based on some random process.[14] Since the second step can reduce the expansion of a set $T$ by at most $2|T|$, and since our setting of parameters implies that $\beta d > 2$, it suffices to show that $\Pr_{H'}[H'$ is $(n^{1-2\beta}, (1 - 2\beta)d)$-expanding$] \geq 1 - 1/(n \ln n)$.

To see this, recall that $G$ is $(r, d' = (1 - \beta)d)$-expanding and therefore, for every $i, j$, the random variable $\phi_i^j(G)$ is $(r, d', 0)$-random. Moreover, the permutations $\phi_i^j$ are sampled independently at random, and therefore $H' = \bigcup_{i,j} \phi_i^j(G)$ is a $(r, d', 0)$-random $(n, \kappa mn)$-hypergraph. Observe that our parameters satisfy the requirements of Lemma 2 (i.e., $4/\beta^2 \leq d' \leq n^{\beta/4}$ and $\kappa mn \leq n^{\beta^2 d'/4}$). By applying the lemma with $t = 0$, we conclude that $H'$ is $(n^{1-\beta}, (1 - \beta)^2 d)$-expanding (and thus also $(n^{1-2\beta}, (1 - 2\beta)d)$-expanding), except with failure probability of at most $n^{-(\beta^2/4)(1-\beta)d}$. The latter quantity is upper-bounded by $1/(n \ln n)$ since $\frac{4nm^3 \ln n}{\varepsilon^2} \leq n^{(\beta^2/4)(1-\beta)d}$. This completes the proof of the first part of Lemma 3.

We proceed with the proof of the second item of Lemma 3. Let $S = (\ell^*, i_2, \ldots, i_d)$ be the last hyperedge of $G$. Let $S'$ denote the $d-1$ tuple $(i_2, \ldots, i_d)$ and let $P_m : \{0, 1\}^d \rightarrow \{0, 1\}$ be the predicate computed by the last output of $f_{G,P}$. We assume (WLOG) that the first input of $P_m$ is sensitive and so it can be written as $P_m(w_1, \ldots, w_d) = w_1 \oplus Q(w_2, \ldots, w_d)$ for some $(d - 1)$-ary predicate $Q$.

The algorithm $B$ is a variant of the inversion algorithms given in [App13]. The input is a hypergraph $H = \bigcup_{j \in [n], i \in [\kappa]} G_i^j$ and a string $y \in \{0, 1\}^{\kappa \cdot n \cdot m}$ such that $y = f_{H, P'}(x)$. Let $y$ be parsed as $(y_i^j)_{j \in [n], i \in [\kappa]}$ where each $y_i^j = f_{G_i^j, P}(x)$. For each $j \in [n]$ and $i \in [\kappa]$, the algorithm $B$ runs $A_G$ on input

---

[13] The choice of the last bit unpredictability is without loss of generality since we can permute the order of the bits of $f_{G,P}$ (see [App13]).

[14] Specifically, sample a random index $\tau \in [n]$, and for every sub-hypergraph $H'_{i,j}$ and hyperedge $S \in H'_{i,j}$ swap the node $\phi_i^j(\ell^*)$ with the node $\tau$, except for the first entry in the last hyperedge of $H_{i,j}$ which $\phi_i^j(\ell^*)$ is replaced by $j$.

$y_i^j[1, \ldots, m-1]$ and gets a prediction bit $e_i^j$. Let $\sigma_i^j$ be the inverse permutation of $\pi_i^j$, and $x_i^j = \sigma_i^j(x)$; then, we get that $y_i^j = f_{\sigma_i(G_i^j), P}(x_i^j)$. By construction, this means that $y_i^j[1, \ldots, m-1] = f_{G, P}(x_i^j)[1, \ldots, m-1]$ and so $A_G$ attempts to predict the value $P_m(x_i^j[S]) = x_i^j[\ell^*] \oplus Q(x_i^j[S'])$. Note that the bit $y_i^j[m]$ equals to $x_i^j[\sigma_i^j(j)] \oplus Q(x_i^j[S'])$, and so

$$P_m(x_i^j[S]) \oplus y_i^j[m] = x_i^j[\ell^*] \oplus x_i^j[\sigma_i^j(j)] = x[\pi_i^j(\ell^*)] \oplus x[j] = x[\tau] \oplus x[j].$$

Assuming that $x[\tau]$ is known (indeed, we can either guess it or try both values), the above equation provides an estimation for $x[j]$. Since our predictor may err, this estimation is "noisy", i.e., it equals to $x[j]$ only with probability $\frac{1}{2} + \Omega(\varepsilon_p)$. After collecting $\kappa$ such votes (and arguing that these votes are "independent enough") we eventually recover the input $x$ bit by bit by deciding on the majority of the votes for each $x[j]$. We proceed by formally describing the algorithm $B$ (Fig. 2).

---

- **Input:** A hypergraph $H = \bigcup_{j \in [n], i \in [\kappa]} G_i^j$ and $y_i^j = f_{G_i^j, P}(x)$.
- Initialize $v_1, \ldots, v_n$ to 0.
- For $j \in [n]$ and $i \in [\kappa]$:
    1. Compute $e_i^j := A_G(y_i^j[1, \ldots, m-1])$, and let $b_i^j = e_i^j \oplus y_i^j[m]$.
    2. If $b_i^j = 1$, then increase $v_j$ by 1, otherwise decrease $v_j$ by 1.
- For $j \in [n]$, set $z_j$ to 1 if $v_j > 0$, otherwise set it to 0. Let $s_0 = z_1 \cdots z_n$ and $s_1 = \overline{z_1} \cdots \overline{z_n}$.
- **Output:** $s_0$ if $y = f_{H, P'}(s_0)$ and $s_1$ if $y = f_{H, P'}(s_1)$. Otherwise, output $\perp$.

---

**Fig. 2.** The inverter $B$

We now prove that $B$ inverts $f_{H, P'}$ well. Let $\mathrm{wt}(x)$ be the hamming weight of $x \in \{0, 1\}^n$ and for $w \in [n]$, let $X_w = \{x \in \{0, 1\}^n | \mathrm{wt}(x) = w\}$. Call $x$ *good* if $A_G$ predicts with advantage $\varepsilon_p/2$ the last bit of $f_{G, P}(x')$ for $x' \xleftarrow{R} X_{\mathrm{wt}(x)}$, i.e.,

$$\Pr_{x' \xleftarrow{R} X_{\mathrm{wt}(x)}, y = f_{G, P}(x')} [A_G(y[1 \ldots m-1]) = y[m]] - 1/2 \geq \varepsilon_p/2.$$

We let Good denote the set of good $x$'s and show that this set is $\varepsilon_p/2$-dense.

*Claim.* $\Pr_{x \xleftarrow{R} \{0, 1\}^n} [x \in \mathrm{Good}] \geq \varepsilon_p/2.$

*Proof.* Recall that our predictor $A_G$ has an advantage of $\varepsilon_p$ when it is invoked on $f_{G, P}(x')$ where $x' \xleftarrow{R} U_n$. Note that we can sample a uniform vector $x' \xleftarrow{R} \{0, 1\}^n$ by first selecting $x \xleftarrow{R} U_n$ and then selecting $x' \xleftarrow{R} X_{\mathrm{wt}(x)}$. Hence, the claim follows from Markov's inequality.                                                    $\square$

Now fix a good $x$. Let $S_n$ denote the set of all permutations from $[n]$ to $[n]$. Observe that sampling $x' \xleftarrow{R} X_{\mathrm{wt}(x)}$ is equivalent to taking a random permutation $\sigma \xleftarrow{R} S_n$ and computing $x' = \sigma(x)$. Hence, it holds that

$$\Pr_{\sigma \xleftarrow{R} S_n, y = f_{G,P}(\sigma(x))} [A_G(y[1 \ldots m-1]) = y[m]] - 1/2 \geq \varepsilon_p/2.$$

By an averaging argument, we get that there exists an index $\tau_x \in [n]$ such that

$$\Pr_{\sigma \xleftarrow{R} \{\pi \in S_n | \pi(\tau_x) = \ell^*\}, y = f_{G,P}(\sigma(x))} [A_G(y[1 \ldots m-1]) = y[m]] - 1/2 \geq \varepsilon_p/2.$$

Next, we show that the algorithm $B$ recovers $x$ with probability at least $\frac{1}{2}$ when invoked with a good input $x$ and with a hypergraph $H$ generated under condition that $\tau = \tau_x$. Since $\tau$ is generated uniformly at random this implies that $\Pr_H[B(H, f_{H,P'}(x)) = x] \geq 1/(2n)$.

*Claim.* For every good $x$, it holds that $\Pr_H[B(H, f_{H,P'}(x)) = x | \tau = \tau_x] \geq \frac{1}{2}$.

*Proof.* We assume that $x[\tau] = 0$ and show that, with high probability, $s_0$ is likely to be $x$. (A similar argument shows that when $x[\tau] = 1$, $s_1$ is likely to be $x$). We prove that for each $j \in [n]$ the value $z_j$ equals to $x[j]$ with probability $1 - 1/(2n)$. The theorem then follows by applying a union bound over all $n$ indices.

Fix some index $j \in [n]$. Call a vote $b_i^j$ good if it is equal to $x[j]$. Our goal is to show that with high probability a majority of the votes are good. Observe that in each iteration $i \in [\kappa]$, the predictor $A_G$ is invoked on $y_i^j[1, \ldots, m-1] = f_{G,P}(x_i^j)[1, \ldots, m-1]$ where $x_i^j = \sigma_i^j(x)$ and that the vote $b_i^j$ is good if the predictor succeeds in predicting $P_m(x_i^j[S])$. Since the permutations $\sigma_i^j$'s (that are the inverses of $\pi_i^j$'s) are independent and are uniform subject to $\sigma_i^j(\tau) = \ell^*$, and since $x$ is good, each call to the predictor succeeds independently with probability $\frac{1}{2} + \varepsilon_p/2$. Hence, by an additive Chernoff bound, the majority of the votes are good except with probability $\exp(-2\kappa \cdot (\varepsilon_p/2)^2) = \exp(-2 \ln n) < 1/(2n)$.  □

This completes the proof of Lemma 3.  □

# 4 PRF Constructions

We describe a general template for constructing pseudorandom functions. The template is parameterized with a predicate family $\mathcal{P} = \{P_d\}$ where $P_d$ is a $d$-ary predicate[15] and two (possibly keyed) algorithms: *mapper* $M$ and *extractor* $E$. Let $n \in \mathbb{N}$ denote the security parameter and let $d = d(n)$ be a locality parameter. Given an input $x \in \{0,1\}^n$ and a uniformly chosen key $k \in \{0,1\}^n$ we define the output of the function as follows. First, we use the mapper $M$ to map $x$ to an $(n, n/(d \log n), d)$-hypergraph $G_x$. Second, given the key $k$ we

---

[15] The construction can be easily generalized to handle non-uniform hypergraphs and/or different predicates $d$-ary predicates for each output.

compute a pseudorandom string $y = f_{G_x,P}(k)$, where $P = P_d$. Finally, we apply a randomness extractor $E$ to $y$ in order to produce the final output. (The keys of $E$ and $M$ are appended to the key $k$ and are treated as part of the key of the construction.) The main intuition behind this template is that if the hypergraph $G_x$ has good expanding properties, the string $y$ contains enough pseudoentropy which once extracted via $E$ looks pseudorandom.

In the following we describe several instantiations of the template by choosing different $M$ and $E$.

*Notation Switch.* Through this section, the symbol $x$ denotes a query to the PRF while $k$ denotes the PRF's key. Due to the structure of our construction, this means that the input to the function $f_{G,P}$ is denoted by $k$ (the key) and the hypergraph $G$ is computed based on the input $x$. (Unlike the notation used in Sect. 3.)

## 4.1   Instantiation $F_1$

The first instantiation $F_1$ can be seen as a "plain" instantiation of the template, where the inputs are mapped to the hypergraphs directly and no extractor is applied in the end (Fig. 3).

---

- **Parameters:** Let $\mathcal{K} = \{0,1\}^n$ be the key space, $\mathcal{X} = \{0,1\}^n$ be the input space, and $\mathcal{Y} = \{0,1\}^{n/(d \log n)}$ be the output space of $F_1$. Let $d = \Theta(\log n)$ and let $P \in \mathcal{P}$ be a $d$-ary predicate.
- **Mapper $M$:** The input $x$ is parsed into $n/(\log n)$ indices, then each consecutive group of $d$ indices is interpreted as a hyperedge of the hypergraph $G$.
- **Extractor $E$:** No extractor is applied in the end.
- **Code of $F_1$:** The function $F_1 : \mathcal{K} \times \mathcal{X} \to \mathcal{Y}$ is defined as $F_1(k,x) := f_{M(x),P}(k)$.

---

**Fig. 3.** Instantiation $F_1$

**Theorem 2.** *Let $n$ be the security parameter. For every $q = n^{o(\log n)}$, every $t(n), \varepsilon(n)$, and every constant $\beta \in (0,1)$ the function $F_1$ is a $(q, t, \varepsilon + n^{-\Omega(\log n)})$ weak PRF under assumption $\mathsf{EPRG}(\mathcal{P}, n \cdot q, \beta, t, \varepsilon)$.*

*Proof.* Fix some constant $\beta$ and let $d = \Theta(\log n)$. Let $x_1, \ldots, x_q$ be $q = n^{o(\log n)}$ random strings from $\{0,1\}^n$ asked by the adversary. For $i \in [q]$, let $G_i = M(x_i)$. Since the $x_i$'s are uniformly distributed, the hypergraph $H := \bigcup_{i=1}^{q} G_i$ is a $(n^{1-\beta}, 0)$-random $(n, m, d)$-hypergraph with $m = qn/(d \log n) < n^{d\beta^2/4}$. Hence, by Lemma 2 (with imperfectness parameter $t = 0$), $H$ is $(n^{1-\beta}, (1 - \beta)d)$-expanding except with probability $\varepsilon_{\mathrm{EXP}} = n^{-\Omega(\log n)}$. (The condition $4/\beta^2 \leq d \leq n^{\beta/4}$ required for Lemma 2 holds since $\beta$ is constant and $d = \Theta(\log n)$.)

The theorem follows by noting that conditioned on $H$ being $(n^{1-\beta}, (1-\beta)d)$-expanding, the EPRG$(\mathcal{P}, n \cdot q, \beta, t, \varepsilon)$ assumption implies that the random variable $V = (F_1(k, x_i))_{i=1}^{q}$, induced by a uniformly chosen $k \in \{0,1\}^n$, is $(t, \varepsilon)$-pseudorandom.

*Remark 1.* We note that the theorem extends to the case where $\log_n q + 1 < \beta^2 d/4$.

**Corollary 2.** *Suppose that* EOWF(XOR-MAJ) *holds. Then, there exists a weak PRF* $F_1 : \{0,1\}^n \times \{0,1\}^n \to \{0,1\}^{n/\log^2 n}$ *which is computable in linear time of* $O(n)$ *on a RAM machine with* $O(\log n)$ *word size, or by a boolean circuit of size* $\tilde{O}(n)$*. Moreover, for every fixed key* $k$*, the function* $F_1(k, \cdot)$ *can be computed by a depth-3* $\mathbf{AC^0}$ *circuit.*

*Proof.* By Corollary 1, EOWF(XOR-MAJ) implies EPRG(XOR-MAJ), which in turn, implies, by Theorem 2, that $F_1$ is a weak PRF.

Observe that the computation of $F_1$ consists of two steps. (1) Access the key $k$ in the $n/\log n$ addresses specified by the input $x$ and retrieve the corresponding content. Namely, for $1 \le i \le \ell$ where $\ell = n/\log n$, output the bits $z_i = k[x[(i-1)\log n + 1 : i\log n]]$ where $x[i : i+j]$ denotes the address represented by the substring $(x[i] \cdots x[i+j])$ under the standard binary representation. (2) Partition the bits $z_1, \ldots, z_\ell$ to $d$-size $\ell/d$ blocks, and compute for each block $1 \le i \le \ell/d$ the bit $y_i = \text{XOR-MAJ}_d(z_{(i-1)d+1}, \ldots, z_{id})$.

**Time.** On a RAM machine with $\log n$ word size, the first step is implemented in time $O(n)$ (these are just accesses to an array) and the second step takes $O(n/\log n)$ time.

**Size.** In Appendix A we show that the first step can be implemented by a circuit of quasilinear size $O(n \log^2 n \log \log n)$. In the second part, each computation of $z_i$ consists of computing two symmetric functions (XOR and Majority) over $d/2$-long inputs. The classical result of [MP75] (see also [Weg87]) shows that every $d$-ary symmetric predicate can computed by a linear-size circuit (of size $O(d)$) and so the overall complexity of the second step is linear in $n$.

**Depth.** Fix some key $k$. Observe that both the first part and the second part of the computation have logarithmic locality (each bit $z_i$ depends on at most $O(\log n)$ bits of $x$ and each $y_i$ depends on at most $O(\log n)$ bits of the $z_i$'s). Observe that any such function can be computed by a polynomial size DNF (OR of AND's) and a polynomial size CNF (AND of OR's). Hence, the overall computation can be naively computed by a depth-4 circuit. In fact, by using DNF for the first part and CNF for the second part we can collapse the two middle layers of OR gates and implement $F_1$ by a depth-3 $\mathbf{AC^0}$ circuit.     □

We note that, under strong EOWF(XOR-MAJ), $F_1$ achieves security against adversaries of almost-exponential size ($\exp(n^{1-\beta})$ for every $\beta > 0$) who make polynomially many queries (or even slightly super-polynomial number of queries

$q$) with quasipolynomial distinguishing advantage of $\varepsilon = n^{-\Omega(\log n)}$. As mentioned in the introduction, the quasipolynomial value of $\varepsilon$ is inherent for $\mathbf{AC^0}$ constructions.

We also remark that one can extend the output length of $F_1$ to $\{0,1\}^n$ by stretching the output using a pseudorandom generator $G : \{0,1\}^{n/\log n} \rightarrow \{0,1\}^n$. Using fast constructions of PRGs (e.g., [App13]) one can do this while keeping the efficiency guarantees stated in the theorem.

### 4.2 Instantiation $F_2$

The second instantiation $F_2$ is a modification of $F_1$, where an extractor is applied in the end. As explained in the introduction, this allows us to reduce the distinguishing advantage $\varepsilon$ (Fig. 4).

---

- **Parameters:** Let $\mathcal{K} = \mathcal{K}_f \times \mathcal{K}_e = \{0,1\}^n \times \{0,1\}^{O(n)}$ be the key space, $\mathcal{X} = \{0,1\}^n$ be the input space, and $\mathcal{Y} = \{0,1\}^{n/(2d\log n)}$ be the output space of $F_2$. Let $P \in \mathcal{P}$ be a $d$-ary predicate.
- **Mapper $M$:** As in $F_1$, $M(x)$ parses $x$ as $(n, n/(d\log n), d)$-hypergraph.
- **Extractor:** Let $\mathrm{Ext} : \mathcal{K}_e \times \{0,1\}^{n/(d\log n)} \rightarrow \{0,1\}^{n/(2d\log n)}$ be a strong $(\ell, \varepsilon_{\mathrm{Ext}})$-extractor where $\ell = 0.9 \cdot n/(d\log n)$ and $\varepsilon_{\mathrm{Ext}} = 2^{-\Omega(n)}$).
- **Code of $F_2$:** The function $F_2 : \mathcal{K} \times \mathcal{X} \rightarrow \mathcal{Y}$ is defined as $F_2((k,s),x) := \mathrm{Ext}_s(f_{M_2(x),P}(k))$, where $(k,s) \in \mathcal{K}_f \times \mathcal{K}_e$.

---

**Fig. 4.** Instantiation $F_2$

Our goal is to provide a tight security reduction from breaking $F_2$ to the EPRG assumption. For this, we will have to rely on the security of EPRG over a predicate family $\mathcal{P}_\beta$ containing all predicates which can be obtained by selecting some $d$-ary predicate $P \in \mathcal{P}$ and arbitrarily fixing at most $\beta d$ of its inputs. Although the security of EOWF with respect to $\mathcal{P}_\beta$ may seem like a strong assumption, we will later show that natural candidates for EOWF already satisfy it.

**Theorem 3.** *Let $n$ be the security parameter. Let $\beta$ be a constant in $(0,1)$, $q = q(n)$ and $d = d(n)$ such that $4/\beta^2 \leq d \leq n^{\beta/4}$ and $q \leq n^{d\beta^2/4-1}$. Let $t = t(n)$, $\varepsilon = \varepsilon(n)$ be arbitrary functions. Then, the function $F_2$ is a $(q, t, \varepsilon + n^{-\Omega(dn^{1-\beta})} + q \cdot 2^{-\Omega(n)})$ weak PRF, under assumption $\mathsf{EPRG}(\mathcal{P}_\beta, n \cdot q, \beta, t, \varepsilon)$.*

*Proof.* Let $x_1, \ldots, x_q$ be $q$ random strings from $\{0,1\}^n$ asked by the adversary. For $i \in [q]$, let $G_i = M_2(x_i)$. Consider the $(n, m, d)$-hypergraph $H := \bigcup_{i=1}^q G_i$ where $m = nq/(d\log n) < n^{d\beta^2/4}$. Since the $G_i$'s are random, the hypergraph $H$ is $(2r, 0)$-random for $r = n^{1-\beta}$. By applying Lemma 2 with $t = r = n^{1-\beta}$ and $d$-uniform hypergraphs, we conclude that, except with probability $\varepsilon_{\mathrm{EXP}} = n^{-\Omega(dr)}$, the hypergraph $H$ is $r$-imperfect $(r, (1-\beta)d)$-expander.

From now on we fix a sequence of queries $(x_1, \ldots, x_q)$ which leads to such an imperfect expander $H$. It suffices to prove that, for a uniformly chosen $(k, s) \in \mathcal{K}$, the random variable $V := (F_2((k, s), x_i))_{i=1}^q$ is $(t, \varepsilon + q \cdot \varepsilon_{\text{Ext}})$-pseudorandom for $\varepsilon_{\text{Ext}} = 2^{-\Omega(n)}$.

By construction, $V$ can be rewritten as $\overline{\text{Ext}}_s(f_{H,P}(k))$ where $\overline{\text{Ext}}_s(y_1, \ldots, y_q) := (\text{Ext}_s(y_1), \ldots, \text{Ext}_s(y_q))$. First, we show that the distribution of $f_{H,P}(k)$ is computationally indistinguishable from a generalized bit-fixing source (the proof is deferred to the full version [AR16]).

**Lemma 4.** *Let $G$ be a $[n, m, d]$-hypergraph which is $n^{1-\beta}$-imperfect $(n^{1-\beta}, (1 - \beta)d)$-expander for some constant $\beta \in (0, 1)$. Then, given that the assumption $\mathsf{EPRG}(\mathcal{P}_\beta, m, \beta, t, \varepsilon)$ holds, the random variable $f_{G,P}(U_n)$ is $(t, \varepsilon)$-computationally indistinguishable from a generalized $(m - n^{1-\beta})$ bit-fixing source.*

It follows that $f_{H,P}(k)$ is $(t, \varepsilon)$-computationally indistinguishable from some generalized $(\frac{qn}{d \log n} - r)$ bit-fixing source $Y$. We therefore conclude that $V = \overline{\text{Ext}}_s(f_{H,P}(k))$ is $(t, \varepsilon)$-indistinguishable from $\overline{\text{Ext}}_s(Y)$. By Lemma 1 (Sect. 2), the latter distribution is $(q \cdot \varepsilon_{\text{Ext}})$-statistically indistinguishable from uniform. Hence, conditioned on $H$ being an almost expander, $V$ must be $(t, \varepsilon + q\varepsilon_{\text{Ext}})$-indistinguishable from uniform. Overall, we conclude that for $q$ random queries, $V$ is $(t, \varepsilon + q\varepsilon_{\text{Ext}} + \varepsilon_{\text{EXP}})$-pseudorandom, as required.  $\square$

**Corollary 3.** *Suppose that strong $\mathsf{EPRG}(\text{XOR-TH})$ holds. Then, there exists a weak PRF $F_2 : \{0, 1\}^{O(n)} \times \{0, 1\}^n \to \{0, 1\}^{n/2 \log^2 n}$ which is $(q, t = \exp(n^{1-\beta}), \varepsilon = \exp(-n^{1-\beta}))$ for every polynomial $q$ and every constant $\beta$, and can be computed in linear time of $O(n)$ on a RAM machine with $O(\log n)$ word size, and by a boolean circuit of size $\tilde{O}(n)$. Moreover, for every fixed key $k$, the function $F_2(k, \cdot)$ can be computed by an $\mathbf{MOD_2 \circ AC^0}$ circuit.*

*Proof.* Instantiate $F_2$ with $\mathcal{P} = \text{XOR-MAJ}$ and observe that $\mathcal{P}_\beta = \text{XOR-TH}$ for sufficiently small $\beta$ (e.g., every $\beta < 1/6$). By Theorem 3, the strong $\mathsf{EPRG}(\text{XOR-TH})$ assumption implies that, for every polynomial $q$ and constant $\beta > 0$, $F_2$ is $(q, t = \exp(n^{1-\beta}), \varepsilon = \exp(-n^{1-\beta}))$ weak PRF.

The efficiency analysis is identical to the analysis of $F_1$ except that we need to add the complexity of the extractor. Ishai et al. [IKOS08, Theorem 3.3] constructed a strong $(0.9 \cdot N, 2^{-\Omega(N)})$-extractor for $N$-bit sources outputting an $(N/2)$-bit string using a seed of length $O(N)$ that can be computed by a linear function (over the binary field) whose circuit is of size $O(N)$. By employing this extractor we get a linear-time implementation in the RAM model and quasilinear-size circuit implementation. Furthermore, since the extractor is a linear function it can be implemented by a single layer of XOR gates and so the overall computation is in $\mathbf{MOD_2 \circ AC^0}$.  $\square$

## 4.3   Instantiation $F_3$

The third instantiation, $F_3$, is a modification of $F_2$, where the input $x$ is mapped to a hypergraph using an $(n, 2^{-\Omega(n)})$-bitwise independent generator $M : \mathcal{K}_m \times$

> - **Parameters:** Let $\mathcal{K} = \mathcal{K}_f \times \mathcal{K}_m \times \mathcal{K}_e = \{0,1\}^n \times \{0,1\}^{2n} \times \{0,1\}^n$ be the key space, $\mathcal{X} = \{0,1\}^n$ be the input space, and $\mathcal{Y} = \{0,1\}^{n/(2d\log n)}$ be the output space of $F_3$. Let $P$ be some $d$-ary predicate chosen from $\mathcal{P}$.
> - **Mapper** $M$: Let $M : \mathcal{K}_m \times \mathcal{X} \to \mathcal{X}$ be a $(n, 2^{-\Omega(n)})$-biased generator and let $\sigma \xleftarrow{R} \mathcal{K}_m$ be its key. We parse the $n$-bit output of $M$ as an $(n, n/(d\log n), d)$-hypergraph.
> - **Extractor:** Let $\mathrm{Ext} : \mathcal{K}_e \times \{0,1\}^{n/(d\log n)} \to \{0,1\}^{n/(2d\log n)}$ be a strong $(0.9 \cdot n/(d\log n), 2^{-\Omega(n)})$-extractor.
> - **Code of $F_3$:** The function $F_3 : \mathcal{K} \times \mathcal{X} \to \mathcal{Y}$ is defined as $F_3((k,\sigma,s), x) := \mathrm{Ext}_s(f_{M_\sigma(x),P}(k))$.

**Fig. 5.** Instantiation $F_3$

$\mathcal{X} \to \mathcal{X}$. An efficient construction of such a $(n, 2^{-\Omega(n)})$-bias generator (with $\mathcal{K}_m = \{0,1\}^{2n}$) is presented in Theorem 5 (Fig. 5).

**Theorem 4.** *Let $n$ be the security parameter. Let $\beta$ be a constant in $(0,1)$, $q = q(n)$ and $d = d(n)$ such that $4/\beta^2 \le d \le n^{\beta/4}$ and $q \le n^{d\beta^2/4-1}$. Let $t = t(n)$, $\varepsilon = \varepsilon(n)$ be arbitrary functions. Then, the function $F_3$ is a non-adaptive $(q, t, \varepsilon + n^{-\Omega(dn^{1-\beta})} + q \cdot 2^{-\Omega(n)})$-PRF, under assumption $\mathsf{EPRG}(\mathcal{P}_\beta, n \cdot q, \beta, t, \varepsilon)$.*

*Proof.* Fix a sequence of $q$ distinct non-adaptive queries $x_1, \ldots, x_q$. For $i \in [q]$, let $G_i := M_\sigma(x_i)$. Since $M$ is $(n, 2^{-\Omega(n)})$-biased, the hypergraph $H := \bigcup_{i=1}^q G_i$ is $(\ell, 2^{-\Omega(n)})$-random hypergraph for $\ell = n/(d\log n) \ge 2n^{1-\beta}$. Recall also that $H$ has at most $n \cdot q \le n^{d\beta^2/4}$ hyperedges and $d$ is chosen such that $4/\beta^2 \le d \le n^{\beta/4}$. By applying Lemma 2 with $t = r = n^{1-\beta}$ and $d$-uniform hypergraphs, we conclude that, except with probability $\varepsilon_{\mathrm{EXP}} = n^{-\Omega(dr)}$, the hypergraph $H$ is $r$-imperfect $(r, (1-\beta)d)$-expander (where the probability is taken over $\sigma \xleftarrow{R} \mathcal{K}_m$).

From now on we fix a good $\sigma$ which leads to such an imperfect expander $H$. It suffices to prove that, for a uniformly chosen $(k, s)$, the random variable $V := (F_3((k, \sigma, s), x_i))_{i=1}^q$ is $(t, \varepsilon + q \cdot \varepsilon_{\mathrm{Ext}})$-pseudorandom for $\varepsilon_{\mathrm{Ext}} = 2^{-\Omega(n)}$. By construction, $V$ can be rewritten as $\overline{\mathrm{Ext}}_s(f_{H,P}(k))$ where $\overline{\mathrm{Ext}}_s(y_1, \ldots, y_q)$ stands for $(\mathrm{Ext}_s(y_1), \ldots, \mathrm{Ext}_s(y_q))$. Lemma 4 shows that the random variable $f_{H,P}(k)$ is $(t, \varepsilon)$-computationally close to some generalized $(qn/(d\log n) - r)$ bit-fixing source $Y$, and Lemma 1 shows that $\overline{\mathrm{Ext}}_s(Y)$ is $q \cdot \varepsilon_{\mathrm{Ext}}$-close to uniform. The theorem follows.                                          $\square$

In Theorem 5 we show that there exists a $(n, 2^{-\Omega(n)})$-bias generator $M : \{0,1\}^{2n} \times \{0,1\}^n \to \{0,1\}^n$ which can be computed in quasilinear time $\tilde{O}(n)$ or by a $\mathbf{TC^0}$ circuit (i.e., a constant-depth circuit with unbounded fan-in AND, OR and Majority gates). The following corollary follows.

**Corollary 4.** *Suppose that $\mathsf{EOWF(XOR\text{-}MAJ)}$ holds. Then, there exists a non-adaptive PRF $F_3 : \{0,1\}^{3n} \times \{0,1\}^n \to \{0,1\}^{n/\log^2 n}$ which is computable by a* *boolean circuit of size $\tilde{O}(n)$. Moreover, for every fixed key $k$, the function $F_3(k, \cdot)$ can be computed by a $\mathbf{TC^0}$ circuit.*

*Proof.* Let $\mathcal{P}$ = XOR-MAJ and observe that $\mathcal{P}_\beta$ = XOR-TH for sufficiently small $\beta$ (e.g., every $\beta < 1/6$). By Corollary 1, EOWF(XOR-TH) implies EPRG(XOR-TH), which in turn, implies, by Theorem 4, that $F_3$ is a non-adaptive PRF.

The efficiency analysis is identical to the analysis of $F_2$ except that we need to add the complexity of $M$ which can be computed in quasilinear time $\tilde{O}(n)$ or by a $\mathbf{TC^0}$ circuit (See Theorem 5).

Under the strong EPRG(XOR-TH) assumption, a logarithmic $d$ implies that $F_3$ is $(q, t = \exp(n^{1-\beta}), \varepsilon = \exp(-n^{1-\beta}))$ secure for every polynomial $q$ and every constant $\beta$. For polynomial locality $d = n^\delta$, for some constant $\delta > 0$, we get $q = \exp(n^{\Omega(\delta)})$, $t = \exp(n^{1-\Omega(\delta)})$ and $\varepsilon = \exp(-n^{1-\Omega(\beta)})$.

**A Bitwise Independent Generator Construction.** We now construct an efficient generator that is $(t, \varepsilon)$-bitwise independent in the regime of $t = n$ and negligible $\varepsilon$.

**Theorem 5.** *Let $k_0, k_1$ be two keys chosen uniformly from $\mathrm{GF}(2^n)$. For $x \in \mathrm{GF}(2^n)$, define the generator $\mathcal{V}_{k_0,k_1}(x) := \frac{k_1}{k_0+x}$. Then, $\mathcal{V}$ is $(d, d \cdot 2^{d/2+1-n})$-bitwise independent for any $d \leq 2^n$. Furthermore, the generator $\mathcal{V}$ can be computed by a circuit of quasilinear size $O(n \log^2 n \log \log n)$ and by a $\mathbf{TC^0}$ circuit.*

*Proof.* We observe that in order to prove that $\mathcal{V}$ is $(d, d \cdot 2^{d/2+1-n})$-bitwise independent, it is sufficient to prove that $\mathcal{V}$ is $(d, \frac{d}{2^{n-1}})$-linear-fooling over $\mathrm{GF}(2^n)$. Indeed, we know that $(t, \varepsilon)$-linear-fooling over $\mathrm{GF}(2^n)$ implies $(t, \varepsilon)$-bias over $\mathrm{GF}(2)$ [Tzu09, Theorem 4.5], which in turn implies $(t, 2^{t/2} \cdot \varepsilon)$-bitwise independence [NN93, Corollary 2.1].

We now turn to showing that $\mathcal{V}$ is $(d, \frac{d}{2^{n-1}})$-linear-fooling over $\mathrm{GF}(2^n)$ for any $d \leq 2^n$. The proof is based on the work of [MV12, Theorem 3.5]. We prove that $\mathcal{V}$ is $(d, \frac{d}{2^{n-1}})$-linear-fooling over $\mathrm{GF}(2^n)$, i.e., for any distinct $a_1, \ldots, a_d \in \mathrm{GF}(2^n)$, any $d$ constants $b_1, \ldots, b_t$ from $\mathrm{GF}(2^n)$ (that are not all equal to zero), we have that

$$\Delta\left(\sum_{i=1}^d b_i \mathcal{V}_{k_0,k_1}(a_i) \; ; \; U_{\mathrm{GF}(2^n)}\right) \leq \frac{d}{2^{n-1}}.$$

After letting $p(x)$ denote the polynomial $\sum_{i=1}^d \frac{b_i}{x+a_i} = \sum_{i=1}^d b_i(x + a_i)^{2^n - 2}$, we get that $\sum_{i=1}^d b_i \mathcal{V}_{k_0,k_1}(a_i)$ can be rewritten as $k_1 \cdot p(k_0)$. Observe that conditioned on $p(k_0) \neq 0$, we have that $k_1 \cdot p(k_0)$ is uniformly distributed over $\mathrm{GF}(2^n)$. Hence, it suffices to show that $p(x)$ has at most $2d - 1$ distinct roots. First, we define auxiliary polynomials:

$$\bar{p}(x) := p(x) \cdot \prod_{j=1}^d (a_j + x) = \sum_{i=1}^d \left[ b_i(x + a_i)^{2^n - 1} \prod_{j \neq i} (a_j + x) \right],$$

and

$$\overline{p}_*(x) := \sum_{i=1}^{d} b_i \prod_{j \neq i} (a_j + x).$$

Observe that any root $y$ of $p(x)$ is also a root of $\overline{p}(x)$. Moreover, note that for any $y \notin \{a_1, \ldots, a_d\}$ we have that $\overline{p}(y) = \overline{p}_*(y)$ (since $y^{2^n-1} = 1$ for any non-zero $y$). Hence, the only possible roots of $p(x)$ are the roots of $\overline{p}_*(x)$ and $\{a_1, \ldots, a_d\}$. This means that in order to show that $p(x)$ has at most $2d - 1$ distinct roots, it is sufficient to show that $\overline{p}_*(x)$ has at most $d - 1$ distinct roots. Because $\overline{p}_*(x)$ is a degree $d - 1$ polynomial, this will always be the case unless $\overline{p}_*(x)$ is identically zero. This is ruled out by observing that $\overline{p}_*(a_i) \neq 0$, where $i$ is chosen such that $b_i \neq 0$. Indeed, $\overline{p}_*(a_i) = b_i \prod_{j \neq i}(a_j + a_i)$ which is non-zero because $a_1, \ldots, a_d$ are distinct.

(**Complexity of** $\mathcal{V}$) Finally, we turn to the analysis of the circuit complexity of $\mathcal{V}$. The complexity of $\mathcal{V}$ equals to the complexity of the division and summation circuits (dividing $k_1$ by $k_0 + x$). As stated in [MV12] this can be done by a $\mathbf{TC^0}$ circuit or by a circuit of size $O(n \log^2 n \log \log n)$ using the techniques of [GvzGPS00].

**Acknowledgement.** We thank Adam Klivans and Shai Shalev-Shwartz for helpful discussions.

## A    Array Multi-access in Quasilinear Time

We consider the following functionality. Given $\ell = n/\log n$ indices of length $\log n$ each $I[1], \ldots, I[\ell]$ and a data vector $K \in \{0,1\}^n$ output $K[I[1]], \ldots, K[I[\ell]]$. We will show that this can done by $O(n \log^2 n \log \log n)$-size circuit. We assume that the input indices are sorted which is without loss of generality since $t$ elements of bit-length $b = \log n$ can be sorted by a circuit of size $O(b\ell \log \ell) = O(n \log n)$ (e.g., using a sorting network [AKS83] where comparison is implemented via Parallel Prefix Computation [LF80]). Instead of describing an $O(n \log^2 n \log \log n)$-size circuit, we describe a Turing Machine $M$ that solves the problem in time $T = O(n \log n)$ using a constant number of tapes. The latter can be simulated by a circuit of size $O(T \log T)$ (e.g., by turning the computation $M$ into an oblivious Turing machine $M'$ of complexity $O(T \log T)$ [PF79] and then moving to a circuit of size $O(T \log T)$). We sketch the description of the machine $M$. The machine $M$ places the indices $I$ on one tape, the data $K$ on another tape and places the output on a special output tape. During its run, $M$ maintains two counters $i$ and $j$ which are initialized to 1. At each step, $M$ checks if the index $I[i]$ equals to $j$ if this is the case then $K[j]$ is written to the current position in the output tape. Also, the head of the output tape is moved one step and the head of the index tape is moved to the next index. In case of inequality, the head of the data tape is moved forward by one step, and the counter $j$ is increased by one. Since each step costs $O(\log n)$ operations and there are at most $n$ steps, the overall complexity is $O(n \log n)$.

# References

[ABG+14] Akavia, A., Bogdanov, A., Guo, S., Kamath, A., Rosen, A.: Candidate weak pseudorandom functions in $AC^0$ $MOD_2$. In: Naor, M. (ed.) Innovations in Theoretical Computer Science, ITCS 2014, Princeton, NJ, USA, 12–14 January 2014, pp. 251–260. ACM (2014)

[ABR12] Applebaum, B., Bogdanov, A., Rosen, A.: A dichotomy for local small-bias generators. In: Cramer, R. (ed.) TCC 2012. LNCS, vol. 7194, pp. 600–617. Springer, Heidelberg (2012). doi:10.1007/978-3-642-28914-9_34

[ABW10] Applebaum, B., Barak, B., Wigderson, A.: Public-key cryptography from different assumptions. In: Schulman, L.J. (ed.) Proceedings of the 42nd ACM Symposium on Theory of Computing, STOC 2010, Cambridge, Massachusetts, USA, 5–8 June 2010, pp. 171–180. ACM (2010)

[AHI05] Alekhnovich, M., Hirsch, E.A., Itsykson, D.: Exponential lower bounds for the running time of DPLL algorithms on satisfiable formulas. J. Autom. Reasoning 35(1–3), 51–72 (2005)

[AKS83] Ajtai, M., Komlós, J., Szemerédi, E.: An o(n log n) sorting network. In: Johnson, D.S., Fagin, R., Fredman, M.L., Harel, D., Karp, R.M., Lynch, N.A., Papadimitriou, C.H., Rivest, R.L., Ruzzo, W.L., Seiferas, J.I. (eds.) Proceedings of the 15th Annual ACM Symposium on Theory of Computing, Boston, Massachusetts, USA, 25–27 April 1983, pp. 1–9. ACM (1983)

[AL15] Applebaum, B., Lovett, S.: Algebraic attacks against random local functions, their countermeasures. In: Electronic Colloquium on Computational Complexity (ECCC), STOC 2016, vol. 22, p. 172 (2015, to appear)

[Ale03] Alekhnovich, M.: More on average case vs approximation complexity. In: 44th Symposium on Foundations of Computer Science (FOCS 2003), Cambridge, MA, USA, Proceedings, 11–14 October 2003, pp. 298–307. IEEE Computer Society (2003)

[App13] Applebaum, B.: Pseudorandom generators with long stretch, low locality from random local one-way functions. SIAM J. Comput. 42(5), 2008–2037 (2013). Preliminary version in STOC 2012

[App14] Applebaum, B.: Bootstrapping obfuscators via fast pseudorandom functions. In: Sarkar, P., Iwata, T. (eds.) ASIACRYPT 2014. LNCS, vol. 8874, pp. 162–172. Springer, Heidelberg (2014). doi:10.1007/978-3-662-45608-8_9

[App15] Applebaum, B.: Cryptographic hardness of random local functions - survey. In: Electronic Colloquium on Computational Complexity (ECCC), vol. 22, p. 27 (2015)

[AR16] Applebaum, B., Raykov, P.: Fast pseudorandom functions based on expander graphs. In: Electronic Colloquium on Computational Complexity (ECCC), vol. 23, p. 82 (2016). Full version of this paper

[BFKL93] Blum, A., Furst, M., Kearns, M., Lipton, R.J.: Cryptographic primitives based on hard learning problems. In: Stinson, D.R. (ed.) CRYPTO 1993. LNCS, vol. 773, pp. 278–291. Springer, Heidelberg (1994). doi:10.1007/3-540-48329-2_24

[BH08] Brodsky, A., Hoory, S.: Simple permutations mix even better. Random Struct. Algorithms 32(3), 274–289 (2008)

[BH15] Berman, I., Haitner, I.: From non-adaptive to adaptive pseudorandom functions. J. Cryptol. 28(2), 297–311 (2015)

[BMR10] Boneh, D., Montgomery, H.W., Raghunathan, A.: Algebraic pseudoran-dom functions with improved efficiency from the augmented cascade. In: Al-Shaer, E., Keromytis, A.D., Shmatikov, V. (eds.) Proceedings of the 17th ACM Conference on Computer and Communications Security, CCS 2010, Chicago, Illinois, USA, 4–8 October 2010, pp. 131–140. ACM (2010)

[BPR12] Banerjee, A., Peikert, C., Rosen, A.: Pseudorandom functions and lat-tices. In: Pointcheval, D., Johansson, T. (eds.) EUROCRYPT 2012. LNCS, vol. 7237, pp. 719–737. Springer, Heidelberg (2012). doi:10.1007/978-3-642-29011-4_42

[BQ12] Bogdanov, A., Qiao, Y.: On the security of Goldreich's one-way function. Comput. Complexity 21(1), 83–127 (2012)

[BR13] Bogdanov, A., Rosen, A.: Input locality and hardness amplification. J. Cryptol. 26(1), 144–171 (2013)

[CEMT14] Cook, J., Etesami, O., Miller, R., Trevisan, L.: On the one-way function candidate proposed by Goldreich. ACM Trans. Comput. Theor. 6(3), 1401–1435 (2014)

[CGH+85] Chor, B., Goldreich, O., Håstad, J., Friedman, J., Rudich, S., Smolensky, R.: The bit extraction problem of t-resilient functions (prelim-inary version). In: 26th Annual Symposium on Foundations of Computer Science, Portland, Oregon, USA, 21–23 October 1985, pp. 396–407. IEEE Computer Society (1985)

[DI05] Damgård, I., Ishai, Y.: Constant-round multiparty computation using a black-box pseudorandom generator. In: Shoup, V. (ed.) CRYPTO 2005. LNCS, vol. 3621, pp. 378–394. Springer, Heidelberg (2005). doi:10.1007/11535218_23

[DLS14] Daniely, A., Linial, N., Shalev-Shwartz, S.: From average case complexity to improper learning complexity. In: Shmoys, D.B. (ed.) Symposium on Theory of Computing, STOC 2014, New York, NY, USA, 31 May – 03 June 2014, pp. 441–448. ACM (2014)

[FPV15] Feldman, V., Perkins, W., Vempala, S.: On the complexity of random sat-isfiability problems with planted solutions. In: Servedio, R.A., Rubinfeld, R. (eds.) Proceedings of the Forty-Seventh Annual ACM on Symposium on Theory of Computing, STOC 2015, Portland, OR, USA, 14–17 June 2015, pp. 77–86. ACM (2015)

[GGM86] Goldreich, O., Goldwasser, S., Micali, S.: How to construct random func-tions. J. ACM 33(4), 792–807 (1986)

[GL89] Goldreich, O., Levin, L.A.: A hard-core predicate for all one-way func-tions. In: Johnson, D.S. (ed.) Proceedings of the 21st Annual ACM Sym-posium on Theory of Computing, Seattle, Washigton, USA, 14–17 May 1989, pp. 25–32. ACM (1989)

[Gol00] Goldreich, O.: Candidate one-way functions based on expander graphs. In: Electronic Colloquium on Computational Complexity (ECCC), vol. 7, no. 90 (2000)

[Gow96] Gowers, W.T.: An almost m-wise independent random permutation of the cube. Comb. Probab. Comput. 5(2), 119–130 (1996)

[GVW12] Gorbunov, S., Vaikuntanathan, V., Wee, H.: Functional encryption with bounded collusions via multi-party computation. In: Safavi-Naini, R., Canetti, R. (eds.) CRYPTO 2012. LNCS, vol. 7417, pp. 162–179. Springer, Heidelberg (2012). doi:10.1007/978-3-642-32009-5_11

[GvzGPS00] Gao, S., von Zur Gathen, J., Panario, D., Shoup, V.: Algorithms for exponentiation in finite fields. J. Symb. Comput. 29(6), 879–889 (2000)

[HILL99]  Håstad, J., Impagliazzo, R., Levin, L.A., Luby, M.: A pseudorandom generator from any one-way function. SIAM J. Comput. **28**(4), 1364–1396 (1999). Preliminary versions in STOC 1989 and STOC 1990

[HMMR05] Hoory, S., Magen, A., Myers, S., Rackoff, C.: Simple permutations mix well. Theor. Comput. Sci. **348**(2–3), 251–261 (2005)

[IKOS08]  Ishai, Y., Kushilevitz, E., Ostrovsky, R., Sahai, A.: Cryptography with constant computational overhead. In: Dwork, C. (ed.) Proceedings of the 40th Annual ACM Symposium on Theory of Computing, Victoria, British Columbia, Canada, May 17–20, 2008, pp. 433–442. ACM (2008)

[Kha93]   Kharitonov, M.: Cryptographic hardness of distribution-specific learning. In: Kosaraju, S.R., Johnson, D.S., Aggarwal, A. (eds.) Proceedings of the Twenty-Fifth Annual ACM Symposium on Theory of Computing, San Diego, CA, USA, 16–18 May 1993, pp. 372–381. ACM (1993)

[KS09]    Klivans, A.R., Sherstov, A.A.: Cryptographic hardness for learning intersections of halfspaces. J. Comput. Syst. Sci. **75**(1), 2–12 (2009)

[LF80]    Ladner, R.E., Fischer, M.J.: Parallel prefix computation. J. ACM **27**(4), 831–838 (1980)

[LMN93]   Linial, N., Mansour, Y., Nisan, N.: Constant depth circuits, fourier transform, and learnability. J. ACM **40**(3), 607–620 (1993)

[LW09]    Lewko, A.B., Waters, B.: Efficient pseudorandom functions from the decisional linear assumption and weaker variants. In: Al-Shaer, E., Jha, S., Keromytis, A.D. (eds.) Proceedings of the 2009 ACM Conference on Computer and Communications Security, CCS 2009, Chicago, Illinois, USA, 9–13 November 2009, pp. 112–120. ACM (2009)

[MP75]    Muller, D.E., Preparata, F.P.: Bounds to complexities of networks for sorting and for switching. J. ACM **22**(2), 195–201 (1975)

[MST03]   Mossel, E., Shpilka, A., Trevisan, L.: On e-biased generators in NC0. In: 44th Symposium on Foundations of Computer Science (FOCS 2003), Cambridge, MA, USA, Proceedings, 11–14 October 2003, pp. 136–145. IEEE Computer Society (2003)

[MV12]    Miles, E., Viola, E.: Substitution-permutation networks, pseudorandom functions, and natural proofs. In: Safavi-Naini, R., Canetti, R. (eds.) CRYPTO 2012. LNCS, vol. 7417, pp. 68–85. Springer, Heidelberg (2012). doi:10.1007/978-3-642-32009-5_5

[NN93]    Naor, J., Naor, M.: Small-bias probability spaces: efficient constructions and applications. SIAM J. Comput. **22**(4), 838–856 (1993)

[NR95]    Naor, M., Reingold, O.: Synthesizers and their application to the parallel construction of psuedo-random functions. In: 36th Annual Symposium on Foundations of Computer Science, Milwaukee, Wisconsin, 23–25 October 1995, pp. 170–181. IEEE Computer Society (1995)

[NR97]    Naor, M., Reingold, O.: Number-theoretic constructions of efficient pseudo-random functions. In: 38th Annual Symposium on Foundations of Computer Science, FOCS 1997, Miami Beach, Florida, USA, 19–22 October 1997, pp. 458–467. IEEE Computer Society (1997)

[NRR00]   Naor, M., Reingold, O., Rosen, A.: Pseudo-random functions and factoring (extended abstract). In: Yao, F.F., Luks, E.M. (eds.) Proceedings of the Thirty-Second Annual ACM Symposium on Theory of Computing, Portland, OR, USA, 21–23 May 2000, pp. 11–20. ACM (2000)

[OW14]    O'Donnell, R., Witmer, D., Goldreich's, P.R.G.: Evidence for near-optimal polynomial stretch. In: IEEE 29th Conference on Computational

Complexity, CCC 2014, Vancouver, BC, Canada, June 11–13, 2014, pp. 1–12. IEEE (2014)

[PF79] Pippenger, N., Fischer, M.J.: Relations among complexity measures. J. ACM **26**(2), 361–381 (1979)

[PW88] Pitt, L., Warmuth, M.K.: Reductions among prediction problems on the difficulty of predicting automata. In: Proceedings: Third Annual Structure in Complexity Theory Conference, Georgetown University, Washington, D.C., USA, 14–17 June 1988, pp. 60–69. IEEE Computer Society (1988)

[RR97] Razborov, A.A., Rudich, S.: Natural proofs. J. Comput. Syst. Sci. **55**(1), 24–35 (1997)

[Tzu09] Tzur, Y.: Notions of weak pseudorandomness and $GF(2^n)$-polynomials. Master's thesis, Weizmann Institute of Science (2009)

[Val84] Valiant, L.G.: A theory of the learnable. In: DeMillo, R.A. (ed.) Proceedings of the 16th Annual ACM Symposium on Theory of Computing, Washington, DC, USA, 30 April–2 May1984, pp. 436–445. ACM (1984)

[Weg87] Wegener, I.: The Complexity of Boolean Functions. Teubner/Wiley, Stuttgart (1987)

[Yao82] Yao, A.C.: Theory and applications of trapdoor functions (extended abstract). In: 23rd Annual Symposium on Foundations of Computer Science, Chicago, Illinois, USA, 3–5 November 1982, pp. 80–91. IEEE Computer Society (1982)

# 3-Message Zero Knowledge Against Human Ignorance

Nir Bitansky[1]($\boxtimes$), Zvika Brakerski[2], Yael Kalai[3], Omer Paneth[4],
and Vinod Vaikuntanathan[1]

[1] MIT, Cambridge, USA
nirbitan@csail.mit.edu
[2] Weizmann, Rehovot, Israel
[3] Microsoft Research, Cambridge, USA
[4] Boston University, Boston, USA

**Abstract.** The notion of Zero Knowledge has driven the field of cryptography since its conception over thirty years ago. It is well established that two-message zero-knowledge protocols for NP do not exist, and that four-message zero-knowledge arguments exist under the minimal assumption of one-way functions. Resolving the precise round complexity of zero-knowledge has been an outstanding open problem for far too long.

In this work, we present a three-message zero-knowledge argument system with soundness against uniform polynomial-time cheating provers. The main component in our construction is the recent delegation protocol for RAM computations (Kalai and Paneth, TCC 2016B and Brakerski, Holmgren and Kalai, ePrint 2016). Concretely, we rely on a three-message variant of their protocol based on a *key-less* collision-resistant hash functions secure against uniform adversaries as well as other standard primitives.

More generally, beyond uniform provers, our protocol provides a natural and meaningful security guarantee against real-world adversaries, which we formalize following Rogaway's "human-ignorance" approach (VIETCRYPT 2006): in a nutshell, we give an explicit uniform reduction from any adversary breaking the soundness of our protocol to finding collisions in the underlying hash function.

N. Bitansky—Research supported in part by DARPA Safeware Grant, NSF CAREER Award CNS-1350619, CNS-1413964 and by the NEC Corporation.

Z. Brakerski—Supported by the Israel Science Foundation (Grant No. 468/14), the Alon Young Faculty Fellowship, Binational Science Foundation (Grant No. 712307) and Google Faculty Research Award.

O. Paneth—Supported by the Simons award for graduate students in Theoretical Computer Science and an NSF Algorithmic foundations grant 1218461.

V. Vaikuntanathan—Research supported in part by DARPA Grant number FA8750-11-2-0225, NSF CAREER Award CNS-1350619, NSF Grant CNS-1413964 (MACS: A Modular Approach to Computer Security), Alfred P. Sloan Research Fellowship, Microsoft Faculty Fellowship, NEC Corporation and a Steven and Renee Finn Career Development Chair from MIT.

M. Hirt and A. Smith (Eds.): TCC 2016-B, Part I, LNCS 9985, pp. 57–83, 2016.
DOI: 10.1007/978-3-662-53641-4_3

# 1   Introduction

The fascinating notion of zero knowledge, conceived over thirty years ago by Goldwasser, Micali and Rackoff [GMR89], has been the source of a great many ideas that revolutionized cryptography, including the simulation paradigm and passive-to-active security transformations [GMW91,FLS99,Bar01,IKOS09].

A central and persistent open question in the theory of zero knowledge is that of round complexity (also called message complexity), which refers to the number of messages that the prover and the verifier must exchange in a zero-knowledge protocol. The seminal work of Goldreich, Micali and Wigderson [GMW91] showed the first computational zero-knowledge *proof* system for all of NP. Their protocol required a polynomial (in the security parameter) number of rounds (in order to achieve an exponentially small soundness error). Feige and Shamir [FS89] show a *four-round* computational zero-knowledge *argument* system [BCC88] for all of NPbased on algebraic assumptions.[1] The assumption was reduced to the minimal assumption of one-way functions by Bellare, Jakobsson and Yung [BJY97].

In terms of lower bounds, Goldreich and Oren [GO94] showed that *three rounds* are necessary for non-trivial zero knowledge (arguments as well as proofs) against non-uniform adversarial verifiers. Zero knowledge in the presence of verifiers with non-uniform advice has by now become the gold standard as it is often essential for secure composition (see, e.g., [GK96b]).

This state of affairs leaves behind a question that has been open for far too long:

*What is the minimal round-complexity of zero knowledge?*

By the works of [FS89,BJY97], the answer is at most 4 while Goldreich and Oren tell us that the answer is at least 3. So far, all constructions of three-message computational zero-knowledge argument systems for NPwere based on strong "auxiliary-input knowledge assumptions" [HT98,BP04b,CD09,BP12,BCC+14]. The plausibility of these assumptions was questioned already around their introduction [HT98] and they were recently shown to be false assuming the existence of *indistinguishability obfuscation* [BCPR14,BM14]. In summary, finding a three-message zero-knowledge argument (under reasonable, falsifiable assumptions) matching the Goldreich-Oren lower bound remains wide open.

**Why is Three-Message Zero Knowledge so Interesting.** Aside from its significance to the theory of zero knowledge, the question of three-message zero knowledge is also motivated by its connections to two fundamental notions in cryptography, namely *non-black-box security proofs* and *verifiable computation*.

In order to make sense of this, let us tell you the one other piece of the zero knowledge story. An important dimension of zero-knowledge proofs is whether the zero-knowledge simulator treats the (adversarial) verifier as a black-box or

---

[1] While zero-knowledge *proofs* [GMW91] provide soundness against computationally unbounded cheating provers, zero-knowledge *arguments* [BCC88] are weaker in that they provide soundness only against computationally bounded cheating provers.

not. In all the protocols referenced above (with the exception of the ones based on "auxiliary-input knowledge assumptions"), the simulator treats the verifier as a black box. Goldreich and Krawczyk [GK96b] show that in any three-message zero-knowledge protocol for a language outside BPP, the simulator *must* make non-black-box use of the verifier's code. In other words, any future three-message zero-knowledge protocol has to "look different" from the ones referenced above.

The pioneering work of Barak [Bar01] demonstrated that barriers of this kind can sometimes be circumvented via non-black-box simulation. However, Barak's technique, and all other non-black-box techniques developed thus far, have only led to protocols with at least four messages [BP13, COP+14].

A bottleneck to reducing the round-complexity of Barak's protocol is the reliance on four-message *universal arguments* [BG08], a notion that enables fast verification of NPcomputations. Accordingly, developments in round-efficient systems for verifiable computation may very well lead to corresponding developments in three-message zero knowledge. In fact, strong forms of verifiable computation have recently proven instrumental in producing novel non-black-box simulation techniques, such as in the context of constant-round concurrent protocols [CLP13b, CLP15]. It is natural, then, to wonder whether these and related developments help us construct three-message zero-knowledge argument systems.

**On Uniform (and Bounded Non-uniform) Verifiers.** Bitansky, Canetti, Paneth and Rosen [BCPR14] study three-message protocols satisfying a relaxed notion of zero knowledge. Instead of requiring the zero knowledge guarantee against all non-uniform verifiers, they only consider verifiers that have an a-priori bounded amount of non-uniformity (but may still run for an arbitrary polynomial time). This includes, in particular, zero-knowledge against uniform verifiers. They demonstrate a three-message zero-knowledge protocol against verifiers with bounded non-uniformity based on the verifiable delegation protocol of Kalai, Raz, and Rothblum [KRR14].

Notably, restricting attention to verifiers with bounded uniformity comes with a great compromise. For once, the zero knowledge property is not preserved under sequential composition. More broadly, such protocols may not provide a meaningful security guarantee against real-world adversaries. As a concrete example, the zero knowledge property of the protocol in [BCPR13] crucially relies on the fact that messages sent by the verifier can be simulated by a Turing machine with a short description, shorter than the protocol's communication. However, this assumption may not hold for real-world adversaries, which can certainly have access to arbitrarily long strings with no apparent short description.

## 1.1   Our Results

In this work, we construct a three-message argument for NPthat is zero knowledge against fully non-uniform verifiers and sound against provers with a-priori bounded (polynomial amount of) non-uniformity. The main component in our construction is a verifiable delegation protocol for RAM computations recently constructed by Kalai and Paneth [KP15] and improved by Brakerski, Holmgren and Kalai [BHK16]. Concretely, we rely on a three-message variant of the [BHK16] protocol based on *keyless* collision-resistant hash functions secure against adversaries with bounded non-uniformity and slightly super-polynomial running time, and a (polynomially-secure) computational private information retrieval (PIR) scheme, as well as other more standard cryptographic assumptions.

In contrast to the setting of verifiers with bounded non-uniformity, our protocol remains secure under sequential composition. Furthermore, our protocol provides a natural and meaningful security guarantee against real-world adversaries, which we formalize following Rogaway's "human ignorance approach" [Rog06], described in greater detail below.

**Rogaway's "Human Ignorance" Approach and Real-World Security.** A more informative way of describing the soundness of our protocol is by the corresponding security reduction. We construct a zero-knowledge *argument* system, meaning that the soundness of the protocol is computational. That is, any prover that breaks the soundness of our protocol, regardless of how non-uniform it is, can be *uniformly* turned into a collision finder for an underlying hash function. In other words, there is a *uniform* algorithm called collision-finder who finds collisions in the hash function given oracle access to the soundness-breaker.

In our protocol, the hash function must already be determined before the first message is sent, thus requiring that we rely on a fixed (key-less) function as opposed to a function family as is normally the case when dealing with collision-resistant hash functions. Clearly, a fixed hash function cannot be collision-resistant against non-uniform adversaries (as such an adversary can have a collision for the function hard-wired as part of its non-uniform advice). However, as argued by Rogaway, a *uniform* reduction from finding collisions in such a function to breaking the security of a protocol is sufficient to argue the real-world security of the protocol. Briefly, the rationale is that an adversarial algorithm that breaks the security of the protocol (with or without non-uniform advice) can be turned into an explicit algorithm that finds collisions in the hash function (with the *same* non-uniform advice). Indeed, for common constructions of hash functions, such as SHA-3, collisions (while they surely exist) are simply not known.

Our main result can accordingly be stated as follows.

**Informal Theorem 1.1** [See Theorem 3.1]. *Assuming the existence of a computational private information retrieval (PIR) scheme, a circuit-private 1-hop homomorphic encryption scheme, and a non-interactive commitment scheme, there exists a three-message argument for NP with a uniform reduction $\mathcal{R}$ (described in the proof of Theorem 3.1) running in quasi-polynomial time, such that, for every non-uniform PPT adversary $\mathcal{A}$, if $\mathcal{A}$ breaks the soundness of the protocol instantiated with a keyless hash function $\mathcal{H}$, then $\mathcal{R}^{\mathcal{A}}$ outputs a collision in $\mathcal{H}$. The protocol is zero knowledge against non-uniform probabilistic polynomial-time (PPT) verifiers.*

All the cryptographic primitives (except the key-less hash function) can be instantiated from the learning with errors (LWE) assumption [Reg09].

**Asymptotic Interpretations.** As discussed above, implementing our protocol with a key-less hash function such as SHA-3 guarantees security against "ignorant" adversaries that are unable to find hash collisions. This class of adversaries may include all the adversaries we care about in practice, however, since functions like SHA-3 do not provide any asymptotic security, we cannot use standard asymptotic terminology to define the class of "SHA-ignorant adversaries".

We formalize the security of our protocol and hash function in conventional asymptotic terms. For any asymptotic hash family $\mathcal{H} = \{\mathcal{H}_n\}_{n \in \mathbb{N}}$, we can accordingly think of the class of adversaries that are $\mathcal{H}$-ignorant. Trying to capture more natural classes of adversaries, we focus on the subclass of adversaries with bounded non-uniformity. It may be reasonable to assume that an asymptotic keyless hash function is indeed collision-resistant against this class as long as the corresponding non-uniform advice is shorter than the hash input length. Therefore, the result for adversaries with bounded non-uniformity stated above follows as a corollary of our explicit reduction.

**The Global Common Random String Model and Resettable Security.** Another direct corollary of our result is that assuming (the standard notion of) keyed collision-resistant hash-function families, there is a 3-message zero-knowledge protocol that is sound against fully non-uniform provers in the *global (or non-programable) common random string model* [Pas03, CDPW07] or in the global hash model [CLP13a]. As observed in [Pas03], both the Goldreich-Oren lower bound and the Goldreich-Krawczyk black-box lower bound hold even in these models.

Another property of our protocol is that it can be made resettably sound [BGGL01] via the (round-preserving) transformation of Barak, Goldreich, Goldwasser and Lindell [BGGL01]. This holds for the three-message version of the protocol (against provers with bounded uniformity, or alternatively, against non-uniform provers in the global random string model).

## 1.2  Our Techniques

We now give an overview of the main ideas behind the new protocol.

**Barak's Protocol.** As explained above, three-message zero-knowledge can only be achieved via *non-black-box* simulation (and the Goldreich-Krawczyk lower bound, in fact, holds even when considering uniform provers). Thus, a natural starting point is the non-black-box simulation technique of Barak [Bar01], which we outline next. Following the Feige-Lapidot-Shamir paradigm [FLS99], the prover and verifier in Barak's protocol first execute a *trapdoor generation preamble*: the verifier sends a key $h$ for a collision-resistant hash function, the prover responds with a commitment cmt, and then, the verifier sends a random challenge $u$. The preamble defines a "trapdoor statement" asserting that there exists a program $\Pi$ such that cmt is a commitment to $h(\Pi)$ and $\Pi(\mathsf{cmt})$ outputs

$u$. Intuitively, no cheating prover is able to commit to a code that predicts the random $u$ ahead of time, and thus cannot obtain a witness (a program $\Pi$) for the trapdoor statement. In contrast, a simulator that is given the code of the (malicious) verifier, can commit to it in the preamble and use it as the witness for the trapdoor statement.

In the second stage of the protocol, the prover and the verifier engage in a witness-indistinguishable (WI) protocol intended to convince the verifier that either the real statement or the trapdoor statement is true, without revealing to the verifier which is the case. Here, since the trapdoor statement corresponds to a computation $\Pi(\mathsf{cmt})$ that may be longer than the honest verifier's run-time, a standard WI system is insufficient. This difficulty is circumvented using the 4-message universal arguments mentioned before, where verification time is independent of the statement being proven.

Overall, Barak's protocol is executed in six messages. In the first message, the verifier sends a key for a collision-resistant hash function, which effectively serves both as the first message (out of three) of the preamble and as the first message (out of four) of the universal argument to come. Then, the two remaining messages of the preamble are sent, following by the remaining three messages of a WI universal argument.[2]

**Squashing Barak's Protocol.** To achieve a three-message protocol, we will squash Barak's protocol. Using a keyless hash function, we can eliminate the first verifier message (which, in Barak's protocol, consists of a key for a collision-resistant hash function). It is just this step that restricts our soundness guarantee to only hold against provers that are unable to find collisions in the key-less hash function (e.g., provers with bounded non-uniformity). This leaves us with a *five*-message protocol, which is still worse than what is achievable using black-box techniques. The bulk of the technical contribution of this work is devoted to the task of squashing this protocol into only *three* messages.

Having eliminated the verifier's first message, we are now left with a 2-message preamble followed by a 3-message WI universal argument. A natural next step is to attempt executing the preamble and the WI argument in parallel. The main problem with this idea is that in Barak and Goldreich's universal arguments, the statement must be fixed before the first prover message is computed. However, in the protocol described, the trapdoor statement is only fixed once the entire preamble has been executed.

We observe that, paradoxically, while the trapdoor statement is only fixed after the preamble has been executed, *the witness for this statement is fixed before the protocol even starts!* Indeed, the witness for the trapdoor statement is simply the verifier's code. It is therefore sufficient to replace Barak and Goldreich's universal argument with a 3-message verifiable delegation protocol that has the following structure: the first prover message depends on the witness alone, the verifier's message fixes the statement, and the third and last prover response includes the proof (which already depends on both the statement and witness).

---

[2] Barak's original construction, in fact, consists of seven messages, but can be squashed into six by using an appropriate WI system (see, e.g., [OV12]).

**Verifiable Memory Delegation.** To obtain a verifiable delegation scheme with the desired structure, we consider the notion of verifiable memory delegation [CKLR11]. In memory delegation, the prover and verifier interact in two phases. In the offline phase, the verifier sends a large memory string $m$ to the prover, saving only a short digest of $m$. In the online phase, the verifier sends a function $f$ to the prover and the prover responds with the output $f(m)$ together with a proof of correctness. The time to verify the proof is independent of the memory size and the function running time.

In our setting, we think of the memory as the witness and of the delegated function as verifying that its input is a valid witness for a specified statement (encoded in the function). One important difference between the settings of verifiable memory delegation and ours is that in the former, the offline phase is executed by the verifier, but in our setting, the prover may adversarially choose any digest (which may not even correspond to any memory string). We therefore rely on memory delegation schemes that remain secure for an adversarially chosen digest. We observe that the verifiable delegation protocols for RAM computations of [KP15, BHK16] yield exactly such a memory delegation scheme, and when implemented using a keyless hash function this delegation scheme is secure against the class of adversaries that cannot find collisions in the hash function (e.g. adversaries with bounded non-uniformity).

Fulfilling the above plan encounters additional hurdles. The main such hurdle is the fact that the verifiable delegation schemes of [KP15, BHK16] are not witness-indistinguishable. We ensure witness-indistinguishability by leveraging special properties of the Lapidot-Shamir WI protocol [LS90a, OV12], and 1-hop homomorphic encryption [GHV10] (similar ideas were used in [BCPR14]).

## 1.3   More Related Work

We mention other related works on round-efficient zero knowledge.

**On Zero-Knowledge Proof Systems.** In this work we show a 3-message *argument* system for NP. If one requires a *proof* system instead, with soundness against unbounded provers, Goldreich and Kahan [GK96a] showed a 5-round (black-box) zero-knowledge proof system for NP. On the other hand, Katz [Kat12], extending the result of Goldreich and Krawczyk [GK96b], shows that, assuming the polynomial hierarchy does not collapse, zero-knowledge protocols for an NP-complete language require at least 5 rounds if the simulator only makes black-box use of the verifier's code. The question of 3-round and 4-round zero-knowledge proof systems for NP(necessarily with non-black-box simulation) still remains wide open.

**On Quasi-Polynomial Time Simulation.** Barak and Pass [BP04a] show a 1-round *weak* zero-knowledge argument for NPwith soundness against uniform polynomial-time provers, based on non-standard assumptions. (One of their assumptions is the existence of a key-less collision-resistant hash function against uniform adversaries with sub-exponential running time.) Their notion of *weak* zero knowledge allows for a quasi-polynomial-time simulator. The fact that the simulator can run longer than (any possible) cheating prover means that the simulator can (and does) break the soundness of the protocol. This has the effect that the round-complexity lower bounds referenced above do not apply in this model. Furthermore, such a protocol may leak information that cannot be simulated in polynomial time (but only in quasi-polynomial time).

**Organization.** In Sect. 2, we give the basic definitions used throughout the paper, including the modeling of adversaries and reductions, the definition of keyless hash functions, and memory delegation. In Sect. 3, we describe and analyze the new protocol.

## 2   Definitions and Tools

In this section, we define the adversarial model we work in, zero-knowledge protocols against restricted classes of provers (e.g., ones with bounded non-uniformity), as well as the tools used in our construction.

### 2.1   Modeling Adversaries, Reductions, and Non-uniformity

In this section, we recall the notion of (black-box) reductions, and address two general classes of adversaries touched in this paper. Commonly in crypto, we consider (uniform) polynomial time reductions between different non-uniform polynomial-time adversaries. In this paper, we will sometimes consider more general types of reductions, e.g. uniform reductions that run in slightly super-polynomial time, as well as different classes of adversaries, e.g. uniform PPT adversaries, or adversaries with bounded non-uniformity. In such cases, we will be explicit about the concrete classes of reductions and adversaries involved.

**Rogaway's "Human Ignorance" Approach to Reductions.** As discussed in the introduction, the most informative way of describing the soundness of our protocol is by the corresponding security reduction from collision-resistance to soundness. Rogaway [Rog06] suggests a framework for formalizing such statements. In this work, however, for the sake of simpler exposition, we do not fully follow Rogaway's framework; we explain the differences next.

While Rogaway's approach gives a meaningful result even for non-asymptotic hash functions such as SHA-3 in terms of concrete security, our security definitions are still formalized in asymptotic terms. We parameterize the security definitions by the class of adversaries. Our main theorem states that for every class of adversaries $\mathbb{A}$, the soundness of the protocol against adversaries in $\mathbb{A}$ can be reduced to the security of the hash function against the same class of adversaries.

We note that the security of our protocol is based on other primitives except keyless collision-resistant hash. In our theorems, we do not emphasize the reduction to these primitives; rather, we simply restrict our result only to classes of adversaries that are unable to break the security of these primitives (most naturally non-uniform polynomial time adversaries).

**Reductions.** For two classes of adversaries $\mathbb{R}, \mathbb{A}$, we denote by $\mathbb{R}^{\mathbb{A}}$ the class of adversaries $\mathcal{R}^{\mathcal{A}} = \{\mathcal{R}_n^{\mathcal{A}_n}\}_{n \in \mathbb{N}}$ where $\mathcal{R}_n$ makes calls to $\mathcal{A}_n$.[3]

**The class $\mathbb{P}$ of non-uniform PPT adversaries.** A general class of adversaries considered in this paper are non-uniform probabilistic polynomial-time Turing machines, or in short non-uniform PPT, which we denote by $\mathbb{P}$. Any such adversary $\mathcal{A} \in \mathbb{P}$ is modeled as a sequence $\{\mathcal{A}_n\}_{n \in \mathbb{N}}$, where $n$ is the security parameter, and where the description and running time of $\mathcal{A}_n$ are polynomially bounded in $n$.

For a super-polynomial $\gamma(n) = n^{\omega(1)}$, we denote by $\mathbb{P}_\gamma$ the class of non-uniform probabilistic adversaries whose description and running time are polynomial in $\gamma(n)$.

**The class $\mathbb{B}$ of PPT adversaries with bounded non-uniformity.** We shall also consider the class $\mathbb{B}_\beta \subset \mathbb{P}$ of adversaries with bounded non-uniformity $O(\beta)$. Concretely, for a fixed function $\beta(n) \leq n^{O(1)}$, the class $\mathbb{B}_\beta$ consists of all non-uniform adversaries $\mathcal{A} \in \mathbb{P}$ whose description $|\mathcal{A}_n|$ is bounded by $O(\beta(n))$, *but their running time could be an arbitrary polynomial*. Abusing notation, we denote by $\mathbb{B}_0$ the class of *uniform* PPT *adversaries*.

For a super-polynomial function $\gamma(n) = n^{\omega(1)}$, we denote by $\mathbb{B}_{\beta,\gamma}$ the class of non-uniform probabilistic adversaries whose description is bounded by $O(\beta(n))$ (or the class of uniform probabilistic adversaries if $\beta = 0$) and running time is polynomial in $\gamma(n)$.

## 2.2 Zero Knowledge Arguments of Knowledge Against Provers with Bounded Non-uniformity

The standard definition of zero knowledge [GMR89, Gol04] considers general non-uniform provers (and verifiers). We define soundness (or argument of knowledge) more generally against provers from a given class $\mathbb{A} \subset \mathbb{P}$. In particular, we will be interested in strict subclasses of $\mathbb{P}$, such as adversaries with bounded non-uniformity.

In what follows, we denote by $\langle P \leftrightarrows V \rangle$ a protocol between two parties $P$ and $V$. For input $w$ for $P$, and common input $x$, we denote by $\langle P(w) \leftrightarrows V \rangle(x)$ the output of $V$ in the protocol. For honest verifiers this output will be a single bit indicating acceptance (or rejection), whereas we assume (without loss of generality) that malicious verifiers outputs their entire view. Throughout, we assume that honest parties in all protocols are uniform PPT algorithms.

**Definition 2.1.** *A protocol* $\langle P \leftrightarrows V \rangle$ *for an NP relation* $\mathcal{R}_\mathcal{L}(x, w)$ *is a zero knowledge argument of knowledge against provers in class* $\mathbb{A} \subset \mathbb{P}$ *if it satisfies:*

1. **Completeness:** *For any* $n \in \mathbb{N}, x \in \mathcal{L} \cap \{0,1\}^n, w \in \mathcal{R}_\mathcal{L}(x)$:

$$\Pr\left[\langle P(w) \leftrightarrows V \rangle(x) = 1\right] = 1.$$

---

[3] In this paper, we shall explicitly address different classes of black-box reductions. One can analogously define non-black-box reductions.

2. **Computational zero knowledge:** *For every non-uniform PPT verifier* $V^* = \{V_n^*\}_{n \in \mathbb{N}} \in \mathbb{P}$, *there exists a (uniform) PPT simulator $\mathcal{S}$ such that:*

$$\{\langle P(w) \leftrightarrows V_n^*(x)\rangle\}_{\substack{(x,w) \in \mathcal{R}_\mathcal{L} \\ |x|=n}} \approx_c \{\mathcal{S}(V_n^*, x)\}_{\substack{(x,w) \in \mathcal{R}_\mathcal{L} \\ |x|=n}}.$$

3. **Argument of knowledge:** *There is a uniform PPT extractor $\mathcal{E}$, such that for any noticeable function $\varepsilon(n) = n^{-O(1)}$, any prover $P^* = \{P_n^*\}_{n \in \mathbb{N}} \in \mathbb{A}$, any security parameter $n \in \mathbb{N}$, and any $x \in \{0,1\}^n$ generated by $P_n^*$ prior to the interaction:*

$$\text{if } \Pr\left[\langle P_n^* \leftrightarrows V\rangle(x) = 1\right] \geq \varepsilon(n),$$

$$\text{then } \Pr\left[\begin{matrix} w \leftarrow \mathcal{E}^{P_n^*}(1^{1/\varepsilon(n)}, x) \\ w \notin \mathcal{R}_\mathcal{L}(x) \end{matrix}\right] = \text{negl}(n).$$

## 2.3  Collision-Resistant Hashing

We define the notion of a keyless hash function that is collision resistant against a class $\mathbb{A} \subseteq \mathbb{P}_\gamma$ of adversaries. In particular, the definition may be realizable only for strict subclasses of $\mathbb{P}_\gamma$, such as the class $\mathbb{B}_{\beta,\gamma}$ of adversaries with bounded non-uniformity and $\text{poly}(\gamma(n))$ running time (where the description length of the adversary, namely $\beta$, will be shorter than the length of the input to the hash).

**Definition 2.2.** *Let $n < \ell(n) \leq n^{O(1)}$. A polynomial-time computable function*

$$\mathcal{H} = \{\mathcal{H}_n\}_{n \in \mathbb{N}}, \mathcal{H}_n : \{0,1\}^{\ell(n)} \to \{0,1\}^n,$$

*is collision resistant against adversaries in $\mathbb{A}$ if for any $\mathcal{A} = \{\mathcal{A}_n\}_{n \in \mathbb{N}} \in \mathbb{A}$, and every $n \in \mathbb{N}$*

$$\Pr\left[\begin{matrix} x, y \leftarrow \mathcal{A}_n; \\ \mathcal{H}_n(x) = \mathcal{H}_n(y) \end{matrix}\right] = \text{negl}(n).$$

*where the probability is over the coins of $\mathcal{A}_n$.*

**Instantiation.** Common constructions of keyless hash functions such as SHA-3 have a fixed output length and therefore do not directly provide a candidate for an asymptotic hash function as in Definition 2.2. One way to obtain candidates for an asymptotic hash function is to start with a family $\mathcal{H}'$ of (keyed) hash-functions

$$\mathcal{H}' = \{\mathcal{H}'_{n,k}\}_{n \in \mathbb{N}, k \in \{0,1\}^n}, \mathcal{H}'_{n,k} : \{0,1\}^{\ell(n)} \to \{0,1\}^n,$$

and fix a uniform polynomial time algorithm $K$ that given a security parameter $1^n$ outputs a key $k \in \{0,1\}^n$. The keyless hash $\mathcal{H}$ is then given by

$$\mathcal{H}_n = \mathcal{H}'_{n,K(1^n)}.$$

For $\mathcal{H}_n$ to be a good candidate collision resistant hash against adversaries in $\mathbb{B}_\beta$, we should make sure that $\beta = o(\ell)$, the family $\mathcal{H}'$ is collision resistant, and the algorithm $K$ behaves "sufficiently like a random oracle". For example we can choose an algorithm $K$ that uses a hash function like SHA-3 (or a version of it that can hash strings of arbitrary length) as a random oracle to output sufficiently many random bits.

## 2.4    Memory Delegation with Public Digest

A two-message memory delegation scheme [CKLR11] allows a client to delegate a large memory to an untrusted server, saving only a short digest of the memory. The client then selects a deterministic computation to be executed over the memory and delegates the computation to the server. The server responds with the computation's output as well as a short proof of correctness that can be verified by the client in time that is independent of that of the delegated computation and the size of the memory.

The notion of memory delegation we consider differs from that of [CKLR11] in the following ways.

- **Read-only computation.** We do not consider computations that update the memory. In particular, the digest of the delegated memory is computed once and does not change as a result of the computations.
- **Soundness.** We define soundness more generally for servers from a given class $\mathbb{A} \subset \mathbb{P}$. Whereas soundness is usually required against the class of all non-uniform PPT adversaries $\mathbb{P}$, we will also be interested in strict subclasses of $\mathbb{P}$, such as adversaries with bounded non-uniformity.
- **Soundness for slightly super-polynomial computations.** We require soundness to hold even for delegated computations running in slightly super-polynomial time.
- **Public digest.** We require that the digest of the memory can be computed non-interactively, and can be made public and used by any client to delegate computations over the same memory without compromising soundness. In particular, the client is not required to save any secret state when delegating the memory.
  Importantly, we do not assume that the party computing the digest is honest. We require that no efficient adversary can produce valid proofs for two different outputs for the same computation with respect to the same digest, even if the digest and computation are adversarially chosen.[4]
- **First message independent of function being delegated.** The first message of the delegation scheme (denoted below by $q$) depends only on the security parameter, and does not depend on the public digest or on the function being delegated.

Concretely, a two-message memory delegation scheme with public digest consists of four polynomial-time algorithms:

- $d \leftarrow \mathsf{Digest}(1^n, D)$ is a deterministic algorithm that takes a security parameter $1^n$ and memory $D$ and outputs a digest $d \in \{0, 1\}^n$.
- $(q, \tau) \leftarrow \mathsf{Query}(1^n)$ is a probabilistic algorithm that outputs a query $q$ and a secret state $\tau$. We assume w.l.o.g that the secret state $\tau$ is simply the random coins used by $\mathsf{Query}$.

---

[4] Soundness with respect to an adversarial digest can be defined in a stronger way, for example, requiring knowledge of the memory corresponding to the digest. However, this stronger requirement is not necessary for our application.

- $\pi \leftarrow \mathsf{Prov}(1^t, \mathcal{M}, D, q)$ is a deterministic algorithm that takes a description of a Turing machine $\mathcal{M}$ and a bound $t$ on the running time of $\mathcal{M}(D)$ and outputs a proof $\pi \in \{0,1\}^n$.
- $\mathsf{b} \leftarrow \mathsf{Ver}(d, \tau, \mathcal{M}, t, y, \pi)$ is a deterministic algorithm that takes a computation output $y$ and outputs an acceptance bit $\mathsf{b}$.

**Definition 2.3 (Memory Delegation with Public Digest).** *Let $\gamma(n)$ be a super-polynomial function such that $n^{\omega(1)} = \gamma(n) < 2^n$. A two-message memory delegation scheme* $(\mathsf{Digest}, \mathsf{Query}, \mathsf{Prov}, \mathsf{Ver})$ *for $\gamma$-time computations with public digest against provers in a class $\mathbb{A} \subset \mathbb{P}$ satisfies the following.*

- **Completeness.** *For every security parameter $n \in \mathbb{N}$, every Turing machine $\mathcal{M}$ and every memory $D \in \{0,1\}^*$ such that $\mathcal{M}(D)$ outputs $y$ within $t \leq 2^n$ steps:*

$$
\Pr\left[1 = \mathsf{Ver}(d, \tau, \mathcal{M}, t, y, \pi) \;\middle|\; \begin{array}{l} d \leftarrow \mathsf{Digest}(1^n, D) \\ (q, \tau) \leftarrow \mathsf{Query}(1^n) \\ \pi \leftarrow \mathsf{Prov}(1^t, \mathcal{M}, D, q) \end{array}\right] = 1.
$$

- **Soundness.** *For every adversary $\mathcal{A} = \{\mathcal{A}_n\}_{n \in \mathbb{N}} \in \mathbb{A}$, there exists a negligible function $\mathsf{negl}(\cdot)$ such that for every security parameter $n \in \mathbb{N}$,*

$$
\Pr\left[\begin{array}{l} t \leq \gamma(n) \\ y \neq y' \\ 1 = \mathsf{Ver}(d, \tau, \mathcal{M}, t, y, \pi) \\ 1 = \mathsf{Ver}(d, \tau, \mathcal{M}, t, y', \pi') \end{array} \;\middle|\; \begin{array}{l} (\mathcal{M}, t, d, y, y') \leftarrow \mathcal{A}_n \\ (q, \tau) \leftarrow \mathsf{Query}(1^n) \\ (\pi, \pi') \leftarrow \mathcal{A}_n(q) \end{array}\right] = \mathsf{negl}(n).
$$

**Instantiation.** A memory delegation scheme satisfying Definition 2.3 can be obtained based on the delegation schemes for RAM computations of Kalai and Paneth [KP15] and that of Brakerski, Holmgren and Kalai [BHK16] with slight adaptations.[5] Below we describe the required adaptations. We focus on the scheme of [BHK16] that can be instantiated based on polynomially-secure PIR.

- **Remove public parameters.** The scheme of [BHK16] has public parameters that are generated honestly before the memory is delegated. These parameters consist of the description of a hash function chosen randomly from a family of collision-resistant hash functions. Here we remove the public parameters and instead use a keyless collision resistant hash against adversaries from a restricted class $\mathbb{A}$. (E.g., $\mathbb{A}$ can be the class of adversaries with $\beta$-bounded non-uniformity $\mathbb{B}_\beta$.) The security of our modified scheme against provers from $\mathbb{A}$ follows the same argument as in [BHK16], who show a uniform black-box reduction from a cheating prover to an adversary that finds collisions.
- **Soundness for slightly super-polynomial computations.** While the scheme of [BHK16] has completeness even for exponentially long delegated

---

[5] We note that we cannot use here the memory delegation scheme of [CKLR11, KRR14] since the soundness of their scheme assumes that the digest is honestly generated.

computations, soundness is only proved when the delegated computation is polynomial time. Here we require soundness even against slightly super-polynomial time $\gamma = n^{\omega(1)}$. In the [BHK16] reduction the running time of the adversary breaking the hash is proportional to the running time of the delegated computation. Therefore, soundness for slightly super-polynomial computations follows by the same argument, assuming a slightly stronger collision-resistance against adversaries from $\mathbb{B}_{0,\gamma}^{\mathbb{A}}$ who can run in time $\gamma$ and use A as a black box.

Recall that $\mathbb{B}_{0,\gamma}^{\mathbb{A}}$ is the class of uniform probabilistic machines running in time $\gamma(n)^{O(1)}$ and given oracle access to an adversary in A. Brakerski, Holmgren and Kalai prove that there is a $\gamma(n)^{O(1)}$-time uniform reduction from breaking the soundness of their scheme to breaking any underlying hash function, assuming the existence of a (polynomially secure) computational PIR scheme.

**Theorem 2.1 [BHK16].** *For any* $\mathbb{A} \subset \mathbb{P}$ *and (possibly super-polynomial) function* $\gamma(\cdot)$, *assuming collision-resistant hash functions against adversaries in* $\mathbb{B}_{0,\gamma}^{\mathbb{A}}$ *and a computational PIR scheme, there exists a two-message memory delegation scheme for* $\gamma$-*time computations with public digest against provers in* A.

## 2.5 Witness Indistinguishability with First-Message-Dependent Instances

We define 3-message WI proofs of knowledge where the choice of statement and witness may depend on the first message in the protocol. In particular, the first message is generated independently of the statement and witness. Also, while we do allow the content of the message to depend on the length $\ell$ of the statement, the message length should be of fixed to $n$ (this allows to also deal with statements of length $\ell > n$). The former requirement was formulated in several previous works (see, e.g., [HV16]) and the latter requirement was defined in [BCPR14].

**Definition 2.4 (WIPOK with first-message-dependent instances).** *Let* $\langle P \leftrightarrows V \rangle$ *be a 3-message argument for* $\mathcal{L}$ *with messages* $(\mathsf{wi}_1, \mathsf{wi}_2, \mathsf{wi}_3)$; *we say that it is a WIPOK with first-message-dependent instances if it satisfies:*

1. **Completeness with first-message-dependent** instances: *For any instance choosing function* $X$, *and* $\ell, n \in \mathbb{N}$,

$$\Pr \left[ V(x, \mathsf{wi}_1, \mathsf{wi}_2, \mathsf{wi}_3; r') = 1 \middle| \begin{array}{l} \mathsf{wi}_1 \leftarrow P(1^n, \ell; r) \\ (x, w) \leftarrow X(\mathsf{wi}_1) \\ x \in \mathcal{L}, w \in \mathcal{R}_{\mathcal{L}}(x) \\ \mathsf{wi}_2 \leftarrow V(\ell, \mathsf{wi}_1; r') \\ \mathsf{wi}_3 \leftarrow P(x, w, \mathsf{wi}_1, \mathsf{wi}_2; r) \end{array} \right] = 1,$$

*where* $r, r' \leftarrow \{0, 1\}^{\mathrm{poly}(n)}$ *are the randomness used by* $P$ *and* $V$.

*The honest prover's first message* $\mathsf{wi}_1$ *is of length* $n$, *independent of the length* $\ell$ *of the statement* $x$.

2. **Adaptive witness-indistinguishability:** *For any polynomial $\ell(\cdot)$, non-uniform PPT verifier $V^* = \{V_n^*\}_{n \in \mathbb{N}} \in \mathbb{P}$ and all $n \in \mathbb{N}$:*

$$\Pr\left[ V_n^*(x, \mathsf{wi}_1, \mathsf{wi}_2, \mathsf{wi}_3) = b \;\middle|\; \begin{array}{l} \mathsf{wi}_1 \leftarrow P(1^n, \ell(n); r) \\ x, w_0, w_1, \mathsf{wi}_2 \leftarrow V_n^*(\mathsf{wi}_1) \\ \mathsf{wi}_3 \leftarrow P(x, w_b, \mathsf{wi}_1, \mathsf{wi}_2; r) \end{array} \right] \leq \frac{1}{2} + \mathrm{negl}(n),$$

*where $b \leftarrow \{0,1\}$, $r \leftarrow \{0,1\}^{\mathrm{poly}(n)}$ is the randomness used by $P$, $x \in \mathcal{L} \cap \{0,1\}^{\ell(n)}$ and $w_0, w_1 \in \mathcal{R}_\mathcal{L}(x)$.*

3. **Adaptive proof of knowledge:** *there is a uniform PPT extractor $\mathcal{E}$ such that for any polynomial $\ell(\cdot)$, all large enough $n \in \mathbb{N}$, and any deterministic prover $P^*$:*

$$\text{if } \Pr\left[ V(\mathsf{tr}; r') = 1 \;\middle|\; \begin{array}{l} \mathsf{wi}_1 \leftarrow P^* \\ \mathsf{wi}_2 \leftarrow V(\ell(n), \mathsf{wi}_1; r') \\ x, \mathsf{wi}_3 \leftarrow P^*(\mathsf{wi}_1, \mathsf{wi}_2) \\ \mathsf{tr} = (x, \mathsf{wi}_1, \mathsf{wi}_2, \mathsf{wi}_3) \end{array} \right] \geq \varepsilon,$$

$$\text{then } \Pr\left[ \begin{array}{l} V(\mathsf{tr}; r') = 1 \\ w \leftarrow \mathcal{E}^{P^*}(1^{1/\varepsilon}, \mathsf{tr}) \\ w \notin \mathcal{R}_\mathcal{L}(x) \end{array} \;\middle|\; \begin{array}{l} \mathsf{wi}_1 \leftarrow P^* \\ \mathsf{wi}_2 \leftarrow V(\ell(n), \mathsf{wi}_1; r') \\ x, \mathsf{wi}_3 \leftarrow P^*(\mathsf{wi}_1, \mathsf{wi}_2) \\ \mathsf{tr} = (x, \mathsf{wi}_1, \mathsf{wi}_2, \mathsf{wi}_3) \end{array} \right] \leq \mathrm{negl}(n),$$

where $x \in \{0,1\}^{\ell(n)}$, and $r' \leftarrow \{0,1\}^{\mathrm{poly}(n)}$ is the randomness used by $V$.

**Instantiation.** Protocols with first-message-dependent instances follow directly from the WIPOK protocol constructed in [BCPR14], assuming ZAPs and non-interactive commitments (there, the first message is taken from a fixed distribution that is completely independent of the instance).

Next, we sketch how such a protocol can be constructed without ZAPs, but assuming keyless collision-resistant hash functions, thus collapsing to an argument of knowledge against adversaries that cannot break the hash (which will anyhow be the class of interest in our zero-knowledge protocol in Sect. 3).

**The Lapidot-Shamir protocol.** As observed in [OV12], the Lapidot-Shamir variant of the 3-message (honest verifier) zero-knowledge protocol for Hamiltonicity [LS90a] is such that the first and second messages only depend on the size of the instance $|x| = \ell$, but not on the instance and witness themselves. The protocol, in particular, supports instances up to size $\ell$ that depend on the prover's first message. However, the size of the first message $\mathsf{wi}_1$ in the protocol is $|\mathsf{wi}_1| > \ell$. We, on the other hand, would like to allow the instance $x$ to be of an arbitrary polynomial size in $|\mathsf{wi}_1|$, and in particular such that $|\mathsf{wi}_1| < \ell$.

We now sketch a simple transformation from any such protocol where, in addition, the verifier's message is independent of the first prover message, into a protocol that satisfies the required first-message dependence of instances. Indeed, the verifier message in the Lapidot-Shamir protocol is simply a uniformly random string, and hence the transformation can be applied here.

**The Transformation.** Let $\ell(n) > n$ be any polynomial function and let $\mathcal{H}$ be a keyless collision-resistant hash function from $\{0,1\}^{\ell(n)}$ to $\{0,1\}^n$. In the new protocol $(P_{\mathsf{new}}, V_{\mathsf{new}})$, the prover computes the first message $\mathsf{mes}_1$ for instances of length $\ell(n)$. Then, rather than sending $\mathsf{mes}_1$ in the clear, the prover $P_{\mathsf{new}}$ sends $y = \mathcal{H}_n(\mathsf{mes}_1) \in \{0,1\}^n$. The verifier proceeds as in the previous protocol $(P, V)$ (note that $\mathsf{mes}_1$ is not required for it to compute $\mathsf{mes}_2$). Finally the prover $P_{\mathsf{new}}$ answers as in the original protocol, and also sends $\mathsf{mes}_1$ in the clear. The verifier $V_{\mathsf{new}}$ accepts, if it would in the original protocol and $\mathsf{mes}_1$ is a preimage of $y$ under $\mathcal{H}_n$.

We first note that now the size of the instance $\ell$ can be chosen to be an arbitrary polynomial in the length $n = |\mathsf{wi}_1|$ of the first WI message. In addition, we note that the protocol is still WI, as the view of the verifier $V_{\mathsf{new}}$ in the new protocol can be perfectly simulated from the view of the verifier $V$ in the old protocol, by hashing the first message on its own.

Finally, we observe that any prover $P_{\mathsf{new}}^*$ that convinces the verifier in the new protocol of accepting with probability $\varepsilon$, can be transformed into a prover $P^*$ that convinces the verifier of the original protocol, or to a collision-finder. Indeed, the prover $P^*$ would first run $P_{\mathsf{new}}^*$ until the last message, i.e., until it obtains a valid preimage $\mathsf{mes}_1$ of $y$. Then it would proceed interacting with $V$ using $\mathsf{mes}_1$ as its first message, and using $P_{\mathsf{new}}^*$ to emulate the third message. By the collision resistance of $\mathcal{H}$ the prover $P_{\mathsf{new}}^*$ indeed cannot make the verifier $V_{\mathsf{new}}$ accept with respect to two different perimages $\mathsf{mes}_1, \mathsf{mes}_1'$, except with negligible probability. Thus the prover $P^*$ convinces $V$ with probability $\varepsilon - \mathsf{negl}(n)$.

## 2.6 1-Hop Homomorphic Encryption

A *1-hop homomorphic encryption scheme* [GHV10] allows a pair of parties to securely evaluate a function as follows: the first party encrypts an input, the second party homomorphically evaluates a function on the ciphertext, and the first party decrypts the evaluation result. (We do not require any compactness of post-evaluation ciphertexts.)

**Definition 2.5.** *A scheme* $(\mathsf{Enc}, \mathsf{Eval}, \mathsf{Dec})$, *where* $\mathsf{Enc}, \mathsf{Eval}$ *are probabilistic and* $\mathsf{Dec}$ *is deterministic, is a semantically-secure, circuit-private, 1-hop homomorphic encryption scheme if it satisfies the following properties:*

- **Perfect correctness:** *For any* $n \in \mathbb{N}$, $x \in \{0,1\}^n$ *and circuit* $C$:

$$\Pr \begin{bmatrix} (\mathsf{ct}, \mathsf{sk}) \leftarrow \mathsf{Enc}(x) \\ \widehat{\mathsf{ct}} \leftarrow \mathsf{Eval}(\mathsf{ct}, C) \\ \mathsf{Dec}_{\mathsf{sk}}(\widehat{\mathsf{ct}}) = C(x) \end{bmatrix} = 1.$$

*where the probability is over the coin tosses of* $\mathsf{Enc}$ *and* $\mathsf{Eval}$.
- **Semantic security:** *For any non-uniform PPT* $\mathcal{A} = \{\mathcal{A}_n\}_{n \in \mathbb{N}} \in \mathbb{P}$, *every* $n \in \mathbb{N}$, *and any pair of inputs* $x_0, x_1 \in \{0,1\}^{\mathsf{poly}(n)}$ *of equal length,*

$$\Pr_{\substack{b \leftarrow \{0,1\} \\ (\mathsf{ct}, \cdot) \leftarrow \mathsf{Enc}(x_b)}} [\mathcal{A}_n(\mathsf{ct}) = b] \leq \frac{1}{2} + \mathsf{negl}(n).$$

– **Circuit privacy:** *The randomized evaluation procedure,* Eval, *should not leak information on the input circuit $C$. This should hold even for malformed ciphertexts. Formally, let* $\mathcal{E}(x) = \mathsf{Supp}(\mathsf{Enc}(x))$ *be the set of all legal encryptions of $x$, let* $\mathcal{E}_n = \cup_{x\in\{0,1\}^n}\mathcal{E}(x)$ *be the set legal encryptions for strings of length $n$, and let $\mathcal{C}_n$ be the set of all circuits on $n$ input bits. There exists a (possibly unbounded) simulator $\mathcal{S}_{1\mathsf{hop}}$ such that:*

$$\{C, \mathsf{Eval}(c, C)\}_{\substack{n\in\mathbb{N}, C\in\mathcal{C}_n \\ x\in\{0,1\}^n, c\in\mathcal{E}(x)}} \approx_c \left\{C, \mathcal{S}_{1\mathsf{hop}}(c, C(x), 1^{|C|})\right\}_{\substack{n\in\mathbb{N}, C\in\mathcal{C}_n \\ x\in\{0,1\}^n, c\in\mathcal{E}(x)}}$$

$$\{C, \mathsf{Eval}(c, C)\}_{\substack{n\in\mathbb{N} \\ C\in\mathcal{C}_n, c\notin\mathcal{E}_n}} \approx_c \left\{C, \mathcal{S}_{1\mathsf{hop}}(c, \perp, 1^{|C|})\right\}_{\substack{n\in\mathbb{N} \\ C\in\mathcal{C}_n, c\notin\mathcal{E}_n}}.$$

**Instantiation.** 1-hop homomorphic encryption schemes can be instantiated based on any two-message two-party computation protocol secure against semi-honest adversaries; in particular, using Yao's garbled circuits and an appropriate 2-message oblivious transfer protocol, which can be based on the Decisional Diffie-Hellman assumption, the Quadratic Residuosity assumption, or the learning with errors assumption [Yao86, GHV10, NP01, AIR01, PVW08, HK12].

# 3   The Protocol

In this section, we construct a 3-message ZK argument of knowledge based on 2-message memory delegation schemes. More precisely, we show that for any class of adversaries $\mathbb{A} \subseteq \mathbb{P}$, given a delegation scheme that is sound against $\mathbb{B}_1^{\mathbb{A}}$, the protocol is an argument of knowledge against $\mathbb{A}$. For simplicity we focus on classes $\mathbb{A}$ that are closed under uniform reductions; namely $\mathbb{B}_1^{\mathbb{A}} \subseteq \mathbb{A}$. These will indeed capture the adversary classes of interest for this work. We start by listing the ingredients used in the protocol, as well as introducing relevant notation.

**Ingredients and notation:**

– A two-message memory delegation scheme (Digest, Query, Prov, Ver) for $\gamma$-bounded computations, sound against provers in $\mathbb{A} \subseteq \mathbb{P}$, for a class $\mathbb{A}$ closed under uniform reductions as in Definition 2.3.
– A semantically secure and circuit-private, 1-hop homomorphic encryption scheme (Enc, Eval, Dec) as in Definition 2.5.
– A 3-message WIPOK for NP with first-message-dependent instances as in Definition 2.4. We denote its messages by $(\mathsf{wi}_1, \mathsf{wi}_2, \mathsf{wi}_3)$.
– A non-interactive perfectly-binding commitment scheme Com.
– For some $\mathsf{wi}_1$, cmt, denote by $\mathcal{M}_{\mathsf{wi}_1,\mathsf{cmt}}$ a Turing machine that given memory $D = V^*$ parses $V^*$ as a Turing machine, runs $V^*$ on input $(\mathsf{wi}_1, \mathsf{cmt})$, parses the result as $(u, \mathsf{wi}_2, q, \widehat{\mathsf{ct}}_r)$, and outputs $u$.
– Denote by $\mathcal{V}_{\mathsf{param}}$ a circuit that has the string param hard-coded and operates as follows. Given as input a verification state $\tau$ for the delegation scheme:
  • parse $\mathsf{param} = (\mathsf{wi}_1, \mathsf{cmt}, q, u, d, t, \pi)$,
  • return 1 ("accept") if either of the following occurs:

 * the delegation verifier accepts: $\mathsf{Ver}(d, \tau, \mathcal{M}_{\mathsf{wi}_1,\mathsf{cmt}}, t, u, \pi) = 1$,
 * the query is inconsistent: $q \neq \mathsf{Query}(1^n; \tau)$.

In words, $\mathcal{V}_{\mathsf{param}}$, given the verification state $\tau$, first verifies the proof $\pi$ that "$\mathcal{M}_{\mathsf{wi}_1,\mathsf{cmt}}(D) = (u, \cdots)$" where $D$ is the database corresponding to the digest $d$. In addition, it verifies that $q$ is truly consistent with the coins $\tau$. If the query is consistent, but the proof is rejected $\mathcal{V}_{\mathsf{param}}$ also rejects.

- Denote by $\mathbf{1}$ a circuit of the same size as $\mathcal{V}_{\mathsf{param}}$ that always returns 1.

We now describe the protocol in Fig. 1.

**Theorem 3.1.** *Given a 2-message memory delegation scheme for $\gamma$-bounded computations sound against provers in $\mathbb{A}$, a semantically-secure, circuit-private,*

---

**Protocol 1**

**Common Input:** an instance $x \in \mathcal{L} \cap \{0,1\}^n$, for security parameter $n$.
$P$: a witness $w \in \mathcal{R}_{\mathcal{L}}(x)$.

1. $P$ computes
   - $\mathsf{wi}_1$, the first message of the WIPOK for statements of length $\ell_\Psi(n)$, where $\ell_\Psi$ is the length of the statement $\Psi$ defined in Step 3 below,
   - $\mathsf{cmt} \leftarrow \mathsf{Com}(0^n, 0^{\log \gamma(n)})$, a commitment to the all zero string,
   and sends $(\mathsf{wi}_1, \mathsf{cmt})$.
2. $V$ computes
   - $\mathsf{wi}_2$, the second message of the WIPOK.
   - $(\tau, q) \leftarrow \mathsf{Query}(1^n)$, verification state (w.l.o.g the coins of $\mathsf{Query}$) and query,
   - $(\mathsf{ct}_\tau, \mathsf{sk}) \leftarrow \mathsf{Enc}_{\mathsf{sk}}(\tau)$, an encryption of the verification state,
   - $u \leftarrow \{0,1\}^n$, a uniformly random string,
   and sends $(u, \mathsf{wi}_2, q, \mathsf{ct}_\tau)$.
3. $P$ computes
   - $\widehat{\mathsf{ct}} \leftarrow \mathsf{Eval}(\mathbf{1}, \mathsf{ct}_\tau)$, an evaluation of the constant one function,
   - $\mathsf{wi}_3$, the third WIPOK message for the statement $\Psi = \Psi_1(x) \vee \Psi_2(\mathsf{wi}_1, \mathsf{cmt}, q, u, \mathsf{ct}_\tau, \widehat{\mathsf{ct}})$ of length $\ell_\Psi(n)$ given by:

$$\left\{ \exists w \,\middle|\, (x,w) \in \mathcal{R}_{\mathcal{L}} \right\} \bigvee$$

$$\left\{ \exists \begin{array}{l} d, \pi, r_{\mathsf{cmt}} \in \{0,1\}^n \\ t \leq \gamma(n) \end{array} \,\middle|\, \begin{array}{l} \mathsf{cmt} = \mathsf{Com}(d, t; r_{\mathsf{cmt}}) \\ \mathsf{param} = (\mathsf{wi}_1, \mathsf{cmt}, q, u, d, t, \pi) \\ \widehat{\mathsf{ct}} = \mathsf{Eval}(\mathcal{V}_{\mathsf{param}}, \mathsf{ct}_\tau) \end{array} \right\},$$

   using the witness $w \in \mathcal{R}_{\mathcal{L}}(x)$ for $\Psi_1$,
   and sends $(\widehat{\mathsf{ct}}, \mathsf{wi}_3)$.
4. $V$ verifies the WIPOK proof $(\mathsf{wi}_1, \mathsf{wi}_2, \mathsf{wi}_3)$ for the statement $\Psi$ and that $\mathsf{Dec}_{\mathsf{sk}}(\widehat{\mathsf{ct}}) = 1$.

---

**Fig. 1.** A 3-message ZK argument of knowledge against prover in $\mathbb{A}$.

*1-hop homomorphic encryption scheme, a 3-message WIPOK with first-message-dependent instances, and a non-interactive perfectly-binding commitment scheme. The corresponding Protocol 1 (Fig. 1) is a zero-knowledge argument of knowledge against provers in* $\mathbb{A}$.

**Overview of proof.** For simplicity, let us focus on showing that the protocol is sound and zero knowledge. (Showing it is an argument of knowledge follows a similar reasoning.) We start with soundness. Assuming that $x \notin \mathcal{L}$, in order to pass the WIPOK with respect to an evaluated cipher $\widehat{\mathsf{ct}}$ that decrypts to 1, the prover must know a digest $d \in \{0, 1\}^n$, a time bound $t \leq \gamma(n)$, and proof $\pi \in \{0, 1\}^n$, such that $\mathcal{V}_{\mathsf{param}}(\tau) = 1$. This, by definition, means that $(d, t, \pi)$ are such that the delegation verifier $\mathsf{Ver}$ is convinced that the digest $d$ corresponds to a machine $V^*$ such that $V^*(\mathsf{wi}_1, \mathsf{cmt}) = u$. Intuitively, this implies that the prover managed to commit to a program that predicts the random string $u$ before it was ever sent, which is unlikely. Formally, we show that such a prover can be used to break the underlying delegation scheme. Here we will also rely on the semantic security of the encryption scheme to claim that the encrypted verification state $\tau$ is hiding. Since the delegation scheme is sound against provers in $\mathbb{A}$, we shall only get soundness against such provers.

To show ZK, we construct a non-black-box simulator following the simulator of Barak [Bar01]. At high-level, the simulator uses the code of the (malicious) verifier $V^*$ as the memory for the delegation scheme, and completes the WIPOK using *the trapdoor branch* $\Psi_2$ of the statement $\Psi = \Psi_1 \vee \Psi_2$. The *trapdoor witness* is basically $(d, t, \pi)$, where $d$ is the digest corresponding to $V^*$, $t \approx |V^*|$ and $\pi$ is the corresponding delegation proof that $V^*(\mathsf{wi}_1, \mathsf{cmt}) = u$, which is now true by definition. By the perfect completeness of the delegation scheme, we know that as long as the verifier honestly encrypts some randomness $\tau$ as the private state, and gives a query $q$ that is consistent with $\tau$, the delegation verifier $\mathsf{Ver}$ will accept the corresponding proof. Thus, the circuit privacy of homomorphic evaluation (which holds also if the verifier produces a malformed ciphertext) would guarantee indistinguishability from a real proof, where the prover actually evaluates the constant 1 circuit.

A detailed proof follows. We first prove in Sect. 3.1 that the protocol is an argument of knowledge. Then we prove in Sect. 3.2 that the protocol is zero knowledge.

## 3.1  Proving that the Protocol Is an Argument of Knowledge

In this section, we show that the protocol is an argument of knowledge against provers in $\mathbb{A}$.

**Proposition 3.1.** *Protocol 1 (Fig. 1) is an argument of knowledge against provers in* $\mathbb{A}$.

*Proof.* We show that there exists a uniform PPT extractor $\mathcal{E} \in \mathbb{B}_1$ and a uniform PPT reduction $\mathcal{R} \in \mathbb{B}_1$, such that for any prover $P^* = \{P_n^*\}_{n \in \mathbb{N}} \in \mathbb{A}$ that generates $x_n \in \{0,1\}^n$ and convinces $V$ of accepting $x_n$ with non-negligible probability $\varepsilon(n)$, one of the following holds:

- $\mathcal{E}^{P_n^*}(1^{1/\varepsilon(n)}, x_n)$ outputs $w \in \mathcal{R}_{\mathcal{L}}(x_n)$ with probability $\varepsilon(n)^2/4 - \mathrm{negl}(n)$,[6] or
- $\mathcal{R}^{P_n^*}$ breaks the soundness of the delegation scheme with probability $n^{-O(1)}$.

We start by describing the extractor. Throughout the description (and following proof), we will often omit $n$, when it is clear from the context.

The witness extractor $\mathcal{E}^{P_n^*}(1^{1/\varepsilon(n)}, x_n)$ operates as follows:

1. Derives from $P^*$ a new prover $P_{\mathsf{wi}}^*$ for the WIPOK as follows. $P_{\mathsf{wi}}^*$ emulates the role of $P^*$ in the WIPOK; in particular, it would (honestly) sample $(\tau, (\mathsf{sk}, \mathsf{ct}_\tau), u)$ on its own to compute the second verifier message $(\mathsf{wi}_2, q, \mathsf{ct}_\tau, u)$ that $P^*$ receives.
2. Chooses the random coins $r$ for the new prover $P_{\mathsf{wi}}^*$, and samples a transcript $\mathsf{tr} = (\Psi, \mathsf{wi}_1, \mathsf{wi}_2, \mathsf{wi}_3)$ of an execution with the honest WIPOK verifier $V_{\mathsf{wi}}$.
3. Applies the WIPOK extractor $\mathcal{E}_{\mathsf{wi}}$ on the transcript $\mathsf{tr}$, with oracle access to $P_{\mathsf{wi}}^*$, and extraction parameter $2/\varepsilon$. That is, computes $w \leftarrow \mathcal{E}_{\mathsf{wi}}^{P_{\mathsf{wi}}^*(r)}(1^{2/\varepsilon}, \mathsf{tr})$.
4. Outputs $w$.

Our strategy will be to show the required reduction $\mathcal{R}$, such that if the extractor fails to extract with the required probability, then the reduction breaks the underlying delegation scheme. Thus from hereon, we assume that for some noticeable function $\eta(n) = n^{-O(1)}$, with probability at most $\varepsilon^2/4 - \eta$ the extracted witness $w$ is in $\mathcal{R}_{\mathcal{L}}(x)$. Rather than already describing the reduction $\mathcal{R}$, we shall first establish several claims regarding the extraction procedure and the consequences of extraction failure. These will motivate our concrete construction of the reduction $\mathcal{R}$.

We start by noting that an execution of $P_{\mathsf{wi}}^*(r)$ with the honest WIPOK verifier $V_{\mathsf{wi}}$ induces a perfectly emulated execution of $P^*$ with the honest verifier $V$. Thus, we know that $V$, and in particular $V_{\mathsf{wi}}$, accepts in such an execution with probability $\varepsilon(n) \geq n^{-O(1)}$.

**Good coins $r$.** We say that random coins $r$ for $P_{\mathsf{wi}}^*$ are good if with probability at least $\varepsilon/2$ over the coins of the WIPOK verifier $V_{\mathsf{wi}}$, the induced execution of $P^*$ with $V$ is such that the zero-knowledge verifier $V$ accepts. By a standard averaging argument, at least an $(\varepsilon/2)$-fraction of the coins $r$ for $P_{\mathsf{wi}}^*$ are good.

Recall that every execution of $\mathcal{E}_{\mathsf{wi}}$ induces a choice $r$ for $P_{\mathsf{wi}}^*$, a WIPOK transcript $\mathsf{tr} = (\Psi, \mathsf{wi}_1, \mathsf{wi}_2, \mathsf{wi}_3)$, and values $(\mathsf{cmt}, q, u, \mathsf{ct}_\tau, \widehat{\mathsf{ct}})$ exchanged in the induced interaction between the zero-knowledge prover $P^*$ and the zero-knowledge verifier $V$. These values, in turn, determine the formula

$$\Psi = \Psi_1(x) \vee \Psi_2(\mathsf{wi}_1, \mathsf{cmt}, q, u, \mathsf{ct}_\tau, \widehat{\mathsf{ct}}).$$

---

[6] We note that the extraction probability can then be amplified to $1 - \mathrm{negl}(n)$ by standard repetition.

We next claim that for any good $r$, such an extraction procedure outputs a witness for $\Psi$ and simultaneously the homomorphic evaluation result $\widehat{ct}$ decrypts to one (under the secret key sk sampled together with $ct_\tau$), with non-negligible probability.

**Claim 3.2 (Extraction for good $r$).** *For any* good *$r$ for $P^*_{wi}$, it holds that $w$ satisfies the induced statement $\Psi$ and $\mathsf{Dec}_{sk}(\widehat{ct}) = 1$ with probability $\varepsilon(n)/2 - \mathrm{negl}(n)$ over a transcript* tr, *and coins for $\mathcal{E}_{wi}$.*

**Proof of Claim 3.2.** Fix some good coins $r$. Since the coins $r$ are good, the WIPOK verifier $V_{wi}$ is convinced by $P^*_{wi}$ with probability at least $\varepsilon/2$, meaning that $V_{wi}$ accepts and in addition $\mathsf{Dec}_{sk}(\widehat{ct}) = 1$. We claim that when this occurs then, except with probability $\mathrm{negl}(n)$, the extractor $\mathcal{E}_{wi}$, also outputs a valid witness $w$ for $\Psi$. This follows directly from the extraction guarantee of the WIPOK. □

Now, relying on the fact that overall the extractor fails to output a witness for $x$, we deduce that with non-negligible probability, the extracted witness satisfies the trapdoor statement $\Psi_2$.

**Claim 3.3 (Extracting a trapdoor witness).** *In a random execution of the extractor, the extracted witness $w$ satisfies the trapdoor statement, namely $\Psi_2(wi_1, cmt, q, u, ct_\tau, \widehat{ct})$, and in addition $\mathsf{Dec}_{sk}(\widehat{ct}) = 1$, with probability at least $\eta(n) - \mathrm{negl}(n)$ over the choice of $r$ for $P^*_{wi}$, a transcript* tr, *and coins for $\mathcal{E}_{wi}$.*

**Proof of Claim 3.3.** First, by the $(\varepsilon/2)$-density of good $r$'s and Claim 3.2, we deduce that in a random execution the extracted $w$ satisfies the statement $\Psi = \Psi_1 \vee \Psi_2$, and in addition $\mathsf{Dec}_{sk}(\widehat{ct}) = 1$, with probability at least $\varepsilon^2/4 - \mathrm{negl}(n)$. Combining this with the fact that $w \in \mathcal{R}_{\mathcal{L}}(x)$ with probability at most $\varepsilon^2/4 - \eta$, the claim follows. □

Next, recall that by the definition of $\Psi_2$, whenever $w$ is a witness for $\Psi_2$, it holds that

$$w = (d, \pi, t, r_{cmt}) : \begin{array}{l} d, \pi \in \{0,1\}^n, t \leq \gamma(n) \\ \widehat{ct} = \mathsf{Eval}(\mathcal{V}_{param}, ct_\tau) \\ param = (wi_1, cmt, q, u, d, t, \pi) \\ cmt = \mathsf{Com}(d, t; r_{cmt}) \end{array}.$$

Furthermore, by the definition of $\mathcal{V}_{param}$ and the perfect completeness of the 1-hop homomorphic encryption,

$$\mathsf{Dec}_{sk}(\widehat{ct}) = \mathcal{V}_{param}(\tau) = \mathsf{Ver}(d, \tau, \mathcal{M}_{wi_1, cmt}, t, u, \pi).$$

We can thus deduce that, with probability $\eta$, the witness $w = (d, \pi, t, r_{cmt})$ extracted by $\mathcal{E}$ is such that: (a) $\mathsf{Ver}(d, \tau, \mathcal{M}_{wi_1, cmt}, t, u, \pi) = 1$, and (b) $cmt = \mathsf{Com}(d, t; r_{cmt})$.

**An equivalent experiment that hides the secret verification state $\tau$.** We now consider an augmented extraction procedure $\mathcal{E}_{aug} \in \mathbb{B}_1$ that behaves exactly as the original extractor $\mathcal{E}$, except that, when $P_{\mathsf{wi}}^*$ emulates $P^*$, it does not sample an encryption $\mathsf{ct}_\tau$ of the secret verification state $\tau$, but rather it samples an encryption $\mathsf{ct}_0$ of $0^{|\tau|}$. We claim that in this alternative experiment, the above two conditions (a) and (b) still hold with the same probability up to a negligible difference.

**Claim 3.4 (Convincing probability in alternative experiment).** *With probability $\eta - \mathsf{negl}(n)$, the witness $w = (d, \pi, t, r_{\mathsf{cmt}})$ extracted by $\mathcal{E}_{\mathsf{aug}}$ is such that:*
*(a) $\mathsf{Ver}(d, \tau, \mathcal{M}_{\mathsf{wi}_1, \mathsf{cmt}}, t, u, \pi) = 1$, and (b) $\mathsf{cmt} = \mathsf{Com}(d, t; r_{\mathsf{cmt}})$.*

**Proof sketch of Claim 3.4.** This claim follows from the semantic security of the 1-hop homomorphic encryption scheme. Indeed, if the above was not the case, we can distinguish between an encryption of $\tau$ and one of $0^{|\tau|}$. For this, note that the first experiment with $\mathsf{ct}_\tau$ (respectively, the second with $\mathsf{ct}_0$) can be perfectly emulated given $\tau$ and the ciphertext $\mathsf{ct}_\tau$ (respectively, $\mathsf{ct}_0$), and in addition the above two conditions (a) and (b) can be tested efficiently.        □

**The reduction $\mathcal{R}$ to the soundness of delegation.** We are now ready to describe the reduction $\mathcal{R}$ that breaks the soundness of the delegation scheme. In what follows, we view the randomness $r$ for $P_{\mathsf{wi}}^*$ as split into $r = (r_1, \tau, u, r_2)$, where $r_1$ is any randomness used to generate the first prover message $(\mathsf{wi}_1, \mathsf{cmt})$, $\tau$ is the randomness for $\mathsf{Query}$ and $u$ is the random string both used to emulate the second verifier message, and $r_2$ are any additional random coins used by $P_{\mathsf{wi}}^*$. The reduction $\mathcal{R}^{P_n^*}(1^{1/\varepsilon(n)}, x_n)$ breaks the delegation scheme as follows:[7]

1. Samples $r^* = (r_1^*, \tau^*, u^*, r_2^*)$ uniformly at random.
2. Runs $\mathcal{E}_{\mathsf{aug}}^{P^*}(1^{1/\varepsilon}, x)$ using $r^*$ as the randomness for $P_{\mathsf{wi}}$. Let $(\mathsf{cmt}^*, \mathsf{wi}_1^*)$ be the corresponding first prover message (which is completely determined by the choice of $r_1^*$), and let $w^* = (d^*, \pi^*, t^*, r_{\mathsf{cmt}}^*)$ be the witness output by the extractor.
3. Samples $u, u' \leftarrow \{0,1\}^n$ uniformly at random.
4. Declares $d^*$ as the digest, $\mathcal{M}_{\mathsf{wi}_1^*, \mathsf{cmt}^*}$ as the machine to be evaluated over the memory, $t^*$ the bound on its running time, and $(u, u')$ as the two outputs for the attack.
5. Given a delegation query $q$, $\mathcal{R}$ generates two proofs $\pi$ and $\pi'$ for $u$ and $u'$ respectively as follows:
   (a) Samples $r = (r_1^*, \perp, u, r_2)$ and $r' = (r_1^*, \perp, u', r_2')$, where in both $r_1^*$ is the same randomness sampled before, $(u, u')$ are the random strings sampled before, and $(r_2, r_2')$ are uniformly random strings.
   (b) Runs $\mathcal{E}_{\mathsf{aug}}^{P^*}(1^{1/\varepsilon}, x)$ once with respect to $r$ and another time with respect to $r'$, with one exception—the prover $P_{\mathsf{wi}}^*$ constructed by $\mathcal{E}_{\mathsf{aug}}^{P^*}$ does not emulate on its own the delegation query in the verifier's message, but

---

[7] Here we give the reduction $(1^{1/\varepsilon(n)}, x_n)$ for the sake of simplicity and clarity of exposition. Recall that $x_n$ is generated by $P_n^*$. Also, $\varepsilon$ can be approximated by sampling. Thus the reduction can (uniformly) obtain these two inputs from $P^*$.

rather it uses the external query $q$ that $\mathcal{R}$ is given. The two executions of $\mathcal{E}_{aug}^{P^*}$ then produce witnesses $w = (d, \pi, t, r_{cmt})$ and $w' = (d', \pi', t', r'_{cmt})$.

(c) Output $(\pi, \pi')$.

We first note that the running time of $\mathcal{R}$ is polynomial in $n$ and in the running of $\mathcal{E}_{aug}$, which is in turn polynomial in the running time of $P^*$ and in $1/\varepsilon(n) = n^{O(1)}$. Thus it is overall polynomial in $n$.

To complete the proof, we show that $\mathcal{R}$ breaks the scheme with noticeable probability.

**Claim 3.5.** $u \neq u'$ and $\pi$ and $\pi'$ both convince the delegation verifier with probability $\Omega(\eta(n)^5)$.

**Proof of Claim 3.5.** Throughout, let us denote by $G$ the event that the witness $w = (d, \pi, t, r_{cmt})$ extracted by $\mathcal{E}_{aug}$ is such that: (a) $\mathsf{Ver}(d, \tau, \mathcal{M}_{wi_1, cmt}, t, u, \pi) = 1$, and (b) $\mathsf{cmt} = \mathsf{Com}(d, t; r_{cmt})$. We will call $r_1^*$ good$_1$, if with probability $\eta/2$ (over all other randomness), $G$ occurs. Then by Claim 3.4 and averaging, with probability $\eta/2 - \mathsf{negl}(n)$ over a choice of a random $r_1^*$, it is good$_1$. Next, for a fixed $r_1^*$ and $\tau$, we will say that $\tau$ is $r_1^*$-good, if with probability $\eta/4$ over a choice of random $(u, r_2^*)$, $G0$ occurs. Then, by averaging, for any good$_1$ $r_1^*$, with probability $\eta/4 - \mathsf{negl}(n)$ over a choice of a random $\tau$, it is $r_1^*$-good.

We are now ready to lower bound the probability that $\mathcal{R}$ breaks the delegation scheme. This is based on the following assertions:

1. In Step 1, with probability $\eta/2 - \mathsf{negl}(n)$, $\mathcal{R}$ samples a good$_1$ $r_1^*$.
2. Conditioned on $r_1^*$ being good$_1$:
   (a) In Step 2, with probability $\eta/2$, $G$ occurs. In particular, the extracted $(d^*, t^*, r_{cmt}^*)$ are valid in the sense that $\mathsf{cmt}^* = \mathsf{Com}(d^*, t^*; r_{cmt}^*)$, $\mathsf{cmt}^*$ is the commitment generated in the first prover message (determined by the choice of $r_1^*$).
   (b) In Step 5, with probability $\eta/4 - \mathsf{negl}(n)$, the coins $\tau$ chosen by the delegation $\mathsf{Query}$ algorithm (inducing the query $q$) are $r_1^*$-good.
   (c) Conditioned on the coins $\tau$ of $\mathsf{Query}$ being $r_1^*$-good:
      i. In Step 5, with probability $\eta/4$, the event $G$ occurs. Thus the extracted $(d, t, r_{cmt}, \pi)$ are valid in the sense that $\mathsf{cmt}^* = \mathsf{Com}(d, t; r_{cmt})$, as well as $\mathsf{Ver}(d, \tau, \mathcal{M}_{wi_1^*, cmt^*}, t, u, \pi) = 1$. Recall that $(wi_1^*, cmt^*)$ are generated in the first prover message (and are determined by the choice of $r_1^*$).
      ii. The same holds independently for the second random output $u'$.
3. In Step 3, with probability $1 - 2^{-n}$, the outputs $u, u'$ sampled by $\mathcal{R}$ are distinct.
4. If $\mathsf{cmt}^* = \mathsf{Com}(d^*, t^*; r_{cmt}^*) = \mathsf{Com}(d, t; r_{cmt}) = \mathsf{Com}(d', t'; r'_{cmt})$, then $(d, t) = (d', t') = (d^*, t^*)$.

The first two assertions follow directly from the definitions and averaging arguments made above. The third assertion follows from the collision probability of two random strings of length $n$. The last assertion follows from the fact that the commitment Com is perfectly binding.

It is left to note that if all of the above occur, then $\mathcal{R}$ manages to produce accepting proofs $(\pi, \pi')$ for two different outcomes $(u, u')$ with respect to the same digest $d^*$ and machine $\mathcal{M}_{\mathsf{wi}_1^*,\mathsf{cmt}^*}$; thus, it breaks soundness. This happens with probability

$$\left(\frac{\eta}{2} - \mathrm{negl}(n)\right) \cdot \frac{\eta}{2} \cdot \left(\frac{\eta}{4} - \mathrm{negl}(n)\right) \cdot \left(\frac{\eta}{4}\right)^2 - 2^{-n} = \Omega(\eta^5).$$

This completes the proof of Claim 3.5.                                     □
This completes the proof of Proposition 3.1.

## 3.2   Proving that the Protocol Is Zero Knowledge

In this section, we prove

**Proposition 3.2.** *Protocol 1 (Fig. 1) is* ZK *against non-uniform PPT verifiers.*

*Proof.* We describe a universal ZK simulator $\mathcal{S}$ that given the code of any non-uniform PPT $V^* = \{V_n^*\}_{n\in\mathbb{N}}$, a polynomial bound $t(n) = n^{O(1)}$ on its running time (or more precisely the time required for a universal machine to run it), and $x \in \mathcal{L}$, simulates the view of $V$. We shall assume $V^*$ is deterministic; this is w.l.o.g as we can always sample random coins for $V^*$ and hardwire them into its non-uniform description. Throughout, we often omit the security parameter $n$ when clear from the context.

The simulator $\mathcal{S}(V_n^*, t(n), x)$, where $|x| = n$, operates as follows:

1. Generates the first message $(\mathsf{wi}_1, \mathsf{cmt})$ as follows:
   (a) Samples a first message $\mathsf{wi}_1 \in \{0,1\}^n$ of the WIPOK.
   (b) Computes a digest $d = \mathsf{Digest}(1^n, V^*)$ of the verifier's code.
   (c) Computes a commitment $\mathsf{cmt} = \mathsf{Com}(d, t; r_{\mathsf{cmt}})$ to the digest $d$ and $V^*$'s running time $t$, using random coins $r_{\mathsf{cmt}} \leftarrow \{0,1\}^n$. Here $t$ is interpreted as string in $\{0,1\}^{\log \gamma(n)}$. This is possible, for all large enough $n$, as $t(n) = n^{O(1)} \ll n^{\omega(1)} = \gamma(n)$.
2. Runs the verifier to obtain $(\mathsf{wi}_2, q, u, \mathsf{ct}_\tau) \leftarrow V^*(\mathsf{wi}_1, \mathsf{cmt})$.
3. Computes the third message $(\widehat{\mathsf{ct}}, \mathsf{wi}_3)$ as follows:
   (a) Computes a proof $\pi = \mathsf{Prov}(1^t, \mathcal{M}_{\mathsf{wi}_1,\mathsf{cmt}}, V^*, q)$ that the digested code of $V^*$ outputs $u$.
   (b) Samples $\widehat{\mathsf{ct}} \leftarrow \mathsf{Eval}(\mathcal{V}_{\mathsf{param}}, \mathsf{ct}_\tau)$, for $\mathsf{param} = (\mathsf{wi}_1, \mathsf{cmt}, q, u, d, t, \pi)$.
   (c) Computes the third WIPOK message $\mathsf{wi}_3$ for the statement $\Psi = \Psi_1(x) \vee \Psi_2(\mathsf{wi}_1, \mathsf{cmt}, q, u, \mathsf{ct}_\tau, \widehat{\mathsf{ct}})$ given by:

$$\left\{\exists w \;\middle|\; \begin{array}{c}(x,w) \\ \in \mathcal{R}_\mathcal{L}\end{array}\right\} \vee \left\{\begin{array}{c}\exists \, d, \pi, r_{\mathsf{cmt}} \in \{0,1\}^n \\ t \le \gamma(n)\end{array} \;\middle|\; \begin{array}{l}\widehat{\mathsf{ct}} = \mathsf{Eval}(\mathcal{V}_{\mathsf{param}}, \mathsf{ct}_\tau) \\ \mathsf{param} = (\mathsf{wi}_1, \mathsf{cmt}, q, u, d, t, \pi) \\ \mathsf{cmt} = \mathsf{Com}(d, t; r_{\mathsf{cmt}})\end{array}\right\},$$

using the witness $(d, \pi, r_{\mathsf{cmt}}, t)$ for the trapdoor statement $\Psi_2$.

(d) Outputs the view $(\mathsf{wi}_1, \mathsf{cmt}, \widehat{\mathsf{ct}}, \mathsf{wi}_3)$ of $V^*$.

We now show that the view generated by $\mathcal{S}$ is computationally indistinguishable from the view of $V^*$ in an execution with the honest prover $P$. We do this by exhibiting a sequence of hybrids.

**Hybrid 1:** The view $(\mathsf{wi}_1, \mathsf{cmt}, \widehat{\mathsf{ct}}, \mathsf{wi}_3)$ is generated by $\mathcal{S}$.

**Hybrid 2:** Instead of generating $\mathsf{wi}_3$ using the witness $(d, \pi, r_{\mathsf{cmt}}, t)$ for $\Psi_2$, it is generated using a witness $w$ for $\Psi_1 = \{x \in \mathcal{L}\}$. By the adaptive witness-indistinguishability of the WIPOK system, this hybrid is computationally indistinguishable from Hybrid 1.

**Hybrid 3:** Instead of generating $\mathsf{cmt}$ as a commitment $\mathsf{cmt} = \mathsf{Com}(d, t; r_{\mathsf{cmt}})$ to $(d, t)$, it is generated as a commitment to $0^{n+\log \gamma(n)}$. Note that in this hybrid the commitment's randomness $r_{\mathsf{cmt}}$ is not used anywhere, but in the generation of $\mathsf{cmt}$. Thus, by the computational hiding of the commitment, this hybrid is computationally indistinguishable from Hybrid 2.

**Hybrid 4:** The view $(\mathsf{wi}_1, \mathsf{cmt}, \widehat{\mathsf{ct}}, \mathsf{wi}_3)$ is generated in an interaction of $V^*$ with the honest prover $P$. The difference from Hybrid 3 is in that $\widehat{\mathsf{ct}}$ is sampled from $\mathsf{Eval}(1, \mathsf{ct}_\tau)$ instead of $\mathsf{Eval}(\mathcal{V}_{\mathsf{param}}, \mathsf{ct}_\tau)$. First, note that by the perfect completeness of the delegation scheme, for any $\tau \in \{0,1\}^n$, $\mathcal{V}_{\mathsf{param}}(\tau) = 1(\tau) = 1$. Indeed, by definition we know that

$$\mathcal{M}_{\mathsf{wi}_1, \mathsf{cmt}}(V^*) = V^*(\mathsf{wi}_1, \mathsf{cmt})[1] = u,$$

and this output is produced after at most $t$ steps. Thus, assuming that $q = \mathsf{Query}(1^n; \tau)$, the delegation verifier accepts; namely, $\mathsf{Ver}(d, \tau, \mathcal{M}_{\mathsf{wi}_1, \mathsf{cmt}}, t, u, \pi) = 1$, and by definition $\mathcal{V}_{\mathsf{param}}(\tau) = 1$. Also, if $q \neq \mathsf{Query}(1^n; \tau)$, then $\mathcal{V}_{\mathsf{param}}(\tau) = 1$ by definition.

By the circuit privacy of the 1-hop homomorphic encryption, the above guarantees indistinguishability whenever $\mathsf{ct}_\tau$ is a well-formed ciphertext since

$$\mathsf{Eval}(\mathcal{V}_{\mathsf{param}}, \mathsf{ct}_\tau) \approx_c \mathcal{S}_{\mathsf{1hop}}(\mathsf{ct}_\tau, \mathcal{V}_{\mathsf{param}}(\tau), |\mathcal{V}_{\mathsf{param}}|) \equiv$$
$$\mathcal{S}_{\mathsf{1hop}}(\mathsf{ct}_\tau, 1(\tau), |1|) \approx_c \mathsf{Eval}(1, \mathsf{ct}_\tau).$$

Also, for any malformed ciphertext $\mathsf{ct}^*$ it holds that

$$\mathsf{Eval}(\mathcal{V}_{\mathsf{param}}, \mathsf{ct}^*) \approx_c \mathcal{S}_{\mathsf{1hop}}(\mathsf{ct}^*, \perp, |\mathcal{V}_{\mathsf{param}}|) \equiv \mathcal{S}_{\mathsf{1hop}}(\mathsf{ct}^*, \perp, |1|) \approx_c \mathsf{Eval}(1, \mathsf{ct}^*).$$

It follows that Hybrid 4 is computationally indistinguishable from Hybrid 3.

This completes the proof of Proposition 3.2.

**Acknowledgments.** We thank Ran Canetti, Shai Halevi and Hugo Krawczyk for helpful comments and for pointing out the connection to [Rog06].

# References

[AIR01] Aiello, B., Ishai, Y., Reingold, O.: Priced oblivious transfer: how to sell digital goods. In: Pfitzmann, B. (ed.) EUROCRYPT 2001. LNCS, vol. 2045, pp. 119–135. Springer, Heidelberg (2001). doi:10.1007/3-540-44987-6_8

[Bar01] Barak, B.: How to go beyond the black-box simulation barrier. In: FOCS, pp. 106–115 (2001)

[BCC88] Brassard, G., Chaum, D., Crépeau, C.: Minimum disclosure proofs of knowledge. J. Comput. Syst. Sci. 37(2), 156–189 (1988)

[BCC+14] Bitansky, N., Canetti, R., Chiesa, A., Goldwasser, S., Lin, H., Rubinstein, A., Tromer, E.: The hunting of the SNARK. IACR Cryptology ePrint Archive, 2014:580 (2014)

[BCPR13] Bitansky, N., Canetti, R., Paneth, O., Rosen, A.: More on the impossibility of virtual-black-box obfuscation with auxiliary input. IACR Cryptology ePrint Archive, 2013:701 (2013)

[BCPR14] Bitansky, N., Canetti, R., Paneth, O., Rosen, A.: On the existence of extractable one-way functions. In: Symposium on Theory of Computing, STOC 2014, New York, NY, USA, May 31–June 03 2014, pp. 505–514 (2014)

[BG08] Barak, B., Goldreich, O.: Universal arguments and their applications. SIAM J. Comput. 38(5), 1661–1694 (2008)

[BGGL01] Barak, B., Goldreich, O., Goldwasser, S., Lindell, Y.: Resettably-sound zero-knowledge and its applications. In: FOCS, pp. 116–125 (2001)

[BHK16] Brakerski, Z., Holmgren, J., Kalai, Y., Non-interactive ram, batch np delegation from any pir. Cryptology ePrint Archive, Report 2016/459 (2016). http://eprint.iacr.org/

[BJY97] Bellare, M., Jakobsson, M., Yung, M.: Round-optimal zero-knowledge arguments based on any one-way function. In: Fumy, W. (ed.) EUROCRYPT 1997. LNCS, vol. 1233, pp. 280–305. Springer, Heidelberg (1997). doi:10.1007/3-540-69053-0_20

[BM14] Brzuska, C., Mittelbach, A.: Indistinguishability obfuscation versus multi-bit point obfuscation with auxiliary input. In: Sarkar, P., Iwata, T. (eds.) ASIACRYPT 2014. LNCS, vol. 8874, pp. 142–161. Springer, Heidelberg (2014). doi:10.1007/978-3-662-45608-8_8

[BP04a] Barak, B., Pass, R.: On the possibility of one-message weak zero-knowledge. In: Naor, M. (ed.) TCC 2004. LNCS, vol. 2951, pp. 121–132. Springer, Heidelberg (2004). doi:10.1007/978-3-540-24638-1_7

[BP04b] Bellare, M., Palacio, A.: The knowledge-of-exponent assumptions and 3-round zero-knowledge protocols. In: Franklin, M. (ed.) CRYPTO 2004. LNCS, vol. 3152, pp. 273–289. Springer, Heidelberg (2004). doi:10.1007/978-3-540-28628-8_17

[BP12] Bitansky, N., Paneth, O.: From the impossibility of obfuscation to a new non-black-box simulation technique. In: FOCS (2012)

[BP13] Bitansky, N., Paneth, O.: On the impossibility of approximate obfuscation and applications to resettable cryptography. In: STOC, pp. 241–250 (2013)

[CD09] Canetti, R., Dakdouk, R.R.: Towards a theory of extractable functions. In: Reingold, O. (ed.) TCC 2009. LNCS, vol. 5444, pp. 595–613. Springer, Heidelberg (2009). doi:10.1007/978-3-642-00457-5_35

[CDPW07] Canetti, R., Dodis, Y., Pass, R., Walfish, S.: Universally composable security with global setup. In: Vadhan, S.P. (ed.) TCC 2007. LNCS, vol. 4392, pp. 61–85. Springer, Heidelberg (2007). doi:10.1007/978-3-540-70936-7_4

[CKLR11] Chung, K.-M., Kalai, Y.T., Liu, F.-H., Raz, R.: Memory delegation. In: Rogaway, P. (ed.) CRYPTO 2011. LNCS, vol. 6841, pp. 151–168. Springer, Heidelberg (2011). doi:10.1007/978-3-642-22792-9_9

[CLP13a] Canetti, R., Lin, H., Paneth, O.: Public-coin concurrent zero-knowledge in the global hash model. In: Sahai, A. (ed.) TCC 2013. LNCS, vol. 7785, pp. 80–99. Springer, Heidelberg (2013). doi:10.1007/978-3-642-36594-2_5

[CLP13b] Chung, K.-M., Lin, H., Pass, R.: Constant-round concurrent zero knowledge from p-certificates. In: FOCS (2013)

[CLP15] Chung, K.-M., Lin, H., Pass, R.: Constant-round concurrent zero-knowledge from indistinguishability obfuscation. In: Gennaro, R., Robshaw, M. (eds.) CRYPTO 2015. LNCS, vol. 9215, pp. 287–307. Springer, Heidelberg (2015). doi:10.1007/978-3-662-47989-6_14

[COP+14] Chung, K.-M., Ostrovsky, R., Pass, R., Venkitasubramaniam, M., Visconti, I.: 4-round resettably-sound zero knowledge. In: Lindell, Y. (ed.) TCC 2014. LNCS, vol. 8349, pp. 192–216. Springer, Heidelberg (2014). doi:10.1007/978-3-642-54242-8_9

[FLS99] Feige, U., Lapidot, D., Shamir, A.: Multiple noninteractive zero knowledge proofs under general assumptions. SIAM J. Comput. 29(1), 1–28 (1999)

[FS89] Feige, U., Shamir, A.: Zero knowledge proofs of knowledge in two rounds. In: Brassard, G. (ed.) CRYPTO 1989. LNCS, vol. 435, pp. 526–544. Springer, Heidelberg (1990). doi:10.1007/0-387-34805-0_46

[GHV10] Gentry, C., Halevi, S., Vaikuntanathan, V.: i-Hop homomorphic encryption and rerandomizable yao circuits. In: CRYPTO, pp. 155–172 (2010)

[GK96a] Goldreich, O., Kahan, A.: How to construct constant-round zero-knowledge proof systems for NP. J. Cryptol. 9(3), 167–190 (1996)

[GK96b] Goldreich, O., Krawczyk, H.: On the composition of zero-knowledge proof systems. SIAM J. Comput. 25(1), 169–192 (1996)

[GMR89] Goldwasser, S., Micali, S., Rackoff, C.: The knowledge complexity of interactive proof systems. SIAM J. Comput. 18(1), 186–208 (1989)

[GMW91] Goldreich, O., Micali, S., Wigderson, A.: Proofs that yield nothing but their validity for all languages in NP have zero-knowledge proof systems. J. ACM 38(3), 691–729 (1991)

[GO94] Goldreich, O., Oren, Y.: Definitions and properties of zero-knowledge proof systems. J. Cryptol. 7(1), 1–32 (1994)

[Gol04] Goldreich, O.: Foundations of Cryptography: Basic Applications, vol. 2. Cambridge University Press, New York (2004)

[HK12] Halevi, S., Kalai, Y.T.: Smooth projective hashing and two-message oblivious transfer. J. Cryptol. 25(1), 158–193 (2012)

[HT98] Hada, S., Tanaka, T.: On the existence of 3-round zero-knowledge protocols. In: Krawczyk, H. (ed.) CRYPTO 1998. LNCS, vol. 1462, pp. 408–423. Springer, Heidelberg (1998). doi:10.1007/BFb0055744

[HV16] Hazay, C., Venkitasubramaniam, M.: On the power of secure two-party computation. Cryptology ePrint Archive, Report 2016/074 (2016). http://eprint.iacr.org/

[IKOS09] Ishai, Y., Kushilevitz, E., Ostrovsky, R., Sahai, A.: Zero-knowledge proofs from secure multiparty computation. SIAM J. Comput. 39(3), 1121–1152 (2009)

[Kat12] Katz, J.: Which languages have 4-round zero-knowledge proofs? J. Cryptol. 25(1), 41–56 (2012)

[KP15] Kalai, Y.T., Paneth, O.: Delegating ram computations. Cryptology ePrint Archive, Report 2015/957 (2015). http://eprint.iacr.org/

3-Message Zero Knowledge Against Human Ignorance

[KRR14] Kalai, Y.T., Raz, R., Rothblum, R.D.: How to delegate computations: the power of no-signaling proofs. In: Symposium on Theory of Computing, STOC 2014, New York, NY, USA, May 31–June 03 2014, pp. 485–494 (2014)

[LS90a] Lapidot, D., Shamir, A.: Publicly verifiable non-interactive zero-knowledge proofs. In: Menezes, A.J., Vanstone, S.A. (eds.) CRYPTO 1990. LNCS, vol. 537, pp. 353–365. Springer, Heidelberg (1991). doi:10.1007/3-540-38424-3_26

[NP01] Naor, M., Pinkas, B.: Efficient oblivious transfer protocols. In: SODA, pp. 448–457 (2001)

[OV12] Ostrovsky, R., Visconti, I.: Simultaneous resettability from collision resistance. Electronic Colloquium on Computational Complexity (ECCC) (2012)

[Pas03] Pass, R.: Simulation in quasi-polynomial time, and its application to protocol composition. In: Biham, E. (ed.) EUROCRYPT 2003. LNCS, vol. 2656, pp. 160–176. Springer, Heidelberg (2003). doi:10.1007/3-540-39200-9_10

[PVW08] Peikert, C., Vaikuntanathan, V., Waters, B.: A framework for efficient and composable oblivious transfer. In: Wagner, D. (ed.) CRYPTO 2008. LNCS, vol. 5157, pp. 554–571. Springer, Heidelberg (2008). doi:10.1007/978-3-540-85174-5_31

[Reg09] Regev, O.: On lattices, learning with errors, random linear codes, and cryptography. J. ACM 56(6), 34.1–34.40 (2009)

[Rog06] Rogaway, P.: Formalizing human ignorance. In: Nguyen, P.Q. (ed.) VIETCRYPT 2006. LNCS, vol. 4341, pp. 211–228. Springer, Heidelberg (2006). doi:10.1007/11958239_14

[Yao86] Yao, A.C.-C.: How to generate and exchange secrets (extended abstract). In: FOCS, pp. 162–167 (1986)

# The GGM Function Family Is a Weakly One-Way Family of Functions

Aloni Cohen[1(✉)] and Saleet Klein[2]

[1] MIT, Cambridge, MA, USA
aloni@mit.edu
[2] Tel Aviv University, Tel Aviv, Israel
saleetklein@mail.tau.ac.il

**Abstract.** We give the first demonstration of the cryptographic hardness of the Goldreich-Goldwasser-Micali (GGM) function family when the secret key is exposed. We prove that for any constant $\epsilon > 0$, the GGM family is a $1/n^{2+\epsilon}$-weakly one-way family of functions, when the lengths of secret key, inputs, and outputs are equal. Namely, any efficient algorithm fails to invert GGM with probability at least $1/n^{2+\epsilon}$ – *even when given the secret key*.

Additionally, we state natural conditions under which the GGM family is strongly one-way.

## 1 Introduction

Pseudorandom functions (PRFs) are fundamental objects in general and in cryptography in particular. A pseudorandom function ensemble is a collection of (efficient) functions $\mathcal{F} = \{f_s\}_{s \in \{0,1\}^*}$ indexed by a *secret key* $s \in \{0,1\}^*$ with the dual properties that (1) given the secret key $s$, $f_s$ is efficiently computable and (2) without knowledge of the secret key, no probabilistic polynomial-time algorithm can distinguish between oracle access to a random function from the ensemble and access to a random oracle. The security property of PRFs depends on the absolute secrecy of the key, and no security is guaranteed when the secret key is revealed. Pseudorandom functions have found wide use: in cryptography to construct private-key encryption and digital signatures [Gol04], in computational learning theory for proving negative results [Val84], and in computational complexity to demonstrate the inherent limits of using natural proofs to prove circuit lower-bounds [RR97].

The first construction of pseudorandom function families starting from any one-way functions came in 1986 by Goldreich, Goldwasser, and Micali [GGM86]. Assuming only that a function is hard to invert, the construction amplifies the secrecy of a short random secret key into an exponentially-long, randomly-accessible sequence of pseudorandom values. For about 10 years, this was the only known method to construct provably secure PRFs, even from specific number-theoretic assumptions. Almost 30 years later, it remains the only generic approach to construct PRFs from any one-way function.

M. Hirt and A. Smith (Eds.): TCC 2016-B, Part I, LNCS 9985, pp. 84–107, 2016.
DOI: 10.1007/978-3-662-53641-4_4

Almost three decades after its conception, we are continuing to discover surprising power specific to the GGM pseudorandom function family. The basic ideas of this construction were used in constructions of broadcast encryption schemes in the early 90s [FN94]. Additionally, these same ideas were to construct function secret sharing schemes for point functions, leading to 2-server computationally-secure PIR schemes with poly-logarithmic communication [BGI15]. More recently, Zhandry exhibited the first quantum-secure PRF by demonstrating that the (classical) GGM ensemble (instantiated with a quantum-secure pseudorandom generator) is secure even against quantum adversaries [Zha12]. In [BW13, BGI14, KPTZ13], the notion of constrained pseudorandom functions was introduced. The "constrained keys" for these PRFs allow a user to evaluate the function on special subsets of the domain while retaining pseudorandomness elsewhere. The GGM ensemble (and modifications thereof) is a constrained PRF for the family of prefix-constraints (including point-puncturing), and GGM yields the simplest known construction of constrained PRFs. This family of constraints is powerful enough to enable many known applications of these families for program obfuscation [SW14].

In this work, we give the first demonstration that the GGM family enjoys some measure of security even when the secret key is revealed to an attacker. In this setting, pseudorandom functions do not necessarily guarantee *any security*. For example, the Luby-Rackoff family of pseudorandom permutations [LR88] are efficiently invertible given knowledge of the secret key. This suggests that we must examine *specific* constructions of pseudorandom functions to see if security is retained when the secret key is revealed. In this work, we ask the following question:

> *What security, if any, does the GGM ensemble provide when the secret key is known?*

A version of this question was posed and addressed by Goldreich[1] in 2002 [Gol02]. Goldreich casts the question from the angle of correlation intractability. Informally, a function ensemble $\{f_s\}_{s \in \{0,1\}^*}$ is correlation intractable if – even given the function description $s$ – it is computationally infeasible to find an input $x$ such that $x$ and $f_s(x)$ satisfy some "sparse" relation. Correlation intractability was formalized in [CGH04], which proved that no such family exists for $|x| \geq |s|$.

In [Gol02], Goldreich proves that the GGM ensemble is not correlation intractable, even for $|x| < |s|$, in a very strong sense. Goldreich constructs a pseudorandom generator $G^{(0)}$ which, when used to instantiate the GGM ensemble, allows an adversary with knowledge of the secret key $s$ to efficiently find preimages $x \in f_s^{-1}(0^n)$. This allows the inversion of $f_s$ for a specific image $0^n$, but not necessarily for random images.

---

[1] And posed much earlier by Micali and by Barak: see Acknowledgments of [Gol02].

## 1.1   Our Contributions

In this work, we prove that the length-preserving[2] GGM ensemble is a weakly one-way family of functions. This means that any efficient algorithm $\mathcal{A}$, when given a random secret key $s$ and $f_s(x)$ for a random input $x$, must fail to invert with non-negligible probability.

Moreover, we prove that if either a random function in $\mathcal{F}_G$ is "regular" in the sense that each image has a polynomially-bounded number of pre-images, or is "nearly surjective" in a sense made precise below, then the length-preserving GGM ensemble is strongly one-way. Formally:

**Theorem 1.** *Let* $\mathcal{F}_G = \{f_s\}_{s \in \{0,1\}^*}$ *be the length-preserving GGM function ensemble with pseudorandom generator* $G$, *where* $f_s : \{0,1\}^{|s|} \to \{0,1\}^{|s|}$. *Then for every constant* $\epsilon > 0$, $\mathcal{F}_G$ *is a* $1/n^{2+\epsilon}$*-weakly one-way collection of functions. That is, for every probabilistic polynomial-time algorithm* $\mathcal{A}$, *for every constant* $\epsilon > 0$, *and all sufficiently large* $n \in \mathbb{N}$,

$$\Pr_{\substack{s \leftarrow U_n \\ x \leftarrow U_n}} [\mathcal{A}(s, f_s(x)) \in f_s^{-1}(f_s(x))] < 1 - \frac{1}{n^{2+\epsilon}} \qquad (1)$$

*where* $U_n$ *is the uniform distribution over* $\{0,1\}^n$.

**Theorem 2.** *Let* $\mathcal{F}_G$ *be the GGM ensemble with pseudorandom generator* $G$. $\mathcal{F}_G$ *is a strongly one-way collection of functions if either of the following hold:*

(a) *There exists a negligible function* negl($\cdot$) *such that for all sufficiently large* $n \in \mathbb{N}$

$$\mathop{\mathbb{E}}_{s \leftarrow U_n} \left[ \frac{|\mathsf{Img}(f_s)|}{2^n} \right] \geq 1 - \mathsf{negl}(n) \qquad (2)$$

(b) *There exists a polynomial* $B$ *such that for all sufficiently large* $n \in \mathbb{N}$ *and for all* $s, y \in \{0,1\}^n$

$$\left| f_s^{-1}(y) \right| \leq B(n) \qquad (3)$$

*Remark 1.* The conditions of Theorem 2 are very strong conditions. Whether a pseudorandom generator $G$ exists which makes the induced GGM ensemble satisfy either condition is an interesting and open question. The possibility of such a generator is open even for the stronger requirement that for every secret key $s$, $f_s$ is a permutation.

*Remark 2.* The length-preserving restriction can be somewhat relaxed to the case when $|x| = |s| \pm O(\log |s|)$, affecting the weakly one-way parameter. A partial result holds when $|x| > |s| + \omega(\log |s|)$, and nothing is currently known if $|x| < |s| - \omega(\log |s|)$. See the full version for further discussion.

---

[2] We consider the secret keys, inputs, and outputs to be of the same lengths. See Remark 2.

## 1.2  Overview of Proof

Let's go into the land of wishful thinking and imagine that for each secret key $s \in \{0,1\}^n$, every string $y \in \{0,1\}^n$ occurs exactly once in the image of $f_s$; that is, suppose that the GGM ensemble $\mathcal{F}_G$ is a family of permutations. In this case we can prove that the GGM family is strongly one-way (in fact, this is a special case of Theorem 2).

The assumption that $\mathcal{F}_G$ is a permutation implies the following two facts.[3]

- *Fact 1:* For each secret key $s \in \{0,1\}^n$, the distributions $f_s(U_n)$ and $U_n$ are identical.
- *Fact 2:* For each string $y \in \{0,1\}^n$, there are exactly two pairs $(b, x) \in \{0,1\} \times \{0,1\}^n$ such that $G_b(x) = y$, where $G$ is the PRG underlying the GGM family, and $G_0(x)$ and $G_1(x)$ are the first and second halves of $G(x)$ respectively.

We may now prove that the GGM ensemble is strongly one-way in two steps:

- *Step 1:* Switch the adversary's input to uniformly random.
- *Step 2:* Construct a distinguisher for the PRG.

*Step 1.* For a PPT algorithm $\mathcal{A}$, let $1/\alpha(n)$ be $\mathcal{A}$'s probability of successfully inverting $y$ with secret key $s$; namely:

$$\Pr_{\substack{s \leftarrow U_n \\ y \leftarrow f_s(U_n)}} [\mathcal{A}(s, y) \in f_s^{-1}(y)] = \frac{1}{\alpha(n)}$$

By Fact 1, $\mathcal{A}$ has exactly the same success probability if $y$ is sampled uniformly from $\{0,1\}^n$:

$$\Pr_{\substack{s \leftarrow U_n \\ y \leftarrow U_n}} [\mathcal{A}(s, y) \in f_s^{-1}(y)] = \frac{1}{\alpha(n)}$$

*Step 2.* We now construct a PPT algorithm $\mathcal{D}$ that has advantage $1/2\alpha(n) -$ negl$(n)$ in distinguishing outputs from the PRG $G$ from random strings (i.e., $U_{2n}$ and $G(U_n)$). By the security of $G$, this implies that $1/\alpha(n) = $ negl$(n)$, completing the proof.

The distinguisher $\mathcal{D}$ is defined as follows:

**Input:** $(y_0, y_1)$   // a sample from either $G(U_n)$ or $U_{2n}$
Sample a secret key $s \leftarrow U_n$ and a bit $b \leftarrow U$;
Compute $x \leftarrow \mathcal{A}(s, y_b)$;
Let $\tilde{x} = x \oplus 0^{n-1}1$   // $\tilde{x}$ differs from $x$ only at the last bit;
if $f_s(x) = y_b$ and $f_s(\tilde{x}) = y_{1-b}$ then
| Output 1;     // Guess ''PRG''
else
| Output 0;     // Guess ''random''
end

**Algorithm 1.** The PRG distinguisher $\mathcal{D}$

---

[3] While these are indeed facts in the land of wishful thinking, they are not generally true. In this overview we wish to highlight only the usefulness of these facts, and believe that their proofs (though elementary), do not further this goal.

Notice that if $\mathcal{D}$ outputs 1, then either $(y_0, y_1)$ or $(y_1, y_0)$ is in $\mathsf{Img}(G)$. If $(y_0, y_1)$ was sampled uniformly from $U_{2n}$, then this happens with probability at most $2^{n+1}/2^{2n}$. Therefore,

$$\Pr[\mathcal{D}(U_{2n}) = 1] \leq 1/2^{n-1}.$$

Now we use Fact 2 from above. There are only 2 possible $x$'s that $\mathcal{A}$ could have output in agreement with $f_s(x)$; if $(y_0, y_1)$ was sampled from $G(U_n)$ and $f_s(x) = y_b$ (which happens with probability $1/\alpha(n)$), then with probability at least $1/2$: $f_s(\tilde{x}) = y_{1-b}$. Therefore,

$$\Pr[\mathcal{D}(G(U_n)) = 1] \geq 1/2\alpha(n),$$

completing the proof of this special case.

Leaving the land of wishful thinking, the proof that the GGM ensemble is weakly one-way follows exactly the same two steps as the special case proved above, but the facts we used are not true in general. We carry out Step 1 in the Input Switching Proposition (Proposition 1): we more carefully analyze the relationship between the distributions $f_s(U_n)$ and $U_n$, losing a factor of $1 - 1/n^{2+\epsilon}$ in the adversary's probability of successfully inverting. We carry out Step 2 in the Distinguishing Lemma (Lemma 2): we analyze the success probability of the distinguisher (the same one as above) by more carefully reasoning about the number of preimages for a value $y$.

**Organization.** Section 2 contains standard definitions and the notation used throughout this work. Section 3 contains the proof of Theorem 1, leaving the proof of the crucial Combinatorial Lemma (Lemma 1) to Sect. 4. Theorem 2 is proved in Sect. 5, and Sect. 6 concludes.

# 2    Preliminaries

## 2.1    Notation

For two strings $a$ and $b$ we denote by $a\|b$ their concatenation. For a bit string $x \in \{0,1\}^n$, we denote by $x[i]$ its $i$-th bit, and by $x[i:j]$ (for $i < j$) the sequence $x[i]\|x[i+1]\|\cdots\|x[j]$. We abbreviate 'probabilistic polynomial time' as 'PPT'.

For a probability distribution $D$, we use $\mathsf{Supp}(D)$ to denote the support of $D$. We write $x \leftarrow D$ to mean that $x$ is a sample from the distribution $D$. By $U_n$, we denote the uniform distribution over $\{0,1\}^n$, and omit the subscript when $n = 1$. For a probabilistic algorithm $A$, we let $A(x)$ denote a sample from the probability distribution induced over the outputs of $A$ on input $x$, though we occasionally abuse notation and let $A(x)$ denote the distribution itself. For a function $f : X \to Y$ and a distribution $D$ over $X$, we denote by $f(D)$ the distribution $(f(x))_{x \leftarrow D}$ over $Y$.

**Definition 1 (Computationally   Indistinguishable).** *Two   ensembles* $\{X_n\}_{n\in\mathbb{N}}$, $\{Y_n\}_{n\in\mathbb{N}}$ *are computationally indistinguishable if for every probabilistic polynomial-time algorithm* $\mathcal{A}$*, every polynomial* $p(\cdot)$*, and all sufficiently large* $n \in \mathbb{N}$

$$|\Pr\left[\mathcal{A}\left(X_n\right)=1\right]-\Pr\left[\mathcal{A}\left(Y_n\right)=1\right]| \leq \frac{1}{p(n)}$$

*We write* $X_n \approx_c Y_n$ *to denote that* $\{X_n\}_{n\in\mathbb{N}}$ *and* $\{Y_n\}_{n\in\mathbb{N}}$ *are computationally indistinguishable.*

**Definition 2 (Multiset).** *A* multi-set $M$ *over a set* $S$ *is a function* $M : S \to \mathbb{N}$*. For each* $s \in S$*, we call* $M(s)$ *the* multiplicity *of* $s$*. We say* $s \in M$ *if* $M(s) \geq 1$*, and denote the* size *of* $M$ *by* $|M| = \sum_S M(s)$*. For two multi-sets* $M$ *and* $M'$ *over* $S$*, we define their intersection* $M \cap M'$ *to be the multiset* $(M \cap M')(s) = \min[M(s), M'(s)]$ *containing each element with the smaller of the two multiplicities.*

## 2.2   Standard Cryptographic Notions, and the GGM Ensemble

**Definition 3 (One-way collection of functions; adapted from [Gol04]).** *A collection of functions* $\{f_s : \{0,1\}^{|s|} \to \{0,1\}^*\}_{s\in\{0,1\}^*}$ *is called* strongly (weakly) one-way *if there exists a probabilistic polynomial-time algorithm* Eval *such that the following two conditions hold:*

– Efficiently computable: *On input* $s \in \{0,1\}^*$*, and* $x \in \{0,1\}^{|s|}$*, algorithm* Eval *always outputs* $f_s(x)$*.*
– Strongly one-way: *For every polynomial* $w(\cdot)$*, for every probabilistic polynomial-time algorithm* $\mathcal{A}$ *and all sufficiently large* $n$*,*

$$\Pr_{\substack{s\leftarrow U_n\\x\leftarrow U_n}}\left[\mathcal{A}(s,f_s(x)) \in f_s^{-1}(f_s(x))\right] < \frac{1}{w(n)} \tag{4}$$

– Weakly one-way: *There exists a polynomial* $w(\cdot)$ *such that for every probabilistic polynomial-time algorithm* $\mathcal{A}$ *and all sufficiently large* $n$*,*

$$\Pr_{\substack{s\leftarrow U_n\\x\leftarrow U_n}}\left[\mathcal{A}(s,f_s(x)) \in f_s^{-1}(f_s(x))\right] < 1 - \frac{1}{w(n)} \tag{5}$$

*In this case, the collection is said to be* $1/w(n)$*-weakly one-way.*

We emphasize that in weakly one-way definition the polynomial $w(n)$ bounds the success probability of *every* efficient adversary. Additionally, weakly one-way collections can be easily amplified to achieve (strongly) one-way collections [Gol04].

We will use the following notation.

**Definition 4 (Inverting Advantage).** *For an adversary $\mathcal{A}$ and distribution $D$ over $(s, y) \in \{0,1\}^n \times \{0,1\}^n$, we define the inverting advantage of $\mathcal{A}$ on distribution $D$ as*

$$\mathsf{Adv}_{\mathcal{A}}(D) = \Pr_{(s,y) \leftarrow D} \left[ \mathcal{A}(s, y) \in f_s^{-1}(y) \right] \tag{6}$$

**Definition 5 (Pseudo-random generator).** *An efficiently computable function $G : \{0,1\}^n \to \{0,1\}^{2n}$ is a (length-doubling) pseudorandom generator (PRG), if $G(U_n)$ is computationally indistinguishable from $U_{2n}$. Namely for any PPT $\mathcal{D}$*

$$\left| \Pr[\mathcal{D}(G(U_n)) = 1] - \Pr[\mathcal{D}(U_{2n}) = 1] \right| = \mathsf{negl}(n)$$

**Definition 6 (GGM function ensemble [GGM86]).** *Let $G$ be a deterministic algorithm that expands inputs of length $n$ into string of length $2n$. We denote by $G_0(s)$ the $|s|$-bit-long prefix of $G(s)$, and by $G_1(s)$ the $|s|$-bit-long suffix of $G(s)$ (i.e., $G(s) = G_0(s) \| G_1(s)$). For every $s \in \{0,1\}^n$ (called the secret key), we define a function $f_s^G : \{0,1\}^n \to \{0,1\}^n$ such that for every $x \in \{0,1\}^n$,*

$$f_s^G(x[1], \ldots, x[n]) = G_{x[n]}(\cdots (G_{x[2]}(G_{x[1]}(s))) \cdots) \tag{7}$$

*For any $n \in \mathbb{N}$, we define $F_n$ to be a random variable over $\{f_s^G\}_{s \in \{0,1\}^n}$. We call $\mathcal{F}_G = \{F_n\}_{n \in \mathbb{N}}$ the GGM function ensemble instantiated with generator $G$.*
*We will typically write $f_s$ instead of $f_s^G$.*

The construction is easily generalized to the case when $|x| \neq n$. Though we define the GGM function ensemble as the case when $|x| = n$, it will be useful to consider the more general case.

### 2.3   Statistical Distance

For two probability distributions $D$ and $D'$ over some universe $X$, we recall two equivalent definitions of their statistical distance $\mathrm{SD}(D, D')$:

$$\mathrm{SD}(D, D') := \frac{1}{2} \sum_{x \in X} |D(x) - D'(x)| = \max_{S \subseteq X} \sum_{x \in S} D(x) - D'(x)$$

For a collection of distributions $\{D(p)\}$ with some parameter $p$, and a distribution $P$ over the parameter $p$, we write

$$(p, D(p))_P$$

to denote the distribution over pairs $(p, x)$ induced by sampling $p \leftarrow P$ and subsequently $x \leftarrow D(p)$.[4] It follows from the definition of statistical distance (see appendix) that for distributions $P$, $D(P)$, and $D'(P)$:

$$\mathrm{SD}\left( (p, D(p))_P, (p, D'(p))_P \right) = \mathop{\mathbb{E}}_{p \leftarrow P} \left[ \mathrm{SD}(D(p), D'(p)) \right] \tag{8}$$

---

[4] For example, the distribution $(x, \mathsf{Bernoulli}(x))_{\mathsf{Uniform}[0,1]}$ is the distribution over $(x, b)$ by drawing the parameter $x$ uniformly from $[0, 1]$, and subsequently taking a sample $b$ from the Bernoulli distribution with parameter $x$.

The quantity $|\mathsf{Img}(f)|$ is related to the statistical distance between the uniform distribution $U_n$ and the distribution $f(U_n)$. For any $f : \{0,1\}^n \to \{0,1\}^n$,

$$\mathsf{SD}(f(U_n), U_n) = 1 - \frac{|\mathsf{Img}(f)|}{2^n} \tag{9}$$

This identity can be easily shown by expanding the definition of statistical distance, or by considering the histograms of the two distributions and a simple counting argument. See the appendix for a proof.

## 2.4 Rényi Divergences

Similar to statistical distance, the Rényi divergence is a useful tool for relating the probability of some event under two distributions. Whereas the statistical distance yields an additive relation between the probabilities in two distributions, the Rényi divergence yields a multiplicative relation. The following is adapted from Sect. 2.3 of [BLL+15].

For any two discrete probability distributions $P$ and $Q$ such that $\mathrm{Supp}(P) \subseteq \mathrm{Supp}(Q)$, we define *the power of the* Rényi *divergence* (of order 2) by

$$R\left(P \| Q\right) = \left( \sum_{x \in \mathrm{Supp}(Q)} \frac{P(x)^2}{Q(x)} \right). \tag{10}$$

An important fact about Rényi divergence is that for an abitrary event $E \subseteq \mathrm{Supp}(Q)$

$$Q(E) \geq \frac{P(E)^2}{R\left(P \| Q\right)}. \tag{11}$$

## 3 The weak one-wayness of GGM

We now outline the proof of Theorem 1: that the GGM function ensemble is $1/n^{2+\epsilon}$-weakly one-way. The proof proceeds by contradiction, assuming that there exists a PPT $\mathcal{A}$ which inverts on input $(s, y)$ with $> 1 - 1/n^{2+\epsilon}$ probability, where $s$ is a uniform secret key and $y$ is sampled as a uniform image of $f_s$.

At a high level there are two steps. The first step (captured by the Input Switching Proposition below) is to show that the adversary successfully inverts with some non-negligible probability, even when $y$ is sampled uniformly from $\{0,1\}^n$, instead of as a uniform image from $f_s$. The second step (captured by the Distinguishing Lemma below) will then use the adversary to construct a distinguisher for the PRG underlying the GGM ensemble. The proof of Input Switching Proposition (Proposition 1) depends on the Combinatorial Lemma proved in Sect. 4. Together, these suffice to prove Theorem 1.

## 3.1   Step 1: The Input Switching Proposition

As discussed in the overview, our goal is to show that for any adversary that inverts with probability $> 1 - 1/n^{2+\epsilon}$ on input distribution $(s, y) \leftarrow (s, f_s(U_n))_{s \leftarrow U_n}$ will invert with non-negligible probability on input distribution $(s, y) \leftarrow (U_n, U_n)$. For convenience, we name these distributions:

- $D_{\mathsf{owf}}$: This is $\mathcal{A}$'s input distribution in the weakly one-way function security game in Definition 3. Namely,

$$D_{\mathsf{owf}} = (s, f_s(U_n))_{s \leftarrow U_n}$$

- $D_{\mathsf{rand}}$: This is our target distribution (needed for Step 2), in which $s$ and $y$ are drawn uniformly at random. Namely,

$$D_{\mathsf{rand}} = (U_n, U_n)$$

**Proposition 1 (Input Switching Proposition).** *For every constant $\epsilon > 0$ and sufficiently large $n \in \mathbb{N}$*

$$\mathsf{Adv}_{\mathcal{A}}(D_{\mathsf{owf}}) > 1 - 1/n^{2+\epsilon} \quad \Longrightarrow \quad \mathsf{Adv}_{\mathcal{A}}(D_{\mathsf{rand}}) > 1/\mathsf{poly}(n) \qquad (12)$$

It suffices to show that for every constant $\epsilon > 0$ and sufficiently large $n \in \mathbb{N}$

$$|\mathsf{Adv}_{\mathcal{A}}(D_{\mathsf{owf}}) - \mathsf{Adv}_{\mathcal{A}}(D_{\mathsf{rand}})| < 1 - 1/n^{2+\epsilon} - 1/\mathsf{poly}(n) \qquad (13)$$

If $\mathsf{SD}(D_{\mathsf{owf}}, D_{\mathsf{rand}}) < 1 - 1/n^2$, then the above follows immediately (even for an unbounded adversary).[5] If instead $\mathsf{SD}(D_{\mathsf{owf}}, D_{\mathsf{rand}}) \geq 1 - 1/n^2$, we must proceed differently.[6]

What if instead $y$ is sampled as a random image from $f_{s'}$, where $s'$ is a *totally independent* seed? Namely, consider the following distribution over $(s, y)$:

- $D_{\mathsf{mix}}$: This is the distribution in which $y$ is sampled as a uniform image from $f_{s'}$ and $s, s'$ are independent secret keys.

$$D_{\mathsf{mix}} = (s, f_{s'}(U_n))_{s, s' \leftarrow U_n \times U_n}$$

In order to understand the relationship between $\mathsf{Adv}_{\mathcal{A}}(D_{\mathsf{owf}})$ and $\mathsf{Adv}_{\mathcal{A}}(D_{\mathsf{mix}})$ we define our final distributions, parameterized by an integer $k \in [0, n-1]$. These distributions are related to $D_{\mathsf{owf}}$ and $D_{\mathsf{mix}}$, but instead of sampling $(s, s')$ from $U_n \times U_n$, they are sampled from $(G(f_r(U_k)))_{r \leftarrow U_n}$. If $k = 0$, we define $f_r(U_k) = r$.

---

[5] Whether this indeed holds depends on the PRG used to instantiate the GGM ensemble. We do not know if such a PRG exists.

[6] If there exists a PRG, then there exists a PRG such that $\mathsf{SD}(D_{\mathsf{owf}}, D_{\mathsf{rand}}) = 1 - \mathbb{E}_{s \leftarrow U_n}[|\mathsf{Img}(f_s)|/2^n] \geq 1 - 1/n^2$. For example, if the PRG only uses the first $n/2$ bits of its input, then $|\mathsf{Img}(f_s)| < 2^{n/2+1}$.

- $D_0^k$: Like $D_{\mathsf{owf}}$ but the secret key is $s = G_0(\hat{s})$ where $\hat{s}$ is sampled as $\hat{s} \leftarrow (f_r(U_k))_{r \leftarrow U_n}$. Namely,

$$D_0^k = (s, f_s(U_n))_{\substack{r \leftarrow U_n; \ \hat{s} \leftarrow f_r(U_k) \\ s = G_0(\hat{s})}}$$

- $D_1^k$: Like $D_{\mathsf{mix}}$, but the secret keys are $s = G_0(\hat{s})$ and $s' = G_1(\hat{s})$ where $\hat{s}$ is sampled as $\hat{s} \leftarrow (f_r(U_k))_{r \leftarrow U_n}$. Namely,

$$D_1^k = (s, f_{s'}(U_n))_{\substack{r \leftarrow U_n; \ \hat{s} \leftarrow f_r(U_k) \\ (s, s') = (G_0(\hat{s}), G_1(\hat{s}))}}$$

*Claim (Indistinguishability of Distributions).* For every $k \in [0, n-1]$,

(a) $D_{\mathsf{owf}} \approx_c D_0^k$,    (b) $D_1^k \approx_c D_{\mathsf{mix}}$,    (c) $D_{\mathsf{mix}} \approx_c D_{\mathsf{rand}}$

*Proof (Indistinguishability of Distributions).* By essentially the same techniques as in [GGM86], the pseudorandomness of the PRG implies that for any $k \le n$, the distribution $f_{U_n}(U_k)$ is computationally indistinguishable from $U_n$. Claim (c) follows immediately. By the same observation, $D_0^k \approx_c D_0^0$ and $D_1^k \approx_c D_1^0$. Finally, by the pseudorandomness of the PRG, $D_{\mathsf{owf}} \approx_c D_0^0$ and $D_1^0 \approx D_{\mathsf{mix}}$. This completes the proofs of (a) and (b).

The above claim and the following lemma (proved in Sect. 4) allow us to complete the proof of the Input Switching Proposition (Proposition 1).

**Lemma 1 (Combinatorial Lemma).** *Let* $D_{\mathsf{owf}}$, $D_0^k$, $D_1^k$, $D_{\mathsf{mix}}$ *and* $D_{\mathsf{rand}}$ *be defined as above. For every constant* $\epsilon' > 0$ *and every* $n \in \mathbb{N}$,

- *either there exists* $k^* \in [0, n-1]$ *such that*

$$SD\left(D_0^{k^*}, D_1^{k^*}\right) \le 1 - \frac{1}{n^{2+\epsilon'}} \tag{L.1}$$

- *or*

$$SD\left(D_{\mathsf{owf}}, D_{\mathsf{rand}}\right) < \frac{2}{n^{\epsilon'/2}} \tag{L.2}$$

We now prove (13) and thereby complete the proof of Input Switching Proposition (Proposition 1). Fix a constant $\epsilon > 0$ and $n \in \mathbb{N}$. Apply the Combinatorial Lemma (Lemma 1) with $\epsilon' = \epsilon/2$. In the case that (L.2) is true,

$$\left|\mathsf{Adv}_\mathcal{A}(D_{\mathsf{owf}}) - \mathsf{Adv}_\mathcal{A}(D_{\mathsf{rand}})\right| \le SD(D_{\mathsf{owf}}, D_{\mathsf{rand}}) < \frac{2}{n^{\epsilon/4}}$$

In the case that (L.1) is true, we use the Triangle Inequality. Let $k^* \in [0, n-1]$ be as guaranteed by (L.1):

$$\left|\mathsf{Adv}_\mathcal{A}(D_{\mathsf{owf}}) - \mathsf{Adv}_\mathcal{A}(D_{\mathsf{rand}})\right|$$
$$\le \left|\mathsf{Adv}_\mathcal{A}(D_{\mathsf{owf}}) - \mathsf{Adv}_\mathcal{A}(D_0^{k^*})\right| + \left|\mathsf{Adv}_\mathcal{A}(D_0^{k^*}) - \mathsf{Adv}_\mathcal{A}(D_1^{k^*})\right|$$
$$\quad + \left|\mathsf{Adv}_\mathcal{A}(D_1^{k^*}) - \mathsf{Adv}_\mathcal{A}(D_{\mathsf{mix}})\right| + \left|\mathsf{Adv}_\mathcal{A}(D_{\mathsf{mix}}) - \mathsf{Adv}_\mathcal{A}(D_{\mathsf{rand}})\right|$$
$$\le \mathsf{negl}(n) + \left(1 - \frac{1}{n^{2+\epsilon/2}}\right) + \mathsf{negl}(n) + \mathsf{negl}(n)$$
$$\le 1 - \frac{1}{n^{2+\epsilon/4}} + \mathsf{negl}(n)$$

## 3.2    Step 2: The Distinguishing Lemma

As discussed in the overview, in this step we show that any efficient algorithm $\mathcal{A}$ that can invert $f_s$ on uniformly random values $y \in \{0,1\}^n$ with probability $\geq 1/\alpha(n)$ can be used to distinguish the uniform distribution from uniform images of the PRG $G$ underlying the GGM ensemble with probability $\geq 1/\mathsf{poly}(\alpha(n))$. Formally, we prove the following lemma:

**Lemma 2 (Distinguishing Lemma).** *Let $G$ be a PRG and $\mathcal{F}_G$ the corresponding GGM ensemble. For all PPT algorithms $\mathcal{A}$ and polynomials $\alpha(n)$, there exists a PPT distinguisher $\mathcal{D}$ which for all $n \in \mathbb{N}$:*

$$\mathsf{Adv}_{\mathcal{A}}(U_n \times U_n) \geq \frac{1}{\alpha(n)}$$

$$\implies \left| \Pr\left[\mathcal{D}\left(G\left(U_n\right)\right) = 1\right] - \Pr\left[\mathcal{D}\left(U_{2n}\right) = 1\right] \right| \geq \left(\frac{1}{4\alpha(n)}\right)^5 - \mathsf{negl}(n)$$

*Proof.* Let $\mathcal{A}$ be a PPT algorithm such that for some polynomial $\alpha(n)$

$$\mathsf{Adv}_{\mathcal{A}}(U_n \times U_n) \geq \frac{1}{\alpha(n)} \tag{14}$$

The distinguisher $\mathcal{D}$ is defined as follows:

> **Input:** $(y_0, y_1)$    // a sample from either $G(U_n)$ or $U_{2n}$
> Sample a secret key $s \leftarrow U_n$ and a bit $b \leftarrow U$;
> Compute $x \leftarrow \mathcal{A}(s, y_b)$;
> Let $\tilde{x} = x \oplus 0^{n-1}1$    // $\tilde{x}$ differs from $x$ only at the last bit;
> **if** $f_s(x) = y_b$ *and* $f_s(\tilde{x}) = y_{1-b}$ **then**
> |    Output 1;        // Guess ``PRG''
> **else**
> |    Output 0;        // Guess ``random''
> **end**

**Algorithm 2.** The PRG distinguisher $\mathcal{D}$

Next we show that the distinguisher $\mathcal{D}$ outputs 1 given input sampled uniformly with only negligible probability, but outputs 1 with some non-negligible probability given input sampled from $G(U_n)$. This violates the security of the PRG, contradicting assumption (14).

Observe that if $\mathcal{D}$ outputs 1, then either $(y_0, y_1)$ or $(y_1, y_0)$ is in $\mathsf{Img}(G)$. If $(y_0, y_1)$ was sampled uniformly from $U_{2n}$, then this happens with probability at most $2^{n+1}/2^{2n}$. Therefore,

$$\Pr[\mathcal{D}(U_{2n}) = 1] = \mathsf{negl}(n) \tag{15}$$

We prove that

$$\Pr[\mathcal{D}(G(U_n)) = 1] \geq \left(\frac{1}{4\alpha(n)}\right)^5 \tag{16}$$

At a very high level, the intuition is that for most $(y_0, y_1) \in \mathsf{Img}(G)$, there are not too many $y_1'$ for which either $(y_0, y_1') \in \mathsf{Img}(G)$ or $(y_1', y_0) \in \mathsf{Img}(G)$ (similarly for $y_0'$ and $y_1$). After arguing that $\mathcal{A}$ must invert even on such "thin" $y$'s, the chance that $y_{1-b}' = y_{1-b}$ is significant. We now formalize this high level intuition.

We define the function $G_* : \{0,1\} \times \{0,1\}^n \to \{0,1\}^n$

$$G_*(b, y) = G_b(y)$$

**Definition 7 ($\theta$-thin, $\theta$-fat).** *An element $y \in \mathsf{Img}(G_*)$ is called $\theta$-thin under $G$ if $|G_*^{-1}(y)| \leq \theta$. Otherwise, it is called $\theta$-fat. Define the sets*

$$\mathsf{Thin}_\theta := \{y \in \mathsf{Img}(G_*) \ : \ y \ is \ \theta - thin\}$$

$$\mathsf{Fat}_\theta := \{y \in \mathsf{Img}(G_*) \ : \ y \ is \ \theta - fat\}$$

*Note that $\mathsf{Thin}_\theta \sqcup \mathsf{Fat}_\theta = \mathsf{Img}(G_*)$*

We define an ensemble of distributions $\{Z_n\}$, where each $Z_n$ is the following distribution over $(s, y_0, y_1, b) \in \{0,1\}^n \times \{0,1\}^n \times \{0,1\}^n \times \{0,1\}$:

$$Z_n = (U_n, G_0(r), G_1(r), U)_{r \leftarrow U_n}. \tag{17}$$

Additionally, for every $x \in \{0,1\}^n$, we define $\widetilde{x}$ to be $x$ with its last bit flipped, namely

$$\widetilde{x} = x \oplus 0^{n-1}1.$$

We begin by expanding $\Pr[\mathcal{D}(G(U_n)) = 1]$.

$$\Pr[\mathcal{D}(G(U_n)) = 1]$$

$$= \Pr_{(s,y_0,y_1,b) \leftarrow Z_n} [f_s(x) = y_b \wedge f_s(\widetilde{x}) = y_{1-b} \mid x \leftarrow \mathcal{A}(s, y_b)]$$

$$\geq \Pr_{(s,y_0,y_1,b) \leftarrow Z_n} [y_b \in \mathsf{Thin}_\theta] \tag{18}$$

$$\cdot \Pr_{(s,y_0,y_1,b) \leftarrow Z_n} \left[ f_s(x) = y_b \ \middle| \ \begin{array}{c} x \leftarrow \mathcal{A}(s, y_b) \\ y_b \in \mathsf{Thin}_\theta \end{array} \right] \tag{19}$$

$$\cdot \Pr_{(s,y_0,y_1,b) \leftarrow Z_n} \left[ f_s(\widetilde{x}) = y_{1-b} \ \middle| \ \begin{array}{c} x \leftarrow \mathcal{A}(s, y_b) \\ y_b \in \mathsf{Thin}_\theta \ \wedge \ f_s(x) = y_b \end{array} \right] \tag{20}$$

To show that $\Pr[\mathcal{D}(G(U_n)) = 1]$ is non-negligible, it's enough to show that (18), (19), and (20) are each non-negligible.

**The first term** can be lower-bounded by

$$\Pr_{(s,y_0,y_1,b) \leftarrow Z_n} [y \in \mathsf{Thin}_\theta] \geq \frac{1}{2\alpha(n)} - \frac{1}{\theta} \tag{21}$$

To see why, first recall that by hypothesis $\mathsf{Adv}_{\mathcal{A}}(U_n \times U_n) \geq \frac{1}{\alpha(n)}$. If $y \notin \mathsf{Img}(f_s)$, then of course $\mathcal{A}(s, y)$ cannot output a preimage of $y$. Therefore $2^n/\alpha(n) \leq |\mathsf{Img}(f_s)| \leq |\mathsf{Img}(G_*)|$. On the other hand, because each $\theta$-fat $y$ must have at least $\theta$ preimages, and the domain of $G_*$ is of size $2^{n+1}$, there cannot be too many $\theta$-fat $y$'s:

$$|\mathsf{Fat}_\theta| \leq \frac{2^{n+1}}{\theta} \tag{22}$$

Recalling that $\mathsf{Img}(G_*) = \mathsf{Thin}_\theta \sqcup \mathsf{Fat}_\theta$:

$$\Pr_{y \leftarrow G_U(U_n)}[y \in \mathsf{Thin}] = \frac{|\{(b, x) \; : \; G_b(x) \in \mathsf{Thin}_\theta\}|}{2^{n+1}}$$

$$\geq \frac{|\mathsf{Thin}_\theta|}{2^{n+1}}$$

$$= \frac{1}{2\alpha(n)} - \frac{1}{\theta}$$

**The second term** can be lower-bounded by:

$$\Pr_{(s, y_0, y_1, b) \leftarrow Z_n} \left[ f_s(x) = y_b \; \middle| \; \begin{array}{l} x \leftarrow \mathcal{A}(s, y_b) \\ y_b \in \mathsf{Thin}_\theta \end{array} \right] \geq \left( \frac{1}{4\alpha(n)} \right)^3 \tag{23}$$

We now provide some intuition for the proof of the above, which is included in the appendix in full. In the course of that argument, we will set $\theta = 4\alpha(n)$.

**Definition 8 (q-good).** *For any $q \in [0, 1]$, an element $y \in \{0, 1\}^n$ is called $q$-good with respect to $\theta$ if it is both $\theta$-thin and $\mathcal{A}$ finds some preimage of $y$ for a uniformly random secret key $s$ with probability at least $q$. Namely,*

$$\mathsf{Good}_q := \left\{ y \in \mathsf{Thin}_\theta \; : \; \Pr_{s \leftarrow U_n}[\mathcal{A}(s, y) \in f_s^{-1}(y)] > q \right\}$$

The marginal distribution of $y_b$ where $(s, y_0, y_1, b) \leftarrow Z_n$ is $G_U(U_n)$. To make the notation more explicit, we use the latter notation for the intuition below. In this notation, (23) can be written

$$\Pr_{\substack{s \leftarrow U_n \\ y \leftarrow G_U(U_n)}} [\mathcal{A}(s, y) \in f_s^{-1}(y) \mid y \in \mathsf{Thin}_\theta] \geq \left( \frac{1}{4\alpha(n)} \right)^3$$

The proof of the above inequality boils down to two parts. First, we show that, by the definition of $\theta$-thin:

$$\Pr_{\substack{s \leftarrow U_n \\ y \leftarrow U_n}} [y \in \mathsf{Good}_q \mid y \in \mathsf{Thin}_\theta] \geq \theta \cdot \Pr_{\substack{s \leftarrow U_n \\ y \leftarrow G_U(U_n)}} [y \in \mathsf{Good}_q \mid y \in \mathsf{Thin}_\theta]$$

Second, we must lower-bound the latter quantity. At a high level, this second step follows from the fact that most of the $y \in \{0, 1\}^n$ are $\theta$-thin. By assumption, $\mathcal{A}$ inverts with decent probability when $y \leftarrow U_n$, and therefore must invert with some not-too-much-smaller probability when conditioning on the event $y \in \mathsf{Thin}_\theta$.

**The third term** can be lower-bounded by:

$$\Pr_{(s,y_0,y_1,b)\leftarrow Z_n}\left[f_s(\widetilde{x}) = y_{1-b} \ \middle| \ \begin{array}{l} x \leftarrow \mathcal{A}(s, y_b) \\ y_b \in \mathsf{Thin}_\theta \ \wedge \ f_s(x) = y_b \end{array}\right] \geq \frac{1}{\theta} \quad (24)$$

To see why, suppose that indeed $y_b \in \mathsf{Thin}_\theta$ and $f_s(x) = y_b$. Because $y_b$ is $\theta$-thin, there are at most $\theta$-possible values of $y'_{1-b} := f_s(\widetilde{x})$, where $\widetilde{x} = x \oplus 0^{n-1}1$. The true $y_{1-b}$ is hidden from the adversary's view, and takes each of the possible values with probability at least $1/\theta$. Thus the probability that $y_{1-b} = y'_{1-b}$ is as above.

Finally, letting $\theta = 4\alpha(n)$ as required to lower-bound the second term and putting it all together implies that

$$\Pr\left[\mathcal{D}(G(U_n)) = 1\right] > \left(\frac{1}{2\alpha(n)} - \frac{1}{\theta}\right) \cdot \left(\frac{1}{4\alpha(n)}\right)^3 \cdot \frac{1}{\theta} \quad (25)$$

$$\geq \left(\frac{1}{4\alpha(n)}\right)^5 \quad (26)$$

This completes the proof of Lemma 2.

## 4    The Combinatorial Lemma

In the proof of the Input Switching Proposition (Proposition 1), we defined the following distributions over $(s, y) \in \{0,1\}^n \times \{0,1\}^n$, for $k \in [0, n-1]$. If $k = 0$, we define $f_r(U_k) = r$.

$$D_{\mathsf{owf}} = (s, f_s(U_n))_{s\leftarrow U_n}$$

$$D_0^k = \left(G_0(\hat{s}), f_{G_0(\hat{s})}(U_n)\right)_{r\leftarrow U_n; \ \hat{s}\leftarrow f_r(U_k)}$$

$$D_1^k = \left(G_0(\hat{s}), f_{G_1(\hat{s})}(U_n)\right)_{r\leftarrow U_n; \ \hat{s}\leftarrow f_r(U_k)}$$

$$D_{\mathsf{mix}} = (s, f_{s'}(U_n))_{s,s'\leftarrow U_n \times U_n}$$

$$D_{\mathsf{rand}} = (U_n, U_n)$$

We define two additional distributions:

$$\widehat{D}_0^k = \left(\hat{s}, f_{G_0(\hat{s})}(U_n)\right)_{r\leftarrow U_n; \ \hat{s}\leftarrow f_r(U_k)}$$

$$\widehat{D}_1^k = \left(\hat{s}, f_{G_1(\hat{s})}(U_n)\right)_{r\leftarrow U_n; \ \hat{s}\leftarrow f_r(U_k)}$$

We restate the lemma stated and used in the proof Input Switching Proposition.

**Lemma 1 (Combinatorial Lemma).** *Let* $D_{\mathsf{owf}}$, $D_0^k$, $D_1^k$, $D_{\mathsf{mix}}$ *and* $D_{\mathsf{rand}}$ *be defined as above. For every constant* $\epsilon' > 0$ *and every* $n \in \mathbb{N}$,

– *either there exists* $k^* \in [0, n-1]$ *such that*

$$\mathrm{SD}\left(D_0^{k^*}, D_0^{k^*}\right) \leq 1 - \frac{1}{n^{2+\epsilon'}} \quad (\text{L.1})$$

– *or*

$$\mathrm{SD}\left(D_{\mathsf{owf}}, D_{\mathsf{rand}}\right) < \frac{2}{n^{\epsilon'/2}} \tag{L.2}$$

We will prove something slightly stronger, namely that either (L.1*) or (L.2) holds, where (L.1*) is:

$$\mathrm{SD}\left(\widehat{D}_0^{k^*}, \widehat{D}_1^{k^*}\right) \leq 1 - \frac{1}{n^{2+\epsilon'}} \tag{L.1*}$$

To see why (L.1*) implies (L.1), observe that for every $k$, given a sample from $\widehat{D}_0^k$ (resp. $\widehat{D}_1^k$) it is easy to generate a sample from $D_0^k$ (resp. $D_1^k$). Thus an (unbounded) distinguisher for the former pair of distributions implies an (unbounded) distinguisher with at least the same advantage for the latter pair.[7]

*Remark 3.* By (8) and (9), $\mathrm{SD}(D_{\mathsf{owf}}, D_{\mathsf{rand}}) = 1 - \mathbb{E}_{s \leftarrow U_n}[\mathsf{Img}(f_s)/2^n]$. Using (L.1*) and this interpretation of (L.2), the lemma informally states that either:

– There is a level $k^*$ such that for a random node $\hat{s}$ on the $k^*$th level, the subtrees induced by the left child $G_0(\hat{s})$ and the right child $G_1(\hat{s})$ are not too dissimilar.
– The image of $f_s$ is in expectation, a very large subset of the co-domain.

Finally, it is worth noting that the proof of this lemma is purely combinatorial and nowhere makes use of computational assumptions. As such, it holds for and GGM-like ensemble instantiated with arbitrary length-doubling function $G$.

*Proof (Combinatorial Lemma).* Fix $n \in \mathbb{N}$ and a secret key $s \in \{0,1\}^n$. Recall that for a multi-set $M$, $M(x)$ is the multiplicity of the element $x$ in $M$.

For every $k \in [0, n-1]$ and $v \in \{0,1\}^k$ (letting $\{0,1\}^0 = \{\varepsilon\}$, where $\varepsilon$ is the empty string), we define two multi-sets over $\{0,1\}^n$ ('$L$' for 'leaves') which together contain all the leaves contained in the subtree with prefix $v$ of the GGM tree rooted at $s$.

$$\begin{aligned}
L_{v,0}^s &= \{f_s(x) : x = v\|0\|t\}_{t \in \{0,1\}^{n-k-1}} \\
L_{v,1}^s &= \{f_s(x) : x = v\|1\|t\}_{t \in \{0,1\}^{n-k-1}}
\end{aligned} \tag{27}$$

Define $I_v^s := L_{v,0}^s \cap L_{v,1}^s$ to be their intersection.

For each $v \in \{0,1\}^k$, we define a set $B_v^s$ of "bad" inputs $x$ to the function $f_s$. For each $y \in I_v^s$, there are at least $I_v^s(y)$-many distinct $x_0$ (respectively, $x_1$) such that $f_s(x_0) = y$ and $x_0 = v\|0\|t$ begins with the prefix $v\|0$ (respectively, $v\|1$). Assign arbitrarily $I_v^s(y)$-many such $x_0$ and $x_1$ to the set $B_v^s$. By construction,

$$|B_v^s| = 2|I_v^s| \tag{28}$$

Let $B^s = \bigcup_{k=0}^{n-1} \bigcup_{v \in \{0,1\}^k} B_v^s$, and let $Q^s := \{0,1\}^n \backslash B^s$ be the set of "good" inputs.

---

[7] This essentially a data-processing inequality.

Observe that $f_s$ is injective on $Q^s$. To see why, consider some $x \in Q^s$, and let $x' \neq x$ be such that $f_s(x) = f_s(x') = y$ if one exists. Suppose that the length of their longest common prefix $v$ is maximal among all such $x'$. By the maximality of the prefix $v$, $x$ must be in $B_v^s$. Therefore,

$$|\mathsf{Img}(f_s)| \geq |Q^s| \tag{29}$$

To reduce clutter we define the following additional notation: for every secret key $r \in \{0,1\}^n$ and level $\ell \in [n]$ we define

$$\Delta_{\mathsf{mix}}(r;\ell) = \mathsf{SD}(f_{G_0(r)}(U_\ell); f_{G_1(r)}(U_\ell))$$

Informally, $\Delta_{\mathsf{mix}}(r;\ell)$ is the difference between the left and right subtrees rooted at $r$ of depth $\ell$. For all $\ell < n$ and $r \in \{0,1\}^n$:

$$\Delta_{\mathsf{mix}}(r;\ell) \geq \Delta_{\mathsf{mix}}(r;n) \tag{30}$$

This can be seen by expanding the definitions, or by considering the nature of the distributions as follows. The GGM construction implies that if two internal nodes have the same label, then their subtrees exactly coincide. Thus, the fraction of nodes at level $n$ that coincide on trees rooted at $G_0(r)$ and $G_1(r)$ is at least the fraction of nodes at level $\ell$ that coincide.

For every secret key $s \in \{0,1\}^n$, $k \in [0, n-1]$, and $v \in \{0,1\}^k$, it holds that:

$$\Delta_{\mathsf{mix}}(f_s(v); n - k - 1) = 1 - \frac{|I_v^s|}{2^{n-k-1}} \tag{31}$$

Rearranging (31) and using (30) with $\ell = n - k$, we have that

$$\frac{|I_v^s|}{2^{n-k-1}} \leq 1 - \Delta_{\mathsf{mix}}(f_s(v); n) \tag{32}$$

*Claim.* For $\epsilon > 0$, $n \in \mathbb{N}$, if $\mathsf{SD}(\widehat{D}_0^{k^*}, \widehat{D}_1^{k^*}) \leq 1 - \frac{1}{n^{2+\epsilon'}}$ (i.e., if (L.1*) is false), then

$$1 - \mathop{\mathbb{E}}_{s \leftarrow U_n}\left[\frac{|Q^s|}{2^n}\right] = \mathop{\mathbb{E}}_{s \leftarrow U_n}\left[\frac{|B^s|}{2^n}\right] < \frac{2}{n^{\epsilon/2}} \tag{33}$$

See proof below. This claim implies (L.2) as follows, completing the proof:

$$\mathsf{SD}(D_{\mathsf{owf}}, D_{\mathsf{rand}}) = 1 - \mathop{\mathbb{E}}_{s \leftarrow U_n}\left[\frac{|\mathsf{Img}(f_s)|}{2^n}\right] \leq 1 - \mathop{\mathbb{E}}_{s \leftarrow U_n}\left[\frac{|Q^s|}{2^n}\right] < 1 - \frac{2}{n^{\epsilon/2}} \tag{34}$$

*Proof (of Claim).* We can now bound the expected size of $|B^s|$ as follows.

$$\mathop{\mathbb{E}}_{s\leftarrow U_n}\left[\frac{|B^s|}{2^n}\right] \tag{35}$$

$$= \mathop{\Pr}_{\substack{s\leftarrow U_n \\ x\leftarrow U_n}}[x\in B^s]$$

$$\leq \sum_{k=0}^{n-1}\sum_{v\in\{0,1\}^k}\mathop{\Pr}_{s,x}[x\in B_v^s] \qquad \text{by the definition of } B^s$$

$$= \sum_{k=0}^{n-1}\mathop{\Pr}_{s,x}\left[x\in B_{x[1:k]}^s\right]$$

$$\leq \sum_{k=0}^{n-1}T\cdot\mathop{\Pr}_{s,x}\left(\frac{|B_{x[1:k]}^s|}{2^{n-k}}\leq T\right)+\mathop{\Pr}_{s,x}\left(\frac{|B_{x[1:k]}^s|}{2^{n-k}}>T\right) \qquad \text{for any } 0\leq T\leq 1$$

$$\leq \sum_{k=0}^{n-1}T+\mathop{\Pr}_{s,x}\left(\frac{|I_{x[1:k]}^s|}{2^{n-k-1}}>T\right) \qquad \text{by (28)}$$

Fix constant $\epsilon>0$. Suppose (L.1*) is false; namely, for all $k\in[0,n-1]$,

$$\text{SD}\left(\widehat{D}_0^{k*},\widehat{D}_1^{k*}\right) = \mathop{\mathbb{E}}_{\substack{r\leftarrow U_n \\ \hat{s}\leftarrow f_r(U_k)}}\left[\Delta_{\text{mix}}(\hat{s};n)\right] > 1-\frac{1}{n^{2+\epsilon}} \tag{36}$$

By Markov's Inequality, for any $\tau>0$:

$$\mathop{\Pr}_{\substack{r\leftarrow U_n \\ \hat{s}\leftarrow f_r(U_k)}}\left[1-\Delta_{\text{mix}}(\hat{s};n)>\frac{\tau}{n^{2+\epsilon}}\right] < \frac{1}{\tau} \tag{37}$$

Observe that the distributions $(f_s(x[1:k]))_{\substack{s\leftarrow U_n \\ x\leftarrow U_n}}$ and $(\hat{s})_{\substack{r\leftarrow U_n \\ \hat{s}\leftarrow f_r(U_k)}}$ are identical. Therefore, by inequality (32) and the above Markov bound:

$$\mathop{\Pr}_{\substack{s\leftarrow U_n \\ x\leftarrow U_n}}\left(\frac{|I_{x[1:k]}^s|}{2^{n-k-1}}>T\right) \leq \mathop{\Pr}_{\substack{s\leftarrow U_n \\ x\leftarrow U_n}}\left(1-\Delta_{\text{mix}}(f_s(x[1:k]);n)>T\right) \leq \frac{1}{Tn^{2+\epsilon}} \tag{38}$$

Continuing the series of inequalities from (35):

$$\leq \sum_{k=0}^{n-1}\left(T+\frac{1}{Tn^{2+\epsilon}}\right) \qquad \text{by (32)}$$

$$\leq n\frac{\tau}{n^{2+\epsilon}}+n\frac{1}{\tau} \qquad \text{for } T=\frac{\tau}{n^{2+\epsilon}}, \text{by (37)}$$

$$= \frac{2}{n^{\epsilon/2}} \qquad \text{for } \tau=n^{1+\epsilon/2}$$

This completes the proof of the claim.

## 5    When Is GGM Strongly One-Way?

Theorem 2 shows that under some natural – albeit strong – conditions, the GGM function ensemble is strongly one-way. Whether pseudorandom generators $G$ exist that induce these conditions in the GGM ensemble is, as yet, unknown.

**Theorem 2.** *Let $\mathcal{F}_G$ be the GGM ensemble with pseudorandom generator $G$. $\mathcal{F}_G$ is a strongly one-way collection of functions if either of the following hold:*

(a)  *There exists a negligible function $\mathsf{negl}(\cdot)$ such that for all sufficiently large $n$*

$$\mathop{\mathbb{E}}_{s \leftarrow U_n}\left[\frac{|\mathsf{Img}(f_s)|}{2^n}\right] \geq 1 - \mathsf{negl}(n) \tag{39}$$

(b)  *There exists a polynomial $\beta(\cdot)$ such that for all sufficiently large $n$ and for all $s, y \in \{0,1\}^n$*

$$\left|f_s^{-1}(y)\right| \leq \beta(n) \tag{40}$$

*Remark 4.* These two conditions have some overlap, but neither is contained in the other. Additionally, a weaker – but somewhat more abstruse – condition than (b) also suffices: namely, that $\sum_{s,y}\left(\frac{|f_s^{-1}(y)|}{2^n}\right)^2$ is bounded above by some polynomial. This quantity is related to the collision entropy of the distribution $(s, f_s(U_n))_{s \leftarrow U_n}$.

*Proof (Theorem 2).* Suppose $\mathcal{F}_G$ satisfies one of the conditions of Theorem 2. Further suppose towards contradiction that there exists a probabilistic polynomial-time $\mathcal{A}$ and a polynomial $w(\cdot)$, such that for infinitely-many $n \in \mathbb{N}$

$$\mathsf{Adv}_{\mathcal{A}}\big((s, f_s(U_n))_{s \leftarrow U_n}\big) \geq \frac{1}{w(n)} \tag{41}$$

By the Distinguishing Lemma, to derive a contradiction it suffices to prove for some polynomial $\alpha(\cdot)$ related to $w$

$$\mathsf{Adv}_{\mathcal{A}}(U_n \times U_n) > \frac{1}{\alpha(n)} \tag{42}$$

**Case (a):** Applying Eqs. (8) and (9) to the assumption on $\mathbb{E}_{s \leftarrow U_n}\left[\frac{\mathsf{Img}(f_s)}{2^n}\right]$ yields

$$\mathsf{SD}\left((s, f_s(U_n))_{U_n}, (U_n, U_n)\right) \leq \mathsf{negl}(n) \tag{43}$$

It follows immediately that (42) holds for $1/\alpha(n) = 1/w(n) - 1/\mathsf{poly}(n)$, for any polynomial $\mathsf{poly}$ (e.g. for $1/\alpha(n) = 1/2w(n)$).

**Case (b):** For this case, we use the facts about Rényi divergence from the Preliminaries and follow that notation closely. Let $P = D_{\mathsf{owf}} = (s, f_s(U_n))_{s \leftarrow U_n}$ and $Q = D_{\mathsf{rand}} = U_{2n}$ be probability distributions over $\{0,1\}^{2n}$.

*Claim.* $R(P\|Q) \leq \beta(n)^2$.

*Proof (of Claim).*

$$R(P\|Q) = \sum_{(s,y)\in\{0,1\}^{2n}} \frac{P(s,y)^2}{Q(s,y)}$$

$$= 2^{2n} \sum_{s,y} P(s,y)^2$$

$$= 2^{2n} \sum_{s,y} \left(\frac{1}{2^n} \cdot \Pr_P[y|s]\right)^2$$

$$= \sum_{s,y} \Pr_P[y|s]^2$$

$$= \sum_{s,y} \left(\frac{|f_s^{-1}(y)|}{2^n}\right)^2$$

$$\le \beta(n)^2$$

Let the event

$$E = \left\{(s,y) \in \{0,1\}^n \times \{0,1\}^n : \Pr_{\mathcal{A}}[\mathcal{A}(s,y) \in f_s^{-1}(y)] > \frac{1}{2w(n)}\right\}$$

be the set of pairs $(s,y)$ on which $\mathcal{A}$ successfully inverts with probability at least $1/2w(n)$. By an averaging argument:

$$\frac{1}{w(n)} < \mathsf{Adv}_{\mathcal{A}}(P) = \Pr_{(s,y)\leftarrow P}[\mathcal{A}(s,y) \in f_s^{-1}(y)]$$

$$= \Pr_P[\mathcal{A}(s,y) \in f_s^{-1}(y) \wedge E]$$

$$+ \Pr_P[\mathcal{A}(s,y) \in f_s^{-1}(y) \wedge \neg E]$$

$$\le \Pr_P[E] + \Pr[\mathcal{A}(s,y) \in f_s^{-1}(y) \mid \neg E]$$

$$\le P(E) + \frac{1}{2w(n)}$$

Using (11) from the Preliminaries (i.e., $Q(E) \ge \frac{P(E)^2}{R(P\|Q)}$), we get that

$$P(E) > \frac{1}{2w(n)} \quad \Longrightarrow \quad Q(E) > \frac{1}{4w(n)^2 B(n)^2} \tag{44}$$

From the definition of event $E$, it follows that the condition in (42) holds, completing the proof:

$$\mathsf{Adv}_{\mathcal{A}}(Q) = \Pr_{(s,y)\leftarrow U_{2n}}[\mathcal{A}(s,y) \in f_s^{-1}(y)] > \frac{Q(E)}{2w(n)} > \frac{1}{8w(n)^3 B(n)^2} \tag{45}$$

# 6 Conclusion

In this work, we demonstrated that the length-preserving Goldreich-Goldwasser-Micali function family is weakly one-way. This is the first demonstration that the family maintains some cryptographic hardness even when the secret key is exposed.

**Open Questions.** Two interesting open questions suggest themselves.

1. Is GGM strongly one-way for all pseudorandom generators, or does there exist a generator for which the induced GGM ensemble can be inverted some non-negligible fraction of the time? A positive answer to this question would be very interesting and improve upon this work; a negative answer would be a spiritual successor to [Gol02].
2. In the absence of a positive answer to the above, do there exist pseudorandom generators for which the induced GGM ensemble is strongly one-way? In particular, do there exist generators that satisfy the requirements of Theorem 2?

**Acknowledgments.** We would like to thank Shafi Goldwasser, Ran Canetti, and Alon Rosen for their encouragement throughout this project. We would additionally like to thank Justin Holmgren for discussions about the proof of Lemma 1, and Krzysztof Pietrzak, Nir Bitansky, Vinod Vaikuntanathan, Adam Sealfon, and anonymous reviewers for their helpful feedback.

This work was done in part while the authors were visiting the Simons Institute for the Theory of Computing, supported by the Simons Foundation and by the DIMACS/Simons Collaboration in Cryptography through NSF grant CNS-1523467. Aloni Cohen was supported in part by the NSF GRFP, along with NSF MACS - CNS-1413920, DARPA IBM - W911NF-15-C-0236, and Simons Investigator Award Agreement Dated 6-5-12. Saleet Klein was supported in part by ISF grant 1536/14, along with ISF grant 1523/14, and the Check Point Institute for Information Security. Both authors were supported by the MIT-Israel Seed Fund.

# A Appendix

*Proof of (8):*

$$\mathrm{SD}\left((p, D(p))_P, (p, D'(p))_P\right)$$

$$= \frac{1}{2} \sum_{(p,x)\in\mathrm{Supp}(P)\times X} \left| \Pr_{(p,D(p))_P}(p,x) - \Pr_{(p,D'(p))_P}(p,x) \right|$$

$$= \sum_{p\in\mathrm{Supp}(P)} \Pr_P(p) \cdot \frac{1}{2} \sum_{x\in X} \left| \Pr_{D(p)}(x) - \Pr_{D'(p)}(x) \right|$$

$$= \sum_{p\in\mathrm{Supp}(P)} \Pr_P(p) \cdot \mathrm{SD}\left(D(p), D'(p)\right)$$

$$= \mathop{\mathbb{E}}_{p\leftarrow P}\left[\mathrm{SD}\left(D(p), D'(p)\right)\right]$$

*Proof of* (9):

$$\mathrm{SD}(f(U_n), U_n) = \frac{1}{2} \sum_{\alpha \in \{0,1\}^n} \left| \Pr[f(U_n) = \alpha] - \Pr[U_n = \alpha] \right|$$

$$= \frac{1}{2} \sum_{\alpha} \left| \frac{|f^{-1}(\alpha)|}{2^n} - \frac{1}{2^n} \right|$$

$$= \frac{1}{2} \left( \sum_{\alpha \in \mathsf{Img}(f)} \left| \frac{|f^{-1}(\alpha)|}{2^n} - \frac{1}{2^n} \right| + \sum_{\alpha \notin \mathsf{Img}(f)} \frac{1}{2^n} \right)$$

$$= \frac{1}{2} \left( 1 - \frac{|\mathsf{Img}(f)|}{2^n} + 1 - \frac{|\mathsf{Img}(f)|}{2^n} \right)$$

$$= 1 - \frac{|\mathsf{Img}(f)|}{2^n}$$

*Proof of Inequality* (23): Recall the following definition.

**Definition 8 ($q$-good).** *For any $q \in [0,1]$, an element $y \in \{0,1\}^n$ is called $q$-good with respect to $\theta$ if it is both $\theta$-thin and $\mathcal{A}$ finds some preimage of $y$ for a uniformly random secret key $s$ with probability at least $q$. Namely,*

$$\mathsf{Good}_q := \left\{ y \in \mathsf{Thin}_\theta : \Pr_{s \leftarrow U_n} [\mathcal{A}(s,y) \in f_s^{-1}(y)] > q \right\}$$

We begin with two observations:

- The distribution over $y_b$ is equivalent to the distribution $(G_b(x))_{(b,x) \leftarrow U \times U_n}$. The number of pairs $(b,x)$ such that $G_b(x) \in \mathsf{Good}_q$ is at least $|\mathsf{Good}_q|$, while the number of pairs $(b,x)$ such that $G_b(x) \in \mathsf{Thin}_\theta$ is at most $\theta |\mathsf{Thin}_\theta|$. Therefore:

$$\Pr_{\substack{s \leftarrow U_n \\ y \leftarrow G_U(U_n)}} [y \in \mathsf{Good}_q \mid y \in \mathsf{Thin}_\theta]$$

$$= \Pr_{(b,x) \leftarrow U \times U_n} [G_b(x) \in \mathsf{Good}_q \mid G_b(x) \in \mathsf{Thin}_\theta]$$

$$\geq \frac{1}{\theta} \cdot \frac{|\mathsf{Good}_q|}{|\mathsf{Thin}_\theta|}$$

$$= \frac{1}{\theta} \cdot \Pr_{s,y \leftarrow U_n} [y \in \mathsf{Good}_q \mid y \in \mathsf{Thin}_\theta]$$

- By definition of $\mathsf{Good}_q$:

$$\Pr_{\substack{s \leftarrow U_n \\ y \leftarrow G_U(U_n)}} [\mathcal{A}(s,y) \in f_s^{-1}(y) \mid y \in \mathsf{Good}_q] > q \tag{46}$$

Combining the above

$$\Pr_{\substack{s \leftarrow U_n \\ y \leftarrow G_U(U_n)}} \left[ \mathcal{A}(s, y) \in f_s^{-1}(y) \mid y \in \mathsf{Thin}_\theta \right]$$

$$\geq \Pr_{\substack{s \leftarrow U_n \\ y \leftarrow G_U(U_n)}} \left[ y \in \mathsf{Good}_q \mid y \in \mathsf{Thin}_\theta \right] \cdot \Pr_{\substack{s \leftarrow U_n \\ y \leftarrow G_U(U_n)}} \left[ \mathcal{A}(s, y) \in f_s^{-1}(y) \mid y \in \mathsf{Good}_q \right]$$

$$\geq \frac{q}{\theta} \cdot \Pr_{s, y \leftarrow U_n} \left[ y \in \mathsf{Good}_q | y \in \mathsf{Thin}_\theta \right] \tag{47}$$

If we show that

$$\Pr_{s, y \leftarrow U_n} \left[ y \in \mathsf{Good}_q | y \in \mathsf{Thin}_\theta \right] \geq \frac{1}{\alpha(n)} - \frac{2}{\theta} - q \tag{48}$$

then selecting $\theta = 4\alpha(n)$ and $q = 1/4\alpha(n)$, the value of (47) is bounded below by

$$\Pr_{\substack{s \leftarrow U_n \\ y \leftarrow G_U(U_n)}} \left[ \mathcal{A}(s, y) \in f_s^{-1}(y) \mid y \in \mathsf{Thin}_\theta \right] \geq \frac{q}{\theta} \cdot \Pr_{s, y \leftarrow U_n} \left[ y \in \mathsf{Good}_q | y \in \mathsf{Thin}_\theta \right]$$

$$\geq \left( \frac{1}{4\alpha(n)} \right)^3$$

The following proves inequality (48) and completes the proof of (23).

$$\frac{1}{\alpha(n)} < \Pr_{\substack{s \leftarrow U_n \\ y \leftarrow U_n}} \left[ \mathcal{A}(s, y) \in f_s^{-1}(y) \right] \qquad \text{by (14)}$$

$$= \Pr_{\substack{s \leftarrow U_n \\ y \leftarrow U_n}} \left[ \mathcal{A}(s, y) \in f_s^{-1}(y) \wedge y \in \mathsf{Thin}_\theta \right]$$

$$+ \Pr_{\substack{s \leftarrow U_n \\ y \leftarrow U_n}} \left[ \mathcal{A}(s, y) \in f_s^{-1}(y) \wedge y \notin \mathsf{Thin}_\theta \right]$$

$$\leq \Pr_{\substack{s \leftarrow U_n \\ y \leftarrow U_n}} \left[ \mathcal{A}(s, y) \in f_s^{-1}(y) \mid y \in \mathsf{Thin}_\theta \right] + \Pr_{y \leftarrow U_n} \left[ y \notin \mathsf{Thin}_\theta \right]$$

$$\leq \Pr_{\substack{s \leftarrow U_n \\ y \leftarrow U_n}} \left[ \mathcal{A}(s, y) \in f_s^{-1}(y) \mid y \in \mathsf{Thin}_\theta \right] + \frac{2^{n+1}/\theta}{2^n} \qquad \text{by (22)}$$

$$\implies \frac{1}{\alpha(n)} - \frac{2}{\theta} < \Pr_{\substack{s \leftarrow U_n \\ y \leftarrow U_n}} [\mathcal{A}(s, y) \in f_s^{-1}(y) \mid y \in \mathsf{Thin}_\theta]$$

$$= \Pr_{\substack{s \leftarrow U_n \\ y \leftarrow U_n}} [y \in \mathsf{Good}_q \mid y \in \mathsf{Thin}_\theta]$$

$$\cdot \Pr_{\substack{s \leftarrow U_n \\ y \leftarrow U_n}} [A(s, y) \in f_s^{-1}(y) \mid y \in \mathsf{Good}_q]$$

$$+ \Pr_{\substack{s \leftarrow U_n \\ y \leftarrow U_n}} [y \notin \mathsf{Good}_q \mid y \in \mathsf{Thin}_\theta]$$

$$\cdot \Pr_{\substack{s \leftarrow U_n \\ y \leftarrow U_n}} [A(s, y) \in f_s^{-1}(y) \mid y \in \mathsf{Thin}_\theta \backslash \mathsf{Good}_q]$$

$$\leq \Pr_{\substack{s \leftarrow U_n \\ y \leftarrow U_n}} [y \in \mathsf{Good}_q \mid y \in \mathsf{Thin}_\theta] + q$$

The final inequality is by the definition of $\mathsf{Thin}_\theta \backslash \mathsf{Good}_q$.

# References

[BGI14] Boyle, E., Goldwasser, S., Ivan, I.: Functional signatures and pseudorandom functions. In: Krawczyk, H. (ed.) PKC 2014. LNCS, vol. 8383, pp. 501–519. Springer, Heidelberg (2014). doi:10.1007/978-3-642-54631-0_29

[BGI15] Boyle, E., Gilboa, N., Ishai, Y.: Function secret sharing. In: Oswald, E., Fischlin, M. (eds.) EUROCRYPT 2015. LNCS, vol. 9057, pp. 337–367. Springer, Heidelberg (2015). doi:10.1007/978-3-662-46803-6_12

[BLL+15] Bai, S., Langlois, A., Lepoint, T., Stehlé, D., Steinfeld, R.: Improved security proofs in lattice-based cryptography: using the Rényi divergence rather than the statistical distance. In: Iwata, T., Cheon, J.H. (eds.) ASIACRYPT 2015. LNCS, vol. 9452, pp. 3–24. Springer, Heidelberg (2015). doi:10.1007/978-3-662-48797-6_1

[BW13] Boneh, D., Waters, B.: Constrained pseudorandom functions and their applications. In: Sako, K., Sarkar, P. (eds.) ASIACRYPT 2013. LNCS, vol. 8270, pp. 280–300. Springer, Heidelberg (2013). doi:10.1007/978-3-642-42045-0_15

[CGH04] Canetti, R., Goldreich, O., Halevi, S.: The random oracle methodology, revisited. J. ACM (JACM) 51(4), 557–594 (2004)

[FN94] Fiat, A., Naor, M.: Broadcast encryption. In: Stinson, D.R. (ed.) CRYPTO 1993. LNCS, vol. 773, pp. 480–491. Springer, Heidelberg (1994). doi:10.1007/3-540-48329-2_40

[GGM86] Goldreich, O., Goldwasser, S., Micali, S.: How to construct random functions. J. ACM (JACM) 33(4), 792–807 (1986)

[Gol02] Goldreich, O.: The GGM construction does not yield correlation intractable function ensembles (2002)

[Gol04] Goldreich, O.: Foundations of Cryptography: Basic Applications, vol. 2. Cambridge University Press, Cambridge (2004)

[KPTZ13] Kiayias, A., Papadopoulos, S., Triandopoulos, N., Zacharias, T.: Delegatable pseudorandom functions and applications. In: 2013 ACM SIGSAC Conference on Computer and Communications Security, CCS 2013, Berlin, Germany, 4–8 November 2013, pp. 669–684 (2013)

[LR88]  Luby, M., Rackoff, C.: How to construct pseudorandom permutations from pseudorandom functions. SIAM J. Comput. **17**(2), 373–386 (1988)

[RR97]  Razborov, A.A., Rudich, S.: Natural proofs. J. Comput. Syst. Sci. **55**(1), 24–35 (1997)

[SW14]  Sahai, A., Waters, B.: How to use indistinguishability obfuscation, deniable encryption, and more. In: Symposium on Theory of Computing, STOC 2014, 31 May–3 June 2014, pp. 475–484. ACM, New York (2014)

[Val84]  Valiant, L.G.: A theory of the learnable. Commun. ACM **27**(11), 1134–1142 (1984)

[Zha12]  Zhandry, M.: How to construct quantum random functions. In: 2012 IEEE 53rd Annual Symposium on Foundations of Computer Science (FOCS), pp. 679–687. IEEE (2012)

# On the (In)Security of SNARKs in the Presence of Oracles

Dario Fiore[1]([✉]) and Anca Nitulescu[2]

[1] IMDEA Software Institute, Madrid, Spain
dario.fiore@imdea.org
[2] CNRS, ENS, INRIA, PSL, Paris, France
anca.nitulescu@ens.fr

**Abstract.** In this work we study the feasibility of knowledge extraction for succinct non-interactive arguments of knowledge (SNARKs) in a scenario that, to the best of our knowledge, has not been analyzed before. While prior work focuses on the case of adversarial provers that may receive (statically generated) *auxiliary information*, here we consider the scenario where adversarial provers are given *access to an oracle*. For this setting we study if and under what assumptions such provers can admit an extractor. Our contribution is mainly threefold.

First, we formalize the question of extraction in the presence of oracles by proposing a suitable proof of knowledge definition for this setting. We call SNARKs satisfying this definition O-SNARKs. Second, we show how to use O-SNARKs to obtain formal and intuitive security proofs for three applications (homomorphic signatures, succinct functional signatures, and SNARKs on authenticated data) where we recognize an issue while doing the proof under the standard proof of knowledge definition of SNARKs. Third, we study whether O-SNARKs exist, providing both negative and positive results. On the negative side, we show that, assuming one way functions, there do not exist O-SNARKs in the standard model for every signing oracle family (and thus for general oracle families as well). On the positive side, we show that when considering signature schemes with appropriate restrictions on the message length O-SNARKs for the corresponding signing oracles exist, based on classical SNARKs and assuming extraction with respect to specific distributions of auxiliary input.

## 1 Introduction

**Succinct Arguments.** Proof systems [GMR89] are fundamental in theoretical computer science and cryptography. Extensively studied aspects of proof systems are the expressivity of provable statements and the efficiency. Related to efficiency, it has been shown that statistically-sound proof systems are unlikely to allow for significant improvements in communication [BHZ87, GH98, GVW02, Wee05]. When considering proof systems for NP this means that, unless some complexity-theoretic collapses occur, in a statistically sound proof system any prover has to communicate, roughly, as much information as the size of the

M. Hirt and A. Smith (Eds.): TCC 2016-B, Part I, LNCS 9985, pp. 108–138, 2016.
DOI: 10.1007/978-3-662-53641-4_5

NP witness. The search of ways to beat this bound motivated the study of *computationally-sound* proof systems, also called *argument systems* [BCC88]. Assuming existence of collision-resistant hash functions, Kilian [Kil92] showed a four-message interactive argument for NP. In this protocol, membership of an instance $x$ in an NP language with NP machine $M$ can be proven with communication and verifier's running time bounded by $p(\lambda, |M|, |x|, \log t)$, where $\lambda$ is a security parameter, $t$ is the NP verification time of machine $M$ for the instance $x$, and $p$ is a *universal* polynomial. Argument systems of this kind are called *succinct*.

**Succinct Non-interactive Arguments.** Starting from Kilian's protocol, Micali [Mic94] constructed a *one-message* succinct argument for NP whose soundness is set in the random oracle model. The fact that one-message succinct arguments are unlikely to exist for hard-enough languages in the plain model motivated the consideration of *two-message* non-interactive arguments, in which the verifier generates its message (a common reference string, if this can be made publicly available) ahead of time and independently of the statement to be proved. Such systems are called *succinct non-interactive arguments* (SNARGs) [GW11]. Several SNARGs constructions have been proposed [CL08, Mie08, Gro10, BCCT12, Lip12, BCC+14, GGPR13, BCI+13, PHGR13, BSCG+13, BCTV14] and the area of SNARGs has become popular in the last years with the proposal of constructions which gained significant improvements in efficiency. Noteworthy is that all such constructions are based on non-falsifiable assumptions [Nao03], a class of assumptions that is likely to be inherent in proving the security of SNARGs (without random oracles), as shown by Gentry and Wichs [GW11].

Almost all SNARGs are also *arguments of knowledge*—so called SNARKs [BCCT12, BCC+14]. Intuitively speaking, this property (which replaces soundness) says that every prover producing a convincing proof must "know" a witness. On the one hand, proof of knowledge turns out to be useful in many applications, such as delegation of computation where the untrusted worker contributes its own input to the computation, or recursive proof composition [Val08, BCCT13]. On the other hand, the formalization of proof of knowledge in SNARKs is a delicate point. Typically, the concept that the prover "must know" a witness is expressed by assuming that such knowledge can be efficiently extracted from the prover by means of a so-called *knowledge extractor*. In SNARKs, extractors are inherently non-black-box and proof of knowledge requires that for every adversarial prover $\mathcal{A}$ generating an accepting proof $\pi$ there must be an extractor $\mathcal{E}_{\mathcal{A}}$ that, given the same input of $\mathcal{A}$, outputs a valid witness.

**Extraction with Auxiliary Input.** Unfortunately, stated as above, proof of knowledge is insufficient for being used in many applications. The problem is that, when using SNARKs in larger cryptographic protocols, adversarial provers may get additional information which can contribute to the generation of adversarial proofs. To address this problem, a stronger, and more useful, definition of proof of knowledge requires that for any adversary $\mathcal{A}$ there is an extractor $\mathcal{E}_{\mathcal{A}}$ such that, for any honestly generated crs and any polynomial-size auxiliary input $aux$, whenever $\mathcal{A}(\text{crs}, aux)$ returns an accepting proof, $\mathcal{E}_{\mathcal{A}}(\text{crs}, aux)$ outputs a valid witness. This type of definition is certainly more adequate when using SNARKs in larger cryptographic protocols, but it also introduces other subtleties. As first discussed in [HT98], extraction in the presence of arbitrary auxiliary input can be problematic, if not implausible. Formal evidence of this issue has been recently given in [BCPR14, BP15]. Bitansky et al. [BCPR14] show that, assuming indistinguishability obfuscation, there do not exist extractable one-way functions (and thus SNARKs) with respect to arbitrary auxiliary input of unbounded polynomial length. Boyle and Pass [BP15] generalize this result showing that assuming collision-resistant hash functions and differing-input obfuscation, there is a fixed auxiliary input distribution for which extractable one-way functions do not exist.

## 1.1   Extraction in the Presence of Oracles

In this work we continue the study on the feasibility of extraction by looking at a scenario that, to the best of our knowledge, has not been explicitly analyzed before. We consider the case in which adversarial provers run in interactive security experiments where they are given *access to an oracle*. For this setting we study if and under what assumptions such provers can admit an extractor.

Before giving more detail on our results, let us discuss a motivation for analyzing this scenario. To keep the presentation simple, here we give a motivation via a hypotetical example; more concrete applications are discussed later.

A CASE STUDY APPLICATION. Consider an application where Alice gets a collection of signatures generated by Bob, and she has to prove to a third party that she owns a valid signature of Bob on some message $m$ such that $P(m) = 1$. Let us say that this application is secure if Alice, after asking for signatures on several messages, cannot cheat letting the third party accept for a false statement (i.e., $P(m) = 0$, or $P(m) = 1$ but Alice did not receive a signature on $m$). If messages are large and one wants to optimize bandwidth, SNARKs can be a perfect candidate solution for doing such proofs,[1] i.e., Alice can generate a proof of knowledge of $(m, \sigma)$ such that "$(m, \sigma)$ verifies with Bob's public key and $P(m) = 1$".

AN ATTEMPT OF SECURITY PROOF. Intuitively, the security of this protocol should follow easily from the proof of knowledge of the SNARK and the unforgeability of the signature scheme. However, somewhat surprisingly, the proof becomes quite subtle. Let us consider a cheating Alice that always outputs a proof for a statement in the language.[2] If Alice is still cheating, then it must be that she is using a signature on a message that she did not query – in other words a forgery. Then one would like to reduce such a cheating Alice to a forger

---

[1] Further motivation can be to keep the privacy of $m$ by relying on zero-knowledge SNARKs.

[2] The other case of statements not in the language can be easily reduced to the soundness of the SNARK.

for the signature scheme. To do this, one would proceed as follows. For any Alice one defines a forger that, on input the verification key vk, generates the SNARK crs, gives (crs, vk) to Alice, and simulate's Alice's queries using its own signing oracle. When Alice comes with the cheating proof, the forger would need an extractor for Alice in order to obtain the forgery from her. However, even if we see Alice as a SNARK prover with auxiliary input vk, Alice does not quite fit the proof of knowledge definition in which adversaries have no oracles. To handle similar cases, one typically shows that for every, interactive, Alice there is a non-interactive algorithm $\mathcal{B}$ that runs Alice simulating her oracles (i.e., $\mathcal{B}$ samples the signing key) and returns the same output. The good news is that for such $\mathcal{B}$ one can claim the existence of an extractor $\mathcal{E}_{\mathcal{B}}$ as it fits the proof of knowledge definition. The issue is though that $\mathcal{E}_{\mathcal{B}}$ expects the same input of $\mathcal{B}$, which includes the secret signing key. This means that our candidate forger mentioned above (which does not have the secret key) cannot run $\mathcal{E}_{\mathcal{B}}$.

APPLICATIONS THAT NEED EXTRACTION WITH ORACLES. Besides the above example, this issue can show up essentially in every application of SNARKs in which adversaries have access to oracles with a secret state, and one needs to run an extractor during an experiment (e.g., a reduction) where the secret state of the oracle is not available. For instance, we recognize this issue while trying to formally prove the security of a "folklore" construction of homomorphic signatures based on SNARKs and digital signatures that is mentioned in several papers (e.g., [BF11, GW13, CF13, GVW15]). The same issue appears in a generic construction of SNARKs on authenticated data in [BBFR15] (also informally discussed in [BCCT12]), where the security proof uses the existence of an extractor for the oracle-aided prover, but without giving particular justification. A similar issue also appears in the construction of succinct functional signatures of [BGI14]. To be precise, in [BGI14] the authors provide a (valid) proof but under a stronger definition of SNARKs in which the adversarial prover and the extractor are independent PPT machines without *common* auxiliary input: a notion for which we are not aware of standard model constructions. In contrast, if one attempts to prove the succinct functional signatures of [BGI14] using the standard definition of SNARKs, one incurs the same issues illustrated above, i.e., the proof would not go through.

In this work we address this problem by providing both negative and positive results to the feasibility of extraction in the presence of oracles. On one hand, our negative results provide an explanation of why the above proofs do not go through so easily. On the other hand, our positive results eventually provide some guidelines to formally state and prove the security of the cryptographic constructions mentioned above (albeit with various restrictions).

## 1.2 An Overview of Our Results

**Defining SNARKs in the Presence of Oracles.** As a first step, we formalize the definition of non-black-box extraction in the presence of oracles by proposing a notion of *SNARKs in the presence of oracles* (O-SNARKs, for short). In a nutshell, an O-SNARK is like a SNARK except that adaptive proof of knowledge must hold with respect to adversaries that have access to an oracle $\mathcal{O}$ sampled from some oracle family $\mathbb{O}$.[3] Slightly more in detail, we require that for any adversary $\mathcal{A}^{\mathcal{O}}$ with access to $\mathcal{O}$ there is an extractor $\mathcal{E}_{\mathcal{A}}$ such that, whenever $\mathcal{A}^{\mathcal{O}}$ outputs a valid proof, $\mathcal{E}_{\mathcal{A}}$ outputs a valid witness, by running on the same input of $\mathcal{A}$, plus the transcript of oracle queries-answers of $\mathcal{A}$.

**Existence of O-SNARKs.** Once having defined their notion, we study whether O-SNARKs exist and under what assumptions. Below we summarize our results.

O-SNARKs in the random oracle model. As a first positive result, we show that the construction of Computationally Sounds (CS) proofs of Micali [Mic00] yields an O-SNARK for *every* oracle family, in the random oracle model. This result follows from the work of Valiant [Val08] which shows that Micali's construction already allows for extraction. More precisely, using the power of the random oracle model, Valiant shows a *black-box* extractor. This powerful extractor can then be used to build an O-SNARK extractor that works for any oracle family.

Insecurity of O-SNARKs for every oracle family. Although the above result gives a candidate O-SNARK, it only works in the random oracle model, and it is tailored to one construction [Mic00]. It is therefore interesting to understand whether extraction with oracles is feasible in the *standard model*. And it would also be interesting to see if this is possible based on the classical SNARK notion. Besides its theoretical interest, the latter question has also a practical motivation since there are several efficient SNARK constructions proposed in the last years that one might like to use in place of CS proofs. Our first result in this direction is that assuming existence of one way functions (OWFs) there do not exist O-SNARKs for NP with respect to *every* oracle family. More precisely, we show the following:

**Theorem 1 (Informal).** *Assume OWFs exist. Then for any polynomial $p(\cdot)$ there is an unforgeable signature scheme $\Sigma_p$ such that any candidate O-SNARK, that is correct and succinct with proofs of length bounded by $p(\cdot)$, cannot satisfy adaptive proof of knowledge with respect to signing oracles corresponding to $\Sigma_p$.*

The above result shows the existence of an oracle family for which O-SNARKs do not exist. A basic intuition behind it is that oracles provide additional auxiliary input to adversaries and, as formerly shown in [BCPR14,BP15], this can create issues for extraction. In fact, to obtain our result we might also have designed an oracle that simply outputs a binary string following a distribution with respect to which extraction is impossible due to [BCPR14,BP15]. However, in this case the result should additionally assume the existence of indistinguishability (or differing-input) obfuscation. In contrast, our result shows that such impossibility holds by only assuming existence of OWFs, which is a much weaker assumption.

---

[3] The notion is parametrized by the family $\mathbb{O}$, i.e., we say $\Pi$ is an O-SNARK for $\mathbb{O}$.

In addition to ruling out existence of O-SNARKs for general oracles, our theorem also rules out their existence for a more specific class of oracle families – *signing oracles* – that is motivated by the three applications mentioned earlier.[4] Its main message is thus that one cannot assume existence of O-SNARKs that work with *any* signature scheme. This explains why the security proofs of the primitives considered earlier do not go through, if one wants to base it on an arbitrary signature scheme.

EXISTENCE OF O-SNARKS FOR SPECIFIC FAMILIES OF SIGNING ORACLES. We study ways to circumvent our impossibility result for signing oracles of Theorem 1. Indeed, the above result can be interpreted as saying that there exist (perhaps degenerate) signature schemes such that there are no O-SNARKs with respect to the corresponding signing oracle family. This is not ruling out that O-SNARKs may exist for specific signature schemes, or – even better – for specific classes of signature schemes. We provide the following results:

1. Hash-and-sign signatures, where the hash is a random oracle, yield "safe oracles", i.e., oracles for which any SNARK is an O-SNARK for that oracle, in the ROM.
2. Turning to the standard model setting, we show that any classical SNARK is an O-SNARK for signing oracles if the message space of the signature scheme is properly bounded, and O-SNARK adversaries query "almost" the entire message space. This positive result is useful in applications that use SNARKs with signing oracles, under the condition that adversaries make signing queries on almost all messages.

NON-ADAPTIVE O-SNARKS. Finally, we consider a relaxed notion of O-SNARKs in which adversaries are required to declare in advance (i.e., before seeing the common reference string) all the oracle queries. For this weaker notion we show that, in the standard model, every SNARK (for arbitrary auxiliary inputs) is a non-adaptive O-SNARK.

**Applications of O-SNARKs.** A nice feature of the O-SNARK notion is that it lends itself to easy and intuitive security proofs in all those applications where one needs to execute extractors in interactive security games with oracles. We show that by replacing SNARKs with O-SNARKs (for appropriate oracle families) we can formally prove the security of the constructions of homomorphic signatures, succinct functional signatures and SNARKs on authenticated data that we mentioned in the previous section. By combining these O-SNARK-based constructions with our existence results mentioned earlier we eventually reach conclusions about the possible secure instantiations of these constructions. The first option is to instantiate them by using Micali's CS proofs as an O-SNARK: this solution essentially yields secure instantiations in the random oracle model

---

[4] We do believe that many more applications along the same line – proving knowledge of valid signatures – are conceivable. Two recent examples which considered our work in such a setting are [DLFKP16, NT16].

that work with a specific proof system [Mic00] (perhaps not the most efficient one in practice). The second option is to instantiate them using hash-and-sign signatures, apply our result on hash-and-sign signatures mentioned above, and then conjecture that replacing the random oracle with a suitable hash function preserves the overall security.[5] Third, one can instantiate the constructions using a classical SNARK scheme $\Pi$ and signature scheme $\Sigma$, and then conjecture that $\Pi$ is also an O-SNARK with respect to the family of signing oracles corresponding to $\Sigma$. Compared to the first solution, the last two ones have the advantage that one could use some of the recently proposed efficient SNARKs (e.g., [PHGR13, BSCG+13]); on the other hand, these solutions have the drawback that security is based only on a heuristic argument. Finally, as a fourth option we provide security proofs of these primitives under a weak, non-adaptive, notion where adversaries declare all their queries in advance. Security in this weaker model can be proven assuming *non-adaptive O-SNARKs*, and thus classical SNARKs. The advantage of this fourth option is that one obtains a security proof for these instantiations based on clear – not newly crafted – assumptions, although under a much weaker security notion. Finally, worth noting is that we cannot apply the positive result on O-SNARK for signing oracles to the O-SNARK-based constructions of homomorphic signatures, functional signatures and SNARKs on authenticated data that we provide, and thus conclude their security under classical SNARKs. The inapplicability is due to the aforementioned restriction of our result, for which adversaries have to query almost the entire message space.[6]

**Interpretation of Our Results.** In line with recent work [BCPR14, BP15] on the feasibility of extraction in the presence of auxiliary input, our results indicate that additional care must be taken when considering extraction *in the presence of oracles*. While for auxiliary input impossibility of extraction is known under obfuscation-related assumptions, in the case of oracles we show that extraction becomes impossible even by only assuming one-way functions. Our counterexamples are of artificial nature and do not rule out the feasibility of extraction in the presence of "natural, benign" oracles. Nevertheless, our impossibility results provide formal evidence of why certain security proofs do not go through, and bring out important subtle aspects of security proofs. Given the importance of provable security and considered the increasing popularity of SNARKs in more practical scenarios, we believe these results give a message that is useful to protocol designers and of interest to the community at large.

---

[5] The need of this final heuristic step is that hash-and-sign signatures use a random oracle in verification and in our applications the SNARK is used to prove knowledge of valid signatures, i.e., one would need a SNARK for $\mathsf{NP}^{\mathcal{O}}$.

[6] The exact reason is rather technical and requires to see the precise definitions and constructions of these primitives first. For the familiar reader, the intuition is that in these primitives/constructions an adversary that queries almost the entire message space of the underlying signature scheme becomes able to trivially break their security.

## 1.3   Organization

The paper is organized as follows. In Sect. 2 we recall notation and definitions used in the rest of our work. Section 3 introduces the notion of O-SNARKs, Sect. 4 includes positive and negative results about the existence of O-SNARKs, and in Sect. 5 we give three applications where our new notion turns out to be useful. For lack of space, additional definitions and detailed proofs are deferred to the full version [FN16].

## 2   Preliminaries

**Notation.** We denote with $\lambda \in \mathbb{N}$ the security parameter. We say that a function $\epsilon(\lambda)$ is *negligible* if it vanishes faster than the inverse of any polynomial in $\lambda$. If not explicitly specified otherwise, negligible functions are negligible with respect to $\lambda$. If $S$ is a set, $x \xleftarrow{\$} S$ denotes the process of selecting $x$ uniformly at random in $S$. If $\mathcal{A}$ is a probabilistic algorithm, $x \xleftarrow{\$} \mathcal{A}(\cdot)$ denotes the process of running $\mathcal{A}$ on some appropriate input and assigning its output to $x$. For binary strings $x$ and $y$, we denote by $x|y$ their concatenation and by $x_i$ the $i$-th bit of $x$. For a positive integer $n$, we denote by $[n]$ the set $\{1, \ldots, n\}$. For a random-access machine $M$ we denote by $\#M(x, w)$ the number of execution steps needed by $M$ to accept on input $(x, w)$.

**The Universal Relation and NP Relations.** We recall the notion of universal relation from [BG08], here adapted to the case of non-deterministic computations.

**Definition 1.** *The* universal relation *is the set $\mathcal{R}_\mathcal{U}$ of instance-witness pairs $(y, w) = ((M, x, t), w)$, where $|y|, |w| \leq t$ and $M$ is a random-access machine such that $M(x, w)$ accepts after running at most $t$ steps. The* universal language $\mathcal{L}_\mathcal{U}$ *is the language corresponding to $\mathcal{R}_\mathcal{U}$.*

For any constant $c \in \mathbb{N}$, $\mathcal{R}_c$ denotes the subset of $\mathcal{R}_\mathcal{U}$ of pairs $(y, w) = ((M, x, t), w)$ such that $t \leq |x|^c$. $\mathcal{R}_c$ is a "generalized" NP relation that is decidable in some fixed time polynomial in the size of the instance.

### 2.1   Succinct Non-interactive Arguments

In this section we provide formal definitions for the notion of succinct non-interactive arguments of knowledge (SNARKs).

**Definition 2 (SNARGs).** *A succinct non-interactive argument (SNARG) for a relation $\mathcal{R} \subseteq \mathcal{R}_\mathcal{U}$ is a triple of algorithms $\Pi = (\mathsf{Gen}, \mathsf{Prove}, \mathsf{Ver})$ working as follows*

$\mathsf{Gen}(1^\lambda, T) \rightarrow \mathsf{crs}$: *On input a security parameter $\lambda \in \mathbb{N}$ and a time bound $T \in \mathbb{N}$, the generation algorithm outputs a common reference string $\mathsf{crs} = (\mathsf{prs}, \mathsf{vst})$ consisting of a public prover reference string $\mathsf{prs}$ and a verification state $\mathsf{vst}$.*

Prove(prs, $y, w$) → $\pi$: *Given a prover reference string* prs, *an instance* $y = (M, x, t)$ *with* $t \leq T$ *and a witness* $w$ *s.t.* $(y, w) \in \mathcal{R}$, *this algorithm produces a proof* $\pi$.

Ver(vst, $y, \pi$) → $b$: *On input a verification state* vst, *an instance* $y$, *and a proof* $\pi$, *the verifier algorithm outputs* $b = 0$ *(reject) or* $b = 1$ *(accept).*

*and satisfying* completeness, succinctness, *and* (adaptive) soundness:

– **Completeness.** *For every time bound* $T \in \mathbb{N}$, *every valid* $(y, w) \in \mathcal{R}$ *with* $y = (M, x, t)$ *and* $t \leq T$, *there exists a negligible function* negl *such that*

$$\Pr\left[ \text{Ver(vst, } y, \pi) = 0 \ \middle| \ \begin{array}{c} (\text{prs, vst}) \leftarrow \text{Gen}(1^\lambda, T) \\ \pi \leftarrow \text{Prove(prs, } y, w) \end{array} \right] \leq \text{negl}(\lambda)$$

– **Succinctness.** *There exists a fixed polynomial* $p(\cdot)$ *independent of* $\mathcal{R}$ *such that for every large enough security parameter* $\lambda \in \mathbb{N}$, *every time bound* $T \in \mathbb{N}$, *and every instance* $y = (M, x, t)$ *such that* $t \leq T$, *we have*

• Gen *runs in time* $\begin{cases} p(\lambda + \log T) & \text{for a fully-succinct SNARG} \\ p(\lambda + T) & \text{for a pre-processing SNARG} \end{cases}$

• Prove *runs in time*

$$\begin{cases} p(\lambda + |M| + |x| + t + \log T) & \text{for fully-succinct SNARG} \\ p(\lambda + |M| + |x| + T) & \text{for pre-processing SNARG} \end{cases}$$

• Ver *runs in time* $p(\lambda + |M| + |x| + \log T)$
• *a honestly generated proof has size* $|\pi| = p(\lambda + \log T)$.

– **Adaptive Soundness.** *For every non-uniform* $\mathcal{A}$ *of size* $s(\lambda) = \text{poly}(\lambda)$ *there is a negligible function* $\epsilon(\lambda)$ *such that for every time bound* $T \in \mathbb{N}$,

$$\Pr\left[ \begin{array}{c} \text{Ver(vst, } y, \pi) = 1 \\ \wedge \ y \notin \mathcal{L}_{\mathcal{R}} \end{array} \ \middle| \ \begin{array}{c} (\text{prs, vst}) \leftarrow \text{Gen}(1^\lambda, T) \\ (y, \pi) \leftarrow \mathcal{A}(\text{prs}) \end{array} \right] \leq \epsilon(\lambda)$$

The notion of SNARG can be extended to be an argument of knowledge (a SNARK) by replacing soundness by an appropriate proof of knowledge property.

**Definition 3 (SNARKs [BCC+14]).** *A succinct non-interactive argument of knowledge (SNARK) for a relation* $\mathcal{R} \subseteq \mathcal{R}_\mathcal{U}$ *is a triple of algorithms* $\Pi = (\text{Gen}, \text{Prove}, \text{Ver})$ *that constitutes a SNARG (as per Definition 2) except that soundness is replaced by the following property:*

– **Adaptive Proof of Knowledge.** *For every non-uniform prover* $\mathcal{A}$ *of size* $s(\lambda) = \text{poly}(\lambda)$ *there exists a non-uniform extractor* $\mathcal{E}_\mathcal{A}$ *of size* $t(\lambda) = \text{poly}(\lambda)$ *and a negligible function* $\epsilon(\lambda)$ *such that for every auxiliary input* aux $\in \{0, 1\}^{poly(\lambda)}$, *and every time bound* $T \in \mathbb{N}$,

$$\Pr\left[ \begin{array}{c} \text{Ver(vst, } y, \pi) = 1 \\ \wedge \\ (y, w) \notin \mathcal{R} \end{array} \ \middle| \ \begin{array}{c} (\text{prs, vst}) \leftarrow \text{Gen}(1^\lambda, T) \\ (y, \pi) \leftarrow \mathcal{A}(\text{prs, aux}) \\ w \leftarrow \mathcal{E}_\mathcal{A}(\text{prs, aux}) \end{array} \right] \leq \epsilon(\lambda)$$

*Furthermore, we say that* $\Pi$ *satisfies* $(s, t, \epsilon)$-*adaptive proof of knowledge if the above condition holds for concrete values* $(s, t, \epsilon)$.

*Remark 1 (Publicly verifiable vs. designated verifier).* If security (adaptive PoK) holds against adversaries that have also access to the verification state vst (i.e., $\mathcal{A}$ receives the whole crs) then the SNARK is called *publicly verifiable*, otherwise it is *designated verifier*. For simplicity, in the remainder of this work all definitions are given for the publicly verifiable setting; the corresponding designated-verifier variants are easily obtained by giving to the adversary only the prover state prs.

*Remark 2 (About extraction and auxiliary input).* First, we stress that in the PoK property the extractor $\mathcal{E}_{\mathcal{A}}$ takes exactly the same input of $\mathcal{A}$, including its random tape. Second, the PoK definition can also be relaxed to hold with respect to auxiliary inputs from specific distributions (instead of arbitrary ones). Namely, let $\mathcal{Z}$ be a probabilistic algorithm (called the auxiliary input generator) that outputs a string $aux$, and let compactly denote this process as $aux \leftarrow \mathcal{Z}$. Then we say that adaptive proof of knowledge holds for $\mathcal{Z}$ if the above definition holds for auxiliary inputs sampled according to $\mathcal{Z}$ – $aux \leftarrow \mathcal{Z}$ – where $\mathcal{Z}$ is also a non-uniform polynomial-size algorithm. More formally, we have the following definition.

**Definition 4 ($\mathcal{Z}$-auxiliary input SNARKs).** *$\Pi$ is called a $\mathcal{Z}$-auxiliary input SNARK if $\Pi$ is a SNARK as in Definition 3 except that adaptive proof of knowledge holds for auxiliary input $aux \leftarrow \mathcal{Z}$.*

For ease of exposition, in our proofs we compactly denote by AdPoK($\lambda, T, \mathcal{A}, \mathcal{E}_{\mathcal{A}}, \mathcal{Z}$) the adaptive proof of knowledge experiment executed with adversary $\mathcal{A}$, extractor $\mathcal{E}_{\mathcal{A}}$ and auxiliary input generator $\mathcal{Z}$. See below its description:

AdPoK($\lambda, T, \mathcal{A}, \mathcal{E}_{\mathcal{A}}, \mathcal{Z}$)
  $aux \leftarrow \mathcal{Z}(1^\lambda)$; crs$\leftarrow$Gen($1^\lambda, T$)
  $(y, \pi) \leftarrow \mathcal{A}(\text{crs}, aux)$  $w \leftarrow \mathcal{E}_{\mathcal{A}}(\text{crs}, aux)$
  **if** Ver(crs, $y, \pi$) $= 1 \wedge (y, w) \notin \mathcal{R}$ **return** 1
  **else return** 0

We say that $\Pi$ satisfies adaptive proof of knowledge for $\mathcal{Z}$-auxiliary input if for every non-uniform $\mathcal{A}$ of size $s(\lambda) = \text{poly}(\lambda)$ there is a non-uniform extractor of size $t(\lambda) = \text{poly}(\lambda)$ and a negligible function $\epsilon(\lambda)$ such that for every time bound $T$ we have $\Pr[\text{AdPoK}(\lambda, T, \mathcal{A}, \mathcal{E}_{\mathcal{A}}, \mathcal{Z}) \Rightarrow 1] \leq \epsilon$. Furthermore, $\Pi$ has $(s, t, \epsilon)$-adaptive proof of knowledge for $\mathcal{Z}$-auxiliary input if the above condition holds for concrete $(s, t, \epsilon)$.

SNARKs FOR NP. A SNARK for the universal relation $\mathcal{R}_{\mathcal{U}}$ is called a universal SNARK. SNARKs for NP are instead SNARKs in which the verification algorithm Ver takes as additional input a constant $c > 0$, and adaptive proof of knowledge is restricted to hold only for relations $\mathcal{R}_c \subset \mathcal{R}_{\mathcal{U}}$. More formally,

**Definition 5 (SNARKs for NP).** *A SNARK for NP is a tuple of algorithms $\Pi = (\text{Gen}, \text{Prove}, \text{Ver})$ satisfying Definition 3 except that the adaptive proof of knowledge property is replaced by the following one:*

- **Adaptive Proof of Knowledge for** NP. *For every non-uniform polynomial-size prover* $\mathcal{A}$ *there exists a non-uniform polynomial-size extractor* $\mathcal{E}_\mathcal{A}$ *such that for every large enough* $\lambda \in \mathbb{N}$, *every auxiliary input* $aux \in \{0,1\}^{poly(\lambda)}$, *and every time bound* $T \in \mathbb{N}$, *and every constant* $c > 0$,

$$\Pr\left[\begin{array}{c} \mathsf{Ver}_c(\mathsf{vst}, y, \pi) = 1 \\ \wedge \\ (y, w) \notin \mathcal{R}_c \end{array} \middle| \begin{array}{c} \mathsf{crs}\leftarrow\mathsf{Gen}(1^\lambda, T) \\ (y, \pi)\leftarrow\mathcal{A}(\mathsf{crs}, aux) \\ w\leftarrow\mathcal{E}_\mathcal{A}(\mathsf{crs}, aux) \end{array}\right] \leq negl(\lambda)$$

In the case of fully-succinct SNARKs for NP, it is not necessary to provide a time bound as one can set $T = \lambda^{\log \lambda}$. In this case we can write $\mathsf{Gen}(1^\lambda)$ as a shorthand for $\mathsf{Gen}(1^\lambda, \lambda^{\log \lambda})$.

# 3    SNARKs in the Presence of Oracles

In this section we formalize the notion of extraction in the presence of oracles for SNARKs. We do this by proposing a suitable adaptive proof of knowledge definition, and we call a SNARK satisfying this definition a *SNARK in the presence of oracles* (O-SNARK, for short). As we shall see, the advantage of O-SNARKs is that this notion lends itself to easy and intuitive security proofs in all those applications where one needs to execute extractors in interactive security games with oracles (with a secret state). Below we provide the definition while the existence of O-SNARKs is discussed in Sect. 4.

## 3.1    O-SNARKs: SNARKs in the Presence of Oracles

Let $\mathbb{O} = \{\mathcal{O}\}$ be a family of oracles. We denote by $\mathcal{O}\leftarrow\mathbb{O}$ the process of sampling an oracle $\mathcal{O}$ from the family $\mathbb{O}$ according to some (possibly probabilistic) process. For example, $\mathbb{O}$ can be a random oracle family, i.e., $\mathbb{O} = \{\mathcal{O} : \{0,1\}^\ell \to \{0,1\}^L\}$ for all possible functions from $\ell$-bits strings to $L$-bits strings, in which case $\mathcal{O}\leftarrow\mathbb{O}$ consists of choosing a function $\mathcal{O}$ uniformly at random in $\mathbb{O}$. As another example, $\mathbb{O}$ might be the signing oracle corresponding to a signature scheme, in which case the process $\mathcal{O}\leftarrow\mathbb{O}$ consists of sampling a secret key of the signature scheme according to the key generation algorithm (and possibly a random tape for signature generation in case the signing algorithm is randomized).

For any oracle family $\mathbb{O}$, we define an O-SNARK $\Pi$ for $\mathbb{O}$ as follows.

**Definition 6 ($\mathcal{Z}$-auxiliary input O-SNARKs for $\mathbb{O}$).** *We say that $\Pi$ is a $\mathcal{Z}$-auxiliary input O-SNARK for the oracle family $\mathbb{O}$, if $\Pi$ satisfies the properties of completeness and succinctness as in Definition 3, and the following property of adaptive proof of knowledge for $\mathbb{O}$:*

- **Adaptive Proof of Knowledge for** $\mathbb{O}$. *Consider the following experiment for security parameter $\lambda \in \mathbb{N}$, time bound $T \in \mathbb{N}$, adversary $\mathcal{A}$, extractor $\mathcal{E}_\mathcal{A}$, auxiliary input generator $\mathcal{Z}$ and oracle family $\mathbb{O}$:*

O-AdPoK($\lambda, T, \mathcal{A}, \mathcal{E}_\mathcal{A}, \mathcal{Z}, \mathbb{O}$)
   $aux \leftarrow \mathcal{Z}(1^\lambda)$;  $\mathcal{O} \leftarrow \mathbb{O}$;  crs $\leftarrow$ Gen($1^\lambda, T$)
   $(y, \pi) \leftarrow \mathcal{A}^\mathcal{O}$(crs, $aux$)  $w \leftarrow \mathcal{E}_\mathcal{A}$(crs, $aux$, qt)
   if Ver(crs, $y, \pi$) $= 1 \wedge (y, w) \notin \mathcal{R}$ return 1
   else return 0

where qt $= \{q_i, \mathcal{O}(q_i)\}$ *is the transcript of all oracle queries and answers made and received by $\mathcal{A}$ during its execution.*

$\Pi$ *satisfies adaptive proof of knowledge with respect to oracle family $\mathbb{O}$ and auxiliary input from $\mathcal{Z}$ if for every non-uniform oracle prover $\mathcal{A}^\mathcal{O}$ of size $s(\lambda) = \mathsf{poly}(\lambda)$ making at most $Q(\lambda) = \mathsf{poly}(\lambda)$ queries there exists a non-uniform extractor $\mathcal{E}_\mathcal{A}$ of size $t(\lambda) = \mathsf{poly}(\lambda)$ and a negligible function $\epsilon(\lambda)$ such that for every time bound $T$, $\Pr[\text{O-AdPoK}(\lambda, T, \mathcal{A}, \mathcal{E}_\mathcal{A}, \mathcal{Z}, \mathbb{O}) \Rightarrow 1] \leq \epsilon(\lambda)$. Furthermore, we say that $\Pi$ satisfies $(s, t, Q, \epsilon)$-adaptive proof of knowledge with respect to oracle family $\mathbb{O}$ and auxiliary input from $\mathcal{Z}$ if the above condition holds for concrete values $(s, t, Q, \epsilon)$.*

### 3.2 Non-adaptive O-SNARKs

In this section we define a relaxation of O-SNARKs in which the adversary is non-adaptive in making its queries to the oracle. Namely, we consider adversaries that first declare all their oracle queries $q_1, \ldots, q_Q$ and then run on input the common reference string as well as the queries' outputs $\mathcal{O}(q_1), \ldots, \mathcal{O}(q_Q)$. More formally,

**Definition 7 ($\mathcal{Z}$-auxiliary input non-adaptive O-SNARKs for $\mathbb{O}$).** *We say that $\Pi$ is a $\mathcal{Z}$-auxiliary input non-adaptive O-SNARK for the oracle family $\mathbb{O}$, if $\Pi$ satisfies the properties of completeness and succinctness as in Definition 3, and the following property of non-adaptive queries proof of knowledge for $\mathbb{O}$:*

– **Non-adaptive Proof of Knowledge for $\mathbb{O}$.** *Consider the following experiment for security parameter $\lambda \in \mathbb{N}$, time bound $T \in \mathbb{N}$, adversary $\mathcal{A} = (\mathcal{A}_1, \mathcal{A}_2)$, extractor $\mathcal{E}_\mathcal{A}$, auxiliary input generator $\mathcal{Z}$ and oracle family $\mathbb{O}$:*

O-NonAdPoK($\lambda, T, \mathcal{A}, \mathcal{E}_\mathcal{A}, \mathcal{Z}, \mathbb{O}$)
   $(q_1, \ldots, q_Q, st) \leftarrow \mathcal{A}_1(1^\lambda)$
   $aux \leftarrow \mathcal{Z}(1^\lambda)$;  $\mathcal{O} \leftarrow \mathbb{O}$;  crs $\leftarrow$ Gen($1^\lambda, T$)
   qt $= (q_1, \mathcal{O}(q_1), \ldots, q_Q, \mathcal{O}(q_Q))$
   $(y, \pi) \leftarrow \mathcal{A}_2(st, \text{crs}, aux, \text{qt})$  $w \leftarrow \mathcal{E}_\mathcal{A}$(crs, $aux$, qt)
   if Ver(crs, $y, \pi$) $= 1 \wedge (y, w) \notin \mathcal{R}$ return *1*
   else return *0*

where st *is simply a state information shared between $\mathcal{A}_1$ and $\mathcal{A}_2$.*

$\Pi$ *satisfies non-adaptive proof of knowledge with respect to oracle family $\mathbb{O}$ and auxiliary input from $\mathcal{Z}$ if for every non-uniform prover $\mathcal{A} = (\mathcal{A}_1, \mathcal{A}_2)$*

of size $s(\lambda) = \mathsf{poly}(\lambda)$ *making at most* $Q(\lambda) = \mathsf{poly}(\lambda)$ *non-adaptive queries there exists a non-uniform extractor* $\mathcal{E}_\mathcal{A}$ *of size* $t(\lambda) = \mathsf{poly}(\lambda)$ *and a negligible function* $\epsilon(\lambda)$ *such that for every time bound* $T$, $\Pr[\mathsf{O\text{-}NonAdPoK}(\lambda, T, \mathcal{A}, \mathcal{E}_\mathcal{A}, \mathcal{Z}, \mathbb{O}) \Rightarrow 1] \leq \epsilon(\lambda)$. *Furthermore, we say that* $\Pi$ *satisfies* $(s, t, Q, \epsilon)$-*non-adaptive proof of knowledge with respect to oracle family* $\mathbb{O}$ *and auxiliary input from* $\mathcal{Z}$ *if the above condition holds for concrete values* $(s, t, Q, \epsilon)$.

It is also possible to define a stronger variant of the above definition in which $\mathcal{A}_1$ is given (adaptive) oracle access to $\mathcal{O}$, whereas $\mathcal{A}_2$ has no access to $\mathcal{O}$, except for the query transcript obtained by $\mathcal{A}_1$. It is not hard to see that the result given in the following paragraph works under this intermediate definition as well.

**Existence of Non-adaptive O-SNARKs from SNARKs.** Below we prove a simple result showing that non-adaptive O-SNARKs follow directly from classical SNARKs for which the proof of knowledge property holds for arbitrary auxiliary input distributions.

**Theorem 2.** *Let* $\mathbb{O}$ *be any oracle family. If* $\Pi$ *is a SNARK satisfying* $(s, t, \epsilon)$-*adaptive PoK (for arbitrary auxiliary input), then* $\Pi$ *is a non-adaptive O-SNARK for* $\mathbb{O}$ *satisfying* $(s, t, Q, \epsilon)$-*non-adaptive PoK.*

For lack of space we only provide an intuition of the proof, which is given in detail in the full version. The idea is that the second stage adversary $\mathcal{A}_2$ of non-adaptive O-SNARKs is very much like a classical SNARK adversary that makes no queries and receives a certain auxiliary input which contains the set of oracle queries chosen by $\mathcal{A}_1$ with corresponding answers. The fact that the auxiliary input includes the set of queries chosen by $\mathcal{A}_1$, which is an arbitrary adversary, implies that the SNARK must support arbitrary, *not necessarily benign*, auxiliary inputs (i.e., it is not sufficient to fix an auxiliary input distribution that depends only on the oracle family $\mathbb{O}$).

# 4   On the Existence of O-SNARKs

In this section we study whether O-SNARKs exist and under what assumptions. In the following sections we give both positive and negative answers to this question. For lack of space, a positive existence result about O-SNARKs for (pseudo)random oracles is given in the full version [FN16].

## 4.1   O-SNARKs in the ROM from Micali's CS Proofs

In this section we briefly discuss how the construction of CS proofs of Micali [Mic00] can be seen as an O-SNARK for *any* oracle family, albeit in the random oracle model. To see this, we rely on the result of Valiant [Val08] who shows that Micali's construction is a "CS proof of knowledge" in the random oracle model. The main observation is in fact that Valiant's proof works by showing a *black-box* extractor working for any prover.

**Proposition 1.** *Let* $\mathbb{O}$ *be any oracle family and* RO *be a family of random oracles. Let* $\Pi_{\mathsf{Mic}}$ *be the CS proof construction from [Mic00]. Then* $\Pi_{\mathsf{Mic}}$ *is an O-SNARK for* $(\mathsf{RO}, \mathbb{O})$*, in the random oracle model.*

*Proof (Sketch).* Let $\mathcal{E}^{\mathsf{RO}}$ be Valiant's black-box extractor[7] which takes as input the code of the prover and outputs a witness $w$. For any adversary $\mathcal{A}^{\mathsf{RO},\mathcal{O}}$ we can define its extractor $\mathcal{E}_{\mathcal{A}}$ as the one that, on input the query transcript qt of $\mathcal{A}$, executes $w \leftarrow \mathcal{E}^{\mathsf{RO}}(\mathcal{A})$ by simulating all the random oracle queries of $\mathcal{E}^{\mathsf{RO}}$ using qt, and finally outputs the same $w$. The reason why qt suffices to $\mathcal{E}_{\mathcal{A}}$ for simulating random oracle queries to $\mathcal{E}^{\mathsf{RO}}$ is that Valiant's extractor $\mathcal{E}^{\mathsf{RO}}$ makes exactly the same queries of the prover.

## 4.2   Impossibility of O-SNARKs for Every Family of Oracles

In this section we show that, *in the standard model*, there do not exist O-SNARKs with respect to *every* family of oracles. We show this under the assumption that universal one-way hash functions (and thus one-way functions [Rom90]) exist. To show the impossibility, we describe an oracle family in the presence of which any candidate O-SNARK that is correct and succinct cannot satisfy adaptive proof of knowledge with respect to that oracle family. Our impossibility result is shown for designated-verifier O-SNARKs, and thus implies impossibility for publicly verifiable ones as well (since every publicly verifiable O-SNARK is also designated-verifier secure). More specifically, we show the impossibility by means of a signing oracle family. Namely, we show a secure signature scheme $\Sigma_p$ such that every correct and succinct O-SNARK $\Pi$ cannot satisfy adaptive proof of knowledge in the presence of the signing oracle corresponding to $\Sigma_p$. Interestingly, such a result not only shows that extraction cannot work for general families of oracles, but also for families of signing oracles, a class which is relevant to several applications.

For every signature scheme $\Sigma = (\mathsf{kg}, \mathsf{sign}, \mathsf{vfy})$ we let $\mathbb{O}_{\Sigma}$ be the family of oracles $\mathcal{O}(m) = \mathsf{sign}(\mathsf{sk}, m)$, where every family member $\mathcal{O}$ is described by a secret key sk of the signature scheme, i.e., the process $\mathcal{O} \leftarrow \mathbb{O}_{\Sigma}$ corresponds to obtaining sk through a run of $(\mathsf{sk}, \mathsf{vk}) \xleftarrow{\$} \mathsf{kg}(1^{\lambda})$. For the sake of simplicity, we also assume that the oracle allows for a special query, say $\mathcal{O}(`vk')$,[8] whose answer is the verification key vk.

**Theorem 3.** *Assume that one-way functions exist. Then for every polynomial $p(\cdot)$ there exists a* UF-CMA-*secure signature scheme $\Sigma_p$ such that every candidate designated-verifier O-SNARK $\Pi$ for* NP*, that is correct and succinct with proofs of length bounded by $p(\cdot)$, does not satisfy adaptive proof of knowledge with respect to $\mathbb{O}_{\Sigma_p}$.*

---

[7] The CS proofs of knowledge definition used by Valiant considers adversaries that are non-adaptive in choosing the statement. However it easy to see that the construction and the proof work also for the adaptive case.

[8] Here $vk$ is an arbitrary choice; any symbol not in $\mathcal{M}$ would do so. Introducing the extra query simplifies the presentation, otherwise vk should be treated as an auxiliary input from a distribution generated together with the oracle sampling.

AN INTUITION OF THE RESULT. Before delving into the details of the proof, we provide the main intuition of this result. This intuition does not use signature schemes but includes the main ideas that will be used in the signature counterexample. Given a UOWHF function family $\mathcal{H}$, consider the NP binary relation $\tilde{R}_{\mathcal{H}} = \{((h,x),w) : h \in \mathcal{H}, h(w) = x\}$, let $\Pi$ be a SNARK for NP and consider $p(\cdot)$ the polynomial for which $\Pi$ is succinct. The idea is to show an oracle family $\tilde{\mathbb{O}}$ and an adversary $\bar{A}$ for which there is no extractor unless $\mathcal{H}$ is not a universal one-way family. For every polynomial $p(\cdot)$, the oracle family contains oracles $\mathcal{O}_p$ that given a query $q$, interpret $q$ as the description of a program $\mathcal{P}(\cdot,\cdot)$, samples a random member of the hash family $h \xleftarrow{\$} \mathcal{H}$, a random $w$, computes $x = h(w)$, and outputs $(h,x)$ along with $\pi \leftarrow \mathcal{P}((h,x),w)$. If $\mathcal{P}(\cdot,\cdot) = \mathsf{Prove}(\mathsf{prs},\cdot,\cdot)$, then the oracle is simply returning an hash image with a proof of knowledge of its (random) preimage. The adversary $\bar{A}^{\mathcal{O}_p}$ is the one that on input $\mathsf{prs}$, simply asks one query $q = \mathcal{P}(\cdot,\cdot) = \mathsf{Prove}(\mathsf{prs},\cdot,\cdot)$, gets $((h,x),\pi) \leftarrow \mathcal{O}_p(q)$ and outputs $((h,x),\pi)$. Now, the crucial point that entails the non-existence of an extractor is that, provided that the input $w$ is sufficiently longer than $\pi$, every valid extractor for such $\bar{A}$ that outputs a valid $w'$ immediately implies a collision $(w,w')$ for $h$.[9] Finally, to prevent adversarially chosen $\mathcal{P}$ from revealing too much information, we require the oracle to check the length of $\pi$, and the latter is returned only if $|\pi| \leq p(\lambda)$.

*Proof (Proof of Theorem 3).* The proof consists of two main steps. First, we describe the construction of the signature scheme $\Sigma_p$ based on any other UF-CMA-secure signature scheme $\hat{\Sigma}$ with message space $\mathcal{M} = \{0,1\}^*$ (that exists assuming OWFs [Lam79, Rom90]), and show that $\Sigma_p$ is UF-CMA-secure. $\Sigma_p$ uses also an UOWHF family $\mathcal{H}$. Second, we show that, when considering the oracle family $\mathbb{O}_{\Sigma_p}$ corresponding to the signature scheme $\Sigma_p$, a correct $\Pi$ with succinctness $p(\cdot)$ cannot be an O-SNARK for $\mathbb{O}_{\Sigma_p}$, i.e., we show an efficient O-SNARK adversary $\mathcal{A}_p^{\mathcal{O}}$ (with access to a $\Sigma_p$ signing oracle $\mathcal{O}(\cdot) = \mathsf{sign}(\mathsf{sk},\cdot)$), for which there is no extractor unless $\mathcal{H}$ is not one-way.

**The Counterexample Signature Scheme $\Sigma_p$.** Let $\hat{\Sigma}$ be any UF-CMA-secure scheme with message space $\mathcal{M} = \{0,1\}^*$. Let $\mathbb{H} = \{\mathcal{H}\}_\lambda$ be a collection of function families $\mathcal{H} = \{h : \{0,1\}^{L(\lambda)} \to \{0,1\}^{\ell(\lambda)}\}$ where each $\mathcal{H}$ is an universal one-way hash family with $L(\lambda) \geq p(\lambda) + \ell(\lambda) + \lambda$. Let $M_{\mathcal{H}}((h,x),w)$ be the machine that on input $((h,x),w)$ accepts iff $h(w) = x$, and $\mathcal{R}_{\mathcal{H}}$ be the NP relation consisting of all pairs $(y,w)$ such that, for $y = (M_{\mathcal{H}},(h,x),t)$, $M_{\mathcal{H}}((h,x),w)$ accepts in at most $t$ steps.

The scheme $\Sigma_p$ has message space $\mathcal{M} = \{0,1\}^*$; its algorithms work as follows:

$\mathsf{kg}(1^\lambda)$: Run $(\widehat{\mathsf{vk}},\widehat{\mathsf{sk}}) \leftarrow \hat{\Sigma}.\mathsf{kg}(1^\lambda)$, set $\mathsf{vk} = \widehat{\mathsf{vk}}$, $\mathsf{sk} = \widehat{\mathsf{sk}}$.
$\mathsf{sign}(\mathsf{sk},m)$: Signing works as follows

---

[9] This relies on the fact that sufficiently many bits of $w$ remain unpredictable, even given $\pi$.

- generate $\hat{\sigma} \leftarrow \widehat{\Sigma}.\mathsf{sign}(\widehat{\mathsf{sk}}, m)$;
- sample $h \xleftarrow{\$} \mathcal{H}$ and $w \xleftarrow{\$} \{0,1\}^{L(\lambda)}$;
- compute $x = h(w)$, $t = \#M_{\mathcal{H}}((h,x),w)$, and set $y = (M_{\mathcal{H}}, (h,x), t)$;
- interpret $m$ as the description of program $\mathcal{P}(\cdot, \cdot)$ and thus run $\pi \leftarrow \mathcal{P}(y, w)$;
- if $|\pi| \leq p(\lambda)$, set $\pi' = \pi$, else set $\pi' = 0$;
- output $\sigma = (\hat{\sigma}, h, x, \pi')$.

$\mathsf{vfy}(\mathsf{vk}, m, \sigma)$: Parse $\sigma = (\hat{\sigma}, h, x, \pi')$ and return the output of $\widehat{\Sigma}.\mathsf{vfy}(\widehat{\mathsf{vk}}, m, \hat{\sigma})$.

It is trivial to check that, as long as $\widehat{\Sigma}$ is a UF-CMA-secure scheme, $\Sigma_p$ is also UF-CMA-secure. Moreover, remark that the scheme $\Sigma_p$ does not depend on the specific O-SNARK construction $\Pi$ but only on the universal polynomial $p(\cdot)$ bounding its succinctness.

**Impossibility of O-SNARKs for $\mathbb{O}_{\Sigma_p}$.** To show that $\Pi$ is not an O-SNARK for $\mathbb{O}_{\Sigma_p}$ (under the assumption that $\mathcal{H}$ is universally one-way), we prove that there is an adversary $\mathcal{A}_p^{\mathcal{O}}$ such that every candidate extractor $\mathcal{E}$ fails in the adaptive proof of knowledge game.

**Lemma 1.** *If $\mathcal{H}$ is universally one way then every $\Pi$ for NP that is correct and succinct with proofs of length $p(\cdot)$ is not a designated-verifier O-SNARK for $\mathbb{O}_{\Sigma_p}$.*

*Proof.* Let $\mathcal{A}_p^{\mathcal{O}}$ be the following adversary: on input prs, encode the Prove algorithm of $\Pi$ with hardcoded prs as a program $\mathcal{P}(\cdot, \cdot) := \mathsf{Prove}(\mathsf{prs}, \cdot, \cdot)$; let $q$ be $\mathcal{P}$'s description, and make a single query $\sigma = (\hat{\sigma}, h, x, \pi') \leftarrow \mathcal{O}(q)$; return $(y, \pi')$ where $y = (M_{\mathcal{H}}, (h, x), t)$ is appropriately reconstructed. We show that for every polynomial-size extractor $\mathcal{E}$ it holds

$$\Pr[\mathsf{O}\text{-}\mathsf{AdPoK}(\lambda, \mathcal{A}_p, \mathcal{E}, \mathbb{O}_{\Sigma_p}) \Rightarrow 0] \leq \nu_{\mathcal{H}}(\lambda) + 2^{-\lambda}$$

where $\nu_{\mathcal{H}}(\lambda) = \mathbf{Adv}_{\mathcal{B},\mathcal{H}}^{UOWHF}(\lambda)$ is the advantage of any adversary $\mathcal{B}$ against $\mathcal{H}$'s universal one-wayness. This means that there is no extractor unless $\mathcal{H}$ is not an universal one-way family.

We proceed by contradiction assuming the existence of a polynomial-size extractor $\mathcal{E}$ such that the above probability is greater than some non-negligible $\epsilon$. We show how to build an adversary $\mathcal{B}$ that breaks universal one-wayness of $\mathcal{H}$ with non-negligible probability.

$\mathcal{B}$ first chooses an hash input $w \xleftarrow{\$} \{0,1\}^{L(\lambda)}$, and then receives an instance $h$ of $\mathcal{H}$. Next, $\mathcal{B}$ generates $(\mathsf{prs}, \mathsf{vst}) \leftarrow \mathsf{Gen}(1^\lambda)$ and $(\widehat{\mathsf{vk}}, \widehat{\mathsf{sk}}) \leftarrow \widehat{\Sigma}.\mathsf{kg}(1^\lambda)$, and runs $\mathcal{A}_p^{\mathcal{O}}(\mathsf{prs})$ simulating the oracle $\mathcal{O}$ on the single query $q := \mathcal{P}(\cdot, \cdot) = \mathsf{Prove}(\mathsf{crs}, \cdot, \cdot)$ asked by $\mathcal{A}_p$. In particular, to answer the query $\mathcal{B}$ uses the secret key $\widehat{\mathsf{sk}}$ to generate $\hat{\sigma}$, and computes $x = h(w)$ using the function $h$ received from its challenger, and the input $w$ chosen earlier. Notice that such a simulation can be done perfectly in a straightforward way, and that $\mathcal{A}_p$'s output is the pair $(y, \pi)$ created by $\mathcal{B}$. Next, $\mathcal{B}$ runs the extractor $w' \leftarrow \mathcal{E}(\mathsf{prs}, \mathsf{qt} = (\mathcal{P}(\cdot, \cdot), (\hat{\sigma}, h, x, \pi)))$, and outputs $w'$.

By correctness of $\Pi$ it holds that the pair $(y, \pi)$ returned by $\mathcal{A}_p$ satisfies $\mathsf{Ver}(\mathsf{vst}, y, \pi) = 1$. Thus, by our contradiction assumption, with probability $\geq \epsilon(\lambda)$, $\mathcal{E}$ outputs $w'$ such that $(y, w') \in \mathcal{R}_\mathcal{H}$. Namely, $h(w') = x = h(w)$. To show that this is a collision, we argue that, information-theoretically, $w' \neq w$ with probability $\geq 1 - 1/2^\lambda$. This follows from the fact that $w$ is randomly chosen of length $L(\lambda) \geq p(\lambda) + \ell(\lambda) + \lambda$ and the only information about $w$ which is leaked to $\mathcal{E}$ is through $\pi$ and $x = h(w)$, an information of length at most $p(\lambda) + \ell(\lambda)$. Therefore there are at least $\lambda$ bits of entropy in $w$, from which $\Pr[w' = w] \leq 2^{-\lambda}$ over the random choice of $w$. Hence, $\mathcal{B}$ can break the universal one-wayness of $\mathcal{H}$ with probability $\geq \epsilon(\lambda) - 2^{-\lambda}$.                                    □

### 4.3   O-SNARKs for Signing Oracles from SNARKs in the Random Oracle Model

In this section we show that it is possible to "immunize" *any* signature scheme in such a way that any classical SNARK is also an O-SNARK for the signing oracle corresponding to the transformed scheme. The idea is very simple and consists into applying the hash-then-sign approach using a hash function that will be modeled as a random oracle. A limitation of this result is that, since the verification algorithm uses a random oracle, in all those applications where the SNARK is used to prove knowledge of valid signatures, one would need a SNARK for $\mathsf{NP}^\mathcal{O}$. Hence, the best one can do is to conjecture that this still works when replacing the random oracle with a suitable hash function.

Let us now state formally our result. To this end, for any signature scheme $\Sigma$ and polynomial $Q(\cdot)$ we define $\mathcal{Z}_{Q,\Sigma}$ as the distribution on tuples $\langle \mathsf{vk}, m_1, \sigma_1, \ldots, m_Q, \sigma_Q \rangle$ obtained by running the following probabilistic algorithm:

$\mathcal{Z}_{Q,\Sigma}(1^\lambda)$
```
let Q = Q(λ); (sk, vk)←kg(1^λ)
M̃ ←$ MsgSample(M, Q) ; let M̃ = {m_1, ..., m_Q}
for i = 1 to Q do : σ_i←sign(sk, m_i)
return ⟨vk, {m_i, σ_i}_{i=1}^Q⟩
```

where $\mathsf{MsgSample}(\mathcal{M}, Q)$ is an algorithm that returns $Q$ distinct messages, each randomly chosen from $\mathcal{M}$. The proof of the following theorem appears in the full version.

**Theorem 4.** *Let $\Sigma$ be a UF-CMA-secure signature scheme, and $\mathcal{H}$ be a family of hash functions modeled as a random oracle. Let $\mathcal{U}_n$ be the uniform distribution over strings of length $n$, and $\mathcal{Z}_{Q,\Sigma}$ be the distribution defined above, where $Q$ is any polynomial in the security parameter. Then there exists a signature scheme $\Sigma_\mathcal{H}$ such that every $(\mathcal{Z}, \mathcal{U}, \mathcal{Z}_{\Sigma,Q})$-auxiliary input SNARK $\Pi$ is a $\mathcal{Z}$-auxiliary input O-SNARK for $(\mathbb{O}_\mathcal{H}, \mathbb{O}_{\Sigma_\mathcal{H}})$ where $\mathbb{O}_\mathcal{H}$ is a random oracle.*

### 4.4   O-SNARKs for Signing Oracles from SNARKs

In this section we give a positive result showing that any SNARK $\Pi$ is an O-SNARK for the signing oracle of signature scheme $\Sigma$ if: (i) the message space of $\Sigma$ is appropriately bounded (to be polynomially or at most superpolynomially large); (ii) $\Pi$ tolerates auxiliary input consisting of the public key of $\Sigma$ plus a collection of signatures on randomly chosen messages; (iii) one considers O-SNARK adversaries that query the signing oracle on almost the entire message space. Furthermore, in case of superpolynomially large message spaces, one needs to assume sub-exponential hardness for $\Pi$.

The intuition behind this result is to simulate the O-SNARK adversary by using a (non-interactive) SNARK adversary that receives the public key and a set of signatures on (suitably chosen) messages as its auxiliary input. If these messages exactly match[10] those queried by the O-SNARK adversary, the simulation is perfect. However, since the probability of matching exactly all the $Q = \text{poly}(\lambda)$ queries may decrease exponentially in $Q$ (making the simulation meaningless), we show how to put proper bounds so that the simulation can succeed with probability depending only on the message space size.

More formally, our result is stated as follows. Let $\Sigma$ be a signature scheme with message space $\mathcal{M}$, and let $Q := Q(\cdot)$ be a function of the security parameter. Let $\mathcal{Z}_{Q,\Sigma}$ be the following auxiliary input distribution

$\mathcal{Z}_{Q,\Sigma}(1^\lambda)$
   let $Q = Q(\lambda)$; $(\text{sk},\text{vk}) \leftarrow \text{kg}(1^\lambda)$
   $\tilde{\mathcal{M}} \xleftarrow{\$} \text{MsgSample}(\mathcal{M}, Q)$ ; let $\tilde{\mathcal{M}} = \{m_1, \ldots, m_Q\}$
   for $i = 1$ to $Q$ do : $\sigma_i \leftarrow \text{sign}(\text{sk}, m_i)$
   return $\langle \text{vk}, \{m_i, \sigma_i\}_{i=1}^Q \rangle$

where $\text{MsgSample}(\mathcal{M}, Q)$ is a probabilistic algorithm that returns a subset $\tilde{\mathcal{M}} \subseteq \mathcal{M}$ of cardinality $Q$ chosen according to some strategy that we discuss later. At this point we only assume a generic strategy such that $\delta(|\mathcal{M}|, Q) = \Pr[\text{MsgSample}(\mathcal{M}, Q) = \mathcal{M}^*]$ for any $\mathcal{M}^* \subseteq \mathcal{M}$ of cardinality $Q$. The proof is in the full version.

**Theorem 5.** *Let $\Sigma$ be a signature scheme with message space $\mathcal{M}$, let $\mathbb{O}_\Sigma$ be the associated family of signing oracles, and let $\mathcal{Z}_{Q,\Sigma}$ be as defined above. If $\Pi$ is a $\mathcal{Z}_{Q,\Sigma}$-auxiliary input SNARK satisfying $(s, t, \epsilon)$-adaptive PoK, then $\Pi$ is an O-SNARK for $\mathbb{O}_\Sigma$ satisfying $(s', t', Q, \epsilon')$-adaptive PoK, where $\epsilon' = \epsilon/\delta(|\mathcal{M}|, Q)$, $s' = s - O(Q \cdot \log |\mathcal{M}|)$, and $t' = t$.*

**Implications of Theorem 5.** The statement of Theorem 5 is parametrized by values $|\mathcal{M}|, Q$ and the function $\delta(|\mathcal{M}|, Q)$, which in turn depends on the query guessing strategy. As for the $\text{MsgSample}(\mathcal{M}, Q)$ algorithm, let us consider the one that *samples a random subset* $\tilde{\mathcal{M}} \subseteq \mathcal{M}$ of cardinality $Q$. For this algorithm we have $\delta(|\mathcal{M}|, Q) = \frac{1}{\binom{|\mathcal{M}|}{Q}}$. Notice that $\delta(|\mathcal{M}|, Q)$ is governing the success probability of our reduction, and thus we would like this function not to become

---

[10] We note that the proof requires an exact match and it is not sufficient that the O-SNARK adversary's queries are a subset of the sampled messages. A more precise explanation of this fact is given at the end of the proof in the full version.

negligible. However, since $Q = \mathsf{poly}(\lambda)$ is a parameter under the choice of the adversary, it might indeed be the case that $\delta(|\mathcal{M}|, Q) \approx 2^{-Q} \approx 2^{-\lambda}$, which would make our reduction meaningless. To avoid this bad case, we restrict our attention to adversaries for which $Q = |\mathcal{M}| - c$ for some constant $c \geq 1$, i.e., adversaries that ask for signatures on the entire message but a constant number of messages. For this choice of $Q$ we indeed have that $\delta(|\mathcal{M}|, Q) = \frac{1}{|\mathcal{M}|^c}$ depends only on the cardinality of $|\mathcal{M}|$. This gives us

**Corollary 1.** *Let $\Sigma$ be a signature scheme with message space $\mathcal{M}$ where $|\mathcal{M}| = \mathsf{poly}(\lambda)$ (resp. $|\mathcal{M}| = \lambda^{\omega(1)}$), and let $Q = |\mathcal{M}| - c$ for constant $c \in \mathbb{N}$. If $\Pi$ is a polynomially (resp. sub-exponentially) secure $\mathcal{Z}_{Q,\Sigma}$-auxiliary input SNARK, then $\Pi$ is an O-SNARK for $\mathbb{O}_\Sigma$ (for adversaries making $Q$ queries).*

# 5    Applications of O-SNARKs

In this section we show three applications of O-SNARKs for building homomorphic signatures [BF11], succinct functional signatures [BGI14], and SNARKs on authenticated data [BBFR15].

Generally speaking, our results show constructions of these primitives based on a signature scheme $\Sigma$ and a succinct non-interactive argument $\Pi$, and show their security by assuming that $\Pi$ is an O-SNARK for signing oracles corresponding to $\Sigma$. Once these results are established, we can essentially reach the following conclusions about the possible secure instantiations of these constructions. First, one can instantiate them by using Micali's CS proofs as O-SNARK (cf. Sect. 4.1): this solution essentially yields secure instantiations in the random oracle model that work with a specific proof system (perhaps not the most efficient one in practice). Second, one can instantiate them with a classical SNARK and a hash-and-sign signature scheme (cf. Sect. 4.3), and conjecture that replacing the random oracle with a suitable hash function preserves the overall security. Third, one can instantiate the constructions using a classical SNARK construction $\Pi$ and signature scheme $\Sigma$, and then conjecture that $\Pi$ is an O-SNARK with respect to the family of signing oracles corresponding to $\Sigma$. Compared to the first solution, the last two ones have the advantage that one could use some of the recently proposed efficient SNARK schemes (e.g., [PHGR13, BSCG+13]); on the other hand these solutions have the drawback that the security of the instantiations would be heavily based on a heuristic argument. Finally, a fourth option that we provide are security proofs of these primitives which consider only non-adaptive adversaries (i.e., adversaries that declare all their queries in advance). In this case we can prove security based on non-adaptive O-SNARKs, and thus based on classical SNARKs (applying our Theorem 2). The advantage of this fourth option is that one obtains a security proof for these instantiations based on classical, not new, assumptions, although the proof holds only for a much weaker security notion.

## 5.1  Homomorphic Signatures

As first application of O-SNARKs we revisit a "folklore" construction of homomorphic signatures from SNARKs. This construction has been mentioned several times in the literature (e.g., [BF11, GW13, CF13, CFW14, GVW15]) and is considered as the 'straightforward' approach for constructing this primitive. In this section we formalize this construction, and notice that its security proof is quite subtle as one actually incurs the extraction issues that we mentioned in the introduction. Namely, one needs to run an extractor in an interactive security game in the presence of a signing oracle. Here we solve this issue by giving a simple proof based on our notion of O-SNARKs (for families of signing oracles).

**Definition of Homomorphic Signatures.** We begin by recalling the definition of homomorphic signatures. The definition below can be seen as the public key version of the notion of homomorphic message authenticators for labeled programs of Gennaro and Wichs [GW13].

Labeled Programs [GW13]. A labeled program consists of a tuple $\mathcal{P} = (F, \tau_1, \ldots \tau_n)$ such that $F : \mathcal{M}^n \to \mathcal{M}$ is a function on $n$ variables (e.g., a circuit), and $\tau_i \in \{0,1\}^\ell$ is the label of the $i$-th variable input of $F$. Let $F_{id} : \mathcal{M} \to \mathcal{M}$ be the canonical identity function and $\tau \in \{0,1\}^\ell$ be a label. We consider $\mathcal{I}_\tau = (F_{id}, \tau)$ as the identity program for input label $\tau$. Given $t$ labeled programs $\mathcal{P}_1, \ldots \mathcal{P}_t$ and a function $G : \mathcal{M}^t \to \mathcal{M}$, the composed program $\mathcal{P}^*$ is the one obtained by evaluating $G$ on the outputs of $\mathcal{P}_1, \ldots \mathcal{P}_t$, and is compactly denoted as $\mathcal{P}^* = G(\mathcal{P}_1, \ldots \mathcal{P}_t)$. The labeled inputs of $\mathcal{P}^*$ are all distinct labeled inputs of $\mathcal{P}_1, \ldots \mathcal{P}_t$, i.e., all inputs with the same label are grouped together in a single input of the new program.

**Definition 8 (Homomorphic Signatures for Labeled Programs).** *A homomorphic signature scheme* HomSig *is a tuple of probabilistic, polynomial-time algorithms* (HomKG, HomSign, HomVer, HomEval) *that work as follows*

HomKG($1^\lambda$) *takes a security parameter* $\lambda$ *and outputs a public key* VK *and a secret key* SK. *The public key* VK *defines implicitly a message space* $\mathcal{M}$, *the label space* $\mathcal{L}$, *and a set* $\mathcal{F}$ *of admissible functions.*

HomSign(SK, $\tau, m$) *takes a secret key* SK, *a (unique) label* $\tau \in \mathcal{L}$ *and a message* $m \in \mathcal{M}$, *and it outputs a signature* $\sigma$.

HomEval(VK, $F, (\sigma_1, \ldots \sigma_n)$) *takes a public key* VK, *a function* $F \in \mathcal{F}$ *and a tuple of signatures* $(\sigma_1, \ldots \sigma_n)$. *It outputs a new signature* $\sigma$.

HomVer(VK, $\mathcal{P}, m, \sigma$) *takes a public key* VK, *a labeled program* $\mathcal{P} = (F, (\tau_1 \ldots \tau_n))$ *with* $F \in \mathcal{F}$, *a message* $m \in \mathcal{M}$, *and a signature* $\sigma$. *It outputs either 0 (reject) or 1 (accept).*

*and satisfy* authentication correctness, evaluation correctness, succinctness, *and* security, *as described below.*

- **Authentication Correctness.** *Informally, we require that signatures generated by* HomSign(SK, $\tau, m$) *verify correctly for* $m$ *as the output of the identity program* $\mathcal{I} = (F_{id}, \tau)$.

- **Evaluation Correctness.** *Intuitively, we require that running the evaluation algorithm on signatures $(\sigma_1, \ldots \sigma_n)$, where $\sigma_i$ is a signature for $m_i$ on label $\tau_i$, produces a signature $\sigma$ which verifies for $F(m_1, \ldots m_n)$.*
- **Succinctness.** *For every large enough security parameter $\lambda \in \mathbb{N}$, there is a polynomial $p(\cdot)$ such that for every $(\mathsf{SK}, \mathsf{VK}) \leftarrow \mathsf{HomKG}(1^\lambda)$ the output size of $\mathsf{HomSign}$ and $\mathsf{HomEval}$ is bounded by $p(\lambda)$ for any choice of their inputs.*
- **Security.** *A homomorphic signature scheme $\mathsf{HomSig}$ is secure if for every PPT adversary $\mathcal{A}$ there is a negligible function $\epsilon$ such that $\Pr[\mathbf{Exp}_{\mathcal{A}, \mathsf{HomSig}}^{\mathsf{HomSig\text{-}UF}}(\lambda) = 1] \leq \epsilon(\lambda)$ where the experiment $\mathbf{Exp}_{\mathcal{A}, \mathsf{HomSig}}^{\mathsf{HomSig\text{-}UF}}(\lambda)$ is described in the following:*

Key generation: *Run $(\mathsf{VK}, \mathsf{SK}) \leftarrow \mathsf{HomKG}(1^\lambda)$ and give $\mathsf{VK}$ to $\mathcal{A}$.*

Signing queries: *$\mathcal{A}$ can adaptively submit queries of the form $(\tau, m)$, where $\tau \in \mathcal{L}$ and $m \in \mathcal{M}$. The challenger initializes an empty list $T$ and proceeds as follows:*

  * *If $(\tau, m)$ is the first query with label $\tau$, then the challenger computes $\sigma \leftarrow \mathsf{HomSign}(\mathsf{SK}, \tau, m)$, returns $\sigma$ to $\mathcal{A}$ and updates the list of queries $T \leftarrow T \cup \{(\tau, m)\}$.*
  * *If $(\tau, m) \in T$ (i.e., the adversary had already queried the tuple $(\tau, m)$), then the challenger replies with the same signature generated before.*
  * *If $T$ contains a tuple $(\tau, m_0)$ for some different message $m_0 \neq m$, then the challenger ignores the query.*

  *Note that each label $\tau$ can be queried only once.*

Forgery: *After the adversary is done with the queries of the previous stage, it outputs a tuple $(\mathcal{P}^*, m^*, \sigma^*)$. Finally, the experiment outputs 1 iff the tuple returned by the adversary is a forgery (as defined below).*
  *Forgeries are tuples $(\mathcal{P}^* = (F^*, (\tau_1^*, \ldots \tau_n^*)), m^*, \sigma^*)$ such that $\mathsf{HomVer}(\mathsf{VK}, \mathcal{P}^*, m^*, \sigma^*) = 1$ and they satisfy one the following conditions:*

  * *Type 1 Forgery: There is $i \in [n]$ such that $(\tau_i^*, \cdot) \notin T$ (i.e., no message $m$ has ever been signed w.r.t. label $\tau_i^*$ during the experiment).*
  * *Type 2 Forgery: All labels $\tau_i^*$ have been queried—$\forall i \in [n], (\tau_i^*, m_i) \in T$—but $m^* \neq F^*(m_1, \ldots m_n)$ (i.e., $m^*$ is not the correct output of the labeled program $\mathcal{P}^*$ when executed on the previously signed messages).*

A homomorphic signature scheme can also be required to be *context-hiding* [BF11]. Intuitively this property says that signatures on outputs of functions do not reveal information about the inputs. The formal definition is recalled in the full version.

**Homomorphic Signatures from O-SNARKs.** To build the homomorphic signature we use a regular signature scheme $\Sigma$ and a fully-succinct O-SNARK $\Pi$ for NP. The resulting scheme is homomorphic for all functions $F$ whose running time is upper bounded by some fixed polynomial $t_\mathcal{F}(\cdot)$, and the scheme is 1-hop, i.e., it is not possible to apply $\mathsf{HomEval}$ on signatures obtained from other executions of $\mathsf{HomEval}$.[11]

---

[11] Previous work hinted the possibility of achieving multi-hop homomorphic signatures by using SNARKs with recursive composition. However, given the issues we already

DEFINING THE MACHINE $M_{\Sigma,F}$. Let $\Sigma$ be a signature scheme, and $F$ be the description of a function $F : \mathcal{X}^n \to \mathcal{X}$ where $\mathcal{X}$ is some appropriate domain (e.g., $\mathcal{X} = \{0,1\}^\mu$). Then $M_{\Sigma,F}(x, w)$ is the random-access machine that works as follows. It takes inputs $(x, w)$ where values $x$ are of the form $x = (\mathsf{vk}, m, \tau_1, \ldots, \tau_n)$ where $\mathsf{vk}$ is a public key of the scheme $\Sigma$, $m \in \mathcal{X}$ is a message and $\tau_i \in \{0,1\}^\ell$ are labels, for $1 \leq i \leq n$. The values $w$ are instead tuples $w = (m_1, \sigma_1, \ldots, m_n, \sigma_n)$ where for every $i \in [n]$, $m_i \in \mathcal{X}$ is a message and $\sigma_i$ is a signature of the scheme $\Sigma$. On input such a pair $(x, w)$, $M_{\Sigma,F}(x, w)$ accepts iff

$$m = F(m_1, \ldots, m_n) \ \wedge \ \mathsf{vfy}(\mathsf{vk}, \tau_i | m_i, \sigma_i) = 1, \forall i = 1, \ldots, n$$

Associated to such machine there is also a polynomial time bound $t_{\Sigma,F}(k) = k^{e_{\Sigma,F}}$, such that $M_{\Sigma,F}$ rejects if it does more than $t_{\Sigma,F}(|x|)$ steps. Finally, we note that given a polynomial bound $t_F(k) = k^{e_F}$ on the running time of every $F$ supported by the scheme, a polynomial bound $t_\Sigma(k) = k^{e_\Sigma}$ on the running time of $\Sigma$'s verification algorithm, and values $n, \mu, \ell$, one can efficiently deduce the constant exponent $e_{\Sigma,F}$ for the time bound $t_{\Sigma,F}(|x|) = |x|^{e_{\Sigma,F}}$.

We call $\mathcal{R}_\Sigma$ the NP binary relation consisting of all pairs $(y, w)$ such that, parsing $y = (M_{\Sigma,F}, x, t)$, $M_{\Sigma,F}(x, w)$ accepts in at most $t$ steps and $t \leq t_{\Sigma,F}(|x|)$.

THE CONSTRUCTION. Let $\Sigma = (\mathsf{kg}, \mathsf{sign}, \mathsf{vfy})$ be a signature scheme and $\Pi = (\mathsf{Gen}, \mathsf{Prove}, \mathsf{Ver})$ be a fully-succinct O-SNARK for NP. The homomorphic signature scheme $\mathsf{HomSig}[\Sigma, \Pi]$ is defined as follows.

$\mathsf{HomKG}(1^\lambda)$: Run $(\mathsf{sk}, \mathsf{vk}) \leftarrow \mathsf{kg}(1^\lambda)$ and $\mathsf{crs} \leftarrow \mathsf{Gen}(1^\lambda)$. Define $\mathsf{SK} = \mathsf{sk}$ and $\mathsf{VK} = (\mathsf{vk}, \mathsf{crs})$. Let the message be $\mathcal{M} = \{0,1\}^\mu$ and the label space be $\mathcal{L} = \{0,1\}^\ell$. Output $(\mathsf{SK}, \mathsf{VK})$.

$\mathsf{HomSign}(\mathsf{SK}, \tau, m)$: Run $\sigma \leftarrow \mathsf{sign}(\mathsf{sk}, \tau | m)$. Output $\bar{\sigma} = (signature, (\tau, m, \sigma))$.

$\mathsf{HomEval}(\mathsf{VK}, m, F, (\bar{\sigma}_1, \ldots, \bar{\sigma}_n))$: Parse every $\bar{\sigma}_i = (signature, (\tau_i, m_i, \sigma_i))$, compute $m = F(m_1, \ldots, m_n)$, reconstruct an instance $y = (M_{\Sigma,F}, x, t)$ where $x = (\mathsf{vk}, m, \tau_1, \ldots, \tau_n)$ and $t = |x|^{e_{\Sigma,F}}$, and the witness $w = (m_1, \sigma_1, \ldots, m_n, \sigma_n)$. Finally, run $\pi \leftarrow \mathsf{Prove}(\mathsf{crs}, y, w)$ and output $\bar{\sigma} = (proof, \pi)$.

$\mathsf{HomVer}(\mathsf{VK}, \mathcal{P} = (F, (\tau_1, \ldots \tau_n)), m, \bar{\sigma})$: Parse the signature $\bar{\sigma} = (flag, \cdot)$ and output the bit $b$ computed as follows:
If $\bar{\sigma} = (signature, (\tau, m, \sigma))$ and $\mathcal{P} = \mathcal{I} = (F_{id}, \tau)$ run $\mathsf{vfy}(\mathsf{vk}, \tau | m, \sigma) \to b$.
If $\bar{\sigma} = (proof, \pi)$ run $\mathsf{Ver}_{e_{\Sigma,F}}(\mathsf{crs}, y, \pi) \to b$ where $y = (M_{\Sigma,F}, x = (\mathsf{vk}, m, \tau_1, \ldots, \tau_n), |x|^{e_{\Sigma,F}})$.
Recall that in a SNARK for NP, $\mathsf{Ver}_c$ is given a constant $c > 0$ and only works for relation $\mathcal{R}_c$.

In what follows we show that the scheme above is a homomorphic signature. Correctness follows from the correctness of $\Sigma$ and $\Pi$, while succinctness is implied by that of $\Pi$. More precise arguments are given in the full version.

---

notice in using classical SNARKs, it is unclear to us whether such a multi-hop construction would allow for a proof.

**Security.** As in Sect. 4.2, for every signature scheme $\Sigma = (\mathsf{kg}, \mathsf{sign}, \mathsf{vfy})$ we denote by $\mathbb{O}_\Sigma$ the family of oracles $\mathcal{O}(m) = \mathsf{sign}(\mathsf{sk}, m)$ (where the verification key is returned as output of a special query $\mathcal{O}('vk')$). We show the security of the scheme $\mathsf{HomSig}[\Sigma, \Pi]$ via the following theorem. The proof is in the full version.

**Theorem 6.** *Let $\Sigma$ be a signature scheme. If $\Pi$ is an O-SNARK for $\mathbb{O}_\Sigma$, and $\Sigma$ is UF-CMA-secure, then $\mathsf{HomSig}[\Sigma, \Pi]$ is a secure homomorphic signature scheme.*

**Non-adaptive Security.** Alternatively, one can modify the previous proof to show that the scheme has security against homomorphic signature adversaries that make non-adaptive signing queries, assuming the weaker assumption that $\Pi$ is a *non-adaptive* O-SNARK (see Definition 7). In particular, combining this change with the result of Theorem 2 one obtains the following:

**Theorem 7.** *If $\Pi$ is a SNARK, and $\Sigma$ is a UF-CMA-secure signature scheme, then $\mathsf{HomSig}[\Sigma, \Pi]$ is secure against adversaries that make non-adaptive signing queries.*

*Remark 3 (On the applicability of Corollary 1).* We note that we cannot combine the positive result of Corollary 1 with Theorem 6 to conclude that the security of the homomorphic signature scheme holds under classical SNARKs. The inapplicability of Corollary 1 is due to its restriction for which adversaries have to query almost the entire message space. By looking at the HomSig construction (and the definition of homomorphic signatures too) one can note that an adversary who queries almost the entire message space of the underlying signature scheme can trivially break the security (for example he could obtain signatures on two distinct messages under the same label).

**Insecurity of HomSig.** In the full version we show that the generic homomorphic signature construction HomSig is not (adaptive) secure for an arbitrary choice of the signature scheme $\Sigma$. This insecurity result does not contradict our Theorem 6, but is closely related with (and confirms) our impossibility of O-SNARKs for any signing oracles (Theorem 3).

## 5.2   Succinct Functional Signatures

As second application of O-SNARKs we revisit the construction of succinct functional signatures of Boyle, Goldwasser, and Ivan [BGI14]. In [BGI14] this construction is proven secure using a notion of SNARKs which significantly differs from the standard one [BCC+14]. To the best of our knowledge, there are no known instantiations of SNARKs under this definition, in the standard model (and is not clear whether it is possible to find some). On the other hand, if one wants to prove the security of this construction using the classical SNARK definition, the security proof incurs the same subtleties related to running an extractor in the presence of a signing oracle.

In this section, we revisit the construction of [BGI14], and we prove its security using O-SNARKs. Interestingly, this proof differs a little from the one of homomorphic signature as here we have to consider O-SNARKs for *multiple* signing oracles.

**Definition 9 (Functional Signatures [BGI14]).** *A functional signature scheme* FS *for a message space* $\mathcal{M}$ *and function family* $\mathcal{F} = \{f : \mathcal{D}_f \to \mathcal{M}\}$ *is a tuple of probabilistic, polynomial-time algorithms* (FS.Setup, FS.KeyGen, FS.Sign, FS.Ver) *that work as follows*

FS.Setup($1^\lambda$) *takes a security parameter* $\lambda$ *and outputs a master verification key* mvk *and a master secret key* msk.

FS.KeyGen(msk, $f$) *takes the master secret key* msk *and a function* $f \in \mathcal{F}$ *(represented as a circuit) and it outputs a signing key* $\mathrm{sk}_f$ *for* $f$.

FS.Sign(mvk, $f$, $\mathrm{sk}_f$, $m$) *takes as input a function* $f \in \mathcal{F}$, *a signing key* $\mathrm{sk}_f$, *and a message* $m \in \mathcal{D}_f$, *and it outputs* $(f(m), \sigma)$.

FS.Ver(mvk, $m^*$, $\sigma$) *takes as input the master verification key* mvk, *a message* $m^* \in \mathcal{M}$ *and a signature* $\sigma$, *and outputs either* 1 *(accept) or* 0 *(reject)*.

*and satisfy* correctness, unforgeability, *and* function privacy *as described below*.

- **Correctness.** *A functional signature scheme is correct if the following holds with probability 1:*

$$\forall f \in \mathcal{F}, \ \forall m \in \mathcal{D}_f, \ (\mathrm{msk}, \mathrm{mvk}) \leftarrow \mathrm{FS.Setup}(1^\lambda), \ \mathrm{sk}_f \leftarrow \mathrm{FS.KeyGen}(\mathrm{msk}, f),$$

$$(m^*, \sigma) \leftarrow \mathrm{FS.Sign}(\mathrm{mvk}, f, \mathrm{sk}_f, m), \mathrm{FS.Ver}(\mathrm{mvk}, m^*, \sigma) = 1$$

- **Unforgeablity.** *A functional signature scheme is unforgeable if for every PPT adversary* $\mathcal{A}$ *there is a negligible function* $\epsilon$ *such that* $\Pr[\mathbf{Exp}_{\mathcal{A},\mathrm{FS}}^{\mathrm{FS-UF}}(\lambda) = 1] \leq \epsilon(\lambda)$ *where the experiment* $\mathbf{Exp}_{\mathcal{A},\mathrm{FS}}^{\mathrm{FS-UF}}(\lambda)$ *is described in the following:*

  Key generation: *Generate* (msk, mvk) $\leftarrow$ FS.Setup($1^\lambda$), *and gives* mvk *to* $\mathcal{A}$.

  Queries: *The adversary is allowed to adaptively query a key generation oracle* $\mathcal{O}_{\mathrm{key}}$ *and a signing oracle* $\mathcal{O}_{\mathrm{sign}}$, *that share a dictionary* $D$ *indexed by tuples* $(f, i) \in \mathcal{F} \times \mathbb{N}$, *whose entries are signing keys. For answering these queries, the challenger proceeds as follows:*

  - $\mathcal{O}_{\mathrm{key}}$ $(f, i)$:
    * *If* $(f, i) \in D$ *(i.e., the adversary had already queried the tuple* $(f, i)$*), then the challenger replies with the same key* $\mathrm{sk}_f^i$ *generated before.*
    * *Otherwise, generate a new* $\mathrm{sk}_f^i \leftarrow \mathrm{FS.KeyGen}(\mathrm{msk}, f)$, *add the entry* $(f, i) \to \mathrm{sk}_f^i$ *in* $D$, *and return* $\mathrm{sk}_f^i$.
  - $\mathcal{O}_{\mathrm{sign}}$ $(f, i, m)$:
    * *If there is an entry for the key* $(f, i)$ *in* $D$, *then the challenger generates a signature on* $f(m)$ *using this key, i.e.,* $\sigma \leftarrow \mathrm{FS.Sign}(\mathrm{mvk}, f, \mathrm{sk}_f^i, m)$.

* * Otherwise, generate a new key $\mathsf{sk}_f^i \leftarrow \mathsf{FS.KeyGen}(\mathsf{msk}, f)$, add an entry $(f, i) \rightarrow \mathsf{sk}_f^i$ to $D$, and generate a signature on $f(m)$ using this key, i.e., $\sigma \leftarrow \mathsf{FS.Sign}(\mathsf{mvk}, f, \mathsf{sk}_f^i, m)$.

Forgery: After the adversary is done with its queries, it outputs a pair $(m^*, \sigma)$, and the experiment outputs 1 iff the following conditions hold

* * $\mathsf{FS.Ver}(\mathsf{mvk}, m^*, \sigma) = 1$.
* * there does not exist $m$ such that $m^* = f(m)$ for any $f$ which was sent as a query to the $\mathcal{O}_{\mathsf{key}}$ oracle.
* * there does not exist a pair $(f, m)$ such that $(f, m)$ was a query to the $\mathcal{O}_{\mathsf{sign}}$ oracle and $m^* = f(m)$.

- **Function privacy.** Intuitively, function privacy requires that the distribution of signatures on a message $m$ that are generated via different keys $\mathsf{sk}_f$ should be computationally indistinguishable, even given the secret keys and master signing key. See [BGI14] or the full version for a more formal definition.

**Definition 10 (Succinct Functional Signatures).** A functional signature scheme is called succinct if there exists a polynomial $s(\cdot)$ such that, for every security parameter $\lambda \in \mathbb{N}$, $f \in \mathcal{F}$, $m \in \mathcal{D}_f$, it holds with probability 1 over $(\mathsf{mvk}, \mathsf{msk}) \leftarrow \mathsf{FS.Setup}(1^\lambda)$, $\mathsf{sk}_f \leftarrow \mathsf{FS.KeyGen}(\mathsf{msk}, f)$, $(f(m), \sigma) \leftarrow \mathsf{FS.Sign}(\mathsf{sk}_f, m)$ that $|\sigma| \leq s(\lambda, |f(m)|)$. In particular, the size of $\sigma$ is independent of the function's size, $|f|$, and the function's input size, $|m|$.

**Succinct Functional Signatures from O-SNARKs.** In the following we show a construction for message space $\mathcal{M}$ and family of functions $\mathcal{F} = \{f : \mathcal{D}_f \rightarrow \mathcal{M}\}$ whose running time is bounded by some fixed polynomial $t_\mathcal{F}(|m|)$. To build the scheme, we use two UF-CMA-secure signature schemes, $\Sigma_0 = (\mathsf{kg}_0, \mathsf{sign}_0, \mathsf{vfy}_0)$ for message space $\mathcal{M}_0$ and $\Sigma' = (\mathsf{kg}', \mathsf{sign}', \mathsf{vfy}')$ for message space $\mathcal{D}$, together with a fully succinct zero-knowledge O-SNARK $\Pi = (\mathsf{Gen}, \mathsf{Prove}, \mathsf{Ver})$ for the NP language $L$ defined below. While in [BGI14] a single signature scheme is used, we prefer to use two different ones as this allows for a more precise statement since we will need to apply different restrictions to $\mathcal{M}_0$ and $\mathcal{D}$ to obtain a precise proof.

DEFINING THE RELATION $\mathcal{R}_L$. Let $M_L$ be a random-access machine as defined below, and $t_L(k) = k^{e_L}$ be a polynomial. $\mathcal{R}_L$ is the binary relation consisting of all pairs $(y, w)$ such that, parsing $y = (M_L, x, t)$, $M_L(x, w)$ accepts in at most $t$ steps and $t \leq t_L(|x|)$. The values $x$ are of the form $x = (m^*, \mathsf{mvk}_0)$ where $\mathsf{mvk}_0$ is a public key of the scheme $\Sigma_0$, and $m^* \in \mathcal{M}$ is a message. The values $w$ are instead tuples $w = (m, f, \mathsf{vk}', \sigma_{\mathsf{vk}'}, \sigma_m)$ such that $m \in \mathcal{D}_f$ with $\mathcal{D}_f \subset \mathcal{D}$, and $\sigma_{\mathsf{vk}'}, \sigma_m$ are signatures for the schemes $\Sigma_0$ and $\Sigma'$ respectively. On input such a pair $(x, w)$, $M_L(x, w)$ is the random-access machine that accepts iff the following conditions (1), (2) and (3) hold:

(1) $m^* = f(m)$
(2) $\mathsf{vfy}'(\mathsf{vk}', m, \sigma_m) = 1$
(3) $\mathsf{vfy}_0(\mathsf{mvk}_0, f|\mathsf{vk}', \sigma_{\mathsf{vk}'}) = 1$

Given polynomial bounds on the running times of verification algorithms $\mathsf{vfy'}$ and $\mathsf{vfy_0}$, and a (fixed) bound $t_{\mathcal{F}}(\cdot)$ on the size and running time of every $f \in \mathcal{F}$, one can deduce a polynomial time bound $t_L(|x|) = |x|^{e_L}$ for the machine $M_L$.

THE CONSTRUCTION. Using the signature schemes $\Sigma_0, \Sigma'$ and a fully-succinct zero-knowledge O-SNARK $\Pi$ for NP, we construct the functional signature scheme $\mathsf{FS}[\Sigma_0, \Sigma', \Pi] = (\mathsf{FS.Setup}, \mathsf{FS.KeyGen}, \mathsf{FS.Sign}, \mathsf{FS.Ver})$ as follows:

$\mathsf{FS.Setup}(1^\lambda)$: This probabilistic algorithm takes a security parameter $\lambda$ and outputs a master verification key mvk and a master secret key msk:
Generate $(\mathsf{msk_0}, \mathsf{mvk_0}) \leftarrow \mathsf{kg_0}(1^\lambda)$, $\mathsf{crs} \leftarrow \mathsf{Gen}(1^\lambda)$. Set the master secret key $\mathsf{msk} = \mathsf{msk_0}$, and the master verification key $\mathsf{mvk} = (\mathsf{mvk_0}, \mathsf{crs})$.

$\mathsf{FS.KeyGen}(\mathsf{msk}, f)$: This algorithm takes the master secret key msk and a function $f \in \mathcal{F}$ (represented as a circuit) and it outputs a signing key $\mathsf{sk}_f$ for $f$.
Generate a new key pair $(\mathsf{sk'}, \mathsf{vk'}) \leftarrow \mathsf{kg'}(1^\lambda)$ for the scheme $\Sigma'$, compute $\sigma_{\mathsf{vk'}} \leftarrow \mathsf{sign_0}(\mathsf{msk_0}, f|\mathsf{vk'})$, and let the certificate $c$ be $c = (f, \mathsf{vk'}, \sigma_{\mathsf{vk'}})$. Finally output $\mathsf{sk}_f = (\mathsf{sk'}, c)$.

$\mathsf{FS.Sign}(\mathsf{mvk}, f, \mathsf{sk}_f, m)$: The algorithm takes as input a function $f \in \mathcal{F}$, a signing key $\mathsf{sk}_f$, and a message $m \in \mathcal{D}_f$, and it outputs $(f(m), \pi)$ where $\pi$ represents a signature on $f(m)$.
Parse $\mathsf{sk}_f$ as $(\mathsf{sk'}, c = (f, \mathsf{vk'}, \sigma_{\mathsf{vk'}}))$, generate $\sigma_m \leftarrow \mathsf{sign'}(\mathsf{sk'}, m)$, set $y = (M_L, x, t)$ with $x = (\mathsf{mvk_0}, f(m))$, $t = |x|^{e_L})$, and $w = (m, f, \mathsf{vk'}, \sigma_{\mathsf{vk'}}, \sigma_m)$. Run $\pi \leftarrow \mathsf{Prove}(\mathsf{crs}, y, w)$ and output $(m^* = f(m), \pi)$.

$\mathsf{FS.Ver}(\mathsf{mvk}, m^*, \pi)$: This algorithms takes as input the master verification key mvk, a message $m^* \in \mathcal{M}$ and a signature $\pi$, and outputs either 1 (accept) or 0 (reject):
Parse $\mathsf{mvk} = (\mathsf{mvk_0}, \mathsf{crs})$ and set $y = (M_L, x, t)$ with $x = (\mathsf{mvk_0}, m^*)$ and $t = |x|^{e_L}$. Then output the same bit returned by $\mathsf{Ver}_{e_L}(\mathsf{crs}, y, \pi)$.

**Correctness.** It is not hard to see that as long as $\Sigma_0, \Sigma'$ and $\Pi$ are correct, then FS is also correct.

**Succinctness.** Intuitively, a functional signature is succinct if the size of any signature depends only on the size of functions' outputs (and the security parameter). In the above construction this property immediately follows from the succinctness of $\Pi$.

**Unforgeability.** We prove the security of FS under the unforgeability of schemes $\Sigma_0$ and $\Sigma'$ and using the notion of O-SNARKs for a specific family of oracles $\mathbb{O}_{m\Sigma,Q}$ that we define below.

$\mathbb{O}_{m\Sigma,Q}$ is parametrized by the algorithms of the signature schemes $\Sigma_0, \Sigma'$ and by a polynomial $Q = Q(\lambda)$. Every member $\mathcal{O}$ of $\mathbb{O}_{m\Sigma,Q}$ is described by a set of secret keys $\mathsf{msk_0}, \mathsf{sk'_1}, \ldots, \mathsf{sk'_Q}$ (i.e., the process of sampling $\mathcal{O} \leftarrow \mathbb{O}$ consists of running $(\mathsf{mvk_0}, \mathsf{msk_0}) \overset{\$}{\leftarrow} \mathsf{kg_0}(1^\lambda)$ and $(\mathsf{vk'_i}, \mathsf{sk'_i}) \overset{\$}{\leftarrow} \mathsf{kg'_1}(1^\lambda), \forall i \in [Q])$. The oracle $\mathcal{O}$ works as follows:

$$\mathcal{O}(i, \text{'}vk'\text{'}) = \begin{cases} \mathsf{mvk_0} & \text{If } i = 0, \\ \mathsf{vk'_i} & \text{otherwise.} \end{cases} \qquad \mathcal{O}(i, \text{'}sk'\text{'}) = \begin{cases} \perp & \text{If } i = 0, \\ \mathsf{sk'_i} & \text{otherwise.} \end{cases}$$

$$\mathcal{O}(i,m) = \begin{cases} (\mathsf{Cnt}, \mathsf{sign}_0(\mathsf{msk}_0, m|\mathsf{vk}'_{\mathsf{Cnt}})), \mathsf{Cnt} \leftarrow \mathsf{Cnt}+1 & \text{If } i=0 \text{ and } \mathsf{Cnt} \leq Q, \\ \bot & \text{If } i=0 \text{ and } \mathsf{Cnt} > Q, \\ \mathsf{sign}'(\mathsf{sk}'_i, m) & \text{otherwise.} \end{cases}$$

For the sake of simplicity we compactly denote $\mathcal{O}_0(\cdot) = \mathcal{O}(0,\cdot)$ and $\mathcal{O}'_i(\cdot) = \mathcal{O}(i,\cdot)$ for all $i > 0$. From the above description, note that oracle $\mathcal{O}_0$ is stateful and we assume it starts with $\mathsf{Cnt} = 1$.

Finally, we point out that for some technical reasons that we mention in Remark 5 at the end of this section, it is not possible to use the notion of O-SNARK for a *single* signing oracle to prove the security of the functional signature scheme. This is the reason why we explicitly considered O-SNARKs for this more complex family of multiple signing oracles.

**Theorem 8.** *If $\Pi$ is an O-SNARK for $\mathbb{O}_{m\Sigma,Q}$ for every $Q = \mathsf{poly}(\lambda)$, and $\Sigma_0, \Sigma'$ are UF-CMA-secure, then $\mathsf{FS}[\Sigma_0, \Sigma', \Pi]$ is an unforgeable functional signature.*

Our proof consists of the following two steps:

1. We show that for every successful $\mathcal{A}_{\mathsf{FS}}$ against the unforgeability of FS there exists an O-SNARK adversary $\tilde{A}$ for an oracle from $\mathbb{O}_{m\Sigma,Q}$ such that $\tilde{A}$ outputs a valid proof with the same (non-negligible) probability of success of $\mathcal{A}_{\mathsf{FS}}$. By the adaptive proof of knowledge for $\mathbb{O}_{m\Sigma,Q}$ we then obtain that for such $\tilde{A}$ there exists a suitable extractor $\mathcal{E}_{\tilde{A}}$ that outputs a valid witness with all but negligible probability.
2. From the previous point, considering adversary $\tilde{A}$ and the corresponding extractor, we can partition adversary-extractor pairs in two types: (1) those that yield a witness $w$ containing a pair $(f, \mathsf{vk}')$ that was never signed before, and (2) those that yield $w$ containing $(f, \mathsf{vk}')$ that was signed before. We show that adversaries of type (1) can be used to break the security of the signature scheme $\Sigma_0$, whereas adversaries of type (2) can be used to break the security of $\Sigma'$.

For lack of space the complete proof appears in the full version, where we also show that the scheme has function privacy.

**Non-adaptive Unforgeability.** Similarly to the homomorphic signature case, it is possible to show that the functional signature scheme achieves security against (functional signature) adversaries that make *non-adaptive* signing queries (i.e., all queries are declared at the beginning of the game). This weaker security can be proven assuming that $\Pi$ is a *non-adaptive* O-SNARK (see Definition 7). Combining this change with the result of Theorem 2 we obtain the following:

**Theorem 9.** *If $\Pi$ is a SNARK and $\Sigma_0, \Sigma'$ are UF-CMA-secure signature schemes, then $\mathsf{FS}[\Sigma_0, \Sigma', \Pi]$ is a functional signature where unforgeability holds against adversaries that make non-adaptive signing queries.*

*Remark 4 (On the applicability of Corollary 1).* For the same reasons discussed in Remark 3, it is not possible to apply the result of Corollary 1 to conclude the that the (adaptive) security of the functional signature scheme holds under classical SNARKs.

*Remark 5 (On the use of multiple signing oracles).* In order to prove the security of the functional signature scheme, one might be tempted to use the notion of O-SNARK with a single signing oracle. Precisely, one might use O-SNARKs for $\mathbb{O}_{\Sigma_0}$ when making a reduction to $\Sigma_0$ and O-SNARKs for $\mathbb{O}_{\Sigma'}$ when making a reduction to $\Sigma'$. Unfortunately, this approach does not work for an intricate technical reason that we explain here. Intuitively, assume that one wants to build an O-SNARK adversary $\tilde{\mathcal{A}}$ that has access to a single signing oracle, say from $\mathbb{O}_{\Sigma_0}$. Then the secret keys needed to simulate all the other oracles have to be given to $\tilde{\mathcal{A}}$ as part of its auxiliary input ($\tilde{\mathcal{A}}$ needs them to simulate $\mathcal{A}_{\mathsf{FS}}$). At this point the issue is that such secret keys in fact give an efficient way to compute a witness for several $y$ in the relation $\mathcal{R}_L$. Therefore, if the extractor gets these secret keys as auxiliary information, we then have no guarantee that, while doing a reduction to the unforgeability of the signature scheme, the extractor will output a witness of the form we expect.

## 5.3   SNARKs on Authenticated Data

As another application of O-SNARKs we consider the generic construction of SNARKs on authenticated data that is given in [BBFR15]. Since this construction is very similar to the homomorphic signature scheme that we present in Sect. 5.1, we only provide an informal discussion of this application. In [BBFR15] Backes et al. introduce the notion of SNARKs on authenticated data to capture in an explicit way the possibility of performing (zero-knowledge) proofs about statements that are authenticated by third parties, i.e., to prove that $(x, w) \in \mathcal{R}$ for some $x$ for which there is a valid signature. While the main focus of that work is on a concrete construction based on quadratic arithmetic programs, the authors also show a generic construction based on SNARKs and digital signatures. Roughly speaking, this construction consists in letting the prover use a SNARK to prove a statement of the form "$\exists x, w, \sigma : (x, w) \in \mathcal{R} \wedge \mathsf{vfy}(\mathsf{vk}, \tau | x, \sigma) = 1$", for some public label $\tau$ of the statement. The formalization of their model is rather similar to that of homomorphic signatures in this paper (e.g., they also use labels). Noticeable differences are that their construction uses pre-processing SNARKs for arithmetic circuit satisfiability, and that to handle several functions they use different SNARK instantiations (one per function).

In [BBFR15] the security proof of this generic construction is only sketched, and in particular they use the existence of an extractor for an adversary that interacts with a signing oracle without providing a particular justification on its existence. With a more careful look, it is possible to see that this security proof incurs the same issue of extraction in the presence of oracles. Using the same techniques that we developed in this paper for the homomorphic signature scheme,[12] it is possible to prove the security of that generic construction using O-SNARKs for signing oracles (or non-adaptive security based on classical SNARKs). In conclusion, for this construction one can either conjecture that a

---

[12] The only major difference is that one has to consider a specification of our definitions to the case of pre-processing SNARKs.

specific SNARK scheme (e.g., [PHGR13]) is secure in the presence of oracles, or, more conservatively, argue only the non-adaptive security of the primitive under the existence of classical SNARKs.

**Acknowledgements.** We would like to thank Manuel Barbosa and Bogdan Warinschi for valuable discussions on this work, and the anonymous reviewers of Crypto 2016 and TCC 2016-B for their useful comments and suggestions. This work was partially supported by the European Union's Horizon 2020 Research and Innovation Programme under grant agreement 688722 (NEXTLEAP), the Spanish Ministry of Economy under project reference TIN2015-70713-R (DEDETIS) and a Juan de la Cierva fellowship to Dario Fiore, by the Madrid Regional Government under project N-Greens (ref. S2013/ICE-2731), and by the European Research Council under the European Community's Seventh Framework Programme (FP7/2007-2013 Grant Agreement no. 339563 CryptoCloud).

# References

[BBFR15]  Backes, M., Barbosa, M., Fiore, D., Reischuk, R.M.: ADSNARK: nearly practical and privacy-preserving proofs on authenticated data. In: 2015 IEEE Symposium on Security and Privacy, pp. 271–286. IEEE Computer Society Press (2015)

[BCC88]  Brassard, G., Chaum, D., Crépeau, C.: Minimum disclosure proofs of knowledge. J. Comput. Syst. Sci. **37**(2), 156–189 (1988)

[BCC+14]  Bitansky, N., Canetti, R., Chiesa, A., Goldwasser, S., Lin, H., Rubinstein, A., Tromer, E.: The hunting of the SNARK. Cryptology ePrint Archive, Report 2014/580 (2014). http://eprint.iacr.org/2014/580

[BCCT12]  Bitansky, N., Canetti, R., Chiesa, A., Tromer, E.: From extractable collision resistance to succinct non-interactive arguments of knowledge, and back again. In: Goldwasser, S. (ed.) ITCS 2012, pp. 326–349. ACM, January 2012

[BCCT13]  Bitansky, N., Canetti, R., Chiesa, A., Tromer, E.: Recursive composition and bootstrapping for SNARKS and proof-carrying data. In: Boneh, D., Roughgarden, T., Feigenbaum, J. (eds.) 45th ACM STOC, pp. 111–120. ACM Press, June 2013

[BCI+13]  Bitansky, N., Chiesa, A., Ishai, Y., Paneth, O., Ostrovsky, R.: Succinct non-interactive arguments via linear interactive proofs. In: Sahai, A. (ed.) TCC 2013. LNCS, vol. 7785, pp. 315–333. Springer, Heidelberg (2013). doi:10.1007/978-3-642-36594-2_18

[BCPR14]  Bitansky, N., Canetti, R., Paneth, O., Rosen, A.: On the existence of extractable one-way functions. In: Shmoys, D.B. (ed.) 46th ACM STOC, pp. 505–514. ACM Press, May/June 2014

[BCTV14]  Ben-Sasson, E., Chiesa, A., Tromer, E., Virza, M.: Scalable zero knowledge via cycles of elliptic curves. In: Garay, J.A., Gennaro, R. (eds.) CRYPTO 2014. LNCS, vol. 8617, pp. 276–294. Springer, Heidelberg (2014). doi:10.1007/978-3-662-44381-1_16

[BF11]  Boneh, D., Freeman, D.M.: Homomorphic signatures for polynomial functions. In: Paterson, K.G. (ed.) EUROCRYPT 2011. LNCS, vol. 6632, pp. 149–168. Springer, Heidelberg (2011). doi:10.1007/978-3-642-20465-4_10

[BG08]    Barak, B., Goldreich, O.: Universal arguments and their applications. SIAM J. Comput. **38**(5), 1661–1694 (2008)

[BGI14]   Boyle, E., Goldwasser, S., Ivan, I.: Functional signatures and pseudorandom functions. In: Krawczyk, H. (ed.) PKC 2014. LNCS, vol. 8383, pp. 501–519. Springer, Heidelberg (2014). doi:10.1007/978-3-642-54631-0_29

[BHZ87]   Boppana, R.B., Hastad, J., Zachos, S.: Does co-NP have short interactive proofs? Inf. Process. Lett. **25**(2), 127–132 (1987)

[BP15]    Boyle, E., Pass, R.: Limits of extractability assumptions with distributional auxiliary input. In: Iwata, T., Cheon, J.H. (eds.) ASIACRYPT 2015. LNCS, vol. 9453, pp. 236–261. Springer, Heidelberg (2015). doi:10.1007/978-3-662-48800-3_10

[BSCG+13] Ben-Sasson, E., Chiesa, A., Genkin, D., Tromer, E., Virza, M.: SNARKs for C: verifying program executions succinctly and in zero knowledge. In: Canetti, R., Garay, J.A. (eds.) CRYPTO 2013. LNCS, vol. 8043, pp. 90–108. Springer, Heidelberg (2013). doi:10.1007/978-3-642-40084-1_6

[CF13]    Catalano, D., Fiore, D.: Practical homomorphic MACs for arithmetic circuits. In: Johansson, T., Nguyen, P.Q. (eds.) EUROCRYPT 2013. LNCS, vol. 7881, pp. 336–352. Springer, Heidelberg (2013). doi:10.1007/978-3-642-38348-9_21

[CFW14]   Catalano, D., Fiore, D., Warinschi, B.: Homomorphic signatures with efficient verification for polynomial functions. In: Garay, J.A., Gennaro, R. (eds.) CRYPTO 2014. LNCS, vol. 8616, pp. 371–389. Springer, Heidelberg (2014). doi:10.1007/978-3-662-44371-2_21

[CL08]    Crescenzo, G., Lipmaa, H.: Succinct NP proofs from an extractability assumption. In: Beckmann, A., Dimitracopoulos, C., Löwe, B. (eds.) CiE 2008. LNCS, vol. 5028, pp. 175–185. Springer, Heidelberg (2008). doi:10.1007/978-3-540-69407-6_21

[DLFKP16] Delignat-Lavaud, A., Fournet, C., Kohlweiss, M., Parno, B.: Cinderella: Turning shabby x. 509 certificates into elegant anonymous credentials with the magic of verifiable computation. In: IEEE Symposium on Security and Privacy (2016)

[FN16]    Fiore, D., Nitulescu, A.: On the (in)security of SNARKs in the presence of oracles. Cryptology ePrint Archive, Report 2016/112 (2016)

[GGPR13]  Gennaro, R., Gentry, C., Parno, B., Raykova, M.: Quadratic span programs and succinct NIZKs without PCPs. In: Johansson, T., Nguyen, P.Q. (eds.) EUROCRYPT 2013. LNCS, vol. 7881, pp. 626–645. Springer, Heidelberg (2013). doi:10.1007/978-3-642-38348-9_37

[GH98]    Goldreich, O., Håstad, J.: On the complexity of interactive proofs with bounded communication. Inf. Process. Lett. **67**(4), 205–214 (1998)

[GMR89]   Goldwasser, S., Micali, S., Rackoff, C.: The knowledge complexity of interactive proof systems. SIAM J. Comput. **18**(1), 186–208 (1989)

[Gro10]   Groth, J.: Short pairing-based non-interactive zero-knowledge arguments. In: Abe, M. (ed.) ASIACRYPT 2010. LNCS, vol. 6477, pp. 321–340. Springer, Heidelberg (2010). doi:10.1007/978-3-642-17373-8_19

[GVW02]   Goldreich, O., Vadhan, S., Wigderson, A.: On interactive proofs with a laconic prover. Comput. Complex. **11**(1–2), 1–53 (2002)

[GVW15]   Gorbunov, S., Vaikuntanathan, V., Wichs, D.: Leveled fully homomorphic signatures from standard lattices. In: Servedio, R.A., Rubinfeld, R. (eds.) 47th ACM STOC, pp. 469–477. ACM Press, June 2015

[GW11]  Gentry, C., Wichs, D.: Separating succinct non-interactive arguments from all falsifiable assumptions. In Fortnow, L.P. Vadhan, S. (eds.) 43rd ACM STOC, pp. 99–108. ACM Press, June 2011

[GW13]  Gennaro, R., Wichs, D.: Fully homomorphic message authenticators. In: Sako, K., Sarkar, P. (eds.) ASIACRYPT 2013. LNCS, vol. 8270, pp. 301–320. Springer, Heidelberg (2013). doi:10.1007/978-3-642-42045-0_16

[HT98]  Hada, S., Tanaka, T.: On the existence of 3-round zero-knowledge protocols. In: Krawczyk, H. (ed.) CRYPTO 1998. LNCS, vol. 1462, pp. 408–423. Springer, Heidelberg (1998). doi:10.1007/BFb0055744

[Kil92]  Kilian, J.: A note on efficient zero-knowledge proofs and arguments (extended abstract). In 24th ACM STOC, pp. 723–732. ACM Press, May 1992

[Lam79]  Lamport, L.: Constructing digital signatures from a one-way function. Technical report SRI-CSL-98, SRI International Computer Science Laboratory, October 1979

[Lip12]  Lipmaa, H.: Progression-free sets and sublinear pairing-based non-interactive zero-knowledge arguments. In: Cramer, R. (ed.) TCC 2012. LNCS, vol. 7194, pp. 169–189. Springer, Heidelberg (2012). doi:10.1007/978-3-642-28914-9_10

[Mic94]  Micali, S.: CS proofs (extended abstracts). In: 35th FOCS, pp. 436–453. IEEE Computer Society Press, November 1994

[Mic00]  Micali, S.: Computationally sound proofs. SIAM J. Comput. **30**(4), 1253–1298 (2000)

[Mie08]  Mie, T.: Polylogarithmic two-round argument systems. J. Math. Crypt. **2**(4), 343–363 (2008)

[Nao03]  Naor, M.: On cryptographic assumptions and challenges. In: Boneh, D. (ed.) CRYPTO 2003. LNCS, vol. 2729, pp. 96–109. Springer, Heidelberg (2003). doi:10.1007/978-3-540-45146-4_6

[NT16]  Naveh, A., Tromer, E.: Photoproof: cryptographic image authentication for any set of permissible transformations. In: IEEE Symposium on Security and Privacy (2016)

[PHGR13]  Parno, B., Howell, J., Gentry, C., Raykova, M.: Pinocchio: nearly practical verifiable computation. In: 2013 IEEE Symposium on Security and Privacy, pp. 238–252. IEEE Computer Society Press, May 2013

[Rom90]  Rompel, J.: One-way functions are necessary and sufficient for secure signatures. In 22nd ACM STOC, pp. 387–394. ACM Press, May 1990

[Val08]  Valiant, P.: Incrementally verifiable computation or proofs of knowledge imply time/space efficiency. In: Canetti, R. (ed.) TCC 2008. LNCS, vol. 4948, pp. 1–18. Springer, Heidelberg (2008). doi:10.1007/978-3-540-78524-8_1

[Wee05]  Wee, H.: On round-efficient argument systems. In: Caires, L., Italiano, G.F., Monteiro, L., Palamidessi, C., Yung, M. (eds.) ICALP 2005. LNCS, vol. 3580, pp. 140–152. Springer, Heidelberg (2005). doi:10.1007/11523468_12

# Leakage Resilient One-Way Functions: The Auxiliary-Input Setting

Ilan Komargodski[✉]

Weizmann Institute of Science, Rehovot, Israel
ilan.komargodski@weizmann.ac.il

**Abstract.** Most cryptographic schemes are designed in a model where perfect secrecy of the secret key is assumed. In most physical implementations, however, some form of information leakage is inherent and unavoidable. To deal with this, a flurry of works showed how to construct basic cryptographic primitives that are resilient to various forms of leakage.

Dodis et al. (FOCS '10) formalized and constructed leakage resilient one-way functions. These are one-way functions $f$ such that given a random image $f(x)$ and leakage $g(x)$ it is still hard to invert $f(x)$. Based on any one-way function, Dodis et al. constructed such a one-way function that is leakage resilient assuming that an attacker can leak any lossy function $g$ of the input.

In this work we consider the problem of constructing leakage resilient one-way functions that are secure with respect to *arbitrary computationally hiding* leakage (a.k.a auxiliary-input). We consider both types of leakage — selective and adaptive — and prove various possibility and impossibility results.

On the negative side, we show that if the leakage is an adaptively-chosen arbitrary one-way function, then it is *impossible* to construct leakage resilient one-way functions. The latter is proved both in the random oracle model (without any further assumptions) and in the standard model based on a strong vector-variant of DDH. On the positive side, we observe that when the leakage is chosen ahead of time, there are leakage resilient one-way functions based on a variety of assumption.

## 1 Introduction

The holy grail of cryptography is designing systems that remain secure in the presence of adversarial behavior. For this, one has to specify (1) a cryptographic primitive of interest (e.g. an encryption scheme or a signature scheme), and (2) a model that captures the power of a potential adversary and what it means for it to break the system.

I. Komargodski—Supported in part by a Levzion fellowship, by grants from the Israel Science Foundation grant no. 1255/12, BSF and from the I-CORE Program of the Planning and Budgeting Committee and the Israel Science Foundation (grant no. 4/11).

© International Association for Cryptologic Research 2016
M. Hirt and A. Smith (Eds.): TCC 2016-B, Part I, LNCS 9985, pp. 139–158, 2016.
DOI: 10.1007/978-3-662-53641-4_6

One of the most common assumptions is that secret keys are perfectly secret and are completely unknown to an adversary. However, in many physical implementations some information does leak due to various side-channel attacks, reuse of randomness, and more.

This deficiency raised the necessity to build a theory of security against classes of side-channel attacks. Starting with the works of [14,27,30], a flurry of works in which different classes of side channel attacks have been defined and different cryptographic primitives have been designed to provably withstand these attacks (see, for example, [1,2,6–9,14–16,18,20–24,27,30,31,34]).

We consider the problem of constructing the most basic cryptographic primitive, a one-way function, in a setting where an adversary obtains side-channel information (this notion was first formalized by [3,17]). A one-way function $f$ is an efficiently computable function such that given $f(x)$ for a random input $x$, any efficient adversary cannot find an $x'$ such that $f(x) = f(x')$. A leakage resilient one-way function $f$ is a one-way function such that given $f(x)$ as above and $g(x)$, where $g$ is adversarially chosen, it is still hard to invert $f$ and recover such an $x'$.

To obtain some sort of security, one clearly has to restrict the adversary to choose $g$ from some collection of functions that do not trivially reveal $x$ by themselves. Indeed, if $g$ is the identity function, no leakage resilient function $f$ exists. Thus, several assumptions on the power of the adversary have been considered. Already in the work of Canetti et al. [14], the authors showed how to obtain a leakage-resilient one-way function assuming that the attacker can leak an arbitrary but sufficiently small subset of the bits of the input. However, this may be overly restrictive as it provides no guarantees if the attacker can learn the XOR of all the input bits. This issue was addressed in several works (see, for example, [3,15,17]) showing that there exists a leakage-resilient one-way function assuming that the attacker can leak any lossy function of the input, namely, any function whose image size is significantly smaller than the domain size. The leakage-resilience in both settings is proven based on the existence of any one-way function which is the weakest assumption possible. For completeness, we provide a proof of the following theorem in Appendix A.

**Theorem 1 ([3,17], Informal).** *Assuming that one-way functions exist, there exists a one-way function $f$, such that for any adversarially-chosen lossy function $g$, given $f(x)$ and $g(x)$ for a random $x$, it is computationally hard to invert $f$.*

Motivated by the positive results for a wide class of leakage functions, we study the question of designing leakage-resilient one-way functions that are secure with respect to *arbitrary computationally hiding* leakage function. We model this by allowing the leakage to be an arbitrary one-way function, even such that fully determine the input.[1] We consider both an *adaptive* notion of security in which the leakage function is adversarially chosen (from a restricted

---

[1] This setting is sometimes referred to as the *auxiliary-input* setting (see, for example, [16,18,26]).

pre-defined collection) after $f$ is fixed, and a selective notion in which the leakage is chosen ahead of time, before $f$ is.

## 1.1 Our Contributions

**Adaptively-chosen leakage.** We show that if the leakage can be an arbitrary one-way function, then there cannot be a leakage resilient one-way function $f$. More precisely, we show that for every one-way function $f$, there exists a one-way function $g$ (that depends on $f$) such that when one gets both $f(x)$ and $g(x)$, it is easy to invert $f$.

We prove this result in two ways: in the random oracle model and in the standard model based on a strong vector-variant of DDH. Specifically, we first show that if the leakage function has access to a random oracle O, then we can construct an oracle-aided function $g^O$ which is one-way and $g^O(x)$ together with $f(x)$ allow to recover $x$. For the result in the standard model, we rely on multi-bit point obfuscators that exist based on a strong vector-variant of the DDH assumption [5,13]; see Sect. 2.3 and Theorem 8.

**Theorem 2 (Informal).** *Let* O *be a random oracle. For every one-way function $f$, there is a one-way function $g^O$ such that for every $x$ given $f(x)$ and $g(x)$ it is easy to recover $x$.*

**Theorem 3 (Informal).** *Assuming multi-bit point obfuscators, for every one-way function $f$, there is a one-way function $g$ such that for every $x$ given $f(x)$ and $g(x)$ it is easy to recover $x$.*

*Moreover, such multi-bit point obfuscators can be constructed from a strong vector-variant of the DDH assumption.*

**Selectively-chosen leakage.** We show that if the leakage function $g$ is fixed ahead of time, then there exists a leakage resilient one-way function $f$ for $g$ from various assumptions. To this end, we observe that one-wayness with respect to selectively-chosen leakage is tightly related to extracting polynomially-many hard-core bits.

**Theorem 4 (Informal).** *For every leakage one-way function $g$, a hardcore function for $g$ that outputs polynomially-many hard-core bits is a leakage-resilient one-way function for $g$.*

If $g$ is a sub-exponentially hard one-way function, then extracting polynomially-many hard-core bits is possible due to Goldreich and Levin [25] (and any pseudorandom generator). Bellare, Stepanovs, and Tessaro [4] (see also the follow-up work of Brzuska and Mittelbach [11]) were the first to show how to extract *any* polynomial number of hard-core bits from *any* one-way function. Their construction is based on obfuscation. More recently, Zhandry [36] obtained the same result based on exponentially-hard DDH.

Thus, instantiating Theorem 4 with the variety of known methods for extracting polynomially-many hard-core bits from $g$, we obtain a leakage-resilient one-way function for $g$, whose security is based either on one-way functions, on obfuscation, on exponential hardness of DDH, and more.

## 1.2  Overview of Our Techniques

In Theorem 2 the underlying idea is very simple. We assume a random oracle O and assume that there exists a leakage resilient one-way function $f$, where the leakage is any one-way function. We define a leakage function $g(x) = \mathsf{O}(f(x)) \oplus x$. Recovering $x$ given $f(x)$ and $g(x)$ is easy by first applying O to $f(x)$ and then XORing the result with $g(x)$. The non-trivial part is showing that this function $g$ is also one-way.

Roughly speaking, our analysis uses the fact that any adversary trying to invert $g(x)$ will have to query the oracle at the point $f(x)$. Otherwise, all it sees are uniform strings from which it cannot infer anything about a possible pre-image. It is left to show that $f(x)$ is sufficiently random so that it cannot be guessed by any polynomial-time adversary with non-negligible probability. Indeed, since $f$ by itself is a one-way function, its image distribution has super-logarithmic min-entropy which satisfied our requirement.

For Theorem 3, our construction is based on multi-bit point obfuscators MBPO and can be seen as an instantiation of the above idea in the standard model. The leakage function, on input $x$, will output a multi-bit point obfuscation of the multi-bit point function that maps $f(x)$ to $x$, denoted by $g(x) = \mathsf{MBPO}(I_{f(x) \to x})$. One obstacle is that an obfuscator is a *probabilistic* procedure, and thus cannot be used directly in our setting. Hence, we use *public-coin* multi-bit point obfuscators, which are obfuscators that output their internal random coins. This allows us to define a leakage function which has hard-wired random coins for the use of the point obfuscator. Specifically, we hardwire into $g$ random coins $r$ and define $g_r(x) = \mathsf{MBPO}(I_{f(x) \to x}; r)$. We show that $g_r$, with very high probability, is a one-way function using the security of the obfuscator.[2]

We observe that such a multi-bit point obfuscator exists based on the strong vector-variant of DDH of Bitansky and Canetti [5] given in Sect. 2.3.[3]

## 2  Preliminaries

In this section we present the notation and basic definitions that are used in this work. For an integer $n \in \mathbb{N}$ we denote by $[n]$ the set $\{1, \ldots, n\}$. For a distribution $X$ we denote by $x \leftarrow X$ the process of sampling a value $x$ from the distribution $X$. Similarly, for a set $\mathcal{X}$ we denote by $x \leftarrow \mathcal{X}$ the process of sampling a value $x$ from the uniform distribution over $\mathcal{X}$. For a randomized function $f$ and an input $x \in \mathcal{X}$, we denote by $y \leftarrow f(x)$ the process of sampling a value $y$ from the distribution $f(x)$. A function $\mathsf{neg} : \mathbb{N} \to \mathbb{R}$ is *negligible* if for every constant $c > 0$ there exists an integer $N_c$ such that $\mathsf{neg}(\lambda) < \lambda^{-c}$ for all $\lambda > N_c$. For two strings $x \in \{0,1\}^n$ and $y \in \{0,1\}^m$ we denote by $x\|y$ the string concatenation of $x$ and $y$.

---

[2] Theorem 2 can also be proved by first showing how to use a random oracle to construct a multi-bit point obfuscator. We thank a reviewer for pointing this out.

[3] We emphasize we do not require security with respect to auxiliary-input, which was shown to be a problematic assumption [10].

Two sequences of random variables $X = \{X_\lambda\}_{\lambda \in \mathbb{N}}$ and $Y = \{Y_\lambda\}_{\lambda \in \mathbb{N}}$ are *computationally indistinguishable* if for any probabilistic polynomial-time algorithm $\mathcal{A}$ there exists a negligible function $\mathsf{neg}(\cdot)$ such that $|\mathbf{Pr}[\mathcal{A}(1^\lambda, X_\lambda) = 1] - \mathbf{Pr}[\mathcal{A}(1^\lambda, Y_\lambda) = 1]| \leq \mathsf{neg}(\lambda)$ for all sufficiently large $\lambda \in \mathbb{N}$.

## 2.1  Min-Entropy

The min-entropy of a distribution $X$ over $\{0,1\}^n$ is defined by

$$\mathsf{H}_\infty(X) = - \min_{x \in \{0,1\}^n} \log_2 \mathbf{Pr}[X = x].$$

## 2.2  One-Way Functions

**Definition 1 (One-way functions).** *A function $f\colon \{0,1\}^* \to \{0,1\}^*$ is said to be one-way if the following two conditions hold:*

1. *There exists a polynomial-time algorithm $A$ such that $A(x) = f(x)$ for every $x \in \{0,1\}^*$.*
2. *For every probabilistic polynomial-time algorithm $B$ there exists a negligible function $\mathsf{neg}(\cdot)$ such that*

$$\mathsf{ADV}_{f,B}^{\mathsf{OWF}} = \mathbf{Pr}[B(1^n, f(x)) \in f^{-1}(f(x))] \leq \mathsf{neg}(n),$$

*where the probability is taken uniformly over all possible $x \in \{0,1\}^n$ and the internal randomness of $B$.*

The following claim will be useful.

**Claim 5.** *Let $f\colon \{0,1\}^n \to \{0,1\}^m$ be a one-way function where $m = m(n)$ is a polynomial. It holds that $\mathsf{H}_\infty(f(X)) \geq \omega(\log n)$.*

*Proof.* Since $f$ is a one-way function, it must be that on a random $x \in \{0,1\}^n$, it is hard to a preimage for $f(x)$. Assume, towards contradiction, that $\mathsf{H}_\infty(f(X)) = O(\log n)$. That is,

$$\min_{y \in \{0,1\}^m} \log_2 \frac{1}{\mathbf{Pr}_{x \in \{0,1\}^n}[f(x) = y]} = O(\log n).$$

Thus, there exists a $y^* \in \{0,1\}^m$ for which $\mathbf{Pr}_{x \in \{0,1\}^n}[f(x) = y^*] \geq 1/p(n)$ for some polynomial $p(\cdot)$.

Define an adversary $\mathcal{A}$ that given a random image $y = f(x)$ outputs a uniformly random $x'$. This adversary wins if both $f(x) = y^*$ and $f(x') = y^*$. Since $x$ and $x'$ are chosen independently and uniformly at random, we have that

$$\mathbf{Pr}_{x' \in \{0,1\}^n}[\mathcal{A}(f(x')) \in f^{-1}(y)] \geq \mathbf{Pr}_{x,x' \in \{0,1\}^n}[f(x) = y^* \text{ and } f(x') = y^*]$$

$$= (\mathbf{Pr}_{x \in \{0,1\}^n}[f(x) = y^*])^2 \geq 1/(p(n))^2.$$

That is, $\mathcal{A}$ will successfully invert $y$ with non-negligible probability, contradiction the one-wayness of $f$.

We extend the definition of a one-way function to *oracle-aided* one-way functions. Roughly speaking, an oracle-aided function $f^O$ is an oracle-aided one-way function if there is an oracle-aided efficient algorithm that computes $f^O$ on every point, and given an image of $f^O$ on a random preimage, any efficient algorithm (that has oracle access to $O$) cannot find the preimage.

**Definition 2 (Oracle aided one-way function).** *Let $O$ be an oracle. A function $f^O$ that has oracle access to $O$ is said to be oracle aided one-way if the following two conditions hold:*

1. *There exists an oracle-aided polynomial-time algorithm $A^O$ such that $A^O(x) = f^O(x)$ for every $x \in \{0,1\}^*$.*
2. *For every oracle-aided probabilistic polynomial-time algorithm $B^O$ and $n \in \mathbb{N}$,*

$$\mathsf{ADV}^{\mathsf{OOWF}}_{f,B} = \mathbf{Pr}[B^O(1^n, f^O(x)) \in (f^O)^{-1}(f^O(x))] < \mathsf{neg}(n),$$

*where the probability is taken uniformly over all possible $x \in \{0,1\}^n$ and the internal randomness of $B$.*

### 2.3    Point Obfuscations

A point function $I_x \colon \{0,1\}^n \to \{0,1\}$ returns 1 on input $x \in \{0,1\}^n$ and 0 on all other inputs. A point obfuscator is an obfuscator that gets a point function $I_x$ as input (in some canonical form in which $x$ is explicit) and outputs a circuit with the same functionality but where $x$ is computationally hidden.

**Definition 3 (Point obfuscator).** *A point obfuscator $\mathsf{PO}(\cdot)$ is a probabilistic polynomial-time algorithm that gets as input a point function $I_x$, where $x \in \{0,1\}^n$, and outputs a circuit $C$ such that*

1. *For all $x$, the circuit $C \leftarrow \mathsf{PO}(I_x)$ is functionally equivalent to $I_x$.*
2. *For any probabilistic polynomial-time algorithm $\mathcal{A}$, there is an probabilistic polynomial-time simulator $S$ and a negligible function $\mathsf{neg}(\cdot)$, such that for all $x \in \{0,1\}^n$ and $n \in \mathbb{N}$,*

$$\mathsf{ADV}^{\mathsf{PO}}_{\mathcal{A},\mathcal{D}} = |\mathop{\mathbf{Pr}}_{\mathcal{A},\mathsf{PO}}[\mathcal{A}(\mathsf{PO}(I_x)) = 1] - \mathop{\mathbf{Pr}}_{S}[S^{I_x}(1^n)) = 1]| \le \mathsf{neg}(n).$$

*Moreover, a point obfuscator is called public coin if it publishes all internal coin tosses as part of its output.*

In [12], Canetti provided a construction that satisfies Definition 3 assuming a strong variant of the DDH assumption. The construction of Canetti is given next.

**Construction 6 ([12]'s point obfuscator).** *Let $\mathcal{G} = \{\mathbb{G}_n\}_{n \in \mathbb{N}}$ be a group ensemble with uniform and efficient representation and operations, where each $\mathbb{G}_n$ is a group of prime order $p_n \in (2^{n-1}, 2^n)$. The public coin point obfuscator $\mathsf{PO}$ for points in the domain $\mathbb{Z}_{p_n}$ is defined as follows: $\mathsf{PO}(I_x)$ samples a random generator $r \leftarrow \mathbb{G}_n^*$ of $\mathbb{G}_n$ and outputs $r, r^x$. Evaluation of the obfuscation at point $z$ is done by checking whether $r^x = r^z$.*

A multi-bit point function $I_{x \to y} \colon \{0,1\}^n \to \{0,1\}^m$ is a function that returns $y \in \{0,1\}^m$ on input $x \in \{0,1\}^n$ and $\bot$ on all other inputs. A multi-bit point function obfuscator, given a multi-bit point function in some canonical form in which $x$ and $y$ are explicit, outputs a circuit with the same functionality but where $x$ and $y$ are computationally hidden.

**Definition 4 (Multi-bit point obfuscator).** *A multi-bit point obfuscator* MBPO *is a probabilistic polynomial-time algorithm that gets as input a multi-bit point function $I_{x \to y}$, where $x \in \{0,1\}^n$ and $y \in \{0,1\}^m$, and outputs a circuit $C$ such that*

1. *For all $x \in \{0,1\}^n$ and $y \in \{0,1\}^m$, the circuit $C \leftarrow \mathsf{MBPO}(I_{x \to y})$ is functionally equivalent to the function $I_{x \to y}$.*
2. *For any probabilistic polynomial-time algorithm $\mathcal{A}$, there is a probabilistic polynomial-time simulator $S$ and a negligible function $\mathsf{neg}(\cdot),$[4] such that for all $n \in \mathbb{N}$, $x \in \{0,1\}^n$, and $y \in \{0,1\}^m$ and*

$$\mathsf{ADV}^{\mathsf{MBPO}}_{\mathcal{A},\mathcal{D}} = \big| \Pr_{\mathcal{A},\mathsf{MBPO}}[\mathcal{A}(\mathsf{MBPO}(I_{x \to y})) = 1] - \Pr_{S}[S^{I_{x \to y}}(1^{n+m}) = 1] \big|$$

$$\leq \mathsf{neg}(n).$$

*Moreover, a multi-bit point obfuscator is called **public coin** if it publishes all internal coin tosses as part of its output.*

One way to obtain a multi-bit point obfuscator was suggested by Canetti and Dakdouk [13]. Specifically, they showed that a *composable* point obfuscator gives rise to a multi-bit point obfuscator.

**Definition 5 (Composable point obfuscator).** *A point obfuscator $\mathsf{PO}(\cdot)$ is said to be $t$-composable if for any probabilistic polynomial-time algorithm $\mathcal{A}$, there is a probabilistic polynomial-time simulator $S$ and a negligible function $\mathsf{neg}(\cdot)$ such that for any $x_1, \ldots, x_t$ it holds that*

$$\mathsf{ADV}^{t\text{-}\mathsf{PO}}_{\mathcal{A},\mathcal{D}} = \big| \Pr_{\mathcal{A},\mathsf{PO}}[\mathcal{A}(\mathsf{PO}(I_{x_1}), \ldots, \mathsf{PO}(I_{x_t})) = 1] - \Pr_{S}[S^{I_{x_1}, \ldots, I_{x_t}}(1^{t \cdot n})) = 1] \big|$$

$$\leq \mathsf{neg}(n).$$

Canetti and Dakdouk [13] showed how to use an $m$-composable point function obfuscator PO to obtain a multi-bit point function that supports outputs (i.e. $y$ values) of length $m$. Specifically, they suggested the following construction.

---

[4] We note that for our application of the multi-bit point obfuscator, it is enough to consider the seemingly relaxed notion of virtual grey-box (VGB) multi-bit point obfuscators, where the simulator has a polynomial bound on the number of queries to its oracle, but is otherwise unlimited. We use the stronger definition which is implied by the weaker one [5, Proposition 7.3].

**Construction 7 ([13]'s multi-bit point obfuscator).** *Let* PO *be a point obfuscator for the domain* $\{0,1\}^n$. *Given a point* $x \in \{0,1\}^n$ *and value* $y = y_1 \ldots y_m \in \{0,1\}^m$, *sample* $s \leftarrow \{0,1\}^n$ *uniformly at random and let*

$$
a_i = \begin{cases} x & \text{if } i = 0 \text{ or } y_i = 1, \\ s & \text{otherwise.} \end{cases}
$$

*Now, the obfuscation of* $I_{x,y}$ *is*

$$
\mathsf{MBPO}(I_{x \to y}) = \mathsf{PO}(I_{a_0}), \ldots, \mathsf{PO}(I_{a_m}), \tag{1}
$$

*and in order to evaluate* $\mathsf{MBPO}(I_{x \to y})$ *on input* $z$ *one first checks if* $z = a_0 = x$ *(by evaluating the first obfuscated circuit). If not (namely,* $z \neq a_0$*), then it outputs* $\perp$. *Otherwise (namely, if* $z = a_0$*), it evaluated all other point obfuscations to find all coordinates in which* $z = a_i = x$ *and outputs* $y_1 \ldots y_m$, *where* $y_i = 1$ *if* $a_1 = z = x$ *(and* 0 *otherwise). Notice that if* PO *is public coin then so is* MBPO.

Bitansky and Canetti [5] showed that under the $(m+1)$-strong vector DDH assumption (defined next), the point obfuscator of Canetti from Theorem 6 is $(m+1)$-composable and thus can be used to get a multi-bit point function. We further observe that since Canetti's point obfuscator is public coin (see Theorem 6), it follows that Canetti and Dakdouk's multi-bit point obfuscator is public coin. We begin with the assumption and then state the theorem.

**Definition 6 (Well spread distribution).** *A distribution* $\mathcal{X}_n$ *over* $\{0,1\}^n$ *is* well-spread *if it is efficiently and uniformly samplable, and it has super-logarithmic min-entropy. Namely,* $\mathsf{H}_\infty(\mathcal{X}_n) \geq \omega(\log n)$.

*Let* $m = m(n)$ *be a polynomial. An ensemble of distributions* $\mathcal{X}_n^{(1)}, \ldots, \mathcal{X}_n^{(m)}$ *(each over* $\{0,1\}^n$*) is* coordinate-wise well-spread *if for each* $i \in [m]$, $\mathcal{X}_n^{(i)}$ *is well-spread.*

**Assumption 8 ($m$-strong vector DDH [5]).** *Let* $m = \mathsf{poly}(n)$. *There exists a group ensemble* $\mathcal{G} = \{\mathbb{G}_n\}_{n \in \mathbb{N}}$, , *where each* $\mathbb{G}_n$ *is a group of prime order* $p_n$ *with uniform and efficient representation and operations, such that for any coordinate-wise well-spread distribution ensemble* $\mathcal{X} = \{\mathcal{X}_n = (\mathcal{X}_n^{(1)}, \ldots, \mathcal{X}_n^{(m)})\}_{n \in \mathbb{N}}$ *over vectors in* $\mathbb{Z}_{p_n}^m$ *the following two ensembles are computationally indistinguishable:*[5]

$$
((g_1, g_1^{a_1}), \ldots, (g_m, g_m^{a_m})), \text{ where } g_1, \ldots, g_m \leftarrow \mathbb{G}_n^* \text{ and } (a_1, \ldots, a_m) \leftarrow \mathcal{X}_n
$$

*and*

$$
((g_1, g_1^{a_1}), \ldots, (g_m, g_m^{a_m})), \text{ where } g_1, \ldots, g_m \leftarrow \mathbb{G}_n^* \text{ and } (a_1, \ldots, a_m) \leftarrow \mathbb{Z}_{p_n}^m.
$$

---

[5] There is a variant for this definition which bears more similarities to DDH, generalizes the assumption of Canetti [12], and it is equivalent to the definition we presented as long as $m \geq 2$. See [5] for more information.

Now we are ready to state the resulting theorem of [5] from Theorem 7 with the underlying Theorem 8.[6]

**Theorem 9.** *Assume the $(m + 1)$-strong vector DDH assumption. Then, the construction from Eq. 1 is a* public coin *multi-bit point obfuscator for multi-bit point functions that output $m$ bits.*

# 3   Definition of Leakage Resilient One-Way Functions

Here we define leakage resilient one-way functions. Intuitively, a one-way function $f$ is leakage resilient for leakage function $g$ if given $f(x)$ and $g(x)$ it is hard to recover an $x'$ such that $f(x') = f(x)$, where $x$ is chosen uniformly at random. Our actual definition is a relaxation and a generalization of the above informal description: (1) we allow $f$ to be sampled from a collection of functions, and (2) we let $g$ come from an a-priori fixed collection of leakage functions.

More precisely, a leakage resilient one-way function collection $\mathcal{F} = \{f : \{0,1\}^n \to \{0,1\}^*\}$ is defined with respect to a collection of leakage functions $\mathcal{L} = \{g : \{0,1\}^n \to \{0,1\}^*\}$. $\mathcal{F}$ is said to be leakage resilient one-way if given $f \leftarrow \mathcal{F}$ it is hard to invert $f(x)$ on a random image even given $f$ and $g(x)$ for any adaptively chosen $g \in \mathcal{L}$ (namely, the choice of $g$ can depend on $f$).

**Definition 7 (Leakage resilient one-way function).** *Let $\mathcal{F} = \{f : \{0,1\}^n \to \{0,1\}^*\}$ be a collection of functions associated with an efficient probabilistic sampler $\mathsf{Gen}_{\mathcal{F}}(1^n)$ that outputs a function $f \in \mathcal{F}$ together with an efficient (deterministic) algorithm for evaluating $f$.*

*The function collection $\mathcal{F}$ is a* leakage resilient one-way function collection *for a collection of functions $\mathcal{L} = \{g : \{0,1\}^n \to \{0,1\}^*\}$ if for every probabilistic polynomial-time algorithms $\mathcal{A} = (\mathcal{A}_0, \mathcal{A}_1)$, there exists a negligible function $\mathsf{neg}(\cdot)$ such that for every $n \in \mathbb{N}$ it holds that*

$$\mathsf{ADV}^{\mathsf{lrOWF}}_{\mathcal{A},\mathcal{F},\mathcal{L}} = \mathbf{Pr}[\mathsf{EXP}_{\mathcal{A},\mathcal{F},\mathcal{L}}(n) = 1] \le \mathsf{neg}(n),$$

*where the random variable $\mathsf{EXP}_{\mathcal{A},\mathcal{F},\mathcal{L}}(n)$ is defined via the following experiment:*

1. *$f \leftarrow \mathsf{Gen}_{\mathcal{F}}(1^n)$.*
2. *$(g, \mathsf{state}) \leftarrow \mathcal{A}_0(1^n, f)$, where $g \in \mathcal{L}$.*
3. *$x^* \leftarrow \{0,1\}^n$ (chosen uniformly at random and independently of $f$ and $g$).*
4. *$x \leftarrow \mathcal{A}_1(f, f(x^*), g(x^*), \mathsf{state})$.*
5. *If $f(x) = f(x^*)$, then output 1, and otherwise output 0.*

---

[6] It may seem odd that Definitions 3 and 4 are stated in a "worst-case" language, while Theorem 8 is stated in an "average-case" language. However, notice that the former are definitions that are given in a simulation-based language while the latter is an indistinguishability-based one. It is known that for (multi-bit) point functions all of these variants are equivalent (see [5, Theorem 5.1 and Proposition 7.3] for a proof).

*If $\mathcal{L}$ consists of one fixed leakage function $g$,[7] then we say that $f$ is a* selective
*leakage resilient one-way function for $\mathcal{L}$. Otherwise, it is called an* adaptive
*leakage resilient one-way function.*

**One vs. a collection of leakage resilient functions.** One may also be interested in a single one-way function $f\colon \{0,1\}^n \to \{0,1\}^*$ which is leakage resilient.
In this case, Item 1 in the definition of the experiment $\mathsf{EXP}_{\mathcal{A},\mathcal{F},\mathcal{L}}(n)$ can be
ignored. We chose to present and work with a definition which allows $f$ to be
chosen from a family as it is more general and since some of our results actually
require having $f$ be chosen from a collection.

**Adaptive vs. selective security.** Our definition captures both adaptive and
selective (i.e. non-adaptive) choice of the leakage. Indeed, if the collection $\mathcal{L}$
consists of a single function $g$, then we can choose the leakage resilient collection
$\mathcal{F}$ *knowing* the leakage $g$ ahead of time (we think of this as the *selective* setting).
On the other hand, if the collection $\mathcal{L}$ contains more functions, we view the
security requirement as an adaptive one, since one has to design the collection
$\mathcal{F}$ without knowing in advance which $g \in \mathcal{L}$ will be chosen by an adversary.
To exemplify an extreme case of the last point, consider the case in which $\mathcal{L}$ is
the set of *all* one-way functions. Then, when designing $\mathcal{F}$, one has very little
information about the leakage.

**What kind of leakage makes sense?** It does not make sense to allow $g \in \mathcal{L}$
to output $x$, as in this case there is no leakage resilient one-way function family
$\mathcal{L}$. This means that every $g \in \mathcal{L}$ has to introduce some hardness for inverting
$x$ from $g(x)$ (when $x$ is a uniform input). (This is a standard and necessary
assumption.) There are several interesting settings for the leakage collection $\mathcal{L}$,
for example:

1. All one-way functions.
2. All sub-exponentially hard one-way functions.
3. All functions whose image size is significantly smaller than the domain size.
4. An arbitrary single one-way function.

The notion in Item 3 was studied earlier (see, for example, [3, 17] and implicitly in [2, 29]) and was proven to be achievable from any one-way function. For
completeness we present the construction and proof in Appendix A. In the main
body, we study all other notions.

## 4    Impossibility of Adaptive Leakage Resilient One-Way Functions

In this section we prove our negative results. We show that without non-trivial
limitation on the leakage collection $\mathcal{L}$, there cannot be a leakage resilient one-
way functions. Specifically, we show that if the leakage collection $\mathcal{L}$ consists of

---

[7] Recall that $f$ and $g$ receive the same input so defining $g$ to be some sort of a universal
circuit and thereby obtaining a huge family of functions is useless.

all one-way functions, there cannot be a leakage resilient one-way function for $\mathcal{L}$. In particular, the leakage can be chosen after the leakage resilient function is chosen and depend on it.

In Sect. 4.1 we prove this in the random oracle model, where functions have access to a random oracle (and without any further cryptographic assumptions). In Sect. 4.2 we provide a construction in the standard model whose security relies on any public-coin multi-bit point obfuscator.

### 4.1 Impossibility in the Random Oracle Model

The following theorem shows that there cannot be a leakage resilient one-way function family $\mathcal{F}$ if the leakage function can depend on the function $f$ chosen from $\mathcal{F}$ and if it has oracle access to a random oracle.

**Theorem 10.** *Let* $O: \{0,1\}^* \rightarrow \{0,1\}^n$ *be a random oracle. Let* $\mathcal{L}^O = \{g: \{0,1\}^n \rightarrow \{0,1\}^*\}$ *be the collection of all oracle-aided one-way functions. There is no leakage-resilient one-way function family* $\mathcal{F} = \{f: \{0,1\}^n \rightarrow \{0,1\}^*\}$ *for the collection* $\mathcal{L}^O$.

*Proof.* Assume towards contradiction that such a function $f: \{0,1\}^n \rightarrow \{0,1\}^*$ exists, where $f \in \mathcal{F}$. We shall define an oracle-aided one-way function $g \in \mathcal{L}^O$ for which

$$\Pr_{x \leftarrow \{0,1\}^n} [\mathcal{A}(1^n, f(x), g(x)) = x] = 1. \qquad (2)$$

This will contradict the assumption that $f$ is leakage-resilient one-way.

Let $g: \{0,1\}^n \rightarrow \{0,1\}^*$ be the following function:

$$g(x) = O(f(x)) \oplus x.$$

We show that Eq. (2) holds and that $g$ is indeed in $\mathcal{L}^O$. Given $y = f(x)$ and $y' = g(x)$ on a uniform $x \in \{0,1\}^n$, $\mathcal{A}$ can recover $x$ as follows. Apply the random oracle $O$ on $y$ to get $O(y) = O(f(x))$ and XOR the output with $y'$. By the definition of $g$, the output must be $x$.

We are left with showing that $g$ is in $\mathcal{L}^O$, that is, it is one-way. Fix $n \in \mathbb{N}$ and let $\mathcal{A}$ be any $q(n)$-query inverter. For $y \in \{0,1\}^*$ and $i \in [q(n)]$ let $Q_i(y)$ be the random variable corresponding to the $i$-th query made by $\mathcal{A}$ to $O$ when $\mathcal{A}$ is given as input the string $y$. Let us denote by $\mathsf{Suc}_i(y)$ the event that the $i$-th query of $\mathcal{A}$ to the random oracle defines a preimage. Namely,

$$\mathsf{Suc}_i(y) = 1 \iff \exists x' \in \{0,1\}^n: Q_i(y) = f(x') \text{ and } O(f(x')) \oplus x' = y$$

Therefore,

$$\Pr[\mathcal{A}^O(y) \in f^{-1}(y)] \leq \Pr[\mathsf{Suc}_1(y) = 1]$$
$$+ \sum_{i=1}^{q(n)} \Pr[\mathsf{Suc}_{i+1}(y) = 1 \mid \mathsf{Suc}_1(y), \ldots, \mathsf{Suc}_i(y) = 0],$$

where $y = g(x)$ and the probabilities are taken over the choice of O and over the choice of $x \in \{0,1\}^n$.

To bound the probability of the event $\mathsf{Suc}_1(y) = 1$, notice that

$$\mathbf{Pr}[\mathsf{Suc}_1(y) = 1] \leq \mathbf{Pr}[Q_1(y) = f(x)] + \mathbf{Pr}[\mathsf{Suc}_1(y) = 1 \mid Q_1(y) \neq f(x)]$$

**Claim 11.** $\mathbf{Pr}[Q_1(y) = f(x)] \leq \mathsf{neg}(n)$.

*Proof.* Recall that $Q_1(y)$ is the *first* query that $\mathcal{A}$ makes to O. Since $x$ is random and O maps every input to a random output, in the view of $\mathcal{A}$, $f(x)$ is distributed uniformly in the distribution of images of $f$. Since $\mathsf{H}_\infty(f(X)) \geq \omega(\log n)$ (see Theorem 5), it holds that $\mathbf{Pr}[Q_1(y) = f(x)] \leq \mathsf{neg}(n)$.

**Claim 12.** $\mathbf{Pr}[\mathsf{Suc}_1(y) = 1 \mid Q_1(y) \neq f(x)] = 1/2^n$.

*Proof.* Note that

$$\mathbf{Pr}[\mathsf{Suc}_1(y) = 1 \mid Q_1(y) \neq f(x)]$$
$$\leq \mathbf{Pr}[\mathsf{O}(Q_1(y)) = z \oplus \mathsf{O}(f(x)) \oplus x \text{ and } z \in f^{-1}(Q_1(y)) \mid Q_1(y) \neq f(x)].$$

Since $Q_1(y) \neq f(x)$, then the value $\mathsf{O}(Q_1(y))$ is completely uniform over $\{0,1\}^n$ and independent of $\mathsf{O}(f(x))$. Therefore, the probability that indeed $\mathsf{O}(Q_1(y)) \oplus z = \mathsf{O}(f(x)) \oplus x$, where $z \in f^{-1}(Q_1(y))$, is $1/2^n$.

We use a similar argument to bound the probability that $\mathsf{Suc}_{i+1}(y) = 1$ conditioned on $\mathsf{Suc}_1(y) \ldots, \mathsf{Suc}_i(y) = 0$. Specifically, we bound the expression

$$\mathbf{Pr}[\mathsf{Suc}_{i+1}(y) = 1 \mid \mathsf{Suc}_1(y), \ldots, \mathsf{Suc}_i(y) = 0]$$
$$\leq \mathbf{Pr}[Q_{i+1}(y) = f(x) \mid \mathsf{Suc}_1(y), \ldots, \mathsf{Suc}_i(y) = 0]$$
$$+ \mathbf{Pr}[\mathsf{Suc}_{i+1}(y) = 1 \mid \mathsf{Suc}_1(y), \ldots, \mathsf{Suc}_i(y) = 0 \text{ and } Q_{i+1}(y) \neq f(x)]$$

Notice that $\mathsf{Suc}_1(y), \ldots, \mathsf{Suc}_i(y) = 0$ implies that $Q_1(y) \ldots, Q_i(y) \neq f(x)$. Thus, the view of $\mathcal{A}$ is that $f(x)$ is uniformly distributed in the distribution of images of $f$ except the points $Q_1(y) \ldots, Q_i(y)$ (some of which may not even be valid images). Namely, for $\mathcal{A}$ the value $f(x)$ is uniformly distribution w.r.t the distribution in which one samples a random $x' \leftarrow \{0,1\}^n$, computes $f(x')$ and outputs $f(x')$ conditioned on $f(x') \notin \{Q_1(y) \ldots, Q_i(y)\}$ (otherwise, we sample $x'$ again). This distribution has super-logarithmic min-entropy, namely,

$$\mathsf{H}_\infty(f(X) \mid f(X) \notin \{Q_1(y) \ldots, Q_i(y)\}) \geq \mathsf{H}_\infty(f(X)) - \log i$$
$$\geq \omega(\log n),$$

where the last inequality follows from Theorem 5 and since $i \leq q(n)$ is a polynomial in $n$. Therefore, as in Theorem 11, we get that

$$\mathbf{Pr}[Q_{i+1}(y) = f(x) \mid \mathsf{Suc}_1(y), \ldots, \mathsf{Suc}_i(y) = 0] \leq \mathsf{neg}(n).$$

Given that $\mathsf{Suc}_1(y), \ldots, \mathsf{Suc}_i(y) = 0$ and $Q_{i+1}(y) \neq f(x)$, we have that $Q_{1+1}(y)$ is completely uniform over $\{0,1\}^n$ and independent of $\mathsf{O}(f(x))$ and all previous queries $O(Q_1(y)), \ldots, O(Q_i(y))$ (we assume, without loss of generality, that all queries to $O$ are distinct). Therefore, the probability that $O(Q_{i+1}(y)) \oplus z = O(f(x)) \oplus x$, where $z \in f^{-1}(Q_{i+1}(y))$, is $1/2^n$. Thus, as in Theorem 12, we have that

$$\mathbf{Pr}[\mathsf{Suc}_{i+1}(y) = 1 \mid \mathsf{Suc}_1(y), \ldots, \mathsf{Suc}_i(y) = 0 \text{ and } Q_{i+1}(y) \neq f(x)] = 1/2^n.$$

In conclusion, since $q(n)$ is a polynomial, we get that

$$\mathbf{Pr}[\mathcal{A}^O(y) \in f^{-1}(y)] \leq \sum_{i=0}^{q(n)} \mathbf{Pr}[\mathsf{Suc}_{i+1}(y) = 1 \mid \mathsf{Suc}_1(y), \ldots, \mathsf{Suc}_i(y) = 0]$$

$$\leq \sum_{i=0}^{q(n)} (\mathsf{neg}(n) + 1/2^n) \leq \mathsf{neg}(n).$$

### 4.2  Impossibility in the Standard Model

The following theorem shows that there cannot be a leakage resilient one-way function family $\mathcal{F}$ if the leakage function can depend on the function $f$ chosen from $\mathcal{F}$.

**Theorem 13.** *Let $\mathcal{L} = \{g \colon \{0,1\}^n \to \{0,1\}^*\}$ be the collection of all one-way functions. Assuming a public-coin multi-bit point obfuscator, there is no leakage resilient one-way function collection $\mathcal{F} = \{f \colon \{0,1\}^n \to \{0,1\}^m\}$ for the collection $\mathcal{L}$.*

*Proof.* Assume towards contradiction that such a function $f \colon \{0,1\}^n \to \{0,1\}^m$ in $\mathcal{F}$ exists. We shall construct a function $g \in \mathcal{L}$ (depending on $f$) and show that for any $x \in \{0,1\}^n$, $f(x)$ together with $g(x)$ reveal $x$. Our building block is a public-coin multi-bit point obfuscator MBPO. Assume that MBPO takes as input a pair of strings $(x,y) \in \{0,1\}^m \times \{0,1\}^n$ and randomness of length $\lambda$. Let $r \leftarrow \{0,1\}^\lambda$ be a uniformly random string. We define $g_r \colon \{0,1\}^n \to \{0,1\}^m$ that outputs, on input $x$, a multi-bit point obfuscation of the function $I_{f(x) \to x}$. Namely,

$$g_r(x) = \mathsf{MBPO}(I_{f(x) \to x}; r) \tag{3}$$

For correctness, we argue that given $f(x)$ and $g_r(x)$ together it is easy to recover $x$. Indeed, one can just plug in $f(x)$ into the output of $g_r(x)$, namely into $\mathsf{MBPO}(I_{f(x) \to x}; r)$. By the correctness of the multi-bit point obfuscator it follows that the output of this operation has to be $x$.

For security we have to prove that $g_r(x)$ is a one-way function. Namely, given $g_r$ and $g_r(x)$ on a uniformly random $x$, one cannot recover any $x'$ such that $g_r(x') = g_r(x)$. First, we observe that by the (perfect) correctness of MBPO it holds that for every $x' \neq x$, it cannot be that $\mathsf{MBPO}(I_{f(x) \to x}; r) = \mathsf{MBPO}(I_{f(x') \to x'}; r)$. Thus, $g_r$ is injective. It is left to show that given $g_r(x)$

any computationally bounded adversary cannot recover $x$ with non-negligible probability.

We consider an even easier task for $\mathcal{A}$ of just outputting the first bit of $x$. By the security of MBPO, we have that for every such adversary $\mathcal{A}$, if there exists a polynomial $p$ such that

$$\mathbf{Pr}[\mathcal{A}(\mathsf{MBPO}(I_{f(x)\to x}; r)) = x_1] \geq 1/2 + 1/p(n),$$

then there is an efficient simulator $S$ such that

$$\mathbf{Pr}[S^{I_{f(x)\to x}}(1^n) = x_1] \geq 1/2 + 1/p(n) - \mathsf{neg}(n).$$

However, since $I_{f(x)\to x}$ outputs $\perp$ on all inputs which are not $f(x)$, and since the distribution $f(x)$ has super-logarithmic min entropy (see Theorem 5), any efficient simulator will never query the oracle on $f(x)$ and thus will get no information about $x$. Hence, it is impossible for it to guess with non-negligible advantage the first bit of $x$.

# 5    Possibility of Selective Leakage Resilient One-Way Functions

In both impossibility results (Theorems 10 and 13) we used the fact that the leakage functions can be chosen adaptively and depend on $f$. In contrast, the following theorem shows that if we limit the choice of the leakage to be independent of $f$, a leakage resilient one-way function exists based on various assumptions.

The high level idea is that if the leakage $g$ is fixed ahead of time, we can still extract from the input (for $f$ and $g$) enough pseudorandom bits that will ensure one-wayness.

**Theorem 14.** Let $g\colon \{0,1\}^n \to \{0,1\}^m$ be a fixed leakage one-way function. Then, there is a leakage-resilient one-way function $f : \{0,1\}^n \to \{0,1\}^*$ for $\mathcal{L} = \{g\}$ assuming that polynomially-many hardcore bits can be extracted from $g$.

Instantiating the theorem with known results we obtain the following corollaries:

1. if $g$ is sub-exponentially secure (with known hardness), then $f$ can be based on any one-way function.
2. if $g$ is a one-way function (with known hardness), then $f$ can be based on any exponentially-secure one-way function.
3. if $g$ is a injective one-way function, then $f$ can be based on indistinguishability obfuscation [4].
4. if $g$ is a one-way function, then $f$ can be based on indistinguishability obfuscation and auxiliary-input point obfuscators [11].
5. if $g$ is a one-way function, then $f$ can be based on exponential hardness of DDH [36].

**Proof of Theorem 14.** Let $g$ be the leakage function and let $\mathcal{H} = \{h \colon \{0,1\}^n \to \{0,1\}^{2n}\}$ be a family of hardcore function for any one-way function that output polynomially-many hard-core bits. Note that letting the range be $2n$ is without loss of generality since from any polynomial number of hardcore bits we can use a (standard) PRG and obtain the desired length. The leakage resilient one-way function $f$ is defined as follows. We sample a random hard-core function from $\mathcal{H}$ and let

$$f_H(x) = H(x)$$

We argue that $f_H(x)$ is a one-way function even given $g(x)$, where $g$ is a one-way function. For this we use the definition of a hard-core function which says that the distribution

$$(H, H(x), g(x))$$

is computationally indistinguishable from

$$(H, r, g(x)),$$

where $x \leftarrow \{0,1\}^n$, $H \leftarrow \mathcal{H}$, and $r \leftarrow \{0,1\}^{2n}$ are chosen independently uniformly at random. Now, since $r$ is of length $2n$, with all but exponentially small probability, it holds that there is no preimage $x'$ for $f_H$ for which $f_H(x') = r$. Thus, since $g$ is one-way as well, any polynomial-time adversary cannot find a preimage.

## 6   Future Directions

In this work we introduced and studied leakage resilient one-way functions with arbitrary computationally-hiding leakage. We showed that the natural adaptive definition is impossible to achieve in the random oracle model and in the standard model based on a (non-standard) computation assumption. We further observed that the non-adaptive variant is very related to hardcore functions and in some sense is dual to it.

It is interesting to base the impossibility result on other assumptions (any one-way function, DDH or even based on indistinguishability obfuscation). Also, extracting polynomially-many hardcore bits from any one-way function based on better assumptions is also an interesting problem.

**Acknowledgements.** We thank Zvika Brakerski, Moni Naor, Gil Segev, and Eylon Yogev for many fruitful discussions on the subject of this paper. We thank the reviewers for their useful comments.

## A   One-Way Functions Resilient for Bounded Leakage

In both impossibility results (Theorems 10 and 13) we used the fact that the leakage functions can output enough information to allow anyone to invert the original one-way function. In contrast, the theorem below shows that if we limit the image size of the functions in the leakage collection $\mathcal{L}$, a leakage resilient one-way function exists assuming one-way functions exist.

We start with a definition of a lossy function (as defined by [33]). This will capture our restriction on the amount of information the output of the leakage must "lose".

**Definition 8 ($(n, \ell)$-lossy function).** *A function $f\colon \{0,1\}^n \to \{0,1\}^n$ is said to be $(n, \ell)$-lossy if its image $\{f(x) \mid x \in \{0,1\}^n\}$ has size at most $2^{n-\ell}$ for every $x \in \{0,1\}^n$.*

Roughly speaking, the parameter $\ell$ captures the number of information bits $f$ loses about a typical input $x$. We note that it is enough for us to relax the definition of a lossy function and only require that it has bounded image size on all but a negligible fraction of the $x$'s. We use the stronger requirement for simplicity.

The construction will rely on universal one-way hash functions (UOWHFs) that were introduced by Naor and Yung [32]. The main feature of UOWHFs is that given an element $x$ in the domain, it is computationally hard to find a *different* domain element $x' \neq x$ which collides with $x$. Naor and Yung showed how to use UOWHFs to construct digital signatures. Besides this application, they showed how to construct them using any injective one-way function. Later, Rompel [35] showed how to construct UOWHFs from any one-way function (see also [28]).

In the following definition we define a weak variant of UOWHFs in which the initial domain element is a uniform random input (rather than an adversarially chosen input).[8] The goal of the adversary is then to find a collision with that random input.

**Definition 9 ((Weak) universal one-way hash functions).** *Let $p(n) = n^{1/c}$ be a polynomial where $c \in \mathbb{N}$ is a constant. A collection of functions $\{F_h\colon \{0,1\}^n \to \{0,1\}^{p(n)}\}$ mapping strings of length $n$ to strings of length $p(n)$ is a (collection of) **universal one-way hash functions** if it is described by a pair of efficient algorithms $\mathsf{UOWHF} = (\mathsf{Gen}, F)$ with the following properties.*

1. *$\mathsf{Gen}$ is a probabilistic algorithm that is given as input the unary value of $n$, and it outputs a function index $h$.*
2. *For every function index $h$ in the image of $\mathsf{Gen}$, $F_h$ is given as input $x \in \{0,1\}^n$ and it outputs a string of length $p(n)$.*
3. *For every probabilistic polynomial-time adversary $\mathcal{A}$, there exists a negligible function $\mathsf{neg}(\cdot)$, such that*

$$\mathbf{Pr}[x' \leftarrow \mathcal{A}(h, x, F_h(x))\colon x \neq x' \text{ and } F_h(x') = F_h(x)] \leq \mathsf{neg}(n),$$

*where the probability is over the choice of $x \leftarrow \{0,1\}^n$, the choice of $h \leftarrow \mathsf{Gen}(1^n)$, and the internal randomness of $\mathcal{A}$.*

---

[8] This is sometimes called a *second pre-image resistant function*.

**Theorem 15 ([3,17]).** *Let $k = n^{1/c}$ for a constant $c \in \mathbb{N}$ and let $\kappa = \omega(\log n)$. Let $\mathcal{F} = \{F_h : \{0,1\}^n \to \{0,1\}^{k-\kappa}\}$ be a family of universal one-way hash functions mapping strings of length $n$ to strings of length $k - \kappa$ described by (Gen, $F$). Let $\mathcal{L} = \{g : \{0,1\}^n \to \{0,1\}^n\}$ be the collection of all $(n,k)$-lossy functions. Then, $\mathcal{F}$ is a leakage resilient one-way function collection for $\mathcal{L}$ .*

The proof uses the notion of **average min-entropy** defined by Dodis et al. [19] which captures the remaining unpredictability of $X$ conditioned on the value of $Y$. Roughly speaking, the average min-entropy of $X$ given $Y$ is the logarithm of the average probability of the most likely value of $X$ given $Y$. That is,

$$\tilde{H}_\infty(X \mid Y) = -\log\left(\mathop{\mathbf{E}}_{y \leftarrow Y}[2^{-H_\infty(X|Y=y)}]\right).$$

The following property of average min-entropy was shown by Dodis et al. [19].

**Lemma 1 ([19, Lemma 2.2]).** *Let $X$ and $Y$ be two random variables. Then,*

*1. For any $\delta > 0$, it holds that*

$$\mathop{\mathbf{Pr}}_{y \leftarrow Y}[H_\infty(X \mid Y = y) \geq \tilde{H}_\infty(X \mid Y) - \log(1/\delta)] \geq 1 - \delta.$$

*2. If $Y$ has at most $2^k$ possible values, then $\tilde{H}_\infty(X \mid Y) \geq H_\infty(X) - k$.*

**Proof of Theorem 15.** We assume towards contradiction that the statement is false. Namely, there exists a function $g \colon \{0,1\}^n \to \{0,1\}^{n-k}$ for which there exists an adversary $\mathcal{A}$ such that for $x^* \leftarrow \{0,1\}^n$ chosen uniformly at random given

$$h, F_h(x^*), g(x^*),$$

where $h \leftarrow$ Gen$(1^n)$, $\mathcal{A}$ is able to recover any $x$ such that $F_h(x^*) = F_h(x)$ with non-negligible probability $1/p(n)$. We use this adversary $\mathcal{A}$ and construct an adversary $\mathcal{B}$ that breaks the security of the universal one-way hash function.

Let $h, x^*, F_h(x^*)$ be a challenge for the universal one-way hash function, where $h \leftarrow$ Gen$(1^n)$ and $x^* \leftarrow \{0,1\}^n$ is chosen uniformly and independently. Our adversary $\mathcal{B}$ will first simulate the choice of $g$ and compute $g(x^*)$. Then, it runs the inverter $\mathcal{A}$ on input $(h, F_h(x^*), g(x^*))$ and obtains a preimage $x$. Finally, $\mathcal{B}$ outputs $x$ as its guess for the collision. We now argue that this adversary indeed breaks the security of the UOWHF. First, it is clear by the correctness of the adversary $\mathcal{A}$ that $F_h(x) = F_h(x^*)$. We are left to argue that $x \neq x^*$ with non-negligible probability.

Roughly speaking, the idea is that since $x^*$ is chosen uniformly at random, given only $F_h(x^*)$ and $g(x^*)$, whose image size altogether $\ll 2^n$, there is not enough information regarding the real $x^*$ that maps to $F_h(x^*)$ and $g(x^*)$. Namely, we will show that with high probability over the choice of $x^*$, there could be many consistent $x$'s that map to the same output. The inverted cannot distinguish between them and thus will output the real $x^*$ with very small probability. We formalize this intuition next.

Fix the function index $h \leftarrow \mathsf{Gen}(1^n)$ and leakage function $g$ (that might depend $h$). Since $F_h(x^*)$ and $g(x^*)$ have together at most $2^{k-\kappa} \cdot 2^{n-k} = 2^{n-\kappa}$ possible outputs and $x^* \leftarrow \{0,1\}^n$ is uniform and independent of $h$, by item 2 of Lemma 1 we have that

$$\tilde{\mathsf{H}}_\infty(x^* \mid h, F_h(x^*), g(x^*)) \geq \mathsf{H}_\infty(x^* \mid h) - (n - \kappa) = \kappa.$$

By item 1 of Lemma 1, we get that for any $\delta > 0$, it holds that

$$\Pr_{x^* \leftarrow \{0,1\}^n}[\mathsf{H}_\infty(x^* \mid h, F_h(x^*), g(x^*)) \geq \tilde{\mathsf{H}}_\infty(x^* \mid h, F_h(x^*), g(x^*)) - \log(1/\delta)]$$

$$\geq 1 - \delta.$$

Therefore,

$$\Pr_{x^* \leftarrow \{0,1\}^n}[\mathsf{H}_\infty(x^* \mid h, F_h(x^*), g(x^*)) \geq \kappa - \log(1/\delta)] \geq 1 - \delta. \tag{4}$$

Let $\delta = 1/2^{\kappa/2}$. Then, with all but a negligible probability over the choice of $x^*$, it holds that

$$\mathsf{H}_\infty(x^* \mid h, F_h(x^*), g(x^*)) \geq \kappa - \kappa/2 = \kappa/2.$$

Therefore, since $\kappa = \omega(\log n)$, by the definition of min-entropy $\Pr[x^* \leftarrow \mathcal{A}(h, F_h(x^*), g(x^*))] \leq \mathsf{neg}(\cdot)$. In conclusion, the adversary $\mathcal{B}$ is able to find a collision with non-negligible probability:

$$\Pr[x \leftarrow \mathcal{B}(h, F_h(x^*), g(x^*)): x \neq x^* \text{ and } F_h(x^*) = F_h(x), g(x^*) = g(x)]$$
$$= \Pr[x \leftarrow \mathcal{A}(h, F_h(x^*), g(x^*)): F_h(x^*) = F_h(x), g(x^*) = g(x)]$$
$$- \Pr[x^* \leftarrow \mathcal{A}(h, F_h(x^*), g(x^*))] \geq$$
$$1/p(n) - \mathsf{neg}(n) \geq 1/(2p(n)).$$

# References

1. Akavia, A., Goldwasser, S., Vaikuntanathan, V.: Simultaneous hardcore bits and cryptography against memory attacks. In: Reingold, O. (ed.) TCC 2009. LNCS, vol. 5444, pp. 474–495. Springer, Heidelberg (2009). doi:10.1007/978-3-642-00457-5_28
2. Alwen, J., Dodis, Y., Wichs, D.: Leakage-resilient public-key cryptography in the bounded-retrieval model. In: Halevi, S. (ed.) CRYPTO 2009. LNCS, vol. 5677, pp. 36–54. Springer, Heidelberg (2009). doi:10.1007/978-3-642-03356-8_3
3. Alwen, J., Dodis, Y., Wichs, D.: Survey: leakage resilience and the bounded retrieval model. In: Kurosawa, K. (ed.) ICITS 2009. LNCS, vol. 5973, pp. 1–18. Springer, Heidelberg (2010). doi:10.1007/978-3-642-14496-7_1

4. Bellare, M., Stepanovs, I., Tessaro, S.: Poly-many hardcore bits for any one-way function and a framework for differing-inputs obfuscation. In: Sarkar, P., Iwata, T. (eds.) ASIACRYPT 2014. LNCS, vol. 8874, pp. 102–121. Springer, Heidelberg (2014). doi:10.1007/978-3-662-45608-8_6

5. Bitansky, N., Canetti, R.: On strong simulation and composable point obfuscation. J. Cryptol. **27**(2), 317–357 (2014)

6. Boyle, E., Goldwasser, S., Jain, A., Kalai, Y.T.: Multiparty computation secure against continual memory leakage. In: Proceedings of the 44th Symposium on Theory of Computing Conference, STOC, pp. 1235–1254 (2012)

7. Boyle, E., Segev, G., Wichs, D.: Fully leakage-resilient signatures. J. Cryptol. **26**(3), 513–558 (2013)

8. Brakerski, Z., Goldwasser, S.: Circular and leakage resilient public-key encryption under subgroup indistinguishability. In: Rabin, T. (ed.) CRYPTO 2010. LNCS, vol. 6223, pp. 1–20. Springer, Heidelberg (2010). doi:10.1007/978-3-642-14623-7_1

9. Brakerski, Z., Kalai, Y.T., Katz, J., Vaikuntanathan, V.: Overcoming the hole in the bucket: public-key cryptography resilient to continual memory leakage. In: 51th Annual IEEE Symposium on Foundations of Computer Science, FOCS, pp. 501–510 (2010)

10. Brzuska, C., Mittelbach, A.: Indistinguishability obfuscation versus multi-bit point obfuscation with auxiliary input. In: Sarkar, P., Iwata, T. (eds.) ASIACRYPT 2014. LNCS, vol. 8874, pp. 142–161. Springer, Heidelberg (2014). doi:10.1007/978-3-662-45608-8_8

11. Brzuska, C., Mittelbach, A.: Using indistinguishability obfuscation via UCEs. In: Sarkar, P., Iwata, T. (eds.) ASIACRYPT 2014. LNCS, vol. 8874, pp. 122–141. Springer, Heidelberg (2014). doi:10.1007/978-3-662-45608-8_7

12. Canetti, R.: Towards realizing random oracles: hash functions that hide all partial information. In: Kaliski, B.S. (ed.) CRYPTO 1997. LNCS, vol. 1294, pp. 455–469. Springer, Heidelberg (1997). doi:10.1007/BFb0052255

13. Canetti, R., Dakdouk, R.R.: Obfuscating point functions with multibit output. In: Smart, N. (ed.) EUROCRYPT 2008. LNCS, vol. 4965, pp. 489–508. Springer, Heidelberg (2008). doi:10.1007/978-3-540-78967-3_28

14. Canetti, R., Dodis, Y., Halevi, S., Kushilevitz, E., Sahai, A.: Exposure-resilient functions and all-or-nothing transforms. In: Preneel, B. (ed.) EUROCRYPT 2000. LNCS, vol. 1807, pp. 453–469. Springer, Heidelberg (2000). doi:10.1007/3-540-45539-6_33

15. Davì, F., Dziembowski, S., Venturi, D.: Leakage-resilient storage. In: Garay, J.A., Prisco, R. (eds.) SCN 2010. LNCS, vol. 6280, pp. 121–137. Springer, Heidelberg (2010). doi:10.1007/978-3-642-15317-4_9

16. Dodis, Y., Goldwasser, S., Tauman Kalai, Y., Peikert, C., Vaikuntanathan, V.: Public-key encryption schemes with auxiliary inputs. In: Micciancio, D. (ed.) TCC 2010. LNCS, vol. 5978, pp. 361–381. Springer, Heidelberg (2010). doi:10.1007/978-3-642-11799-2_22

17. Dodis, Y., Haralambiev, K., López-Alt, A., Wichs, D.: Cryptography against continuous memory attacks. In: 51th Annual IEEE Symposium on Foundations of Computer Science, FOCS, pp. 511–520 (2010)

18. Dodis, Y., Kalai, Y.T., Lovett, S.: On cryptography with auxiliary input. In: Proceedings of the 41st Annual ACM Symposium on Theory of Computing, STOC, pp. 621–630 (2009)

19. Dodis, Y., Ostrovsky, R., Reyzin, L., Smith, A.D.: Fuzzy extractors: how to generate strong keys from biometrics and other noisy data. SIAM J. Comput. **38**(1), 97–139 (2008)

20. Dodis, Y., Pietrzak, K.: Leakage-resilient pseudorandom functions and side-channel attacks on Feistel networks. In: Rabin, T. (ed.) CRYPTO 2010. LNCS, vol. 6223, pp. 21–40. Springer, Heidelberg (2010). doi:10.1007/978-3-642-14623-7_2

21. Dodis, Y., Sahai, A., Smith, A.: On perfect and adaptive security in exposure-resilient cryptography. In: Pfitzmann, B. (ed.) EUROCRYPT 2001. LNCS, vol. 2045, pp. 301–324. Springer, Heidelberg (2001). doi:10.1007/3-540-44987-6_19

22. Dziembowski, S., Pietrzak, K.: Leakage-resilient cryptography. In: 49th Annual IEEE Symposium on Foundations of Computer Science, FOCS, pp. 293–302 (2008)

23. Faust, S., Kiltz, E., Pietrzak, K., Rothblum, G.N.: Leakage-resilient signatures. In: Micciancio, D. (ed.) TCC 2010. LNCS, vol. 5978, pp. 343–360. Springer, Heidelberg (2010). doi:10.1007/978-3-642-11799-2_21

24. Faust, S., Rabin, T., Reyzin, L., Tromer, E., Vaikuntanathan, V.: Protecting circuits from computationally bounded and noisy leakage. SIAM J. Comput. **43**(5), 1564–1614 (2014)

25. Goldreich, O., Levin, L.A.: A hard-core predicate for all one-way functions. In: Proceedings of the 21st Annual ACM Symposium on Theory of Computing, STOC, pp. 25–32 (1989)

26. Goldwasser, S., Kalai, Y.T.: On the impossibility of obfuscation with auxiliary input. In: 46th Annual IEEE Symposium on Foundations of Computer Science, FOCS, pp. 553–562 (2005)

27. Ishai, Y., Sahai, A., Wagner, D.: Private circuits: securing hardware against probing attacks. In: Boneh, D. (ed.) CRYPTO 2003. LNCS, vol. 2729, pp. 463–481. Springer, Heidelberg (2003). doi:10.1007/978-3-540-45146-4_27

28. Katz, J., Koo, C.: On constructing universal one-way hash functions from arbitrary one-way functions. IACR Cryptology ePrint Archive, p. 328 (2005)

29. Katz, J., Vaikuntanathan, V.: Signature schemes with bounded leakage resilience. In: Matsui, M. (ed.) ASIACRYPT 2009. LNCS, vol. 5912, pp. 703–720. Springer, Heidelberg (2009). doi:10.1007/978-3-642-10366-7_41

30. Micali, S., Reyzin, L.: Physically observable cryptography. In: Naor, M. (ed.) TCC 2004. LNCS, vol. 2951, pp. 278–296. Springer, Heidelberg (2004). doi:10.1007/978-3-540-24638-1_16

31. Naor, M., Segev, G.: Public-key cryptosystems resilient to key leakage. SIAM J. Comput. **41**(4), 772–814 (2012)

32. Naor, M., Yung, M.: Universal one-way hash functions and their cryptographic applications. In: Proceedings of the 21st Annual ACM Symposium on Theory of Computing, STOC, pp. 33–43 (1989)

33. Peikert, C., Waters, B.: Lossy trapdoor functions and their applications. SIAM J. Comput. **40**(6), 1803–1844 (2011)

34. Pietrzak, K.: A leakage-resilient mode of operation. In: Joux, A. (ed.) EUROCRYPT 2009. LNCS, vol. 5479, pp. 462–482. Springer, Heidelberg (2009). doi:10.1007/978-3-642-01001-9_27

35. Rompel, J.: One-way functions are necessary and sufficient for secure signatures. In: Proceedings of the 22nd Annual ACM Symposium on Theory of Computing, STOC, pp. 387–394 (1990)

36. Zhandry, M.: The magic of ELFs. In: Robshaw, M., Katz, J. (eds.) CRYPTO 2016. LNCS, vol. 9814, pp. 479–508. Springer, Heidelberg (2016). doi:10.1007/978-3-662-53018-4_18

# Simulating Auxiliary Inputs, Revisited

Maciej Skórski[(✉)]

University of Warsaw, Warsaw, Poland
maciej.skorski@mimuw.edu.pl

**Abstract.** For any pair $(X, Z)$ of correlated random variables we can think of $Z$ as a randomized function of $X$. If the domain of $Z$ is small, one can make this function computationally efficient by allowing it to be only approximately correct. In folklore this problem is known as *simulating auxiliary inputs*. This idea of simulating auxiliary information turns out to be a very usefull tool, finding applications in complexity theory, cryptography, pseudorandomness and zero-knowledge. In this paper we revisit this problem, achieving the following results:

  (a) We present a novel boosting algorithm for constructing the simulator. This boosting proof is of independent interest, as it shows how to handle "negative mass" issues when constructing probability measures by shifting distinguishers in descent algorithms. Our technique essentially fixes the flaw in the TCC'14 paper "How to Fake Auxiliary Inputs".

  (b) The complexity of our simulator is better than in previous works, including results derived from the uniform min-max theorem due to Vadhan and Zheng. To achieve $(s, \epsilon)$-indistinguishability we need the complexity $O\left(s \cdot 2^{5\ell}\epsilon^{-2}\right)$ in time/circuit size, which improve previous bounds by a factor of $\epsilon^{-2}$. In particular, with we get meaningful provable security for the EUROCRYPT'09 leakage-resilient stream cipher instantiated with a standard 256-bit block cipher, like AES256.

Our boosting technique utilizes a two-step approach. In the first step we shift the current result (as in gradient or sub-gradient descent algorithms) and in the separate step we fix the biggest non-negative mass constraint violation (if applicable).

**Keywords:** Simulating auxiliary inputs · Boosting · Leakage-resilient cryptography · Stream ciphers · Computational indistinguishability

The full (and updated) version of this paper is available at the Cryptology ePrint archive and the arXiv archive (http://arxiv.org/abs/1503.00484).

M. Skorski—Supported by the National Science Center, Poland (2015/17/N/ST6/03564).

M. Hirt and A. Smith (Eds.): TCC 2016-B, Part I, LNCS 9985, pp. 159–179, 2016.
DOI: 10.1007/978-3-662-53641-4_7

# 1    Introduction

## 1.1    Simulating Correlated Information

**Informal Problem Statement.** Let $(X, Z) \in \mathcal{X} \times \mathcal{Z}$ be a pair of correlated random variables. We can think of $Z$ as a *randomized* function of $X$. More precisely, consider the randomized function $h : \mathcal{X} \to \mathcal{Z}$, which for every $x$ outputs $z$ with probability $\Pr[Z = z | X = x]$. By definition it satisfies

$$(X, h(X)) \overset{d}{=} (X, Z) \tag{1}$$

however the function $h$ is *inefficient* as we need to hardcode the conditional probability table of $Z|X$. It is natural to ask, if this limitation can be overcome

**Q1**: Can we represent $Z$ as an *efficient* function of $X$?

Not surprisingly, it turns out that a positive answer may be given only in computational settings. Note that replacing the equality in Eq. (1) by closeness in the total variation distance (allowing the function $h$ to make some mistakes with small probability) is not enough[1]. This discussion leads to the following reformulated question

**Q1'**: Can we *efficiently simulate* $Z$ as a function of $X$?

**Why It Matters?** Aside from being very foundational, this question is relevant to many areas of computer science. We will not discuss these applications in detail, as they are well explained in [JP14]. Below we only mention where such a generic simulator can be applied, to show that this problem is indeed well-motivated.

(a) Complexity Theory. From the simulator one can derive Dense Model Theorem [RTTV08], Impagliazzo's hardcore lemma [Imp95] and a version of Szemeredis Regularity Lemma [FK99].
(b) Cryptography. The simulator can be applied for settings where $Z$ models short leakage from a secret state $X$. It provides tools for improving and simplifying proofs in leakage-resilient cryptography, in particular for leakage-resilient stream ciphers [JP14].
(c) Pseudorandomness. Using the simulator one can conclude results called chain rules [GW11], which quantify pseudorandomness in conditioned distributions. They can be also applied to leakage-resilient cryptography.
(d) Zero-knowledge. The simulator can be applied to represent the text exchanged in verifier-prover interactions $Z$ from the common input $X$ [CLP15].

Thus, the simulator may be used as a tool to unify, simplify and improve many results. Having briefly explained the motivation we now turn to answer the posed question, leaving a more detailed discussion of some applications to Sect. 1.6.

---

[1] Indeed, consider the simplest case $\mathcal{Z} = \{0, 1\}$, define $X$ to be uniform over $\mathcal{X} = \{0, 1\}^n$, and take $Z = f(X)$ where $f$ is a function which is 0.5-hard to predict by circuits exponential in $n$, Then $(X, h(X))$ and $(X, Z)$ are at least $\frac{1}{4}$-away in total variation.

## 1.2    Problem Statement

The problem of simulating auxiliary inputs in the computational setting can be defined precisely as follows

> Given a random variables $X \in \{0,1\}^n$ and correlated $Z \in \{0,1\}^\ell$, what is the minimal complexity $s_h$ of a (randomized) function $h$ such that the distributions of $h(X)$ and $Z$ are $(\epsilon, s)$-indistinguishable given $X$, that is
>
> $$|\mathbb{E}\,D(X, h(X)) - \mathbb{E}\,D(X, Z)| < \epsilon$$
>
> holds for all (deterministic) circuits D of size $s$?

The indistinguishability above is understood with respect to deterministic circuits. However it doesn't really matter for distinguishing two distributions, where randomized and deterministic distinguishers are equally powerful[2].

It turns out that it is relatively easy[3] to construct a simulator $h$ with a polynomial blowup in complexity, that is when

$$s_h = \mathrm{poly}\left(s, \epsilon^{-1}, 2^\ell\right).$$

However, more challenging is to minimize the dependency on $\epsilon^{-1}$. This problem is especially important for cryptography, where security definitions require the advantage $\epsilon$ to be possibly small. Indeed, for meaningful security $\epsilon = 2^{-80}$ or at least $\epsilon = 2^{-40}$ it makes a difference whether we lose $\epsilon^{-2}$ or $\epsilon^{-4}$. We will see later how much inefficient bounds here may affect provable security of stream ciphers.

## 1.3    Related Works

**Original Work of Jetchev and Pietrzak (TCC'14).** The authors showed that $Z$ can be "approximately" computed from $X$ by an "efficient" function h.

**Theorem 1 ([JP14], corrected).** *For every distribution $(X, Z)$ on $\{0,1\}^n \times \{0,1\}^\ell$ and every $\epsilon$, $s$, there exists a "simulator" $h : \{0,1\}^n \to \{0,1\}^\ell$ such that*

*(a) $(X, h(X))$ and $(X, Z)$ are $(\epsilon, s)$-indistinguishable*
*(b) h is of complexity $s_h = O\left(s \cdot 2^{4\ell}\epsilon^{-4}\right)$*

The proof uses the standard min-max theorem. In the statement above we correct two flaws. One is a missing factor of $2^\ell$. The second (and more serious) one is the (corrected) factor $\epsilon^{-4}$, claimed incorrectly to be $\epsilon^{-2}$. The flaws are discussed in Appendix A.

---

[2] If two distributions can be distinguished by a randomized circuit, we can fix a specific choice of coins to achieve at least the same advantage.

[3] We briefly sketch the idea of the proof: note first that it is easy to construct a simulator for every single distinguisher. Having realized that, we can use the min-max theorem to switch the quantifiers and get one simulator for all distinguishers.

**Vadhan and Zheng (CRYPTO'13).** The authors derived a version of Theorem 1 but with incomparable bounds

**Theorem 2 ([VZ13]).** *For every distribution* $X, Z$ *on* $\{0,1\}^n \times \{0,1\}^\ell$ *and every* $\epsilon$, $s$, *there exists a "simulator"* $h : \{0,1\}^n \to \{0,1\}^\ell$ *such that*

*(a)* $(X, h(X))$ *and* $(X, Z)$ *are* $(s, \epsilon)$*-indistinguishable*
*(b)* $h$ *is of complexity* $s_h = O\left(s \cdot 2^\ell \epsilon^{-2} + 2^\ell \epsilon^{-4}\right)$

The proof follows from a general regularity theorem which is based on their uniform min-max theorem. The additive loss of $O\left(2^\ell \epsilon^{-4}\right)$ appears as a consequence of a sophisticated weight-updating procedure. This error is quite large and may dominate the main term for many settings (whenever $s \ll \epsilon^{-2}$).

As we show later, Theorems 1 and 2 give in fact comparable security bounds when applied to leakage-resilient stream ciphers (see Sect. 1.6)

## 1.4  Our Results

We reduce the dependency of the simulator complexity $s_h$ on the advantage $\epsilon$ to only a factor of $\epsilon^{-2}$, from the factor of $\epsilon^{-4}$.

**Theorem 3 (Our Simulator).** *For every distribution* $X, Z$ *on* $\{0,1\}^n \times \{0,1\}^\ell$ *and every* $\epsilon$, $s$, *there exists a "simulator"* $h : \{0,1\}^n \to \{0,1\}^\ell$ *such that*

*(a)* $(X, h(X))$ *and* $(X, Z)$ *are* $(s, \epsilon)$*-indistinguishable*
*(b)* $h$ *is of complexity* $s_h = O\left(s \cdot 2^{5\ell} \log(1/\epsilon)\epsilon^{-2}\right)$

Below in Table 1 we compare our result to previous works.

**Table 1.** The complexity of simulating $\ell$-bit auxiliary information given required indistinguishability strength, depending on the proof technique. For simplicity, terms polylog$(1/\epsilon)$ are omitted.

| Author | Technique | Advantage | Size | Cost of simulating |
|---|---|---|---|---|
| [JP14] (Theorem 1) | Min-Max | $\epsilon$ | $s$ | $s_h = O\left(s \cdot 2^{4\ell} \epsilon^{-4}\right)$ |
| [VZ13] (Theorem 2) | Complicated boosting | | | $s_h = O\left(s \cdot 2^\ell / \epsilon^2 + 2^\ell \epsilon^{-4}\right)$ |
| **This paper** (Theorem 3) | Simple boosting | | | $s_h = O\left(s \cdot 2^{5\ell} \epsilon^{-2}\right)$ |

Our result is slightly worse in terms of dependency $\ell$, but outperforms previous results in terms of dependency on $\epsilon^{-1}$. However, the second dependency is more crucial for cryptographic applications. Note that the typical choice is sub-logarithmic leakage, that is $\ell = o\left(\log \epsilon^{-1}\right)$ is asymptotic settings[4] (see for example [CLP15]). Stated in non-asymptotic settings this assumption translates

---

[4] This is a direct consequence of the fact that we want $\ell$ to fit poly-preserving reductions.

to $\ell < c \log \epsilon^{-1}$ where $c$ is a small constant (for example $c = \frac{1}{12}$ see [Pie09]). In these settings, we outperform previous results.

To illustrate this, suppose we want to achieve security $\epsilon = 2^{-60}$ simulating just one bit from a 256-bit input. As it follows from Table 1, previous bounds are useless as they give the complexity bigger than $2^{256}$ which is the worst complexity of all boolean functions over the chosen domain. In settings like this, only our bound can be applied to conclude meaningful results. For more concrete examples of settings where our bounds are even only meaningful, we refer to Table 2 in Sect. 1.6.

## 1.5   Our Techniques

Our approach utilizes a simple boosting technique: as long as the condition (a) in Theorem 3 fails, we can use the distinguisher to improve the simulator. This makes our algorithm constructive with respect to distinguishers obtained from an oracle[5], similarly to other boosting proofs [JP14, VZ13]. In short, if for a "candidate" solution $h$ there exists D such that

$$\mathbb{E}\, D(X, Z) - \mathbb{E}\, D(X, h(X)) > \epsilon$$

then we construct a new solution $h'$ using D and $h$, according to the equation[6]

$$\Pr[h'(x) = z] = \Pr[h(x) = z] + \gamma \cdot \mathsf{Shift}\,(D(x, z)) + \mathsf{Corr}(x, z)$$

where

(a) The parameter $\gamma$ is a *fixed step* chosen in advance (its optimal value depends on $\epsilon$ and $\ell$ and is calculated in the proof.)
(b) $\mathsf{Shift}\,(D(x, z))$ is a *shifted* version of D, so that $\sum_z \mathsf{Shift}\,(D(x, z)) = 0$. This restriction correspond to the fact that we want to preserve the constraint $\sum_z h(x, z) = 1$. More precisely, $\mathsf{Shift}\,(D(x, z)) = D(x, z) - \mathbb{E}_{z' \leftarrow U_\ell} D(x, z)$
(c) $\mathsf{Corr}(x, z)$ is a *correction term* used to fix (some of) possibly negative weights.

The procedure is being repeated in a loop, over and over again. The main technical difficulty is to show that it eventually stops after not so many iterations.

Note that in every such a step the complexity cost of the shifting term is $O\left(2^\ell \cdot \mathsf{size}(D)\right)$[7]. The correction term, in our approach, does a search over $z$ looking for the biggest negative mass, and redistributes it over the remaining points. Intuitively, it works because the total negative mass is getting smaller with every step. See Algorithm 1 for a pseudo-code description of the algorithm and the rest of Sect. 3 for a proof.

---

[5] The oracle evaluates the distance of the given candidate solution and the simulated distribution, answering with a distinguisher if the distance is smaller than required.
[6] As we already mentioned, we can assume that D is deterministic without loss of generality. Then all the terms in the equation are well-defined.
[7] By definition, it requires computing the average of $D(x, \cdot)$ over $2^\ell$ elements.

## 1.6   Applications

**Better Security for the EUROCRYPT'09 Stream Cipher.** The first construction of leakage-resilient stream cipher was proposed by Dziembowski and Pietrzak in [DP08]. On Fig. 1 below we present a simplified version of this cipher [Pie09], based on a weak pseudorandom function (wPRF).

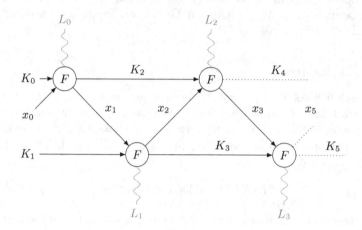

**Fig. 1.** The EUROCRYPT'09 stream cipher (adaptive leakage). $F$ denotes a weak pseudorandom function. By $K_i$ and $x_i$ we denote, respectively, values of the secret state and keystream bits. Leakages are denotted in gray with $L_i$.

Jetchev and Pietrzak in [JP14] showed how to use the simulator theorem to simplify the security analysis of the EUROCRYPT'09 cipher. The cipher security depends on the complexity of the simulator as explained in Theorem 1 and Remark 2. We consider the following setting:

- number of rounds $q = 16$,
- $F$ instantiated with AES256 (as in [JP14])
- cipher security we aim for $\epsilon' = 2^{-40}$
- $\lambda = 3$ bits of leakage per round

The concrete bounds for $(q, \epsilon', s')$-security of the cipher (which roughly speaking means that $q$ consecutive outputs is $(s', \epsilon')$-pseudorandom, see Sect. 2 for a formal definition) are given in Table 2 below. We ommit calculations as they are merely putting parameters from Theorems 1, 2 and 3 into Remark 2 and assuming that AES as a weak PRF is $(\epsilon, s)$-secure for any pairs $s/\epsilon \approx 2^k$ (following the similar example in [JP14]).

More generally, we can give the following comparison of security bounds for different wPRF-based stream ciphers, in terms of time-sccess ratio. The bounds in Table 3 follow from the simple lemma in Sect. 4, which shows how the time-success ratio changes under explicit reduction formulas.

**Table 2.** The security of the EUROCRYPT'09 stream cipher, instantiated with AES256 as a weak PRF of rouhgly $k = 256$ bits of security. In this settngs only our new bounds provide non-trivial bounds.

| Analysis/authors | wPRF security | Leakage | Advantage $\epsilon'$ | Size $s'$ |
|---|---|---|---|---|
| [JP14] (Theorem 1) | 256 | $\lambda = 3$ | $2^{-40}$ | 0 |
| [VZ13] (Theorem 2) | | | | 0 |
| **This paper** (Theorem 3) | | | | $2^{66}$ |

**Table 3.** Different bounds for wPRF-based leakage-resilient stream ciphers. $k$ is the security level of the underlying wPRF. The value $k'$ is the security level for the cipher, understood in terms of time-success ratio. the numbers denote: (1) The EUROCRYPT'09 cipher, (2) The CSS'10/CHESS'12 cipher, (3) The CT-RSA'13 cipher.

| Cipher | Analysis | Proof techniques | Security level | Comments |
|---|---|---|---|---|
| (1) | [Pie09] | Pseudoentropy chain rules | $k' \ll \frac{1}{8}k$ | Large number of blocks |
| (1) | [JP14] | Aux. Inputs Simulator (corr.) | $k' \approx \frac{k}{6} - \frac{5}{6}\lambda$ | |
| (1) | [VZ13] | Aux. Inputs Simulator | $k' \approx \frac{k}{6} - \frac{1}{3}\lambda$ | |
| (1) | **This work** | Aux. Inputs Simulator | $k' \approx \frac{k}{4} - \frac{4}{3}\lambda$ | |
| (2) | [FPS12] | Pseudoentropy chain rules | $k' \approx \frac{k}{5} - \frac{3}{5}\lambda$ | Large public seed |
| (3) | [YS13] | Square-friendly apps. | $k' \approx \frac{k}{4} - \frac{3}{4}\lambda$ | Only in minicrypt |

## 1.7  Organization

In Sect. 2 we discuss basic notions and definitions. The proof of Theorem 3 appears in Sect. 3.

## 2  Preliminaries

### 2.1  Notation

By $\mathbb{E}_{y \leftarrow Y} f(y)$ we denote an expectation of $f$ under $y$ sampled according to the distribution $Y$.

### 2.2  Basic Notions

*Indistinguishability.* Let $\mathcal{V}$ be a finite set, and $\mathcal{D}$ be a class of deterministic $[0,1]$-valued functions on $\mathcal{V}$. For any two real functions $f_1, f_2$ on $\mathcal{V}$, we say that $f_1, f_2$ are $(\mathcal{D}, \epsilon)$-indistinguishable if

$$\forall D \in \mathcal{D}: \quad \left| \sum_{x \in \mathcal{V}} D(x) \cdot f_1(x) - \sum_{x \in \mathcal{V}} D(x) \cdot f_2(x)) \right| \leqslant \epsilon$$

Note that the domain $V$ depends on the context. If $X_1, X_2$ are two probability distributions, we say that they are $(s, \epsilon)$-indistinguishable if their probability mass functions are indistinguishable, that is when

$$\left| \sum_{x \in V} \mathrm{D}(x) \cdot \Pr[X_1 = x] - \sum_{x \in V} \mathrm{D}(x) \cdot \Pr[X_2 = x] \right| \leqslant \epsilon$$

for all $\mathrm{D} \in \mathcal{D}$. If $\mathcal{D}$ consists of all circuits of size $s$ we say that $f_1, f_2$ are $(s, \epsilon)$-indistinguishable.

*Remark 1.* This an extended notion of indistinguishability, borrowed from [TTV09], which captures not only probability measures but also real-valued functions. A good intuition is provided by the following observation [TTV09]: think of functions over $V$ as $|V|$-dimensional vectors then $\epsilon \geqslant |\sum_{x \in V} \mathrm{D}(x) \cdot f_1(x) - \sum_{x \in V} \mathrm{D}(x) \cdot f_2(x)| = |\langle f_1 - f_2, \mathrm{D} \rangle|$ means that $f_1$ and $f_2$ are *nearly orthogonal* for all test functions in $\mathcal{D}$.

*Distinguishers.* In the definition above we consider deterministic distinguishers, as this is required by our algorithm. However, being randomized doesn't help in distinguishing, as any randomized-distinguisher achieving advantage $\epsilon$ when run on two fixed distributions can be converted into a deterministic distinguishers of the same size and advantage (by fixing one choice of coins). Moreover, any real-valued distinguisher can be converted, by a boolean threshold, into a boolean one with at least the same advantage [FR12].

*Relative Complexity.* We say that a function $h$ has complexity at most $T$ relative to the set of functions $\mathcal{D}$ if there are functions $\mathrm{D}_1, \ldots, \mathrm{D}_T$ such $h$ can be computed by combining them using at most $T$ of the following operations: (a) multiplication by a constant, (b) application of a boolean threshold function, (c) sum, (d) product.

## 2.3    Stream Ciphers Definitions

We start with the definition of weak pseudorandom functions, which are *computationally indistinguishable* from random functions, when queried on random inputs and fed with uniform secret key.

**Definition 1 (Weak pseudorandom functions).** *A function* $\mathrm{F} : \{0,1\}^k \times \{0,1\}^n \to \{0,1\}^m$ *is an* $(\epsilon, s, q)$-*secure weak PRF if its outputs on* $q$ *random inputs are indistinguishable from random by any distinguisher of size* $s$, *that is*

$$\left| \Pr\left[ \mathrm{D}\left( (X_i)_{i=1}^q, \mathrm{F}((K, X_i)_{i=1}^q) = 1 \right] - \Pr\left[ \mathrm{D}\left( (X_i)_{i=1}^q, (R_i)_{i=1}^q \right) = 1 \right] \right| \leqslant \epsilon$$

*where the probability is over the choice of the random* $X_i \leftarrow \{0,1\}^n$, *the choice of a random key* $K \leftarrow \{0,1\}^k$ *and* $R_i \leftarrow \{0,1\}^m$ *conditioned on* $R_i = R_j$ *if* $X_i = X_j$ *for some* $j < i$.

Stream ciphers generate a keystream in a recursive manner. The security requires the output stream should be indistinguishable from uniform[8].

**Definition 2 (Stream ciphers).** *A stream-cipher* $\mathsf{SC} : \{0,1\}^k \to \{0,1\}^k \times \{0,1\}^n$ *is a function that, when initialized with a secret state* $S_0 \in \{0,1\}^k$, *produces a sequence of output blocks* $X_1, X_2, \ldots$ *computed as*

$$(S_i, X_i) := \mathsf{SC}(S_{i-1}).$$

*A stream cipher* $\mathsf{SC}$ *is* $(\epsilon, s, q)$-*secure if for all* $1 \leqslant i \leqslant q$, *the random variable* $X_i$ *is* $(s, \epsilon)$-*pseudorandom given* $X_1, \ldots, X_{i-1}$ *(the probability is also over the choice of the initial random key* $S_0$).

Now we define leakage resilient stream ciphers, following the "only computation leaks" assumption.

**Definition 3 (Leakage-resilient stream ciphers).** *A leakage-resilient stream-cipher is* $(\epsilon, s, q, \lambda)$-*secure if it is* $(\epsilon, s, q)$-*secure as defined above, but where the distinguisher in the* $j$-*th round gets* $\lambda$ *bits of arbitrary deceptively chosen leakage about the secret state accessed during this round. More precisely, before* $(S_j, X_j) := \mathsf{SC}(S_{j-1})$ *is computed, the distinguisher can choose any leakage function* $f_j$ *with range* $\{0,1\}^\lambda$, *and then not only get* $X_j$, *but also* $\Lambda_j := f_j(\hat{S}_{j-1})$, *where* $\hat{S}_{j-1}$ *denotes the part of the secret state that was modified (i.e., read and/or overwritten) in the computation* $\mathsf{SC}(S_{j-1})$.

## 2.4 Security of Leakage-Resilient Stream Ciphers

Best provable secure constructions of leakage-resilient streams ciphers are based on so called weak PRFs, primitives which look random when queried on random inputs [Pie09, FPS12, JP14, DP10, YS13]. The most recent (TCC'14) analysis is based on a version of Theorem 1.

**Theorem 4 (Proving Security of Stream Ciphers [JP14]).** *If* $F$ *is a* $(\epsilon_F, s_F, 2)$-*secure weak PRF then* $\mathsf{SC}^F$ *is a* $(\epsilon', s', q, \lambda)$-*secure leakage resilient stream cipher where*

$$\epsilon' = 4q\sqrt{\epsilon_F 2^\lambda}, \quad s' = \Theta(1) \cdot \frac{s_F \epsilon'^4}{2^{4\lambda}}.$$

*Remark 2 (The exact complexity loss).* An inspection of the proof in [JP14] shows that $s_F$ equals the complexity of the simulator $h$ in Theorem 1, with circuits of size $s'$ as distingusihers and $\epsilon$ replaced by $\epsilon'$.

---

[8] We note that in a more standard notion the entire stream $X_1, \ldots, X_q$ is indistinguishable from random. This is implied by the notion above by a standard hybrid argument, with a loss of a multiplicative factor of $q$ in the distinguishing advantage.

## 2.5    Time-Success Ratio

The running time (circuit size) $s$ and success probability $\epsilon$ of attacks (practical and theoretical) against a particular primitive or protocol may vary. For this reason Luby [LM94] introduced the time-success ratio $\frac{t}{\epsilon}$ as a universal measure of security. This model is widely used to analyze provable security, cf. [BL13] and related works.

**Definition 4 (Security by Time-Success Ratio [LM94]).** *A primitive P is said to be $2^k$-secure if for every adversary with time resources (circuit size in the nonuniform model) $s$, the success probability in breaking P (advantage) is at most $\epsilon < s \cdot 2^{-k}$. We also say that the time-success ratio of P is $2^k$, or that is has k bits of security.*

For example, AES with a 256-bit random key is believed to have 256 bits of security as a *weak* PRF[9].

## 3    Proof of Theorem 3

For technical convenience, we attempt to efficiently approximate the conditional probability function $g(x, z) = \Pr[Z = z | X = x]$ rather than building the sampler directly. Once we end with building an efficient approximation $h(x, z)$, we transform it into a sampler $h_{\mathsf{sim}}$ which outputs $z$ with probability $h(x, z)$ (this transformation yields only a loss of $2^\ell \log(1/\epsilon)$). We are going to prove the following fact

> For every function $g$ on $\mathcal{X} \times \mathcal{Z}$ which is a $\mathcal{X}$-conditional probability mass function over $Z$ (that is $g(x, z) \geqslant 0$ for all $x, z$ and $\sum_z g(x, z) = 1$ for every $x$), and for every class $\mathcal{D}$ closed under complements[10] there exists $h$ such that
> (a) $h$ is a $\mathcal{X}$-conditional probability mass function over $Z$
> (b) $h$ is of complexity $s_h = O(2^{4\ell}\epsilon^{-2})$ with respect to $\mathcal{D}$
> (c) $(X, Z)$ and $(X, h_{\mathsf{sim}}(X))$ are indistinguishable, which in terms of $g$ and $h$ means

$$\left| \sum_z \mathbb{E}_{x \sim X} \left[ \mathrm{D}(x, z) \cdot (g(x, z) - h(x, z)) \right] \right| \leqslant \epsilon \tag{2}$$

The sketch of the construction is shown in Algorithm 1. Here we would like to point out two things. First, we stress that we do not produce a strictly positive function; what our algorithm guarantees, is that the total negative mass is *small*. We will see later that this is enough. Second, our algorithm performs essentially same operations for every $x$, which is why its complexity depends only on $\mathcal{Z}$.

We denote for shortness $\overline{\mathrm{D}}(x, z) = \mathrm{D}(x, z) - \mathbb{E}_{z' \leftarrow U_Z} \mathrm{D}(x, z')$ for any D (the "shift" transformation).

---

[9] We consider the security of AES256 as a weak PRF, and not a standard PRF, because of non-uniform attacks which show that no PRF with a $k$-bit key can have $s/\epsilon \approx 2^k$ security [DTT09], at least unless we additionally require $\epsilon \gg 2^{-k/2}$.

[10] This is a standard assumption in indistinguishability proofs. We can always extend the class by adding $-\mathrm{D}$ for every $\mathrm{D} \in \mathcal{D}$, which increases the complexity only by 1.

---

**Algorithm 1.** Construct the Auxiliary Inputs Simulator

---

**input** : Function $g : \{0,1\}^n \times \{0,1\}^\ell \to [0,1]$, accuracy paramter $\epsilon > 0$, class
$\mathcal{D}$, step $\gamma$

**output**: Function $h$ which is $\epsilon$-indistinguishable from $g$ under $\mathcal{D}$, add up to 1
for every $x$, and with total negative mass smaller $\gamma |\mathcal{Z}|^3$

1  $t \leftarrow 0$
2  $h^0(x,z) \leftarrow \frac{1}{|\mathcal{Z}|}$ for every $x$ and $z$
3  **while** *exists* $D \in \mathcal{D}$ *such that* $\mathbb{E}_{x \sim X} \left[ \sum_z \overline{D}(x,z) \cdot \left( g(x,z') - h^t(x,z') \right) \right] \geqslant \epsilon$ **do**
   /* while the simulator is not good enough */
4  |   $D^{t+1} \leftarrow D$
5  |   **for** $z' \in \mathcal{Z}$ **do**    /* improve the simulator towards the distinguisher
   |   direction */
6  |   |   $h^{t+1}(x,z') \leftarrow h^t(x,z') + \gamma \cdot \overline{D^{t+1}}(x,z')$
7  |   $t \leftarrow t+1$
8  |   $m \leftarrow 0$
9  |   **for** $z' \in \mathcal{Z}$ **do**           /* locate the biggest negative point mass */
10 |   |   **if** $h^t(x,z') < m$ **then**
11 |   |   |   $m \leftarrow h^t(x,z')$
12 |   |   |   $z^- \leftarrow z'$
13 |   $h^t(x,z^-) = 0$    /* cut the biggest negative mass */ **for** $z' \in \mathcal{Z}$ **do**
14 |   |   $h^t(x,z') \leftarrow h^t(x,z') + \frac{m}{|\mathcal{Z}|-1}$    /* redestribute the cut mass */

15 **return** $h^t(x,z)$

---

*Proof.* Consider the functions $h^t$. Define $\tilde{h}^{t+1}(x,z) \stackrel{def}{=} h^t(x,z) + \gamma \cdot \overline{D}^{t+1}(x,z)$.
According to Algorithm 1, we have

$$h^{t+1}(x,z) = h^t(x,z) + \gamma \cdot \overline{D}^{t+1}(x,z) + \theta^{t+1}(x,z) \qquad (3)$$

with the correction term $\theta^t(x,z)$ that be computed recursively as (see Line 13
in Algorithm 1)

$$\theta^t(x,z) = 0$$

$$\theta^t(x,z) = \begin{cases} -\min\left( h^t(x,z) + \gamma \cdot \overline{D}^{t+1}(x,z), 0 \right), & \text{if } z = z^t_{\min}(x) \\ \frac{\min\left( h^t(x,z^t_{\min}(x)) \right) + \gamma \cdot \overline{D}^{t+1}(x,z^t_{\min}(x)), 0 \right)}{\#\mathcal{Z}-1} & \text{if } z \neq z^t_{\min}(x) \end{cases} \quad t = 0,1,\ldots$$

$$(4)$$

where $z^t_{\min}(x)$ is one of the points $z$ minimizing $h^t(x,z) + \gamma \cdot \overline{D}^{t+1}(x,z)$ (chosen
and fixed for every $t$) . In particular

$$h^t(x,z^t_{\min}(x))) + \gamma \cdot \overline{D}^{t+1}(x,z^t_{\min}(x)) < 0 \iff \exists z : \ h^t(x,z) + \gamma \cdot \overline{D}^{t+1}(x,z) < 0$$

$$(5)$$

<u>Notation</u>: for notational convenience we indenify the functions $D^t(x,z)$, $\overline{D}^t(x,z)$,
$\theta^t(x,z)$, $\tilde{h}^t(x,z)$ and $h^t(x,z)$ with matrices where $x$ are columns and $z$ are rows.

That is $h_x^t$ denotes the $|\mathcal{Z}|$-dimensional vector with entries $h^t(x,z)$ for $z \in \mathcal{Z}$ and similarly for other functions $D^t(x,z)$, $\overline{D}^t(x,z)$, $\theta^t(x,z)$, $\tilde{h}^t(x,z)$.

*Claim 1 (Complexity of Algorithm 1).* $T$ executions of the "while loop" can be realized with time $O\left(T \cdot |\mathcal{Z}| \cdot \text{size}(\mathcal{D})\right)$ and memory $O(|\mathcal{Z}|)$.[11]

This claim describes precisely resources required to compute the function $h^T$ for every $T$. In order to bound $T$, we define the energy function as follows:

*Claim 2 (Energy function).* Define the auxiliary function

$$\Delta^t = \sum_{i=0}^{t-1} \mathbb{E}_{x \sim X} \left[ \overline{D}_x^{i+1} \cdot \left(g_x - h_x^i\right) \right].\tag{6}$$

Then we have $\Delta^t = E_1 + E_2$ where

$$E_1 = \tfrac{1}{\gamma} \mathbb{E}_{x \sim X} \left[ \left(h_x^t - h_x^0\right) \cdot g_x + \tfrac{1}{2} \sum_{i=0}^{t-1} \left(h_x^{i+1} - h_x^i\right)^2 - \tfrac{1}{2}\left(\left(h_x^t\right)^2 - \left(h_x^0\right)^2\right) \right]$$
$$E_2 = \tfrac{1}{\gamma} \mathbb{E}_{x \sim X} \left[ -\sum_{i=0}^{t-1} \theta_x^{i+1} \cdot \left(g_x - h_x^{i+1}\right) - \sum_{i=0}^{t-1} \theta_x^{i+1} \cdot \left(h_x^{i+1} - h_x^i\right) \right]\tag{7}$$

Note that all the symbols represent vectors and multiplications, including squares, should be understood as scalar products. The proof is based on simple algebraic manipulations and appears in Appendix B.

*Remark 3 (Technical issues and intuitions).* To upper-bound the formulas in Eq. (7), we need the following important properties

(a) *Boundedness of correction terms,* that is ideally $|\theta^i(x.z)| = O(\text{poly}(|\mathcal{Z}|) \cdot \gamma)$.
(b) *Acute angle between the correction and the error,* that is $\theta_x^i \cdot \left(g_x - h_x^i\right) \geqslant 0$.

Below we present an outline of the proof, discussing more technical parts in the appendix.

**Proof Outline.** Indeed, with these assumptions we prove an upper bound on the energy function, namely

$$E_1 + E_2 \leqslant O\left(\text{poly}(|\mathcal{Z}|) \cdot \left(t\gamma + \gamma^{-1}\right)\right),\tag{8}$$

which follows from the properties (a) and (b) above (they are proved in Claims 4 and 3 below, and the inequality on $E_1 + E_2$ is derived in Claim 5). Note that, except a factor $\text{poly}(|\mathcal{Z}|)$, our formula (not the proof, though) is identical to the bound used in [TTV09] (see Claim 3.4 in the eprint version). Indeed, our theorem is, to some extent, an extension to the main result in [TTV09] to cover the conditional case, where $|\mathcal{X}| > 1$. The main difference is that we show how to simulate a short leakage $|Z|$ given $X$, whereas [TTV09] shows how to simulate

---

[11] The RAM model.

$Z$ alone, under the assumption that the distribution of $Z$ is dense in the uniform distribution (the min-entropy gap being small)[12].

Since the bound above is valid for any step $t$, and since on the other hand we have $t\epsilon \leqslant \Delta^t$ after $t$ steps of the algorithm, we achieve a contradiction (to the number of steps) setting $\gamma = \epsilon/\text{poly}(|\mathcal{Z}|)$. Indeed, suppose that $t\epsilon \leqslant A|\mathcal{Z}|^B(\gamma^{-1} + t\gamma)$ for some positive constants $A, B$. Since the step size $\gamma$ can be chosen arbitrarily, we can set $\gamma = \frac{\epsilon}{2A|\mathcal{Z}|^B}$ which yields $\frac{t\epsilon}{2} \leqslant \frac{2A^2|\mathcal{Z}|^B}{\epsilon}$ or $t \leqslant 4A^2|\mathcal{Z}|^B\epsilon^{-2}$, which means that the algorithm terminates after at most $T = \text{poly}(|\mathcal{Z}|)\epsilon^{-2}$ steps. Our proof goes exactly this way, except some extra optimization do obtain better exponent $A$.

We stress that it outputs only a *signed measure*, not a probability distribution yet. However, because of property (a) the negative mass is only of order $\text{poly}(|\mathcal{Z}|)\epsilon$ and the function we end with can be simply rescaled (we replace negative masses by 0 and normalize the function dividing by a factor $1 - m$ where $m$ is the total negative mass). With this transformation, we keep the expected advantage $O(\epsilon)$ and lose an extra factor $O(|\mathcal{Z}|)$ in the complexity. We can then. Finally, we need to remember that we construct only a probability distribution function, not a sampler. Transforming it into a sampler yields an overhead of $O(\mathcal{Z})$. This discussion shows that it is possible to build a sampler of complexity $\text{poly}(|\mathcal{Z}|)\epsilon^{-2}$ with respect to $\mathcal{D}$. A more careful inspection of the proof shows that we can actually achieve the claimed bound $|\mathcal{Z}|^5\epsilon^{-2}$ (see Remark 4 at the end of the proof).

**Technical Discussion.** We note that condition (b) somehow means that mass cuts should go in the right direction, as it is much simpler to prove that Algorithm 1 terminates when there are no correction terms $\theta^t$; thus we don't want to go in a wrong direction and ruin the energy gain. Concrete bounds on properties (a) and (b) are given in Claims 3 and 4.

In Algorithm 1 in every round we shift only one negative point mass (see Line 13). However, since this point mass is chosen to be as big as possible and since $h^{t+1}$ and $h^t$ differ only by a small term $\gamma \cdot \overline{\text{D}}^{t+1}$ except the mass shift $\theta^{t+1}$, one can expect that we have the negative mass under control. Indeed, this is stated precisely in Claim 3 below.

*Claim 3 (The total negative mass is small).* Let

$$\text{NegativeMass}(h^t(x, \cdot)) = -\sum_z \min(h^t(x, z), 0)$$

be the total negative mass in $h^t(x, z)$ as the function of $z$. Then we have

$$\text{NegativeMass}(h^t(x, \cdot) < |\mathcal{Z}|^3\gamma. \tag{9}$$

---

[12] It's not possible to extend the result from [TTV09] directly, the issue is that the constraint on the marginal distribution are not preserved. That's why [JP14] and this paper require much more extra work.

for every $x$ and every $t$. In fact, for all $x, z$ and $t$ we have the following stronger bound

$$\max_z \left| \min(h^t(x,z),0) \right| < |\mathcal{Z}|\gamma.$$

The proof is based on a recurrence relation that links $\mathsf{NegativeMass}(h^{t+1}(x,\cdot))$ with $\mathsf{NegativeMass}(h^t(x,\cdot))$, and appears in Appendix C.

*Claim 4 (The angle formed by the correction and the difference vector is acute).* For every $x,t$ we have $\mathsf{Angle}\left(\theta_x^{t+1}, g_x - h_x^{t+1}\right) \in \left[-\frac{\pi}{2}, \frac{\pi}{2}\right].$

The proof appears in Appendix D.

Having established Claims 3 and 4 we are now in position to prove a concrete bound in Eq. (8). To this end, we give upper bounds on $E_1$ and $E_2$, defined in Eq. (7), separately.

*Claim 5 (Algorithm 1 terminates after a small number of steps).* The energy function in Claim 2 can be bounded as follows

$$E_1 \leqslant \gamma^{-1}\left(1 + 2|\mathcal{Z}|^2\gamma + |\mathcal{Z}|t\gamma^2 + |\mathcal{Z}|^3 t\gamma^2\right), \quad E_2 \leqslant 2|\mathcal{Z}|^2 t\gamma.$$

In particular, we conclude that with $\gamma = \frac{\epsilon}{8|\mathcal{Z}|^4}$ the algorithm terminates after at most $t = O(|\mathcal{Z}|^3)\epsilon^{-2}$ steps.

First, note that by Claim 4 we have $-\sum_{i=0}^{t-1}\theta_x^{i+1}\cdot\left(g_x - h_x^{i+1}\right) \leqslant 0$. Second, by definition of the sequence $(h^i)_i$ we have $-\sum_{i=0}^{t-1}\theta_x^{i+1}\cdot\left(h_x^{i+1} - h_x^i\right) = -\sum_{i=0}^{t-1}\theta_x^{i+1}\cdot$ $\theta_x^{i+1} - \sum_{i=0}^{t-1}\gamma\theta_x^{i+1}\cdot\overline{\mathrm{D}}_x^{i+1}$ which is at most $2|\mathcal{Z}|^3 t\gamma^2$, because of Eq. (9) (the sum of absolute correction terms $\sum_z|\theta^{i+1}(x,z)|$ is, by definition, twice the total negative mass, and $|\overline{\mathrm{D}}^{i+1}(x,z)| \leqslant 1$). This proves that

$$E_2 \leqslant \frac{1}{\gamma}\cdot 2|\mathcal{Z}|^3 t\gamma^2 = 2|\mathcal{Z}|^3 t\gamma.$$

To bound $E_1$, note that we have to bounds two non-negative terms, namely $\frac{1}{2}\sum_i\left(h_x^{i+1} - h_x^i\right)^2$ and $\left(h_x^t - h_x^0\right)\cdot g_x$. As for the first one, we have

$$\left(h_x^{i+1} - h_x^i\right)^2 = \left(\gamma\overline{\mathrm{D}}_x^{i+1} + \theta_x^{i+1}\right)^2 \leqslant 2(\gamma\overline{\mathrm{D}}_x^{i+1})^2 + 2\left(\theta_x^{i+1}\right)^2,$$

where the inequality follows by the Cauchy-Schwarz inequality[13]. We trivially have $\left(\overline{\mathrm{D}}_x^{i+1}\right)^2 \leqslant |\mathcal{Z}|$ (because of $|\overline{\mathrm{D}}(x,z)| \leqslant 1$). By the definition of correction terms in Eq. (4) we have $\left(\theta_x^{i+1}\right)^2 = \sum_z(\theta^{i+1}(x,z))^2 < 2(\theta^{i+1}(x,z_0))^2$, where $\theta^{i+1}(x,z_0)$ is the smallest negative mass, which is at most $(2|\mathcal{Z}|^3\gamma)^2$ by Eq. (9) . Thus, we have $\left(h_x^{i+1} - h_x^i\right)^2 \leqslant 2|\mathcal{Z}|\gamma^2 + 8|\mathcal{Z}|^6\gamma^2$. To bund $\left(h_x^t - h_x^0\right)\cdot g_x$ note that $-h_x^0\cdot g_x \leqslant 0$ and that $h_x^t\cdot g_x \leqslant \max_z|h^t(x,z)|$ (because $g(x,z) \geqslant 0$

---
[13] Or cam be concluded from the parallelogram identity $(x+y)^2 + (x-y)^2 = x^2 + y^2$.

and $\sum_x g(x, z) = 1)$ which means $h_x^t \cdot g_x \leqslant 1 + 2\mathsf{NegativeMass}(h_x^t)$ (as $\sum_z \max(h^t(x, z), 0) = 1 - \sum_z \min(h^t(x, z), 0) = 1 + \mathsf{NegativeMass}(h_x^t)$ and $-\sum_z \min(h^t(x, z), 0) = \mathsf{NegativeMass}(h_x^t)$ by $\sum_z \max(h^t(x, z) = 1$ and the definition of the total negative mass). This allows us to estimate $E_1$ as follows

$$E_1 \leqslant \gamma^{-1}\left(1 + 2|\mathcal{Z}|^3\gamma + |\mathcal{Z}|t\gamma^2 + 4|\mathcal{Z}|^6 t\gamma^2\right)$$

After $t$ steps, the energy is at least $t\epsilon$. On the other hand, it at most $E_1 + E_2$. Since $|\mathcal{Z}|, |\mathcal{Z}|^3 \leqslant |\mathcal{Z}|^6$, we obtain

$$t\epsilon < \gamma^{-1} + 2|\mathcal{Z}|^3 + 7|\mathcal{Z}|^6 t\gamma$$

Since this is true for any positive $\gamma$, we choose $\gamma = \frac{\epsilon}{14|\mathcal{Z}|^6}$, which gives us (slightly weaker than claimed)

$$t < 32|\mathcal{Z}|^6\epsilon^{-2}.$$

*Remark 4 (Optimized bounds).* By the second part of Claim 3 we have $|\theta^i(x, z)| < |\mathcal{Z}|\gamma$ for every $x, z$ and $i$. An inspection of the discussion above shows that this allows us to improve the bounds on $E_1, E_2$

$$E_1 \leqslant \gamma^{-1}\left(1 + 2|\mathcal{Z}|^2\gamma + |\mathcal{Z}|t\gamma^2 + |\mathcal{Z}|^2 t\gamma^2\right), \quad E_2 \leqslant 2|\mathcal{Z}|^2 t\gamma$$

Setting $\gamma = \frac{\epsilon}{8|\mathcal{Z}|^2}$ we get $E_1 + E_2 \leqslant 20|\mathcal{Z}|^2\epsilon^{-1}$ and $t \leqslant 20|\mathcal{Z}|^2\epsilon^{-2}$.

This finishes the proof of the claim.

From Claim 5 we conclude that after $t = O\left(|\mathcal{Z}|^2\epsilon^{-2}\right)$ steps we end up with a function $h = h^t$ that is $(s, \epsilon)$-indistinguishable from $g$, because the algorithm terminated (and, clearly, has the complexity at most $O\left(|\mathcal{Z}|^3\epsilon^{-2}\right)$ relative to circuits of size $s$ (including an overhead of $O(|\mathcal{Z}|)$ to compute $\overline{\mathsf{D}}$ from D). To finish the proof, we need to solve two issues

*Claim 6 (From the signed measure to the probability measure).* Let $h^t$ be the output of the algorithm. Define the probability distribution

$$h(x, z) = \frac{\max(h^t(x, z), 0)}{\sum_{z'} \max(h^t(x, z'), 0)}$$

for every $x, z$. Then $h^t(x, \cdot)$ and $h(x, \cdot)$ are $O(\epsilon)$-statistically close for every $x$.

To prove the claim, we note that $\sum_{z'} \max(h^t(x, z'), 0)$ equals $1 + \beta$ where $\beta = \mathsf{NegativeMass}(h^t(x, \cdot)$. Thus we have $|h(x, z) - h^t(x, z)| \leqslant |h^t(x, z)| \cdot \frac{\beta}{1+\beta}$. Since $\sum_{z'} |h^t(x, z')| = \sum_{z'} \max(h^t(x, z'), 0) - \sum_{z'} \min(h^t(x, z'), 0) = 1 + 2\beta$, we get $\sum |h(x, z) - h^t(x, z)| = O(\beta)$ which is $O(\epsilon)$ by Claim 3 for $\gamma$ defined as in Claim 5.

Recall that we have constructed an approximating probability measure $h$ for the probability mass function $g$, which is not a sampler yet. However, we can fix it by rejection sampling, as shown below.

*Claim 7 (From the pmf to the sampler).* There exists a (probabilistic) function $h_{\mathsf{sim}} : \mathcal{X} \to \mathcal{Z}$ which calls $h(x, z)$ (defined as above) at most $O(|\mathcal{Z}| \log(1/\epsilon))$ times and for every $x$ the distribution of its output is $\epsilon$-close to $h(x, \cdot)$ for every $x$.

The proof goes by a simple rejection sampling argument: we sample a point $z \leftarrow \mathcal{Z}$ at random and reject with probability $h(x, z)$. The rejection probability in one turn is $\frac{1}{|\mathcal{Z}|}$. If we repeat the experiment $|\mathcal{Z}| \log(1/\epsilon)|$ then the probability of rejection in every round is only $\epsilon$. On the other hand, conditioned on the opposite event, we get the distribution identical to $h(x, \cdot)$. So the distance is at most $\epsilon$ as claimed. note that

The last two claims prove that the distribution of $h_{sim}(x)$ is $(s, O(\epsilon))$-close to $h_x^t = h^t(x, \cdot)$, for every $x$. Since $h^t$, as a function of $x, z$ is $(s, \epsilon)$-close to $g$, and $g$ is the conditional distribution of $Z|X$, we obtain

$$X, h_{sim}(X) \approx^{s, O(\epsilon)} X, Z$$

and the complexity of the final sampler $h_{sim}(X)$ is $O(|\mathcal{Z}|^5 \epsilon^{-2})$

## 4    Time-Success Ratio Under Algebraic Transformations

In Theorem 1 below we provide a quantitative analysis of how the time-success ratio changes under concrete formulas in security reductions.

**Lemma 1 (Time-success ratio for algebraic transformations).** *Let $a, b, c$ and $A, B, C$ be positive constants. Suppose that $P'$ is secure against adversaries $(s', \epsilon')$, whenever $P$ is secure against adversaries $(s, \epsilon)$, where*

$$\begin{aligned} s' &= s \cdot c\epsilon^C - b\epsilon^{-B} \\ \epsilon' &= a\epsilon^A. \end{aligned} \tag{10}$$

*In addition, suppose that the following condition is satisfied*

$$A \leqslant C + 1. \tag{11}$$

*Then the following is true: if $P$ is $2^k$-secure, then $P'$ is $2^{k'}$-secure (in the sense of Definition 4) where*

$$k' = \begin{cases} \frac{A}{B+C+1}k + \frac{A}{B+C+1}(\log c - \log b) - \log a, & b \geqslant 1 \\ \frac{A}{C+1}k + \frac{A}{C+1}\log c - \log a, & b = 0 \end{cases} \tag{12}$$

The proof is elementary though not immediate. It can be found in [Skó15].

*Remark 5 (On the technical condition(11)).* This condition is satisfied in almost all applications, at in the reduction proof typically $\epsilon'$ cannot be better (meaning higher exponent) than $\epsilon$. Thus, quite often we have $A \leqslant 1$.

## A    More on the Flaw in [JP14]

In the original setting we have $\mathcal{Z} = \{0,1\}^\lambda$. In the proof of the claimed better bound $O\left(s \cdot 2^{3\lambda}\epsilon^{-2}\right)$ there is a mistake on page 18 (eprint version), when the authors enforce a signed measure to be a probability measure by a mass shifting argument. The number $M$ defined there is in fact a function of $x$ and is hard to compute, whereas the original proof amuses that this is a constant independent of $x$. During iterations of the boosting loop, this number is used to modify distinguishers class step by step, which drastically blows up the complexity (exponentially in the number of steps, which is already polynomial in $\epsilon$). In the min-max based proof giving the bound $O\left(s \cdot 2^{3\lambda}\epsilon^{-4}\right)$ a fixable flaw is a missing factor of $2^\lambda$ in the complexity (page 16 in the eprint version), which is because what is constructed in the proof is only a probability mass function, not yet a sampler [Pie15].

## B    Proof of Claim 2

We can rewrite Eq. (6) as

$$
\Delta^t = \frac{1}{\gamma} \mathbb{E}_{x \sim X} \left[ \sum_{i=0}^{t-1} \left( \left( h_x^{i+1} - h_x^i \right) - \theta_x^{i+1} \right) \cdot \left( g_x - h_x^i \right) \right]
$$

$$
= \frac{1}{\gamma} \mathbb{E}_{x \sim X} \left[ \sum_{i=0}^{t-1} \left( h_x^{i+1} - h_x^i \right) \cdot \left( g_x - h_x^i \right) - \sum_{i=0}^{t-1} \theta_x^{i+1} \cdot \left( g_x - h_x^i \right) \right] \quad (13)
$$

First, note that

$$
\sum_{i=0}^{t-1} \left( h_x^{i+1} - h_x^i \right) \cdot \left( g_x - h_x^i \right)
$$

$$
= \left( h_x^t - h_x^0 \right) \cdot g_x - \sum_{i=0}^{t-1} h_x^i \cdot \left( h_x^{i+1} - h_x^i \right)
$$

$$
= \left( h_x^t - h_x^0 \right) \cdot g_x + \frac{1}{2} \sum_{i=0}^{t-1} \left( h_x^{i+1} - h_x^i \right) \cdot \left( h_x^{i+1} - h_x^i \right) +
$$

$$
- \frac{1}{2} \sum_{i=0}^{t-1} \left( h_x^{i+1} + h_x^i \right) \cdot \left( h_x^{i+1} - h_x^i \right)
$$

$$
= \left( h_x^t - h_x^0 \right) \cdot g_x + \frac{1}{2} \sum_{i=0}^{t-1} \left( h_x^{i+1} - h_x^i \right)^2 - \frac{1}{2} \left( \left( h_x^t \right)^2 - \left( h_x^0 \right)^2 \right)
$$

$$
(14)
$$

As to the second term in Eq. (13), we observe that

$$
- \sum_{i=0}^{t-1} \theta_x^{i+1} \cdot \left( g_x - h_x^i \right) = - \sum_{i=0}^{t-1} \theta_x^{i+1} \cdot \left( g_x - h_x^{i+1} \right) - \sum_{i=0}^{t-1} \theta_x^{i+1} \cdot \left( h_x^{i+1} - h_x^i \right) \quad (15)
$$

# C    Proof of Claim 3

*Proof (Proof of Claim 3).* We start by comparing the total negative mass in the functions $h^{t+1} = h^t + \overline{D}^{t+1} + \theta^{t+1}$ and $h^t$. Suppose first that $\tilde{h}^t(x, z_0) < 0$ where $z_0 = z^t_{\min}(x)$. Since $\sum_{z \neq z_0} \tilde{h}^{t+1} = 1 - \tilde{h}^{t+1}(x, z_0)$, there exists $z_1$ such that $\tilde{h}^{t+1}(x, z_1) \geqslant \frac{1 - \tilde{h}^{t+1}(x, z_0)}{|\mathcal{Z}| - 1} > 0$. Combining this with Eq. (4) we obtain

$$h^{t+1}(x, z_1) = \tilde{h}^{t+1}(x, z_1) + \frac{\tilde{h}^{t+1}(x, z_0)}{|\mathcal{Z}| - 1} \geqslant \frac{1}{|\mathcal{Z}| - 1} \tag{16}$$

These observations together with Eq. (3) give us

$$
\begin{aligned}
\sum_{z \in \mathcal{Z}} \min\left(h^{t+1}(x, z), 0\right) &= \sum_{z \in \mathcal{Z}} \min\left(\tilde{h}^{t+1}(x, z) + \theta^{t+1}(x, z), 0\right) \\
&= \sum_{z \in \mathcal{Z} \setminus \{z_0, z_1\}} \min\left(\tilde{h}^{t+1}(x, z) + \frac{\tilde{h}^{t+1}(x, z_0)}{|\mathcal{Z}| - 1}, 0\right) \\
&\geqslant \sum_{z \in \mathcal{Z} \setminus \{z_0, z_1\}} \min\left(\tilde{h}^{t+1}(x, z), 0\right) + (|\mathcal{Z}| - 2) \cdot \frac{\tilde{h}^{t+1}(x, z_0)}{|\mathcal{Z}| - 1} \\
&= \sum_{z \in \mathcal{Z}} \min\left(\tilde{h}^{t+1}(x, z), 0\right) + (|\mathcal{Z}| - 2) \cdot \frac{\tilde{h}^{t+1}(x, z_0)}{|\mathcal{Z}| - 1} - \tilde{h}^{t+1}(x, z_1) \\
&= \sum_{z \in \mathcal{Z}} \min\left(\tilde{h}^{t+1}(x, z), 0\right) + \min\left(\frac{\tilde{h}^{t+1}(x, z_0)}{|\mathcal{Z}| - 1}, 0\right)
\end{aligned}
\tag{17}
$$

where the inequality line follows from $\tilde{h}^{t+1}(x, z_0) < 0$ and Eq. (16). But by the definition of $z_0 = z^t_{\min}(x)$ we have $\tilde{h}^{t+1}(x, z_0) = \min_z \tilde{h}^{t+1}(x, z)$. Since this value is negative, we get

$$\tilde{h}^{t+1}(x, z_0) \leqslant \frac{1}{|\mathcal{Z}| - 1} \cdot \sum_{z \in \mathcal{Z}} \min\left(\tilde{h}^{t+1}(x, z), 0\right) \tag{18}$$

Combining Eqs. (17) and (18) we obtain

$$-\sum_{z \in \mathcal{Z}} \min\left(h^{t+1}(x, z), 0\right) \leqslant -\left(1 - \frac{1}{(|\mathcal{Z}| - 1)^2}\right) \sum_{z \in \mathcal{Z}} \min\left(\tilde{h}^{t+1}(x, z), 0\right). \tag{19}$$

Since $|h^{t+1}(x, z) - \tilde{h}^t(x, z)| \leqslant \gamma$ by Eq. (3), we get the following recursion

$$-\sum_{z \in \mathcal{Z}} \min\left(h^{t+1}(x, z), 0\right) \leqslant -\left(1 - \frac{1}{(|\mathcal{Z}| - 1)^2}\right) \sum_{z \in \mathcal{Z}} \min\left(h^t(x, z), 0\right) + |\mathcal{Z}|\gamma \tag{20}$$

which can be rewritten as

$$\mathsf{NegativeMass}\left(h^{t+1}(x, \cdot)\right) < \left(1 - \frac{1}{|\mathcal{Z}|^2}\right) \mathsf{NegativeMass}\left(h^t(x, \cdot)\right) + |\mathcal{Z}|\gamma. \tag{21}$$

which is in addition trivially true if $\tilde{h}^{t+1}(x,z) \geqslant 0$ for all $z$. Since we have NegativeMass $\left(h^0(x,\cdot)\right) = 0$, expanding this recursion till $t = 0$ gives an upper bound $|\mathcal{Z}|\gamma \cdot \sum_{j \leqslant t+1} \left(1 - |\mathcal{Z}|^{-2}\right)^j$ which is smaller than by $|\mathcal{Z}|^3 \gamma$ by the convergence of the geometric series. This finishes the proof of the first part.

To prove the second part, recall that by the definition of $z_0$ we have $\tilde{h}^{t+1}(x, z_0) = \min_z \tilde{h}^{t+1}(x, z)$. Suppose that $\tilde{h}^{t+1}(x, z_0) < 0$ (that is, there is a negative mass in $\tilde{h}^{t+1}(x, \cdot)$). Now, by the definition of $h^{t+1}$, we get

$$\max_z \left|\min(h^{t+1}(x,z), 0)\right| = \max_{z \neq z_0} \left|\min(h^{t+1}(x,z), 0)\right|$$

$$= \max_{z \neq z_0} \left|\min\left(\tilde{h}^{t+1}(x,z) + \frac{|\tilde{h}^{t+1}(x, z_0)|}{|\mathcal{Z}| - 1}, 0\right)\right|.$$

Suppose that $\tilde{h}^{t+1}(x, z) + \frac{|\tilde{h}^{t+1}(x,z_0)|}{|\mathcal{Z}|-1} \leqslant 0$ for some $z$. Then, by the definition of $z_0$, we also have

$$0 \geqslant \tilde{h}^{t+1}(x, z) + \frac{|\tilde{h}^{t+1}(x, z_0)|}{|\mathcal{Z}| - 1}$$

$$\geqslant \tilde{h}^{t+1}(x, z_0) + \frac{|\tilde{h}^{t+1}(x, z_0)|}{|\mathcal{Z}| - 1}$$

$$= -\left(1 - \frac{1}{|\mathcal{Z}| - 1}\right)\left|\tilde{h}^{t+1}(x, z_0)\right|.$$

From this we conclude that for *any* $z$ we have

$$\min\left(\tilde{h}^{t+1}(x, z) + \frac{|\tilde{h}^{t+1}(x, z_0)|}{|\mathcal{Z}| - 1}, 0\right) \geqslant -\left(1 - \frac{1}{|\mathcal{Z}| - 1}\right)\left|\tilde{h}^{t+1}(x, z_0)\right|.$$

and thus

$$\max_{z \neq z_0} \left|\min\left(\tilde{h}^{t+1}(x, z) + \frac{|\tilde{h}^{t+1}(x, z_0)|}{|\mathcal{Z}| - 1}, 0\right)\right| \leqslant \left(1 - \frac{1}{|\mathcal{Z}| - 1}\right)\left|\tilde{h}^{t+1}(x, z_0)\right|$$

which means that (still assuming that $\tilde{h}^{t+1}(x, z_0) < 0$)

$$\max_z \left|\min(h^{t+1}(x,z), 0)\right| \leqslant \left(1 - \frac{1}{|\mathcal{Z}| - 1}\right) \max_z \left|\min\left(\tilde{h}^{t+1}(x, z), 0\right)\right|.$$

Note that $0 \geqslant \min\left(\tilde{h}^{t+1}(x, z), 0\right) \geqslant \min\left(h^t(x, z), 0\right) - \gamma$ by the definition of $h^{t+1}$ and $\tilde{h}^{t+1}$. Then

$$\max_z \left|\min(h^{t+1}(x,z), 0)\right| \leqslant \left(1 - \frac{1}{|\mathcal{Z}| - 1}\right) \max_z \left|\min(h^t(x, z), 0)\right| + \gamma.$$

Note that this inequality is true even if $\tilde{h}^{t+1}(x, z_0) = 0$, that is $\tilde{h}^{t+1}(x, z) \geqslant 0$ for all $z$ as then $h^{t+1}(x, z) \geqslant 0$ for all $z$. By expanding this recursion, and noticing that $\min(h^0(x, z), 0) = 0$ for all $x, z$ by definition, we get

$$\max_z \left| \min(h^{t+1}(x, z), 0) \right| \leqslant \gamma \sum_{j=0}^{t} \left( 1 - \frac{1}{|\mathcal{Z}| - 1} \right)^j < |\mathcal{Z}| \gamma.$$

# D   Proof of Claim 4

*Proof.* If $\theta^{t+1}(x, z) = 0$ then there is nothing to prove. Suppose that $\theta^{t+1}(x, z) < 0$. Let $z_0 = z_{\min}^t(x)$. According to Eq. (4) we have $\theta^{t+1}(x, z_0) = -\tilde{h}^{t+1}(x, z_0)$ and $\theta^{t+1}(x, z) = \frac{\tilde{h}^{t+1}(x, z_0)}{\#\mathcal{Z} - 1}$ for $z \neq z_0$. Therefore

$$\theta_x^{t+1} \cdot \left( g_x - \tilde{h}_x^{t+1} \right) = -\tilde{h}^{t+1}(x, z_0) \left( g(x, z_0) - \tilde{h}^{t+1}(x, z_0) \right)$$

$$+ \sum_{z \neq z_0} \frac{\tilde{h}^{t+1}(x, z_0)}{|\mathcal{Z}| - 1} \cdot \left( g(x, z) - \tilde{h}^{t+1}(x, z) \right)$$

$$= -\tilde{h}^{t+1}(x, z_0) \left( g(x, z_0) - \tilde{h}^{t+1}(x, z_0) \right)$$

$$- \frac{\tilde{h}^{t+1}(x, z_0)}{|\mathcal{Z}| - 1} \left( g(x, z_0) - \tilde{h}^{t+1}(x, z_0) \right) \tag{22}$$

and

$$-\theta_x^{t+1} \cdot \theta_x^{t+1} = -\tilde{h}^{t+1}(x, z_0) \cdot \tilde{h}^{t+1}(x, z_0) \left( 1 + \frac{1}{|\mathcal{Z} - 1|} \right). \tag{23}$$

Putting Eqs. (22) and (23) together we obtain

$$\theta_x^{t+1} \cdot \left( g_x - h_x^{t+1} \right) = \theta_x^{t+1} \cdot \left( g_x - \tilde{h}_x^{t+1} \right) - \theta_x^{t+1} \cdot \theta_x^{t+1}$$

$$= - \left( 1 + \frac{1}{|\mathcal{Z}| - 1} \right) \tilde{h}^{t+1}(x, z_0) \cdot g(x, z_0)$$

which is positive because $\tilde{h}^{t,r}(x, z_0) < 0$ and $g(x, z_0) \geqslant 0$. This proves Claim 4.

# References

[BL13] Buldas, A., Laanoja, R.: Security proofs for hash tree time-stamping using hash functions with small output size. In: Boyd, C., Simpson, L. (eds.) ACISP. LNCS, vol. 7959, pp. 235–250. Springer, Heidelberg (2013). doi:10.1007/978-3-642-39059-3_16

[CLP15] Chung, K.-M., Lui, E., Pass, R.: From weak to strong zero-knowledge and applications. In: Dodis, Y., Nielsen, J.B. (eds.) TCC 2015. LNCS, vol. 9014, pp. 66–92. Springer, Heidelberg (2015). doi:10.1007/978-3-662-46494-6_4

[DP08] Dziembowski, S., Pietrzak, K.: Leakage-resilient cryptography. In: Proceedings of the 49th Annual IEEE Symposium on Foundations of Computer Science, Washington, DC, USA, FOCS 2008, pp. 293–302. IEEE Computer Society (2008)

[DP10]  Dodis, Y., Pietrzak, K.: Leakage-resilient pseudorandom functions and side-channel attacks on Feistel networks. In: Rabin, T. (ed.) CRYPTO 2010. LNCS, vol. 6223, pp. 21–40. Springer, Heidelberg (2010). doi:10.1007/978-3-642-14623-7_2

[DTT09]  De, A., Trevisan, L., Tulsiani, M.: Non-uniform attacks against one-way functions and prgs. In: Electronic Colloquium on Computational Complexity (ECCC), vol. 16, p. 113 (2009)

[FK99]  Frieze, A.M., Kannan, R.: Quick approximation to matrices and applications. Combinatorica 19(2), 175–220 (1999)

[FPS12]  Faust, S., Pietrzak, K., Schipper, J.: Practical leakage-resilient symmetric cryptography. In: Prouff, E., Schaumont, P. (eds.) CHES 2012. LNCS, vol. 7428, pp. 213–232. Springer, Heidelberg (2012). doi:10.1007/978-3-642-33027-8_13

[FR12]  Fuller, B., Reyzin, L.: Computational entropy and information leakage. Cryptology ePrint Archive, report 2012/466 (2012). http://eprint.iacr.org/

[GW11]  Gentry, C., Wichs, D.: Separating succinct non-interactive arguments from all falsifiable assumptions. In: Fortnow, L., Vadhan, S.P. (eds.) STOC, pp. 99–108. ACM (2011)

[Imp95]  Impagliazzo, R.: Hard-core distributions for somewhat hard problems. In: 36th Annual Symposium on Foundations of Computer Science, pp. 538–545. IEEE (1995)

[JP14]  Jetchev, D., Pietrzak, K.: How to fake auxiliary input. In: Lindell, Y. (ed.) TCC 2014. LNCS, vol. 8349, pp. 566–590. Springer, Heidelberg (2014)

[LM94]  Luby, M.G., Michael, L.: Pseudorandomness and Cryptographic Applications. Princeton University Press, Princeton (1994)

[Pie09]  Pietrzak, K.: A leakage-resilient mode of operation. In: Joux, A. (ed.) EUROCRYPT 2009. LNCS, vol. 5479, pp. 462–482. Springer, Heidelberg (2009). doi:10.1007/978-3-642-01001-9_27

[Pie15]  Pietrzak, K.: Private communication, May 2015

[RTTV08]  Reingold, O., Trevisan, L., Tulsiani, M., Vadhan, S.: Dense subsets of pseudorandom sets. In: Proceedings of the 49th Annual IEEE Symposium on Foundations of Computer Science, Washington, DC, USA, FOCS 2008, pp. 76–85. IEEE Computer Society (2008)

[Skó15]  Skórski, M.: Time-advantage ratios under simple transformations: applications in cryptography. Cryptography and Information Security in the Balkans - Second International Conference, BalkanCryptSec: Koper, Slovenia, 3–4 September 2015. Revised Selected Papers, pp. 79–91 (2015)

[TTV09]  Trevisan, L., Tulsiani, M., Vadhan, S.: Regularity, boosting, and efficiently simulating every high-entropy distribution. In: Proceedings of the 24th Annual IEEE Conference on Computational Complexity, Washington, DC, USA, CCC 2009, pp. 126–136. IEEE Computer Society (2009)

[VZ13]  Vadhan, S., Zheng, C.J.: A uniform min-max theorem with applications in cryptography. In: Canetti, R., Garay, J.A. (eds.) CRYPTO 2013. LNCS, vol. 8042, pp. 93–110. Springer, Heidelberg (2013). doi:10.1007/978-3-642-40041-4_6

[YS13]  Yu, Y., Standaert, F.-X.: Practical leakage-resilient pseudorandom objects with minimum public randomness. In: Dawson, E. (ed.) CT-RSA 2013. LNCS, vol. 7779, pp. 223–238. Springer, Heidelberg (2013). doi:10.1007/978-3-642-36095-4_15

# Unconditional Security

# Pseudoentropy: Lower-Bounds for Chain Rules and Transformations

Krzysztof Pietrzak[1,2]([✉]) and Maciej Skórski[1,2]

[1] IST Austria, Klosterneuburg, Austria
krzpie@gmail.com
[2] University of Warsaw, Warsaw, Poland

**Abstract.** Computational notions of entropy have recently found many applications, including leakage-resilient cryptography, deterministic encryption or memory delegation. The two main types of results which make computational notions so useful are (1) Chain rules, which quantify by how much the computational entropy of a variable decreases if conditioned on some other variable (2) Transformations, which quantify to which extend one type of entropy implies another.

Such chain rules and transformations typically lose a significant amount in quality of the entropy, and are the reason why applying these results one gets rather weak quantitative security bounds. In this paper we for the first time prove lower bounds in this context, showing that existing results for transformations are, unfortunately, basically optimal for non-adaptive black-box reductions (and it's hard to imagine how non black-box reductions or adaptivity could be useful here.)

A variable $X$ has $k$ bits of HILL entropy of quality $(\epsilon, s)$ if there exists a variable $Y$ with $k$ bits min-entropy which cannot be distinguished from $X$ with advantage $\epsilon$ by distinguishing circuits of size $s$. A weaker notion is Metric entropy, where we switch quantifiers, and only require that for every distinguisher of size $s$, such a $Y$ exists.

We first describe our result concerning transformations. By definition, HILL implies Metric without any loss in quality. Metric entropy often comes up in applications, but must be transformed to HILL for meaningful security guarantees. The best known result states that if a variable $X$ has $k$ bits of Metric entropy of quality $(\epsilon, s)$, then it has $k$ bits of HILL with quality $(2\epsilon, s \cdot \epsilon^2)$. We show that this loss of a factor $\Omega(\epsilon^{-2})$ in circuit size is necessary. In fact, we show the stronger result that this loss is already necessary when transforming so called deterministic real valued Metric entropy to randomised boolean Metric (both these variants of Metric entropy are implied by HILL without loss in quality).

The chain rule for HILL entropy states that if $X$ has $k$ bits of HILL entropy of quality $(\epsilon, s)$, then for any variable $Z$ of length $m$, $X$ conditioned on $Z$ has $k - m$ bits of HILL entropy with quality $(\epsilon, s \cdot \epsilon^2 / 2^m)$. We show that a loss of $\Omega(2^m / \epsilon)$ in circuit size necessary here. Note that this still leaves a gap of $\epsilon$ between the known bound and our lower bound.

K. Pietrzak—Supported by the European Research Council consolidator grant (682815-TOCNeT).

M. Skórski—Supported by the National Science Center, Poland (2015/17/N/ST6/03564).

M. Hirt and A. Smith (Eds.): TCC 2016-B, Part I, LNCS 9985, pp. 183–203, 2016.
DOI: 10.1007/978-3-662-53641-4_8

# 1    Introduction

There exist various information theoretic notions of entropy that quantify the "uncertainty" of a random variable. A variable $X$ has $k$ bits of Shannon entropy if it cannot be compressed below $k$ bits. In cryptography we mostly consider min-entropy, where we say that $X$ has $k$ bits of min-entropy, denoted $\mathbf{H}_\infty(X) = k$, if for any $x$, $\Pr[X = x] \leq 2^{-k}$.

In a cryptographic context, we often have to deal with variables that only appear to have high entropy to computationally bounded observers. The most important case is pseudorandomness, where we say that $X \in \{0,1\}^n$ is pseudo-random, if it cannot be distinguished from the uniform distribution over $\{0,1\}^n$.

More generally, we say that $X \in \{0,1\}^n$ has $k \leq n$ bits of HILL pseudoen-tropy [12], denoted $\mathbf{H}_{\epsilon,s}^{\mathsf{HILL}}(X) = k$ if it cannot be distinguished from some $Y$ with $\mathbf{H}_\infty(Y) = k$ by any circuit of size $s$ with advantage $> \epsilon$, note that we get pseudorandomness as a special case for $k = n$. We refer to $k$ as the *quantity* and to $(\epsilon, s)$ as the *quality* of the entropy.

A weak notion of pseudoentropy called Metric pseudoentropy [3] often comes up in security proofs. This notion is defined like HILL, but with the quantifiers exchanged: We only require that for every distininguisher there exists a distrib-ution $Y, \mathbf{H}_\infty(Y) = k$ that fools this particular distinguisher (not one such $Y$ to fool them all).

HILL pseudoentropy is named after the authors of the [12] paper where it was introduced as a tool for constructing a pseudorandom generator from any one-way function. Their construction and analysis was subsequently improved in a series of works [11,13,28]. A lower bound on the number of calls to the underlying one-way function was given by [14].[1] More recently HILL pseudoentropy has been used in many other applications like leakage-resilient cryptography [6,17], deterministic encryption [7] and memory delegation [4].

The two most important types of tools we have to manipulate pseudoentropy are chain rules and transformations from one notion into another. Unfortunately, the known transformations and chain rules lose large factors in the quality of the entropy, which results in poor quantitative security bounds that can be achieved using these tools. In this paper we provide lower bounds, showing that unfortunately, the known results are tight (or almost tight for chain rules), at least when considering non-adaptive black-box reductions. Although black-box impossibility results have been overcome by non black-box constructions in the past [2], we find it hard to imagine how non black-box constructions or adaptivity could help in this setting. We believe that relative to the oracles we construct also adaptive reductions are impossible as adaptivity "obviously" is no of use, but proving this seems hard. Our results are summarized in Figs. 1 and 2.

**Complexity of the Adversary.** In order to prove a black-box separation, we will construct an oracle and prove the separation unconditionally relative to this

---

[1] Their $\Omega(n/log(n))$ lower bound matches existing constructions from *regular* one-way functions [10]. For general one-way functions this lower bound is still far of the best construction [28] making $\tilde{\Theta}(n^3)$ calls.

oracle, i.e., assuming all parties have access to it. This then shows that any construction/proof circumventing or separation in the plain model cannot be relativizing, which in particular rules out all black-box constructions [1,16].

In the discussion below we measure the complexity of adversaries only in terms of numbers of oracle queries. Of course, in the actual proof we also bound them in terms of circuit size. For our upper bounds the circuits will be of basically the same size as the number of oracle queries (so the number of oracle queries is a good indication of the actual size), whereas for the lower bounds, we can even consider circuits of exponential size, thus making the bounds stronger (basically, we just require that one cannot hard-code a large fraction of the function table of the oracle into the circuit).

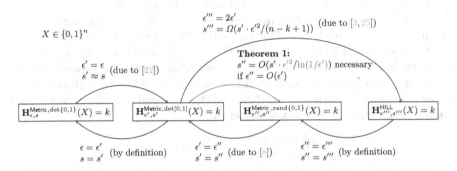

**Fig. 1.** Transformations: our bound comparing to the state of art. Our Theorem 1, stating that a loss of $\epsilon'^2/\ln(1/\epsilon')$ in circuit size is necessary for black-box reductions that show how deterministic implies randomized metric entropy (if the advantage $\epsilon'$ remains in the same order) requires $\epsilon' = 2^{-O(n-k+1)}$ and thus $\ln(1/\epsilon') \in O(n-k+1)$, so there's no contradiction between the transformations from [3,25] and our lower bound (i.e., the blue term is smaller than the red one). (Color figure online)

**Transformations.** It is often easy to prove that a variable $X \in \{0,1\}^n$ has so called Metric pseudoentropy against deterministic distinguishers, denoted $\mathbf{H}^{\text{Metric,det}\{0,1\}}_{\epsilon,s}(X) = k$. Unfortunately, this notion is usually too weak to be useful, as it only states that for every (deterministic, boolean) distinguisher, there exists some $Y$ with $\mathbf{H}_\infty(Y) = k$ that fools this particular distinguisher, but one usually needs a single $Y$ that fools all (randomised) distinguishers, this is captured by HILL pseudoentropy.

Barak et al. [3] show that any variable $X \in \{0,1\}^n$ that has Metric entropy, also has the same amount of HILL entropy. Their proof uses the min-max theorem, and although it perseveres the amount $k$ of entropy, the quality drops from $(\epsilon, s)$ to $(2\epsilon, \Omega(s \cdot \epsilon^2/n))$. A slightly better bound $(2\epsilon, \Omega(s \cdot \epsilon^2/(n+1-k)))$ (where again $k$ is the amount of Metric entropy), was given recently in [25]. The argument uses the min-max theorem and some results on convex approximation in $L_p$ spaces.

In Theorem 1 we show that this is optimal – up to a small factor $\Theta((n - k + 1)/\ln(1/\epsilon))$ – as a loss of $\Omega(\ln(1/\epsilon)/\epsilon^2)$ in circuit size is necessary for any black-box reduction. Note that for sufficiently small $\epsilon \in 2^{-\Omega(n-k+1)}$ our bound even matches the positive result up to a small constant factor.

The high-level idea of our separation is as follows; We construct an oracle $\mathcal{O}$ and a variable $X \in \{0,1\}^n$, such that relative to this oracle $X$ can be distinguished from any variable $Y$ with high min-entropy when we can make one randomized query, but for any deterministic distinguisher A, we can find a $Y$ with high min-entropy which A cannot distinguish from $X$.

To define $\mathcal{O}$, we first choose a uniformly random subset $S \in \{0,1\}^n$ of size $|S| = 2^m$. Moreover we chose a sufficiently large set of boolean functions $D_1(\cdot), \ldots, D_h(\cdot)$ as follows: for every $x \in S$ we set $D_i(x) = 1$ with probability $1/2$ and for every $x \notin S$, $D_i(x) = 1$ with probability $1/2 + \delta$.

Given any $x$, we can distinguish $x \in S$ from $x \notin S$ with advantage $\approx 2\delta$ by quering $D_i(x)$ *for a random $i$*. This shows that $X$ cannot have much more than $\log(|S|) = m$ bits of HILL entropy (in fact, even probabilistic Metric entropy) as any variable $Y$ with $\mathbf{H}_\infty(Y) \geqslant m + 1$ has at least half of its support outside $S$, and thus can be distinguished with advantage $\approx 2\delta/2 = \delta$ with one query as just explained. Concretely (recall that in this informal discussion we measure size simply by the number of oracle queries).

$$\mathbf{H}_{\delta,1}^{\text{Metric,rand}\{0,1\}}(X) \leqslant m + 1$$

On the other hand, if the adversary is allowed $q$ *deterministic* queries, then intuitively, the best he can do is to query $D_1(x), \ldots, D_q(x)$ and guess that $x \in S$ if less than a $1/2 + \delta/2$ fraction of the outputs is 1. But even if $q = 1/\delta^2$, this strategy will fail with constant probability. Thus, we can choose a $Y$ with large support outside $S$ (and thus also high min-entropy) which will fool this adversary. This shows that $X$ does have large Metric entropy against deterministic distinguishers, even if we allow the adversaries to run in time $1/\delta^2$, concretely, we show that

$$\mathbf{H}_{\Theta(\delta),O(1/\delta^2)}^{\text{Metric,det}\{0,1\}}(X) \geqslant n - O(\log(1/\delta))$$

**The Adversary.** Let us stress that we show impossibility in the non-uniform setting, i.e., for any input length, the distinguisher circuit can depend arbitrarily on the oracle. Like in many non-uniform black-box separation results (including [19,22,24,30,31]), the type of adversaries for which we can rigorously prove the lower bound is not completely general, but the necessary restrictions seem "obviously" irrelevant. In particular, given some input $x$ (where we must decide if $x \in S$), we only allow the adversary queries on input $x$. This doesn't seem like a real restriction as the distribution of $D_i(x')$ for any $x' \neq x$ is independent of $x$, and thus seems useless (but such queries can be used to make the success probability of the adversary on different inputs correlated, and this causes a problem in the proof). Moreover, we assume the adversary makes his queries non-adaptively, i.e., it choses the indices $i_1, \ldots, i_q$ before seeing the outputs of the queries $D_{i_1}(x), \ldots, D_{i_q}(x)$. As the distribution of all the $D_i$'s is identical, this doesn't seem like a relevant restriction either.

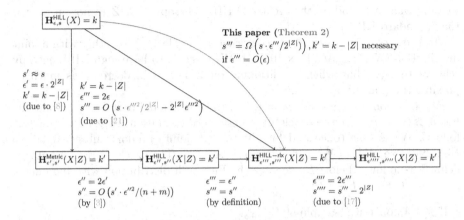

**Fig. 2.** Chain rules: our lower bounds comparing to the state of art. In the literature there are basically three approaches to prove a chain rule for HILL entropy. The first one reduces the problem to an efficient version of the dense model theorem [22], the second one uses the so called auxiliary input simulator [17], and the last one is by a convex optimization framework [21,26]. The last approach yields a chain rule with a loss of $\approx 2^m/\epsilon^2$ in circuit size, where $m$ is the length of leakage $Z$.

**Chain Rules.** Most (if not all) information theoretic entropy notions $H(.)$ satisfy some kind of chain rule, which states that the entropy of a variable $X$, when conditioned on another variable $Z$, can decrease by at most the bitlength $|Z|$ of $Z$, i.e., $H(X|Z) \geqslant H(X) - |Z|$.

Such a chain rule also holds for some computational notions of entropy. For HILL entropy a chain rule was first proven in [6,22] by a variant of the *dense model theorem*, and was improved by Fuller and Reyzin [8]. A different approach using a *simulator* was proposed in [17] and later improved by Vadhan and Zheng [29]. A unified approach, based on convex optimization techniques was proposed recently in [21,26] achieving best bounds so far.

The "dense model theorem approach" [8] proceeds as follows: one shows that if $X$ has $k$ bits of HILL entropy, then $X|Z$ has $k-m$ (where $Z \in \{0,1\}^m$) bits of Metric entropy. In a second step one applies a Metric to HILL transformation, first proven by Barak et al. [3], to argue that $X|Z$ has also large HILL. The first step loses a factor $2^m$ in advantage, the second another $2^{2m}\epsilon^2$ in circuit size. Eventually, the loss in circuit size is $2^{2m}/\epsilon^2$ and the loss in advantage is $2^m$ which measured in terms of the security ratio size/advantage gives a total loss of $2^m/\epsilon^2$.

A more direct "simulator" approach [29] loses only a multiplicative factor $2^m/\epsilon^2$ in circuit size (there's also an additive $1/\epsilon^2$ term) but there is no loss in advantage. The additive term can be improved to only $2^m\epsilon^2$ as shown in [21,26].

In this paper we show that a loss of $2^m/\epsilon$ is necessary. Note that this still is a factor $1/\epsilon$ away from the positive result. Our result as stated in Theorem 2 is a bit stronger as just outlined, as we show that the loss is necessary even if we only want a bound on the "relaxed" HILL entropy of $X|Z$ (a notion weaker than standard HILL).

To prove our lower bound, we construct an oracle $\mathcal{O}(.)$, together with a joint distribution $(X, Z) \in \{0, 1\}^n \times \{0, 1\}^m$. We want $X$ to have high HILL entropy relative to $\mathcal{O}(.)$, but when conditioning on $Z$ it should decrease as much as possible (in quantity and quality).

We first consider the case $m = 1$, i.e., the conditional part $Z$ is just one bit. For $n \gg \ell \gg m = 1$ the oracle $\mathcal{O}(.)$ and the distribution $(X, Z)$ is defined as follows. We sample (once and for all) two (disjoint) random subset $\mathcal{X}_0, \mathcal{X}_1 \subseteq \{0, 1\}^n$ of size $|\mathcal{X}_0| = |\mathcal{X}_1| = 2^{\ell-1}$, let $\mathcal{X} = \mathcal{X}_0 \cup \mathcal{X}_1$. The oracle $\mathcal{O}(.)$ on input $x$ is defined as follows (below $B_p$ denotes the Bernoulli distribution with parameter $p$, i.e., $\Pr[b = 1 : b \leftarrow B_p] = p$).

- If $x \in \mathcal{X}_0$ output a sample of $B_{1/2+\delta}$.
- If $x \in \mathcal{X}_1$ output a sample of $B_{1/2-\delta}$.
- Otherwise, if $x \notin \mathcal{X}$, output a sample of $B_{1/2}$.

Note that our oracle $\mathcal{O}(.)$ is probabilistic, but it can be "derandomized" as we'll explain at the beginning of Sect. 4. The joint distribution $(X, Z)$ is sampled by first sampling a random bit $Z \leftarrow \{0, 1\}$ and then $X \leftarrow \mathcal{X}_Z$.

Given a tuple $(V, Z)$, we can distinguish the case $V = X$ from the case where $V = Y$ for any $Y$ with large support outside of $\mathcal{X}$ ($X$ has min-entropy $\ell$, so let's say we take a variable $Y$ with $\mathbf{H}_\infty(Y|Z) \geq \ell + 1$ which will have at least half of its support outside $\mathcal{X}$) with advantage $\Theta(\delta)$ by quering $\alpha \leftarrow \mathcal{O}(V, Z)$, and outputting $\beta = \alpha \oplus Z$.

- If $(V, Z) = (X, Z)$ then $\Pr[\beta = 1] = 1/2 + \delta$. To see this, consider the case $Z = 0$, then $\Pr[\beta = 1] = \Pr[\alpha = 1] = \Pr[\mathcal{O}(X) = 1] = 1/2 + \delta$.
- If $(V, Z) = (Y, Z)$ then $\Pr[\beta = 1] = \Pr[Y \notin \mathcal{X}](1/2) + \Pr[Y \in \mathcal{X}](1/2 + \delta) \leq 1/2 + \delta/2$.

Therefore $X|Z$ doesn't have $\ell + 1$ bits of HILL entropy

$$\mathbf{H}^{\mathsf{HILL}}_{\delta/2,1}(X|Z) < \ell + 1$$

On the other hand, we claim that $X$ (without $Z$ but access to $\mathcal{O}(.)$) cannot be distinguished from the uniform distribution over $\{0, 1\}^n$ with advantage $\Theta(\delta)$ unless we allow the distinguisher $\Omega(1/\delta)$ oracle queries (the hidden constant in $\Theta(\delta)$ can be made arbitrary large by stetting the hidden constant in $\Omega(1/\delta)$ small enough), i.e.,

$$\mathbf{H}^{\mathsf{HILL}}_{\Theta(\delta),\Omega(1/\delta)}(X) = n \tag{1}$$

To see why (1) holds, we first note that given some $V$, a single oracle query is useless to tell whether $V = X$ or $V = U_n$: although in the case where $V = X \in \mathcal{X}_Z$ the output $\mathcal{O}(X)$ will have bias $\delta$, one can't decide in which direction the bias goes as $Z$ is (unconditionally) pseudorandom. If we're allowed in the order $1/\delta^2$ queries, we can distinguish $X$ from $U_n$ with constant advantage, as with $1/\delta^2$ samples one can distinguish the distribution $B_{1/2+\delta}$ (or $B_{1/2-\delta}$) from $B_{1/2}$ with constant advantage. If we just want $\Theta(\delta)$ advantage, $\Omega(1/\delta)$ samples are necessary, which proves (1). While it is easy to prove that for the coin with bias $\delta$ one needs $O\left(1/\delta^2\right)$ trials to achieve 99% of certainty, finding the number of trials for some confidence level in $o(1)$ as in our case, is more challenging. We solve this problem by a tricky application of *Renyi divergences*[2] The statement of our "coin problem" with precise bounds is given in Lemma 3.

So far, we have only sketched the case $m = 1$. For $m > 1$, we define a random function $\pi : \{0,1\}^n \to \{0,1\}^{m-1}$. The oracle now takes an extra $m-1$ bit string $j$, and for $x \in \mathcal{X}$, the output of $\mathcal{O}(x,j)$ only has bias $\delta$ if $\pi(x) = j$ (and outputs a uniform bit everywhere else). We define the joint distribution $(X,Z)$ by sampling $X \leftarrow \mathcal{X}$, define $Z'$ s.t. $X \in \mathcal{X}_{Z'}$, and set $Z = \pi(X)\|Z'$. Now, given $Z$, we can make one query $\alpha \leftarrow \mathcal{O}(V, Z[1\ldots m-1])$ and output $\beta = \alpha \oplus Z[m]$, where, as before, getting advantage $\delta$ in distinguishing $X$ from any $Y$ with min-entropy $\geq \ell + 1$.

On the other hand, given some $V$ (but no $Z$) it is now even harder to tell if $V = X$ or $V = Y$. Not only don't we know in which direction the bias goes as before in the case $m = 1$ (this information is encoded in the last bit $Z[m]$ of $Z$), but we also don't know on which index $\pi(V)$ (in the case $V = X$) we have to query the oracle to observe any bias at all. As there are $2^{m-1}$ possible choices for $\pi(V)$, this intuitively means we need $2^{m-1}$ times as many samples as before to observe any bias, which generalises (1) to

$$\mathbf{H}^{\mathsf{HILL}}_{\Theta(\delta),\Omega(2^{m-1}/\delta)}(X) = n$$

### 1.1  Some Implications of Our Lower Bounds

**Leakage Resilient Cryptography.** The chain rule for HILL entropy is a main technical tool used in several security proofs like the construction of leakage-resilient schemes [6, 20]. Here, the quantitative bound provided by the chain rule directly translates into the amount of leakage these constructions can tolerate. Our Theorem 2 implies a lower bound on the necessary security degradation for this proof technique. This degradation is, unfortunately, rather severe: even if we just leak $m = 1$ bit, we will lose a factor $2^m/\epsilon$, which for a typical security parameter $\epsilon = 2^{-80}$ means a security degradation of "80 bits".

Let us also mention that Theorem 2 answers a question raised by Fuller and Reyzin [8], showing that for any chain rule the *simultaneous loss* in quality and quantity is necessary,[3]

---

[2] Lower bounds [30,31] also require nontrivial binomial estimates. They were obtained, however by direct and involved calculations.

[3] Their question was about chain rules bounding the worst-case entropy, that is bounding $\mathbf{H}^{\mathsf{HILL}}(X|Z = z)$ for every $z$. Our result, stated simply for average entropy $\mathbf{H}^{\mathsf{HILL}}(X|Z)$, is much more general and applies to qualitatively better chain rules obtained by simulator arguments.

**Faking Auxiliary Inputs.** [17,27,29] consider the question how efficiently one can "fake" auxiliary inputs. Concretely, given any joint distribution $(X, Z)$ with $Z \in \{0,1\}^m$, construct an *efficient* simulator $h$ s.t. $(X, h(X))$ is $(\epsilon, s)$-indistinguishable from $(X, Z)$. For example [29] gives a simulator $h$ of complexity $O\left(2^m \epsilon^2 \cdot s\right)$ (plus additive terms independent of $s$). This result has found many applications in leakage-resilient crypto, complexity theory and zero-knowledge theory. The best known lower bound (assuming exponentially hard OWFs) is $\Omega\left(\max(2^m, 1/\epsilon)\right)$. Since the chain rule for relaxed HILL entropy follows by a simulator argument [17] with the same complexity loss, our Theorem 2 yields a better lower bound $\Omega\left(2^m/\epsilon\right)$ on the complexity of simulating auxiliary inputs.

**Dense Model Theorem.** The computational dense model theorem [22] says, roughly speaking, that dense subsets of pseudorandom distributions are computationally indistinguishable from true dense distributions. It has found applications including differential privacy, memory delegation, graph decompositions and additive combinatorics. It is well known that the worst-case chain rule for HILL-entropy is equivalent to the dense model theorem, as one can think of dense distributions as uniform distributions $X$ given short leakage $Z$. For settings with constant density, which correspond to $|Z| = O\left(1\right)$, HILL and relaxed HILL entropy are equivalent [17]; moreover, the complexity loss in the chain rule is then equal to the cost of transforming Metric Entropy into HILL Entropy. Now our Theorem 1 implies a necessary loss in circuit size $\Omega\left(1/\epsilon^2\right)$ if one wants $\epsilon$-indistinguishability. This way we reprove the tight lower bound due to Zhang [31] for constant densities.

## 2    Basic Definitions

Let $X_1$ and $X_2$ be two distributions over the same finite set. The *statistical distance* of $X_1$ and $X_2$ equals SD $(X_1; X_2) = \frac{1}{2} \sum_x |\Pr[X_1 = x] - \Pr[X_2 = x]|$.

**Definition 1 (Min-Entropy).** *A random variable $X$ has min-entropy $k$, denoted by $\mathbf{H}_\infty(X) = k$, if $\max_x \Pr[X = x] \leq 2^{-k}$.*

**Definition 2 (Average conditional min-Entropy [5]).** *For a pair $(X, Z)$ of random variables, the* average min-entropy *of $X$ conditioned on $Z$ is*

$$\widetilde{\mathbf{H}}_\infty(X|Z) = -\log \mathbb{E}_{z \leftarrow Z}\left[\max_x \Pr[X = x|Z = z]\right] = -\log \mathbb{E}_{z \leftarrow Z}\left[2^{-\mathbf{H}_\infty(X|Z=z)}\right]$$

**Distinguishers.** We consider several classes of distinguishers. With $\mathcal{D}_s^{\text{rand},\{0,1\}}$ we denote the class of randomized circuits of size at most $s$ with boolean output (this is the standard non-uniform class of distinguishers considered in cryptographic definitions). The class $\mathcal{D}_s^{\text{rand},[0,1]}$ is defined analogously, but with real valued output in $[0, 1]$. $\mathcal{D}_s^{\text{det},\{0,1\}}$, $\mathcal{D}_s^{\text{det},[0,1]}$ are defined as the corresponding classes for *deterministic* circuits. With $\Delta^D(X; Y) = |\mathbb{E}_X[D(X)] - \mathbb{E}_Y[D(Y)]|$ we denote $D$'s advantage in distinguishing $X$ and $Y$.

**Definition 3 (HILL pseudoentropy [12,15]).** *A variable $X$ has* HILL *entropy at least $k$ if*

$$\mathbf{H}_{\epsilon,s}^{\mathsf{HILL}}(X) \geq k \iff \exists Y,\ \mathbf{H}_\infty(Y) = k\ \forall D \in \mathcal{D}_s^{\mathrm{rand},\{0,1\}} : \Delta^D(X;Y) \leq \epsilon$$

*For a joint distribution $(X, Z)$, we say that $X$ has $k$ bits* conditonal Hill entropy *(conditionned on $Z$) if*

$$\mathbf{H}_{\epsilon,s}^{\mathsf{HILL}}(X|Z) \geq k$$
$$\iff \exists (Y, Z), \widetilde{\mathbf{H}}_\infty(Y|Z) = k\ \forall D \in \mathcal{D}_s^{\mathrm{rand},\{0,1\}} : \Delta^D((X, Z); (Y, Z)) \leq \epsilon$$

**Definition 4 (Metric pseudoentropy [3]).** *A variable $X$ has* Metric *entropy at least $k$ if*

$$\mathbf{H}_{\epsilon,s}^{\mathsf{Metric}}(X) \geq k \iff \forall D \in \mathcal{D}_s^{\mathrm{rand},\{0,1\}} \exists Y_D,\ \mathbf{H}_\infty(Y_D) = k\ :\ \Delta^D(X; Y_D) \leq \epsilon$$

Metric star entropy *is defined analogousely but using deterministic real valued distinguishers*

$$\mathbf{H}_{\epsilon,s}^{\mathsf{Metric}*}(X) \geq k \iff \forall D \in \mathcal{D}_s^{\mathrm{det},[0,1]} \exists Y_D,\ \mathbf{H}_\infty(Y_D) = k : \Delta^D(X; Y_D) \leq \epsilon$$

**Relaxed Versions of HILL and Metric Entropy.** A weaker notion of conditional HILL entropy allows the conditional part to be replaced by some computationally indistinguishable variable

**Definition 5 (Relaxed HILL pseudoentropy [9,23]).** *For a joint distribution $(X, Z)$ we say that $X$ has* relaxed HILL entropy $k$ *conditioned on $Z$ if*

$$\mathbf{H}_{\epsilon,s}^{\mathsf{HILL}-\mathsf{rlx}}(X|Z) \geq k$$
$$\iff \exists (Y, Z'), \widetilde{\mathbf{H}}_\infty(Y|Z') = k, \forall D \in \mathcal{D}_s^{\mathrm{rand},\{0,1\}},\ :\ \Delta^D((X, Z); (Y, Z')) \leq \epsilon$$

The above notion of *relaxed* HILL satisfies a chain rule whereas the chain rule for the standard definition of conditional HILL entropy is known to be false [18]. One can analogously define relaxed variants of metric entropy, we won't give these as they will not be required in this paper.

**Pseudoentropy Against Different Distinguisher Classes.** For randomized distinguishers, it's irrelevant if the output is boolean or real values, as we can replace any $D \in \mathcal{D}_s^{\mathrm{rand},[0,1]}$ with a $D' \in \mathcal{D}^{\mathrm{rand},\{0,1\}}$ s.t. $\mathbb{E}[D'(X)] = \mathbb{E}[D(X)]$ by setting (for any $x$) $\Pr[D'(x) = 1] = \mathbb{E}[D(x)]$. For HILL entropy (as well as for its relaxed version), it also doesn't matter if we consider randomized or deterministic distinguishers in Definition 3, as we always can "fix" the randomness to an optimal value. This is no longer true for metric entropy,[4] and thus the distinction between metric and metric star entropy is crucial.

---

[4] It might be hard to find a high min-entropy distribution $Y$ that fools a randomized distinguisher $D$, but this task can become easy once $D$'s randomness is fixed.

# 3   A Lower Bound on Metric-to-HILL Transformations

**Theorem 1.** *For every* $n$, $k$, $m$ *and* $\epsilon$ *such that* $n \geqslant k + \log(1/\epsilon) + 4$, $\frac{1}{8} > \epsilon$
*and* $n - 1 \geq m > 6\log(1/\epsilon)$ *there exist an oracle* $\mathcal{O}$ *and a distribution* $X$ *over*
$\{0,1\}^n$ *such that*

$$\mathbf{H}_{\epsilon,T}^{\text{Metric,det}\{0,1\}}(X) \geqslant k \tag{2}$$

*here the complexity* $T$ *denotes any circuit of size* $2^{O(m)}$ *that makes at most* $\frac{\ln(2/\epsilon)}{216\epsilon^2}$
*non-adaptive queries and, simultaneously,*

$$\mathbf{H}_{2\epsilon,T'}^{\text{Metric,rand}\{0,1\}}(X) \leqslant m + 1 \tag{3}$$

*where the distinguishers size* $T'$ *is only* $O(n)$ *and the query complexity is* 1.

Let $S$ be a random subset of $\{0,1\}^n$ of size $2^m$, where $m \leqslant n - 1$, and let
$D_1, \ldots, D_h$ be boolean functions drawn independently from the following distribution $D$: $D(x) = 1$ on $S$ with probability $p$ if $x \in S$ and $D(x) = 1$ with
probability $q$ if $x \in S^c$, where $p > q$ and $p + q = 1$. Denote $X = U_S$. We will
argue that the metric entropy against a probabilistic adversary who is allowed
one query is roughly $m$ with advantage $\Omega(p-q)$. But the metric entropy against
non-adaptive deterministic adversary who can make $t$ queries of the form $D_i(x)$
is much bigger, even if $t = O\left((p-q)^{-2}\right)$. Let us sketch an informal argument
before we give the actual proof. We need to prove two facts:

(i) There is a probabilistic adversary $\mathsf{A}^*$ such that with high probability over
   $X, D_1, \ldots, D_h$ we have $\Delta^{\mathsf{A}^*}(X, Y) = \Omega(p-q)$ for all $Y$ with $\mathbf{H}_\infty(Y) \geqslant m+1$.
(ii) For every deterministic adversary $\mathsf{A}$ making at most $t = O\left((p-q)^{-2}\right)$
   non-adaptive queries, with high probability over $X, D_1, \ldots, D_h$ we have
   $\Delta^{\mathsf{A}}(X;Y) = 0$ for some $Y$ with $\mathbf{H}_\infty(Y) = n - \Theta(1)$.

To prove (i) we observe that the probabilistic adversary can distinguish between
$S$ and $S^c$ by comparing the bias of ones. We simply let $\mathsf{A}^*$ forward its input to
$D_i$ for a randomly chosen $i$, i.e.,

$$\mathsf{A}^*(x) = D_i(x), \quad i \leftarrow [1, \ldots, h]$$

With extremely high probability we have $\Pr[\mathsf{A}^*(x) = 1] \in [p - \delta, p + \delta]$ if $x \in S$
and $\Pr[\mathsf{A}^*(x) = 1] \in [q - \delta, q + \delta]$ if $x \notin S$ for some $\delta \ll p - q$ (by a Chernoff
bound, $\delta$ drops exponentially fast in $h$, so we just have to set $h$ large enough).
We have then $\Pr[\mathsf{A}^*(X) = 1] \geqslant p + \delta$ and $\Pr[\mathsf{A}^*(Y) = 1] \leqslant 1/2 \cdot (p + q + 2\delta)$
for every $Y$ of min-entropy at least $m + 1$ (since then $\Pr[Y \in S] \leqslant 1/2$). This
yields $\Delta^{\mathsf{A}^*}(X;Y) = (p - q)/2$. In order to prove (ii) one might intuitively argue
that the best a $t$-query deterministic adversary can do to contradict to (ii), is to
guess whether some value $x$ has bias $p$ or $q = 1 - p$, by taking the majority of $t$
samples

$$\mathsf{A}(x) = \text{Maj}(D_1(x), \ldots, D_t(x))$$

But even if $t = \Theta(1/(p-q)^2)$, majority will fail to predict the bias with constant probability. This means there exists a variable $Y$ with min-entropy $n - \Theta(1)$ such that $\Pr[A(Y) = 1] = \Pr[A(X) = 1]$. The full proof gives quantitative forms of (i) and (ii), showing essentially that "majority is best" and appears in Appendix A.

## 4    Lower Bounds on Chain Rules

For any $n \gg \ell \gg m$, we construct a distribution $(X, Z) \in \{0,1\}^n \times \{0,1\}^m$ and an oracle $\mathcal{O}(.)$ such that relative to this oracle, $X$ has very large HILL entropy but the HILL entropy of $X|Z$ is much lower in quantity and quality: for arbitrary $n \gg \ell \gg m$ (where $|Z| = m$, $X \in \{0,1\}^n$), the quantity drops from $n$ to $\ell - m + 2$ (it particular, by much more than $|Z| = m$), even if we allow for a $2^m/\epsilon$ drop in quality.

**Theorem 2 (A lower bound on the chain rule for $H^{\text{HILL}-\text{rlx}}$).** *There exists a joint distribution $(X, Z)$ over $\{0,1\}^n \times \{0,1\}^m$, and an oracle $\mathcal{O}$ such that, relative to $\mathcal{O}$, for any $(\ell, \delta)$ such that $\frac{n}{2} - \frac{\log(1/\delta)}{2} > m$ and $\ell > m + 6\log(1/\delta)$, we have*

$$H^{\text{HILL}}_{\delta/2,T}(X) = n \tag{4}$$

*where[5] $T > c \cdot 2^m/\delta$ with some absolute constant $c$ but*

$$H^{\text{HILL}-\text{rlx}}_{\delta/2,T'}(X|Z) < \ell + 1 \tag{5}$$

*where $T'$ captures a circuit of size only $O(n)$ making only 1 oracle query.*

*Remark 1 (On the technical restrictions).* Note that the assumptions on $\ell$ and $\delta$ are automatically satisfied in most interesting settings, as typically we assume $m \ll n$ and $\log(1/\delta) \ll n$.

*Remark 2 (A strict separation).* The theorem also holds if we insist on a larger distinguishing advantage after leakage. Concretely, allowing for more than just one oracle query, the $\delta/2$ advantage in (5) can be amplified to $C\delta$ for any constant $C$ assuming $\delta$ is small enough to start with (see Remark 4 in the proof).

The full proof appears in Appendix B. The heart of the argument is a lower bound on the query complexity for the corresponding "coin problem": we need to distinguish between $T$ random bits, and the distribution where we sample equally likely $T$ independent bits $B_p$ or $T$ independent bits $B_q$ where $p = \frac{1}{2} + \delta$ and $q = 1 - p$. (see Appendix C for more details). The rest of the proof is based on a standard concentration argument, using extensively Chernoff Bounds.

---

[5] The class of adversaries here consists of all circuits with the total number of gates, including oracle gates, at most $T$. Theorem 2 is also true when the circuit size $s$ is much bigger than the total number of oracle gates $T$ (under some assumption on $s$, $\ell$, $\epsilon$). For simplicity, we do not state this version.

## 5    Open Problems

As shown in Fig. 2, there remains a gap between the best proofs for the chain-rule, which lose a factor $\epsilon^2/2^{|Z|}$ in circuit size, and the required loss of $\epsilon/2^{|Z|}$ we prove in this paper. Closing this bound by either improving the proof for the chain-rule or give an improved lower bound remains an intriguing open problem.

Our lower bounds are only proven for adversaries that make their queries non-adaptively. Adaptive queries don't seem to help against our oracle, but rigorously proving this fact seems tricky.

Finally, the lower bounds we prove on the loss of circuit size assume that the distinguishing advantage remains roughly the same. There exist results which are not of this form, in particular – as shown in Fig. 2 – the HILL to Metric transformation from [8] only loses in distinguishing advantage, not in circuit size (i.e., we have $s \approx s'$). Proving lower bounds and giving constructions for different circuit size vs. distinguishing advantage trade-offs leave many challenges for future work.

## A    Proof of Theorem 1

### A.1    Majority Is Best

We prove two statements which are quantitative forms of (i) and (ii) discussed after the statement of Theorem 1. First we show that the probabilistic adversary $A^*$ easily distinguishes $X$ from all $Y$ of high min-entropy.

**Claim 1 (Probabilistic Metric Entropy of $X$ is small).** *Let $A^*$ be a probabilistic adversary who on input $x$ samples a random $i \in [1, \ldots, h]$, then queries for $D_i(x)$ and outputs the response. Then for any $\delta \leqslant (p - q)/3$ we have*

$$\Pr[\forall Y : \mathbf{H}_\infty (Y) \geqslant m + 1, \ \Delta^{A^*} (X; Y) \geqslant (p - q)/3] \geqslant 1 - 2^{\max(n-1, m+1)} \exp(-h\delta^2). \quad (6)$$

*Remark 3 (The complexity of the probabilistic distinguisher).* We can chose $h$ in Claim 1 to be $2^n$, then $A^*$ is of size $O(n)$ and makes only one query.

Consider now a deterministic adversary $A$ who on input $x$ can make at most $t$ queries learning $D_i(x)$ for $t$ different $i \in [1, \ldots, h]$. We claim that

**Claim 2 (Deterministic Metric Entropy is big).** *Suppose that we have $n \geqslant k + \log(1/\epsilon) + 4$ and $\delta = \frac{\epsilon^2}{2+2\epsilon}$. Then for every nonadaptive adversary $A$ which makes $t \leqslant \frac{\ln(2/\epsilon)}{6(p-q)^2}$ queries we have*

$$\Pr_{X, D_1, \ldots, D_h} [\exists Y : \mathbf{H}_\infty (Y) \geqslant k, \ \Delta^A(X; Y) \leqslant \epsilon] \geqslant 1 - 4\exp(-2^m \delta^2). \quad (7)$$

Setting $p - q = 6\epsilon$ we see that Eq. (2) follows from Claim 1 and Eq. (3) follows from Eq. (7) combined with the union bound over all distinguishers. Note that the right hand side of Eq. (7) converges to 1 with the rate *doubly* exponential in $m$, so we can even afford taking a union bound over all distinguishers of size exponential in $m$.

*Proof (of Claim 1).* By a Chernoff bound[6] and the union bound

$$\Pr_{X,D_1,\ldots,D_h}[\forall x \in S^c : \Pr[A^*(x) = 1] \leqslant q + \delta] \geqslant 1 - 2^{n-1}\exp(-2\delta^2 h) \qquad (8)$$

similarly

$$\Pr_{X,D_1,\ldots,D_h}[\forall x \in S : |\Pr[A^*(x) = 1] - p| \leqslant \delta] \geqslant 1 - 2^m \cdot 2\exp(-2\delta^2 h). \qquad (9)$$

The advantage of $A^*$, with probability $1 - 2^{n-1}\exp(-2h\delta^2)$, equals

$$\Delta^{A^*}(X;Y) \geqslant (p - \delta) - (p + \delta)\Pr[Y \in S] - (q + \delta)\Pr[Y \in S^c]$$
$$\geqslant p - q - (p - q)\Pr[Y \in S] - 2\delta.$$

Since by the assumption we have $\Pr[Y \in S] \leqslant \frac{1}{2}$, Eq. (6) follows.

*Proof (of Claim 2).* The adversary $A$ non-adaptively queries for $D_i(x)$ values for $t$ distinct $i$'s and then outputs a bit, this bit is thus computed by a function of the form

$$f\left(x, D_{i_1(x)}(x), \ldots, D_{i_t(x)}(x)\right), \qquad (10)$$

for some fixed boolean function $f : \{0,1\}^n \times \{0,1\}^t \to \{0,1\}$. We start by simplifying the event (7) using the following proposition, which gives an alternative characterization of the deterministic metric entropy.

**Lemma 1 ([3,25]).** *Let $D$ be a boolean deterministic function on $\{0,1\}^n$. Then there exists $Y$ of min-entropy at least $k$ such that $\Delta^D(X;Y) \leqslant \epsilon$ if and only if*

$$\mathbb{E}\,D'(X) \leqslant 2^{n-k}\,\mathbb{E}\,D'(U) + \epsilon \qquad (11)$$

*holds for $D' \in \{D, 1 - D\}$*

Since $|S^c| \geqslant 2^{n-1}$, we have $\mathbb{E}\,D(U) \geqslant \mathbb{E}_{x \leftarrow S^c}D(x)/2$ for any function $D$. Therefore, by Lemma 1, the inequality (7) will be proved if we show that the following inequality holds:

$$\Pr_{X,D_1,\ldots,D_h}\left[\forall A' \in \{A, 1 - A\} : \mathbb{E}_{x \leftarrow S}A'(x) \leqslant 2^{n-k-1}\mathbb{E}_{x \leftarrow S^c}A'(x) + \epsilon\right] \geqslant 1 - 4\exp(-2^m\delta^2) \qquad (12)$$

By the union bound, it is enough to show that for $A' \in \{A, 1 - A\}$ we have

$$\Pr_{X,D_1,\ldots,D_h}\left[\mathbb{E}_{x \leftarrow S}A'(x) \leqslant 2^{n-k-1}\mathbb{E}_{x \leftarrow S^c}A'(x) + \epsilon\right] \geqslant 1 - 2\exp(-2^m\delta^2) \qquad (13)$$

In the next step we simplify the expressions $\mathbb{E}_{x \leftarrow S}A'(x)$ and $\mathbb{E}_{x \leftarrow S^c}A'(x)$. The following fact is a direct consequence of the Chernoff bound.

---

[6] We use the following version: let $X_i$ for $i = 1,\ldots,N$ be independent random variables such that $X_i \in [a_i, b_i]$. Then for any positive $t$ we have $\Pr_{X_1,\ldots,X_N}\left[\sum_{i=1}^N X_i - \mathbb{E}\left[\sum_{i=1}^N X_i\right] \geqslant t\right] \leqslant \exp\left(\frac{2t^2}{\sum_{i=1}^N (b_i - a_i)^2}\right)$.

**Proposition 1.** *For any function $f \in \{0,1\}^n \times \{0,1\}^t \to [0,1]$ we have*

$$\left| \mathbb{E}_{x \leftarrow S} f\left(x, D_{i_1(x)}(x), \ldots, D_{i_t(x)}(x)\right) - \mathbb{E} f(U_n, B_p^1, \ldots, B_p^t) \right| \leqslant \delta \qquad (14)$$

$$\left| \mathbb{E}_{x \leftarrow S^c} f\left(x, D_{i_1(x)}(x), \ldots, D_{i_t(x)}(x)\right) - \mathbb{E} f(U_n, B_q^1, \ldots, B_q^t) \right| \leqslant \delta \qquad (15)$$

*with probability $1 - 2\exp(-2 \cdot 2^m \delta^2)$ over the choice of $X$ and $D_1, \ldots, D_h$.*

For any $\mathbf{r} = (\mathbf{r}_1, \mathbf{r}_2, \ldots, \mathbf{r}_t) \in [0,1]^t$, and any (deterministic or randomized) function $f \in \{0,1\}^t \to [0,1]$ we denote $\mathbb{E}_{\mathbf{r}} f = \mathbb{E} f(B_{\mathbf{r}_1}, \ldots, B_{\mathbf{r}_t})$. It is enough to show that if $\mathbf{r}, \mathbf{r}'$ are both chosen from $\{p,q\}^t$ then we have

$$\mathbb{E}_{\mathbf{r}} f + \delta \leqslant 2^{n-k-1} \max(\mathbb{E}_{\mathbf{r}'} f - \delta, 0) + \epsilon. \qquad (16)$$

This inequality will follow by the following lemma (applied to $f$ in the proposition but considered as a function of $\{0,1\}^t$ randomized with the first $n$ input bits).

**Lemma 2.** *Suppose that $p, q > 0$ are such that $p + q = 1$. Let $f : \{0,1\}^t \to [0,1]$ be an arbitrary function and let $\mathbf{r}, \mathbf{r}' \in \{p,q\}^t$. Then for any $c > 0$ we have*

$$\mathbf{E}_{\mathbf{r}} f \leqslant \exp\left(\frac{(c+1)(p-q)^2}{q} \cdot t\right) \cdot \mathbf{E}_{\mathbf{r}'} f + \exp(-2c^2(p-q)^2 t).$$

*Proof.* The idea of the proof is to show that for most values of $z$ the ratio $\Pr[B_{\mathbf{r}} = z]/\Pr[B_{\mathbf{r}'} = z]$ is bounded. We have

$$\Pr[B_{\mathbf{r}} = z]/\Pr[B_{\mathbf{r}'} = z]$$
$$= (p/q)^{\#\{i:z_i=1,\ \mathbf{r}_i > \mathbf{r}'_i\} - \#\{i:z_i=1,\ \mathbf{r}_i < \mathbf{r}'_i\}} \cdot (q/p)^{\#\{i:z_i=0,\ \mathbf{r}_i > \mathbf{r}'_i\} - \#\{i:z_i=0,\ \mathbf{r}_i < \mathbf{r}'_i\}}$$
$$= (p/q)^{\#\{i:z_i=1,\ \mathbf{r}_i > \mathbf{r}'_i\} - \#\{i:z_i=0,\ \mathbf{r}_i > \mathbf{r}'_i\} - \#\{i:z_i=1,\ \mathbf{r}_i < \mathbf{r}'_i\} + \#\{i:z_i=0,\ \mathbf{r}_i < \mathbf{r}'_i\}} \qquad (17)$$
$$= (p/q)^{\sum_{i=1}^t (2z_i - 1) \cdot \mathrm{sgn}(\mathbf{r}_i - \mathbf{r}'_i)} \qquad (18)$$

The random variables $\xi_i = (2z_i - 1) \cdot \mathrm{sgn}(\mathbf{r}_i - \mathbf{r}'_i)$ for $i = 1, \ldots, t$, where $z$ is sampled from $B_{\mathbf{r}}$, are independent with the expectations $\mathbb{E}\xi_i = (2\mathbf{r}_i - 1)\mathrm{sgn}(\mathbf{r}_i - \mathbf{r}'_i) \leqslant p - q$. By the Chernoff bound for any $c > 0$ we get

$$\Pr_{z \leftarrow B_{\mathbf{r}}} \left[ \sum_{i=1}^t (2z_i - 1) \cdot \mathrm{sgn}(\mathbf{r}_i - \mathbf{r}'_i) \geqslant (p-q)t + c(p-q)t \right] \leqslant \exp(-2c^2(p-q)^2 t). \qquad (19)$$

Therefore,

$$\mathbb{E}_{\mathbf{r}} f \leqslant (p/q)^{(c+1)(p-q)t} \mathbb{E}_{\mathbf{r}'} f + 2\exp(-2c^2(p-q)^2 t) \qquad (20)$$

and the claim follows by observing that $p/q = 1 + (p-q)/q \leqslant \exp((p-q)/q)$.

From Lemma 2 it follows that Eq. (16) is satisfied with

$$\delta \leqslant \frac{\epsilon}{2\exp\left((c+1)(p-q)^2 \cdot t/q\right) + 2} \qquad (21)$$

provided that

$$\exp(-2c^2(p-q)^2 \cdot t) \leqslant \epsilon/2 \tag{22}$$

$$\exp\left((c+1)(p-q)^2 \cdot t/q\right) \leqslant 2^{n-k-1} \tag{23}$$

It is easy to see that Eqs. (23) and (22) are satisfied if and only if

$$\frac{\ln(2/\epsilon)}{2c^2(p-q)^2} \leqslant t \leqslant (n-k-3)\ln 2 \cdot \frac{q}{(c+1)(p-q)^2}.$$

This inequality can be satisfied if and only if

$$\epsilon \geqslant 2 \cdot 2^{(k-n+3) \cdot \frac{2qc^2}{c+1}}.$$

If we set $t = \frac{\ln(2/\epsilon)}{2c^2(p-q)^2}$ then Eq. (21) becomes

$$\delta \leqslant \frac{\epsilon}{(2/\epsilon)^{\frac{c+1}{2qc^2}} + 2}$$

Choosing $c$ so that $\frac{2qc^2}{c+1} = 1$ we see that it is enough to assume $\epsilon \geqslant 2 \cdot 2^{k-n+3}$, any $\delta$ such that $\delta \leqslant \frac{\epsilon^2}{2+2\epsilon}$ and $t \approx \frac{\ln(2/\epsilon)}{6(p-q)^2}$ (the constant 6 is sightly bigger than the exact value, but if Claim 2 holds true for some $t$ then also for $t' < t$). This finishes the proof of Claim 2.

# B   Proof of Theorem 2

**A Remark on the Oracle.** For convenience, the oracle $\mathcal{O} : \{0,1\}^n \to \{0,1\}$ we use in the proof is probabilistic, in the sense that it flips some random coins before answering a query (in particular, making the same query twice might give different outputs). We remark that, as the adversaries considered are probabilistic, one can replace this oracle with a deterministic one $\mathcal{O}_{\mathrm{det}}$ by assigning to every possible query $x$ a $2^L$ tuple $(x,r), r \in \{0,1\}^L$ of queries (for some sufficiently large $L$), where the output for $\mathcal{O}_{\mathrm{det}}((x,r))$ is sampled according to $\mathcal{O}(x)$ for every $r$. We can emulate the output distribution $\mathcal{O}(x)$ by querying $\mathcal{O}((x,r))$ for a random $r$. On the other hand, for a random $x$, even an exponential size distinguisher will not be able to distinguish $\mathcal{O}_{\mathrm{def}}((x,\cdot))$ from an oracle which, when queried on input $(x,r)$ for the first time, samples the output according to the distribution of $\mathcal{O}(x)$.[7]

*Proof (of Theorem 2).* We first describe how we construct the distribution $(X, Z)$ and the oracle $\mathcal{O}$.

---

[7] This can be shown along the lines of the proof that a random exponential size subset is unconditionally pseudorandom against exponential size distinguishers, see Goldreich's book "Foundations of Cryptography – Basic Techniques", Proposition 3.2.3.

**Construction details.** We chose at random two disjoint sets $\mathcal{X}_0, \mathcal{X}_1 \subset \{0,1\}^n$ of size $2^\ell$ and define $\mathcal{X} = \mathcal{X}_0 \cup \mathcal{X}_1$. Let $\pi : \{0,1\}^n \to \{0,1\}^{m-1}$ be a random function. The oracle $\mathcal{O}$ on input $(x, j) \in \mathcal{X} \times \{0,1\}^{m-1}$ outputs a sample of $B_{1/2}$ (i.e., a uniformly random bit), except if $x \in \mathcal{X}$ and $\pi(x) = j$, in this case the output bit has bias $\delta$; If $x \in \mathcal{X}_0$, the oracle outputs a sample of $B_{1/2-\delta}$, and otherwise, if $x \in \mathcal{X}_1$, a sample of $B_{1/2+\delta}$. We define the joint distribution $(X, Z)$ by sampling $Z' \leftarrow \{0,1\}, X \leftarrow \mathcal{X}_{Z'}$ and setting $Z = \pi(X) \| Z'$ (note that $X$ is uniform in $\mathcal{X}$)

**Adversaries.** The adversary on input $x \in \{0,1\}^n$ makes $T$ non-adaptive queries $(x, j_1(x)), \ldots, (x, j_T(x))$ to the oracle. We denote $\mathcal{O}$'s response with $R(x) = \left(R^i(x, j_i(x))\right)_{i=1}^T$. The adversary's final output $f(x, R(x))$ is computed by a boolean function $f : \{0,1\}^n \times \{0,1\}^T \to \{0,1\}$.

**Formal proof.** Let $R(x) = (R^1(x, j_1(x)), \ldots, R^T(x, j_T(x)))$ be the sequences of the oracle's responses and Let $B(x) = (B_{1/2}^1, \ldots, B_{1/2}^T)$ be independent random bits. For every $x$ the number of *useful* responses, that is indexes $i$ such that $R^i(x, j_i(x))$ is biased, is defined to be

$$T(x) = \sum_{i=1}^{T} [j_i(x) = \pi(x)] \tag{24}$$

On average we have $\mathbb{E}_{\mathcal{O}(\cdot)} T(x) = T/2^{m-1}$. We claim that the adversary actually learns basically nothing about $\mathcal{X}$: the sequence of oracle outptus is close to the sequence of unbiased bits. We start by showing that $\mathcal{X}$ is pseudorandom for our adversary.

**Claim 3** ($X$ **is pseudorandom, even given oracle responses**). *For any $f$ and $\epsilon > 0$ we have*

$$|\mathbb{E}_{x \leftarrow \mathcal{X}} f(x, R(x)) - \mathbb{E}_{x \leftarrow U_n} f(x, R(x))| \leq \epsilon + O\left(\delta^2 T / 2^m\right) \tag{25}$$

*with error probability at most $O\left(\exp\left(-\Omega\left(2^{n-m}\right)\right) + \exp\left(-\Omega\left(2^\ell \epsilon^2\right)\right)\right)$.*

*Proof.* By Lemma 3 and the definition of $\mathcal{O}$, for every $x \in \mathcal{X}$ we obtain

$$|\mathbb{E}f(x, R(x)) - \mathbb{E}f(x, B(x))| = \begin{cases} O\left(T(x)\delta^2\right), & x \in \mathcal{X} \\ 0, & x \notin \mathcal{X} \end{cases} \tag{26}$$

for every boolean function $f$ and some absolute constant hidden under big-Oh. Thus

$$\left| \mathbb{E}_{x \leftarrow \mathcal{X}} f(x, R(x)) - \mathbb{E}_{x \leftarrow \mathcal{X}} f(x, B(x)) \right| = O\left( \mathbb{E}_{x \leftarrow \mathcal{X}} T(x)\delta^2 \right) \tag{27}$$

Note that the random variables $f(x, R(x))$ for different values of $x$ are independent and similarly $f(x, B(x))$ for different values of $x$ are independent. Since the set $\mathcal{X}$ is chosen at random by the Hoeffding-Chernoff bound we obtain that with probability $1 - 2\exp\left(-\Omega\left(2^{\ell}\epsilon^2\right)\right)$ over $\mathcal{O}$ the following holds:

$$\left|\underset{x \leftarrow \mathcal{X}}{\mathbb{E}} f(x, B(x)) - \underset{x \leftarrow U_n}{\mathbb{E}} f(x, B(x))\right| \leqslant \epsilon \tag{28}$$

Combining Eqs. (27) and (28) we obtain (with probability $1 - 2\exp\left(-\Omega\left(2^{\ell}\epsilon^2\right)\right)$ over $\mathcal{O}$).

$$\left|\underset{x \leftarrow \mathcal{X}}{\mathbb{E}} f(x, R(x)) - \underset{x \leftarrow U_n}{\mathbb{E}} f(x, B(x))\right| \leqslant \epsilon + O\left(\underset{x \leftarrow \mathcal{X}}{\mathbb{E}} T(x)\delta^2\right) \tag{29}$$

By Eq. (26) we have

$$\left|\underset{x \leftarrow U_n}{\mathbb{E}} f(x, R(x)) - \underset{x \leftarrow U_n}{\mathbb{E}} f(x, B(x))\right| \leqslant O\left(\underset{x \leftarrow U_n}{\mathbb{E}} T(x)\delta^2\right). \tag{30}$$

Now Eqs. (29) and (30) imply

$$\left|\underset{x \leftarrow \mathcal{X}}{\mathbb{E}} f(x, R(x)) - \underset{x \leftarrow U_n}{\mathbb{E}} f(x, R(x))\right| \leqslant \epsilon + O\left(\underset{x \leftarrow U_n}{\mathbb{E}} T(x)\delta^2\right). \tag{31}$$

The random variables $T(x)$ for different $x$ are independent, bounded by $T$ and have the first moment $\underset{\mathcal{O}}{\mathbb{E}}(T(x)) = T/2^{m-1}$. By the multiplicative Chernoff bound with probability $1 - 2\exp\left(-\Omega\left(2^{n-m}\right)\right)$ over $\mathcal{O}$ it holds that $\underset{x \leftarrow U_n}{\mathbb{E}} T(x) < 2 \cdot T/2^{m-1}$. This implies Eq. (25) with error probability at most

$$P_{\mathrm{err}} = O\left(\exp\left(-\Omega\left(2^{n-m}\right)\right) + \exp\left(-\Omega\left(2^{\ell}\epsilon^2\right)\right)\right).$$

**Claim 4.** *There exists a distinguisher* $\mathsf{D} : \{0,1\}^n \times \{0,1\}^m \to \{0,1\}$ *which calls the oracle* $\mathcal{O}$ *one time and such that for any joint distribution* $Y, Z'$ *over* $\{0,1\}^n \times \{0,1\}^m$ *with entropy* $\widetilde{\mathbf{H}}_{\infty}(Y|Z') \geqslant \ell + 1$ *it holds that*

$$\mathbb{E}\,\mathsf{D}(X, Z) - \mathbb{E}\,\mathsf{D}(Y, Z') \geqslant \frac{\delta}{2}$$

*with probability* $1 - 2\exp(-\Omega\left(2^{\ell}\delta^2\right))$.

*Remark 4 (Amplified distinguisher).* Assuming that $T$ is sufficiently large, we can modify $\mathsf{D}$ by taking the majority vote over $T$ queries on $\mathcal{O}(x, z)$. This will boost the distinguishing advantage from $\delta/2$ to $C\delta$ where $C$ can be an arbitrary constant (for sufficiently small $\delta$).

*Proof (of Claim 4).* The distinguisher $\mathsf{D}$ simply calls the oracle $\mathcal{O}$ on the pair $(x, z)$. The probability that $\mathsf{D}$ outputs 1 on input $(Y, Z')$ is at most (the probabilities below are over the choice of $\mathcal{O}$ and $Y, Z'$)

$$
\begin{aligned}
\Pr\left(\mathsf{D}(Y, Z') = 1\right) &= \operatorname*{E}_{z \leftarrow Z'} \Pr\left(\mathsf{D}(Y|_{Z'=z}, z) = 1\right) \\
&= \operatorname*{E}_{z \leftarrow Z'} \left[\Pr\left(\mathsf{D}(Y, z) = 1 \wedge Y \notin \mathcal{X} \mid Z' = z\right)\right] \\
&\quad + \operatorname*{E}_{z \leftarrow Z'} \left[\Pr\left(\mathsf{D}(Y, z) = 1 \wedge Y \in \mathcal{X} \mid Z' = z\right)\right] \\
&= \frac{1}{2} + \delta \cdot \operatorname*{E}_{z \leftarrow Z'} \left[\Pr\left(Y \in \mathcal{X} \mid Z' = z\right)\right] \\
&\leqslant \frac{1}{2} + \delta \operatorname*{E}_{z \leftarrow Z'} \left[|\mathcal{X}| \cdot 2^{-\mathbf{H}_\infty(Y|Z'=z)}\right] \\
&= \frac{1}{2} + \delta \cdot |\mathcal{X}| \cdot 2^{-\tilde{\mathbf{H}}_\infty(Y|Z')}
\end{aligned}
$$

which is at most $\frac{1}{2} + \frac{\delta}{2}$. On the other hand we have $\Pr(\mathsf{D}(X, Z) = 1) = \frac{1}{2} + \delta$. From this we see that the advantage is $\delta$ on average - but we need stronger concentration guarantees. Note that $\Pr(\mathsf{D}(X, Z) = 1) = \sum_{x \in S} \Pr[X = x] \cdot \mathsf{D}(x, i(x))$ can be viewed as a sum of independent random variables. By the Chernoff-Hoeffding bound we get

$$
\operatorname*{Pr}_{\mathcal{O}}\left[\Pr(\mathsf{D}(X, Z) = 1) \geqslant \frac{1}{2} + \delta - \frac{\delta}{8}\right] \geqslant 1 - \exp(-\Omega\left(2^\ell \delta^2\right)))
$$

Similarly, $\Pr(\mathsf{D}(Y, Z') = 1) = \sum_{x,z} \Pr[Y = x, Z' = z] \cdot \mathsf{D}(x, z')$. Since

$$
\begin{aligned}
\sum_{x,z} \Pr[Y = x, Z' = z]^2 &= \sum_z \sum_x \Pr[Z' = z]^2 \Pr[Y = x|Z' = z]^2 \\
&\leqslant \sum_z \Pr[Z' = z] 2^{-\mathbf{H}_\infty(Y|Z'=z)} \\
&\leqslant 2^{-\tilde{\mathbf{H}}_\infty(Y|Z)},
\end{aligned}
$$

the Chernoff-Hoeffding bound implies

$$
\operatorname*{Pr}_{\mathcal{O}}\left[\Pr(\mathsf{D}(Y', Z) = 1) \leqslant \frac{1}{2} + \frac{\delta}{2} + \frac{\delta}{8}\right] \geqslant 1 - \exp(-\Omega\left(2^\ell \delta^2\right)) \tag{32}
$$

and the result follows. We set $\epsilon = \frac{\delta}{3}$ and $T = c \cdot 2^m/\epsilon$. Now Claim 4 directly implies Eq. (5) whereas Eq. (4) follows, when $c$ is sufficiently small, from Claim 3 by a union bound; To see this, note that the right hand side of (32) is doubly exponentially close (in $\ell$) to 1, and recall that $\ell > m + 6\log(1/\delta)$. So we can take a union bound over all $O(\exp(T))$ circuits $\mathsf{D}$ of size $T$ and deduce that with high probability the left hand side of (32) hold for all of them.

## C    Proof of Lemma 3

**Lemma 3 (Lower bounds on the coin problem).** *Fix $\delta \in (0, 1/2)$ and define $p = \frac{1}{2} + \delta$ and $q = 1 - p$. Consider the following two experiments:*

(a) *We flip a fair coin, and depending on the result we toss $T$ times a biased coin $B_p$ (probability of the head is $p$) or toss $T$ times a coin $B_q$ (probability of the head is $q$). The output is the result of these $T$ flips.*

(b) *We flip $T$ times a fair coin and output the results.*

*Then one cannot distinguish (a) from (b) better than with advantage $O\left(T\delta^2\right)$.*

*Remark 5.* We give a simple proof based on calculating Renyi divergences. This result can be also derived by more sophisticaed techniques from Fourier analysis (the generalized XOR lemma).

Before we give the proof, let's recall some basic facts about *Pearson Chi-Squared Distance*. For any two distributions $P, Q$ over the same space, their Chi-Squared distance defined by

$$D_{\chi^2}(P \parallel Q) = \sum_x Q(x) \left(\frac{P(x)}{Q(x)} - 1\right)^2 = \sum_x \frac{P(x)^2}{Q(x)^2} - 1 \tag{33}$$

Now let $U_1, \ldots, U_n$ be independent uniform bits, $X_1, \ldots, X_n$ be i.i.d. bits where 1 appears with probability $p = \frac{1}{2} + \delta$ and $Y_1, \ldots, Y_n$ be i.i.d. bits where 1 appears with probability $q = 1 - p = \frac{1}{2} - \delta$. We want to estimate the distance between $U = U_1, \ldots, U_n$ and $Z$ distributed as an equally weighted combination of $X = X_1, \ldots, X_n$ and $Y = Y_1, \ldots, Y_n$. We think of $\delta$ as a fixed parameter and $n$ as a growing number. Our statement will easily follow by combining the following two claims

**Claim 5.** *With $U$ and $Z$ as above, and for $n = O\left(\delta^{-2}\right)$, it holds that*

$$D_{\chi^2}\left(U; Z\right) = O\left(n^2\delta^4\right) \tag{34}$$

**Claim 6.** *For any $R$ and uniform $U$*

$$\mathrm{SD}(R \parallel U) \leqslant \sqrt{D_{\chi^2}(R \parallel U)}, \tag{35}$$

Indeed, combining these claims we obtain $\mathrm{SD}(Z \parallel U) = O(n\delta^2)$ when $n = O\left(\delta^{-2}\right)$. Since the left-hand side is bounded by 1, this is true also when $n > c\delta^{-2}$ for some absolute constant $c$ and the result follows.

*Proof (of Claim 5).* We have

$$D_{\chi^2}\left(\frac{1}{2}P_{X_1,\ldots,X_n} + \frac{1}{2}P_{Y_1,\ldots,Y_n} \parallel P_{U_1} \cdot \ldots \cdot P_{U_n}\right)$$

$$= 2^n \cdot \sum_{z_1,\ldots,z_n} \left(\frac{1}{2}P_{X_1}(z_1) \cdot \ldots \cdot P_{X_n}(z_n) + \frac{1}{2}P_{Y_1}(z_1) \cdot \ldots \cdot P_{Y_n}(z_n)\right)^2 - 1$$

$$= \frac{1}{4} \cdot 2^n \prod_i \left(\sum_z P_{X_i}(z)^2\right) + \frac{1}{4} \cdot 2 \cdot 2^n \prod_i \left(\sum_z P_{X_i}(z)P_{Y_i}(z)\right)$$

$$+\frac{1}{4} \cdot 2^n \prod_i \left( \sum_z P_{Y_i}(z)^2 \right) - 1$$

$$= \frac{1}{4} \left( (1 + 4\delta^2)^n + 2(1 - 4\delta^2)^n + (1 + 4\delta^2)^n - 4 \right)$$

$$(36)$$

and the result follows by the Taylor expansion $(1 + u)^n = 1 + nu + O(n^2 u^2)$ where $nu = O(1)$ applied to $u = 4\delta^2$. The bound is valid as long as $n = O\left(\delta^{-2}\right)$.

*Proof (of Claim 6).* This inequality follows immediately from the Cauchy-Schwarz inequality and the definition of $D_{\chi^2}$.

# References

1. Baker, T., Gill, J., Solovay, R.: Relativizations of the p=?np question. SIAM J. Comput. **4**(4), 431–442 (1975)
2. Barak, B.: How to go beyond the black-box simulation barrier. In: 42nd FOCS, pp. 106–115. IEEE Computer Society Press, October 2001
3. Barak, B., Shaltiel, R., Wigderson, A.: Computational analogues of entropy. In: 11th International Conference on Random Structures and Algorithms, pp. 200–215 (2003)
4. Chung, K.-M., Kalai, Y.T., Liu, F.-H., Raz, R.: Memory delegation. In: Rogaway, P. (ed.) CRYPTO 2011. LNCS, vol. 6841, pp. 151–168. Springer, Heidelberg (2011)
5. Dodis, Y., Reyzin, L., Smith, A.: Fuzzy extractors: how to generate strong keys from biometrics and other noisy data. In: Cachin, C., Camenisch, J.L. (eds.) EUROCRYPT 2004. LNCS, vol. 3027, pp. 523–540. Springer, Heidelberg (2004)
6. Dziembowski, S., Pietrzak, K.: Leakage-resilient cryptography. In: 49th FOCS, pp. 293–302. IEEE Computer Society Press, October 2008
7. Fuller, B., O'Neill, A., Reyzin, L.: A unified approach to deterministic encryption: new constructions and a connection to computational entropy. In: Cramer, R. (ed.) TCC 2012. LNCS, vol. 7194, pp. 582–599. Springer, Heidelberg (2012)
8. Fuller, B., Reyzin, L.: Computational entropy and information leakage. Cryptology ePrint Archive, report 2012/466 (2012). http://eprint.iacr.org/
9. Gentry, C., Wichs, D.: Separating succinct non-interactive arguments from all falsifiable assumptions. In: Fortnow, L., Vadhan, S.P. (eds.) 43rd ACM STOC, pp. 99–108. ACM Press, June 2011
10. Goldreich, O., Krawczyk, H., Luby, M.: On the existence of pseudorandom generators. SIAM J. Comput. **22**(6), 1163–1175 (1993)
11. Haitner, I., Reingold, O., Vadhan, S.P.: Efficiency improvements in constructing pseudorandom generators from one-way functions. In: Schulman, L.J. (ed.) 42nd ACM STOC, pp. 437–446. ACM Press, June 2010
12. Håstad, J., Impagliazzo, R., Levin, L.A., Luby, M.: A pseudorandom generator from any one-way function. SIAM J. Comput. **28**(4), 1364–1396 (1999)
13. Holenstein, T.: Pseudorandom generators from one-way functions: a simple construction for any hardness. In: Halevi, S., Rabin, T. (eds.) TCC 2006. LNCS, vol. 3876, pp. 443–461. Springer, Heidelberg (2006)
14. Holenstein, T., Sinha, M.: Constructing a pseudorandom generator requires an almost linear number of calls. In: 53rd FOCS, pp. 698–707. IEEE Computer Society Press, October 2012

15. Hsiao, C.-Y., Lu, C.-J., Reyzin, L.: Conditional computational entropy, or toward separating pseudoentropy from compressibility. In: Naor, M. (ed.) EUROCRYPT 2007. LNCS, vol. 4515, pp. 169–186. Springer, Heidelberg (2007)
16. Impagliazzo, R., Rudich, S.: Limits on the provable consequences of one-way permutations. In: Goldwasser, S. (ed.) CRYPTO 1988. LNCS, vol. 403, pp. 8–26. Springer, Heidelberg (1990)
17. Jetchev, D., Pietrzak, K.: How to fake auxiliary input. In: Lindell, Y. (ed.) TCC 2014. LNCS, vol. 8349, pp. 566–590. Springer, Heidelberg (2014)
18. Krenn, S., Pietrzak, K., Wadia, A.: A counterexample to the chain rule for conditional HILL entropy. In: Sahai, A. (ed.) TCC 2013. LNCS, vol. 7785, pp. 23–39. Springer, Heidelberg (2013)
19. Lu, C.-J., Tsai, S.-C., Wu, H.-L.: On the complexity of hard-core set constructions. In: Arge, L., Cachin, C., Jurdziński, T., Tarlecki, A. (eds.) ICALP 2007. LNCS, vol. 4596, pp. 183–194. Springer, Heidelberg (2007)
20. Pietrzak, K.: A leakage-resilient mode of operation. In: Joux, A. (ed.) EUROCRYPT 2009. LNCS, vol. 5479, pp. 462–482. Springer, Heidelberg (2009)
21. Pietrzak, K., Skórski, M.: The chain rule for HILL pseudoentropy, revisited. In: Lauter, K., Rodríguez-Henríquez, F. (eds.) LatinCrypt 2015. LNCS, vol. 9230, pp. 81–98. Springer, Heidelberg (2015)
22. Reingold, O., Trevisan, L., Tulsiani, M., Vadhan, S.P.: Dense subsets of pseudorandom sets. In: 49th FOCS, pp. 76–85. IEEE Computer Society Press, October 2008
23. Reyzin, L.: Some notions of entropy for cryptography. In: Fehr, S. (ed.) ICITS 2011. LNCS, vol. 6673, pp. 138–142. Springer, Heidelberg (2011)
24. Simon, D.R.: Findings collisions on a one-way street: can secure hash functions be based on general assumptions? In: Nyberg, K. (ed.) EUROCRYPT 1998. LNCS, vol. 1403, pp. 334–345. Springer, Heidelberg (1998)
25. Skorski, M.: Metric pseudoentropy: characterizations, transformations and applications. In: Lehmann, A., Wolf, S. (eds.) Information Theoretic Security. LNCS, vol. 9063, pp. 105–122. Springer, Heidelberg (2015)
26. Skorski, M.: A better chain rule for hill pseudoentropy - beyond bounded leakage. In: Information Theoretic Security - 9th International Conference, ICITS 2016 (2016)
27. Skorski, M.: Simulating auxiliary information, revisited. In: TCC 2016-B (2016)
28. Vadhan, S.P., Zheng, C.J.: Characterizing pseudoentropy and simplifying pseudorandom generator constructions. In: Karloff, H.J., Pitassi, T. (eds.) 44th ACM STOC, pp. 817–836. ACM Press, May 2012
29. Vadhan, S., Zheng, C.J.: A uniform min-max theorem with applications in cryptography. In: Canetti, R., Garay, J.A. (eds.) CRYPTO 2013, Part I. LNCS, vol. 8042, pp. 93–110. Springer, Heidelberg (2013)
30. Watson, T.: Advice lower bounds for the dense model theorem. TOCT 7(1), 1 (2014)
31. Zhang, J.: On the query complexity for showing dense model. Electron. Colloquium Comput. Complexity (ECCC) 18, 38 (2011)

# Oblivious Transfer from Any Non-trivial Elastic Noisy Channel via Secret Key Agreement

Ignacio Cascudo[1]([⊠]), Ivan Damgård[2], Felipe Lacerda[2], and Samuel Ranellucci[2]

[1] Department of Mathematics, Aalborg University, Aalborg, Denmark
ignacio@math.aau.dk
[2] Department of Computer Science, Aarhus University, Aarhus, Denmark
{ivan,lacerda,samuel}@cs.au.dk

**Abstract.** A $(\gamma, \delta)$-elastic channel is a binary symmetric channel between a sender and a receiver where the error rate of an honest receiver is $\delta$ while the error rate of a dishonest receiver lies within the interval $[\gamma, \delta]$. In this paper, we show that from *any* non-trivial elastic channel (i.e., $0 < \gamma < \delta < \frac{1}{2}$) we can implement oblivious transfer with information-theoretic security. This was previously (Khurana et al., Eurocrypt 2016) only known for a subset of these parameters. Our technique relies on a new way to exploit protocols for information-theoretic key agreement from noisy channels. We also show that information-theoretically secure commitments where the receiver commits follow from any non-trivial elastic channel.

**Keywords:** Oblivious transfer · Elastic channels · Key agreement · Commitments

## 1 Introduction

In this paper we consider oblivious transfer (OT), a well known two-party cryptographic primitive. In oblivious transfer, a sender has two messages and a receiver chooses to learn one of them. The receiver gains no information about the other message, while the sender does not know which of the messages the receiver has learned. Oblivious transfer is an important primitive because it is sufficient for information-theoretic secure computation [Kil88].

However, information-theoretic secure computation and therefore oblivious transfer are well known to be impossible if sender and receiver communicate in the plain model, without additional resources. Therefore, several alternative models have been studied where information-theoretically secure oblivious transfer is possible because we assume additional resources.

One such assumption is the existence of a noisy channel between the sender and the receiver. It was shown in [CK88] that binary symmetric channels are in fact enough to realize oblivious transfer. A binary symmetric channel is one where each bit sent is flipped with a certain probability, known as the error rate of the channel. More efficient constructions, and different variants of noisy

© International Association for Cryptologic Research 2016
M. Hirt and A. Smith (Eds.): TCC 2016-B, Part I, LNCS 9985, pp. 204–234, 2016.
DOI: 10.1007/978-3-662-53641-4_9

channels, were provided in subsequent papers, such as [BCS96, Cré97, DKS99, DFMS04, CMW05, CS06, PDMN11, IKO+11].

In particular, it was realized that it is problematic to assume that we are given a noisy channel with known and fixed parameters, such that the OT protocol we construct is allowed to depend on the parameter values. One reason for this is that it can be very hard to reliably estimate the parameters of a real channel. Another, more serious problem is that by fixing the parameters we are implicitly assuming that the adversary cannot change them. This is clearly unrealistic, and was the main motivation for introducing *unfair noisy channels* (UNC) in [DKS99]. In this model, the channel is a binary symmetric channel where, however, an adversary who corrupts one of the two parties can also choose the error rate to be within some range $[\gamma, \delta]$. For $\delta \geq 2\gamma(1-\gamma)$, the channel is easily seen to be trivial (it can be simulated from noiseless communication). It was shown in [DKS99] that information-theoretically secure oblivious transfer follows from UNC for a certain subset of the possible non-trivial parameter choices, while information-theoretically secure commitments follow from any non-trivial UNC.

*Elastic channels* (EC), a relaxation of unfair noisy channels, have been introduced in [KMS16]. For an EC, the noise can only be reduced by an adversary who corrupts the receiver. More precisely, given $0 \leq \gamma < \delta \leq 1/2$, a $(\gamma, \delta)$-elastic channel is one where the communication between the sender and an honest receiver has error rate $\delta$, but a dishonest receiver may reduce this to be in the interval $[\gamma, \delta]$. Clearly, in this setting, $\delta = 1/2$ would correspond to a channel where all information is lost for the honest receiver, while $\gamma = 0$ would yield a channel where a dishonest receiver has full information about the messages sent by the sender. We cannot implement oblivious transfer in either case, and hence these channels are deemed trivial.

It was shown in [KMS16] that commitments where the sender commits follow from any non-trivial EC, and that oblivious transfer follow from EC for a certain subset of parameters, which is larger than in the case of an UNC. More specifically, they show that $\delta \leq \ell(\gamma)$ where $\ell(\gamma) := \left(1 + (4\gamma(1-\gamma))^{-1/2}\right)^{-1}$ is sufficient.

It is of course interesting that going from UNC to EC allows a larger range of parameters from which we can get OT. However, for both channels, we are still left with a "grey area" of parameter values where we do not know if OT is possible. One might say that we still do not know if an EC is *fundamentally and qualitatively different* from a UNC as far as OT is concerned. Moreover, for commitments, we know that we can have the sender commit, but since an EC is asymmetric w.r.t what corrupted senders and receivers can do, it is not clear that we can get commitments where the receiver commits for any non-trivial EC.

*Our Contribution.* In this paper, we make progress on the above questions. First, we close the gap left open in [KMS16] and show that information theoretically secure oblivious transfer follows from any non-trivial EC. Along the way, we also construct commitments where the receiver commits, from any non-trivial EC.

Our main technical contribution is a new way to exploit a certain type of key agreement protocol towards implementing OT. More specifically, we consider a key agreement protocol between two parties (Alice and Bob) in the following model: Alice can send messages to Bob through a binary symmetric channel $C$ with error rate $\delta$, and the adversary Eve will receive what Alice sends via an independent binary symmetric channel with error rate $\gamma' \in [\gamma, \delta]$. On top of this, Alice and Bob may also communicate via a public error-free channel. Several key agreement protocols exist in this model [Mau93]. The main idea is to use the public channel to identify transmissions where Alice and Bob are more likely to agree on what was sent on the noisy channel. Because Eve's channel is independent, this may create a situation where Eve has a disadvantage compared to Bob, even if her noise rate is initially smaller.

In this work, we consider key agreement protocols that are secure in the usual sense: Alice and Bob agree on the output, and Eve gets essentially no information on the key. But in addition, we require an extra property we call *emulatability*: We can replace Bob by a "fake" Bob', who gets *no information* on what Alice sends on the noisy channel (but Eve gets information with error rate $\gamma'$ as usual). Still, Bob' can complete the conversation on the public channel such that neither Alice nor Eve can distinguish Bob' from Bob. As we explain later, we can modify the key agreement protocol presented in [Mau93, Sect. 5] so that it is emulatable. We show that an oblivious transfer protocol secure against semi-honest adversaries can be constructed from *any* emulatable key agreement protocol. Furthermore, by using information-theoretic commitments where the committing party is the receiver (which can be constructed from any non-trivial EC, as we will show) we can upgrade our protocol to achieve security against a malicious receiver too. Finally, we show how to achieve security against a malicious sender in the case where our emulatable key agreement protocol is the one mentioned above.

*Technical Overview.* To give an intuition of how our protocol works, consider first the case of semi-honest security where a semi-honest receiver reduces the error rate to the minimal value $\gamma$ (which is without loss of generality).

We turn an emulatable key agreement (KA) protocol as described above into an OT protocol as follows. The sender and the receiver engage in two independent instances (indexed respectively by 0 and 1) of the key agreement protocol above. In both cases, the sender from the OT protocol takes the role of Alice in the KA protocol, while the receiver does the following: in the instance of the KA protocol corresponding to his selection bit $b$, he acts as Bob would, while in the other instance, he acts as Bob' (so in particular his actions are independent of what he receives from the sender on the EC). Finally Alice sends her messages $m_0, m_1$ one-time padded respectively with $k_0$ and $k_1$, each of these keys obtained in the corresponding key agreement protocol.

Now, an honest receiver will learn $m_b$ as he should, which follows from correctness of the KA protocol. Second, a corrupt sender cannot learn the choice bit $b$. This follows from the emulatability property of the KA protocol: the sender cannot distinguish in which of the two instances she is interacting with the real

Bob. Finally, a corrupt receiver cannot learn $m_{1-b}$. This follows from the fact that, in the instance of the KA corresponding to $1 - b$, the view of the receiver is the same as the view of Eve, namely he sees everything Alice sends with error rate $\gamma$, and he sees the public discussion (the fact that he generates that discussion himself by running Bob' makes no difference). One can then show that emulatability implies that this view is distributed identically to the case where Eve watches Alice interact with Bob, and the usual definition of key agreement security guarantees that this is independent from the exchanged key $k_{1-b}$.

Security in the malicious case is more involved. First, we need to ensure that the malicious receiver follows the protocol. It turns out to be sufficient that the receiver proves that for one of the KA instances, the messages he sends on the public channel are generated by Bob', of course without revealing which one. To this end we can use the fact that commitments where the committing party is the receiver also follow from any EC (see below) and, via a known reduction, zero-knowledge proofs on committed values follow as well. Thus, we are doing something very similar to the GMW compiler. As a result we get a protocol that is secure against a semi-honest sender and a malicious receiver.

To further protect against a malicious sender, we execute many instances of the OT. The receiver checks the statistics of what he receives on the EC and discards instances that are too far from what he expects to see from an honest sender. This creates a protocol where the sender will (at least sometimes) have non-trivial uncertainty about the choice bit. We can now use standard techniques to clean this up to get a secure OT.

As for our construction on receiver commitments from any non-trivial EC, we observe that the commitment protocol from [DKS99] (that was designed for a UNC) can be modified to work for an EC. All we essentially have to do is to choose the parameters correctly. On the one hand, handling an EC is harder because $\delta$ and $\gamma$ are much further apart than for a UNC, however, on the other hand an EC is easier because one party has to live with the large noise rate even if he is corrupt. Intuitively, the observation is that these two issues balance each out so that (almost) the same protocol still works.

*Outline.* In Sect. 2 we define the basic functionalities we will deal with for the remainder of the paper, namely oblivious transfer and the elastic channel. In Sect. 3, we introduce the notion of emulatable key agreement, as well as a protocol that implements it. Emulatable key agreement is used in Sect. 4 to implement an OT protocol that is secure against semi-honest adversaries. This protocol is then used in Sect. 5 in the construction of a protocol secure against a malicious receiver. Finally, in Sect. 6 we present a construction that builds upon the one of Sect. 5 to obtain security against malicious adversaries.

## 2   Preliminaries

### 2.1   Security Model

We prove our protocols secure in the Universal Composability framework introduced in [Can01]. This model is explained in Appendix A.

## 2.2  Oblivious Transfer

Oblivious transfer is a two-party primitive where one party (the sender) inputs two messages and the other party (the receiver) chooses to receive one—and only one—of them. Crucially, the sender does not learn the receiver's choice, and the receiver does not learn the message it did not choose. This primitive is formalized in the figure below. Note that the description includes an adversary $\mathcal{A}$, which can corrupt parties.

---

**Functionality $\mathcal{F}_{\text{OT}}$ (Oblivious transfer)**

$\mathcal{F}_{\text{OT}}$ runs with two parties: a sender and a receiver.

**Send:** Upon receiving (send, sid, $m_0, m_1$) from the sender: store (sid, $m_0, m_1$) and send (sent, sid) to $\mathcal{A}$.

**Receipt:** Upon receiving (choice, sid, $b$) from the receiver: if a message of the form (sid, $m_0, m_1$) has been stored, send (receipt, $m_b$) to the receiver.

---

## 2.3  Elastic Channel

A $(\gamma, \delta)$-elastic channel, as introduced in [KMS16], is a binary symmetric channel with crossover probability $\delta$ where a receiver that has been corrupted by the adversary can choose to reduce the crossover probability to a level $\nu$ with $\gamma \leq \nu \leq \delta$. In the functionality below, we define a more general version where the channel is composed by $\ell$ binary symmetric channels (all with crossover probability $\nu$).

---

**Functionality $\mathcal{F}_{\text{EC}}(\gamma, \delta)$ (Elastic channel)**

$\mathcal{F}_{\text{EC}}$ runs with parties $P_1, P_2$ and eavesdropper $\mathcal{A}$ as follows:

**Initialization:** $\nu \leftarrow \delta$

**Noise:** Upon receiving (noise, $\bar{\nu}$) from $\mathcal{A}$, if the receiver is corrupt and $\gamma \leq \bar{\nu} \leq \delta$ then set $\nu \leftarrow \bar{\nu}$.

**Send:** On (send, sid, $m$) from the sender, where $m \in \{0, 1\}^\ell$, produce $\bar{m}$ by flipping each bit of $m$ independently with probability $\nu$. Then send the message (sent, sid) to $\mathcal{A}$ and the message (sent, sid, $\bar{m}$) to the receiver.

---

# 3  Emulatable Key Agreement

Key agreement is the problem where two parties, Alice and Bob, want to establish a common key (a random element from $\{0, 1\}^\ell$) so that an eavesdropper Eve has no information about this key. In other words, the goal is to implement the following functionality $\mathcal{F}_{\mathsf{KA}}$.[1]

---

### Functionality $\mathcal{F}_{\mathsf{KA}}$ (Key agreement)

$\mathcal{F}_{\mathsf{KA}}$ runs with security parameter $u$, parties $P_1, P_2$ and eavesdropper $\mathcal{A}$ as follows:

**Establish:** Upon receiving $(\mathtt{establish}, \mathrm{sid}, P_i, P_{3-i})$ from $P_i$ (where $i \in \{1, 2\}$), store $(\mathrm{sid}, P_i, P_{3-i})$ and send $(\mathrm{sid}, P_i, P_{3-i})$ to $\mathcal{A}$. If the tuple $(\mathrm{sid}, P_{3-i}, P_i)$ had also been stored, choose $k \leftarrow_R \{0, 1\}^t$, send $(\mathtt{sent}, 1^t)$ to $\mathcal{A}$ and send $(\mathtt{key}, \mathrm{sid}, k)$ to $P_1, P_2$.

---

In this section, we consider the scenario in which Alice can communicate to Bob via a wiretap channel $\mathcal{F}_\mathsf{C}$ where each bit is flipped (independently) with probability $\delta$. Eve can obtain another noisy version of this communication, where each bit is flipped with probability $\gamma$ and this noise is independent from Bob's. Furthermore, there is a feedback public channel $\mathcal{F}_{\mathsf{Pub}}$ through which Alice and Bob can communicate.

---

### Functionality $\mathcal{F}_\mathsf{C}$ (Wiretap channel)

$\mathcal{F}_\mathsf{C}$ runs with parameters $\gamma, \delta \in (0, 1/2)$, message size $\ell$, parties $P_1, P_2$ and eavesdropper $\mathcal{A}$ as follows:

**Send:** Upon receiving $(\mathtt{send}, \mathrm{sid}, P_1, P_2, m)$ where $m \in \{0, 1\}^\ell$:
1. Produce $\bar{m}$ by flipping each bit of $m$ independently with probability $\delta$. Furthermore, produce $\tilde{m}$ by flipping each bit of $m$ independently with probability $\gamma$.
2. Send $(\mathtt{sent}, \mathrm{sid})$ to $P_1$, $(\mathtt{receipt}, \mathrm{sid}, \bar{m})$ to $P_2$ and $(\mathtt{receipt}, \mathrm{sid}, \tilde{m})$ to $\mathcal{A}$.

---

### Functionality $\mathcal{F}_{\mathsf{Pub}}$ (Public channel)

$\mathcal{F}_{\mathsf{Pub}}$ runs with message size $\ell$, parties $P_1, P_2$ and eavesdropper $\mathcal{A}$ as follows:

**Send:** Upon receiving $(\mathtt{send}, \mathrm{sid}, P_i, P_j, m)$ where $m \in \{0, 1\}^\ell$, send $(\mathtt{sent}, \mathrm{sid})$ to $P_i$ and $(\mathtt{receipt}, \mathrm{sid}, m)$ to $P_j$ and $\mathcal{A}$.

---

In this setting, we are interested in key agreement protocols with an additional property that we call *emulatability*. A key agreement protocol $\pi$ is emulatable if, in addition to implementing the key agreement functionality as it should,

---

[1] In the remainder of this section, we interchangeably call the parties Alice, Bob, Eve or respectively $P_1, P_2, \mathcal{A}$.

the role of Bob can be simulated by some entity $\mathcal{E}$, the emulator, that learns *no information* about the messages transmitted through $\mathcal{F}_C$, other than their lengths, and neither Alice nor Eve can distinguish whether Alice is interacting with Bob or with $\mathcal{E}$.

We formalize this below. We first define a functionality $\mathcal{F}_{DC}$ that models a dummy channel whose task is to erase every information sent through the channel $\mathcal{F}_C$ except for the length of the messages.

---

**Functionality $\mathcal{F}_{DC}$ (Dummy channel)**

$\mathcal{F}_{DC}$ runs with message size $\ell$ and parties $P_1, P_2$ as follows:

**Send:** Upon receiving (send, sid, $m$) from $P_1$ where $m \in \{0,1\}^\ell$: If no such command has already been sent, send (sent, sid, $|1|^\ell$) to $P_2$. Otherwise, ignore the command.

---

**Definition 1.** *A key agreement protocol $\pi$ between Alice and Bob using a wiretap channel $\mathcal{F}_C$ and a public channel $\mathcal{F}_{Pub}$ is emulatable if:*

1. *It realizes the functionality $\mathcal{F}_{KA}$. That is, there exists a simulator $S$ such that for all eavesdroppers $\mathcal{A}$,*

$$\pi \diamond \mathcal{F}_C \diamond \mathcal{F}_{Pub} \equiv_{\mathcal{A}} \mathcal{F}_{KA} \diamond S.$$

2. *There exists an emulator $\mathcal{E}$ such that the following happens: suppose that we consider the protocol $\pi'$ where we replace Bob by $\mathcal{F}_{DC} \diamond \mathcal{E}$, i.e., $\mathcal{E}$ is linked to $\mathcal{F}_C$ via the dummy channel $\mathcal{F}_{DC}$, and Alice acts as in protocol $\pi$, while in both cases the eavesdropper $\mathcal{A}$ receives information from $\mathcal{F}_C$ and $\mathcal{F}_{Pub}$. Then from the point of view of Alice and all eavesdroppers $\mathcal{A}$, the protocol executions of $\pi$ and $\pi'$ are indistinguishable.*

*That is, we have*

$$\pi \diamond \mathcal{F}_C \diamond \mathcal{F}_{Pub} \equiv_{Alice, \mathcal{A}} \pi' \diamond \mathcal{F}_C \diamond \mathcal{F}_{Pub}.$$

We will need the following property later on.

**Proposition 1.** *Suppose that a key agreement protocol $\pi$ is emulatable. Then for any eavesdropper $\mathcal{A}$, if Alice is executing the protocol $\pi'$ with the emulator $\mathcal{E}$ as in the definition, $\mathcal{A}$ obtains no information about Alice's output.*

This is because, if $\mathcal{A}$ could obtain any information about Alice's output in the execution of $\pi'$, then either she would be able to obtain information about Alice's output in the execution of $\pi$ (contradicting property 1 of emulatability) or she would be able to distinguish $\pi$ and $\pi'$ (contradicting property 2).

## 3.1   The Emulatable Key Agreement Protocol

We now describe an emulatable key agreement protocol for a wiretap channel $\mathcal{F}_C$ with $\gamma < \delta$, that is, for which the channel to the eavesdropper Eve is more reliable than the channel to Bob.

This is a small modification of a key agreement protocol from [Mau93, Sect. 5]. For each $\gamma, \delta$, the protocol specifies numbers $s, \ell, n \in \mathbb{N}$, to be determined below. In addition, $\ell = 2m + 1$ is an odd number. The protocol consists of three phases: advantage distillation, information reconciliation and privacy amplification.

The goal of the advantage distillation step is to create a conceptual channel between Alice and Bob which is more reliable than the one between Alice and Eve. In our protocol, this step proceeds as follows. Alice samples $n$ random bits $b_i$ and encodes each bit $b$ as a bitstring $v$ in $\{0,1\}^\ell$ by selecting uniformly at random a set $\mathcal{J} \subseteq \{1, \dots, \ell\}$ of size $m$ and setting the $j$-th coordinate of $v$ to be $1 - b$ if $j \in \mathcal{J}$ and $b$ if $j \notin \mathcal{J}$. Note that this means $m + 1$ of the bits of the encoding equal $b$ while the other $m$ bits equal $1 - b$. Now, if for a given sent bit $b$, Bob receives a message of the form $(c, c, \dots, c)$ for some $c$, we say Bob accepts the bit $b$ and $c$ is his guess about $b$. Now Alice creates the bitstring $b_{i_1} b_{i_2} \dots b_{i_s}$ given by the first $s$ bits accepted by Bob and Bob creates the bitstring $c_{i_1} c_{i_2} \dots c_{i_s}$ of his guesses. They both disregard the remaining bits. Alternatively, one can see Alice's encoding process as first encoding her bit with the repetition code and then introducing errors in exactly $m$ positions. As we discuss in Sect. 3.2, the protocol is similar to that in [Mau93, Sect. 5], except that the global error introduced here is of fixed weight $m$, rather than flipping each bit with certain probability. In Sect. 3.2 below, we discuss why we need this to introduce this modification. Yet, from the point of view of advantage distillation, the intuition why this protocol works is the same as in [Mau93]: namely, even though Eve has more information over messages sent over the wiretap channel than Bob has, she has less information about the ones accepted by Bob; in other words, the probability that Bob decodes those bits correctly is higher than that of Eve's. We formalize this later.

The information reconciliation step is carried out over the public channel. After this step, Alice and Bob will share a common bitstring with overwhelming probability, and Eve is still guaranteed to have some uncertainty about it. In the description below, we use the information reconciliation protocol in [BS94], where Alice sends the evaluation on her bitstring of a hash function chosen from a 2-universal family with an appropriate range size. Then Bob corrects his bitstring by finding the closest bitstring to it which is consistent with this evaluation.

Alice and Bob can then apply privacy amplification to obtain a random string about which Eve has *no* information. This can also be done by having only Alice send information over the public channel. The fact that both the information reconciliation and privacy amplification steps involve only Alice sending information over the public channel is important to guarantee the emulatability property.

We note that the information reconciliation step may in general not be computationally efficient for Bob; however, in fact any information reconciliation protocol can be used, as long as it is non-interactive. One efficient option is to employ a fuzzy extractor, as in [DORS08, Sect. 8.1], in order to execute both steps.

This description is formalized below. (For simplicity, we omit the description of the "establish" step introduced in the functionality of Sect. 3.)

---

### Protocol $\pi_{\mathsf{KA}}$ (Emulatable key agreement)

**Parameters:**

- $\sigma$: security parameter.
- $\ell := \ell(\gamma, \delta)$: an odd natural number which only depends on $\gamma$ and $\delta$.
- $m := \frac{\ell-1}{2}$.
- $s \in \omega(\sigma)$.
- $t \in \Theta(\sigma)$.
- $n > \lceil s/(\delta(1-\delta)^m)\rceil$.
- $0 < \epsilon < \frac{1}{2} - \delta$: a small constant.

Let $\mathfrak{h}$ denote the binary entropy function and $\mathcal{H}_1 : \{0,1\}^s \to \{0,1\}^{s \cdot \mathfrak{h}(\delta+\epsilon)+\sigma}$, $\mathcal{H}_2 : \{0,1\}^s \to \{0,1\}^t$ be 2-universal families of hash functions.

*Advantage distillation:*

**Alice:**

Select $b_1, \ldots, b_n \in_R \{0,1\}$.

For $i \in \{1, \ldots, n\}$:

1. Select a set $\mathcal{J}_i \subseteq \{1, \ldots, \ell\}$ of size $m$ uniformly at random.

2. Set $v_i$ to be the bitstring in $\{0,1\}^\ell$ such that $v_i[j] = 1 - b_i$ for $j \in \mathcal{J}_i$ and $v_i[j] = b_i$ for $j \notin \mathcal{J}_i$ where $v_i[j]$ denotes the $j$-th coordinate of $v_i$.

3. Send (**send**, Alice, Bob, $\mathrm{sid}_i$, $v_i$) to $\mathcal{F}_{\mathsf{C}}$.

**Bob:**

For $i \in \{1, \ldots, n\}$, await (**receipt**, Alice, Bob, $\mathrm{sid}_i$, $\bar{v}_i$) from $\mathcal{F}_{\mathsf{C}}$.

Construct the set $\mathcal{I} \subseteq \{1, \ldots, n\}$ consisting of the indices $i$ for which $\bar{v}_i = c_i^\ell$ for some $c_i \in \{0,1\}$

Encode the set $\mathcal{I}$ as a bit string $u$ and send (**send**, sid, Bob, Alice, $u$) to $\mathcal{F}_{\mathsf{Pub}}$.

**Alice:**

Await (**sent**, sid, Bob, Alice, $u$) from $\mathcal{F}_{\mathsf{Pub}}$.

---

**Alice ↔ Bob:**

Alice sets $X^s = (b_{i_1}, b_{i_2}, \ldots, b_{i_s})$ and Bob sets $Y^s = (c_{i_1}, c_{i_2}, \ldots, c_{i_s})$, where $i_1, \ldots, i_s$ are the first $s$ indices in $\mathcal{I}$.

*Information reconciliation and privacy amplification:*

**Alice:**

Sample $h_1 \in_R \mathcal{H}_1$, $h_2 \in_R \mathcal{H}_2$, send $(\mathsf{send}, \mathrm{sid}, \mathrm{Alice}, \mathrm{Bob}, h_1, h_1(X^s), h_2)$ to $\mathcal{F}_{\mathsf{Pub}}$.

Output $h_2(X^s)$.

**Bob:**

Await $(\mathsf{send}, \mathrm{sid}, \mathrm{Alice}, \mathrm{Bob}, h_1, h_1(X^s), h_2)$ from $\mathcal{F}_{\mathsf{Pub}}$.

Find the closest (in the Hamming metric) bitstring $\widetilde{X}^s$ to $Y^s$ satisfying $h_1(\widetilde{X}^s) = h_1(X^s)$.

Output $h_2(\widetilde{X}^s)$.

---

In order to prove that our protocol is indeed an emulatable key agreement protocol, we introduce the following notation. Let $X$ denote a variable with the uniform distribution over $\{0,1\}$. Let $Y$ and $Z$ be the random variables that describe respectively the output bit $c$ of Bob and the received bitstring of Eve (which is an element in $\{0,1\}^\ell$) when Alice samples a bit $b$ according to $X$, encodes it as in our protocol, sends it through the wiretap channel *and Bob accepts*. An important point to make is that, since the noise of Bob and Eve are independent, the probability distribution of $Z$ would be the same if we removed the conditioning on Bob accepting the bit. We have the following theorem.

**Theorem 1.** *The protocol $\pi_{\mathsf{KA}}$ is an emulatable key agreement protocol.*

We use the following lemma which intuitively means that, as $\ell$ grows, the probability that Eve receives a bitstring where most bits are 0 approaches $1/2$ if Alice encoded a 0 (naturally an analogous statement holds if Alice encoded a 1). The proof of the lemma can be found in Appendix B.

**Lemma 1.** *For $i \in \{0,1\}$, let $S_i \subseteq \{0,1\}^\ell$ be the set of all bitstrings where most bits are $i$. Then $\Pr[Z \in S_i | X = i] \to 1/2$ as $\ell \to \infty$.*

*Proof (of Theorem 1).* The detailed proof is in Appendix C. Here we give a sketch.

First we argue about the correctness of the protocol. It is not difficult to see that, for each index $i$, Bob accepts the corresponding bit with probability $p_{\mathsf{accept}} = (\delta(1-\delta))^m$. Furthermore, condition to Bob having accepted a bit, the probability that he decodes it correctly is again exactly $1 - \delta$, i.e., the advantage distillation step creates another conceptual noisy channel where the noise parameter is still $\delta$, the same as in the original noisy channel.

Since we set $n$ slightly larger than $\lceil s/p_{\mathsf{accept}} \rceil$, for large enough parameters Bob will, with very high probability, accept at least $s$ bits, of which roughly $\delta \cdot s$ will be incorrect. By the results on information reconciliation in [BS94] our choice of $\mathcal{H}_1$ guarantees that Bob corrects to the right string in the information reconciliation step, and hence that they output the same key at the end of the protocol.

Next, we consider privacy. Let $X$ and $Z$ be as above. We can use Lemma 1 in order to establish that $H_\infty(X|Z) \to 1$ as $\ell \to \infty$. We can then select $\ell$ large enough so that $H_\infty(X^s|Z^s, h_1, h_1(X^s)) \geq t + 2\sigma$ (see the full proof for details), and apply the leftover hash lemma to conclude that conditioned on everything seen by Eve during the protocol, the distribution of $h_2(X^s)$ is $2^{-\sigma}$-close to the uniform distribution over $\{0,1\}^t$.

Finally, to show that the protocol is emulatable, we have to construct an emulator $\mathcal{E}$ that satisfies Property 2 in Definition 1. Note the only information sent by Bob is the description of the set $\mathcal{I}$ of indices for which Bob accepted Alice's message. Hence, this can be emulated by sampling a random index set $\mathcal{I} \subseteq \{0,1\}^n$, where each index belongs to $\mathcal{I}$ with independent probability $p_{\mathsf{accept}}$.

## 3.2    On the Emulatability of Other Key Agreement Protocols

Protocol $\pi_{\mathsf{KA}}$ described above is based on the protocol given in [Mau93, Sect. 5]. As a matter of fact, several key agreement protocols for noisy channels are described in [Mau93] and subsequent works. However, they are either not emulatable (or, at least, it seems difficult to show they are) or they do not work for all non-trivial sets of parameters $(\gamma, \delta)$.

First, [Mau93, Sect. 5], considers a slightly different scenario, in which there is only a public channel available for communication but on the other hand at the beginning of the protocol Alice, Bob and Eve have noisy versions (respectively $r_A$, $r_B$ and $r_E$) of a common string $r$, where each bit is independently flipped with probabilities $\epsilon_A$, $\epsilon_B$ and $\epsilon_E$ for Alice, Bob and Eve respectively. Then having Alice mask a message (by xoring it with $r_A$) and send it through the public channel, induces a conceptual noisy channel where the input of Alice is $m$, and the outputs of Bob and Eve are $m \oplus r_A \oplus r_B$ and $m \oplus r_A \oplus r_E$ respectively. In the protocol proposed in [Mau93, Sect. 5] Alice encodes random bits with a repetition code and sends the information over the conceptual channel. From this point, the protocol proceeds as ours (Bob accepts the bits corresponding to codewords and they execute information reconciliation and privacy amplification on the resulting string). It can be shown that any parameters $0 < \epsilon_A, \epsilon_B, \epsilon_E < 1/2$ lead to a secure key agreement protocol.

In our scenario, the players do not start with noisy versions of a common string, but have a $(\gamma, \delta)$-wiretap channel. We can reproduce the situation above in our scenario as follows: in order to send the message $m$, Alice flips each bit independently with probability $\epsilon_A > 0$ and sends the result through the $(\gamma, \delta)$-wiretap channel. This would be an equivalent situation of the above where $\epsilon_B = \gamma$ and $\epsilon_E = \delta$, and therefore it would lead to a secure key agreement protocol. However, the protocol would not be emulatable: the reason is that the probability that Bob accepts a given instance depends on the exact number of bitflips introduced by Alice. However, because this artificial noise has been introduced by Alice and not by the channel, this information is known by Alice; on the other hand, the number of bitflips in a given instance cannot be determined precisely by the emulator, even though it knows $\epsilon_A$. Hence, regardless of how we define the emulator, Alice will be able to distinguish when she is interacting with it or with Bob.

If Alice does not introduce this artificial noise (i.e., if $\epsilon_A = 0$), then there *is* an emulator that can reproduce Bob's answer in every case, but the range of $(\gamma, \delta)$ for which this protocol is a secure key agreement protocol does not include all possible $0 < \gamma, \delta < 1/2$ and, in fact, it can be seen is exactly the very same range of parameters $(\gamma, \delta)$ for which [KMS16] shows the existence of an OT protocol for a $(\gamma, \delta)$-elastic noisy channel (i.e. those pairs satisfying $\delta \leq \left(1 + (4\gamma(1 - \gamma))^{-1/2}\right)^{-1}$).

In our protocol, we solve these problems by having Alice introduce artificial noise, but making this noise be of a fixed Hamming weight $m$. This solves the problem with the existence of the emulator, while still preserving the security of the protocol.

Finally, we also need to mention that a simpler protocol for key agreement in our wire-tap channel scenario is presented in [Mau93, Proposition 1]. The protocol first creates a conceptual channel from Bob to Alice in which Alice has more information about Bob's message than Eve does. This protocol works for all non-trivial wire-tap channel noise parameters. However, the information reconciliation and privacy amplification steps cannot be performed in such a way that it is only Alice who sends information (because in this case these steps are going to correct Alice's knowledge of Bob's string). Then, it is unclear whether this protocol can be made emulatable, because we would also need the emulator to simulate the information sent by Bob in these steps, and this does not appear to be straightforward.

## 4  Semi-honest Protocol

Now we present an OT protocol over the elastic channel $\mathcal{F}_{EC}(\gamma, \delta)$ for semi-honest adversaries. We show that such an oblivious transfer protocol can be constructed from any emulatable key agreement protocol that works in the setting of Sect. 3 (where Alice, Bob and Eve are connected by a wiretap channel $\mathcal{F}_C$ with the noise parameters being $\delta$ for Bob and $\gamma$ for Eve).

The idea of the protocol is for sender and receiver to engage in two separate subprotocols. In one, they run the emulatable key agreement protocol with the sender acting as Alice and the receiver acting as Bob. In the other subprotocol, the sender follows again the key agreement protocol as Alice, whereas the receiver runs the emulator, according to Definition 1. The choice bit $c$ determines whether the receiver will follow the protocol or act as the emulator. Here, the elastic channel is used as a conceptual wiretap channel $\mathcal{F}_C$, where an honest receiver gets the output of the legitimate (noisier) channel, whereas an adversarial receiver gets the output of the less noisy channel.

To see why the protocol is secure, we note that since the key agreement protocol is emulatable, the sender does not know whether she is interacting with Bob (that is, whether she is engaging in the actual key agreement protocol) or with the emulator. Hence, she does not learn any information about the choice bit $c$. This guarantees the receiver's privacy.

On the other hand, by definition the emulator can generate the transcript for the key agreement protocol without knowing anything about the exchanged key. Therefore in this case the receiver has no information about the key output by Alice at the end of the key agreement protocol.

This proof sketch is formalized in Theorem 2, below.

---

**Protocol $\pi_{\mathsf{OTSH}}$ (Semi-honest oblivious transfer)**

Let $\pi_{\mathsf{KA}}$ be an emulatable key agreement protocol, as stated in Definition 1. We denote the sender's input as $m_0, m_1$ and denote the receiver's input as $c$.

**Sender $\leftrightarrow$ Receiver:**

Sender and receiver execute two copies $\pi_0, \pi_1$ of $\pi_{\mathsf{KA}}$, where the sender behaves in both as Alice. In $\pi_c$, the receiver acts as Bob and in $\pi_{1-c}$, it acts as the emulator $\mathcal{E}$ prescribed by $\pi'_{\mathsf{KA}}$.

**Receiver:**

On completion of $\pi_0, \pi_1$, record the output of $\pi_c$ as $k$.

**Sender:**

Await $k_0, k_1$ from $\pi_0, \pi_1$.

Set $\bar{m}_i := m_i \oplus k_i$ for $i = 0, 1$.

Send $(\mathsf{send}, \mathsf{sid}_0, \bar{m}_0)$ and $(\mathsf{send}, \mathsf{sid}_1, \bar{m}_1)$ to $\mathcal{F}_{\mathsf{Pub}}$.

**Receiver:**

Await $(\mathsf{sent}, \mathsf{sid}_0, \bar{m}_0)$, $(\mathsf{sent}, \mathsf{sid}_1, \bar{m}_1)$ from $\mathcal{F}_{\mathsf{Pub}}$.

Output $m_c := \bar{m}_c \oplus k$.

---

**Theorem 2.** *The protocol $\pi_{\mathsf{OTSH}}$ realizes $\mathcal{F}_{\mathsf{OT}}$. That is, there exists a simulator $\mathcal{S}$ such that*

$$\pi_{\mathsf{OTSH}} \diamond \mathcal{F}_{\mathsf{EC}} \diamond \mathcal{F}_{\mathsf{Pub}} \equiv_{\mathcal{Z}} \mathcal{F}_{\mathsf{OT}} \diamond \mathcal{S}$$

*for all semi-honest environments $\mathcal{Z}$.*

*Proof.* For each activation, the environment $\mathcal{Z}$ chooses $m_0, m_1, c$. When interacting with the protocol, $\mathcal{Z}$ receives $m'_c$, and when interacting with $\mathcal{F}_{\mathsf{OT}}$, it receives $m_c$. We note first that since $\pi_c$ is an instance of $\pi_{KA}$, which implements $\mathcal{F}_{KA}$, we have $m'_c = m_c$. All that remains to be shown is that there exists a simulator for $\mathcal{F}_{OT}$ that can reproduce the view of the environment.

First, assume $P_1$ is corrupted, so that $\mathcal{Z}$ gets access to $P_1$'s internal state. During the real execution, it gets access to $k_0, k_1$ (through $P_1$), $\bar{m}_0, \bar{m}_1$ plus the leakage from $\pi_0$ and $\pi_1$ (through the adversary $\mathcal{A}$, which interacts with $\mathcal{F}_{EC}$ and $\mathcal{F}_{Pub}$). At the end of the execution, it gets $P_2$'s output, which is given by $m_c$.

In the ideal process, the simulator $\mathcal{S}$ corrupts $P_1$, so that it gets access to $m_0, m_1$. $\mathcal{S}$ proceeds as follows. First, it executes two copies of $\mathcal{F}_{KA} \diamond \mathcal{S}'$, where $\mathcal{S}'$ is the simulator for the key agreement protocol. By assumption, this internal simulator replicates the leakage from $\pi_0$ and $\pi_1$, which is relayed to $\mathcal{Z}$. Additionally, at the end of $\mathcal{F}_{KA}$'s execution, $\mathcal{S}$ gets two random keys, which we denote by $k''_0, k''_1$. It then computes $\bar{m}_i = m_i \oplus k''_i$ for $i = 1, 2$ and sends both to $\mathcal{Z}$. Finally, it sends $m_0, m_1$ to $\mathcal{F}_{OT}$, which will then send $m_c$ to $P_2$. It is easy to see that $\mathcal{F}_{OT} \diamond \mathcal{S}$ provides $\mathcal{Z}$ with the same view as in the real protocol.

Now assume $P_2$ is corrupted. Throughout the real execution, $\mathcal{Z}$ gets access to $k_c$ (through $P_2$), $\bar{m}_0, \bar{m}_1, m_c$ plus the leakage from $\pi_0$ and $\pi_1$ (through the eavesdropper $\mathcal{A}$). In the ideal process, $\mathcal{S}$ gets $c$ by corrupting $P_2$. It proceeds as follows. It runs one copy of $\mathcal{F}_{KA} \diamond \mathcal{S}'$, obtaining a random key $k''_c$, and relays $c$ to $\mathcal{F}_{OT}$. After $P_2$ receives $m_c$ from $\mathcal{F}_{OT}$, $\mathcal{S}$ computes $\bar{m}_c = m_c \oplus k''_c$ and sends it to $\mathcal{Z}$. Clearly, $m_c$ and $\bar{m}_c$ have the same distribution as in the real execution.

Finally, we look at the leakage from the execution of $\pi_{1-c}$ (executing the instance of $\pi'_{KA}$ with the emulator $\mathcal{E}$). Due to Proposition 1, $\pi_{1-c}$ gives no information on $k_{1-c}$ to the eavesdropper $\mathcal{A}$. Therefore $\bar{m}_{1-c}$ gives no additional information to $\mathcal{Z}$. Moreover, since the execution of $\mathcal{E}$ only depends on the outputs of the dummy channel $\mathcal{F}_{DC}$, its view provides $\mathcal{Z}$ with no additional information, even given the rest of $\mathcal{Z}$'s view. The view of $\mathcal{Z}$ is therefore the same in both scenarios.

## 5  OT Protocol Secure Against a Malicious Receiver

In this section, we make our protocol secure against a malicious receiver. Note that in our semi-honest protocol, we rely on the fact that the players will engage in two instances of an emulatable key agreement protocol, where the receiver will play the role of Bob in one of them and the emulator in the other. Of course, if the receiver is malicious, he will not necessarily adopt this behaviour. We will use standard techniques to solve this problem. Namely, we want to use the paradigm introduced in [GMW86]: we will make the receiver prove in zero knowledge that he is acting as in the semi-honest protocol.

To do this, we will need that the receiver can commit to bits. Recall that in [KMS16] it was shown that commitments where the sender commits follow from any non-trivial EC, but since an EC is asymmetric, it is not clear that this allows the receiver to commit. Therefore, we solve this problem first.

## 5.1 Receiver Commitment from Any Non-trivial EC

The solution in a nutshell is to observe that the commitment protocol from [DKS99] will work for receiver commitments on any non-trivial EC, if we slightly tune some of the parameters.

First, note that we can reverse the direction of the EC, by simply having the sender send a random bit $x$ on the EC, the receiver chooses a bit $b$ to *send* and sends $x \oplus b$ back on the public channel. This is clearly a noisy channel in the opposite direction. In this subsection we will rename the sender and call him the verifier $V$, while the receiver will be called the committer $C$. What we just constructed is a "reversed EC" where the $C$ sends and $V$ receives. $V$ always receives with noise rate $\delta$, but $C$ can reduce his noise rate to $\gamma$ if he is corrupted (and hence get a better idea of what $V$ received). The goal is now to build an unconditionally secure commitment scheme based on such a channel.

In fact, we show that, under a careful choice of parameters, the commitment protocol from [DKS99] already works with no change. A complete description of the protocol, as well as an intuition for why it is secure, is provided in Appendix D.

## 5.2 From Commitment to Security Against a Malicious Receiver

Recall that the GMW compiler [GMW86] transforms a semi-honestly secure protocol into a maliciously secure one by using the following three steps: in the first step, each party commits to his input; in the second step, each party is forced to commit to a random tape, where it is important that the tape is hidden from the other party and is chosen at random. This is done by having the party that is committing to a random tape commit to a random value. The other party then sends a random string. The tape is then defined to be the xor of both strings. This technique is known as coin-tossing in the well. Finally, in the third step, each player follows the protocol with the committed inputs and their committed tape and whenever they send a message, they also prove in zero-knowledge that this is the correct message given their committed input, their committed random tape and the transcript of the protocol.

In this section, we are only interested in achieving security against a malicious receiver, so we apply the compiler to the receiver only. This results in the following approach: In the first step, the receiver will commit to his choice of input $c$; this also indicates the instance of the key agreement protocol where he will play the role of Bob. In the second step, the receiver will be forced to commit to a random tape $t$ for the emulator using coin-tossing in the well. Then the sender and receiver will run an augmented version of the semi-honest protocol. Each instance of the key agreement protocol will be associated to an index $b$. Each time a receiver sends a message, the receiver also proves in zero-knowledge: "Either the given instance of key agreement has index $b = c$ or the message was produced by following the description of the emulator with random tape $t$".

There is, however, one difficulty: In [GMW86], the commitments were computational. It was therefore possible to prove statements about committed values directly. For a black-box information-theoretically secure commitment, it is not directly possible to prove statements that involve the committed values. To fix this problem, we use a commitment scheme which can indeed be used for any number of zero-knowledge proofs. This is the commitment scheme from [CvdGT95] which was later proven UC-secure in [Est04]. As shown in [CvdGT95], this commitment scheme can be constructed in a black-box manner from any commitment schemes. Although this commitment scheme only allows proofs of xor relationships directly, one can use techniques such as [BCC88] to prove arbitrary statements involving the committed values.

---

**Functionality $\mathcal{F}_{\text{COMZK}}$** (Commitment with zero-knowledge)

$\mathcal{F}_{\text{COMZK}}$ runs with two parties: a sender and a receiver.

**Commit:** On receiving $(\text{commit}, \text{cid}, m)$ from the sender:
If such a command has already been sent, ignore the message. Otherwise, record $(\text{cid}, m)$ and send $(\text{committed}, \text{cid})$ to $\mathcal{A}$ and to the receiver.

**Reveal:** On receiving $(\text{reveal}, \text{cid})$ from the sender:
If no pair $(\text{cid}, m)$ was recorded then ignore the message. Otherwise, send $(\text{open}, \text{cid}, m)$ to $\mathcal{A}$ and to the receiver.

**Proof:** On receiving $(\text{prove}, x, \text{cid}_1, \ldots, \text{cid}_n, R)$ from the sender:
Check that for each $\text{cid}_i$, there exists a $m_i$ such that the pair $(\text{cid}_i, m_i)$ has been recorded. If this is not the case then ignore the command. Let $w = (m_1, \ldots, m_n)$. Check that $(x, w) \in R$. If this is not the case then ignore the command. Otherwise, send $(\text{proven}, x, \text{cid}_1, \ldots, \text{cid}_n, R)$ to the receiver and $\mathcal{A}$.

---

**Protocol $\pi_{\text{OTMR}}$** (Oblivious transfer–malicious receiver)

We denote $b$ as the index of the key agreement instance. We denote $m_0, m_1$ as the sender's input and $c$ denotes the receiver's input. We denote $\mathcal{E}(t, r)$, the next message function of the emulator given transcript $t$ and random tape $r$. If the emulator is awaiting a message for a given transcript $t$, we let $\mathcal{E}(t, r) = \bot$. We define the following two relationships: $R_1$ and $R_2$.

$$R_1(a, (b, c)) := \begin{cases} 1 & a = b \oplus c \\ 0 & \text{otherwise} \end{cases}$$

$$R_2((t, m, b), (r, c)) := \begin{cases} 1 & \text{if } b = c \\ 1 & \mathcal{E}(t; r) = m \\ 0 & \text{otherwise} \end{cases}$$

**Receiver:**

$r_1 \in_R \{0,1\}^k$

Send $(\mathtt{commit}, \mathtt{cid}, c)$, $(\mathtt{commit}, \mathtt{rid}_1, r_1)$ to $\mathcal{F}_{\mathtt{COMZK}}$

**Sender:**

Await $(\mathtt{committed}, \mathtt{cid})$, $(\mathtt{committed}, \mathtt{rid}_1)$ from $\mathcal{F}_{\mathtt{COMZK}}$

$r_2 \in_R \{0,1\}^k$

Send $r_2$ to the receiver.

**Receiver:**

$r \leftarrow r_1 \oplus r_2$ (random tape)

Send $(\mathtt{commit}, \mathtt{rid}, r)$ to $\mathcal{F}_{\mathtt{COMZK}}$ (commit to the random tape)

Send $(\mathtt{prove}, r_2, \mathtt{rid}_1, \mathtt{rid}, R_1)$ to $\mathcal{F}_{\mathtt{COMZK}}$ (prove that the commited value associated to rid is indeed a commitment to the random tape)

**Sender:**

Await $(\mathtt{committed}, \mathtt{rid})$ and $(\mathtt{proven}, r_2, \mathtt{rid}_1, \mathtt{rid}, R_1)$ from $\mathcal{F}_{\mathtt{COMZK}}$.

**Sender $\leftrightarrow$ Receiver:**

Sender and receiver run $\pi_{\mathtt{OTSH}}$ as defined in Section 4 where the sender inputs $m_0, m_1$ and the receiver inputs $c$ with the following modification:

Whenever a receiver would send a message $m$ in the semi-honest protocol, let $b$ be the instance of the key agreement protocol they are executing, and let $t$ be the transcript up to that point for that instance of the key agreement protocol. The receiver sends $m$ to the sender and also sends the command $(\mathtt{prove}, (t, m, b), \mathtt{rid}, \mathtt{cid}, R_2)$ to $\mathcal{F}_{\mathtt{COMZK}}$. Whenever the sender receives a message $m$ from the receiver, he awaits that $\mathcal{F}_{\mathtt{COMZK}}$ send him $(\mathtt{proven}, (t, m, b), \mathtt{rid}, \mathtt{cid}, R_2)$ before proceeding.

**Theorem 3.** $\pi_{\mathtt{OTMR}}$ *securely realizes* $\mathcal{F}_{\mathtt{OT}}$ *in the* $\mathcal{F}_{\mathtt{EC}}$-*hybrid model against an environment that can only semi-honestly corrupt the sender.*

This theorem follows directly from the construction of XOR commitments from [CvdGT95, Est04], the security of the GMW compiler [GMW86] and the security of the zero-knowledge protocol from [BCC88, Kil92].

## 6 Secure Protocol

In this section we consider our oblivious transfer protocol $\pi_{\mathtt{OTMR}}$ from Sect. 5, which is secure against a semi-honest sender and a malicious receiver and we show that, if $\pi_{\mathtt{OTMR}}$ is implemented with the key agreement protocol from Sect. 3.1, we can transform $\pi_{\mathtt{OTMR}}$ into a protocol $\pi_{\mathtt{OT}}$ secure against an malicious sender too.

Note that in the aforementioned key agreement protocol, the sender is supposed to send through the channel several bitstrings of length $\ell$ and Hamming weight either $m$ or $m + 1$, where $\ell = 2m + 1$. From now on, we refer to bitstrings of weights $m$ and $m + 1$ as codewords, while the rest will be non-codewords. A problem that arises when using this key agreement protocol as a basis for our oblivious transfer protocol, is that an active sender could use non-codewords to bias the distribution of indices and learn the receiver's choice. For example, if she sends the all-one bitstring, this index will be accepted by the receiver with higher probability if he is playing the role of Bob, than it will if he is playing the role of the emulator.

We will prevent an active sender from using non-codewords in her advantage by combining cut-and-choose techniques, a typicality test and an OT-combiner. The protocol works essentially as follows: the sender and receiver will start to run $N$ instances of $\pi_{\mathrm{OTMR}}$ in parallel. Right after the sender has sent the intended codewords through the channel $\mathcal{F}_{\mathrm{EC}}$ in all these instances, the receiver will then choose half of those instances and request the sender to open her view (i.e., to reveal the information that she sent through the channel). The receiver now runs a typicality test on those instances: he counts the number of differences between what the sender claims to have sent and what he received for those instances. If this distance is higher than what would be typically expected from the noisy channel then the receiver aborts. If the test passes then it is guaranteed that, except with negligible probability, there is at least one unopened instance where no bad codeword was sent.

The sender and receiver now apply a $(1, N/2)$ OT-combiner on the half of the instances of $\pi_{\mathrm{OTMR}}$ that have not been opened; in general, a $(t, n)$ OT-combiner [HKN+05] is a primitive which given (black-box) access to $n$ OT candidates, implements a secure OT as long as $t$ of them are secure; in our case, our candidates are the unopened instances of $\pi_{\mathrm{OTMR}}$ and we use a simple XOR-based OT-combiner which only needs to be secure against a malicious sender (all the candidates are already guaranteed to be secure against a malicious receiver). Since the sender has behaved well in at least one of these instances, we achieve a secure oblivious transfer protocol by applying this combiner.

The sender could also try to cheat in the public channel part of the key agreement protocol by sending some inconsistent information (for example in the information reconciliation step) to see the aborting behaviour of the receiver; however, we have the receiver abort in the global protocol if he sees at least one inconsistency in some instance of the protocol. Given the properties of the combiner the only way to obtain information about the receiver's input bit is that the sender cheats in one of the key agreement protocols of every unopened instance and the receiver never aborts, which happens if the sender guesses each of the $b_i$'s for the unopened instances and in turn this happens with probability $2^{-N/2}$ (in fact, we could make her cheating probability even lower by having the receiver abort if he detects inconsistent information in the opened instances).

## 6.1  Protocol

The protocol $\pi_{OT}$ is described below.

---

### Protocol $\pi_{OT}$ (Oblivious transfer)

The protocol involves two players: the sender and the receiver. The sender provides inputs $m_0, m_1 \in \{0,1\}$ and receives no output. The receiver provides $c \in \{0,1\}$ and outputs $m_c$. Fix $\kappa$ a security parameter for $\pi_{OT}$.

For the protocol $\pi_{OTMR}$ secure against a malicious receiver from previous section instantiated with security parameter $x$, denote $W(x)$ the expected number of bits flipped during such protocol if the noise parameter is not changed (that is, $\delta$ times the number of bits sent through the elastic channel). Now define the following parameters:

$$Q(x) := \frac{32}{(1 - 2\delta)^2} W(x).$$

$$\sigma := \min\{x \in \mathbb{Z} : x - \log Q(x) - \log \kappa \geq \kappa\}.$$

$$N := \kappa Q(\sigma).$$

$$\tau := W(\sigma) + \frac{1 - 2\delta}{2}.$$

and we instantiate $\pi_{OTMR}$ with security parameter $\sigma$.

(Note that, once $\kappa$ is fixed, $\sigma$ is well defined because $Q(x)$ is polynomial in $x$ and hence $x - \log Q(x) \geq \kappa + \log \kappa$ for sufficiently large $x$.)

**Sender:**
  Sample $\Delta \in_R \{0,1\}$.
  Sample $w_0^1, \ldots, w_0^N \in_R \{0,1\}$.
  Sample $w_1^1, \ldots, w_1^N \in_R \{0,1\}$.
  Let $\Delta_i := w_0^i \oplus w_1^i \oplus \Delta$, $i = 1, \ldots, N$.

**Receiver:**
  Sample $b_1, \ldots, b_N \in_R \{0,1\}$.

**Sender $\leftrightarrow$ Receiver:**
  Sender and receiver run $N$ instances of the protocol $\pi_{OTMR}$ as defined in Section 5. Let $(w_0^i, w_1^i)$ be the sender's input and $b_i$ be the receiver's input in the $i$th instance. If at some point in one of the instances the sender sends any information through the public channel that the receiver detects as invalid (such as incorrectly formed $h_1$, $h_2$, or a value $v$ that is not of the form $h_1(X^s)$ for any string $X^s$), then the receiver waits until all instances are completed and then aborts.
  Moreover, the sender records the bits that she sends through the elastic channel in each of the instances as $X = \{(i, j, x_{i,j}) \mid 1 \leq i \leq N, 1 \leq j \leq B\}$. The receiver records the noisy version of bits that he receives from each instance as $Y = \{(i, j, y_{i,j}) \mid 1 \leq i \leq N, 1 \leq j \leq B\}$.

---

**Receiver:**

Choose $T \in_R \{I \mid I \subseteq \{1, \ldots, N\}, |I| = N/2\}$.

Send $T$ to receiver.

Set $\mathcal{L} := \{1, \ldots, N\} \setminus T$.

**Sender:**

Await $T$.

If $|T| \neq N/2$ then abort.

Set $\mathcal{L} := \{1, \ldots, N\} \setminus T$, send $S := \{(i, \Delta_i) \mid i \in \mathcal{L}\}$ and $\tilde{X} := \{(i, j, x_{i,j}) \in X \mid i \in T\}$ to the receiver.

**Receiver:**

Await $\tilde{X}$ and $S$.

Check that $\tilde{X}$ indeed corresponds to a set of bits that the sender should have sent in $\pi_{\mathsf{OTMR}}$ (i.e., that the appropriate parts of $\tilde{X}$ correspond to codewords). If not, abort.

Check that

$$\sum_{i \in T, 1 \leq j \leq B} |x_{i,j} - y_{i,j}| \leq \frac{\tau N}{2}.$$

If it fails, then abort.

Let $b := \bigoplus_{i \in \mathcal{L}} b_i$. Send $d := b \oplus c$ to the sender.

**Sender:**

Let $w_0 := \bigoplus_{i \in \mathcal{L}} w_0^i$, $w_1 := w_0 \oplus \Delta$. Send $(v_0, v_1) := (m_0 \oplus w_d, m_1 \oplus w_{1 \oplus d})$ to the receiver.

**Receiver:**

Let $w := \bigoplus_{i \in \mathcal{L}} w_{b_i}^i \oplus (b_i \wedge \Delta_i)$. Output $w \oplus v_c$.

In protocol $\pi_{\mathsf{OT}}$, the parameters $N$ (the number of instances of $\pi_{\mathsf{OTMR}}$ that will be run), $\sigma$ (the security parameter of $\pi_{\mathsf{OTMR}}$) and $\tau$ (a threshold parameter for the test, which is $W(\sigma)$ plus a small offset) are defined so that we have the following guarantees:

1. The probability that at least one instance of $\pi_{\mathsf{OTMR}}$ is broken by a dishonest receiver is smaller than $2^{-\kappa}$: Indeed, each individual instance can be broken with probability at most $2^{-\sigma}$, and it is easy to see that with our choice of parameters, it holds that $N \cdot 2^{-\sigma} \leq 2^{-\kappa}$.

2. The probability that an honest sender passes the typicality test is at least $1 - 2^{-\kappa}$: see proof in Appendix E.1.
3. If a malicious sender sends at least one non-valid codeword in at least $N/2 - \kappa$ instances of $\pi_{\mathrm{OTMR}}$ from the testing set, then she passes the typicality test with probability at most $2^{-\kappa}$: see proof in Appendix E.1.

Note that the third property prevents a malicious sender to cheat except with probability $2^{-\kappa}$. Indeed, in order for a malicious sender to cheat successfully, she would need to break each of the $N/2$ instances of $\pi_{\mathrm{OTMR}}$ from the evaluation set, and for that she would need to send at least one bad codeword in each of those instances. By 3., in order to pass the test she needs to send all the correct codewords in at least $\kappa$ instances of the testing set. But since she does not know which instances will be selected for the evaluation set and which for the testing set, then the probability that none of these (at least) $\kappa$ correct instances end up in the evaluation set is at most $2^{-\kappa}$.

With all these remarks in mind, we can show (Appendix E) that

**Theorem 4.** $\pi_{\mathrm{OT}}$ *securely realizes* $\mathcal{F}_{\mathrm{OT}}$ *in* $\mathcal{F}_{\mathrm{EC}}$*-hybrid model.*

**Acknowledgments.** Part of this work was carried out while Ignacio Cascudo was with Aarhus University. The authors acknowledge support from the Danish National Research Foundation and The National Science Foundation of China (under the grant 61361136003) for the Sino-Danish Center for the Theory of Interactive Computation and from the Center for Research in Foundations of Electronic Markets (CFEM), supported by the Danish Strategic Research Council. In addition, Ignacio Cascudo acknowledges support from the Danish Council for Independent Research, grant no. DFF-4002-00367, Ivan Damgrd was also supported by the advanced ERC grant MPCPRO and Samuel Ranellucci was supported by European Research Council Starting Grant 279447. We thank Jesper Buus Nielsen, Maciej Obremski and the anonymous reviewers for their helpful comments.

# A     Universal Composability

The Universal Composability security framework, introduced in [Can01], is based on the simulation paradigm. Roughly, the idea is to compare the execution of the actual protocol (the real world) with an idealized scenario (the ideal world) in which the computations are carried out by a trusted third party (the ideal functionality) which receives inputs from and hands in outputs to the players. The goal is to show that these two worlds are indistinguishable. In order to formalize this goal, we introduce a party called the environment $\mathcal{Z}$, whose task is to distinguish between both worlds. Furthermore, in the ideal world, we introduce a simulator $\mathcal{S}$, its task being to simulate any action of the adversary in the real protocol and thereby to make the two views indistinguishable for any environment. More precisely, in the real world execution of protocol $\pi$, with the adversary $\mathcal{A}$ and environment $\mathcal{Z}$, the environment provides input and receives output from both $\mathcal{A}$ and $\pi$. Call $\mathrm{Real}_{\mathcal{A},\pi,\mathcal{Z}}$ the view of $\mathcal{Z}$ in this execution. In the ideal world $\mathcal{Z}$ provides input and receives output from $\mathcal{S}$ and the ideal functionality $\mathcal{F}$. Call $\text{Ideal}_{\mathcal{S},\mathcal{F},\mathcal{Z}}$ the view of $\mathcal{Z}$ in the ideal execution. We can proceed to define what it means for a protocol to be secure.

**Definition 2.** *A protocol $\pi$ UC-implements a functionality $\mathcal{F}$ against a certain class of adversaries $\mathcal{C}$ if for every adversary $\mathcal{A} \in \mathcal{C}$ there exists a simulator $\mathcal{S}$ such that for every environment $\mathcal{Z}$, $\text{Real}_{\mathcal{A},\pi,\mathcal{Z}} \approx \text{Ideal}_{\mathcal{S},\mathcal{F},\mathcal{Z}}$.*

The cornerstone of the universal composability framework is the composition theorem, which works as follows. Denote by $\pi \diamond G$ a protocol $\pi$ that during its execution makes calls to an ideal functionality $G$. The composition proof shows that if $\pi_f \diamond G$ securely implements $\mathcal{F}$ and if $\pi_g$ securely implements $G$ then $\pi_f \diamond \pi_g$ securely implements $\mathcal{F}$. This provides modularity in construction of protocols and simplifies proofs dramatically. It is also shown that proving security against a dummy adversary, one who acts as a communication channel, is sufficient for proving general security.

# B   Proof of Lemma 1

Clearly, we only need to cover the case $i = 0$. First, note that $\Pr[Z \in S_0|X = 0] \geq 1/2$ since $\gamma < 1/2$. Let $\overline{X}$ denote the random variable describing the encoding of $b = 0$ by Alice, i.e., $\overline{X}$ has the uniform distribution over the set of bitstrings in $\{0,1\}^\ell$ of weight exactly $m$ or $m + 1$. We observe that since Eve's noise is independent and identically distributed for each bit sent through the wiretap channel, for every string $\overline{x} \in \{0,1\}^\ell$ of weight $m$, $\Pr[Z \in S_0|X = 0] = \Pr[Z \in S_0|\overline{X} = \overline{x}]$. So we now compute $\Pr[Z \in S_0|\overline{X} = \overline{x}]$ for $\overline{x} = 010101\ldots010$.

For $i = 1,\ldots,m$, let $V_i$ be the random variable that takes value 1 if $Z = z$ and $z_{2i-1} = z_{2i} = 1$, the value $-1$ if $z_{2i-1} = z_{2i} = 0$ and the value 0 if $z_{2i-1} \neq z_{2i}$. Then clearly $\Pr[Z \in S_0|\overline{X} = \overline{x}] \leq \Pr[\sum_{i=1}^m V_i \leq 0]$.

Now note that $V_i$ are independent identically distributed variables such that $\Pr[V_i = 1] = \Pr[V_i = -1] = p$ and $\Pr[V_i = 0] = 1 - 2p$ where $p = \gamma(1 - \gamma)$. Hence $\Pr[\sum_{i=1}^m V_i < 0] = \Pr[\sum_{i=1}^m V_i > 0]$ and clearly (using for example the central limit theorem) $\Pr[\sum_{i=1}^m V_i = 0] \to 0$ as $\ell$ (and consequently $m$) grows. Therefore

$$1/2 \leq \Pr[Z \in S_0|X = 0] \leq \Pr[\sum_{i=1}^m V_i \leq 0] \to 1/2$$

and consequently $\Pr[Z \in S_0|X = 0] \to 1/2$.

# C   Proof of Theorem 1

We first argue that, if we set the parameters adequately, the protocol is correct and secure, i.e., with overwhelming probability Alice and Bob have a common string at the end of the protocol about which Eve has a negligible amount of information.

Remember that for each index $i$, Alice encodes $b_i$ as a bitstring containing $m+1$ bits equal to $b_i$ and $m$ bits equal to $(1-b_i)$ and Bob accepts if he receives $c_i^\ell$. Hence, the probability that Bob accepts $i$, i.e., the probability that an index $i$ is in $\mathcal{I}$ is $p_{\text{accept}} = \delta^m (1-\delta)^{m+1} + \delta^{m+1}(1-\delta)^m = (\delta(1-\delta))^m$.

On the other hand, conditioned on Bob accepting index $i$, the probability that $c_i \neq b_i$ is

$$\frac{\delta^{m+1}(1-\delta)^m}{(\delta(1-\delta))^m} = \delta.$$

Furthermore these probabilities are independent from each $i$, so the advantage distillation step creates another conceptual noisy channel where Alice communicates $s$ bits to Bob and the noise parameter is still $\delta$ (independently of how large $\ell$ is).

Hence if we set $n$ slightly larger than $\lceil s/p_{\text{accept}} \rceil$ for large enough parameters, Bob will, with very high probability, accept at least $s$ bits, of which roughly $\delta \cdot s$ will be incorrect. By the results on information reconciliation in [BS94], if $h_1$ is chosen from the 2-universal family of hash functions $\mathcal{H}_1$, then Bob can correct to the right string $X^s$ with very high probability given his original string, $h_1$ and $h_1(X^s)$. Hence both Alice and Bob will compute the same value $h_2(X^s)$ with high probability and hence the protocol is correct.

As for privacy, remember that $X$ denotes the uniform distribution over $\{0,1\}$ and $Z$ the variable that represents Eve's output when Alice chooses $b$ according to $X$, encodes it, and sends it through the channel. Then

$$H_\infty(X|Z) = \sum_{z \in \{0,1\}^\ell} \Pr[Z = z] \cdot (-\log(\max_{b \in \{0,1\}} \Pr[X = b|Z = z])).$$

Now the maximum of $\Pr(X = b|Z = z)$ is reached for $b = 0$ if $z \in S_0$ and for $b = 1$ if $z \in S_1$, where $S_i$ is defined as in Lemma 1. On the other hand, for every $z \in S_0$, we have $z' := (1, \ldots, 1) - z \in S_1$ and clearly, $\Pr[X = 0|Z = z] = \Pr[X = 1|Z = z']$. Hence we can write

$$H_\infty(X|Z) = \sum_{z \in S_0} 2 \cdot \Pr[Z = z] \cdot (-\log \Pr[X = 0|Z = z]).$$

Now, clearly $\sum_{z \in S_0} 2 \cdot Pr[Z = z] = 1$ and $-\log$ is a convex function. This means we can apply Jensen's inequality to get

$$H_\infty(X|Z) \geq -\log\left(\sum_{z \in S_0} 2 \cdot \Pr[Z = z]\Pr[X = 0|Z = z]\right).$$

Now we use that $\Pr[Z = z]\Pr[X = 0|Z = z] = \Pr[X = 0]\Pr[Z = z|X = 0] = \frac{1}{2}\Pr[Z = z|X = 0]$, so after summing over $z \in S_0$ we obtain:

$$H_\infty(X|Z) \geq -\log \Pr[Z \in S_0|X = 0] \to 1$$

as $\ell \to \infty$ because of Lemma 1. Since $\delta + \epsilon < 1/2$, for large enough $\ell$, we have $\mathfrak{h}(\delta + \epsilon) < H_\infty(X|Z)$ (remember $\mathfrak{h}(\cdot)$ denotes the binary entropy function).

Now let $X^s, Y^s$ denote the random variables denoting the $s$ bits outputted by Alice and Bob respectively and let $Z^s$ be the variable representing the $s$ bitstrings outputted by Eve. Then clearly $H_\infty(X^s|Z^s) = sH_\infty(X|Z) > s\mathfrak{h}(\delta + \epsilon) + t + 3\sigma$ since $t, \sigma = o(s)$ and therefore $H_\infty(X^s|Z^s, h_1, h_1(X^s)) \geq H_\infty(X^s|Z^s) - s\mathfrak{h}(\delta + \epsilon) - \sigma > t + 2\sigma$.

Now, the leftover hash lemma guarantees that conditioned on everything seen by Eve during the protocol, the distribution of $h_2(X^s)$ is $2^{-\sigma}$-close to the uniform distribution over $\{0, 1\}^t$.

To show that the protocol is emulatable, we have to construct an emulator $\mathcal{E}$ that satisfies Property 2 in Definition 1. We note that the only information Bob sends to Eve is the description of the set $\mathcal{I}$ of indices for which Bob accepted Alice's message. We can construct an emulator for Bob thus. After $\mathcal{E}$ receives a message from the dummy channel $\mathcal{F}_{DC}$, it samples a random index set $\mathcal{I} \subseteq \{0, 1\}^n$, where each index is chosen according to a Bernoulli distribution with parameter $p_{accept}$—the index is included in $\mathcal{I}$ if the trial succeeds. $\mathcal{E}$ then sends a description of $\mathcal{I}$ to Alice via $\mathcal{F}_{Pub}$. It is clear that such an emulator satisfies Property 2.

# D  Commitment Protocol for ECs from [DKS99]

In this section we describe the commitment protocol from [DKS99] and show that, under the adequate choice of parameters, it is a receiver commitment protocol for any $(\gamma, \delta)$-elastic noisy channels.

We define some constants as follows. $d_0$ is defined by $\delta = \gamma(1 - d_0) + d_0(1 - \gamma)$. That is, $d_0$ is such that adding noise with rate $\gamma$ and then noise with rate $d_0$ produces total noise rate $\delta$. This means that $d_0 = (\delta - \gamma)/(1 - 2\gamma)$, and from it follows trivially that since $\delta < 1/2$, we have $\delta > d_0$. We can therefore choose constants $d_1, d$ and $d^*$ such that $d_0 < d_1 < d^* < d < \delta$. Finally, we define $\delta' = \gamma(1 - d_1) + d_1(1 - \gamma)$. Note that since $d_1 > d_0$ we have $\delta' > \delta$.

Furthermore, we define $\ell$ to be the logarithm of the number of elements in a Hamming ball of radius $d$, and likewise $\ell^*$ the logarithm of number of elements in a Hamming ball of radius $d^*$.

We will need three families of universal hash functions $\mathcal{H}, \mathcal{H}_1, \mathcal{H}_2$ that are $64k$-wise independent and map from $\{0, 1\}^k$ to $\{0, 1\}, \{0, 1\}^{\ell^*}, \{0, 1\}^{\ell - \ell^*}$, respectively.

Finally, remember that, as explained in Sect. 5.1, we reverse the direction of the elastic channel, so the protocol that we describe next uses a noisy channel with noise rate $\delta$ where the committer $C$ sends information and the verifier $V$ receives, but where it is $C$ who can alter the noise rate and reduce it to $\gamma$.

---

**Protocol** Commit

**C:**

Sample $X \in_R \{0,1\}^k$, send (**send**, sid, $C, V, X$) to $\mathcal{F}_{\mathsf{EC}}$.

**V:**

Await (**send**, sid, $C, V, X'$) from $\mathcal{F}_{\mathsf{EC}}$

Sample $h_1 \in_R \mathcal{H}_1$, send (**send**, sid$_1$, $V, C, h_1$) to $\mathcal{F}_{\mathsf{Pub}}$.

**C:**

Await (**send**, sid$_1$, $V, C, h_1$) from $\mathcal{F}_{\mathsf{Pub}}$.

Set $y_1 := h_1(X)$, send (**send**, sid$_2$, $C, V, y_1$) to $\mathcal{F}_{\mathsf{Pub}}$.

**V:**

Await (**send**, sid$_2$, $C, V, y_1$) from $\mathcal{F}_{\mathsf{Pub}}$.

Sample $h_2 \in_R \mathcal{H}_2$, send (**send**, sid$_3$, $V, C, h_2$) to $\mathcal{F}_{\mathsf{Pub}}$.

**C:**

Await (**send**, sid$_3$, $V, C, h_2$) from $\mathcal{F}_{\mathsf{Pub}}$.

Sample $h \in_R \mathcal{H}$, set $y_2 := h_2(X)$ and $b := h(X)$.

Send (**send**, sid$_4$, $C, V, y_2$) and (**send**, sid$_5$, $C, V, h$) to $\mathcal{F}_{\mathsf{Pub}}$.

Output $b$.

**V:**

Await (**send**, sid$_4$, $C, V, y_2$) and (**send**, sid$_5$, $C, V, h$) from $\mathcal{F}_{\mathsf{Pub}}$.

---

**Protocol** Open

We define $\Delta$ as the Hamming distance.

**C:**

Send (**send**, sid, $C, V, X$) to $\mathcal{F}_{\mathsf{Pub}}$.

**V:**

Await (**sent**, sid, $C, V, X$) from $\mathcal{F}_{\mathsf{Pub}}$.

Check that $y_1 = h_1(X)$, $y_2 = h_2(X)$ and $\Delta(X, X') \leq \delta' k$. If either condition is false, then abort.

Output $b := h(X)$.

---

We have defined our constants slightly differently from what was done in [DKS99], but $d_0$ is defined in the same way, and the rest of the constants satisfy the same inequalities. It therefore turns out that exactly the same proofs can be used to show this version secure. We will not repeat the proofs here, but give some intuition why the protocol is secure. We let $\Delta$ denote the Hamming distance, and by negligible we mean negligible as a function of $k$.

**Both parties are honest.** In this case we expect $X'$ to be at distance $\delta k$ from $X$. Since $\delta' > \delta$, the probability that the distance is greater than $\delta'k$ is negligible, so $V$ will accept the opening.

$C$ **is corrupt.** We want to argue that there is only one string $C$ can convincingly open after commitment time. Suppose first that $C$ tries to claim a string $X^*$ with $\Delta(X^*, X) > d^*k$. Then note than in his view, the received string $X'$ is expected to be such that $\delta(X, X') = \gamma k$. So we expect that $\Delta(X^*, X') > (d^*(1 - \gamma) + \gamma(1 - d^*))k > \delta'k$ because $d^* > d_1$. So $V$ would reject with overwhelming probability in this case. This means that $X^*$ must be in a Hamming ball with radius $d^*$ and center in $X$. But by sending $h_1(X), h_2(X)$, $C$ reveals $\ell$ bits of information on $X$. Since $\ell > \ell^*$ this is more than required to identify uniquely an string in a ball of radius $\ell^*$, so there is only one string that can be opened.

$V$ **is corrupt.** We want to argue that $V$ has essentially no information about $h(X)$ before opening. Note that in $V$'s view $X$ is in a Hamming ball with radius $\delta$ and center in $X'$. Via the hashing $V$ gets only $\ell$ bits of information, and since $d < \delta$, one can show that there are exponentially many candidates left for $X$, even after hashing. Now by a standard privacy amplification argument, it follows that the expected information $V$ has on $h(X)$ is negligible.

# E    Proof of Security of $\pi_{\mathsf{OT}}$

## E.1    Statements About the Typicality Test

We will need to establish some statements about the typicality test from our protocol.

Define $\mathsf{X}[\mu]$ to be a binomial variable with expectation $\mu$. By abuse of notation, we denote by $\sum_{i=1}^{N} \mathsf{X}[\mu]$ the variable defined by sampling $N$ independent random variables with expectation $\mu$ and adding the result.

**Probability that an Honest Sender Passes the Typicality Test.** We show that the honest sender passes the typicality test with probability at least $1 - 2^{-\kappa}$. Let $T = \sum_{i=1}^{N/2} \mathsf{X}[W(\sigma)]$. An honest receiver does *not* pass the typicality test if and only if $T \geq \tau N/2$. Now let $\mu = \mathsf{E}[T] = \frac{N}{2}W(\sigma)$ and $\beta = \frac{1}{2W(\sigma)}$. We can apply Chernoff's bound to see that

$$\Pr[T \geq \tau N/2] = \Pr[T \geq (1 + \beta)\mu] \leq e^{-\mu\beta^2/4} \leq 2^{-\kappa}.$$

**Probability that a Malicious Sender Breaks the Typicality Test.** We show that if a malicious sender cheats in $N/2 - \kappa$ instances of the testing set, she passes the typicality test with probability at most $2^{-\kappa}$. Note that in order for the sender to send something different from a codeword in a given instance, at least one of the bits she sent does not correspond to the bit she communicates when she sends $\tilde{X}$. Now note that, for a given bit $x_{i,j}$ communicated by the sender when she sends $\tilde{X}$, if this bit was indeed correct, then $x_{i,j} \neq y_{i,j}$ with probability $\delta$, while if she sent $1 - x_{i,j}$ instead, then $x_{i,j} \neq y_{i,j}$ with probability $1 - \delta$. Note that the difference between these probabilities is $1 - 2\delta$. This means that, in expectation, if the sender assumes the cheating behaviour we just described, the distance between the bitstrings $(x_{i,j})$ and $(y_{i,j})$ will grow by an additive factor of $(1 - 2\delta)(N/2 - \kappa)$ with respect to the case where the sender would be honest. We want to show that in these conditions, the malicious sender will fail the test with high probability. That is, again defining $T = \sum_{i=1}^{N/2} X[W(\sigma)]$, we need to show:

$$\Pr\left[T \leq \frac{\tau N}{2} - (1 - 2\delta)\left(\frac{N}{2} - \kappa\right)\right] \leq 2^{-\kappa}.$$

Let $\mu = \mathsf{E}[T] = \frac{N}{2}W(\sigma)$ and $\beta = \frac{(1-2\delta)(N-4\kappa)}{2NW(\sigma)}$. Chernoff's bound then says

$$\Pr\left[T \leq (1 - \beta)\mu\right] \leq e^{-\mu\beta^2/2}.$$

Now it is easy to see that, for the values of $\mu$ and $\beta$ detailed above, $(1-\beta)\mu = \frac{\tau N}{2} - (1 - 2\delta)\left(\frac{N}{2} - \kappa\right)$ (so this probability is indeed what we want to bound) and that $e^{-\mu\beta^2/2} \leq 2^{-\kappa}$.

In the rest of the section we will prove Theorem 4.

## E.2    Correctness

If both players are honest, then the protocol is correct with probability at least $1 - 2^{-\kappa}$. Indeed, with at least this probability the honest sender passes the typicality test and the protocol is completed. Then, note that:

$$w_{b_i}^i \oplus (b_i \wedge \Delta_i) = \begin{cases} w_0^i & \text{, if } b_i = 0 \\ w_1^i \oplus \Delta_i = w_0^i \oplus \Delta & \text{, if } b_i = 1 \end{cases}$$

Hence $w = w_0$ if there is an even number of $i \in \mathcal{L}$ such that $b_i = 1$, i.e., if $b = 0$, and $w = w_0 \oplus \Delta = w_1$ if $b = 1$. In other words, $w = w_b$.

On the other hand $v_c = m_c \oplus w_{c \oplus d} = m_c \oplus w_b$.

Therefore the output of the receiver equals $w \oplus v_c = m_c$, so the protocol outputs the correct value.

## E.3    Security Against a Malicious Receiver

*Simulation.* The simulator $\mathcal{S}$ for $\pi_{\mathsf{OT}}$ will first proceed by running $N$ instances $\mathcal{S}_1, \ldots, \mathcal{S}_N$ of the simulator for $\pi_{\mathsf{OTMR}}$. Upon receiving (choice, $b_i$) from the environment, it will record it and send a random $w_i$. If any of the simulators aborts, then the simulator aborts. In the next step, it awaits the test set $T$ from $\mathcal{Z}$. Now, the simulator must send a $\tilde{X}$ such that the view of $\mathcal{Z}$ for the test instances is the same as in the real world.

Each of the views produced by the simulators are statistically indistinguishable (within $2^{-\sigma}$) from real instances of the OT protocol. Therefore, there must be a distribution $D$ for $\tilde{X}$ that depends only on the transcript between the simulators and $\mathcal{Z}$ that is $(1/2^\kappa)$-close to one which would be produced in the real world.

Indeed, if this was not the case, since

$$\frac{N}{2^\sigma} = \frac{\kappa Q(\sigma)}{2^\sigma} = \frac{1}{2^{\sigma - \log Q(\sigma) - \log \kappa}} \le \frac{1}{2^\kappa},$$

then $\mathcal{Z}$ would be able to distinguish with probability larger than $1/2^\sigma$ between a run of the simulated malicious-receiver OT and a run with the malicious-receiver OT protocol with the elastic channel for at least one of the $N$ instances, which contradicts the security of $\pi_{\mathsf{OTMR}}$.

> $\mathcal{S}$ samples $\tilde{X} \in_R D$. $\mathcal{S}$ sets $\mathcal{L} = \{1, \ldots, N\} \setminus T$. $\mathcal{S}$ samples $\Delta_i \in_R \{0,1\}$, for $i \in \mathcal{L}$ and sets $S = \{(i, \Delta_i) \mid i \in \mathcal{L}\}$.
> $\mathcal{S}$ sends $S, \tilde{X}$ to $\mathcal{Z}$. $\mathcal{S}$ computes $w := \bigoplus_{i \in \mathcal{L}} w_i \oplus (\Delta_i \wedge b_i)$ and $b := \bigoplus_{i \in \mathcal{L}} b_i$. $\mathcal{S}$

awaits that the environment inputs $d$. $\mathcal{S}$ samples a random $x \in_R \{0,1\}$, sets $c = b \oplus d$ and sends $(\mathsf{choice}, c)$ to $\mathcal{F}_{\mathsf{OT}}$. Upon receiving $(\mathsf{receipt}, m)$, $\mathcal{S}$ sets $u_0 = m \oplus w, u_1 = x$ and sends $(v_0, v_1) := (u_d, u_{1 \oplus d})$ to $\mathcal{Z}$.

*Indistinguishability.* This follows from the fact that the given robust OT-combiner is universally composable and that the underlying OT protocol is secure against a malicious receiver.

## E.4  Security Against a Malicious Sender

*Simulation.* The simulator $\mathcal{S}$ employs the following strategy. First, for each instance of OT, $\mathcal{S}$ runs an instance $\mathcal{S}_i$ of the simulator for the protocol $\pi_{\mathsf{OTMR}}$ (for the semi-honest sender) for as long as $\mathcal{Z}$ does not send invalid codewords for that instance. If any $\mathcal{S}_i$ aborts, then $\mathcal{S}$ aborts.

When, for a given instance, $\mathcal{Z}$ sends an invalid codeword, $\mathcal{S}$ takes the simulator $\mathcal{S}_i$, samples a $b_i$ at random and samples a receiver $R_i$ whose input is $b_i$ and whose view is consistent with what has been sent by the environment for that instance.

From this point on, instead of running the simulator for the given instance, $\mathcal{S}$ runs $R_i$ and whenever $\mathcal{Z}$ sends a message which is meant to be communicated through the elastic channel, $\mathcal{S}$ simulates the channel and sends the result to $R_i$.

Once the instances of OT (both simulated and run with honest receiver) have completed, $\mathcal{S}$ samples a random test set $T$ and sends it to $\mathcal{Z}$. $\mathcal{S}$ awaits $\tilde{X}, S$ from the environment. $\mathcal{S}$ simulates the typicality test. $\mathcal{S}$ takes each instance of OT for the test that is still run by the simulator for the test cases and replaces it with a receiver in the same way that was described above. Then once $\mathcal{S}$ has produced the given views, $\mathcal{S}$ takes these views and runs the typicality test. If the test fails, the simulator aborts.

$\mathcal{S}$ denotes the set of instances $\mathcal{I}$ that were only run by simulators and were not part of the test set. Let $\mathcal{J}$ be the set of instances that were run by the receivers and were not part of the test set. The simulators provided the values $\{(w_0^i, w_1^i) \mid i \in \mathcal{I}\}$ and the receivers provided the values $\{w_{b_j}^j \mid j \in \mathcal{J}\}$.

$\mathcal{S}$ samples a $u \in \mathcal{I}$ and, for each $i \in \mathcal{I}$, selects a random $b_i$. $\mathcal{S}$ selects $b = \bigoplus_{i \in \mathcal{L}, i \neq u} b_i$, $w := \bigoplus_{i \in \mathcal{L}, i \neq u} w_{b_i}^i \oplus (\Delta_i \wedge b_i)$. $\mathcal{S}$ sets $m_0' := w \oplus w_0^u$, $m_1' := w \oplus w_1^u \oplus \Delta_u$.

$\mathcal{S}$ samples a random $r$ and sends $d = b \oplus r$ to $\mathcal{Z}$. $\mathcal{S}$ awaits $v_0, v_1$ from $\mathcal{Z}$. $\mathcal{S}$ sets $m_0 := w \oplus v_{b \oplus r} \oplus m_{b \oplus r}'$ and $m_1 := w \oplus v_{b \oplus r \oplus 1} \oplus m_{b \oplus r \oplus 1}'$. $\mathcal{S}$ sends (send, sid, $m_0, m_1$) to $\mathcal{F}_{\text{OT}}$.

*Indistinguishability.* The real-world instances of OT where the sender did not send bad codewords are indistinguishable from the ideal-world instances run by local simulators. This follows from the security of $\pi_{\text{OTSH}}$ against semi-honest adversaries.

Next, we consider, the instances of OT where the sender sent bad codewords. These are also indistinguishable from instances run by the simulator because, on seeing a bad codeword, the simulator replaces the local simulator with a receiver $R_i$, with random input $b_i$, that acts as in the real world (including the communication between the sender and this receiver, which is simulated by imitating the behaviour of the channel). Furthermore, the receiver is constructed so that it is consistent with what had been previously sent through the channel and the given choice of inputs.

The last step of our simulation needs, however, to make sure that $\mathcal{I}$ is non-empty, i.e., that there is at least one instance of the evaluation set where $\mathcal{Z}$ sends only correct codewords. But notice that, as we have shown before, if $\mathcal{Z}$ would send a non-codeword in each instance, it would result (except with probability $2^{-\kappa}$) in an abort due to the typicality test.

# References

[BCC88] Brassard, G., Chaum, D., Crépeau, C.: Minimum disclosure proofs of knowledge. J. Comput. Syst. Sci. **37**(2), 156–189 (1988)

[BCS96] Brassard, G., Crépeau, C., Santha, M.: Oblivious transfers and intersecting codes. IEEE Trans. Inf. Theory **42**(6), 1769–1780 (1996)

[BS94] Brassard, G., Salvail, L.: Secret-key reconciliation by public discussion. In: Helleseth, T. (ed.) EUROCRYPT 1993. LNCS, vol. 765, pp. 410–423. Springer, Heidelberg (1994). doi:10.1007/3-540-48285-7_35

[Can01] Canetti, R.: Universally composable security: a new paradigm for cryptographic protocols. In: Proceedings of 42nd IEEE Symposium on Foundations of Computer Science, pp. 136–145. IEEE (2001)

[CK88]    Crépeau, C., Kilian, J.: Achieving oblivious transfer using weakened security assumptions (Extended Abstract). In: 29th Annual Symposium on Foundations of Computer Science, White Plains, New York, USA, 24–26 October 1988, pp. 42–52 (1988)

[CMW05]   Crépeau, C., Morozov, K., Wolf, S.: Efficient unconditional oblivious transfer from almost any noisy channel. In: Blundo, C., Cimato, S. (eds.) SCN 2004. LNCS, vol. 3352, pp. 47–59. Springer, Heidelberg (2005). doi:10.1007/978-3-540-30598-9_4

[Cré97]   Crépeau, C.: Efficient cryptographic protocols based on noisy channels. In: Fumy, W. (ed.) EUROCRYPT 1997. LNCS, vol. 1233, pp. 306–317. Springer, Heidelberg (1997). doi:10.1007/3-540-69053-0_21

[CS06]    Crépeau, C., Savvides, G.: Optimal reductions between oblivious transfers using interactive hashing. In: Proceedings of Advances in Cryptology - EUROCRYpPT, 25th Annual International Conference on the Theory and Applications of Cryptographic Techniques, St. Petersburg, Russia, May 28–June 1, pp. 201–221 (2006)

[CvdGT95] Crépeau, C., Graaf, J., Tapp, A.: Committed oblivious transfer and private multi-party computation. In: Coppersmith, D. (ed.) CRYPTO 1995. LNCS, vol. 963, pp. 110–123. Springer, Heidelberg (1995). doi:10.1007/3-540-44750-4_9

[DFMS04]  Damgård, I., Fehr, S., Morozov, K., Salvail, L.: Unfair noisy channels and oblivious transfer. In: Naor, M. (ed.) TCC 2004. LNCS, vol. 2951, pp. 355–373. Springer, Heidelberg (2004). doi:10.1007/978-3-540-24638-1_20

[DKS99]   Damgård, I., Kilian, J., Salvail, L.: On the (im)possibility of basing oblivious transfer and bit commitment on weakened security assumptions. In: Stern, J. (ed.) EUROCRYPT 1999. LNCS, vol. 1592, pp. 56–73. Springer, Heidelberg (1999). doi:10.1007/3-540-48910-X_5

[DORS08]  Dodis, Y., Ostrovsky, R., Reyzin, L., Smith, A.D.: Fuzzy extractors: how to generate strong keys from biometrics and other noisy data. SIAM J. Comput. 38(1), 97–139 (2008)

[Est04]   Estren, G.: Universally composable committed oblivious transfer and multi-party computation assuming only basic black-box primitives. Ph.D. thesis, McGill University (2004)

[GMW86]   Goldreich, O., Micali, S., Wigderson, A.: How to prove all NP statements in zero-knowledge and a methodology of cryptographic protocol design (Extended Abstract). In: Odlyzko, A.M. (ed.) CRYPTO 1986. LNCS, vol. 263, pp. 171–185. Springer, Heidelberg (1987). doi:10.1007/3-540-47721-7_11

[HKN+05]  Harnik, D., Kilian, J., Naor, M., Reingold, O., Rosen, A.: On robust combiners for oblivious transfer and other primitives. In: Proceedings of Advances in Cryptology - EUROCRYpPT, 24th Annual International Conference on the Theory and Applications of Cryptographic Techniques, Aarhus, Denmark, pp. 96–113, 22–26 May 2005

[IKO+11]  Ishai, Y., Kushilevitz, E., Ostrovsky, R., Prabhakaran, M., Sahai, A., Wullschleger, J.: Constant-rate oblivious transfer from noisy channels. In: Rogaway, P. (ed.) CRYPTO 2011. LNCS, vol. 6841, pp. 667–684. Springer, Heidelberg (2011). doi:10.1007/978-3-642-22792-9_38

[Kil88]   Kilian, J.: Founding cryptography on oblivious transfer. In: Proceedings of the Twentieth Annual ACM Symposium on Theory of Computing, pp. 20–31. ACM (1988)

[Kil92]  Kilian, J.: A note on efficient zero-knowledge proofs and arguments. In: Proceedings of the Twenty-Fourth Annual ACM Symposium on Theory of Computing, pp. 723–732. ACM (1992)

[KMS16]  Khurana, D., Maji, H.K., Sahai, A.: Secure computation from elastic noisy channels. In: Proceedings of Advances in Cryptology - EUROCRYpPT - 35th Annual International Conference on the Theory and Applications of Cryptographic Techniques, Vienna, Austria, Part II, pp. 184–212, 8–12 May 2016

[Mau93]  Maurer, U.M.: Secret key agreement by public discussion from common information. IEEE Trans. Inf. Theory 39(3), 733–742 (1993)

[PDMN11]  Pinto, A.C.B., Dowsley, R., Morozov, K., Nascimento, A.C.A.: Achieving oblivious transfer capacity of generalized erasure channels in the malicious model. IEEE Trans. Inf. Theory 57(8), 5566–5571 (2011)

# Simultaneous Secrecy and Reliability Amplification for a General Channel Model

Russell Impagliazzo[1]([✉]), Ragesh Jaiswal[2], Valentine Kabanets[3],
Bruce M. Kapron[4], Valerie King[4], and Stefano Tessaro[5]

[1] University of California, San Diego, San Diego, USA
russell@cs.ucsd.edu
[2] Indian Institute of Technology Delhi, New Delhi, India
rjaiswal@cse.iitd.ac.in
[3] Simon Fraser University, Burnaby, Canada
kabanets@cs.sfu.ca
[4] University of Victoria, Victoria, Canada
bmkapron@uvic.ca, val@cs.uvic.ca
[5] University of California, Santa Barbara, Santa Barbara, USA
tessaro@cs.ucsb.edu

**Abstract.** We present a general notion of channel for cryptographic purposes, which can model either a (classical) physical channel or the consequences of a cryptographic protocol, or any hybrid. We consider *simultaneous secrecy and reliability amplification* for such channels. We show that simultaneous secrecy and reliability amplification is not possible for the most general model of channel, but, at least for some values of the parameters, it is possible for a restricted class of channels that still includes both standard information-theoretic channels and keyless cryptographic protocols.

Even in the restricted model, we require that for the original channel, the failure chance for the attacker must be a factor $c$ more than that for the intended receiver. We show that for any $c > 4$, there is a one-way protocol (where the sender sends information to the receiver only) which achieves simultaneous secrecy and reliability. From results of Holenstein and Renner (*CRYPTO'05*), there are no such one-way protocols for $c < 2$. On the other hand, we also show that for $c > 1.5$, there are two-way protocols that achieve simultaneous secrecy and reliability.

We propose using similar models to address other questions in the theory of cryptography, such as using noisy channels for secret agreement, trade-offs between reliability and secrecy, and the equivalence of various notions of oblivious channels and secure computation.

## 1 Introduction

Modern cryptography has its roots in the work of Shannon [35], using channels as the model of communication where some secrecy is attainable [9,39]. A cryptographic protocol can also be interpreted as implicitly defining a *computational* channel, where the loss of information is merely computational. For example,

© International Association for Cryptologic Research 2016
M. Hirt and A. Smith (Eds.): TCC 2016-B, Part I, LNCS 9985, pp. 235–261, 2016.
DOI: 10.1007/978-3-662-53641-4_10

consider a channel sending a message $m$ as the pair consisting of a public key $pk$, and an encryption $c$ of $m$ under $pk$. If the encryption scheme provides some form of (even weak) security, a computationally bounded adversarial observer of the channel output will only learn partial information about $m$, even though information-theoretically the channel may well uniquely define its input.

In some circumstances, it may not even be clear whether the limitation is computational or informational. For example, an adversary may not be able to perfectly tune in to a low-power radio broadcast. This might appear an information-theoretic limitation, but improved algorithms to interpolate signals or to predict interference due to atmospheric conditions could also improve the adversary's ability to eavesdrop.

In this work, we introduce a model of computation that combines information-theoretic and computational limitations. Specifically, we present a general notion of channel for cryptographic purposes, which can model either a (classical) physical channel or the consequences of a cryptographic protocol, or any hybrid.

We require our model to satisfy the following properties:

- [**Agnostic**] It should not matter *why* an adversary is limited. Protocols designed exploiting an adversary's weakness should remain secure whether that weakness is due to limited information, computational ability, or any other reason.
- [**Composable**] We should be able to safely combine a protocol that achieves one goal from an assumption, and a second protocol that achieves a second goal from the first, into one that achieves the second goal from the original assumption.
- [**Functional**] The assumptions underlying our protocols should concern what the parties *can do*, rather than concerning what they or the channels through which they communicate *are*. In particular, we should be able to use this to evaluate the danger of side information, and enhanced functionality should not threaten secrecy properties.
- [**Combining reliability and secrecy**] Instead of viewing reliability of a channel and its secrecy as separate issues, our model should combine the two in a seamless way. We want to study how enhancing secrecy might impact reliability, and vice versa. In other words, we view reliability as equally necessary for the overall secrecy.

In this paper, we focus on the *simultaneous secrecy and reliability amplification* for such channels. We start with a channel where the intended receiver gets the transmitted bit except with some probability and the attacker can guess the transmitted bit except with a somewhat higher probability. We wish to use the channel to define one where the receiver gets the transmitted bit almost certainly while only negligible information is leaked to the attacker. We show that simultaneous secrecy and reliability amplification is not possible for the most general model of channel, but, at least for some values of the parameters, it is possible for a restricted class of channels that still includes both standard information-theoretic channels and keyless cryptographic protocols.

Note that, traditionally, error-correction and encryption have been thought of in communications theory as separate layers, with one performed first and then the other on top. However, when one wants to leverage the secrecy of an unreliable channel, it does not seem possible to separate the two. Using an error-correcting code prior to secrecy considerations could totally eliminate even the partial secrecy, and amplifying secrecy could make the channel totally unreliable. (In some sense, our solution alternates primitive error-correction stages with secrecy amplification stages, but we need several rounds of each nested carefully.)

## 1.1  Our Results

We propose a very general model of channel with state, which makes few assumptions about the way the channel is constructed or the computational resources of the users and attackers. In the present paper, such a channel is used for communication between Alice and Bob, with an active attacker Eve. The channel has certain reliability and secrecy guarantees, ensuring that Bob receives a bit sent to him by Alice with sufficiently higher probability than Eve (see Sect. 2).

We show (in Sect. 3) how secrecy and reliability of such channels can be simultaneously amplified with efficient protocols (using one-way communication only), provided that the original channel has a constant-factor gap (at least 4) between its secrecy and reliability (i.e., Eve is 4 times more likely to make a mistake on a random bit sent by Alice across the channel than Bob is on any given bit sent by Alice). We prove (in Sect. 4) that some constant-factor gap (the factor 2) is necessary for any one-way protocol. Finally, we present (in Sect. 5) an efficient two-way communication protocol for amplifying secrecy and reliability, assuming the original channel has the factor 1.5 gap between secrecy and reliability.

For our one-way protocol in Sect. 3, we tighten a result of Halevi and Rabin [16] on the secrecy analysis of a repetition protocol. If the eavesdropper has probability at most $1 - \alpha$ of guessing a bit sent across the channel from Alice to Bob, then the eavesdropper has probability at most $1 - (2\alpha)^n/2$ of learning the bit, if this bit is sent across the channel $n$ times. This improves upon the analysis of [16], who showed $1 - \alpha^n$ probability for the eavesdropper.

Our two-way protocol in Sect. 5 applies to secret-key agreement between two parties both in the information-theoretic and complexity-theoretic setting, extending the results of Holenstein and Renner [19] on one-way protocols.

## 1.2  Related Work

Our results exhibit both technical and conceptual similarities with the rich line of works on secrecy amplification for cryptographic primitives and protocols. A number of them developed amplification results for both soundness and correctness of specific two-party protocols [1,4,5,15–17,20,32,33,37]. Different from our work, however, these consider settings where one of two parties is corrupt, and secrecy for the other party is desired. Here, we envision a scenario with two honest parties, Alice and Bob, communicating in presence of a malicious third party, Eve. Previously, this was only considered in works on secrecy and correctness amplification for public-key encryption and key agreement [11,18,19,26]. We note that our framework is far more general than these previous works.

Following Shannon's impossibility result showing that perfect secrecy requires a secret key as large as the plaintext [35] (see also [10]), there has been a large body of research in information-theoretic cryptography. This line of work shows that perfect secrecy is possible, if one assumes that physical communication channels are noisy. One such model of a noisy communication channel is Wyner's wiretap channel of [39], generalized by [9], and extensively studied since (see [25] for a survey). A number of both possibility and impossibility results were shown for various models of noisy channels, see, e.g., [6–8,12,21,29–31,38].

Different formalizations of secrecy in the information-theoretic setting were studied by [2,22,23,36]. In particular, Bellare et al. [2] consider the wiretap channel and relate the information-theoretic notion of secrecy (traditionally used in information-theoretic cryptography) to the semantic secrecy in the spirit of [14] (used in complexity-theoretic cryptography).

We remark that in the information-theoretic approach to cryptography, the focus is usually on what the channel *is*: for example, a channel between Alice and Bob, with eavesdropper Eve, is modeled as a triple of correlated random variables $A, B, E$, with certain assumptions on the joint distribution of these variables. Then the question is studied what such a channel can be used for, and how efficiently (e.g., at what rate). In contrast, our main focus is on the *utilization* of the channel, i.e., what the channel can be used for. For example, if a channel can be used for somewhat secret and reliable transmission of information, we would like to know if that channel can be used to construct a new channel for totally secret and reliable transmission.

Below we provide a more detailed comparison between our work and the most closely related previous work.

*Comparison with [19].* Perhaps the most closely related to the present paper is the work by Holenstein and Renner [19] that considers the task of secret-key agreement in the information-theoretic setting, where two honest parties, Alice and Bob, have access to some correlated randomness such that the eavesdropper, Eve, has only partial information on that randomness. In particular, [19] consider a special case where the random variables of Alice and Bob, $A$ and $B$, are binary and have correlation at least $\alpha$ (i.e., $A$ and $B$ are equal with probability at least $(1 + \alpha)/2$), whereas with probability at least $1 - \beta$, the random variable $E$ of Eve contains no information on $A$. One of the main results of [19] shows that secret key agreement, using one-way communication from Alice to Bob, is possible when $\alpha^2 > \beta$, and impossible otherwise. Holenstein and Renner also observe that one-way secret-key agreement for such random variables is equivalent to the task of black-box circuit polarization, introduced by Sahai and Vadhan [34] in the context of statistical zero knowledge. The impossibility result for one-way secret-key agreement in [19] implies that the parameters for circuit polarization achieved by Sahai and Vadhan [34] are in fact optimal for such black-box protocols.

The setting of binary random variables $A$, $B$, $E$ in [19] is similar to the channel model we consider. Their condition on $A$ and $B$ being correlated corresponds to channel's reliability, and the condition on $E$ sometimes having no information on $A$ corresponds to channel's secrecy. We use the impossibility result of [19] (almost directly) to argue the need of a constant-factor (factor 2) separation between reliability and secrecy of channels for the case of one-way protocols. However, our one-way channel protocol (for the case of factor 4 separation between reliability and secrecy) is for a more general, not necessarily information-theoretic, setting. Moreover, we go beyond the one-way communication, and describe an efficient two-way protocol that works for the case where the constant-factor gap between reliability and secrecy of a channel is smaller (factor 1.5) than the gap required by one-way protocols. This yields a new protocol that works both for the information-theoretic setting (as in [19]), and for the complexity-theoretic setting, using the results of [18].

*Comparison with [30].* Maurer [30] considered the information-theoretic setting of a channel between Alice and Bob, with eavesdropper Eve, where the channel from Alice to Bob is symmetric noisy channel with the noise parameter $\epsilon$, and the channel from Alice to Eve is an independent symmetric noisy channel with the noise parameter $\delta$. Using the earlier work by [9], Maurer shows that Alice and Bob can securely agree on a secret in this setting, provided $\epsilon < \delta$. Surprisingly, Maurer also shows that secret-key agreement between Alice and Bob is still possible even if $\epsilon \geq \delta$, by using a two-way protocol (where Bob also sends messages to Alice over the public channel)! Like Maurer, we also use a two-way protocol to overcome the limitations of one-way protocols. The difference is that our setting is more general than his information-theoretic setting (of two independent noisy channels). For example, in Maurer's setting, it is easy to see that Eve has less information than Alice about the bit Bob receives, which is not always true in our setting (unless $\alpha > 2\beta$). However, his results raise the question of what additional reasonable conditions on our channel model could be used to reduce the gap between secrecy and reliability that one needs to assume. One natural condition is that Eve has a small probability of learning a random bit sent from Bob to Alice (in addition to the existent secrecy condition that Eve has small probability of learning a random bit sent from Alice to Bob). We leave the study of this channel model with "symmetric secrecy" for future research.

*Comparison with [27].* The framework of *constructive cryptography* by Maurer [27] also deals with reductions between channels, using the formalism from the abstract cryptography framework [28]. In constructive cryptography, the main goal is to capture traditional security goals (like secrecy and authenticity) in terms of channel transformations. Contrary to our framework, channels in constructive cryptography are described *exactly* through ideal functionalities, in the same spirit as in Canetti's UC framework [3]. Maurer's framework in fact also allows the definition of *classes* of channels (as we consider here), but this feature appears to be mostly definitional, as we are not aware of any results that would apply to the context of our work.

## 1.3  Our Techniques

We use fairly standard tools such as the direct-product and XOR protocols, relying on the proof techniques in [13,24]. We also use the repetition protocol, whose secrecy in the cryptographic setting was first analyzed in [16]. We generalize and improve their analysis (see Theorem 14), getting better secrecy ($(2\alpha)^n/2$ instead of $\alpha^n$), which is crucial for our applications. While the techniques we ended up using in this paper are standard, finding the right techniques to use for our applications was nontrivial, and involved considering many other standard techniques that turned out to be inapplicable to our setting. For example, error-correcting codes are an obvious approach to amplifying reliability. But it is still very unclear how such codes affect secrecy. Also, many of the ways we apply standard techniques are delicate. The XOR protocol we use is standard, but fails dramatically if one reverses the order in which the messages are sent. There seems to be a subtle and intricate interplay between the contradictory requirements of secrecy and reliability that we want to achieve simultaneously.

## 2  The Model and Axioms

### 2.1  Channels

The following is a definition of a one-way channel that communicates information from a user Alice to a user Bob. An attacker Eve is capable of launching possibly active attacks, and can gain some information about communicated messages. We can generalize such a channel to one allowing two-way communication or multi-party channels. Note that while we do capture a variety of classical physical systems with this definition, we do not necessarily capture quantum channels or protocols, because we assume that computation does not change the system's state. We could generalize further, but it's already getting pretty complicated.

**Definition 1 (Channel).** *A* one-way channel *from user Alice to user Bob with attacker Eve has the following components:*

1. SECURITY PARAMETER: $k \in \mathbb{N}$;
2. STATES: *for each* $k$, *a countable set of possible* underlying states, $\Sigma_k \subseteq \{0,1\}^*$;
3. ATTACKS: *for each* $k$, *a countable set of possible attacks* $\Gamma_k \subseteq \{0,1\}^*$;
4. TRANSITION FUNCTION: *for each* $k$, *a probabilistic transition function* $\delta_k$ *which takes as input the current state* $s \in \Sigma_k$, *an attack* $\gamma \in \Gamma_k$ *from Eve, and a* transmitted bit $b$ *from Alice, and produces a probability distribution* $\delta_k(s, \gamma, b)$ *on the* updated state $s' \in \Sigma_k$ *and received message* $b' \in \{0,1\}$;
5. EVE'S VIEW FUNCTION: *a function* $v_E(s)$ *from states to strings, giving the* visible part of the state *for Eve;*
6. RESOURCE LIMITS: *a set* $F$ *of probabilistic functions from strings to strings, computable within the* computational limits *of the adversary. We assume* $F$ *is closed under polynomial-time (in the lengths of strings and the secrecy*

*parameter) Turing reductions, and under fixing as advice any single bit, visible state or action.*[1]

*Remark 2.* For our application of secret and reliable information transmission from Alice to Bob in the presence of an active evesedropper Eve, we can assume that Alice and Bob, as trusted parties, do not need to keep track of the channel state. This simplifies our definition of channel above. However, for other tasks (e.g., Oblivious Transfer, bit flipping over the phone, secure multiparty computation), we need to include in our model Alice's and Bob's view functions of the channel state, $v_A(s)$ and $v_B(s)$, respectively. This would match the standard information-theoretic view of such a channel as a triple of correlated random variables $A$ (for Alice), $B$ (for Bob), and $E$ (for Eve).

Our main results only apply to limited classes of channels that we call *transparent* and *semi-transparent*.

**Definition 3 (Transparency).** *A channel of Definition 1 is called* transparent *if it satisfies the following additional properties:*

- $v_E(s) = s$ *(i.e., all of the state is visible to the attacker), and*
- *for every $k \in \mathbb{N}$, $\delta_k \in F$ (i.e., the attacker can simulate the channel).*

*A channel of Definition 1 is called* semi-transparent *if it satisfies the following additional properties:*

- $v_E(s) = s$ *(i.e., all of the state is visible to the attacker), and*
- *for every $k \in \mathbb{N}$, computing the new state under $\delta_k$ is in $F$ (i.e., the attacker can simulate the channel as far as the information they get, but not necessarily the output).*

*Remark 4.* The utility of transparency condition on the channel is that it enables the eavesdropper Eve to simulate the channel forward, by taking control of a virtual Alice. In fact, as was pointed out to us by Daniele Micciancio [personal communication, 2015], given an arbitrary channel that can be simulated forward, one can define a new, equivalent channel that is transparent; the converse is also true. So transparency is equivalent to being simulatable forward.

Transparent channels include any memoryless channel with computationally unbounded (information-theoretic) attackers, and any two-party protocol where there are no secret inputs for either party before the protocol starts.

---

[1] If a channel is such that the state description rapidly grows (say, squares) after each use, then after very few uses, the adversary that is allowed polynomial time in the size of the state will get to use exponential-time computation for her attacks. A standard cryptographic channel will unlikely be secure in this case. However, it is up to the designer of the channel to ensure that it remains secure, with respect to polynomial-time adversaries (which will probably force the designer to make sure that the state description does not grow too fast with respect to $k$).

**Definition 5 ($\alpha$-Secrecy and $\beta$-Reliability).** *Let $1/2 > \alpha > \beta \geq 0$ be constants (or functions of the security parameter). A channel is called $\alpha$-secret and $\beta$-reliable if it satisfies the following axioms:*

– **Secrecy Axiom:** *For all but finitely many $k \in \mathbb{N}$, $\forall f \in F$, $\forall s \in \Sigma_k$, $\forall \gamma \in \Gamma_k$, and for $b \in_U \{0,1\}$ uniformly chosen,*

$$\Pr_{(s',b')=\delta_k(s,\gamma,b)} [f(v_E(s')) = b] \leq 1 - \alpha.$$

– **Reliability Axiom:** *$\forall k \in \mathbb{N}$, $\forall s \in \Sigma_k$, $\forall \gamma \in \Gamma_k$, and $\forall b \in \{0,1\}$,*

$$\Pr_{(s',b')=\delta_k(s,\gamma,b)} [b' = b] \geq 1 - \beta.$$

These conditions are met by the (non-transparent) channel that works as follows. Initially the state is the empty string. The intended receiver always gets the sent bit. The eavesdropper is allowed exponential computation time, and has two attacks: "defer" or "break". If "defer" is chosen, the eavesdropper learns nothing at the time (the visible state contains no bits), but the current bit sent is appended to the channel state. If "break" is chosen, with probability $1 - 2\alpha$, the channel state is updated as normal but becomes visible to the eavesdropper; with probability $2\alpha$, the channel state is erased (becomes the empty string).

The first example provably shows that secrecy amplification cannot be based solely on the above axioms. Consider any protocol to send a bit secretly from Alice to Bob, using the channel above. Eve can use the strategy of using "defer" until the last bit is sent, and attacking the last bit with "break". With probability $1 - 2\alpha$, Eve learns the entire conversation between Alice and Bob. By simulating all possible random choices used by Alice and Bob, and seeing which ones are consistent with the conversation, Eve can learn the secret.

To see where non-transparency could actually prevent secrecy amplification in the cryptographic setting, consider a channel that simulates the following private-key protocol. Alice and Bob share a secret key $\kappa$, and to send a message, Alice sends $E_\kappa(m)$ and a weak commitment $C(\kappa)$ to Bob. If an eavesdropper can break the secrecy of the commitment scheme with some small probability $\alpha$, then no matter how the scheme is used repeatedly and combined, the attacker will learn the key with probability at least $\alpha$. In general, protocols that assume prior shared information such as a private key will not be transparent, because the attacker cannot simulate a run of the protocol without this shared information. We will show that for *transparent* channels this problem does *not* arise.

## 2.2  Examples

We give some examples of both channels in an information-theoretic setting and computational setting. Our results hold for channels that are some hybrid of the two as well, but these two extremes are the most familiar, so will serve as intuition. In general, we'll be using complexity-theoretic methods when proving possibility results, and prove impossibility results using information-theoretic means, so we will be shifting back and forth between the two.

## Information-Theoretic Channels

**Noise vs. erasure:** One interesting channel is a joint symmetric binary noise and erasure channel, where, when Alice sends $b$, Bob receives the bit $b'$ which is equal to $b$ with probability $1 - \beta$ and equal to $1 - b$ otherwise. Eve receives (i.e., the new state equals) the bit $b$ with probability $1 - 2\alpha$ and the message $\perp$ otherwise.[2] There might or might not be correlation between Eve's erasures and Bob's noise. The channel is memoryless, in that the current state does not actually affect the transition function. Any memoryless channel is equivalent to a transparent one in the information-theoretic setting, since we might as well replace the state with the visible state and Eve can always simulate the fixed transition function.

**Noise attacks:** An active Eve might be able to control the noise of the channel, but not gain any information about the bit sent. For example, say attacks are numbers $\gamma$ between 0 and $\beta$. Bob receives a bit $b'$ with binary symmetric noise $\gamma$, and Eve receives (i.e., the new state is) $b' \oplus b$, whether or not Bob got the bit sent. This channel gives Eve no information about the bit sent, but allows her to attack reliability. Again it is memoryless, hence transparent.

**Arbitrary memoryless channels:** We can embed conventional results about secrecy capacity of channels in our model. Consider any fixed distribution on triples $(A, B, E)$, where we view a single use of a device as giving Alice information $A$, Bob $B$ and the attacker $E$, and Alice and Bob can communicate in the clear as well. Using the device $K$ times gives a sequence of $K$ values of these variables $A_1, ..., A_K, B_1, ..., B_K$, and $E_1, ..., E_K$ from the same joint distribution. At some point, after using the device and sending some messages, Bob will output a guess as to the bit Alice meant to send him. The new state would be the $K$ tuple of values $E_1, ..., E_K$, and the messages sent in the clear. While the sequence $A$ and $B$ are used, and help determine the output, we don't include them in the state (because they will not be used in future transmissions), and since Alice and Bob are trusted participants, there is no reason to keep track of their side information, rather than just the secret they agree on. The system is memoryless, and hence transparent.

## Complexity-Theoretic Channels

**Private key encryption:** If Alice and Bob use a secret key and send messages using a private key encryption, then the state would be both the key and the messages sent in the clear, but the visible state for Eve would just be the messages sent in the clear. So this type of protocol is not transparent, since including the key in the visible state would render it useless.

**Noisy trapdoor function with fixed public key:** Say Bob creates a trapdoor function with probabilistic encryption and noisy decryption, and Alice always sends bits with Bob's fixed public key. Then the state of the channel

---

[2] Note that Eve can guess the bit with probability 1/2 when she receives $\perp$. So the probability of her knowing the bit $b$ is $1 - 2\alpha + (1/2) \cdot (2\alpha) = 1 - \alpha$.

is the public key and the encryption of the bit sent. This channel is semi-transparent, because Eve can simulate the new state (only the encryption of the bit is changed), but cannot necessarily simulate whether Bob will get the bit correctly without Bob's secret key. If there is feed-back from Bob to Alice, Eve might be able to simulate a chosen cyphertext attack on the encryption function.

**Noisy trapdoor function with fresh public keys:** On the other hand, using the same encryption function but with a fresh key every message, the channel becomes fully transparent. Eve can simulate the channel and Bob's received bit by generating her own keys and using them. Chosen cyphertext attacks become a non-issue, so protocols using feedback are fine.

## 2.3   Virtual Channels and Protocol Channels

A protocol using a channel defines a new, *virtual channel*. The inputs to this virtual channel are strategies for the participants and attacker, using the old channel. The virtual channel's states accumulate the protocol history, that is the sequence of observable states during the protocol, together with any messages sent in the clear. The transition function simulates the protocol with the given strategies to obtain the history.

A *protocol channel* fixes the inputs from Alice and Bob in the virtual channel to specific strategies of Alice and Bob.

**Definition 6 (Amplifying secrecy and reliability).** *For $\alpha' > \alpha > \beta > \beta'$, secrecy and reliability amplification from $(\alpha, \beta)$ to $(\alpha', \beta')$ means defining a protocol which guarantees that, for any (transparent) channel satisfying $\alpha$-secrecy and $\beta$-reliability, the protocol channel satisfies $\alpha'$-secrecy and $\beta'$-reliability.*

We note that by construction, states of a protocol channel have the same degree of visibility as states of the underlying channel. Furthermore, since transitions of the protocol channel simulate the strategies of the participants, we conclude the following.

**Lemma 7.** *If a channel is transparent, and the legitimate users' strategies are in $F$, then the protocol channel is also transparent, regardless of whether the protocol uses one-way or two-way communication. If a channel is semi-transparent, and the legitimate users' strategies are in $F$, then the protocol channel is also semi-transparent, provided that the protocol uses one-way (from Alice to Bob) communication only.*

Thus, protocol constructions or secrecy and reliability amplifications which assume the axiom of transparency will always be *composable*. In other words, we can have a series of protocols built on top of channels. The protocols will only utilize the channels as black boxes and so not require any knowledge of how the underlying channel works. They will have the property that if the channel is transparent, $\alpha$-secret and $\beta$-reliable, then the protocol is $\alpha'$-secret and $\beta'$-reliable. Then we can use the protocol as the channel in any way of converting $\alpha'$-secret and $\beta'$-reliable channels into $\alpha''$-secret and $\beta''$-reliable ones. The same is true also for *one-way* protocols using *semi-transparent* channels.

# 3    Secrecy and Reliability Amplification for One-Way Protocols

The main result of this section is the following.

**Theorem 8.** *For any non-negligible $\epsilon$ and any $1/2 > \alpha > 4\beta > 0$, there is a one-way protocol for secrecy and reliability amplification from $(\alpha, \beta)$ to $(1/2 - \epsilon, 2^{-k})$.*

The required protocol will rely on the Direct-Product protocol, the Parity protocol, and the Repetition protocol that we discuss next.

## 3.1    Direct-Product Protocols

The direct product is one of the fundamental constructions in complexity and the theory of cryptography. Direct product theorems state that if one instance of a problem is unlikely to be solved, then two independent instances are even less likely to be both solved. There are many proofs of direct product theorems that apply to a wide variety of models and circumstances. Modern proofs utilize connections to coding theory, hard-core sets, and so on. However, these proofs do not seem to work in our setting. What does work is one of the oldest techniques in direct products, estimates of conditional probabilities, used, for example, by Levin [24].

Direct product constructions generally decrease reliability but enhance secrecy. The simplest direct product constructions just concatenate the various solutions. We'll analyze such a protocol, but it will not be immediate how to translate the result about concatenating secrets into one where the secrets are combined into a single bit.

Consider the following **Direct-Product Protocol**:

Alice sends $n$ independent random bits $b_n, \ldots, b_1$ (we number them in reverse order to make an inductive argument cleaner) through the channel.

We compare the probability that Bob receives all $n$ bits with the probability that Eve can guess all $n$ bits. First, for Bob's probability of receiving all $n$ bits, we can use that the reliability axiom holds for each state of the channel. Conditioned on any event for the first $i$ bits, and in particular, conditional on Bob receiving the first $i$ bits correctly, the probability of his receiving the $i$th bit correctly is at least $1 - \beta$. Therefore, the probability that he receives all $n$ bits correctly is at least $(1 - \beta)^n$.

Next, we use the method of conditional probabilities, due to Levin, to bound the probability that Eve can guess all $n$ bits.

**Theorem 9 (Direct-Product Theorem for Channels).** *For any non-negligible function $\epsilon$ of the secrecy parameter, and any polynomially bounded $n$, the probability that Eve can guess all $n$ bits is at most $(1 - \alpha)^n + n\epsilon$.*

*Proof.* Consider the distribution on the information available to Eve by an attack. An attack on the protocol will be determined by two functions $A$ which receives a list of states and determines the next attack $a$ on the channel, and $f$ which after the protocol ends outputs the guess $B_n...B_1$. The protocol under this strategy will evolve as follows:

1. The protocol starts in some state $s_{n+1}$. Let the initial history $H_{n+1}$ be the list containing only $s_{n+1}$.
2. For each $i$ from $n$ to $1$:
   (a) Alice picks a random bit $b_i \in \{0, 1\}$.
   (b) Eve picks channel attack $a_i = A(H_{i+1})$.
   (c) The new state $s_i$ and the bit $b'_i$ received by Bob are given by $(s_i, b'_i) = \delta_k(s_{i+1}, a_i, b_i)$.
   (d) Append $s_i$ to $H_{i+1}$ to get an updated history $H_i$.
3. Eve guesses $B_n, \ldots, B_1 = f(H_1)$.

Note that given any $H_i$, Eve can simulate the rest of the process to produce $H_1$ according to the correct conditional distribution, using randomly generated bits $b_{i-1}, \ldots, b_1$ (since $\delta_k \in F$). (This is where we use transparency.)

Let Success$_i$ be the event that $B_i = b_i, \ldots, B_1 = b_1$. The theorem will follow from the next claim for $i = n$.

*Claim.* For any $1 \leq i \leq n$ and history $H_{i+1}$, $\Pr[\text{Success}_i \mid H_{i+1}] \leq (1-\alpha)^i + i\epsilon$.

*Proof (of Claim).* Our proof is by induction on $i$. The $i = 1$ case is just the secrecy property of the channel at state $s_2$. Fix $H_{i+1}$. Consider the following attack on a single bit $b_i$ sent on the channel at state $s_{i+1}$:

> Eve uses attack $a_i$, bit $b_i$ is sent by Alice, and the channel arrives in state $s_i$. Then she repeatedly simulates the conditional distribution on histories starting from $H_i$ as given above, until either Success$_{i-1}$ or the number of simulations reaches $T = (1/\epsilon)\ln(1/\epsilon)$. If the former, she outputs $B_i$ as her guess for $b_i$, otherwise, the simulations time out without success, she outputs no guess.

By transparency of the channel and its $\alpha$-secrecy, we get that

$$\Pr[B_i = b_i \mid H_{i+1}] \leq (1 - \alpha). \tag{1}$$

Next, $\Pr[B_i = b_i \mid H_i]$ is $\Pr[\text{Success}_i \mid H_i, \text{Success}_{i-1}]$ times the probability of not timing out, which is $1-(1-\Pr[\text{Success}_{i-1}|H_i])^T$. In particular, if $\Pr[\text{Success}_i \mid H_i] \geq \epsilon$, so is $\Pr[\text{Success}_{i-1} \mid H_i]$ and the probability of not timing out is at least $1 - (1 - \epsilon)^T \geq 1 - e^{-\epsilon T} = 1 - \epsilon$ by our choice of $T$. Then

$$\Pr[B_i = b_i \mid H_i] \geq \frac{\Pr[\text{Success}_i \mid H_i]}{\Pr[\text{Success}_{i-1} \mid H_i]} - \epsilon$$

$$\geq \frac{\Pr[\text{Success}_i \mid H_i]}{(1 - \alpha)^{i-1} + (i - 1)\epsilon} - \epsilon,$$

where the last inequality is by the induction hypothesis applied to $H_{i-1}$. So we get

$$\Pr[\text{Success}_i \mid H_i] \leq (1 - \alpha)^{i-1} \cdot \Pr[B_i = b_i \mid H_i] + i\epsilon. \tag{2}$$

If $\Pr[\text{Success}_i \mid H_i] < \epsilon$, then Eq. (2) holds for trivial reasons. Finally, averaging over $H_i$ in Eq. (2) and then using the inequality of Eq. (1), concludes the proof.

### 3.2 Parity Protocols

Next we want to use our direct-product protocol to get a single bit message across the channel. Before showing a protocol that works (under some circumstances), we give an illuminating example of a tempting protocol that fails.

**Naive Parity Protocol.** Consider the naive parity protocol for sending a bit $b$ from Alice to Bob:

> Alice sends random bits $b_n, \ldots, b_1$ as above, and then sends $b \oplus b_n \oplus \cdots \oplus b_1$. Bob's guess at $b$ is the parity of all the bits he receives.

We are not sure whether this protocol boosts secrecy, but it actually fails miserably when it comes to reliability. In fact, there are channels where this protocol is much worse than random guessing from Bob's point of view!

**Theorem 10.** *For any $1/2 > \beta > 0$, there is a transparent $1/2$-secret and $\beta$-reliable channel such that the naive parity protocol above yields the protocol channel with reliability $1 - (1 - \beta)^n$.*

*Proof.* Indeed, consider a channel where Eve decides whether each bit is sent with symmetric noise $\beta$ or with no noise, and learns nothing about the bit sent, only the noise. In other words, the channel has two states, 0 and 1, and there are two attacks, 0 and 1. A coin $\eta$ of bias $\beta$ is flipped by the channel, and the new state is $\eta$ (regardless of the bit sent or the attack). The bit received by Bob is $b \oplus a\eta$, i.e., is flipped if Eve picks attack 1 and the noise is 1, and is not otherwise. One can think of Alice and Bob as communicating by low power radio, and Eve can make the channel noisy by broadcasting at the same time, but can only tell if she disrupted the signal, not what the message was.

This channel has secrecy $1/2$ and $\beta$-reliability. But if Alice and Bob use the parity protocol, Eve can use attack 1 (keep the channel noisy) until $\eta = 1$, and then set $a = 0$ after that. Bob only gets the correct bit if $\eta$ is never 1, so with probability $(1 - \beta)^n$. 

So the reliability of the naive parity protocol goes totally out the window!

**Modified Parity Protocol.** Next we show a modification of this protocol that amplifies secrecy of a given channel, albeit at the price of possibly worsening its reliability somewhat. This will be later combined with another protocol that will significantly improve reliability while somewhat worsening secrecy. By carefully choosing the parameters of the protocols in this combination, we will be able to achieve both secrecy and reliability amplification for a given $\alpha$-secret and $\beta$-reliable channel, provided that $\alpha > 4\beta$.

The modified parity protocol sends the parity of a random subset of bits $b_n, \ldots, b_1$, rather than all of them. Consider the **Parity Protocol**:

> To send a given bit $b$ to Bob, Alice uses the channel to send random bits $b_n, \ldots, b_1$, and then, in the clear, sends random bits $r_n, \ldots, r_1$, followed by $b \oplus (\oplus_{i=1}^{n} b_i r_i)$. Bob receives bits $b'_n, \ldots, b'_1$ through the channel, and outputs $(b \oplus (\oplus_{i=1}^{n} b_i r_i)) \oplus (\oplus_{i=1}^{n} b'_i r_i)$.

**Theorem 11.** *Given any $\alpha$-secret and $\beta$-reliable transparent channel, the Parity Protocol above yields the protocol channel that is $\alpha'$-secret and $\beta'$-reliable for $\alpha' \approx (1 - e^{-\alpha n/2})/2$ and $\beta' \approx (1 - e^{-\beta n})/2$.*

*Proof.* The probability that Bob receives all $n$ bits is $(1 - \beta)^n$, and then he correctly recovers $b$ with probability 1 over the choice of random bits $r_n, \ldots, r_1$. Otherwise, Bob's string $b'_n \ldots b'_1$ is different from the string $b_n \ldots b_1$, but the two strings have the same inner product modulo 2 with the random string $r_n \ldots r_1$, with probability $1/2$ over the choice of $r_n, \ldots, r_1$. Thus, Bob's overall chance of guessing $b$ correctly is $(1 + (1 - \beta)^n)/2$, which means that the protocol is about $(1/2)(1 - e^{-\beta n})$-reliable.

On the other hand, if Eve can guess $b$ with conditional probability $1/2 + \gamma_b$ after $b = b_n, \ldots, b_1$ are sent, using the algorithm of Goldreich and Levin [13], varying over choices of bits $r$, she can guess the entire vector $b$ with probability $c \cdot \gamma_b^2$, for some constant $c > 0$. Set $\gamma = \mathbf{Exp}_b[\gamma_b]$. We conclude that if Eve can guess $b$ with probability $1/2+\gamma$, then she can recover the entire $b$ with probability at least $c \cdot \mathbf{Exp}_b[\gamma_b^2]$, which by Jensen's Inequality is at least $c \cdot (\mathbf{Exp}_b[\gamma_b])^2 = c \cdot \gamma^2$.

Finally, using the Direct-Product Theorem for Channels, Theorem 9, we must have $c \cdot \gamma^2 \leq (1 - \alpha)^n + n\epsilon$ for any non-negligible $\epsilon$, or $\gamma \leq \sqrt{c} \cdot (1 - \alpha)^{n/2} + \epsilon'$ for any such $\epsilon'$. So secrecy is roughly $1/2(1 - e^{-\alpha n/2})$.

While both secrecy and reliability in the above protocol are close to $1/2$, a multiplicative difference in $\alpha$ vs. $\beta$ has become an exponent in the advantage over random guessing, with the factor of 2 lost in the process.

*Remark 12.* Note that order matters in the protocol. Although sending $b_n, \ldots, b_1$ then $r_n, \ldots, r_1$ is the same information as sending $r$ first then $b$, the reverse order would be subject to the same attack as the naive parity protocol above.

## 3.3   Repetition Protocol

Here we get a protocol for improving reliability. It is the following **Repetition Protocol**:

> To transmit a given bit $b$ to Bob, Alice sends this $b$ over the channel $n$ times. Bob takes the majority value of the received bits.

This protocol is somewhat dual to direct product: here reliability is enhanced at the price of secrecy dropping substantially. In fact, it is not clear that any secrecy would remain. In the cryptographic setting, Halevi and Rabin [16] showed that at least $\alpha^n$ secrecy remains. We generalize and improve their result, showing that the repetition protocol has at least $(2\alpha)^n/2$ secrecy.

First, we analyze reliability using familiar probabilistic tools.

**Theorem 13.** *The Repetition Protocol applied to a $\beta$-reliable channel yields a channel with reliability $\beta' \leq e^{-(1-2\beta)^2 n/8}$.*

*Proof.* We need to show that, for any attack on the Repetition Protocol over a $\beta$-reliable channel, the probability that Bob fails to output $b$ is at most $e^{-(1-2\beta)^2 n/8}$. Let $b'_n, \ldots, b'_1$ be the bits received by Bob. Look at the quantity that adds $\beta$ each time bit $b'_i = b$ and subtracts $(1 - \beta)$ if the bit received is incorrect. By the definition of $\beta$-reliability, this quantity is a sub-martingale, with the difference bounded by 1. Bob only returns the wrong bit if there are more incorrect bits received than correct bits, in which case this quantity is at most $\beta n/2 - (1-\beta)n/2 = -(1-2\beta)n/2$. By Azuma's inequality, the probability of this is at most $e^{-((1-2\beta)n/2)^2/(2n))}$, as claimed.

Next we show:

**Theorem 14.** *For any parameters $\alpha$ and $n$ (with $n$ polynomially bounded in the security parameter, and $(2\alpha)^n$ non-negligible), the $n$-bit Repetition Protocol over an $\alpha$-secret transparent channel has secrecy at least $(2\alpha)^n/2$.*

*Proof.* As in the proof of Theorem 9, fixing functions $A$ and $f$ that describe Eve's attack, the process can be described as follows:

1. Alice picks a random bit $r$ (to be sent over the channel $n$ times).
2. The protocol starts in some state $s_{n+1}$. Let the initial history $H_{n+1}$ be the list containing only $s_{n+1}$.
3. For each $i$ from $n$ to 1:
   (a) Eve picks channel attack $a_i = A(H_{i+1})$.
   (b) The new state and bit Bob receives is $(s_i, b'_i) = \delta_k(s_{i+1}, a_i, r)$.
   (c) Append $s_i$ to $H_{i+1}$ to get an updated history $H_i$.
4. Eve guesses $R = f(H_1)$.

Consider starting from partial history $H_{i+1}$, picking a new random bit $r_1$ and simulating the protocol from then on sending $r_1$ for the $i$ remaining bits to be sent. The theorem will follow from the next claim when $i = n$.

*Claim.* For every $1 \leq i \leq n$, $\Pr[R \neq r_1 \mid H_{i+1}] \geq (2\alpha)^i/2$.

*Proof.* The proof is by induction on $i$. For $i = 1$, this is exactly the definition of $\alpha$-secrecy. Consider the following attack on a single bit $r_1$ sent on the channel at state $s_{i+1}$:

Eve uses attack $a_i$ and $r_1$ is sent by Alice, and the channel arrives in state $s_i$. Then she picks a new random bit $r_2$ and simulates the repetition protocol starting from $H_i$, with Alice sending $r_2$ each time. If the simulation returns an $R \neq r_2$, Eve guesses $R$. Otherwise, Eve repeats the simulation for a fresh random bit $r_2$. (Note that the expected number of repetitions is at most $2(2\alpha)^{-i}$, by the induction hypothesis, which is feasible by assumption).

By $\alpha$-secrecy, the described strategy must fail with probability at least $\alpha$, i.e.,

$$\Pr[R \neq r_1 \mid R \neq r_2, H_{i+1}] \geq \alpha. \tag{3}$$

Now fix any history $H_i$ and bit $r_1$. For the $R$ returned by Eve in the above strategy, the probability that $R \neq r_1$ is the conditional probability

$$\Pr[R \neq r_1 \mid R \neq r_2, H_i] = \frac{\Pr[R = \neg r_1 = \neg r_2 \mid H_i]}{\Pr[R \neq r_2 \mid H_i]}.$$

By induction, for each $H_i$ the denominator of this expression is at least $(2\alpha)^{i-1}/2$. So for each $H_i$ and $r_1$, we have

$$((2\alpha)^{i-1}/2) \cdot \Pr[R \neq r_1 \mid R \neq r_2, H_i] \leq \Pr[r_1 = r_2, R \neq r_1 \mid H_i].$$

Averaging both sides over $H_i$, we get

$$((2\alpha)^{i-1}/2) \cdot \Pr[R \neq r_1 \mid R \neq r_2, H_{i+1}] \leq \Pr[r_2 = r_1, R \neq r_1 \mid H_{i+1}]. \tag{4}$$

Finally, applying Eq. (3) to the left-hand side of Eq. (4), we get

$$\begin{aligned} ((2\alpha)^{i-1}/2) \cdot \alpha &\leq \Pr[r_2 = r_1, R \neq r_1 \mid H_{i+1}] \\ &= \Pr[r_2 = r_1] \cdot \Pr[R \neq r_1 \mid r_2 = r_1, H_{i+1}] \\ &= (1/2) \cdot \Pr[R \neq r_1 \mid r_2 = r_1, H_{i+1}], \end{aligned}$$

and so $\Pr[R \neq r_1 \mid r_2 = r_1, H_{i+1}] \geq (2\alpha)^{i-1}(2\alpha)/2 = (2\alpha)^i/2$. Observe that the last probability is for the process where, starting at $H_{i+1}$, the same bit $r_1$ is sent $i$ times. This is exactly the probability in the statement of our claim (for the repetition protocol starting at $H_{i+1}$).

This completes the proof of the theorem.

## 3.4   Assembling the Pieces for One-Way Protocols

Here we show how to combine the two building blocks we just used: the Parity protocol and the repetition protocol. Let $\alpha > 4(1 + 2\delta)\beta$. We re-state the main theorem of this section.

**Theorem 15.** *For any non-negligible $\epsilon$ and any $1/2 > \alpha > 4\beta > 0$, there is a one-way protocol for secrecy and reliability amplification from $(\alpha, \beta)$ to $(1/2 - \epsilon, 2^{-k})$.*

*Proof.* First, we can use the following protocol to make $\alpha$ and $\beta$ suitably small without changing their ratios:

With probability $p$, Alice uses the channel to send a random bit $b$, otherwise she sends $b$ in the clear. This protocol is $\alpha' = p\alpha$ secret and $\beta' = p\beta$ reliable.

Since $1 - \alpha' \approx e^{-\alpha'}$ for small $\alpha'$, we can pick $p$ small enough so that $(1 - \alpha') < e^{-\alpha(1-\delta)}$. Then we use the Parity protocol of Theorem 11 with $n = \log k$ to define a channel that has secrecy at least

$$(1/2) \cdot \left(1 - (1 - \alpha')^{n/2}\right) \geq (1/2) \cdot \left(1 - k^{-(\alpha/2)(1-\delta)}\right)$$
$$\geq (1/2) \cdot \left(1 - k^{-2\beta(1+\delta)}\right),$$

and reliability at least $(1/2) \cdot \left(1 - e^{-\beta n}\right) = (1/2) \cdot \left(1 - k^{-\beta}\right)$.

We use the repetition protocol on this channel for $N = k^{2\beta(1+\delta/2)}$ repetitions. By Theorem 14, the resulting channel has secrecy at least $(1/2) \cdot (1 - k^{-\beta\delta})$ and, by Theorem 14, reliability at most $e^{-k^{-2\beta} N/8} = e^{-(1/8)k^{\beta\delta}}$, which tends to 0 exponentially fast with $k$. We can use the Parity protocol with $n = k$ on this protocol, to get one that is $(1/2 - \epsilon)$-secret for arbitrary non-negligible $\epsilon$, and still has exponentially small reliability. If we want, we can then use repetition on this protocol for any polynomial number of times to keep the advantage of an adversary negligible, while making the reliability as good as desired.

*Remark 16.* The above shows a one-way protocol when $\alpha > 4\beta$. The factor of 4 can be thought of as two factors of two. The first one is due to the quadratic dependence of list size on the advantage when list decoding the Hadamard code (cf. the proof of Theorem 11 above). The second factor of 2 is because repeating a message through a symmetric channel takes quadratic time in the advantage, whereas for an erasure channel, the advantage grows linearly (cf. the proof of Theorem 17 below).

## 4   Impossibility Results for One-Way Protocols

Here, we show that a constant factor difference of two between $\alpha$ and $\beta$ is *necessary*. To get our negative result, we will look at a particular channel; of course, it follows that if no protocol exists for this channel, then no protocol exists for an unknown channel. Our particular channel is stateless, and is

- SYMMETRIC $\beta$-NOISE CHANNEL FOR BOB: each bit sent over the channel is flipped with probability $\beta$, and is unchanged with probability $1 - \beta$,
- $2\alpha$-ERASURE CHANNEL FOR EVE: each bit sent over the channel is erased with probability $2\alpha$ (with Eve getting a special symbol '?'), and is unchanged with probability $1 - 2\alpha$.

In addition, we allow Eve to have unlimited computational power.

We prove the following result, using the techniques of Holenstein and Renner [19].

**Theorem 17.** *If $\alpha \leq 2\beta - 2\beta^2$, then no one-way protocol for the above channel has reliability .01 and secrecy .49.*

*Proof.* We use the techniques of Holenstein and Renner [19] who showed that the same relationship between secrecy and reliability parameters is necessary for any information-theoretic one-way protocol for secret key agreement. Let a random variable $B$ denote the bit to be sent. Let $X_1, \ldots, X_n$ be the distribution on bits Alice sends through the channel, and let $V$ be the distribution on messages she sends in the clear. Let $Y_1, \ldots, Y_n$ be the bits Bob receives, and $Z_1, \ldots, Z_n$ be the information Eve receives.

Let $H$ be the entropy function. Let $B'$ bet the Boolean random variable that is 1 iff Bob correctly guesses the bit $B$, given $V$ and $Y_1, \ldots, Y_n$. Since, given $V, Y_1, \ldots, Y_n$, Bob guesses $B$ correctly with probability at least .99, we get $H(B' \mid V, Y_1, \ldots, Y_n) \leq H(.99)$. On the other hand, note that $V$ and $Y_1, \ldots, Y_n$ determine Bob's guess at $B$, and so if we know $B$, then we also know $B'$, and vice versa. It follows that $H(B \mid V, Y_1, \ldots, Y_n) = H(B' \mid V, Y_1, \ldots, Y_n) \leq H(.99) \approx 0$. By a similar reasoning for Eve, we get that $H(B \mid V, Z_1, \ldots, Z_n) \geq H(.49) \approx 1$.

Consider $H(B \mid V, Y_1, ..Y_i, Z_{i+1}...Z_n)$. When $i = n$, this is close to 0, and when $i = 0$, close to 1. So there must exist an index $i$, $0 \leq i \leq n$, such that

$$H(B \mid V, Y_1, \ldots, Y_i, Z_{i+1}, \ldots, Z_n) < H(B \mid V, Y_1, \ldots, Y_{i-1}, Z_i, \ldots, Z_n).$$

Then by an averaging argument, there must exist values for $V$, $Y_1, \ldots, Y_{i-1}$ and $Z_{i+1}, \ldots, Z_n$, so that in the conditional distribution, we have

$$H(B \mid Y_i) < H(B \mid Z_i). \tag{5}$$

Note that, because the protocol is one-way, conditioning on these values does not change the conditional distributions of $Y_i$ or $Z_i$ as functions of $X_i$ (the bit sent)[3]. It will possibly change both the distributions of $B$ and $X_i$ to arbitrary distributions.

By Eq. (5), and using the entropy chain rule twice, we get

$$
\begin{aligned}
0 &> H(B \mid Y_i) - H(B \mid Z_i) \\
&= H(B, Y_i) - H(Y_i) - H(B, Z_i) + H(Z_i) \\
&= H(B) + H(Y_i \mid B) - H(Y_i) - H(B) - H(Z_i \mid B) + H(Z_i) \\
&= H(Y_i \mid B) - H(Y_i) - H(Z_i \mid B) + H(Z_i).
\end{aligned}
$$

---

[3] In contrast, consider a 2-way protocol where Bob, after receiving his $n$ bits over the channel, sends Alice a message in the clear stating whether all his received bits are the same. Then fixing the value of Bob's message to Alice *will change* the distribution of $Y_i$ as a function of $X_i$. So the argument in the present theorem does not apply to this 2-way protocol. (In fact, we use such a 2-way protocol in Sect. 5 in order to overcome the "factor-2 barrier" for one-way protocols given by the present theorem).

Next we analyze each of the four summands in the last equation above.

Let $q$ be the conditional probability that $B = 1$, and let $p_1$ be the conditional probability that $X_i = 1$ if $B = 1$, and $p_0$ be the conditional probability that $X_i = 1$ if $B = 0$. Then the overall probability that $X_i = 1$ is

$$p := qp_1 + (1 - q)p_0.$$

Note that $Y_i$ is equal to $X_i$ with probability $1 - \beta$, and to $\neg X_i$ otherwise. It follows that

$$H(Y_i) = H(p(1 - 2\beta) + \beta). \tag{6}$$

Next, given $B = 1$, $Y_i$ is distributed as first flipping a coin with probability $p_1$ to determine $X_1$, then a coin with probability $\beta$, and finally taking the parity. So we have

$$H(Y_i \mid B = 1) = H(p_1(1 - 2\beta) + \beta),$$

and similarly,

$$H(Y_i \mid B = 0) = H(p_0(1 - 2\beta) + \beta).$$

Combining the two conditional entropies, we conclude

$$H(Y_i \mid B) = q \cdot H(p_1(1 - 2\beta) + \beta) + (1 - q) \cdot H(p_0(1 - 2\beta) + \beta), \tag{7}$$

Finally, $Z_i$ reveals whether the bit is erased, a random event with probability $2\alpha$ no matter what, and then, with probability $1 - 2\alpha$, it reveals the value of $X_i$. Thus, $H(Z_i) = H(2\alpha) + (1 - 2\alpha) \cdot H(X_i)$, and the same for any conditional distribution. So we get

$$H(Z_i) = H(2\alpha) + (1 - 2\alpha) \cdot H(p), \tag{8}$$

and

$$H(Z_i \mid B) = H(2\alpha) + (1 - 2\alpha) \cdot (q \cdot H(p_1) + (1 - q) \cdot H(p_0)). \tag{9}$$

Combining Eqs. (6)–(9), we get

$$0 > (H(Y_i \mid B) - H(Y_i)) - (H(Z_i \mid B) - H(Z_i))$$
$$= q \cdot H(p_1(1 - 2\beta) + \beta) + (1 - q) \cdot H(p_0(1 - 2\beta) + \beta) - H(p(1 - 2\beta) + \beta)$$
$$- (H(2\alpha) + (1 - 2\alpha) \cdot (q \cdot H(p_1) + (1 - q) \cdot H(p_0)) - H(2\alpha) - (1 - 2\alpha) \cdot H(p)).$$

Rearranging the terms in the last expression, we can write it as

$$q \cdot (H(p_1(1 - 2\beta) + \beta) - (1 - 2\alpha) \cdot H(p_1))$$
$$+ (1 - q) \cdot (H(p_0(1 - 2\beta) + \beta) - (1 - 2\alpha) \cdot H(p_0))$$
$$- (H(p(1 - 2\beta) + \beta) - (1 - 2\alpha) \cdot H(p))$$
$$= q \cdot F(p_1) + (1 - q) \cdot F(p_0) - F(p),$$

for the function $F(x) := H(x \cdot (1 - 2\beta) + \beta) - (1 - 2\alpha) \cdot H(x)$. Thus, we have

$$q \cdot F(p_1) + (1 - q) \cdot F(p_0) - F(p) < 0,$$

which is equivalent (recalling that $p = qp_1 + (1 - q)p_0$) to

$$F(qp_1 + (1 - q)p_0) > q \cdot F(p_1) + (1 - q) \cdot F(p_0). \tag{10}$$

Observe that Eq. (10) states that the function $F$ at a convex combination of two points is greater than the convex combination of its values at those two points. This condition is violated if $F$ is a convex function on the interval $[0, 1]$. So, to complete our proof by contradiction, it suffices to show

*Claim.* The function $F(x)$ defined above is convex on $[0, 1]$.

*Proof (of Claim).* We use the convexity criterion for twice differentiable functions: such a function is convex over an interval iff its second derivative is nonnegative on that interval. We can change the binary logs to natural logs, since that just multiplies $F$ by a positive constant factor. For the ln-based entropy function $h(x) = -x \ln x - (1 - x) \ln(1 - x)$, its first derivative is $h'(x) = -\ln x + \ln(1 - x)$, and its second derivative is $h''(x) = -1/x - 1/(1 - x)$.

Similarly, for the linear function $L(x) := x(1 - 2\beta) + \beta$, one can easily verify that

$$(h(L(x)))' = (1 - 2\beta) \cdot (\ln(1 - L(x)) - \ln(L(x))),$$

and

$$(h(L(x)))'' = (1 - 2\beta)^2 \cdot \left( -\frac{1}{1 - L(x)} - \frac{1}{L(x)} \right).$$

Using these expressions for the second derivatives of $h(x)$ and $h(L(x))$, we get

$$F''(x) = (H(L(x)))'' - (1 - 2\alpha) \cdot H''(x)$$

$$= (1 - 2\beta)^2 \cdot \left( -\frac{1}{1 - L(x)} - \frac{1}{L(x)} \right) + (1 - 2\alpha) \cdot \left( \frac{1}{x} + \frac{1}{1 - x} \right)$$

$$= -(1 - 2\beta)^2 \cdot \frac{1}{L(x) \cdot (1 - L(x))} + (1 - 2\alpha) \cdot \frac{1}{x(1 - x)}.$$

We want to show that $F''(x) \geq 0$ for all $x \in [0, 1]$, i.e., that

$$\frac{1 - 2\alpha}{x(1 - x)} \geq \frac{(1 - 2\beta)^2}{L(x) \cdot (1 - L(x))}.$$

Note that $L(x) = x(1 - 2\beta) + (1/2)(2\beta)$, and so $L(x)$ is always between $x$ and $1/2$ (no matter which side of $1/2$ the point $x$ is). Since the function $x(1 - x)$ is symmetric around $1/2$, and achieves its maximum at the point $1/2$, we conclude that $L(x)(1 - L(x)) \geq x(1 - x)$. Thus it suffices to show

$$\frac{1 - 2\alpha}{x(1 - x)} \geq \frac{(1 - 2\beta)^2}{x(1 - x)},$$

equivalent to $1 - 2\alpha \geq (1 - 2\beta)^2$. The latter is equivalent to $\alpha \leq 2\beta - 2\beta^2$, which is our assumption on the $\alpha$ and $\beta$.

This completes the proof of the theorem.

## 5    Breaking the Factor of Two Barrier with Two-Way Protocols

By the lower bound of Theorem 17, we know that it is impossible to amplify secrecy and reliability of a given $\alpha$-secret and $\beta$-reliable channel when $\alpha < 2\beta$, if we use *one-way* communication only. Here we show that a *two-way* communication protocol exists that works even for $\alpha < 2\beta$, as long as $\alpha > (3/2)\beta$.

Our main result of the section is the following.

**Theorem 18.** *For any non-negligible $\epsilon$ and for any $1/2 > \alpha > 1.5 \cdot \beta > 0$, there is a two-way protocol for secrecy and reliability amplification from $(\alpha, \beta)$ to $(1/2 - \epsilon, 2^{-k})$.*

We will need a simple variant on the repetition protocol where Bob communicates one bit in the clear. Like the repetition protocol, this variant will reduce both secrecy and reliability exponentially. But, if $\alpha > 1.5\beta$, the exponent that secrecy decreases by will be larger than that for Bob's failure chance. So the ratio between them will improve with the number of repetitions. We can then pick the number of repetitions to be such that the ratio is greater than 4, and use this protocol as the channel in the one-way protocol from Theorem 15.

The variant protocol is **Repetition with Feedback**:

1. Alice uses the channel to send $b$ to Bob $n$ times.
2. If Bob receives the same bit $b'$ each time, he sends the message "Consistent" to Alice in the clear and uses $b'$ as his output. Otherwise he sends the message "Inconsistent" to Alice in the clear.
3. If Bob sends "Inconsistent", Alice sends $b$ in the clear, and Bob uses that as his output.

We show the following.

**Theorem 19.** *Let $\alpha, \beta, n$ be any parameters such that $n$ is poly-bounded in the security parameter, and $(2(\alpha - \beta))^n$ is non-negligible. The n-bit Repetition with Feedback protocol applied to an $\alpha$-secret and $\beta$-reliable transparent channel yields a new $\alpha'$-secret and $\beta'$-reliable channel, for $\alpha' \geq (2(\alpha - \beta))^n/2$ and $\beta' \leq \beta^n$.*

*Proof.* RELIABILITY: First we argue reliability of the new channel. We need to show that for any attack on the Repetition with Feedback Protocol over a $\beta$-reliable channel, the probability that Bob fails to output $b$ is at most $\beta^n$. Indeed, Bob gets $b$ unless he receives the same bit $b'$ each of $n$ times, and $b' \neq b$. Thus, the protocol only fails if the channel fails $n$ times in a row, which happens with probability at most $\beta^n$.

SECRECY: Next we argue secrecy of the new channel. We need to show that no attack on the $n$-bit Repetition with Feedback protocol using an $\alpha$-secret and $\beta$-reliable transparent channel can predict a random bit $b$ sent by the protocol with better than $1 - (2(\alpha - \beta))^n/2$ probability of success. As before, fixing functions $A$ and $f$ that describe Eve's attack, the process can be described as:

1. Alice picks a random bit $r$.
2. The protocol starts in some state $s_{n+1}$. Let the initial history $H_{n+1}$ be the list containing only $s_{n+1}$.
3. For each $i$ from $n$ to $1$:
   (a) Eve picks channel attack $a_i = A(H_{i+1})$.
   (b) The new state and bit Bob receives is $(s_i, b'_i) = \delta_k(s_{i+1}, a_i, r)$ .
   (c) Append $s_i$ to $H_{i+1}$ to get an updated history $H_i$.
4. If all $b'_i$ are equal (according to Bob's message in the clear), Eve guesses $R = f(H_1, \text{"Consistent"})$. Otherwise she learns $b$ when it is sent in the clear.

The intuition is that, even if we revealed the secret to Eve whenever Bob fails to get the secret, the channel would remain $(\alpha - \beta)$-secret, because failure happens with probability at most $\beta$. We could then apply the analysis of the repetition protocol to this altered channel.

Define random variable $R = f(H_1, \text{"Consistent"})$, even if the bits received are possibly inconsistent. Consider starting from partial history $H_{i+1}$, picking a new random bit $r_1$ and simulating the protocol from then on sending $r_1$ for the $i$ remaining bits to be sent, and verifying that $b'_i = r_1$ each time. The theorem will follow form the next claim for $i = n$ (which shows that with probability at least $(2(\alpha - \beta))^n/2$, Bob gets $b$ all $n$ times, sends "Consistent", and Eve outputs $R \neq b$).

*Claim.* For each $1 \leq i \leq n$, $\Pr[R \neq r_1, \wedge_{1 \leq j \leq i}(b'_j = r_1) \mid H_{i+1}] \geq (2(\alpha - \beta))^i/2$.

*Proof (of Claim).* Our proof is by induction on $i$. For $i = 1$, this follows from $\alpha$-secrecy and $\beta$-reliability: the probability that $R \neq r_1$ is at least $\alpha$, and the probability that $b'_1 \neq r_1$ is at most $\beta$, so the probability that $R \neq r_1 = b'_1$ is at least $\alpha - \beta$. Consider the following strategy for Eve to predict a single bit $r_1$ sent on the channel at state $s_{i+1}$:

> Eve uses $a_i$ as her attack when Alice sends $r_1$, and the channel arrives in state $s_i$. Then she picks a new random bit $r_2$ and simulates the repetition protocol with feedback starting from $H_i$, with Alice sending $r_2$ each time (including simulating the bit Bob receives). If the simulation returns an $R \neq r_2$ and Bob receives $r_2$ each time, Eve guesses $R$. Otherwise, Eve

repeats the simulation for a fresh random bit $r_2$. (Note that the expected number of repetitions is at most $2(2(\alpha-\beta))^{-i}$, by the induction hypothesis, which is feasible by assumption).

Denote by Success$_i$ the event that Bob receives $r_2$ each of the last $i$ times. Fix any history $H_i$, together with $r_1$. The probability that, for the $R$ returned by Eve in the above strategy, $R \neq r_1$ is

$$\Pr[R \neq r_1 \mid R \neq r_2, H_i, \text{Success}_{i-1}] = \frac{\Pr[R = \neg r_1 = \neg r_2, \text{Success}_{i-1} \mid H_i]}{\Pr[R \neq r_2, \text{Success}_{i-1} \mid H_i]}.$$

By induction, for each such $H_i$, the denominator of this expression is at least $(2(\alpha - \beta))^{i-1}/2$. So for each $H_i$ where $b_i' = r_1$,

$$\frac{(2(\alpha - \beta))^{i-1}}{2} \cdot \Pr[R \neq r_1 \mid R \neq r_2, H_i, \text{Success}_{i-1}]$$
$$\leq \Pr[r_2 = r_1, R \neq r_1, \text{Success}_{i-1} | H_i].$$

Note that $H_i$ already determines (although Eve doesn't know which way) whether Bob received $r_1$, i.e., whether $b_i' = r_1$. For those histories where this did happen, the conditional probability that $R \neq r_1$ and Bob receives $r_1$ is the same as just the first clause, and for the others, it is 0. So either way we get

$$\frac{1}{2} \cdot (2(\alpha - \beta))^{i-1} \cdot \Pr[R \neq r_1, b_i' = r_1 \mid R \neq r_2, H_i, \text{Success}_{i-1}]$$
$$\leq \Pr[r_2 = r_1, R \neq r_1, b_i' = r_1, \text{Success}_{i-1} \mid H_i].$$

Then we can average both sides over all $H_i$, to get

$$\frac{1}{2} \cdot (2(\alpha - \beta))^{i-1} \cdot \Pr[R \neq r_1, b_i' = r_1 \mid R \neq r_2, H_{i+1}, \text{Success}_{i-1}]$$
$$\leq \Pr[r_2 = r_1, R \neq r_1, b_i' = r_1, \text{Success}_{i-1} \mid H_{i+1}].$$

By $\alpha$-secrecy and $\beta$-reliability, the probability on the left-hand side of the inequality above is at least $\alpha - \beta$. The probability on the right-hand side is $1/2$ (the probability that $r_2 = r_1$), times the probability that $R \neq r_1$ and Success$_i$ when $r_1$ is sent $i$ times starting at $H_{i+1}$. The latter probability is exactly the probability in the statement of the claim. Thus, we get

$$\Pr[R \neq r_1, \wedge_{1 \leq j \leq i}(b_j' = r_1) \mid H_{i+1}] \geq \frac{1}{2} \cdot (2(\alpha - \beta))(2(\alpha - \beta))^{i-1}.$$

This completes the proof of the theorem.

As a corollary, we get the desired proof of the main result of this section.

*Proof (of Theorem 18).* Given $\alpha > 1.5\beta$, we first use the Repetition with Feedback protocol for an appropriate number of times to get a new protocol channel with $\alpha'$-secrecy and $\beta'$-reliability for $\alpha' > 4\beta'$. Then we use the protocol of Theorem 15 on this protocol channel.

*Tightness of the Analysis of the Repetition with Feedback Protocol.* In our analysis of the Repetition with Feedback protocol, the ratio of secrecy to reliability improves with $n$ when $2(\alpha - \beta) > \beta$, i.e., when $\alpha > 1.5\beta$. In other cases, it makes things worse, rather than better. We now show this analysis is actually tight.

Consider the channel where, with probability $2\beta$, Eve and Bob both receive a random bit $b'$. In addition, Eve receives A, denoting that this is the case in question. With probability $2(\alpha - \beta)$, Bob receives the correct bit $b$, and Eve receives just the message B, saying that this is the case. With the remaining probability $1 - 2\alpha$, Bob receives the correct bit $b$, and Eve also receives $b$ and the message C.

In the repetition with feedback, if the messages Bob receives are consistent, and C has occurred, Eve knows with certainty one bit Bob received and hence that bit must have been received all $n$ times. If the messages Bob receives are consistent, and A occurred, then Eve and Bob get the same random bit $b'$ all $n$ times.

If Bob's messages are inconsistent, the secret is sent in the clear and Eve gets it. Eve fails to get the secret when either *(i)* case B happens all $n$ times, and thereafter Eve does not guess the random bit sent by Alice, or *(ii)* case A happens all $n$ times, and the random bit $b'$ is different from Alice's bit. Thus the overall failure probability for Eve is at most $(2(\alpha - \beta))^n/2 + \beta^n$.

# 6    Conclusions and Open Problems

In this paper, we considered just the simplest issue in secure communication, the transmission of secret information from one party to another. Even here, there are unexpected complications arising from the joint consideration of secrecy and reliability. We gave non-trivial constructions of secure protocols that under some circumstances are guaranteed to amplify both secrecy and reliability to within negligible amounts of the ideal.

However, our results raise more questions than they answer. We hope that these will be addressed in future work, and that future work will consider similar models for more complex issues in secure communications. We suggest the following tasks to consider for the case of trusted parties: authentication, covert channels (steganography), and traffic analysis. For the case of untrusted parties, it will be interesting to use an appropriate channel model to argue about: coin flipping, oblivious transfer, multi-party computation, and broadcast.

It would also be very interesting to study channel models with weaker restrictions on transparency. For example, can one generalize our channel model to include the quantum-computational setting?

**Acknowledgments.** We thank Yevgeny Dodis, Noah Stevens-Davidowitz, Giovanni di Crescenzo, Daniele Micciancio, Thomas Holenstein and Steven Rudich for helpful comments and discussions. Russell Impagliazzo's work was partially supported by the Simons Foundation and NSF grant CCF-121351; this work was done [in part] while Russell Impagliazzo was visiting the Simons Institute for the Theory of Computing, supported by the Simons Foundation and by the DIMACS/Simons Collaboration in

Cryptography through NSF grant #CNS-1523467. Valentine Kabanets was partially supported by the NSERC Discovery grant. Bruce Kapron's work was supported in part by the NSERC Discovery Grant "Foundational Studies in Privacy and Security". Stefano Tessaro was partially supported by NSF grants CNS-1423566, CNS-1553758, CNS-1528178, IIS-1528041 and the Glen and Susanne Culler Chair.

# References

1. Bellare, M., Impagliazzo, R., Naor, M.: Does parallel repetition lower the error in computationally sound protocols? In: Proceedings of the 38th IEEE Annual Symposium on Foundations of Computer Science, FOCS 1997, pp. 374–383 (1997)

2. Bellare, M., Tessaro, S., Vardy, A.: Semantic security for the wiretap channel. In: Safavi-Naini, R., Canetti, R. (eds.) CRYPTO 2012. LNCS, vol. 7417, pp. 294–311. Springer, Heidelberg (2012). doi:10.1007/978-3-642-32009-5_18

3. Canetti, R.: Universally composable security: a new paradigm for cryptographic protocols. In: 42nd Annual Symposium on Foundations of Computer Science, FOCS 2001, Las Vegas, Nevada, USA, 14–17 October 2001, pp. 136–145. IEEE Computer Society (2001)

4. Chung, K.-M., Liu, F.-H.: Parallel repetition theorems for interactive arguments. In: Micciancio, D. (ed.) TCC 2010. LNCS, vol. 5978, pp. 19–36. Springer, Heidelberg (2010). doi:10.1007/978-3-642-11799-2_2

5. Chung, K.-M., Pass, R.: Tight parallel repetition theorems for public-coin arguments using KL-divergence. In: Dodis, Y., Nielsen, J.B. (eds.) TCC 2015, Part II. LNCS, vol. 9015, pp. 229–246. Springer, Heidelberg (2015). doi:10.1007/978-3-662-46497-7_9

6. Crépeau, C.: Efficient cryptographic protocols based on noisy channels. In: Fumy, W. (ed.) EUROCRYPT 1997. LNCS, vol. 1233, pp. 306–317. Springer, Heidelberg (1997). doi:10.1007/3-540-69053-0_21

7. Crépeau, C., Kilian, J.: Achieving oblivious transfer using weakened security assumptions. In: 29th Annual Symposium on Foundations of Computer Science, 1988, pp. 42–52, October 1988

8. Crépeau, C., Morozov, K., Wolf, S.: Efficient unconditional oblivious transfer from almost any noisy channel. In: Blundo, C., Cimato, S. (eds.) SCN 2004. LNCS, vol. 3352, pp. 47–59. Springer, Heidelberg (2005). doi:10.1007/978-3-540-30598-9_4

9. Csiszar, I., Körner, J.: Broadcast channels with confidential messages. IEEE Trans. Inf. Theory **24**(3), 339–348 (1978)

10. Dodis, Y.: Shannon impossibility, revisited. In: Smith, A. (ed.) ICITS 2012. LNCS, vol. 7412, pp. 100–110. Springer, Heidelberg (2012). doi:10.1007/978-3-642-32284-6_6

11. Dwork, C., Naor, M., Reingold, O.: Immunizing encryption schemes from decryption errors. In: Cachin, C., Camenisch, J.L. (eds.) EUROCRYPT 2004. LNCS, vol. 3027, pp. 342–360. Springer, Heidelberg (2004). doi:10.1007/978-3-540-24676-3_21

12. Garg, S., Ishai, Y., Kushilevitz, E., Ostrovsky, R., Sahai, A.: Cryptography with one-way communication. In: Gennaro, R., Robshaw, M. (eds.) CRYPTO 2015. LNCS, vol. 9216, pp. 191–208. Springer, Heidelberg (2015). doi:10.1007/978-3-662-48000-7_10

13. Goldreich, O., Levin, L.A.: A hard-core predicate for all one-way functions. In: Proceedings of the Twenty-First Annual ACM Symposium on Theory of Computing, pp. 25–32 (1989)

14. Goldwasser, S., Micali, S.: Probabilistic encryption. J. Comput. Syst. Sci. **28**(2), 270–299 (1984)
15. Haitner, I.: A parallel repetition theorem for any interactive argument. In: Proceedings of the 50th IEEE Annual Symposium on Foundations of Computer Science, FOCS 2009, pp. 241–250 (2009)
16. Halevi, S., Rabin, T.: Degradation and amplification of computational hardness. In: Canetti, R. (ed.) TCC 2008. LNCS, vol. 4948, pp. 626–643. Springer, Heidelberg (2008). doi:10.1007/978-3-540-78524-8_34
17. Håstad, J., Pass, R., Wikström, D., Pietrzak, K.: An efficient parallel repetition theorem. In: Micciancio, D. (ed.) TCC 2010. LNCS, vol. 5978, pp. 1–18. Springer, Heidelberg (2010). doi:10.1007/978-3-642-11799-2_1
18. Holenstein, T.: Key agreement from weak bit agreement. In: Proceedings of the 37th Annual ACM Symposium on Theory of Computing, STOC 2005, pp. 664–673 (2005)
19. Holenstein, T., Renner, R.: One-way secret-key agreement and applications to circuit polarization and immunization of public-key encryption. In: Shoup, V. (ed.) CRYPTO 2005. LNCS, vol. 3621, pp. 478–493. Springer, Heidelberg (2005). doi:10. 1007/11535218_29
20. Holenstein, T., Schoenebeck, G.: General hardness amplification of predicates and puzzles. In: Ishai, Y. (ed.) TCC 2011. LNCS, vol. 6597, pp. 19–36. Springer, Heidelberg (2011). doi:10.1007/978-3-642-19571-6_2
21. Ishai, Y., Kushilevitz, E., Ostrovsky, R., Prabhakaran, M., Sahai, A., Wullschleger, J.: Constant-rate oblivious transfer from noisy channels. In: Rogaway, P. (ed.) CRYPTO 2011. LNCS, vol. 6841, pp. 667–684. Springer, Heidelberg (2011). doi:10. 1007/978-3-642-22792-9_38
22. Iwamoto, M., Ohta, K.: Security notions for information theoretically secure encryptions. In: 2011 IEEE International Symposium on Information Theory Proceedings (ISIT), pp. 1777–1781, July 2011
23. Iwamoto, M., Ohta, K., Shikata, J.: Security formalizations and their relationships for encryption and key agreement in information-theoretic cryptography. CoRR, abs/1410.1120 (2014)
24. Levin, L.A.: One-way functions and pseudorandom generators. Combinatorica **7**(4), 357–363 (1987)
25. Liang, Y., Poor, H.V., Shamai (Shitz), S.: Information theoretic security. Found. Trends Commun. Inf. Theory **5**(45), 355–580 (2008)
26. Lin, H., Tessaro, S.: Amplification of chosen-ciphertext security. In: Johansson, T., Nguyen, P.Q. (eds.) EUROCRYPT 2013. LNCS, vol. 7881, pp. 503–519. Springer, Heidelberg (2013). doi:10.1007/978-3-642-38348-9_30
27. Maurer, U.: Constructive cryptography – a new paradigm for security definitions and proofs. In: Mödersheim, S., Palamidessi, C. (eds.) TOSCA 2011. LNCS, vol. 6993, pp. 33–56. Springer, Heidelberg (2012). doi:10.1007/978-3-642-27375-9_3
28. Maurer, U., Renner, R.: Abstract cryptography. In: ICS, pp. 1–21. Tsinghua University Press (2011)
29. Maurer, U.M.: Perfect cryptographic security from partially independent channels. In: Proceedings of the Twenty-Third Annual ACM Symposium on Theory of Computing, STOC 1991, pp. 561–571. ACM, New York (1991)
30. Maurer, U.M.: Secret key agreement by public discussion from common information. IEEE Trans. Inf. Theory **39**(3), 733–742 (1993)
31. Ueli, M.: Information-theoretic cryptography. In: Wiener, M. (ed.) CRYPTO 1999. LNCS, vol. 1666, pp. 47–65. Springer, Berlin Heidelberg (1999). doi:10.1007/ 3-540-48405-1_4

32. Pass, R., Venkitasubramaniam, M.: An efficient parallel repetition theorem for Arthur-Merlin games. In: Proceedings of the 39th Annual ACM Symposium on Theory of Computing, STOC 2007, pp. 420–429 (2007)

33. Pietrzak, K., Wikström, D.: Parallel repetition of computationally sound protocols revisited. In: Vadhan, S.P. (ed.) TCC 2007. LNCS, vol. 4392, pp. 86–102. Springer, Heidelberg (2007). doi:10.1007/978-3-540-70936-7_5

34. Sahai, A., Vadhan, S.P.: A complete promise problem for statistical zero-knowledge. In: 38th Annual Symposium on Foundations of Computer Science, FOCS 1997, Miami Beach, Florida, USA, 19–22 October 1997, pp. 448–457. IEEE Computer Society (1997)

35. Shannon, C.E.: Communication theory of secrecy systems. Bell Syst. Tech. J. **28**, 656–715 (1949)

36. Shikata, J.: Formalization of information-theoretic security for key agreement, revisited. In: 2013 IEEE International Symposium on Information Theory Proceedings (ISIT), pp. 2720–2724, July 2013

37. Wullschleger, J.: Oblivious-transfer amplification. In: Naor, M. (ed.) EUROCRYPT 2007. LNCS, vol. 4515, pp. 555–572. Springer, Heidelberg (2007). doi:10.1007/978-3-540-72540-4_32

38. Wullschleger, J.: Oblivious transfer from weak noisy channels. In: Reingold, O. (ed.) TCC 2009. LNCS, vol. 5444, pp. 332–349. Springer, Heidelberg (2009). doi:10.1007/978-3-642-00457-5_20

39. Wyner, A.D.: The wire-tap channel. Bell Syst. Tech. J. **54**, 1355–1387 (1975)

# Proof of Space from Stacked Expanders

Ling Ren[✉] and Srinivas Devadas

Massachusetts Institute of Technology, Cambridge, MA, USA
{renling,devadas}@mit.edu

**Abstract.** Recently, proof of space (PoS) has been suggested as a more egalitarian alternative to the traditional hash-based proof of work. In PoS, a prover proves to a verifier that it has dedicated some specified amount of space. A closely related notion is memory-hard functions (MHF), functions that require a lot of memory/space to compute. While making promising progress, existing PoS and MHF have several problems. First, there are large gaps between the desired space-hardness and what can be proven. Second, it has been pointed out that PoS and MHF should require a lot of space not just at some point, but throughout the entire computation/protocol; few proposals considered this issue. Third, the two existing PoS constructions are both based on a class of graphs called superconcentrators, which are either hard to construct or add a logarithmic factor overhead to efficiency. In this paper, we construct PoS from stacked expander graphs. Our constructions are simpler, more efficient and have tighter provable space-hardness than prior works. Our results also apply to a recent MHF called Balloon hash. We show Balloon hash has tighter space-hardness than previously believed and consistent space-hardness throughout its computation.

## 1 Introduction

Proof of work (PoW) has found applications in spam/denial-of-service countermeasures [13,22] and in the famous cryptocurrency Bitcoin [36]. However, the traditional hash-based PoW does have several drawbacks, most notably poor resistance to application-specific integrated circuits (ASIC). ASIC hash units easily offer $\sim 100\times$ speedup and $\sim 10,000\times$ energy efficiency over CPUs. This gives ASIC-equipped adversaries a huge advantage over common desktop/laptop users. Recently, proof of space (PoS) [11,24] has been suggested as a potential alternative to PoW to address this problem. A PoS is a protocol between two parties, a prover and a verifier. Analogous to (but also in contrast to) PoW, the prover generates a cryptographic proof that it has invested a significant amount of memory or disk space (as opposed to computation), and the proof should be easy for the verifier to check. It is believed that if an ASIC has to access a large external memory, its advantage over a CPU will be small, making PoS more egalitarian than PoW.

Somewhat unfortunately, two competing definitions of "proof of space" have been proposed [11,24] with very different security guarantees and applications. Adding to the confusion are other closely related and similar-sounding notions

© International Association for Cryptologic Research 2016
M. Hirt and A. Smith (Eds.): TCC 2016-B, Part I, LNCS 9985, pp. 262–285, 2016.
DOI: 10.1007/978-3-662-53641-4_11

such as memory-hard functions (MHF) [39], proof of secure erasure (PoSE) [40], provable data possession (PDP) [12] and proof of retrievability (PoR) [30]. A first goal of this paper is to clarify the connections and differences between these notions. Section 2 will give detailed comparisons. For now, we give a short summary below and in Fig. 1.

PoSE ≈ MHF $\xrightarrow{\substack{\text{+ efficient} \\ \text{verification}}}$ PoTS $\xrightarrow{\substack{\text{+ repeated} \\ \text{audits}}}$ PoPS $\xleftarrow{\substack{\text{if w/o the initial} \\ \text{long message}}}$ PDP/PoR

**Fig. 1.** Relation between PoSE/MHF, PoTS, PoPS and PDP/PoR.

As its name suggests, a memory-hard function (MHF) is a function that requires a lot of memory/space to compute. Proof of secure erasure (PoSE) for the most part is equivalent to MHF. Proof of space by Ateniese et al. [11] extends MHF with efficient verification. That is, a verifier only needs a small amount of space and computation to check a prover's claimed space usage. Proof of space by Dziembowski et al. [24] further gives a verifier the ability to repeatedly audit a prover and check if it is still storing a large amount of data. The key difference between the two proofs of space lies in whether the proof is for transient space or persistent space. We shall distinguish these two notions of space and define them separately as *proof of transient space (PoTS)* and *proof of persistent space (PoPS)*. PDP and PoR solve a very different problem: availability check for a user's outsourced storage to an untrusted server. They do force the server to use a lot of persistent space but do not meet the succinctness (short input) requirement since a large amount of data needs to be transferred initially. Since PoPS is the strongest among the four related primitives (MHF, PoSE, PoTS, and PoPS), the end goal of this paper will be a PoPS with improved efficiency and space-hardness. Along the way, our techniques and analysis improve MHF/PoSE and PoTS as well.

Let us return to the requirements for a MHF $f$. Besides being space-hard, it must take short inputs. This is to rule out trivial solutions that only take long and incompressible input $x$ of size $|x| = N$. Such an $f$ trivially has space-hardness since $N$ space is needed to receive the input, but is rather uninteresting. We do not have many candidate problems that satisfy both requirements. Graph pebbling (also known as pebble games) in the random oracle model is the only candidate we know of so far. Although the random oracle model is not the most satisfactory assumption from a theoretical perspective, it has proven useful in practice and has become standard in this line of research [8,11,20,24,26,28, 31]. Following prior work, we also adopt the graph pebbling framework and the random oracle model in this work.

A pebble game is a single-player game on a directed acyclic graph (DAG). The player's goal is to put pebbles on certain vertices. A pebble can be placed on a vertex if it has no predecessor or if all of its predecessors have pebbles on them. Pebbles can be removed from any vertex at any time. The number of pebbles on the graph models the space usage of an algorithm.

Pebble games on certain graphs have been shown to have high space complexity or sharp space-time trade-offs. The most famous ones are stacked superconcentrators [32,38], which have been adopted in MHF [28], PoSE [31] and PoS [11,24]. However, bounds in graph pebbling are often very loose, especially for stacked superconcentrators [32,38]. This translates to large gaps between the desired memory/space-hardness and the provable guarantees in MHF and PoS (Sect. 1.1). Furthermore, MHFs and PoS need other highly desired properties that have not been studied in graph pebbling before (Sect. 1.2). The main contribution of this paper is to close these unproven or unstudied gaps while maintaining or even improving efficiency.

We illustrate these problems in the next two subsections using MHF as an example, but the analysis and discussion apply to PoS as well. We use "memory-hard" and "space-hard" interchangeably throughout the paper.

## 1.1   Gaps in Provable Memory Hardness

The most strict memory-hardness definition for a MHF $f$ is that for any $x$, $f(x)$ can be efficiently computed using $N$ space, but is impossible to compute using $N-1$ space. Here, "impossible to compute" means the best strategy is to take a random guess in the output space of $f(\cdot)$ (by the random oracle assumption). Achieving this strict notion of memory-hardness is expensive. Aside from the trivial solution that sets input size to $|x| = N$, the best known construction has $O(N^2)$ time complexity for computing $f$ [26]. The quadratic runtime makes this MHF impractical for large space requirements.

All other MHFs/PoSE and PoS in the literature have quasilinear runtime, i.e., $N \cdot \text{polylog} N$, by adopting much more relaxed notions of memory-hardness. One relaxation is to introduce an unproven gap [24,31]. For example, in the PoSE by Karvelas and Kiayias [31], while the best known algorithm to compute $f$ needs $N$ space, it can only be shown that computing $f$ using less than $N/32$ space is impossible. No guarantees can be provided if an adversary uses more than $N/32$ but less than $N$ space.

The other way to relax memory-hardness is to allow space-time trade-offs, and it is usually combined with unproven gaps. Suppose the best known algorithm (to most users) for a MHF takes $S$ space and $T$ time. These proposals hope to claim that any algorithm using $S' = S/q$ space should run for $T'$ time, so that the time penalty $T'/T$ is "reasonable". If the time penalty is linear in $q$, it corresponds to a lower bound on $S'T' = \Omega(ST)$, as scrypt [39] and Catena-BRG [28] did. Notice that the hidden constant in the bound leaves an unproven gap. Other works require the penalty to be superlinear in $q$ [15,28] or exponential in some security parameter [11,20,33], but the penalty only kicks in when $S'$ is below some threshold, e.g., $N/8$, again leaving a gap.

We believe an exponential penalty is justifiable since it corresponds to the widely used computational security in cryptography. However, an $ST$ lower bound and a large unproven gap are both unsatisfactory. Recall that the motivation of MHF is ASIC-resistance. With an $ST$ bound, an attacker is explicitly allowed to decrease space usage, at the cost of a proportional increase in computation. Then, an adversary may be able to fit $S/100$ space in an ASIC, and get in return a speedup or energy efficiency gain well over 100.

A large unproven gap leaves open the possibility that an adversary may gain an unfair advantage over honest users, and fairness is vital to applications like voting and cryptocurrency. A more dramatic example is perhaps PoSE [31]. With an unproven gap of 32, a verifier can only be assured that a prover has wiped $1/32$ fraction of its storage, which can hardly be considered a "proof of erasure". The authors were well aware of the problem and commented that this gap needs to be very small for a PoSE to be useful [31], yet they were unable to tighten it. Every MHF, PoSE or PoS with quasilinear efficiency so far has a large unproven gap (if it has a provable guarantee at all).

In fact, MHFs have been broken due to the above weaknesses. Scrypt [39], the most popular MHF, proved an asymptotic $ST$ lower bound. But an $ST$ bound does not prevent space-time trade-offs, and the hidden constants in the bounds turned out to be too small to provide meaningful guarantees [16]. As a result, ASICs for scrypt are already commercially available [1]. The lesson is that space-hardness is one of the examples where exact security matters. PoS proposals so far have not received much attention from cryptanalysis, but the loose hidden constants in prior works are equally concerning. Therefore, we will be explicit about every constant in our constructions, and also make best efforts to analyze hidden constants in prior works (in Tables 1, 2 and 3).

## 1.2  Consistent Memory Hardness

In a recent inspiring paper, Alwen and Serbinenko pointed out an overlooked weakness in all existing MHFs' memory-hardness guarantees [8]. Again, the discussion below applies to PoS. The issue is that in current definitions, even if a MHF $f$ is proven to require $N$ space in the most strict sense, it means $N$ space is needed *at some point* during computation. It is possible that $f$ can be computed by an algorithm that has a short memory-hard phase followed by a long memory-easy phase. In this case, an adversary can carry out the memory-hard phase on a CPU and then offload the memory-easy phase to an ASIC, defeating the supposed ASIC-resistance.

Alwen and Serbinenko argue, and we fully agree, that a good MHF should require a lot of memory not just at some point during its computation, but throughout the majority of its computation. However, we think the solution they presented has limitations. Alwen and Serbinenko suggested lower bounding a MHF's cumulative complexity (CC), the sum of memory usage in all steps of an algorithm [8]. For example, if the algorithm most users adopt takes $T$ time and uses $S$ space at every time step, its CC is $ST$. If we can lower bound the CC of any algorithm for this MHF to $ST$, it rules out an algorithm that runs for $T$ time, uses $S$ space for a few steps but very little space at other steps. A CC bound is thus an improved version of an $ST$ bound, and this is also where the problem is. Like an $ST$ bound, CC explicitly allows proportional space-time trade-offs: algorithms that run for $qT$ time and use $S/q$ space for any factor $q$. Even when combined with a strict space lower bound of $S$, it still does not rule out an algorithm that runs for $qT$ time, uses $S$ space for a few steps but $S/q$ space at all other steps. We have discussed why a proportional space-time trade-off or a long memory-easy phase can be harmful, and CC allows both.

Instead, we take a more direct approach to this problem. Recall that our goal is to design a MHF that consistently uses a lot of memory during its computation. So we will simply lower bound the number of time steps during the computation with high space usage. If this lower bound is tight, we say a MHF has consistent memory-hardness.

Another difference between our approach and that of [8] is the computation model. Alwen and Serbinenko assumed their adversaries possess infinite parallel processing power, and admirably proved lower bounds for their construction against such powerful adversaries. But their construction is highly complicated and the bound is very loose. We choose to stay with the sequential model or limited parallelism for two reasons. First, cumulative/consistent memory-hardness and parallelism are two completely independent issues and should not be coupled. Consistent (cumulative) memory-hardness is extremely important in the sequential model. Mixing it with the parallel model gives the wrong impression that it only becomes a problem when an adversary has infinite parallelism. Second, massive parallelism seems unlikely for MHFs in the near future. Even if parallel computation is free in ASICs, to take advantage of it, an adversary also needs proportionally higher memory bandwidth (at least in our construction). Memory bandwidth is a scarce resource and is *the* major bottleneck in parallel computing, widely known as the "memory wall" [10]. It is interesting to study the infinitely parallel model from a theoretical perspective as memory bandwidth may become cheap in the future. But at the time being, it is not worth giving up practical and provably secure solutions in the sequential model.

## 1.3   Our Results

We construct PoTS and PoPS from stacked expanders. Our constructions are conceptually simpler, more efficient and have tighter space-hardness guarantees than prior works [11,24]. We could base our space-hardness on a classical result by Paul and Tarjan [37], but doing so would result in a large unproven gap. Instead, we carefully improve the result by Paul and Tarjan to make the gap arbitrarily small. We then introduce the notion of consistent memory-hardness and prove that stacked expanders have this property.

These results lead to better space-hardness guarantees for our constructions. For our PoTS, we show that no computationally bounded adversary using $\gamma N$ space can convince a verifier with non-negligible probability, where $\gamma$ can be made arbitrarily close to 1. The prover also needs close to $N$ space not just at some point in the protocol, but consistently throughout the protocol. In fact, the honest strategy is very close to the theoretical limits up to some tight constants. For PoPS, we show that an adversary using a constant fraction of $N$ persistent space (e.g., $N/3$) will incur a big penalty. It is a bit unsatisfactory that we are unable to further tighten the bound and have to leave a small gap. But our result still represents a big improvement over the only previous PoPS [24] whose gap is as large as $2 \times 256 \times 25.3 \times \log N$.

Our tight and consistent memory-hardness results can be of independent interest. Independent of our work, Corrigan-Gibbs et al. recently used stacked expanders to build a MHF called Balloon hash [20]. They invoked Paul and Tarjan [37] for space-hardness and left an unproven gap of 8. (Their latest space-hardness proof no longer relies on [37], but the gap remains the same.) Our results show that Balloon hash offers much better space-hardness than previously believed. Our work also positively answers several questions left open by Corrigan-Gibbs et al. [20]: Balloon hash is consistently space-hard, over time and under batching.

## 2    Related Work

***MHF***. It is well known that service providers should store hashes of user passwords. This way, when a password hash database is breached, an adversary still has to invert the hash function to obtain user passwords. However, ASIC hash units have made the brute force attack considerably easier. This motivated memory-hard functions (MHF) as better password scramblers. Percival [39] proposed the first MHF, scrypt, as a way to derive keys from passwords. Subsequent works [3,15,20,28,33] continued to study MHFs as key derivation functions, password scramblers, and more recently as proof of work. In the recent Password Hashing Competition [27], the winner Argon2 [15] and three of the four "special recognitions"—Catena [28], Lyra2 [3] and yescrypt [41]—claimed memory-hardness.

The most relevant MHF to our work is Balloon hash [20], which also adopted stacked expanders. We adopt a technique from Balloon hash to improve our space-hardness. Our analysis, in turn, demonstrates better space-hardness for Balloon hash and positively answers several open questions regarding its consistent memory-hardness [20]. We also develop additional techniques to obtain PoS.

***Attacking MHF***. MHFs have been classified into data-dependent ones (dMHF) and data-independent ones (iMHF), based on whether a MHF's memory access pattern depends on its input [20,28]. Catena and Balloon hash are iMHF, and the rest are dMHF. Some consider dMHFs less secure for password hashing due to cache timing attacks.

Most MHF proposals lack rigorous analysis, and better space-time trade-offs (in the traditional sequential model) have been shown against them [16, 20]. The only two exceptions are Catena-DBG and Balloon, both of which use graph pebbling. Alwen and Blocki considered adversaries with infinite parallel processing power, and showed that such a powerful attacker can break any iMHF, including Catena-DBG and Balloon [5,6].

*MBF*. Prior to memory-hard functions, Dwork et al. [21,23] and Abadi et al. [2] proposed memory-bound functions (MBF). The motivation of MBF is also ASIC-resistance, but the complexity metric there is the number of cache misses. A MHF may not be memory-bound since its memory accesses may hit in cache most of the time. A MBF has to be somewhat memory-hard to spill from cache, but it may not consume too much memory beyond the cache size. A construction that enjoys both properties should offer even better resistance to ASICs, but we have not seen efforts in this direction.

*PoSE*. Proof of secure erasure (PoSE) was first studied by Perito and Tsudik [40] as a way to wipe a remote device. Assuming a verifier knows a prover (a remote device) has exactly $N$ space, any protocol that forces the prover to use $N$ space was considered a PoSE [40]. This includes the trivial solution where the verifier sends the prover a large random file of size $N$, and then asks the prover to send it back. Since this trivial solution is inefficient and uninteresting, for the rest of the paper when we say PoSE, we always mean communication-efficient PoSE [31], where the prover receives a short challenge but needs a lot of space to generate a proof. A reader may have noticed that space-hardness and short input are exactly the same requirements we had earlier for MHFs. Thus, we can think of PoSE as an application of MHFs, with one small caveat in the definition of space-hardness. We have mentioned that a proportional space-time trade-off or a large unproven gap is undesirable for MHFs; for PoSE, they are unacceptable. On the flip side, PoSE does not need consistent space-hardness.

*PoTS and memory-hard PoW*. Two independent works named their protocols "proofs of space" [11,24]. The key difference is whether the proof is for transient space or persistent space. Ateniese et al. [11] corresponds to a proof of transient space (PoTS). It enables efficient verification of a MHF with polylog($N$) verifier space and time. If we simply drop the efficient verification method and have the verifier redo the prover's work, PoTS reduces to PoSE/MHF.

Two recent proposals Cuckoo Cycle [48] and Equihash [17] aim to achieve exactly the same goal as PoTS, and call themselves memory-hard proof of work. This is also an appropriate term because a prover in PoTS has to invest both space and time, usually $N$ space and $N \cdot$ polylog($N$) computation. Cuckoo Cycle and Equihash are more efficient than Ateniese et al. [11] but do not have security proofs. An attack on Cuckoo Cycle has already been proposed [9].

*PoPS*. Dziembowski et al. [24] is a proof of persistent space (PoPS). Compared to Ateniese et al. [11], it supports "repeated audits". The protocol has two stages. In the first stage, the prover generates some data of size $N$, which we call *advice*. The prover is supposed to store the advice persistently throughout the second stage. In the second stage, the verifier can repeatedly audit the prover and check if it is still storing the advice. All messages exchanged between the two parties and the verifier's space/time complexity in both stages should be polylog($N$). If the prover is audited only once, PoPS reduces to PoTS.

It is worth pointing out that an adversary can always discard the advice and rerun setup when audited. To this end, the space-hardness definition is somewhat relaxed (see Sect. 6 for details). PoPS also attempts to address the other drawback of PoW: high energy cost. It allows an honest prover who faithfully stores the advice to respond to audits using little computation, hence consuming little dynamic energy. Whether these features are desirable depends heavily on the application. Proof of Space-Time [35] is a recent proposal that resembles PoPS but differs in the above two features. In their protocol, an honest prover needs to access the entire $N$-sized advice or at least a large fraction to pass each audit. In return, they argue that the penalty they guarantee for a cheating prover is larger.

***PDP and PoR.*** Provable data possession (PDP) [12] and proof of retrievability (PoR) [30] allow a user who outsources data to a server to repeatedly check if the server is still storing his/her data. If a verifier (user) outsources large and incompressible data to a prover (server), PDP and PoR can achieve the space-hardness goal of both PoTS and PoPS. However, transmitting the initial data incurs high communication cost. In this aspect, PoS schemes [11,24] are stronger as they achieve low communication cost. PDP and PoR are stronger in another aspect: they can be applied to arbitrary user data while PoS populates prover/server memory only with random bits. In summary, PDP and PoR solve a different problem and are out of the scope of this paper.

***Graph pebbling.*** Graph pebbling is a powerful tool in computer science, dating back at least to 1970s in studying Turing machines [19,29] and register allocation [46]. More recently, graph pebbling has found applications in various areas of cryptography [11,23–26,28,31,34,47].

***Superconcentrators.*** The simplest superconcentrator is perhaps the butterfly graph, adopted in MHF/PoSE [28,31] and PoTS [11], but it has a logarithmic factor more vertices and edges than linear superconcentrators or expanders. Linear superconcentrators, adopted in PoPS [24], on the other hand, are hard to construct and recursively use expanders as building blocks [4,18,44,45]. Thus, it is expected that superconcentrator-based MHFs and PoS will be more complicated and less efficient than expander-based ones (under comparable space-hardness).

## 3 Pebble Games on Stacked Expanders

### 3.1 Graph Pebbling and Labelling

A pebble game is a single-player game on a directed acyclic graph (DAG) $G$ with a constant maximum in-degree $d$. A vertex with no incoming edges is called a *source* and a vertex with no outgoing edges is called a *sink*. The player's goal is to put pebbles on certain vertices of $G$ using a sequence of *moves*. In each move, the player can place one pebble and remove an arbitrary number of pebbles (removing pebbles is free in our model). The player's moves can be represented as a sequence of transitions between pebble placement configurations on the graph, $\mathbf{P} = (P_0, P_1, P_2 \cdots, P_T)$. If a pebble exists on a vertex $v$ in a configuration $P_i$, we say $v$ is *pebbled* in $P_i$. The starting configuration $P_0$ does not have to be empty; vertices can be pebbled in $P_0$. The pebble game rule is as follows: to transition from $P_i$ to $P_{i+1}$, the player can *pebble* (i.e., place a pebble on) one vertex $v$ if $v$ is a source or if all predecessors of $v$ are pebbled in $P_i$, and then *unpebble* (i.e., remove pebbles from) any subset of vertices. We say a sequence **P** pebbles a vertex $v$ if there exists $P_i \in \mathbf{P}$ such that $v$ is pebbled in $P_i$. We say a sequence **P** pebbles a set of vertices if **P** pebbles every vertex in the set.

A pebble game is just an abstraction. We need a concrete computational problem to enforce the pebble game rules. Prior work has shown that the graph labelling problem with a random oracle $\mathcal{H}$ implements pebble games. In graph labelling, vertices are numbered, and each vertex $v_i$ is associated with a label $h(v_i) \in \{0,1\}^\lambda$ where $\lambda$ is the output length of $\mathcal{H}$.

$$h(v_i) = \begin{cases} \mathcal{H}(i,x) & \text{if } v_i \text{ is a source} \\ \mathcal{H}(i,h(u_1),h(u_2),\cdots,h(u_d)) & \text{otherwise, } u_1 \text{ to } u_d \text{ are } v_i's \text{ predecessors} \end{cases}$$

Clearly, any legal pebbling sequence gives a graph labelling algorithm. It has been shown that the converse is also true for PoSE/MHF [23, 26, 31] and PoTS [11], via a now fairly standard "ex post facto" argument. The equivalence has not been shown for PoPS due to subtle issues [24], but there has been recent progress in this direction [7]. We refer readers to these papers and will not restate their results.

Given the equivalence (by either a proof or a conjecture), we can use metrics of the underlying pebble games to analyze higher-level primitives. Consider a pebble sequence $\mathbf{P} = (P_0, P_1, P_2 \cdots, P_T)$. Let $|P_i|$ be the number of pebbles on the graph in configuration $P_i$. We define the space complexity of a sequence $S(\mathbf{P}) = \max_i(|P_i|)$, i.e., the maximum number of pebbles on the graph at any step. It is worth noting that space in graph labelling is measured in "label size" $\lambda$ rather than bits.

We define the time complexity of a sequence $T(\mathbf{P})$ to be the number of transitions in **P**. $T(\mathbf{P})$ equals the number of random oracle $\mathcal{H}$ calls, because we only allow one new pebble to be placed per move. This corresponds to the sequential model. We can generalize to limited parallelism, say $q$-way parallelism, by allowing up to $q$ pebble placements per move. But we do not consider infinite parallelism in this paper as discussed in Sect. 1.2.

For a more accurate timing model in graph labelling, we assume the time to compute a label is proportional to the input length to $\mathcal{H}$, i.e., the in-degree of the vertex. Another way to look at it is that we can transform a graph with maximum in-degree $d$ into a graph with maximum in-degree 2 by turning each vertex into a binary tree of up to $d$ leaves.

To capture consistent space-hardness, we define $M_{S'}(\mathbf{P}) = |\{i : |P_i| \geq S'\}|$, i.e., the number of configurations in **P** that contain at least $S'$ pebbles. Consider a pebble game that has a legal sequence **P**. If there exist some $S' < S(\mathbf{P})$ and $T' < T(\mathbf{P})$, such that any legal sequence $\mathbf{P}'$ for that same pebble game has $M_{S'}(\mathbf{P}') \geq T'$, we say the pebble game is consistently memory-hard. The distance between $(S', T')$ and $(S(\mathbf{P}), T(\mathbf{P}))$ measures the quality of consistent memory-hardness.

## 3.2   Bipartite Expanders

Now we introduce bipartite expanders, the basic building blocks for our constructions, and review classical results on their efficient randomized constructions.

**Definition 1.** *An $(n, \alpha, \beta)$ bipartite expander $(0 < \alpha < \beta < 1)$ is a directed
bipartite graph with $n$ sources and $n$ sinks such that any subset of $\alpha n$ sinks are
connected to at least $\beta n$ sources.*

Prior work has shown that bipartite expanders for any $0 < \alpha < \beta < 1$
exist given sufficiently many edges. We adopt the randomized construction by
Chung [18]. This construction gives a $d$-regular bipartite expander, i.e., there
are $d$ outgoing edges from each source and $d$ incoming edges to each sink. It
simply connects the $dn$ outgoing edges of the sources and the $dn$ incoming edges
of the sinks according to a random permutation. Given a permutation $\Pi$ on
$\{0, 1, 2, \cdots, dn - 1\}$, if $\Pi(i) = j$, add an edge from source $(i \mod n)$ to sink
$(j \mod n)$.

**Theorem 1.** *Chung's construction yields an $(n, \alpha, \beta)$ bipartite expander
$(0 < \alpha < \beta < 1)$ for sufficiently large $n$ with overwhelming probability if*

$$d > \frac{H_b(\alpha) + H_b(\beta)}{H_b(\alpha) - \beta H_b(\frac{\alpha}{\beta})}$$

*where $H_b(\alpha) = -\alpha \log_2 \alpha - (1 - \alpha) \log_2(1 - \alpha)$ is the binary entropy function.*

The theorem has been proven by Bassalygo [14] and Schöning [44], but both
proofs were quite involved. We give a proof using a simple counting argument.

*Proof.* There are $(dn)!$ permutations in total. We analyze how many permutations are "bad", i.e., do not yield an expander. A bad permutation must connect
some subset $U$ of $\alpha n$ sinks to a subset $V$ of $\beta n$ sources. There are $\binom{n}{\alpha n}\binom{n}{\beta n}$ combinations. Within each combination, there are $\binom{d\beta n}{d\alpha n}(d\alpha n)!$ ways to connect $U$ to
$V$. There are $(dn - d\alpha n)!$ ways to connect the rest of edges (those not incident
to $U$). The probability that we hit a bad permutation is

$$\Pr(\Pi \text{ is bad}) = \binom{n}{\alpha n}\binom{n}{\beta n}\binom{d\beta n}{d\alpha n}(d\alpha n)!(dn - d\alpha n)!/(dn)!$$

$$= \binom{n}{\alpha n}\binom{n}{\beta n}\binom{d\beta n}{d\alpha n}\bigg/\binom{dn}{d\alpha n}$$

Using Robbins' inequality for Stirling's approximation $\sqrt{2\pi n}(n/e)^n e^{\frac{1}{12n+1}} <
n! < \sqrt{2\pi n}(n/e)^n e^{\frac{1}{12n}}$ [43], we have $\log_2 \binom{n}{\alpha n} = n H_b(\alpha) - \frac{1}{2}\log_2 n + o(1)$. Thus,

$$\log_2 \Pr(\Pi \text{ is bad}) = n[H_b(\alpha) + H_b(\beta) + d\beta H_b(\alpha/\beta) - dH_b(\alpha)] - \log_2 n + o(1).$$

If $H_b(\alpha) + H_b(\beta) + d\beta H_b(\alpha/\beta) - dH_b(\alpha) < 0$, or equivalently the bound on $d$
in the theorem statement holds, then $\Pr(\Pi \text{ is bad})$ decreases exponentially as $n$
increases.                                                                              $\square$

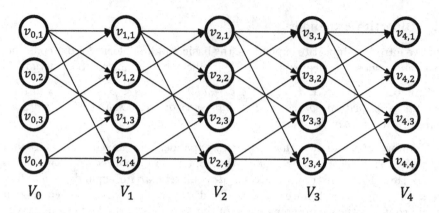

**Fig. 2.** A stacked bipartite expander $G_{(4,4,\frac{1}{4},\frac{1}{2})}$.

Pinsker [42] used a different randomized construction, which independently selects $d$ predecessors for each sink. Pinsker's construction requires $d > \frac{H_b(\alpha)+H_b(\beta)}{-\alpha\log_2\beta}$ [20], which is a slightly worse bound than Theorem 1. But Pinsker's construction is arguably simpler than Chung's because it only needs a random function as opposed to a random permutation.

### 3.3   Pebble Games on Stacked Bipartite Expanders

Construct $G_{(n,k,\alpha,\beta)}$ by stacking $(n,\alpha,\beta)$ bipartite expanders. $G_{(n,k,\alpha,\beta)}$ has $n(k+1)$ vertices, partitioned into $k+1$ sets each of size $n$, $V = \{V_0, V_1, V_2, \cdots, V_k\}$. All edges in $G_{(n,k,\alpha,\beta)}$ go from $V_{i-1}$ to $V_i$ for some $i$ from 1 to $k$. For each $i$ from 1 to $k$, $V_{i-1}$ and $V_i$ plus all edges between them form an $(n,\alpha,\beta)$ bipartite expander. The bipartite expanders at different layers can but do not have to be the same. $G_{(n,k,\alpha,\beta)}$ has $n$ sources, $n$ sinks, and the same maximum in-degree as the underlying $(n,\alpha,\beta)$ bipartite expander. Figure 2 is an example of $G_{(4,4,\frac{1}{4},\frac{1}{2})}$ with in-degree 2.

Obviously, simply pebbling each expander in order results in a sequence $\mathbf{P}$ that pebbles $G_{(n,k,\alpha,\beta)}$ using $S(\mathbf{P}) = 2n$ space in $T(\mathbf{P}) = n(k+1)$ moves. Paul and Tarjan [37] showed that $G_{(n,k,\frac{1}{8},\frac{1}{2})}$ has an exponentially sharp space-time trade-off. Generalized to $(n,\alpha,\beta)$ expanders, their result was the following:

**Theorem 2 (Paul and Tarjan [37]).** *If $\mathbf{P}$ pebbles any subset of $2\alpha n$ sinks of $G_{(n,k,\alpha,\beta)}$, starting with $|P_0| \leq \alpha n$ and using $S(\mathbf{P}) \leq \alpha n$ space, then $T(\mathbf{P}) \geq \lfloor\frac{\beta}{2\alpha}\rfloor^k$.*

This theorem forms the foundation of Balloon hash. We could base our PoTS/PoPS protocols on it. However, the space-hardness guarantee we get will be at most $n/4$. We need $\frac{\beta}{2\alpha} \geq 2$ to get an exponential time penalty, so $\alpha n < \beta n/4 < n/4$.

Through a more careful analysis, we show a tighter space-time trade-off for stacked bipartite expanders, which will lead to better space-hardness for our PoTS/PoPS protocols as well as Balloon hash. We improve Theorem 2 by considering only initially unpebbled sinks. Let $\gamma = \beta - 2\alpha > 0$ for the rest of the paper.

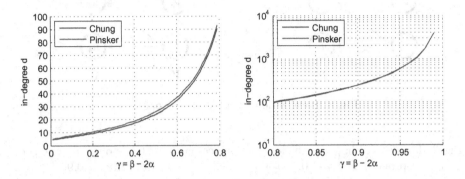

**Fig. 3.** Minimum in-degree $d$ to achieve a given $\gamma = \beta - 2\alpha$.

**Theorem 3.** *If* $\mathbf{P}$ *pebbles any subset of* $\alpha n$ *initially unpebbled sinks of* $G_{(n,k,\alpha,\beta)}$, *starting with* $|P_0| \leq \gamma n$ *and using* $S(\mathbf{P}) \leq \gamma n$ *space, then* $T(\mathbf{P}) \geq 2^k \alpha n$.

*Proof.* For the base case $k = 0$, $G_{(n,0,\alpha,\beta)}$ is simply a collection of $n$ isolated vertices with no edges. Each vertex is both a source and a sink. The theorem is trivially true since the $\alpha n$ initially unpebbled sinks have to be pebbled.

Now we show the inductive step for $k \geq 1$ assuming the theorem holds for $k - 1$. In $G_{(n,k,\alpha,\beta)}$, sinks are in $V_k$. The $\alpha n$ to-be-pebbled sinks in $V_k$ are connected to at least $\beta n$ vertices in $V_{k-1}$ due to the $(n, \alpha, \beta)$ expander property. Out of these $\beta n$ vertices in $V_{k-1}$, at least $\beta n - \gamma n = 2\alpha n$ of them are unpebbled initially in $P_0$ since $|P_0| \leq \gamma n$. These $2\alpha n$ vertices in $V_{k-1}$ are unpebbled sinks of $G_{(n,k-1,\alpha,\beta)}$. Divide them into two groups of $\alpha n$ each in the order they are pebbled in $\mathbf{P}$ for the first time. $\mathbf{P}$ can be then divided into two parts $\mathbf{P} = (\mathbf{P}_1, \mathbf{P}_2)$ where $\mathbf{P}_1$ pebbles the first group ($\mathbf{P}_1$ does not pebble any vertex in the second group) and $\mathbf{P}_2$ pebbles the second group. Due to the inductive hypothesis, $T(\mathbf{P}_1) \geq 2^{k-1} \alpha n$. The starting configuration of $\mathbf{P}_2$ is the ending configuration of $\mathbf{P}_1$. At the end of $\mathbf{P}_1$, there are at most $\gamma n$ pebbles on the graph, and the second group of $\alpha n$ vertices are all unpebbled. So we can invoke the inductive hypothesis again, and have $T(\mathbf{P}_2) \geq 2^{k-1} \alpha n$. Therefore, $T(\mathbf{P}) = T(\mathbf{P}_1) + T(\mathbf{P}_2) \geq 2^k \alpha n$. □

Theorem 3 lower bounds the space complexity of any feasible pebbling strategy for stacked bipartite expanders to $\gamma n$, where $\gamma = \beta - 2\alpha$. If we increase $\beta$ or decrease $\alpha$, $\gamma$ improves but the in-degree $d$ also increases due to Theorem 1. For each $\gamma = \beta - 2\alpha$, we find the $\alpha$ and $\beta$ that minimize $d$, and plot them in

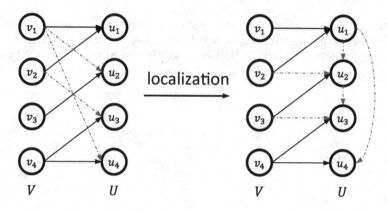

**Fig. 4.** Localization for a bipartite expander.

Fig. 3. The curves show the efficiency vs. space-hardness trade-offs our constructions can provide. For $\gamma < 0.7$, $d$ is reasonably small. Beyond $\gamma = 0.9$, $d$ starts to increase very fast. We recommend parameterizing our constructions around $0.7 \leq \gamma \leq 0.9$.

However, even if $\gamma$ is close to 1, we still have a gap of 2 as our simple pebbling strategy for stacked bipartite expanders needs $2n$ space. To address this gap, we adopt the localization technique in Balloon hash [20].

### 3.4    Localization of Bipartite Expanders

Localization [20] is a transformation on the edges of a bipartite expander. Consider an $(n, \alpha, \beta)$ bipartite expander with sources $V = \{v_1, v_2, \cdots v_n\}$ and sinks $U = \{u_1, u_2, \cdots u_n\}$. The localization operation first adds an edge $(v_i, u_i)$ for all $i$ (if it does not already exist), and then replaces each edge $(v_i, u_j)$ where $i < j$ with $(u_i, u_j)$. Figure 4 highlights the removed and the added edges in red. Pictorially, it adds an edge for each horizontal source-sink pair, and replaces each "downward diagonal" edge with a corresponding "downward vertical" edge. This adds at most one incoming edge for each vertex in $U$.

Let $LG_{(n,k,\alpha,\beta)}$ be a stack of localized expanders, i.e., the resulting graph after localizing the bipartite expander at every layer of $G_{(n,k,\alpha,\beta)}$. $LG_{(n,k,\alpha,\beta)}$ can be efficiently pebbled using $n$ space, by simply pebbling each layer in order and within each layer from top to bottom. Once $v_{k,i}$ is pebbled, $v_{k-1,i}$ can be unpebbled because no subsequent vertices depend on it. A vertex $v_{k,j} \in V_k$ that originally depended on $v_{k-1,i}$ is either already pebbled (if $j \leq i$), or has its dependency changed to $v_{k,i}$ by the localization transformation.

When we localize a bipartite expander, the resulting graph is no longer bipartite. The expanding property, however, is preserved under a different definition. After localization, the graph has $n$ sources and $n$ non-sources (the original sinks). Any subset $U'$ of $\alpha n$ non-sources collectively have $\beta n$ sources as ancestors ($v$ is an ancestor of $u$ if there is a path from $v$ to $u$). Crucially, the paths between them are vertex-disjoint outside $U'$. This allows us to prove the same result in Theorem 3 for stacked localized expanders.

**Lemma 1.** *Let $U'$ be any subset of $\alpha n$ sinks of an $(n, \alpha, \beta)$ bipartite expander, and $V'$ be the set of sources connected to $U'$ (we have $|V'| \geq \beta n$). After localization, there exist $\beta n$ paths from $V'$ to $U'$ that are vertex-disjoint outside $U'$.*

*Proof.* After localization, vertices in $V'$ fall into two categories. A vertex $v_i \in V'$ may still be an immediate predecessor to some $u \in U'$, which obviously does not share any vertex outside $U'$ with a path starting from any $v_j$ ($j \neq i$). If $v_i$ is not an immediate predecessor, then the path $v_i \rightarrow u_i \rightarrow u$ must exist for some $u \in U'$, because there was an edge $(v_i, u)$ in the original bipartite expander. Any other $v_j \in V'$ ($j \neq i$) is either an immediate predecessor or uses $u_j$ as the intermediate hop in its path to $U'$. In either case, $v_i$ does not share any source or intermediate-hop with any other $v_j$.   $\square$

**Theorem 4.** *Let $\gamma = \beta - 2\alpha > 0$. If $\mathbf{P}$ pebbles any subset $U'$ of $\alpha n$ initially unpebbled vertices in the last layer $V_k$ of $LG_{(n,k,\alpha,\beta)}$, starting with $|P_0| \leq \gamma n$ and using $S(\mathbf{P}) \leq \gamma n$ space, then $T(\mathbf{P}) \geq 2^k \alpha n$.*

*Proof.* The proof remains unchanged from Theorem 3 as long as we show that $\mathbf{P}$ still needs to pebble $2\alpha n$ initially unpebbled vertices in $V_{k-1}$.

A path from $v$ to $u$ is "initially pebble-free", if no vertex on the path, including $v$ and $u$, is pebbled in $P_0$. Due to the pebble game rule, if a vertex $v \in V_{k-1}$ has an initially pebble-free path to some $u \in U'$, then it needs to be pebbled before $u$ can be pebbled. Since $V_{k-1}$ and $V_k$ form a localized expander, due to Lemma 1, there exist at least $\beta n$ ancestors in $V_{k-1}$ whose paths to $U'$ are vertex-disjoint outside $U'$. Since vertices in $U'$ are initially unpebbled, pebbles in $P_0$ can only be placed on the vertex-disjoint parts of these paths. Therefore, each pebble can block at most one of these paths. Since $|P_0| \leq \gamma n$, there must be at least $\beta n - \gamma n = 2\alpha n$ vertices in $V_{k-1}$ that have initially pebble-free paths to $U'$, and they have to be pebbled by $\mathbf{P}$.   $\square$

We now have tight space lower bounds for pebble games on stacked localized expanders. $LG_{(n,k,\alpha,\beta)}$ can be efficiently pebbled with $n$ space but not with $\gamma n$ space, where $\gamma$ can be set close to 1. Next, we show that pebble games on localized stacked expanders are also consistently space-hard.

### 3.5   Consistent Space Hardness

**Theorem 5.** *Let $0 < \eta < \gamma = \beta - 2\alpha$. If $\mathbf{P}$ pebbles any subset of $\alpha n$ initially unpebbled vertices in the last layer of $LG_{(n,k,\alpha,\beta)}$, starting with $|P_0| \leq \eta n$, and using $T(\mathbf{P}) \leq 2^{k_0} \alpha n$ moves, then*

$$M_{\eta n}(\mathbf{P}) \geq \begin{cases} 0 & k < k_0 \\ 2^{k - k_0} & k_0 \leq k \leq k_1 \\ (k - k_1 + 1)(\gamma - \eta)n & k > k_1 \end{cases}$$

*where $k_1 = k_0 + \lceil \log_2 (\gamma - \eta)n \rceil$.*

*Proof.* We will derive a lower bound $M_k$ for $M_{\eta n}(\mathbf{P})$ where $k$ is the number of layers in $LG_{(n,k,\alpha,\beta)}$. Similar to the proof of Theorem 3, there are $(\beta - \eta)n$ initially unpebbled vertices in $V_{k-1}$ that have to be pebbled by $\mathbf{P}$. Let $U$ be the set of these $(\beta - \eta)n$ vertices. Again, we sort $U$ according to the time they are first pebbled. We divide $\mathbf{P}$ into three parts $\mathbf{P} = (\mathbf{P}_1, \mathbf{P}_2, \mathbf{P}_3)$. $\mathbf{P}_1$ pebbles the first $\alpha n$ vertices in $U \subset V_{k-1}$. $\mathbf{P}_1$ starts from the same initial configuration as $\mathbf{P}$ and has fewer moves than $\mathbf{P}$, so we have $M_{\eta n}(\mathbf{P}_1) \geq M_{k-1}$.

We define $\mathbf{P}_2$ to include all consecutive configurations immediately after $\mathbf{P}_1$ until (and including) the first configuration whose space usage is below $\eta n$. $\mathbf{P}_3$ is then the rest of $\mathbf{P}$. By definition, every $P_i \in \mathbf{P}_2$, except the last one, satisfies $|P_i| > \eta n$. The last configuration in $\mathbf{P}_2$ is also the starting configuration of $\mathbf{P}_3$, and its space usage is below $\eta n$. It is possible that $T(\mathbf{P}_2) = 1$ or $T(\mathbf{P}_3) = 0$, if the space usage after $\mathbf{P}_1$ immediately drops below $\eta n$ or never drops below $\eta n$.

Now we have two cases based on $T(\mathbf{P}_2)$. If $T(\mathbf{P}_2) > (\gamma - \eta)n$, we have $M_k > M_{k-1} + (\gamma - \eta)n$. If $T(\mathbf{P}_2) \leq (\gamma - \eta)n$, then $\mathbf{P}_3$ has to pebble at least $\alpha n$ vertices in $U$, because $\mathbf{P}_1$ and $\mathbf{P}_2$ combined have pebbled no more than $\alpha n + (\gamma - \eta)n = (\beta - \alpha - \eta)n$ vertices in $U$. And $\mathbf{P}_3$ starts with no more than $\eta n$ pebbles and has fewer moves than $\mathbf{P}$, so $M_{\eta n}(\mathbf{P}_3) \geq M_{k-1}$. In this case, we have $M_k \geq 2M_{k-1}$. Combining the two cases, we have the following recurrence

$$M_k \geq \min(M_{k-1} + (\gamma - \eta)n, 2M_{k-1}).$$

For a base case of this recurrence, we have $M_{k_0} \geq 1$, because Theorem 4 says any pebbling strategy that never uses $\eta n$ space needs at least $2^{k_0} \alpha n$ moves. Solving the recurrence gives the result in the theorem.    $\square$

For a tight bound on the entire sequence, we further chain the vertices in $LG_{(n,k,\alpha,\beta)}$ by adding an edge $(v_{i,j}, v_{i,j+1})$ for every $0 \leq i \leq k$ and $1 \leq j \leq n-1$. (We can prove a looser bound without the chaining technique.) This forces any sequence to pebble all vertices in the same order as the simple strategy.

**Corollary 1.** *Any sequence $\mathbf{P}$ that pebbles the chained stacked localized expanders $LG_{(n,k,\alpha,\beta)}$ starting from an empty initial configuration in $T(\mathbf{P}) \leq 2^{k_0} \alpha n$ steps has $M_{(\beta-3\alpha)n}(\mathbf{P}) \geq n(k - k_1)$ where $k_1 = k_0 + \lceil \log_2(\alpha n) \rceil$.*

*Proof.* Set $\eta = \beta - 3\alpha$. Theorem 5 shows that beyond the first $k_1$ layers, it is expensive to ever reduce space usage below $\eta n$. Doing so on layer $k > k_1$ would require at least $(k - k_1 + 1)\alpha n > \alpha n$ steps with $\eta n$ space usage to pebble the next $\alpha n$ vertices. The penalty keeps growing with the layer depth. So a better strategy is to maintain space usage higher than $\eta n$ for every step past layer $k_1$. There are at least $n(k - k_1)$ steps past layer $k_1$, and hence the theorem holds. $\square$

The simple strategy maintains $n$ space for $nk$ steps, i.e., the entire duration except for the first $n$ steps which fill memory. Corollary 1 is thus quite tight as $n(k - k_1)$ and $\eta n$ can be very close to $nk$ and $n$ with proper parameters.

## 4   Improved Analysis for Balloon Hash MHF

A memory-hard function (MHF) is a function $f$ that (i) takes a short input, and (ii) requires a specified amount of, say $N$, space to compute efficiently. To our knowledge, all MHF proposals first put the input through a hash function $\mathcal{H}$ so that $f(\mathcal{H}(\cdot))$ can take an input of any size, and $f(\cdot)$ only deals with a fixed input size $\lambda = |\mathcal{H}(\cdot)|$. $\lambda$ is considered short since it does not depend on $N$. There is no agreed upon criterion of memory-hardness. As discussed in Sect. 1.1, we adopt the exponential penalty definition.

**Definition 2 (MHF).** *Let $k$ be a security parameter, $N$ be the space requirement, and $N'$ be the provable space lower bound. A memory-hard function $y = f(x)$, parameterized $k$, $N$ and $N'$, has the following properties:*

   **(non-triviality)** *the input size $|x|$ does not depend on $N$,*
   **(efficiency)** *$f$ can be computed using $N$ space in $T = \mathrm{poly}(k, N)$ time,*
   **(memory-hardness)** *no algorithm can compute $f$ using less than $N'$ space in $2^k$ time with non-negligible probability.*

   The graph labelling problem on a hard-to-pebble graph immediately gives a MHF. Table 1 lists the running time $T$ and the provable space lower bound $N'$ for all existing MHFs with strict memory-hardness or exponential penalty (though the former two did not use the term MHF). All of them are based on graph pebbling. DKW [26] has perfect memory-hardness but requires a quadratic runtime. The other three have quasilinear runtime but large gaps in memory-hardness. The single-buffer version of Balloon hash (Balloon-SB) [20] used stacked localized expanders. Using the analysis in the Balloon hash paper, the space lower bound $N'$ for Balloon-SB is at most $N/4$ no matter how much we sacrifice runtime. Our improved analysis shows that Balloon-SB enjoys tighter space-hardness as well as consistent space-hardness. Theorem 4 shows that Balloon-SB with $T = dkN$ achieves $N' = \gamma N$, where $\gamma$ can be made arbitrarily small. The relation between $\gamma$ and $d$ is shown in Fig. 3. In addition, Corollary 1 gives a tight bound on consistent memory-hardness. This gives positive answers to two open questions left in the Balloon hash paper [20]: Balloon hash is space-hard over time and under batching.

**Table 1.** Comparison of MHFs with strict space-hardness or exponential penalty.

|       | DKW [26] | KK [31]           | Catena-DBG [28] | Balloon [20] | Balloon + our analysis |
|-------|----------|-------------------|-----------------|--------------|------------------------|
| $T$   | $N^2$    | $2N(\log_2 N)^2$  | $2kN \log_2 N$  | $7kN$        | $dkN$                  |
| $N'$  | $N$      | $N/32$            | $N/20$          | $N/8$        | $\gamma N$             |

KK [31] reported $T = \Theta(N \log_2 N)$ due to a miscalculation on the number of vertices. We generalized Catena-DBG [28] to exponential penalty though the designers of Catena recommended tiny security parameters like $k = 2$ for better efficiency.

# 5    Proof of Transient Space from Stacked Expanders

## 5.1    Definition

We use notation $(y_v, y_p) \leftarrow \langle \mathsf{V}(x_v), \mathsf{P}(x_p) \rangle$ to denote an interactive protocol between a verifier $\mathsf{V}$ and a prover $\mathsf{P}$. $x_v$, $x_p$, $y_v$, $y_p$ are $\mathsf{V}$'s input, $\mathsf{P}$'s input, $\mathsf{V}$'s output and $\mathsf{P}$'s output, respectively. We will omit $(x_v)$ or $(x_p)$ if a party does not take input. We will omit $y_p$ if $\mathsf{P}$ does not have output. For example, $\{0,1\} \leftarrow \langle \mathsf{V}, \mathsf{P} \rangle$ means neither $\mathsf{V}$ nor $\mathsf{P}$ takes input, and $\mathsf{V}$ outputs one bit indicating if it accepts (output 1) or rejects (output 0) $\mathsf{P}$'s proof. Both $\mathsf{P}$ and $\mathsf{V}$ can flip coins and have access to the same random oracle $\mathcal{H}$.

**Definition 3 (PoTS).** *Let $k$, $N$ and $N'$ be the same as in Definition 2. A proof of transient space is an interactive protocol $\{0,1\} \leftarrow \langle \mathsf{V}, \mathsf{P} \rangle$ that has the following properties:*

> **(succinctness)** *all messages between $\mathsf{P}$ and $\mathsf{V}$ have size $\mathsf{poly}(k, \log N)$,*
> **(efficient verifiability)** *$\mathsf{V}$ uses $\mathsf{poly}(k, \log N)$ space and time,*
> **(completeness)** *$\mathsf{P}$ uses $N$ space, runs in $\mathsf{poly}(k, N)$ time, and $\langle \mathsf{V}, \mathsf{P} \rangle = 1$,*
> **(space-hardness)** *there does not exist $\mathsf{A}$ that uses less than $N'$ space, runs in $2^k$ time, and makes $\langle \mathsf{V}, \mathsf{A} \rangle = 1$ with non-negligible probability.*

The definition above is due to Ateniese et al. [11]. Metrics for a PoTS include message size, prover runtime, verifier space/runtime, and the gap between $N$ and $N'$. The first three measure efficiency and the last one measures space-hardness. We also care about consistent space-hardness as defined in Sect. 3.5.

## 5.2    Construction

We adopt the Merkle commitment framework in Ateniese et al. [11] and Dziembowski et al. [24] to enable efficient verification. At a high level, the prover computes a Merkle commitment $C$ that commits the labels of all vertices in $LG_{(n,k,\alpha,\beta)}$ using the same random oracle $\mathcal{H}$. The verifier then checks if $C$ is "mostly correct" by asking the prover to open the labels of some vertices. The opening of label $h(v)$ is the path from the root to the leaf corresponding to $v$ in the Merkle tree. To compute a commitment $C$ that is "mostly correct", a prover cannot do much better than pebbling the graph following the rules, which we have shown to require a lot of space consistently. We say "a vertex" instead of "the label of a vertex" for short. For example, "commit/open a vertex" means "commit/open the label of a vertex".

Computing a Merkle tree can be modeled as a pebble game on a binary tree graph. It is not hard to see that a complete binary tree with $n$ leaves can be efficiently pebbled with roughly $\log_2 n$ space ($\lceil \log_2 n \rceil + 1$ to be precise) in $n$ moves. So $\mathsf{P}$ can compute the commitment $C$ using $N = n + \log_2 n + k \approx n$ space. The strategy is as follows: pebble $V_0$ using $n$ space, compute Merkle commitment $C_0$ for all vertices in $V_0$ using additional $\log_2 n$ space, discard the Merkle tree except the root, and then pebble $V_1$ rewriting $V_0$, compute $C_1$, discard the rest of the Merkle tree, and continue like this. Lastly, $C_1$ to $C_k$ are committed into a single Merkle root $C$.

After receiving $C$, V randomly selects $l_0$ vertices, and for each vertex $v$ asks P to open $v$, and all predecessors of $v$ if $v$ is not a source. Note that P did not store the entire Merkle tree but was constantly rewriting parts of it because the entire tree has size $nk \gg n$. So P has to pebble the graph for a second time to reconstruct the $l_0(d+1)$ paths/openings V asked for. This is a factor of 2 overhead in prover's runtime.

**Table 2.** Efficiency and space-hardness of PoTS.

|  | prover runtime | verifier space/time and message size | $N'$ |
|---|---|---|---|
| ABFG [11] | $12kN\log_2 N$ | $6\delta^{-1}k^2(\log_2 N)^2$ | $(\frac{1}{6} - \delta)N$ |
| This paper | $2(d+1)kN$ | $(d+1)\delta^{-1}k^2\log_2 N$ | $(\gamma - \delta)N$ |

Given the labels of all the predecessors of $v$ (or if $v$ is a source), V can check if $h(v)$ is correctly computed. If any opening or $h(v)$ is incorrect, V rejects. If no error is found, then $C$ is "mostly correct". We say a label $h(v_i)$ is a "fault" under $C$ if it is not correctly computed either as $h(i, x)$ or from $v_i$'s predecessors' labels under $C$. A cheating prover is motivated to create faults using pseudorandom values, because these faulty labels are essentially free pebbles that are always available but take no space. Dziembowski et al. [24] called them red pebbles and pointed out that a red pebble is no more useful than a free normal pebble because a normal pebble can be removed and later placed somewhere else. In other words, any sequence **P** that starts with $|P_0| = s_0$ initial pebbles and uses $m$ red pebbles and $s$ normal pebbles can be achieved by some sequence **P'** that starts with $|P_0'| = s_0 + m$ initial pebbles and uses 0 red pebbles and $s + m$ normal pebbles. We would like to bound the number of faults, which translate to a bounded loss in provable space-hardness.

If we want to lower bound the number of faults to $\delta n$ ($\delta < 1$) with overwhelming probability, we can set $l_0 = \frac{k|V|}{\delta n} = \delta^{-1}k^2$. Then, any commitment $C$ with $\delta n$ faults passes the probabilistic checking with at most $(1 - \frac{\delta n}{|V|})^{l_0} < e^{-k}$. Again, $k$ is our security parameter. With at most $\delta n$ faults, P needs to pebble at least $n - \delta n$ sinks ($> \alpha n$ with a proper $\delta$). By Theorem 4 and accounting for faults, a cheating prover needs at least $N' = (\gamma - \delta)n \approx (\gamma - \delta)N$ space to avoid exponential time.

### 5.3 Efficiency and Space-Hardness

Table 2 gives the efficiency and space-hardness of our construction, and compares with prior work using stacked butterfly superconcentrators [11]. Our prover runtime is $2(d+1)Nk$ where 2 is due to pebbling the graph twice, and $d+1$ is due to the in-degree of our graph plus hashing in Merkle tree. Message size includes

Merkle openings for the $l_0 = \delta^{-1}k^2$ challenges and their predecessors. The verifier has to check all these Merkle openings, so its space/time complexity are the same as message size. The efficiency of ABFG [11] can be calculated similarly using the fact that stacked butterfly superconcentrators have $2kN \log N$ vertices with in-degree 2. To match their space-hardness, which cannot be improved past $N' = \frac{1}{6}N$ with existing proof techniques, we only need in-degree $d = 9$. To match their efficiency, we set $d = 6 \log_2 N$, which we approximate as 150. That gives our construction very tight space-hardness at $N' = (\gamma - \delta)N$ with $\gamma = 0.85$. Furthermore, Corollary 1 gives a tight bound on consistent memory-hardness. Adjusting for faults, an adversary needs $n(k - k_1)$ steps whose space usage is at least $(\beta - 3\alpha - \delta)n$.

For simplicity, we used a single security parameter $k$. But in fact, the term $k^2$ in message size and verifier complexity should be $kk'$ where $k'$ is a statistical security parameter. $k'$ can be set independently from our graph depth $k$, which captures computational security. The same applies to the DFKP construction in Table 3.

# 6   Proof of Persistent Space from Stacked Expanders

## 6.1   Definition

**Definition 4 (PoPS).** *Let $k$ be a security parameter, $N$ be the space/advice requirement, $N_0'$ and $N_1'$ be two space lower bound parameters. A proof of persistent space is a pair of interactive protocols $(C, y) \leftarrow \langle V_0, P_0 \rangle$ and $\{0, 1\} \leftarrow \langle V_1(C), P_1(y) \rangle$ that have the following properties:*

**(succinctness)** *all messages between $P_0$ and $V_0$, and between $P_1$ and $V_1$ have size $\mathsf{poly}(k, \log N)$,*
**(efficient verifiability)** *$V_0$ and $V_1$ use $\mathsf{poly}(k, \log N)$ space and time,*
**(completeness)** *$P_0$ and $P_1$ satisfy the following*
  *– $P_0$ uses $N$ space, runs in $\mathsf{poly}(k, N)$ time, and outputs $y$ of size $N$,*
  *– $P_1$ uses $N$ space, runs in $\mathsf{poly}(k, \log N)$ time, and $\langle V_1(C), P_1(y') \rangle = 1$,*
**(space-hardness)** *there do not exist $A_0$ and $A_1$ such that*
  *– $A_0$ uses $\mathsf{poly}(k, N)$ space, runs in $\mathsf{poly}(k, N)$ time, and $\langle V_0, A_0 \rangle = (C, y')$ where $|y'| < N_0'$,*
  *– $A_1$ takes $y'$ as input, uses $N_1'$ space, runs in $2^k$ time, and makes $\langle V_1(C), A_1(y') \rangle = 1$ with non-negligible probability.*

$(C, y) \leftarrow \langle V_0, P_0 \rangle$ represents the setup stage. P outputs advice $y$ of size $N$, which is supposed to be stored persistently. V (through interaction with P) outputs a verified commitment $C$. $\{0, 1\} \leftarrow \langle V_1(C), P_1(y) \rangle$ represents one audit. The inputs of two parties are their respective output from the setup stage, and in the end V either accepts or rejects. It is implied that an audit $V_1$ has to use random challenges. Otherwise, it is easy to find $A_1$ that takes as input and also outputs the correct response to a fixed audit.

Efficiency metrics (message size, prover runtime, verifier space/runtime) are defined similarly to PoTS but now take into account both stages of the protocol.

The space-hardness definition and metric become a little tricky. Since the focus here is persistent space or advice size, one natural definition is to require that no polynomial adversary $A_1$ with advice size $|y'| < N_0'$ can convince V with non-negligible probability. Unfortunately, this ideal space-hardness definition is not achievable given the succinctness requirement, and we will describe an adversary who can violate it. In the setup phase, $A_0$ behaves in the same way as an honest $P_0$ except that it outputs the transcript (all the messages combined) between $A_0$ and $V_0$ as the cheating advice $y'$. Due to succinctness, the transcript size $|y'| = \text{poly}(k, \log N)$ is much shorter than any reasonable $N_0'$. In an audit, $A_1$ can rerun $P_0$ by simulating $V_0$, using the recorded transcript $y'$, to obtain the advice $y$ that $P_0$ would have produced, and then go on to run an honest $P_1$ to pass the audit.

Given the impossibility of ideal space-hardness, multiple alternative definitions have been proposed. Dziembowski et al. [24] gave two definitions. The first one requires $A_1$ to use $O(N)$ space. The second one requires $A_1$ to use $O(N)$ runtime, which is strictly weaker than the first one because $O(N)$ transient space implies $O(N)$ runtime. Proof of Space-Time [35] introduces a conversion rate between space and computation, and requires the sum of the two resources spent by $A_1$ to be within a constant factor of the sum spent by $P_0$. In this paper, we adopt the first definition of Dziembowski et al. [24] because it forces a prover to use space (either transient or persistent). In contrast, the latter two definitions explicitly allow a cheating prover to use tiny space and compensate with computation.

Under our space-hardness definition, if a cheating prover discards persistent advice in an attempt to save space, he/she will find himself/herself repeatedly refilling that space he/she attempts to save. If $(N_0', N_1')$ are close to $(N, N)$, a rational prover will choose to dedicate persistent space for the advice. We would like to be explicit that such a PoPS relies on a prover's cost of persistent space relative to computation and transient space, and very importantly the frequency of audits.

## 6.2   Construction

The setup phase is basically the PoTS protocol we presented in Sect. 5. P computes a Merkle commitment $C$, and V makes sure $C$ is "mostly correct" through a probabilistic check. At the end of the setup phase, an honest P stores the labels of all sinks $V_k$ and the Merkle subtree for $V_k$ as advice. Any vertices in $V_i$ for $i < k$ are no longer needed. V now can also discard $C$ and use $C_k$ which commits $V_k$ from this point onward. Since an honest P has to store the Merkle tree, it makes sense to use a different random oracle $\mathcal{H}_1$ with smaller output size for the Merkle commitment. If $|\mathcal{H}(\cdot)|$ is reasonably larger than $|\mathcal{H}_1(\cdot)|$, then the labels in the graph dominate, and the advice size is thus roughly $n$. Using the same random oracle results in an additional factor of 2 loss in space-hardness.

**Table 3.** Efficiency and space-hardness of PoPS.

| | | Prover runtime | Verifier space/runtime and message size | $N_0'\ N_1'$ |
|---|---|---|---|---|
| DFKP [24] | Setup | $3N$ | $3\delta^{-1}k(\log_2 N)^2$ | $(\frac{1}{3} \times \frac{1}{256 \times 25.3} - \delta)\frac{N}{\log_2 N}$ |
| | Audit | $0$ | $k \log_2 N$ | $(\frac{2}{3} \times \frac{1}{256 \times 25.3} - \delta)\frac{N}{\log_2 N}$ |
| This paper | Setup | $2(d+1)kN$ | $(d+1)\delta^{-1}k^2 \log_2 N$ | $(\frac{1}{3}\gamma - \delta)N$ |
| | Audit | $0$ | $k \log_2 N$ | $(\frac{2}{3}\gamma - \delta)N$ |

In the audit phase, V asks P to open $l_1$ randomly selected sinks. The binding property of the Merkle tree forces P to pebble these sinks, possibly with the help of at most $\delta n$ faults. But due to the red pebble argument, we can focus on the case with no faults first and account for faults later.

There is still one last step from Theorem 4 to what we need. Theorem 4 says any subset of $\alpha n$ initially unpebbled sinks are hard to pebble, but we would hope to challenge P on $l_1 \ll \alpha n$ sinks. Therefore, we need to show that a significant fraction of sinks are also hard to pebble individually.

**Theorem 6.** *Let $\gamma = \beta - 2\alpha$. Starting from any initial configuration $P_0$ of size $|P_0| \le \frac{1}{3}\gamma n$, less than $\alpha n$ initially unpebbled sinks of $G_{(n,k,\alpha,\beta)}$ can be pebbled individually using $\frac{2}{3}\gamma n$ space in $2^k$ moves.*

*Proof.* Suppose for contradiction that there are at least $\alpha n$ such sinks. Consider a strategy that pebbles these sinks one by one, never unpebbles $P_0$, and restarts from $P_0$ after pebbling each sink. This strategy pebbles a subset of $\alpha n$ initially unpebbled sinks, starting with $|P_0| < \frac{1}{3}\gamma n < \gamma n$, using at most $\frac{1}{3}\gamma n + \frac{2}{3}\gamma n = \gamma n$ pebbles in at most $2^k \alpha n$ moves. This contradicts Theorem 4. □

At most $\frac{1}{3}\gamma n$ pebbles may be initially pebbled in $P_0$, so no more than $(\frac{1}{3}\gamma + \alpha)n < \frac{1}{2}n$ individual sinks can be pebbled using $\frac{2}{3}\gamma n$ space in $2^k$ moves by Theorem 6. With more than half of the sinks being hard to pebble individually, we can set $l_1 = k$. The probability that no hard-to-pebble sink is included in the challenge is at most $2^{-k}$. Lastly, accounting for faults, no computationally bounded P using $N_0' = (\frac{1}{3}\gamma - \delta)n$ advice and $N_1' = (\frac{2}{3}\gamma - \delta)n$ space can pass an audit. The choice of constants $\frac{1}{3}$ and $\frac{2}{3}$ are arbitrary. The theorem holds for any pair of constants that sum to 1.

## 6.3  Efficiency and Space-Hardness

We compare with prior work [24] based on recursively stacked linear superconcentrators [38] in Table 3. The efficiency and (consistent) space-hardness of the setup phase are the same as our PoTS. In the audit phase, the prover sends Merkle openings for $k$ sinks to the verifier to check. If the prover stores less than $N_0' = (\frac{1}{3}\gamma - \delta)N$ advice, it needs at least $N_1' = (\frac{2}{3}\gamma - \delta)N$ space to pass an audit. This also sets a lower bound of $N_1' - N_0' = (\frac{1}{3}\gamma - \delta)N$ on prover's time to pass the audit, as it needs to fill its space to $N_1'$. Consistent space-hardness is not well defined for audits as an honest prover needs very little time to respond to audits.

The DFKP construction [24] treats the labels of all vertices as advice. This optimizes runtime but leaves a very large (even asymptotic) gap in space-hardness. It is possible for them to run audits only on the sinks, essentially generating less advice using the same graph and runtime. This will improve space-hardness up to $N_0' = N_1' = (\frac{1}{2} \times \frac{1}{256} - \delta)N$ while increasing runtime by a factor of $25.3 \log N$. There is currently no known way to remove the remaining gap of 512. We did not count the cost of including vertex ID in random oracle calls (for both our construction and theirs) since vertex ID is small compared to labels. This explains the different prover runtime we report in Table 3 compared to [24]. For completeness, we mention that the second construction of DFKP [24] and proof of Space-Time [35] are not directly comparable to our construction or the first DFKP construction because they provide very different efficiency and/or space-hardness guarantees.

## 7    Conclusion and Future Work

We derived tight space lower bounds for pebble games on stacked expanders, and showed that a lot of space is needed not just at some point, but throughout the pebbling sequence. These results gave MHF (Balloon hash) and PoTS with tight and consistent space-hardness. We also constructed a PoPS from stacked expanders with much better space-hardness than prior work.

While the space-hardness gap for Balloon hash and our PoTS can be made arbitrarily small, pushing it towards the limit would lead to very large constants for efficiency. How to further improve space-hardness for MHF and PoS remains interesting future work. It is also interesting to look for constructions that maintain consistent space-hardness under massive or even infinite parallelism.

At the moment, PoTS and PoPS are still far less efficient than PoW in terms of proof size and verifier complexity. A PoW is a single hash, while a PoS consists of hundreds (or more) of Merkle paths. The challenge remains in constructing practical PoTS/PoPS with tight and consistent space-hardness.

## References

1. Zoom Hash Scrypt ASIC. http://zoomhash.com/collections/asics. Accessed: 20 May 2016
2. Abadi, M., Burrows, M., Manasse, M., Wobber, T.: Moderately hard, memory-bound functions. ACM Trans. Internet Technol. 5(2), 299–327 (2005)
3. Almeida, L.C., Andrade, E.R., Barreto, P.S.L.M., Marcos, A., Simplicio Jr., M.A.: Lyra: password-based key derivation with tunable memory and processing costs. J. Crypt. Eng. 4(2), 75–89 (2014)
4. Alon, N., Capalbo, M.: Smaller explicit superconcentrators. In: Proceedings of the Fourteenth Annual ACM-SIAM Symposium on Discrete Algorithms, pp. 340–346. Society for Industrial and Applied Mathematics (2003)
5. Alwen, J., Blocki, J.: Efficiently computing data-independent memory-hard functions. Cryptology ePrint Archive, Report 2016/115 (2016)

6. Alwen, J., Blocki, J.: Towards practical attacks on argon2i and balloon hashing. Cryptology ePrint Archive, Report 2016/759 (2016)

7. Alwen, J., Chen, B., Kamath, C., Kolmogorov, V., Pietrzak, K., Tessaro, S.: On the complexity of scrypt and proofs of space in the parallel random oracle model. In: Fischlin, M., Coron, J.-S. (eds.) EUROCRYPT 2016. LNCS, vol. 9666, pp. 358–387. Springer, Heidelberg (2016). doi:10.1007/978-3-662-49896-5_13

8. Alwen, J., Serbinenko, V.: High parallel complexity graphs and memory-hard functions. In: Proceedings of the Forty-Seventh Annual ACM on Symposium on Theory of Computing, pp. 595–603. ACM (2015)

9. Andersen, D.G.: Exploiting time-memory tradeoffs in cuckoo cycle (2014). https://www.cs.cmu.edu/~dga/crypto/cuckoo/analysis.pdf. Accessed Aug 2016

10. Asanovic, K., Bodik, R., Catanzaro, B.C., Gebis, J.J., Husbands, P., Keutzer, K., Patterson, D.A., Plishker, W.L., Shalf, J., Williams, S.W.: The landscape of parallel computing research: a view from berkeley. Technical Report UCB/EECS-2006-183, EECS Department, University of California, Berkeley (2006)

11. Ateniese, G., Bonacina, I., Faonio, A., Galesi, N.: Proofs of space: when space is of the essence. In: Abdalla, M., De Prisco, R. (eds.) SCN 2014. LNCS, vol. 8642, pp. 538–557. Springer, Heidelberg (2014)

12. Ateniese, G., Burns, R., Curtmola, R., Herring, J., Kissner, L., Peterson, Z., Song, D.: Provable data possession at untrusted stores. In: Proceedings of the 14th ACM Conference on Computer and Communications Security, pp. 598–609. ACM (2007)

13. Back, A.: Hashcash-a denial of service counter-measure (2002)

14. Leonid Alexandrovich Bassalygo: Asymptotically optimal switching circuits. Problemy Peredachi Informatsii **17**(3), 81–88 (1981)

15. Biryukov, A., Dinu, D., Khovratovich, D.: Fast and tradeoff-resilient memory-hard functions for cryptocurrencies and password hashing (2015)

16. Biryukov, A., Khovratovich, D.: Tradeoff cryptanalysis of memory-hard functions. Cryptology ePrint Archive, Report 2015/227 (2015)

17. Biryukov, A., Khovratovich, D.: Equihash: asymmetric proof-of-work based on the generalized birthday problem. In: NDSS (2016)

18. Chung, F.R.K.: On concentrators, superconcentrators, generalizers, and nonblocking networks. Bell Syst. Techn. J. **58**(8), 1765–1777 (1979)

19. Cook, S.A.: An observation on time-storage trade off. In: Proceedings of the Fifth Annual ACM Symposium on Theory of Computing, pp. 29–33. ACM (1973)

20. Corrigan-Gibbs, H., Boneh, D., Schechter, S.: Balloon hashing: a provably memory-hard function with a data-independent access pattern. Cryptology ePrint Archive, Report 2016/027 (2016)

21. Dwork, C., Goldberg, A., Naor, M.: On memory-bound functions for fighting spam. In: Boneh, D. (ed.) CRYPTO 2003. LNCS, vol. 2729, pp. 426–444. Springer, Heidelberg (2003). doi:10.1007/978-3-540-45146-4_25

22. Dwork, C., Naor, M.: Pricing via processing or combatting junk mail. In: Brickell, E.F. (ed.) CRYPTO 1992. LNCS, vol. 740, pp. 139–147. Springer, Heidelberg (1993)

23. Dwork, C., Naor, M., Wee, H.M.: Pebbling and proofs of work. In: Shoup, V. (ed.) CRYPTO 2005. LNCS, vol. 3621, pp. 37–54. Springer, Heidelberg (2005)

24. Dziembowski, S., Faust, S., Kolmogorov, V., Pietrzak, K.: Proofs of space. In: Gennaro, R., Robshaw, M. (eds.) CRYPTO 2015. LNCS, vol. 9216, pp. 585–605. Springer, Heidelberg (2015)

25. Dziembowski, S., Kazana, T., Wichs, D.: Key-evolution schemes resilient to space-bounded leakage. In: Rogaway, P. (ed.) CRYPTO 2011. LNCS, vol. 6841, pp. 335–353. Springer, Heidelberg (2011)

26. Dziembowski, S., Kazana, T., Wichs, D.: One-time computable self-erasing functions. In: Ishai, Y. (ed.) TCC 2011. LNCS, vol. 6597, pp. 125–143. Springer, Heidelberg (2011)
27. Forler, C., List, E., Lucks, S., Wenzel, J.: Overview of the candidates for the password hashing competition (2015)
28. Forler, C., Lucks, S., Wenzel, J.: Catena: a memory-consuming password-scrambling framework. Cryptology ePrint Archive, Report 2013/525 (2013)
29. Hopcroft, J., Paul, W., Valiant, L.: On time versus space and related problems. In: 16th Annual Symposium on Foundations of Computer Science, pp. 57–64. IEEE (1975)
30. Juels, A., Kaliski Jr., B.S.: PORs: proofs of retrievability for large files. In: Proceedings of the 14th ACM Conference on Computer and Communications Security, pp. 584–597. ACM (2007)
31. Karvelas, N.P., Kiayias, A.: Efficient proofs of secure erasure. In: Abdalla, M., De Prisco, R. (eds.) SCN 2014. LNCS, vol. 8642, pp. 520–537. Springer, Heidelberg (2014)
32. Lengauer, T., Tarjan, R.E.: Asymptotically tight bounds on time-space trade-offs in a pebble game. J. ACM **29**(4), 1087–1130 (1982)
33. Lerner, S.D.: Strict memory hard hashing functions (preliminary v0. 3, 01-19-14)
34. Mahmoody, M., Moran, T., Vadhan, S.: Publicly verifiable proofs of sequential work. In: Proceedings of the 4th Conference on Innovations in Theoretical Computer Science, pp. 373–388. ACM (2013)
35. Moran, T., Orlov, I.: Proofs of space-time and rational proofs of storage. Cryptology ePrint Archive, Report 2016/035 (2016)
36. Nakamoto, S.: Bitcoin: a peer-to-peer electronic cash system (2008)
37. Paul, W.J., Tarjan, R.E.: Time-space trade-offs in a pebble game. Acta Informatica **10**(2), 111–115 (1978)
38. Paul, W.J., Tarjan, R.E., Celoni, J.R.: Space bounds for a game on graphs. Math. Syst. Theory **10**(1), 239–251 (1976)
39. Percival, C.: Stronger key derivation via sequential memory-hard functions (2009)
40. Perito, D., Tsudik, G.: Secure code update for embedded devices via proofs of secure erasure. In: Gritzalis, D., Preneel, B., Theoharidou, M. (eds.) ESORICS 2010. LNCS, vol. 6345, pp. 643–662. Springer, Heidelberg (2010)
41. Peslyak, A.: yescrypt - a password hashing competition submission (2014). https://password-hashing.net/submissions/specs/yescrypt-v2.pdf. Accessed Aug 2016
42. Pinsker, M.S.: On the complexity of a concentrator. In: 7th International Telegraffic Conference, vol. 4 (1973)
43. Robbins, H.: A remark on Stirling's formula. Am. Math. Monthly **62**(1), 26–29 (1955)
44. Schöning, U.: Better expanders and superconcentrators by Kolmogorov complexity. In: SIROCCO, pp. 138–150 (1997)
45. Schöning, U.: Smaller superconcentrators of density 28. Inf. Process. Lett. **98**(4), 127–129 (2006)
46. Sethi, R.: Complete register allocation problems. SIAM J. Comput. **4**(3), 226–248 (1975)
47. Smith, A., Zhang, Y.: Near-linear time, leakage-resilient key evolution schemes from expander graphs. Cryptology ePrint Archive, Report 2013/864 (2013)
48. Tromp, J.: Cuckoo cycle: a memory-hard proof-of-work system (2014)

# Perfectly Secure Message Transmission in Two Rounds

Gabriele Spini[1,2,3]([⊠]) and Gilles Zémor[1]

[1] Institut de Mathématiques de Bordeaux, UMR 5251, Université de Bordeaux,
351 Cours de la Libération, 33400 Talence, France
[2] Mathematical Institute, Leiden University, Leiden, The Netherlands
[3] CWI Amsterdam, Amsterdam, The Netherlands
spini@cwi.nl

**Abstract.** In the model that has become known as "Perfectly Secure Message Transmission" (PSMT), a sender Alice is connected to a receiver Bob through $n$ parallel two-way channels. A computationally unbounded adversary Eve controls $t$ of these channels, meaning she can acquire and alter any data that is transmitted over these channels. The sender Alice wishes to communicate a secret message to Bob *privately* and *reliably,* i.e. in such a way that Eve will not get any information about the message while Bob will be able to recover it completely.

In this paper, we focus on protocols that work in two transmission rounds for $n = 2t + 1$. We break from previous work by following a conceptually simpler blueprint for achieving a PSMT protocol. We reduce the previously best-known communication complexity, i.e. the number of transmitted bits necessary to communicate a 1-bit secret, from $O(n^3 \log n)$ to $O(n^2 \log n)$. Our protocol also answers a question raised by Kurosawa and Suzuki and hitherto left open: their protocol reaches optimal transmission rate for a secret of size $O(n^2 \log n)$ bits, and the authors raised the problem of lowering this threshold. The present solution does this for a secret of $O(n \log n)$ bits.

**Keyword:** Perfectly Secure Message Transmission

## 1 Introduction

The problem of Perfectly Secure Message Transmission (PSMT for short) was introduced by Dolev et al. in [2] and involves two parties, a sender Alice and a receiver Bob, who communicate over $n$ parallel channels in the presence of an adversary Eve. Eve is computationally unbounded and controls $t \leq n$ of the channels, meaning that she can read and overwrite any data sent over the channels under her control. The goal of PSMT is to design a protocol that allows

G. Spini—Supported by the Algant-Doc doctoral program, www.algant.eu.
G. Zémor—Supported by the "Investments for the Future" Programme IdEx Bordeaux – CPU (ANR-10-IDEX- 03-02).

M. Hirt and A. Smith (Eds.): TCC 2016-B, Part I, LNCS 9985, pp. 286–304, 2016.
DOI: 10.1007/978-3-662-53641-4_12

Alice to communicate a secret message to Bob *privately* and *reliably,* i.e. in such a way that Eve will not be able to acquire any information on the message, while Bob will always be able to completely recover it.

Two factors influence whether PSMT is possible and how difficult it is to achieve, namely the number $t$ of channels corrupted and controlled by Eve, and the number $r$ of transmission rounds, where a transmission round is a phase involving only one-way communication (either from Alice to Bob, or from Bob to Alice).

It was shown in Dolev et al.'s original paper [2] that for $r = 1$, i.e. when communication is only allowed from Alice to Bob, PSMT is possible if and only if $n \geq 3t + 1$. It was also shown in [2] that for $r \geq 2$, i.e. when communication can be performed in two or more rounds, PSMT is possible if and only if $n \geq 2t + 1$, although only a very inefficient way to do this was proposed. A number of subsequent efforts were made to improve PSMT protocols, notably in the most difficult case, namely for two rounds and when $n = 2t + 1$. The following two quantities, called *communication complexity* and *transmission rate,* were introduced and give a good measure of the efficiency of a PSMT protocol. They are defined as follows:

$$\text{Communication complexity} := \text{total number of bits transmitted to}$$
$$\text{communicate a single-bit secret,}$$

$$\text{Transmission rate} := \frac{\text{total number of bits transmitted}}{\text{bit-size of the secret}}.$$

Focusing exclusively on the case $n = 2t + 1$, Dolev et al. [2] presented a PSMT protocol for $r = 3$ with transmission rate $O(n^5)$: for $r = 2$ a protocol was presented with non-polynomial rate.

Sayeed and Abu-Amara [8] were the first to propose a two-round protocol with a polynomial transmission rate of $O(n^3)$. They also achieved communication complexity of $O(n^3 \log n)$. Further work by Agarwal et al. [1] improved the transmission rate to $O(n)$ meeting, up to a multiplicative constant, the lower bound of [10]. However, this involved exponential-time algorithms for the participants in the protocol. The current state-of-the art protocol is due to Kurosawa and Suzuki [4,5]. It achieves $O(n)$ transmission rate with a polynomial-time effort from the participants. All these protocols do not do better than $O(n^3 \log n)$ for the communication complexity.

We contribute to this topic in the following ways. We present a constructive protocol for which only polynomial-time, straightforward computations are required of the participants, that achieves the improved communication complexity of $O(n^2 \log n)$. In passing, we give an affirmative answer to an open problem of Kurosawa and Suzuki (at the end of their paper [5]) that asks whether it is possible to achieve the optimal transmission rate $O(n)$ for a secret of size less than $O(n^2 \log n)$ bits. We do this for a secret of $O(n \log n)$ bits.

Just as importantly, our solution is conceptually significantly simpler than previous protocols. Two-round PSMT involves Bob initiating the protocol by first sending an array of symbols $(x_{ij})$ over the $n$ parallel channels, where the first index $i$ means that symbol $x_{ij}$ is sent over the $i$-th channel. All previous proposals relied on arrays $(x_{ij})$ with a lot of structure, with linear relations between symbols that run both along horizontal (constant $j$) and vertical (constant $i$) lines. In contrast, we work with an array $(x_{ij})$ consisting of completely independent rows $\mathbf{x}^{(j)} = (x_{1j}, x_{2j}, \ldots, x_{nj})$ that are simply randomly chosen words of a given Reed-Solomon code. In its simplest, non-optimized, form, the PSMT protocol we present only involves simple syndrome computations from Alice, and one-time padding the secrets it wishes to transfer with the image of linear forms applied to corrupted versions of the codewords $\mathbf{x}^{(j)}$ it has received from Bob. Arguably, the method could find its way into textbooks as relatively straightforward applications of either secret-sharing or wiretap coset-coding techniques. In its optimized form, the protocol retains sufficient simplicity to achieve a transmission rate $5n + o(n)$, compared to the previous record of $6n + o(n)$ of [3] obtained by painstakingly optimizing the $25n + o(n)$ transmission rate of [5].

In the next Section we give an overview of our method and techniques.

## 2    Protocol Overview

The procedure takes as input the number $n = 2t + 1$ of channels between Alice and Bob and the number $\ell$ of secret messages to be communicated; we assume that the messages lie in a finite field $\mathbb{F}_q$. First, a code $\mathcal{C}$ that will be the basic communication tool is selected; $\mathcal{C}$ is a linear block code of length $n$ over $\mathbb{F}_q$, dimension $t + 1$ and minimum distance $t + 1$. It furthermore has the property that the knowledge of $t$ symbols of any of its codewords $\mathbf{x}$ leaves $\mathbf{h}\mathbf{x}^T$ completely undetermined, where $\mathbf{h}$ is a vector produced together with $\mathcal{C}$ at the beginning of the protocol. The code $\mathcal{C}$ can be a Reed-Solomon code.

Since we require at most two rounds of communication, Bob starts the procedure; he chooses a certain number of random and independent codewords $\mathbf{x}$, and communicates them by sending the $i$-th symbol of each codeword over the $i$-th channel. This is a first major difference from previous papers, notably [5], where codewords are communicated in a more complicated "horizontal-and-vertical" fashion; our construction is thus conceptually simpler and eliminates techniques introduced by early papers [8] which marked substancial progress at the time but also hindered the development of more efficient protocols when they survived in subsequent work.

As a result of this first round of communication, Alice receives a corrupted version $\mathbf{y} = \mathbf{x} + \mathbf{e}$ for each codeword $\mathbf{x}$ sent by Bob. As in previous PSMT protocols, Alice then proceeds by broadcast, meaning every symbol she physically sends to Bob, she sends $n$ times, once over every channel $i$. In this way privacy is sacrificed, since Eve can read everything Alice sends, but reliability is ensured, since Bob recovers every transmitted symbol by majority decoding.

A secret message consisting of a single symbol $s \in \mathbb{F}_q$ is encoded by Alice as $s + \mathbf{h}\mathbf{y}^T$ for some received vector $\mathbf{y}$. In other words, $s$ is one-time padded with the quantity $\mathbf{h}\mathbf{y}^T$ and this is broadcast to Bob. Notice that at this point, revealing $s + \mathbf{h}\mathbf{y}^T$ to Eve gives her zero information on $s$. This is because she can have intercepted at most $t$ symbols of the codeword $\mathbf{x}$: therefore the element $\mathbf{hx}^T$ is completely unknown to her by the above property of $\mathcal{C}$ and $\mathbf{h}$, and the mask $\mathbf{hy}^T = \mathbf{hx}^T + \mathbf{he}^T$ is unknown to her as well.

Now broadcasting the quantity $s + \mathbf{hy}^T$ is not enough by itself to convey the secret $s$ to Bob, because Bob also does not have enough information to recover the mask $\mathbf{hy}^T$. To make the protocol work, Alice needs to give Bob extra information that tells Eve nothing she doesn't already know.

This extra information comes in two parts. The first part is simply the syndrome $\sigma(\mathbf{y}) = \mathbf{Hy}^T$ of $\mathbf{y}$, where $\mathbf{H}$ is a parity-check matrix of $\mathcal{C}$; notice that this data is indeed useless to Eve, who already knows it given that $\mathbf{Hy}^T = \mathbf{Hx}^T + \mathbf{He}^T = \mathbf{He}^T$ where $\mathbf{e}$ is chosen by herself.

The second part makes use of the fact that during the first phase, Bob has not sent a single codeword $\mathbf{x}$ to Alice, but a batch of codewords $\mathcal{X}$ and Alice has received a set $\mathcal{Y}$ of vectors made up of the corrupted versions $\mathbf{y} = \mathbf{x} + \mathbf{e}$ of the codewords $\mathbf{x}$. Alice will sacrifice a chosen subset of these vectors $\mathbf{y}$ and reveal them completely to Bob and Eve by broadcast. Note that this does not yield any information on the unrevealed vectors $\mathbf{y}$ since Bob has chosen the codewords $\mathbf{x}$ of $\mathcal{X}$ randomly and independently. At this point we apply an idea that originates in [5]: the chosen revealed subset of $\mathcal{Y}$ is called in [5] a *pseudo-basis* of $\mathcal{Y}$. To compute a pseudo-basis of $\mathcal{Y}$, Alice simply computes all syndromes $\sigma(\mathbf{y})$ for $\mathbf{y} \in \mathcal{Y}$, and choses a minimal subset of $\mathcal{Y}$ whose syndromes generate linearly all syndromes $\sigma(\mathbf{y})$ for $\mathbf{y} \in \mathcal{Y}$. A pseudo-basis of $\mathcal{Y}$ could alternatively be called a *syndrome-spanning* subset of $\mathcal{Y}$. Now elementary coding-theory arguments imply that the syndrome function $\sigma$ is injective on the subspace generated by the set of *all* errors $\mathbf{e}$ that Eve applies to all Bob's codewords $\mathbf{x}$ (Lemma 2 and Proposition 1). Therefore a pseudo-basis of $\mathcal{Y}$ gives Bob access to the whole space spanned by Eve's errors and allows him, *for any non-revealed* $\mathbf{y} = \mathbf{x} + \mathbf{e}$, to recover the error $\mathbf{e}$ from the syndrome $\sigma(\mathbf{y}) = \sigma(\mathbf{e})$.

The above protocol is arguably "the right way" of exploiting the pseudo-basis idea of Kurosawa and Suzuki, by which we mean it is the simplest way of turning it into a two-round PSMT protocol. We shall present optimised variants that achieve the communication complexity and transmission rate claimed in the Introduction. Our final protocol involves two additional ideas; the first involves a more efficient broadcasting scheme than pure repetition: this idea was also used by Kurosawa and Suzuki. The second idea is new and involves using a decoding algorithm for the code $\mathcal{C}$.

The rest of the paper is organised as follows. In Sect. 3 we recall the coding theory that we need to set up the protocol. In particular, Sect. 3.1 introduces the code $\mathcal{C}$ and the vector $\mathbf{h}$ with the desired properties. Section 3.2 introduces Kurosawa and Suzuki's pseudo-basis idea, though we depart somewhat from their original description to fit our syndrome-coding approach to PSMT.

In Sect. 4 we describe in a formal way the protocol sketched above, and we compute its communication cost; it will turn out that this construction has a communication complexity of $O\left(n^3 \log n\right)$ and a transmission rate of $O\left(n^2\right)$.

Section 5 is devoted to improving the efficiency of the protocol; specifically, Sect. 5.1 introduces generalized broadcast, Sects. 5.2 and 5.3 show how to lower the cost of transmitting the pseudo-basis, while Sect. 5.4 presents a way to improve the efficiency of the last part of the protocol. A key aspect of this section is that Alice must make extensive use of a *decoding* algorithm for linear codes, a new feature compared to previous work on the topic.

Finally, in Sect. 6 we implement these improvements and compute the cost of the resulting protocol, reaching a communication complexity of $O\left(n^2 \log n\right)$ and a transfer rate of $5n + o(n)$; we also show in this section that optimal transfer rate is achieved for a secret of $O(n \log n)$ bits. Section 7 gives concluding remarks.

# 3    Setting and Techniques

## 3.1    Error-Correcting Codes for Communication

We will use the language of Coding Theory, for background, see e.g. [6]. Let us briefly recall that when a linear code over the finite field $\mathbb{F}_q$ is defined as $\mathcal{C} = \{\mathbf{x} \in \mathbb{F}_q^n, \mathbf{H}\mathbf{x}^T = 0\}$, the $r \times n$ matrix $\mathbf{H}$ is called a *parity-check matrix* for $\mathcal{C}$ and the mapping

$$\sigma : \mathbb{F}_q^n \to \mathbb{F}_q^r$$
$$\mathbf{x} \mapsto \mathbf{H}\mathbf{x}^T$$

is referred to as the *syndrome map*. Recall also that a code of parameters (length, dimension, minimum Hamming distance) $[n, k, d]$ is said to be Maximum Distance Separable or *MDS*, if $d + k = n + 1$. Particular instances of MDS codes are Reed-Solomon codes, which exist whenever the field size $q$ is equal to or larger than the length $n$. In a secret-sharing context, Reed-Solomon codes are equivalent to Shamir's secret-sharing scheme [9], and they have been used extensively to construct PSMT protocols. We could work from the start with Reed-Solomon codes, equivalently Shamir's scheme, but prefer to use more general MDS codes, not purely for generality's sake, but to stay unencumbered by polynomial evaluations and to highlight that we have no need for anything other than Hamming distance properties. In Sect. 6, we will need our MDS codes to come with a decoding algorithm and will have to invoke Reed-Solomon codes specifically: we will only need to know of the existence of a polynomial-time algorithm though, and will not require knowledge of any specifics.

We will need an MDS code $\mathcal{C}$ that will be used to share randomness, together with a vector $\mathbf{h}$ such that the value of $\mathbf{h}\mathbf{x}^T$ is completely undetermined for a codeword $\mathbf{x} \in \mathcal{C}$ even when $t$ symbols of $\mathbf{x}$ are known. The linear combination given by $\mathbf{h}$ will then be used to create the masks that hide the secrets.

The following Lemma states the existence of such a pair $(\mathcal{C}, \mathbf{h})$: it is a slightly non-standard use of Massey's secret sharing scheme [7]. It is implicit that we suppose $q > n$, so that MDS codes exist for all dimensions and length up to $n + 1$.

**Lemma 1.** *For any $n$ and any $t < n$ there exists an MDS code $\mathcal{C}$ of parameters $[n, t+1, n-t]$ and a vector $\mathbf{h} \in \mathbb{F}_q^n$ such that given a random codeword $\mathbf{x} \in \mathcal{C}$, the scalar product $\mathbf{h}\mathbf{x}^T$ is completely undetermined even when $t$ symbols of $\mathbf{x}$ are known.*

*Proof.* Let $\mathcal{C}'$ be an MDS code of parameters $[n+1, t+1, n-t+1]$; notice that such a code exists for any $n$ and $t \le n$ [6]. Let $\mathcal{C}$ be the code obtained from $\mathcal{C}'$ by puncturing at its last coordinate, i.e.

$$\mathcal{C} := \left\{ \mathbf{x} \in \mathbb{F}_q^n : \exists x \in \mathbb{F}_q \text{ with } (\mathbf{x}, x) \in \mathcal{C}' \right\}$$

The minimum distance of $\mathcal{C}$ is at most one less than that of $\mathcal{C}'$, and $\mathcal{C}$ is MDS of parameters $[n, t+1, n-t]$ as requested. Now let $\mathbf{H}'$ be a parity-check matrix of $\mathcal{C}'$; since $\mathcal{C}'$ has minimum distance $n-t+1 > 1$, there is at least one row of $\mathbf{H}'$ whose last symbol is non-zero, i.e. such a row is of the form

$$(\mathbf{h}, \alpha) \in \mathbb{F}_q^{n+1} \text{ with } \mathbf{h} \in \mathbb{F}_q^n, 0 \ne \alpha \in \mathbb{F}_q.$$

We claim that the pair $(\mathcal{C}, \mathbf{h})$ is of the desired type: indeed, let $\mathbf{x}$ be a random codeword of $\mathcal{C}$. Then there exists a (unique) codeword $\mathbf{x}'$ of $\mathcal{C}'$ such that $\mathbf{x}' = (\mathbf{x}, x)$; now $\mathbf{h}\mathbf{x}^T = -\alpha x$, i.e. the knowledge of $\mathbf{h}\mathbf{x}^T$ is equivalent to the knowledge of $x$, given that $\alpha$ is non-zero.

Now since $\mathcal{C}'$ has dimension $t+1$, for any $t$ known symbols $x_{i_1}, \cdots, x_{i_t}$ of $\mathbf{x}$ and any $\tilde{x} \in \mathbb{F}_q$, there exists exactly one $\tilde{\mathbf{x}}' \in \mathcal{C}'$ such that $\tilde{\mathbf{x}}'_{i_j} = x_{i_j}$ for any $j$ and such that $\tilde{\mathbf{x}}'_{n+1} = \tilde{x}$. Hence the claim holds. $\qquad\square$

As stated above, this lemma will guarantee the privacy of our protocols; conversely, we can achieve reliable (although not private) communication via the following remark: Alice and Bob can *broadcast* a symbol by sending it over all the channels; since Eve only controls $t < n/2$ of them, the receiver will be able to correct any error introduced by Eve with a simple majority choice. Broadcast thus guarantees reliability by sacrificing privacy.

### 3.2 Pseudo-Bases or Syndrome-Spanning Subsets

The second fundamental building block of our paper is the notion of *pseudo-basis*, introduced by Kurosawa and Suzuki [5]. The concept stems from the following intuition: assume that Bob communicates a single codeword $\mathbf{x}$ of an MDS code $\mathcal{C}$ to Alice by sending each of its $n$ symbols over the corresponding channel. Eve intercepts $t$ of these symbols, thus $\mathcal{C}$ must have dimension at least $t+1$ if we want to prevent her from learning $\mathbf{x}$; but this means that the minimum distance of $\mathcal{C}$ cannot exceed $n+1-(t+1) = t+1$, which is not enough for Alice to correct an arbitrary pattern of up to $t$ errors that Eve can introduce.

If, however, we repeat the process for several different $\mathbf{x}^{(i)}$, then Alice and Bob have an important advantage: they know that all the errors introduced by Eve *always lie in the same subset of $t$ coordinates*. Kurosawa and Suzuki propose the following strategy to exploit this knowledge: Alice can compute a pseudo-basis (a subset with special properties) of the received vectors; she can then transmit it to Bob, who will use this special structure of the errors to determine their support.

The key is the following simple lemma:

**Lemma 2.** *Let $C$ be a linear code of parameters $[n, k, d]_q$, and let $\mathbf{H}$ be a parity-check matrix of $C$; let $E$ be a linear subspace of vectors of $\mathbb{F}_q^n$ such that the Hamming weight $w_H(\mathbf{e})$ of $\mathbf{e}$ satisfies $w_H(\mathbf{e}) < d$ for any $\mathbf{e} \in E$.*

*We then have that the following map is injective:*

$$\sigma_{|E} : E \to \mathbb{F}_q^{n-k}$$

$$\mathbf{e} \mapsto \mathbf{H}\mathbf{e}^T$$

*Proof.* Simply notice that $\ker\left(\sigma_{|E}\right) = \{\mathbf{0}\}$: indeed, $\ker\left(\sigma_{|E}\right) \subseteq C$; but by assumption all elements of $E$ have weight smaller than $d$, so that $\ker\left(\sigma_{|E}\right) = \{\mathbf{0}\}$. □

We can now introduce the concept of pseudo-basis; for the rest of this section, we assume that a linear code $C$ of parameters $[n, k, d]_q$ has been chosen, together with a parity-check matrix $\mathbf{H}$ and associated syndrome map $\sigma$.

**Definition 1 (Pseudo-Basis [5]).** *Let $\mathcal{Y}$ be a set of vectors of $\mathbb{F}_q^n$; a pseudo-basis of $\mathcal{Y}$ is a subset $\mathcal{W} \subseteq \mathcal{Y}$ such that $\sigma(\mathcal{W})$ is a basis of the syndrome subspace $\langle \sigma(\mathcal{Y}) \rangle$.*

*Notice that a pseudo-basis has thus cardinality at most $n - k$, and that it can be computed in time polynomial in $n$.*

The following property formalizes the data that Bob can acquire after he obtains a pseudo-basis of the words received by Alice:

**Proposition 1 ([5]).** *Let $\mathcal{X}, \mathcal{E}, \mathcal{Y}$ be three subsets:*

$$\mathcal{X} := \left\{\mathbf{x}^{(1)}, \cdots, \mathbf{x}^{(r)}\right\} \subseteq C,$$

$$\mathcal{E} := \left\{\mathbf{e}^{(1)}, \cdots, \mathbf{e}^{(r)}\right\} \subseteq \mathbb{F}_q^n \quad \text{such that } \#\bigcup\left(support\left(\mathbf{e}^{(j)}\right) : j = 1, \cdots, r\right) < d,$$

$$\mathcal{Y} := \left\{\mathbf{y}^{(1)}, \cdots, \mathbf{y}^{(r)}\right\} \subseteq \mathbb{F}_q^n \quad \text{with } \mathbf{y}^{(j)} = \mathbf{x}^{(j)} + \mathbf{e}^{(j)} \text{ for every } j$$

*Then, given knowledge of $\mathcal{X}$ and a pseudo-basis of $\mathcal{Y}$, we can compute $\mathbf{e}^{(j)}$ from its syndrome $\sigma(\mathbf{e}^{(j)})$, for any $1 \le j \le r$.*

*Proof.* The hypothesis on the supports of the elements of $\mathcal{E}$ implies that the subspace $E = \langle \mathcal{E} \rangle$ satisfies the hypothesis of Lemma 2 and the syndrome function is therefore injective on $\langle \mathcal{E} \rangle$. Given the pseudo-basis $\left\{\mathbf{y}^{(i)} : i \in I\right\}$, we can decompose any syndrome $\sigma(\mathbf{e}^{(j)})$ as

$$\sigma(\mathbf{e}^{(j)}) = \sum_{i \in I} \lambda_i \sigma(\mathbf{y}^{(i)}) = \sum_{i \in I} \lambda_i \sigma(\mathbf{e}^{(i)})$$

$$= \sigma \left( \sum_{i \in I} \lambda_i \mathbf{e}^{(i)} \right)$$

which yields

$$\mathbf{e}^{(j)} = \sum_{i \in I} \lambda_i \mathbf{e}^{(i)}$$

by injectivity of $\sigma$ on $E$. □

*Remark 1.* Since the syndrome map induces a one-to-one mapping from $E$ to $\sigma(E)$, we also have that $\{\mathbf{y}^{(i)} : i \in I\}$ is a pseudo-basis of $\mathcal{Y}$ if and only if $\{\mathbf{e}^{(i)} : i \in I\}$ is a basis of $E = \langle \mathcal{E} \rangle$.

The reader should now have a clear picture of how the pseudo-basis will be used to obtain shared randomness: Bob will select a few codewords $\mathbf{x}^{(1)}, \cdots, \mathbf{x}^{(r)}$ in an MDS code of distance at least $t + 1$, then communicate them to Alice by sending the $i$-th symbol of each codeword over channel $i$; Alice will be able to compute a pseudo-basis of the received words, a clearly non-expensive computation, then communicate it to Bob. Bob will then be able to determine any error introduced by Eve just from its syndrome as just showed in Proposition 1.

The following section gives all the details.

## 4    A First Protocol

We now present the complete version of our first communication protocol, following the blueprint of Sect. 2.

---

**Protocol 1.** *The protocol allows Alice to communicate $\ell$ secret elements $s^{(1)}, \cdots, s^{(\ell)}$ of $\mathbb{F}_q$ to Bob, where $q$ is an arbitrary integer with $q > n$. The protocol takes as input an MDS code $C$ of parameters $[n, t+1, t+1]_q$ and a vector $\mathbf{h}$ of length $n$ as in Lemma 1.*

I. *Bob chooses $t + \ell$ uniformly random and independent codewords $\mathbf{x}^{(1)}, \cdots, \mathbf{x}^{(t+\ell)}$ of $C$ and communicates them to Alice by sending the $i$-th symbol of each codeword over the $i$-th channel.*

II. *Alice receives the corrupted versions $\mathbf{y}^{(1)} = \mathbf{x}^{(1)} + \mathbf{e}^{(1)}, \cdots, \mathbf{y}^{(t+\ell)} = \mathbf{x}^{(t+\ell)} + \mathbf{e}^{(t+\ell)}$; she then proceeds with the following actions:*

    *(i) She computes a pseudo-basis $\{\mathbf{y}^{(i)} : i \in I\}$ for $I \subset \{1, \cdots, t+\ell\}$ of the received values and broadcasts to Bob $\left(i, \mathbf{y}^{(i)} : i \in I\right)$.*

---

> (ii) She then considers the first $\ell$ words that do not belong to the pseudo-basis; to ease the notation, we will re-name them $\mathbf{y}^{(1)}, \cdots, \mathbf{y}^{(\ell)}$. For each secret $s^{(j)}$ to be communicated she broadcasts to Bob the following two elements:
>   – $\mathbf{H}\left(\mathbf{y}^{(j)}\right)^{T}$, the syndrome of $\mathbf{y}^{(j)}$;
>   – $s^{(j)} + \mathbf{h}\left(\mathbf{y}^{(j)}\right)^{T}$.
>
> III. Proposition 1 guarantees that for any $j$, $1 \leq j \leq \ell$, Bob can compute the error vector $\mathbf{e}^{(j)}$ and hence reconstruct $\mathbf{y}^{(j)} = \mathbf{x}^{(j)} + \mathbf{e}^{(j)}$ from his knowledge of $\mathbf{x}^{(j)}$. He can therefore open the mask $\mathbf{h}\left(\mathbf{y}^{(j)}\right)^{T}$ and obtain the secret $s^{(j)}$.

**Proposition 2.** *The above protocol allows for private and reliable communication of $\ell$ elements of $\mathbb{F}_q$.*

*Proof.* As a first remark, notice that since the pseudo-basis has cardinality at most $t$ as remarked in Definition 1, Alice has enough words to mask her $\ell$ secret messages, since the total number of words is equal to $t + \ell$. We can now prove that the protocol is private and reliable:

– *Privacy:* Eve can intercept at most $t$ coordinates of each codeword sent over the channels in the first step; the codewords corresponding to the pseudo-basis are revealed in step II-(i), but this information is useless since the words are chosen independently and those belonging to the pseudo-basis are no longer used. For any $\mathbf{y}^{(j)}$ that does not belong to the pseudo-basis, the syndrome $\mathbf{H}\left(\mathbf{y}^{(j)}\right)^{T}$ is also transmitted, but Eve already knows it since $\mathbf{H}\left(\mathbf{y}^{(j)}\right)^{T} = \mathbf{H}\left(\mathbf{x}^{(j)} + \mathbf{e}^{(j)}\right)^{T} = \mathbf{H}\left(\mathbf{e}^{(j)}\right)^{T}$, where $\mathbf{e}^{(j)}$ denotes the error she introduced herself on $\mathbf{x}^{(j)}$.

   Hence thanks to Lemma 1, Eve has no information on any $\mathbf{h}\left(\mathbf{y}^{(j)}\right)^{T}$, so that privacy holds.
– *Reliability:* Eve can disrupt the communication only at step I, since all the following ones only use broadcasts. Proposition 1 then ensures that Bob can recover the vectors $\mathbf{y}^{(j)}$ from their syndromes and the corresponding codeword $\mathbf{x}^{(j)}$. From there he can compute and remove the mask $\mathbf{h}\left(\mathbf{y}^{(j)}\right)^{T}$ without error. □

We now compute the communication complexity and transmission rate of this first protocol, underlining the most expensive parts:

*Communication complexity:* we can set $\ell := 1$.

– Step I requires transmitting $t + 1$ codewords over the channels, thus requiring a total of $O\left(n^2\right)$ symbols to be transmitted.
– Step II-(i) requires broadcasting up to $t$ words of $\mathbb{F}_q^n$, thus giving a total of $O\left(n^3\right)$ symbols to be transmitted.

- Finally, step II-(ii) requires broadcasting a total of $t+1$ symbols (a size-$t$ syndrome and the masked secret), thus giving a total of $O\left(n^2\right)$ elements to be transmitted.

Hence since we can assume that $q = O(n)$, we get a total communication complexity of

$$O\left(n^3 \log n\right)$$

bits to be transmitted to communicate a single-bit secret.

*Tranfer rate:* optimal rate is achieved for $\ell = \Omega\left(n\right)$.

- Step I requires transmitting $t+\ell$ codewords, for a total of $O\left(n^2 + n\ell\right)$ symbols.
- Step II-(i) remains unchanged from the single-bit case, and thus requires transmitting $O\left(n^3\right)$ symbols.
- Finally, step II-(ii) requires broadcasting a total of $\ell(t+1)$ symbols ($\ell$ size-$t$ syndromes and the masked secrets), thus giving a total of $O\left(n^2\ell\right)$ symbols;

To sum up, the overall transmission rate is equal to

$$\frac{O\left(n^2 + n\ell + n^3 + n^2\ell\right)}{\ell} = O\left(n^2\right).$$

It is immediately seen that the main bottleneck for communication complexity is step II-(i), i.e. the communication of the pseudo-basis, while for transmission rate it is step II-(ii), i.e. the communication of the masked secrets and of the syndromes. We address these issues in the following sections.

# 5   Improvements to the Protocol

We discuss in this section some key improvements to the protocol; Sect. 5.1 presents the key technique of generalized broadcast, Sects. 5.2 and 5.3 show a new way to communicate the pseudo-basis (the main bottleneck for communication complexity) and Sect. 5.4 a new way to communicate the masked secret and the information to open the masks (bottleneck for transmission rate).

## 5.1   Generalized Broadcast

Our improvements on the two bottlenecks showed in Sect. 4 rely on the fundamental technique of generalized broadcast, which has been highlighted in the paper by Kurosawa and Suzuki [5].

The intuition is the following: we want to choose a suitable code $\mathcal{C}_{\text{BCAST}}$ for perfectly reliable transmission, i.e. we require that if any word $\mathbf{x} \in \mathcal{C}_{\text{BCAST}}$ is communicated by sending each symbol $\mathbf{x}_i$ over the $i$-th channel, then $\mathbf{x}$ can always be recovered in spite of the errors introduced by the adversary. In the general situation, since Eve can introduce up to $t$ errors, $\mathcal{C}_{\text{BCAST}}$ must have minimum distance $2t + 1 = n$, and hence dimension 1; for instance, $\mathcal{C}_{\text{BCAST}}$ can be a repetition code, yielding the broadcast protocol of Sect. 3.1.

Now assume that at a certain point of the protocol, Bob gets to know the position of $m$ channels under Eve's control; then the communication system between the two has been improved: instead of $n$ channels with $t$ errors, we have $n$ channels with $m$ erasures and $t - m$ errors (since Bob can ignore the symbols received on the $m$ channels under Eve's control that he has identified). We can thus expect that reliable communication between Alice and Bob (i.e., broadcast) can be performed at a lower cost by using a code with smaller distance and greater dimension; the following lemma formalizes this intuition.

**Lemma 3 (Generalized Broadcast).** *Let $m \leq t$ and let $\mathcal{C}_m$ be an MDS code of parameters $[n, m+1, n-m]_q$; assume that Bob knows the location of $m$ channels controlled by Eve. Then Alice can communicate with perfect reliability $m+1$ symbols $x_1, \cdots, x_{m+1}$ of $\mathbb{F}_q$ to Bob in the following way: she first takes the codeword $\mathbf{c} \in \mathcal{C}_m$ which encodes $(x_1, \cdots, x_{m+1})$, then sends each symbol of $\mathbf{c}$ through the corresponding channel; Eve cannot prevent Bob from completely recovering the message.*

*We refer to this procedure as $m$-generalized broadcast.*

*Proof.* Notice that $\mathbf{c}$ is well-defined since $\mathcal{C}_m$ has dimension $m+1$. Now since Bob knows the location of $m$ channels that are under Eve's control, he can replace the symbols of $\mathbf{c}$ received via these channels with erasure marks $\perp$, and consider the truncated codeword $\tilde{\mathbf{c}}$ lacking these symbols. Now $\tilde{\mathbf{c}}$ belongs to the punctured code obtained from $\mathcal{C}_m$ by removing $m$ coordinates, which has minimum distance $(n - m) - m \geq 2(t - m) + 1$; it can thus correct up to $t - m$ errors, which is exactly the maximum number of errors that Eve can introduce (since she controls at most $t - m$ of the remaining channels). Once he has obtained the shortened codeword $\tilde{\mathbf{c}}$, he can then recover the complete one since $\mathcal{C}_m$ can correct from $m$ erasures, given that it has minimum distance $n - m \geq m$. $\square$

Hence if Alice knows that Bob has identified at least $m$ channels under Eve's control, she can divide the cost of a broadcast by a factor $m$ (since the above method requires to transmit $n$ symbols of $\mathbb{F}_q$ to communicate $m + 1$ symbols of $\mathbb{F}_q$).

In the following sections we will make use of Lemma 3 to improve the efficiency of the protocol.

## 5.2    Improved Transmission of the Pseudo-Basis: A Warm-Up

We present here a new method of communicating the pseudo-basis, which is a straightforward implementation of the generalized broadcasting technique.

The key point is the following observation:

**Lemma 4.** *Let $\mathcal{W} = (\mathbf{y}^{(i)} : i \in I)$ be a pseudo-basis of the set of received vectors; then if Bob knows $m$ elements of $\mathcal{W}$, he knows at least $m$ channels that have been forged by Eve.*

*Proof.* By subtracting the original codeword from an element of the pseudo-basis, Bob knows the corresponding error; furthermore, these errors form a basis of the entire error space (Remark 1). Now if Bob knows $m$ elements of the pseudo-basis, he then knows $m$ of these errors, which necessarily affect at least $m$ coordinates since they are linearly independent. The claim then follows.    □

The sub-protocol consisting of the transmission of the pseudo-basis by Alice is simply the following:

---

**Protocol 2.** *Alice wishes to communicate to Bob a pseudo-basis $\mathcal{W}$ of cardinality $w$.*

*For any $i = 1, \cdots, w$, she then uses $(i-1)$-generalized broadcast to communicate the $i$-th element of the pseudo-basis to Bob.*

---

Lemmas 3 and 4 ensure that this technique is secure; we now compute its cost:

- Each element of the pseudo-basis is a vector of $\mathbb{F}_q^n$;
- using $m$-generalized broadcast to communicate $n$ elements of $\mathbb{F}_q$ requires communicating $\left\lceil \frac{n}{m+1} \right\rceil n$ field elements;
- hence Protocol 2 requires communicating the following number of elements of $\mathbb{F}_q$:

$$\sum_{i=1}^{w} \left\lceil \frac{n}{i} \right\rceil n = O\left( n^2 \sum_{i=1}^{w} \frac{1}{i} \right) = O\left( n^2 \log n \right)$$

which means that we have reduced to $O\left( n^2 \log^2 n \right)$ the total communication complexity.

This complexity is still one logarithmic factor short of our goal; in the next section we show a more advanced technique that allows to bring down the cost to $O\left( n^2 \right)$ field elements.

## 5.3   Improved Transmission of the Pseudo-Basis: The Final Version

In this section we show a more advanced technique to communicate the pseudo-basis. The key idea is the following: denote by $w$ the size of the pseudo-basis; if Alice can find a received word $\mathbf{y}$ which is subject to an error of weight $cw$ for some constant $c$ and sends it to Bob, then Bob will learn the position of at least $cw$ corrupted channels. Alice will thus be able to use $cw$-generalized broadcast as in Lemma 3 to communicate the elements of the pseudo-basis (which amount to $wn$ symbols); since $cw$-generalized broadcast of a symbol has a cost of $O(n/cw)$, the total cost of communicating the pseudo-basis will thus be $(wn) \cdot O(n/cw) = O\left( n^2 \right)$.

We thus devise an algorithm that allows Alice to find a word $\mathbf{y}$ subject to at least $m = \Omega(w)$ errors (for instance, such condition is met if $\mathbf{y}$ is subject to $\Omega(t)$ errors, since $w \leq t$). Notice that such a word $\mathbf{y}$ may not exist among the received words $\{\mathbf{y}^{(i)}\}$, therefore we will look for a linear combination of the $\mathbf{y}(i)$ with this property.

As mentioned in Sects. 2 and 3, Alice will make extensive use of a decoding algorithm. Recall that a code of distance $d$ can be uniquely decoded from up to $\lfloor (d-1)/2 \rfloor$ errors, and that in the case of Reed-Solomon codes, such decoding can be performed in time polynomial in $n$ [6]; this means that for any Reed-Solomon code $\mathcal{C}$ there exists an algorithm that takes as input a word $\mathbf{y} \in \mathbb{F}_q^n$ and outputs a decomposition $\mathbf{y} = \mathbf{x} + \mathbf{e}$ with $\mathbf{x} \in \mathcal{C}$ and $w_\mathrm{H}(\mathbf{e}) \leq \lfloor (d-1)/2 \rfloor$ (if such a decomposition does not exist, the algorithm outputs an error message $\perp$).

---

**Protocol 3.** *Alice has received the words* $\mathbf{y}^{(1)}, \cdots, \mathbf{y}^{(r)}$ *and has computed a pseudo-basis* $\{\mathbf{y}^{(i)} : i \in I\}$ *of them; denote by $w$ its cardinality. Alice proceeds with the following actions:*

- *she uses Algorithm 1 below to find a "special word"* $\mathbf{y}$, *with coefficients* $(\mu_i : i \in I)$ *such that* $\mathbf{y} = \sum_{i \in I} \mu_i \mathbf{y}^{(i)}$. *She then communicates to Bob the triplet* $(I, (\mu_i : i \in I), \mathbf{y})$ *by using ordinary broadcast.*
- *Finally, she communicates the pseudo-basis of the received values by using $m$-generalized broadcast, where* $m := \min(w, t/3)$, *$w$ being the cardinality of the pseudo-basis.*

---

Before describing the algorithm formally and proving its validity, we sketch the idea. Alice has computed a pseudo-basis $\{\mathbf{y}^{(i)} : i \in I\}$. For $i \in I$, she applies the decoding algorithm to $\mathbf{y}^{(i)} = \mathbf{x}^{(i)} + \mathbf{e}^{(i)}$. If the decoding algorithm fails, it means that $\mathbf{y}^{(i)}$ is at a large Hamming distance from any codeword, in particular from Bob's codeword $\mathbf{x}^{(i)}$, and the single $\mathbf{y}^{(i)}$ is the required linear combination. If the decoding algorithm succeeds for every $i$, Alice obtains decompositions

$$\mathbf{y}^{(i)} = \tilde{\mathbf{x}}^{(i)} + \tilde{\mathbf{e}}^{(i)}$$

where $\tilde{\mathbf{x}}^{(i)}$ is some codeword. Alice must be careful, because she has no guarantee that the codeword $\tilde{\mathbf{x}}^{(i)}$ coincides with Bob's codeword $\mathbf{x}^{(i)}$, and hence that $\tilde{\mathbf{e}}^{(i)}$ coincides with Eve's error vector $\mathbf{e}^{(i)}$. What Alice then does is look for a linear combination $\sum_i \mu_i \tilde{\mathbf{e}}^{(i)}$ that has Hamming weight at least $t/3$ and at most $2t/3$. If she is able to find one, then a simple Hamming distance argument guarantees that the corresponding linear combination of Eve's original errors $\sum_i \mu_i \mathbf{e}^{(i)}$ also has Hamming weight at least $t/3$. If Alice is unable to find such a linear combination, then she falls back on constructing one that has weight not more than $2t/3$ and at least the cardinality $w$ of the pseudo-basis. This will yield an alternative form of the desired result. We now describe this formally.

**Algorithm 1.** *Alice has a pseudo-basis* $\left(\mathbf{y}^{(i)} : i = 1, \cdots, w\right)$ *(indices have been changed to simplify the notation); the algorithm allows Alice to identify a word* $\mathbf{y}$ *subject to at least* $m := \min(w, t/3)$ *errors introduced by Eve.*

*In the following steps, whenever we say that the output of the algorithm is a word* $\mathbf{y}^{(i)}$, *we implicitly assume that the algorithm also outputs the index* $i$; *more generally, whenever the algorithm outputs a linear combination* $\sum_i \mu_i \mathbf{y}^{(i)}$ *of the words in the pseudo-basis, we assume that it also outputs the coefficient vector* $(\mu_1, \cdots, \mu_w)$ *of the linear combination.*

1. *Alice uses a unique decoding algorithm to decode the elements of the pseudo-basis; if the algorithm fails for a given word* $\mathbf{y}^{(i)}$ *(i.e., it doesn't output a codeword having distance at most* $t/2$ *from* $\mathbf{y}^{(i)}$*), then Algorithm 1 stops and outputs* $\mathbf{y}^{(i)}$.

2. *If the decoding algorithm worked for every* $i$, *Alice gets a decomposition* $\mathbf{y}^{(i)} = \tilde{\mathbf{x}}^{(i)} + \tilde{\mathbf{e}}^{(i)}$ *with* $\tilde{\mathbf{x}}^{(i)} \in \mathcal{C}$ *and* $w_H\left(\tilde{\mathbf{e}}^{(i)}\right) \leq t/2$ *for every* $i$; *notice that it is not guaranteed that the* $\tilde{\mathbf{x}}^{(i)}$ *coincide with the codewords* $\mathbf{x}^{(i)}$ *originally chosen by Bob.*
   *If any of the* $\tilde{\mathbf{e}}^{(i)}$ *has weight greater than* $t/3$, *the algorithm stops and outputs* $\mathbf{y}^{(i)}$.

3. *Define* $\tilde{\mathbf{f}}^{(1)} := \tilde{\mathbf{e}}^{(1)}$ *and* $\tilde{\mathbf{y}}^{(1)} := \mathbf{y}^{(1)}$. *For any* $i = 2, \cdots, w$, *proceed with the following actions:*
   - *let* $\lambda^{(i)}$ *be a non-zero element of* $\mathbb{F}_q$ *such that* $\tilde{\mathbf{f}}_j^{(i-1)} + \lambda^{(i)} \tilde{\mathbf{e}}_j^{(i)} \neq 0$ *for any coordinate* $j \in \{1, 2, \ldots, n\}$ *for which* $\tilde{\mathbf{f}}_j^{(i-1)} \neq 0$.
   - *let* $\tilde{\mathbf{f}}^{(i)} := \tilde{\mathbf{f}}^{(i-1)} + \lambda^{(i)} \tilde{\mathbf{e}}^{(i)}$ *and* $\tilde{\mathbf{y}}^{(i)} := \tilde{\mathbf{y}}^{(i-1)} + \lambda^{(i)} \mathbf{y}^{(i)}$;
     *if* $w_H\left(\tilde{\mathbf{f}}^{(i)}\right) > t/3$, *stop and output* $\tilde{\mathbf{y}}^{(i)}$.

4. *Output* $\tilde{\mathbf{y}}^{(w)}$.

We can now prove that this algorithm allows Alice to find the desired codeword, which naturally implies that Protocol 3 indeed allows for reliable communication of the pseudo-basis:

**Proposition 3.** *Algorithm 1 allows Alice to find a word* $\mathbf{y}$ *subject to an error introduced by Eve of weight at least* $m := \min(w, t/3)$.

*Proof.* The following observation is the key point of the algorithm:

**Lemma 5.** *Let* $\mathbf{y} = \mathbf{x} + \mathbf{e} = \tilde{\mathbf{x}} + \tilde{\mathbf{e}}$ *for* $\mathbf{x}, \tilde{\mathbf{x}} \in \mathcal{C}$. *Then if* $\tilde{\mathbf{e}}$ *satisfies* $w_H(\tilde{\mathbf{e}}) \leq 2t/3$, *we have that* $w_H(\mathbf{e}) \geq \min \{w_H(\tilde{\mathbf{e}}), t/3\}$.

*Proof.* The claim is trivial if $\mathbf{e} = \tilde{\mathbf{e}}$; hence assume that $\mathbf{e} \neq \tilde{\mathbf{e}}$. Notice that $\mathbf{e} - \tilde{\mathbf{e}} = \tilde{\mathbf{x}} - \mathbf{x}$; hence since $d_{\min}(\mathcal{C}) = t + 1$, we have that

$$t + 1 \leq w_H(\mathbf{e} - \tilde{\mathbf{e}}) \leq w_H(\mathbf{e}) + w_H(\tilde{\mathbf{e}}) \leq w_H(\mathbf{e}) + \frac{2t}{3}$$

Hence we have that $w_H(\mathbf{e}) \geq t/3$, so that the claim is proved.   □

We now analyze the algorithm step-by-step:

1. if decoding fails for a word $\mathbf{y}^{(i)}$, then it is guaranteed that the error introduced by Eve on it has weight bigger than $t/2 > m$ (otherwise, the unique decoding algorithm would succeed since $d_{\min}(\mathcal{C}) = t + 1$).
2. since by assumption $w_H(\tilde{\mathbf{e}}^{(i)}) \leq t/2 \leq 2t/3$, if we also have $t/3 \leq w_H(\tilde{\mathbf{e}}^{(i)})$, then thanks to Lemma 5 the output $\mathbf{y}^{(i)}$ is of the desired type.
3. Since the algorithm did not abort at step 2, all elements $\tilde{\mathbf{e}}^{(i)}$ have weight at most $t/3$.
   First notice that if the algorithm did not produce $\tilde{\mathbf{f}}^{(i-1)}$ as output, then $\tilde{\mathbf{f}}^{(i)}$ is well-defined: indeed, we have that $w_H\left(\tilde{\mathbf{f}}^{(i-1)}\right) \leq t/3$; this means that $\lambda^{(i)}$ is well-defined, since it is an element of $\mathbb{F}_q$ that has to be different from 0 and from at most $t/3 < n - 1$ elements.
   Now if the algorithm outputs $\tilde{\mathbf{f}}^{(i)}$, then necessarily $w_H\left(\tilde{\mathbf{f}}^{(i-1)}\right) \leq t/3$ (otherwise the algorithm would have stopped before computing $\tilde{\mathbf{f}}^{(i)}$); furthermore, by assumption we have that $w_H\left(\tilde{\mathbf{e}}^{(i)}\right) \leq t/3$, so that $w_H\left(\tilde{\mathbf{f}}^{(i)}\right) \leq 2t/3$ and we can apply Lemma 5, so that the output is of the desired type.
4. Notice that for any $i = 1, \cdots, w$, we have that $\tilde{\mathbf{f}}^{(i)}$ has maximal weight among elements of the vector space $\langle \tilde{\mathbf{e}}^{(1)}, \cdots, \tilde{\mathbf{e}}^{(i)} \rangle$ (the condition on $\lambda^{(i)}$ ensures that this condition is met at each step). Hence since the elements $\{\tilde{\mathbf{e}}^{(1)}, \cdots, \tilde{\mathbf{e}}^{(w)}\}$ are linearly independent (because their syndromes are linearly independent, since $(\mathbf{y}^{(1)}, \ldots, \mathbf{y}^{(w)})$ is a pseudo-basis), we have that $w_H\left(\tilde{\mathbf{f}}^{(i)}\right) \geq i$ for any $i$.
   In particular, we have that $w_H\left(\tilde{\mathbf{f}}^{(w)}\right) \geq w$; hence since $w_H\left(\tilde{\mathbf{f}}^{(w)}\right) \leq 2t/3$ as remarked above, we have that the output $\tilde{\mathbf{y}}^{(w)}$ is of the desired type.   □

*Remark 2.* Protocol 3 requires Alice to use ordinary broadcast to communicate a single vector of $\mathbb{F}_q^n$ (hence transmitting $n^2$ elements of $\mathbb{F}_q$), then to use $m$-generalized broadcast with $m \geq \min\{w, t/3\}$ to communicate $w \leq t$ vectors of $\mathbb{F}_q^n$ (hence transmitting at most $3n^2$ elements of $\mathbb{F}_q$). We thus get a total of at most $4n^2$ elements of $\mathbb{F}_q$ to be transmitted.

Furthermore, Algorithm 1 has running time polynomial in $n$, as long as the code $\mathcal{C}$ has a unique-decoding algorithm of polynomial running time as well. As already remarked, such algorithms exist for instance for Reed-Solomon codes.

We study the second bottleneck of the original protocol in the next section.

## 5.4   The Improved Communication of the Masked Secrets

We present in this section the second key improvement to the protocol: after the pseudo-basis is communicated, we devise a way to lower the cost of transmitting to Bob the masked secrets and the information to open the masks. We aim at a cost linear in the number $\ell$ of secrets to be transmitted (while it was quadratic in Protocol 1). As in Sect. 5.3, Alice makes use of a unique decoding algorithm.

**Protocol 4.** *The protocol is performed once the pseudo-basis has been communicated to Bob; we thus assume that Bob knows the global support $S := \cup_i support\left(\mathbf{e}^{(i)}\right)$ of the errors affecting the elements $\mathbf{y}^{(i)}$ (cf. Remark 1). We assume that Alice wishes to communicate $\ell$ secret elements $s^{(1)}, \cdots, s^{(\ell)}$ of $\mathbb{F}_q$ to Bob, and that $\ell$ codewords $\mathbf{x}^{(1)}, \cdots, \mathbf{x}^{(\ell)}$ of $C$ have been sent by Bob to Alice (who has received $\mathbf{y}^{(1)}, \cdots, \mathbf{y}^{(\ell)}$) and have not been disclosed in other phases.*

- *Alice uses a unique decoding algorithm to decode $\mathbf{y}^{(i)}$, so that for every $i$ she obtains (if decoding was successful) a decomposition $\mathbf{y}^{(i)} = \tilde{\mathbf{x}}^{(i)} + \tilde{\mathbf{e}}^{(i)}$ with $\tilde{\mathbf{x}}^{(i)} \in C$ and $w_H\left(\tilde{\mathbf{e}}^{(i)}\right) \leq t/2$.*
  *For every $i = 1, \cdots, \ell$ she then communicates the following elements to Bob:*
  - *the syndrome $\mathbf{H}\left(\mathbf{y}^{(i)}\right)^T$ via $t/2$-generalized broadcast;*
  - *the elements $z_1^{(i)}, z_2^{(i)}$ of $\mathbb{F}_q$ by ordinary broadcast, where*

  $$z_1^{(i)} := \quad s^{(i)} + \mathbf{h}\left(\mathbf{y}^{(i)}\right)^T$$

  $$z_2^{(i)} := \begin{cases} s^{(i)} + \mathbf{h}\left(\tilde{\mathbf{x}}^{(i)}\right)^T & \text{if decoding succeeded,} \\ 0 & \text{otherwise.} \end{cases}$$

- *Bob can then obtain each secret $s^{(i)}$ in a different way depending on the size of the global support $S$ of the errors:*
  - *if $|S| \geq t/2$, he uses the knowledge of the syndrome of $\mathbf{y}^{(i)}$ and of the support of the error to compute $\mathbf{y}^{(i)}$, so that he can compute $z_1^{(i)} - \mathbf{h}\left(\mathbf{y}^{(i)}\right)^T$ as well.*
  - *if $|S| < t/2$, he ignores the syndrome that has been communicated to him, and computes $z_2^{(i)} - \mathbf{h}\left(\mathbf{x}^{(i)}\right)^T$.*

We now prove that this protocol works and is secure:

**Proposition 4.** *The above protocol allows for private and reliable communication of $\ell$ elements of $\mathbb{F}_q$.*

*Proof.* We check Privacy and Reliability.

*Privacy:* we have already observed in Proposition 2 that Eve has no information on $\mathbf{hy}^T$ (we drop the index $(i)$ to simplify notation), so that $z_1$ perfectly hides the secret. Now notice that if $\mathbf{y}$ can be decoded, then $z_2 = s + \mathbf{hx}^T = z_1 - \mathbf{h}\tilde{\mathbf{e}}^T$; hence to conclude, it suffices to prove that Eve already knows whether $\mathbf{y}$ can be decoded or not, and that she knows $\tilde{\mathbf{e}}$ if $\mathbf{y}$ can be decoded. We prove this claim in the following lemma:

**Lemma 6.** *Let $\mathbf{x}$ be a codeword sent by Bob to Alice, and let $\mathbf{y} = \mathbf{x} + \mathbf{e}$ be the received vector. Then Eve knows whether $\mathbf{y}$ can be decoded (i.e. $\mathbf{y} = \tilde{\mathbf{x}} + \tilde{\mathbf{e}}$ as above) or not; furthermore, if $\mathbf{y}$ can be decoded, then she knows $\tilde{\mathbf{e}}$.*

*Proof.* By definition, $\tilde{\mathbf{e}}$ is a vector of minimum weight (and of weight at most $t/2$) such that $\mathbf{y} - \tilde{\mathbf{e}}$ belongs to $\mathcal{C}$; notice that the last condition is equivalent to require that $\mathbf{e} - \tilde{\mathbf{e}}$ belongs to $\mathcal{C}$. Now these requirements uniquely determine $\tilde{\mathbf{e}}$: indeed, if by contradiction $\mathbf{e} - \mathbf{e}' \in \mathcal{C}$ for another $\mathbf{e}'$, then $\mathbf{e}' - \tilde{\mathbf{e}}$ would belong to $\mathcal{C}$, a contradiction since $w_H(\mathbf{e}' - \tilde{\mathbf{e}}) \leq t/2 + t/2 < d_{\min}(\mathcal{C})$.

Hence $\tilde{\mathbf{e}}$ is uniquely determined by $\mathbf{e}$ and $\mathcal{C}$: Eve can thus compute it from the data in her possession. Notice that, in particular, she knows whether $\tilde{\mathbf{e}}$ exists or not, i.e. whether decoding of $\mathbf{y}$ is possible or not. □

*Reliability:* we have two possible cases:

- if $|\mathcal{S}| \geq t/2$, then Bob is able to acquire the syndrome $\mathbf{H}\mathbf{y}^T$ of $\mathbf{y}$ via $t/2$-generalized broadcast (cf. Lemma 3); thus as remarked in Proposition 2, he can recover $\mathbf{y}$ and open the mask to get the secret.
- if $|\mathcal{S}| < t/2$, then Bob knows that Alice has correctly decoded $\mathbf{y}$, since Eve introduced less than $d_{\min}/2$ errors; thus $\tilde{\mathbf{x}} = \mathbf{x}$ so that $z_2 - \mathbf{h}\mathbf{x}^T = (s + \mathbf{h}\tilde{\mathbf{x}}^T) - \mathbf{h}\mathbf{x}^T = s$.

  Notice that in this case Bob will have failed to decode the $t/2$-generalized broadcast but he will simply ignore the elements received in this way. □

*Remark 3.* Notice that we could further improve the efficiency of this protocol by requiring Alice to use $w$-generalized broadcast (instead of regular one) to communicate the elements $z_1^{(i)}$ and $z_2^{(i)}$, where $w$ is the size of the pseudo-basis; this, however, would not reduce the order of magnitude of the total cost.

## 6    The Improved Protocol

The improved protocol simply implements the new techniques of Sects. 5.3 and 5.4.

---

**Protocol 5.** *The protocol allows Alice to communicate $\ell$ secret elements $s^{(1)}, \cdots, s^{(\ell)}$ of $\mathbb{F}_q$ to Bob, where $q$ is an arbitrary integer with $q > n$. The protocol takes as input an MDS code $\mathcal{C}$ of parameters $[n, t+1, t+1]_q$ and a vector $\mathbf{h}$ of length $n$ as in Lemma 1.*

I. *Bob chooses $t + \ell + 1$ uniformly random and independent codewords $\mathbf{x}^{(1)}, \cdots, \mathbf{x}^{(t+\ell+1)}$ of $\mathcal{C}$ and sends them over the channels to Alice.*

II. *Alice receives the corrupted versions $\mathbf{y}^{(1)}, \cdots, \mathbf{y}^{(t+\ell+1)}$, and she computes a pseudo-basis $\{\mathbf{y}^{(i)} : i \in I\}$ of the received values; she then proceeds with the following actions:*

---

> (i) She uses Protocol 3 to communicate the pseudo-basis to Bob.
> (ii) She then uses the remaining words to communicate to Bob the masked secrets and the data to retrieve them as in the first part of Protocol 4.
> III. Upon receiving the pseudo-basis, Bob proceeds to compute the global support $S$ of the error space; he can then obtain each secret $s^{(i)}$ as specified in the corresponding part of Protocol 4.

Notice that privacy and reliability of the protocol follow from the previous discussions; we now analyze the complexity of the protocol:

*Communication complexity:* we can set $\ell := 1$.

- Step I requires transmitting $t+2$ words of $\mathbb{F}_q^n$ over the channels, thus requiring a total of $O\left(n^2\right)$ symbols to be transmitted.
- Step II-(i) requires transmitting $O\left(n^2\right)$ elements of $\mathbb{F}_q$ as shown in Remark 2.
- Finally, step II-(ii) requires using $t/2$-generalized broadcast to communicate $n$ symbols, and standard broadcast to communicate 2 symbols, thus giving a total of $O(n)$ elements to be transmitted.

Hence since we can assume that $q = O(n)$, we get a total communication complexity of

$$O\left(n^2 \log n\right)$$

bits to be transmitted to communicate a single-bit secret.

*Transfer rate:* optimal rate is achieved for $\ell = \Omega(n)$.

- Step I requires transmitting $t + \ell + 1 = \ell + O(n)$ codewords, for a total of $n\ell + O(n^2)$ symbols.
- Step II-(i) remains unchanged from the single-bit case, and thus requires transmitting $O\left(n^2\right)$ symbols.
- Finally, step II-(ii) uses $t/2$-generalized broadcast to communicate $\ell t$ elements of $\mathbb{F}_q$ and standard broadcast to communicate $2\ell$ elements of $\mathbb{F}_q$, so that the overall cost is equal to $4n\ell$ symbols to be transmitted.

To sum up, the overall transmission rate is equal to

$$\frac{5n\ell + O\left(n^2\right)}{\ell} = 5n + O\left(n^2/\ell\right).$$

Furthermore, by using Reed-Solomon codes (instead of arbitrary MDS ones), we then have that Protocol 5 has computational cost polynomial in $n$ for both Alice and Bob.

# 7    Concluding Remarks

We have presented a two-round PSMT protocol that has polynomial computational cost for both sender and receiver, and that achieves transmission rate linear in $n$ and communication complexity in $O\left(n^2 \log n\right)$; we believe that our protocol is conceptually simpler compared to previous work and fully harnesses the properties of the pseudo-basis.

As proved in [10], the transfer rate is asymptotically optimal; furthermore, our protocol has a low multiplicative constant of 5.

Conversely, it remains open whether the $O\left(n^2 \log n\right)$ communication complexity is optimal or not; the only known lower bound on this parameter is still $O(n)$, as the one for transfer rate [10]. We believe that a communication complexity lower than $O\left(n^2\right)$ is unlikely to be achievable, at least not without a completely different approach to the problem.

**Acknowledgments.** The authors would like to thank Serge Fehr and Ronald Cramer for their useful comments and suggestions.

# References

1. Agarwal, S., Cramer, R., Haan, R.: Asymptotically optimal two-round perfectly secure message transmission. In: Dwork, C. (ed.) CRYPTO 2006. LNCS, vol. 4117, pp. 394–408. Springer, Heidelberg (2006). doi:10.1007/11818175_24
2. Dolev, D., Dwork, C., Waarts, O., Yung, M.: Perfectly secure message transmission. J. ACM **40**(1), 17–47 (1993)
3. Griggio, J.: Perfectly secure message transmission protocols with low communication overhead and their generalization. Master thesis (2012). http://algant.eu/documents/theses/griggio.pdf
4. Kurosawa, K., Suzuki, K.: Truly efficient 2-round perfectly secure message transmission scheme. In: Smart, N. (ed.) EUROCRYPT 2008. LNCS, vol. 4965, pp. 324–340. Springer, Heidelberg (2008). doi:10.1007/978-3-540-78967-3_19
5. Kurosawa, K., Suzuki, K.: Truly efficient 2-round perfectly secure message transmission scheme. IEEE Trans. Inf. Theory **55**(11), 5223–5232 (2009)
6. MacWilliams, F., Sloane, N.: The Theory of Error Correcting Codes. North-Holland mathematical library. North-Holland Publishing Company (1977)
7. Massey, J.L.: Some applications of coding theory in cryptography. In: Codes, Ciphers: Cryptography and Coding IV, pp. 33–47 (1995)
8. Sayeed, H.M., Abu-Amara, H.: Efficient perfectly secure message transmission in synchronous networks. Inf. Comput. **126**(1), 53–61 (1996)
9. Shamir, A.: How to share a secret. Commun. ACM **22**(11), 612–613 (1979)
10. Srinathan, K., Narayanan, A., Pandu Rangan, C.: Optimal perfectly secure message transmission. In: Franklin, M. (ed.) CRYPTO 2004. LNCS, vol. 3152, pp. 545–561. Springer, Heidelberg (2004). doi:10.1007/978-3-540-28628-8_33

# Foundations of Multi-Party Protocols

# Almost-Optimally Fair Multiparty Coin-Tossing with Nearly Three-Quarters Malicious

Bar Alon$^{(\boxtimes)}$ and Eran Omri

Department of Computer Science, Ariel University, Ariel, Israel
`alonbar08@gmail.com, omrier@ariel.ac.il`

**Abstract.** An $\alpha$-fair coin-tossing protocol allows a set of mutually distrustful parties to generate a uniform bit, such that no efficient adversary can bias the output bit by more than $\alpha$. Cleve [STOC 1986] has shown that if half of the parties can be corrupted, then, no $r$-round coin-tossing protocol is $o(1/r)$-fair. For over two decades the best known $m$-party protocols, tolerating up to $t \geq m/2$ corrupted parties, were only $O\left(t/\sqrt{r}\right)$-fair. In a surprising result, Moran, Naor, and Segev [TCC 2009] constructed an $r$-round two-party $O(1/r)$-fair coin-tossing protocol, i.e., an optimally fair protocol. Beimel, Omri, and Orlov [Crypto 2010] extended the result of Moran et al. to the *multiparty setting* where strictly fewer than 2/3 of the parties are corrupted. They constructed a $2^{2^k}/r$-fair $r$-round $m$-party protocol, tolerating up to $t = \frac{m+k}{2}$ corrupted parties.

Recently, in a breakthrough result, Haitner and Tsfadia [STOC 2014] constructed an $O\left(\log^3(r)/r\right)$-fair (almost optimal) three-party coin-tossing protocol. Their work brought forth a combination of novel techniques for coping with the difficulties of constructing fair coin-tossing protocols. Still, the best coin-tossing protocols for the case where more than 2/3 of the parties may be corrupted (and even when $t = 2m/3$, where $m > 3$) were $\theta\left(1/\sqrt{r}\right)$-fair. We construct an $O\left(\log^3(r)/r\right)$-fair $m$-party coin-tossing protocol, tolerating up to $t$ corrupted parties, whenever $m$ is constant and $t < 3m/4$.

## 1 Introduction

Secure multiparty computation allows a set of mutually distrustful parties to perform a computational task, while guaranteeing some security properties to hold. Examples of desirable security properties of a secure protocol are correctness, privacy, and *fairness* (roughly, the requirement that either all parties receive their respective outputs, or none do). When a strict majority of honest parties can be guaranteed, protocols for secure computation (see, e.g., [9,19]) provide full security, i.e., they provide all the security properties mentioned above (and others), including fairness. When there is no honest majority, however, this is no longer the case, and full security (specifically, full fairness) is not achievable in

---

Research supported by ISF grant 544/13.

M. Hirt and A. Smith (Eds.): TCC 2016-B, Part I, LNCS 9985, pp. 307–335, 2016.
DOI: 10.1007/978-3-662-53641-4_13

general. As was shown by Cleve [14], this is already evident for the elementary (no input) task of coin-tossing.

The coin-tossing functionality, introduced by Blum [12], allows a set of parties to agree on a uniformly chosen bit. Cleve [14] showed that this functionality cannot be computed with complete fairness without a strict honest majority. He proved that for any $r$-round two-party coin-tossing protocol, there exists an (efficient) adversary that can bias the output of the honest party by $\Omega(1/r)$. Cleve's impossibility naturally generalizes to the multiparty setting with no honest majority and has ramifications to general secure computation, implying that any function that implies coin-tossing (e.g., the XOR function) cannot be computed with full fairness without an honest majority. The question of optimal fairness for the coin-tossing functionality seems to be crucial towards understanding general secure and fair multiparty computation.

On the positive end, Averbuch et al. [6], Cleve [14] showed how to compute the coin-tossing functionality with partial fairness, limiting the bias of any adversary to $O(1/\sqrt{r})$. For over two decades, these constructions were believed to be optimal. This belief was supported by the work of Cleve and Impagliazzo [15], showing that in a model, where commitments are available only as black-box (and no other assumptions are made), the bias of any coin-tossing protocol is $\Omega(1/\sqrt{r})$. In a breakthrough result, Moran, Naor, and Segev [30] showed that the $\Omega(1/r)$-bias lowerbound of Cleve is tight for the case of two-party coin-tossing. They constructed an $r$-round two-party coin-tossing protocol with bias $O(1/r)$. The protocol of Moran et al. follows the special-round paradigm[1], previously appearing in [22,27].

Beimel, Omri, and Orlov [8] constructed (via the special-round paradigm) an optimal $O(1/r)$-bias protocol for any constant number of parties, whenever strictly less than a 2/3-fraction of the parties are malicious. More accurately, for their construction to yield an $O(1/r)$ bound on the bias of their protocol, it suffices that the gap between the number of corrupted parties and the number of honest parties is constant (rather than the total number of parties).

Still, the question whether optimal $O(1/r)$-coin-tossing was possible when the set of malicious parties may consist of two-thirds or more of the parties remained open. Specifically, even the case of three-party optimally-fair coin-tossing, where two of the parties may be corrupted remained unsettled. Answering the question regarding the three party case seemed to require new techniques and a novel understanding of coin-tossing protocols. In another breakthrough result, Haitner and Tsfadia [24] constructed an $O\left(\log^3(r)/r\right)$-fair (almost optimal) three-party coin-tossing protocol. Their work, indeed, offers some profound insight into the difficulties of constructing coin-tossing protocols, and brings forth a combination of novel techniques for coping with these difficulties. However, while it may be tempting to expect that the solution for the three-party case (and, specifically, that of [24]) will soon lead to a solution for fair coin-tossing for any (constant) number of parties, this has not been the case so far.

---

[1] The idea is to randomly and secretly choose a special round in which the parties unknowingly get the output of the computation.

## 1.1  Our Results

Our main contribution is a multiparty coin-tossing protocol that has small bias whenever the number of parties is constant fewer than $3/4$ of them are corrupted.

**Theorem 1 (informal).** *Assume that oblivious transfer protocols exist. Let $m$ and $t$ be constants (in the security parameter $n$) such that $m/2 \leq t < 3m/4$, and let $r = r(n)$ be an integer. There exists an $r$-round $m$-party coin-tossing protocol tolerating up to $t$ corrupted parties that has bias $O(2^{2^m} \log^3(r)/r)$.*

The formal statements and proofs implying Theorem 1 are given in Sect. 3, a warmup construction illustrating the ideas behind the general construction is given in Sect. 1.4. The $2^{2^m}$ factor in the upperbound on the bias of our construction is due the fact that in each round, the adversary sees defense values for many corrupted subsets. For this reason, we require $m$ to be constant.

## 1.2  Additional Related Work

Partially fair coin-tossing is an example of $1/p$-secure computation. Informally, a protocol is $1/p$-secure if it emulates the ideal functionality within $1/p$ distance. The formal definition of $1/p$-secure computation appears in Sect. 2.3.1. $1/p$-security *with abort* was suggested by Katz [27]. Gordon and Katz [21] defined $1/p$-security and constructed 2-party $1/p$-secure protocols for every functionality whose size of either the domain or the range of the functionality is polynomial (in the security parameter). Beimel et al. [7] studied multiparty $1/p$-secure protocols for general functionalities. The main result in [7] is constructions of $1/p$-secure protocols that are resilient against *any* number of corrupted parties, provided that the number of parties is constant and that the size of the range of the functionality is at most polynomial in the security parameter $n$. The bias of the coin-tossing protocol resulting from [7] is $O(1/\sqrt{r})$.

The impossibility result of Cleve [14] made many researchers believe that no interesting functions can be computed with full fairness without an honest majority. A surprising result by Gordon et al. [22] showed that there are even functions containing embedded XOR that can be computed with fairness. This led to a line of works, investigating complete fairness in secure multiparty computation without an honest majority [2,3,29]. Recently, Asharov et al. [4] gave a full characterization of fairness secure two-party computation of Boolean functions.

Coin-tossing is an interesting and useful task even in weaker models, e.g., secure-with-abort coin-tossing – where honest parties are not requested to output a bit upon a premature abort by the adversary, and weak coin-tossing – where each party has an a priori desire for the output bit. Indeed, the latter type of coin-tossing was the one formulated by Blum [11], who suggested a fully secure weak (and actually, secure with abort) coin-tossing protocol based on the existence of one-way functions ([25,31]). His protocol is also a $1/4$-secure implementation of the fair coin-tossing functionality. Conversely, the existence of secure-with-abort protocols imply the existence of one-way functions [10,23,28]. For the cryptographic complexity of optimally-fair coin-tossing, [16,17] gave some evidence that one-way functions may not suffice.

## 1.3   Our Techniques

Towards explaining the ideas behind our protocol, we give a brief overview of the constructions of [8,24,30]. We restrict our discussion to the fail-stop model, where corrupted parties follow the prescribed protocol, unless choosing to prematurely abort at some point in the execution. Indeed, the core difficulties in constructing fair coin-tossing protocols stand in this model as well. Specifically, an $r$-round multiparty coin-tossing protocol in the fail-stop model can be adapted to the malicious setting by adding signatures to each message (or by applying the GMW compiler [19]).

### 1.3.1   The Protocol of Moran et al. [30].

The protocol of Moran, Naor, and Segev [30] is a two-party $r$-round coin-tossing protocol with optimal bias $1/4r$. That is, their protocol matches the lowerbound of Cleve [14] (up to a factor 2). The basic idea of the protocol is that in each round $i$, each of the parties is given an independently chosen uniform bit, which will be its output, in case the other party aborts. This is done until some special round $i^*$. From round $i^*$ and on, both parties get the same bit $c$. Finally, $i^*$ is chosen uniformly from $[r]$ and is kept secret from the parties. The security of the protocol relies on the inability of the adversary to guess the value of $i^*$ with probability higher than $1/r$. We next give a slightly more detailed overview of the MNS protocol restricted to fail-stop adversaries.

*A skeleton for two-party coin-tossing protocols.* We start by describing the skeleton for the two-party protocol of [30]. Indeed, this is a more generic skeleton and can be used to describe any two-party coin-tossing protocol $(A, B)$.

*The preliminary phase of the protocol.* In this phase, the parties jointly compute defense values for each of the $r$ rounds of interaction. Denote the defense value assigned to A for round $i \in [r]$ by $a_i$ and the value assigned to B for round $i$ by $b_i$ (in the MNS protocol, these defense values are actually bits). At the end of this preliminary phase, the parties do not learn these defense values, but rather hold a share in a 2-out-of-2 secret sharing scheme (separately, for each defense value). Denote by $a_i[P]$ and $b_i[P]$ the shares of $a_i$ and $b_i$ (respectively) held by party P.

*Interaction rounds.* In round $i$, party A reveals $b_i[A]$ and party B reveals $a_i[B]$. Specifically, in round $i$, party A learns $a_i$ and party B learns $b_i$. The role of these defense values is to define the output of an honest party, upon a premature abort of the other party. For example, if party A aborts in round $i$ (not allowing B to learn $b_i$), then B halts and outputs $b_{i-1}$. If an abort never occurs, then parties output $a_r = b_r$.

*The MNS instantiation of the two-party skeleton.* We now specify how the defense values are selected in the protocol of [30]. The parties jointly select a special round number $i^* \in \{1, \ldots, r\}$, uniformly at random, and select bits $a_1, \ldots, a_{i^*-1}, b_1, \ldots, b_{i^*-1}$, independently, uniformly at random. Then, they uniformly select a bit $w \in \{0, 1\}$ and set $a_i = b_i = w$ for all $i^* \le i \le r$.

The security of the protocol follows from the fact that, unless the adversary aborts in round $i^*$, it cannot bias the output of the protocol. This is true, since before round $i^*$ the view of the adversary is independent of the prescribed output bit $w$, and hence, given that the adversary aborts *before* round $i^*$, the output of the honest party is a uniform bit. On the other hand, after round $i^*$ is completed, the output of the honest party is fixed. Hence, aborting in any round after $i^*$ is equivalent to never aborting at all, therefore, given that the adversary aborts *after* round $i^*$, the output of the honest party is also a uniform bit. Finally, the view of any of the parties up to round $i \le i^*$ is independent of the value of $i^*$, hence, any adversary corrupting a single party can guess $i^*$ with probability at most $1/r$.

### 1.3.2 The Protocols of Haitner and Tsfadia [24].

Haitner and Tsfadia [24] constructed a three-party $r$-round coin-tossing protocol with close to optimal bias $O\left(\log^3 r/r\right)$. Towards achieving this goal, Haitner and Tsfadia [24] first constructed several new two-party fair coin-tossing protocols with bias $O\left(\log^3 r/r\right)$. Evidently, the bias of these protocols does not match the Cleve [14] lowerbound (as does the MNS protocol), however, the techniques and insight introduced in these constructions make them interesting even before considering the final three-party construction, for which they serve as a building block. In fact, most of the techniques that enable the three-party construction of [24] come up already in their two-party protocols.

Before describing the protocols of [24], let us first highlight some of the ideas underlying them. We stress that none of their protocols follows the special round paradigm. Alternatively, their protocols have the value of the game (i.e., the expected outcome in an honest continuation of the current state) gradually shift from being $1/2$ (or some other $\alpha \in [0, 1]$, for that matter) to being either 0 or 1. This is done by having the parties run in the background – jointly and hidden from each of them – a protocol with a gradually shifting and publicly known game value (in this case, a weighted variant of the majority protocol of [6,14]). Let $O_i$ be the game value in round $i$.

One of the core observations underlying all the constructions of Haitner and Tsfadia [24] is that letting the defense value $a_i$ be a bit sampled according to $O_i$, fully protects A in case of an abort by B in round $i$. More importantly, if the gap between $O_i$ and $O_{i-1}$ is typically $O\left(1/\sqrt{r}\right)$, then $a_i$ does not reveal too much information about the current value of $O_i$ to A. Finally, Haitner and Tsfadia [24] show that $a_i$ can be instantiated, not only as a bit, but also as a description of a full execution of a two-party protocol with output and (defense values) sampled according to $O_i$ (where, this form of $a_i$ still does not reveal too much information about the current value of $O_i$ to A). Going from here to their construction of a three-party coin-tossing protocol is fairly natural.

We next describe the two-party protocols of Haitner and Tsfadia [24]. We do so using the skeleton for two-party protocols described in Sect. 1.3.1. That is, we explain how the defense values $a_i, b_i$ for each round $i$ are selected. We note that Haitner and Tsfadia [24] did not present their protocols in this exact manner, but rather divided each interaction round $i$ into two steps. The first step is exactly the one described in the above skeleton, i.e., where A learns $a_i$ and B learns $b_i$. In the second step of round $i$, the parties reconstruct a value $x_i$ that describes the expected value of the game $O_i$. This extra step is not necessary for the correctness of the protocol, and hence, does not affect the security of the protocol (since any attack on the protocol not using $x_i$ can also be applied to the protocol that gives $x_i$).

*The basic two-party protocol of* [24]. We now specify how the defense values are selected in the basic two-party protocol of [24] (parametrized by $\alpha \in [0,1]$), such that the common output bit is 1 with probability $\alpha$. The basic idea is to sample $O(r^2)$ bits (i.e., elements from $\{-1,1\}$) i.i.d., such that the sum of all bits is positive with probability $\alpha$. The prescribed output of the protocol is 1 if the sum of all bits is positive, and 0 otherwise. Towards revealing this output (gradually, in $r$ rounds), let $\delta_i$ be the value of the game, conditioned on the value of the first $\sum_{k=r-i+1}^{r} k$ bits. Note that $\delta_0 = \alpha$ and that in each round $i$, the value of $\delta_i$ is computed conditioned on less and less *new* bits (i.e., bits that were not used to compute $\delta_{i-1}$). The defense value given to each of the parties in round $i$ is simply a sample from $\delta_i$.

Slightly more formally, let $\varepsilon \in \left[-\frac{1}{2}, \frac{1}{2}\right]$ be such that the sum of $r(r+1)/2$ elements from $\{-1,1\}$ is positive with probability $\alpha$, where each element is 1 with probability $1/2 + \varepsilon$. Let $x_i$ be the sum of $r - i + 1$ elements from $\{-1,1\}$, where each element takes the value of 1 with probability $1/2 + \varepsilon$. Let $\delta_i$ be the expected game value in round $i$, that is, $\delta_i$ is the probability that the sum of $\sum_{k=1}^{r-i} k$ elements from $\{-1,1\}$, is at least $\sum_{k=1}^{i} x_k$. The bits $a_i$ and $b_i$ are independently sampled according to $\delta_i$, i.e., $a_i = 1$ (and $b_i = 1$) w.p. $\delta_i$.

For some intuition on the security of the protocol, consider the case where party A, wishing to bias the output of party B, receives a defense value $a_i$ before party B receives its defense value $b_i$. If A chooses to abort, then B is instructed to output $a_{i-1}$, which was sampled according to $\delta_{i-1}$. Indeed, if A could see $\delta_i$ before deciding whether to abort or not, it could bias the output of B by $\Omega(1/\sqrt{r})$. The crux of the analysis is to show that this is not the case when A only receives a sample from $\delta_i$. Towards this end, Haitner and Tsfadia [24] bound, on expectation, the gap between $\delta_{i-1}$ and $\widehat{\delta_{i-1}}$, defined to be the value of the game, conditioned on the value of the first $\sum_{k=r-i+1}^{r} k$ bits and on the value of $a_i$.

*The three party protocol of* [24]. The construction of [24] for three parties follows a very similar rationale to the above protocol. That is, in each round $i$ every single party, as well as, every pair of parties obtain a defense value that should behave as a sample from $\delta_i$. A pair of parties cannot simply be given a single bit, since one of them may be corrupt. Rather, they should be given a two-party protocol similar to the above, with their defenses set with parameter $\alpha = \delta_i$. A problem arises here, since the simple application of the above idea would require giving the adversary information based on $\Omega(r^3)$ bits sampled according to the appropriate $\varepsilon$ value. This would be devastating to the security of the protocol, as it would allow the adversary to reveal $\delta_i$. To tackle this problem, [24] came up with a derandomized version of the above two-party protocol. They were then able to show that sending the shares for this protocol as the defense values for pairs of parties does not reveal too much about $\delta_i$ to the adversary. We next describe the derandomized two-party protocol of Haitner and Tsfadia [24].

*The two-party derandomized protocol of* [24]. We now specify how the defense values are selected in the derandomized version of the protocol of [24], such that the common output bit is 1 with probability $\alpha$. Let $\varepsilon \in \left[-\frac{1}{2}, \frac{1}{2}\right]$ be such that the sum of $r(r+1)/2$ elements from $\{-1, 1\}$ is positive with probability $\alpha$, where each element is 1 with probability $1/2 + \varepsilon$. For $j \in \{a, b\}$, let $S^j$ be a set of size $r(r+1)$, over $\{-1, 1\}$, where each element takes the value of 1 with probability $1/2 + \varepsilon$. Let $x_i$ be the sum of $r - i + 1$ elements from $\{-1, 1\}$, where each element takes the value of 1 with probability $1/2 + \varepsilon$. Let $\delta_i^j$ be the expected game value in round $i$, according to the set $S^j$, that is, $\delta_i^j$ is the probability that the sum of the elements in a randomly chosen subset of $S^j$, of size $\sum_{k=1}^{r-i} k$, is at least $\sum_{k=1}^{i} x_k$. The bit $a_i$ (respectively $b_i$) is sampled according to $\delta_i^a$ (respectively $\delta_i^b$), i.e., $a_i = 1$ (respectively $b_i = 1$) with probability $\delta_i^a$ (respectively $\delta_i^b$).

The security of the various constructions of [24] is proved via a series of bounds on weighted Binomial games. In Sect. 2, we recall these results, and in Sect. 3 we use them to prove the security of our construction.

### 1.3.3 Reducing Many-Party Coin-Tossing to Few-Party Coin-Tossing.

Reducing multiparty coin-tossing protocols for the setting without an honest majority to 2-party protocols is quite straightforward. Indeed, the impossibility of [14] is generalized from the two-party setting to the many party setting via such a reduction. In this section, we show that sometimes the other direction is also possible.

**The Protocol of Beimel et al.** [8]. The protocol of Beimel, Omri, and Orlov [8] extends the results of [30] to the multiparty model, where fewer than 2/3 of the parties are corrupted. The bias of their protocol is proportional to $1/r$ and doubly exponential in the gap between the number of corrupted parties $t$ and the number of honest parties $h$ in the protocol ($m = h + t$). In particular, for a constant number of parties $m$, where fewer than $2m/3$ are corrupted, [8] present an $r$-round $m$-party coin-tossing protocol with an optimal bias of $O(1/r)$. Interestingly, their protocol has an $O(1/r)$-bias even when the number of parties $m$ is non-constant, as long as the $t - h$ is constant. In the following description, however, we present a simplified version of the protocol of [8], which requires $t$ (rather than $t - h$) to be constant in order to achieve an $O(1/r)$-bias.

While not presented this way, the result of Beimel et al. [8] is achieved via a generic reduction to (a certain type of) two-party protocols. They use a few layers of secret sharing schemes to allow for each subset $J$ of parties, containing an honest majority (i.e., $h \leq |J| < 2h$, hence if all the parties outside of $J$ abort the execution, then there is an honest majority in $J$) to obtain a defense value, i.e., a bit $d_i^J$. For each round $i$ and for each such $J$, the value of $d_i^J$ is shared in an *inner* secret sharing scheme with threshold $h$-out-of-$|J|$. The idea is that the shares of this inner secret sharing scheme (of $d_i^J$) should be revealed to the parties of $J$ at round $i$ of the execution. Namely, each party in $J$ should get one of the (inner scheme) shares of $d_i^J$ in round $i$.

To make sure that the above shares are not revealed to any subset before round $i$, and at the same time, that the execution of the protocol proceeds, as long as, the set of remaining active parties does not contain an honest majority, the shares (of the inner scheme) for round $i$ are shared in an *outer* secret sharing scheme with threshold $(t+1)$-out-of-$m$. As a result, the adversary can never learn anything about the shares of the $i$'th inner scheme without the help of honest parties. In addition, to halt the computation in round $i$, the adversary must instruct at least $h$ parties to abort the computation.

Now, given a two-party protocol according to the above skeleton, and with the additional property that $a_i$ and $b_i$ are sampled from the same distribution $D_i$ and that it is possible to sample many such samples, completing the reduction is done by selecting the defense values $d_i^J$ from the distribution $D_i$.

If the following extra property holds, then the resulting many-party protocol would be $\alpha$-fair as long as $t < 2m/3$. The extra property that we need to require is that if the adversary in the 2-party protocol is given $2^{2^m}$ defense values, sampled from $D_i$ (and the honest party gets a single one), it will not be able to bias the 2-party protocol by more than $\alpha$.[2]

### 1.3.4 Applying the Reduction of Beimel et al. to the Protocols of Haitner and Tsfadia.

In this work, we use secret sharing schemes, in a manner similar to [8], to reduce an $m$-party coin-tossing with $t < 3m/4$ malicious to the 3-party construction of [24]. We do so in two steps. First, we apply the above (simplified version of the) reduction of [8] to the (derandomized) two-party protocol of [24] to obtain an *auxiliary* $\hat{m}$-party coin-tossing protocol, tolerating $\hat{t} < 2\hat{m}/3$ corruptions. Then, we use the auxiliary protocol, as a building block in the construction of the *final* $m$-party protocol that tolerates $t < 3m/4$ corruptions. More specifically, the auxiliary protocol, parametrized by some $\varepsilon \in [0, 1]$, is used as defense values for subsets of parties for the case that at least $m/4$ corrupted parties abort the execution of the final protocol.

---

[2] Beimel et al. [8] use a slightly more involved technique to distribute defense values to the different subsets of parties, allowing several subsets to be assigned the same output bit, while maintaining the guarantee that the adversary cannot bias the output of the honest parties without guessing the value of the special round $i^*$.

We next give an overview of both constructions. In Sect. 1.4, we exemplify the constructions for the case that $m = 7$ and $t = 5$; in Sect. 1.4.1, we instantiate the auxiliary protocol for the case of five parties with up to three corruptions, and in Sect. 1.4.2, we use this construction to instantiate the final protocol for the case of seven parties with up to five corruptions. In the following, let $\hat{h} = \hat{m} - \hat{t}$ and $h = m - t$ be lowerbounds on the number of honest parties in the respective protocols. In our discussion the auxiliary protocol will be used with $\hat{m}$ being the number of active parties remaining after some corrupted parties have prematurely aborted the execution of the final $m$-party protocol. Specifically, we will have $\hat{h} = h$, since honest parties never prematurely abort the computation.

Both the basic and the final protocols use two layers of (threshold) secret sharing schemes. For each round $i$ and for each *protected* subset of parties $J$ (we specify below which subsets are called protected for each construction), the defense value for the set $J$ in round $i$ is $d_i^J$. This defense value is shared among the parties of $J$ in an appropriate secret sharing scheme (actual parameters for each construction are specified below). This is called the *inner* secret sharing scheme. For each round $i$, all the shares of all parties in the inner secret sharing schemes for round $i$ are shared in an $(\tilde{t} + 1)$-out-of-$\hat{m}$ threshold secret sharing scheme, where $\hat{m}$ and $\tilde{t}$ are the number of parties and the bound on the number of corruptions in the respective construction. This is called the *outer* secret sharing scheme.

The idea behind the outer secret sharing scheme is to provide two guarantees. First, the adversary is never able to reconstruct the secrets without the participation of honest parties (which will only participate in the appropriate round). Second, the adversary is only able to prevent the reconstruction of the secret of the outer scheme (for round $i$) by instructing at least $\tilde{h} = \hat{m} - \tilde{t}$ corrupted parties to abort before completing the reconstruction. Hence, the protocol proceeds normally as long as more than $\tilde{t}$ parties are active. We stress that the adversary is indeed able to instruct $\tilde{h}$ parties to abort in the process of reconstruction of the secret of the outer secret sharing scheme, hence, seeing all the shares of corrupted parties for round $i$, while not allowing honest parties to see their shares of the inner scheme. Furthermore, since the adversary is rushing, it can actually decide whether to do so or not – after seeing the shares of all honest parties.

In addition to the above, assume that $\tilde{t} < \frac{b\hat{m}}{b+1}$ for some natural $b > 1$, and assume that at least $\tilde{h}$ corrupted parties aborted (which is the case if the secret of the outer scheme cannot be reconstructed). Let $J$ be the set of the remaining parties and let $t_J$ be the number of corrupted parties in $J$. Since the number of honest parties in $J$ remains the same as before, i.e., at least $h > \frac{\hat{m}}{b+1}$, it follows that $t_J < |J| - \frac{\hat{m}}{b+1}$. By assumption $|J| \leq \tilde{t} < \frac{b\hat{m}}{b+1}$, it follows that $t_J < |J| - \frac{\hat{m}}{b+1} < \frac{(b-1)\cdot|J|}{b}$. Thus, if $h$ parties abort the execution of the final construction, then less than 2/3 of the remaining parties are corrupted, and if $\hat{h}$ parties abort the execution of the auxiliary construction, then most of the remaining parties are honest.

We now explain what protected subsets are and how the parameters for the inner secret sharing schemes are chosen for each of the two constructions. We begin with the final construction. Protected subsets of parties are subsets $J$ that are assigned a defense value $d_i^J$ in each round $i$. These should include all subsets that are liable to become the set of active parties, after a premature abort by at least $h$ parties. Since the number of aborting (corrupted) parties may be anything between $h$ and $t$, we should let protected subsets be all subsets of parties $J$, such that $h \leq |J| \leq t$.[3]

To determine the parameters for the inner secret sharing scheme, consider the case that $a \geq h$ corrupted parties have aborted in round $i$, hence the set of active parties $J$ is of size $m - a$. Let $t_J$ be the number of corrupted parties in $J$, then $t_J \leq t - a$. Therefore, using a $(t - m + |J| + 1)$-out-of-$|J|$, we require at least $t - a + 1 = t - m + |J| + 1$ parties of $J$ for the reconstruction of $d_i^J$. This ensures that the adversary was never able to reconstruct $d_{i-1}^J$ (which is the defense value that the parties in $J$ will use). Very similar reasoning are used for the auxiliary construction, where a subset of parties is protected if it of any size between $\hat{h}$ and $2\hat{h} - 1$, and the threshold of the inner secret sharing scheme is set to $\hat{h}$-out-of-$|J|$.

It is left to specify what are the defense values $d_i^J$, which are the secrets that are shared in the inner secret sharing schemes. Roughly speaking these values are selected in the auxiliary and in the final constructions in a very similar manner to that of the derandomized two-party and the three-party protocols of [24] (respectively). In a bit more detail, in these protocols, there is a value $\delta_i$ representing the expected value of the game, and the defense values for all protected subsets describe a way to reveal a sample a bit according to $\delta_i$.

In the final protocol, a defense value is an instantiation of the auxiliary protocol, such that the output bit is 1 with probability $\delta_i$. To be more precise, $d_i^J$ is the set of shares in the outer secret sharing of the instantiation of the auxiliary protocol to be executed by the parties of $J$, in case all other parties abort the computation. The exact same information can also be encapsulated into a set of $O(r^2)$ elements from $\{-1, 1\}$ taking the value with probability $1/2 + \varepsilon$, where $\varepsilon = \varepsilon(\delta_i) \in \left[-\frac{1}{2}, \frac{1}{2}\right]$ is such that the sum of $r(r+1)/2$ elements from $\{-1, 1\}$ is positive with probability $\delta_i$, whenever each element is 1 with probability $1/2 + \varepsilon$. Indeed, this fact will allow us to use the vector game lemma of [24] (see Lemma 2) to bound the bias that the adversary can inflict by seeing the defense values of all corrupted protected sets. The proof of security of the final protocol is obtained by combining the above bound with a bound on the bias of the auxiliary protocol.

We now specify how the defense values are selected in the auxiliary protocol. Let $J$ be a protected subset of parties, the parties of $J$ jointly hold a set $S^J$ of size $r(r+1)$, over $\{-1, 1\}$, where each element takes the value of 1 with probability $1/2 + \varepsilon$. Recall that $x_i$ be the sum of $r - i + 1$ elements from $\{-1, 1\}$, where each element takes the value of 1 with probability $1/2 + \varepsilon$. Let $\delta_i^J$ be the expected game value in round $i$, according to the set $S^J$, that is, $\delta_i^J$ is the probability that the

---

[3] Actually, in our construction, we only call subsets $J$, such that $2h - 1 \leq |J| \leq t$. This suffices, since if a smaller subset of active parties is left, it can use the defense value of its lexicographically first superset of size $2h - 1$.

sum of the elements in a randomly chosen subset of $S^J$, of size $\sum_{k=1}^{r-i} k$, is at least $\sum_{k=1}^{i} x_k$. The bit $b_i^J$ is sampled according to $\delta_i^J$, i.e., $b_i^J = 1$ with probability $\delta_i^J$. To prove the security of this protocol, we introduce an extended version of the Hypergeometric game (Lemma 3), presented in [24]. More specifically, we show that even when the adversary sees a (constant) number of independent samples, each from a different set, it cannot bias the output by much.

### 1.4 A Warm-Up Construction – A Seven-Party Protocol Tolerating up to Five Corrupted Parties

Following the overview of our constructions, given in Sect. 1.3.4 in this section, we show how to instantiate our final construction for the case of 7 parties, where at most 5 are corrupted. In Sect. 1.4.1, we instantiate the auxiliary protocol for 5 parties with at most 3 corruptions, and in Sect. 1.4.2 we use it to instantiate the final protocol for 7 parties with at most 5 corruptions. In the following, let $\varepsilon \in \left[-\frac{1}{2}, \frac{1}{2}\right]$, and for $i \in \{0, \ldots, r\}$ let $s_i = \sum_{k=1}^{r-i} k$.

### 1.4.1 A Five-Party Protocol Tolerating up to Three Corrupted Parties. 
We now describe the algorithm $HG(\varepsilon, 5, 3)$, generating shares for 5 parties with 3 corrupted parties. This is a specific instantiation of the more general functionality described in Algorithm 5. Let $\mathcal{B}in_{n,\varepsilon}$ denote the binomial distribution over $\{-1, 1\}$ (i.e., a sum of $n$ samples from $\{-1, 1\}$, each taking the value of 1 with probability $\frac{1}{2} + \varepsilon$).

**Selecting defenses:**

1. For every $J \subset [5]$ of size 3, let $S^J$ be a set with $2s_0$ elements from $\{-1, 1\}$, each taking the value of 1 with probability $\frac{1}{2} + \varepsilon$.
2. For every $i \in [r]$ let $\hat{x}_i \leftarrow \mathcal{B}in_{r-i+1,\varepsilon}$.
3. For every $i \in \{0, \ldots, r\}$ and every $J \subseteq [5]$ of size 3:
   (a) Let $A_i^J$ be a random subset of $S^J$ of size $s_i$.
   (b) Let $\hat{d}_i^J$ be 1 if $\sum_{k=1}^{i} \hat{x}_k + \sum_{a \in A_i^J} a \geq 0$, and 0 otherwise.

**Sharing the values:**

- For every $i \in \{0, 1 \ldots, r\}$, $J \subset [5]$ of size 3, and $j \in J$, let $d_i^J[j]$ be the share of party $P_j$ of the secret $d_i^J$, in a 2-out-of-3 secret sharing.
- For every $i \in [r]$, $J \subset [5]$ of size 3, and for every $j' \in J$, let $d_i^J[j', j]$ be the share of party $P_j$ of the secret $d_i^J[j']$, in a 4-out-of-5 secret sharing, such that party $P_{j'}$ is required in order to recover $d_i^J[j']$ (See Construction 4).

*Interaction rounds.* The interaction of the parties proceeds in $r$ rounds. In round $i \in [r]$, party $P_j$ broadcasts $d_i^J[j', j]$, for every $J \subset [5]$ of size 3, and for every $j' \in J$.

If a single party aborts the execution, then the remaining 4 parties can continue with the protocol. If two or three parties abort the execution, then the remaining parties reconstruct $d_{i'}^J$, where $J$ is lexicographically first set of size 3, which contains all the indices of the active parties, and $i'$ is the maximum $i$ for which the parties have enough shares to reconstruct. The honest parties output that bit.

If after $r$ rounds, there are at least 4 active parties, then the parties reconstruct the last joint defense for the lexicographically first subset of them, and the honest parties output that bit.

*Security.* By the properties of the two layers of secret sharing, in each round the adversary learns a constant number of defense values, which are sampled according to the appropriate Hypergeometric distribution. Roughly speaking, the security of the above protocol is reduced to an extended version of the Hypergeometric game considered by [24], with a constant number of samples. The proof of security of the general construction, as well as, the froof of the bound for the extended Hypergeometric game are given in the full version of the paper [1].

### 1.4.2   The Seven-Party Protocol.

We are now ready to describe our 7 party protocol. We first describe the share generator. Given $x_1 \ldots x_i$, for some $i \in [r]$ we let $\delta_i(x_1 \ldots x_i)$ be the probability that then sum of $s_i$ uniform $\{-1, 1\}$ bits is at least $-\sum_{k=1}^i x_k$. We call $\delta_i$ the expected outcome of the protocol in round $i$. In the following we let $\mathcal{B}in_n := \mathcal{B}in_{n,0}$.

**Selecting defenses:**

1. For every $i \in [r]$, let $x_i \leftarrow \mathcal{B}in_{r-i+1}$.
2. Let $\varepsilon_i \in \left[-\frac{1}{2}, \frac{1}{2}\right]$ be such that, the expected outcome of an honest execution with parameter $\varepsilon = \varepsilon_i$ of the 5-party protocol from Sect. 1.4.1 is $\delta_i(x_1 \ldots x_i)$.
3. For every $J \subset [7]$, such that $4 \leq |J| \leq 5$, let $d_i^J \leftarrow HG(\varepsilon_i, |J|, |J| - 2)$.
4. For every $J \subset [7]$, such that $2 \leq |J| \leq 3$, let $d_i^J$ be a bit, sampled with probability $\delta_i(x_1 \ldots x_i)$.

**Sharing the values:**

– For every $i \in [r]$ and $J \subset [7]$, such that $4 \leq |J| \leq 5$, let $d_i^J[j]$ be the share of party $P_j$ of the secret $d_i^J$, in a $(|J| - 1)$-out-of-$|J|$ secret sharing.
– For every $i \in [r]$, $J \subset [7]$, such that $4 \leq |J| \leq 5$, and for every $j' \in J$, let $d_i^J[j', j]$ be the share of party $P_j$ of the secret $d_i^J[j']$, in a 6-out-of-7 secret sharing, such that party $P_{j'}$ is required in order to recover $d_i^J[j']$ (See Construction 4).
– For every $i \in [r]$ and $J \subset [7]$, such that $2 \leq |J| \leq 3$, let $d_i^J[j]$ be the share of party $P_j$ of the secret $d_i^J$, in a 2-out-of-$|J|$ secret sharing.

*Interaction rounds.* The interaction of the parties proceeds in $r$ rounds. In round $i \in [r]$ party $P_j$ broadcasts $d_i^J[j', j]$, for every $J \subset [7]$, such that $3 \leq |J| \leq 5$, and for every $j' \in J$.

If a single party aborts the execution, then the remaining 6 parties can continue with the protocol (they can do so by the properties of the 6-out-of-7 secret sharing scheme). If more parties abort the execution, then the remaining active parties reconstruct $d_{i'}^J$, where $J$ is the lexicographic first set containing all their indices, and $i'$ is the maximum $i$ for which the parties have enough shares to reconstruct. If more than three parties remain, then they execute the five party protocol from Sect. 1.4.1. Otherwise, there is an honest majority, and hence, the remaining parties reconstruct $d_{i'}^J$, which is a bit.

If after $r$ rounds, there are at least 5 active parties, then each pair reconstruct its last common defense (Note that either all of these defenses are equal to 1 or all of them are equal to 0).

*Security.* In each round $i \in [r]$, the adversary learns an $O\left(r^2\right)$ bits sampled according to $\varepsilon_i$. If only one party aborts the execution, then the remaining parties can still continue, as the secret sharing is a 6-out-of-7. Hence the adversary must instruct at least two parties to abort. In case at least two parties abort at round $i$, the remaining active parties can reconstruct the defense from the round $i-1$. They then, execute the protocol described in Sect. 1.4.1. As this is the Vector game considered by [24], the adversary does not gain much advantage from aborting after seeing the above $O\left(r^2\right)$ bits samples (assuming that the remaining parties run the defense protocol honestly). Of course, we cannot assume that they do, however, combining the above with the security of the 5-party protocol, we get that in total, the adversary's gain remains small.

### 1.5   Organization

In Sect. 2, we provide some notations and definitions that we use in this work, and recall some bounds on online Binomial games from [24]. In Sect. 3 we present our main construction and provide a proof for Theorem 1.

## 2   Preliminaries

### 2.1   Notation

We use calligraphic letters to denote sets, uppercase for random variables, and lowercase for values. All logarithms considered here are in base two. For $n \in \mathbb{N}$, let $[n] = \{1, 2 \ldots n\}$. Given a random variable (or a distribution) $X$, we write $x \leftarrow X$ to indicate that $x$ is selected according to $X$. The support of a distribution $D$ over a finite set $S$, denoted $\mathrm{Supp}(D)$, is defined as $\{s \in S \mid D(s) > 0\}$. For a random variable $X$ and a natural number $n$ we let $X^n = \left(X^{(1)}, X^{(2)}, \ldots, X^{(n)}\right)$, where the $X^{(i)}$'s are i.i.d. copies of $X$.

Let $n \in \mathbb{N}$ and $\varepsilon \in \left[-\frac{1}{2}, \frac{1}{2}\right]$. Let $Ber(\varepsilon)$ be the Bernoulli distribution over $\{-1, 1\}$, taking 1 with probability $\frac{1}{2} + \varepsilon$. Define the Binomial distribution $Bin_{n,\varepsilon}$, by $Bin_{n,\varepsilon}(k) = \Pr\left[\sum_{i=1}^n x_i = k\right]$ where $x_i$ are i.i.d according to $Ber(\varepsilon)$. Let

$\widehat{Bin}_{n,\varepsilon}(k) = \Pr_{x \leftarrow Bin_{n,\varepsilon}}[x \geq k] = \sum_{t \geq k} Bin_{n,\varepsilon}(t)$. For $\varepsilon = 0$ we will simply write $Bin_n$ and $\widehat{Bin}_n$.

Define the Hypergeometric distribution $\mathcal{HG}_{n,w,m}$, by $\mathcal{HG}_{n,w,m}(k) = \Pr_{S \subseteq \mathcal{S}, |S|=m}\left[\sum_{s \in S} s = k\right]$, where $S$ is chosen uniformly, $\mathcal{S}$ is a set of size $n$, whose members are from $\{-1, 1\}$, and it holds that $\sum_{s \in \mathcal{S}} s = w$. Let $\widehat{\mathcal{HG}}_{n,w,m}(k) = \Pr_{x \leftarrow \mathcal{HG}_{n,w,m}}[x \geq k] = \sum_{t \geq k} \mathcal{HG}_{n,w,m}(t)$. For $i \in \{0, 1, \ldots n\}$ let $s_i(n) = \sum_{k=1}^{n-i} k = \frac{(n-i+1)(n-i)}{2}$. When $n$ is clear from the context we write $s_i$. For a set $S$ we let $w(S) = \sum_{s \in S} s$.

We make use of the following facts.

**Fact 2 (Hoeffding's inequality for $\{-1, 1\}$).** *Let $n, t \in \mathbb{N}$ and let $\varepsilon \in \left[-\frac{1}{2}, \frac{1}{2}\right]$. Then*

$$\Pr_{x \leftarrow Bin_{n,\varepsilon}}[|x - 2\varepsilon n| \geq t] \leq 2e^{-\frac{t^2}{2n}}.$$

**Fact 3 (Hoeffding's inequality for the hypergeometric distribution).** *Let $m \leq n \in \mathbb{N}$ and let $w \in \mathbb{Z}$ satisfying $|w| \leq n$. Then*

$$\Pr_{x \leftarrow \mathcal{HG}_{n,w,m}}[|x - \mu| \geq t] \leq e^{-\frac{t^2}{2m}},$$

*where $\mu = \mathbb{E}_{x \leftarrow \mathcal{HG}_{n,w,m}}[x] = \frac{mw}{n}$*

## 2.2    Coin-Tossing Protocols

A multiparty coin-tossing protocol with $m$ parties is defined using $m$ probabilistic polynomial-time Turing machines $p_1, \ldots, p_m$ having the security parameter $1^n$ as their only input. The coin-tossing computation proceeds in rounds, in each round, the parties broadcast and receive messages on a broadcast channel. The number of rounds in the protocol is typically expressed as some polynomially-bounded function $r$ in the security parameter. At the end of protocol, the (honest) parties should hold a common bit $w$. We denote by $\text{CoinToss}_\varepsilon()$ the ideal functionality that gives the honest parties the same bit $w$, distributed according to $\varepsilon$, that is, $\Pr[w = 1] = 1/2 + \varepsilon$ and $\Pr[w = 0] = 1/2 - \varepsilon$. We let $\text{CoinToss}()$ be $\text{CoinToss}_0()$.

In this work we consider a malicious static computationally-bounded adversary, i.e., a non-uniform that runs in a polynomial-time. The adversary is allowed to corrupt some subset of the parties. That is, before the beginning of the protocol, the adversary corrupts a subset of the parties that may deviate arbitrarily from the protocol, and thereafter the adversary sees the messages sent to the corrupt parties and controls the messages sent by the corrupted parties. Still, for the most of the technical discussion of the paper, we only discuss fail-stop adversaries. A fail-stop adversary acts completely honestly (i.e., as required by the prescribed protocol), with the only difference that it can abort the computation at any point in the execution of the protocol. We, then, use standard techniques ([8,19]) to turn a coin-tossing protocol in the fail-stop model into a coin-tossing protocol (with the same fairness and round-complexity) in the malicious model. The honest parties follow the instructions of the protocol.

The parties communicate in a synchronous network, using only a broadcast channel. The adversary is rushing, that is, in each round the adversary hears the messages sent by the honest parties before broadcasting the messages of the corrupted parties for this round (thus, the messages broadcast by corrupted parties can depend on the messages of the honest parties broadcast in this round).

## 2.3   Security Definitions for Multiparty Protocols

The security of multiparty computation protocols is defined using the real vs. ideal paradigm. In this paradigm, we consider the real-world model, in which protocols are executed. We then formulate an ideal model for executing the task at hand. This ideal model involves a trusted party whose functionality captures the security requirements of the task. Finally, we show that the real-world protocol "emulates" the ideal-world protocol: For any real-life adversary $\mathcal{A}$ there should exist an ideal-model adversary $\mathcal{S}$ (also called simulator) such that the global output of an execution of the protocol with $\mathcal{A}$ in the real-world model is distributed similarly to the global output of running $\mathcal{S}$ in the ideal model. In the coin-tossing protocol, the parties do not have inputs. Thus, to simplify the definitions, we define secure computation without inputs (except for the security parameters).

*The Real Model.* Let $\Pi$ be an $m$-party protocol computing $\mathcal{F}$. Let $\mathcal{A}$ be a non-uniform probabilistic polynomial time adversary with auxiliary input aux, corrupting a subset $\mathcal{C}$ of the parties. Let $REAL_{\Pi,\mathcal{A}(\text{aux})}(1^n)$ be the random variable consisting of the view of the adversary (i.e., its random input and the messages it got) and the output of the honest parties, following an execution of $\Pi$, where each party $p_j$ begins by holding the input $1^n$.

*The Ideal Model.* The basic ideal model we consider is a model without abort. Specifically, there are parties $\{p_1, \ldots, p_m\}$, and an adversary $\mathcal{S}$ who has corrupted a subset $I$ of them. An ideal execution for the computing $\mathcal{F}$ proceeds as follows:

**Inputs:** Party $p_j$ holds a security parameter $1^n$. The adversary $\mathcal{S}$ has some auxiliary input aux.

**Trusted party sends outputs:** The trusted party computes $\mathcal{F}(1^n)$ with uniformly random coins and sends the appropriate outputs to the parties.

**Outputs:** The honest parties output whatever they received from the trusted party, the corrupted parties output nothing, and $\mathcal{S}$ outputs an arbitrary probabilistic polynomial-time computable function of its view.

Let $IDEAL_{\mathcal{F},\mathcal{S}(\text{aux})}(1^n)$ be the random variable consisting of the output of the adversary $\mathcal{S}$ in this ideal world execution and the output of the honest parties in the execution.

In this work we consider a few formulations of the ideal-world, and consider composition of a few protocols, all being executed in the same real-world, however, each secure with respect to a different ideal-world. We prove the security of the resulting protocol, using the hybrid model techniques of Canetti [13].

### 2.3.1    1/p-Indistinguishability and 1/p-Secure Computation

As explained in the introduction, the ideal functionality CoinToss() cannot be implemented when there is no honest majority. We use $1/p$-secure computation, defined by [20,27], to capture the divergence from the ideal world. This notion applies to general secure computation. We start with some notation.

A function $\mu(\cdot)$ is *negligible* if for every positive polynomial $q(\cdot)$ and all sufficiently large $n$ it holds that $\mu(n) < 1/q(n)$. A *distribution ensemble* $X = \{X_{a,n}\}_{a\in\{0,1\}^*, n\in\mathbb{N}}$ is an infinite sequence of random variables indexed by $a \in \{0,1\}^*$ and $n \in \mathbb{N}$.

**Definition 1 (Statistical Distance and 1/p-indistinguishability).** *We define the* statistical distance *between two random variables $A$ and $B$ as the function*

$$\mathrm{SD}(A,B) = \frac{1}{2}\sum_{\alpha}\Big|\Pr[A=\alpha] - \Pr[B=\alpha]\Big|.$$

*For a function $p(n)$, two distribution ensembles $X = \{X_{a,n}\}_{a\in\{0,1\}^*, n\in\mathbb{N}}$ and $Y = \{Y_{a,n}\}_{a\in\{0,1\}^*, n\in\mathbb{N}}$ are computationally $1/p$-indistinguishable, denoted $X \stackrel{1/p}{\approx} Y$, if for every non-uniform polynomial-time algorithm $D$ there exists a negligible function $\mu(\cdot)$ such that for every $n$ and every $a \in \{0,1\}^*$,*

$$\Big|\Pr[D(X_{a,n}) = 1] - \Pr[D(Y_{a,n})) = 1]\Big| \le \frac{1}{p(n)} + \mu(n).$$

Two distribution ensembles are *computationally indistinguishable*, denoted $X \stackrel{\mathrm{C}}{\equiv} Y$, if for every $c \in N$ they are computationally $\frac{1}{n^c}$-indistinguishable.

We next define the notion of $1/p$-secure computation [7,20,27]. The definition uses the standard real/ideal paradigm [13,18], except that we consider a completely fair ideal model (as typically considered in the setting of honest majority), and require only $1/p$-indistinguishability rather than indistinguishability.

**Definition 2 (perfect 1/p-secure computation).** *An $m$-party protocol $\Pi$ is said to perfectly $(t, 1/p)$-secure compute a functionality $\mathcal{F}$ if for every non-uniform adversary $\mathcal{A}$ in the real model, corrupting up to $t$ of the parties, there exists a polynomial-time adversary $\mathcal{S}$ in the ideal model, corrupting the same parties as $\mathcal{A}$, such that for every $n \in \mathbb{N}$ and for every $\mathrm{aux} \in \{0,1\}^*$*

$$\mathrm{SD}(\mathrm{IDEAL}_{\mathcal{F},\mathcal{S}(\mathrm{aux})}(1^n), \mathrm{REAL}_{\Pi,\mathcal{A}(\mathrm{aux})}(1^n)) \le \frac{1}{p(n)}.$$

**Definition 3 (1/p-secure computation [7,20,27]).** *Let $p = p(n)$ be a function. An m-party protocol $\Pi$ is said to be $(t, 1/p)$-securely compute a functionality $\mathcal{F}$ if for every non-uniform probabilistic polynomial-time adversary $\mathcal{A}$ in the real model, corrupting up to $t$ of the parties, there exists a non-uniform probabilistic polynomial-time adversary $\mathcal{S}$ in the ideal model, corrupting the same parties as $\mathcal{A}$, such that the following two distribution ensembles are computationally $1/p(n)$-indistinguishable*

$$\left\{ \text{IDEAL}_{\mathcal{F},\mathcal{S}(\text{aux})}(1^n) \right\}_{\text{aux}\in\{0,1\}^*, n\in\mathbb{N}} \overset{1/p}{\approx} \left\{ \text{REAL}_{\Pi,\mathcal{A}(\text{aux})}(1^n) \right\}_{\text{aux}\in\{0,1\}^*, n\in\mathbb{N}} .$$

We next define the notion of *secure computation* and notion of *bias of a coin-tossing protocol* by using the previous definition.

**Definition 4 (secure computation).** *An m-party protocol $\Pi$ t-securely computes a functionality $\mathcal{F}$, if for every $c \in N$, the protocol $\Pi$ is $(t, 1/n^c)$-securely compute the functionality $\mathcal{F}$.*

**Definition 5 ($\varepsilon$-coin-toss).** *We say that a protocol is a $\varepsilon$-coin-toss protocol with bias $1/p$, tolerating up to $t$ corruptions, if it is a $(t, 1/p)$-secure protocol for the functionality $\text{CoinToss}_\varepsilon()$.*

**Definition 6 (coin tossing).** *We say that a protocol is a coin-tossing protocol with bias $1/p$, tolerating up to $t$ corruptions, if it is a $(t, 1/p)$-secure protocol for the functionality $\text{CoinToss}()$.*

### 2.4 Security with Identifiable Abort

We use here a variant of secure computation with abort, where upon abort, at least one cheating party is identified to all honest parties. This definition was first formally stated by Aumann and Lindell [5], and was also considered in [7,8,26], (in the first two, it was called security with abort and cheat detection).

Roughly speaking, our definition requires that one of two events is possible: If at least one party deviates from the prescribed protocol, then the adversary obtains the outputs of these parties (but nothing else), and all honest parties are notified by the protocol that these parties have aborted. Otherwise, the protocol terminates normally, and all parties receive their outputs. Again, we consider the restricted case where parties hold no private inputs. The formal definition is omitted for lack of space, and will appear in the full version of the paper [1].

### 2.5 Cryptographic Tools

We next informally describe two cryptographic tools that we use in our protocols.

*Signature Schemes.* A signature on a message proves that the message was created by its presumed sender, and its content was not altered. A signature scheme is a triple (Gen, Sign, Ver) containing the key generation algorithm Gen, which gets as input a security parameter $1^n$ and outputs a pair of keys, the signing key $K_S$ and the verification key $K_v$, the signing algorithm Sign, and the verifying algorithm Ver. We assume that it is infeasible to produce signatures without holding the signing key.

*Secret-Sharing Schemes.* An $\alpha$-out-of-$m$ secret-sharing scheme is a mechanism for sharing data among a set of parties such that every set of parties of size $\alpha$ can reconstruct the secret, while any smaller set knows nothing about the secret. In this paper, we use Shamir's $\alpha$-out-of-$m$ secret-sharing scheme [33]. In this scheme, the shares of any $\alpha - 1$ parties are uniformly distributed and independent of the secret. Furthermore, given at most such $\alpha - 1$ shares and a secret $s$, one can *efficiently* complete them to $m$ shares of the secret $s$. Using this scheme, [8] presented a way to construct a secret sharing scheme *with respect to a certain party*. We use that in our construction as well.

**Construction 4.** *Let $s$ be some secret taken from some finite field $\mathbb{F}$. We share $s$ among $m$ parties with respect to a special party $p_j$ in an $\alpha$-out-of-$m$ secret-sharing scheme as follows:*

1. *Choose shares $\left(s^{(1)}, s^{(2)}\right)$ of the secret $s$ in a two-out-of-two secret-sharing scheme, that is, select $s^{(1)} \in \mathbb{F}$ uniformly at random and compute $s^{(2)} = s - s^{(1)}$. Denote these shares by $\mathrm{mask}_j(s)$ and $\mathrm{comp}(s)$, respectively.*
2. *Generate shares $\left(\lambda^{(1)}, \ldots, \lambda^{(j-1)}, \lambda^{(j+1)}, \ldots, \lambda^{(m)}\right)$ of the secret $\mathrm{comp}(s)$ in an $(\alpha - 1)$-out-of-$(m - 1)$ Shamir's secret-sharing scheme. For each $\ell \neq j$, denote $\mathrm{comp}_\ell(s) = \lambda^{(\ell)}$.*

**Output:**

- *The share of party $p_j$ is $\mathrm{mask}_j(s)$. We call this share, $p_j$'s masking share.*
- *The share of each party $p_\ell$, where $\ell \neq j$, is $\mathrm{comp}_\ell(s)$. We call this share, $p_\ell$'s complement share.*

In the above, the secret $s$ is shared among the parties in $P$ in a secret-sharing scheme such that any set of size at least $\alpha$ that contains $p_j$ can reconstruct the secret. In addition, similarly to the Shamir secret-sharing scheme, the following property holds: for any set of $\beta < \alpha$ parties (regardless if the set contains $p_j$), the shares of these parties are uniformly distributed and independent of the secret. Furthermore, given such $\beta < \alpha$ shares and a secret $s$, one can *efficiently* complete them to $m$ shares of the secret $s$ and *efficiently* select uniformly at random one vector of shares competing the $\beta$ shares to $m$ shares of the secret $s$.

## 2.6    Claims and Definitions from [24]

The following definitions and propositions are taken verbatim from [24] and they will serve us as well. Given a partial view of a fail-stop adversary, we are interested in the expected outcome of the parties, conditioned on this view and the adversary making no further aborts.

**Definition 7 (view value).** *Let $\pi$ be a protocol in which the honest parties always output the same bit value. For a partial view $v$ of the parties in a fail-stop execution of $\pi$, let $C_\pi(v)$ denote the parties full view in an honest execution of $\pi$ conditioned on $v$ (i.e. all parties that do not abort in $v$ act honestly in $C_\pi(v)$). Let $\Delta_\pi(v) = E_{v' \leftarrow C_\pi(v)}[out(v')]$, where $out(v')$ is the common output of the non-aborting parties in $v'$.*

A protocol is unbiased, if no fail-stop adversary can bias the common output of the honest parties by too much.

**Definition 8 ($(t, \alpha)$-unbiased protocol).** *Let $\pi$ be an $m$-party, $r$-round protocol, in which the honest parties always output the same bit value. We say that $\pi$ is $(t, \alpha)$-unbiased, if the following holds for every fail-stop adversary $\mathcal{A}$ controlling the parties indexed by a subset $\mathcal{C} \subset [m]$ of size at most $t$. Let $V$ be $\mathcal{A}$'s view in a random execution of $\pi$, and let $I_j$ be the index of the $j$'th round in which $\mathcal{A}$ sent an abort message (set to $r + 1$ if no abort occurred). Let $V_i$ be the prefix of $V$ at the end of the $i$'th round, letting $V_0$ be the view consisting of only the random coins of $\mathcal{A}$, and let $V_i^-$ be the prefix of $V_i$ with the $i$'th round abort message (if any) removed. Then,*

$$\mathop{E}_{V} \left[ \left| \sum_{j \in |\mathcal{C}|} \left( \Delta(V_{I_j}) - \Delta(V_{I_j}^-) \right) \right| \right] \le \alpha$$

*where $\Delta = \Delta_\pi$ according to Definition 7.*

The following is an alternative characterization of fair coin-tossing protocols (against fail-stop adversaries).

**Lemma 1 ([24, Lemma 2.18]).** *Let $n \in \mathbb{N}$ be a security parameter and let $\pi$ be a $(t, \alpha)$-unbiased coin-tossing protocol with $\alpha(n) \le \frac{1}{2} - \frac{1}{p(n)}$, for some polynomial $p$. Then $\pi$ is a $(t, \alpha(n) + neg(n))$-secure coin tossing protocol against fail-stop adversaries.*

The following lemmata and propositions assume that the protocol is of a specific form. More concretely, let $\varepsilon \in \left[ -\frac{1}{2}, \frac{1}{2} \right]$, $f$ be a randomized function (that may depend on $\varepsilon$), and let $\pi_{\varepsilon, f}$ be an $r$-round $m$-party coin-tossing protocol, such that, before any interaction takes place, every party learns $D_0$, which is sampled according to the current game value, and for every round $i \in [r]$, every party first learns a defense $D_i = f(i, Y_i)$, and then the coin $X_i$, where $X_i \leftarrow \mathcal{B}in_{r-i+1,\varepsilon}$, $Y_i = \sum_{k=1}^{i} X_k$. We let $V_{\pi_{\varepsilon, f}}$ denote the adversary's view in a random execution of $\pi_{\varepsilon, f}$. We further assume that adversary never aborts after seeing $X_i$.

**Lemma 2 (Vector Game** [24, Lemma 4.5]**).** *Let* $c \in \mathbb{N}$ *and let* $r \in \mathbb{N}$ *be the number of rounds. Let* $f : [r] \times \mathbb{Z} \to \{-1, 1\}^{c \cdot r^2}$ *be a randomized function that on input* $(i, y)$ *outputs* $c \cdot r^2$ *elements from* $\{-1, 1\}$*, each takes the value of 1 with probability* $\mathcal{B}er(\varepsilon)$*, where* $\varepsilon \in \left[-\frac{1}{2}, \frac{1}{2}\right]$ *satisfies* $\widehat{Bin}_{s_0, \varepsilon}(0) = \widehat{Bin}_{s_i}(-y)$*. Then:*

$$\mathop{\mathrm{E}}_{V_{\pi_{0,f}}} \left[ \left| \Delta\left(V_{\pi_{0,f}}\right) - \Delta\left(V_{\pi_{0,f}}^{-}\right) \right| \right] = O\left(\frac{\log^3 r}{r}\right).$$

**Lemma 3 (Hypergeometric Game** [24, Lemma 4.4]**).** *Let* $w \in \mathbb{Z}$, $\varepsilon \in \left[-\frac{1}{2}, \frac{1}{2}\right]$ *and let* $r \in \mathbb{N}$ *be the number of rounds. Let* $f : [r] \times \mathbb{Z} \to \{0, 1\}$ *be a randomized function that on input* $(i, y)$ *outputs 1 with probability* $\widehat{HG}_{2s_0, w, s_i}(-y)$ *and 0 otherwise. Assuming that* $|w| \le c \cdot \sqrt{\log r \cdot s_0}$*, for some constant* $c$*, then:*

$$\mathop{\mathrm{E}}_{V_{\pi_{\varepsilon,f}}} \left[ \left| \Delta\left(V_{\pi_{\varepsilon,f}}\right) - \Delta\left(V_{\pi_{\varepsilon,f}}^{-}\right) \right| \right] = O\left(\frac{\log^3 r}{r}\right).$$

**Lemma 4 (Ratio Lemma** [24, Lemma 4.10]**).** *Let* $r \in \mathbb{N}$ *be the number of rounds, and let* $\varepsilon \in \left[-\frac{1}{2}, \frac{1}{2}\right]$*. In the following we let* $Y_0 = 0$*. Let*

$$\mathcal{X}_i := \left\{ x \in \mathrm{Supp}(X_i) : |x| \le 4\sqrt{\log r \cdot (r - i + 1)} \right\}$$

*and*

$$\mathcal{Y}_i := \left\{ y' \in \mathrm{Supp}(Y_{i-1}) : |y' + 2\varepsilon \cdot s_{i-1}| \le 4\sqrt{\log r \cdot s_{i-1}} \right\}.$$

*Assume* $|\varepsilon| \le 2\sqrt{\frac{\log r}{s_0}}$ *and that for every* $i \in [r - \lfloor \log^{2.5} r \rfloor]$ *and* $y \in \mathcal{Y}_i$*, there exists a set* $\mathcal{D}_{i,y}$ *such that for every* $x \in \mathcal{X}_i$*, and every* $d \in \mathcal{D}_{i,y} \cap \mathrm{Supp}(f(i, y + X_i) \mid Y_{i-1} = y, X_i \in \mathcal{X}_i)$*, it holds that:*

$$\Pr[f(i, y + X_i) \notin \mathcal{D}_{i,y} \mid Y_{i-1} = y] \le \frac{1}{r^2}$$

*and*

$$\left| 1 - \frac{\Pr[f(i, y + X_i) = d \mid Y_{i-1} = y \wedge X_i = x]}{\Pr[f(i, y + X_i) = d \mid Y_{i-1} = y \wedge X_i \in \mathcal{X}_i]} \right| \le c \cdot \sqrt{\frac{\log r}{r - i}} \cdot \left(1 + \frac{|x|}{\sqrt{r - i + 1}}\right),$$

*for some constant* $c$*. Then:*

$$\mathop{\mathrm{E}}_{V_{\pi_{\varepsilon,f}}} \left[ \left| \Delta\left(V_{\pi_{\varepsilon,f}}\right) - \Delta\left(V_{\pi_{\varepsilon,f}}^{-}\right) \right| \right] = O\left(\frac{\log^3 r}{r}\right).$$

**Proposition 1 (**[24, Proposition 4.6]**).** *For every randomized functions* $f, g$*, and for every* $\varepsilon \in \left[-\frac{1}{2}, \frac{1}{2}\right]$*, it holds that*

$$\mathop{\mathrm{E}}_{V_{\pi_{\varepsilon,g \circ f}}} \left[ \left| \Delta\left(V_{\pi_{\varepsilon,g \circ f}}\right) - \Delta\left(V_{\pi_{\varepsilon,g \circ f}}^{-}\right) \right| \right] \le \mathop{\mathrm{E}}_{V_{\pi_{\varepsilon,f}}} \left[ \left| \Delta\left(V_{\pi_{\varepsilon,f}}\right) - \Delta\left(V_{\pi_{\varepsilon,f}}^{-}\right) \right| \right]$$

**Proposition 2** ([24, Proposition 4.7]). *Let $\varepsilon \in \left[-\frac{1}{2}, \frac{1}{2}\right]$ and $f$ be some randomized function. If $\Pr[Y_r \geq 0] \notin \left[\frac{1}{r^2}, 1 - \frac{1}{r^2}\right]$, where $r \in \mathbb{N}$ is the number of rounds, then*

$$\underset{V_{\pi_\varepsilon,f}}{\mathrm{E}} \left[ \left| \Delta\left(V_{\pi_\varepsilon,f}\right) - \Delta\left(V_{\pi_\varepsilon,f}^-\right) \right| \right] \leq \frac{2}{r}.$$

## 2.7 An Extension of the Hypergeometric Game

In this section we introduce an extended version of the Hypergeometric game (Lemma 3), presented in [24]. More specifically, we let the adversary see a constant number of independent samples, each from a different set. Furthermore, we augment the view of the adversary with all of these sets.

**Lemma 5.** *Let $\xi \in \mathbb{N}$ be some constant, let $\mathbf{w} = (w_1 \ldots, w_\xi) \in \mathbb{Z}^\xi$, let $\varepsilon \in \left[-\frac{1}{2}, \frac{1}{2}\right]$, and let $r \in \mathbb{N}$ be the number of rounds. For $k \in [\xi]$, let $h_k : [r] \times \mathbb{Z} \to \{0, 1\}$ be a randomized function that on input $(i, y)$ outputs 1 with probability $\widehat{\mathcal{HG}}_{2s_0, w_k, s_i}(-y)$ and 0 otherwise. Assuming that for every $k \in [\xi]$, it holds that $|w_k| \leq c\sqrt{\log r \cdot s_0}$, for some constant $c$, then:*

$$\underset{V_{\pi_\varepsilon,h}}{\mathrm{E}} \left[ \left| \Delta\left(V_{\pi_\varepsilon,h}\right) - \Delta\left(V_{\pi_\varepsilon,h}^-\right) \right| \right] = O\left(2^\xi \cdot \frac{\log^3 r}{r}\right),$$

*where $h(i, y) = (h_1(i, y), \ldots, h_\xi(i, y))$.*

The proof of Lemma 5 is deferred to the full version of this paper [1].

# 3 The Multiparty Protocol

In this section, we describe our construction and prove Theorem 1. This result is formally restated in Sect. 3.3 (as Corollary 1) and proved therein.

In Sect. 3.1, we describe a construction of an $m$-party coin-tossing protocol tolerating up to 2/3 corruptions. In Sect. 3.2, we describe the main construction of an $m$-party almost optimally fair coin-tossing protocol tolerating up to 3/4 corruptions.

## 3.1 A Coin-Tossing Protocol for $t < 2m/3$

The following algorithm, is an extension of the two-party share generator, presented in [24], to the multiparty case.

**Algorithm 5 (MultipartyShareGen$_{<2/3}$ – $HG(\varepsilon, m, t)$).** *Let $r \in \mathbb{N}$ be the number of rounds.*

**Input:** *Number of rounds $r$, $\varepsilon = \varepsilon(n) \in \left[-\frac{1}{2}, \frac{1}{2}\right]$, the number of parties $m$, and an upper bound $t$ on the number of corrupted parties. Denote $h = m - t$. Observe that a subset $J \subset [m]$ of size $2h - 1$, containing all honest parties has an honest majority.*

**Selecting coins and defenses:**

1. *For every $J \subset [m]$ of size $2h - 1$:*
   (a) *Let $S^J$ be a set with $2s_0$ elements from $\{-1, 1\}$, where each element is sampled according to $\mathcal{B}er(\varepsilon)$.*
   (b) *Let $A_0^J$ be a random subset of $S^J$ of size $s_0$.*
   (c) *Let $d_0^J$ be 1 if $\sum\limits_{a \in A_0^J} a \geq 0$, and 0 otherwise.*

2. *For $i = 1$ to $r$:*
   (a) *Sample $x_i \leftarrow \mathcal{B}in_{r-i+1,\varepsilon}$.*
   (b) *For every $J \subset [m]$ of size $2h - 1$, we let $A_i^J$ be a random subset of $S^J$ of size $s_i$.*
   (c) *For every $J \subset [m]$ of size $2h - 1$, let $d_i^J$ be 1 if $\sum\limits_{k=1}^{i} x_k + \sum\limits_{a \in A_i^J} a \geq 0$, and 0 otherwise.*

**Sharing the values:**

1. *For $i \in [r]$, let $x_i[j]$ be a share of $x_i$ in a $(t+1)$-out-of-$m$ secret sharing.*
2. *For $i \in \{0, \ldots, r\}$, $j \in [m]$, and $J \subset [m]$ of size $2h - 1$, let $d_i^J[j]$ be a share of $d_i^J$ in a $h$-out-of-$(2h - 1)$ secret sharing.*
3. *For $i \in [r]$, $j \in [m]$, $J \subset [m]$ of size $2h - 1$, and $j' \in J$, let $d_i^J[j', j]$ be a share of $d_i^J[j']$ in a $(t+1)$-out-of-$m$ secret sharing, such that party $P_{j'}$ is required in order to recover $d_i^J[j']$. This can be done with Construction 4.*

**Output:** *Party $P_j$ receives $d_i^{J'}[j', j]$, $d_0^J[j]$, $x_i[j]$ for all $i \in [r]$, $J, J' \subset [m]$ of size $2h - 1$, $j \in J$, and $j' \in J'$.*

**Protocol 6 (Multiparty$_{<2/3}$ Coin-Toss).** *Let $r \in \mathbb{N}$ be the number of rounds. Let $\hat{m}$, and $\hat{t}$ be two constants where $\hat{m}$ denotes the number of parties, and $\hat{t}$ is an upper bound on the number of corrupted parties.*

**Common input:** *Number of rounds $r$ and output distribution parameter $\varepsilon$ (jointly reconstructable, possibly unknown to parties).*

**Private inputs:** *The private inputs of the parties were given to them by an oracle computing $\mathrm{HG}(\varepsilon, \hat{m}, \hat{t})$ as defined in Algorithm 5. The input of party $P_j$ for $j \in [\hat{m}]$ is $\boldsymbol{x}_j, \boldsymbol{d}_j$, where*

$$\boldsymbol{x}_j = (x_1[j], \ldots, x_r[j]) \text{ and } \boldsymbol{d}_j = (D_0[j], D_1[j], \ldots D_r[j]),$$

*where*

$$D_i[j] = \left\{ d_i^J[j', j] \mid J \subset [\hat{m}] \wedge |J| = 2h - 1 \wedge j' \in J \right\}, \text{ for } i \in [r]$$

*and*

$$D_0[j] = \left\{ d_0^J[j] \mid J \subset [\hat{m}] \wedge |J| = 2h - 1 \wedge j \in J \right\}.$$

**Interaction rounds:** *For $i = 1$ to $r$:*
(a) *Each party $P_j$ sends $d_i^J[j', j]$ to $P_{j'}$ for every $j' \neq j$ and $J \subset [\hat{m}]$ of size $2h - 1$, such that $j' \in J$.*
(b) *The parties reconstruct $x_i$.*

**Output:** *The honest parties output 1 if $\sum_{i=1}^{r} x_i \geq 0$, and outputs 0 otherwise.*

**In case of abort:** *Let $J \subset [\hat{m}]$ be the set of remaining parties. If $|J| \geq \hat{t} + 1$, then the parties in $J$ go on with the execution of the protocol. Otherwise, they reconstruct and output $d_i^{J'}$, for the lexicographically first $J' \subset [\hat{m}]$ of size $2h - 1$, such that $J \subseteq J'$, and for the largest $i$ for which they have all of the corresponding shares (for the parties of $J$).*

## 3.2  A Coin-Tossing Protocol for $t < 3m/4$

**Algorithm 7 (MultipartyShareGen$_{<3/4}$).** *Let $r \in \mathbb{N}$ be the number of rounds. Let $m$ be a constant representing the number of parties, and let $t$ be a constant which is a bound on the number of corrupted parties. We denote $h = m - t$ (i.e., a lower bound on the number of honest parties). In the following, we call a subset $J \subset [m]$ protected if $2h - 1 \leq |J| \leq t$.*

**Input:** *Number of rounds $r$.*
**Selecting coins and defenses:**
    *For $i = 1$ to $r$:*
1. *Sample $x_i \leftarrow \mathcal{B}in_{r-i+1}$.*
2. *Let $\varepsilon_i \in \left[-\frac{1}{2}, \frac{1}{2}\right]$ be such that $\widehat{\mathcal{B}in}_{s_i,\varepsilon}\left(-\sum_{k=1}^{i} x_k\right) = \widehat{\mathcal{B}in}_{s_0,\varepsilon_i}(0)$.*
3. *For every protected $J \subset [m]$, sample $d_i^J \leftarrow \mathrm{HG}(\varepsilon_i, |J|, t - m + |J|)$.*
**Sharing the values:**
1. *For $i \in [r]$, let $x_i[j]$ be a share of $x_i$ in a $(t+1)$-out-of-$m$ secret sharing.*
2. *For $i \in [r]$, $j \in [m]$, and a protected $J \subset [m]$, let $d_i^J[j]$ be a share of $d_i^J$ in a $(t - m + |J| + 1)$-out-of-$|J|$ secret sharing.*
3. *For $i \in [r]$, $j \in [m]$, a protected $J \subset [m]$, and $j' \in J$, let $d_i^J[j', j]$ be a share of $d_i^J[j']$ in a $(t+1)$-out-of-$m$ secret sharing, such that party $P_{j'}$ is required in order to recover $d_i^J[j']$. This can be done with Construction 4.*
**Output:** *Party $P_j$ receives $d_i^J[j', j]$, $x_i[j]$ for every $i \in [r]$, $J \subset [m]$, and $j' \in J$.*

We are now ready to describe the actual multiparty coin-tossing protocol. We remark that the protocol is defined in the fail-stop model, where corrupted parties must follow the prescribed protocol, unless they decide to prematurely abort the execution at some point. This is done for the sake of simplicity of presentation and compiling the following protocols so that they tolerate any malicious behavior is done by standard techniques, using signatures.

### Protocol 8 (Multiparty$_{<3/4}$ Coin-Toss).

**Common input:** *Number of rounds $r$.*
**Preprocessing:** *Parties run a secure with identifiable abort implementation of Algorithm 7 to obtain their respective outputs. If an abort occurred during the execution, then the remaining parties restart the protocol without the aborting parties.*
**Interaction rounds:** *For $i = 1$ to $r$:*

(a) *Each party $P_j$ sends $d_i^J[j',j]$ to $P_{j'}$ for every $j' \neq j$ and every protected $J \subset [m]$ such that $j' \in J$.*

(b) *The parties reconstruct $x_i$.*

**Output:** *The honest parties output 1 if $\sum\limits_{i=1}^{r} x_i \geq 0$, and outputs 0 otherwise.*

**In case of abort:** *Let $J \subset [m]$ be the set of remaining active parties. If $|J| \geq t+1$, then the parties in $J$ continue with the execution of the protocol. Assume that $|J| \leq t$. If the abort happened before the execution of Algorithm 7, then the parties run a secure with identifiable abort implementation of Algorithm 5 to obtain their respective outputs, and they execute Protocol 6. If the abort happened during the interaction rounds, then the parties execute Protocol 6 with $d_i^{J'}[j]$ as the private input for $P_j$, for the lexicographic first $J' \subset [m]$ such that $J \subseteq J'$, and for the largest $i$ for which they have all of the corresponding shares.[4]*

### 3.3   Stating the Main Results

**Theorem 9.** *Let $m$ and $t$ be two constants such that $t < 3m/4$. Assuming $OT$ exists, then for every $r \in \mathbb{N}$, Protocol 8 is an $r$-round $m$-party $O\left(2^{2^m} \cdot \frac{\log^3 r}{r}\right)$-secure coin-tossing protocol tolerating any fail-stop adversary that corrupts up to $t$ parties, in the $\left(\text{MultipartyShareGen}_{<3/4}, \text{MultipartyShareGen}_{<2/3}\right)$-hybrid model (guaranteeing security with identifiable abort).*

**Corollary 1.** *Let $n$ be the security parameter, and let $m$ and $t$ be two constants, such that $t < 3m/4$. Assuming $OT$ exists, then for every polynomial $r = r(n)$, there exists an $r$-round $m$-party $O\left(2^{2^m} \cdot \frac{\log^3 r}{r}\right)$-secure coin-tossing protocol, against any PPT adversary corrupting up to $t$ parties.*

In order to prove Theorem 9 we first need to show that Protocol 6 is secure. The security of Protocol 6 by itself does not suffice, as in Protocol 8 after an abort, the adversary's view contains some additional information, and so, the following Lemma states that the additional information won't help him to bias the outcome.

**Lemma 6.** *Let $\varepsilon \in \left[-\frac{1}{2}, \frac{1}{2}\right]$, and let $\hat{m}$ and $\hat{t}$ be two constants, such that $\hat{t} < 2\hat{m}/3$. Then for every $r \in \mathbb{N}$, Protocol 6 is an $r$-round $\hat{m}$-party $\left(\hat{t}, O\left(2^{2^m} \cdot \frac{\log^3 r}{r}\right)\right)$-unbiased $\varepsilon$-coin-toss protocol tolerating any fail-stop adversary, corrupting up to $\hat{t}$ parties. Moreover, the above holds even when the adversary gets $\varepsilon$ as an auxiliary input.*

---

[4] Note that in the case where $|J| \leq 2h - 1$, there is an honest majority, and so, in MultipartyShareGen$_{<3/4}$ we could have given them a common bit reconstruct with full security. We decided to instruct the parties to execute Protocol 6 for the sake of simplicity.

The proof of Lemma 6 is deferred to the final version of this paper [1]. We now use it in combination with the results of [24] to prove Theorem 9.

*Proof of (Theorem 9).* Assume without loss of generality that $r \equiv 1 \mod 4$ (otherwise, we set the number of rounds to be the largest $r' < r$ such that $r' \equiv 1 \mod 4$). Hence, $s_i(r)$ is odd, and the output of the parties in an honest execution (without aborts) is a uniform bit. We also assume that $r$ is larger than some constant, which will be determined by the analysis, as otherwise the theorem holds trivially.

Let $\mathcal{A}$ be a fail-stop adversary and let $\mathcal{C} \subset [m]$ be the set of parties that $\mathcal{A}$ corrupts. By assumption, it holds that $|\mathcal{C}| < 3m/4$. Let $V$ be the view of the adversary $\mathcal{A}$ in a random execution of Protocol 8. For a round $I \in [r] \times \{(a), (b)\}$ in the outer protocol, let $V_I$ be the view of the adversary in round $I$ and let $V_I^-$ be it's view without the abort (if happened). We show that the protocol is $\left(t, O\left(2^{2^m} \cdot \frac{\log^3 r}{r}\right)\right)$-unbiased, i.e., we show that:

$$\underset{V}{\mathbb{E}} \left[ |\Delta(V) - \Delta(V^-)| \right] = O\left(2^{2^m} \cdot \frac{\log^3 r}{r}\right). \tag{1}$$

Applying Lemma 1 to Eq. (1) yields that the protocol is $\left(|\mathcal{C}|, O\left(2^{2^m} \cdot \frac{\log^3 r}{r}\right) + neg(n)\right)$-secure. We next prove the correctness of Eq. (1).

We need to analyze the gain of the adversary by prematurely aborting the execution of the protocol. Recall that to prematurely abort the execution of the outer protocol, the adversary needs to instruct at least $m - t$ parties to abort. Otherwise, the remaining active parties are instructed to go on as usual, and indeed, by the properties of the secret sharing scheme, they are able to go through with reconstructing their appropriate secrets. Namely, upon receiving (in Step a of round $i$) shares $d_i^J[j, j']$ from at least $t$ parties $P_{j'}$, party $P_j$ is able to reconstruct $d_i^J[j]$ (using its own share of it). Similarly, upon receiving (in Step b of round $i$) shares $x_i[j']$ from at least $t$ parties $P_{j'}$, party $P_j$ is able to reconstruct $x_i$.

Assume an abort occurred before the interaction rounds. Moreover, we assume that at most $t$ parties remain active. Then by the description of the protocol, the parties are instructed to run a secure with identifiable abort implementation of Protocol 6, and there is no bias in the samples. Then $\Delta(V) = \Delta(V^-) = \frac{1}{2}$, which yields no advantage to the adversary.

Assume an abort occurred during the interaction rounds. Let $I = (i, (\cdot))$ be the first round for which there is an abort and there are at most $t$ active parties remaining. We define two adversaries $\mathcal{A}_{(a)}$ and $\mathcal{A}_{(b)}$ as follows: $\mathcal{A}_{(a)}$ and $\mathcal{A}_{(b)}$ act exactly as does $\mathcal{A}$, until round $I$, in which $\mathcal{A}$ decided to abort. If $I = (i, (a))$, then $\mathcal{A}_{(a)}$ aborts at $(i, (a))$, and $\mathcal{A}_{(b)}$ completes the execution honestly without aborting. If $I = (i, (b))$, then $\mathcal{A}_{(a)}$ completes the execution honestly without aborting, and $\mathcal{A}_{(b)}$ aborts at $(i, (b))$. Let $V_I^{(a)}$ and $V_I^{(b)}$ be the view of $\mathcal{A}_{(a)}$ and $\mathcal{A}_{(b)}$, respectively.

**Assume that $I = (i, (a))$:**

The view of the adversary $\mathcal{A}_{(a)}$ consists of:

$$\{x_1, x_2 \ldots x_{i-1}\} \text{ and } D_i^{\mathcal{C}},$$

where

$$D_i^{\mathcal{C}} = \left\{ d_k^{\mathcal{C}'} : |\mathcal{C}' \cap \mathcal{C}| > t - m + |\mathcal{C}'| \wedge k \leq i \right\},$$

is the set of all the defenses that the adversary can see up to and including round $i$. In addition, the adversary $\mathcal{A}_{(a)}$ sees many shares that are useless to it. Specifically, the adversary $\mathcal{A}_{(a)}$ holds shares of two different types. The first type are shares of the elements in its view that $\mathcal{A}_{(a)}$ completely reconstructed (i.e., those specified above). This type of shares are useless to $\mathcal{A}_{(a)}$, as they were chosen independently of all other information. The second type are shares of the defense values of other sets that $\mathcal{A}_{(a)}$ cannot reconstruct (since it sees at most $t$ such shares). This type of shares are useless to $\mathcal{A}_{(a)}$ by the properties of secret sharing schemes. We thus, disregard these two types of shares, and continue with the analysis as if the view of $\mathcal{A}_{(a)}$ consists only of the random coins and of $D_i^{\mathcal{C}}$. Formally, the view of the adversary $\mathcal{A}_{(a)}$ may contain only part of $D_i^{\mathcal{C}}$, however, an adversary with more information can always emulate one with less information by simply disregarding parts of its view.

Each $d_k^{\mathcal{C}'}$ is a vector, which consists of $O(r^2)$ elements from $\{-1, 1\}$, where the elements are sampled according to $Ber(\varepsilon_k)$, where $\varepsilon_k$ satisfies $\widehat{Bin}_{s_0, \varepsilon_k}(0) = \widehat{Bin}_{s_k}\left(-\sum_{l=1}^{k} x_l\right)$. As $D_i^{\mathcal{C}}$ has $O(r^2)$ bits in total, Lemma 2 tell us that:

$$\mathop{E}_{V_I^{(a)}} \left[ \left| \Delta\left(V_I^{(a)}\right) - \Delta\left(V_I^{(a)-}\right) \right| \right] = O\left(\frac{\log^3 r}{r}\right), \tag{2}$$

**Assume that $I = (i, (b))$:**

The view of the adversary $\mathcal{A}_{(b)}$ consists of:

$$\{x_1, x_2 \ldots x_i\} \text{ and } D_i^{\mathcal{C}},$$

As in the previous case, we disregard the other shares that $\mathcal{A}_{(b)}$ sees. Since the defenses are sampled independently, given $x_i$, and since the expectation of each $d_i^{\mathcal{C}'}$ is exactly the game value given $x_1, \ldots x_i$, the adversary gains nothing by aborting in this rounds.

Combining the two cases yields the bound on the maximum bias $\mathcal{A}$ can do in round $I$:

$$\mathop{E}_{V_I} \left[ \left| \Delta(V_I) - \Delta(V_I^-) \right| \right]$$

$$= \mathop{E}_{V_I^{(a)}} \left[ \left| \Delta\left(V_I^{(a)}\right) - \Delta\left(V_I^{(a)-}\right) \right| \right] + \mathop{E}_{V_I^{(b)}} \left[ \left| \Delta\left(V_I^{(b)}\right) - \Delta\left(V_I^{(b)-}\right) \right| \right]$$

$$= O\left(\frac{\log^3 r}{r}\right).$$

In order the conclude the proof of security we need to show that the remaining corrupted parties can't bias the outcome by more than $O\left(\frac{\log^3 r}{r}\right)$. Let $J \subset [m]$ be the set of the remaining parties, let $\hat{m} = |J|$, and let $h = m - t$ be a lower bound on the number of honest parties. Since, at least $h$ parties aborted, it follows that there are at most $\hat{t} := \hat{m} - h$ corrupted parties in $J$. By assumption, $t < \frac{3m}{4}$, and hence, $\hat{t} < \hat{m} - \frac{m}{4}$. Since $\hat{m} \leq t < \frac{3m}{4}$, it holds that $\hat{t} < \hat{m} - \frac{\hat{m}}{3} = \frac{2\hat{m}}{3}$. Therefore by Lemma 6 it holds that:

$$\mathop{\mathbb{E}}_{V_{\text{inner}}} \left[ |\Delta\left(V_{\text{inner}}\right) - \Delta\left(V_{\text{inner}}^-\right)| \right] = O\left(2^{2^m} \cdot \frac{\log^3 r}{r}\right), \tag{3}$$

where $V_{\text{inner}}$ is the view of $\mathcal{A}$ in Protocol 6, with $\varepsilon_i$ included. Note that Lemma 6 assumes that the adversary's view contains only $\varepsilon_i$ as auxiliary input. However, Eq. (3) still holds, as the rest of the view is independent of $V_I$, and give no information to the adversary. □

### 3.3.1  Proof of Corollary 1. We next sketch the proof of Corollary 1.

*Proof Sketch of Corollary 1.* We adjust Protocols 6 and 8, so that each message that any of the parties ever needs to send is signed, and all other parties verify this signature upon receiving this messages. If at some point in the execution, party $P$ broadcasts a message that is not properly signed, then, all parties treat this as if $P$ has aborted the computation and is no longer active. This is done similarly to the way presented in [8]. Towards this end, Algorithm 7 is changed so that for every round $i$, every two parties $P_j, P_{j'}$, and every appropriate subset $J$, both $x_i[j]$ and $d_i^J[j]$ are signed. In addition, let $\sigma(i, J, j')$ be the signature attributed to $d_i^J[j']$, then, $d_i^J[j', j]$ is redefined to be a share of $\left(d_i^J[j'], \sigma(i, J, j')\right)$ in a $(t + 1)$-out-of-$m$ secret sharing, such that party $P_{j'}$ is required in order to recover $d_i^J[j']$. Finally, $d_i^J[j', j]$ is also signed.

We further modify Algorithm 7 so that for every $i \in [r]$, the computation of $\varepsilon_i$ (see Item 2) can be done efficiently, similarly to the way done in [24]. Observe that $\varepsilon_i$ is only used to sample $O\left(r^2\right)$ independent bits, hence it can be efficiently estimated with $\tilde{\varepsilon}_i$, such that the statistical difference between the samples is bounded by $\frac{1}{r^2}$. It follows that the adjusted Protocol 8 is a $r$-round, $m$-party $O\left(2^{2^m} \cdot \frac{\log^3 r}{r} + \frac{r}{r^2}\right)$-secure coin-tossing protocol against any PPT adversary.

Finally, similarly to [8], the modified (efficient) functionality is replaced by a secure with identifiable abort protocol that runs in a constant number of rounds. As explained in [8], this can be done using (a variation on) the protocol of [32].

**Acknowledgements.** We are grateful to Iftach Haitner and Amos Beimel for useful conversations.

## References

1. Alon, B., Omri, E.: Almost-optimally fair multiparty coin-tossing with nearly three-quarters malicious (2016). http://omrier.wixsite.com/eran-omri/almost-opt-fair-multiparty-coin-tos. Full version of this paper

2. Asharov, G.: Towards characterizing complete fairness in secure two-party computation. In: Lindell, Y. (ed.) TCC 2014. LNCS, vol. 8349, pp. 291–316. Springer, Heidelberg (2014)

3. Asharov, G., Lindell, Y., Rabin, T.: A full characterization of functions that imply fair coin tossing and ramifications to fairness. In: Sahai, A. (ed.) TCC 2013. LNCS, vol. 7785, pp. 243–262. Springer, Heidelberg (2013)

4. Asharov, G., Beimel, A., Makriyannis, N., Omri, E.: Complete characterization of fairness in secure two-party computation of Boolean functions. In: Dodis, Y., Nielsen, J.B. (eds.) TCC 2015, Part I. LNCS, vol. 9014, pp. 199–228. Springer, Heidelberg (2015)

5. Aumann, Y., Lindell, Y.: Security against covert adversaries: efficient protocols for realistic adversaries. In: Vadhan, S.P. (ed.) TCC 2007. LNCS, vol. 4392, pp. 137–156. Springer, Heidelberg (2007). doi:10.1007/978-3-540-70936-7_8

6. Averbuch, B., Blum, M., Chor, B., Goldwasser, S., Micali, S.: How to implement Bracha's $O(\log n)$ Byzantine agreement algorithm (1985, Unpublished manuscript)

7. Beimel, A., Lindell, Y., Omri, E., Orlov, I.: $1/p$-secure multiparty computation without honest majority and the best of both worlds. In: Rogaway, P. (ed.) CRYPTO 2011. LNCS, vol. 6841, pp. 277–296. Springer, Heidelberg (2011)

8. Beimel, A., Omri, E., Orlov, I.: Protocols for multiparty coin toss with dishonest majority. In: Rabin, T. (ed.) CRYPTO 2010. LNCS, vol. 6223, pp. 538–557. Springer, Heidelberg (2010)

9. Ben-Or, M., Goldwasser, S., Wigderson, A.: Completeness theorems for non-cryptographic fault-tolerant distributed computation (extended abstract). In: Proceedings of the 29th Annual Symposium on Foundations of Computer Science (FOCS), pp. 1–10 (1988)

10. Berman, I., Haitner, Tentes, A.: Coin flipping of any constant bias implies one-way functions. In: Symposium on Theory of Computing, STOC 2014, New York, NY, USA, 31 May - 03 June 2014, pp. 398–407 (2014)

11. Blum, M.: Coin flipping by telephone. In: Advances in Cryptology - CRYPTO 1981, pp. 11–15 (1981)

12. Blum, M.: Coin flipping by telephone a protocol for solving impossible problems. SIGACT News 15(1), 23–27 (1983)

13. Canetti, R.: Security and composition of multiparty cryptographic protocols. J. Cryptol. 13(1), 143–202 (2000)

14. Cleve, R.: Limits on the security of coin flips when half the processors are faulty. In: Proceedings of the 18th Annual ACM Symposium on Theory of Computing (STOC), pp. 364–369 (1986)

15. Cleve, R., Impagliazzo, R.: Martingales, collective coin flipping and discrete control processes (1993, Manuscript)

16. Dachman-Soled, D., Lindell, Y., Mahmoody, M., Malkin, T.: On the black-box complexity of optimally-fair coin tossing. In: Ishai, Y. (ed.) TCC 2011. LNCS, vol. 6597, pp. 450–467. Springer, Heidelberg (2011)

17. Dachman-Soled, D., Mahmoody, M., Malkin, T.: Can optimally-fair coin tossing be based on one-way functions? In: Lindell, Y. (ed.) TCC 2014. LNCS, vol. 8349, pp. 217–239. Springer, Heidelberg (2014)

18. Goldreich, O.: Foundations of Cryptography: Volume 2, Basic Applications. Cambridge University Press, New York (2009)

19. Goldreich, O., Micali, S., Wigderson, A.: How to play any mental game or a completeness theorem for protocols with honest majority. In: STOC 19, pp. 218–229 (1987)

20. Gordon, S.D., Katz, J.: Partial fairness in secure two-party computation. In: Gilbert, H. (ed.) EUROCRYPT 2010. LNCS, vol. 6110, pp. 157–176. Springer, Heidelberg (2010)

21. Gordon, S.D., Katz, J.: Partial fairness in secure two-party computation. J. Cryptol. **25**(1), 14–40 (2012)

22. Gordon, S.D., Hazay, C., Katz, J., Lindell, Y.: Complete fairness in secure two-party computation. In: Proceedings of the 40th Annual ACM Symposium on Theory of Computing (STOC), pp. 413–422 (2008)

23. Haitner, I., Omri, E.: Coin flipping with constant bias implies one-way functions. In: Proceedings of the 52nd Annual Symposium on Foundations of Computer Science (FOCS), pp. 110–119 (2011)

24. Haitner, I., Tsfadia, E.: An almost-optimally fair three-party coin-flipping protocol. In: Symposium on Theory of Computing, STOC 2014, New York, NY, USA, 31 May - 03 June 2014, pp. 408–416 (2014). http://www.cs.tau.ac.il/~iftachh/papers/3PartyCF/QuasiOptimalCF_Full.pdf

25. Haitner, I., Nguyen, M., Ong, S.J., Reingold, O., Vadhan, S.: Statistically hiding commitments and statistical zero-knowledge arguments from any one-way function. SIAM J. Comput. **39**(3), 1153–1218 (2009)

26. Ishai, Y., Ostrovsky, R., Zikas, V.: Secure multi-party computation with identifiable abort. In: Garay, J.A., Gennaro, R. (eds.) CRYPTO 2014, Part II. LNCS, vol. 8617, pp. 369–386. Springer, Heidelberg (2014)

27. Katz, J.: On achieving the "best of both worlds" in secure multiparty computation. In: STOC07, pp. 11–20 (2007)

28. Maji, H.K., Prabhakaran, M., Sahai, A.: On the computational complexity of coin flipping. In: Proceedings of the 51st Annual Symposium on Foundations of Computer Science (FOCS), pp. 613–622 (2010)

29. Makriyannis, N.: On the classification of finite Boolean functions up to fairness. In: Abdalla, M., De Prisco, R. (eds.) SCN 2014. LNCS, vol. 8642, pp. 135–154. Springer, Heidelberg (2014)

30. Moran, T., Naor, M., Segev, G.: An optimally fair coin toss. In: Reingold, O. (ed.) TCC 2009. LNCS, vol. 5444, pp. 1–18. Springer, Heidelberg (2009)

31. Naor, M.: Bit commitment using pseudorandomness. J. Cryptol. **4**(2), 151–158 (1991). Preliminary version in CRYPTO 1989

32. Pass, R.: Bounded-concurrent secure multi-party computation with a dishonest majority. In: Proceedings of the 36th Annual ACM Symposium on Theory of Computing (STOC), pp. 232–241 (2004)

33. Shamir, A.: How to share a secret. Commun. ACM **22**(11), 612–613 (1979)

# Binary AMD Circuits from Secure Multiparty Computation

Daniel Genkin[1,2(✉)], Yuval Ishai[1,3], and Mor Weiss[1]

[1] Technion, Haifa, Israel
{danielg3,yuvali,morw}@cs.technion.ac.il
[2] Tel Aviv University, Tel Aviv, Israel
[3] UCLA, Los Angeles, USA

**Abstract.** An *AMD circuit* over a finite field $\mathbb{F}$ is a randomized arithmetic circuit that offers the "best possible protection" against additive attacks. That is, the effect of every additive attack that may blindly add a (possibly different) element of $\mathbb{F}$ to every internal wire of the circuit can be simulated by an ideal attack that applies only to the inputs and outputs.

Genkin et al. (STOC 2014, Crypto 2015) introduced AMD circuits as a means for protecting MPC protocols against active attacks, and showed that every arithmetic circuit $C$ over $\mathbb{F}$ can be transformed into an equivalent AMD circuit of size $O(|C|)$ with $O(1/|\mathbb{F}|)$ simulation error. However, for the case of the binary field $\mathbb{F} = \mathbb{F}_2$, their constructions relied on a tamper-proof output decoder and could only realize a weaker notion of security.

We obtain the first constructions of fully secure binary AMD circuits. Given a boolean circuit $C$ and a statistical security parameter $\sigma$, we construct an equivalent binary AMD circuit $C'$ of size $|C| \cdot \mathrm{polylog}(|C|, \sigma)$ (ignoring lower order additive terms) with $2^{-\sigma}$ simulation error. That is, the effect of toggling an arbitrary subset of wires can be simulated by toggling only input and output wires.

Our construction combines in a general way two types of "simple" honest-majority MPC protocols: protocols that only offer security against *passive* adversaries, and protocols that only offer *correctness* against active adversaries. As a corollary, we get a conceptually new technique for constructing active-secure two-party protocols in the OT-hybrid model, and reduce the open question of obtaining such protocols with constant computational overhead to a similar question in these simpler MPC models.

**Keywords:** Algebraic Manipulation Detection · AMD circuits · Secure multiparty computation

## 1   Introduction

In this paper we give the first construction of boolean circuits which are secure against attacks that can toggle an arbitrary subset of the wires, in the sense that

© International Association for Cryptologic Research 2016
M. Hirt and A. Smith (Eds.): TCC 2016-B, Part I, LNCS 9985, pp. 336–366, 2016.
DOI: 10.1007/978-3-662-53641-4_14

every such attack is equivalent to attacking *only* the inputs and outputs of the circuit. We begin with a short overview of the problem and related background.

An *Algebraic Manipulation Detection* (AMD) code [3] over a finite field $\mathbb{F}$ is a randomized coding scheme that offers the best possible protection against *additive* attacks, namely attacks that can blindly add a fixed (but possibly different) element from $\mathbb{F}$ to every entry of the codeword. Since an attacker can destroy all information by adding a random field element to every symbol, the best one can hope for is to *detect* errors with high probability, rather than correct them.

An analogous goal of protecting *computations* against additive attacks was recently considered by Genkin et al. [11]. This goal is captured by the notion of an *AMD circuit*, a randomized arithmetic circuit which offers the best possible protection against additive attacks that may add a (possibly different) field element to every wire. Since the adversary can legitimately attack input and output wires, the best one can hope for is to limit the adversary to these inevitable attacks. That is, in an AMD circuit the effect of every additive attack that may apply to all internal wires in the circuit can be simulated by an ideal attack that applies only to the inputs and outputs. Combining such AMD circuits with a standard AMD code, one can also protect the inputs and outputs by employing small tamper-proof input encoder and output decoder.

The study of AMD circuits in [11] was motivated by the observation that in the simplest information-theoretic MPC protocols from the literature, that were only designed to offer protection against *passive* (i.e., semi-honest) adversaries, the effect of every *active* (malicious) adversary corresponds precisely to an additive attack on the circuit being evaluated. Thus, a useful paradigm for tackling the difficult goal of protecting against active attacks is to apply such a simple passive-secure protocol to an AMD-encoded computation. This paradigm seems quite promising even from a concrete efficiency perspective [10,13].

The main result of [11] is that every arithmetic circuit $C$ over $\mathbb{F}$ can be transformed into an equivalent AMD circuit of size $O(|C|)$ with $O(1/|\mathbb{F}|)$ simulation error. This provides poor security guarantees over small fields, and in fact the construction used to achieve this can be completely broken when applied over the binary field $\mathbb{F} = \mathbb{F}_2$. (The natural approach of using an arithmetic circuit over a large extension field does not apply here, because the computation of field multiplications is also subject to attacks.) For the binary case, an alternative construction from [11] relies on the use of a tamper-proof output decoder and can only realize a weaker notion of security that allows for arbitrary correlations between the input and the event an attack is detected.

The goal of this work is to remedy this state of affairs and provide fully secure AMD circuits over small fields, with a primary focus on the binary case. Binary AMD circuits can be viewed as standard (randomized) boolean circuits (over the full basis) that are subject to arbitrary *toggling* attacks: the adversary may choose to toggle the values of an arbitrary subset of the wires. This seems quite natural even from a pure fault tolerance perspective and can be viewed as a strict generalization of the classical "random noise" fault model considered by von Neumann [20], Dobrushin and Ortyukov [7], and Pippenger [18]. Such a toggling attack model may not be too far from some real-life scenarios like faults introduced by faulty hardware or cosmic radiation.

In the context of applications to MPC, the binary case is important because it enables us to apply the AMD circuits methodology also to natural protocols that are cast in the *OT-hybrid model*. These include the simple passive-secure version of the GMW protocol [12]. In contrast, the MPC applications in [11] for the case of dishonest majority could only apply to arithmetic extensions of the GMW protocol that employ an arithmetic extension of OT denoted by OLE.[1] Replacing OLE by OT is particularly attractive in light of efficient OT extension techniques [14,17] that do not apply to OLE.

We obtain the first constructions of fully secure binary AMD circuits. Given a boolean circuit $C$ and a statistical security parameter $\sigma$, we construct an equivalent binary AMD circuit $\hat{C}$ of size $|C| \cdot \text{polylog}(|C|, \sigma)$ (ignoring lower order additive terms) with $2^{-\sigma}$ simulation error. That is, the effect of toggling an arbitrary subset of wires can be simulated by toggling only input and output wires.

Our construction combines in a general way two types of "simple" honest-majority MPC protocols: protocols that only offer security against *passive* adversaries, and protocols that only offer *correctness* against active adversaries. It proceeds according to the following steps. First, we use the correct-only MPC protocol to convert a relatively simple AMD circuit that provides only *constant correctness* (i.e., any "potentially harmful" attack is detected with some positive probability) into one that offers full correctness (i.e., attacks are detected except with $2^{-\sigma}$ probability). However, this notion of correctness is not enough, mainly because it does not rule out correlations between the input and the event an attack is detected. We eliminate such correlations generically by distributing the computation using a passive-secure MPC protocol. The analysis of this step crucially relies on a recent lemma due to Bogdanov et al. [1] that uses the degree of approximating the OR function by real-valued polynomials to upper bound its best-case advantage in distinguishing between two distributions that are $t$-wise indistinguishable.

As a byproduct, we get a conceptually different technique for constructing active-secure two-party protocols in the OT-hybrid model from these simpler building blocks. This technique is appealing because in a sense it counters the common wisdom that "security" is more than a combination of "correctness" and "secrecy." Indeed, our construction shows a *general* way to obtain full security (for MPC protocols in the OT-hybrid model) by only combining one MPC protocol that guarantees *correctness* and another that only guarantees *secrecy*, namely security in the presence of passive attacks. Moreover, the "correct-only MPC" component can be instantiated by a trivial protocol in which each party performs the entire computation locally. (To get the asymptotic efficiency

---

[1] An Oblivious Linear-function Evaluation (OLE) over a field $\mathbb{F}$ takes a field element $x \in \mathbb{F}$ from Receiver and a pair $(a, b) \in \mathbb{F}^2$ from Sender and delivers $ax + b$ to Receiver. In the case of binary fields, OLE can be realized via a single call to standard (bit-) OT.

mentioned above, we need to apply more sophisticated correct-only MPC proto-cols that offer better efficiency.) This can be compared with the IPS compiler [16], which also provides a general way of obtaining active-secure protocols in the OT-hybrid model, but requires an honest-majority MPC protocol that provides *active security* (which is strictly stronger than relying on active correctness and passive security).

In addition to its conceptual appeal, our new methodology also sheds new light on an intriguing open question about the complexity of secure computation [15]: Are there active-secure two-party protocols that achieve *constant computational overhead*? In other words, does the asymptotic multiplicative cost of protecting against active adversaries have to grow with the level of security? This question is open even when allowing a trusted source of correlated randomness, and in particular it is open in the OT-hybrid model. The best known protocols [6] have a polylogarithmic overhead in the security parameter (a result that we can match using binary AMD circuits). Our work reduces this question to the same open question in arguably simpler models. Indeed, while our construction involves some additional ad-hoc components (on top of the two types of MPC protocols discussed above) the additional cost they incur depends only on the input and output sizes, and not on the size of the computation. Furthermore, our construction also employs AMD codes to encode the entire protocol transcript, but these can be implemented with constant computational overhead (see Claim 18 and Corollary 1 in Sect. 6).

## 1.1 Our Results and Techniques

We now provide more details about our results, and the underlying techniques (summarized in Fig. 1 below). We begin by defining the notion of *additive correctness*, which allows the evaluation of a function $f : \mathbb{F}^n \to \mathbb{F}^k$ in the presence of an additive attack[2] on the circuit computing $f$.

**Definition 1 (Additive correctness; cf. full version of [11], Definition 4.1).** *Let $\epsilon > 0$. We say that a randomized circuit $\widehat{C} : \mathbb{F}^n \to \mathbb{F}^t \times \mathbb{F}^k$ is an $\epsilon$-additively-correct implementation of a function $f : \mathbb{F}^n \to \mathbb{F}^k$ if the following holds:*

- **Completeness.** *For all $\mathbf{x} \in \mathbb{F}^n$ it holds that $\Pr[\widehat{C}(\mathbf{x}) = (0^t, f(\mathbf{x}))] = 1$.*
- **Additive correctness.** *For any additive attack $\mathbf{A}$ there exists $\mathbf{a}^{\mathsf{in}} \in \mathbb{F}^n$, and $\mathbf{a}^{\mathsf{out}} \in \mathbb{F}^k$, such that for every input $\mathbf{x}$ it holds that $\Pr[\widehat{C}^{\mathbf{A}}(\mathbf{x}) \notin \mathsf{ERR} \cup \{(0^t, f(\mathbf{x} + \mathbf{a}^{\mathsf{in}}) + \mathbf{a}^{\mathsf{out}})\}] \leq \epsilon$, where $C^{\mathbf{A}}$ is the circuit obtained by subjecting $C$ to the additive attack $\mathbf{A}$, and $\mathsf{ERR} = (\mathbb{F}^t \setminus \{0^t\}) \times \mathbb{F}^k$.*

*We say that $\widehat{C}$ is an $\epsilon$-additively-correct implementation of a circuit $C$ if $\widehat{C}$ is an $\epsilon$-additively-correct implementation of the function $f_C$ computed by $C$.*

Previous works [10,11] constructed additively correct implementations for arithmetic circuits over any finite field $\mathbb{F}$, with constant overhead, and $\epsilon = O(1/|\mathbb{F}|)$. In particular, for $\mathbb{F} = \mathbb{F}_2$ the error is constant.

---

[2] For a formal definition of additive attacks, see Definition 3.

### 1.1.1    Correctness Amplification via Correct-Only MPC

For any function $f$, and security parameter $\sigma$, we show the first $2^{-\sigma}$-additively-correct implementation of $f$, with polylogarithmic blowup:

**Theorem 1 (Cf. Theorem 11).** *For any depth-$d$ arithmetic circuit $C : \mathbb{F}^n \to \mathbb{F}^k$, and any security parameter $\sigma$, there exists a $2^{-\sigma}$-additively-correct implementation $\widehat{C}$ of $C$, where $|\widehat{C}| = |C| \cdot polylog(|C|, \sigma) + poly(n, k, d, \sigma)$.*

To prove Theorem 1, we present a general method of amplifying additive correctness based on "correct-only" MPC protocols. Such protocols enable a single client, aided by $m$ servers, to evaluate an arithmetic circuit $C$ on its input, while guaranteeing correctness of the computation in the presence of an active adversary that corrupts a constant fraction of the servers. Moreover, the only interaction between the client and servers is in the first and last rounds.

More specifically, for $m$ servers, and some constant $c$, let $\pi$ be a $d$-round $cm$-correct MPC protocol, namely correctness holds even if $cm$ servers are corrupted. Let InpEnc, OutDec denote the functions used by the client in the first and last rounds (respectively) to compute its messages to the servers, and its output (respectively). Let NextMSG denote the function used by the servers to compute their messages in each round of the protocol. The naive approach towards implementing the circuit $\widehat{C}$ using $\pi$ is to implement every sub-circuit (namely, each of NextMSG, InpEnc, and OutDec) using an $\epsilon$-additively-correct implementation. This naive approach fails because an additive attack may influence the computation of *all* NextMSG *functions*, which corresponds to *actively corrupting all servers in* $\pi$, whereas the correctness of the protocol only holds when at most $cm$ servers are corrupted. Consequently, additive attacks on $\widehat{C}$ can be divided into two categories:

1. **"Small" Attacks.** The sub-circuits of $\widehat{C}$ that these attacks influence correspond to at most $cm$ servers of $\pi$, so by the $cm$-correctness of $\pi$, such attacks cannot affect the output.
2. **"Large" Attacks.** The sub-circuits of $\widehat{C}$ that these attacks affect correspond to *more* than $cm$ servers of $\pi$. Since each sub-circuit (computing NextMSG) is implemented using an $\epsilon$-additively-correct implementation, then except with probability $\epsilon^{cm}$ at least one of these attacks is detected, or their effect on the computations in the sub-circuits is equivalent to additive attacks on the inputs and outputs of the sub-circuits.

Additionally, we notice that any additive attack on $\pi$ consists of sub-attacks of one of three types:

1. **Attacks on communication channels.** These attacks only affect the messages that parties receive in $\pi$, but do not modify the NextMSG functions. By encoding all messages sent in the protocol using an AMD encoding scheme (and altering InpEnc, NextMSG, OutDec to operate on AMD codewords) we can guarantee that such attacks are detected with high probability.
2. **Attacks on NextMSG functions.** These attacks arbitrary modify the NextMSG function of the corresponding server, but (as noted above) can be

protected against by replacing all NextMSG functions with their $\epsilon$-additively-correct implementations.

3. **Attacks on client functions.** Since $\pi$ is correct only as long as the client is honest, such attacks may arbitrarily affect the outputs. Therefore, to guarantee that such attacks are detected except with negligible probability, InpEnc and OutDec should be replaced with their $2^{-\sigma}$-additively-correct implementation. The crucial point here is that since $|\mathsf{InpEnc}| + |\mathsf{OutDec}|$ is polynomial in the inputs and outputs, but otherwise independent of $|C|$, then *any* efficient $2^{-\sigma}$-additively-correct implementation will do, and the resultant overhead would still be polylog$(m\,|C|)$. (We show an example of a $2^{-\sigma}$-additively-correct implementation in Appendix A.)

Consequently, we implement the circuit $\widehat{C}$ using $\pi$ as follows. We first replace the NextMSG functions of $\pi$ with the functions NextMSG$'$ that operate on AMD codewords, and replace NextMSG$'$ with its $\epsilon$-additively correct implementation, $\widehat{\mathsf{NextMSG}'}$, such that $\left|\widehat{\mathsf{NextMSG}'}\right| = O\left(|\mathsf{NextMSG}'|\right)$, and $\epsilon$ is constant. Additionally, we replace InpEnc (resp., OutDec) with the function InpEnc$'$ (resp., OutDec$'$) which outputs (resp., takes as input) AMD codewords, and replace InpEnc$'$, OutDec$'$ with their $2^{-\sigma}$-additively correct implementations $\widehat{\mathsf{InpEnc}}, \widehat{\mathsf{OutDec}}$. Thus, $|\widehat{C}| = |\widehat{\mathsf{InpEnc}}| + |\widehat{\mathsf{OutDec}}| + \sum_{i=1}^{m}\sum_{j=1}^{d}|\widehat{\mathsf{NextMSG}_i^j}|$. We use an efficient correct-only MPC protocol $\pi$ (e.g., a slightly simplified version of [6]) to guarantee that when $m = \sigma$, the multiplicative computational overhead is only polylog$(\sigma, |C|)$. (Since we would like the overhead to be sublinear in $\sigma$, we cannot use a trivial correct-only MPC protocol for evaluating $C$ on input $x$.) For this choice of $\pi$, $|\mathsf{InpEnc}| + |\mathsf{OutDec}| = \mathrm{poly}(n, k)$, so $|\widehat{\mathsf{InpEnc}}| + |\widehat{\mathsf{OutDec}}| = \mathrm{poly}(n, k)$. Similarly, $\sum_{i=1}^{\sigma}\sum_{j=1}^{d}|\mathsf{NextMSG}_i^j| = |C| \cdot \mathrm{polylog}(|C|, \sigma) + \mathrm{poly}(n, k, d, \sigma)$, so $\sum_{i=1}^{\sigma}\sum_{j=1}^{d}|\widehat{\mathsf{NextMSG}_i^j}| = |C| \cdot \mathrm{polylog}(|C|, \sigma) + \mathrm{poly}(n, k, d, \sigma)$. (See Sect. 4 for a more complete discussion.)

### 1.1.2 From Correctness to Security via Passive-Secure MPC

Additive correctness (as guaranteed by Theorem 1) does not rule out the possibility that the probability of ERR (due to set flags) is correlated with the inputs of $\widehat{C}$. Thus, additive attacks on additively-correct circuits may leak information about the inputs to $\widehat{C}$, making additive correctness insufficient for applications to secure multiparty computation (as described in, e.g., [11]) that require that no such correlations exist. This stronger property is achieved by the following *additive security* property which, intuitively, guarantees that any additive attack on $\widehat{C}$ is equivalent (up to a small statistical distance) to an additive attack on the inputs and outputs of the function that $\widehat{C}$ computes. Formally,

**Definition 2 (Additively-secure implementation).** *Let $\epsilon > 0$. We say that a randomized circuit $C : \mathbb{F}^n \to \mathbb{F}^k$ is an $\epsilon$-additively-secure implementation of a function $f : \mathbb{F}^n \to \mathbb{F}^k$ if the following holds.*

- **Completeness.** *For every $x \in \mathbb{F}^n$, $\Pr[C(x) = f(x)] = 1$.*

– **Additive-attack security.** *For any additive attack $\mathcal{A}$ there exist $\mathbf{a}^{\text{in}} \in \mathbb{F}^n$, and a distribution $\mathcal{A}^{\text{Out}}$ over $\mathbb{F}^k$, such that for every $\mathbf{x} \in \mathbb{F}^n$, $\text{SD}(C^{\mathbf{A}}(\mathbf{x}), f(\mathbf{x} + \mathbf{a}^{\text{in}}) + \mathcal{A}^{\text{out}}) \leq \epsilon$.*

As in the case of additive correctness, previous works [10,11] constructed additively-secure implementations for arithmetic circuits over any finite field $\mathbb{F}$, with constant overhead, and $\epsilon = O(1/|\mathbb{F}|)$. Unfortunately, their results and techniques are of little use in the binary case, since the error is too large. We present the first additively-secure circuits with negligible error probability over the binary field. Formally:

**Theorem 2 (Cf. Theorem 14).** *For any depth-d arithmetic circuit $C : \mathbb{F}^n \to \mathbb{F}^k$, and security parameter $\sigma$, there exists a $2^{-\sigma}$-additively-secure implementation $\widehat{C}$ of $C$, where $|\widehat{C}| = |C| \cdot polylog(|C|, \sigma) + poly(n, k, d, \sigma)$.*

As in Sect. 1.1.1, the high-level idea is to implement $C$ using an $m$-party protocol (in the standard model, namely not in the server-client model), where the functions computed by the parties are replaced with additively-correct implementations that operate over AMD encodings. However, since our main concern now is privacy, and not correctness, we use *passive-secure* protocols which *only guarantee privacy* against a constant fraction $c$ of passively-corrupted parties. This privacy guarantee allows us to decouple the probability of ERR of the additively *correct* circuits from their inputs, resulting in additively *secure* circuits.

More specifically, the input of the circuit $C$ is shared between the parties using an additive secret-sharing, and the $d$-round passive-secure protocol $\pi$ computes the functionality that reconstructs the input from the shares, evaluates $C$, and outputs an additive secret-sharing of the output. The privacy property of $\pi$, together with the secrecy property of the secret-sharing scheme, guarantee that the joint view of a constant fraction of passively-corrupted parties reveals no information about the inputs, or outputs, of the computation. As in Sect. 1.1.1, $\widehat{C}$ is obtained from $\pi$ by first replacing all NextMSG functions with the functions NextMSG′ that operate on AMD encodings, and then implementing each NextMSG′ using a $2^{-\sigma}$-additively-correct implementation $\widehat{\text{NextMSG}′}$ with constant overhead (such as the one from Theorem 1). As $\widehat{C}$ should emulate $C$ (rather than output a secret sharing of the output of $C$), the output is reconstructed from the outputs of the parties in $\pi$ by summing their shares, and is then combined with the flags generated by *all* the additively-correct implementations, such that if *any* of the flags were set then the output of $\widehat{C}$ is random.

Using a union-bound over the additive-correctness property of the additively-correct implementations, except with probability at most $|C| \cdot 2^{-\sigma}$ any additive attack on the $\widehat{\text{NextMSG}′}$ functions either sets a flag, or is equivalent to an attack on the inputs and outputs of NextMSG′. Except for the inputs, and output, of $\widehat{C}$, the inputs and outputs of the NextMSG′ functions are protected by the AMD encoding scheme, so by the additive soundness of the AMD encoding scheme, any attack (except for an attack on the inputs, and output, of $\widehat{C}$) will set a flag with overwhelming probability. Thus, the only additive attacks that do not set a flag (with overwhelming probability) are attacks on the inputs and outputs of $\widehat{\mathsf{C}}$, which are equivalent to attacks on the inputs and outputs of $C$. Thus, with overwhelming probability the execution of $\pi$ is correct *even in the presence of additive attacks*.

It remains to show that the probability of setting a flag in $\widehat{\mathsf{C}}$, thus causing the output to be random, is input independent. We use the fact that the probability that a subset of $\widehat{\mathsf{NextMSG}}'$ implementations set their flags depends only on their joint inputs and outputs, and distinguish between two types of attacks.

1. **"Small" attacks.** These attacks attempt to corrupt less than $cm$ parties. Therefore, the probability that a flag is set depends only on the inputs and outputs of these parties which, by the privacy of $\pi$, and the secrecy of the secret-sharing scheme, is independent of the inputs of $\widehat{\mathsf{C}}$.
2. **"Large" attacks.** These attacks attempt to corrupt more than $cm$ parties, and so we can no longer use the privacy of $\pi$. However, notice that in this case the output of $\widehat{\mathsf{C}}$ is random if and only if at least one additively-correct implementation set a flag (regardless of the identity or number of flags that were set). That is, the output is random if and only if the OR of the flags is 1. Using a recent lemma of [1] (stated as Lemma 1 below), the correlation of the OR with the input is negligible, because the OR is computed over a large fraction of the flags.

As for the size of $\widehat{\mathsf{C}}$, notice that $|\widehat{\mathsf{C}}| = \sum_{i=1}^{m} \sum_{j=1}^{d} |\widehat{\mathsf{NextMSG}_i^j}|$. To obtain the small overhead guaranteed by Theorem 2, we use a $cm$ private (for some constant $c > 0$), $m$-party protocol of [6] in which the total circuit size of all the $\mathsf{NextMSG}$ functions is $|C| \cdot \mathrm{polylog}(|C|, m) + \mathrm{poly}(m, n, k, d, \log |C|)$. Setting $m = \mathrm{poly}(\sigma)$, $\sum_{i=1}^{\sigma} \sum_{j=1}^{d} |\widehat{\mathsf{NextMSG}_i^j}| = |C| \cdot \mathrm{polylog}(|C|, \sigma) + \mathrm{poly}(n, k, d, \sigma)$, and so if all the $\mathsf{NextMSG}_i^j$ are generated using Theorem 1, $|\widehat{\mathsf{C}}| = |C| \cdot \mathrm{polylog}(|C|, \sigma) + \mathrm{poly}(n, k, d, \sigma)$. (See Sect. 5 for a more detailed analysis.)

## 1.2 On the Difference Between Additive Correctness and Additive Security

As noted in Sect. 1.1.2, Definition 1 is weaker than Definition 2. In particular, the correctness guarantee of Definition 1 is insufficient for many MPC applications, since the probability of ERR (due to set flags) might be correlated with the inputs, and consequently reveal information regarding the inputs of $\widehat{\mathsf{C}}$. As we now show, such correlations exist in many natural constructions of additively correct implementations (and, in particular, in all additivity correct constructions discussed in this paper as well as the constructions in [10,11]).

As a typical example of correlations between inputs and the probability of ERR created by additive attacks, consider the simpler case of an AMD code. Specifically, consider the code which encodes a field element $x \in \mathbb{F}$ as $(x, v_1, \cdots, v_\sigma, r_1, \cdots, r_\sigma)$, where $v_1, \cdots, v_\sigma \in_R \mathbb{F}$ are uniformly random, and $r_i = v_i \cdot x$ for all $1 \leq i \leq \sigma$. To decode $(x, v_1, \cdots, v_\sigma, r_1, \cdots, r_\sigma)$, the decoder

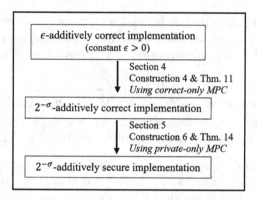

**Fig. 1.** Additive security from weak additive correctness (both steps use AMD codes)

verifies that $x \cdot v_i = r_i$ for all $1 \leq i \leq \sigma$. Consider the additive attack that adds the same arbitrary constant $\delta \neq 0$ to all the $v_i$'s. If $x = 0$ then $r_i = 0$ for every $1 \leq i \leq \sigma$, thus the test $0 \cdot (v_i + \delta) = 0$ passes for all $i$, and decoding succeeds. However, if $x \neq 0$ then every $x \cdot v_i = r_i$ test fails except with probability $1/|\mathbb{F}|$. Since decoding succeeds only if *all* tests succeed, decoding fails in this case with probability at least $1 - 1/|\mathbb{F}|^\sigma$.

Overall, this attack leaks information regarding the value of $x$ because if $x = 0$ then the decoder aborts with probability zero, whereas if $x \neq 0$ then the decoder aborts with probability almost 1. Similar attacks apply to all additively-correct constructions presented in this paper, thus requiring the transformation of Sect. 5.

## 2   Preliminaries

In the following, $\mathbb{F}$ will denote a finite field, $n$ usually denotes the input length, $k$ usually denotes the output length, $d, s$ denote depth and size, respectively (e.g., of circuits, as defined below), and $m$ is used to denote the number of parties. Vectors will be denoted by boldface letters (e.g., **a**). If $\mathcal{D}$ is a distribution then $X \leftarrow \mathcal{D}$, or $X \in_R \mathcal{D}$, denotes sampling $X$ according to the distribution $\mathcal{D}$. Given two distributions $X, Y$, $\mathsf{SD}(X, Y)$ denotes the statistical distance between $X, Y$.

The following lemma regarding $k$-wise indistinguishable distributions over $\{0, 1\}^n$ will be used to construct additively-secure circuits.

**Lemma 1 (Cf. Claim 3.9 in [1]).** *Let $n, k$ be positive integers, and $\mathcal{X}, \mathcal{Y}$ be $k$-wise indistinguishable distributions over $\{0, 1\}^n$. Then*

$$|\Pr[(x_1, \cdots, x_n) \leftarrow \mathcal{X} : \vee_{i=1}^n x_i = 1] - \Pr[(y_1, \cdots, y_n) \leftarrow \mathcal{Y} : \vee_{i=1}^n y_i = 1]| \leq 2^{-\Omega(k/\sqrt{n})}.$$

**Additive Attacks.** We follow the terminology of [10].

**Definition 3 (Additive attack).** *An additive attack* **A** *on a circuit* $C$ *is a fixed vector of field elements which is independent from the inputs and internal values of* $C$. **A** *contains an entry for every wire, and every output gate, of* $C$, *and has the following effect on the evaluation of the circuit. For every wire* $\omega$ *connecting gates* a *and* b *in* $C$, *the entry of* **A** *that corresponds to* $\omega$ *is added to the output of* a, *and the computation of the gate* b *uses the derived value. Similarly, for every output gate* o, *the entry of* **A** *that corresponds to the wire in the output of* o *is added to the value of this output.*

**Notation 3.** *For a (possibly randomized) circuit* $C$ *and for a gate* $g$ *of* $C$, *we denote by* $g_{\mathbf{x}}$ *the distribution of the output value of* $g$ *(defined in a natural way) when* $C$ *is evaluated on an input* $\mathbf{x}$.

**Notation 4.** *Let* $C$ *be a (possibly randomized) circuit, and* **A** *be an additive attack on* $C$. *We denote by* $\mathbf{A}_{c,c'}$ *the attack* **A** *restricted to the wire connecting the gates* $c, c'$ *of* $C$. *Similarly we denote by* $\mathbf{A}^{\mathsf{out}}$ *the restriction of* **A** *to all the outputs of* $C$.[3]

**Encoding Schemes.** An encoding scheme E over a set $\Sigma$ of symbols (called "the alphabet") is a pair (Enc, Dec) of algorithms, where the *encoding algorithm* Enc is a PPT algorithm that given a message $x \in \Sigma^n$ outputs an encoding $\hat{x} \in \Sigma^{\hat{n}}$ for some $\hat{n} = \hat{n}(n)$; and the *decoding algorithm* Dec is a deterministic algorithm, that given an $\hat{x}$ of length $\hat{n}$ in the image of Enc, outputs an $x \in \Sigma^n$. Moreover, $\Pr[\mathsf{Dec}(\mathsf{Enc}(x)) = x] = 1$ for every $x \in \Sigma^n$. We will assume that when $n > 1$, Enc encodes every symbol of $x$ separately, and in particular $\hat{n}(n) = n \cdot \hat{n}(1)$.

**Parameterized Encoding Schemes.** We consider encoding schemes in which the encoding and decoding algorithms are given an additional input $1^t$, which is used as a security parameter. Concretely, the encoding length depends also on $t$ (and not only on $n$), i.e., $\hat{n} = \hat{n}(n, t)$, and for every $t$ the resultant scheme is an encoding scheme (in particular, for every $x \in \Sigma^n$ *and every* $t \in \mathbb{N}$, $\Pr[\mathsf{Dec}(\mathsf{Enc}(x, 1^t), 1^t) = x] = 1$). We call such schemes *parameterized*. We will only consider parameterized encoding schemes, and therefore when we say "encoding scheme" we mean a *parameterized* encoding scheme.

**Algebraic Manipulation Detection (AMD) Encoding Schemes.** Informally, AMD encoding schemes over a finite field $\mathbb{F}$ guarantee that additive attacks on codewords are detected by the decoder with some non-zero probability:

**Definition 4 (AMD encoding scheme, [3,11]).** *Let* $\mathbb{F}$ *be a finite field,* $n \in \mathbb{N}$ *be an input length parameter,* $t \in \mathbb{N}$ *be a security parameter, and* $\epsilon(n,t) : \mathbb{N} \times \mathbb{N} \to \mathbb{R}^+$. *An* $(n, t, \epsilon(n, t))$*-algebraic manipulation detection (AMD) encoding scheme* (Enc, Dec) *over* $\mathbb{F}$ *is an encoding scheme with the following guarantees.*

- *Perfect completeness. For every* $\mathbf{x} \in \mathbb{F}^n$, $\Pr[\mathsf{Dec}(\mathsf{Enc}(\mathbf{x}, 1^t), 1^t) = (0, \mathbf{x})] = 1$.

---

[3] Note that $\mathbf{A}_{c,c'}$ is a single field element whereas $\mathbf{A}^{\mathsf{out}}$ is a vector of field elements.

- **Additive soundness.** *For every* $0^{\hat{n}(n,t)} \neq \mathbf{a} \in \mathbb{F}^{\hat{n}(n,t)}$, *and every* $\mathbf{x} \in \mathbb{F}^n$,
  $\Pr\left[\mathsf{Dec}\left(\mathsf{Enc}\left(\mathbf{x}, 1^t\right) + \mathbf{a}, 1^t\right) \notin \mathsf{ERR}\right] \leq \epsilon(n,t)$ *where* $\mathsf{ERR} = (\mathbb{F}\backslash\{0\}) \times \mathbb{F}^n$, *and the probability is over the randomness of* $\mathsf{Enc}$.

**Remark 1.** *It will sometime be useful to represent* $(\mathsf{Enc}, \mathsf{Dec})$ *as families of arithmetic circuits (instead of polynomial-time algorithms) that are parameterized by the security parameter* $t$. *That is,* $(\mathsf{Enc} = \{\mathsf{Enc}_n\}, \mathsf{Dec} = \{\mathsf{Dec}_n\})$ *are families of arithmetic circuits over* $\mathbb{F}$, *where* $\mathsf{Enc}_n : \mathbb{F}^n \to \mathbb{F}^{\hat{n}}$ *is randomized, and* $\mathsf{Dec}_n : \mathbb{F}^{\hat{n}} \to \mathbb{F} \times \mathbb{F}^n$ *is deterministic. (Here, the security parameter* $t$ *is "hard-wired" into the circuits.) Somewhat abusing notation, we use* $\mathsf{Enc}, \mathsf{Dec}$ *to denote both the families of circuits, and the circuits* $\mathsf{Enc}_n, \mathsf{Dec}_n$ *for a specific* $n$, *omitting the subscript (when* $n$ *is clear from the context).*

We will sometimes need AMD codes with a stronger *robustness* guarantee which, roughly speaking, guarantees additive correctness *even in the presence of additive attacks on the internal wires of the encoding procedure*, where the ideal additive attack on the output is independent of the additive attack:

**Definition 5 (Robust AMD encoding schemes).** *Let* $\mathbb{F}$ *be a finite field,* $n \in \mathbb{N}$ *be an input length parameter,* $\hat{n} \in \mathbb{N}$ *be an output length parameter,* $t \in \mathbb{N}$ *be a security parameter, and* $\epsilon(n,t) : \mathbb{N} \times \mathbb{N} \to \mathbb{R}^+$. *We say that an encoding scheme* $(\mathsf{Enc}, \mathsf{Dec})$ *over* $\mathbb{F}$ *is an* $(n, \hat{n}, t, \epsilon(n,t))$-*robust AMD encoding scheme, if it is an* $(n, t, \epsilon(n,t))$-*AMD encoding scheme in which the additive soundness property is replaced with the following additive robustness property. Let* $\mathsf{Enc} : \mathbb{F}^n \to \mathbb{F}^{\hat{n}}$, $\mathsf{Dec} : \mathbb{F}^{\hat{n}} \to \mathbb{F} \times \mathbb{F}^n$, *then for any additive attack* $\mathbf{A}$ *on* $\mathsf{Enc}$ *there exists an ideal attack* $\mathbf{a}^{\mathsf{in}} \in \mathbb{F}^n$ *such that for any* $\mathbf{b} \in \mathbb{F}^{\hat{n}}$, *and any* $\mathbf{x} \in \mathbb{F}^n$, *it holds that* $\Pr\left[\mathsf{Dec}\left(\mathsf{Enc}^{\mathbf{A}}\left(\mathbf{x}, 1^t\right) + \mathbf{b}, 1^t\right) \notin \mathsf{ERR} \cup \left\{\left(0, \mathbf{x} + \mathbf{a}^{\mathsf{in}}\right)\right\}\right] \leq \epsilon$, *where* $\mathsf{ERR} = (\mathbb{F}\backslash\{0\}) \times \mathbb{F}^n$, *and the probability is over the randomness of* $\mathsf{Enc}$.

**Secure Multiparty Computation.** We recall a few standard definitions that will be used in subsequent sections.

We view an MPC protocol $\pi$ as a collection of $\mathsf{NextMSG}$ functions. The protocol proceeds in rounds, where in round $j$, the description of $\pi$ contains a *next message function* $\mathsf{NextMSG}_i^j$ of round $j$ for party $P_i$, defined as follows. $\mathsf{NextMSG}_i^j$ takes as input all the messages $m_i^{j-1}$ that $P_i$ received before round $j$, its input $x_i$, and its randomness $r_i$; and outputs the messages that $P_i$ sends in round $j$. If $j$ is the last round of $\pi$, then for every party $P_i$, $\mathsf{NextMSG}_i^j$ outputs the output of $P_i$ in $\pi$.

**The Client-Server Model.** The client-server model (see [2,4,5] for a more detailed discussion) is a refinement of the standard MPC model in which each party has one of two possible roles: *clients* hold inputs and receive outputs; and *servers* have no inputs and receive no outputs, but may participate in the computation. Notice that every protocol in the client-server model can be converted to a protocol in the standard MPC model by asking every party to emulate a single server and a single client (assuming the protocol has the same number of clients and servers). See Fig. 2.

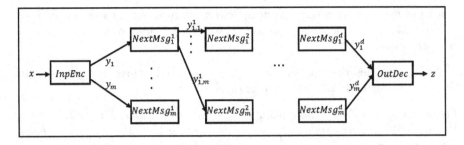

**Fig. 2.** MPC protocol with a single client and $m$ servers

In the following, we assume that the protocol consists of a single input client, a single output client, and $m_S$ servers. We call such protocols $m_S$-server protocols. We use the simulation-based paradigm, and say that a protocol $\pi$ in the client-server model is $(s, \epsilon)$-*secure* ($(s, \epsilon)$-private) if it is secure (up to distance $\epsilon$) against all active (passive) adversaries corrupting at most $s$ servers, and no clients. We assume that the description of a protocol in the client-server model consists of the following:

1. **Input Encoding.** A description of a function InpEnc whose input is the input of the input client, and whose output is the messages that the input client sends to the servers.

2. **Circuit Evaluation.** For every server $S_i$, and every round $j$, a description of a function $\mathsf{NextMSG}_i^j$ which specifies the messages that $S_i$ sends to all the servers (to the output client) in round $j$ (in the last round).

3. **Output Decoding.** A description of a function OutDec whose input is the messages sent to the output client (from the servers) in the last round, and whose output is the output of $\pi$.

We will use a relaxed notion of security, which we call *correct-only* MPC. Intuitively, it guarantees output correctness even in the presence of an active adversary that corrupts a "small" subset of the servers. This notion relaxes the standard security notion because it does not guarantee input privacy. We formalize correct-only MPC as follows, where for a protocol $\pi$, and an adversary Adv, $\pi^{\mathsf{Adv}}(\mathbf{x})$ denotes the outputs (of the clients) in an execution of $\pi$ on inputs $\mathbf{x}$ in the presence of Adv.

**Definition 6.** *Let $f : X \to Y$ be a function, and $\pi$ be a single client, $m_S$-server protocol. We say that $\pi$ $(t, \epsilon)$-correctly computes $f$ if for every active adversary Adv corrupting a set $T, |T| \leq t$ of servers, and every client input $\mathbf{x} \in X$, it holds that $\Pr\left[f(\mathbf{x}) \neq \pi^{\mathsf{Adv}}(\mathbf{x})\right] \leq \epsilon$.*

*We say that $\pi$ $t$-correctly computes $f$ if it $(t, \epsilon)$-correctly computes $f$ for $\epsilon = 0$.*

**Remark 2.** *Notice that any protocol $\pi$ for $t$-correctly computing $f$ in the client-server model can be assumed to be deterministic without loss of generality. This* *is because the adversary* Adv *has no effect on the randomness used by the input clients. Therefore, any $\pi$ can be de-randomized by fixing its randomness to some arbitrary value.*

Next, we describe a simple replication-based $m$-server protocol for $(\lceil m/2 \rceil - 1)$-correctly computing a function $f$.

**Theorem 5.** *Let $\mathbb{F}$ be a finite field. Then for every arithmetic circuit $C : \mathbb{F}^n \to \mathbb{F}^k$, and $m \in \mathbb{N}$, there exists an $m$-server protocol for $(\lceil m/2 \rceil - 1)$-correctly computing $f$. Moreover, the computational complexity (in field operations) of $\pi$ is $|C| \cdot m$.*

*Proof.* The input client replicates the input $\mathbf{x}$ among all the servers, who locally compute $\mathbf{z}_i \leftarrow C(\mathbf{x})$ and send $\mathbf{z}_i$ to the output client, who outputs $\mathrm{maj}\{\mathbf{z}_1, \cdots, \mathbf{z}_m\}$. □

We will use the following theorem regarding the existence of correct-only MPC protocols.

**Theorem 6 (Implicit in [6]).** *Let $\sigma$ be a security parameter, $m \in \mathbb{N}$, $\mathbb{F}$ be a finite field, and $C : \mathbb{F}^n \to \mathbb{F}^k$ be a depth-$d$ arithmetic circuit. Then there exists a $d$-round, $m$-server protocol $\pi$ that $m/10$-correctly computes $C$, where:*

- *The total circuit size of the input encoding function* InpEnc, *and the output decoding function* OutDec, *is $poly(n, k, m)$.*
- *The total circuit size of all the* NextMSG *functions is $|C| \cdot polylog(|C|, \sigma) + poly(m, d, n, k, \log |C|)$.*
- *In each round of $\pi$, the messages sent by each party contain in total at most $poly(n, k, \log |C|)$ field elements.*

## 3    Circuit Transformations

In this section we describe a few circuit transformations which will be used in Sects. 4 and 5 to construct additively-correct and additively-secure circuits. At a high level, these transformations replace a given circuit $C$ over field $\mathbb{F}$ with a new circuit that operates on AMD encodings. We first describe a randomized gadget that combines and amplifies error flags. This gadget will be used in the following constructions to combine error flags obtained from AMD decoding of several codewords.

**Construction 1.** *Let $n_f \in \mathbb{N}$ be an input length parameter, and $\sigma \in \mathbb{N}$ be a security parameter. The* flag combining *gadget* $\mathcal{F}_{\mathsf{comb}} : \mathbb{F}^{n_f} \to \mathbb{F}^\sigma$, *on input $f_1, \cdots, f_{n_f} \in \mathbb{F}$, operates as follows.*

1. *Generates $n_f$ random vectors $\mathbf{r}_1, \cdots, \mathbf{r}_{n_f} \in_R \mathbb{F}^\sigma$.*
2. *Outputs $\mathbf{f} \leftarrow \sum_{i=1}^{n_f} \mathbf{r}_i \cdot f_i$.*

**Observation 7.** *If $(f_1, \cdots, f_{n_f}) \neq \mathbf{0}$ then $\mathbb{F}_{\mathsf{comb}}(f_1, \cdots, f_{n_f}) \neq \mathbf{0}$ except with probability at most $2^{-\sigma}$.*

Next, we describe a circuit transformation $\mathcal{T}_{\mathsf{inter}}$ that will be used to replace intermediate rounds in secure protocols. Intuitively, given a circuit $C$, the transformed circuit $\mathcal{T}_{\mathsf{inter}}(C)$ takes AMD encodings of the inputs of $C$, decodes them, uses the flag combining gadget $\mathcal{F}_{\mathsf{comb}}$ of Construction 1 to combine the error flags generated during decoding, evaluates the circuit $C$, and outputs AMD encodings of the output, concatenated with the combined error flag.

**Construction 2.** *Given a circuit $C : \mathbb{F}^n \to \mathbb{F}^k$, and an AMD encoding scheme* $(\mathsf{Enc}, \mathsf{Dec})$ *that outputs encodings of length $\hat{n}(n)$, the circuit $\mathcal{T}_{\mathsf{inter}}(C) : \mathbb{F}^{\hat{n}(n)} \to \mathbb{F}^\sigma \times \mathbb{F}^{\hat{n}(k)}$, on input $(\mathbf{x}_1, \cdots, \mathbf{x}_n)$, operates as follows.*

1. *For every $1 \le i \le n$, computes $(f_i, \mathbf{x}_i') \leftarrow \mathsf{Dec}(\mathbf{x}_i)$.*
2. *Computes $(y_1, \cdots, y_k) \leftarrow C(\mathbf{x}_1', \cdots, \mathbf{x}_n')$.*
3. *Computes $\mathbf{f} \leftarrow \mathcal{F}_{\mathsf{comb}}(f_1, \cdots, f_n)$.*
4. *Outputs $(\mathbf{f}, \mathsf{Enc}(y_1), \cdots, \mathsf{Enc}(y_k))$.*

Finally, we describe a circuit transformation $\mathcal{T}_{\mathsf{fin}}$ that will be used to replace the output generation rounds. This transformation differs from the transformation $\mathcal{T}_{\mathsf{inter}}$ of Construction 2 only in the fact that it does not encode the outputs.

**Construction 3.** *Given a circuit $C : \mathbb{F}^n \to \mathbb{F}^k$, and an AMD encoding scheme* $(\mathsf{Enc}, \mathsf{Dec})$ *that outputs encodings of length $\hat{n}(n)$, the circuit $\mathcal{T}_{\mathsf{fin}}(C) : \mathbb{F}^{\hat{n}(n)} \to \mathbb{F}^\sigma \times \mathbb{F}^k$, on input $(\mathbf{x}_1, \cdots, \mathbf{x}_n)$, operates as follows.*

1. *Performs Steps 1–3 of Construction 2, and let $(y_1, \cdots, y_k), \mathbf{f}$ denote the outputs of Steps 2 and 3, respectively.*
2. *Outputs $(\mathbf{f}, y_1, \cdots, y_k)$.*

Finally, we will use the following notation.

**Notation 8.** *Given a circuit $C : \mathbb{F}^n \to \mathbb{F}^k$, and an AMD encoding scheme* $(\mathsf{Enc}, \mathsf{Dec})$ *that outputs encodings of length $\hat{n}(n)$, we use $(\mathsf{Enc} \circ C) : \mathbb{F}^n \to \mathbb{F}^{\hat{n}(k)}$ to denote the circuit that on input $\mathbf{x} \in \mathbb{F}^n$, computes $(y_1, \cdots, y_k) \leftarrow C(\mathbf{x})$, and outputs $(\mathsf{Enc}(y_1), \cdots, \mathsf{Enc}(y_k))$.*

## 4    Efficient Additive Correctness Using Correct-Only MPC

In this section we construct a $2^{-\sigma}$-additively-correct circuit with $\mathrm{polylog}(|C|, \sigma)$ overhead. Specifically, for every depth-$d$ arithmetic circuit $C : \mathbb{F}^n \to \mathbb{F}^k$ we construct a $2^{-\sigma}$-additively correct implementation $\widehat{C}$, where $\left|\widehat{C}\right| = |C| \cdot \mathrm{polylog}(|C|, \sigma) + \mathrm{poly}(n, k, d, \sigma)$, thus proving Theorem 1.

Recall that when $\widehat{C}$ is constructed from a correct-only MPC protocol $\pi$ then each attack on $\widehat{C}$ can be divided into three "parts". The first "part" attacks connecting wires between sub-circuits of $\widehat{C}$ (these sub-circuits are $\mathsf{InpEnc}$, $\mathsf{OutDec}$ and $\mathsf{NextMSG}$), and we protect against such attacks by having these sub-circuits operate on AMD codewords. The second "part" attacks the NextMSG functions, and we protect against such attacks by replacing NextMSG with its $\epsilon$-additively correct implementation. Thus, every such attack either affects only few NextMSG functions, in which case the correctness of $\pi$ guarantees that it does not affect the outputs; or it affects many NextMSG functions, in which case $\epsilon$-additive correctness guarantees that (except with negligible probability) the attack is either detected, or corresponds to an additive attack on the inputs and outputs of NextMSG. (Additive attacks on the inputs and outputs correspond to the first type of attacks, namely attacks on the connecting wires, which are detected by the AMD encoding scheme.) The third and final "part" attacks the clients, and we protect against such attacks by replacing InpEnc, OutDec with their $2^{-\sigma}$-additively-correct implementations (e.g., Construction 9 and Appendix A). This is formalized in the following construction, and described in Fig. 3.

**Construction 4.** *Let $\mathbb{F}$ be a finite field, $C : \mathbb{F}^n \to \mathbb{F}^k$ be an arithmetic circuit over $\mathbb{F}$, $\sigma$ be a security parameter, and $\pi$ be a $d$-round, $\sigma$-correct $m$-server protocol for computing $C$ using only point-to-point channels. We assume (without loss of generality) that every message sent in $\pi$ consists of exactly $s$ field elements, for some $s \in \mathbb{N}$. Let $(\mathsf{Enc}, \mathsf{Dec})$ be an $(s, \sigma, 2^{-\sigma})$-AMD encoding scheme that outputs encodings of length $\hat{n}(s)$. The circuit $\widehat{C}$ will use the following ingredients.*

1. **Input Encoding.** *Let $h$ denote the number of messages sent by the input client in the first round, namely $\mathsf{InpEnc} : \mathbb{F}^n \to (\mathbb{F}^s)^h$. Let $\widehat{\mathsf{InpEnc}} : \mathbb{F}^n \to \mathbb{F}^{t'} \times (\mathbb{F}^{\hat{n}(s)})^h$ denote the $2^{-\sigma}$-additively correct implementation, with $t'$ flags, of the circuit $(\mathsf{Enc} \circ \mathsf{InpEnc}) : \mathbb{F}^n \to (\mathbb{F}^{\hat{n}(s)})^h$ (as defined in Notation 8).*

2. **Message Generation.** *For every $1 \leq i \leq m$, and $2 \leq j \leq d-1$, let $g$ $(h)$ denote the number of messages received (sent) by the $i$'th server in round $j-1$ $(j)$.[4] That is, $\mathsf{NextMSG}_i^j : (\mathbb{F}^s)^g \to (\mathbb{F}^s)^h$. Let $\widehat{\mathsf{NextMSG}_i^j} : (\mathbb{F}^{\hat{n}(s)})^g \to \mathbb{F}^t \times \mathbb{F}^\sigma \times (\mathbb{F}^{\hat{n}(s)})^h$ denote the $\epsilon$-additively correct implementation, with $t$ flags, of the circuit $\mathcal{T}_{\mathsf{inter}}\left(\mathsf{NextMSG}_i^j\right) : (\mathbb{F}^{\hat{n}(s)})^g \to \mathbb{F}^\sigma \times (\mathbb{F}^{\hat{n}(s)})^h$ (see Construction 2).*

3. **Output Generation.** *Let $g$ denote the number of messages received by the output client in the final round, namely $\mathsf{OutDec} : (\mathbb{F}^s)^g \to \mathbb{F}^k$. Let $\widehat{\mathsf{OutDec}} : (\mathbb{F}^{\hat{n}(s)})^g \to \mathbb{F}^{t''} \times \mathbb{F}^\sigma \times \mathbb{F}^k$ denote the $2^{-\sigma}$-additively correct implementation, with $t''$ flags, of the circuit $\mathcal{T}_{\mathsf{fin}}(\mathsf{OutDec}) : (\mathbb{F}^{\hat{n}(s)})^g \to \mathbb{F}^\sigma \times \mathbb{F}^k$ (see Construction 3).*

4. **Circuit Construction.** *The circuit $\widehat{C}$, on input $\mathbf{x} \in \mathbb{F}^n$:*

   (a) *Emulates $\pi$, with $x$ as the input of the client, and where $\widehat{\mathsf{InpEnc}}$, $\widehat{\mathsf{NextMSG}_i^j}$ and $\widehat{\mathsf{OutDec}}$ of Steps 1–3 above (connected in the natural way) replace $\mathsf{InpEnc}$, $\mathsf{NextMSG}_i^j$ and $\mathsf{OutDec}$. That is, for every round $1 \leq j \leq d$, if server $S_i$ sends a message to server $S_{i'}$, then the corresponding output of $\widehat{\mathsf{NextMSG}_i^j}$ is wired to the corresponding input of $\widehat{\mathsf{NextMSG}_{i'}^{j+1}}$. Denote the output of the client in the above execution by $\mathbf{z}$.*

---

[4] We assume each server transfers its internal state from one round to the next by sending a message to itself.

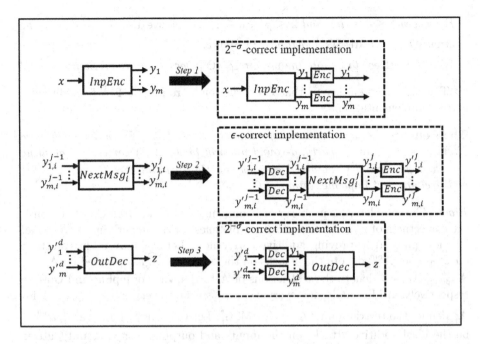

**Fig. 3.** Components of Construction 4

(b) *For every $1 \leq i \leq m$, and every $1 \leq j \leq d$, let $f'^j_{i,1}, \cdots, f'^j_{i,t}$ be the first $t$ outputs of $\widehat{\mathsf{NextMSG}^j_i}$, and let $f^j_{i,1}, \cdots, f^j_{i,\sigma}$ be the next $\sigma$ outputs of $\widehat{\mathsf{NextMSG}^j_i}$. (The $f'^j_{i,w}$'s are the flags of the $\epsilon$-correct implementation, and the $f^j_{i,w}$'s are the flags generated during the AMD decoding.)*

(c) *Let $f'^1_1, \cdots, f'^1_{t'}$ be the first $t'$ outputs of $\widehat{\mathsf{InpEnc}}$. (These are the flags of the $2^{-\sigma}$-correct implementation.)*

(d) *Let $f'^d_1, \cdots, f'^d_{t''}$ be the first $t''$ outputs of $\widehat{\mathsf{OutDec}}$ and let $f^d_1, \cdots, f^d_\sigma$ be the next $\sigma$ outputs of $\widehat{\mathsf{OutDec}}$. (The $f'^d_i$'s are the flags of the $2^{-\sigma}$-correct implementation, and the $f^d_i$'s are the flags generated during the AMD decoding.)*

(e) *For every $1 \leq w' \leq \sigma$, compute $f''_{w'} \leftarrow \sum_{i=1}^m \sum_{j=2}^{d-1} \left( \sum_{w=1}^t f'^j_{i,w} \cdot r_{i,j,w,w'} + \sum_{w=1}^\sigma f^j_{i,w} \cdot r_{i,j,t+w,w'} \right) + \sum_{w=1}^{t''} f'^d_w \cdot r_{1,d,w,w'} + \sum_{w=1}^\sigma f^d_w \cdot r_{1,d,t+w,w'} + \sum_{w=1}^{t'} f'^1_w \cdot r_{1,1,w,w'}$ where $r_{i,j,w,w'} \in_R \mathbb{F}$.*

(f) *Output $\mathbf{z} + \sum_{w=1}^\sigma f''_w \cdot \mathbf{r}'_w$ where $\mathbf{r}'_w \in_R \mathbb{F}^k$.*

We now analyze the properties of Construction 4. The following notation will be useful.

**Notation 9.** *We denote the ingredients of Construction 4 as follows.*

- *We use $\mathsf{InpEnc}'$ to denote the circuit $(\mathsf{Enc} \circ \mathsf{InpEnc})$ obtained in Step 1.*

- For every $1 \leq i \leq m$, and $2 \leq j \leq d-1$, we use $\mathsf{NextMSG}_i'^j$ to denote the circuit $\mathcal{T}_{\mathsf{inter}}\left(\mathsf{NextMSG}_i^j\right)$ obtained in Step 2.
- We use $\mathsf{OutDec}'$ to denote the circuit $\mathcal{T}_{\mathsf{fin}}(\mathsf{OutDec})$ obtained in Step 3.

The next theorem shows that Construction 4 produces a $2^{-\Omega(\sigma)}$-additively-correct implementation.

**Theorem 10.** *Let $\sigma$ be a security parameter, $C : \mathbb{F}^n \to \mathbb{F}^k$ be an arithmetic circuit, and $\pi$ be an $m$-party, $d$-round protocol for $(\sigma, 2^{-\sigma})$-correctly computing $C$. Then the circuit $\widehat{C}$ obtained by applying Construction 4 to $C$ is a $2^{-\Omega(\sigma)}$-additively-correct implementation of $C$.*

*Proof.* The completeness property of $\widehat{C}$ immediately follows from Construction 4, the correctness of $\pi$, and the perfect completeness of the underlying AMD code. We now proceed to proving additive correctness. Let $\mathbf{A}$ be an additive attack on $\widehat{C}$, and let $\mathbf{A}^{\mathsf{out}}$ denote the attacks on the outputs of $\widehat{C}$ as specified by $\mathbf{A}$. Let $\mathbf{A}_{\mathsf{InpEnc}}, \mathbf{A}_{\mathsf{OutDec}}$ denote the restrictions of $\mathbf{A}$ to the wires of $\widehat{\mathsf{InpEnc}}$ and $\widehat{\mathsf{OutDec}}$ respectively. Additionally, for every $1 \leq i \leq m$ and every $2 \leq j \leq d-1$ let $\mathbf{A}_i^j$ denote the restriction of $\mathbf{A}$ to $\widehat{\mathsf{NextMSG}_i^j}$. Let $(\mathbf{a}^{\mathsf{in},1}, \mathbf{a}^{\mathsf{out},1})$ and $(\mathbf{a}^{\mathsf{in},d}, \mathbf{a}^{\mathsf{out},d})$ be the ideal additive attacks on the inputs and outputs of $\widehat{\mathsf{InpEnc}}$ and $\widehat{\mathsf{OutDec}}$ corresponding to $\mathbf{A}_{\mathsf{InpEnc}}, \mathbf{A}_{\mathsf{OutDec}}$. Similarly, for every $1 \leq i \leq m$ and every $2 \leq j \leq d-1$, let $\mathbf{a}_i^{\mathsf{in},j}$, and $\mathbf{a}_i^{\mathsf{out},j}$ be the ideal additive attacks on the inputs and outputs of $\widehat{\mathsf{NextMSG}_i^j}$ corresponding to $\mathbf{A}_i^j$. Define $\mathbf{a}^{\mathsf{in}} = \mathbf{a}^{\mathsf{in},1}$ and $\mathbf{a}^{\mathsf{out}} = \mathbf{a}^{\mathsf{out},d} + \mathbf{A}^{\mathsf{out}}$. We claim that for every input $\mathbf{x}$ it holds that

$$\Pr[\widehat{C}^{\mathbf{A}}(\mathbf{x}) \notin \mathsf{ERR} \cup \{(0^\sigma, C(\mathbf{x} + \mathbf{a}^{\mathsf{in}}) + \mathbf{a}^{\mathsf{out}})\}] \leq 2^{-\Omega(\sigma)}$$

where $\mathsf{ERR} = (\mathbb{F}^\sigma \backslash \{0^\sigma\}) \times \mathbb{F}^k$.

Indeed, let $\mathbf{x} \in \mathbb{F}^n$ be an input to $\widehat{C}$, and define $P_{\mathsf{bad}}$ as the event that $\widehat{C}^{\mathbf{A}}(\mathbf{x}) \notin \mathsf{ERR} \cup \{(0^\sigma, C(\mathbf{x} + \mathbf{a}^{\mathsf{in}}) + \mathbf{a}^{\mathsf{out}})\}$, namely

$$\Pr[\widehat{C}^{\mathbf{A}}(\mathbf{x}) \notin \mathsf{ERR} \cup \{(0^\sigma, C(\mathbf{x} + \mathbf{a}^{\mathsf{in}}) + \mathbf{a}^{\mathsf{out}})\}] = \Pr[P_{\mathsf{bad}}].$$

Next, denote by $P_f$ the event that

$$\bigwedge_{i=1}^{m} \bigwedge_{j=2}^{d-1} \bigwedge_{w=1}^{t} (f_{i,w,\mathbf{x}}'^{j\mathbf{A}} = f_{i,w,\mathbf{x}}^{j\mathbf{A}} = 0) \bigwedge_{w=1}^{t'} f_w'^1 = 0 \bigwedge_{w=1}^{t''} f_w'^d = 0.$$

Notice that by construction of $\widehat{C}$ we obtain that

$$\Pr[\widehat{C}^{\mathbf{A}}(\mathbf{x}) \notin \mathsf{ERR} \cup \{(0^\sigma, C(\mathbf{x} + \mathbf{a}^{\mathsf{in}}) + \mathbf{a}^{\mathsf{out}})\}] \leq 2^{-\Omega(\sigma)} + \Pr[P_{\mathsf{bad}} \wedge P_f].$$

We proceed by defining the event $P_{\mathsf{OK}}^{1,1}$ as $\widehat{\mathsf{InpEnc}}^{\mathbf{A}}(\mathbf{x}) \in \mathsf{ERR} \cup \{\mathsf{InpEnc}(\mathbf{x} + \mathbf{a}^{\mathsf{in},1}) + \mathbf{a}^{\mathsf{out},1}\}$ and $P_{\mathsf{OK}}^{d,d}$ as $\widehat{\mathsf{OutDec}}^{\mathbf{A}}(\mathbf{y}_{\mathbf{x}}^{\mathbf{A}}) \in \mathsf{ERR} \cup \{\mathsf{OutDec}(\mathbf{y}_{\mathbf{x}}^{\mathbf{A}} + \mathbf{a}^{\mathsf{in},d}) + \mathbf{a}^{\mathsf{out},d}\}$, where $\mathbf{y}_\mathbf{x}^\mathbf{A}$ is the random variable corresponding to the messages received by the client from the servers during the last round of $\pi$ inside $\widehat{\mathsf{C}}^\mathbf{A}(\mathbf{x})$. We notice that by the $2^{-\sigma}$-correctness of $\widehat{\mathsf{InpEnc}}$ and $\widehat{\mathsf{OutDec}}$ it holds that

$$\Pr\left[P_{\mathsf{bad}} \wedge P_f\right] \leq 2^{-\Omega(\sigma)} + \Pr\left[P_{\mathsf{bad}} \wedge P_f \wedge P_{\mathsf{OK}}^{1,1} \wedge P_{\mathsf{OK}}^{d,d}\right].$$

Next, for every round $2 \leq j \leq d-1$ and party $1 \leq i \leq m$, denote by $\mathsf{In}_i^j$ the set of servers which send messages to the $i$th server during the $j$th round, and denote by $\mathbf{a}_{i,i'}^{\mathsf{in},j}$ the ideal additive attacks on the inputs of $\widehat{\mathsf{NextMSG}}_i^j$ which correspond to the message received by server $i$ from server $i'$ during the $j$th round. Similarly, denote by $\mathsf{Out}_i^j$ the set of servers to which the $i$th server sends messages during the $j$th round, and denote by $\mathbf{a}_{i,i'}^{\mathsf{out},j}$ the ideal additive attacks on the outputs of $\widehat{\mathsf{NextMSG}}_i^j$ which correspond to the message sent by server $i$ to server $i'$ during the $j$th round. In addition, we assume without loss of generality that the client sends a message to all the servers during the first round, and receives a message from all the servers during the last round. Finally, for every server $1 \leq i \leq m$, we denote by $\mathbf{a}_i^{\mathsf{out},1}$ the restriction of $\mathbf{a}^{\mathsf{out},1}$ to the messages that the client sends to the $i$th server during the first round and by $\mathbf{a}_i^{\mathsf{in},d}$ the restriction of $\mathbf{a}^{\mathsf{in},d}$ to the messages that the client receives from the $i$th server during the $d$th round. Finally, we denote by $\mathbf{a}_i^{\mathsf{in},2}$ the messages received by the $i$th server from the client, and we denote by $\mathbf{a}_{i'}^{\mathsf{out},d-1}$ the messages sent by the $i'$th server to the client.

For any $1 \leq i, i' \leq m$ and $2 \leq j \leq d-1$, we say that a tuple $(i', i, j)$ is *problematic* if one of the following three conditions hold.

1. **Input Corruption.** It holds that $\mathbf{a}_i^{\mathsf{in},2} + \mathbf{a}_i^{\mathsf{out},1} \neq 0$ and $i' = j = 1$.
2. **Intermediate Corruption.** It holds that $\mathbf{a}_{i,i'}^{\mathsf{in},j} + \mathbf{a}_{i',i}^{\mathsf{out},j-1} \neq 0$.
3. **Output Corruption.** It holds that $\mathbf{a}_{i'}^{\mathsf{out},d-1} + \mathbf{a}_{i'}^{\mathsf{in},d} \neq 0$ and $i = j = d$.

Next, we define the set $\mathcal{A} = \{(i', i, j) : \text{the tuple } (i', i, j) \text{ is problematic}\}$ and we split the proof into two cases.

- **Case 1: $|\mathcal{A}| > \sigma$.** Intuitively, in this case a large portion of $\widehat{\mathsf{C}}$ was corrupted. We show that in this case $\widehat{\mathsf{C}}$ will almost always abort the computation by setting at least one of the flags to a non zero value, namely the probability of an incorrect output (i.e., not in $\mathsf{ERR} \cup \{(0^\sigma, C(\mathbf{x} + \mathbf{a}^{\mathsf{in}}) + \mathbf{a}^{\mathsf{out}})\}$) is low.

  We denote the random variables describing the messages exchanged during the evaluation of $\widehat{\mathsf{C}}^\mathbf{A}$ on input $\mathbf{x}$ as follows: for every $1 \leq i \leq m$ and $2 \leq j \leq d-2$, $\widehat{y}_{i,i',\mathbf{x}}^{\mathbf{A},j}$ corresponds to the message sent by the $i$th server to the $i'$th server in round $j$; $\widehat{y}_{i,\mathbf{x}}^{\mathbf{A},1}$ corresponds to the messages sent by the client to the $i$th server in the first round; and $\widehat{y}_{i,\mathbf{x}}^{\mathbf{A},d-1}$ corresponds to the message sent by the $i$th server to the client in round $d-1$.

  Next, for any $1 \leq i \leq m$ and $2 \leq j \leq d-1$ denote by $P_{\mathsf{OK}}^{i,j}$ the event that $\widehat{\mathsf{NextMSG}}_i^{\mathbf{A},j}\left(\left(\widehat{y}_{i',i,\mathbf{x}}^{\mathbf{A},j-1}\right)_{i' \in \mathsf{In}_i^j}\right)$ is in $\mathsf{ERR} \cup$

$$\left\{ \left( 0^t, \text{NextMSG}_i'^j \left( \left( \widehat{y}_{i',i,\mathbf{x}}^{\mathbf{A},j-1} \right)_{i' \in \text{In}_i^j} + \mathbf{a}_i^{\text{in},j} \right) + \mathbf{a}_i^{\text{out},j} \right) \right\}, \quad \text{where} \quad \text{ERR} =$$

$(\{\mathbb{F}^t\} \backslash \{0^t\}) \times \mathbb{F}^{o_i^j}$, and $o_i^j$ is the output length of $\text{NextMSG}_i'^j$.

Next, notice that for every tuple $(i', i, j)$ the randomness of $\widehat{\text{NextMSG}}_i^{\mathbf{A},j}$ is independent from the randomness of $\widehat{\text{NextMSG}}_{i'}^{\mathbf{A},j-1}$. Thus, it holds that $\Pr\left[ P_{\text{OK}}^{i',j-1} \wedge P_{\text{OK}}^{i,j} \right] \geq (1 - \epsilon)^2$, yielding $\Pr\left[ \overline{P_{\text{OK}}^{i',j-1} \wedge P_{\text{OK}}^{i,j}} \right] \leq 1 - (1 - \epsilon)^2$. Next, across all the problematic tuples in $\mathcal{A}$ we obtain that $\Pr\left[ P_{\text{bad}} \wedge P_f \wedge P_{\text{OK}}^{1,1} \wedge P_{\text{OK}}^{d,d} \right]$ is at most

$$\left( 1 - (1 - \epsilon)^2 \right)^\sigma + \Pr\left[ \begin{array}{c} P_{\text{bad}} \wedge P_f \wedge P_{\text{OK}}^{1,1} \wedge P_{\text{OK}}^{d,d} \wedge \\ \left( \exists(i', i, j) \in \mathcal{A} : (P_{\text{OK}}^{i',j-1} \wedge P_{\text{OK}}^{i,j}) \right) \end{array} \right].$$

Finally, the fact that $P_{\text{OK}}^{i',j-1} \wedge P_{\text{OK}}^{i,j}$ for some problematic tuple $(i', i, j) \in \mathcal{A}$ implies that there is a non-zero additive attack on the wires between server $i'$ (or the client in case $j = 1$) and server $i$ (again, or the client in case $j = d$) during the $j$th round. Thus, by the additive soundness of $(\text{Enc}, \text{Dec})$ we obtain that except with probability $2^{-\sigma}$, $(f_{i,1}^j, \cdots, f_{i,o}^j) \neq 0$, namely $P_f$ does not hold. Consequently,

$$\Pr\left[ \begin{array}{c} P_{\text{bad}} \wedge P_f \wedge P_{\text{OK}}^{1,1} \wedge P_{\text{OK}}^{d,d} \wedge \\ \left( \exists(i', i, j) \in \mathcal{A} : (P_{\text{OK}}^{i',j-1} \wedge P_{\text{OK}}^{i,j}) \right) \end{array} \right] \leq 2^{-\Omega(\sigma)}.$$

– **Case 2: $|\mathcal{A}| \leq \sigma$.** Notice that having less than $\sigma$ problematic tuples implies that for the protocol $\pi$ inside $\widehat{\mathsf{C}}$, the additive attack $\mathbf{A}$ only corrupted less than $\sigma$ parties. In this case we get that except with probability $2^{-\sigma}$, the protocol $\pi$ manages to correctly compute $C$. Thus, in this case

$$\Pr\left[ P_{\text{bad}} \wedge P_f \wedge P_{\text{OK}}^{1,1} \wedge P_{\text{OK}}^{d,d} \right] \leq 2^{-\Omega(\sigma)}. \qquad \square$$

We show that for an appropriate choice of parameters, Construction 4 is a $2^{-\sigma}$-additively correct implementation. This is formalized in the next Theorem.

**Theorem 11.** *For any depth-$d$ arithmetic circuit $C : \mathbb{F}^n \rightarrow \mathbb{F}^k$, and any security parameter $\sigma$, there exists a $2^{-\Omega(\sigma)}$-additively-correct implementation $\widehat{\mathsf{C}}$ of $C$ where $|\widehat{\mathsf{C}}| = |C| \cdot polylog(|C|, \sigma) + poly(n, k, d, \sigma)$.*

We first state several results regarding AMD encoding schemes, which will be used in the proof.

Asymptotically optimal constructions of AMD encoding schemes have been presented by [3,8]. In fact, [3] consider a slightly weaker definition of AMD codes which guarantees that $\Pr[\text{Dec}(\text{Enc}(\mathbf{x}) + \mathbf{a}) \notin \text{ERR} \cup \{(0, \mathbf{x})\}] \leq \epsilon$, allowing for ERR on some inputs and correct output on others (see Definition 7 below). However, their construction actually possesses the stronger security property of Definition 4.

**Theorem 12 (Implicit in [3], Corollary 1).** *For any $n, \sigma \in \mathbb{N}$, and field $\mathbb{F}$, there exists a pair of families of circuits $(\mathsf{Enc}, \mathsf{Dec})$ over $\mathbb{F}$ that is an $(n, \sigma, \frac{1}{|\mathbb{F}|^\sigma})$-AMD encoding scheme with encodings of length $n + \sigma$. Moreover, the size of $\mathsf{Enc}$ and $\mathsf{Dec}$ is $\widetilde{O}(n + \sigma)$.*

**Theorem 13 (Implicit in [10]).** *There exists a constant $\epsilon \in (0, 1)$ such that for any field $\mathbb{F}$ and arithmetic circuit $C : \mathbb{F}^n \to \mathbb{F}^k$ there exist a circuit $\widehat{C} : \mathbb{F}^n \to \mathbb{F} \times \mathbb{F}^k$ which is an $\epsilon$-additively-correct implementation of $C$. Moreover, $\left|\widehat{C}\right| = O\left(|C|\right)$.*

*Proof (of Theorem 11).* Apply Construction 4 to $C$ using an AMD code of Theorem 12, the $\epsilon$-additively-correct construction from Theorem 13 and the $\sigma$-server protocol $\pi$ from Theorem 6. To obtain the $2^{-\sigma}$-additively-correct implementation of $\widehat{\mathsf{InpEnc}}$ and $\widehat{\mathsf{OutDec}}$ used in Steps 1 and 3 of Construction 4, we use an additively-correct circuit compiler $\mathsf{Comp}^{\mathsf{In}}$ that on input a circuit $C$ outputs a circuit $\widehat{C}$ such that $\left|\widehat{C}\right| = \sigma \cdot |C|$ (e.g., Construction 9 of Appendix A). Since $\pi$ $(\sigma/10)$-correctly computes $C$ we obtain that $\widehat{C}$ is a $2^{-\Omega(\sigma)}$-additively-correct implementation of $C$.

Next, we proceed to analyze the size of $\widehat{C}$. By the construction of $\widehat{C}$ we have that $|\widehat{C}| = |\widehat{\mathsf{InpEnc}}| + |\widehat{\mathsf{OutDec}}| + \sum_{i=1}^{\sigma} \sum_{j=1}^{d} |\widehat{\mathsf{NextMSG}_i^j}|$. From Theorem 6 we obtain that $|\mathsf{InpEnc}| + |\mathsf{OutDec}|$ is $\mathrm{poly}(n, k, \sigma)$. Thus, when $\mathsf{InpEnc}$ and $\mathsf{OutDec}$ are implemented using Construction 9 (Appendix A, $|\widehat{\mathsf{InpEnc}}| + |\widehat{\mathsf{OutDec}}|$ is also $\mathrm{poly}(n, k, \sigma)$. We now proceed to analyze $\sum_{i=1}^{\sigma} \sum_{j=1}^{d} |\widehat{\mathsf{NextMSG}_i^j}|$.

We begin by noticing that in each round of $\pi$, each server sends messages containing a total of $\mathrm{poly}(n, k, \log |C|)$ field elements. Thus, by having $\mathsf{NextMSG}'$ encode every message sent during the execution of $\pi$ with the AMD codes from Theorem 12 we obtain that the circuit size of every $\mathsf{NextMSG}'$ function increases by an additive term which is $\mathrm{poly}(n, k, \log |C|, \sigma)$ compared to $\mathsf{NextMSG}$. Next, since the overall circuit size of all the $\mathsf{NextMSG}$ functions is $|C| \cdot \mathrm{polylog}(|C|, \sigma) + \mathrm{poly}(\sigma, d, n, k, \log |C|)$ and since $|\widehat{\mathsf{NextMSG}}| = O(\mathsf{NextMSG}')$ we obtain that the total circuit size of all the $\widehat{\mathsf{NextMSG}}$ circuits inside $\widehat{C}$ is also $|C| \cdot \mathrm{polylog}(|C|, \sigma) + \mathrm{poly}(\sigma, d, n, k, \log |C|)$. $\qquad \square$

**Remark 3.** *The proof of Theorem 11 uses an ad-hoc "feasibility" construction to achieve $\mathrm{polylog}(\sigma)$ overhead. However, it is possible to improve the simplicity, and concrete efficiency, of the construction by replacing the feasibility construction with simpler gadgets implementing the input encoder and output decoder. We now outline a more direct construction (which matches the complexity of Theorem 11). We begin by observing that for the protocol of Theorem 6, we can assume (without loss of generality) that $\mathsf{InpEnc}(\mathbf{x}) = (\mathbf{x}, \cdots, \mathbf{x})$, and $\mathsf{OutDec}(\mathbf{y}_1, \cdots, \mathbf{y}_m)$ outputs $(0^\sigma, \mathbf{y}_1)$ if $\mathbf{y}_1 = \cdots = \mathbf{y}_m$, otherwise it outputs a random value in $(\mathbb{F}^\sigma \backslash \{0^\sigma\}) \times \mathbb{F}^k$. Next, we implement $\widehat{\mathsf{InpEnc}}$ and $\widehat{\mathsf{OutDec}}$ directly using the following simple gadgets.*

- **Implementing InpEnc.** *We define* $\widehat{\mathsf{InpEnc}}(\mathbf{x}) = (\mathsf{Enc}(\mathbf{x}), \cdots, \mathsf{Enc}(\mathbf{x}))$, *where* Enc *is the encoding procedure of a* $2^{-\sigma}$-*robust AMD code (as in Definition 5). The stronger robustness property guarantees the existence of a single consistent value such that (with high probability) every server either decodes to it, or aborts.*
- **Implementing OutDec.** *We modify each server to compute a MAC value of its outputs. In addition, C is evaluated in the clear:* $\mathbf{z} \leftarrow C(\mathbf{x})$, *and the output* $\mathbf{z}$ *is MACed to obtain* $\tilde{\mathbf{z}}$. *Finally,* $\widehat{\mathsf{OutDec}}$ *contains a gadget that compares all MACed outputs of the servers to* $\tilde{\mathbf{z}}$, *and outputs* $\mathbf{z}$ *if the test passes, otherwise it outputs* ERR.[5]

# 5   From Additive Correctness to Additive Security via Passive-Secure MPC

In this section we combine additively-correct circuits with passive-secure MPC protocols to construct binary additively-secure circuits with a negligible error, thus proving Theorem 2.

Recall that (as described in Sect. 1.1.2) we construct the additively-secure implementation $\widehat{C}$ of $C$ from a passive-secure MPC protocol $\pi$. More specifically, the inputs of parties in $\pi$ are additive secret-shares of the input of $C$, and $\pi$ evaluates the function that: (1) reconstructs the input from the secret shares; (2) evaluates $C$; and (3) outputs an additive secret-sharing of the output.

Consequently, every additive attack on $\widehat{C}$ can be divided into two "parts". The first "part" targets the wires connecting different sub-circuits NextMSG of $\widehat{C}$, and we protect against such attacks by having these sub-circuits operate on AMD codewords. The second "part" modifies the internal computations of the NextMSG functions, and we protect against such attacks by replacing each NextMSG with its $2^{-\sigma}$-additively-correct implementation. Thus, the resultant $\widehat{C}$ is a $2^{-\Omega(\sigma)}$-additively correct implementation of $C$, where every attack is with overwhelming probability either "harmless" (namely, corresponds to an additive attack on the inputs and output of $C$), or causes the output to be random. Moreover, as we argued in Sect. 1.1.2, the probability that the output is random is independent of the inputs.

We start by defining the circuit $C_{\mathsf{AUG}}$, which implements the functionality computed by $\pi$ (namely, emulates $C$ on secret shares).

**Construction 5.** *Let* $C : \mathbb{F}^n \to \mathbb{F}^k$ *be an arithmetic circuit, and* $m \in \mathbb{N}$. *The circuit* $C_{\mathsf{AUG}}$, *on inputs* $(\mathbf{x}_1, \cdots, \mathbf{x}_m) \in (\mathbb{F}^n)^m$, *performs the following.*

1. *Computes* $\mathbf{x} \leftarrow \sum_{i=1}^{m} \mathbf{x}_i$, *and* $\mathbf{y} \leftarrow C(\mathbf{x})$. *(This step reconstructs the input to* $C$ *from the secret shares, and evaluates* $C$.)

---

[5] To implement $\widehat{\mathsf{OutDec}}$ without leaking information regarding the outputs of $C$, we compare *only the MAC tags* generated by the servers (and not *the actual outputs*). This necessitates an additional evaluation of $C$ (in the clear) to generate the output.

2. *Generates* $\mathbf{y}_1, \cdots, \mathbf{y}_{m-1} \in \mathbb{F}^n$ *uniformly at random, and compute* $\mathbf{y}_m \leftarrow$ $\mathbf{y} - \sum_{i=1}^{m-1} \mathbf{y}_i$. *(*$\mathbf{y}_1, \cdots, \mathbf{y}_m$ *is an additive secret sharing of the output* $\mathbf{y}$.*)*
3. *Outputs* $(\mathbf{y}_1, \cdots, \mathbf{y}_m)$.

Next, we use $C_{\mathsf{AUG}}$ to construct the circuit $\widehat{C}$, see also Fig. 4.

**Construction 6.** *Let* $C : \mathbb{F}^n \to \mathbb{F}^k$ *be an arithmetic circuit over a finite field* $\mathbb{F}$, $\sigma$ *be a security parameter, and* $\pi$ *be a d-round, t-private, m-party protocol for computing the circuit* $C_{\mathsf{AUG}}$ *of Construction 5, using only point-to-point channels. We assume (without loss of generality) that every message sent in* $\pi$ *consists of exactly s field elements, for some* $s \in \mathbb{N}$. *Let* $(\mathsf{Enc}, \mathsf{Dec})$ *be an* $(s, \sigma, 2^{-\sigma})$-*AMD encoding scheme that outputs encodings of length* $\hat{n}(s)$, *and* $\mathsf{Dec}$ *outputs* $\sigma$ *flags during decoding. The circuit* $\widehat{C}$ *will use the following ingredients.*

1. **Protecting the first round.** *For every* $1 \le i \le m$, *assume that party* $P_i$ *sends h messages in the first round, namely* $\mathsf{NextMSG}_i^1 : \mathbb{F}^n \to (\mathbb{F}^s)^h$. *Let* $\widehat{\mathsf{NextMSG}}_i^1 : \mathbb{F}^n \to \mathbb{F}^t \times (\mathbb{F}^{\hat{n}(s)})^h$ *be the* $2^{-\sigma}$-*additively correct implementation, with t flags, of the circuit* $(\mathsf{Enc} \circ \mathsf{NextMSG}_i^1) : \mathbb{F}^n \to (\mathbb{F}^{\hat{n}(s)})^h$ *(see Construction 3).*

2. **Protecting middle rounds.** *For every* $1 \le i \le m$, *and* $2 \le j \le d-1$, *assume that in round* $j-1$ *(j)* $P_i$ *receives (sends)* $g$ *(h) messages, namely* $\mathsf{NextMSG}_i^j : (\mathbb{F}^s)^g \to (\mathbb{F}^s)^h$. *Let* $\widehat{\mathsf{NextMSG}}_i^j : (\mathbb{F}^{\hat{n}(s)})^g \to \mathbb{F}^t \times \mathbb{F}^\sigma \times (\mathbb{F}^{\hat{n}(s)})^h$ *be the* $2^{-\sigma}$-*additively correct implementation, with t flags, of the circuit* $\mathcal{T}_{\mathsf{inter}}\left(\mathsf{NextMSG}_i^j\right) : (\mathbb{F}^{\hat{n}(s)})^g \to \mathbb{F}^\sigma \times (\mathbb{F}^{\hat{n}(s)})^h$ *(see Construction 2).*

3. **Protecting the last round.** *For every* $1 \le i \le m$ *assume that* $P_i$ *receives* $g$ *messages in the final round, namely* $\mathsf{NextMSG}_i^d : (\mathbb{F}^s)^g \to \mathbb{F}^k$. *Let* $\widehat{\mathsf{NextMSG}}_i^d : (\mathbb{F}^{\hat{n}(s)})^g \to \mathbb{F}^t \times \mathbb{F}^\sigma \times \mathbb{F}^k$ *be the* $2^{-\sigma}$-*additively correct implementation, with t flags, of the circuit* $\mathcal{T}_{\mathsf{fin}}\left(\mathsf{NextMSG}_i^d\right) : (\mathbb{F}^{\hat{n}(s)})^g \to \mathbb{F}^t \times \mathbb{F}^k$ *(see Construction 3).*

4. **Circuit construction.** *The circuit* $\widehat{C}$ *on input* $\mathbf{x}$ *performs the following.*
   (a) *Generate* $\mathbf{x}_1, \cdots, \mathbf{x}_{m-1} \in \mathbb{F}^n$ *uniformly at random and compute* $\mathbf{x}_m \leftarrow \mathbf{x} - \sum_{i=1}^{m-1} \mathbf{x}_i$.
   (b) *Emulates* $\pi$ *with* $\mathbf{x}_i$ *as the input of party* $P_i$, *where the* $\widehat{\mathsf{NextMSG}}_i^j$ *described in Steps Steps 1–3 (connected in the natural way) replace the* $\mathsf{NextMSG}_i^j$. *That it, for every round* $1 \le j \le d-1$, *if party* $P_i$ *sends a message to party* $P_{i'}$, *we wire the corresponding output of* $\widehat{\mathsf{NextMSG}}_i^j$ *to the corresponding input of* $\widehat{\mathsf{NextMSG}}_{i'}^{j+1}$.
   (c) *Let* $\mathbf{z}_i$ *denote* $P_i$'s *output in the above execution. Compute* $\mathbf{z} \leftarrow \sum_{i=1}^m \mathbf{z}_i$.
   (d) *For every* $1 \le i \le m$, *and* $2 \le j \le d$, *let* $f_{i,1}'^j, \cdots, f_{i,t}'^j$ *denote the first t outputs of* $\widehat{\mathsf{NextMSG}}_i^j$, *and* $f_{i,1}'^j, \cdots, f_{i,\sigma}'^j$ *denote the* $t+1$ *to* $t+\sigma$ *outputs of* $\widehat{\mathsf{NextMSG}}_i^j$. *(The* $f_{i,w}'^j$'s *are the flags of the* $2^{-\sigma}$-*correct implementation, and the* $f_{i,w}^j$'s *are the flags generated during the AMD decoding.)*

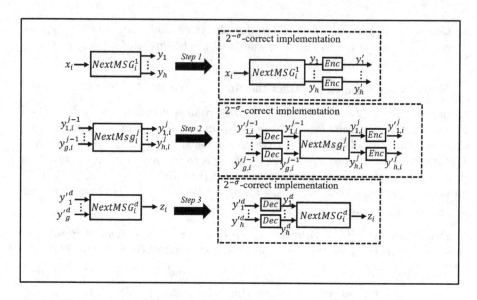

**Fig. 4.** Components of Construction 6

(e) *For every* $1 \leq i \leq m$ *let* $\widehat{f_{i,1}'^d, \cdots, f_{i,t}'^d}$ *denote the first $t$ outputs of* $\mathsf{NextMSG}_i^1$. *(These are the flags of the $2^{-\sigma}$-correct implementation.)*

(f) *For every* $1 \leq w' \leq \sigma$, *compute* $f_{w'}'' \leftarrow \sum_{i=1}^{m} \sum_{j=2}^{d}$ $\left( \sum_{w=1}^{t} f_{i,w}'^j \cdot r_{i,j,w} + \sum_{w=1}^{\sigma} f_{i,w}^j \cdot r_{t+i,j,w} \right) + \sum_{i=1}^{m} \sum_{w=1}^{t} f_{i,w}'^1 \cdot r_{i,d,w}$, *where* $r_{i,j,w} \in_R \mathbb{F}$.

(g) *Output* $\mathbf{z} + \sum_{w=1}^{m} f_w'' \cdot \mathbf{r}_w'$, *where* $\mathbf{r}_w' \in_R \mathbb{F}^k$.

We show that any additive attack on $\widehat{C}$ is either equivalent to an additive attack on the inputs and output of $C$, or will sets flags inside $\widehat{C}$ to non-zero values. Moreover, the probability that a flag is set depends only on the additive attack, and is almost independent of the input. This is captured by the next theorem.

**Theorem 14.** *For any depth-$d$ arithmetic circuit $C : \mathbb{F}^n \to \mathbb{F}^k$, and security parameter $\sigma$, there exists a $2^{-\Omega(\sigma)}$-additively-secure implementation $\widehat{C}$ of $C$, where $|\widehat{C}| = |C| \cdot polylog(|C|, \sigma) + poly(n, k, d, \sigma)$. Moreover, $\widehat{C}$ can be constructed from $C$ in $poly(|C|, \sigma, m)$ time.*

The proof of Theorem 14, which follows the outline presented in Sect. 1.1.2, is deferred to the full version. Here, we only outline the main points and subtle issues in the proof. We first show that with overwhelming probability any additive attack on $\widehat{C}$ either sets error flags in $\widehat{C}$, or is equivalent to an additive attack on its inputs and output. This is proved in two steps: first, using the additive correctness property of the $\mathsf{NextMSG}_i^j$ sub-circuits, except with negligible probability additive attacks on the internal wires of every $\mathsf{NextMSG}_i^j$ can be "pushed" to an additive attack on its inputs and outputs. Second, we examine the additive attacks obtained in this manner between every pair of adjacent $\mathsf{NextMSG}_{i'}^{j-1}$ and $\mathsf{NextMSG}_i^j$ sub-circuits. If all these attacks cancel out, then the output of $\widehat{\mathsf{C}}$ is correct. Otherwise, the additive-security property of the AMD code protecting the communication channels between the $\mathsf{NextMSG}$ sub-circuits guarantees that with overwhelming probability an error flag will be set, causing $\widehat{\mathsf{C}}$ to abort.

Next, we prove that the probability of abort is almost independent of the inputs of $\widehat{\mathsf{C}}$. As before, we first "push" additive attacks on the $\mathsf{NextMSG}_i^j$ sub-circuits to additive attacks on their inputs and outputs. We then traverse the layers of $\widehat{\mathsf{C}}$ from the inputs to the output. In each layer $j$, a flag can be raised either by a $\mathsf{NextMSG}_i^j$ sub-circuit (which corresponds to the computation performed by a single party $P_i$), or by the AMD decoding performed in $\mathsf{NextMSG}_i^j$. In either case, the event that a flag is set depends only on the view of $P_i$ which, by the $t$-privacy of $\pi$ (and of the additive secret sharing of the input), guarantees that the distributions of the flags when evaluating $\widehat{\mathsf{C}}$ on two different inputs $\mathbf{x}, \mathbf{x}'$ are $t$-wise indistinguishable. Since a *single* set flag suffices to cause an abort, the "OR lemma" (Lemma 1) guarantees that the probability of abort is independent of the inputs to $\widehat{\mathsf{C}}$.

## 6 Constant-Overhead AMD Codes and Their Applications to Constant-Overhead MPC

In this section we use AMD codes to relate the open question of constructing actively-secure two-party protocols with constant computational overhead to the simpler questions of constructing passively-secure honest-majority MPC protocols, and correct-only honest-majority MPC protocols, with constant computational overhead. This is done by combining our constructions from Sects. 4 and 5 with a (relaxed) AMD encoding scheme that has constant overhead.

More formally, we say that a secure implementation of a circuit $C$ (e.g., an additively-secure implementation of $C$, or a secure protocol for evaluating $C$) has *constant computational overhead* if its circuit size is $O(|C|) + \mathrm{poly}(\log |C|, \sigma, d, n, k)$ where $\sigma$ is the security parameter, $d$ is the circuit depth, and $n, k$ are the input and output lengths, respectively. (The circuit size of a protocol $\pi$ is the total circuit size of all the $\mathsf{NextMSG}$ functions of $\pi$.)

We first construct relaxed AMD encoding schemes with constant overhead, namely the size of the encoding and decoding circuits is linear in the message length. At a high level, relaxed AMD encoding schemes, first considered by [3], have a weaker soundness guarantee: as long as the output is correct with high probability, (non-zero) additive attacks are allowed to pass unnoticed. This should be contrasted with (standard) AMD codes, in which *every* additive attack is guaranteed to be detected (with high probability).

**Definition 7 (Relaxed AMD encoding scheme [3]).** *Let $\mathbb{F}$ be a finite field, $n \in \mathbb{N}$ be an input length parameter, $t \in \mathbb{N}$ be a security parameter, and $\epsilon(n,t):$ $\mathbb{N} \times \mathbb{N} \to \mathbb{R}^+$. An $(n, t, \epsilon(n,t))$-relaxed AMD encoding scheme* (Enc, Dec) *over $\mathbb{F}$ is an encoding scheme with the following properties.*

- *Perfect completeness. For every* $\mathbf{x}$ $\in$ $\mathbb{F}^n$, $\Pr\left[\mathsf{Dec}\left(\mathsf{Enc}\left(\mathbf{x}, 1^t\right), 1^t\right) = (0, \mathbf{x})\right] = 1$.
- *Relaxed additive soundness. For every* $0^{\hat{n}(n,t)} \neq \mathbf{a} \in \mathbb{F}^{\hat{n}(n,t)}$, *and every* $\mathbf{x} \in \mathbb{F}^n$, $\Pr\left[\mathsf{Dec}\left(\mathsf{Enc}\left(\mathbf{x}, 1^t\right) + \mathbf{a}, 1^t\right) \notin \mathsf{ERR} \cup \{(0, \mathbf{x})\}\right] \leq \epsilon(n, t)$ *where* $\mathsf{ERR} = (\mathbb{F}\backslash\{0\}) \times \mathbb{F}^n$, *and the probability is over the randomness of* Enc.

Roughly speaking, we construct a constant-overhead AMD encoding scheme by composing a linearly encodable and decodable AMD encoding scheme with constant additive soundness, with a linearly encodable error-correcting code with constant rate and relative distance. We will need the following notion of an $[n, k, d]$-error-correcting code.

**Definition 8.** *We say that a pair* (Enc $: \mathbb{F}^k \to \mathbb{F}^n$, Dec $: \mathbb{F}^n \to \mathbb{F}^k$) *of deterministic circuits is an* $[n, k, d]$-*error-correcting code* (ECC) *over $\mathbb{F}$ if any $\mathbf{x}, \mathbf{y} \in \mathbb{F}^k$ it holds that* $\Pr\left[\mathsf{Dec}(\mathsf{Enc}(\mathbf{x})) = \mathbf{x}\right] = 1$ *and that* $|\{i : (\mathsf{Enc}(x))_i \neq (\mathsf{Enc}(y))_i\}| \geq d$.

The following theorem is due to Spielman [19] (see also [9]):

**Theorem 15.** *There exist constants $d_1 > 1$, and $d_2 > 0$, such that for any field $\mathbb{F}$, and any $k \in \mathbb{N}$, there exists a pair of circuits* (Enc$_k$, Dec$_k$) *which is a* $[\lfloor d_1 k \rfloor, k, \lceil d_2 k \rceil]$-*ECC over $\mathbb{F}$. Moreover, the size of* Enc$_k$ *is $O(k)$.*

We can now construct an AMD encoding scheme with constant overhead.

**Construction 7.** *Let $n$ be a positive integer, $\mathbb{F}$ be a finite field, and* (Enc$_n$, Dec$_n$) *be an $[n', n, d]$-ECC over $\mathbb{F}$. In addition, let* (Enc$_{amd} : \mathbb{F} \to \mathbb{F}^k$, Dec$_{amd} : \mathbb{F}^k \to \mathbb{F} \times \mathbb{F}$ *be a $(1, t, \epsilon(t))$-AMD encoding scheme. Consider the circuits* Enc $: \mathbb{F}^n \to \mathbb{F}^{n+k\cdot n'}$ *and* Dec $: \mathbb{F}^{n+k\cdot n'} \to \mathbb{F} \times \mathbb{F}^n$ *which are defined as follows.*

- *The circuit* Enc *on input $\mathbf{x} \in \mathbb{F}^n$ performs the following:*
  1. *Computes $\mathbf{x}' \leftarrow$ Enc$_n(\mathbf{x})$ and for all $1 \leq i \leq n'$ computes $\widehat{x}_i \leftarrow$ Enc$_{amd}(x_i')$.*
  2. *Outputs $(\mathbf{x}, \widehat{\mathbf{x}})$.*
- *The circuit* Dec *on input $(\mathbf{x}, \widehat{\mathbf{x}})$ performs the following:*
  1. *Computes $\mathbf{x}' \leftarrow$ Enc$_n(\mathbf{x})$.*
  2. *For all $1 \leq i \leq n'$ computes $(f_i, y_i') \leftarrow$ Dec$_{amd}(\widehat{x}_i)$ and $f_i' \leftarrow x_i' - y_i'$.*
  3. *In case there exists an $1 \leq i \leq n'$ such that $f_i \neq 0$ or $f_i' \neq 0$, outputs $(1, 0^n)$. Otherwise, outputs $(0, \mathbf{x})$.*

**Theorem 16.** *For any positive integer $n$, the pair of circuits* Enc, Dec *of Construction 7 is an $(n, t, \epsilon(t)^d)$-relaxed AMD encoding scheme.*

*Proof.* The correctness property follows directly from the construction. We now prove the relaxed additive soundness property. Let $\mathbf{x} \in \mathbb{F}^n$ be an input to Enc, and $\mathbf{A} = (\mathbf{a}, \mathbf{b}) \in \mathbb{F}^n \times \mathbb{F}^{kn'}$ be an additive attack on the outputs of Enc. We consider two possible cases.

1. $\mathbf{a} = \mathbf{0}$. In this case, the additive attack does not attempt to alter the value $\mathbf{x}$ passed from Enc to Dec, so $\Pr[\mathsf{Dec}(\mathsf{Enc}(\mathbf{x}, 1^t) + (\mathbf{a}, \mathbf{b}), 1^t) \notin \mathsf{ERR} \cup \{(0, \mathbf{x})\}] = 0$.

2. $a_i \neq 0$ for some $1 \leq i \leq n$. In this case, let $\mathcal{I} = \{i : (\mathsf{Enc}_n(\mathbf{x} + \mathbf{a}))_i \neq (\mathsf{Enc}_n(\mathbf{x}))_i\}$. For an additive attack to successfully cause Dec to output some $\tilde{\mathbf{x}} \neq \mathbf{x}$, it must be the case that $x_i'^{\mathbf{A}} = y_i'^{\mathbf{A}}$ for every $i \in \mathcal{I}$, where $x_i'^{\mathbf{A}} = (\mathsf{Enc}_n(\mathbf{x} + \mathbf{a}))_i$, and $y_i'^{\mathbf{A}} = \mathsf{Dec}_{amd}(\hat{x}_i + \mathbf{b}_i) = \mathsf{Dec}_{amd}((\mathsf{Enc}_{amd}((\mathsf{Enc}_n(\mathbf{x}))_i)) + \mathbf{b}_i)$ (the right equality follows from the definition of $\hat{x}$). For every $i \in \mathcal{I}$, if $\mathbf{b}_i = 0$ then by the correctness of $(\mathsf{Enc}_{amd}, \mathsf{Dec}_{amd})$, $\mathsf{Dec}_{amd}(\mathsf{Enc}_{amd}((\mathsf{Enc}_n(\mathbf{x}))_i)) = \mathsf{Dec}_{amd}(\mathsf{Enc}_{amd}((\mathsf{Enc}_n(\mathbf{x}))_i)) = (\mathsf{Enc}_n(\mathbf{x}))_i \neq (\mathsf{Enc}_n(\mathbf{x} + \mathbf{a}))_i$ (the rightmost equality holds since $i \in \mathcal{I}$), so Dec outputs ERR (with probability 1); otherwise the additive soundness of $(\mathsf{Enc}_{amd}, \mathsf{Dec}_{amd})$ guarantees that $f_i \neq 0$ only with probability $\epsilon(t)$. Moreover, the relative distance property of the ECC guarantees that $|\mathcal{I}| \geq d$. Consequently, $\Pr[\mathsf{Dec}(\mathsf{Enc}(\mathbf{x}, 1^t) + (\mathbf{a}, \mathbf{b}), 1^t) \notin \mathsf{ERR} \cup \{(0, \mathbf{x})\}] \leq \epsilon(t)^d$. $\qquad \square$

Instantiating Construction 7 with the ECC of Theorem 15, we obtain the following result.

**Theorem 17.** *For any positive integer $n$ there exists an $(n, t, 2^{-\Omega(n)})$-relaxed AMD encoding scheme with encoding and decoding circuits of size $\Theta(n)$.*

Theorem 17 can be used to relate the open question of constructing actively-secure two-party protocols with constant computational overhead to the simpler questions of constructing passively-secure honest-majority MPC protocols, and correct-only honest-majority MPC protocols, with constant computational overhead. We first show that actively secure 2-party MPC protocols in the OT-hybrid model, with constant computational overhead, can be constructed from additively-secure circuits with constant computational overhead. Formally,

**Claim 18.** *Assume that any boolean circuit $C$ admits an additively-secure implementation $\widehat{C}$ with constant computational overhead. Then there exists an actively secure 2-party protocol $\pi$ for evaluating $cC$ in the OT-hybrid model with constant computational overhead.*

*Proof (sketch).* The work of [11] observed that the effect of an active attack on an arithmetic version of the *passively-secure* GMW protocol [12] $\pi_{\mathsf{GMW}}$ (in the OLE-hybrid model) corresponds to an additive attack on the underlaying circuit being evaluated. This observation holds in the binary case as well (where $\pi'$ is executed in the OT-hybrid model). Thus, given an additively-secure implementation $\widehat{C}$ of $C$ with constant computational overhead, one can construct an actively secure 2-party protocol $\pi$ for evaluating $C$ in the OT-hybrid model, with constant computational overhead, simply by running $\pi_{\mathsf{GMW}}$ on $\widehat{C}$. $\qquad \square$

The following corollary reduces the task of constructing actively-secure 2-party protocols in the OT-hybrid model, with constant computational overhead, to the following simpler tasks:

1. Constructing passively-secure 2-party protocols in the OT-hybrid model with constant computational overhead.
2. Constructing correct-only (as per Definition 6) 2-party protocols in the OT-hybrid model with constant computational overhead.

**Corollary 1.** *If there exist both correct-only MPC protocols, and passively secure MPC protocols, with constant computational overhead, then there is a secure 2-party protocol in the OT-hybrid model with constant computational overhead.*

*Proof (sketch).* Let $\pi_1, \pi_2$ be correct-only, and passively secure, protocols (resp.) with constant computational overhead. The protocol $\pi$ for evaluating a circuit $C$ is obtained by applying Claim 18 to the circuit $\widehat{C}_{sec}$ constructed below.

1. Construct an additively-correct implementation $\widehat{C}_{corr}$ of $C$ (as per Definition 1) with constant computational overhead using $\pi_1$, Construction 4, and the relaxed AMD codes of Theorem 17.
2. Construct an additively-secure implementation $\widehat{C}_{sec}$ of $C$ (as per Definition 2) with constant computational overhead using $\pi_2$, Construction 6, and the relaxed AMD codes of Theorem 17.

By repeating the analysis of Constructions 4 and 6 while replacing the protocol from [6] with $\pi_1, \pi_2$, we obtain that $\pi$ has constant computational overhead. Regarding the security of $\pi$, the only difference from the analysis in Sects. 4 and 5 is that $\pi$ employs a *relaxed* AMD encoding scheme (whereas Constructions 4 and 6 used (standard) AMD encoding schemes). However, since AMD codes are used in these constructions only to protect the communication channels of $\pi_1, \pi_2$, then relaxed additive soundness suffices for the analysis since it guarantees that no attack can alter the values of these messages. □

**Acknowledgments.** The first author is a member of the Check Point Institute for Information Security and was supported by ERC starting grant 259426; by the Blavatnik Interdisciplinary Cyber Research Center; by the Israeli Centers of Research Excellence I-CORE program (center 4/11); by the Leona M. & Harry B. Helmsley Charitable Trust; and by NATO's Public Diplomacy Division in the Framework of "Science for Peace".

The second author was supported by ERC starting grant 259426, ISF grant 1709/14, BSF grant 2012378, a DARPA/ARL SAFEWARE award, NSF Frontier Award 1413955, NSF grants 1228984, 1136174, 1118096, and 1065276. This material is based upon work supported by the Defense Advanced Research Projects Agency through the ARL under Contract W911NF-15-C-0205. The views expressed are those of the author and do not reflect the official policy or position of the Department of Defense, the National Science Foundation, or the U.S. Government.

The third author was supported by ERC starting grant 259426 and a Check Point Institute for Information Security grant for graduate students and post-doctoral fellows.

## A    Additive Correctness Without a Decoder: Feasibility

In this section we construct a $2^{-\Omega(\sigma)}$ additively-correct circuit compiler $\mathsf{Comp}^{\mathsf{ln}}$ which on input a circuit $C$ outputs a circuit $\widehat{C}$ such that $\left|\widehat{C}\right| = \sigma \cdot |C|$.

We present a method of amplifying security of additively-correct constructions through repetition. The natural approach is for the compiled circuit $\widehat{C}$
to contain $\sigma$ copies of an $\epsilon$-additively-correct implementation $\widehat{C}_\epsilon$ that are all
evaluated on the input $x$. This approach raises two issues. First, an additive
attack $\mathbf{A}$ on $\widehat{C}$ consists of $\sigma$ additive attacks $\mathbf{A}_1, \cdots, \mathbf{A}_\sigma$ on the $\sigma$ copies of $\widehat{C}_\epsilon$.
For each copy $\widehat{C}_{\epsilon,i}$, the additive-security of $\widehat{C}_\epsilon$ guarantees that there exists an
"ideal" additive attack on the inputs and outputs of $\widehat{C}_{\epsilon,i}$ such that except with
probability $\epsilon$, the output of $\widehat{C}_{\epsilon,i}$ under the additive attack $\mathbf{A}_i$ equals its output
under the corresponding ideal additive attack. However, if different copies are
evaluated under *different* additive attacks then *the corresponding ideal additive
attacks may also be different*. This in effect causes different copies to be evaluated *on different inputs*. To overcome this, before compiling $C$ we first modify it
to take inputs encoded using a robust AMD encoding scheme. Since such codes
guarantee additive correctness with an ideal additive attack that is independent
of the additive attack on the outputs of the encoder, this guarantees the existence of a *single* additive attack that *simultaneously* corresponds to the additive
attacks on *all* copies.

The second issue is that $\widehat{C}$ should verify that all copies have the same output,
and this should be *performed in the presence of additive attacks*. Therefore,
before compiling $C$ we first transform it into a circuit that MACs its output.
Thus, the test comparing the MACs of two inconsistent outputs will fail, *even if
it is performed under an additive attack*. These alterations of $C$ are summarized
in the following construction of $C_{\mathsf{AUG}}$.

**Construction 8 ($C_{\mathsf{AUG}}, C_{\mathsf{MAC}}$).** *Let $C : \mathbb{F}^n \to \mathbb{F}^k$ be an arithmetic circuit over
a finite field $\mathbb{F}$, $\sigma \in \mathbb{N}$ be a length parameter, and $(\mathsf{Enc}, \mathsf{Dec})$ be an $(n + \sigma, l, \sigma, \epsilon)$-
robust AMD encoding scheme (as in Definition 5). The circuit $C_{\mathsf{AUG}} : \mathbb{F}^l \to
\mathbb{F}^\sigma \times \mathbb{F}^k \times (\mathbb{F}^\sigma)^k$, on input $\mathbf{x}' \in \mathbb{F}^l$, performs the following.*

1. *Compute $(f, (\mathbf{u}, \mathbf{x})) \leftarrow \mathsf{Dec}(\mathbf{x}')$, where $f \in \mathbb{F}$, $\mathbf{u} = (u_1, \cdots, u_\sigma) \in \mathbb{F}^\sigma$, and
   $\mathbf{x} \in \mathbb{F}^n$. (Intuitively, $\mathbf{x}$ is the input to the original circuit, and $\mathbf{u}$ will be used
   to MAC the outputs.)*
2. *Computes $\mathbf{z} \leftarrow C(\mathbf{x})$, where $\mathbf{z} \in \mathbb{F}^k$.*
3. *For all $1 \leq i \leq k$, computes $(z'_{i,1}, \cdots, z'_{i,\sigma}) \leftarrow (u_1 \cdot z_i, \cdots, u_\sigma \cdot z_i)$. (This step
   MACs each output coordinate $z_i$.)*
4. *Outputs $\left( f \cdot \mathbf{s}, \mathbf{z}, (z'_{1,1}, \cdots, z'_{1,\sigma}), \cdots, (z'_{k,1}, \cdots, z'_{k,\sigma}) \right)$ where $\mathbf{s} \in_R \mathbb{F}^\sigma$.*

*The circuit $C_{\mathsf{MAC}} : \mathbb{F}^n \times \mathbb{F}^\sigma \to \mathbb{F}^k \times (\mathbb{F}^\sigma)^k$ is obtained from $C$ in a similar
manner, except that its input is $(\mathbf{u}, \mathbf{x})$ ("in the clear"), and so it does not perform
the input decoding of Step 1 above, and does not output a list of flags.*

**Construction 9.** *Let $\mathbb{F}$ be a finite field, $\sigma \in \mathbb{N}$ be a security parameter, $n \in \mathbb{N}$
be an input length parameter, and $k, k' \in \mathbb{N}$ be output length parameters. Let $C :
\mathbb{F}^n \to \mathbb{F}^k$ be an arithmetic circuit over $\mathbb{F}$, and $(\mathsf{Enc}, \mathsf{Dec})'$ be an $(n + \sigma, l, \sigma, 2^{-\sigma})$-
additively robust AMD encoding scheme. Let $C_{\mathsf{MAC}}$ and $C_{\mathsf{AUG}}$ denote the circuits
obtained by applying Construction 8 to $C$. Notice that the inputs to $C_{\mathsf{MAC}}$ are*

$\mathbf{x} \in \mathbb{F}^n$ and a MAC key $\mathbf{u} \in \mathbb{F}^\sigma$, and its output is in $\mathbb{F}^{k+\sigma k}$, whereas the inputs to $C_{\mathsf{AUG}}$ are robust-AMD encodings of $\mathbf{u}, \mathbf{x}$, and its output is in $\mathbb{F}^{\sigma+k+\sigma k}$. Let $\widehat{C}_{\mathsf{AUG}}$ be an $\epsilon$-additively-correct implementation of $C_{\mathsf{AUG}}$ with $t$ flags. The randomized circuit $\widehat{C} : \mathbb{F}^n \to \mathbb{F}^\sigma \times \mathbb{F}^k$, on input $\mathbf{x} \in \mathbb{F}^n$, operates as follows.

1. Generates a random MAC key $\mathbf{u} \in \mathbb{F}^\sigma$. ($\mathbf{u}$ will be used to MAC the outputs of $C$ in $C_{\mathsf{AUG}}$.)
2. Computes $\mathbf{z}_{\mathsf{MAC}} \leftarrow C_{\mathsf{MAC}}(\mathbf{x}, \mathbf{u})$, and we denote $\mathbf{z}_{\mathsf{MAC}} = (\mathbf{z}, (\widetilde{z}_{1,1}, \cdots, \widetilde{z}_{1,\sigma}), \cdots, (\widetilde{z}_{k,1}, \cdots, \widetilde{z}_{k,\sigma}))$. (This step evaluates $C$ once directly, and MACs the outputs.)
3. Computes $(\mathbf{x}', \mathbf{u}') \leftarrow \mathsf{Enc}((\mathbf{x}, \mathbf{u}))$. (This step encodes the inputs to $C_{\mathsf{AUG}}$.)
4. For all $1 \leq i \leq \sigma$, computes $(\mathbf{f}_i, \mathbf{y}_i) \leftarrow \widehat{C}_{\mathsf{AUG},i}(\mathbf{x}', \mathbf{u}')$, where $\widehat{C}_{\mathsf{AUG},1}, \cdots, \widehat{C}_{\mathsf{AUG},\sigma}$ denote $\sigma$ separate copies of $\widehat{C}_{\mathsf{AUG}}$.
5. For all $1 \leq i \leq \sigma$, interprets $\mathbf{y}_i$ as $\left( (f'_{i,1}, \cdots, f'_{i,\sigma}), \mathbf{z}'_i, (z'_{i,1,1}, \cdots, z'_{i,1,\sigma}), \cdots, (z'_{i,k,1}, \cdots, z'_{i,k,\sigma}) \right)$. (The output of each copy $\widehat{C}_{\mathsf{AUG},i}$ is interpreted as $\sigma$ flags $f'_{i,1}, \cdots, f'_{i,t}$ indicating whether the decoding of $\mathbf{x}', \mathbf{u}'$ succeeded, a $k$-length output, and $\sigma$ MACs for every output coordinate.)
6. For all $1 \leq i, j \leq \sigma$, computes $f''_{i,j} \leftarrow \sum_{l=1}^{k}(\widetilde{z}_{l,i} - z'_{j,l,i})r_{i,j,l}$ where all the $r_{i,j,l}$ are generated uniformly from $\mathbb{F}$. (This step compares the MACed outputs computed by the $\epsilon$-additively-correct implementations in Step 3, with the MACed output computed directly in Step 2. Specifically, $f''_{i,j}$ compares the $i$'th MAC of the $j$'th copy, to the $i$'th MAC of Step 2.)
7. For all $1 \leq i \leq \sigma$, computes $f''_i \leftarrow \sum_{j=1}^{\sigma} f''_{i,j}r_{i,j} + \sum_{j=1}^{\sigma} r'_{i,j}f'_{j,i}$, where all the $r_{i,j}$ and $r'_{i,j}$ are generated uniformly from $\mathbb{F}$. (This step checks that all copies agree on the $i$'th MAC, and in addition, that the decoding of the inputs in all copies succeeded.)
8. For all $1 \leq i \leq \sigma$ compute $g_i \leftarrow \sum_{j=1}^{\sigma}(\sum_{u=1}^{t} f_{j,u}\widetilde{r}_{i,j,u} + f''_j \widetilde{r}'_{i,j,u})$ where all the $\widetilde{r}_{i,j}$ and $\widetilde{r}'_{i,j}$ are generated uniformly from $\mathbb{F}$. (This step checks that the computation in the $i$'th $\epsilon$-additively-correct implementation succeeded, and in addition, that the input decoding in all copies succeeded, and they all agree on all MACs.)
9. Output $((g_1, \cdots, g_\sigma), \mathbf{z})$.

In the full version of the paper we prove the following:

**Theorem 19.** For any field $\mathbb{F}$, arithmetic circuit $C : \mathbb{F}^n \to \mathbb{F}^k$ and security parameter $\sigma$, the circuit $\widehat{C}$ obtained by applying Construction 9 to $C$ is a $2^{-\Omega(\sigma)}$-additively-correct implementation of $C$. Moreover, $|\widehat{C}| = poly(\sigma, |C|)$.

# References

1. Bogdanov, A., Ishai, Y., Viola, E., Williamson, C.: Bounded indistinguishability and the complexity of recovering secrets. In: Robshaw, M., Katz, J. (eds.) CRYPTO 2016. LNCS, vol. 9816, pp. 593–618. Springer, Heidelberg (2016). doi:10.1007/978-3-662-53015-3_21

2. Cramer, R., Damgård, I.B., Ishai, Y.: Share conversion, pseudorandom secret-sharing and applications to secure computation. In: Kilian, J. (ed.) TCC 2005. LNCS, vol. 3378, pp. 342–362. Springer, Heidelberg (2005)

3. Cramer, R., Dodis, Y., Fehr, S., Padró, C., Wichs, D.: Detection of algebraic manipulation with applications to robust secret sharing and fuzzy extractors. In: Smart, N.P. (ed.) EUROCRYPT 2008. LNCS, vol. 4965, pp. 471–488. Springer, Heidelberg (2008)

4. Damgård, I.B., Ishai, Y.: Constant-round multiparty computation using a black-box pseudorandom generator. In: Shoup, V. (ed.) CRYPTO 2005. LNCS, vol. 3621, pp. 378–394. Springer, Heidelberg (2005)

5. Damgård, I.B., Ishai, Y.: Scalable secure multiparty computation. In: Dwork, C. (ed.) CRYPTO 2006. LNCS, vol. 4117, pp. 501–520. Springer, Heidelberg (2006)

6. Damgård, I., Ishai, Y., Krøigaard, M.: Perfectly secure multiparty computation and the computational overhead of cryptography. In: Gilbert, H. (ed.) EUROCRYPT 2010. LNCS, vol. 6110, pp. 445–465. Springer, Heidelberg (2010)

7. Dobrushin, R., Ortyukov, E.: Upper bound on the redundancy of self-correcting arrangements of unreliable functional elements. Problems Inf. Transm. 23(2), 203–218 (1977)

8. Dodis, Y., Katz, J., Reyzin, L., Smith, A.: Robust fuzzy extractors and authenticated key agreement from close secrets. In: Dwork, C. (ed.) CRYPTO 2006. LNCS, vol. 4117, pp. 232–250. Springer, Heidelberg (2006)

9. Druk, E., Ishai, Y.: Linear-time encodable codes meeting the Gilbert-Varshamov bound and their cryptographic applications. In: ITCS 2014, pp. 169–182. ACM (2014)

10. Genkin, D., Ishai, Y., Polychroniadou, A.: Efficient multi-party computation: from passive to active security via secure SIMD circuits. In: Gennaro, R., Robshaw, M. (eds.) CRYPTO 2015. LNCS, vol. 9216, pp. 721–741. Springer, Heidelberg (2015)

11. Genkin, D., Ishai, Y., Prabhakaran, M., Sahai, A., Tromer, E.: Circuits resilient to additive attacks with applications to secure computation. In: STOC 2014, pp. 495–504 (2014). Full version in Cryptology ePrint Archive: Report 2015/154

12. Goldreich, O., Micali, S., Wigderson, A.: How to play any mental game or a completeness theorem for protocols with honest majority. In: STOC 1987, pp. 218–229. ACM (1987)

13. Ikarashi, D., Kikuchi, R., Hamada, K., Chida, K.: Actively private and correct MPC scheme in $t \leq n/2$ from passively secure schemes with small overhead. IACR Cryptology ePrint Archive 2014:304 (2014)

14. Ishai, Y., Kilian, J., Nissim, K., Petrank, E.: Extending oblivious transfers efficiently. In: Boneh, D. (ed.) CRYPTO 2003. LNCS, vol. 2729, pp. 145–161. Springer, Heidelberg (2003)

15. Ishai, Y., Kushilevitz, E., Ostrovsky, R., Sahai, A.: Cryptography with constant computational overhead. In: STOC 2008, pp. 433–442 (2008)

16. Ishai, Y., Prabhakaran, M., Sahai, A.: Founding cryptography on oblivious transfer – efficiently. In: Wagner, D. (ed.) CRYPTO 2008. LNCS, vol. 5157, pp. 572–591. Springer, Heidelberg (2008)

17. Keller, M., Orsini, E., Scholl, P.: Actively secure OT extension with optimal overhead. In: Gennaro, R., Robshaw, M. (eds.) CRYPTO 2015. LNCS, vol. 9215, pp. 724–741. Springer, Heidelberg (2015). doi:10.1007/978-3-662-47989-6_35

18. Pippenger, N.: On networks of noisy gates. In: FOCS 1985, pp. 30–38. IEEE (1985)

19. Spielman, D.A.: Linear-time encodable and decodable error-correcting codes. IEEE Trans. Inf. Theor. **42**(6), 1723–1731 (1996)
20. von Neumann, J.: Probabilistic logics and synthesis of reliable organisms from unreliable components. In: Shannon, C., McCarthy, J. (eds.) Automata Studies, pp. 43–98. Princeton University Press, Princeton (1956)

# Composable Security in the Tamper-Proof Hardware Model Under Minimal Complexity

Carmit Hazay[1], Antigoni Polychroniadou[2],
and Muthuramakrishnan Venkitasubramaniam[3]([✉])

[1] Bar-Ilan University, Ramat Gan, Israel
carmit.hazay@biu.ac.il
[2] Aarhus University, Aarhus, Denmark
antigoni@cs.au.dk
[3] University of Rochester, Rochester, NY, USA
muthuv@cs.rochester.edu

**Abstract.** We put forth a new formulation of tamper-proof hardware in the Global Universal Composable (GUC) framework introduced by Canetti et al. in TCC 2007. Almost all of the previous works rely on the formulation by Katz in Eurocrypt 2007 and this formulation does not fully capture tokens in a concurrent setting. We address these shortcomings by relying on the GUC framework where we make the following contributions:

1. We construct secure Two-Party Computation (2PC) protocols for general functionalities with optimal round complexity and computational assumptions using stateless tokens. More precisely, we show how to realize arbitrary functionalities in the two-party setting with GUC security in two rounds under the minimal assumption of One-Way Functions (OWFs). Moreover, our construction relies on the underlying function in a black-box way. As a corollary, we obtain feasibility of Multi-Party Computation (MPC) with GUC-security under the minimal assumption of OWFs. As an independent contribution, we identify an issue with a claim in a previous work by Goyal, Ishai, Sahai, Venkatesan and Wadia in TCC 2010 regarding the feasibility of UC-secure computation with stateless tokens assuming collision-resistant hash-functions (and the extension based only on one-way functions).

2. We then construct a 3-round MPC protocol to securely realize arbitrary functionalities with GUC-security starting from any semi-honest secure MPC protocol. For this construction, we require the so-called one-many commit-and-prove primitive introduced in the original work of Canetti, Lindell, Ostrovsky and Sahai in STOC 2002 that is round-efficient and black-box in the underlying commitment. Using specially designed "input-delayed" protocols we realize this primitive (with a 3-round protocol in our framework) using stateless tokens and one-way functions (where the underlying one-way function is used in a black-box way).

**Keywords:** Secure computation · Tamper-proof hardware · Round complexity · Minimal assumptions

© International Association for Cryptologic Research 2016
M. Hirt and A. Smith (Eds.): TCC 2016-B, Part I, LNCS 9985, pp. 367–399, 2016.
DOI: 10.1007/978-3-662-53641-4_15

# 1    Introduction

Secure Multi-Party Computation (MPC) enables a set of parties to mutually run a protocol that computes some function $f$ on their private inputs, while preserving two important properties: *privacy* and *correctness*. The former implies data confidentiality, namely, nothing leaks by the protocol execution but the computed output, while, the later requirement implies that no corrupted party or parties can cause the output to deviate from the specified function. It is by now well known how to securely compute any efficient functionality [4,29,50,58] under the stringent simulation-based definitions (following the ideal/real paradigm). These traditional results prove security in the stand-alone model, where a *single* set of parties run a *single* execution of the protocol. However, the security of most cryptographic protocols proven in the stand-alone setting does not remain intact if many instances of the protocol are executed concurrently [6,8,48]. The strongest (but also the most realistic) setting for concurrent security, known as *Universally Composable* (UC) security [6] considers the execution of an unbounded number of concurrent protocols in an arbitrary and adversarially controlled network environment. Unfortunately, stand-alone secure protocols typically fail to remain secure in the UC setting. In fact, without assuming some *trusted help*, UC-security is impossible to achieve for most tasks [8,10,48]. Consequently, UC-secure protocols have been constructed under various *trusted setup* assumptions in a long series of works; see [2,7,13,22,41,42,46] for few examples.

One such setup assumption and the focus of this work is the use of tamper-proof hardware tokens. The first work to model tokens in the UC framework was by Katz in [42] who introduced the $\mathcal{F}_{\mathrm{WRAP}}$-functionality to capture such tokens and demonstrated feasibility of realizing general functionalities with UC-security. Most of the previous works in the tamper proof hardware [15,17,22,25,30,42,46] rely on this formulation. As we explain next, this formulation does not provide adequate composability guarantees. We begin by mentioning that any notion of composable security in an interactive setting should allow for multiple protocols to co-exist in the same system and interact with each other. We revisit the following desiderata put forth by Canetti, Lin and Pass [11] for any notion of composable security:

**Concurrent multi-instance security:** The security properties relating to local objects (including data and tokens) of the analyzed protocol itself should remain valid even when multiple instances of the protocol are executed concurrently and are susceptible to coordinated attacks against multiple instances. Almost all prior works in the tamper proof model do not specifically analyze their security in a concurrent setting. In other words, they only discuss UC-security of a single instance of the protocol. In particular, when executing protocols in the concurrent setting with tokens, an adversary could in fact transfer a token received from one execution to another and none of the previous works that are based on the $\mathcal{F}_{\mathrm{WRAP}}$-functionality accommodate transfers.

**Modular analysis:** Security of the larger overall protocols must be deducible from the security properties of its components. In other words, composing protocols should preserve security in a modular way. One of the main motivations and features in the UC-framework is the ability to analyze a protocol locally in isolation while guaranteeing global security. This does not only enable easier design but identifies the required security properties. The current framework proposed by Katz [42] does not allow for such a mechanism.

**Environmental friendliness:** Unknown protocols in the system should not adversely affect the security of the analyzed protocol. Prior UC-formulation of tamper proof tokens are not "fully" environment friendly as tokens cannot be transferred to other unknown protocols. Furthermore, prior works in the $\mathcal{F}_{\mathrm{WRAP}}$-hybrid do not explicitly prove multi-instance security in the presence of an environment (i.e., they do not realize the multi-versions of the corresponding complete functionality).

The state-of-affairs regarding tamper-proof tokens leads us to ask the following question.

*Does there exist a UC-formulation of tamper-proof hardware tokens that guarantee strong composability guarantees and allows for modular design?*

Since the work of [42], the power of hardware tokens has been explored extensively in a long series of works, especially in the context of achieving UC-security (for example, [15,17,23,25,26,30,51]). While the work of Katz [42] assumed the stronger stateful tokens, the work of Chandran, Goyal and Sahai [15] was the first to achieve UC-security using only stateless tokens. In this work we will focus only on the weaker stateless token model. In the tamper-proof model with stateless tokens, as we argue below, the issue of minimal assumptions and round-complexity have been largely unaddressed. The work of Chandran et al. [15] gives an $O(\kappa)$-round protocol (where $\kappa$ is the security parameter) based on enhanced trapdoor permutations. Following that, Goyal et al. [30] provided an (incorrect) $O(1)$-round construction based on Collision-Resistant Hash Functions (CRHFs). The work of Choi et al. [17], extending the techniques of [23,30], establishes the same result and provide a five-round construction based on CRHFs.

All previous constructions require assumptions stronger than one-way functions (OWFs), namely either trapdoor permutations or CRHFs. Thus as a first question, we investigate the minimal assumptions required for token-based secure computation protocols. The works of [17,30] rely on CRHFs for realizing statistically-hiding commitment schemes. Towards minimizing assumptions, both these works consider the variant of their protocol where they replace the construction of the statistically-hiding commitment scheme based on CRHFs to the one based on one-way functions [33] to obtain UC-secure protocols under minimal assumptions (See Theorem 3 in [30] and Footnote 7 in [17]). While analyzing the proof of this variant in the work of [30], we found a flaw[1] in the original construction based on CRHFs. We present a concrete attack that breaks the

---

[1] In private communication, the authors have acknowledged this flaw and are in the process of updating their result. We remark that we point out a flaw *only* in one particular result, namely, realizing the UC-secure oblivious transfer functionality based on CRHFs and stateless tokens.

security of their construction in Sect. 3. More recently, the authors of [17] have conveyed in private communication that the variant that naively replaces the commitment in their protocol is in fact vulnerable to covert attacks. They have since retracted this result (see the updated eprint version [16]). Given the state of affairs, our starting point is to address the following fundamental question regarding tokens that remains open.

*Can we construct tamper-proof UC-secure protocols using stateless tokens assuming only one-way functions?*

A second important question that we address here is:

*What is the round complexity of UC-secure two-party protocols using stateless tokens assuming only one-way functions?*

We remark here that relying on black-box techniques, it would be impossible to achieve non-interactive secure computation even in the tamper proof model as any such approach would be vulnerable to a residual function attack.[2] This holds even if we allow an initial token exchange phase, where the two parties exchange tokens (that are independent of their inputs). Hence, the best we could hope for is two rounds.[3]

**(G)UC-secure protocols in the multi-party setting.** In the UC framework, it is possible to obtain UC-secure protocols in the MPC setting by first realizing the UC-secure oblivious transfer functionality (UC OT) in the two-party setting and then combining it with general compilation techniques (e.g., [12,40,44,47] to obtain UC-secure multi-party computation protocols. First, we remark that specifically in the stateless tamper-proof tokens model, prior works fail to consider multi-versions of the OT-functionality while allowing transferrability of tokens which is important in an MPC setting.[4] As such, none of the previous works explicitly study the round complexity of multi-party protocols in the tamper proof model (with stateless tokens), we thus initiate this study in this work and address the following question.

*Can we obtain round-optimal multi-party computation protocols with GUC-security in the tamper proof model?*

---

[2] Intuitively, this attack allows the recipient of the (only) message to repeatedly evaluate the function on different inputs for a fixed sender's input.

[3] Note that in the plain model, without trusted setup, Katz and Ostrovsky [43] showed that five rounds are necessary and sufficient for general 2PC functionalities. Garg et al. [28] revisit the lower bound of [43] and showed that four rounds are necessary and sufficient for realizing general 2PC functionalities in the simultaneous message exchange model where both parties can simultaneously exchange messages in each round.

[4] We remark that the work of [17] considers multiple sessions of OT between a single pair of parties. However, they do not consider multiple sessions between multiple pairs of parties which is required to realize UC-security in the multiparty setting.

**Unidirectional token exchange.** Consider the scenario where companies such as Amazon or Google wish to provide an *email spam-detection* service and users of this service want to keep their emails private (so as to not have unwanted advertisements posted based on the content of their emails). In such a scenario, it is quite reasonable to assume that Amazon or Google have the infrastructure to create tamper-proof hardware tokens in large scale while the clients cannot be expected to create tokens on their own. Most of the prior works assume (require) that both parties have the capability of constructing tokens. When relying on non-black-box techniques, the work of [17] shows how to construct UC-OT using a single stateless token and consequently requires only one of the parties to create the token. The work of Moran and Segev in [51] on the other hand shows how to construct UC-secure two-party computation via a black-box construction where tokens are required to be passed only in one direction, however, they require the stronger model of stateful tokens. It is desirable to obtain a black-box construction when relying on stateless tokens. Unfortunately, the work of [17] shows that this is impossible in the fully concurrent setting. More precisely, they show that UC-security is impossible to achieve for general functionalities via a black-box construction using stateless tokens if only one of the parties is expected to create tokens. In this work, we therefore wish to address the following question:

> *Is there a meaningful security notion that can be realized in a client-server setting relying on black-box techniques using stateless tokens where tokens are created only by the server?*

## 1.1  Our Results

As our first contribution, we put forth a formulation of the tamper-proof hardware as a "global" functionality that provides strong composability guarantees. Towards addressing the various shortcomings of the composability guarantees of the UC-framework, Canetti et al. [7] introduced the Global Universal Composability (GUC) framework which among other things allows to consider global setup functionalities such as the common reference string model, and more recently the global random oracle model [9]. In this work, we put forth a new formulation of tokens in the GUC-framework that will satisfy all our desiderata for composition. Furthermore, in our formulation, we will be able to invoke the GUC composition theorem of [7] in a modular way. A formal description of the $\mathcal{F}_{\text{gWRAP}}$-functionality can be found in Fig. 2 and more detailed discussion is presented in the next section.

In the two-party setting we resolve both the round complexity and computational complexity required to realize GUC-secure protocols in the stronger $\mathcal{F}_{\text{gWRAP}}$-hybrid stated in the following theorem:

**Theorem 1.1 (Informal).** *Assuming the existence of OWFs, there exists a two-round protocol that GUC realizes any (well-formed) two-party functionality in the global tamper proof model assuming stateless tokens. Moreover it only makes black-box use of the underlying OWF.*

As mentioned earlier, any (black-box) non-interactive secure computation protocol is vulnerable to a residual function attack assuming stateless tokens. Therefore, the best round complexity we can hope for assuming (stateless) tamper-proof tokens is two which our results shows is optimal. In concurrent work [24], Dottling et al. show how to obtain UC-secure two-party computation protocol relying on one-way functions via non-black-box techniques.

As mentioned before, we also identify a flaw in a prior construction that attempted to construct a UC-secure protocols in the stateless tamper-proof model from OWFs. We describe a concrete attack on this protocol in Sect. 3. On a high-level, the result of Goyal et al. first constructs a "Quasi oblivious transfer" protocol based on tokens that admits one-sided simulation and one-sided indistinguishability. Next, they provide a transformation from Quasi-OT to full OT. We demonstrate that the transformation in the second step is insecure by constructing an adversary that breaks its security. The purpose of presenting the flaw is to illustrate a subtlety that arises when arguing security in the token model. While there are mechanisms that could potentially facilitate compiling a Quasi-OT to full OT in the token model, we do not pursue this approach for two reasons. First, fixing this issue will still result in a protocol that requires statistically-hiding commitments and in light of the vulnerability of the [17] protocol, it is unclear if we can simply rely on one-way functions for the statistically-hiding commitment scheme to obtain a construction under minimal assumptions. Second, even if this construction is secure, it would yield only a $O(\kappa)$-round protocol [33]. Instead, we directly construct a round-optimal construction based on OWFs using a more modular, and in our opinion, simpler construction.

In the multi-party setting, our first theorem follows as a corollary of our results from the two-party setting.

**Theorem 1.2.** *Assuming the existence of OWFs, there exists a $O(d_f)$-round protocol that GUC realizes any multi-party (well formed) functionality $f$ in the global tamper proof model assuming stateless tokens, where $d_f$ is the depth of any circuit implementing $f$.*

Next, we improve the round-complexity of our construction to obtain the following theorem:

**Theorem 1.3.** *Assuming the existence of OWFs and stand-alone semi-honest MPC in the OT-hybrid, there exists a three-round protocol that GUC realizes any multi-party (well formed) functionality in the global tamper proof model assuming stateless tokens.*

We remark that our construction is black-box in the underlying one-way function but relies on the code of the MPC protocol in a non-black-box way. It is conceivable that one can obtain a round-optimal construction if we do not require it to be black-box in the underlying primitives and leave it as future work.

Finally, in the client-server setting, we prove the following theorem in the full version [34]:

**Theorem 1.4 (Informal).** *Assuming the existence of one-way functions, there exists a two-round protocol that securely realizes any two-party functionality assuming stateless tokens in a client-server setting, where the tokens are created only by the server. We also provide an extension where we achieve UC-security against malicious clients and sequential and parallel composition security against malicious servers.*

In more detail, we provide straight-line (UC) simulation of malicious clients and standard rewinding-based simulation against malicious servers. Our protocols guarantee security of the servers against arbitrary malicious coordinating clients and protects every individual client executing sequentially or in parallel against a corrupted server. We believe that this is a reasonable model in comparison to the Common Reference String (CRS) model where both parties require a trusted entity to sample the CRS. Furthermore, it guarantees meaningful concurrent security that is otherwise not achievable in the plain model in two rounds.

## 1.2   Our Techniques

Our starting point for our round optimal secure two-party computation is the following technique from [30] for an extractable commitment scheme.

Roughly speaking, in order to extract the receiver's input, the sender chooses a function $F$ from a pseudorandom function family that maps $\{0,1\}^m$ to $\{0,1\}^n$ bits where $m >> n$, and incorporates it into a token that it sends to the receiver. Next, the receiver commits to its input $b$ by first sampling a random string $u \in \{0,1\}^m$ and querying the PRF token on $u$ to receive the value $v$. It sends as its commitment the string $\mathsf{com}_b = (\mathsf{Ext}(u;r) \oplus b, r, v)$ where $\mathsf{Ext}(\cdot,\cdot)$ is a strong randomness extractor. Now, since the PRF is highly compressing, it holds with high probability that conditioned on $v$, $u$ has very high min-entropy and therefore $\mathsf{Ext}(u;r) \oplus b, r$ statistically hides $b$. Furthermore, it allows for extraction as the simulator can observe the queries made by the sender to the token and observe that queries that yields $v$ to retrieve $u$. This commitment scheme is based on one-way functions but is only extractable. To obtain a full-fledged UC-commitment from an extractable commitment we can rely on standard techniques (See [35,56] for a few examples). Instead, in order to obtain round-optimal constructions for secure two-party computation, we extend this protocol directly to realize the UC oblivious transfer functionality. A first incorrect approach is the following protocol. The parties exchange two sets of PRF tokens. Next, the receiver commits to its bit $\mathsf{com}_b$ using the approach described above, followed by the sender committing to its input $(\mathsf{com}_{s_0}, \mathsf{com}_{s_1})$ along with an OT token that implements the one-out-of-two string OT functionality. More specifically, it stores two strings $s_0$ and $s_1$, and given a single bit $b$ outputs $s_b$. Specifically, the code of that token behaves as follows:

- On input $b^*, u^*$, the token outputs $(s_b, \mathsf{decom}_{s_b})$ only if $\mathsf{com}_b = (\mathsf{Ext}(u^*;r) \oplus b^*, r, v)$ and $\mathsf{PRF}(u^*) = v$. Otherwise, the token aborts.

The receiver then runs the token to obtain $s_b$ and verifies if $\mathsf{decom}_{s_b}$ correctly decommits $\mathsf{com}_{s_b}$ to $s_b$. This simple idea is vulnerable to an input-dependent abort attack, where the token aborts depending on the value $b^*$. The work of [30] provides a combiner to handle this particular attack which we demonstrate is flawed. We describe the attack in Sect. 3. We instead will rely on a combiner from the recent work of Ostrovsky, Richelson and Scafuro [54] to obtain a two-round GUC-OT protocol.

**GUC-secure multi-party computation protocols.** In order to demonstrate feasibility, we simply rely on the work of [40] who show how to achieve GUC-secure MPC protocols in the OT-hybrid. By instantiating the OT with our GUC-OT protocol, we obtain MPC protocols in the tamper proof model assuming only one-way functions. While this protocol minimizes the complexity assumptions, the round complexity would be high. In this work, we show how to construct a 3-round MPC protocol. Our starting point is to take any semi-honest MPC protocol in the stand-alone model and compile it into a malicious one using tokens following the paradigm in the original work of Canetti et al. [12] and subsequent works [48,55]. Roughly, the approach is to define a commit-and-prove GUC-functionality $\mathcal{F}_{\mathrm{CP}}$ and compile the semi-honest protocol using this functionality following a GMW-style compilation.

We will follow an analogous approach where we directly construct a full-fledged $\mathcal{F}_{\mathrm{CP}}^{1:M}$-functionality that allows a single prover to commit to a string and then prove multiple statements on the commitment simultaneously to several parties. In the token model, realizing this primitive turns out to be non-trivial. This is because we need the commitment in this protocol to be straight-line extractable and the proof to be about the value committed. Recall that, the extractable commitment is based on a PRF token supplied by the receiver of the commitment (and the verifier in the zero-knowledge proof). The prover cannot attest the validity of its commitment (via an NP-statement) since it does not know the code (i.e. key) of the PRF. Therefore, any commit and prove scheme in the token model necessarily must rely on a zero-knowledge proof that is black-box in the underlying commitment scheme. In fact, in the seminal work of Ishai et al. [39] they showed how to construct such protocols that have been extensively used in several works where the goal is to obtain constructions that are black-box in the underlying primitives. Following this approach and solving its difficulties that appear in the tamper-proof hardwire model, we can compile a $T$-round semi-honest secure MPC protocol to a $O(T)$-round protocol. Next, to reduce the rounds of the computation we consider the approach of Garg et al. [27] who show how to compress the round complexity of any MPC protocol to a two-round GUC-secure MPC protocol in the CRS model using obfuscation primitives.

In more detail, in the first round of the protocol in [27], every party commits to its input along with its randomness. The key idea is the following compiler used in the second round: it takes any (interactive) underlying MPC protocol, and has each party obfuscate their "next-message" function in that protocol, providing one obfuscation for each round. To ensure correctness, zero-knowledge proofs are used to validate the actions of each party w.r.t the commitments made in the first step. Such a mechanism is also referred to as a commit-and-prove strategy. This enables each party to independently evaluate the obfuscation one by one, generating messages of the underlying MPC protocol and finally obtain the output. The observation here is that party $P_i$'s next-message circuit for round $j$ in the underlying MPC protocol depends on its private input $x_i$ and randomness $r_i$ (which are hard-coded in the obfuscation) and on input the transcript of the communication in the first $j - 1$ rounds outputs its message for the next round.

To incorporate this approach in the token model, we can simply replace the obfuscation primitives with tokens. Next, to employ zero-knowledge proofs via a black-box construction, we require a zero-knowledge protocol that allows commitment of a witness via tokens at the beginning of the protocol and then in a later step prove a statement about this witness where the commitment scheme is used in a "black-box" way. A first idea here would be to compile using the zero-knowledge protocol of [39] that facilitate such a commit-and-prove paradigm. However, as we explain later this would cost us in round-complexity. Instead we will rely on so-called input-delayed proofs [45] that have recently received much attention [20,21,36]. In particular, we will rely on the recent work of [36] who shows how to construct the so-called "input-delay" commit-and-prove protocols which allow a prover to commit a string in an initial commit phase and then prove a statement regarding this string at a later stage where the input statement is determined later. However, their construction only allows for proving one statement regarding the commitment. One of our technical contributions is to extend this idea to allow multiple theorems and further extend it so that a single prover can prove several theorems to multiple parties simultaneously. This protocol will be 4-round and we show how to use this protocol in conjunction with the Garg et al.'s round collapsing technique.

## 1.3   Related Work

In recent and independent work, using the approach of [9], Nilges [49,53] consider a GUC-like formulation of the tokens for the two-party setting where the parties have fixed roles. The focus in [49,53] was to obtain a formulation that accommodates reusability of a single token for several independent protocols in the UC-setting for the specific two-party case. In contrast to our work, they do not explicitly model or discuss adversarial transferability of the tokens. In particular they do not discuss in the multi-party case, which is the main motivation behind our work.

Another recent work by Boureanu, Ohkubo and Vaudenay [5] studies the limit of composition when relying on tokens. In this work, they prove that EUC (or GUC)-security is impossible to achieve for most functionalities if tokens can be transferred in a restricted framework. More precisely, their impossibility holds, if the tokens themselves do not "encode" the session identifier in any way. Our work, circumvents this impossibility result by precisely allowing the tokens generated (by honest parties) to encode the session identifier in which they have to be used.

# 2  Modeling Tamper-Proof Hardware in the GUC Framework

In this section we describe our model and give our rationale for our approach. We provide a brief discussion on the Universal Composability (UC) framework [6], UC with *joint state* [14] (JUC) and Generalized UC [7] (GUC). For more details, we refer the reader to the original works and the discussion in [9].

*Basic UC.* Introduced by Canetti in [6], the Universal Composability (UC) framework provides a framework to analyse security of protocols in complex network environments in a modular way. One of the fundamental contributions of this work was to give a definition that will allow to design protocols and demonstrate security by "locally" analyzing a protocol but guaranteeing security in a *concurrent* setting where security of the protocol needs to be intact even when it is run concurrently with many instances of arbitrary protocols. Slightly more technically, in the UC-framework, to demonstrate that a protocol $\Pi$ securely realizes an ideal functionality $\mathcal{F}$, we need to show that for any adversary $\mathcal{A}$ in the real world interacting with protocol $\Pi$ in the presence of arbitrary environments $\mathcal{Z}$, there exists an ideal adversary $\mathcal{S}$ such that for any environment $\mathcal{Z}$ the view of an interaction with $\mathcal{A}$ is indistinguishable from the view of an interaction with the ideal functionality $\mathcal{F}$ and $\mathcal{S}$.

Unfortunately, soon after its inception, a series of impossibility results [8,10,48] demonstrated that most non-trivial functionalities cannot be realized in the UC-framework. Most feasibility results in the UC-framework relied on some sort of trusted setup such as the common reference string (CRS) model [8], tamper-proof model [42] or relaxed security requirements such as super-polynomial simulation [3,55,57]. When modeling trusted setup such as the CRS model, an extension of the UC-framework considers the $\mathcal{G}$-hybrid model where "all" real-world parties are given access to an ideal setup functionality $\mathcal{G}$. In order for the basic composition theorem to hold in such a $\mathcal{G}$-hybrid model, two restrictions have to be made. First, the environment $\mathcal{Z}$ cannot access the ideal setup functionality *directly*; it can only do so indirectly via the adversary. In some sense, the setup $\mathcal{G}$ is treated as "local" to a protocol instance. Second, two protocol instances of the same or different protocol cannot share "state" for the UC-composition theorem to hold. Therefore, a setup model such as the CRS in the UC-framework necessitates that each protocol uses its own local setup. In other words, an independently sampled reference string for every protocol instance. An alternative approach that was pursued in a later work was to realize a multi-version of a functionality and proved security of the multi-version using a single setup. For example, the original feasibility result of Canetti, Lindell, Ostrovsky and Sahai [12] realized the $\mathcal{F}_{\text{MCOM}}$-functionality which is the multi-version of the basic commitment functionality $\mathcal{F}_{\text{COM}}$ in the CRS model.

*JUC.* Towards accommodating a global setup such as the CRS for multiple protocol instances, Canetti and Rabin [14] introduced the Universal Composition with Joint State (JUC) framework. Suppose we want to analyze several instances of protocol $\Pi$ with an instance $\mathcal{G}$ as common setup, then at the least, each instance of the protocol must share some state information regarding $\mathcal{G}$ (e.g., the reference string in the CRS model). The JUC-framework precisely accommodates such a scenario, where a new composition theorem is proven, that allows for composition of protocols that share some state. However, the JUC-model for the CRS setup would only allow the CRS to be accessible to a pre-determined set of protocols and in particular still does not allow the environment to directly access the CRS.

*GUC.* For most feasibility results in the (plain) CRS model both in the UC and JUC framework, the simulator $\mathcal{S}$ in the ideal world needed the ability to "program" the CRS. In particular, it is infeasible to allow the environment to access the setup reference string. As a consequence, we can prove security only if the reference string is privately transmitted to the protocols that we demand security of and cannot be made *publicly* accessible. The work of Canetti, Pass, Dodis and Walfish [7] introduced the Generalized UC-framework to overcome this shortcoming in order to model the CRS as a global setup that is publicly available. More formally, in the GUC-framework, a global setup $\mathcal{G}$ is accessible by any protocol running in the system and in particular allows direct access by the environment. This, in effect, renders all previous protocols constructed in the CRS model not secure in the GUC framework as the simulator loses the programmability of the CRS. In fact, it was shown in [7] that the CRS setup is insufficient to securely realize the ideal commitment functionality in the GUC-framework. More generally, they show that any setup that simply provides only "public" information is not sufficient to realize GUC-security for most non-trivial functionalities. They further demonstrated a feasibility in the Augmented CRS model, where the CRS contains signature keys, one for each party and a secret signing key that is not revealed to the parties, except if it is corrupt, in which case the secret signing key for that party is revealed.

As mentioned before, the popular framework to capture the tamper-proof hardware is the one due to [42] who defined the $\mathcal{F}_{\text{WRAP}}$-functionality in the UC-framework. In general, in the token model, the two basic advantages that the simulator has over the adversary is "observability" and "programmability". Observability refers to the ability of the simulator to monitor all queries made by an adversary to the token and programmability refers to the ability to program responses to the queries in an online manner. In the context of tokens, both these assumptions are realistic as tamper-proof tokens do provide both these abilities in a real-world. However, when modeling tamper proof hardware tokens in the UC-setting, both these properties can raise issues as we discuss next.

Apriori, it is not clear why one should model the tamper proof hardware as a global functionality. In fact, the tokens are local to the parties and it makes the case for it *not* to be globally accessible. Let us begin with the formulation by Katz [42] who introduced the $\mathcal{F}_{\text{WRAP}}$-functionality (see Fig. 1 for the stateless variant). In the real world the creator or sender of a token specifies the code to be incorporated in a token by sending the description of a Turing machine $M$ to the ideal functionality. The ideal functionality then emulates the code of $M$ to the receiver of the token, only allowing black-box access to the input and output tapes of $M$. In the case of stateful tokens, $M$ is modeled as an interactive Turing machine while for stateless tokens, standard Turing machines would suffice. Slightly more technically, in the UC-model, parties are assigned unique identifiers PID and sessions are assigned identifiers sid. In the tamper proof model, to distinguish tokens, the functionality accepts an identifier mid when a token is created. More formally, when one party $PID_i$ creates a token with program $M$ with token identifier mid and sends it to another party $PID_j$ in session sid, then the $\mathcal{F}_{\mathrm{WRAP}}$ records the tuple $(PID_i, PID_j, \mathrm{mid}, M)$. Then whenever a party with identifier $PID_j$ sends a query $(\mathrm{Run}, \mathrm{sid}, PID_i, \mathrm{mid}, x)$ to the $\mathcal{F}_{\mathrm{WRAP}}$-functionality, it first checks whether there is a tuple of the form $(\cdot, PID_j, \mathrm{mid}, \cdot)$ and then runs the machine $M$ in this tuple if one exists.

---

**Functionality $\mathcal{F}_{\mathrm{WRAP}}^{\mathrm{Stateless}}$**

Functionality $\mathcal{F}_{\mathrm{WRAP}}^{\mathrm{Stateless}}$ is parameterized by a polynomial $p(\cdot)$ and an implicit security parameter $\kappa$.

**Create.** Upon receiving $(\mathrm{Create}, \mathrm{sid}, PID_i, PID_j, \mathrm{mid}, M)$ from S, where $M$ is a Turing machine, do:

    1. Send $(\mathrm{Create}, PID_i, PID_j, \mathrm{mid})$ to R.
    2. Store $(PID_i, PID_j, \mathrm{mid}, M)$.

**Execute.** Upon receiving $(\mathrm{Run}, \mathrm{sid}, PID_i, \mathrm{mid}, x)$ from R, find the unique stored tuple $(PID_i, PID_j, \mathrm{mid}, M)$. If no such tuple exists, do nothing. Run $M(x)$ for at most $p(\kappa)$ steps, and let out be the response (out $= \bot$ if $M$ does not halt in $p(k)$ steps). Send $(PID_i, PID_j, \mathrm{mid}, \mathrm{out})$ to R.

---

**Fig. 1.** The ideal functionality for stateless tokens [42].

In the UC-setting (or JUC), to achieve any composability guarantees, we need to realize the multi-use variants of the specified functionality and then analyze the designed protocol in a concurrent man-in-the-middle setting. In such a multi-instance setting, it is reasonable to assume that an adversary that receives a token from one honest party in a left interaction can forward the token to another party in a right interaction. Unfortunately, the $\mathcal{F}_{\mathrm{WRAP}}$-functionality does not facilitate such a transfer.

Let us modify $\mathcal{F}_{\mathrm{WRAP}}$ to accommodate transfer of tokens by adding a special "transfer" query that allows a token in the possession of one party to be transferred to another party. Since protocols designed in most works do not explicitly prove security in a concurrent man-in-the-middle setting, such a modification renders the previous protocols designed in $\mathcal{F}_{\mathrm{WRAP}}$ insecure. For instance, consider the commitment scheme discussed in the introduction based on PRF tokens. Such a scheme would be insecure as an adversary can simply forward the token from the receiver in a right interaction to the sender in a left interaction leading to a malleable commitment.

In order to achieve security while allowing transferability we need to modify the tokens themselves in such a way to be not useful in an execution different from where it is supposed to be used. If every honestly generated token admits only queries that are prefixed with the correct session identifier then transferring the tokens created by one honest party to another honest party will be useless as honest parties will prefix their queries with the right session and the honestly generated tokens will fail to answer on incorrect session prefixes. This is inspired by an idea in [9], where they design GUC-secure protocols in the Global Random Oracle model [9]. As such, introducing transferrability naturally requires protocols to address the issue of non-malleability.

While this modification allows us to model transferrability, it still requires us to analyze protocols in a concurrent man-in-the-middle setting. In order to obtain a more modular definition, where each protocol instance can be analyzed in isolation we need to allow the token to be transferred from the adversary to the environment. In essence, we require the token to be somewhat "globally" accessible and this is the approach we take.

## 2.1  The Global Tamper-Proof Model

A natural first approach would be to consider the same functionality in the GUC-framework and let the environment to access the $\mathcal{F}_{\mathrm{WRAP}}$-functionality. This is reasonable as an environment can have access to the tokens via auxiliary parties to whom the tokens were transferred to. However, naively incorporating this idea would deny "observability" and "programmability" to the simulator as all adversaries can simply transfer away their tokens the moment they receive them and let other parties make queries on their behalf. Indeed, one can show that the impossibility result of [17] extends to this formulation of the tokens (at least if the code of the token is treated in a black-box manner).[5] A second approach would be to reveal to the simulator all queries made to the token received by the adversary even if transferred out to any party. However, such a formulation would be vulnerable to the following transferring attack. If an adversary received a token from one session, it can send it as its token to an honest party in another session and now observe all queries made by the honest party to the token. Therefore such a formulation of tokens is incorrect.

Our formulation will accommodate transferrability while still guaranteeing observability to the simulator. In more detail, we will modify the definition of $\mathcal{F}_{\mathrm{WRAP}}$ so that it will reveal to the simulator all "illegitimate" queries made to the token by any other party. This approach is analogous to the one taken by Canetti, Jain and Scafuro [9] where they model the Global Random Oracle

---

[5] Informally, the only advantage that remains for the simulator is to see the code of the tokens created by the adversary. This essentially reduces to the case where tokens are sent only in one direction and is impossible due to a result of [17] when the code is treated as a black-box.

Model and are confronted by a similar issue; here queries made to a globally accessible random oracle via auxiliary parties by the environment must be made available to the simulator while protecting the queries made by the honest party. In order to define "legitimate" queries we will require that all tokens created by an honest party, by default, will accept an input of the form $(\mathsf{sid}, x)$ and will respond with the evaluation of the embedded program $M$ on input $x$, only if $\mathsf{sid} = \overline{\mathsf{sid}}$, where $\overline{\mathsf{sid}}$ corresponds to the session where the token is supposed to be used, i.e. the session where the honest party created the token. Furthermore, whenever an honest party in session $\overline{\mathsf{sid}}$ queries a token it received on input $x$, it will prefix the query with the correct session identifier, namely issue the query $(\overline{\mathsf{sid}}, x)$. An *illegitimate query* is one where the sid prefix in a query differs from the session identifier from which the party is querying from. Every illegitimate query will be recorded by our functionality and will be disclosed to the party whose session identifier is actually $\mathsf{sid}$.

More formally, the $\mathcal{F}_{\mathrm{gWRAP}}$-functionality is parameterized by a polynomial $p(\cdot)$ which is the time bound that the functionality will exercise whenever it runs any program. The functionality admits the following queries:

**Creation Query:** This query allows one party S to create and send a token to another party R by sending the query $(\mathsf{Create}, \mathsf{sid}, \mathsf{S}, \mathsf{R}, \mathsf{mid}, M)$ where $M$ is the description of the machine to be embedded in the token, mid is a unique identifier for the token and sid is the session identifier. The functionality records $(\mathsf{R}, \mathsf{sid}, \mathsf{mid}, M)$.[6]

**Transfer Query:** We explicitly provide the ability for parties to transfer tokens to other parties that were not created by them (e.g., received from another session). Such a query will only be used by the adversary in our protocols as honest parties will always create their own tokens. When a transfer query of the form $(\mathsf{transfer}, \mathsf{sid}, \mathsf{S}, \mathsf{R}, \mathsf{mid})$ is issued, the tuple $(\mathsf{S}, \overline{\mathsf{sid}}, \mathsf{mid}, M)$ is erased and a new tuple $(\mathsf{R}, \mathsf{sid}, \mathsf{mid}, M)$ is created where $\overline{\mathsf{sid}}$ is the identifier of the session where it was previously used.

**Execute Query:** To run a token the party needs to provide an input in a particular format. All honest parties will provide the input as $x = (\mathsf{sid}, x')$ and the functionality will run $M$ on input $x$ and supply the answer. In order to achieve non-malleability, we will make sure in all our constructions that tokens generated by honest parties will respond to a query only if it contains the correct sid.

**Retrieve Query:** This is the important addition to our functionality following the approach taken by [9]. $\mathcal{F}_{\mathrm{gWRAP}}$-functionality will record all illegitimate queries made to a token. Namely for a token recorded as the tuple $(\mathsf{R}, \overline{\mathsf{sid}}, \mathsf{mid}, M)$ an illegitimate query is of the form $(\mathsf{sid}, x)$ where $\mathsf{sid} \neq \overline{\mathsf{sid}}$ and such a query will be recorded in a set $\mathcal{Q}_{\mathsf{sid}}$ that will be made accessible to the receiving party corresponding to sid.

---

[6] We remark here that the functionality does not explicitly store the PID of the creator of the token. We made this choice since the simulator in the ideal world will create tokens for itself which will serve as a token created on behalf of an honest party.

A formal description of the ideal functionality $\mathcal{F}_{\text{gWRAP}}$ is presented in Fig. 2. We emphasize that our formulation of the tamper-proof model will now have the following benefits:

1. It overcomes the shortcomings of the $\mathcal{F}_{\text{WRAP}}$-functionality as defined in [42] and used in subsequent works. In particular, it allows for transferring tokens from one session to another while retaining "observability".
2. Our model allows for designing protocols in the UC-framework and enjoys the composition theorem as it allows the environment to access the token either directly or via other parties.
3. Our model explicitly rules out "programmability" of tokens. We remark that it is (potentially) possible to explicitly provide a mechanism for programmability in the $\mathcal{F}_{\text{gWRAP}}$-functionality. We chose to not provide such a mechanism so as to provide stronger composability guarantees.
4. In our framework, we can analyze the security of a protocol in isolation and guarantee concurrent multi-instance security directly using the GUC-composition theorem. Moreover, it suffices to consider a "dummy" adversary that simply forwards the environment everything (including the token).

An immediate consequence of our formulation is that it renders prior works such as [15,23,24,42] that rely on the programmability of the token insecure in our model. The works of [17,30] on the other hand can be modified and proven secure in the $\mathcal{F}_{\text{gWRAP}}$-hybrid as they do not require the tokens to be programmed.

We now provide the formal definition of UC-security in the Global Tamper-Proof model.

**Definition 2.1 (GUC security in the global tamper-proof model).** *Let $\mathcal{F}$ be an ideal functionality and let $\pi$ be a multi-party protocol. Then protocol $\pi$ GUC realizes $\mathcal{F}$ in $\mathcal{F}_{\text{gWRAP}}$-hybrid model, if for every uniform PPT hybrid-model adversary $\mathcal{A}$, there exists a uniform PPT simulator $\mathcal{S}$, such that for every non-uniform PPT environment $\mathcal{Z}$, the following two ensembles are computationally indistinguishable,*

$$\left\{ \mathbf{View}_{\pi,\mathcal{A},\mathcal{Z}}^{\mathcal{F}_{\text{gWRAP}}}(\kappa) \right\}_{\kappa \in \mathbb{N}} \overset{c}{\approx} \left\{ \mathbf{View}_{\mathcal{F},\mathcal{S},\mathcal{Z}}^{\mathcal{F}_{\text{gWRAP}}}(\kappa) \right\}_{\kappa \in \mathbb{N}}.$$

# 3  Issue with *Over Extraction* in Oblivious Transfer Combiners [30]

In the following we identify an issue that affects one of the feasibility results in [30, Sect. 5]. More precisely, this result establishes that UC security for general functionalities is feasible in the tamper-proof hardware model in $O(\kappa)$-round assuming only OWFs (or $O(1)$-round based on CRHFs) based on stateless tokens. The issue arises as a result of *over extraction* where a fully-secure OT protocol is constructed from a weaker variant and the simulation extracts values for sender's inputs even on certain executions where the receiver aborts.

---

**Functionality $\mathcal{F}_{\mathrm{gWRAP}}$**

Parameters: Polynomial $p(\cdot)$.

**Create.** Upon receiving (Create, sid, S, R, mid, $M$) from S, where $M$ is a Turing machine, do:

    1. Send (Receipt, sid, S, R, mid) to R.

    2. Store (R, sid, mid, $M$).

**Execute.** Upon receiving (Run, sid, mid, $x$) from R, find the unique stored tuple (R, sid, mid, $M$). If such a tuple does not exist, do nothing. Otherwise, interpret $x = (\overline{\mathrm{sid}}, x')$ and run $M(x)$ for at most $p(\kappa)$ steps, and let out be the response (out $= \perp$ if $M$ does not halt in $p(k)$ steps). Send (sid, R, mid, out) to R.

    **Handling Illegal Queries:** If sid $\neq \overline{\mathrm{sid}}$, then add $(x', \mathrm{out}, \mathrm{mid})$ to the list $\mathcal{Q}_{\overline{\mathrm{sid}}}$ that is initialized to be empty.

**Transfer.** Upon receiving (transfer, sid, S, R, mid) from S, find the unique stored tuple (S, $\overline{\mathrm{sid}}$, mid, $M$). If no such tuple exists, do nothing. Otherwise,

    1. Send (Receipt, sid, S, R, mid) to R.

    2. Store (R, sid, mid, $M$). Erase (S, $\overline{\mathrm{sid}}$, mid, $M$).

**Retrieve Queries:** Upon receiving a request (retreive, sid) from a party R, return the list $\mathcal{Q}_{\mathrm{sid}}$ of illegitimate queries.

---

**Fig. 2.** The global stateless token functionality.

The term over extraction has been studied before in the context of commitment schemes where a scheme with over extraction is constructed as an intermediate step towards achieving full security [31,56].

On a high-level, in the work of [30], they first construct an OT protocol with milder security guarantees. More precisely, a QuasiOT protocol achieves UC-security against a malicious receiver and straight-line extraction against malicious sender. However, the scheme is not fully secure as a malicious sender can cause an input-dependent abort for an honest receiver. Towards amplifying the security, [30] consider the following protocol:

1. The sender with input $(s_0, s_1)$ and receiver with input $b$ interact in $n$ executions of QuasiOTs. The sender picks $z_1, \ldots, z_n$ and $\Delta$ at random and sets the inputs to the $i^{th}$ QuasiOT instance as $(z_i, z_i + \Delta)$. The receiver on the other hand chooses bits $b_1, \ldots, b_n$ at random subject to the sum being its input $b$.
2. If the first step completes, the sender sends $(s_0' = s_0 + \sum_i z_i, s_1' = s_1 + \sum_i z_i + \Delta)$ to the receiver. The receiver computes its output as $s_b' + \sum_i w_i$ where $w_i$ is the output of the receiver in the $i^{th}$ QuasiOT.

This protocol remains secure against a malicious receiver. However, an issue arises with a malicious sender. To simulate a malicious sender in this protocol, [30] rely on the straight-line extractor of the $n$ QuasiOTs by sampling two sets of random $(b_1, \ldots, b_n)$, one set summing up to 0 and another set summing up to 1 and computing what the receiver outputs in the two cases. As we demonstrate below such a strategy leads to failure in the simulation. More precisely, consider the following malicious sender strategy.

- Pick $z_1, z_2, \ldots, z_{n-1}$ and $\Delta$ at random.
- The inputs of the first $n-1$ tokens are set to $z_1, z_1 + \Delta, \ldots, z_{n-1}, z_{n-1} + \Delta$.
- Let $z_1 + \ldots + z_{n-1} = a$ and $z_1 + \ldots + z_{n-1} + \Delta = b$.
- The inputs to the $n$-th token are some fixed values $c$ (when $b_n = 0$) and $d$ (when $b_n = 1$), where $c + d \neq \Delta$.

Next, the sender modifies the code of the tokens used in the QuasiOT protocol so that the first $n-1$ QuasiOTs never abort. The $n$-th instantiations however is made to abort whenever the input $b_n$, the receiver's input is 1. Let $s_0 = 0$ and $s_1 = 1$ (we remark that we are not concerned about the actual inputs of the sender, but focus on what the receiver learns). We next examine the honest receiver's output in both the real and ideal worlds. First, in the real world the honest receiver learns an output only if $b_n = 0$ (since the $n$-th token aborts whenever $b_n = 1$). We consider two cases:

Case 1: The receiver's input is $b = 0$. Then $b_n = 0$ with probability $1/2$, and $b_n = 1$ with probability $1/2$. Moreover, when $b_n = 0$, the sum of the outputs obtained by the receiver is $a + c$. This is because when $b_n = 0$, then, $b_1 + \ldots + b_{n-1} = 0$, and the receiver learns $a$ as the sum of the outputs in the first $n-1$ QuasiOTs and $c$ from the $n$-th QuasiOT. On the other hand, if $b_n = 1$ then the receiver aborts in the $n$-th QuasiOT and therefore aborts.

Case 2: The receiver's input is $b = 1$. Similarly, in this case the receiver will learn $b + c$ with probability $1/2$ and aborts with probability $1/2$.

In the ideal world, the simulator runs first with a random bit-vector and extracts its inputs in the QuasiOTs by monitoring the queries to the corresponding PRF tokens. Next, it generates two bit-vectors $b_i$'s and $b_i'$'s that add up to 0 and 1, respectively, and computes the sums of the sender's input that correspond to these bits. Then the distribution of these sums can be computed as follows:

Case 1: In case that $\sum b_i = 0$, then $b_n = 0$ with probability $1/2$, and $b_n = 1$ with probability $1/2$. In the former case the receiver learns $a + c$, whereas in the latter case it learns $b + d$.

Case 2: In case that $\sum b_i' = 1$, then with probability $1/2$, $b_n' = 0$ and with probability $1/2$, $b_n' = 1$. In the former case the receiver learns $b + c$, whereas in the latter it learns $a + d$.

Note that this distribution is different from the real distribution, where the receiver never learns $b + d$ or $a + d$ since the token will always abort and not reveal $d$. We remark that in our example the abort probability of the receiver is independent of its input as proven in Claim 17 in [30], yet the distribution of what it learns is different.

On a more general note, our attack presents the subtleties that need to be addressed with the "selective" abort strategy. Recent works by Ciampi et al. [18,19] have identified subtleties in recent construction of non-malleable commitments [32] where selective aborts were not completely addressed.

# 4    Two-Round Token-Based GUC Oblivious Transfer

In this section we present our main protocol that implements GUC OT in two rounds. We first construct a three-round protocol and then show in [34] how to obtain a two-round protocol by exchanging tokens just once in a setup phase. Recall that the counter example to the [30] protocol shows that directly extracting the sender's inputs does not necessarily allow us to extract the sender's inputs correctly, as the tokens can behave maliciously. Inspired by the recently developed protocol from [54] we consider a new approach here for which the sender's inputs are extracted directly by monitoring the queries it makes to the PRF tokens and using additional checks to ensure that the sender's inputs can be verified.

*Protocol intuition.* As a warmup consider the following sender's algorithm that first chooses two random strings $x_0$ and $x_1$ and computes their shares $[x_b] = (x_b^1, \ldots, x_b^{2\kappa})$ for $b \in \{0, 1\}$ using the $\kappa + 1$-out-of-$2\kappa$ Shamir secret-sharing scheme. Next, for each $b \in \{0, 1\}$, the sender commits to $[x_b]$ by first generating two vectors $\alpha_b$ and $\beta_b$ such that $\alpha_b \oplus \beta_b = [x_b]$, and then committing to these vectors. Finally, the parties engage in $2\kappa$ parallel OT executions where the sender's input to the $j$th instance are the decommitments to $(\alpha_0[j], \beta_0[j])$ and $(\alpha_1[j], \beta_1[j])$. The sender further sends $(s_0 \oplus x_0, s_1 \oplus x_1)$. Thus, to learn $s_b$, the receiver needs to learn $x_b$. For this, it enters the bit $b$ for $\kappa + 1$ or more OT executions and then reconstructs the shares for $x_b$, followed by reconstructing $s_b$ using these shares. Nevertheless, this reconstruction procedure works only if there is a mechanism that verifies whether the shares are consistent.

To resolve this issue, Ostrovsky et al. made the observation that the Shamir secret-sharing scheme has the property for which there exists a linear function $\phi$ such that any vector of shares $[x_b]$ is valid if and only if $\phi(x_b) = 0$. Moreover, since the function $\phi$ is linear, it suffices to check whether $\phi(\alpha_b) + \phi(\beta_b) = 0$. Nevertheless, this check requires from the receiver to know the entire vectors $\alpha_b$ and $\beta_b$ for its input $b$. This means it would have to use $b$ as the input to all the $2\kappa$ OT executions, which may lead to an input-dependent abort attack. Instead, Ostrovsky et al. introduced a mechanism for checking consistency indirectly via a *cut-and-choose* mechanism. More formally, the sender chooses $\kappa$ pairs of vectors that add up to $[x_b]$. It is instructive to view them as matrices $A_0, B_0, A_1, B_1 \in \mathbb{Z}_p^{\kappa \times 2\kappa}$ where for every row $i \in [\kappa]$ and $b \in \{0, 1\}$, it holds that $A_b[i, \cdot] \oplus B_b[i, \cdot] = [x_b]$. Next, the sender commits to each entry of each matrix separately and sets as input to the $j$th OT the decommitment information of the entire column $((A_0[\cdot, j], B_0[\cdot, j]), (A_1[\cdot, j], B_1[\cdot, j]))$. Upon receiving the information for a particular column $j$, the receiver checks if for all $i$, $A_b[i, j] \oplus B_b[i, j]$ agree on the same value. We refer to this as the *shares consistency check*.

Next, to check the validity of the shares, the sender additionally sends vectors $[z_1^b], \ldots, [z_\kappa^b]$ in the clear along with the sender's message where it commits to the entries of $A_0, A_1, B_0$ and $B_1$ such that $[z_i^b]$ is set to $\phi(A_0[i, \cdot])$. Depending on the challenge message, the sender decommits to $A_0[i, \cdot]$ and $A_1[i, \cdot]$ if $c_i = 0$ and $B_0[i, \cdot]$ and $B_1[i, \cdot]$ if $c_i = 1$. If $c_i = 0$, then the receiver checks whether

$\phi(A_b[i, \cdot]) = [z_i^b]$, and if $c_i = 1$ it checks whether $\phi(B_b[i, \cdot]) + z_i^b = 0$. This check ensures that except for at most $s \in \omega(\log \kappa)$ of the rows $(A_b[i, \cdot], B_b[i, \cdot])$ satisfy the condition that $\phi(A_b[i, \cdot]) + \phi(B_b[i, \cdot]) = 0$ and for each such row $i$, $A_b[i, \cdot] + B_b[i, \cdot]$ represents a valid set of shares for both $b = 0$ and $b = 1$. This check is denoted by the *shares validity check*. In the final protocol, the sender sets as input in the $j$th parallel OT, the decommitment to the entire $j$th columns of $A_0$ and $B_0$ corresponding to the receiver's input 0 and $A_1$ and $B_1$ for input 1. Upon receiving the decommitment information on input $b_j$, the receiver considers a column "good" only if $A_{b_j}[i, j] + B_{b_j}[i, j]$ add up to the same value for every $i$. Using another cut-and-choose mechanism, the receiver ensures that there are sufficiently many good columns which consequently prevents any input-independent behavior. We refer this to the shares-validity check.

*Our oblivious transfer protocol.* We obtain a two-round oblivious transfer protocol as follows. The receiver commits to its input bits $b_1, \ldots, b_{2\kappa}$ and the challenge bits for the share consistency check $c_1, \ldots, c_\kappa$ using the PRF tokens. Then, the sender sends all the commitments *a la* [54] and $2\kappa + \kappa$ tokens, where the first $2\kappa$ tokens provide the decommitments to the columns, and the second set of $\kappa$ tokens give the decommitments of the rows for the shares consistency check. The simulator now extracts the sender's inputs by retrieving its queries and we are able to show that there cannot be any input dependent behavior of the token if it passes both the shares consistency check and the shares validity check.

We now describe our protocol $\Pi_{\mathrm{OT}}^{\mathrm{GUC}}$ with sender S and receiver R using the following building blocks: let (1) Com be a non-interactive perfectly binding commitment scheme, (2) let $\mathcal{SS} = (\mathsf{Share}, \mathsf{Recon})$ be a $(\kappa+1)$-out-of-$2\kappa$ Shamir secret-sharing scheme over $\mathbb{Z}_p$, together with a linear map $\phi : \mathbb{Z}_p^{2\kappa} \rightarrow \mathbb{Z}_p^{\kappa-1}$ such that $\phi(v) = 0$ iff $v$ is a valid sharing of some secret, (3) $F, F'$ be two families of pseudorandom functions that map $\{0,1\}^{5\kappa} \rightarrow \{0,1\}^\kappa$ and $\{0,1\}^\kappa \rightarrow \{0,1\}^{p(\kappa)}$, respectively (4) H denote a hardcore bit function and (5) $\mathsf{Ext} : \{0,1\}^{5\kappa} \times \{0,1\}^d \rightarrow \{0,1\}$ denote a randomness extractor where the source has length $5\kappa$ and the seed has length $d$. See Protocol 1 for the complete description.

**Protocol 1.** *Protocol* $\Pi_{\mathrm{GUC}}^{\mathrm{OT}}$ - *GUC OT with stateless tokens.*

- **Inputs:** S *holds two strings* $s_0, s_1 \in \{0,1\}^\kappa$ *and* R *holds a bit* $b$. *The common input is* sid.
- **The protocol:**
    1. **S $\rightarrow$ R:** S *chooses* $3\kappa$ *random PRF keys* $\{\gamma_l\}_{[l \in 3\kappa]}$ *for family* $F$. *Let* $\mathsf{PRF}_{\gamma_l} : \{0,1\}^{5\kappa} \rightarrow \{0,1\}^\kappa$ *denote the pseudorandom function.* S *creates token* $\mathsf{TK}_{\mathsf{S}}^{\mathsf{PRF},l}$ *sending* $(\mathsf{Create}, \mathsf{sid}, \mathsf{S}, \mathsf{R}, \mathsf{mid}_l, M_1)$ *to* $\mathcal{F}_{\mathrm{gWRAP}}$ *where* $M_1$ *is the functionality of the token that on input* $(\overline{\mathsf{sid}}, x)$ *outputs* $\mathsf{PRF}_{\gamma_l}(x)$ *for all* $l \in [3\kappa]$; *For the case where* $\overline{\mathsf{sid}} \neq \mathsf{sid}$ *the token aborts;*
    2. **R $\rightarrow$ S:** R *selects a random subset* $T_{1-b} \subset [2\kappa]$ *of size* $\kappa/2$ *and defines* $T_b = [2\kappa]/T_{1-b}$. *For every* $j \in [2\kappa]$, R *sets* $b_j = \beta$ *if* $j \in T_\beta$. R *samples uniformly at random* $c_1, \ldots, c_\kappa \leftarrow \{0,1\}$. *Finally,* R *sends*

(a) $(\{\mathsf{com}_{b_j}\}_{j\in[2\kappa]}, \{\mathsf{com}_{c_i}\}_{i\in[\kappa]})$ *to* S *where*

$$\forall\, j \in [2\kappa], i \in [\kappa]\quad \mathsf{com}_{b_j} = (\mathsf{Ext}(u_j) \oplus b_j, v_j) \quad and \quad \mathsf{com}_{c_i} = (\mathsf{Ext}(u_i') \oplus c_i, v_i')$$

$u_j, u_i' \leftarrow \{0,1\}^{5\kappa}$ *and* $v_j, v_i'$ *are obtained by sending respectively* $(\mathsf{Run}, \mathsf{sid}, \mathsf{mid}_j, u_j)$ *and* $(\mathsf{Run}, \mathsf{sid}, \mathsf{mid}_{2\kappa+i}, u_i')$.

(b) R *generates the tokens* $\{\mathsf{TK}_\mathsf{R}^{\mathsf{PRF},l'}\}_{l'\in[8\kappa^2]}$ *which are analogous to the* PRF *tokens* $\{\mathsf{TK}_\mathsf{S}^{\mathsf{PRF},l}\}_{l\in[3\kappa]}$ *by sending* $(\mathsf{Create}, \mathsf{sid}, \mathsf{R}, \mathsf{S}, \mathsf{mid}_{l'}, M_2)$ *to* $\mathcal{F}_{\mathsf{gWRAP}}$ *for all* $l' \in [8\kappa^2]$.

3. **S → R**: S *picks two random strings* $x_0, x_1 \leftarrow \mathbb{Z}_p$ *and secret shares them using* $\mathcal{SS}$. *In particular,* S *computes* $[x_b] = (x_b^1, \dots, x_b^{2\kappa}) \leftarrow \mathsf{Share}(x_b)$ *for* $b \in \{0,1\}$. S *commits to the shares* $[x_0], [x_1]$ *as follows. It picks random matrices* $A_0, B_0 \leftarrow \mathbb{Z}_p^{\kappa\times2\kappa}$ *and* $A_1, B_1 \leftarrow \mathbb{Z}_p^{\kappa\times2\kappa}$ *such that* $\forall i \in [\kappa]$:

$$A_0[i, \cdot] + B_0[i, \cdot] = [x_0], \quad A_1[i, \cdot] + B_1[i, \cdot] = [x_1].$$

S *computes two matrices* $Z_0, Z_1 \in \mathbb{Z}_p^{\kappa\times\kappa-1}$ *and sends them in the clear such that:*

$$Z_0[i, \cdot] = \phi(A_0[i, \cdot]), Z_1[i, \cdot] = \phi(A_1[i, \cdot]).$$

S *sends:*

(a) *Matrices* $(\mathsf{com}_{A_0}, \mathsf{com}_{B_0}, \mathsf{com}_{A_1}, \mathsf{com}_{B_1})$ *to* R, *where,*

$$\forall i \in [\kappa],\; j \in [2\kappa], \beta \in \{0,1\}\quad \mathsf{com}_{A_\beta[i,j]} = (\mathsf{Ext}(u^{A_\beta[i,j]}) \oplus A_\beta[i,j], v^{A_\beta[i,j]})$$
$$\mathsf{com}_{B_\beta[i,j]} = (\mathsf{Ext}(u^{B_\beta[i,j]}) \oplus B_\beta[i,j], v^{B_\beta[i,j]})$$

*where* $(u^{A_\beta[i,j]}, u^{B_\beta[i,j]}) \leftarrow \{0,1\}^{5\kappa}$ *and* $(v^{A_\beta[i,j]}, v^{B_\beta[i,j]})$ *are obtained by sending* $(\mathsf{Run}, \mathsf{sid}, \mathsf{mid}_{[i,j,\beta]}, u^{A_\beta[i,j]})$ *and* $(\mathsf{Run}, \mathsf{sid}, \mathsf{mid}_{2\kappa^2+[i,j,\beta]}, u^{B_\beta[i,j]})$, *respectively, to the token* $\mathsf{TK}_\mathsf{R}^{\mathsf{PRF},[i,j,\beta]}$ *where* $[i,j,\beta]$ *is an encoding of the indices* $i, j, \beta$ *into an integer in* $[2\kappa^2]$.

(b) $C_0 = s_0 \oplus x_0$ *and* $C_1 = s_1 \oplus x_1$ *to* R.

(c) *For all* $j \in [2\kappa]$, S *creates a token* $\mathsf{TK}_j$ *sending* $(\mathsf{Create}, \mathsf{sid}, \mathsf{S}, \mathsf{R}, \mathsf{mid}_{3\kappa+j}, M_3)$ *to* $\mathcal{F}_{\mathsf{gWRAP}}$ *where* $M_3$ *is the functionality that on input* $(\overline{\mathsf{sid}}, b_j, \mathsf{decom}_{b_j})$, *aborts if* $\overline{\mathsf{sid}} \neq \mathsf{sid}$ *or if* $\mathsf{decom}_{b_j}$ *is not verified correctly. Otherwise it outputs* $(A_{b_j}[\cdot, j], \mathsf{decom}_{A_{b_j}[\cdot,j]}, B_{b_j}[\cdot, j], \mathsf{decom}_{B_{b_j}[\cdot,j]})$.

(d) *For all* $i \in [\kappa]$, S *creates a token* $\widehat{\mathsf{TK}}_i$ *sending* $(\mathsf{Create}, \mathsf{sid}, \mathsf{S}, \mathsf{R}, \mathsf{mid}_{5\kappa+i}, M_4)$ *to* $\mathcal{F}_{\mathsf{gWRAP}}$ *where* $M_4$ *is the functionality that on input* $(\overline{\mathsf{sid}}, c_i, \mathsf{decom}_{c_i})$ *aborts if* $\overline{\mathsf{sid}} \neq \mathsf{sid}$ *or if* $\mathsf{decom}_{c_i}$ *is not verified correctly. Otherwise it outputs,*

$$(A_0[i, \cdot], \mathsf{decom}_{A_0[i,\cdot]}, A_1[i, \cdot], \mathsf{decom}_{A_1[i,\cdot]}), \; if\; c = 0$$
$$(B_0[i, \cdot], \mathsf{decom}_{B_0[i,\cdot]}, B_1[i, \cdot], \mathsf{decom}_{B_1[i,\cdot]}), \; if\; c = 1$$

4. **Output Phase:** *For all* $j \in [2\kappa]$, *R sends* $(\mathsf{Run}, \mathsf{sid}, \mathsf{mid}_{3\kappa+j}, (b_j, \mathsf{decom}_{b_j}))$ *and receives*

$$(A_{b_j}[\cdot, j], \mathsf{decom}_{A_{b_j}[\cdot, j]}, B_{b_j}[\cdot, j], \mathsf{decom}_{B_{b_j}[\cdot, j]}).$$

*For all* $i \in [\kappa]$, *R sends* $(\mathsf{Run}, \mathsf{sid}, \mathsf{mid}_{5\kappa+i}, (c_i, \mathsf{decom}_{c_i}))$ *and receives*

$$(A_0[\cdot, i], A_1[\cdot, i]) \text{ or } (B_0[\cdot, i], B_1[\cdot, i]).$$

(a) SHARES VALIDITY CHECK PHASE: *For all* $i \in [\kappa]$, *if* $c_i = 0$ *check that* $Z_0[i, \cdot] = \phi(A_0[i, \cdot])$ *and* $Z_1[i, \cdot] = \phi(A_1[i, \cdot])$. *Otherwise, if* $c_i = 1$ *check that* $\phi(B_0[i, \cdot]) + Z_0[i, \cdot] = 0$ *and* $\phi(B_1[i, \cdot]) + Z_1[i, \cdot] = 0$. *If the tokens do not abort and all the checks pass, the receiver proceeds to the next phase.*

(b) SHARES CONSISTENCY CHECK PHASE: *For each* $b \in \{0, 1\}$, *R randomly chooses a set* $T_b$ *for which* $b_j = b$ *of* $\kappa/2$ *coordinates. For each* $j \in T_b$, *R checks that there exists a unique* $x_b^j$ *such that* $A_b[i, j] + B_b[i, j] = x_b^j$ *for all* $i \in [\kappa]$. *If so,* $x_b^j$ *is marked as consistent. If the tokens do not abort and all the shares obtained in this phase are consistent, R proceeds to the reconstruction phase. Else it abort.*

(c) OUTPUT RECONSTRUCTION: *For* $j \in [2\kappa]/T_{1-b}$, *if there exists a unique* $x_b^j$ *such that* $A_b[i, j] + B_b[i, j] = x_b^j$, *mark share* $j$ *as a* good *column. If R obtains less than* $\kappa+1$ good *shares, it aborts. Otherwise, let* $x_b^{j_1}, \ldots, x_b^{j_{\kappa+1}}$ *be any set of* $\kappa + 1$ *consistent shares. R computes* $x_b \leftarrow \mathsf{Recon}(x_b^{j_1}, \ldots, x_b^{j_{\kappa+1}})$ *and outputs* $s_b = C_b \oplus x_b$.

Next, we state the following theorem, the proof can be found in [34].

**Theorem 4.1.** *Assume the existence of one-way functions, then protocol* $\Pi_{\mathrm{GUC}}^{\mathrm{OT}}$ *GUC realizes* $\mathcal{F}_{\mathrm{OT}}$ *in the* $\mathcal{F}_{\mathrm{gWRAP}}$*-hybrid.*

*Proof overview.* On a high-level, when the sender is corrupted our simulation proceeds analogously to the simulation from [54] where the simulator generates the view of the malicious sender by honestly generating the receiver's messages and then extracting all the values committed to by the sender. Nevertheless, while in [54] the authors rely on extractable commitments and extract the sender's inputs via rewinding, we directly extract its inputs by retrieving the queries made by the malicious sender to the $\{\mathsf{TK}_{\mathsf{R}}^{\mathsf{PRF}, i}\}_i$ tokens. The proof of correctness follows analogously. More explicitly, the share consistency check ensures that for any particular column that the receiver obtains, if the sum of the values agree on the same bit, then the receiver extracts the correct share of $[x_b]$ with high probability. Note that it suffices for the receiver to obtain $\kappa+1$ good columns for its input $b$ to extract enough shares to reconstruct $x_b$ since the shares can be checked for validity. Namely, the receiver chooses $\kappa/2$ indices $T_b$ and sets its input for these OT executions as $b$. For the rest of the OT executions, the receiver sets its input as $1 - b$. Denote this set of indices by $T_{1-b}$. Then, upon receiving the sender's response to its challenge and the OT responses, the receiver first performs the shares consistency check. If this check passes, it performs the shares validity check for all columns, both with indices in $T_{1-b}$ and for the indices in a random subset of size $\kappa/2$ within $T_b$. If one of these checks do not pass, the receiver aborts. If both checks pass, it holds with high probability that the decommitment information for $b = 0$ and $b = 1$ are correct in all but $s \in \omega(\log n)$ indices. Therefore, the receiver will extract $[x_b]$ successfully both when its input $b = 0$ and $b = 1$. Furthermore, it is ensured that if the two checks performed by the receiver pass, then a simulator can extract both $x_0$ and $x_1$ correctly by simply extracting the sender's input to the OT protocol and following the receiver's strategy to extract.

On the other hand, when the receiver is corrupted, our simulation proceeds analogous to the simulation in [54] where the simulator generates the view of the malicious receiver by first extracting the receiver's input $b$ and then obtaining $s_b$ from the ideal functionality. It then completes the execution following the honest sender's code with $(s_0, s_1)$, where $s_{1-b}$ is set to random. Moreover, while in [54] the authors rely on a special type of interactive commitment that allows the extraction of the receiver's input via rewinding, we instead extract this input directly by retrieving the queries made by the malicious receiver to the $\{\mathsf{TK}_S^{\mathsf{PRF},l}\}_{l \in [3\kappa]}$ tokens. The proof of correctness follows analogously. Informally, the idea is to show that the receiver can learn $\kappa + 1$ or more shares for either $x_0$ or $x_1$ but not both. In other words there exists a bit $b$ for which a corrupted receiver can learn at most $\kappa$ shares relative to $s_{1-b}$. Thus, by replacing $s_{1-b}$ with a random string, it follows from the secret-sharing property that obtaining at most $\kappa$ shares keeps $s_{1-b}$ information theoretically hidden.

*On relying on one-way functions.* In this protocol the only place where one-way permutations are used is in the commitments made by the sender in the second round of the protocol via a non-interactive perfectly-binding commitment. This protocol can be easily modified to rely on statistically-binding commitments which have two-round constructions based on one-way functions [52]. Specifically, since the sender commits to its messages only in the second-round, the receiver can provide the first message of the two-round commitment scheme along with the first message of the protocol.

## 5 Three-Round Token-Based GUC Secure Multi-party Computation

In this section, we show how to compile an arbitrary round semi-honest protocol $\Pi$ to a three-round protocol using stateless tokens. As discussed in the introduction, the high-level of our approach is borrowing the compressing round idea from [27] which proceeds in three steps. In the first step, all parties commit to their inputs via an extractable commitment and then in the second step, each party provides a token to emulate their actions with respect to $\Pi$ given the commitments. Finally, each party runs the protocol $\Pi$ locally and obtains the result of the computation. For such an approach to work, it is crucial that an adversary, upon receiving the tokens, is not be able to "rewind" the computation and launch a resetting attack. This is ensured via zero-knowledge proofs that are provided in each round. In essence, the zero-knowledge proofs validates the actions of each party with respect to the commitments made in the first step. Such a mechanism is also referred to as a commit-and-prove strategy. In Sect. 5.1, we will present a construction of a commit-and-prove protocol in the $\mathcal{F}_{\text{gWRAP}}$-hybrid and then design our MPC protocol using this protocol. We then take a modular approach by describing our MPC protocol in an idealized version of the commit-and-prove functionality analogous to [12] and then show how to realize this functionality. As we mentioned before we then rely on the approach of [27] to compress the rounds of our MPC protocol compiled with our commit and prove protocol in 3 rounds. This is presented in the full version [34].

## 5.1   One-Many Commit-and-Prove Functionality

The commit and prove functionality $\mathcal{F}_{\text{CP}}$ introduced in [12] is a generalization of the commitment functionality and is core to constructing protocols in the GUC-setting. The functionality parameterized by an NP-relation $\mathcal{R}$ proceeds in two stages: The first stage is a commit phase where the receiver obtains a commitment to some value $w$. The second phase is a prove phase where the functionality upon receiving a statement $x$ from the committer sends $x$ to the receiver along with the value $\mathcal{R}(x, w)$. We will generalize the $\mathcal{F}_{\text{CP}}$-functionality in two ways. First, we will allow for asserting multiple statements on a single committed value $w$ in the $\mathcal{F}_{\text{gWRAP}}$-hybrid. Second, we will allow a single party to assert the statement to many parties. In an MPC setting this will be useful as each party will assert the correctness of its message to all parties in each step. Our generalized functionality can be found in Fig. 3 and is parameterized by an NP relation $\mathcal{R}$ and integer $m \in \mathbb{N}$ denoting the number of statements to be proved.

To realize this functionality, we will rely on the so-called input-delayed proofs [20, 21, 36, 45]. In particular, we rely on the recent work of Hazay and Venkitasubramaniam [36], who showed how to obtain a 4-round commit-and-prove protocol where the underlying commitment scheme and one-way permutation are used in a black-box way, and requires the statement only in the last round. Below, we extend their construction and design a protocol $\Pi_{\text{CP}}$ that securely realizes functionality $\mathcal{F}_{\text{CP}}^{1:M}$, and then prove the following theorem.

**Theorem 5.1.** *Assuming the existence of one-way functions, then protocol $\Pi_{\text{CP}}$ securely realizes the $\mathcal{F}_{\text{CP}}^{1:M}$-functionality in the $\mathcal{F}_{\text{gWRAP}}$-hybrid.*

**Realizing $\mathcal{F}_{\text{CP}}^{1:M}$ in the $\mathcal{F}_{\text{gWRAP}}$ -Hybrid.** In the following section we extend ideas from [36] in order to obtain a one-many commit-and-prove protocol with negligible soundness using a specialized randomized encodings (RE) [1, 38], where the statement is only known at the last round. Loosely speaking, RE allows to represent a "complex" function by a "simpler" randomized function. Given a

---

**Functionality $\mathcal{F}_{CP}^{1:M}$**

Functionality $\mathcal{F}_{CP}^{1:M}$ is parameterized by an NP-relation $\mathcal{R}$, an integer $m$ and an implicit security parameter $\kappa$, and runs with set of parties $\mathcal{P} = \{P_1, \ldots, P_n\}$.

**Commit Phase:** Upon receiving a message (commit, sid, $\mathcal{P}, w$) from $P_i$, where $w \in \{0,1\}^{\kappa}$, record the tuple (sid, $P_i, \mathcal{P}, w, 0$) and send (receipt, $P_i, \mathcal{P}$, sid) to all parties in $\mathcal{P}$.

**Prove Phase:** Upon receiving a message (prove, sid, $\mathcal{P}, x$) from $P_i$, where $w \in \{0,1\}^{poly(\kappa)}$, find the record (sid, $P_i, \mathcal{P}, w, ctr_{sid}$). If no such record is found or $ctr_{sid} \geq m$ then ignore. Otherwise, send (proof, sid, $\mathcal{P}, (x, \mathcal{R}(x, w))$) to all parties in $\mathcal{P}$. Replace the tuple (sid, $P_i, \mathcal{P}, w, ctr_{sid}$) with (sid, $P_i, \mathcal{P}, w, ctr_{sid} + 1$).

---

**Fig. 3.** The one-many multi-theorem commit and prove functionality [12].

string $w_0 \in \{0,1\}^n$, the [36] protocol considers a randomized encoding of the following function:

$$f_{w_0}(x, w_1) = (\mathcal{R}(x, w_0 \oplus w_1), x, w_1)$$

where $\mathcal{R}$ is the underlying NP relation and the function has the value $w_0$ hardwired in it. The RE we consider needs to be secure against *adaptively chosen inputs* and *robust*. Loosely speaking, an RE is secure against adaptive chosen inputs if both the encoding and the simulation can be decomposed into offline and online algorithms and security should hold even if the input is chosen adaptively after seeing the offline part of the encoding. Moreover, an offline/online RE is said to be robust if no adversary can produce an offline part following the honest encoding algorithm and a (maliciously generated) online part that evaluates to a value outside the range of the function. Then the ZK proof follows by having the prover generate the offline phase of the randomized encoding for this functionality together with commitments to the randomness $r$ used for this generation and $w_1$. Next, upon receiving a challenge bit $ch$ from the verifier, the prover completes the proof as follows. In case $ch = 0$, then the prover reveals $r$ and $w_1$ for which the verifier checks the validity of the offline phase. Otherwise, the prover sends the online part of the encoding and a decommitment of $w_1$ for which the verifier runs the decoder and checks that the outcome is $(1, x, w_1)$.

A concrete example based on garbled circuits [58] implies that the offline part of the randomized encoding is associated with the garbled circuit, where the randomness $r$ can be associated with the input key labels for the garbling. Moreover, the online part can be associated with the corresponding input labels that enable to evaluate the garbled circuit on input $x, w_1$. Clearly, a dishonest prover cannot provide both a valid garbling and a set of input labels that evaluates the circuits to 1 in case $x$ is a false statement. Finally, adaptive security is achieved by employing the construction from [37] (see [36] for a discussion regarding the robustness of this scheme).

We discuss next how to extend Theorem 5.5 from [36] by adding the one-many multi-theorem features. In order to improve the soundness parameter of their ZK proof Hazay and Venkitasubramaniam repeated their basic proof sufficiently many times in parallel, using fresh witness shares each time embedding the [39] approach in order to add a mechanism that verifies the consistency of the shares. Consider a parameter $N$ to be the number of repetitions and let $m$ denote the number of proven theorems. Our protocol employs two types of commitments schemes: (1) Naor's commitment scheme [52] denoted by Com. (2) Token based extractable commitment scheme in the $\mathcal{F}_{\text{gWRAP}}$-hybrid denoted by $\text{Com}_{\text{gWRAP}}$ and defined as follows. First, the receiver $R$ in the commitment scheme will prepare a token that computes a PRF under a randomly chosen key $k$ and send it to the committer in an initial setup phase, incorporated with the session identifier sid. Such that on input $(x, \text{sid})$ the token outputs PRF evaluated on the input $x$. More, precisely, the receiver on input sid creates a token $\text{TK}^{\text{PRF}_k}$ with the following code:

– On input $(x, \widetilde{\text{sid}})$: If $\widetilde{\text{sid}} = \text{sid}$ output $\text{PRF}_k(x)$. Otherwise, output $\bot$.

Then, to commit to a bit $b$, the committer $C$ first queries the token $\text{TK}^{\text{PRF}_k}$ on input $(u, \text{sid})$ where $u \in \{0,1\}^{5\kappa}$ is chosen at random and sid is the session identifier. Upon receiving the output $v$ from the token, it sends $(\text{Ext}(u) \oplus b, v)$ where Ext is a randomness extractor as used in Sect. 4. We remark here that if the tokens are exchanged initially in a token exchange phase, then the commitment scheme is non-interactive.

**Protocol 2.** *Protocol $\Pi_{\text{CP}}$ - one-many commit-and-prove protocol.*

- **Input:** *The prover holds a witness $w$, where the prover is a designated party $P_\tau$ for some $\tau \in [n]$.*
- **The Protocol:**
  1. *Each party $P_k$ for $k \neq \tau$ plays the role of the verifier and picks random $m$ $t$-subsets $I_j^k$ of $[N]$ for each $j \in [m]$ and $k \in [n-1]$ where $m$ is the number of proven statements. It also picks $t$ random challenge bits $\{ch_{i,j}^k\}_{i \in I_j^k}$ and commits to them using $\text{Com}_{\text{gWRAP}}^k$. It further sends the first message of the Naor's commitment scheme.*
  2. *The prover then continues as follows:*
     (a) *It first generates $N \times m \times (n-1)$ independent XOR sharings of the witness $w$, say*
     $$\{w_{i,j,k}^0, w_{i,j,k}^1\}_{(i \times j \times k) \in [N \times m \times (n-1)]}.$$
     (b) *Next, for each $j \in [m]$ and $k \in [n-1]$, it generates the views of $2N$ parties $P_{i,j,k}^0$ and $P_{i,j,k}^1$ for all $i \in [N]$ executing a $t$-robust $t$-private MPC protocol, where $P_{i,j,k}^b$ has input $w_{i,j,k}^b$, that realizes the functionality that checks if $w_{i,j,k}^0 \oplus w_{i,j,k}^1$ are all equal. Let $V_{i,j,k}^b$ be the view of party $P_{i,j,k}^b$.*
     (c) *Next, for each $j \in [m]$ and $k \in [n-1]$, it computes $N$ offline encodings of the following set of functions:*
     $$f_{w_{i,j,k}^0, V_{i,j,k}^0}(x_j, w_{i,j,k}^1, V_{i,j,k}^1) = (b, x_j, w_{i,j,k}^1, V_{i,j,k}^1)$$
     *where $b = 1$ if and only if $\mathcal{R}(x_j, w_{i,j,k}^0 \oplus w_{i,j,k}^1)$ holds and the views $V_{i,j,k}^0$ and $V_{i,j,k}^1$ are consistent with each other.*

(d) *Finally, the prover broadcasts to all parties the set containing*

$$\{((f^{\mathrm{off}}_{w^0_{i,j,k}, V^0_{i,j,k}}(r_{i,j,k}), \mathsf{Com}(r_{i,j,k}), \mathsf{Com}(w^0_{i,j,k}), \mathsf{Com}(w^1_{i,j,k}),$$

$$\mathsf{Com}(V^0_{i,j,k}), \mathsf{Com}(V^1_{i,j,k}))\}_{(i \times j \times k) \in [N \times m \times (n-1)]}.$$

*Moreover, let* $\mathrm{decom}_{r_{i,j,k}}, \mathrm{decom}_{w^0_{i,j,k}}, \mathrm{decom}_{w^1_{i,j,k}}, \mathrm{decom}_{V^0_{i,j,k}}, \mathrm{decom}$ $V^1_{i,j,k}$ *be the respective decommitment information of the above commitments. Then for every* $k \in [n-1]$, $P_i$ *commits to the above decommitment information with respect to party* $P_k$ *and all* $(i \times j) \in [N] \times [m]$, *using* $\mathsf{Com}_{\mathrm{gWRAP}}$.

3. *The verifier decommits to all its challenges.*
4. *For every index* $(i,j)$ *in the* $t$ *subset the prover replies as follows:*
   - *If* $ch^i_{j,k} = 0$ *then it decommits to* $r_{i,j,k}$, $w^0_{i,j,k}$ *and* $V^0_{i,j,k}$. *The verifier then checks if the offline part was constructed correctly.*
   - *If* $ch^i_{j,k} = 1$ *then it sends* $f^{\mathrm{on}}_{w^0_{i,j,k}, V^0_{i,j,k}}(r_{i,j,k}, x_j, w^1_{i,j,k}, V^1_{i,j,k})$ *and decommits* $w^1_{i,j,k}$ *and* $V^1_{i,j,k}$. *The verifier then runs the decoder and checks if it obtains* $(1, x_j, w^1_{i,j,k}, V^1_{i,j,k})$.

*Furthermore, from the decommitted views* $V^{ch^i_{j,k}}_{i,j,k}$ *for every index* $(i,j)$ *that the prover sends, the verifier checks if the MPC-in-the-head protocol was executed correctly and that the views are consistent.*

**Theorem 5.2.** *Assuming the existence of one-way functions, then protocol* $\Pi_{\mathrm{CP}}$ *GUC realizes* $\mathcal{F}^{1:M}_{\mathrm{CP}}$ *in the* $\mathcal{F}_{\mathrm{gWRAP}}$-*hybrid.*

**Proof.** Let $\mathcal{A}$ be a malicious PPT real adversary attacking protocol $\Pi_{\mathrm{CP}}$ in the $\mathcal{F}_{\mathrm{gWRAP}}$-hybrid model. We construct an ideal adversary $\mathcal{S}$ with access to $\mathcal{F}^{1:M}_{\mathrm{CP}}$ which simulates a real execution of $\Pi_{\mathrm{CP}}$ with $\mathcal{A}$ such that no environment $\mathcal{Z}$ can distinguish the ideal process with $\mathcal{S}$ and $\mathcal{F}_{\mathrm{gWRAP}}$-hybrid from a real execution of $\Pi_{\mathrm{CP}}$ with $\mathcal{A}$ in the $\mathcal{F}_{\mathrm{gWRAP}}$-hybrid. $\mathcal{S}$ starts by invoking a copy of $\mathcal{A}$ and running a simulated interaction of $\mathcal{A}$ with environment $\mathcal{Z}$, emulating the honest party. We describe the actions of $\mathcal{S}$ for every corruption case.

*Simulating the communication with* $\mathcal{Z}$: Every message that $\mathcal{S}$ receives from $\mathcal{Z}$ it internally feeds to $\mathcal{A}$ and every output written by $\mathcal{A}$ is relayed back to $\mathcal{Z}$. In case the adversary $\mathcal{A}$ issues a transfer query on any token $(\mathsf{transfer}, \cdot)$, $\mathcal{S}$ relays the query to the $\mathcal{F}_{\mathrm{gWRAP}}$.

*Party* $P_\tau$ *is not corrupted.* In this scenario the adversary only corrupts a subset of parties $\mathcal{I}$ playing the role of the verifiers in our protocol. The simulator proceeds as follows.

1. Upon receiving a commitment $\mathsf{Com}^k_{\mathrm{gWRAP}}$ from a corrupted party $P_k$, the simulator extracts the $m$ committed $t$-subsets $I^k_j$ and the challenge bits $\{ch^k_{i,j}\}_{i \in I^k_j}$ for all $j \in [m]$, by retrieving the queries made to the tokens.
2. For each $j \in [m]$ and $k \in [\mathcal{I}]$, the simulator generates the views of $2N$ parties $P^0_{i,j,k}$ and $P^1_{i,j,k}$ for all $i \in [N]$ emulating the simulator of the $t$-robust $t$-private MPC protocol underlying in the real proof, where the set of corrupted parties for the $(j,k)^{th}$ execution is fixed to be $I^k_j$ extracted above. Let $V^b_{i,j,k}$ be the view of party $P^b_{i,j,k}$.

3. Next, for each $j \in [m]$ and $k \in [\mathcal{I}]$, the simulator computes $N$ offline encodings as follows.
   - For every index $i$ in the $t$ subset $I_j^k$ the simulator replies as follows:
     - If $ch_{i,j}^k = 0$, then the simulator broadcasts the following honestly generated message: $f_{w_{i,j,k}^0, V_{i,j,k}^0}^{\text{off}}(r_{i,j,k})$, $\text{Com}(r_{i,j,k})$, $\text{Com}(w_{i,j,k}^0)$, $\text{Com}(0)$, $\text{Com}(V_{i,j,k}'^0)$, $\text{Com}(V_{i,j,k}'^1)$. where $V_{i,j,k}'^0 = 0$ and $V_{i,j,k}'^1 = V_{i,j,k}^1$ if the matched challenge bit equals one, and vice versa.
     - Else, if $ch_{i,j}^k = 1$, then the simulator invokes the simulator for the randomized encoding and broadcasts the following message:
     
     $$\{\mathcal{S}_{w_{i,j,k}^0, V_{i,j,k}^0}^{\text{off}}(r_{i,j,k}), \text{Com}(0), \text{Com}(0), \text{Com}(w_{i,j,k}^1),$$
     $$\text{Com}(V_{i,j,k}'^0), \text{Com}(V_{i,j,k}'^1)\}_{(i \times j \times k) \in [N \times m \times (n-1)]}$$
     
     where $w_{i,j,k}^1$ is a random string and $V_{i,j,k}'^0 = 0$ and $V_{i,j,k}'^1 = V_{i,j,k}^1$ if the matched challenge bit equals one, and vice versa.
   - For every index $i$ not in the $t$ subset $I_j^k$ the simulator broadcasts
   
   $$f_{w_{i,j,k}^0, V_{i,j,k}^0}^{\text{off}}(r_{i,j,k}), \text{Com}(r_{i,j,k}), \text{Com}(w_{i,j,k}^0), \text{Com}(0), \text{Com}(0), \text{Com}(0).$$
   
   The simulator correctly commits to the decommitments information with respect to the honestly generated commitments (namely, as the honest prover would have done) using $\text{Com}_{\text{gWRAP}}$. Else, it commits to the zero string.
4. Upon receiving the decommitment information from the adversary, the simulator aborts if the adversary decommits correctly to a different set of messages than the one extracted above by the simulator.
5. Else, $\mathcal{S}$ completes the protocol by replying to the adversary as the honest prover would do.

Note that the adversary's view is modified with respect to the views it obtains with respect to the underlying MPC and both types of commitments. Indistinguishability follows by first replacing the simulated views of the MPC execution with a real execution. Namely the simulator for this hybrid game commits to the real views. Indistinguishability follows from the privacy of the protocol. Next, we modify the fake commitments into real commitments computed as in the real proof. The reduction for this proof follows easily as the simulator is not required to open these commitments.

*Party $P_\tau$ is corrupted.* In this scenario the adversary corrupts a subset of parties $\mathcal{I}$ playing the role of the verifiers in our protocol as well as the prover. The simulator for this case follows the honest verifier's strategy $\{P_k\}_{k \notin [\mathcal{I}]}$, with the exception that it extracts the prover's witness by extracting one of the witness' pairs. Recall that only the decommitment information is committed via the extractable commitment scheme $\text{Com}_{\text{gWRAP}}$. Since a commitment is made using tokens from every other party and there is at least one honest party, the simulator can extract the decommitment information and from that extract the real value. We point out that in general extracting out shares from only one-pair could cause the problem of "over-extraction" where the adversary does not necessarily commit to shares of the same string in each pair. In our protocol this is not an issue because in conjunction with committing to these shares, it also commits to the views of an MPC-in-the-head protocol which verifies that all shares are correct. Essentially, the soundness argument follows by showing that if an adversary deviates, then with high-probability the set $\mathcal{I}$ will include a party with an "inconsistent view". This involves a careful argument relying on the so-called $t$-robustness of the underlying MPC-in-the-head protocol. Such an argument is presented in [36] to get negligible soundness from constant soundness and this proof can be naturally extended to our setting (our protocol simply involves more repetitions but the MPC-in-the-head views still ensure correctness of all repetition simultaneously).

As for straight-line extraction, the argument follows as for the simpler protocol. Namely, when simulating the verifier's role the simulator extracts the committed values within the forth message of the prover. That is, following a similar procedure of extracting the committed message via obtaining the queries to the token, it is sufficient to obtain two shares of the witness as the robustness of the MPC protocol ensures that all the pairs correspond to the same witness. ∎

## 5.2  Warmup: Simple MPC Protocol in the $\mathcal{F}_{\mathrm{CP}}^{1:\mathrm{M}}$-Hybrid

We next describe our MPC protocol in the $\mathcal{F}_{\mathrm{CP}}^{1:\mathrm{M}}$-hybrid. On a high-level, we follow GMW-style compilation [29] of a semi-honest secure protocol $\Pi$ to achieve malicious security using the $\mathcal{F}_{\mathrm{CP}}^{1:\mathrm{M}}$-functionality. Without loss of generality, we assume that in each round of the semi-honest MPC protocol $\Pi$, each party broadcasts a single message that depends on its input and randomness and on the messages that it received from all parties in all previous rounds. We let $m_{i,j}$ denote the message sent by the $i^{th}$ party in the $j^{th}$ round in the protocol $\Pi$. We define the function $\pi_i$ such that $m_{i,t} = \pi_i(x_i, r_i, (M_1, \dots, M_{t-1}))$ where $m_{i,t}$ is the $t^{th}$ message generated by party $P_i$ in protocol $\Pi$ with input $x_i$, randomness $r_i$ and where $M_r$ is the message sent by all parties in round $i$ of $\Pi$. We leave the complete construction to the full version [34].

*Protocol description.* Our protocol $\Pi_{\mathrm{MPC}}$ proceeds as follows:

**Round 1.** In the first round, the parties commit to their inputs and randomness. More precisely, on input $x_i$, party $P_i$ samples random strings $r_{i,1}, r_{i,2}, \dots, r_{i,n}$ and sends $(\mathsf{commit}, \mathsf{sid}, \mathcal{P}, \overline{w})$ to $\mathcal{F}_{\mathrm{CP}}^{1:\mathrm{M}}$ and $\overline{w} = (x, R_i)$ where $R_i = (r_{i,1}, r_{i,2}, \dots, r_{i,n})$.

**Round 2.** $P_i$ broadcasts shares $\overline{R}_i = R_i - \{r_{i,i}\}$ and sends $(\mathsf{prove}, P_i, \mathcal{P}, \overline{R}_i)$. Let $M_0 = (\overline{R}_1, \dots, \overline{R}_n)$.

**Round $2 + \delta$.** Let $M_{\delta-1}$ be the messages broadcast by all parties in rounds $3, 4, \dots, 2+(\delta-1)$ and let $m_{i,\delta} = \pi_i(x_i, r_i, (M_1, \dots, M_{\delta-1}))$ where $r_i = \oplus_j r_{j,i}$. $P_i$ broadcasts $m_{i,\delta}$ and sends to $\mathcal{F}_{\mathrm{CP}}^{1:\mathrm{M}}$ the message $(\mathsf{prove}, P_i, \mathcal{P}, \overline{M}_{t-1} : m_{i,\delta})$ where $\overline{M}_{\delta-1} = (M_0, M_1, \dots, M_{\delta-1})$.

The NP-relation $\mathcal{R}$ used to instantiate the $\mathcal{F}_{CP}^{1:M}$ functionality will include:

1. $(M_0, R_i)$ : if $M_0$ contains $\overline{R}_i$ as its $i^{th}$ component where $\overline{R}_i = R_i - \{r_{i,i}\}$ and $R_i = \{r_{i,1}, \ldots, r_{i,n}\}$.
2. $((\overline{M}_{\delta-1}, m_{i,\delta}), (x_i, R_i))$ : if $(M_0, R_i) \in \mathcal{R}$ and $m_{i,\delta} = \pi_i(x_i, r_i, (M_1, \ldots, M_{\delta-1}))$ where $r_i = \oplus_{j \in [n]} r_{j,i}$, $\overline{M}_{\delta-1} = (M_0, M_1 \ldots, M_{\delta-1})$ and $R_i = \{r_{i,1}, \ldots, r_{i,n}\}$.

**Theorem 5.3.** *Let $f$ be any deterministic polynomial-time function with $n$ inputs and a single output. Assume the existence of one-way functions and an $n$-party semi-honest MPC protocol $\Pi$. Then the protocol $\Pi_{MPC}$ GUC realizes $\mathcal{F}_f$ in the $\mathcal{F}_{CP}^{1:M}$-hybrid.*

**Proof.** Let $\mathcal{A}$ be a malicious PPT real adversary attacking protocol $\Pi_{MPC}$ in the $\mathcal{F}_{CP}^{1:M}$-hybrid model. We construct an ideal adversary $\mathcal{S}$ with access to $\mathcal{F}_f$ which simulates a real execution of $\Pi_{MPC}$ with $\mathcal{A}$ such that no environment $\mathcal{Z}$ can distinguish the ideal process with $\mathcal{S}$ interacting with $\mathcal{F}_f$ from a real execution of $\Pi_{MPC}$ with $\mathcal{A}$ in the $\mathcal{F}_{CP}^{1:M}$-hybrid. $\mathcal{S}$ starts by invoking a copy of $\mathcal{A}$ and running a simulated interaction of $\mathcal{A}$ with environment $\mathcal{Z}$, emulating the honest party. We describe the actions of $\mathcal{S}$ for every corruption case.

*Simulating honest parties:* Let $\mathcal{I}$ be the set of parties corrupted by the adversary $\mathcal{A}$. This means $\mathcal{S}$ needs to simulate all messages from parties in $\mathcal{P}/\mathcal{I}$. $\mathcal{S}$ emulates the $\mathcal{F}_{CP}^{1:M}$ functionality for $\mathcal{A}$ as follows. For every $P_j \in \mathcal{P}/\mathcal{I}$ it sends the commitment message (receipt, $P_j, \mathcal{P}$, sid) to all parties $P_i \in \mathcal{I}$. Next, for every message (commit, sid, $P_i, \mathcal{P}, \overline{w}_i$) received from $\mathcal{A}$, it records $\overline{w}_i = (x_i, r_{i,1}, \ldots, r_{i,n})$. Upon receiving this message on behalf of every $P_i \in \mathcal{I}$, the simulator $\mathcal{S}$ sends $x_i$ on behalf of every $P_i \in \mathcal{I}$ to $\mathcal{F}_f$ and obtains the result of the computation output. Then using the simulator of the semi-honest protocol $\Pi$, it generates random tapes $r_i$ for every $P_i \in \mathcal{I}$ and messages $m_{j,\delta}$ for all honest parties $P_j \in \mathcal{P}/\mathcal{I}$ and all rounds $\delta$. Next, it sends $\overline{R}_j$ on behalf of the honest parties $P_j \in \mathcal{P}/\mathcal{I}$ so that for every $P_i \in \mathcal{I}$, $r_i = \oplus r_{j,i}$. This is possible since there is at least one party $P_j$ outside $\mathcal{I}$ and $\mathcal{S}$ can set $r_{j,i}$ so that it adds to $r_i$. Next, in round $2 + \delta$, it receives the messages from $P_i \in \mathcal{I}$ and supplies messages from the honest parties according to the simulation of $P_i$. Along with each message it receives the prove message that the parties in $\mathcal{I}$ send to $\mathcal{F}_{CP}^{1:M}$. $\mathcal{S}$ simply honestly emulates $\mathcal{F}_{CP}^{1:M}$ for these messages. For messages that the honest parties send to $\mathcal{F}_{CP}^{1:M}$, $\mathcal{S}$ simply sends the receipt message to all parties in $\mathcal{I}$.

Indistinguishability of the simulation follows from the following two facts:

- Given an input $x_i$ and random tape $r_i$ for every $P_i \in \mathcal{I}$ and the messages from the honest parties, there is a unique emulation of the semi-honest protocol $\Pi$ where all the messages from parties $P_i$ if honestly generated are deterministic.
- Since the simulation is emulating the $\mathcal{F}_{CP}^{1:M}$ functionality, the computation immediately aborts if a corrupted party $P_i$ deviates from the deterministic strategy. ∎

**Acknowledgements.** We thank Yuval Ishai, Amit Sahai, and Vipul Goyal for fruitful discussions regarding token-based cryptography. The first author acknowledges support from the Israel Ministry of Science and Technology (grant No. 3-10883) and support by the BIU Center for Research in Applied Cryptography and Cyber Security in conjunction with the Israel National Cyber Bureau in the Prime Minister's Office. The second author was also supported by the Danish National Research Foundation; the National Science Foundation of China (grant no. 61061130540) for the Sino-Danish CTIC; the CFEM supported by the Danish Strategic Research Council. In addition, this work was done in part while visiting the Simons Institute for the Theory of Computing, supported by the Simons Foundation and by the DIMACS/Simons Collaboration in Cryptography through NSF grant CNS-1523467. The third author was supported by Google Faculty Research Grant and NSF Award CNS-1526377.

# References

1. Applebaum, B., Ishai, Y., Kushilevitz, E.: Cryptography in $NC^0$. In: FOCS, pp. 166–175 (2004)
2. Barak, B., Canetti, R., Nielsen, J.B., Pass, R.: Universally composable protocols with relaxed set-up assumptions. In: FOCS, pp. 186–195 (2004)
3. Barak, B., Sahai, A.: How to play almost any mental game over the net - concurrent composition via super-polynomial simulation. In: FOCS, pp. 543–552 (2005)
4. Beaver, D.: Foundations of secure interactive computing. In: Feigenbaum, J. (ed.) CRYPTO 1991. LNCS, vol. 576, pp. 377–391. Springer, Heidelberg (1992). doi:10. 1007/3-540-46766-1_31
5. Boureanu, I., Ohkubo, M., Vaudenay, S.: The limits of composable crypto with transferable setup devices. In: CCS, pp. 381–392 (2015)
6. Canetti, R.: Universally composable security: a new paradigm for cryptographic protocols. In: FOCS, pp. 136–145 (2001)
7. Canetti, R., Dodis, Y., Pass, R., Walfish, S.: Universally composable security with global setup. In: Vadhan, S.P. (ed.) TCC 2007. LNCS, vol. 4392, pp. 61–85. Springer, Heidelberg (2007). doi:10.1007/978-3-540-70936-7_4
8. Canetti, R., Fischlin, M.: Universally composable commitments. In: Kilian, J. (ed.) CRYPTO 2001. LNCS, vol. 2139, pp. 19–40. Springer, Heidelberg (2001). doi:10. 1007/3-540-44647-8_2
9. Canetti, R., Jain, A., Scafuro, A.: Practical UC security with a global random oracle. In: CCS, pp. 597–608 (2014)
10. Canetti, R., Kushilevitz, E., Lindell, Y.: On the limitations of universally composable two-party computation without set-up assumptions. J. Cryptology **19**(2), 135–167 (2006)
11. Canetti, R., Lin, H., Pass, R.: Adaptive hardness and composable security in the plain model from standard assumptions. In: FOCS, pp. 541–550 (2010)
12. Canetti, R., Lindell, Y., Ostrovsky, R., Sahai, A.: Universally composable two-party and multi-party secure computation. In: STOC (2002)
13. Canetti, R., Pass, R., Shelat, A.: Cryptography from sunspots: how to use an imperfect reference string. In: FOCS, pp. 249–259 (2007)
14. Canetti, R., Rabin, T.: Universal composition with joint state. In: Boneh, D. (ed.) CRYPTO 2003. LNCS, vol. 2729, pp. 265–281. Springer, Heidelberg (2003). doi:10. 1007/978-3-540-45146-4_16

15. Chandran, N., Goyal, V., Sahai, A.: New constructions for UC secure computation using tamper-proof hardware. In: Smart, N.P. (ed.) EUROCRYPT 2008. LNCS, vol. 4965, pp. 545–562. Springer, Heidelberg (2008). doi:10.1007/978-3-540-78967-3_31

16. Choi, S.G., Katz, J., Schröder, D., Yerukhimovich, A., Zhou, H.-S.: (Efficient) universally composable oblivious transfer using a minimal numberof stateless tokens. *IACR Cryptology ePrint Archive*, 2013:840 (2013)

17. Choi, S.G., Katz, J., Schröder, D., Yerukhimovich, A., Zhou, H.-S.: (Efficient) universally composable oblivious transfer using a minimal number of stateless tokens. In: Lindell, Y. (ed.) TCC 2014. LNCS, vol. 8349, pp. 638–662. Springer, Heidelberg (2014). doi:10.1007/978-3-642-54242-8_27

18. Ciampi, M., Ostrovsky, R., Siniscalchi, L., Visconti, I.: Concurrent non-malleable commitments (and more) in 3 rounds. In: Robshaw, M., Katz, J. (eds.) CRYPTO 2016. LNCS, vol. 9816, pp. 270–299. Springer, Heidelberg (2016). doi:10.1007/978-3-662-53015-3_10

19. Michele, C., Rafail, O., Luisa, S., Ivan, V.: On round-efficient non-malleable protocols. *IACR Cryptology ePrint Archive*, 2016:621 (2016)

20. Ciampi, M., Persiano, G., Scafuro, A., Siniscalchi, L., Visconti, I.: Improved or-composition of sigma-protocols. In: TCC, pp. 112–141 (2016)

21. Ciampi, M., Persiano, G., Scafuro, A., Siniscalchi, L., Visconti, I.: Online/Offline OR composition of sigma protocols. In: Fischlin, M., Coron, J.-S. (eds.) EUROCRYPT 2016. LNCS, vol. 9666, pp. 63–92. Springer, Heidelberg (2016). doi:10.1007/978-3-662-49896-5_3

22. Dachman-Soled, D., Malkin, T., Raykova, M., Venkitasubramaniam, M.: Adaptive and concurrent secure computation from new adaptive, non-malleable commitments. In: Sako, K., Sarkar, P. (eds.) ASIACRYPT 2013, Part I. LNCS, vol. 8269, pp. 316–336. Springer, Heidelberg (2013). doi:10.1007/978-3-642-42033-7_17

23. Döttling, N., Kraschewski, D., Müller-Quade, J.: Unconditional and composable security using a single stateful tamper-proof hardware token. In: Ishai, Y. (ed.) TCC 2011. LNCS, vol. 6597, pp. 164–181. Springer, Heidelberg (2011). doi:10.1007/978-3-642-19571-6_11

24. Döttling, N., Kraschewski, D., Möller-Quade, J., Nilges, T.: From stateful hardware to resettable hardware using symmetric assumptions. In: ProvSec, pp. 23–42 (2015)

25. Döttling, N., Kraschewski, D., Müller-Quade, J., Nilges, T.: General statistically secure computation with bounded-resettable hardware tokens. In: Dodis, Y., Nielsen, J.B. (eds.) TCC 2015, Part I. LNCS, vol. 9014, pp. 319–344. Springer, Heidelberg (2015). doi:10.1007/978-3-662-46494-6_14

26. Döttling, N., Mie, T., Müller-Quade, J., Nilges, T.: Implementing resettable UC-functionalities with untrusted tamper-proof hardware-tokens. In: Sahai, A. (ed.) TCC 2013. LNCS, vol. 7785, pp. 642–661. Springer, Heidelberg (2013). doi:10.1007/978-3-642-36594-2_36

27. Garg, S., Gentry, C., Halevi, S., Raykova, M.: Two-round secure MPC from indistinguishability obfuscation. In: TCC, pp. 74–94 (2014)

28. Garg, S., Mukherjee, P., Pandey, O., Polychroniadou, A.: The exact round complexity of secure computation. In: Fischlin, M., Coron, J.-S. (eds.) EUROCRYPT 2016. LNCS, vol. 9666, pp. 448–476. Springer, Heidelberg (2016). doi:10.1007/978-3-662-49896-5_16

29. Goldreich, O., Micali, S., Wigderson, A.: How to play any mental game or a completeness theorem for protocols with honest majority. In: STOC, pp. 218–229 (1987)

30. Goyal, V., Ishai, Y., Sahai, A., Venkatesan, R., Wadia, A.: Founding cryptography on tamper-proof hardware tokens. In: Micciancio, D. (ed.) TCC 2010. LNCS, vol. 5978, pp. 308–326. Springer, Heidelberg (2010). doi:10.1007/978-3-642-11799-2_19

31. Goyal, V., Lee, C.-K., Ostrovsky, R., Visconti, I.: Constructing non-malleable commitments: a black-box approach. In: FOCS, pp. 51–60 (2012)

32. Goyal, V., Richelson, S., Rosen, A., Vald, M.: An algebraic approach to nonmalleability. In: 55th IEEE Annual Symposium on Foundations of Computer Science, FOCS 2014, Philadelphia, PA, USA, 18–21 October 2014, pp. 41–50 (2014)

33. Haitner, I., Hoch, J.J., Reingold, O., Segev, G.: Finding collisions in interactive protocols - tight lower bounds on the round and communication complexities of statistically hiding commitments. SIAM J. Comput. 44(1), 193–242 (2015)

34. Carmit, H., Antigoni, P., Muthuramakrishnan, V.: Composable security in the tamper proof hardware model under minimal complexity. IACR Cryptology ePrint Archive 2015:887 (2015)

35. Hazay, C., Venkitasubramaniam, M.: On black-box complexity ofuniversally composable security in the CRS model. In: ASIACRYPT, pp. 183–209 (2015)

36. Hazay, C., Venkitasubramaniam, M.: On the power of secure two-party computation. In: Robshaw, M., Katz, J., Wooten, M.B. (eds.) CRYPTO 2016. LNCS, vol. 9815, pp. 397–429. Springer, Heidelberg (2016). doi:10.1007/978-3-662-53008-5_14

37. Brett, H., Zahra, J., Rafail, O., Alessandra, S., Daniel, W.: Adaptively secure garbled circuits from one-way functions. IACR Cryptology ePrint Archive 2015:1250 (2015)

38. Ishai, Y., Kushilevitz, E. Randomizing polynomials: a new representation with applications to round-efficient secure computation. In: FOCS, pp. 294–304 (2000)

39. Ishai, Y., Kushilevitz, E., Ostrovsky, R., Sahai, A.: Zero-knowledge proofs from secure multiparty computation. SIAM J. Comput. 39(3), 1121–1152 (2009)

40. Ishai, Y., Prabhakaran, M., Sahai, A.: Founding cryptography on oblivious transfer – efficiently. In: Wagner, D. (ed.) CRYPTO 2008. LNCS, vol. 5157, pp. 572–591. Springer, Heidelberg (2008). doi:10.1007/978-3-540-85174-5_32

41. Kalai, Y.T., Lindell, Y., Prabhakaran, M.: Concurrent composition of secure protocols in the timing model. J. Cryptology 20(4), 431–492 (2007)

42. Katz, J.: Universally composable multi-party computation using tamper-proof hardware. In: Naor, M. (ed.) EUROCRYPT 2007. LNCS, vol. 4515, pp. 115–128. Springer, Heidelberg (2007). doi:10.1007/978-3-540-72540-4_7

43. Katz, J., Ostrovsky, R.: Round-optimal secure two-party computation. In: Franklin, M. (ed.) CRYPTO 2004. LNCS, vol. 3152, pp. 335–354. Springer, Heidelberg (2004). doi:10.1007/978-3-540-28628-8_21

44. Kilian, J.: Founding cryptography on oblivious transfer. In: STOC, pp. 20–31 (1988)

45. Lapidot, D., Shamir, A.: Publicly verifiable non-interactive zero-knowledge proofs. In: Menezes, A., Vanstone, S.A. (eds.) CRYPTO 1990. LNCS, vol. 537, pp. 353–365. Springer, Heidelberg (1991). doi:10.1007/3-540-38424-3_26

46. Lin, H., Pass, R., Venkitasubramaniam, M.: A unified framework for concurrent security: universal composability from stand-alone non-malleability. In: STOC, pp. 179–188 (2009)

47. Pass, R., Lin, H., Venkitasubramaniam, M.: A unified framework for UC from only OT. In: Wang, X., Sako, K. (eds.) ASIACRYPT 2012. LNCS, vol. 7658, pp. 699–717. Springer, Heidelberg (2012). doi:10.1007/978-3-642-34961-4_42

48. Lindell, Y.: General composition and universal composability in secure multi-party computation. In: FOCS, pp. 394–403 (2003)

49. Jeremias, M., Jörn, M.-Q., Tobias, N.: Universally composable (non-interactive) two-party computation from untrusted reusable hardware tokens. IACR Cryptology ePrint Archive 2016:615 (2016)

50. Micali, S., Rogaway, P.: Secure computation. In: Feigenbaum, J. (ed.) CRYPTO 1991. LNCS, vol. 576, pp. 392–404. Springer, Heidelberg (1992). doi:10.1007/3-540-46766-1_32

51. Moran, T., Segev, G.: David and Goliath commitments: UC computation for asymmetric parties using tamper-proof hardware. In: Smart, N.P. (ed.) EUROCRYPT 2008. LNCS, vol. 4965, pp. 527–544. Springer, Heidelberg (2008). doi:10.1007/978-3-540-78967-3_30

52. Naor, M.: Bit commitment using pseudorandomness. J. Cryptology 4(2), 151–158 (1991)

53. Nilges, T.: The Cryptographic Strength of Tamper-Proof Hardware. Ph.D. thesis, Karlsruhe Institute of Technology (2015)

54. Ostrovsky, R., Richelson, S., Scafuro, A.: Round-optimal black-box two-party computation. In: Gennaro, R., Robshaw, M. (eds.) CRYPTO 2015. LNCS, vol. 9216, pp. 339–358. Springer, Heidelberg (2015). doi:10.1007/978-3-662-48000-7_17

55. Pass, R.: Simulation in quasi-polynomial time, and its application to protocol composition. In: EUROCRYPT, pp. 160–176 (2003)

56. Pass, R., Wee, H.: Black-box constructions of two-party protocols from one-way functions. In: Reingold, O. (ed.) TCC 2009. LNCS, vol. 5444, pp. 403–418. Springer, Heidelberg (2009). doi:10.1007/978-3-642-00457-5_24

57. Prabhakaran, M., Sahai, A.: New notions of security: achieving universal composability without trusted setup. In: STOC, pp. 242–251 (2004)

58. Yao, A.C.-C.: How to generate and exchange secrets (extended abstract). In: FOCS, pp. 162–167 (1986)

# Composable Adaptive Secure Protocols Without Setup Under Polytime Assumptions

Carmit Hazay[1][✉] and Muthuramakrishnan Venkitasubramaniam[2]

[1] Bar-Ilan University, Ramat Gan, Israel
carmit.hazay@cs.biu.ac.il
[2] University of Rochester, Rochester, NY, USA
muthuv@cs.rochester.edu

**Abstract.** All previous constructions of general multiparty computation protocols that are secure against adaptive corruptions in the concurrent setting either require some form of setup or non-standard assumptions. In this paper we provide the first general construction of secure multi-party computation protocol *without* any setup that guarantees composable security in the presence of an *adaptive adversary* based on standard polynomial-time assumptions. We prove security under the notion of "UC with super-polynomial helpers" introduced by Canetti et al. (FOCS 2010), which is closed under universal composition and implies "super-polynomial-time simulation". Moreover, our construction relies on the underlying cryptographic primitives in a black-box manner.

Next, we revisit the zero-one law for two-party secure functions evaluation initiated by the work of Maji, Prabhakaran and Rosulek (CRYPTO 2010). According to this law, every two-party functionality is either trivial (meaning, such functionalities can be reduced to any other functionality) or complete (meaning, any other functionality can be reduced to these functionalities) in the Universal Composability (UC) framework. As our second contribution, assuming the existence of a simulatable public-key encryption scheme, we establish a zero-one law in the adaptive setting. Our result implies that every two-party non-reactive functionality is either trivial or complete in the UC framework in the presence of adaptive, malicious adversaries.

**Keywords:** UC security · Adaptive secure computation · Coin-tossing · Black-box construction · Extractable commitments · Zero-one law

## 1 Introduction

Secure computation enables a set parties to mutually run a protocol that computes some function $f$ on their private inputs, while preserving a number of

---

C. Hazay—Research supported by the Israel Ministry of Science and Technology (grant No. 3-10883) and by the BIU Center for Research in Applied Cryptography and Cyber Security in conjunction with the Israel National Cyber Bureau in the Prime Minister's Office.

M. Venkitasubramaniam—Research supported by Google Faculty Research Grant and NSF Award CNS-1526377.

© International Association for Cryptologic Research 2016
M. Hirt and A. Smith (Eds.): TCC 2016-B, Part I, LNCS 9985, pp. 400–432, 2016.
DOI: 10.1007/978-3-662-53641-4_16

security properties. Two of the most important properties are privacy and correctness. The former implies data confidentiality, namely, nothing leaks by the protocol execution but the computed output. The later requirement implies that no corrupted party or parties can cause the output to deviate from the specified function. It is by now well known how to securely compute any efficient functionality [Yao86, GMW87, MR91, Bea91, Can01] in various models and under the stringent simulation-based definitions (following the ideal/real paradigm). Security is typically proven with respect to two adversarial models: the semi-honest model (where the adversary follows the instructions of the protocol but tries to learn more than it should from the protocol transcript), and the malicious model (where the adversary follows an arbitrary polynomial-time strategy), and feasibility results are known in the presence of both types of attacks. The initial model considered for secure computation was of a static adversary where the adversary controls a subset of the parties (who are called corrupted) before the protocol begins, and this subset cannot change. In a stronger corruption model the adversary is allowed to choose which parties to corrupt throughout the protocol execution, and as a function of its view; such an adversary is called adaptive.

These feasibility results rely in most cases on stand-alone security, where a *single* set of parties run a *single* execution of the protocol. Moreover, the security of most cryptographic protocols proven in the stand-alone setting does not remain intact if many instances of the protocol are executed concurrently [Lin03]. The strongest (but also the most realistic) setting for concurrent security is known by *Universally Composable* (UC) security [Can01]. This setting considers the execution of an unbounded number of concurrent protocols in an arbitrary and adversarially controlled network environment. Unfortunately, stand-alone secure protocols typically fail to remain secure in the UC setting. In fact, without assuming some *trusted help*, UC security is impossible to achieve for most tasks [CF01, CKL06, Lin03]. Consequently, UC secure protocols have been constructed under various *trusted setup* assumptions in a long series of works; see [BCNP04, CDPW06, KLP07, CPS07, LPV09, DMRV13] for few examples.

**Concurrent Security *Without Any* Setup.** In many situations, having a trusted set-up might be hard or expensive. Designing protocols in the plain model that provide meaningful security in a concurrent setting is thus an important challenge. In this regard, a relaxation of UC security allows the adversary in an ideal execution to run in *super-polynomial time*; this notion is referred to as super-polynomial security (or SPS) [Pas03]. On a high-level, this security notion guarantees that any attack carried out by an adversary running in polynomial time can be mounted in the ideal execution with super-polynomial resources. In many scenarios, such a guarantee is meaningful and indeed several past works have designed protocols guaranteeing this relaxed UC security against static adversaries [Pas03, BS05, LPV09] and adaptive adversaries [BS05, DMRV13, Ven14]. While initial works relied on sub-exponential hardness assumptions, more recent works in the static setting have been constructed based on standard polynomial-time hardness assumptions.

The work of [CLP10], put forth some basic desiderata regarding security notions in a concurrent setting. One of them requires supporting *modular analysis:* Namely, there should be a way to deduce security properties of the overall protocol from the security properties of its components. Quite surprisingly, it was shown in [CDPW06] that most protocols in the UC framework that consider both trusted setups and relaxed models of security, in fact, do not support this.

Towards remedying the drawbacks of SPS security, Prabhakaran and Sahai [PS04] put forth the notion of Angel-based UC security that provides guarantees analogous to SPS security while at the same time supporting modular analysis. In this model, both the adversary and the simulator have access to an oracle, referred to as an "angel" that provides judicious use of super-polynomial resources. In the same work and subsequent effort [MMY06] the authors provided constructions under this security notion relying on non-standard hardness assumptions. Recently, Canetti, Lin and Pass [CLP10] provided the first constructions in this model relying on standard polynomial time assumptions. Moreover, to emphasize the modular analysis requirement, they recast the notion of Angel-based security in the extended UC (EUC) framework of [CDPW06] calling it UC with super-polynomial helpers. While prior approaches relied on non-interactive helpers that were stateless, this work designed a helper that was highly interactive and stateful. Since this work, several follow up works [LP12a, GLP+15, Kiy14] have improved both the round complexity and the computational assumptions. The most recent work due to Kiyoshima [Kiy14] provides a $\widetilde{O}(\log^2 n)$-round protocol to securely realize any functionality in this framework based on semi-honest oblivious transfer protocols where the underlying primitives are used in a black-box manner. In this line of research, the work of Canetti, Lin and Pass [CLP13] distinguishes itself by designing protocols that guarantee a stronger notion of security. More precisely, they extend the angel-based security so that protocols developed in this extended framework additionally preserve security of other protocols running the system (i.e. cause minimal "side-effect"). They refer to such protocols "environment friendly" protocols. However, as observed in the same work, this strong notion inherently requires non-black-box simulation techniques. Moreover, the constructions presented in [CLP13] are non-black-box as well.

While considerable progress has been made in constructing protocols secure against static adversaries, very little is known regarding adaptive adversaries. Specifically, the work of Barak and Sahai [BS05] and subsequent works [DMRV13, Ven14] show how to achieve SPS security under non-standard assumptions. Besides these works, every other protocol that guarantees any meaningful security against adaptive adversaries in a concurrent setting has required setup. The main question left open by previous work regarding adaptive security is:

*Can we realize general functionalities with SPS security in the plain model under standard polynomial time assumptions? and,*

*Can we show adaptively secure angel-based (or EUC-security) under standard hardness assumptions where the underlying primitives are used in a black-box manner?*

We stress that even the works that provide SPS security require non-standard or sub-exponential hardness assumptions and are *non-black-box*, that is, the constructions rely on the underlying assumptions in non-black-box way. A more ambitious goal would be to construct "environment-friendly" protocols [CLP13] and we leave it as future work.

## 1.1   Our Results

In this work we resolve both these questions completely and provide the first realizations of general functionalities under EUC security against malicious, adaptive adversaries (See [CDPW06, CLP10] for a formal definition). More formally, we prove the following theorem:

**Theorem 1.1.** *Assume the existence of a simulatable public-key encryption scheme. Then there exists a sub-exponential time computable (interactive) helper machine $\mathcal{H}$ such that for any "well formed" polynomial-time functionality $\mathcal{F}$, there exists a protocol that realizes $\mathcal{F}$ with $\mathcal{H}$-EUC security, in the plain model secure against malicious, adaptive adversaries. Furthermore, the protocol makes only black-box use of the underlying encryption scheme.*

We recall here that simulatable public-key encryption (PKE), introduced by Damgard and Nielsen [DN00], allows to obliviously sample the public key/ciphertext without the knowledge of the corresponding secret key/plaintext.

As far as we know, this is the first construction based on polynomial-time hardness assumptions of a secure multi-party computation that achieves any non-trivial notion of concurrent security against adaptive adversaries without any trusted-set up (in the plain model) and without assuming an honest majority. Also, the construction supports *modular analysis* and relies on the underlying scheme in a black-box way. In essence, our protocol provides the *strongest* possible security guarantees in the plain model.

**A Zero-One Law for Adaptive Security.** In [PR08], Prabhakaran and Rosulek initiated the study of the "cryptographic complexity" of two-party secure computation tasks in the UC framework. Loosely speaking, in their framework a functionality $\mathcal{F}$ *UC-reduces* to another functionality $\mathcal{G}$ if there is a UC secure protocol for $\mathcal{F}$ in the $\mathcal{G}$-hybrid, i.e., using ideal access to $\mathcal{G}$. Under this notion of a reduction in the presence of static adversaries, Maji et al. in [MPR10] established a *zero-one* law for two-party (non-reactive) functionalities which states that every functionality is either trivial or complete. In this work, we extend their result to the adaptive setting to obtain the following theorem.

**Theorem 1.2 (Informal).** *All non-reactive functionalities are either trivial or complete under UC-reductions in the presence of adaptive adversaries.*

## 1.2 Previous Techniques

All previous approaches for Angel-based UC secure protocols relied on a particular "adaptive hardness" assumption which amounts to guaranteeing security in the presence of an adversary that has adaptive access to a helper function. Indeed, as pursued in the orginal approaches by [PS04, MMY06], complexity leveraging allows for designing such primitives. A major breakthrough was made by Canetti, Lin and Pass [CLP10] that showed that a helper function could be based on standard assumptions. The main technical tool introduced in this work is a new notion of a commitment scheme that is secure against an adaptive chosen commitment attack (CCA security). On a high-level, a tag-based commitment scheme, which are schemes that have additionally a tag as a common input, is said to be CCA-secure if a commitment made with tag id is hiding even if the receiver has access to a (possibly, super-polynomial time) oracle that is capable of "breaking" commitments made using any tag id' $\neq$ id. In the original work, they constructed a $O(n^\epsilon)$-round CCA-secure commitment scheme based on one-way functions (OWFs) [CLP10]. Since then, several followup works have improved this result, culminating in the work of Kiyoshima [Kiy14] who gave a $\tilde{O}(\log^2 n)$-round construction of a CCA-secure commitment scheme based on OWFs while relying on the underlying OWF in a black-box way.[1] We remark here that Angel-based security based on standard polynomial-time assumptions have been constructed only in the static setting. Moreover, all constructions in this line of work, first construct a CCA-secure commitment scheme and then realize a complete UC functionality, such as the commitment or oblivious-transfer functionality using a "decommitment" oracle as the helper functionality.

When considering the adaptive setting, we begin with the observation that any cryptographic primitive in use must be secure in the presence of adaptive corruptions. Namely, we require a simulation that can produce random coins consistent with any honest party during the execution as soon as it is adaptively corrupted. A first attempt would be to enhance a CCA-secure commitment scheme to the adaptive setting. This means there must be a mechanism to equivocate the commitment scheme. It is in fact crucial in all works using CCA-secure commitments that the helper functionality be able to break the commitment and obtain the unique value (if any) that the commitment can be decommitted to. However, equivocal commitments by definition can have commitments that do not have unique decommitments. In essence, standard CCA-secure commitment schemes are necessarily statistically binding (and all previous constructions indeed are statistically binding). Hence, it would be impossible to use any of those schemes in the adaptive setting.

The previous works [DMRV13, Ven14] get around this issue by relying on some sort of setup, namely, a mechanism by which the commitments will be statistically binding in the real world for adversaries, yet can be equivocated in the ideal world by the simulator. The notion of an adaptive instance-dependent

---

[1] We further note that Goyal et al. [GLP+15] gave a $\tilde{O}(\log n)$-round CCA-secure commitment scheme but makes use of the OWF in a non-black-box way.

scheme [LZ11] provides exactly such a primitive. Informally, such commitment schemes additionally take as input an NP-statement and provide the following guarantee: If the statement is true, the commitment can be equivocated using the witness, whereas if the statement is false then the commitment is statistically binding. Moreover, it admits adaptive corruptions where a simulator can produce random coins for an honest committer, revealing a simulated commitment to any value. The work of [BS05] relies on complexity leveraging in order to generate statements that a simulator, in super-polynomial time can break but an adversary, in polynomial time, cannot break. On the other hand, the works of [DMRV13, Ven14] rely on the so called *UC puzzle*, that provides a similar advantage for the simulator while relying on milder assumptions.

A second issue arises in the adaptive setting where any commitment scheme that tolerates concurrent executions (even with fixed roles) and is equivocal, implies some sort of selective opening security. Indeed, the result of Ostrovsky et al. [ORSV13] proves that it is impossible, in general, to construct concurrent commitments secure w.r.t. selective opening attacks. Getting around this lower bound is harder. Previous results [DMRV13, Ven14] get around this lower bound by first constructing a "weaker" commitment scheme in a limited concurrent environment. Namely, they construct an equivocal non-malleable commitment scheme that can simulate any man-in-the-middle adversary receiving "left" commitments made to independent and identically distributed values (via some *a priori* fixed distribution), and is acting as a committer in many "right" interactions. This allows to get around the [ORSV13] lower bound, as Ostrovsky et al. lower bound holds only if the simulator *does not* know the distribution of the commitments received by the adversary. In any case, all previous works fail to achieve the stronger Angel-based UC security, where the helper function is provided to the adversary and the simulator in the real and ideal world respectively are the same.

Given these bottlenecks, it seems unlikely to use a commitment scheme with such a property. In this work, we introduce a new primitive that will allow to both provide the adaptive hardness property as well as admit adaptive corruptions. This primitive is coin-tossing and will additionally require to satisfy an *adaptive hardness guarantee* that we define in the next section. We chose coin-tossing as a primitive as it does not require any inputs from the parties and the output is independent of any "global" inputs of the parties participating in the coin-tossing. Roughly speaking, if a party is adaptively corrupted it is possible to sample a random string as the output and equivocate the interaction to output this string. On the other hand, a commitment scheme will not allow such a mechanism as corrupting a sender requires equivocating the interaction to a particular value (that could potentially depend on a global input).

## 2  Our Main Tool: CCA-Secure Coin-Tossing

The main technical tool used in our construction is a new notion of a coin-tossing protocol that is secure against adaptive chosen coins attack (CCA security).

Cryptographic primitives with an adaptive hardness property has been studied extensively in the case of the encryption schemes (chosen ciphertext attack security), and more recently in the case of commitments [CLP10, KMO14, Kiy14, GLP+15]. We define here an analogous notion for coin-tossing protocols for the stronger case of adaptive corruptions.

A natural approach is to say that a coin-tossing protocol is CCA-secure if the coin-tossing scheme retains its simulatability even if a "Receiver" has access to a "biasing" oracle $\mathcal{O}$ that has the power to bias the protocol outcome of the coin-tossing to any chosen value. Unfortunately, we do not know how to realize such a notion and will instead, consider a weaker "indistinguishability"-based notion (as opposed to simulation based notion) that will be sufficient for our application.

**A motivating example.** We motivate our definition by discussing what security properties are desirable for coin-tossing protocols (in general). Consider a public-key cryptosystem that additionally has a property that a public-key can be obliviously sampled using random coins without knowledge of the secret-key (e.g., dense cryptosystems, simulatable public-key encryption schemes). Furthermore, semantic security holds for a key sampled using the oblivious strategy. Consider a protocol where the parties after engaging in a coin-tossing protocol sample a public-key using the outcome of the coin-tossing. In such a scenario we would like the coin-tossing scheme to ensure that the semantic-security continues to hold if parties encrypt messages using the public-key.

The natural "simulatable" definition requires the coin-tossing to be "simulatable". If we instantiate a simulatable coin-tossing protocol in our motivating application, semantic security of ciphertexts constructed using the public-key sampled from the coin-tossing outcome indeed holds via a simple security reduction. Suppose there exists an adversary that distinguishes an encryption of 0 from 1 when encrypted under a public-key sampled using the coin-tossing. We can use the simulator to construct an adversary that violates the security of the underlying encryption scheme. Consider a simulator that receives as a challenge a uniformly sampled string and a ciphertext generated with the associated public-key. The simulator can internally simulate the coin-tossing to be this sampled string and thereby use the adversary to break the security of the encryption scheme.

A weaker alternative to simulatability is an information-theoretic based definition where the requirement would be that the entropy of the outcome is sufficiently high. However, such a definition will not suffice in our motivating example.[2] This is because we will not be able to "efficiently" reduce a cheating adversary to the violating the security game of the underlying cryptosystem.

Instead, we take a more direct approach where the security for the coin-tossing is defined so that it will be useful in our motivating example. First, we generalize the security game of the underlying encryption scheme in our

---

[2] Unless the cryptosystems have additional properties. For instance, consider dual-mode encryption schemes where there are keys sampled via a high-entropy string and could potentially be statistically hiding.

motivating example to any indistinguishability based primitive. We model such a primitive via a (possibly) interactive challenger $C$ that receives as input a random string $o$ and a private bit $b$. We say that an adversary interacting with $C$ succeeds if when interacting on a randomly chosen $o$ and bit $b$, the adversary can guess $b$ with probability better than a $\frac{1}{2}$. Let $\pi$ be a (two-party) coin-tossing protocol. Our motivating example can be formulated using the following experiment $\mathsf{EXP}_b$ with an adversary $A$:

- $A$ interacts with an honest party using $\pi$ to generate $o$.
- Next, it interacts with a challenger $C$ on input $o$ and bit $b$.

We compare this experiment with a stand-alone experiment $\mathsf{STA}_b$ where an adversary $B$ simply interacts with $C$ on input $b$ and $o$ where $o$ is uniformly sampled. Our security definition of the coin-tossing protocol must preserve the following security property against a challenger $C$: *if the stand-alone game is hard to distinguish, i.e. $\mathsf{STA}_0$ from $\mathsf{STA}_1$, then the experiments $\mathsf{EXP}_0$ from $\mathsf{EXP}_1$ must also be hard to distinguish.* More formally, our definition will (explicitly) give a reduction from any adversary that $A$ distinguishes $\mathsf{EXP}_b$ to a stand-alone adversary $B$ that can distinguish $\mathsf{STA}_b$. Finally, in a CCA-setting, we generalize this definition by requiring that if there exists any oracle adversary $A^{\mathcal{O}}$ with access to a biasing oracle $\mathcal{O}$ that can distinguish $\mathsf{EXP}_0$ from $\mathsf{EXP}_1$, then there exists a stand-alone adversary $B$ (without access to any oracle) that can distinguish $\mathsf{STA}_b$ from $\mathsf{STA}_1$.

Towards formalizing this notion and incorporating adaptive corruptions, we first consider a tag-based coin-tossing protocol between two parties, an *Initiator* $I$ and a *Receiver* $R$ with $l(n)$-bit identities and $m(n)$-bit outcomes. A biasing oracle $\mathcal{O}$ interacts with an adversary $A$ as follows: $\mathcal{O}$ participates with $A$ in many sessions using the protocol where the oracle controls the initiator, using identities of length $l(n)$ that are chosen adaptively by $A$. At the beginning of each session, the adversary produces a coin outcome $c \in \{0,1\}^{m(n)}$ to the oracle where at the end of this session, if the initiator that is initially controlled by the oracle is not (adaptively) corrupted by the adversary, then the outcome of the interaction must result in the *chosen coin c*. If at any point during the interaction the initiator is corrupted, then the oracle simply provides the random-tape of $I$ that is consistent with the partial transcript of the interaction.

We compare an experiment $\mathsf{EXP}_b$ with oracle PPT adversary $A^{\mathcal{O}}$ and a stand-alone experiment $\mathsf{STA}_b$ with adversary $B$. In the man-in-the-middle experiment, an adversary with oracle access to $\mathcal{O}$ interacts with a honest receiver $R$ on identity id to generate an output $o \in \{0,1\}^n$ where $n$ is the security parameter. Then it interacts with a challenger $C$ on common input $(n, o, \mathsf{id})$ and private input $b$ for $C$. The adversary is allowed to corrupt the receiver $R$, challenger $C$ and any of the interactions with $\mathcal{O}$. If the adversary $A$ corrupts either $C$ or $I$ then the output of the experiment is set to $\perp$. If for some identity $\mathsf{id}'$ on which $A$ queries $\mathcal{O}$, it holds that $\mathsf{id}' = \mathsf{id}$, then the output of the experiment is set to $\perp$. Otherwise, the output of the experiment is set to be the output of the adversary.

In the stand-alone experiment $\mathsf{STA}_b$, we consider a PPT adversary $B$ that interacts with $C$ on common input $(n, o)$ and private input $b$ for $C$ where $o$ is uniformly sampled from $\{0,1\}^n$. The output of the experiment is set to be the output of $\mathcal{B}$. Observe that in the stand-alone experiment $\mathcal{B}$ does not get to corrupt $\mathcal{C}$.

Informally, a tag-based coin-tossing scheme $\langle I, R \rangle$ is said to be CCA-secure against a challenger $\mathcal{C}$, if there exists a biasing oracle $\mathcal{O}$ for $\langle I, R \rangle$ such that for every oracle PPT adversary $\mathcal{A}$ and distinguisher $\mathcal{D}$ such that $\mathcal{D}$ distinguishes $\mathsf{EXP}_0$ and $\mathsf{EXP}_1$ with $\mathcal{A}$, then there exist a (stand-alone) PPT $\mathcal{B}$ and distinguisher $\mathcal{D}'$ such that $\mathcal{D}'$ distinguishes $\mathsf{STA}_0$ and $\mathsf{STA}_1$ with $\mathcal{B}$.

In addition to this security requirement we will additionally consider the following definition of CCA-security which simply requires that any adversary with oracle access to a biasing oracle $\mathcal{O}$ can be simulated by a stand-alone PPT machine. In this case, we simply say $\langle I, R \rangle$ is CCA-secure w.r.t $\mathcal{O}$.

Quite surprisingly, we show how to realize such a primitive by relying on a CCA-secure commitment that is secure only against static adversaries. The idea here is that while CCA-secure commitments cannot admit adaptive corruptions, the basic security game ensures that an *unopened* commitment remains hiding in the presence of an adversary having access to a decommitment oracle. We combine such a commitment scheme with the technique of Hazay and Venkitasubramaniam from [HV15] who showed how to construct an adaptive UC-commitment scheme, starting from a public-key encryption scheme (with an oblivious ciphertext generation property) in the CRS model. On a high-level, the protocol can be abstracted as providing a transformation from a extractable (only) commitment scheme (that has a oblivious generation property) to a full adaptively secure UC-commitment. At first, it would be tempting to simply replace the invocations of extractable commitments with a CCA-secure commitment scheme as we only require extraction from these commitments and not equivocation in the simulation. However, this intuition fails in an adaptive setting when considering the fact that we additionally require that the commitment scheme has a oblivious generation property and it is unclear how to construct such a extractable scheme (based on rewinding) to have this property. Nevertheless, we show how to carefully use CCA-secure commitments in the same protocol to obtain a CCA-secure coin-tossing scheme. Next, we show that given a CCA-secure coin-tossing protocol with a biasing oracle $\mathcal{O}$ it is possible to realize the ideal commitment functionality using a helper functionality. Again, we use another variant of the same protocol from [HV15] to accomplish this transformation. Our constructions and proofs of security are highly modular and quite simple. Moreover, all our transformations rely on the underlying primitives in a black-box manner.

Finally, we show that the black-box construction of an $O(n^\epsilon)$-round CCA-secure commitment scheme from Lin and Pass [LP12a] will satisfy the required property to be instantiated in our protocol for the CCA-secure coin-tossing scheme.

We remark here that while the focus of the present work is to achieve *plain* angel-based security, we could achieve the stronger "environment-friendly" property if we instead rely on a *strongly unprovable* CCA-secure commitment scheme

[CLP13] to construct our CCA-secure coin-tossing scheme. We leave this as future work.

## 2.1    A Formal Definition of CCA-Secure Coin-Tossing

We begin with the simpler security requirement of CCA-security w.r.t biasing oracles.

**Definition 1 (CCA-secure coin-tossing).** *Let $\langle I, R \rangle$ be a tag-based coin-tossing scheme with $l(n)$-bit identities, $m(n)$-bit outcomes and $\mathcal{O}$ a biasing oracle for it. We say that $\langle I, R \rangle$ is robust CCA-secure w.r.t. $\mathcal{O}$, if for every PPT adversary $\mathcal{A}$ there exists a simulator $\mathcal{S}$ such that $\{\mathcal{A}^{\mathcal{O}}(n, z)\}_{n \in \mathbb{N}, z \in \{0,1\}^*} \approx \{\mathcal{S}(n, z)\}_{n \in \mathbb{N}, z \in \{0,1\}^*}$*

## 2.2    CCA-Security w.r.t Challengers

Let the random variable $\mathsf{EXP}_b(\langle I, R \rangle, \mathcal{O}, \mathcal{A}, \mathcal{C}, n, z)$ denote the output of the following experiment:

1. On common input $1^n$ and auxiliary input $z$, $\mathcal{A}^{\mathcal{O}}$ chooses an identity id $\in \{0,1\}^{l(n)}$ and first interacts with a honest receiver $R$ using $\langle I, R \rangle$. Let $o$ be the outcome of the execution.
2. Next, it interacts with $\mathcal{C}$ with common input $(n, o)$ and private input $b$ for $\mathcal{C}$.

Finally, the experiment outputs the view of the adversary $\mathcal{A}$ in the experiment and the output is set to $\bot$ unless $\mathcal{A}$ corrupts either $\mathcal{C}$ or $I$ or any of the identities chosen for the interactions of $\mathcal{A}$ with $\mathcal{O}$ is equal to id. Let the random variable $\mathsf{STA}_b(\mathcal{B}, \mathcal{C}, n, z)$ denote the output of $\mathcal{B}$ in an interaction between $\mathcal{B}$ and $\mathcal{C}$ with common input $(n, o)$ where $o$ is uniformly sampled from $\{0,1\}^n$, private input $b$ for $\mathcal{C}$ and auxiliary input $c$ with $\mathcal{B}$.

**Definition 2 (CCA-secure coin-tossing).** *Let $\langle I, R \rangle$ be a tag-based coin-tossing scheme with $l(n)$-bit identities, $m(n)$-bit outcomes and $\mathcal{O}$ a biasing oracle for it. We say that $\langle I, R \rangle$ is CCA-secure w.r.t. $\mathcal{O}$ against a challenger $\mathcal{C}$, if for every PPT adversary $\mathcal{A}$ and distinguisher $\mathcal{D}$, if $\mathcal{D}$ distinguishes the following ensembles with non-negligible probability:*

- *$\{\mathsf{EXP}_0(\langle I, R \rangle, \mathcal{O}, \mathcal{A}, \mathcal{C}, n, z)\}_{n \in \mathbb{N}, z \in \{0,1\}^*}$*
- *$\{\mathsf{EXP}_1(\langle I, R \rangle, \mathcal{O}, \mathcal{A}, \mathcal{C}, n, z)\}_{n \in \mathbb{N}, z \in \{0,1\}^*}$*

*then there exists a stand-alone adversary (that does not have access to $\mathcal{O}$) $\mathcal{B}$ and distinguisher $\mathcal{D}'$ such that $\mathcal{D}'$ distinguishes the following ensembles with non-negligible probability:*

- *$\{\mathsf{STA}_0(\mathcal{B}, \mathcal{C}, n, z)\}_{n \in \mathbb{N}, z \in \{0,1\}^*}$*
- *$\{\mathsf{STA}_1(\mathcal{B}, \mathcal{C}, n, z)\}_{n \in \mathbb{N}, z \in \{0,1\}^*}$*

We highlight that in a real experiment, $o$ is the result of the outcome of a coin-tossing between the adversary acting as the receiver and an honest initiator. However, the game between $\mathcal{B}$ and $\mathcal{C}_b$ is instantiated with a randomly chosen $o$. In essence, the definition says that if a challenge presented by $\mathcal{C}_0$ and $\mathcal{C}_1$ is hard to distinguish for a randomly sampled $o$, then it will be hard to distinguish even if $o$ was sampled according to $\langle I, R \rangle$ with an adversarial receiver $R$ who has access to oracle $\mathcal{O}$.

# 3 Preliminaries

We assume familiarity with basic notions of Turing machines, probabilistic-polynomial time computation and standard security notions of computational indistinguishability, public-key encryption and commitment schemes.

## 3.1 Simulatable PKE

**Definition 3 (Simulatable public-key encryption scheme).** *A $\ell$-bit simulatable encryption scheme* consists of an encryption scheme *(Gen, Enc, Dec)* augmented with *(oGen, oRndEnc, rGen, rRndEnc). Here, oGen and oRndEnc are the oblivious sampling algorithms for public keys and ciphertexts, and rGen and rRndEnc are the respective inverting algorithms, rGen (resp. rRndEnc) takes $r_G$ (resp. $(PK, r_E, m)$) as the trapdoor information. We require that, for all messages $m \in \{0,1\}^\ell$, the following distributions are computationally indistinguishable:*

$$\{\mathsf{rGen}(PK), \mathsf{rRndEnc}(PK, c), PK, c \mid (PK, SK) = \mathsf{Gen}(1^n; r_G), c = \mathsf{Enc}_{PK}(m; r_E)\}$$
$$\text{and } \{\hat{r}_G, \hat{r}_E, \hat{PK}, \hat{c} \mid (\hat{PK}, \bot) = \mathsf{oGen}(1^n; \hat{r}_G), \hat{c} = \mathsf{oRndEnc}_{\hat{PK}}(1^n; \hat{r}_E)\}$$

*It follows from above that a simulatable encryption scheme is also semantically secure.*

## 3.2 CCA-Secure Commitment Schemes

The following is taken verbatim from [CLP10]. Roughly speaking, a commitment scheme is CCA (chosen-commitment-attack) secure if the commitment scheme retains its hiding property even if the receiver has access to a "decommitment oracle". Let $\langle C, R \rangle$ be a tag-based commitment scheme with $l(n)$-bit identities. A decommitment oracle $\mathcal{O}$ of $\langle C, R \rangle$ acts as follows in interaction with an adversary $\mathcal{A}$: it participates with $\mathcal{A}$ in many sessions of the commit phase of $\langle C, R \rangle$ as an honest receiver, using identities of length $n$, chosen adaptively by $\mathcal{A}$. At the end of each session, if the session is accepting and valid, it reveals a decommitment of that session to $\mathcal{A}$. Otherwise, it sends $\bot$. Note that when a session has multiple decommitments, the decommitment oracle only returns one of them. Hence, there might exist many valid decommitment oracles. We remark that we will rely on a slightly weaker oracle, referred to as "committed-value" oracle in [LP12a] that simply extracts the committed value instead of providing the decommitment information. This relaxation is required for the black-box construction in [LP12a] and we will rely on the same definition.

Loosely speaking, a tag-based commitment scheme $\langle C, R \rangle$ is said to be CCA-secure, if there exists a committed-value oracle $\mathcal{O}$ for $\langle C, R \rangle$, such that the hiding property of the commitment holds even with respect to adversaries with access to $\mathcal{O}$. More precisely, let $\mathcal{A}^{\mathcal{O}}$ denote the adversary $\mathcal{A}$ with access to the oracle $\mathcal{O}$. Let $\mathsf{IND}_b(\langle C, R \rangle, \mathcal{O}, \mathcal{A}, n, z)$, where $b \in \{0, 1\}$, denote the output of the following probabilistic experiment: on common input $1^n$ and auxiliary input $z$, $\mathcal{A}^{\mathcal{O}}$ (adaptively) chooses a pair of challenge values $(v_0, v_1) \in \{0, 1\}$, the values to be committed to, and an identity $\mathsf{id} \in \{0, 1\}^{l(n)}$, and receives a commitment to $v_b$ using identity $\mathsf{id}$. Finally, the experiment outputs the output $y$ of $\mathcal{A}^{\mathcal{O}}$, the output $y$ is replaced by $\perp$ if during the execution $\mathcal{A}$ sends $\mathcal{O}$ any commitment using identity $\mathsf{id}$ (that is, any execution where the adversary queries the committed-value oracle on a commitment using the same identity as the commitment it receives, is considered invalid).

**Definition 4 (CCA-secure commitments).** *Let $\langle C, R \rangle$ be a tag-based commitment scheme with $l(n)$-bit identities, and $\mathcal{O}$ a committed-value oracle for it. We say that $\langle C, R \rangle$ is CCA-secure w.r.t. $\mathcal{O}$, if for every PPT $\mathcal{A}$, the following ensembles are computationally indistinguishable:*

$$(i)\{\mathsf{IND}_0(\langle C, R \rangle, \mathcal{O}, \mathcal{A}, n, z)\}_{n \in \mathbb{N}}, \quad (ii)\{\mathsf{IND}_1(\langle C, R \rangle, \mathcal{O}, \mathcal{A}, n, z)\}_{n \in \mathbb{N}}$$

*We say that $\langle C, R \rangle$ is CCA-secure if there exists a committed-value oracle $\mathcal{O}'$, such that, $\langle C, R \rangle$ is CCA-secure w.r.t. $\mathcal{O}'$.*

We extend this definition to include adversaries that can *adaptively* corrupt the committer $C$ in the left interaction and any of the receivers in the interactions with the committed-value oracle. We present this definition in Appendix A. We stress here that the security definition only requires the standard static guarantee of hiding even in the presence of adaptive corruptions. Finally, we will also require a strengthening of the CCA-security commitment scheme called $k$-robustness [CLP10] that preserves the security of arbitrary $k$-round protocols w.r.t any adversary that has access to the committed-value oracle and its adaptive analogue (For a more precise definition, we refer the reader to the full version).

## 4  Black-Box Adaptive UC Secure Protocols with Super-Polynomial Helpers

We consider the model of UC with super-polynomial helpers introduced in [PS04, CLP10]. Informally speaking, in this UC model, both the adversary and the environment in the real and ideal worlds have access to a super-polynomial time functionality that assists the parties. For more details, we refer the reader to [CLP10]. In the original work of [CLP10] as well as subsequent works, only static adversaries were considered. In this work, we consider the stronger adaptive adversary and obtain the following theorem in this model.

**Theorem 4.1.** *Assume the existence of a simulatable public-key encryption scheme. Then, for every $\epsilon > 0$ there exists a super-polynomial time helper functionality $\mathcal{H}$, such that for every well-formed functionality $\mathcal{F}$, there exists a $\tilde{O}(d_{\mathcal{F}} n^{\epsilon})$-round protocol $\Pi$ that $\mathcal{H}$-EUC emulates $\mathcal{F}$ where $d_{\mathcal{F}}$ is the depth of the circuit implementing the functionality $\mathcal{F}$. Furthermore, the protocol uses the underlying encryption scheme in a black-box way.*

We will rely in our proof the following two lemmas.

**Lemma 4.1.** *Assume the existence of a simulatable public-key encryption scheme and a $T_{\mathrm{COIN}}$-round CCA-secure coin-tossing protocol. Then, there exists a super-polynomial time helper functionality $\mathcal{H}$, such that there exists a $O(T_{\mathrm{COIN}})$-round protocol $\Pi$ that $\mathcal{H}$-EUC emulates $\mathcal{F}_{\mathrm{COM}}$ against malicious adaptive adversaries. Furthermore, the protocol uses the underlying encryption scheme in a black-box way.*

**Lemma 4.2.** *Assume the existence of one-way functions, the for every $\epsilon > 0$ there exists a $O(n^{\epsilon})$-round CCA-secure coin-tossing scheme against malicious adaptive adversaries. Furthermore, the protocol uses the underlying primitives in a black-box way.*

First, we prove the theorem assuming the lemmas hold and then prove the lemmas in the following sections. Towards this, we first describe our helper functionality $\mathcal{H}$. The biasing oracle for the CCA-secure coin-tossing scheme provided in Lemma 4.2 will serve as $\mathcal{H}$. This in turn relies on Lin and Pass construction from [LP12a] of a $\tilde{O}(n^{\epsilon})$-round black-box construction of a CCA-secure commitment scheme based on one-way functions. Since one-way functions can be constructed from a simulatable public-key encryption scheme in a black-box way, combining [LP12a] with Lemmas 4.1 and 4.2 we have a $O(n^{\epsilon})$-round protocol that $\mathcal{H}$-EUC that emulates $\mathcal{F}_{\mathrm{COM}}$. We conclude the proof of the theorem by combining the following three results:

1. The work of Choi et al. [CDMW09] provides a $O(T_{\mathrm{OT}})$-round construction that realizes $\mathcal{F}_{\mathrm{OT}}$ in the $\mathcal{F}_{\mathrm{COM}}$-hybrid assuming the existence of a $T_{\mathrm{OT}}$-round stand-alone adaptively-secure semi-honest oblivious-transfer protocol where the underlying protocol is used in a black-box way.
2. The work of Damgard and Nielsen [DN00] provides a black-box construction of a $O(1)$-round stand-alone adaptively-secure semi-honest oblivious-transfer protocol assuming the existence of simulatable public-key encryption schemes.
3. The work of Ishai et al. [IPS08] provides a $O(d_{\mathcal{F}})$-round protocol that realizes any well-formed functionality $\mathcal{F}$ in the $\mathcal{F}_{\mathrm{OT}}$-hybrid, where $d_{\mathcal{F}}$ is the depth of the circuit implementing functionality $\mathcal{F}$.

We rely on the $O(n^{\epsilon})$ construction of CCA-secure commitment of Lin and Pass [LP12a] instead of the more round efficient construction of Kiyoshima [Kiy14] because we additionally need to prove that the commitment is secure in the presence of adaptive adversaries and we are able to achieve this only for the [LP12a] construction. We leave it as future work to improve it with respect to the [Kiy14] construction.

# 5    CCA-Secure Coin-Tossing from CCA-Secure Commitments

In this section, we provide our construction of CCA-secure coin-tossing protocol $\langle I, R \rangle$. The two primitives we will require are CCA-secure commitments and one-way functions. Recall that, standard CCA-secure commitments require that a value committed to, using a tag id, remains hidden even to an adversary who has access to a "decommitment oracle". We will additionally require that if we consider an adversary that can adaptively corrupt receivers in its interactions with the decommitment oracle, the value committed to the adversary is hidden as long as the committer in this interaction is not corrupted. We show that the CCA-secure commitment scheme of [LP12a] satisfies this guarantee in Appendix A. More formally, let $\langle C, R \rangle$ be a CCA-secure commitment scheme and Com be a statistically-binding commitment scheme with pseudorandom commitments. For instance, the 2-round commitment scheme of Naor [Nao91] based on one-way function satisfies this notion. Next, we prove that the scheme from Fig. 1 is CCA-secure and CCA-secure against challengers.

**Theorem 5.1.** *Suppose,* $\langle C, R \rangle$ *is a 0-robust CCA secure commitment scheme in the presence of adaptive adversaries. Then there exists an oracle helper* $\mathcal{O}$ *such that* $\langle I, R \rangle$ *is a CCA-secure coin tossing protocol w.r.t* $\mathcal{O}$.

*Proof.* To demonstrate that our scheme is CCA-secure, we construct a biasing oracle $\mathcal{O}$ and show that given any PPT adversary $\mathcal{A}$, there exists a PPT simulator $\mathcal{S}$ such that:

$$\{\mathcal{A}^{\mathcal{O}}(n, z)\}_{n \in \mathbb{N}, z \in \{0,1\}^*} \approx \{\mathcal{S}(n, z)\}_{n \in \mathbb{N}, z \in \{0,1\}^*}.$$

We provide the description of our biasing oracle $\mathcal{O}$ in Fig. 2. On a high-level, this oracle follows the equivocation strategy analogous to the simulation in [HV15]. In slight more detail, this protocol that is a variant of the protocol in [HV15] allows for the initiator to equivocate $m$ in Stage 3 if for a chosen set $S$ at the beginning of the execution, the outcome of the coin-tossing in Stage 2 can be biased to yield $S$. Our oracle $\mathcal{O}$ will be able to accomplish this by breaking the commitment made by the receiver $R$ in Stage 2 using $\langle C, R \rangle$ in exponential time. Next, given an adversary $\mathcal{A}$, we construct a simulator $\mathcal{S}$ in two steps:

**Step 1:** Suppose $\mathcal{O}'$ is the oracle w.r.t which $\langle C, R \rangle$ is 0-robust. From the description of our oracle $\mathcal{O}$, it follows that every query to $\mathcal{O}$ can be simulated by a PPT algorithm with access to $\mathcal{O}'$. Recall that the only super-polynomial computation made by $\mathcal{O}$ is breaking a commitment made using $\langle C, R \rangle$, which can be done using $\mathcal{O}'$.[3] Therefore, given any adversary $\mathcal{A}$, there exists another oracle adversary $\widehat{\mathcal{A}}$ such that the following distributions are identically distributed:

---

[3] We remark here that typical CCA-secure commitment schemes are statistically binding and such schemes can be easily broken in exponential time. However, the CCA-secure commitment of [LP12a] is not statistically binding. However, as shown in [LP12a] it is "strongly" computationally binding which will suffice.

**Protocol $\pi_{\text{COIN}} = \langle I, R \rangle$.**

Let $1^n$ be the common input to the initiator $I$ and receiver $R$ and the identity of the interaction id $\in \{0,1\}^{l(n)}$.

**Stage 1: Commit Phase:** The receiver sends the first message $\sigma$ of the Naor's commitment scheme. The initiator first picks a random bit $m$ and chooses a random $n$-degree polynomial $p(\cdot)$ over a field $\mathbb{F}[x]$ such that $p(0) = m$. Namely, it randomly chooses $a_i \leftarrow \mathbb{F}$ for all $i \in [n]$ and sets $a_0 = m$, and defines the polynomial $p(x) = a_0 + a_1 x + \cdots + a_n x^n$. The initiator then creates a commitment to $m$ as follows. For every $i = [3n + 1]$, it first picks $b_i \leftarrow \{0,1\}$ at random and then computes:

$$c_i^{b_i} = \text{Com}_\sigma(p(i); t_i) \text{ and } c_i^{1-b_i} = r_i$$

where $r_i, t_i \leftarrow \{0,1\}^n$. The initiator sends $(c_0^0, c_0^1), \ldots, (c_{3n+1}^0, c_{3n+1}^1)$ to the receiver.

**Stage 2: Cut-and-Choose Phase:** The initiator and receiver interact in a coin-tossing protocol to obtain the cut-and-choose set that is carried out as follows.

1. The receiver chooses a random $\sigma_0$ and commits to the initiator using $\langle C, R \rangle$ using identity id.
2. The initiator picks $\sigma_1 \leftarrow \{0,1\}^N$ at random and sends it in the clear to the receiver.
3. The receiver decommits $r_{\sigma_0}$.

Both the initiator and the receiver compute $\sigma = \sigma_0 \oplus \sigma_1$ and use $\sigma$ as the random string to sample a random subset $\Gamma \subset [3n + 1]$ of size $n$. The initiator provides the decommitments for $\{c_i^{b_i}\}_{i \in \Gamma}$ by sending the sequence $\{b_i, p(i), t_i\}_{i \in \Gamma}$. The receiver verifies that all the decommitments are correct and aborts otherwise.

**Stage 3: Coin-Toss Phase:** In the first two stages, the initiator essentially commits to the string $m$. Next they continue with the coin-tossing protocol.

1. The receiver commits to $m'$ using $\langle C, R \rangle$.
2. This is followed by the initiator revealing its input $m$ as follows: Let $\Gamma' = [3n + 1] - \Gamma$. The initiator decommits $\{c_i^{b_i}\}_{i \in \Gamma'}$ to their respective messages. The receiver checks if the decommitments are correct and aborts otherwise. Using the $n$ polynomial evaluations revealed relative to $i \in \Gamma$ and any additional polynomial evaluation that was revealed relative to $\Gamma'$, the receiver reconstructs the polynomial $p(\cdot)$ (via polynomial interpolation of $n + 1$ points). Next, the receiver verifies whether $p(0) = m$, and that for every $i \in [3n + 1]$ the point $p(i)$ is the decrypted value within $c_i^{m_i}$.
3. The receiver decommits $m'$

Finally, the outcome of the coin-tossing is $m \oplus m'$. More formally, out$(\tau)$ where $\tau$ is the transcript of this protocol is set to $m \oplus m'$.

**Fig. 1.** Our CCA-secure coin-tossing protocol $\langle I, R \rangle$.

- $\{\mathcal{A}^{\mathcal{O}}(n, z)\}_{n \in \mathbb{N}, z \in \{0,1\}^*}$
- $\{\widehat{\mathcal{A}}^{\mathcal{O}'}(n, z)\}_{n \in \mathbb{N}, z \in \{0,1\}^*}$

---

**Description of biasing oracle $\mathcal{O}$:** For the protocol described in Figure 1, our biasing oracle $\mathcal{O}$ on input $c$ proceeds as follows:

- In Stage 1, picks a random subset $\widetilde{\Gamma} \subset [3n + 1]$ of size $n$ and two random $n$-degree polynomials $p_0(\cdot)$ and $p_1(\cdot)$ such that $p_0$ and $p_1$ agree on all points $i \in \widetilde{\Gamma}$ and $p_0(0) = 0$ and $p_1(0) = 1$.

  - For every $i \in \widetilde{\Gamma}$ the simulator proceeds as the honest sender would with polynomial $p_0(\cdot)$. Namely, it first picks $b_i \leftarrow \{0, 1\}$ at random and then sets

    $$c_i^{b_i} = \mathsf{Com}(p_0(i); t_i) \text{ and } c_i^{1-b_i} = r_i$$

    where $t_i \leftarrow \{0, 1\}^n$ (we recall that $p_0(i) = p_1(i)$ for all $i \in \widetilde{\Gamma}$).
  - For every $i \in \widetilde{\Gamma'} = [3n + 1] - \widetilde{\Gamma}$, the simulator picks $b_i \leftarrow \{0, 1\}$ at random and then sets

    $$c_i^{b_i} = \mathsf{Com}(p_0(i); t_i^0) \text{ and } c_i^{1-b_i} = \mathsf{Com}(p_1(i); t_i^1)$$

    where $t_i^0, t_i^1 \leftarrow \{0, 1\}^n$ are chosen uniformly at random.
    Finally, it sends $(c_0^0, c_0^1), \dots, (c_{3n+1}^0, c_{3n+1}^1)$ as the Stage 1 message of the initiator.

- In Stage 2, it breaks the commitment made using $\langle C, R \rangle$ and obtains the decommitted value $\sigma_0$. Next, it sets $\sigma_1$ so that $\sigma = \sigma_1 \oplus \sigma_0$ yields the set $\widetilde{\Gamma}$ as the outcome in Stage 2.

- In Stage 3, it breaks the commitment made using $\langle C, R \rangle$ and obtains $m'$. Then it decommits to $m = c \oplus m'$ using the following strategy. Recall first as the initiator it needs to reveal points on a polynomial $p(\cdot)$ and pairs $\{(b_i, t_i)\}_{i \in [3n+1]}$ such that $p(0) = m$ and $c_i^{b_i} = \mathsf{Com}(p(i); t_i)$. Let $\hat{b}_i = b_i \oplus m$ for all $i \in \widetilde{\Gamma'}$, then $\mathcal{S}$ reveals $p_m(\cdot)$, $\{\hat{b}_i, t_i^{\hat{b}_i}, r_i = c_i^{1-m}\}_{i \in \widetilde{\Gamma'}}$.

**Fig. 2.** Biasing oracle $\mathcal{O}$

**Step 2:** Relying on the 0-robustness CCA-security of the $\langle C, R \rangle$ commitment scheme, it follows that given $\widehat{\mathcal{A}}$, there exists a simulator $\mathcal{S}$ such that the following distributions are indistinguishable.

- $\{\widehat{\mathcal{A}}^{\mathcal{O}'}(n, z)\}_{n \in \mathbb{N}, z \in \{0,1\}^*}$
- $\{\mathcal{S}(n, z)\}_{n \in \mathbb{N}, z \in \{0,1\}^*}$

The statement of the theorem now follows using a standard hybrid argument. ■

Next, we proceed to show the stronger security-preserving property of our scheme.

**Theorem 5.2.** *Suppose, $\langle C, R \rangle$ is a $k$-robust CCA-secure commitment scheme in the presence of adaptive adversaries. Then for every $k$-message PPT $C$, $\langle I, R \rangle$ is a CCA-secure coin-tossing scheme w.r.t. the biasing oracle $\mathcal{O}$ against $C$.*

*Proof.* Assume for contradiction there exist an adversary $\mathcal{A}$, sequence $\{z_n\}_{n \in \mathbb{N}}$ and distinguisher $\mathcal{D}$ such that $\mathcal{D}$ distinguishes the following ensembles

- $\{\mathsf{EXP}_0(\langle I, R\rangle, \mathcal{O}, \mathcal{A}, \mathcal{C}, n, z_n)\}_{n\in\mathbb{N}}$
- $\{\mathsf{EXP}_1(\langle I, R\rangle, \mathcal{O}, \mathcal{A}, \mathcal{C}, n, z_n)\}_{n\in\mathbb{N}}$

with non-negligible probability. Namely, it distinguishes with probability $p(n)$ for some polynomial $p(\cdot)$ and infinitely many $n$'s. We need to construct a machine $\mathcal{B}$ and distinguisher $\mathcal{D}'$ that will distinguish $\mathsf{STA}_0$ from $\mathsf{STA}_1$. Let $\mathcal{O}'$ be the committed-value oracle guaranteed by the $k$-robust CCA-security of $\langle C, R\rangle$ in the presence of adaptive adversaries. We will accomplish our goal of constructing $\mathcal{B}$ in two steps.

**Step 1:** First we construct a simulator $\widetilde{S}$ such that the following distributions are distinguishable with non-negligible probability.

- $\{\mathsf{STA}_0(\widetilde{S}^{\mathcal{O}'}, \mathcal{C}, n, z)\}_{n\in\mathbb{N}, z\in\{0,1\}^*}$
- $\{\mathsf{STA}_1(\widetilde{S}^{\mathcal{O}'}, \mathcal{C}, n, z)\}_{n\in\mathbb{N}, z\in\{0,1\}^*}$

**Step 2:** Since $\mathcal{C}$ interacts in at most $k$-messages, we obtain the required $\mathcal{B}$ directly by relying on the $k$-robustness of the CCA-security of $\langle C, R\rangle$ in the presence of an adaptive adversary.

**Step 1: Constructing $\widetilde{S}^{\mathcal{O}'}$.** Fix an $n$ for which $\mathcal{D}$ distinguishes the two ensembles with probability $p = p(n)$. Recall that in the EXP experiment, $\mathcal{A}$ first interacts with an external $R$ and then interacts with $\mathcal{C}_b$.

In a random instance of the $\mathsf{EXP}_b$ experiment, let $T$ be the random variable representing the partial transcript up until the end of Stage 1 in $\mathcal{A}$'s interaction with external $R$. Now, we consider the modified experiment $\widetilde{\mathsf{EXP}}_b^T$ which starts from the partial transcript[4] $T$ and proceeds identically to the $\mathsf{EXP}_b$.

Now, using an averaging argument, we can conclude that with probability at least $p/2$ over partial transcript $\tau_n \leftarrow T$ it holds that $\mathcal{D}$ distinguishes the following two ensembles with probability at least $p/2$s.

- $\{\widetilde{\mathsf{EXP}}_0^{\tau_n}(\langle I, R\rangle, \mathcal{O}, \mathcal{A}, \mathcal{C}, n, z_n)\}_{n\in\mathbb{N}}$
- $\{\widetilde{\mathsf{EXP}}_0^{\tau_n}(\langle I, R\rangle, \mathcal{O}, \mathcal{A}, \mathcal{C}, n, z_n)\}_{n\in\mathbb{N}}$

Now, we are ready to construct $\widetilde{S}$. The high-level approach is as follows: First, we show that, except with non-negligible probability over random executions starting from $\tau_n$, there is a fixed value $m_n$ that the adversary will decommit to in the Stage 3 of its interaction with $R$. We will rely on an information theoretic lemma from [HV15] for this. We state this step in the Claim 5.1 below.

**Claim 5.1.** *There exists a string $m_n$ such that, starting from partial transcript $\tau_n$, the probability that $\mathcal{A}$ successfully decommits to a message different from $m_n$ in Stage 3 is negligible.*

---

[4] This can be achieved by instantiating the adversary with the same random coins and feeding the messages from $T$ and then running the rest of the experiment with fresh randomness.

On a high-level the idea is that given the transcript until end of Stage 1, there is a unique set $S$ that needs to be the outcome of Stage 2 in order for the an initiator to equivocate in Stage 3. This means that if an adversarial initiator can equivocate with non-negligible probability, then it has to bias the coin-tossing in Stage 2 to yield this unique set $S$ with non-negligible probability. Such an adversary can then show to violate the CCA-security of the commitment made using $\langle C, R \rangle$ in Stage 2. We provide a formal proof of the claim at the end of this section.

Next, for a fixed transcript $\tau_n$, we will give $\tau_n, m_n$ and partial view of $\mathcal{A}$ in the execution as the non-uniform advice. Our simulator $\widetilde{S}$ will start an execution with $\mathcal{A}$ from the partial view with transcript $\tau_n$ and will use $m_n$ to bias the outcome of the coin-tossing to $o$ by setting $m' = m_n \oplus o$ in Stage 3 of the execution. Now, we observe that, if $o$ is uniformly distributed, then $m'$ chosen by $\widetilde{S}$ will also be (non-negligibly) close to the uniform distribution given $m_n$ and hence the view of $S$ output with $\mathcal{C}_b$ will be statistically close to the distribution of $\mathcal{A}$ when interacting with $\mathcal{C}_b$ starting from $\tau_n$. This means that if $\mathcal{D}$ distinguishes the view of $\mathcal{A}$ starting from $\tau_n$ in both the experiments, then it will also distinguish the output of $\widetilde{S}^{\mathcal{O}'}$ in the two experiments.

We now construct our simulation $\widetilde{S}$. On input $(1^n, o, (z, \tau_n, m_n, r_n))$, $\widetilde{S}^{\mathcal{O}'}$ internally emulates an execution of $\mathcal{A}(1^n, z; r)$ in the real experiment starting from the partial transcript $\tau_n$. On the left, $\widetilde{S}^{\mathcal{O}'}$ needs to provide messages for the initiator $I$ such that the outcome is $o$ while simultaneously answering all oracle queries to $\mathcal{O}$. This it accomplished by committing to $m' = o \oplus m_n$ in Stage 3. Then if the adversary reveals anything other than the $m_n$, it simply aborts.

**Answering $\mathcal{O}$ Queries.** In any interaction, the oracle $\mathcal{O}$ first receives a coin $c$. In the internal emulation $\widetilde{S}^{\mathcal{O}'}$ obtains $c$ and needs to emulate $\mathcal{O}$. It carries out the actions exactly as $\mathcal{O}$ with the exception that instead of breaking the commitments made using $\langle C, R \rangle$ (as $\mathcal{O}$ does) $\widetilde{S}^{\mathcal{O}'}$ simply forwards it to $\mathcal{O}'$ which breaks them for $S$.

It follows from the construction and Claim 5.1 that the following distributions are statistically close:

- $\{\widetilde{\mathsf{EXP}}_b^{\tau_n}(\langle I, R \rangle, \mathcal{O}, \mathcal{A}, \mathcal{C}, n, z_n)\}_{n \in \mathbb{N}}$
- $\{\mathsf{STA}_b(\widetilde{S}^{\mathcal{O}'}, \mathcal{C}, n, (z_n, \tau_n, m_n, r_n))\}_{n \in \mathbb{N}, z \in \{0,1\}^*}$

and therefore $\mathcal{D}$ distinguishes the distribution $\mathsf{STA}_0(\widetilde{S}^{\mathcal{O}'}, \mathcal{C}, n, (z_n, \tau_n, m_n, r_n))$ from $\mathsf{STA}_1(\widetilde{S}^{\mathcal{O}'}, \mathcal{C}, n, (z_n, \tau_n, m_n, r_n))$ with with probability at least $p/2 - \nu(n) > p/4$ for all sufficiently large $n$'s.

**Step 2: Constructing a Stand-alone $\mathcal{B}$.** In Step 1, we constructed a machine $\widetilde{S}^{\mathcal{O}'}$ that with access to $\mathcal{O}'$ can violate the game. Now to get a stand-alone $\mathcal{B}$, we rely on the $k$-robustness property of $\langle C, R \rangle$ with $\widetilde{S}^{\mathcal{O}'}$ to obtain $\mathcal{B}$. More precisely, using the robustness we have that the following distributions are computationally indistinguishable:

- $\{\mathsf{STA}_b(\widetilde{S}^{\mathcal{O}'}, \mathcal{C}, n, (z_n, \tau_n, m_n, r_n))\}_{n \in \mathbb{N}, z \in \{0,1\}^*}$

- $\{\mathsf{STA}_b(\mathcal{B}, \mathcal{C}, n, (z_n, \tau_n, m_n, r_n))\}_{n \in \mathbb{N}, z \in \{0,1\}^*}$

and therefore $\mathcal{D}$ distinguishes the distribution $\mathsf{STA}_0$ from $\mathsf{STA}_1$ with probability at least $p/4 - \nu(n) > p/8$ for all sufficiently large $n$'s and this completes the proof of the theorem.

To conclude the proof of Theorem 5.2, it only remains to prove Claim 5.1.

*Proof of Claim 5.1.* Assume for contradiction, the adversary $\mathcal{A}$ equivocates with non-negligible probability starting from $\tau_n$. We now show that $\mathcal{A}^{\mathcal{O}'}$ violates the CCA-security of $\langle C, R \rangle$ w.r.t $\mathcal{O}'$, namely, it violates the hiding property of the commitment made using $\langle C, R \rangle$ in Stage 2.

As stated above, we use an information theoretic lemma from [HV15]. On a high-level, the lemma states that for the adversary to be able to equivocate in Stage 3, there exists a unique set $S$ that it must bias the outcome of the coin-toss in Stage 2 so that the resulting set is $S$. On a high-level, we can rely on this lemma, as a malicious initiator that equivocates must bias the outcome to a particular set $S$ and using the set $S$. Then, we can construct an adversary $\widehat{\mathcal{A}}^{\mathcal{O}'}$ that violates the CCA-game for $\langle C, R \rangle$ by simply detecting this set $S$ in the outcome of Stage 2.

More formally, given $\tau_n$, and a partial view of $\mathcal{A}$, let us assume that $\mathcal{A}$ equivocates with probability $\frac{1}{q(n)}$ for some polynomial $q(\cdot)$ and infinitely many $n$.

Before we recall the information theoretic lemma from [HV15], we first explain how our protocol is an instance of the protocol in their work. In [HV15], they construct an adaptively secure UC-commitment in the CRS hybrid where the protocol proceeds as follows:

1. In Stage 1, the committer using the same strategy as the initiator in our Stage 1 commits to a string $m$, where instead of using $\mathsf{Com}_\sigma$, it uses an encryption scheme with oblivious ciphertext generation property (where the public-key for this scheme is placed in the CRS).
2. In Stage 2, the committer and receiver execute a coin-toss where the receiver makes the first move just as in $\langle I, R \rangle$ with the exception that the receiver in the their protocol uses again an encryption scheme (with the public-key in the CRS) instead of a commitment scheme to commit to $\sigma_0$.
3. In the decommitment phase of their protocol, the committer reveals its commitment just as the initiator does in Step 2 of Stage 3 in our protocol.

We remark that in essence, the protocol in [HV15] is used as a subprotocol in our work here where the initiator commits to a string $m$ and then reveals it. The only property they need of the encryption scheme is that it is statistically binding and has the oblivious generation property. In our protocol, the Naor commitment scheme has both these properties. (See our next protocol for such a variant).

**Claim 5.2 Restatement of Claim 5.5 [HV15].** *Let $\tau$ be a fixed partial transcript up until end of Stage 1. Then, except with negligible probability, there exists no two transcripts $\mathsf{trans}_1, \mathsf{trans}_2$ that satisfy the following conditions:*

1. $\text{trans}_1$ and $\text{trans}_2$ are complete and accepting transcripts of $\pi_{\text{COM}}$ with $\tau$ being their prefix.
2. There exists two distinct sets $S_1, S_2$ such that $S_1$ and $S_2$ are the respective outcomes of the coin-tossing phase within $\text{trans}_1$ and $\text{trans}_2$.
3. There exist valid decommitments to two distinct strings in $\text{trans}_1$ and $\text{trans}_2$.

Since the commitment made by our Initiator can be viewed as an instance of their protocol, we can conclude that there exists a unique set $S$ that should be the outcome of the coin-toss in Stage 2 for a malicious initiator to equivocate $m$. Since $\mathcal{A}$ equivocates with probability $\frac{1}{q(n)}$ it holds, there is a set $S$ such that with the probability negligible close[5] to $\frac{1}{q(n)}$, starting from $\tau_n$, the outcome of Stage 2 is $S$. To construct an adversary $\widehat{\mathcal{A}}$ that violates the CCA-security of the underling $\langle C, R \rangle$ scheme, we simply incorporate $\mathcal{A}$ and use as auxiliary input $\tau_n, S$ and the partial view of $\mathcal{A}$. Next, it forwards the $\langle C, R \rangle$ interaction in Stage 2 to an external committer. All queries to the helper oracle $\mathcal{O}$ by $\mathcal{A}$ can be simulated using $\mathcal{H}$ and $\widehat{\mathcal{A}}$ simply uses $\mathcal{H}$ to emulate $\mathcal{O}$. Then it halts the execution right after the adversary in the internal emulation reveals $\sigma_1$. Now, $\widehat{\mathcal{A}}$ simply outputs $\sigma_0 = \sigma \oplus \sigma_1$ where $\sigma$ is the string that maps to the set $S$. This violates the CCA game as with probability close to $\frac{1}{q(n)}$, $\widehat{\mathcal{A}}$ identifies the message committed using $\langle C, R \rangle$. □ ∎

# 6 Realizing $\mathcal{F}_{\text{COM}}$ Using CCA-Secure Coin-Tossing

In this section, we provide our black-box construction of $\mathcal{H}$-EUC secure protocol $\Pi_{\text{COM}}$. Our protocol is a variant of the protocol described in [HV15] where it is shown how to realize $\mathcal{F}_{\text{COM}}$ in the CRS model assuming only public-key encryption that admits oblivious-ciphertext generation with adaptive UC-security. While the [HV15] protocol assumes that every pair of parties share an independently generated CRS, in this work we assume no setup, but will require the stronger simulatable public-key encryption scheme. Assume that $\langle I, R \rangle$ is a CCA-secure coin-tossing scheme and that the public-key encryption scheme (Gen, Enc, Dec) is augmented with algorithms (oGen, oRndEnc, rGen, rRndEnc) which implies a simulatable public-key encryption scheme. Then we start with a formal description of our protocol.

Consider a helper functionality $\mathcal{H}$ that "biases" the coin-tossing in an interaction using $\langle I, R \rangle$ in the same way as the biasing oracle $\mathcal{O}$ does, subject to the condition that player $P_i$ in a protocol instance $sid$ can only query the functionality on interactions that use identity $(P_i, sid)$. More precisely, every party $P_i$ can simultaneously engage with $\mathcal{H}$ in multiple sessions of $\langle I, R \rangle$ as an initiator using identity $P_i$ where the functionality simply forwards all the messages internally to the biasing oracle $\mathcal{O}$, and ensures that the result of the coin-tossing is biased to a prescribed outcome at the end of each session. See Fig. 3 for a formal description

---

[5] The probability is not identically equal to $\frac{1}{q(n)}$ since the commitment scheme is only statistically binding and not perfectly binding.

---

**Functionality $\mathcal{H}$**

**Initialization:** Upon receiving an input $(\mathsf{Init}, P_i, sid, k)$ from party $P_i$ in the protocol instance $sid$, if there is a previously recorded session $(P_i, sid, k)$, ignore this message; otherwise, initialize a session of $\langle I, R \rangle$ with $\mathcal{O}$ using identity $(P_i, sid)$ and record session $(P_i, sid, k)$.

**Accessing O:** Upon receiving an input $(\mathsf{Mesg}, P_i, sid, k, m)$ from party $P_i$ in the protocol instance $sid$, if there is no previously recorded session $(P_i, sid, k)$, ignore the message; otherwise, forward $m$ to $\mathcal{O}$ in the $k^{th}$ session that uses identity $(P_i, sid)$, obtain a reply $m'$, and return $(\mathsf{Mesg}, P_i, sid, k, m')$ to $P_i$.

---

**Fig. 3.** The helper functionality $\mathcal{H}$ (i.e. angel).

---

**Protocol $\Pi_{\mathrm{COM}}$.**

**Sender's Input:** A message $m \in \{0, 1\}$ and a security parameter $1^n$.

**Commitment Phase:**

**Stage 1: Key Generations Phase:** The sender and receiver engage in a protocol using $\langle I, R \rangle$ where the receiver acts as the initiator $I$ and the sender acts as $R$. Let $\mathsf{PK} = \mathsf{oGen}(\mathsf{out}(\tau_{S \to R}))$ where $\tau_{S \to R}$ is the transcript of the interaction.

**Stage 2: Input Encoding Phase:** The sender chooses a random $n$-degree polynomial $p(\cdot)$ over a field $\mathbb{F}[x]$ such that $p(0) = m$. Namely, it randomly chooses $a_i \leftarrow \mathbb{F}$ for all $i \in [n]$ and sets $a_0 = m$, and defines the polynomial $p(x) = a_0 + a_1 x + \cdots + a_n x^n$. The sender then creates a commitment to $m$ as follows. For every $i = [3n + 1]$, it first pick $b_i \leftarrow \{0, 1\}$ at random and then computes: $c_i^{b_i} = \mathsf{Enc}_{\mathsf{PK}}(p_0(i); t_i)$ and $c_i^{1-b_i} = \mathsf{oRndEnc}(\mathsf{PK}, r_i)$ where $r_i, t_i \leftarrow \{0, 1\}^n$. The sender sends $(c_0^0, c_0^1), \ldots, (c_{3n+1}^0, c_{3n+1}^1)$ to the receiver.

**Stage 3: Cut-and-choose Phase:** The sender and receiver engage in a protocol using $\langle I, R \rangle$ where the sender acts as the initiator $I$ and the receiver acts as $R$. Define a subset $\Gamma \subset [3n + 1]$ of size $n$ using the outcome $\mathsf{out}(\tau_{R \to S}))$ where $\tau_{R \to S}$ is the transcript of the interaction. The sender provides the plaintexts encrypted in $\{c_i^{b_i}\}_{i \in \Gamma}$ by sending the sequence $\{b_i, p(i), t_i\}_{i \in \Gamma}$. The receiver verifies that all the decryptions are correct and aborts otherwise.

**Decommitment Phase:** Let $\Gamma' = [3n + 1] - \Gamma$. The sender reveals its input $m$ and all the plaintexts encrypted in $\{c_i^{b_i}\}_{i \in \Gamma'}$. The receiver checks if all the decryptions are correct and aborts otherwise. Using the $n$ polynomial evaluations revealed relative to $i \in \Gamma$ and any additional polynomial evaluation that was revealed relative to $\Gamma'$, the receiver reconstructs the polynomial $p(\cdot)$ (via polynomial interpolation of $n+1$ points). Next, the receiver verifies whether $p(0) = m$, and that for every $i \in [3n + 1]$ the point $p(i)$ is the decrypted value within $c_i^{m_i}$.

---

**Fig. 4.** Protocol $\Pi_{\mathrm{COM}}$ that realizes $\mathcal{F}_{\mathrm{COM}}$ using a CCA-secure coin-tossing protocol $\langle I, R \rangle$

of the functionality. We note here that since $\mathcal{O}$ can be implemented in super-polynomial time, this functionality can also be implemented in super-polynomial time.

**Proof overview:** Recalling that an adversary can adaptively corrupt both parties, for the overview, we present the hardest cases for simulation, which is static corruption of one party followed by the adaptive corruption of the other party.

*Simulating static corruption of receiver and post-execution corruption of sender.* To simulate the messages for a honest sender, the simulator generates random shares for 0 and 1 that agree on a randomly chosen $n$ subset $\widetilde{\Gamma}$ (chosen in advance). It then encrypts these shares in Stage 2 where for each index it randomly positions the shares for 0 and 1. Next, in Stage 3, the simulator biases $\tau_{R \to S}$ using the helper $\mathcal{H}$ so that the subset generated using $\mathsf{out}(\tau_{R \to S})$ is exactly $\widetilde{\Gamma}$. As these shares are common for a sharing of 0 and 1, revealing them in the commit phase will go undetected. Later in the decommit phase, it can chose to reveal shares of 0 or 1 depending on the real message $m$ (to show that the unopened shares were obliviously generated will be done by exploiting the invertible sampling algorithm for the simulatable encryption scheme). The core argument in proving indistinguishability of simulation will be to reduce the hiding property of Stage 2 to the semantic-security of the underlying encryption scheme on a public-key generated using Gen, i.e., the CPA-security of the encryption scheme, where we will rely on the CCA-security game w.r.t challengers for our coin-tossing protocol to achieve this. We discuss this reduction on a high-level below. Before that we remark that the adversary will not be able to use the helper oracle $\mathcal{H}$ to bias the outcome of the coin-tossing in Stage 1 because the helper oracle will not provide access to the biasing oracle on sessions where the party querying the helper is not the responder $R$ of that coin-tossing session.

*Reduction:* The challengers $\mathcal{C}$ for our CCA-game, on input a string $o$ will set $\mathsf{PK} = \mathsf{rGen}(o)$ and for a predetermined message $t$ it proceed as follows: If its private input $b = 0$, $\mathcal{C}$ will output a ciphertext that is an honest encryption of $t$ using Enc. If its private input $b = 1$, $\mathcal{C}$ will obliviously generate a ciphertext using oRndEnc. It will follow from the security guarantees of the simulatable public-key encryption that for a randomly chosen $o$, no (stand-alone) adversary can distinguish the outputs of $\mathcal{C}|_{b=0}$ or $\mathcal{C}|_{b=1}$ even given $o$ (i.e. $\mathsf{STA}_0 \approx \mathsf{STA}_1$).

Now given an adversary $\mathcal{A}$ controlling the receiver in our coin-tossing scheme $\langle I, R \rangle$ we consider a sequence of hybrid experiments where we replace the encryptions in Stage 2 from the honest sender's strategy to the simulated strategy. Namely, obliviously generated ciphertexts $c_j^{1-b_j}$ will be generated using the encryption algorithm. More precisely, we consider a sequence of hybrids $H^1 \dots, H^{3n+1}$ where in the $H^i$ we generate $c_j^{1-b_j}$ for $j = 1, \dots, i$ in Stage 2 according to the simulator's strategy (i.e. encryption of valid messages as opposed to being obliviously generated). Next we show that $H^{i-1}$ and $H^i$ are indistinguishable. The only difference between the two hybrids is in how $c_i^{1-b_i}$ is generated. More precisely, in $H^{i-1}$, $c_i^{1-b_i}$ is generated using oRndEnc and in $H^i$ it is generated using Enc. We now reduce the indistinguishability of the hybrids to the semantic-security of the encryption scheme via the CCA-game of

$\langle I, R \rangle$. Towards this, we consider a challenger $\mathcal{C}$ described above for which the stand-alone game is hard.

Next, consider an oracle adversary $\widetilde{\mathcal{A}}$ that internally incorporates $\mathcal{A}$ and the environment and proceeds as follows: $\widetilde{\mathcal{A}}$ forwards every oracle query made by $\mathcal{A}$ to its oracle and forwards the interaction using $\langle I, R \rangle$ in Stage 1 externally to an honest receiver. $\widetilde{\mathcal{A}}$ then stalls the internal emulation upon having the interaction within $\langle I, R \rangle$ complete, and outputs the view of $\mathcal{A}$ and the outcome of the coin-tossing $o$ from the internal emulation, in the external interaction. Then it interacts with $\mathcal{C}$ that on input $o$ produces a ciphertext. Internally, $\widetilde{\mathcal{A}}$ feeds the ciphertext in place of $c_i^{1-b_i}$ in Stage 2. The rest of the encryptions are honestly generated according to the strategy in $H^i$.

It now follows that if the message $t$ is chosen according to the strategy in $H^i$, then we have that $\mathsf{hyb}^{i-1} = \mathsf{EXP}_1(\langle I, R \rangle, \mathcal{O}, \mathcal{A}, \mathcal{C}, n, z)\}_{n \in \mathbb{N}, z \in \{0,1\}^*}$ and $\mathsf{hyb}^i = \mathsf{EXP}_0(\langle I, R \rangle, \mathcal{O}, \mathcal{A}, \mathcal{C}, n, z)\}_{n \in \mathbb{N}, z \in \{0,1\}^*}$ where $\mathsf{hyb}^{i-1}$ and $\mathsf{hyb}^i$ are the views of the adversary $\mathcal{A}$ in the hybrids $H^{i-1}$ and $H^i$. Therefore, if $\mathsf{hyb}^{i-1}$ and $\mathsf{hyb}^i$ are distinguishable by the CCA-security of $\langle I, R \rangle$ we have that there exists a stand-alone PPT algorithm $B$ for which $\mathsf{STA}_0$ and $\mathsf{STA}_1$ are distinguishable. Recalling that $\mathsf{STA}_0 \approx \mathsf{STA}_1$ by the hiding property of obliviously generated ciphertexts in the underlying encryption scheme and thus we arrive at a contradiction. Therefore, $\mathsf{hyb}^{i-1}$ and $\mathsf{hyb}^i$ must be indistinguishable.

To complete this case, we need to handle post-execution corruption of the sender. This can be achieved exactly as in the decommitment phase which reveals all the randomness used in the commitment phase.

*Simulating static corruption of sender and post-execution corruption of receiver.* For a honest receiver, the simulator first biases the outcome of the coin-tossing in Stage 1, so that PK is a public-key for which it knows the corresponding secret-key. This will allow the simulator to decrypt the ciphertexts provided by the adversary in Stage 2. However, this does not ensure extraction as an adversarial sender can equivocate just as the simulator for honest senders. Showing that there is a unique value that can be extracted requires showing that a corrupted sender cannot successfully predict exactly the $n$ indexes $\Gamma$ from $\{1, \ldots, 3n + 1\}$ that will be chosen in the coin-tossing protocol. Using an information-theoretic argument from [HV15], we know that after an encoding phase, for any adversary to break binding (i.e. equivocate) it must ensure that the coin-tossing phase results in a particular set $\Gamma$. We can reduce the binding property of our scheme to the CCA-security of underlying coin-tossing scheme. First, we observe that the helper functionality cannot be directly used by the adversary to bias the coin-toss as $\mathcal{H}$ will not help in sessions where the identity of the party controlling $\mathcal{I}$ is the party requesting the help. We will infact rely on the CCA-security against challengers to guarantee that the coin-tossing outcome has high-entropy. Once we have established that the adversary cannot bias the coin-toss used to determine the set $\Gamma$, we can obtain extraction by relying on a strategy from [HV15], that can determine the message using the decryptions from Stage 1 and the coin-tossing outcome in Stage 3. Finally, to address post-execution corruption of the receiver we observe that it suffices to generate the messages for the receiver honestly and upon corruption simply provide the random coins of this honest receiver.

**Formal Proof of Correctness of UC-Commitment Protocol:** Let $\mathcal{A}$ be a PPT adversary that attacks Protocol $\Pi_{\text{COM}}$ described in Fig. 4 and recall that simulator $\mathcal{S}$ interacts with the ideal functionality $\mathcal{F}_{\text{COM}}$ and with the environment $\mathcal{Z}$. Then $\mathcal{S}$ starts by invoking a copy of $\mathcal{A}$ and running a simulated (internal) interaction of $\mathcal{A}$ with the environment $\mathcal{Z}$ and parties running the protocol. We fix the following notation. First, the session and sub-session identifiers are respectively denoted by $sid$ and $ssid$. Next, the committing party is denoted $P_i$ and the receiving party $P_j$. $\mathcal{S}$ proceeds as follows:

**Simulating the Communication With $\mathcal{Z}$:** Every message that $\mathcal{S}$ receives from $\mathcal{Z}$ it internally feeds to $\mathcal{A}$ and every output written by $\mathcal{A}$ is relayed back to $\mathcal{Z}$.

**Simulating the Commitment Phase When the Receiver is Statically Corrupted:** In this case $\mathcal{S}$ uses the honest sender's algorithm in Stage 1 and in Stage 2 proceeds as follows. Upon receiving message $(sid, \text{Sen}, \text{Rec})$ from $\mathcal{F}_{\text{COM}}$, the simulator picks a random subset $\widetilde{\gamma} \subset [3n + 1]$ of size $n$ and two random $n$-degree polynomials $p_0(\cdot)$ and $p_1(\cdot)$ such that $p_0$ and $p_1$ agree on all points $i \in \widetilde{\Gamma}$ and $p_0(0) = 0$ and $p_1(0) = 1$.

- For every $i \in \widetilde{\Gamma}$ the simulator proceeds as the honest sender would with polynomial $p_0(\cdot)$. Namely, it first picks $b_i \leftarrow \{0, 1\}$ at random and then sets the following pairs, $c_i^{b_i} = \text{Enc}_{\text{PK}}(p_0(i); t_i)$ and $c_i^{1-b_i} = \text{oRndEnc}(\text{PK}, r_i)$ where $r_i, t_i \leftarrow \{0, 1\}^n$ (we recall that $p_0(i) = p_1(i)$ for all $i \in \widetilde{\Gamma}$).
- For every $i \in \widetilde{\Gamma}' = [3n + 1] - \widetilde{\Gamma}$ the simulator picks $b_i \leftarrow \{0, 1\}$ at random and then uses the points on both polynomials $p_0(\cdot)$ and $p_1(\cdot)$ to calculate the following pairs, namely $c_i^{b_i} = \text{Enc}_{\text{PK}}(p_0(i); t_i^0)$ and $c_i^{1-b_i} = \text{Enc}_{\text{PK}}(p_1(i), t_i^1)$ where $t_i^0, t_i^1 \leftarrow \{0, 1\}^n$ are chosen uniformly at random.

Finally, the simulator sends the pairs $(c_0^0, c_0^1), \ldots, (c_{3n+1}^0, c_{3n+1}^1)$ to the receiver.

Next, in Stage 3, the simulator biases the coin-tossing result so that the set $\Gamma$ that is chosen in this phase is identical to $\widetilde{\Gamma}$. More precisely, produces coins $c$ that will yield $\widetilde{\Gamma}$ in Stage 3 and sends $c$ to $\mathcal{H}$. Next, it forwards the messages the simulator receives from $\mathcal{A}$ controlling $R$ in this interaction using $\langle I, R \rangle$ to $\mathcal{H}$. Recall that the helper function will bias the outcome of this interaction to $c$ (as the identity of this interaction is not equal to any identity made by the $\mathcal{A}$). Finally, the simulator reveals the plaintexts in all the ciphertexts within $\{c_i^{b_i}\}_{i \in \widetilde{\Gamma}}$.

**Simulating the Decommitment Phase Where the Receiver is Statically Corrupted:** Upon receiving a message $(\text{reveal}, sid, m)$ from $\mathcal{F}_{\text{COM}}$, $\mathcal{S}$ generates a simulated decommitment message as follows. Recall first that the simulator needs to reveal points on a polynomial $p(\cdot)$ and pairs $\{(b_i, t_i)\}_{i \in [3n+1]}$ such that

$p(0) = m$ and $c_i^{b_i} = \mathsf{Enc}_{\mathsf{PK}}(p(i); t_i)$. Let $\hat{b}_i = b_i \oplus m$ for all $i \in \widetilde{\Gamma'}$, then $\mathcal{S}$ reveals $p_m(\cdot)$, $\{\hat{b}_i, t_i^{\hat{b}_i}, r_i = \mathsf{rRndEnc}(\mathsf{PK}, t_i^{1-m}, p_{1-m}(i))\}_{i \in \widetilde{\Gamma'}}$.

**Simulating the Commit Phase When the Sender is Statically Corrupted:** Simulating the sender involves extracting the committed value as follows. In Stage 1, $\mathcal{S}$ first samples $(\mathsf{PK}, \mathsf{SK})$ using the Gen algorithm with randomness $r_G$. Then it runs rGen on $r_G$ to obtain $c$ which it forwards to the helper $\mathcal{H}$. Then, it forwards the messages the simulator receives from $\mathcal{A}$ controlling $R$ in this interaction using $\langle I, R \rangle$ to $\mathcal{H}$. Recall that the helper function will bias the outcome of this interaction to $c$. This means that the public-key obtained from the coin-tossing is $\mathsf{PK}$.

The simulation next uses the honest receiver's algorithm in Stages 2 and 3. Let $\Gamma$ be the set obtained from the outcome of the coin-tossing phase. To extract the input, $\mathcal{S}$ chooses an arbitrary index $j \in [3n + 1] - \Gamma$ and reconstructs two polynomials $q(\cdot)$ and $\widetilde{q}(\cdot)$ such that for all $i \in \Gamma$, $q(i) = \widetilde{q}(i) = \beta_i^{b_i}$, $q(j) = \beta_j^0$, $\widetilde{q}(j) = \beta_j^1$ and $q(0), \widetilde{q}(0) \in \{0, 1\}$. It then verifies whether for all $i \in [3n + 1]$, $q(i) \in \{\beta_i^0, \beta_i^1\}$ and $\widetilde{q}(i) \in \{\beta_i^0, \beta_i^1\}$. The following cases arise:

**Case 1:** *Both $q(\cdot)$ and $\widetilde{q}(\cdot)$ satisfy the condition and $\widetilde{q}(0) \neq q(0)$* Then $\mathcal{S}$ halts returning fail. Below we prove that the simulator outputs fail with negligible probability.

**Case 2:** *At most one of $q(\cdot)$ and $\widetilde{q}(\cdot)$ satisfy the condition or $\widetilde{q}(0) = q(0)$.* $\mathcal{S}$ sends $(\mathsf{commit}, sid, q(0))$ to the $\mathcal{F}_{\mathrm{COM}}$ functionality and stores the committed bit $q(0)$. Otherwise, $\mathcal{S}$ sends a default value.

**Case 3:** *Neither $q(\cdot)$ or $\widetilde{q}(\cdot)$ satisfy the condition.* $\mathcal{S}$ sends a default value to the ideal functionality and need not store the committed bit since it will never be decommitted correctly.

**Simulating Adaptive Corruptions:** We remark that we only provide the description of the simulator for static corruption. If any honest party is adaptively corrupted during the simulation, since the simulation is straight-line and admits post-execution corruption, it can directly generate coins even in the middle of the execution.

Below we analyze each of the scenarios above, and show that no environment $\mathcal{Z}$ interacting with $\mathcal{S}$ in the ideal-world is distinguishable from that with $\mathcal{A}$ in the real-world in each of the cases.

**Analysis of Receiver Corruptions:** Our proof follows a sequence of hybrids from the real world execution to the ideal world execution.

**Hybrid $H_0$:** $H_0$ is identical to the real world execution.

**Hybrid $H_1$:** The hybrid experiment $H_1$ proceeds identically to $H_0$ with the exception that a set $\widetilde{\Gamma}$ of size $n$ is chosen at random and the coin-tossing interaction using $\langle I, R \rangle$ in Stage 3 is biased so that the outcome yields $\widetilde{\Gamma}$. Hybrids $H_0$ and $H_1$ are identically distributed except when the oracle $\mathcal{O}$ fails. Since this happens only with negligible probability, the outputs of the two experiments are statistically close.

**Hybrid $H_2$:** We gradually change the ciphertexts generated in Stage 2 from the real committer to the simulation. Indistinguishability of experiment $H_1$ and $H_2$ will rely on the security of the encryption scheme. However, to reduce the indistinguishability to the security game of the simulatable public-key encryption scheme, we will require to bias the PK chosen in Stage 1 to a challenge public-key obtained from the challenger for the encryption security game. We will be able to do this by relying on the security game of our CCA-secure coin-tossing protocol.

More formally, consider a sequence of hybrids $H_1^0, \dots, H_1^{3n+1}$ where in the $H_1^i$ we generate $c_j^{1-b_j}$ for $j = 1, \dots, i$ according to the simulator's strategy (i.e. encryption of valid messages as opposed to being obliviously generated). Now we show that $H_1^{i-1}$ and $H_1^i$ are indistinguishable. The only difference between the two hybrids is in how $c_i^{1-b_i}$ is generated. More precisely, in $H_1^{i-1}$, $c_i^{1-b_i}$ is generated using oRndEnc and in $H_1^i$ it is generated using Enc. We now reduce the indistinguishability of the hybrids to the semantic-security of the encryption scheme via the CCA-game of $\langle I, R \rangle$. Towards this, we give a challenger $\mathcal{C}$ for which the stand-alone game is hard. On a high-level this game will be the semantic-security of the underlying simulatable public-key encryption scheme where the public-key is sampled using rGen on the coin-toss $o$.

**Reduction:** More formally, given a message $t$, define $\mathcal{C}(o, b)$ as the strategy that sets PK $= \mathsf{rGen}(o)$ and outputs a ciphertext that was honest encryption of $t$ using Enc when $b = 0$ and obliviously generated using oRndEnc when $b = 1$.

Next consider an oracle adversary $\tilde{\mathcal{A}}$ that internally incorporates $\mathcal{A}$ and the environment and proceeds as follows: $\tilde{\mathcal{A}}$ forwards every oracle query made by $\mathcal{A}$ to its oracle and forwards the interaction using $\langle I, R \rangle$ in Stage 1 externally to an honest receiver. Let $o$ be the outcome of the interaction in the internal emulation. an encryption of a message using Enc or generates one obliviously. Then it interacts with $\mathcal{C}$ that on input $o$ produces a ciphertext. Internally, $\tilde{\mathcal{A}}$ feeds the ciphertext in place of $c_i^{1-b_i}$ in Stage 2. It now follows that if the message $t$ is chosen according to the strategy in $H_1^i$, then $\mathsf{hyb}_1^{i-1}(n, z) = \mathsf{EXP}_1(\langle I, R \rangle, \mathcal{O}, \mathcal{A}, \mathcal{C}, n, z)\}$ and $\mathsf{hyb}_1^i(n, z) = \mathsf{EXP}_0(\langle I, R \rangle, \mathcal{O}, \mathcal{A}, \mathcal{C}, n, z)\}$ where $\mathsf{hyb}_1^{i-1}$ and $\mathsf{hyb}_1^i$ are the views of the adversary $\mathcal{A}$ in the hybrids $H_1^{i-1}$ and $H_1^i$. Therefore, if $\mathsf{hyb}_1^{i-1}$ and $\mathsf{hyb}_1^i$ are distinguishable by the CCA-security of $\langle I, R \rangle$ we have that there exists a stand-alone $PPT$ algorithm $B$ that distinguish the interaction with $\mathcal{C}_0$ and $\mathcal{C}_1$ for a randomly sampled coin-toss outcome $o$. This violates the semantic-security of the encryption scheme and thus we arrive at a contradiction. Therefore, $\mathsf{hyb}_1^{i-1}$ and $\mathsf{hyb}_1^i$ must be indistinguishable.

**Hybrid $H_3$:** In this hybrid, we follow $H_2$ except that we use the simulation strategy to decommit to the message $m$ received from the $\mathcal{F}_{\mathrm{COM}}$-functionality. Since in $H_2$ the commitment phase has been setup to be equivocated, this follows directly. Again using the CCA-security of $\langle I, R \rangle$ just as we used to argue indistinguishability for hybrids $H_1$ and $H_2$, we can reduce the indistinguishability of $H_2$ and $H_3$ to the security of the underlying simulatable public-key encryption scheme.

Finally, we conclude by observing the $H_3$ is identical to the ideal world experiment.

**Analysis of Sender Corruptions:** Our proof follows a sequence of hybrids from the real world execution to the ideal world execution.

**Hybrid $H_0$:** $H_0$ is identical to the real world execution.

**Hybrid $H_1$:** This experiment proceeds identical to $H_0$ with the exception that we forward the interaction using $\langle I, R \rangle$ in Stage 1 to the oracle $\mathcal{H}$. More precisely, we pick $(PK, SK)$ using the Gen algorithm with randomness $r_G$. Then rGen is invoked on $r_G$ to obtain $c$ which it forwards to the helper $\mathcal{H}$. Recall that $\mathcal{H}$ will bias the coin-toss outcome to $c$ and the resulting public-key agreed upon will be PK. Indistinguishability of $H_1$ and $H_0$ can be reduced directly to the indistinguishability of real and obliviously generated public-keys of the simulatable public-key encryption scheme using the CCA-security of $\langle I, R \rangle$.

**Hybrid $H_2$:** $H_2$ is the same as $H_1$ with the exception that the value committed to by the adversary is extracted using the simulator's strategy and forwarded to $\mathcal{F}_{\text{COM}}$. The only difference between the hybrids $H_1$ and $H_2$ is that in $H_2$ we extract a value for the commitment from the adversarial sender. This means that to argue indistinguishability it suffices to show that the value extracted is correct (i.e. the scheme is binding). We argue this by relying on the information-theoretic lemma proved in [HV15]. In more detail, this lemma shows that at the end of Stage 2, it is possible to define a set $\Gamma$ such that for any adversarial sender to equivocate it needs to bias the outcome of the coin toss in Stage 3 to result in this set $\Gamma$. This coin-toss is decided using our protocol $\langle I, R \rangle$ where the adversarial sender controls the initiator and by relying on CCA-security we argue next that there exists no adversary that can bias the outcome to result in a particular set with non-negligible probability. Suppose for contradiction there exists an adversary $\mathcal{A}$ that can bias the outcome to $\Gamma$ in $H_1$ with non-negligible probability. We now construct an adversary $\mathcal{A}'$ that incorporates $\mathcal{A}$ and internally emulates the hybrid experiment $H_2$ with the exception that it forwards the interaction of $\mathcal{A}$ in Stage 3 to an external honest receiver. Now, consider the challengers $\mathcal{C}$ for the CCA-security game where $\mathcal{C}|_{b=0}$ outputs 1 if the outcome $o$ results in $\Gamma$ and 0 otherwise. $\mathcal{C}|_{b=1}$ outputs 0 irrespective of the outcome. By our assumption on $\mathcal{A}$, this means that $\text{EXP}_0$ and $\text{EXP}_1$ with the adversary $\mathcal{A}'$ are distinguishable because the adversary biases the coin-toss to result in $\Gamma$ with non-negligible probability. However, since a uniformly sampled coin will result in $\Gamma$ with at most negligible probability we have that $\text{STA}_0$ and $\text{STA}_1$ are indistinguishable which is a contradiction. Therefore, we have that the value extracted by our simulator is correct except with non-negligible probability and this concludes the proof. Finally, we conclude by observing the $H_2$ is identical to the ideal world experiment.

# 7   Application: A Zero-One Law for Adaptive Security

We extend the result of [MPR10] and establish a zero-one law under adaptive UC-reduction. More formally, we show that all (non-reactive)[6] functionalities fall into two categories: *trivial* functionalities, those which can be UC-reduced to any other functionality; and *complete* functionalities, to which any other functionality can be UC-reduced.

**Theorem 7.1.** *Assume the existence of simulatable public-key encryption scheme. Then every two-party non-reactive functionality is either trivial or complete in the UC framework in the presence of adaptive, malicious adversaries.*

*Proof.* An important step in proving the zero-one law in [MPR10] was to identify all non-trivial functionalities into one of four categories (i.e. functionalities):

1. $\mathcal{F}_{\mathrm{XOR}}$: This functionality enables simultaneous exchange of information, such as the XOR function.
2. $\mathcal{F}_{\mathrm{CC}}$: This functionality enables to selectively hide one party's input from the other, typically characterized as a cut-and-choose functionality.
3. $\mathcal{F}_{\mathrm{OT}}$: This functionality enables OT of inputs from one party to another.
4. $\mathcal{F}_{\mathrm{COM}}$: This functionality allows information in internal memory to be hidden between rounds, an instance of which is the commitment functionality.

Specifically, it was shown in [MPR10] that every non-trivial functionality $\mathcal{F}$ can realize one of the above four functionalities with information-theoretic security. We are able to demonstrate the zero-one law by proving the following key lemma.

**Lemma 7.1 (Informal).** *Assume the existence of simulatable public-key encryption scheme. Then $\mathcal{F}_{\mathrm{COM}}$ can be realized in the $\mathcal{F}_{\mathrm{COIN}}$-hybrid model in the presence of adaptive, malicious adversaries, using black-box access to the encryption scheme.*

As mentioned before, in order to demonstrate the zero-one law it suffices to show that the four categories of non-trivial functionalities are complete, where it suffices to only consider $\mathcal{F}_{\mathrm{OT}}$, $\mathcal{F}_{\mathrm{XOR}}$ and $\mathcal{F}_{\mathrm{CC}}$ when considering non-reactive functionalities. Recalling that the previous results [IPS08, CDMW09] establish completeness of $\mathcal{F}_{\mathrm{OT}}$ and $\mathcal{F}_{\mathrm{COM}}$, where the latter result additionally requires the existence of stand-alone adaptively secure semi-honest oblivious-transfer protocol, it is thus left to show that the remaining two categories $\mathcal{F}_{\mathrm{CC}}$ and $\mathcal{F}_{\mathrm{XOR}}$ are complete. We note first that combining our lemma with the result of [CDMW09] establishes that $\mathcal{F}_{\mathrm{XOR}}$ is complete. We remark here that simulatable PKE schemes are sufficient to construct adaptive semi-honest OT which is required in the transformation of [CDMW09]. In order to show that $\mathcal{F}_{\mathrm{CC}}$ is complete, we recall that in [MPR10], $\mathcal{F}_{\mathrm{CC}}$ is reduced to another functionality called the $\mathcal{F}_{\mathrm{EXTCOM}}$-functionality for the static corruptions case. Roughly speaking this functionality is a mild variant of the $\mathcal{F}_{\mathrm{COM}}$ functionality that admits

---

[6] Such functionalities are computed in a single round of communication with the functionality.

straight-line extraction without straight-line equivocation. For more details, we refer the reader to the full version or [MPR10]. We argue that the same protocol also realizes $\mathcal{F}_{\mathrm{EXTCOM}}$ in the presence of adaptive corruptions. On a high-level, we are able to accomplish this since $\mathcal{F}_{\mathrm{EXTCOM}}$ does not require equivocation. To complete the picture, we show how to construct a variant of the $\mathcal{F}_{\mathrm{COIN}}$ functionality in the $\mathcal{F}_{\mathrm{EXTCOM}}$-hybrid and argue that this variant suffices to establish that $\mathcal{F}_{\mathrm{EXTCOM}}$ is complete even for the adaptive case. Our constructions make use of the underlying primitives only in a black-box manner.                                      □

# A   Adaptive Extension to CCA-Secure Commitments

In our work, we need to consider the CCA-Security game in the presence of an adaptive adversary $\mathcal{A}$. We recall the CCA-security game for the commitments as introduced in [CLP10]. Roughly speaking, a commitment scheme is CCA-secure if the commitment scheme retains its hiding property even if the receiver has access to a "decommitment oracle". The experiment considers an oracle adversary $\mathcal{A}$ with oracle access to a helper function $\mathcal{H}$ and interacts as the receiver with an honest committer $C$. In our adaptive setting, we will require two additional properties: (1) The adversary will be allowed to corrupt the external committer $C$. However, security is required to hold, i.e. hiding property of the left commitment, only if the committer is not corrupted, and (2) In the interaction between the adversary and the helper oracle, where it interacts as the committer, the adversary will be allowed to corrupt the receiver. In this case, the helper oracle is required to provide random coins for the receiver consistent with the transcript.

The second property does not require any explicit change in the definition of the security as it only alters the semantics of the interaction between $\mathcal{A}$ and $\mathcal{H}$. The first property however needs to be incorporated in the definition which we do next.

**Modifying the $\mathsf{IND}_b$ Random Variable in the Definition.** In the standard definition $\mathsf{IND}_b(\langle C, R \rangle, \mathcal{O}, \mathcal{A}, n, z)$ represents the output of the $\mathcal{A}^{\mathcal{O}}$ in a experiment where it interacts with an honest committer with input $b \in \{0,1\}^n$. This output is set to $\bot$, if the identity of the execution with $C$ is the same as the identity of any interaction of $\mathcal{A}$ with $\mathcal{O}$. We define a new random variable $\overline{\mathsf{IND}}_b(\langle C, R \rangle, \mathcal{O}, \mathcal{A}, n, z)$ which is equal to $\mathsf{IND}_b(\langle C, R \rangle, \mathcal{O}, \mathcal{A}, n, z)$ only if $\mathcal{A}^{\mathcal{O}}$ does not corrupt the honest committer $C$ in the execution. Otherwise it is set to $\bot$.

**Definition 5 (CCA-secure commitments with adaptive adversary).**
*Let $\langle C, R \rangle$ be a tag-based commitment scheme with $l(n)$-bit identities, and $\mathcal{O}$ a committed-value oracle for it. We say that $\langle C, R \rangle$ is CCA-secure w.r.t. $\mathcal{O}$ in the presence of an adaptive adversary, if for every PPT $\mathcal{A}$, the following ensembles are computationally indistinguishable: $\{\overline{\mathsf{IND}}_0(\langle C, R \rangle, \mathcal{O}, \mathcal{A}, n, z)\}_{n \in \mathbb{N}} \approx \{\overline{\mathsf{IND}}_1(\langle C, R \rangle, \mathcal{O}, \mathcal{A}, n, z)\}_{n \in \mathbb{N}}$ We say that $\langle C, R \rangle$ is CCA-secure if there exists a committed-value oracle $\mathcal{O}'$, such that, $\langle C, R \rangle$ is CCA-secure w.r.t. $\mathcal{O}'$.*

**Theorem A.1.** *Assume the existence of one-way functions. Then, for every $\epsilon > 0$, there exists a $O(n^\epsilon)$, there exists a $O(n^\epsilon)$-round commitment scheme that is CCA-secure w.r.t. the committed-value oracle* in the presence of an adaptive adversary *and only relies on black-box access to one-way functions (where $n$ is the security parameter).*

**Proof Sketch:** Lin and Pass [LP12a] gave a black-box construction of a $O(n^\epsilon)$-round CCA-secure commitment scheme $\langle C, R \rangle$. We rely on the same construction for our stronger definition of security in the presence of an adaptive adversary. We provide a high-level proof sketch of its correctness. We begin with a short overview of their proof.

In the proof of standard security of the scheme provided in [LP12a], the idea is to reduce the indistinguishability of the $\mathsf{IND}_b$ experiments to the stand-alone hiding property of a different commitment scheme $\langle \tilde{C}, \tilde{R} \rangle$ (that is a slight variant of $\langle C, R \rangle$). The main part of the proof is to show that given and oracle adversary for $\langle C, R \rangle$ there exists a stand-alone malicious receiver $R^*$ (that does not have access to the oracle) for $\langle \tilde{C}, \tilde{R} \rangle$. On a high-level, $R^*$ will internally incorporate $\mathcal{A}$ and emulate the committed-value oracle for $\mathcal{A}$ while forwarding the left interaction externally to $\tilde{C}$ (which it can do as it is a variant that has a "similar" structure). To emulate the oracle, $R^*$ needs to extract the value committed value which it will accomplish by rewinding the right interactions. Two issues arise:

- Since the left interaction is forwarded to an external committer, $R^*$ needs to be able to rewind the right interactions without rewinding the left. The main idea here that is reminiscent of previous work [DDN03,LPV08] is to identify the so-called *safe-points* where this can be done. In slight more detail, when rewinding from a safe-point the only thing the adversary can do in the left interaction is to request "complete" (3-round witness-indistinguishable) proofs and such a request will be accommodated by the variant $\langle \tilde{C}, \tilde{R} \rangle$.
- There are unbounded-many right interactions and will result in $R^*$ recursively rewinding interactions to extract the committed value in the interactions. In [LP12a], they achieve this by provided several points to rewind from and rely on the [RK99] to ensure that expected running time of the rewindings in each level is polynomial and the recursive depth is at most a constant.

Next, we argue why the same protocol satisfies our stronger definition of security. We begin with the observation that if the adversary $\mathcal{A}$ does not corrupt the left or right interactions, then our definition reduces to the standard CCA-security. We will prove security identically to [LP12a] by reducing it to the stand-alone hiding property of $\langle \tilde{C}, \tilde{R} \rangle$. We will employ the exact rewinding strategy as in [LP12a] for $R^*$ with the following exception: Our definition of *safe-point* will have one additional requirement: A *safe-point* for our scheme is any *safe-point* according to [LP12a] with the added requirement that the adversary corrupts neither the committer in the left-interaction or the receiver of the right interaction (associated with the safe-point) before the 3-round witness indistinguishable (WI) proof associated with the safe-point completes.[7]

We remark that our definition of *safe-point* can modularly replace the definition in [LP12a] and the entire proof goes through. This is because the definition affects only the run-time analysis of the reduction. For the run-time analysis to go through the only requirements are that there are sufficiently many *safe-point*'s and when rewound from a safe-point, it continues to be a *safe-point* with at least the same probability (See Step 1 in Sub-Claim 2 of [LP12b]). The first property holds because, a right receiver needs to be rewound only if $\mathcal{A}$ completes the entire right session without corrupting the right receiver or the left committer. In this event there will be as many safe points according to the definition of [LP12a] as there according to ours. The second property holds because a rewinding will be cancelled only if the point is not safe. This concludes the proof.

# References

[BCNP04] Barak, B., Canetti, R., Nielsen, J.B., Pass, R.: Focs, pp. 186–195 (2004)

[Bea91] Beaver, D.: Foundations of secure interactive computing. In: Feigenbaum, J. (ed.) CRYPTO 1991. LNCS, vol. 576, pp. 377–391. Springer, Heidelberg (1992)

[BS05] Barak, B., Sahai, A.: How to play almost any mental game over the net - concurrent composition via super-polynomial simulation. In: FOCS, pp. 543–552 (2005)

[Can01] Canetti, R.: Universally composable security: a new paradigm for cryptographic protocols. In: FOCS, pp. 136–145 (2001)

[CDMW09] Choi, S.G., Dachman-Soled, D., Malkin, T., Wee, H.: Simple, black-box constructions of adaptively secure protocols. In: Reingold, O. (ed.) TCC 2009. LNCS, vol. 5444, pp. 387–402. Springer, Heidelberg (2009)

[CDPW06] Canetti, R., Dodis, Y., Pass, R., Walfish, S.: Universally composable security with global setup. IACR Cryptology ePrint Archive, 2006:432 (2006)

[CF01] Canetti, R., Fischlin, M.: Universally composable commitments. In: Kilian, J. (ed.) CRYPTO 2001. LNCS, vol. 2139, pp. 19–40. Springer, Heidelberg (2001)

[CKL06] Canetti, R., Kushilevitz, E., Lindell, Y.: On the limitations of universally composable two-party computation without set-up assumptions. J. Cryptology 19(2), 135–167 (2006)

[CLP10] Canetti, R., Lin, H., Pass, R.: Adaptive hardness and composable security in the plain model from standard assumptions. In: FOCS, pp. 541–550 (2010)

[CLP13] Canetti, R., Lin, H., Pass, R.: From unprovability to environmentally friendly protocols. In: FOCS, pp. 70–79 (2013)

[CPS07] Canetti, R., Pass, R., Shelat, A.: Cryptography from sunspots: how to use an imperfect reference string. In: FOCS, pp. 249–259 (2007)

[DDN03] Dolev, D., Dwork, C., Naor, M.: Nonmalleable cryptography. SIAM Rev. 45(4), 727–784 (2003)

---

[7] In [LP12b] the definition of a safe-point is parameterized with the depth of the recursion and our additional requirement naturally extends to the definition.

[DMRV13] Dachman-Soled, D., Malkin, T., Raykova, M., Venkitasubramaniam, M.: Adaptive and concurrent secure computation from new adaptive, non-malleable commitments. In: Sako, K., Sarkar, P. (eds.) ASIACRYPT 2013, Part I. LNCS, vol. 8269, pp. 316–336. Springer, Heidelberg (2013)

[DN00] Damgård, I.B., Nielsen, J.B.: Improved non-committing encryption schemes based on a general complexity assumption. In: Bellare, M. (ed.) CRYPTO 2000. LNCS, vol. 1880, pp. 432–450. Springer, Heidelberg (2000)

[GLP+15] Goyal, V., Lin, H., Pandey, O., Pass, R., Sahai, A.: Round-efficient concurrently composable secure computation via a robust extraction lemma. In: Dodis, Y., Nielsen, J.B. (eds.) TCC 2015, Part I. LNCS, vol. 9014, pp. 260–289. Springer, Heidelberg (2015)

[GMW87] Goldreich, O., Micali, S., Wigderson, A.: How to play any mental game or a completeness theorem for protocols with honest majority. In: STOC, pp. 218–229 (1987)

[HV15] Hazay, C., Venkitasubramaniam, M.: On black-box complexity of universally composable security in the CRS model. In: Iwata, T., Cheon, J.H. (eds.) ASIACRYPT 2015. LNCS, vol. 9453, pp. 183–209. Springer, Heidelberg (2015). doi:10.1007/978-3-662-48800-3_8

[IPS08] Ishai, Y., Prabhakaran, M., Sahai, A.: Founding cryptography on oblivious transfer – efficiently. In: Wagner, D. (ed.) CRYPTO 2008. LNCS, vol. 5157, pp. 572–591. Springer, Heidelberg (2008)

[Kiy14] Kiyoshima, S.: Round-efficient black-box construction of composable multi-party computation. In: Garay, J.A., Gennaro, R. (eds.) CRYPTO 2014, Part II. LNCS, vol. 8617, pp. 351–368. Springer, Heidelberg (2014)

[KLP07] Kalai, Y.T., Lindell, Y., Prabhakaran, M.: Concurrent composition of secure protocols in the timing model. J. Cryptology 20(4), 431–492 (2007)

[KMO14] Kiyoshima, S., Manabe, Y., Okamoto, T.: Constant-round black-box construction of composable multi-party computation protocol. In: Lindell, Y. (ed.) TCC 2014. LNCS, vol. 8349, pp. 343–367. Springer, Heidelberg (2014)

[Lin03] Lindell, Y.: General composition and universal composability in secure multi-party computation. In: FOCS, pp. 394–403 (2003)

[LP12a] Lin, H., Pass, R.: Black-box constructions of composable protocols without set-up. In: Safavi-Naini, R., Canetti, R. (eds.) CRYPTO 2012. LNCS, vol. 7417, pp. 461–478. Springer, Heidelberg (2012)

[LP12b] Lin, H., Pass, R.: Black-box constructions of composable protocols without set-up (full version) (2012). https://www.cs.ucsb.edu/rachel.lin

[LPV08] Lin, H., Pass, R., Venkitasubramaniam, M.: Concurrent non-malleable commitments from any one-way function. In: Canetti, R. (ed.) TCC 2008. LNCS, vol. 4948, pp. 571–588. Springer, Heidelberg (2008)

[LPV09] Lin, H., Pass, R., Venkitasubramaniam, M.: A unified framework for concurrent security: universal composability from stand-alone nonmalleability. In: STOC, pp. 179–188 (2009)

[LZ11] Lindell, Y., Zarosim, H.: Adaptive zero-knowledge proofs and adaptively secure oblivious transfer. J. Cryptology 24(4), 761–799 (2011)

[MMY06] Malkin, T., Moriarty, R., Yakovenko, N.: Generalized environmental security from number theoretic assumptions. In: Halevi, S., Rabin, T. (eds.) TCC 2006. LNCS, vol. 3876, pp. 343–359. Springer, Heidelberg (2006)

[MPR10] Maji, H.K., Prabhakaran, M., Rosulek, M.: A zero-one law for cryptographic complexity with respect to computational UC security. In: Rabin, T. (ed.) CRYPTO 2010. LNCS, vol. 6223, pp. 595–612. Springer, Heidelberg (2010)

[MR91]  Micali, S., Rogaway, P.: Secure computation. In: Feigenbaum, J. (ed.) CRYPTO 1991. LNCS, vol. 576, pp. 392–404. Springer, Heidelberg (1992)

[Nao91]  Naor, M.: Bit commitment using pseudorandomness. J. Cryptology 4(2), 151–158 (1991)

[ORSV13]  Ostrovsky, R., Rao, V., Scafuro, A., Visconti, I.: Revisiting lower and upper bounds for selective decommitments. In: Sahai, A. (ed.) TCC 2013. LNCS, vol. 7785, pp. 559–578. Springer, Heidelberg (2013)

[Pas03]  Pass, R.: Simulation in quasi-polynomial time, and its application to protocol composition. In: Biham, E. (ed.) EUROCRYPT 2003. LNCS, vol. 2656, pp. 160–176. Springer, Heidelberg (2003). doi:10.1007/3-540-39200-9_10

[PR08]  Prabhakaran, M., Rosulek, M.: Cryptographic complexity of multi-party computation problems: classifications and separations. In: Wagner, D. (ed.) CRYPTO 2008. LNCS, vol. 5157, pp. 262–279. Springer, Heidelberg (2008)

[PS04]  Prabhakaran, M., Sahai, A.: New notions of security: achieving universal-composability without trusted setup. In: STOC, pp. 242–251 (2004)

[RK99]  Richardson, R., Kilian, J.: On the concurrent composition of zero-knowledge proofs. In: Stern, J. (ed.) EUROCRYPT 1999. LNCS, vol. 1592, p. 415. Springer, Heidelberg (1999)

[Ven14]  Venkitasubramaniam, M.: On adaptively secure protocols. In: Abdalla, M., De Prisco, R. (eds.) SCN 2014. LNCS, vol. 8642, pp. 455–475. Springer, Heidelberg (2014)

[Yao86]  Yao, A.C.-C.: How to generate and exchange secrets (extended abstract). In: FOCS, pp. 162–167 (1986)

# Adaptive Security of Yao's Garbled Circuits

Zahra Jafargholi$^{(\boxtimes)}$ and Daniel Wichs

Northeastern University, Boston, USA
{zahra,wichs}@ccs.neu.edu

**Abstract.** A garbling scheme is used to garble a circuit $C$ and an input $x$ in a way that reveals the output $C(x)$ but hides everything else. Yao's construction from the 80's is known to achieve *selective security*, where the adversary chooses the circuit $C$ and the input $x$ in one shot. It has remained as an open problem whether the construction also achieves adaptive security, where the adversary can choose the input $x$ after seeing the garbled version of the circuit $C$.

A recent work of Hemenway et al. (CRYPTO'16) modifies Yao's construction and shows that the resulting scheme is adaptively secure. This is done by encrypting the garbled circuit from Yao's construction with a special type of "somewhere equivocal encryption" and giving the key together with the garbled input. The efficiency of the scheme and the security loss of the reduction is captured by a certain pebbling game over the circuit.

In this work we prove that Yao's construction itself is already adaptively secure, where the security loss can be captured by the same pebbling game. For example, we show that for circuits of depth $d$, the security loss of our reduction is $2^{O(d)}$, meaning that Yao's construction is adaptively secure for NC1 circuits without requiring complexity leveraging. Our technique is inspired by the "nested hybrids" of Fuchsbauer et al. (Asiacrypt'14, CRYPTO'15) and relies on a careful sequence of hybrids where each hybrid involves some limited guessing about the adversary's adaptive choices. Although it doesn't match the parameters achieved by Hemenway et al. in their full generality, the main advantage of our work is to prove the security of Yao's construction as is, without any additional encryption layer.

# 1 Introduction

Garbled circuits, introduced by Yao in (oral presentations of) [Yao82,Yao86], can be used to garble a circuit $C$ and an input $x$ in a way that reveals $C(x)$ but hides everything else. Yao's construction is based on one-way functions and achieves a number of desirable properties with countless applications. One of the features of this construction is that a circuit $C$ can be garbled *off-line* in time

Research supported by NSF grants CNS-1347350, CNS-1314722, CNS-1413964. This work was done in part while the authors were visiting the Simons Institute for the Theory of Computing, supported by the Simons Foundation and by the DIMACS/Simons Collaboration in Cryptography through NSF grant CNS-1523467.

M. Hirt and A. Smith (Eds.): TCC 2016-B, Part I, LNCS 9985, pp. 433–458, 2016.
DOI: 10.1007/978-3-662-53641-4_17

proportional to $|C|$ which is presumably large, but an input $x$ can later be garbled very efficiently *on-line* in time only proportional to $|x|$ which is presumably much smaller. We consider the *on-line complexity* (i.e., time to garble the input $x$) as the main measure of efficiency.

*Selective vs. Adaptive Security.* Unfortunately, Yao's construction is only known to satisfy selective security where the adversary must choose the circuit $C$ and the input $x$ to be garbled in one shot. It has remained an open problem whether Yao's construction also achieves the stronger notion of adaptive security where the adversary can choose the input $x$ after seeing the garbled circuit. Adaptive security is especially important in the off-line/on-line setting where the adversary often sees the garbled circuit first and may be able to influence the choice of the input $x$.

*Prior Work on Adaptive Security.* The work of Bellare, Hoang and Rogaway [BHR12a] raised the question of whether Yao's construction or indeed any construction of garbled circuits achieves adaptive security. They showed a simple adaptively secure construction where the on-line complexity is proportional to the *circuit size*, but left it as an open problem to do better.

The work of Applebaum et al. [AIKW13] shows that the on-line complexity in the adaptive setting must at least exceed the *output size* of the circuit. This is in contrast to the selective setting, where Yao's garbling scheme achieves on-line complexity that depends only on the input size and not the output size. However, there is a small variant of Yao's scheme (by giving the mapping of output labels to output bits with the garbled input) which is natural in the adaptive setting and which raises the on-line complexity to also depend on the output size. We refer to this variant as Yao's construction when we consider the adaptive setting and it has remained as an open problem if this variant is adaptively secure.

Another approach to proving adaptive security of Yao's construction is to use *complexity leveraging* where we guess the adversary's choice of $x$ a-priori. A direct approach results in a security loss of $2^n$ where $n$ is the input size to the circuit. In particular, if we insist on polynomial security loss then this approach can only handle circuits with a logarithmic input size.

We mention that there are also other approaches that depart from Yao's construction and/or rely on significantly heavier assumptions than one-way functions. For example [BHR12a] show how to get an optimal solution (in fact one that bypasses the lower-bound of [AIKW13]) in the random oracle model. The work of [BHK13] shows that this solution also works in the standard-model based on non-standard hash-function assumption referred to as UCE. Boneh et al. [BGG+14] implicitly provides an adaptive garbling scheme with low on-line complexity that scales with the depth of the circuit under LWE, while the work of Ananth and Sahai [AS15] shows how to get an essentially optimal schemes assuming indistinguishability obfuscation.

*Work of Hemenway et al. (CRYPTO'16).* The most relevant prior work is a recent result of Hemenway et al. [HJO+15]. This work modifies Yao's construction by encrypting the garbled circuit with a special type of "somewhere equivocal encryption" and giving the key together with the garbled input. The encryption scheme has an "equivocation parameter" which determines its key size and therefore affects the on-line complexity of the garbling. They show that the resulting scheme is adaptively secure where the equivocation parameter needed and the security loss of the reduction are captured by a certain pebbling game over the circuit. In particular, if a circuit with input size $n$ and output size $m$ can be pebbled with $t$ pebbles in $\gamma$ steps then the resulting scheme can be instantiated so as to achieve on-line complexity $O(n + m + t)$ and security loss $\gamma$. Furthermore they show that any circuit of size $q$, width $w$, and depth $d$ can either be pebbled with $t = O(w)$ pebbles in $\gamma = O(q)$ steps or with $t = O(d)$ pebbles in $\gamma = q \cdot 2^{O(d)}$ steps. In particular, this means that (without complexity leveraging):

- For any circuit of width $w$, the on-line complexity can be made $O(w)$.
- For NC1 circuits, the on-line complexity can be just $O(n + m)$.

*Our Results.* In this work we revisit the question of whether Yao's construction itself (without modification) is adaptively secure. We give a new reduction which connects the security of Yao's construction with the same pebbling game as studied by Hemenway et al. [HJO+15]. In particular, we show that for circuits that can be pebbled with $t$ pebbles in $\gamma$ steps, Yao's construction is adaptively secure with a security loss of $\gamma 2^{O(t)}$. For example, since circuits of size $q$ and depth $d$ can be pebbled in $\gamma = q2^{O(d)}$ steps with $t = O(d)$ pebbles we get a security loss of $q2^{O(d)}$. This means that Yao's construction is already adaptively secure for NC1 circuits, without the use of complexity leveraging.[1]

Next we describe our techniques and compare to those of [HJO+15]. On a very high level, the work of [HJO+15] proves security via a sequence of hybrids, where in each hybrid some small number of garbled gates of the Yao garbled circuit are "equivocal" and only needed to be specified by the reduction in the on-line phase after the input $x$ is known. In this work we replace the role of "equivocation" with the careful use of "guessing". Instead of simply guessing the entire input $x$, our reduction consists of a sequence of hybrids where in each hybrid we guess some small number of the wire values in the circuit and abort if the guess is incorrect. We then show how to patch together hybrids that contain different guessed wires (and even a different number of guessed wires) to get a security proof. This approach is reminiscent of the "nested hybrids" technique employed by Fuchsbauer et al. [FKPR14,FJP15] and we believe our abstraction of this technique via pebbling will be useful in other contexts.

## 1.1  Our Techniques

**Yao's Scheme and the Challenge of Adaptive Security** ([HJO+15]). To describe our technical contribution, we must first describe Yao's construction

---

[1] Unfortunately, we cannot get a meaningful analogue of the width based result of [HJO+15] since the security loss would be $2^w$ which exceeds the trivial security loss of $2^n$ obtained by simply guessing the input.

and the difficulty one faces when trying to prove adaptive security. The following discussion is taken essentially verbatim from [HJO+15], following the ideas of Lindell and Pinkas [LP09] who gave the first detailed proof of security for Yao's garbled circuits in the selective security setting.

*Yao's Scheme.* For each wire $w$ in the circuit, we pick two keys $k_w^0, k_w^1$ for a symmetric-key encryption scheme. For each gate in the circuit computing a function $g : \{0, 1\}^2 \to \{0, 1\}$ and having input wires $a, b$ and output wire $c$ we create a *garbled gate* consisting of 4 randomly ordered ciphertexts created as:

$$c_{0,0} = \mathsf{Enc}_{k_a^0}(\mathsf{Enc}_{k_b^0}(k_c^{g(0,0)})) \quad c_{1,0} = \mathsf{Enc}_{k_a^1}(\mathsf{Enc}_{k_b^0}(k_c^{g(1,0)})),$$
$$c_{0,1} = \mathsf{Enc}_{k_a^0}(\mathsf{Enc}_{k_b^1}(k_c^{g(0,1)})) \quad c_{1,1} = \mathsf{Enc}_{k_a^1}(\mathsf{Enc}_{k_b^1}(k_c^{g(1,1)})) \tag{1}$$

where $(\mathsf{Enc}, \mathsf{Dec})$ is a CPA-secure encryption scheme. The garbled circuit $\widetilde{C}$ consists of all of the gabled gates, along with an *output mapping* $\{k_w^0 \to 0, k_w^1 \to 1\}$ which gives the correspondence between the keys and the bits they represent for each output wire $w$. To garble an $n$-bit value $x = x_1 x_2 \cdots x_n$, the garbled input $\widetilde{x}$ consists of the keys $k_{w_i}^{x_i}$ for the $n$ input wires $w_i$.

To evaluate the garbled circuit on the garbled input, it's possible to decrypt (exactly) one ciphertext in each garbled gate and get the key $k_w^{v(w)}$ corresponding to the bit $v(w)$ going over the wire $w$ during the computation $C(x)$. Once the keys for the output wires are computed, it's possible to recover the actual output bits by looking them up in the output mapping.

*Selective Security Simulator.* To prove the selective security of Yao's scheme, we need to define a simulator that gets the output $y = y_1 y_2 \cdots y_m = C(x)$ and must produce $\widetilde{C}, \widetilde{x}$. The simulator picks random keys $k_w^0, k_w^1$ for each wire $w$ just like the real scheme, but it creates the garbled gates as follows:

$$c_{0,0} = \mathsf{Enc}_{k_a^0}(\mathsf{Enc}_{k_b^0}(k_c^0)) \quad c_{1,0} = \mathsf{Enc}_{k_a^1}(\mathsf{Enc}_{k_b^0}(k_c^0)),$$
$$c_{0,1} = \mathsf{Enc}_{k_a^0}(\mathsf{Enc}_{k_b^1}(k_c^0)) \quad c_{1,1} = \mathsf{Enc}_{k_a^1}(\mathsf{Enc}_{k_b^1}(k_c^0)) \tag{2}$$

where all four ciphertext encrypt the same key $k_c^0$. It creates the output mapping $\{k_w^0 \to y_w, k_w^1 \to 1 - y_w\}$ by "programming it" so that the key $k_w^0$ corresponds to the correct output bit $y_w$ for each output wire $w$. This defines the simulated garbled circuit $\widetilde{C}$. To create the simulated garbled input $\widetilde{x}$ the simulator simply gives out the keys $k_w^0$ for each input wire $w$. Note that, when evaluating the simulated garbled circuit on the simulated garbled input, the adversary only sees the keys $k_w^0$ for every wire $w$.

*Selective Security Hybrids.* To prove indistinguishability between the real world and the simulation, there is a series of carefully defined hybrid games that switch the distribution of one garbled gate at a time. Unfortunately, we cannot directly switch a gate from the real distribution (1) to the simulated one (2) and therefore must introduce an intermediate distribution (3) as below:

$$c_{0,0} = \mathsf{Enc}_{k_a^0}(\mathsf{Enc}_{k_b^0}(k_c^{v(c)})) \quad c_{1,0} = \mathsf{Enc}_{k_a^1}(\mathsf{Enc}_{k_b^0}(k_c^{v(c)})),$$
$$c_{0,1} = \mathsf{Enc}_{k_a^0}(\mathsf{Enc}_{k_b^1}(k_c^{v(c)})) \quad c_{1,1} = \mathsf{Enc}_{k_a^1}(\mathsf{Enc}_{k_b^1}(k_c^{v(c)})) \tag{3}$$

where $v(c)$ is the correct value of the bit going over the wire $c$ during the computation of $C(x)$.

Let us give names to the three modes for creating garbled gates that we defined above: (1) is called RealGate mode, (2) is called SimGate mode, and (3) is called InputDepSimGate mode, since the way that it is defined depends adaptively on the choice of the input $x$.

We can switch a gate from RealGate to InputDepSimGate mode if the *predecessor* gates are in InputDepSimGate mode (or we are in the input level). This follows by CPA security of encryption. In particular, we are *not* changing the value contained in ciphertext $c_{v(a),v(b)}$ encrypted under the keys $k_a^{v(a)}, k_b^{v(b)}$ that the adversary obtains during evaluation, but we *can* change the values contained in all of the other ciphertexts since the keys $k^{1-v(a)}, k^{1-v(b)}$ do not appear anywhere inside the predecessor garbled gates as long as they are already in InputDepSimGate mode.

We can also switch a gate from InputDepSimGate to SimGate mode if the *successor* gates are in InputDepSimGate or SimGate mode (or we are at the output level). This is actually an information theoretic step; since the keys $k_c^0, k_c^1$ are used completely symmetrically in the successor gates there is no difference between always encrypting $k_c^{v(c)}$ as in InputDepSimGate mode or encrypting $k_c^0$ as in SimGate. This allows us to first switch every gate from RealGate to InputDepSimGate mode and then from InputDepSimGate to SimGate, proving the selective security of Yao's construction.

*Challenges in Achieving Adaptive Security.* There are two issues in using the above strategy in the adaptive setting: an immediate but easy to fix problem and a more subtle but difficult to overcome problem.

The first immediate issue is that the selective simulator needs to know the output $y = C(x)$ to create the garbled circuit $\widetilde{C}$ and in particular to program the output mapping $\{k_w^0 \rightarrow y_w, k_w^1 \rightarrow 1 - y_w\}$ for the output wires $w$. However, the adaptive simulator does not get the output $y$ until *after* it creates the garbled circuit $\widetilde{C}$. Therefore, we cannot (even syntactically) use the selective security simulator in the adaptive setting. This issue turns out to be easy to fix by modifying the construction to send the output-mapping as part of the garbled input $\widetilde{x}$ in the on-line phase, rather than as part of the garbled circuit $\widetilde{C}$ in the off-line phase. This modification raises on-line complexity to also being linear in the output size of the circuit, which we know to be necessary by the lower bound of [AIKW13]. We refer to this modification as Yao's garbled circuit construction in the adaptive setting. With this modification, the adaptive simulator can program the output mapping after it learns the output $y = C(x)$ in the on-line phase and therefore we get a syntactically meaningful simulation strategy in the adaptive setting.

The second problem is where the true difficulty lies. Although we have a syntactically meaningful simulation strategy, the previous proof of indistinguishability of the real world and the simulation completely breaks down in the adaptive setting. In particular InputDepSimGate mode as specified in Eq. (3) is syntactically undefined in the adaptive setting. Recall that in this mode the garbled gate is created in a way that depends on the input $x$, but in the adaptive setting the input $x$ is chosen adaptively after the garbled circuit is created! Therefore, *although we have a syntactically meaningful simulation strategy for the adaptive setting, we do not have any syntactically meaningful sequence of intermediate hybrids to prove indistinguishability between the real world and the simulated world.*

**Our Solution.** As described above, in the selective setting there is a proof of security via a sequence of hybrids that changes the distribution of gates from RealGate mode to InputDepSimGate mode to SimGate mode. Unfortunately, InputDepSimGate mode does not make sense (even syntactically) in the adaptive setting since it relies on knowing the value on the outgoing wire of that gate, which isn't defined until the input $x$ is given.

To overcome this problem, the work of [HJO+15] encrypted the entire garbled circuit with a somewhere equivocal encryption scheme which allowed the simulator to put dummy values in place of all of the gates in InputDepSimGate mode and only later after the input $x$ was known replace the dummy values with correctly distributed garbled gates by equivocating the encryption.

*Our Idea: Guess and Hope for the Best.* Our idea to overcome this problem is very different. We define hybrid games in the adaptive setting where we guess the value $v(c)$ on the outgoing wire $c$ of every gate in InputDepSimGate mode a-priori and abort if the adversary's adaptive choice of the input $x$ doesn't match our guesses. Note that although the goal is to have the garbled gates in InputDepSimGate mode depend on the input $x$, we choose them independently of $x$ and only abort later if we chose incorrectly. This defines syntactically meaningful hybrid games, but unfortunately the set of guessed wires and even the number of guessed wires is different in each hybrid making it impossible to compare them directly. However, we show that by carefully adding and removing guesses in different parts of the proof and then only comparing hybrids with an equivalent set of guesses, we can patch together this sequence of a-priori incomparable hybrids and give an indistinguishability reduction. Overall, we can take any valid sequence of $\gamma$ hybrid games that would give an indistinguishability proof in the selective setting and translate it into a proof of security in the adaptive setting with a security loss of $\gamma 2^{O(t)}$ where $t$ is the maximum number of gates in InputDepSimGate mode in any hybrid. This idea of "carefully" guessing different components in different hybrids is reminiscent of the *nested hybrids* technique of Fuchsbauer et al. [FKPR14,FJP15].

In comparison to [HJO+15], we rely on "guessing" instead of "equivocating". Whereas [HJO+15] had to modify Yao's scheme and pay for gates in InputDepSimGate mode by increasing the "equivocation parameter" which resulted in larger key size for the somewhere equivocal encryption, we get to keep the scheme unmodified but pay for gates in InputDepSimGate mode in the security loss of our reduction.

*Sequences of Hybrids and Pebbling.* With the above framework, the goal of proving adaptive security reduces to the goal of giving a sequence of hybrids in the selective setting where the number of gates in InputDepSimGate mode in any hybrid is as small as possible. This is the same challenge as faced in the work of [HJO+15] and we can rely on the same idea.

Recall that we need to start with the real world where all gates are in RealGate mode and end with the simulated world where all gates are in SimGate mode. As discussed in the overview of the selective security proof of Yao's garbled circuits, we are allowed to change a gate from RealGate to InputDepSimGate if all of its predecessors are in InputDepSimGate (or it's an input gate) and we are allowed to change InputDepSimGate to SimGate if all of the successors are in InputDepSimGate or SimGate modes (or it's an output gate). A naive sequence of hybrids, corresponding to the proof of selective security of Lindell and Pinkas [LP09], would first change all the gates from RealGate mode to InputDepSimGate mode one level at a time starting from the input level, and then change them all to SimGate mode by again changing one level at a time starting from the input level. However, this requires that there is a hybrid step where all of the gates are in InputDepSimGate mode, while our goal is to minimize the number of such gates. It turns out that one can do much better.

The work of [HJO+15] abstracts the above problem as a pebbling game. We associate RealGate mode with not having a pebble, InputDepSimGate mode with having a *black pebble* and SimGate mode with having a *gray pebble*. The rules of the game go as follows:

- We can place or remove a black pebble on a gate as long as both predecessors of that gate have black pebbles on them (or the gate is an input gate).
- We can replace a black pebble with a gray pebble on a gate as long as all successors of that gate have black or gray pebbles on them (or the gate is an output gate).

The goal of the game is to end up with a gray pebble on every gate while using as few black pebbles as possible at any point in time. It was shown that any circuit of size $q$, width $w$ and depth $d$ can be pebbled in two different ways: either with $t = O(w)$ black pebbles in $\gamma = O(q)$ steps or with $t = O(d)$ black pebbles in $\gamma = q \cdot 2^{O(d)}$ steps.

*Our Parameters.* Using the second pebbling strategy based on depth, we get a security proof of Yao's garbled circuits in the adaptive setting with a security loss of $q2^{O(d)}$ where $q$ is the circuit size and $d$ is the circuit depth. In particular, for NC1 circuits we get a security reduction showing the adaptive security of Yao's garbled circuits without complexity leveraging.

## 2   Preliminaries

*General Notation.*   For a positive integer $n$, we define $[n] := \{1, \ldots, n\}$. We use the notation $x \leftarrow X$ for the process of sampling a value $x$ according to the distribution $X$. We use $U_n$ for uniform distribution over $n$-bit strings. A function $\mu(\cdot)$ is negligible in $x$ if $\mu(x) \leq 1/p(x)$ for any polynomial function $p$ and all sufficiently large $x$. We use $\mathsf{poly}(x)$ to denote the set of all polynomial functions $p(x)$. For an interactive game GAME with an adversary $\mathcal{A}$, we use GAME$_\mathcal{A}$ to denote the outcome of the game played with $\mathcal{A}$.

**Definition 1.** *Two distributions $X$ and $Y$ are $(T, \varepsilon)$-indistinguishable, denote $\mathbf{D}_T[X, Y] = \varepsilon$ if for any probabilistic algorithm $\mathcal{A}$, running in time $T$,*

$$|\Pr[\mathcal{A}(X) = 1] - \Pr[\mathcal{A}(Y) = 1]| \leq \varepsilon.$$

*For two games GAME and GAME$'$ we say they are $(T(\lambda), \varepsilon(\lambda))$-indistinguishable, $\mathbf{D}_{T(\lambda)}[\text{GAME}, \text{GAME}'] = \varepsilon(\lambda)$, if for any adversary $\mathcal{A}$ running in time $T(\lambda)$,*

$$\left|\Pr[\text{GAME}_\mathcal{A} = 1] - \Pr\left[\text{GAME}'_\mathcal{A} = 1\right]\right| \leq \varepsilon(\lambda).$$

*Let games GAME$(\lambda)$ and GAME$'(\lambda)$ be games parameterized by the security parameter $\lambda$. If for any polynomial function $T(\lambda)$, there exists a negligible function $\varepsilon(\lambda)$, such that for all $\lambda$, $\mathbf{D}_{T(\lambda)}[\text{GAME}(\lambda), \text{GAME}'(\lambda)] \leq \varepsilon(\lambda)$, we say the two games are computationally indistinguishable and denote this by GAME$(\lambda) \overset{comp}{\approx}$ GAME$'(\lambda)$.*

*Circuit Notation.*   A boolean circuit $C$ consists of gates $\mathsf{gate}_1, \ldots, \mathsf{gate}_q$ and wires $w_1, w_2, \ldots, w_p$. A gate is defined by the tuple $\mathsf{gate}_i = (g, w_a, w_b, w_c)$, where $g : \{0, 1\}^2 \to \{0, 1\}$ is the function computed by the gate, $w_a, w_b$ are the incoming wires, and $w_c$ is the outgoing wire. Although each gate has a unique outgoing wire $w_c$, this wire can be used as an incoming wire to several different gates and therefore this models a circuit with fan-in 2 and unbounded fan-out. We also allow $w_a = w_b$, for gates with fan-in 1. We denote the number of gates with $q$, input wires with $m$ and output wires with $m$. The total number of wires is $p = n + q$ (since each wire can either be input wire or an outgoing wire of some gate). For convenience, we denote the $n$ input wires by $\mathsf{in}_1, \ldots, \mathsf{in}_n$ and the $m$ output wires by $\mathsf{out}_1, \ldots, \mathsf{out}_m$. We also use reserve $a, b$ and $c$ as labels for input wires to a gate and output wire of the same gate (instead of $w_a, w_b$, and $w_c$). For $x \in \{0, 1\}^n$ we write $C(x)$ to denote the output of evaluating the circuit $C$ on input $x$.

We say $C$ is leveled, if each gate has an associated level and any gate at level $l$ has incoming wires only from gates at level $l - 1$ and outgoing wires only to gates at level $l + 1$. We let the *depth* $d$ denote the number of levels and the *width* $w$ denote the maximum number of gates in any level.

A circuit $C$ is fully specified by a list of gate tuples $\mathsf{gate}_i = (g, a, b, c)$. We use $\Phi_{\mathsf{topo}}(C)$ to refer to the topology of a circuit– which indicates how gates are connected, without specifying the function implement by each gate. In other words, $\Phi_{\text{topo}}(C)$ is the list of *sanitized gate tuples* $\widehat{\text{gate}}_i = (\perp, a, b, c)$ where the function $g$ that the gate implements is removed from the tuple.

# 3  Garbling Scheme and Adaptive Security ([HJO+15])

The bulk of this section defining what garbled circuits are and presenting Yao's construction is taken verbatim from [HJO+15].

## 3.1  Garbling Scheme

We now give a formal definition of a garbling scheme. There are many variants of such definitions in the literature, we use the definition given in [HJO+15] and refer the reader to [BHR12b] for a comprehensive treatment.

**Definition 2.** *A Garbling Scheme is a tuple of PPT algorithms* GC = (GCircuit, GInput, Eval) *such that:*

- $(\widetilde{C}, k) \xleftarrow{\$} \text{GCircuit}(1^\lambda, C)$: *takes as input a security parameter* $\lambda$, *a circuit* $C: \{0,1\}^n \to \{0,1\}^m$, *and outputs the garbled circuit* $\widetilde{C}$, *and key* $k$.
- $\tilde{x} \leftarrow \text{GInput}(k, x)$: *takes as input,s* $x \in \{0,1\}^n$, *and key* $k$ *and outputs* $\tilde{x}$.
- $y = \text{Eval}(\widetilde{C}, \tilde{x})$: *given a garbled circuit* $\widetilde{C}$ *and a garbled input* $\tilde{x}$ *output* $y \in \{0,1\}^m$.

**Correctness.** *There is a negligible function* $\varepsilon$ *such that for any* $\lambda \in \mathbb{N}$, *any circuit* $C$ *and input* $x$ *it holds that* $\Pr[C(x) = \text{Eval}(\widetilde{C}, \tilde{x})] = 1 - \varepsilon(\lambda)$, *where* $(\widetilde{C}, k) \leftarrow \text{GCircuit}(1^\lambda, C)$, $\tilde{x} \leftarrow \text{GInput}(k, x)$.

*Adaptive Security.*

- GC *is* $(T(\lambda), \varepsilon(\lambda))$-*adaptively secure garbling scheme, if there exists a probabilistic polynomial time simulator* Sim = (SimC, SimIn) *such that, for any probabilistic adversary* $\mathcal{A}$, *running in time* $T(\lambda)$,

$$\left| \Pr[\text{Exp}_{\mathcal{A}, \text{GC}, \text{Sim}}^{\text{adaptive}}(1^\lambda, 0) = 1] - \Pr[\text{Exp}_{\mathcal{A}, \text{GC}, \text{Sim}}^{\text{adaptive}}(1^\lambda, 1) = 1] \right| \leq \varepsilon(\lambda).$$

*In other words,* $\mathbf{D}_{T(\lambda)} \left[ \text{Exp}_{\text{GC}, \text{Sim}}^{\text{adaptive}}(1^\lambda, 0), \text{Exp}_{\text{GC}, \text{Sim}}^{\text{adaptive}}(1^\lambda, 1) \right] = \varepsilon(\lambda).$

- GC *is adaptively secure if* $\text{Exp}_{\text{GC}, \text{Sim}}^{\text{adaptive}}(1^\lambda, 0) \stackrel{\text{comp}}{\approx} \text{Exp}_{\text{GC}, \text{Sim}}^{\text{adaptive}}(1^\lambda, 1)$

*where the experiment* $\text{Exp}_{\mathcal{A}, \text{GC}, \text{Sim}}^{\text{adaptive}}(1^\lambda, b)$ *is defined as follows:*

1. *The adversary* $\mathcal{A}$ *specifies* $C$ *and gets* $\widetilde{C}$ *where* $\widetilde{C}$ *is created as follows:*
   - *if* $b = 0$: $(\widetilde{C}, k) \leftarrow \text{GCircuit}(1^\lambda, C)$,
   - *if* $b = 1$: $(\widetilde{C}, ,) \leftarrow \text{SimC}(1^\lambda, \Phi_{\text{topo}}(C))$, *where* $\Phi_{\text{topo}}(C)$ *reveals the topology of* $C$.
2. *The adversary* $\mathcal{A}$ *specifies* $x$ *and gets* $\tilde{x}$ *created as follows:*
   - *if* $b = 0$, $\tilde{x} \leftarrow \text{GInput}(k, x)$,
   - *if* $b = 1$, $\tilde{x} \leftarrow \text{SimIn}(C(x), \text{state})$.
3. *Finally, the adversary outputs a bit* $b'$, *which is the output of the experiment.*

*On-line Complexity.* The time it takes to garble an input $x$, (i.e., time complexity of $\mathsf{GInput}(\cdot, \cdot)$) is the *on-line complexity* of the scheme. Clearly the on-line complexity of the scheme gives a bound on the size of the garbled input $\widetilde{x}$. Ideally, the on-line complexity should be much smaller than the circuit size $|C|$.

*Projective Scheme.* We say a garbling scheme is *projective* if each bit of the garbled input $\widetilde{x}$ only depends on one bit of the actual input $x$. In other words, each bit of the input, is garbled independently of other bits of the input. Projective schemes are essential for two-party computation where the garbled input is transmitted using an oblivious transfer (OT) protocol. Our constructions will be projective.

*Hiding Topology.* A garbling scheme that satisfies the above security definition may reveal the topology of the circuit $C$. However, there is a way to transform any such garbling scheme into one that hides everything, including the topology of the circuit, without a significant asymptotic efficiency loss. More precisely, we rely on the fact that there is a function $\mathsf{HideTopo}(\cdot)$ that takes a circuit $C$ as input and outputs a functionally equivalent circuit $C'$, such that for any two circuits $C_1, C_2$ of equal size, if $C_1' = \mathsf{HideTopo}(C_1)$ and $C_2' = \mathsf{HideTopo}(C_2)$, then $\Phi_{\mathsf{topo}}(C_1') = \Phi_{\mathsf{topo}}(C_2')$. An easy way to construct such function $\mathsf{HideTopo}$ is by setting $C'$ to be a universal circuit, with a hard-coded description of the actual circuit $C$. Therefore, to get a topology-hiding garbling scheme, we can simply use a topology-revealing scheme but instead of garbling the circuit $C$ directly, we garble the circuit $\mathsf{HideTopo}(C)$.

## 3.2  Yao's Garbling Scheme

In this section we describe Yao's garbling scheme and in the next section we give the simulation strategy.

*Construction.* Let $C$ be a leveled boolean circuit with fan-in 2 and unbounded fan-out, with inputs size $n$, output size $m$, depth $d$. Let $q$ denote the number of gates in $C$. Recall that wires are uniquely identified with labels and a circuit $C$ is specified by a list of gate tuples $\mathsf{gate} = (g, a, b, c)$, where $g$ computes the gate and $a, b$ are the input wire labels and $c$ is the output wire label. The topology of the circuit $\Phi_{\mathsf{topo}}(C)$ consists of the sanitized gate tuples $\widehat{\mathsf{gate}} = (\bot, a, b, c)$. For simplicity, we implicitly assume that $\Phi_{\mathsf{topo}}(C)$ is public and known to the circuit evaluator without explicitly including it as part of the garbled circuit $\widetilde{C}$. To simplify the description of our construction, we first describe the procedure for garbling a single gate, that we denote by $\mathsf{GarbleGate}$.

Let $\Gamma = (\mathsf{KeyGen}, \mathsf{Enc}, \mathsf{Dec})$ be a CPA-secure symmetric-key encryption scheme that satisfies the special correctness property, enabling one to recognize successful decryptions (defined in Appendix A.) $\mathsf{GarbleGate}$ is defined as follows.

| GCircuit($1^\lambda, C$) | Eval($\widetilde{C}, \widetilde{x}$) |
|---|---|
| (Wires) for $i \in [p]$, $\sigma \in \{0,1\}$ <br> $k_i^\sigma \leftarrow$ KeyGen($1^\lambda$). <br> (Input wires) $K = (k_{\mathsf{in}_i}^0, k_{\mathsf{in}_i}^1)_{i \in [n]}$. <br> (Gates) For each gate$_i = (g, a, b, c)$ in $C$: <br> $\quad \widetilde{g}_i \leftarrow$ GarbleGate($g, \{k_a^\sigma, k_b^\sigma, k_c^\sigma\}_{\sigma \in \{0,1\}}$). <br> (Output tables) For each output $\ell \in [m]$: <br> $\quad \widetilde{d}_c := [(k_{\mathsf{out}_\ell}^0 \to 0), (k_{\mathsf{out}_\ell}^1 \to 1)]$. <br> (Garbled Circuit) $\widetilde{C} := (\widetilde{g}_1, \ldots, \widetilde{g}_q)$. <br> **Output** $\widetilde{C}$, $k = (K, (\widetilde{d}_\ell)_{\ell \in [m]})$. <br> GInput($x, k$) <br><br> (Select input keys) $K^x = (k_{\mathsf{in}_1}^{x_1}, \ldots, k_{\mathsf{in}_n}^{x_n})$. <br> **Output** $\widetilde{x} = (K^x, (\widetilde{d}_\ell)_{\ell \in [m]})$. | Parse $\widetilde{x} = (K, (\widetilde{d}_\ell)_{\ell \in [m]})$. <br> Evaluate Circuit. <br> $\quad$ Parse $K = (k_{\mathsf{in}_1}, \ldots, k_{\mathsf{in}_n})$. <br> $\quad$ For each level $j = 1, \ldots, d$, <br> $\quad\quad$ For each $\widehat{\mathsf{gate}}_i = (\bot, a, b, c)$ at level $j$: <br> $\quad\quad$ Let $\widetilde{g}_i = [\mathsf{ct}_1, \mathsf{ct}_2, \mathsf{ct}_3, \mathsf{ct}_4]$ and let <br> $\quad\quad$ $k_c' \leftarrow$ Dec$_{k_a}$(Dec$_{k_b}$($\mathsf{ct}_\delta$)) For $\delta \in [4]$ <br> $\quad\quad$ If $k_c' \neq \bot$ then set $k_c := k_c'$. <br> Decoding output. <br> For $\ell \in [m]$: <br> $\quad$ Parse $\widetilde{d}_\ell = [(k_{\mathsf{out}_\ell}^0 \to 0), (k_{\mathsf{out}_\ell}^1 \to 1)]$. <br> $\quad$ Set $y_\ell = b$ iff $k_{\mathsf{out}_\ell} = k_{\mathsf{out}_\ell}^b$. <br> **Output** $y_1, \ldots, y_m$. |

**Fig. 1.** Yao's garbling scheme.

- $\widetilde{g} \leftarrow$ GarbleGate($g, \{k_a^\sigma, k_b^\sigma, k_c^\sigma\}_{\sigma \in \{0,1\}}$): This function computes 4 ciphertexts $\mathsf{ct}_{\sigma_0, \sigma_1}$ : $\sigma_0, \sigma_1 \in \{0,1\}$ as defined below and outputs them in a random order as $\widetilde{g} = [\mathsf{ct}_1, \mathsf{ct}_2, \mathsf{ct}_3, \mathsf{ct}_4]$.

$$\mathsf{ct}_{0,0} \leftarrow \mathsf{Enc}_{k_a^0}(\mathsf{Enc}_{k_b^0}(k_c^{g(0,0)})), \; \mathsf{ct}_{0,1} \leftarrow \mathsf{Enc}_{k_a^0}(\mathsf{Enc}_{k_b^1}(k_c^{g(0,1)}))$$
$$\mathsf{ct}_{1,0} \leftarrow \mathsf{Enc}_{k_a^1}(\mathsf{Enc}_{k_b^0}(k_c^{g(1,0)})), \; \mathsf{ct}_{1,1} \leftarrow \mathsf{Enc}_{k_a^1}(\mathsf{Enc}_{k_b^0}(k_c^{g(1,1)}))$$

### 3.3 Adaptive Simulator

The adaptive security simulator for our garbling scheme is essentially the same as the selective security simulator for Yao's scheme (as in [LP09]), with the only difference that the output table is sent in the on-line phase, and is computed adaptively to map to the correct output.

More specifically, the adaptive simulator (SimC, SimIn) works as follows. In the off-line phase, SimC computes the garbled gates using procedure GarbleSimGate, that generates 4 ciphertexts that encrypt the same output key. More precisely,

- GarbleSimGate($\{k_a^\sigma, k_b^\sigma\}_{\sigma \in \{0,1\}}, k_c'$) takes both keys for input wires $w_a, w_b$ and a single key for the output wire $w_c$, that we denote by $k_c'$. It then output $\widetilde{g}_c = [\mathsf{ct}_1, \mathsf{ct}_2, \mathsf{ct}_3, \mathsf{ct}_4]$ where the ciphertexts, arranged in random order, are computed as follows.

$$\mathsf{ct}_{0,0} \leftarrow \mathsf{Enc}_{k_a^0}(\mathsf{Enc}_{k_b^0}(k_c')) \; \mathsf{ct}_{1,0} \leftarrow \mathsf{Enc}_{k_a^1}(\mathsf{Enc}_{k_b^0}(k_c'))$$
$$\mathsf{ct}_{0,1} \leftarrow \mathsf{Enc}_{k_a^0}(\mathsf{Enc}_{k_b^1}(k_c')) \; \mathsf{ct}_{1,1} \leftarrow \mathsf{Enc}_{k_a^1}(\mathsf{Enc}_{k_b^0}(k_c'))$$

---

**Simulator**
$\mathsf{SimC}(1^\lambda, \Phi_{\mathsf{topo}}(C))$

(Wires) $k_{w_i}^\sigma \leftarrow \mathsf{KeyGen}(1^\lambda)$ for $i \in [p]$, $\sigma \in \{0,1\}$.
(Garbled gates) For each gate $\widetilde{\mathsf{gate}}_i = (\bot, a, b, c)$ in $\Phi_{\mathsf{topo}}(C)$:
$\quad \tilde{g}_i \leftarrow \mathsf{GarbleSimGate}\left(\{k_a^\sigma, k_b^\sigma\}_{\sigma \in \{0,1\}}, k_c^0\right)$.

**Output** $\widetilde{C}$, state $= (\{k_{w_i}^\sigma\})$.

$\mathsf{SimIn}(y, \mathsf{state})$

(Output table) $\widetilde{sd}_\ell \leftarrow \left[\left(k_{\mathsf{out}_\ell}^{y_\ell} \to 0\right), \left(k_{\mathsf{out}_\ell}^{1-y_\ell} \to 1\right)\right]_{\ell \in [m]}$. // ensures $k_{\mathsf{out}_\ell}^0 \to y_\ell$

**Output** $\tilde{x} = ((k_{\mathsf{in}_i}^0)_{i \in [n]}, (\widetilde{sd}_\ell)_{\ell \in [m]})$.

---

**Fig. 2.** Simulator for adaptive security.

The simulator invokes $\mathsf{GarbleSimGate}$ on input $k_c' = k_c^0$.

In the on-line phase, $\mathsf{SimIn}$, on input $y = C(x)$ adaptively computes the output tables so that the evaluator obtains the correct output. This is easily achieved by associating each bit of the output, $y_j$, to the only key encrypted in the output gate $g_{\mathsf{out}_j}$, which is $k_{\mathsf{out}_j}^0$. For the input keys, $\mathsf{SimIn}$ just sends keys $k_{\mathsf{in}_i}^0$ for each $i \in [n]$. The detailed definition of $(\mathsf{SimC}, \mathsf{SimIn})$ is provided in Fig. 2.

## 4    Hybrid Games

Our goal is to show the indistinguishability of the real world and the simulation in the adaptive setting. We do so by first introducing a template that allows us to define various hybrid games and then showing how to patch such games together to get a full security proof.

### 4.1    Template for Defining Hybrid Games

*Garbling Mode/Guessed Wires.* A gate's garbling mode indicates the way it is computed and can be one of the following RealGate, SimGate, InputDepSimGate which corresponds to the distributions outlined in Fig. 3. A *circuit configuration* is consists of two sets. A set the garbling modes for each gate in the circuit (i.e. $\mathsf{mode}_i$, $i \in [q]$) and as set of *guessed wires* $I \subseteq [p]$. We use the pair $((\mathsf{mode}_i)_{i \in [q]}, I)$ to denote a circuit configuration. A circuit configuration is *valid* if the outgoing wire of every gate in InputDepSimGate mode, is contained in the set of guessed wires $I$.

*The Hybrid Game* $\mathsf{Hyb}((\mathsf{mode}_i)_{i \in [q]}, I)$. Every valid circuit configuration defines a hybrid game as specified formally in Fig. 4 and described informally below. The hybrid game consists of a *guessing* step and a *garbling* step. The garbling step

| RealGate | SimGate | InputDepSimGate |
|---|---|---|
| $ct_{0,0} \leftarrow Enc_{k_a^0}(Enc_{k_b^0}(k_c^{g(0,0)}))$ | $ct_{0,0} \leftarrow Enc_{k_a^0}(Enc_{k_b^0}(k_c^0))$ | $ct_{0,0} \leftarrow Enc_{k_a^0}(Enc_{k_b^0}(k_c^{v(c)}))$ |
| $ct_{0,1} \leftarrow Enc_{k_a^0}(Enc_{k_b^1}(k_c^{g(0,1)}))$ | $ct_{0,1} \leftarrow Enc_{k_a^0}(Enc_{k_b^1}(k_c^0))$ | $ct_{0,1} \leftarrow Enc_{k_a^0}(Enc_{k_b^1}(k_c^{v(c)}))$ |
| $ct_{1,0} \leftarrow Enc_{k_a^1}(Enc_{k_b^0}(k_c^{g(1,0)}))$ | $ct_{1,0} \leftarrow Enc_{k_a^1}(Enc_{k_b^0}(k_c^0))$ | $ct_{1,0} \leftarrow Enc_{k_a^1}(Enc_{k_b^0}(k_c^{v(c)}))$ |
| $ct_{1,1} \leftarrow Enc_{k_a^1}(Enc_{k_b^1}(k_c^{g(1,1)}))$ | $ct_{1,1} \leftarrow Enc_{k_a^1}(Enc_{k_b^1}(k_c^0))$ | $ct_{1,1} \leftarrow Enc_{k_a^1}(Enc_{k_b^1}(k_c^{v(c)}))$ |

**Fig. 3.** Garbling Gate modes: RealGate (left), SimGate (center), InputDepSimGate (right). The value $v(c)$ depends on the input $x$ and corresponds to the bit going over the wire $c$ in the computation $C(x)$.

has two procedures: one for creating the garbled circuit $\widetilde{C}$ and one for creating the garbled input $\widetilde{x}$. The initial guessing step, is necessary in order to create gates in InputDepSimGate mode. For any such gate it is essential to know what is the bit on its output wire, (referred to as $v(c)$ in Fig. 3) once the circuit is computed. However the input is not known at the time of circuit garbling. Therefore we guess it! In some hybrid games we also need to guess values on other wires in the circuit. We define a set (called Guess), that stores all these guessed values for the marked wires. Hyb creates the garbled circuit by picking random keys $k_{w_i}^{\sigma}$ for each wire $w_i$. For each gate $i$, $\text{mode}_i \in \{\text{RealGate}, \text{SimGate}, \text{InputDepSimGate}\}$, it creates a garbled gate $\widetilde{g}_i$ according to the corresponding distribution as described in Fig. 3, and using $\text{Guess}(c)$ instead of the unknown $v(c)$. Once Hyb has the input, it checks whether all the guesses were made correctly. If not, the game is over with a fixed and dedicated output (say 0). However if they are correct, it follows the rules below to create the garbled input and map the output wires to $\{0, 1\}$.

- If all of the gates having $\text{in}_i$ as an input wire are in SimGate mode, then $K[i] := k_{\text{in}_i}^0$ else $K[i] := k_{\text{in}_i}^{x_i}$.
- If the unique gate having $\text{out}_\ell$ as an output wire is in SimGate mode, then we give the output map the simulated values $\widetilde{d}_\ell := [(k_{\text{out}_\ell}^{y_\ell} \to 0), (k_{\text{out}_\ell}^{1-y_\ell} \to 1)]$ else the real ones $\widetilde{d}_\ell := [(k_{\text{out}_\ell}^0 \to 0), (k_{\text{out}_\ell}^1 \to 1)]$.

*Real Game and Simulated Game.* By the definition of adaptively secure garbled circuits (Definition 2), the real game $\text{Exp}_{A,GC,Sim}^{\text{adaptive}}(1^\lambda, 0)$ is equivalent to $\text{Hyb}_A^\lambda((\text{mode}_i = \text{RealGate})_{i \in [q]}, \emptyset)$ and the simulated game $\text{Exp}_{A,GC,Sim}^{\text{adaptive}}(1^\lambda, 1)$ is equivalent to $\text{Hyb}_A^\lambda((\text{mode}_i = \text{SimGate})_{i \in [q]}, \emptyset)$. Therefore, the main aim is to show that these hybrids are indistinguishable.

### 4.2   Rules for Indistinguishable Hybrids

We provide rules that allow us to move from one configuration to another and prove that the corresponding hybrid games are indistinguishable. We define two rules that allow us to do this.

**Game** $\text{Hyb}_{\mathcal{A}}^{\lambda}((\text{mode}_i)_{i \in [q]}, I)$

1. (Guesses) For all $w_i \in I$,
   - Let $\text{Guess}(w_i) \leftarrow \{0, 1\}$
2. **Receive $C$ from $\mathcal{A}$**
   **Garble circuit $C$:**
3. (Wires) $k_i^{\sigma} \leftarrow \text{KeyGen}(1^{\lambda})$ for $i \in [p]$, $\sigma \in \{0, 1\}$.
4. (Gates) For each $\text{gate}_i = (g, a, b, c)$ in $C$.
   - If $\text{mode}_i = \text{RealGate}$:
     run $\tilde{g}_i \leftarrow \text{GarbleGate}(g, \{k_a^{\sigma}, k_b^{\sigma}, k_c^{\sigma}\}_{\sigma \in \{0,1\}})$.
   - if $\text{mode}_i = \text{SimGate}$:
     run $\tilde{g}_i \leftarrow \text{GarbleSimGate}(\{k_a^{\sigma}, k_b^{\sigma}\}_{\sigma \in \{0,1\}}, k_c^0)$.
   - If $\text{mode}_i = \text{InputDepSimGate}$:
     run $\tilde{g}_i \leftarrow \text{GarbleSimGate}((k_a^{\sigma}, k_b^{\sigma})_{\sigma \in \{0,1\}}, k_c^{\text{Guess}(c)})$.
5. **Send $\tilde{C}$ to $\mathcal{A}$ and get back $x$**
   **Garble Input $x$:**
6. (Check the guesses) For each $\forall\, i \in I$,
   - Let $v(w_i)$ be the bit on the wire $w_i$ during the computation $C(x)$.
   - if $v(w_i) \neq \text{Guess}(w_i)$ **Output 0** and abort the game.
7. (Output tables) Let $y = C(x)$. For $\ell = 1, \ldots, m$:
   Let $i$ be the index of the gate with output wire $\text{out}_{\ell}$.
   - If $\text{mode}_i \neq \text{SimGate}$, set $\tilde{d}_{\ell} := [(k_{\text{out}_{\ell}}^0 \rightarrow 0), (k_{\text{out}_{\ell}}^1 \rightarrow 1)]$,
   - else, set $\tilde{d}_{\ell} := [(k_{\text{out}_{\ell}}^{y_{\ell}} \rightarrow 0), (k_{\text{out}_{\ell}}^{1-y_{\ell}} \rightarrow 1)]$.
8. (Select input keys) For $\ell = 1, \ldots, n$:
   - If all gates $i$ having $\text{in}_{\ell}$ as an input wire satisfy $\text{mode}_i = \text{SimGate}$,
     then set $K[\ell] := k_{\text{in}_{\ell}}^0$,
   - else set $K[\ell] := k_{\text{in}_{\ell}}^{x_{\ell}}$.
9. **Send $\tilde{x} := (K, \{\tilde{d}_{\ell}\}_{\ell \in [m]})$ to $\mathcal{A}$ and receive $\mathcal{A}$'s output**
10. **Output $\mathcal{A}$'s output**

**Fig. 4.** The hybrid game.

**Definition 3 (Neighboring Hybrids).** *We say two valid hybrids or configurations* $((\text{mode}_i)_{i \in [q]}, I), ((\text{mode}_i')_{i \in [q]}, I)$ *are "neighboring", if the set of guessed wires $I$ is the same in both of them and the garbling modes of all gates except one are the same; i.e. there exists some $j \in [q]$ such that for all $i \neq j$ we have* $\text{mode}_i = \text{mode}_i'$. *We call $\text{gate}_j$ the target gate of the two hybrids or configurations.*

**Definition 4 (Predecessor/Successor/Sibling Gates).** *[HJO+15] Given a circuit $C$ and a gate $j \in [q]$ of the form $\text{gate}_j = (g, w_a, w_b, w_c)$ with incoming wires $w_a, w_b$ and outgoing wire $w_c$:*

- *We define the* predecessors *of $j$, denoted by* $\mathsf{Pred}(j)$, *to be the set of gates whose outgoing wires are either $w_a$ or $w_b$. If $w_a, w_b$ are input wires then* $\mathsf{Pred}(j) = \emptyset$, *else* $|\mathsf{Pred}(j)| = 2$.
- *We define the* successors *of $j$, denoted by* $\mathsf{Succ}(j)$ *to be the set of gates that contain $w_c$ as an incoming wire. If $w_c$ is an output wires then* $\mathsf{Succ}(j) = \emptyset$.
- *We define the* siblings *of $j$, denoted by* $\mathsf{Siblings}(j)$ *to be the set of gates that contain either $w_a$ or $w_b$ as an incoming wire.*

We define $\mathsf{TimeGC}(x)$ to be the time it takes to garble a circuit of size $x$ using Yao's garbling scheme. For convenience, we let mode $\overset{\text{def}}{=} (\mathsf{mode}_i)_{i \in [q]}$ and omit writing the security parameter $\lambda$ in the superscript of the hybrid games, since it is the same for all the games discussed here. For the same reason we use, $\varepsilon$ and $T$ instead of $\varepsilon(\lambda)$ and $T(\lambda)$.

**Indistinguishability Rule 1: RealGate $\leftrightarrow$ InputDepSimGate:** This rule allows us to change the garbling mode of a gate from RealGate to InputDepSimGate. It says that one can move from a circuit configuration (mode, $I$) to neighboring circuit configuration (mode$'$, $I$) where the mode of the target gate changes from RealGate in mode to InputDepSimGate in mode$'$ (and vice versa).

**Lemma 1.** *Let* $\mathsf{Hyb}(\mathsf{mode}, I)$ *and* $\mathsf{Hyb}(\mathsf{mode}', I)$ *be two neighboring hybrids, with target gate$_j$ such that* $\mathsf{mode}_j = \mathsf{RealGate}$ *and* $\mathsf{mode}'_j = \mathsf{InputDepSimGate}$. *In addition, for all $i \in \mathsf{Pred}(j)$:* $\mathsf{mode}_i = \mathsf{InputDepSimGate}$. *Then* $\mathsf{Hyb}(\mathsf{mode}, I)$ *and* $\mathsf{Hyb}(\mathsf{mode}', I)$ *are $(T(\lambda), \varepsilon(\lambda))$-indistinguishable as long as $\Gamma = (\mathsf{KeyGen}, \mathsf{Enc}, \mathsf{Dec})$ is an encryption scheme $(T'(\lambda), \varepsilon(\lambda))$-secure under CPA double encryption as per Definition 6 and $T'(\lambda) = T(\lambda) + \mathsf{TimeGC}(|C|)$.*

*Proof.* Let (mode, $I$) and (mode$'$, $I$) be as in the statement of the lemma, two valid circuit configurations. Towards a contradiction, assume that there exists a adversary $\mathcal{A}$ who runs in time $T$ and distinguishes $H^0 := \mathsf{Hyb}(\mathsf{mode}, I)$ and $H^1 := \mathsf{Hyb}(\mathsf{mode}', I)$. i.e.,

$$\left| \Pr\left[H^0_{\mathcal{A}} = 1\right] - \Pr\left[H^1_{\mathcal{A}} = 1\right] \right| > \varepsilon.$$

We construct an adversary $\mathcal{B}$, running in time $T'$ that breaks the double CPA-security of the encryption scheme $\Gamma = (\mathsf{KeyGen}, \mathsf{Enc}, \mathsf{Dec})$ which is used to garble gates. More specifically, we show that $\mathcal{B}$ wins the chosen double encryption security game (Definition 6) which is implied by CPA security. The formal description of adversary $\mathcal{B}$ is provided in Fig. 5.

Informally, $\mathcal{B}$ –on input (mode, $I$) and target gate $j$– aims to use her CPA-oracle access in $\mathsf{Exp}^{\mathsf{double}}(1^\lambda, b)$ to generate distribution $H^b$. The only difference between $H^0$ and $H^1$ is in the way gate $\tilde{g}_j$ is computed. On a high level, the reduction $\mathcal{B}$ will compute all garbled gates $\tilde{g}_i$ for $i \neq j$, according to experiment $\mathsf{Hyb}(\mathsf{mode}, I)$, and will compute the garbled gate $\tilde{g}_j$ using the ciphertexts obtained as a challenge in the experiment $\mathsf{Exp}^{\mathsf{double}}(1^\lambda, b)$.

**Adversary $\mathcal{B}$ (Reduction)**
**Input mode, $I$ and $j$.**

1. (Guesses) For all $w_i \in I$,
   - Let $\mathsf{Guess}(w_i) \leftarrow U_1$
2. **Receive $C$ from $\mathcal{A}$**
   Garble circuit $C$:
3. (Wires) $k_i^\sigma \leftarrow \mathsf{KeyGen}(1^\lambda)$ for $i \in [p]$, $\sigma \in \{0,1\}$. *except* for the two keys $k_{a^*}^{1-\alpha}, k_{b^*}^{1-\beta}$.
4. (Gates) For each $\mathbf{gate}_i = (g, a, b, c)$ in $C$ except $\mathbf{gate}_j$.
   - If $\mathsf{mode}_i = \mathsf{RealGate}$:
     run $\widetilde{g}_i \leftarrow \mathsf{GarbleGate}(g, \{k_a^\sigma, k_b^\sigma, k_c^\sigma\}_{\sigma \in \{0,1\}})$.
   - if $\mathsf{mode}_i = \mathsf{SimGate}$:
     run $\widetilde{g}_i \leftarrow \mathsf{GarbleSimGate}(\{k_a^\sigma, k_b^\sigma\}_{\sigma \in \{0,1\}}, k_c^0)$.
   - if $\mathsf{mode}_i = \mathsf{InputDepSimGate}$:
     run $\widetilde{g}_i \leftarrow \mathsf{GarbleSimGate}((k_a^\sigma, k_b^\sigma)_{\sigma \in \{0,1\}}, k_c^{\mathsf{Guess}(c)})$
   4·A) Let $\alpha := \mathsf{Guess}(a^*)$, $\beta := \mathsf{Guess}(b^*)$
   4·B) Let $x_0 = k_{c^*}^{g^*(1-\alpha,\beta)}, y_0 = k_{c^*}^{g^*(\alpha,1-\beta)}, z_0 = k_{c^*}^{g^*(1-\alpha,1-\beta)}$ and $x_1 = y_1 = z_1 = k_{c^*}^{g^*(\alpha,\beta)}$.
   4·C) Give $k_{a^*}^\alpha, k_{b^*}^\beta$ and $(x_0, y_0, z_0), (x_1, y_1, z_1)$ to the challenger of $\mathsf{Exp}^{\mathsf{double}}(1^\lambda, b)$. The challenger of $\mathsf{Exp}^{\mathsf{double}}(1^\lambda, b)$ chooses two keys which we implicitly define as $k_{a^*}^{1-\alpha}, k_{a^*}^{1-\beta}$. It gives $B$ the ciphertexts $\mathsf{ct}_x, \mathsf{ct}_y, \mathsf{ct}_z$ and oracle access to $\mathsf{Enc}_{k_{a^*}^{1-\alpha}}(\cdot)$ and $\mathsf{Enc}_{k_{b^*}^{1-\beta}}(\cdot)$.
   4·D) For the gate $j$ :
   - Compute $\mathsf{ct}_{\alpha,\beta} \leftarrow \mathsf{Enc}_{k_{a^*}^\alpha}(\mathsf{Enc}_{k_{b^*}^\beta}(k_{c^*}^{g(\alpha,\beta)}))$.
   - Set $\mathsf{ct}_{1-\alpha,\beta} := \mathsf{ct}_x$, $\mathsf{ct}_{\alpha,1-\beta} := \mathsf{ct}_y$, $\mathsf{ct}_{1-\alpha,1-\beta} := \mathsf{ct}_z$.
   - Let $\widetilde{g}_j$ be a random ordering of $[\mathsf{ct}_{0,0}, \mathsf{ct}_{0,1}, \mathsf{ct}_{1,0}, \mathsf{ct}_{1,1}]$
5. **Send $\widetilde{C}$ to $\mathcal{A}$. Obtain $x$ from $\mathcal{A}$.**

   Garble Input $x$:
6. (Check the guesses) For each $\forall i \in I$,
   - Let $v(w_i)$ be the bit on the wire $w_i$ during the computation $C(x)$.
   - if $v(w_i) \neq \mathsf{Guess}(w_i)$ **Output 0 and abort.**
7. (Output tables) Let $y = C(x)$. For $\ell = 1, \ldots, m$:
   Let $i$ be the index of the gate with output wire $\mathsf{out}_\ell$.
   - If $\mathsf{mode}_i \neq \mathsf{SimGate}$, set $\widetilde{d}_\ell := [(k_{\mathsf{out}_\ell}^0 \to 0), (k_{\mathsf{out}_\ell}^1 \to 1)]$,
   - else, set $\widetilde{d}_\ell := [(k_{\mathsf{out}_\ell}^{y_\ell} \to 0), (k_{\mathsf{out}_\ell}^{1-y_\ell} \to 1)]$.
8. (Select input keys) For $\ell = 1, \ldots, n$:
   - If all gates $i$ having $\mathsf{in}_\ell$ as an input wire satisfy $\mathsf{mode}_i = \mathsf{SimGate}$, then set $K[\ell] := k_{\mathsf{in}_\ell}^0$,
   - else set $K[\ell] := k_{\mathsf{in}_\ell}^{x_\ell}$.
9. **Set $\widetilde{x} := (K, \{\widetilde{d}_\ell\}_{\ell \in [m]})$. Send $\widetilde{x}$ to $\mathcal{A}$ and output whatever $\mathcal{A}$ outputs.**

15

**Fig. 5.** Proof of security for rule 1: the reduction $B$ uses an adversary $\mathcal{A}$ that distinguishes the hybrids to play the chosen double encryption security game (Definition 6) denoted by $\mathsf{Exp}^{\mathsf{double}}$.

In more detail, let $\mathsf{gate}_j = (g^*, a^*, b^*, c^*)$ be the target gate. Recall, the predecessors of $\mathsf{gate}_j$ (with output wires $a^*$ and $b^*$) are in InputDepSimGate mode. Therefore garbling of each gate in $\mathsf{Pred}(j)$, includes encryptions of one wire label only. We call these wires (which are fixed by the bit values guessed in step 1, $\alpha, \beta \in \{0, 1\}$) $k_{a^*}^{\alpha}$ and $k_{b^*}^{\beta}$. Consequently the wire label decrypted during the evaluation of $\mathsf{gate}_j$ is also the same wire label in both games, $k_{c^*}^{g(\alpha,\beta)}$. The difference is $\mathsf{mode}_j = \mathsf{RealGate}$ in $\mathsf{Hyb}(\mathsf{mode}, I)$, meaning, there is another wire label, which was used to garble $\mathsf{gate}_j$ and its ciphertext is one of the four ciphertexts $\mathsf{ct}_s$, $s \in \{0, 1\}^2$. But in $\mathsf{Hyb}(\mathsf{mode}', I)$, garbling mode of $\mathsf{gate}_j$ is InputDepSimGate and the only wire label used is $k_{c^*}^{g(\alpha,\beta)}$. To create the same garbled gate distributions using the challenger of the $\mathsf{Exp}^{\mathsf{double}}(1^\lambda, b)$, the reduction $\mathcal{B}$ –who knows all wire keys $except$ for $k_{a^*}^{1-\alpha}, k_{b^*}^{1-\beta}$– will create $\mathsf{ct}_{\alpha,\beta}$ as an encryption of $k_{c^*}^{g(\alpha,\beta)}$ on its own, but the remaining three ciphertexts $\mathsf{ct}_{\alpha',\beta'}$ will come from the experiment $\mathsf{Exp}^{\mathsf{double}}(1^\lambda, b)$ as either encryptions of different values $k_{c^*}^{g(\alpha',\beta')}$ (real) or of the same value $k_{c^*}^{g(\alpha,\beta)}$[2].

The one subtlety is that the reduction needs to create encryptions under the keys $k_{a^*}^{1-\alpha}, k_{b^*}^{1-\beta}$ to create garbled gates $\widetilde{g}_i$ for gates $i$ that are siblings of gate $j$. It can do that by using the encryption oracles which are given to it as part of the experiment $\mathsf{Exp}^{\mathsf{double}}(1^\lambda, b)$. The formal description of the reduction $\mathcal{B}$ is provided in Fig. 5. Finally notice that $\mathcal{B}$'s running time is, the time it takes to create the garble circuit plus the time it takes to run $\mathcal{A}$, so $T' = T + \mathsf{TimeGC}(|C|)$.

$$\left| \Pr[H_{\mathcal{A}}^0 = 1] - \Pr[H_{\mathcal{A}}^1 = 1] \right|$$
$$\leq \left| \Pr[\mathsf{Exp}_{\mathcal{B}}^{\mathsf{double}}(1^\lambda, 0) = 1] - \Pr[\mathsf{Exp}_{\mathcal{B}}^{\mathsf{double}}(1^\lambda, 1) = 1] \right| \leq \varepsilon.$$

which proves the Lemma.

**Indistinguishability Rule 2. InputDepSimGate $\leftrightarrow$ SimGate:** This rule allows us to change the mode of a gate $j$ from InputDepSimGate to SimGate under the condition that all successor gates $i \in \mathsf{Succ}(j)$ satisfy that $\mathsf{mode}_i \in \{\mathsf{InputDepSimGate}, \mathsf{SimGate}\}$.

**Lemma 2.** *Let* $\mathsf{Hyb}(\mathsf{mode}, I)$ *and* $\mathsf{Hyb}(\mathsf{mode}', I)$ *be two neighboring hybrids, with target* $\mathsf{gate}_j$ *such that* $\mathsf{mode}_j = \mathsf{InputDepSimGate}$ *in* mode *and* $\mathsf{mode}_j = \mathsf{SimGate}$ *in* mode'. *In addition, for all* $i \in \mathsf{Succ}(j)$ *we have* $\mathsf{mode}_i \in \{\mathsf{SimGate}, \mathsf{InputDepSimGate}\}$. *Then for any* $\mathcal{A}$, $\mathsf{Hyb}_{\mathcal{A}}(\mathsf{mode}, I)$ *and* $\mathsf{Hyb}_{\mathcal{A}}(\mathsf{mode}', I')$ *are identically distributed.*

*Proof.* Fix any adversary $\mathcal{A}$. Define $H_0 := \mathsf{Hyb}_{\mathcal{A}}(\mathsf{mode}, I)$ and $H_1 := \mathsf{Hyb}_{\mathcal{A}}(\mathsf{mode}', I)$. The difference between the hybrids is in how the garbled gate $\widetilde{g}_j$ is created:

---

[2] If $a^* = b^*$, ($\mathsf{gate}_j$ has fan-in 1), then $\mathcal{B}$ uses the challenger of the CPA encryption instead of the double-encryption scheme. The reduction considers the CPA challenger's key as $k_{b^*}^{1-\alpha}$, and using appropriate queries garbles $\mathsf{gate}_j$.

- In $H_0$, we set $\widetilde{g}_j \leftarrow \mathsf{GarbleSimGate}((k_{a*}^\sigma, k_{b*}^\sigma)_{\sigma \in \{0,1\}}, k_{c*}^{\mathsf{Guess}(c^*)})$.
- In $H_1$, we set $\widetilde{g}_j \leftarrow \mathsf{GarbleSimGate}((k_{a*}^\sigma, k_{b*}^\sigma)_{\sigma \in \{0,1\}}, k_{c*}^0)$.

If $j$ is not an output gate, and all successor gates $i \in \mathsf{Succ}(j)$ are in {SimGate, InputDepSimGate} modes then the keys $k_{c*}^0$ and $k_{c*}^1$ are treated symmetrically everywhere in the game other than in $\widetilde{g}_j$. Therefore, by symmetry, there is no difference between using $k_{c*}^0$ and $k_{c*}^{\mathsf{Guess}(c^*)}$ in $\widetilde{g}_j$

If $j$ is an output gate then the keys $k_{c*}^0$ and $k_{c*}^1$ are only used in $\widetilde{g}_j$ and in the output map $\widetilde{d}_j$. Therefore, by symmetry, there is no difference between using $k_{c*}^{y_j}$ in $\widetilde{g}_j$ and setting $\widetilde{d}_j := [(k_{\mathsf{out}_j}^0 \to 0), (k_{\mathsf{out}_j}^1 \to 1)]$ (in $H_0$) versus using $k_{c*}^0$ in $\widetilde{g}_j$ and setting $\widetilde{d}_j := [(k_{\mathsf{out}_j}^{y_j} \to 0), (k_{\mathsf{out}_j}^{1-y_j} \to 1)]$ (in $H_1$).

One last difference between the hybrids occurs if some wire $\mathsf{in}_i$ becomes only connected to gates that are in SimGate in $H_1$. In this case, when we create the garbled input $\widetilde{x}$, then in $H_0$ we give $K[i] := k_{\mathsf{in}_i}^{x_i}$ but in $H_1$ we give $K[i] := k_{\mathsf{in}_i}^0$. Since the keys $k_{\mathsf{in}_i}^0, k_{\mathsf{in}_i}^1$ are treated symmetrically everywhere in the game (both in $H_0$ and $H_1$) other than in $K[i]$, there is no difference between setting $K[i] := k_{\mathsf{in}_i}^0$ versus $K[i] := k_{\mathsf{in}_i}^{x_i}$.

**Scaling Indistinguishability.** We now show that by adding guesses we can make the hybrids more indistinguishable, or equivalently, removing guesses makes the hybrids more distinguishable. This lemma is crucial for comparing hybrids with different guesses by scaling the number of guesses up or down to make the comparison possible.

**Lemma 3.** *If* $\mathbf{D}_T\left[\mathsf{Hyb}(\mathsf{mode}, I), \mathsf{Hyb}(\mathsf{mode}', I)\right] = \varepsilon$ *and* $J$ *is a set of wires, disjoint from* $I$ *then*

$$\mathbf{D}_T\left[\mathsf{Hyb}(\mathsf{mode}, I \cup J), \mathsf{Hyb}(\mathsf{mode}', I \cup J)\right] = 2^{-|J|} \cdot \varepsilon.$$

*Proof.* For any probabilistic $T$ bounded adversary $\mathcal{A}$, we have

$$\Pr\left[\mathsf{Hyb}_{\mathcal{A}}(\mathsf{mode}, I \cup J) = 1\right] = 2^{-|J|} \Pr\left[\mathsf{Hyb}_{\mathcal{A}}(\mathsf{mode}, I) = 1)\right]$$

Because with probability $2^{-|J|}$, (the probability of guessing the extra $|J|$ wires correctly) $\mathcal{A}$ playing the game $\mathsf{Hyb}(\mathsf{mode}, I \cup J)$ has the exact same interactions as in game $\mathsf{Hyb}(\mathsf{mode}, I)$ and therefore the same exact outputs. The same holds for $\Pr\left[\mathsf{Hyb}_{\mathcal{A}}(\mathsf{mode}', I \cup J) = 1\right]$ therefore,

$$\left[\Pr\left[\mathsf{Hyb}_{\mathcal{A}}(\mathsf{mode}, I \cup J) = 1\right] - \Pr\left[\mathsf{Hyb}_{\mathcal{A}}(\mathsf{mode}', I \cup J) = 1\right]\right]$$
$$= 2^{-|J|}\left|\Pr\left[\mathsf{Hyb}_{\mathcal{A}}(\mathsf{mode}, I) = 1\right] - \Pr\left[\mathsf{Hyb}_{\mathcal{A}}(\mathsf{mode}', I) = 1\right]\right| \leq 2^{-|J|} \cdot \varepsilon$$

## 5    Pebbling and Sequences of Hybrid Games

In the last section we defined hybrid games parameterized by a configuration (mode, $I$). We also gave 2 rules, which describe ways that allow us to move from one configuration to another in indistinguishable steps. Now our goal is to use the given rules so as to define a *sequence of indistinguishable hybrid games* that takes us from the *real game* $\mathsf{Hyb}((\mathsf{mode}_i = \mathsf{RealGate})_{i\in[q]}, I = \emptyset)$ to the *simulation* $\mathsf{Hyb}((\mathsf{mode}_i = \mathsf{SimGate})_{i\in[q]}, I = \emptyset)$.

*Pebbling Game.* We capture the problem of finding a sequences of hybrid games using a certain type of *pebbling game* on the graph of circuit $C$.

- *Graph of circuit $C$* is obtained by assigning a node to each gate, and a directed edge from node $i$ to node $j$ for each wire going out of gate $i$ and into gate $j$. To make this consistent, we think of each input wire (in) as outgoing wire of an empty (dummy) gate, going into a gate in level 1 of the circuit. Since we are always considering a pebbling on the graph of a circuit, we use words gate/node and wire/edge interchangeably.
- *Pebbles.* Each gate can either have *no pebble*, a *black pebble*, or *a gray pebble* on it (this will correspond to RealGate, InputDepSimGate and SimGate modes respectively). Initially, the circuit starts out with no pebbles on any gate. The game consist of the following possible moves:
  **Pebbling Rule A.** We can place or remove a black pebble on a gate as long as both predecessors of that gate have black pebbles on them (or it is an input gate).
  **Pebbling Rule B.** We can replace a black pebble with a gray pebble on a gate as long as all successors of that gate have black or gray pebbles on them (or the gate is an output gate).
- A *pebbling* of a circuit $C$ starts with no pebbles on the graph and is a sequence of $\gamma$ moves that follow rules A and B and that end up with a gray pebble on every gate. We say that a pebbling uses $t$ black pebbles if this is the maximal number of black pebbles on the circuit at any point in time during the game.
- A *pebble configuration* specifies for each gate, whether it contains no pebble, a gray pebble, or a black pebble.

*From Pebbling to Sequence of Hybrids.* A pebbling in $\gamma$ moves has a sequence of $\gamma+1$ pebble configurations starting with no pebbles and ending with a gray pebble on each gate. Each pebble configuration follows from the preceding one by a

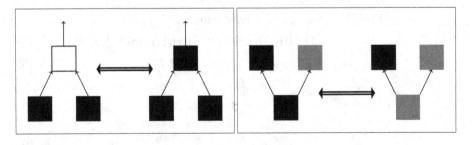

**Fig. 6.** Pebbling rules

move that satisfies pebbling rules A or B. Next we create a sequence of hybrids by defining one hybrid from each pebbling configuration.

- For every gate $i \in [q]$, we set $\mathsf{mode}_i = \mathsf{RealGate}$ if gate $i$ has no pebble, $\mathsf{mode}_i = \mathsf{InputDepSimGate}$ if gate $i$ has a black pebble, and $\mathsf{mode}_i = \mathsf{SimGate}$ if gate $i$ has a gray pebble.
- We set $I$ to be the set of the output wires of the gates with black pebbles.

Therefore a pebbling defines a sequence of hybrids $H_\alpha = \mathsf{Hyb}(\mathsf{mode}^\alpha, I^\alpha)$ for $\alpha = 0, \ldots, \gamma$ where $H_0 = \mathsf{Hyb}((\mathsf{mode}_i^0 = \mathsf{RealGate})_{i \in [q]}, \emptyset)$ is the real game and $H_\gamma = \mathsf{Hyb}((\mathsf{mode}_i^\gamma = \mathsf{SimGate})_{i \in [q]}, \emptyset)$ is the simulated game, and each $H_\alpha$ is induced by the pebbling configuration after $\alpha$ moves. In our next theorem and the following corollary, we prove that the sequence of hybrids obtained from a pebbling, as explained above, shows indistinguishability of the real and simulated games.

**Theorem 1.** *Assume that there is a pebbling of circuit $C$ in $\gamma$ moves, using $t$ black pebbles. Also assume that the encryption scheme $\Gamma = (\mathsf{KeyGen}, \mathsf{Enc}, \mathsf{Dec})$ is $(T + \mathsf{TimeGC}(|C|), \varepsilon)$-secure under CPA double encryption.*

*Then, the sequence of hybrids obtained from such pebbling as described above has the following property. For any $\alpha \in \{0, 1, \cdots, \gamma\}$, $H_\alpha = \mathsf{Hyb}(\mathsf{mode}^\alpha, I^\alpha)$*

$$\mathbf{D}_T\left[\mathsf{Hyb}(\mathsf{mode}^0, I^\alpha), H_\alpha\right] \leq \sum_{i=1}^{\alpha} 2^{r_i - |I^\alpha|} \cdot \varepsilon \leq \alpha 2^{t - |I^\alpha|} \varepsilon$$

*where $r_\alpha = \max\left(|I^{\alpha-1}|, |I^\alpha|\right) \leq t$, for $\alpha \in [\gamma]$.*

*Proof.* We show the claim holds for $\mathsf{mode}_0$ and any configurations; $(\mathsf{mode}^\alpha, I^\alpha)$, $\alpha \in \{0, 1, \cdots \gamma\}$ by induction on the number of pebbling steps taken so far (i.e., $\alpha$). For convenience, let $s_\alpha = |I^\alpha|$ and remember $r_\alpha = \max(s_{\alpha-1}, s_\alpha)$.

**Base case.** Let $\alpha = 0$, $\mathbf{D}_T\left[\mathsf{Hyb}(\mathsf{mode}^0, I^0), H_0\right] = \mathbf{D}_T\left[H_0, H_0\right] = 0$.
**Inductive step.** Assume the claim holds for $\alpha$, we show it holds for $\alpha + 1$.

- If the $\alpha+1$st move in the pebbling game is to add a black pebble: $s_{\alpha+1} = s_\alpha + 1$ and $r_{\alpha+1} = s_{\alpha+1}$

$$\mathbf{D}_T\left[\mathsf{Hyb}(\mathsf{mode}^0, I^{\alpha+1}), \mathsf{Hyb}(\mathsf{mode}^{\alpha+1}, I^{\alpha+1})\right]$$
$$\leq \mathbf{D}_T\left[\mathsf{Hyb}(\mathsf{mode}^0, I^{\alpha+1}), \mathsf{Hyb}(\mathsf{mode}^\alpha, I^{\alpha+1})\right]$$
$$\qquad + \mathbf{D}_T\left[\mathsf{Hyb}(\mathsf{mode}^\alpha, I^{\alpha+1}), \mathsf{Hyb}(\mathsf{mode}^{\alpha+1}, I^{\alpha+1})\right] \tag{4}$$
$$\leq 2^{-1} \cdot \mathbf{D}_T\left[\mathsf{Hyb}(\mathsf{mode}^0, I^\alpha), \mathsf{Hyb}(\mathsf{mode}^\alpha, I^\alpha)\right] + \varepsilon \tag{5}$$
$$\leq 2^{-1} \sum_{i=1}^{\alpha} 2^{r_i - s_\alpha} \cdot \varepsilon + \varepsilon \leq \sum_{i=1}^{\alpha} 2^{r_i - s_\alpha - 1} \cdot \varepsilon + \varepsilon$$
$$\leq \sum_{i=1}^{\alpha} 2^{r_i - s_{\alpha+1}} \cdot \varepsilon + 2^{r_{\alpha+1} - s_{\alpha+1}} \cdot \varepsilon \leq \sum_{i=1}^{\alpha+1} 2^{r_i - s_{\alpha+1}} \cdot \varepsilon \tag{6}$$

Line 4 follows from the previous line by the Triangle Inequality. By Lemmas 1 or 2, $\mathbf{D}_T[\mathsf{Hyb}(\mathsf{mode}^\alpha, I^{\alpha+1}), \mathsf{Hyb}(\mathsf{mode}^{\alpha+1}, I^{\alpha+1})] \leq \varepsilon$. By Lemma 3 we have that $\mathbf{D}_T\left[\mathsf{Hyb}(\mathsf{mode}^0, I^{\alpha+1}), \mathsf{Hyb}(\mathsf{mode}^\alpha, I^{\alpha+1})\right] \leq \mathbf{D}_T[\mathsf{Hyb}(\mathsf{mode}^0, I^\alpha),$ $\mathsf{Hyb}(\mathsf{mode}^\alpha, I^\alpha)]/2$. Combining the two we get Line 5. We use the induction hypothesis to arrive at Line 6. The last line follows by noticing $s_{\alpha+1} = s_\alpha + 1$ and $r_{\alpha+1} = s_{\alpha+1}$.

- If the $\alpha + 1$st move in the pebbling game is to remove a black pebble: $s_{\alpha+1} = s_\alpha - 1$ and $r_{\alpha+1} = s_\alpha$

$$\mathbf{D}_T\left[\mathsf{Hyb}(\mathsf{mode}^0, I^{\alpha+1}), \mathsf{Hyb}(\mathsf{mode}^{\alpha+1}, I^{\alpha+1})\right]$$

$$\leq 2\mathbf{D}_T\left[\mathsf{Hyb}(\mathsf{mode}^0, I^\alpha), \mathsf{Hyb}(\mathsf{mode}^{\alpha+1}, I^\alpha)\right] \tag{7}$$

$$\leq 2\mathbf{D}_T\left[\mathsf{Hyb}(\mathsf{mode}^0, I^\alpha), \mathsf{Hyb}(\mathsf{mode}^\alpha, I^\alpha)\right]$$

$$+ 2\mathbf{D}_T\left[\mathsf{Hyb}(\mathsf{mode}^\alpha, I^\alpha), \mathsf{Hyb}(\mathsf{mode}^{\alpha+1}, I^\alpha)\right] \tag{8}$$

$$\leq 2\left(\sum_{i=1}^\alpha 2^{r_i - s_\alpha} \cdot \varepsilon + \varepsilon\right) \leq \sum_{i=1}^\alpha 2^{r_i - s_\alpha + 1} \cdot \varepsilon + 2\varepsilon$$

$$\leq \sum_{i=1}^\alpha 2^{r_i - s_{\alpha+1}} \cdot \varepsilon + 2^{r_{\alpha+1} - s_{\alpha+1}} \cdot \varepsilon \leq \sum_{i=1}^{\alpha+1} 2^{r_i - s_{\alpha+1}} \cdot \varepsilon \tag{9}$$

Similar to the last case, Line 7 follows from the previous line By Lemma 3. Line 8 follows from the Triangle inequality. By Lemma 1 or 2 and the induction hypothesis we arrive at Line 9. The last line follows by noticing $s_{\alpha+1} = s_\alpha - 1$ and $r_{\alpha+1} = s_\alpha$.

The reason we can apply Lemmas 1 or 2, is that the pebbling game rules (A and B) guarantee that the garbling modes of each two hybrids in our sequence have the necessary properties for applying Lemmas 1 or 2. In addition we created the set $I$ such that it includes all the necessary wires to keep the configuration valid. For more details, see Fig. 7, where we change $\mathsf{gate}_j$'s mode at step $\alpha + 1$, following rule A or B.

**Corollary 1.** *Assume that $\Gamma = (\mathsf{KeyGen}, \mathsf{Enc}, \mathsf{Dec})$ is an encryption scheme which is $(T(\lambda), \varepsilon(\lambda))$-secure under CPA double encryption. If there is a pebbling of circuit $C$ in $\gamma$ moves, using $t$ black pebbles then $\mathsf{Exp}_{\mathsf{GC},\mathsf{Sim}}^{\mathsf{adaptive}}(1^\lambda, 0)$ and $\mathsf{Exp}_{\mathsf{GC},\mathsf{Sim}}^{\mathsf{adaptive}}(1^\lambda, 1)$ are $(T'(\lambda), \varepsilon'(\lambda))$-indistinguishable where*

- $\varepsilon'(\lambda) \leq \sum_{i=1}^\gamma 2^{r_i} \cdot \varepsilon(\lambda) \leq \gamma \cdot 2^t \cdot \varepsilon(\lambda)$
- $T'(\lambda) = T(\lambda) - \mathsf{TimeGC}(|C|)$.

*where $r_i = \max(s_{i-1}, s_i)$ and $s_i$ is the number of black pebbles used at the $i$th pebbling step.*

*Proof.* By definition $\mathsf{Exp}_{\mathsf{GC},\mathsf{Sim}}^{\mathsf{adaptive}}(1^\lambda, 0) = \mathsf{Hyb}^\lambda(\mathsf{mode}^0, I^0)$ and $\mathsf{Exp}_{\mathsf{GC},\mathsf{Sim}}^{\mathsf{adaptive}}(1^\lambda, 1) = \mathsf{Hyb}^\lambda(\mathsf{mode}^\gamma, I^\gamma)$ where $I^0 = I^\gamma = \emptyset$. By Theorem 1 with $\alpha = \gamma$, we have $\mathbf{D}_{T(\lambda)}\left[\mathsf{Hyb}^\lambda(\mathsf{mode}^0, \emptyset), \mathsf{Hyb}^\lambda(\mathsf{mode}^\gamma, \emptyset)\right] \leq \sum_{i=1}^\gamma 2^{r_i} \cdot \varepsilon(\lambda)$ which proves the Corollary.

| Rule | wires in $I^\alpha$, $I^{\alpha+1}$ | $\text{mode}_j^\alpha \to \text{mode}_j^{\alpha+1}$ | Hybrids | Lemma |
|------|-------------------------------------|------------------------------------------------------|---------|-------|
| A | $\text{WPred} \subseteq I^\alpha$, $\text{WPred} \cup \{c^*\} \subseteq I^{\alpha+1}$ | RealGate $\to$ InputDepSimGate | $(\text{mode}^\alpha, I^{\alpha+1})$ $(\text{mode}^{\alpha+1}, I^{\alpha+1})$ | 1 |
|   | $\text{WPred} \cup \{c^*\} \subseteq I^\alpha$, $\text{WPred} \subseteq I^{\alpha+1}$ | InputDepSimGate $\to$ RealGate | $(\text{mode}^\alpha, I^\alpha)$ $(\text{mode}^{\alpha+1}, I^\alpha)$ | |
| B | $\text{WPred} \cup \{c^*\} \subseteq I^\alpha$, $\text{WPred} \subseteq I^{\alpha+1}$ | InputDepSimGate $\to$ SimGate | $(\text{mode}^\alpha, I^\alpha)$ $(\text{mode}^{\alpha+1}, I^\alpha)$ | 2 |
|   | $\text{WPred} \subseteq I^\alpha$, $\text{WPred} \cup \{c^*\} \subseteq I^{\alpha+1}$ | SimGate $\to$ InputDepSimGate | $(\text{mode}^\alpha, I^{\alpha+1})$ $(\text{mode}^{\alpha+1}, I^{\alpha+1})$ | |

**Fig. 7.** From pebbling rules to indistinguishable hybrids. $\text{WPred}$ := $\{$output wires of $\text{Pred}(j)\}$, $\text{WSucc}$ := $\{$output wires of $\text{Succ}(j)\}$.

**Corollary 2.** *If there is a pebbling of circuit $C$ in $\gamma$ moves, using $t$ black pebbles then* GC *is adaptively secure with online complexity*

1. $(m+n)\lambda$, *when* $\Gamma$ *is secure under* CPA *double encryption and* $2^t\gamma = \text{poly}(\lambda)$.
2. $(m + n)\text{poly}(\lambda + \log \gamma + t)$, *when* $\Gamma$ *is sub-exponentially secure under* CPA *double encryption and* $\log(\gamma) + t = \text{poly}(\lambda)$.

*Proof.* The online complexity of the garbling scheme consist of $(m + n)$ secret keys of the scheme $\Gamma$.

For case (1) we only need standard security of $\Gamma$ to survive a polynomial security loss of $2^t\gamma = \text{poly}(\lambda)$. Therefore, we can set the security parameter of $\Gamma$ to $\lambda$, which gives a key size of $\lambda$.

For case (2) we need to survive a security loss of $2^t\gamma = 2^{\text{poly}(\lambda)}$. If the encryption scheme $\Gamma$ is sub-exponentially secure it means that when instantiated with security parameter $\lambda'$ it has security $\varepsilon(\lambda') \leq 2^{-(\lambda')^\nu}$ for some constant $\nu$ and all large enough $\lambda'$. Therefore we need to set $\lambda' = (\lambda + \log(\gamma) + t)^{1/\nu}$ to ensure that $2^t\gamma\varepsilon(\lambda')$ is negligible, which results in a key size of $\lambda' = \text{poly}(\lambda + \log(\gamma) + t)$.

### 5.1 Pebbling Strategies

We now rely on a result of [HJO+15] to instantiate Corollary 2. In particular, it shows that for any circuit with $q$ gates and depth $d$ there is a pebbling strategy which makes at most $\gamma = q \cdot 2^{2d}$ moves and uses $t = 2d$ black pebbles. See Appendix B for the description of the strategy. By instantiating Corollary 2 with the above strategy, we obtain the following corollary.

**Corollary 3.** *Assuming the existence of (standard) one-way functions, Yao's garbling schemes is adaptively secure with on-line complexity $(n + m)\lambda$ for all circuits of depth $d = O(\log \lambda)$.*

*Assuming the existence of sub-exponentially secure one-way functions Yao's garbling schemes is adaptively secure with on-line complexity $(n + m)\text{poly}(\lambda, d)$, for arbitrary circuits of depth $d = \text{poly}(\lambda)$.*

# 6    Conclusions

We show that Yao's garbled circuit construction is already adaptively secure, without the need for any modification, at least when it comes to NC1 circuits. More generally, we give a reduction where the security loss is related (exponentially) to the pebble complexity of the circuit, which can often be much smaller than the input size, and therefore beats the naive reduction that guesses the entire input. It remains as an open problem to improve the reduction further or to give some negative results showing that it cannot be done.

# A    Symmetric-Key Encryption with Special Correctness [LP09]

In our construction of the garbling scheme, we use a symmetric-key encryption scheme $\Gamma = (\mathsf{KeyGen}, \mathsf{Enc}, \mathsf{Dec})$ which satisfies the standard definition of CPA security and an additional *special correctness* property below (this is a simplified and sufficient variant of the property described in from [LP09]). We need this property to ensure the correctness of our garbled circuit construction.

**Definition 5 (Special Correctness).** *A CPA-secure symmetric-key encryption $\Gamma = (\mathsf{KeyGen}, \mathsf{Enc}, \mathsf{Dec})$ satisfies special correctness if there is some negligible function $\varepsilon$ such that for any message $m$ we have:*

$$\Pr[\mathsf{Dec}_{k_2}(\mathsf{Enc}_{k_1}(m)) \neq \bot \ : \ k_1, k_2 \leftarrow \mathsf{KeyGen}(1^\lambda)] \leq \varepsilon(\lambda).$$

*Construction.* Let $F = \{f_k\}$ be a family of pseudorandom functions where $f_k : \{0,1\}^\lambda \to \{0,1\}^{\lambda+s}$, for $k \in \{0,1\}^\lambda$ and $s$ is a parameter denoting the message length. Define $\mathsf{Enc}_k(m) = (r, f_k(r) \oplus m0^\lambda)$ where $m \in \{0,1\}^s$, $r \xleftarrow{\$} \{0,1\}^\lambda$ and $m0^\lambda$ denotes the concatenation of $m$ with a string of 0s of length $\lambda$. Define $\mathsf{Dec}_k(c)$ which parses $c = (r, z)$, computes $w = z \oplus f_k(r)$ and if the last $\lambda$ bits of $w$ are 0's it outputs the first $s$ bits of $w$, else it outputs $\bot$.

It's easy to see that this scheme is CPA secure and that it satisfies the special correctness property.

*Double Encryption Encryption Security.* For convenience, we define a notion of double encryption security, following [LP09]. This notion is implied by standard CPA security but is more convenient to use in our security proof of garbled circuit security.

**Definition 6 (Double-encryption security).** *An encryption scheme $\Gamma = (\mathsf{KeyGen}, \mathsf{Enc}, \mathsf{Dec})$*

*– is $(T(\lambda), \varepsilon(\lambda))$-secure under chosen double encryption if*

$$\mathbf{D}_{T(\lambda)} \left[ \mathsf{Exp}^{\mathsf{double}}(1^\lambda, 0), \mathsf{Exp}^{\mathsf{double}}(1^\lambda, 1) \right] = \varepsilon(\lambda).$$

– *is secure under* chosen double encryption *if*

$$\mathsf{Exp}^{\mathsf{double}}(1^\lambda, 0) \overset{\mathrm{comp}}{\approx} \mathsf{Exp}^{\mathsf{double}}(1^\lambda, 1).$$

– *is sub-exponentially secure if*

$$\exists\, \nu > 0, \forall\, T(\lambda) \in \mathsf{poly}(\lambda) \;\; \mathbf{D}_{T(\lambda)}\left[\mathsf{Exp}^{\mathsf{double}}(1^\lambda, 1), \mathsf{Exp}^{\mathsf{double}}(1^\lambda, 0)\right] \leq \varepsilon(\lambda) = 1/2^{\lambda^\nu}.$$

*where the experiment* $\mathsf{Exp}_{\mathcal{A}}^{\mathsf{double}}$ *is defined as follows.*

**Experiment** $\mathsf{Exp}_{\mathcal{A}}^{\mathsf{double}}(1^\lambda, b)$

1. *The adversary* $\mathcal{A}$ *on input* $1^\lambda$ *outputs two keys* $k_a$ *and* $k_b$ *of length* $\lambda$ *and two triples of messages* $(x_0, y_0, z_0)$ *and* $(x_1, y_1, z_1)$ *where all messages are of the same length.*
2. *Two keys* $k'_a, k'_b \overset{\$}{\leftarrow} \mathsf{KeyGen}(1^\lambda)$ *are chosen.*
3. $\mathcal{A}^{\mathsf{Enc}_{k'_a}(\cdot), \mathsf{Enc}_{k'_b}(\cdot)}$ *is given the challenge ciphertexts* $c_x \leftarrow \mathsf{Enc}_{k_a}(\mathsf{Enc}_{k'_b}(x_b))$, $c_y \leftarrow \mathsf{Enc}_{k'_b}(\mathsf{Enc}_{k_b}(y_b))$, $c_z \leftarrow \mathsf{Enc}_{k'_a}(\mathsf{Enc}_{k'_b}(z_b))$ *as well as* **oracle access to** $\mathsf{Enc}_{k'_a}(\cdot)$ *and* $\mathsf{Enc}_{k'_b}(\cdot)$.
4. $\mathcal{A}$ *outputs* $b'$ *which is the output of the experiment.*

The following lemma is essentially immediate - see [LP09] for a formal proof.

**Lemma 4.** *If* $(\mathsf{KeyGen}, \mathsf{Enc}, \mathsf{Dec})$ *is CPA-secure then it is secure under chosen double encryption with the same security parameter.*

## B  Pebbling Strategy [HJO+15]

This is a recursive strategy defined as follows.

– Pebble($C$):
  For each gate $i$ in $C$ starting with the gates at the top level moving to the bottom level:
  1. RecPutBlack($C, i$)
  2. Replace the black pebble on gate $i$ with a gray pebble.
– RecPutBlack($C, i$): // Let LeftPred($C, i$) and RightPred($C, i$) be the two predecessors of gate $i$ in $C$.
  1. If gate $i$ is an input gate, put a black pebble on $i$ and **return**.
  2. Run RecPutBlack($C$, LeftPred($C, i$)), RecPutBlack($C$, RightPred($C, i$))
  3. Put a black pebble on gate $i$.
  4. Run RecRemoveBlack($C$, LeftPred($C, i$)) and
     RecRemoveBlack($C$, RightPred($C, i$)),
– RecRemoveBlack($C, i$): This is the same as RecPutBlack, except that instead of putting a black pebble on gate $i$, in steps 1 and 3, we remove it.

To analyze the correctness of this strategy, we note the following invariants: if the circuit $C$ is in a configuration where it does not contain any pebbles at any level below that of gate $i$, then (1) the procedure RecPutBlack($C, i$) results in a configuration where a single black pebble is added to gate $i$, but nothing else changes, (2) the procedure RecRemoveBlack($C, i$) results in a configuration where a single black pebble is removed from gate $i$, but nothing else changes. Using these two invariants the correctness of of the entire strategy follows.

To calculate the number of black pebbles used and the number of moves that the above strategy takes to pebble $C$, we use the following simple recursive equations. Let #PebPut($d$) and #PebRem($d$) be the number of black pebbles on gate $i$ and below it used to execute RecPutBlack and RecRemoveBlack on a gate at level $d$, respectively. We have,

$$\#\mathsf{PebPut}(1) = 1, \quad \#\mathsf{PebPut}(d) \leq \max(\#\mathsf{PebPut}(d-1), \#\mathsf{PebRem}(d-1)) + 2$$
$$\#\mathsf{PebRem}(1) = 1, \quad \#\mathsf{PebRem}(d) \leq \max(\#\mathsf{PebPut}(d-1), \#\mathsf{PebRem}(d-1)) + 2$$

Therefore the strategy requires at most $2d$ black pebbles to pebble the circuit.

To calculate the number of moves it takes run Pebble($C$), we use the following recursive equations. Let #Moves($d$) be the number of moves it takes to put a black pebble on, or remove a black pebble from, a gate at level $d$. Then

$$\#\mathsf{Moves}(1) = 1, \quad \#\mathsf{Moves}(d) = 4(\#\mathsf{Moves}(d-1)) + 1$$

Hence, each call of RecPutBlack takes at most $4^d$ moves, and the total number of moves to pebble the circuit is at most $q4^d$. In summary, the above gives us a strategy to pebble any circuit with at most $\gamma = q4^d$ moves and $t = 2d$ black pebbles.

# References

[AIKW13] Applebaum, B., Ishai, Y., Kushilevitz, E., Waters, B.: Encoding functions with constant online rate or how to compress garbled circuits keys. In: Canetti, R., Garay, J.A. (eds.) CRYPTO 2013, Part II. LNCS, vol. 8043, pp. 166–184. Springer, Heidelberg (2013)

[AS15] Ananth, P., Sahai, A.: Functional encryption for turing machines. Cryptology ePrint Archive, Report 2015/776 (2015). http://eprint.iacr.org/

[BGG+14] Boneh, D., Gentry, C., Gorbunov, S., Halevi, S., Nikolaenko, V., Segev, G., Vaikuntanathan, V., Vinayagamurthy, D.: Fully key-homomorphic encryption, arithmetic circuit ABE and compact garbled circuits. In: Nguyen, P.Q., Oswald, E. (eds.) EUROCRYPT 2014. LNCS, vol. 8441, pp. 533–556. Springer, Heidelberg (2014)

[BHK13] Bellare, M., Hoang, V.T., Keelveedhi, S.: Instantiating random oracles via UCEs. In: Canetti, R., Garay, J.A. (eds.) CRYPTO 2013, Part II. LNCS, vol. 8043, pp. 398–415. Springer, Heidelberg (2013)

[BHR12a] Bellare, M., Hoang, V.T., Rogaway, P.: Adaptively secure garbling with applications to one-time programs and secure outsourcing. In: Wang, X., Sako, K. (eds.) ASIACRYPT 2012. LNCS, vol. 7658, pp. 134–153. Springer, Heidelberg (2012)

[BHR12b] Bellare, M., Hoang, V.T., Rogaway, P.: Foundations of garbled circuits. In: Yu, T., Danezis, G., Gligor, V.D. (eds.) ACM CCS 12, pp. 784–796. ACM Press, October 2012

[FJP15] Fuchsbauer, G., Jafargholi, Z., Pietrzak, K.: A quasipolynomial reduction for generalized selective decryption on trees. In: Gennaro, R., Robshaw, M. (eds.) CRYPTO 2015. LNCS, vol. 9215, pp. 601–620. Springer, Heidelberg (2015). doi:10.1007/978-3-662-47989-6_29

[FKPR14] Fuchsbauer, G., Konstantinov, M., Pietrzak, K., Rao, V.: Adaptive security of constrained PRFs. In: Sarkar, P., Iwata, T. (eds.) ASIACRYPT 2014, Part II. LNCS, vol. 8874, pp. 82–101. Springer, Heidelberg (2014)

[HJO+15] Hemenway, B., Jafargholi, Z., Ostrovsky, R., Scafuro, A., Wichs, D.: Adaptively secure garbled circuits from one-way functions. IACR Cryptology ePrint Archive, 2015, p. 1250 (2015)

[LP09] Lindell, Y., Pinkas, B.: A proof of security of Yao's protocol for two-party computation. J. Cryptol. $22(2)$, 161–188 (2009)

[Yao82] Yao, A.C.: Protocols for secure computations (extended abstract). In: 23rd FOCS, pp. 160–164. IEEE Computer Society Press, November (1982)

[Yao86] Yao, A.C.: How to generate and exchange secrets (extended abstract). In: 27th FOCS, pp. 162–167. IEEE Computer Society Press, October 1986

# Round Complexity and Efficiency
# of Multi-party Computation

# Efficient Secure Multiparty Computation with Identifiable Abort

Carsten Baum[1]([✉]), Emmanuela Orsini[2], and Peter Scholl[2]

[1] Department of Computer Science, Aarhus University, Aarhus, Denmark
cbaum@cs.au.dk
[2] Department of Computer Science, University of Bristol, Bristol, UK
{emmanuela.orsini,peter.scholl}@bristol.ac.uk

**Abstract.** We study secure multiparty computation (MPC) in the dishonest majority setting providing security with identifiable abort, where if the protocol aborts, the honest parties can agree upon the identity of a corrupt party. All known constructions that achieve this notion require expensive zero-knowledge techniques to obtain active security, so are not practical.

In this work, we present the first efficient MPC protocol with identifiable abort. Our protocol has an information-theoretic online phase with message complexity $O(n^2)$ for each secure multiplication (where $n$ is the number of parties), similar to the BDOZ protocol (Bendlin et al., Eurocrypt 2011), which is a factor in the security parameter lower than the identifiable abort protocol of Ishai et al. (Crypto 2014). A key component of our protocol is a linearly homomorphic information-theoretic signature scheme, for which we provide the first definitions and construction based on a previous non-homomorphic scheme. We then show how to implement the preprocessing for our protocol using somewhat homomorphic encryption, similarly to the SPDZ protocol (Damgård et al., Crypto 2012).

**Keywords:** Secure multiparty computation · Identifiable abort

## 1 Introduction

Multiparty Computation deals with the problem of jointly computing a function among a set of mutually distrusting parties with some security guarantees such as

Full version available at http://eprint.iacr.org/2016/187.pdf

C. Baum—Part of the work was done while visiting University of Bristol. The author acknowledges support from the Danish National Research Foundation and The National Science Foundation of China (under the grant 61061130540) for the Sino-Danish Center for the Theory of Interactive Computation; and also from the CFEM research center (supported by the Danish Strategic Research Council) and the COST Action IC1306.

E. Orsini—Supported in part by ERC Advanced Grant ERC-2010-AdG-267188-CRIPTO.

P. Scholl—Supported in part by EPSRC via grant EP/I03126X, and in part by the DARPA Brandeis program and the US Navy under contract #N66001-15-C-4070.

© International Association for Cryptologic Research 2016
M. Hirt and A. Smith (Eds.): TCC 2016-B, Part I, LNCS 9985, pp. 461–490, 2016.
DOI: 10.1007/978-3-662-53641-4_18

correctness of the output and privacy of the inputs. MPC has been an interesting topic in cryptography for the last 30 years, but while in the past efficiency was the main bottleneck and MPC was exclusively the subject of academic studies, the situation has steadily improved and now even large circuits can be evaluated with acceptable costs in terms of time and space. A key example of this progress is the recent line of work that began with the BDOZ [BDOZ11] and SPDZ [DPSZ12, DKL+13] protocols. These protocols are based on a secret-sharing approach and can provide active security against a dishonest majority, where any number of the parties may be corrupt.

The SPDZ-style protocols work in the *preprocessing model* (or *offline/online* setting), with an offline phase that generates random correlated data independent of the parties' inputs and the function, and an online phase, in which this correlated randomness is used to perform the actual computation. The key advantage of the preprocessing model in SPDZ lies in the efficiency of the online phase, which only uses information-theoretic techniques.

It is a well-known fact that, in the dishonest majority setting, successful termination of protocols cannot be guaranteed, so these protocols simply *abort* if cheating is detected. It was also shown by Cleve in [Cle86] that, unless an honest majority is assumed, it is impossible to obtain protocols for MPC that provide *fairness* and *guaranteed output delivery*. Fairness is a very desirable property and intuitively means that either every party receives the output, or else no-one does.

In this scenario SPDZ-style protocols, and in general all known efficient MPC protocols that allow dishonest majority, are vulnerable to Denial-of-Service attacks, where one or more dishonest parties can force the protocol to abort, so that honest parties never learn the output. They can even do this *after learning the output*, whilst remaining anonymous to the honest parties, which could be a serious security issue in some applications. This motivates the notion of *MPC with identifiable abort* (ID-MPC) [CL14,IOZ14]. Protocols with identifiable abort either terminate, in which case all parties receive the output of the computation, or abort, such that all honest parties agree on the identity of at least one corrupt party. It is clear that, while this property neither guarantees fairness nor output delivery (as it does not prevent a corrupt party from aborting the protocol by refusing to send messages) at the same time it discourages this kind of behaviour because, upon abort, at least one corrupt party will be detected and can be excluded from future computations.

**Why Efficient ID-MPC is not Trivial.** It is easy to see that the SPDZ protocol is not ID-MPC: Each party holds an additive share $x_i$ of each value $x$ and similarly an additive share $m(x)_i$ of an information-theoretic MAC on $x$. To open a shared value, all parties provide their shares of both the value and its MAC, and then check validity of the MAC. A dishonest party $P_i$ can make the protocol abort by sending a share $x_i^* \neq x_i$ or $m(x)_i^* \neq m(x)_i$. However, since the underlying value $x$ is authenticated, and not the individual shares, $P_i$ is neither committed to $x_i$ nor $m(x)_i$, so other parties cannot identify who caused the abort. At first glance, it seems that the [BDOZ11] protocol might satisfy identifiable abort. In this protocol, instead of authenticating $x$, pairwise MACs are set up so that each party holds a MAC on every other party's share. However, the following counterexample (similar to [Sey12, Sect. 3.6]), depicted in Fig. 1, shows that this is not sufficient.

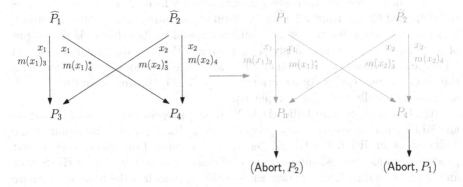

**Fig. 1.** Counterexample for identifiable abort with pairwise MACs

Let the adversary control parties $P_1$ and $P_2$. $P_1$ sends the correct value $x_1$ to both remaining parties $P_3, P_4$, but only the correct MAC $m(x_1)_3$ to $P_3$. To $P_4$, he sends an incorrect MAC $m(x_1)_4^*$. Conversely, $P_2$ will send the incorrect MAC $m(x_2)_3^*$ of his share $x_2$ to $P_3$ and the correct $m(x_2)_4$ to $P_4$. Now both honest parties $P_3, P_4$ can agree that some cheating happened, but as they do not agree on the identity of the corrupt party they are unable to reliably convince each other who cheated. (Note that a corrupt party could also decide to output (Abort, $P_3$), confusing matters even further for the honest $P_2, P_3$.) We conclude that, with an approach based on secret-sharing, special care must be taken so that all honest parties can agree upon the correctness of an opened value.

**Our Contributions.** In this work we present an efficient MPC protocol in the preprocessing model that reactively computes arithmetic circuits over a finite field, providing security with identifiable abort against up to $n - 1$ out of $n$ malicious parties. The online phase relies only on efficient, information-theoretic primitives and a broadcast channel, with roughly the same complexity as the BDOZ protocol [BDOZ11]. The offline phase, which generates correlated randomness, can be instantiated using somewhat homomorphic encryption based on ring-LWE, and allows use of all the relevant optimisations presented in [DPSZ12, DKL+13, BDTZ16].

A first building block towards achieving this goal is our definition of *homomorphic information-theoretic signatures* (HITS). Information-theoretic signature schemes [CR91] cannot have a public verification key (since otherwise an unbounded adversary can easily forge messages), but instead each party holds a private verification key. The main security properties of IT signatures are unforgeability and consistency, meaning that no-one can produce a signature that verifies by one honest party but is rejected by another. Swanson and Stinson [SS11] were the first to formally study and provide security proofs for IT signatures, and demonstrated that many subtle issues can arise in definitions. On the other hand, homomorphic signature schemes [BFKW09,CJL09] feature an additional *homomorphic evaluation* algorithm, which allows certain functions to be applied to signatures. The verification algorithm is then given a signature, a message $m$ and a description of a function $f$, and verifies that $m$ is the output of $f$, applied to some previously signed inputs. We give the first definition of HITS, and the first construction of HITS for affine functions, which is based on the (non-homomorphic) construction from [HSZI00] (proven secure in [SS11]), and has essentially the same complexity.

We then show how to build ID-MPC in the preprocessing model, based on any HITS with some extra basic properties. Our basic protocol is similar to the online phase of SPDZ and BDOZ, based on a correlated randomness setup that produces random shared multiplication triples, authenticated using HITS with an unknown signing key. The downside of this approach is the need for a secure broadcast channel in every round of the protocol. Since broadcast with up to $n-1$ corrupted parties requires $\Omega(n)$ rounds of communication[1] [GKKO07] (assuming a PKI setup), this leads to a round complexity of $\Omega(n \cdot D)$, for depth $D$ arithmetic circuits, and a message complexity of $\Omega(n^3)$ field elements per multiplication gate. The number of broadcast rounds can be reduced to just two—and the total number of rounds to $O(n + D)$—by using an insecure broadcast for each multiplication gate, and then verifying the insecure broadcasts at the end of the protocol in a single round of authenticated broadcast. Additionally, by batching the signature verification at the end of the protocol, we can reduce the message complexity per multiplication to $O(n^2)$. Overall, this gives on online phase that is only around $n$ times slower than the SPDZ protocol, or similar to BDOZ.

In addition, we present a preprocessing protocol that uses somewhat homomorphic encryption to compute the correlated randomness needed for the online phase of our protocol, obtaining security with identifiable abort. The method for creating multiplication triples is essentially the same as [DPSZ12], but creating the additional HITS data is more complex. In addition, we must ensure that the preprocessing protocol has identifiable abort as well.

In the full version of this work, we moreover give two interesting modifications of the scheme: We present an extension of our ID-MPC scheme that implies *verifiable abort*: In the ID-MPC setting, the honest parties agree upon which party is corrupt, but they are not able to convince anyone outside of the computation of this fact. We sketch how our scheme can be modified so that this in fact is possible, using a public bulletin board. We also present an information-theoretic MPC protocol in the preprocessing model that allows use of fields of

---

[1] This is not needed in SPDZ, because a simple 'broadcast with abort' technique can be performed in just two rounds.

substantially smaller size than in our main protocol (with an approach that is similar to the MiniMAC protocol [DZ13]).

**Comparison to Existing Work.** The model of identifiable abort was first explicitly defined in the context of covert security, by Aumann and Lindell [AL10]. Cohen and Lindell [CL14] considered the relationships between broadcast, fairness and identifiable abort, and showed that an MPC protocol with identifiable abort can be used to construct secure broadcast. The classic GMW protocol [GMW87] (and many protocols based on this) satisfies the ID-MPC property, but is highly impractical due to the non-black box use of cryptographic primitives.

The most relevant previous work is by Ishai et al. [IOZ14], who formally studied constructing identifiable abort, and presented a general compiler that transforms any semi-honest MPC protocol in the preprocessing model into a protocol with identifiable abort against malicious adversaries. Their protocol is information-theoretic, and makes use of the 'MPC-in-the-head' technique of Ishai et al. [IKOS07] for proving the correctness of each message in zero-knowledge. Although recent work by Giacomelli et al. [GMO16] shows that this technique can be efficient for certain applications, we show that when applied to ID-MPC as in [IOZ14], the resulting protocol is around $O(\kappa)$ times less efficient than ours, to achieve soundness error $2^{-\kappa}$. Note that Ishai et al. also use IT signatures for authenticating values, similarly to our usage, but without the homomorphic property that allows our protocol to be efficient. For the preprocessing stage, they describe an elegant transformation that converts any protocol for implementing any correlated randomness setup in the OT-hybrid model into one with identifiable abort, which makes black-box use of an OT protocol. Again, unfortunately this method is not particularly practical, mainly because it requires the OT protocol to be secure against an adaptive adversary, which is much harder to achieve than statically secure OT [LZ13].

Several other works have used similar primitives to information-theoretic signatures for various applications. In [CDD+99], a primitive called *IC signatures* is used for adaptively secure multiparty computation. These are very similar to what we use and their construction is linearly homomorphic, but the opening stage requires every party to broadcast values, whereas in our HITS only the sender broadcasts a message. Moreover, IC signatures are required to handle the case of a corrupted dealer (which we do not need, due to trusted preprocessing), and this leads to further inefficiencies. In [IOS12], a *unanimously identifiable commitment scheme* is presented, which is used to construct identifiable secret sharing; this has similarities to a simplified form of IC signatures, but is not linear.

Finally, we note that when the number of parties is constant, it is possible to achieve a relaxed notion of fairness, called *partial fairness*, in the dishonest majority setting by allowing a non-negligible distinguishing probability by the environment [BLOO11].

**Organisation.** In Sect. 2, we describe the model and some basic preliminaries, and also discuss the need for and use of a broadcast channel in our protocols. Section 3 introduces the definition of homomorphic information-theoretic signatures, and Sect. 4 describes our construction. Our information-theoretic ID-MPC protocol in the preprocessing model is presented in Sect. 5, followed by the preprocessing using SHE in Sect. 6. In Sect. 7 we evaluate the efficiency of our protocols, compared with the previous state of the art.

## 2 Preliminaries

### 2.1 Notation

Throughout this work, we denote by $\kappa$ and $\lambda$ the statistical, resp. computational, security parameters, and we use the standard definition of negligible (denoted by $\mathsf{negl}(\kappa)$) and overwhelming function from [Gol01]. We use bold lower case letters for vectors, i.e. $\mathbf{v}$ and refer to the $i$th element of a vector $\mathbf{v}$ as $\mathbf{v}|_i$. The notation $x \leftarrow S$ will be used for the uniform sampling of $x$ from a set $S$, and by $[n]$ we mean the set $\{1, \ldots, n\}$. The $n$ parties in the protocol are denoted as $\mathcal{P} = \{P_1, \ldots, P_n\}$, while the adversary is denoted by $\mathcal{A}$, and has control over a subset $\mathcal{I} \subset [n]$ of the parties. We also sometimes let $\mathcal{P}$ denote the index set $[n]$, depending on the context.

### 2.2 Model

We prove our protocols secure in the universal composability (UC) model of Canetti [Can01], with which we assume the reader has some familiarity. Our protocols assume a single static, active adversary, who can corrupt up to $n - 1$ parties at the beginning of the execution of a protocol, forcing them to behave in an arbitrary manner. We assume a *synchronous* communication model, where messages are sent in rounds, and a *rushing adversary*, who in each round, may receive the honest parties' messages before submitting theirs.

We use the UC definition of *MPC with identifiable abort*, or ID-MPC, from Ishai et al. [IOZ14]. Given any UC functionality $\mathcal{F}$, define $\left[\mathcal{F}\right]_\perp^{\mathsf{ID}}$ to be the functionality with the same behaviour as $\mathcal{F}$, except that at any time the adversary may send a special command (Abort, $P_i$), where $i \in \mathcal{I}$, which causes $\left[\mathcal{F}\right]_\perp^{\mathsf{ID}}$ to output (Abort, $P_i$) to all parties.

**Definition 1 ([IOZ14]).** *Let $\mathcal{F}$ be a functionality and $\left[\mathcal{F}\right]_\perp^{\mathsf{ID}}$ the corresponding functionality with identifiable abort. A protocol $\Pi$ securely realises $\mathcal{F}$ with identifiable abort if $\Pi$ securely realises the functionality $\left[\mathcal{F}\right]_\perp^{\mathsf{ID}}$.*

As noted in [IOZ14], the UC composition theorem [Can01] naturally extends to security with identifiable abort, provided that the higher-level protocol always respects the abort behaviour of any hybrid functionalities.

## 2.3   Broadcast Channel

Our protocols require use of a secure broadcast channel. Since Cohen and Lindell showed that MPC with identifiable abort can be used to construct a broadcast channel [CL14], and it is well known that secure broadcast is possible if and only if there are fewer than $n/3$ corrupted parties, it is not surprising that we assume this (the protocols in [IOZ14] require the same).

In practice, we suggest the broadcast primitive is implemented using *authenticated broadcast*, which exists for any number of corrupted parties, assuming a PKI setup. For example, the classic protocol of Dolev and Strong [DS83] uses digital signatures, and Pfitzmann and Waidner [PW92] extended this method to the information-theoretic setting. Both of these protocols have complexity $O(\ell n^2)$ when broadcasting $\ell$-bit messages. Hirt and Raykov [HR14] presented a protocol that reduces the communication cost to $O(\ell n)$ when $\ell$ is large enough. We are not aware of any works analysing the practicality of these protocols, so we suggest this as an important direction for future research.

## 3   Homomorphic Information-Theoretic Signatures

In this section, we define the notion of *homomorphic information-theoretic signatures* (HITS). It differs slightly from standard cryptographic signatures: First and foremost, in the information-theoretic setting, a signature[2] scheme must have a distinct, private verification key for each verifying party. This is because we define security against computationally unbounded adversaries, hence a verifier could otherwise easily forge signatures. Secondly, allowing homomorphic evaluation of signatures requires taking some additional care in the definitions. To prevent an adversary from exploiting the homomorphism to produce arbitrary related signatures, the verification algorithm must be given a function, and then verifies that the signed message is a valid output of the function on some previously signed messages.

With this in mind, our definition therefore combines elements from the IT signature definition of Swanson and Stinson [SS11], and (computational) homomorphic signature definitions such as [BFKW09, CJL09, GVW15].

**Definition 2 (Homomorphic Information-Theoretic Signature).** *A homomorphic information-theoretic signature (HITS) scheme for the set of verifiers* $\mathcal{P} = \{P_1, \ldots, P_n\}$*, function class* $\mathcal{F}$ *and message space* $\mathcal{M}$*, consists of a tuple of algorithms* (Gen, Sign, Ver, Eval) *that satisfy the following properties:*

$(\mathsf{sk}, \mathbf{vk}) \leftarrow \mathsf{Gen}(1^\kappa, w)$ *takes as input the (statistical) security parameter* $\kappa$ *and an upper bound* $w \in \mathbb{N}$ *on the number of signatures that may be created, and outputs the signing key* $\mathsf{sk}$ *and vector of the parties' (private) verification keys,* $\mathbf{vk} = (\mathsf{vk}_1, \ldots, \mathsf{vk}_n)$.

---

[2] We want to put forward that the name *signature* for the primitive in question can be somewhat misleading, as it shares properties with commitments and MACs. Nevertheless we decided to use the term for historical reasons.

$\sigma \leftarrow \text{Sign}(m, \text{sk})$ *is a deterministic algorithm that takes as input a message* $m \in \mathcal{M}$ *and signing key* $\text{sk}$, *and outputs a signature* $\sigma$.

$\sigma \leftarrow \text{Eval}(f, (\sigma_1, \ldots, \sigma_\ell))$ *homomorphically evaluates the function* $f \in \mathcal{F}$ *on a list of signatures* $(\sigma_1, \ldots, \sigma_\ell)$.

$0/1 \leftarrow \text{Ver}(m, \sigma, f, \text{vk}_j)$ *takes as input a message* $m$, *a signature* $\sigma$, *a function* $f$ *and* $P_j$*'s verification key* $\text{vk}_j$, *and checks that* $m$ *is the valid, signed output of* $f$.

*Remark 1.* HITS schemes can generally be defined to operate over *data sets*. Multiple data sets can be handled by tagging each dataset with a unique identifier and restricting operations to apply only to signatures with the same tag. However, for our application we only require a single dataset, which simplifies the definition.

*Remark 2.* To streamline the definition even more, we consider a setting where there is only one signer, who is honest. This leads to a definition that is conceptually simpler than the IT signature definition of Swanson and Stinson [SS11], which considers a group of users who can all sign and verify each others' messages.

We then define security as follows:

**Definition 3** ($w, \tau$-**security**). *A HITS scheme* ($\text{Gen}, \text{Sign}, \text{Ver}, \text{Eval}$) *is* ($w, \tau$)-*secure for a class of functions* $\mathcal{F}$ *and message space* $\mathcal{M}$ *if it satisfies the following properties:*

**Signing correctness:** *Let* $\ell \leq w$ *and define for* $i \in [\ell]$ *the projection function* $\pi_i(m_1, \ldots, m_\ell) = m_i$. *Then we require that for every pair* $(\text{sk}, \mathbf{vk})$ *output by* $\text{Gen}$, *for any* $(m_1, \ldots, m_\ell) \in \mathcal{M}^\ell$, *and for all* $i \in [\ell], j \in [n]$,

$$\text{Ver}(m_i, \text{Sign}(m_i, \text{sk}), \pi_i, \text{vk}_j) = 1.$$

**Evaluation correctness:** *For every pair* $\text{sk}, \mathbf{vk}$ *output by* $\text{Gen}$, *for every function* $f \in \mathcal{F}$, *for all messages* $(m_1, \ldots, m_\ell) \in \mathcal{M}^\ell$, *and for all* $j \in [n]$,

$$\text{Ver}\left(f(m_1, \ldots, m_\ell), \text{Eval}\left(f, (\text{Sign}(m_1, \text{sk}), \ldots, \text{Sign}(m_\ell, \text{sk}))\right), f, \text{vk}_j\right) = 1.$$

**Unforgeability:** *Let* $\mathcal{I} \subsetneq [n]$ *be an index set of corrupted verifiers, and define the following game between a challenger* $\mathcal{C}$ *and an adversary* $\mathcal{A}$:

1. $\mathcal{C}$ *computes* $(\text{sk}, \mathbf{vk}) \leftarrow \text{Gen}(1^\kappa, w)$ *and sends* $\{\text{vk}_i\}_{i \in \mathcal{I}}$ *to the adversary.*
2. $\mathcal{A}$ *may query* $\mathcal{C}$ *adaptively up to a maximum of* $w$ *times for signatures. Let* $m_1, \ldots, m_{w'}$ *be the list of messages queried to* $\mathcal{C}$.
3. $\mathcal{A}$ *outputs a function* $f \in \mathcal{F}$, *a list of indices* $\{i_1, \ldots, i_\ell\} \subseteq [w']$ *in ascending order, a target message* $m^*$ *and a signature* $\sigma^*$.
4. $\mathcal{A}$ *wins if* $m^* \neq f(m_{i_1}, \ldots, m_{i_\ell})$ *and there exists* $j \in [n] \backslash \mathcal{I}$ *for which*

$$\text{Ver}(m^*, \sigma^*, f, \text{vk}_j) = 1.$$

*A scheme is unforgeable if for any subset of corrupted verifiers $\mathcal{I} \subsetneq [n]$ and for any adversary $\mathcal{A}$,*

$$\Pr[\mathcal{A} \, wins] \leq \tau(|\mathcal{M}|, \kappa).$$

**Consistency:** *The security game for consistency is identical to the unforgeability game, except for the final step (the winning condition), which becomes:*
4. $\mathcal{A}$ wins[3] *if there exist $i, j \in [n] \backslash \mathcal{I}$ such that*

$$\mathsf{Ver}(m^*, \sigma^*, f, \mathsf{vk}_i) = 1 \quad and \quad \mathsf{Ver}(m^*, \sigma^*, f, \mathsf{vk}_j) = 0.$$

*A scheme satisfies* consistency *if for any set $\mathcal{I} \subsetneq [n]$ and for any $\mathcal{A}$ playing the above modified game,*

$$\Pr[\mathcal{A} \, wins] \leq \tau(|\mathcal{M}|, \kappa).$$

Note that evaluation correctness implies signing correctness, but we state two separate properties for clarity. The *consistency* (or *transferability*) property guarantees that a corrupted party cannot create a signature $\sigma$ that will be accepted by one (honest) verifier but rejected by another. In [SS11], a reduction from consistency to unforgeability is given. However, their definition of IT signatures considers a group of users who are all signers and verifiers, any of whom may be corrupted. In our setting, there is a single, honest signer, so consistency is no longer implied and must be defined separately.

Additionally, we require that signatures output by the Eval algorithm do not reveal any information on the input messages $m_1, \ldots, m_\ell$ other than that given by $f(m_1, \ldots, m_\ell)$. This is similar to the concept of *context hiding* [GVW15] in the computational setting, and is captured by the following definition.

**Definition 4 (Evaluation privacy).** *A HITS scheme (Gen, Sign, Eval, Ver) is* evaluation private *if there exists a PPT algorithm* Sim *that, for every* $(\mathsf{sk}, \mathbf{vk}) \leftarrow$ Gen, *for every function $f \in \mathcal{F}$, for all messages $m_1, \ldots, m_\ell$ with $m = f(m_1, \ldots, m_\ell)$, $\sigma_i = \mathsf{Sign}(m_i, \mathsf{sk})$ and $\sigma = \mathsf{Eval}(f, \sigma_1, \ldots, \sigma_\ell)$, computes*

$$\mathsf{Sim}(\mathsf{sk}, m, f) = \sigma.$$

Intuitively, this means that any valid signature that comes from Eval can also be computed without knowing the original inputs to $f$, so is independent of these. This definition is simpler than that of [GVW15], as our signing algorithm is restricted to be deterministic, so we require equality rather than an indistinguishability-based notion.

# 4 Construction of HITS

We now describe our construction of homomorphic information-theoretic signatures. The message space $\mathcal{M}$ is a finite field $\mathbb{F}$. We restrict the function class $\mathcal{F}$

---

[3] There is no requirement that $m^* \neq f(m_{i_1}, \ldots, m_{i_\ell})$.

to be the set of all affine transformations $f : \mathbb{F}^w \to \mathbb{F}$ (where $w$ is the maximum number of signatures that can be produced). The general case of affine functions with fewer than $w$ inputs can be handled by using a default value, $\perp$, for the unused input variables. Note also that the signing algorithm is *stateful*, and must keep track of how many messages have been signed previously.

Gen($1^\kappa, w$): The key generation algorithm is as follows:
1. Sample $\hat{\alpha}_1, \ldots, \hat{\alpha}_n \leftarrow \mathbb{F}$ and $\hat{\beta}_{i,1}, \ldots, \hat{\beta}_{i,n} \leftarrow \mathbb{F}$ for each $i \in [w]$.
2. For each verifier $P_j$, sample $\mathbf{v}_j = (v_{j,1}, \ldots, v_{j,n}) \leftarrow \mathbb{F}^n$ and compute

$$\alpha_j = \sum_{r=1}^n \hat{\alpha}_r \cdot v_{j,r} \quad \text{and} \quad \beta_{j,i} = \sum_{r=1}^n \hat{\beta}_{i,r} \cdot v_{j,r} \quad \text{for } i \in [w].$$

3. Output $\mathsf{sk} = \left( \left\{ \hat{\alpha}_r, \{\hat{\beta}_{i,r}\}_{i \in [w]} \right\}_{r=1}^n \right), \mathsf{vk} = \left( \mathbf{v}_j, \alpha_j, \{\beta_{j,i}\}_{i \in [w]} \right)_{j=1}^n$.

Sign($m, \mathsf{sk}$): To sign the $i$-th message, $m$, (for $i \leq w$) the signer computes the vector

$$\boldsymbol{\sigma}_i = \left( \hat{\alpha}_r \cdot m + \hat{\beta}_{i,r} \right)_{r=1}^n.$$

Eval($f, (\boldsymbol{\sigma}_1, \ldots, \boldsymbol{\sigma}_w)$): Let $f : \mathbb{F}^w \to \mathbb{F}$ be defined by

$$f(x_1, \ldots, x_w) = \mu_1 \cdot x_1 + \cdots + \mu_w \cdot x_w + c,$$

with $\mu_i, c \in \mathbb{F}$. The new signature $\boldsymbol{\sigma}$ is obtained by evaluating $f$, excluding the constant term, over every component of the input signatures:

$$\boldsymbol{\sigma} = \mu_1 \cdot \boldsymbol{\sigma}_1 + \cdots + \mu_w \cdot \boldsymbol{\sigma}_w \in \mathbb{F}^n.$$

Ver($m, \boldsymbol{\sigma}, f, \mathsf{vk}_j$): First use $f$ to compute the additional verification data

$$\beta_j = \sum_{i=1}^w \mu_i \cdot \beta_{j,i} - c \cdot \alpha_j.$$

Then check that

$$\beta_j + \alpha_j \cdot m = \sum_{r=1}^n \boldsymbol{\sigma}|_r \cdot v_{j,r}.$$

If the check passes output 1, otherwise 0.

**Theorem 1.** *Let $\mathbb{F}$ be a finite field, $\mathcal{M} := \mathbb{F}$ and $\mathcal{F}$ be the the set of affine maps from $\mathbb{F}^w$ to $\mathbb{F}$, then the tuple of algorithms* (Gen, Sign, Eval, Ver) *is a $(w, 3/|\mathbb{F}|)$-secure HITS with evaluation privacy.*

*Proof.* See the full version of this work. ☐

As an immediate consequence of the previous theorem, we have:

**Corollary 1.** *Let* $|\mathbb{F}| > 2^\kappa, w = \text{poly}(\kappa)$ *then* (Gen, Sign, Eval, Ver) *is a* $(\text{poly}(\kappa), \text{negl}(\kappa))$*-secure* HITS *with evaluation privacy.*

**Proposition 1.** *Let* $\mathcal{SG}, \mathcal{SK}$ *and* $\mathcal{VK}$ *be the domains of the signatures, signing and verification keys, respectively. Our* HITS *has the following memory sizes:*

$$|\mathcal{SG}| = |\mathbb{F}|^n, \qquad |\mathcal{SK}| = |\mathbb{F}|^{n(w+1)}, \qquad |\mathcal{VF}| = |\mathbb{F}|^{n(w+3)}.$$

In terms of signature size, our scheme is $n$-bits close to the lower bound for MRA-codes, which HITS is a special case of (see the full version of this work for more details). Note that the scheme of [SS11], which is not homomorphic, requires the same memory size as HITS for signatures and signing keys, but has slightly smaller verification keys ($|\mathbb{F}|^{n(w+2)-1}$). We do not know if the signature schemes described in [SS11] and our HITS are optimal in terms of the memory size of the keys, or if the need for larger verification keys in HITS is due to the homomorphic property of the scheme.

## 5   Online Phase for Efficient MPC with Identifiable Abort

In this section we describe our information-theoretic protocol for secure multiparty computation with identifiable abort in the preprocessing model. We assume a set of $n$ parties $\mathcal{P} = \{P_1, \ldots, P_n\}$, and any HITS scheme HITS = (Gen, Sign, Eval, Ver) that satisfies $(w, \text{negl}(\kappa))$-security and evaluation privacy from Sect. 3, and supports homomorphic evaluation of linear functions over a message space of a finite field $\mathbb{F}$.

Our protocol performs reactive computation of arithmetic circuits over $\mathbb{F}$, using correlated randomness from a preprocessing setup, similarly to the BDOZ and SPDZ protocols [BDOZ11, DPSZ12]. Correctness, privacy and identifiable abort are guaranteed by the security properties of HITS. The functionality that we implement is $\mathcal{F}_{\text{MPC}}$, shown in Fig. 2. Note that $\mathcal{F}_{\text{MPC}}$ already contains an explicit command for identifiable abort in the output stage, since it models an unfair execution where the adversary can abort after learning the output. The modified functionality $[\mathcal{F}_{\text{MPC}}]_\perp^{\text{ID}}$ then extends this abort to be possible at any time.

**Authenticated Secret Sharing.** Our protocol is based on authenticated additive secret sharing over the finite field $\mathbb{F}$, and we use the following notation to represent a shared value $a$:

$$[\![a]\!] = \big(a_i, \sigma_{a_i}\big)_{i \in \mathcal{P}},$$

where party $P_i$ holds $a_i \in \mathbb{F}$ and $\sigma_{a_i} = \text{Sign}(a_i, \text{sk})$, such that $\sum_{i \in \mathcal{P}} a_i = a$.

By the linearity of the secret sharing scheme and HITS we can easily define addition of two shares, $[\![z]\!] = [\![x]\!] + [\![y]\!]$, as follows:

1. Compute $z_i = x_i + y_i$.

---

**Functionality $\mathcal{F}_{\mathsf{MPC}}$**

Let $\mathcal{I}$ be the set of indices of corrupt parties.

**Input:** On input $(\mathsf{Input}, P_i, id, x)$ from $P_i$ and $(\mathsf{Input}, P_i, id)$ from all other parties, with $id$ a fresh identifier and $x \in \mathbb{F}$, store $(id, x)$.

**Add:** On input $(\mathsf{Add}, id_1, id_2, id_3)$ from all parties (where $id_1, id_2$ are present in memory), retrieve $(id_1, x)$, $(id_2, y)$ and store $(id_3, x + y)$.

**Multiply:** On input $(\mathsf{Mult}, id_1, id_2, id_3)$ from all parties (where $id_1, id_2$ are present in memory), retrieve $(id_1, x)$, $(id_2, y)$ and store $(id_3, x \cdot y)$.

**Output:** On input $(\mathsf{Output}, id)$ from all parties (where $id$ is present in memory), retrieve $(id, y)$ and send $y$ to the adversary. Wait for the adversary to input either $\mathsf{Deliver}$ or $(\mathsf{Abort}, P_i)$ for some $i \in \mathcal{I}$. If the adversary sends $\mathsf{Deliver}$ then output $y$ to all honest parties, otherwise output $(\mathsf{Abort}, P_i)$.

---

**Fig. 2.** Ideal functionality for reactive MPC in the finite field $\mathbb{F}$.

2. Compute $\sigma_{z_i} = \mathsf{Eval}(f, (\sigma_{x_i}, \sigma_{y_i}))$, where $f(a, b) = a + b$.
3. Output $[\![z]\!] = (z_i, \sigma_{z_i})_{i \in \mathcal{P}}$.

Note that if $\sigma_{x_i}, \sigma_{y_i}$ are already outputs of the $\mathsf{Eval}$ algorithm, then $f$ should instead be defined to include the linear function that was applied to these inputs previously, and $\mathsf{Eval}$ applied to those inputs. However, this is just a technicality and in practice, each homomorphic addition can be computed on-the-fly. We can also define addition or multiplication of shared values by constants, using $\mathsf{Eval}$ in a similar way.[4]

---

$\mathsf{Open}([\![a]\!])$:
1. Every party $P_i \in \mathcal{P}$ broadcasts $(a_i, \sigma_{a_i})$.
2. Each party $P_i$ runs $\mathsf{Ver}(a_j, \sigma_{a_j}, f, \mathsf{vk}_i)$, for each $j \neq i$. If for some $j$ the check fails, $P_i$ outputs $(\mathsf{Abort}, P_j)$, otherwise it outputs $a = \sum_{i \in \mathcal{P}} a_i$.

---

**Fig. 3.** Procedure for opening an authenticated, shared value.

In Fig. 3 we define the basic subprotocol used to open authenticated, shared values. Each time the command $\mathsf{Open}$ is called, parties check the correctness of the opened value using the $\mathsf{Ver}$ algorithm. For each share, the intuition is that if the corresponding signature is verified, then the share is correct with overwhelming probability due to the unforgeability of the scheme; on the contrary,

---

[4] For addition with a constant, only one party (say $P_1$) needs to adjust their share. Signatures stay the same, as the verification algorithm accounts for the constant term in the affine function.

if there exists an index $j \in \mathcal{P} \backslash \mathcal{I}$, where $\mathcal{I}$ denotes the set of corrupt parties, for which the check does not go through, then the same happens for all honest parties, due to the consistency of HITS.

**Preprocessing Requirements.** The preprocessing functionality, $\mathcal{F}_{\mathsf{Prep}}$, is shown in Fig. 4. It generates a set of HITS keys $(\mathsf{sk}, \mathbf{vk})$ and gives each party a verification key, whilst no-one learns the signing key. The functionality then computes two kinds of authenticated data, using $\mathsf{sk}$:

- **Input tuples:** Random shared values $\llbracket r \rrbracket$, such that one party, $P_i$, knows $r$. This is used so that $P_i$ can provide input in the online phase.
- **Multiplication triples:** Random shared triples $\llbracket a \rrbracket, \llbracket b \rrbracket, \llbracket c \rrbracket$, where $a, b \leftarrow \mathbb{F}$ and $c = a \cdot b$.

Note that corrupted parties can always choose their own shares of authenticated values, instead of obtaining random shares from the functionality.

**Protocol.** Our protocol, shown in Fig. 5, is based on the idea of securely evaluating the circuit gate by gate in a shared fashion, using the linearity of the $\llbracket \cdot \rrbracket$-representation for computing all linear gates, preprocessed multiplication triples for multiplication using Beaver's technique [Bea91], and preprocessed input tuples for the inputs.

---

### Functionality $\mathcal{F}_{\mathsf{Prep}}$

Let $\mathcal{I}$ be the set of indices of corrupt parties. The functionality is parametrised by the statistical security parameter, $\kappa$.

**Initialise:**   On input $(\mathsf{Init}, w)$ from all parties, do the following:
1. Compute $(\mathsf{sk}, \mathbf{vk}) \leftarrow \mathsf{Gen}(1^\kappa, w)$.
2. Send $\mathsf{vk}_i$ to party $P_i$ and store $\mathsf{sk}$.

**Macro Bracket:**   On input $(\mathsf{Bracket}, x)$, create the representation $\llbracket x \rrbracket$ as follows:
1. Receive shares $x_i$ for $i \in \mathcal{I}$ from the adversary.
2. Sample random shares $x_i \leftarrow \mathbb{F}$, for $i \notin \mathcal{I}$, subject to the constraint that $x = x_1 + \cdots + x_n$.
3. For $i = 1, \ldots, n$, compute $\sigma_{x_i} = \mathsf{Sign}(x_i, \mathsf{sk})$ and return $\{x_i, \sigma_{x_i}\}_{i \in [n]}$.

**Input:**   On input $(\mathsf{Input}, P_i)$ from all parties, sample a random $r \leftarrow \mathbb{F}$ and run $(\mathsf{Bracket}, r)$ to obtain $\llbracket r \rrbracket$. Output $(r_j, \sigma_{r_j})$ to each party $P_j$, and also $r$ to $P_i$.

**Triple:**   On input $(\mathsf{Triple})$ from all parties, do the following:
1. Sample $a, b \leftarrow \mathbb{F}$ and let $c = a \cdot b$.
2. Run the macro $(\mathsf{Bracket})$ on input $a, b$ and $c$ to obtain $\llbracket a \rrbracket, \llbracket b \rrbracket, \llbracket c \rrbracket$. Output $(a_i, b_i, c_i, \sigma_{a_i}, \sigma_{b_i}, \sigma_{c_i})$ to each party $P_i$.

---

**Fig. 4.** Ideal functionality for the preprocessing phase.

---

**Protocol $\Pi_{\text{Online}}$**

Let $n_M$ be the number of secure multiplications to be performed and $n_I$ the total number of inputs.

**Initialise:** The parties call $\mathcal{F}_{\text{Prep}}$ with $(\text{Init}, n \cdot (3n_M + n_I))$. If $\mathcal{F}_{\text{Prep}}$ outputs $(\text{Abort}, P_i)$, the parties output $(\text{Abort}, P_i)$ and halt.

**Input:** For party $P_i$ to input $x \in \mathbb{F}$, the parties call $\mathcal{F}_{\text{Prep}}$ with $(\text{Input}, P_i)$ to obtain a mask value $[\![r]\!]$, and $P_i$ also obtains $r$:
1. $P_i$ sets $m = r - x$ and broadcasts $m$.
2. All parties locally compute $[\![x]\!] = [\![r]\!] - m$.

**Add:** On input $([\![x]\!], [\![y]\!])$, parties locally compute $[\![x + y]\!] = [\![x]\!] + [\![y]\!]$.

**Multiply:** On input $([\![x]\!], [\![y]\!])$, the parties do the following:
1. Take one multiplication triple $([\![a]\!], [\![b]\!], [\![c]\!])$ from $\mathcal{F}_{\text{Prep}}$, compute $[\![\epsilon]\!] = [\![x]\!] - [\![a]\!]$, $[\![\rho]\!] = [\![y]\!] - [\![b]\!]$.
2. Call $\text{Open}([\![\epsilon]\!])$ and $\text{Open}([\![\rho]\!])$.
3. Locally compute $[\![z]\!] = [\![c]\!] + \epsilon \cdot [\![b]\!] + \rho \cdot [\![a]\!] + \epsilon \cdot \rho$.

**Output:** To output a share $[\![y]\!]$, the parties call $\text{Open}([\![y]\!])$. If for some $i \in \mathcal{P}$ the check fails, output $(\text{Abort}, i)$ and halt, otherwise accept $y$ as a valid output.

---

Fig. 5. Operations for secure function evaluation with identifiable abort.

## 5.1   Security

**Theorem 2.** *In the $\mathcal{F}_{\text{Prep}}$-hybrid model, the protocol $\Pi_{\text{Online}}$ implements $[\mathcal{F}_{\text{MPC}}]_{\perp}^{\text{ID}}$ with statistical security against any static active adversary corrupting up to $n - 1$ parties.*

*Proof.* Let $\mathcal{A}$ be a malicious PPT real adversary attacking the protocol $\Pi_{\text{Online}}$, we construct an ideal adversary $\mathcal{S}$ with access to $\mathcal{F}_{\text{MPC}}$ which simulates a real execution of $\Pi_{\text{Online}}$ with $\mathcal{A}$ such that no environment $\mathcal{Z}$ can distinguish the ideal process with $\mathcal{S}$ and $\mathcal{F}_{\text{MPC}}$ from a real execution of $\Pi_{\text{Online}}$ with $\mathcal{A}$. $\mathcal{S}$ starts by invoking a copy of $\mathcal{A}$ and running a simulated interaction of $\mathcal{A}$ with $\mathcal{Z}$.

After describing the simulator we will argue indistinguishability of the real and ideal worlds. Let $\mathcal{I}$ be the index set of corrupt parties, simulation proceeds as follows:

*Simulating the* **Initialise** *Step.* The simulator $\mathcal{S}$ honestly emulates $\mathcal{F}_{\text{Prep}}$ towards the adversary $\mathcal{A}$. Note that $\mathcal{S}$ knows all the data given to the adversary and the simulated signing key $\text{sk}^*$ of HITS, so can generate a valid signature for any message.

*Simulating the* **Input** *Step.* We distinguish two cases:

- For $i \in \mathcal{P} \backslash \mathcal{I}$, $\mathcal{S}$ emulates towards $\mathcal{A}$ a broadcast of a random value $m \in \mathbb{F}$, and proceeds according to the protocol with the next simulated random input tuple, $r$. Thereafter, $S$ computes $x = r - m$ and stores $x$, the dummy, random input for honest $P_i$.

- For $i \in \mathcal{I}$, $S$ receives from the adversary the message $m$, and retrieves the next random input tuple $r$. It then computes $x = r - m$ and inputs it to the $[\mathcal{F}_{\mathsf{MPC}}]_\perp^{\mathsf{ID}}$.

*Simulating the* **Circuit Evaluation.** For linear gates, the simulator does not have to simulate any message on the behalf of the honest parties. $S$ updates the internal shares and calls the respective procedure in $[\mathcal{F}_{\mathsf{MPC}}]_\perp^{\mathsf{ID}}$.

In a multiplication gate, for each call to Open, $S$ receives all the corrupt shares $(t_j^*, \sigma_{t_j^*})$ from $\mathcal{A}$, and computes and sends the shares and signatures for the dummy honest parties as in the protocol. Let $(t_j, \sigma_{t_j})$ be the values that $S$ expects from the dishonest $P_j$, based on previous computations and the simulated preprocessing data. $S$ checks for all the $(t_j^*, \sigma_{t_j^*})$ received from $\mathcal{A}$ and for all $i \in \mathcal{P} \backslash \mathcal{I}$ that $t_j = t_j^*$ and $\sigma_{t_j} = \sigma_{t_j^*}$. If the check does not pass for some $j \in \mathcal{I}$ then $S$ sends $(\mathsf{Abort}, P_j)$ to $[\mathcal{F}_{\mathsf{MPC}}]_\perp^{\mathsf{ID}}$. Otherwise it proceeds.[5]

*Simulating the* **Output** *Step.* The simulator sends $(\mathsf{Output})$ to the functionality and gets the result $y$ back. Let $y'$ be the output value that the simulator has computed using dummy, random inputs on behalf of the honest parties. Then it picks an honest party $P_{i_0}$, and modifies its share as $y_{i_0}^* = y_{i_0} + (y - y')$, then uses the evaluation privacy algorithm to compute $\sigma_{y_{i_0}^*} = \mathsf{Sim}(\mathsf{sk}^*, y_{i_0}^*, f)$, where $f$ is the same linear function that has been applied to obtain $\sigma_{y_{i_0}}$, and sends the honest shares and signatures to the adversary. It then receives $(y_j^*, \sigma_{y_j^*})_{j \in \mathcal{I}}$ from the adversary, while expecting $y_j, \sigma_{y_j}$. If $y_j = y_j^*$ and $\sigma_{y_j} = \sigma_{y_j^*}$ for all $j \in \mathcal{I}$ then $S$ sends Deliver to the functionality; otherwise it sends $(\mathsf{Abort}, P_j)$ for the lowest $j$ that failed and halts.

*Indistinguishability.* Now we prove that the all the simulated transcripts and the honest parties' outputs are identically distributed to the real transcripts and output in the view of the environment $\mathcal{Z}$, except with probability $\mathsf{negl}(\kappa)$.

During *initialisation*, the simulator honestly runs an internal copy of $\mathcal{F}_{\mathsf{Prep}}$, so the simulation of this step is perfect. In the *input* step, the values $m$ broadcast by honest parties are uniformly random in both cases, as they are masked by a one-time uniformly random value from $\mathcal{F}_{\mathsf{Prep}}$ that is unknown to $\mathcal{Z}$.

In the *multiplication* step, the parties call the command Open. Honest shares and signatures are simulated as in the protocol, using the simulated data from the emulation of $\mathcal{F}_{\mathsf{Prep}}$, and applying the Eval algorithm. The broadcast shares are all uniformly distributed in both worlds, as the shares are always masked by fresh random values from $\mathcal{F}_{\mathsf{Prep}}$, so are perfectly indistinguishable. To argue indistinguishability of the signatures, we need to use the evaluation privacy property.

---

[5] If there is more than invalid share then we always abort with the smallest index where the check fails.

We must prove that

$$\sigma_{t_i} \stackrel{s}{\approx} \sigma_{t_i^*},$$

where $\{\sigma_{t_i^*}\}_{i \notin \mathcal{I}}$ are the simulated ideal-world signatures, and $\{\sigma_{t_i}\}_{i \notin \mathcal{I}}$ are the real-world signatures, for some honest parties' shares $\{t_i\}_{i \notin \mathcal{I}}$.

Since $\sigma_{t_i}$ and $\sigma_{t_i^*}$ are both valid signatures output from Eval, evaluation privacy guarantees that there exists an algorithm Sim such that:

$$\sigma_{t_i} = \mathsf{Sim}(\mathsf{sk}, t_i, g) \quad \text{and} \quad \sigma_{t_i^*} = \mathsf{Sim}(\mathsf{sk}^*, t_i^*, g),$$

where $g$ is the linear function evaluated to get the values $t_i$ and $t_i^*$, and sk and $\mathsf{sk}^*$ are respectively the real-world and ideal-world secret keys. Since $(t_i, \mathsf{sk})$ and $(t_i^*, \mathsf{sk}^*)$ are identically distributed in both the executions, then the same holds for $\sigma_{t_i}$ and $\sigma_{t_i^*}$. Note that it is crucial here that $\mathcal{S}$ computes $\sigma_{t_i^*}$ using Eval and the function $g$, rather than creating a fresh signature using $\mathsf{sk}^*$, otherwise indistinguishability would not hold.

We also must consider the abort behaviour of the Open procedure in the two worlds. If during any opening, $\mathcal{A}$ attempts to open a fake value then it will always be caught in the simulation, whereas it succeeds if it is able to forge a corresponding signature in the real protocol. Hence, if the ideal protocol aborts with the identity of some corrupt party $P_i$, then the same happens in the real world, except with negligible probability due to unforgeability. The consistency property of HITS ensures that if one honest party outputs $(\mathsf{Abort}, P_i)$ in the protocol, then all the honest parties will output the same, except with negligible probability.

Now, if the real or simulated protocol proceeds to the last step, $\mathcal{Z}$ observes the *output* value $y$, and the corresponding honest parties' shares, together with their signatures. The honest shares are consistent with $y$ and the signatures are correctly generated in both worlds. Again, to argue indistinguishability of the signatures we can use the evaluation privacy property of HITS. Hence $\mathcal{Z}$'s view of the honest parties' messages in the last step has the same distribution in the real and hybrid execution.

Finally, we must argue indistinguishability of the outputs in both worlds. In the ideal world, the output $y$ is a correct evaluation on the inputs, so the only way the environment can distinguish is to produce an incorrect output in the real world. This can only happen if a corrupt party sending an incorrect share that is successfully verified. However, as we have seen before, if the adversary attempts to open a fake value, during the input, multiplication or output step, then it will be caught with overwhelming probability, by the unforgeability and consistency properties of HITS. □

## 5.2   An Optimised Protocol

When instantiated with our HITS scheme from Sect. 4, the online phase protocol above requires $O(n^2)$ field elements to be broadcasted per secure multiplication.

Since each authenticated broadcast requires $O(n)$ rounds, this gives a communication complexity of at least $O(n^3)$ field elements per multiplication and $O(D \cdot n)$ rounds overall, where $D$ is the multiplicative depth of the arithmetic circuit. We now describe an optimised variant of our protocol, which reduces the number of rounds to $O(D + n)$ and the communication cost per multiplication to $O(n^2)$.

**Reducing the Number of Broadcasts.** Let $\Pi_{bc}$ be the UC protocol for authenticated broadcast used in the protocol. We make the following assumption about its structure: in the first round of $\Pi_{bc}$, the sender (with input $x$) sends $x$, and nothing more, to all parties.[6] Let the remainder of the protocol be denoted $\Pi'_{bc}$.

We now modify the protocol $\Pi_{\mathsf{Online}}$ so that whenever a party $P_i$ is supposed to broadcast a value $x_i$ in the Open subprotocol, $P_i$ instead sends $x_i$ to all parties, and appends $x_i$ to a list $B_i$. Note that the **Input** stage still requires broadcast, as otherwise it seems difficult for the simulator to extract a corrupted party's input. The **Output** stage is then modified so that first, each party runs $\Pi'_{bc}(B_i)$ to complete the broadcasts that were initialised in the previous rounds. With this change, there are only two broadcast rounds and each multiplication gate requires just one round of communication, reducing the overall number of rounds to $O(D + n)$.

**Batching the Signature Verification.** We can reduce the number of field elements sent during a multiplication to $n - 1$ per party by checking all signatures together in the **Output** stage of the protocol, rather than during the circuit evaluation. This means that during the computation, the parties only send shares without the corresponding signatures. We then check a random linear combination of each parties' signatures just before every output stage.

The complete protocol for the optimised output stage is given in Fig. 6. Since there are only two authenticated broadcast rounds, the number of rounds for computing a depth $D$ circuit with one output gate in the optimised protocol is $O(D + n)$. The total number of field elements sent over the network is no more than[7]

$$n_I \cdot \mathsf{bc}(1) + 2n(n - 1) \cdot n_M + n \cdot \mathsf{bc}(n + 2n_M + 1),$$

where $n_I$ is the total number of private inputs, $n_M$ the number of secure multiplications and $\mathsf{bc}(m)$ the cost of broadcasting $m$ elements using $\Pi_{bc}$. Meanwhile, the storage cost (for the preprocessing data) is $O(n(n_M + n_I))$ per party.

A drawback of this optimization is that in comparison to $\Pi_{\mathsf{Online}}$ a party that sends corrupt signatures $\sigma$ will now only be caught *after $\mathcal{A}$ learns the output.*

---

[6] Almost any broadcast protocol can be easily converted into this form. For example, the Dolev-Strong broadcast [DS83] begins with the sender sending $(x, \mathsf{Sign}(x))$ to all parties; we split this up into one round for $x$ and one round for $\mathsf{Sign}(x)$.

[7] Excluding the cost of $\mathcal{F}_{\mathsf{Rand}}$, which can be implemented using standard techniques such as a hash-based commitment scheme in the random oracle model.

---

**Protocol Output($[\![y]\!]$)**

Let $B_i$ be the list of shares sent by a party $P_i$ in all input and multiplication steps since the last Output invocation.

**OutputCheck:**    Run the following, for each party $P_i$:
1. $P_i$ sends the share $y_i$ to all other parties, and appends $y_i$ to $B_i$.
2. Write $B_i = (a_1, \ldots, a_t)$.
3. Sample $(r_1, \ldots, r_t) \in \mathbb{F}^t$ using $\mathcal{F}_{\mathsf{Rand}}$.
4. $P_i$ computes $\sigma = \mathsf{Eval}(f, (\sigma_{a_1}, \ldots, \sigma_{a_t}))$, where the linear function $f$ is defined as $f(x_1, \ldots, x_t) = r_1 \cdot x_1 + \cdots + r_t \cdot x_t$.
5. $P_i$ invokes $\Pi_{\mathsf{bc}}(\sigma)$ and $\Pi'_{\mathsf{bc}}(a_1, \ldots, a_t)$ to broadcast $\sigma$ and complete the broadcast of the shares $a_1, \ldots, a_t$. If the broadcast fails, output $(\mathsf{Abort}, P_i)$.
6. Every party $P_j$ (for $j \neq i$) computes $f(a_1, \ldots, a_t) = a$ and checks if $\mathsf{Ver}(a, \sigma, f, \mathsf{vk}_j) = 1$. If the check fails, output $(\mathsf{Abort}, P_i)$.

**Output:**    Each party computes $y = \sum_{i=1}^n y_i$ and outputs $y$.

Fig. 6. Output stage of the optimised online protocol.

---

**Functionality $\mathcal{F}_{\mathsf{Rand}}(\mathbb{F})$**

**Random sample:** Upon receiving $(rand; u)$ from all parties, it samples a uniform $r \in \mathbb{F}$ and outputs $(rand, r)$ to all parties.

---

Fig. 7. Functionality $\mathcal{F}_{\mathsf{Rand}}$ that provides random values to all parties.

We stress that this is according to the definition of *identifiable abort* (which does not specify when the abort signal must be sent), but different from $\Pi_{\mathsf{Online}}$ where such behaviour would immediately be detected.

**Security of the Modified Online Phase.** We now argue security of the new online protocol, describing the key differences compared with the previous protocol. In the simulation, the simulator $\mathcal{S}$ now cannot determine whether a corrupt party has sent the correct message during the **Multiply** command, since the signatures are not sent here. Instead, this must be detected in the output stage when the broadcasts and signatures are checked.

In the **OutputCheck** stage, $\mathcal{S}$ first calls the functionality $[\mathcal{F}_{\mathsf{MPC}}]_\perp^{\mathsf{ID}}$ to obtain the result $y$, then adjusts one honest party's share and signature (using the evaluation privacy algorithm) to fix the correct output as before, and sends the honest shares to the adversary. For the remainder of this stage, the simulator acts as in the protocol for the honest parties, computing the random linear combination of signatures using $\mathsf{Eval}$, and then runs the simulator of $\Pi_{\mathsf{bc}}$ for each broadcast. If any broadcast fails for a corrupt sender $P_j$ then $\mathcal{S}$ sends $(\mathsf{Abort}, P_j)$ to $[\mathcal{F}_{\mathsf{MPC}}]_\perp^{\mathsf{ID}}$. If all broadcasts succeed, $\mathcal{S}$ checks the signatures and sends $(\mathsf{Abort}, P_j)$ if the signature of $P_j$ does not verify. Note that an incorrect broadcast can lead to an honest party's signature being incorrect, so it is important that the broadcasts are checked first here.

Indistinguishability of all shares sent up until the **Output** stage follows from uniformity of the preprocessing data, as in the previous protocol. The security of the $\Pi_{bc}$ simulator guarantees indistinguishability of step 5, in particular that all parties agree upon the sets of shares $B_i$ that were sent by each party $P_i$ during the protocol.

If the broadcasts succeed then the honest parties' signatures will always be correctly generated, and the evaluation privacy property of HITS guarantees they are identically distributed. The environment therefore can only distinguish between the worlds by causing the output, $y$, to be incorrect. Suppose a corrupt party $P_i$ broadcasts the values $B_i' = (a_1', \ldots, a_t')$ in the protocol, and $a_j \neq a_j'$ for at least one $j$, where $a_j$ is the original signed value that $P_i$ was supposed to send. Then if the verification in step 5 of the output stage succeeds, the correctness and security properties of HITS guarantee that:

$$\sum_{i=1}^{t}(a_i - a_i') \cdot r_i = 0,$$

It is easy to see that the probability of this check passing is $1/|\mathbb{F}|$, as the values $r_i$ are unknown to the adversary at the time of choosing $a_i'$, so the check prevents an incorrect output with overwhelming probability.

## 6    Preprocessing with Identifiable Abort

This section describes a practical protocol for securely implementing $\mathcal{F}_{Prep}$ with identifiable abort, based on somewhat homomorphic encryption. The protocol is based on the SPDZ preprocessing [DPSZ12, DKL+13], but the cost is around $n^2$ times higher due to the larger amount of preprocessing data needed for the HITS data in our online phase.

We first explain in more detail why the generic preprocessing method of Ishai et al. [IOZ14] does not lead to an efficient protocol. They presented a method to transform any protocol for a correlated randomness setup in the OT-hybrid model into a protocol that is secure with identifiable abort. Although their compiled protocol only requires black-box use of an OT protocol, it is impractical for a number of reasons:

- The protocol to be compiled is assumed to consist only of calls to an ideal OT functionality and a broadcast channel. This means that any pairwise communication must be performed via the OT functionality and so is very expensive.
- The transformation requires first computing an authenticated secret sharing of the required output, and then opening this to get the output. In our case, the output of $\mathcal{F}_{Prep}$ is already secret shared and authenticated, so intuitively, this step seems unnecessary.

- Their security proof requires the underlying OT protocol to be adaptively secure. This is much harder to achieve in practice, and rules out the use of efficient OT extensions [LZ13].

## 6.1  Modified Functionality $\mathcal{F}^*_{\text{Prep}}$

The $\mathcal{F}_{\text{Prep}}$ functionality from Sect. 5 is completely black-box with respect to the HITS scheme used. In this section, we implement preprocessing specifically for the scheme HITS from Sect. 4. We also make one small modification to the initialisation of $\mathcal{F}_{\text{Prep}}$, shown in Fig. 8, which simplifies our preprocessing protocol by not requiring the adversary's verification data, $\mathbf{v}_j$, to be uniformly random. The following proposition shows that the scheme, and therefore online phase, remain secure with this modification.

**Proposition 2.** *The scheme* HITS *remains secure when* Gen *is modified to allow adversarial inputs, as in* $\mathcal{F}^*_{\text{Prep}}$.

*Proof.* This easily follows by inspection of the scheme. Notice that the signing and verification algorithms for honest parties are independent of the values $\{\mathbf{v}_j\}_{j \in \mathcal{I}}$, so changing the distribution of these cannot cause an honest party to accept an invalid signature or reject a valid signature.                                        □

We now show how to use somewhat homomorphic encryption to perform the preprocessing with identifiable abort.

## 6.2  SHE Scheme Requirements

As in SPDZ, we use a *threshold somewhat homomorphic encryption* scheme SHE $= (\text{Gen}, \text{Enc}, \text{Dec}, \boxplus, \boxtimes)$ to generate the preprocessing data. The scheme must have a message space of $\mathbb{F}$, and we represent ciphertexts known to all parties with the notation $\langle x \rangle = \text{Enc}(x)$. The binary operators $\boxplus, \boxtimes$ then guarantee that

$$\langle x + y \rangle = \langle x \rangle \boxplus \langle y \rangle \quad \text{and} \quad \langle x \cdot y \rangle = \langle x \rangle \boxtimes \langle y \rangle,$$

for some suitable choice of randomness in the output ciphertexts. For our purposes, these homomorphic operations only need to support evaluation of circuits

---

On input $(\text{Init}, n_M, n_I)$ from all parties, set $w = n \cdot (3 \cdot n_M + n_I)$ and do the following:

1. Wait for the adversary to input $\mathbf{v}_j \in \mathbb{F}^n$, for each $j \in \mathcal{I}$.
2. Compute $(\text{sk}, \mathbf{vk}) \leftarrow \text{HITS.Gen}(1^\kappa, w)$, except using the provided values $\{\mathbf{v}_j\}_{j \in \mathcal{I}}$ to compute $\{\text{vk}_j\}_{j \in \mathcal{I}}$, instead of sampling fresh values.
3. Send $\text{vk}_i$ to party $P_i$ and store sk.

**Fig. 8.** Initialise command of $\mathcal{F}^*_{\text{Prep}}$.

with polynomially many additions and multiplicative depth 1. As was shown in [DPSZ12, DKL+13], a ring-LWE variant of the BGV scheme [BGV12] is practical for this purpose, and this also allows homomorphic operations to be batched for greater efficiency.

In addition, we require the following interactive protocols that will be used for the preprocessing.

*Zero Knowledge Proof of Plaintext Knowledge.* A protocol $\Pi_{\mathsf{ZKPoK}}$, which is a public-coin zero-knowledge proof of knowledge of the message and randomness that makes up a ciphertext. When used in our preprocessing protocol, all parties will sample the public verifier's messages with a coin-tossing functionality $\mathcal{F}_{\mathsf{Rand}}$ (see Fig. 7), so that the proofs are verified by all parties.

*Distributed Key Generation and Decryption.* The distributed key generation protocol outputs a public key to all parties, and an additively shared secret key. The distributed decryption protocol then allows the parties to decrypt a public ciphertext so that all parties obtain the output. These requirements are modelled in the functionality $\mathcal{F}_{\mathsf{KeyGenDec}}$ (Fig. 9). To achieve security with identifiable abort in our preprocessing protocol, note that the distributed decryption method modelled in $\mathcal{F}_{\mathsf{KeyGenDec}}$ always outputs a *correct* decryption, unlike the method in SPDZ [DPSZ12], which allows a corrupted party to introduce additive errors into the output. The SPDZ method can easily be modified to achieve this, by including a zero-knowledge proof, similar to the $\Pi_{\mathsf{ZKPoK}}$ proof used for ciphertext generation. Efficient zero-knowledge proofs for actively secure LWE-based key generation and distributed decryption were also given in [AJL+12], which can be adapted to the ring-LWE setting.

## 6.3  Basic Subprotocols

In Fig. 10 we describe some basic subprotocols for generating and decrypting ciphertexts. The RandShCtxt subprotocol creates $n$ public ciphertexts encrypting uniformly random shares, where each party holds one share. The ShareDec subprotocol takes a public ciphertext $\langle m \rangle$, encrypting $m$, and performs distributed decryption in such a way that each party learns only a random, additive

---

**Functionality** $\mathcal{F}_{\mathsf{KeyGenDec}}$

**KeyGen($1^\lambda$):** Let $(\mathsf{sk}, \mathsf{pk}) \leftarrow \mathsf{SHE.Gen}(1^\lambda)$. Store $\mathsf{sk}$ and output $\mathsf{pk}$ to all parties.

**DistDec($\langle m \rangle$):** On input a ciphertext $\langle m \rangle$ from all parties, output

$$m = \mathsf{Dec}(\langle m \rangle, \mathsf{sk})$$

to all $P_i$, where $m$ may be a valid message or an invalid ciphertext symbol $\bot$.

---

**Fig. 9.** SHE distributed key generation and decryption functionality.

share of $m$. If the flag new_ctxt is set to 1 then ShareDec additionally outputs a fresh encryption of the message $m$ to all parties. This is used to ensure that SHE only needs to evaluate circuits of multiplicative depth 1. The PrivateDec subprotocol is another variant of distributed decryption that decrypts the ciphertext $\langle x \rangle$ only to $P_i$. Note that the private decryption protocol used in [DPSZ12] is not suitable here, as it involves parties all sending a single message to $P_j$; in the identifiable abort setting, this would allow $P_j$ to claim that an honest party $P_i$ sent an invalid message, as the messages are not verifiable by all parties. To get around this, our PrivateDec protocol only uses broadcasted messages that are verifiable by all parties using the public-coin zero-knowledge proofs.

## 6.4  Creating the Preprocessing Data

The complete preprocessing protocol is shown in Figs. 11 and 12. To create a multiplication triple, each party must obtain random, additive shares $(a_i, b_i, c_i)$ such that $c = a \cdot b$, along with signatures on these shares. Creating shares of triples is essentially identical to the method in [DPSZ12], except we use the correct distributed decryption command of $\mathcal{F}_{\mathsf{KeyGenDec}}$, instead of a possibly faulty method. This means that there is no way the adversary can introduce errors into triples, so we avoid the need for the pairwise sacrificing procedure from [DPSZ12], where half of the triples are wasted to check correctness. The main other difference in our protocol, compared to [DPSZ12], is that we need to setup verification keys for the signature scheme and create signatures on every share, which is more complex than setting up simple MACs.

The setup phase begins by using RandShCtxt to create random, additive shares of the signing key values $\hat{\alpha}_r, \hat{\beta}_{r,i}$, and each party $P_j$'s private verification values $v_{j,r}$, along with ciphertexts encrypting the signing key shares and verification data, in steps 2–3. Next, in steps 4–5, the homomorphism of SHE is used to compute ciphertexts encrypting the signing key, and then ciphertexts encrypting the verification key values $\alpha_j, \beta_{j,i}$ for party $P_j$, for $i \in [w]$. These verification keys are then privately decrypted to each party.

Given encryptions of the signing key, an encrypted share can be authenticated by homomorphic evaluation of the signing algorithm, followed by private decryption of the signature to the relevant party, as seen in the subprotocol Auth (Fig. 11). Recall that in our scheme, a signature on $x_j$ is given by

$$\boldsymbol{\sigma} = \left( \hat{\alpha}_r \cdot x_j + \hat{\beta}_r \right)_{r=1}^{n},$$

where $\hat{\alpha}_r, \hat{\beta}_r$ are uniformly random elements of the secret key. (Note we have dropped the subscript $i$ on $\hat{\beta}$ here.) For party $P_j$ to obtain a signature on the share $x_j$, where all parties already know the ciphertext $\langle x_j \rangle$, all parties homomophically compute

$$\langle \boldsymbol{\sigma}|_r \rangle = (\langle \alpha_r \rangle \boxtimes \langle x_j \rangle) \boxplus \langle \beta_r \rangle,$$

for $r \in [n]$, and use private distributed decryption to output $\boldsymbol{\sigma}$ to $P_j$.

**Subprotocol RandShCtxt():**
1. Each party samples a random share $x_j \in \mathbb{F}$ and computes $\langle x_j \rangle = \mathsf{SHE.Enc}_{\mathsf{pk}}(x_j)$.
2. Each party broadcasts $\langle x_j \rangle$ and runs the protocol $\Pi_{\mathsf{ZKPoK}}$ to prove that $\langle x_j \rangle$ is correctly generated.
3. Each party $P_j$ outputs $x_j, \langle x_1 \rangle, \ldots, \langle x_n \rangle$.

**Subprotocol ShareDec($\langle m \rangle$, new_ctxt):**
1. Run RandShCtxt so that each $P_j$ obtains a share $r_j$ and ciphertexts $\langle r_1 \rangle, \ldots, \langle r_n \rangle$ that encrypt the shares.
2. Homomorphically compute

$$\langle m + r \rangle = \langle m \rangle \boxplus \langle r_1 \rangle \boxplus \cdots \boxplus \langle r_n \rangle.$$

3. Call $\mathcal{F}_{\mathsf{KeyGenDec}}.\mathsf{DistDec}$ to decrypt $\langle m+r \rangle$ so all parties learn $m + r$, where $r = r_1 + \cdots + r_n$.
4. $P_1$ outputs $m_1 = (m + r) - r_1$ and for all $j \neq 1$, $P_j$ outputs $m_j = -r_j$.
5. If new_ctxt $= 1$, each party $P_i$ also outputs $\langle m^* \rangle = \mathsf{SHE.Enc}_{\mathsf{pk}}(m+r) - \langle r \rangle$, where a default, public value is used for the randomness in Enc.

**Subprotocol PrivateDec($\langle x \rangle, P_j$):**
1. $P_j$ samples a random mask $K \leftarrow \mathbb{F}$, broadcasts $\langle K \rangle \leftarrow \mathsf{SHE.Enc}_{\mathsf{pk}}(K)$ and runs $\Pi_{\mathsf{ZKPoK}}$ to prove its correctness.
2. All parties homomorphically compute

$$\langle x + K \rangle = \langle x \rangle \boxplus \langle K \rangle.$$

3. Run $\mathcal{F}_{\mathsf{KeyGenDec}}.\mathsf{DistDec}(\langle x + K \rangle)$ so that all parties obtain the plaintext $x + K$.
4. $P_j$ recovers and outputs $x$.

**Fig. 10.** Subprotocols for the preprocessing protocol using SHE

**Theorem 3.** *The protocol $\Pi_{\mathsf{Prep}}$ (Figs. 11 and 12) securely realises $\mathcal{F}^*_{\mathsf{Prep}}$ (Figs. 8 and 4) with identifiable abort in the $\mathcal{F}_{\mathsf{KeyGenDec}}$-hybrid model, with computational security.*

*Proof.* See the full version of this work. □

## 7   Efficiency Evaluation

We now evaluate the concrete efficiency of our protocol, and compare it with the BDOZ [BDOZ11] and SPDZ [DPSZ12] protocols—which only offer security with abort—and the IOZ protocol [IOZ14], which achieves identifiable abort. First we discuss the complexity of broadcast in the two settings, then we compare the online phases of each protocol, and finally discuss the preprocessing.

---

**Protocol $\Pi_{\text{Prep}}$**

To create $n_M$ triples and $n_I$ input values for $n$ parties, set the parameter $w :=$ $n \cdot (3n_M + n_I)$.

**Setup:**    Creates the verification keys and ciphertexts encrypting the signing key.
1. Run $\mathcal{F}_{\text{KeyGenDec}}.\text{KeyGen}$ to obtain an SHE public key pk.
2. Run $\text{RandShCtxt}()$ $2n$ times, so each party $P_j$ obtains random elements $\hat{\alpha}_r^j, v_{j,r}$, for $r = 1, \ldots, n$, and everyone obtains ciphertexts $\langle \hat{\alpha}_r^j \rangle, \langle v_{j,r} \rangle$ encrypting these.
3. Run $\text{RandShCtxt}()$ $w(n)$ times, so party $P_j$ obtains $\hat{\beta}_{r,i}^j$, for $r \in [n]$ and $i \in [w]$, and everyone gets the corresponding ciphertexts.
4. Homomorphically compute, for $r \in [n]$:

$$\langle \hat{\alpha}_r \rangle = \langle \hat{\alpha}_r^1 \rangle \boxplus \cdots \boxplus \langle \hat{\alpha}_r^n \rangle$$
$$\langle \hat{\beta}_{r,i} \rangle = \langle \hat{\beta}_{r,i}^1 \rangle \boxplus \cdots \boxplus \langle \hat{\beta}_{r,i}^n \rangle \qquad \text{for } i \in [w].$$

5. Now compute the encrypted verification keys, for $j \in [n]$:

$$\langle \alpha_j \rangle = \bigboxplus_{r=1}^{n} (\langle \hat{\alpha}_r \rangle \boxtimes \langle v_{j,r} \rangle) \quad \text{and} \quad s\langle \beta_{j,i} \rangle = \bigboxplus_{r=1}^{n} \left( \langle \hat{\beta}_{j,r} \rangle \boxtimes \langle v_{j,r} \rangle \right)$$

6. Run the subprotocol $\text{PrivateDec}(\langle \alpha_j \rangle, P_j)$ and $\text{PrivateDec}(\langle \beta_{j,i} \rangle, P_j)$ for $i \in [w]$ and $j \in [n]$, so that each party $P_j$ gets their verification key $\text{vk}_j$.
7. All parties store the ciphertexts $\langle \hat{\alpha}_r \rangle, \langle \hat{\beta}_{r,i} \rangle$, for $r \in [n]$ and $i \in [w]$, and their private verification keys, $\text{vk}_j = (\mathbf{v}_j, \alpha_j, \{\beta_{j,i}\}_{i \in [w]})$.

---

**Fig. 11.** Preprocessing protocol with identifiable abort (Setup).

**Cost of Broadcast.** For MPC with identifiable abort, we denote the cost of broadcasting $m$ field elements by $\text{bc}(m)$. To be able to identify a cheater, this must be done using authenticated broadcast, which requires a PKI setup. The classic Dolev-Strong broadcast [DS83] has message complexity $O(mn^2)$, or a more recent protocol by Hirt and Raykov costs $O(mn)$ for large enough messages [HR14]. Note that any authenticated broadcast protocol requires $\Omega(n)$ rounds of communication if up to $n-1$ parties may be corrupt [GKKO07], which is considerably more expensive than the standard abort setting.

When security with (non-unanimous) abort is allowed (here, for SPDZ and BDOZ), a simple "broadcast with abort" protocol suffices [GL05]. Here, the broadcaster sends $x$ to everyone, then all other parties resend $x$ and check they received the same value. This can be further optimised by performing trivial, insecure broadcasts, and then at the end of the protocol, doing a single broadcast of the hash of all sent values to verify correctness [DKL+12]. This means each broadcast costs $O(n)$ messages, with a one-time $O(n^2)$ cost to verify these at the end.

---

**Protocol $\Pi_{\mathsf{Prep}}$ (continued)**

**Subprotocol** $\mathsf{Auth}(x_j, \langle x_j \rangle, P_j)$: Authenticates the share $x_j$ held by party $P_j$, where $\langle x_j \rangle$ is a (public) SHE encryption of $x_j$. Start by initialising a counter $\mathsf{cnt} := 1$.

1. All parties homomorphically compute, for $r \in [n]$,
$$\langle \sigma|_r \rangle = \langle \hat{\alpha}_r \rangle \boxtimes \langle x_j \rangle \boxplus \langle \hat{\beta}_{r,\mathsf{cnt}} \rangle.$$

2. Run $\mathsf{PrivateDec}(\langle \sigma_r \rangle, P_j)$ for $r \in [n]$ so that $P_j$ obtains the signature on $x_j$,
$$\sigma = (\hat{\alpha}_1 \cdot x_j + \hat{\beta}_{1,\mathsf{cnt}}, \ldots, \hat{\alpha}_n \cdot x_j + \hat{\beta}_{n,\mathsf{cnt}}).$$

3. Set $\mathsf{cnt} := \mathsf{cnt} + 1$

**Triple:**   Creates a single, authenticated multiplication triple.
1. The parties run $\mathsf{RandShCtxt}$ twice to obtain additive shares $a_j, b_j$ and public ciphertexts $\langle a_j \rangle, \langle b_j \rangle$ that encrypt the shares.
2. The parties homomorphically compute the ciphertext
$$\langle c \rangle = (\langle a_1 \rangle \boxplus \cdots \boxplus \langle a_n \rangle) \boxtimes (\langle b_1 \rangle \boxplus \cdots \boxplus \langle b_n \rangle).$$

3. The parties run $\mathsf{ShareDec}(\langle c \rangle, 1)$ to obtain decrypted, random shares of $c$ and a fresh ciphertext $\langle c \rangle$.
4. For each $j \in [n]$ and for each symbol $x \in \{a, b, c\}$, run $\mathsf{Auth}(x_j, \langle x_j \rangle, P_j)$ so $P_j$ obtains $\sigma_{x_j} = \mathsf{Sign}(x_j, \mathsf{sk})$.
5. Output the authenticated triple $([\![a]\!], [\![b]\!], [\![c]\!])$.

**Input**$(P_i)$:   Creates a random, authenticated value $[\![r]\!]$, where $r$ is known to $P_i$.
1. Run $\mathsf{RandShCtxt}$ to obtain additive shares $r_j$, and public ciphertexts $\langle r_j \rangle$ that encrypt these shares, for $j \in [n]$.
2. Run $\mathsf{Auth}(r_j, \langle r_j \rangle, P_j)$ for every $j \in [n]$ to obtain $[\![r]\!]$.
3. Compute $\langle r \rangle = \langle r_1 \rangle \boxplus \cdots \boxplus \langle r_n \rangle$ and run $\mathsf{PrivateDec}(\langle r \rangle, P_i)$ so that $P_i$ learns $r$.

---

**Fig. 12.** Preprocessing protocol with identifiable abort (authentication and triple generation).

When opening shared values (such as during multiplication) a more efficient method was described in [DPSZ12], where each party first sends their share to $P_1$, who then computes the sum and sends the result to all parties. This gives a cost of $2(n - 1)$ messages per opening, instead of $n(n - 1)$ for the previous method (again, the actual broadcast is verified at the end of the protocol).

**SPDZ.** In the SPDZ protocol (as in [DKL+13]), an authenticated secret share consists of $n$ additive shares on the secret and $n$ MAC shares, so each party stores two field elements. The preprocessing consists of one authenticated share per input, and three per multiplication triple. In the online phase, each input requires one party to broadcast a single value, for a communication cost of $n - 1$

field elements. A multiplication consists of two openings, each of which requires all parties to broadcast a share at a cost of $2n(n-1)$ messages using the protocol described above.

In the output phase of SPDZ, first the shares are opened, then a random linear combination of the MACs is checked, and finally all broadcasts must be checked. The MAC and broadcast checking methods both have a communication cost in $O(n^2)$.

**BDOZ.** In the BDOZ protocol, each party first obtains a fixed, global MAC key $\alpha_i$. This is fixed for all shared values, so we ignore this cost. For each shared representation $[a]$, party $P_i$ also stores the share $a_i$, $n$ local MAC keys $\beta^i_{a_1}, \ldots, \beta^i_{a_n}$ and $n$ MAC values $m_1(a_i), \ldots, m_n(a_i)$. Each of these are a single field element, so we get a total storage cost of $2n+1$ field elements per party for each authenticated shared value.

If we assume an optimised version of the original protocol, so that all parties open their shares $a_i$ using the SPDZ broadcast and then delay MAC checking until the **Output** stage, then the online communication costs are essentially the same as SPDZ.

**IOZ.** The IOZ online phase takes any semi-honest MPC protocol (with preprocessing), and compiles it to a malicious protocol with identifiable abort, similarly to the GMW paradigm [GMW87]. The compiled protocol has a preprocessing phase that outputs the original semi-honest preprocessing data, authenticated using IT signatures, as well as additional data for zero-knowledge proofs using MPC-in-the-head, which are required for each round of the semi-honest protocol. Using a semi-honest GMW protocol with multiplication triples as a base, the preprocessing data already contains the same number of IT signatures as our protocol, before taking into account the zero-knowledge proofs.

Each zero-knowledge proof requires storing $m$ IT signatures as preprocessing, where $m$ is the number of parties in the MPC-in-the-head method. In [IOZ14], they choose $m = O(\kappa)$ for statistical security level $\kappa$, whereas [GMO16] use $m = 3$, but require repeating the proof $\kappa$ times to get negligible soundness error. Since repeating the proof requires extra preprocessing for each repetition, we obtain a very rough lower bound of storing $\kappa$ signatures (or $\kappa \cdot n$ field elements) per proof with either approach.

For the communication costs, we only take into account the cost for every party to broadcast one proof, plus the (at least) two signatures that are broadcast in the $\Pi_{\mathsf{SCP}}$ protocol of [IOZ14]. According to [GMO16, Sect. 4.2], the proof size is at least $2 \cdot \kappa \cdot \log_2(|\mathbb{F}|)$, for a proof with soundness $2^{-\kappa}$, generously ignoring the size of the circuit representing the NP-relation being proven and other constant factors. If the IOZ version of MPC-in-the-head is used instead, each proof still requires broadcasting $t = O(\kappa)$ field elements in the $\Pi_{\mathsf{1SCP}}$ protocol, so would not have significantly better complexity.

**Table 1.** Comparison of the storage and communication costs of the protocols, measured in number of field elements. $N = n_I + 2n_M$ (where $n_I$ is number of inputs, $n_M$ is number of multiplications), $D$ is the multiplicative depth of the circuit, and $\kappa$ is a statistical security parameter

| Protocol | Prep. storage | | Online Comms. | | | Rounds |
|---|---|---|---|---|---|---|
| | Input | Mult. | Input | Mult. | Output | |
| SPDZ | 2 | 6 | $n-1$ | $4(n-1)$ | $O(n^2)$ | $O(D)$ |
| BDOZ | $2n+1$ | $6n+3$ | $n-1$ | $4(n-1)$ | $O(n^2)$ | $O(D)$ |
| IOZ (at least) | $\kappa n$ | $\kappa n$ | $\kappa \cdot \mathsf{bc}(1)$ | $\kappa n \cdot \mathsf{bc}(1) + 2n \cdot \mathsf{bc}(n)$ | $\kappa n \cdot \mathsf{bc}(1) + 2n \cdot \mathsf{bc}(n)$ | $O(D \cdot n)$ |
| Ours | $2n+1$ | $6n+3$ | $\mathsf{bc}(1)$ | $2n(n-1)$ | $n \cdot \mathsf{bc}(n + 2n_M + 1)$ | $O(D+n)$ |

**Comparison of the Online Phases.** The complexities in Table 1 for our protocol can be derived from the analysis in Sect. 5.2. We have ignored storage costs for the $\mathbf{v}_j, \alpha_j$ parts of the verification keys, as these are independent of the number of signatures. Our protocol is roughly a factor of $n$ times worse than SPDZ in terms of storage and communication cost, and has similar costs to BDOZ, bar the requirement for two rounds of authenticated broadcast. Compared with the IOZ protocol, we improve by at least a multiplicative factor in the security parameter, as well as a greatly reduced number of broadcast rounds.

### 7.1 Preprocessing Cost

For preprocessing, the main factor affecting computation and communication costs in [DPSZ12, DKL+13] is the number of zero-knowledge proofs of correct ciphertext generation that are required, so this is what we measure in our protocol.

The main cost of our preprocessing protocol, compared with [DPSZ12], is to produce the signatures and verification keys for each shared value, instead of MACs as in SPDZ. The **Setup** phase of our protocol (Fig. 11) generates verification keys, whose size depends on the number of signatures. Ignoring any costs independent of the number of signatures, this requires $n$ calls to RandShCtxt for each signature. Each RandShCtxt call requires $n$ zero-knowledge proofs, and since there are $n$ signatures per shared value (one per share) this gives a total of $O(n^3)$ zero-knowledge proofs per multiplication triple or input tuple. This dominates the cost of creating the $n$ signatures for each shared value, which is in $O(n^2)$.

In contrast, SPDZ shared MAC values only require $O(n)$ proofs each, so our protocol requires $O(n^2)$ more proofs than SPDZ in the preprocessing phase. It as an interesting problem to see if this can be reduced, although it seems that with IT signatures a factor of at least $O(n)$ is inherent, due to the signature size.

For comparison, note that the IOZ preprocessing transformation, which is based on any protocol in the OT-hybrid model, uses a verifiable OT protocol which broadcasts a message for every message of the OT protocol, adding an $O(n)$ overhead on top of the OT-hybrid protocol. When accounting for producing the larger amount of preprocessing data needed for the online phase, this gives

488     C. Baum et al.

an overall overhead of $O(n^2)$, the same as ours. However, it seems unlikely that an OT-based protocol for $\mathcal{F}_{Prep}$ could be much more efficient than using SHE, mainly because the need for adaptive security in IOZ prevents the use of efficient OT extensions [LZ13].

# References

[AJL+12] Asharov, G., Jain, A., López-Alt, A., Tromer, E., Vaikuntanathan, V., Wichs, D.: Multiparty computation with low communication, computation and interaction via threshold FHE. In: Pointcheval, D., Johansson, T. (eds.) EUROCRYPT 2012. LNCS, vol. 7237, pp. 483–501. Springer, Heidelberg (2012)

[AL10] Aumann, Y., Lindell, Y.: Security against covert adversaries: efficient protocols for realistic adversaries. J. Cryptology **23**(2), 281–343 (2010)

[BDOZ11] Bendlin, R., Damgård, I., Orlandi, C., Zakarias, S.: Semi-homomorphic encryption and multiparty computation. In: Paterson, K.G. (ed.) EUROCRYPT 2011. LNCS, vol. 6632, pp. 169–188. Springer, Heidelberg (2011)

[BDTZ16] Baum, C., Damgård, I., Toft, T., Zakarias, R.: Better preprocessing for secure multiparty computation. In: Manulis, M., Sadeghi, A.-R., Schneider, S. (eds.) ACNS 2016. LNCS, vol. 9696, pp. 327–345. Springer, Heidelberg (2016). doi:10.1007/978-3-319-39555-5_18

[Bea91] Beaver, D.: Efficient multiparty protocols using circuit randomization. In: Feigenbaum, J. (ed.) CRYPTO 1991. LNCS, vol. 576, pp. 420–432. Springer, Heidelberg (1992)

[BFKW09] Boneh, D., Freeman, D.M., Katz, J., Waters, B.: Signature schemes for network coding: signing a linear subspace. In: Public Key Cryptography - PKC, pp. 68–87 (2009)

[BGV12] Brakerski, Z., Gentry, C., Vaikuntanathan, V.: (leveled) fully homomorphic encryption without bootstrapping. In: ITCS 2012, pp. 309–325 (2012)

[BLOO11] Beimel, A., Lindell, Y., Omri, E., Orlov, I.: $1/p$-secure multiparty computation without honest majority and the best of both worlds. In: Rogaway, P. (ed.) CRYPTO 2011. LNCS, vol. 6841, pp. 277–296. Springer, Heidelberg (2011)

[Can01] Canetti, R.: Universally composable security: a new paradigm for cryptographic protocols. In: FOCS, pp. 136–145 (2001)

[CDD+99] Cramer, R., Damgård, I.B., Dziembowski, S., Hirt, M., Rabin, T.: Efficient multiparty computations secure against an adaptive adversary. In: Stern, J. (ed.) EUROCRYPT 1999. LNCS, vol. 1592, pp. 311–326. Springer, Heidelberg (1999)

[CJL09] Charles, D.X., Jain, K., Lauter, K.E.: Signatures for network coding. IJICoT **1**(1), 3–14 (2009)

[CL14] Cohen, R., Lindell, Y.: Fairness versus guaranteed output delivery in secure multiparty computation. In: Sarkar, P., Iwata, T. (eds.) ASIACRYPT 2014, Part II. LNCS, vol. 8874, pp. 466–485. Springer, Heidelberg (2014)

[Cle86] Cleve, R.: Limits on the security of coin flips when half the processors are faulty (extended abstract). In: STOC 1986, pp. 364–369 (1986)

[CR91] Chaum, David, Roijakkers, Sandra: Unconditionally-Secure Digital Signatures. In: Menezes, Alfred, J., Vanstone, Scott, A. (eds.) CRYPTO 1990. LNCS, vol. 537, pp. 206–214. Springer, Heidelberg (1991). doi:10.1007/3-540-38424-3_15

[DKL+12] Damgård, I., Keller, M., Larraia, E., Miles, C., Smart, N.P.: Implementing AES via an actively/covertly secure dishonest-majority MPC protocol. In: SCN, pp. 241–263 (2012)

[DKL+13] Damgård, I., Keller, M., Larraia, E., Pastro, V., Scholl, P., Smart, N.P.: Practical covertly secure MPC for dishonest majority – or: breaking the SPDZ limits. In: Crampton, J., Jajodia, S., Mayes, K. (eds.) ESORICS 2013. LNCS, vol. 8134, pp. 1–18. Springer, Heidelberg (2013)

[DPSZ12] Damgård, I., Pastro, V., Smart, N., Zakarias, S.: Multiparty computation from somewhat homomorphic encryption. In: Safavi-Naini, R., Canetti, R. (eds.) CRYPTO 2012. LNCS, vol. 7417, pp. 643–662. Springer, Heidelberg (2012)

[DS83] Dolev, D., Strong, H.R.: Authenticated algorithms for byzantine agreement. SIAM J. Comput. 12(4), 656–666 (1983)

[DZ13] Damgård, I., Zakarias, S.: Constant-overhead secure computation of boolean circuits using preprocessing. In: Sahai, A. (ed.) TCC 2013. LNCS, vol. 7785, pp. 621–641. Springer, Heidelberg (2013)

[GKKO07] Garay, J.A., Katz, J., Koo, C.-Y., Ostrovsky, R.: Round complexity of authenticated broadcast with a dishonest majority. In: FOCS 2007, pp. 658–668 (2007)

[GL05] Goldwasser, S., Lindell, Y.: Secure multi-party computation without agreement. J. Cryptology 18(3), 247–287 (2005)

[GMO16] Giacomelli, I., Madsen, J., Orlandi, C.: ZKBoo: faster zero-knowledge for boolean circuits. In: 25th USENIX Security Symposium (USENIX Security 2016), pp. 1069–1083 (2016)

[GMW87] Goldreich, O., Micali, S., Wigderson, A.: How to play any mental game or a completeness theorem for protocols with honest majority. In: STOC 1987, pp. 218–229 (1987)

[Gol01] Goldreich, O.: The Foundations of Cryptography - Basic Techniques, vol. 1. Cambridge University Press, Cambridge (2001)

[GVW15] Gorbunov, S., Vaikuntanathan, V., Wichs, D.: Leveled fully homomorphic signatures from standard lattices. In: STOC 2015, pp. 469–477 (2015)

[HR14] Hirt, M., Raykov, P.: Multi-valued byzantine broadcast: The $t_{jn}$ case. In: Sarkar, P., Iwata, T. (eds.) ASIACRYPT 2014. LNCS, vol. 8874, pp. 448–465. Springer, Heidelberg (2014). doi:10.1007/978-3-662-45608-8_24

[HSZI00] Hanaoka, G., Shikata, J., Zheng, Y., Imai, H.: Unconditionally secure digital signature schemes admitting transferability. In: Okamoto, T. (ed.) ASIACRYPT 2000. LNCS, vol. 1976, pp. 130–142. Springer, Heidelberg (2000). doi:10.1007/3-540-44448-3_11

[IKOS07] Ishai, Y., Kushilevitz, E., Ostrovsky, R., Sahai, A.: Zero-knowledge from secure multiparty computation. In: STOC 2007, pp. 21–30 (2007)

[IOS12] Ishai, Y., Ostrovsky, R., Seyalioglu, H.: Identifying cheaters without an honest majority. In: Cramer, R. (ed.) TCC 2012. LNCS, vol. 7194, pp. 21–38. Springer, Heidelberg (2012)

[IOZ14] Ishai, Y., Ostrovsky, R., Zikas, V.: Secure multi-party computation with identifiable abort. In: Garay, J.A., Gennaro, R. (eds.) CRYPTO 2014, Part II. LNCS, vol. 8617, pp. 369–386. Springer, Heidelberg (2014)

[LZ13] Lindell, Y., Zarosim, H.: On the feasibility of extending oblivious transfer. In: Sahai, A. (ed.) TCC 2013. LNCS, vol. 7785, pp. 519–538. Springer, Heidelberg (2013)

[PW92] Pfitzmann, B., Waidner, M.: Unconditional byzantine agreement for any number of faulty processors. In: STACS 1992, pp. 339–350 (1992)

[Sey12] Seyalioglu, H.A.-J.: Reducing trust when trust is essential. Ph.D. thesis, University of California, Los Angeles 2012. https://escholarship.org/uc/item/7301296m

[SS11] Swanson, C.M., Stinson, D.R.: Unconditionally secure signature schemes revisited. In: Fehr, S. (ed.) ICITS 2011. LNCS, vol. 6673, pp. 100–116. Springer, Heidelberg (2011)

# Secure Multiparty RAM Computation in Constant Rounds

Sanjam Garg[1]([⊠]), Divya Gupta[1], Peihan Miao[1], and Omkant Pandey[2]

[1] University of California, Berkeley, USA
{sanjamg,divyagupta2016,peihan}@berkeley.edu
[2] Stony Brook University, Stony Brook, USA
omkant@gmail.com

**Abstract.** Secure computation of a random access machine (RAM) program typically entails that it be first converted into a circuit. This conversion is unimaginable in the context of big-data applications where the size of the circuit can be exponential in the running time of the original RAM program. Realizing these constructions, without relinquishing the efficiency of RAM programs, often poses considerable technical hurdles. Our understanding of these techniques in the multi-party setting is largely limited. Specifically, the round complexity of all known protocols grows linearly in the running time of the program being computed. In this work, we consider the multi-party case and obtain the following results:

- *Semi-honest model*: We present a constant-round black-box secure computation protocol for RAM programs. This protocol is obtained by building on the new black-box garbled RAM construction by Garg, Lu, and Ostrovsky [FOCS 2015], and constant-round secure computation protocol for circuits of Beaver, Micali, and Rogaway [STOC 1990]. This construction allows execution of multiple programs on the same persistent database.
- *Malicious model*: Next, we show how to extend our semi-honest results to the malicious setting, while ensuring that the new protocol is still constant-round and black-box in nature.

## 1 Introduction

Alice, Bob, and Charlie jointly own a large private database $D$. For instance, the database $D$ can be a concatenation of their individually owned private databases. They want to compute and learn the output of arbitrary dynamically chosen private random access machine (RAM) programs $P_1, P_2, \ldots$, on private

This paper was presented jointly with [25] in proceedings of the 14th IACR Theory of Cryptography Conference (TCC) 2016-B.

Research supported in part from a DARPA/ARL SAFEWARE Award, AFOSR Award FA9550-15-1-0274, NSF CRII Award 1464397 and a research grant from the Okawa Foundation. The views expressed are those of the author and do not reflect the official policy or position of the funding agencies.

M. Hirt and A. Smith (Eds.): TCC 2016-B, Part I, LNCS 9985, pp. 491–520, 2016.
DOI: 10.1007/978-3-662-53641-4_19

inputs $x_1, x_2, \ldots$ and the previously stored database, which gets updated as these programs are executed. Can we do this?

Beginning with the seminal results of Yao [41] and Goldreich, Micali, and Wigderson [17], cryptographic primitives for secure computations are customarily devised for circuits. Using these approaches for random access machine (RAM) programs requires the conversion of the RAM program to a circuit. Using generic transformations [7,38], a program running in time $T$ translates into a circuit of size $O(T^3 \log T)$. Additionally, the obtained circuit must grow at least with the size of the input that includes data, which can be prohibitive for various applications. In particular, this dependence on input length implies an exponential slowdown for binary search. For instance, in the example above, for each program that Alice, Bob, and Charlie want to compute, communication and computational complexities of the protocol need to grow with the size of the database. Using fully homomorphic encryption [13], one can reduce the communication complexity of this protocol, but not the computational cost, which would still grow with the size of the database. Therefore, it is paramount that we realize RAM friendly secure computation techniques, that do not suffer from these inefficiencies.

**Secure computation for RAM programs.** Motivated by the above considerations, various secure computation techniques that work directly for RAM programs have been developed. For instance, Ostrovsky and Shoup [36] achieve general secure RAM computation using oblivious RAM techniques [16,18,35]. Subsequently, Gordon et al. [20] demonstrate an efficient realization based on specialized number-theoretic protocols. However, all these and other follow-up works require round complexity linear in the running time of the program. This changed for the two-party setting with the recent results on garbled RAM [12,14,32] and its black-box variant [11].[1] However, these round-efficient results are limited to the two-party setting.

In this work, we are interested in studying this question in the multiparty setting in the following two natural settings of RAM computations: persistent database setting and the non-persistent database setting. Furthermore, we want constructions that make only a black-box use of the underlying cryptographic primitives.

**Persistent vs. non-persistent database.** In the setting of RAM programs, the ability to store a *persistent* private database that can be computed on multiple times can be very powerful. Traditionally, secure computation on RAM programs is thus studied in two models. In the first model, called the *persistent* database model, one considers execution of many programs on the same database over a time period; the database can be modified by these programs during their execution and these changes persist over time. In this setting, the database

---

[1] We note that several other cutting-edge results [4,6,15,19,31] have been obtained in non-interactive secure computation over RAM programs but they all need to make strong computational assumptions such as [9,10,39]. Additionally they make non-black-box use of the underlying cryptographic primitives.

can be huge and the execution time of each program does not need to depend on the size of the database.

In the non-persistent database setting, one considers only a single program execution. This setting is extremely useful in understanding the underlying difficulties in obtaining a secure solution.

**Black-box vs. non-black-box.** Starting with Impagliazzo-Rudich [26,27], researchers have been very interested in realizing cryptographic goals making just a black-box use of underlying primitive. It has been the topic of many important recent works in cryptography [21,23,29,37,40]. On the other hand, the problem of realizing black-box construction for various primitive is still open, e.g. multi-statement non-interactive zero-knowledge [5,8,24] and oblivious transfer extension [1].[2] From a complexity perspective, black-box constructions are very appealing as they often lead to conceptually simpler and qualitatively more efficient constructions.[3]

Note that putting together Garbled RAM construction of Garg, Lu, Ostrovsky, and Scafuro [12] and the multiparty secure computation protocol of Beaver, Micali, and Rogaway [2] immediately gives a non-black-box protocol for RAM programs with the persistent use of memory. However, motivated by black-box constructions and low round complexity, in this work, we ask:

*Can we realize* constant-round black-box *secure multiparty computation for RAM programs?*

## 1.1 Our Results

In this paper, addressing the above question, we obtain the first constant-round black-box protocols for both the semi-honest and the malicious setting. Specifically, we present the following results:

- **Semi-honest**: We show a constant-round black-box secure computation protocol for RAM programs. This protocol is obtained by building on the new black-box garbled RAM construction by Garg, Lu, and Ostrovsky [11], and constant round secure computation protocol for circuits of Beaver, Micali, and Rogaway [2]. Our construction allows for the execution of multiple programs on the same persistent database. In our construction, for an original database of size $M$, one party needs to maintain a persistent database of size $M \cdot \text{poly}(\log M, \kappa)$. The communication and computational complexities of each program evaluation grow with $T \cdot \text{poly}(\log T, \log M, \kappa)$ where $T$ is the running time of the program and $\kappa$ is the security parameter.

---

[2] Interestingly for oblivious transfer extension we do know black-box construction based on stronger assumptions [28].

[3] Additionally, black-box constructions enable implementations agnostic to the implementation of the underlying primitives. This offers greater flexibility allowing for many optimizations, scalability, and choice of implementation.

- **Malicious**: Next we enhance the security of our construction from semi-honest setting to malicious, while ensuring that the new protocol is still constant-round and black-box. In realizing this protocol we build on the constant round black-box secure computation protocol of Ishai, Prabhakaran, and Sahai [30]. However, this result is only for the setting of the non-persistent database.[4]

Both our constructions only make a black-box use of one-way functions in the OT-hybrid model.

## 1.2   Concurrent and Independent Work

In a concurrent and independent work, Hazay and Yanai [25] consider the question of malicious secure two-party secure RAM computation. They present a constant-round protocol building on the the semi-honest two-party protocols [12,14]. They achieve a similar result as ours in the two-party setting but make a non-black-box use of one-way functions. Moreover, they allow running of multiple programs on a persistent database when all the programs as well as the inputs are known beforehand to the garbler.[5] Finally, the protocol of [25] makes a black-box use of ORAM[6] and only one party needs to store the memory locally. In this work, we can achieve the latter efficiency property in the semi-honest setting but not in the malicious setting.

An independent work of Miao [33] addresses the same problem as [25] but making only a black-box use of one-way functions and for the standard notion of persistent database that allows for programs and inputs of later executions to be chosen dynamically based on previous executions. [33] achieves a constant-round malicious secure two-party computation protocol making a black-box use of one-way functions in the OT-hybrid with the use of random oracle. It builds on the techniques of [3,34].

## 2   Our Techniques

**Semi-honest setting with a single program.** First, we consider the problem of constructing a semi-honest secure protocol for multi-party RAM computation. That is, consider $n$ parties $Q_1, \ldots, Q_n$ and a database $D = D_1 || \ldots || D_n$ such that $Q_i$ holds the database $D_i$. They want to securely evaluate a program $P$ on input $x = x_1, \ldots, x_n$ w.r.t. the database $D$, where $x_i$ is the secret input of party $Q_i$. Recall that our goal is to construct a constant-round protocol that

---

[4] We elaborate on the fundamental issue in extending this result to the persistent database setting at the end of next section.

[5] We note that our malicious secure protocol also achieves this weaker notion of persistent database, but in this paper we only focus on the standard notion of persistent data where later programs and inputs can be chosen dynamically.

[6] Our protocol is non-black-box in the use of ORAM. But, since [11] and our paper use an information theoretic ORAM, we are still black-box in the use of underlying cryptography.

only makes a black-box use of cryptographic primitives such as one-way function and oblivious transfer (OT). Moreover, we require that our protocol should only incur a one-time communication and computational cost that is linear in the size of $D$ up to poly-logarithmic factors. Subsequently, evaluating each new program should require communication and computation that is only linear in the running time of that program up to poly-logarithmic factors.

**High level approach.** Our starting point would be garbled RAM that solves the problem in the two-party setting. Recall that a garbled RAM is the RAM analogue of Yao's garbled circuits [41], and allows for multiple program executions on a persistent database. Recently, Garg et al. [11] gave a construction of a garbled RAM that only makes a black-box use of one-way functions. Given this primitive, a natural approach would be to generate the garbled RAM via a secure computation protocol for circuits. However, since garbled RAM is a cryptographic object and builds on one-way functions, a straight-forward application of this approach leads to an immediate problem of non-black-box use of one-way functions.

**Garbled RAM abstraction.** To handle the above issue regarding non-black-box use of one-way functions, we would massage the garbled RAM construction of [11] such that all calls to one-way functions are performed locally by parties and ensure that the functionality computed by generic MPC is information-theoretic. Towards this goal, we need to understand the structure of the garbled RAM of [11] in more detail. Next, we abstract the garbled RAM of [11] and describe the key aspects of the construction, which avoids the details irrelevant for understanding our work.

The garbled RAM for memory database $D$, program $P$ and input $x$ consists of the garbled memory $\tilde{D}$, the garbled program $\tilde{P}$, and the garbled input $\tilde{x}$. At a high level, the garbled memory $\tilde{D}$ consists of a collection of memory garbled circuits (for reading and writing to the memory) that invoke other garbled circuits depending on the input and the execution. More precisely, the garbled circuits of the memory are connected in a specific manner and each garbled circuit has keys of several other garbled circuits hard-coded inside it and it outputs input labels for the garbled circuit that is to be executed next. Moreover, for some of these garbled circuits, labels for partial inputs are revealed at the time of the generation of garbled RAM, while the others are revealed at run-time. Among the labels revealed at generation time, semantics of some of the labels is public, while the semantics of the others depend on the contents of the database as well as the randomness of the ORAM used.[7] Similarly, the garbled program $\tilde{P}$ consists of a sequence of garbled circuits for CPU steps such that each circuit has input labels of several garbled circuits, from the memory and the program, hard-coded inside itself. Finally, the garbled input consists of the labels for the first circuit in the garbled program that are revealed depending on the input $x$.

---

[7] The labels revealed at generation time are later referred to as the tabled garbled information. For security of garbled RAM, it is crucial to hide the semantics of the labels dependent on the database contents and we will revisit this later in the technical overview.

Our crucial observation about [11] is the following: Though each circuit has hard-coded *secret* labels for several other garbled circuits, the overall structure of the garbled memory as well as garbled program is public. That is, how the garbled circuits are connected in memory and the program as well as the structure of hard-coding is fixed and public, independent of the actual database or the program being garbled. This observation would be useful in two aspects: (1) To argue that the functionality being computed using the generic MPC for circuits is information theoretic. This is crucial in getting a black-box secure computation protocol for RAM. (2) When running more than one program on a persistent database, the basic structure of garbled RAM being public ensures that the cost of garbling additional programs does not grow with the size of the database.

Using above observations, we provide a further simplified formalization of the garbled RAM scheme of [11], where intuitively, we think of the circuits of memory and program as universal circuits that take secret hard-coded labels as additional inputs.[8] The labels for these additional input wires now have to be revealed at the time of garbled RAM generation. For details refer to Sect. 3.2. In light of this, our task is to devise a mechanism to generate all these garbled circuits and (partial) labels in a distributed manner securely. As mentioned above, since these garbled circuits use one-way functions (in generating encrypted gate tables), we cannot generate them naïvely.

**Handling the issue of non-black-box use of one-way functions.** We note that the garbled RAM of [11] makes a black-box use of a circuit garbling scheme and hence, can be instantiated using any secure circuit garbling scheme. This brings us to the next tool we use from the literature, which is the distributed garbling scheme of Beaver et al. [2], referred to as BMR in the following. In BMR, for each wire in the circuit, every party contributes a share of the label such that the wire-label is a concatenation of label shares from all the parties. Moreover, all calls to PRG (for generating encryptions of gate tables) are done locally such that given these PRG outputs and label shares, the generation of a garbled gate-table is information theoretic. This ensures that the final protocol of BMR is black-box in use of one-way functions. Our key observation is that we can instantiate the black-box garbled RAM of [11] with the BMR distributed garbling as the underlying circuit garbling scheme.

Based on what we have described so far, to obtain a constant-round semi-honest secure protocol for RAM, we would do the following: First, we would view the garbled RAM of [11] as a collection of suitable garbled circuits with additional input wires corresponding to the hardcoded secret labels (for simplicity). Next, we would use BMR distributed garbling as the underlying circuit garbling scheme, where each party computes the labels as well as PRG outputs locally. And, finally, we would run a constant-round black-box secure computation protocol for circuits (that is, BMR) to generate all the garbled circuits of the garbled RAM along with labels for partial inputs. In Sect. 5.3, we argue that

---

[8] Though this transformation is not crucial for security, it helps simplify the exposition of our protocol.

the functionality being computed by MPC is information theoretic. Hence, this gives a black-box protocol.

**Subtlety with use of ORAM.** At first, it seems that we can generate all the garbled circuits of the garbled RAM in parallel via the MPC. But, separating the generation of garbled circuits creates a subtle problem in how garbled RAM internally uses oblivious RAM. As mentioned before, some of the labels revealed at the time of garbled RAM generation depend on the database contents and the randomness used for ORAM. For security, the randomness of ORAM is contributed by all the parties and any sub-group of the parties does not learn the semantics of these labels. Therefore, separating the generation of garbled circuits requires all the parties to input the entire database to each garbled circuit generation, which would violate the efficiency requirements. In particular, efficiency of garbling the database would be at least quadratic in its size.

We solve this problem by bundling together the generation of all the garbled circuits under one big MPC. This does not harm security as well as provides the desired efficiency guarantees. More precisely, all the garbled circuits are generated by a single MPC protocol, where all the parties only need to input once the entire database along with all the randomness for the oblivious RAM (as well as their label shares and PRG outputs). We defer the details of this to the main body. There we also describe how we can extend this protocol for the setting of multiple program executions on a persistent database.

**Malicious Setting.** Next, we consider the case of malicious security. Again, to begin with, consider the case of a single program execution. For malicious security, we change the underlying secure computation protocol for generating garbled RAM to be malicious secure instead of just semi-honest secure. This would now ensure that each garbled circuit is generated correctly. Given that this secure computation is correct and malicious secure, the only thing that a malicious adversary can do is choose inputs to this protocol incorrectly or inconsistently. More precisely, as we will see in the main body, it is crucial that the PRG outputs fed into the secure computation protocol are correct. In fact, use of incorrect PRG values can cause honest parties to abort during evaluation of generated garbled RAM. This would be highly problematic for security if the adversary can cause input-dependent abort of honest parties as this is not allowed in ideal world. Note that we cannot use zero-knowledge proofs to ensure the correctness of PRG evaluations as this would lead to a non-black-box use of one-way functions. To get around this hurdle, we prove that the probability that an honest party aborts is independent of honest party inputs and depends only on the PRG values used by the adversary. In fact, given the labels as well as PRG outputs used by our adversary, our simulator can simulate which honest parties would abort and which honest parties would obtain the output.

**The case of persistent data in the malicious setting.** The final question is, can we extend the above approach to handle multiple programs? In the malicious setting, the adversary can choose the inputs for the second program based on the garbled memory that it has access to. Note that the garbled RAM of [11] does not guarantee security when the inputs can be chosen adaptively given the garbled RAM. Recall that the garbled memory of [11] consists of a collection of garbled circuits. In fact, to construct a scheme that satisfies this stronger security guarantee will require a circuit garbling scheme with the corresponding stronger security guarantee. In other words, we would need a circuit garbling scheme that is adaptively secure where the size of garbled input does not grow with the size of the circuit. However, we do not know of any such scheme in the standard model, i.e., without programmable random oracle assumption. Hence, we leave open the question of black-box malicious security for executing multiple RAM programs on a persistent database.

## 3   Preliminaries

We describe garbled RAM formally and give a brief overview of black box garbled RAM construction from [11]. Here we describe an abstraction of their construction which will suffice to describe our protocol for secure multi-party RAM computation as well as its security proof. Parts of this section have been taken verbatim from [11,14]. In the following, let $\kappa$ be the security parameter. For a brief description of RAM model and garbled circuits, refer to the full version of this paper.

### 3.1   Garbled RAM

The garbled RAM [12,14,32] is the extension of garbled circuits to the setting of RAM programs. Here, the memory data $D$ is garbled once and then many different garbled programs can be executed sequentially with the memory changes persisting from one execution to the next.

**Definition 1.** *A secure single-program garbled RAM scheme consists of four procedures* (GData, GProg, GInput, GEval) *with the following syntax:*

- $(\tilde{D}, s) \leftarrow$ GData($1^\kappa, D$): *Given a security parameter $1^\kappa$ and memory $D \in \{0,1\}^M$ as input,* GData *outputs the garbled memory $\tilde{D}$ and a key $s$.*
- $(\tilde{P}, s^{in}) \leftarrow$ GProg($1^\kappa, 1^{\log M}, 1^t, P, s, m$) : *Takes the description of a RAM program $P$ with memory-size $M$ and running-time $t$ as input. It also requires a key $s$ (produced by* GData*) and current time $m$. It then outputs a garbled program $\tilde{P}$ and an input-garbling-key $s^{in}$.*
- $\tilde{x} \leftarrow$ GInput($1^\kappa, x, s^{in}$): *Takes as input $x \in \{0,1\}^n$ and an input-garbling-key $s^{in}$, and outputs a garbled-input $\tilde{x}$.*
- $(y, \tilde{D}') =$ GEval$^{\tilde{D}}(\tilde{P}, \tilde{x})$: *Takes a garbled program $\tilde{P}$, garbled input $\tilde{x}$ and garbled memory data $\tilde{D}$ and outputs a value $y$ along with updated garbled data $\tilde{D}'$. We model* GEval *itself as a RAM program that can read and write to arbitrary locations of its memory initially containing $\tilde{D}$.*

**Efficiency:** *The run-time of* GProg *and* GEval *are* $t \cdot \text{poly}(\log M, \log T, \kappa)$, *which also serves as the bound on the size of the garbled program* $\tilde{P}$. *Here,* $T$ *denotes the combined running time of all programs. Moreover, the run-time of* GData *is* $M \cdot \text{poly}(\log M, \log T, \kappa)$, *which also serves as an upper bound on the size of* $\tilde{D}$. *Finally the running time of* GInput *is* $n \cdot \text{poly}(\kappa)$.

**Correctness:** *For correctness, we require that for any initial memory data* $D \in \{0,1\}^M$ *and any sequence of programs and inputs* $\{P_i, x_i\}_{i \in [\ell]}$, *following holds: Denote by* $y_i$ *the output produced and by* $D_{i+1}$ *the modified data resulting from running* $P_i(D_i, x_i)$. *Let* $(\tilde{D}, s) \leftarrow$ GData$(1^\kappa, D)$. *Also, let* $(\tilde{P}_i, s_i^{in}) \leftarrow$ GProg$(1^\kappa, 1^{\log M}, 1^{t_i}, P_i, s, \sum_{j \in [i-1]} t_j), \tilde{x}_i \leftarrow$ GInput$(1^\kappa, x_i, s_i^{in})$ *and* $(y_i', \tilde{D}_{i+1}) =$ GEval$^{\tilde{D}_i}(\tilde{P}_i, \tilde{x}_i)$. *Then,* $\Pr[y_i = y_i', \text{ for all } i \in \{1, \ldots, \ell\}] = 1$.

**Security:** *For security, we require that there exists a PPT simulator* GramSim *such that for any initial memory data* $D \in \{0,1\}^M$ *and any sequence of programs and inputs* $\{P_i, x_i\}_{i \in [\ell]}$, *the following holds: Denote by* $y_i$ *the output produced and by* $D_{i+1}$ *the modified data resulting from running* $P_i(D_i, x_i)$. *Let* $(\tilde{D}, s) \leftarrow$ GData$(1^\kappa, D), (\tilde{P}_i, s_i^{in}) \leftarrow$ GProg$(1^\kappa, 1^{\log M}, 1^{t_i}, P_i, s, \sum_{j \in [i-1]} t_j)$ *and* $\tilde{x}_i \leftarrow$ GInput$(1^\kappa, x_i, s_i^{in})$, *then*

$$(\tilde{D}, \{\tilde{P}_i, \tilde{x}_i\}_{i \in [\ell]}) \stackrel{\text{comp}}{\approx} \text{GramSim}(1^\kappa, 1^M, \{1^{t_i}, y_i\}_{i \in [\ell]}).$$

## 3.2   Black-Box Garbled RAM of [11]

The work of [11] gives a construction of garbled RAM that only makes a black-box use of one-way functions. In particular, it proves the following theorem.

**Theorem 1 ([11]).** *Assuming the existence of one-way functions, there exists a secure black-box garbled RAM scheme for arbitrary RAM programs satisfying the efficiency, correctness and security properties stated in Definition 1.*

Below, we describe the construction of [11] at a high level. We describe the algorithms (GData, GProg, GInput). In following, in the context of garbled circuits, *labels* refers to one of the labels for an input bit and *keys* refers to both labels (one for 0 and one for 1) corresponding to an input bit.

[11] construct black-box garbled RAM in two steps. First a garbled RAM scheme is constructed under the weaker security requirement of unprotected memory access (UMA2-security) where only the sequence of memory locations being accessed is revealed. Everything else about the program, data and the input is hidden. Next this weaker security guarantee is amplified to get full security by using statistical oblivious RAM that hides the memory locations being accessed.

**Garbled RAM achieving UMA2-security.** Let the corresponding procedures be ($\widetilde{\text{GData}}, \widetilde{\text{GProg}}, \widetilde{\text{GInput}}, \widetilde{\text{GEval}}$).

- $(\tilde{D}, s) \leftarrow \widehat{\mathsf{GData}}(1^\kappa, D)$: $\tilde{D}$ consists of a collection of garbled circuits and a tabled garbled information $\mathsf{Tab}$. The key $s$ corresponds to a PRF key.

**Garbled Circuits.** The collection of garbled circuits is organized as a binary tree of depth $d = O(\log |D|)$ and each node consists of a sequence of *garbled circuits*.[9] For any garbled circuit, its *successor* (resp. predecessor) is defined to be the next (resp. previous) node in the sequence. For a garbled circuit, all garbled circuits in parent (resp. children) node are called parents (resp. children). There are two kinds of circuits: leaf circuits $\mathsf{C}^{\mathsf{leaf}}$ (at the leaves of the tree) and non-leaf circuits $\mathsf{C}^{\mathsf{node}}$ (at the internal nodes of the tree). Intuitively speaking, the leaf nodes carry the actual data.

Each garbled circuit has hard-coded inside it (partial) keys of a set of other garbled circuits. We emphasize that for any circuit $C$, the keys that are hard-coded inside it are fixed (depending on the size of data) and is independent of the actual contents of the data. This would be crucial later.

**Tabled garbled information.** For each node in the tree as described above, the garbled memory consists of a table of information $\mathsf{Tab}(i, j)$, where $(i, j)$ denotes the $j^{th}$ node at $i^{th}$ level from the root. Note that $d$ denotes the depth of the tree. The tabulated information $\mathsf{Tab}(i, j)$ contains labels for partial inputs of the first garbled circuit in the sequence of circuits at node $(i, j)$ (i.e., one label for some of the input bits). As the garbled memory is consumed by executing the garbled circuits, the invariant is maintained that the tabulated information contains partial labels for the first unused garbled circuit at that node.

A crucial point to note is the following: Labels in $\mathsf{Tab}$ entry corresponding to non-leaf nodes, i.e., $\mathsf{C}^{\mathsf{node}}$, depend on the keys on some other garbled circuit. Also, the tabulated information for the leaf nodes depends on actual value in the data being garbled. More precisely, the entry $\mathsf{Tab}(d, j)$ for level $d$ of the leaves contains the partial labels for the first $\mathsf{C}^{\mathsf{leaf}}$ circuit at $j^{th}$ leaf corresponding to value $D[j]$ value.[10]

The keys of the all the garbled circuits are picked to be outputs of a PRF with key $s$ on appropriate inputs.

- $(\tilde{P}, s^{in}) \leftarrow \widehat{\mathsf{GProg}}(1^\kappa, 1^{\log M}, 1^t, P, s, m)$ : The garbled program $\tilde{P}$ consists of a sequence of garbled circuits, called $\mathsf{C}^{\mathsf{step}}$. Again, each garbled circuit has (partial) keys of some circuits of $\tilde{P}$ and $\tilde{D}$ hard-coded inside it. We emphasize that for any circuit $\mathsf{C}^{\mathsf{step}} \in \tilde{P}$, which keys are hard-coded inside it is fixed and is independent of the actual program and actual data.

$s^{in}$ corresponds to the keys of the first circuit in this sequence.

---

[9] Note that for security, any garbled circuit can only be executed once. Hence, to enable multiple reads from the memory, each node consists of a sequence of garbled circuits. Number of garbled circuits at any node is chosen carefully. See [11] for details. For our purpose, we do not need to specify the number of garbled circuits at each node, but it is worth emphasizing that the total number of garbled circuits is $|D| \cdot \mathsf{poly}(\log |D|, \kappa)$.

[10] Note that it is public that $j^{th}$ leaf corresponds to $D[j]$. Later, statistical ORAM is used to hide this correspondence.

- $\tilde{x} \leftarrow \widetilde{\mathsf{GInput}}(1^\kappa, x, s^{in})$: The GInput algorithm uses $x$ as selection bits for the keys provided in $s^{in}$, and outputs $\tilde{x}$ that is just the selected labels.

*Remark 1.* Note that for [11] the labels in Tab for $\mathsf{C}^{\mathsf{node}}$ and hard-coded key values are independent of the actual data $D$ and input $x$. The labels in Tab for $\mathsf{C}^{\mathsf{leaf}}$ depend on data $D$ and labels in $\tilde{x}$ depend on input $x$.

**Garbled RAM Achieving Full Security.** [11] prove the following lemma.

**Lemma 1 [11].** *Given a UMA2-secure garbled RAM scheme* $(\widetilde{\mathsf{GData}}, \widetilde{\mathsf{GProg}}, \widetilde{\mathsf{GInput}}, \widetilde{\mathsf{GEval}})$ *for programs with (leveled) uniform memory access, and a statistical ORAM scheme* $(\mathsf{OData}, \mathsf{OProg})$ *giving (leveled) uniform memory access that protects the access pattern, there exists a fully secure garbled RAM scheme.*

The construction works by first applying ORAM compiler followed by UMA2-secure garbled RAM compiler. More formally,

- $\mathsf{GData}(1^\kappa, D)$: Execute $(D^*) \leftarrow \mathsf{OData}(1^\kappa, D)$ followed by $(\tilde{D}, s) \leftarrow \widetilde{\mathsf{GData}}(1^\kappa, D^*)$. Output $(\tilde{D}, s)$. Note that OData does not require a key as it is a statistical scheme.
- $\mathsf{GProg}(1^\kappa, 1^{\log M}, 1^t, P, \widehat{s}, m)$: Execute $P^* \leftarrow \mathsf{OProg}(1^\kappa, 1^{\log M}, 1^t, P)$ followed by $(\tilde{P}, s^{in}) \leftarrow \widetilde{\mathsf{GProg}}(1^\kappa, 1^{\log M'}, 1^{t'}, P^*, s, m)$ . Output $(\tilde{P}, s^{in})$.
- $\mathsf{GInput}(1^\kappa, x, s^{in})$: Note that $x$ is valid input for $P^*$. Execute $\tilde{x} \leftarrow \widetilde{\mathsf{GInput}}(1^\kappa, x, s^{in})$, and output $\tilde{x}$.
- $\mathsf{GEval}^{\tilde{D}}(\tilde{P}, \tilde{x})$: Execute $y \leftarrow \widetilde{\mathsf{GEval}}^{\tilde{D}}(\tilde{P}, \tilde{x})$ and output $y$.

Note that UMA-2-secure garbled RAM does not hide which leaf nodes of garbled data correspond to which bit of data $D$. In the garbled RAM with full security, this is being hidden due to compilation by ORAM. In the final garbled RAM with full security, which keys are hardwired in each garbled circuit remains public as before. Also, which keys are stored in Tab are public except those for $\mathsf{C}^{\mathsf{leaf}}$ as they correspond to data values. These are determined by the randomness used in ORAM as well as actual data.

**Transformation of Garbled RAM to Remove the Hard-Coding of Keys.** As is clear from the above description, the garbled RAM $(\tilde{D}, \tilde{P}, \tilde{x})$ consists of a collection of garbled circuits of three types: $\mathsf{C}^{\mathsf{leaf}}, \mathsf{C}^{\mathsf{node}}$ and $\mathsf{C}^{\mathsf{step}}$ and a collection of labels given in tabled garbled information of $\tilde{D}$ and $\tilde{x}$. Each of these circuits have (partial) keys of other garbled circuits hard-coded inside them and this structure of hard-coding is public and fixed independent of the actual data, program and randomness of ORAM. In this work, we change the circuits in garbled RAM construction as follows: We consider these hard-coded values as additional inputs whose labels are given out at time of garbled RAM generation. Once we remove the hard-coding of keys from inside these circuits, the structure of these circuits is public and known to all. The remark below summarizes our garbled RAM abstraction.

*Remark 2.* **Garbled RAM abstraction.** The garbled RAM consists of a collection of garbled circuits. The structure of these garbled circuits as well as semantics of each of the input wires are public. For each of these garbled circuits, labels for partial inputs are revealed at the time of garbled RAM generation. Labels which are not revealed become known while evaluating the garbled RAM. Labels which are revealed correspond to one of the following inputs: The ones for which labels were given in Tab, earlier hardcoded keys, or the input $x$.

The semantics of the labels revealed in Tab corresponding to $C^{node}$ is public. But, ORAM hides the semantics of the labels in Tab corresponding to $C^{leaf}$ and these depend on data values in memory at specific locations. Moreover, the mapping of these leaves to locations of memory is also hidden by ORAM and is determined by the randomness used by ORAM.

It is easy to see that the garbed RAM obtained after this transformation is equivalent to the original garbled RAM of [11]. In the following, the garbled RAM will refer to this simpler version of garbled RAM where circuits do not have any hard-coded keys. This transformation would help in ensuring that the secure computation protocols that we describe only make a black-box use of cryptography.

## 4   Our Model

In this section we define the security of secure computation for RAM programs for the case of persistent database. In this work, we consider both semi-honest as well as malicious adversaries. A semi-honest adversary is guaranteed to follow the protocol whereas a malicious adversary may deviate arbitrarily from the protocol specification. In particular, a malicious adversary may refuse to participate in the protocol, may use an input that is different from prescribed input, and may be abort prematurely blocking the honest parties from getting the output. We consider static corruption model, where the adversary may corrupt an arbitrary collection of the parties, and this set is fixed before the start of the protocol. We consider *security with abort* using real world and ideal world simulation based definition. We consider a ideal computation using incorruptible third party to whom parties send their inputs and receive outputs. This ideal scenario is secure by definition. For security, we require that whatever harm the adversary can do by interacting in the real protocol is mimicked by an ideal world adversary. We provide a detailed formal description of our model in our full version.

Consider parties $Q_1, \ldots, Q_n$ holding secret database $D_1, \ldots, D_n$, respectively. They want to compute a sequence of programs $\boldsymbol{P} = (P^{(1)}, \ldots, P^{(\ell)})$ on the persistent database $D = D_1 || \ldots || D_n$. $Q_i$ has secret input $x_i^{(j)}$ for program $P^{(j)}$.

The overall output of the ideal-world experiment consists of all the values output by all parties at the end, and is denoted by $\mathsf{Ideal}_{\mathcal{S}}^{\boldsymbol{P}}(1^\kappa, D, \{\boldsymbol{x}^{(j)}\}_{j \in [\ell]}, z)$, where $z$ is the auxiliary input given to the ideal adversary $\mathcal{S}$ at the beginning. In the case of an semi-honest adversary $\mathcal{S}$, all parties receive the same output from the trusted functionality. In the case of the malicious adversary, $\mathcal{S}$ after receiving the output from the trusted party, chooses a subset of honest parties $\mathcal{J} \subseteq [n] \setminus \mathcal{I}$ and sends $\mathcal{J}$ to the trusted functionality. Here, $\mathcal{I} \subseteq [n]$ denotes the set of corrupt parties. The trusted party sends the output to parties in $\mathcal{J}$ and special symbol $\perp$ to all the other honest parties.

Similarly, the overall output of the real-world experiment consists of all the values output by all parties at the end of the protocol, and is denoted by $\mathsf{Real}^{\pi}_{\mathcal{A}}(1^{\kappa}, D, \{x^{(j)}\}_{j \in [\ell]}, z)$, where $z$ is the auxiliary input given to real world adversary $\mathcal{A}$. Then, security is defined as follows:

**Definition 2 (Security).** *Let* $\boldsymbol{P} = (P^{(1)}, \ldots, P^{(\ell)})$ *be a sequence of well-formed RAM programs and let* $\pi$ *be a n-party protocol for* $\boldsymbol{P}$. *We say that* $\pi$ *securely computes* $\boldsymbol{P}$, *if for every* $\{D_i, x_i^{(1)}, \ldots, x_i^{(\ell)}\}_{i \in [n]}$, *every auxiliary input* $z$, *every real world adversary* $\mathcal{A}$, *there exists an ideal world* $\mathcal{S}$ *such that* $\mathsf{Real}^{\pi}_{\mathcal{A}}(1^{\kappa}, D, \{x^{(j)}\}_{j \in [\ell]}, z) \approx \mathsf{Ideal}^{\boldsymbol{P}}_{\mathcal{S}}(1^{\kappa}, D, \{x^{(j)}\}_{j \in [\ell]}, z)$.

**Efficiency:** We also want the following efficiency guarantees. We consider the following two natural scenarios: Below, $M$ is the size of total database, i.e. $M = |D|$, $t_j$ denotes the running time of $P^{(j)}$ and $T = \max_j t_j$

1. **All parties do computation:** In this case, for all the parties we require the total communication complexity and computation complexity to be bounded by $M \cdot \mathsf{poly}(\log M, \log T, \kappa) + \sum_{j \in [\ell]} t_j \cdot \mathsf{poly}(\log M, \log t_j, \kappa)$ and each party needs to store total database and program of size at most $M \cdot \mathsf{poly}(\log M, \log T, \kappa) + T \cdot \mathsf{poly}(\log M, \log T, \kappa)$. With each additional program to be computed, the additional communication complexity should be $t_j \cdot \mathsf{poly}(\log M, \log t_j, \kappa)$.
2. **Only one party does the computation:** In this case, as before the communication complexity and computation complexity of the protocol is bounded by $M \cdot \mathsf{poly}(\log M, \log T, \kappa) + \sum_{j \in [\ell]} t_j \cdot \mathsf{poly}(\log M, \log t_j, \kappa)$. But, in some cases such as semi-honest security, we can optimize on the space requirements of the parties and all parties do not require space proportional to the total database. The one designated party who does the computation needs to store $M \cdot \mathsf{poly}(\log M, \log T, \kappa) + T \cdot \mathsf{poly}(\log M, \log T, \kappa)$. All the other parties $Q_i$ only need to store their database of size $|D_i|$.[11]

## 5   Semi-honest Multi-party RAM Computation

In this section, we describe the semi-honest secure protocol for RAM computation. We prove the following theorem.

**Theorem 2.** *There exists a constant-round semi-honest secure multiparty protocol for secure RAM computation for the case of persistent database in the OT-hybrid model that makes a black-box use of one-way functions. This protocol satisfies the security and the efficiency requirements of Sect. 4.*

---

[11] During the protocol execution all parties would need space proportional to $M \cdot \mathsf{poly}(\log M, \log T, \kappa) + T \cdot \mathsf{poly}(\log M, \log T, \kappa)$. Looking ahead, this is needed to run a protocol which outputs the garbled RAM to the designated party.

Below, we first describe a secure protocol for the case of a single program execution for the case when all the parties compute the program and hence, need space proportional to the total size of the database. Later, we describe how our protocol can be extended to the case of multiple programs and optimizations of load balancing.

**Protocol Overview.** Our goal is to construct a semi-honest secure protocol for RAM computation. At a high level, in our protocol, the parties will run a multiparty protocol (for circuits) to generate the garbled RAM. We want a constant round protocol for secure computation that only makes a black-box use of one-way functions in the OT-hybrid model. Such a protocol was given by [2]. As already mentioned in technical overview, a naïve use of this protocol to generate the garbled RAM results in a non-black-box use of one way functions. The reason is the following: As explained before, the black-box garbled RAM of [11] consists of a collection of garbled circuits and hence, uses one-way functions inside it. Our main idea is to transform the garbled RAM of [11] in a way that allows each party to compute the one-way functions locally so that the functionality we compute using [2] is non-cryptographic (or, information theoretic). To achieve this, we again use ideas from distributed garbling scheme of [2]. Below we review their main result and the underlying garbling technique.

## 5.1   Distributed Garbling Protocol of [2]

Following result was proven by [2].

**Theorem 3** ([2]). *There exists a constant-round semi-honest secure protocol for secure computation for circuits, which makes black-box calls to PRG in the OT-hybrid model. Let the protocol for a functionality $\mathcal{F}$ be denoted by $\Pi^{\mathcal{F}}_{\mathsf{bmr}}$ and the corresponding simulator be $\mathsf{Sim}_{\mathsf{bmr}}$.*

We describe this protocol next at a high level. Some of the following text has been taken from [22].

Suppose there are $n$ parties $Q_1, \ldots, Q_n$ with inputs $x_1, \ldots, x_n$. The goal is the following: For any circuit $C$, at the end of the garbling protocol, each party holds a garbled circuit $\tilde{C}$ corresponding to $C$ and garbled input $\tilde{x}$ corresponding to $x = x_1, \ldots, x_n$. Then, each party can compute $\tilde{C}$ on $\tilde{x}$ locally. At a high level, the evaluation is as follows: Recall that in a garbled circuit, each wire $w$ has two keys $\mathsf{key}^w_0$ and $\mathsf{key}^w_1$: one corresponding to the bit being 0 and another corresponding to the bit being 1. In the multiparty setting, each wire also has a *wire mask* $\lambda^w$ that determines the correspondence between the two wire keys and the bit value. More precisely, the key $\mathsf{key}^w_b$ corresponds to the bit $b \oplus \lambda^w$.

In the following, let $F$ be a PRF and $G$ be a PRG. The garbling protocol $\Pi^{\mathsf{sh}}_{\mathsf{bmr}}$ is as follows:

- **Stage 1.** Party $Q_i$ picks a seed $s_i$ for the PRF $F$. It generates its shares for keys for all the wires of $C$ and wire masks as follows: Define $(\mathsf{k}^w_0(i), \mathsf{k}^w_1(i), \lambda^w_i) =$

$F_{s_i}(w)$. In the final garbled circuit $\tilde{C}$, key for any wire $w$ will be a concatenation of keys from all the parties. That is, $\mathsf{key}_b^w = \mathsf{k}_b^w(1) \circ \ldots \circ \mathsf{k}_b^w(n)$ and $\lambda^w = \lambda_1^w \oplus \ldots \oplus \lambda_n^w$.

- **Stage 2.** Recall that in a garbled circuit, for any gate, the keys for the output wires are encrypted under the input keys for all the four possible values for the inputs. These encryptions are stored in a garbled table corresponding to each gate. For all garbled circuit constructions, this step of symmetric encryption involves the use of one-way functions. In order to ensure black-box use of one-way functions, each party will make the PRG call locally. The parties locally expand their key parts into large strings that will be used as one-time pads to encrypt the key for the output wire labels. More precisely, $Q_i$ expands the key parts $\mathsf{k}_0^w(i)$ and $\mathsf{k}_1^w(i)$ using PRG $G$ to obtain two new strings, i.e., $(\mathsf{p}_b^w(i), \mathsf{q}_b^w(i)) = G(\mathsf{k}_b^w(i))$, for $b \in \{0,1\}$. Both $\mathsf{p}_b^w(i)$ and $\mathsf{q}_b^w(i)$ have length $n|\mathsf{k}_b^w(i)| = |\mathsf{key}_b^w|$ (enough to encrypt the key for output wire). More precisely, for every gate in $C$, a gate table is defined as follows: Let $\alpha, \beta$ be the two input wires and $\gamma$ be the output wire, and denote the gate operation by $\otimes$. Party $Q_i$ holds the inputs $\mathsf{p}_b^\alpha(i), \mathsf{q}_b^\alpha(i), \mathsf{p}_b^\beta(i), \mathsf{q}_b^\beta(i)$ for $b \in \{0,1\}$ along with shares of masks $\lambda_i^\alpha, \lambda_i^\beta, \lambda_i^\gamma$. The garbled gate table is the following four encryptions:

$$A_g = \mathsf{p}_0^\alpha(1) \oplus \ldots \oplus \mathsf{p}_0^\alpha(n) \oplus \mathsf{p}_0^\beta(1) \oplus \ldots \oplus \mathsf{p}_0^\beta(n)$$
$$\oplus \begin{cases} \mathsf{k}_0^\gamma(1) \circ \ldots \circ \mathsf{k}_0^\gamma(n) \text{ if } \lambda^\alpha \otimes \lambda^\beta = \lambda^\gamma \\ \mathsf{k}_1^\gamma(1) \circ \ldots \circ \mathsf{k}_1^\gamma(n) \text{ otherwise} \end{cases}$$

$$B_g = \mathsf{q}_0^\alpha(1) \oplus \ldots \oplus \mathsf{q}_0^\alpha(n) \oplus \mathsf{p}_1^\beta(1) \oplus \ldots \oplus \mathsf{p}_1^\beta(n)$$
$$\oplus \begin{cases} \mathsf{k}_0^\gamma(1) \circ \ldots \circ \mathsf{k}_0^\gamma(n) \text{ if } \lambda^\alpha \otimes \overline{\lambda^\beta} = \lambda^\gamma \\ \mathsf{k}_1^\gamma(1) \circ \ldots \circ \mathsf{k}_1^\gamma(n) \text{ otherwise} \end{cases}$$

$$C_g = \mathsf{p}_1^\alpha(1) \oplus \ldots \oplus \mathsf{p}_1^\alpha(n) \oplus \mathsf{q}_0^\beta(1) \oplus \ldots \oplus \mathsf{q}_0^\beta(n)$$
$$\oplus \begin{cases} \mathsf{k}_0^\gamma(1) \circ \ldots \circ \mathsf{k}_0^\gamma(n) \text{ if } \overline{\lambda^\alpha} \otimes \lambda^\beta = \lambda^\gamma \\ \mathsf{k}_1^\gamma(1) \circ \ldots \circ \mathsf{k}_1^\gamma(n) \text{ otherwise} \end{cases}$$

$$D_g = \mathsf{q}_1^\alpha(1) \oplus \ldots \oplus \mathsf{q}_1^\alpha(n) \oplus \mathsf{q}_1^\beta(1) \oplus \ldots \oplus \mathsf{q}_1^\beta(n)$$
$$\oplus \begin{cases} \mathsf{k}_0^\gamma(1) \circ \ldots \circ \mathsf{k}_0^\gamma(n) \text{ if } \overline{\lambda^\alpha} \otimes \overline{\lambda^\beta} = \lambda^\gamma \\ \mathsf{k}_1^\gamma(1) \circ \ldots \circ \mathsf{k}_1^\gamma(n) \text{ otherwise} \end{cases}$$

In [2] this garbled table is generated by running a semi-honest secure computation protocol by the parties $Q_1, \ldots, Q_n$. Here, for the secure computation protocol, the private input of party $Q_i$ are $\mathsf{p}_b^\alpha(i), \mathsf{q}_b^\alpha(i), \mathsf{p}_b^\beta(i), \mathsf{q}_b^\beta(i), \mathsf{k}_b^\gamma(i), \lambda_i^\alpha, \lambda_i^\beta, \lambda_i^\gamma$ for $b \in \{0,1\}$. Note that the garbled table is an information theoretic (or, non-cryptographic) function of the private inputs. Hence, the overall protocol is information theoretic in the OT-hybrid model. Moreover, to get the constant round result of [2], it was crucial that the garbled table generation circuit has depth constant (in particular, 2).

- **Stage 3.** The parties also get the garbled input $\tilde{x}$. For a wire $w$ with value $x_w$, let $\Lambda^w = x_w \oplus \lambda_1^w \oplus \ldots \oplus \lambda_n^w$. All parties get $\Lambda^w, \mathsf{key}_{\Lambda^w}^w$. Parties also reveal their masks for each output wire $\lambda_i^o$.

- **Stage 4.** Finally, given the garbled circuit $\tilde{C}$ consisting of all the garbled tables and garbled input $\tilde{x}$, the parties can compute locally as follows: For any wire $w, \rho^w$ denote its correct value during evaluation. It is maintained that for any wire $w$, each party learns the masked value $\Lambda^w$ and the label $\mathsf{key}^w_{\Lambda^w}$ where $\Lambda^w = \lambda^w \oplus \rho^w$. It is clearly true for the input wires. Now, for any gate $g$ with input wires $\alpha, \beta$, each party knows $\Lambda^\alpha, \mathsf{key}^\alpha_{\Lambda^\alpha}, \Lambda^\beta, \mathsf{key}^\beta_{\Lambda^\beta}$. If $(\Lambda^\alpha, \Lambda^\beta) = (0,0)$, decrypt the first row of the garbled gate, i.e., $A_g$, if $(\Lambda^\alpha, \Lambda^\beta) = (0,1)$ decrypt $B_g$, if $(\Lambda^\alpha, \Lambda^\beta) = (1,0)$ decrypt $C_g$, and else if $(\Lambda^\alpha, \Lambda^\beta) = (1,1)$ decrypt $D_g$ and obtain $\mathsf{key}^\gamma = \mathsf{k}^\gamma(1) \circ \ldots \circ \mathsf{k}^\gamma(n)$. Now, party $Q_i$ checks the following: If $\mathsf{k}^\gamma(i) = \mathsf{k}^\gamma_b(i)$ for some $b \in \{0,1\}$, it sets $\Lambda^\gamma = b$. Else, party $Q_i$ aborts. Finally, each parties computes the output using $\lambda^o$ and $\Lambda^o$ for the output wires.

## 5.2   Garbled RAM Instantiated with Distributed Garbling of BMR

The aforementioned garbling protocol implies a special distributed circuit garbling scheme, which we refer to in the following as BMR scheme, denoted by $(\mathsf{GCircuit}_{\mathsf{bmr}}, \mathsf{Eval}_{\mathsf{bmr}}, \mathsf{CircSim}_{\mathsf{bmr}})$. It has the same syntax as the a secure circuit garbling scheme, but with the special labeling structure described above. The scheme has the following properties.

*Black-box use of OWFs.* The scheme only involves a black-box use of one-way functions in the OT-hybrid model.

*Security.* Since the above protocol from [2] is a semi-honest secure computation protocol, the BMR scheme is a secure circuit garbling scheme. That is, it does not reveal anything beyond the output of the circuit $C$ on $x$ to an adversary corrupting a set of parties $\mathcal{I} \subset [n]$. More precisely, we can abstract out the BMR scheme as well as its security as follows: Let us denote the collection of labels used by party $Q_i$ using PRF $s_i$ by $\mathsf{Labels}_i$ and the set of wire masks by $\lambda_i$. Similarly, let $\mathsf{Labels}_\mathcal{I}$ denote $\{\mathsf{Labels}_i\}_{i \in [\mathcal{I}]}$ and $\lambda_\mathcal{I}$ denote $\{\lambda_i\}_{i \in [\mathcal{I}]}$. Then, the following lemma states the security of the BMR scheme.

**Lemma 2 (Security of BMR garbling scheme).** *There exists a PPT simulator $\mathsf{CircSim}_{\mathsf{bmr}}$ such that $\mathsf{CircSim}_{\mathsf{bmr}}(1^\kappa, C, x_\mathcal{I}, \mathsf{Labels}_\mathcal{I}, \lambda_\mathcal{I}, y) \approx (\tilde{C}, \tilde{x})$. Here $(\tilde{C}, \tilde{x})$ correspond to the garbled circuit and the garbled input produced in the real world using $\Pi^{\mathsf{sh}}_{\mathsf{bmr}}$ conditioned on $(\mathsf{Labels}_\mathcal{I}, \lambda_\mathcal{I})$. We denote it by*
$$\mathsf{GCircuit}_{\mathsf{bmr}}(1^\kappa, C, x)\Big|_{(\mathsf{Labels}_\mathcal{I}, \lambda_\mathcal{I})}.$$

The proof of the above lemma follows from the security of [2].

**Distributed garbling scheme of garbled RAM.** Our next step is instantiating the garbled RAM of [11] with the BMR circuit garbling scheme. As mentioned in Lemma 2, the BMR scheme $(\mathsf{GCircuit}_{\mathsf{bmr}}, \mathsf{Eval}_{\mathsf{bmr}}, \mathsf{CircSim}_{\mathsf{bmr}})$ is a secure circuit garbling scheme. And we note that [11] makes a black-box use of a secure circuit garbling scheme $(\mathsf{GCircuit}, \mathsf{Eval}, \mathsf{CircSim})$. Our key observation is that it can be instantiated using the BMR scheme. This would be very useful for our protocol of secure computation for RAM programs. When we instantiate the garbled RAM of [11] with the BMR scheme, the following lemma summarizes the security of the resulting garbled RAM relying upon Lemma 2.

**Lemma 3 (Garbled RAM security with BMR garbling).** *Instantiating the garbled RAM construction of [11] with the BMR circuit garbling scheme* $(\mathsf{GCircuit_{bmr}}, \mathsf{Eval_{bmr}}, \mathsf{CircSim_{bmr}})$ *gives a secure garbled RAM scheme. In particular, the garbler picks* $s_1, \ldots, s_n$ *as the seeds of the PRF to generate the keys of the garbled circuit. Let the* $i^{th}$ *set of keys be* $\mathsf{Labels}_i$*. Denote the resulting scheme by* $\mathsf{Gram_{bmr}}$*. Denote the corresponding simulator for garbled RAM by* $\mathsf{GramSim_{bmr}}$*, which would internally use* $\mathsf{CircSim_{bmr}}$*. Using the security of garbled RAM and the security of BMR scheme, we have the following:*

$$\mathsf{Gram_{bmr}}(1^\kappa, 1^t, D, P, x) \approx_c \mathsf{GramSim_{bmr}}(1^\kappa, 1^t, 1^{|D|}, y),$$

*where* $y$ *denotes the output of* $P(D, x)$*. In fact, using the security property of* $\mathsf{CircSim_{bmr}}$*, we have the following stronger security property. Let* $(x = x_1, \ldots, x_n)$ *and* $D = (D_1||\ldots||D_n)$*. Let* $\mathcal{I} \subset [n]$*. Then*

$$\mathsf{Gram_{bmr}}(1^\kappa, 1^t, D, P, x, \mathsf{Labels}_{\mathcal{I}}, \mathcal{I}) \approx_c \mathsf{GramSim_{bmr}}(1^\kappa, 1^t, 1^{|D|}, (x_{\mathcal{I}}, D_{\mathcal{I}}, \mathsf{Labels}_{\mathcal{I}}), y).$$

Recall that a garbled RAM consists of a collection of garbled circuits along with partial labels. It is easy to see that using the BMR garbling scheme preserves the black-box nature of the garbled RAM construction of [11].

**Removing the hard-coding of keys in the scheme of [11].** As mentioned before, the garbled circuits in garbled RAM of [11] contain hard-coding of keys of other garbled circuits. For ease of exposition, we remove this hard-coding of sensitive information and provide the previously these values as additional inputs.[12] Moreover, labels corresponding to these new inputs would be revealed at the time of garbled RAM generation. More precisely, we would do the following:

Consider a circuit $C$ in the original scheme of [11] which has the partial keys of some circuit $C'$ hardcoded inside it. Since we will be using the BMR garbling scheme, key $\mathsf{key}_b^w$ of any wire $w$ of $C'$ consists of a concatenation of keys $\mathsf{k}_b^w(i)$ such that the party $Q_i$ contributes $\mathsf{k}_b^w(i)$. Wire $w$ will also have a mask $\lambda^w = \lambda_1^w \oplus \ldots \oplus \lambda_n^w$ such that $\lambda_i^w$ is contributed by party $Q_i$. Now, the transformed $C$ will have input wires corresponding to each bit of $\mathsf{key}_0^w$ and $\mathsf{key}_1^w$ and also $\lambda^w$. That is, the circuits along with having keys of some other circuits, will also have masks for the corresponding wires.[13] This is necessary for consistency and correctness of evaluation of the garbled circuits. We further expand the input $\lambda^w$ into $n$ bits as $\lambda_1^w, \ldots, \lambda_n^w$. Finally, input wires of $C$ corresponding to $\mathsf{k}_b^w(i)$ for $b \in \{0, 1\}$ and $\lambda_i^w$ will correspond to input wires of party $Q_i$. Note that the transformed circuit falls in the framework of [2].

---

[12] This would be useful in arguing that the functionality computed under generic MPC is information theoretic as well as arguing efficiency while garbling multiple programs w.r.t. a persistent database.

[13] The circuits described in [11] will be modified naturally to include these mask bits in evaluation. For example, consider a circuit which was originally producing output $\mathsf{qKey}_x$, where $\mathsf{qKey}$ was a collection of keys (two per bit) being selected by string $x$. Now new circuit will output $\mathsf{qKey}_{x \oplus \lambda}$, where $\lambda$ contains the mask bits for all wires in $x$. Hence, if $\mathsf{qKey}$ was given as input, then we also need to provide $\lambda$ as input.

## 5.3   Semi-honest Secure Protocol for RAM Computation

In this section, we describe our constant-round protocol for semi-honest secure
multiparty RAM computation that only makes a black-box use of one-way functions in the OT-hybrid model. The parties will run the semi-honest protocol
from [2] to collectively generate the garbled RAM and compute the garbled
RAM locally to obtain the output. As mentioned before, a naïve implementation of this idea results in a non-black-box use of one-way functions because a
garbled RAM consists of a bunch of garbled circuits (that use PRG inside them).
To overcome this issue, we will use the garbled RAM instantiation based on garbling scheme of [2] described above (see Lemma 3) without the hard-coding of
keys. Here, the main idea is that each party invokes the one-way function locally
and off-the-shelf secure computation protocol is invoked only for an information
theoretic functionality (that we describe below). Moreover, recall that the garbled RAM scheme with full security compiles a UMA-2 secure scheme with a
statistically secure ORAM. It is crucial for black-box nature of our protocol that
the ORAM scheme is statistical and does not use any cryptographic primitives.
Hence, intuitively, since the secure computation protocol of [2] is black-box, the
overall transformation is black-box in use of OWFs.

**The functionality $\mathcal{F}_{\mathsf{Gram}}$.** We begin by describing the functionality $\mathcal{F}_{\mathrm{GRAM}}$
w.r.t. a program $P$ that will be computed by the parties via the constant-round
black-box protocol of [2].

1. **Inputs:** The input of party $Q_i$ consists of $x_i$, database $D_i$, shares of keys for
   all the garbled circuits denoted by $\mathsf{Labels}_i$, all the wire masks denoted by $\lambda_i$,
   the relevant PRG outputs on labels denoted by $\mathsf{PRG}_i$, and randomness for
   ORAM $r_i$.
2. **Output:** Each party gets the garbled RAM $(\tilde{D}, \tilde{P}, \tilde{x})$ for program $P$ promised
   by Lemma 3. The randomness used by ORAM is computed as $\oplus_{i \in [n]} r_i$.

Now we argue that the above described functionality is *information theoretic*.
We first note that garbled RAM consists of a collection of garbled circuits and
the structure of these circuits as well as interconnection of these circuits is known
publicly. This is true because we have removed all the sensitive information that
was earlier hard-coded as additional input to the circuits. Moreover, the circuits
that are being garbled in [11] are information theoretic. This follows from the
fact that [11] only makes a black-box of one-way functions. Secondly, once all the
labels as well as PRG outputs are computed locally by the parties, the garble
table generation is information theoretic (see Sect. 5.1). Thirdly, the circuit to
compute the labels for partial inputs that are revealed at the time of garbled
RAM generation is information theoretic. This is because the values of those
partial inputs can be computed information theoretically from the inputs of the
parties. And finally, we use the fact that the ORAM used is statistical and hence,
information theoretic. Therefore, the overall functionality $\mathcal{F}_{\mathrm{GRAM}}$ is information
theoretic.

**Our protocol.** Consider $n$ parties $Q_1, \ldots, Q_n$ who want to compute a program $P$. The party $Q_i$ holds an input $x_i$ and database $D_i$. Let us denote the semi-honest protocol for RAM computation of $P$ by $\Pi_{\mathsf{RAM}}^{\mathsf{sh}}$ that is as follows:

- **Step 1.** Party $Q_i$ computes the inputs for the functionality $\mathcal{F}_{\mathrm{GRAM}}$ described above. More precisely, party $Q_i$ does the following: It picks a seed $s_i$ for a PRF and randomness $r_i$ for ORAM. It generates its shares for keys to all wires of all the circuits $\mathsf{Labels}_i$ by computing the PRF with key $s_i$ on appropriate inputs. It also picks a random mask for each wire $\lambda_i$. Party $Q_i$ also locally computes the relevant PRG outputs $\mathsf{PRG}_i$ needed for garbled tables (see Sect. 5.1).
- **Step 2.** The parties run the semi-honest secure protocol $\Pi_{\mathsf{bmr}}^{\mathcal{F}_{\mathrm{GRAM}}}$ (provided by Theorem 3) to compute the functionality $\mathcal{F}_{\mathrm{GRAM}}$ described above. Each party will get the garbled RAM $(\tilde{D}, \tilde{P}, \tilde{x})$ as output.
- **Step 3.** Each party runs $\mathsf{GEval}^{\tilde{D}}(\tilde{P}, \tilde{x})$ to obtain the output $y$.

**Correctness.** The correctness of the above protocol follows trivially from the correctness of $\Pi_{\mathsf{bmr}}^{\mathcal{F}_{\mathrm{GRAM}}}$ and correctness of garbled RAM of [11].

**Round complexity.** The round complexity of the above protocol is same as the round complexity of $\Pi_{\mathsf{bmr}}^{\mathcal{F}_{\mathrm{GRAM}}}$. Hence, it is a constant by Theorem 3.

**Black-box use of one-way functions.** This follows from the fact that the protocol $\Pi_{\mathsf{bmr}}^{\mathcal{F}}$ in Theorem 3 only makes a black-box use of one-way functions in the OT-hybrid model and that the functionality $\mathcal{F}_{\mathrm{GRAM}}$ is set up to be information theoretic (as argued above).

**Efficiency.** First of all, [11] guarantees that the number of circuits needed for $\tilde{D}$ is only proportional to $|D|$ up to poly-logarithmic factors, and that the number of circuits needed for $\tilde{P}$ is proportional to the running time of $P$ up to poly-logarithmic factors. The functionality $\mathcal{F}_{\mathrm{GRAM}}$ that generates the garbled RAM ($\tilde{D}$ and $\tilde{P}$) has size linear in the garbled RAM itself. And finally, the communication and computation complexities of $\Pi_{\mathsf{bmr}}^{\mathcal{F}_{\mathrm{GRAM}}}$ grow linearly in the size of $\mathcal{F}_{\mathrm{GRAM}}$. Therefore the entire communication and computation complexities are satisfactory.

Note that for the desired efficiency, it is crucial that we run a single MPC protocol to generate all the garbled circuits instead of running multiple sessions of MPC (each generating one garbled circuit) in parallel. This is because partial labels for some garbled circuits depend on the actual database values and for security it is crucial to hide which circuit corresponds to which index in the database. This security guarantee is achieved by the use of ORAM in the garbled RAM scheme. In our protocol, for security, the randomness of ORAM is contributed by all the parties. Hence, separating the generation of garbled circuits would require all the parties to input the entire database to each garbled circuit generation, and the resulting efficiency of garbling the database would be at least quadratic in its size.

## 5.4   Proof of Semi-Honest Security

In this section, we prove that the above protocol is secure against a semi-honest adversary corrupting a set $\mathcal{I} \subseteq [n]$ of parties. We would rely on the semi-honest security of garbled RAM from [11] when instantiated using the BMR garbling scheme (see Lemma 3) as well as semi-honest security of $\Pi_{\mathsf{bmr}}^{\mathcal{F}}$ (see Theorem 3). We will define our simulator $\mathsf{Sim}_{\mathsf{mpc}}^{\mathsf{sh}}$ and prove that the view computed by $\mathsf{Sim}_{\mathsf{mpc}}^{\mathsf{sh}}$ is indistinguishable from the view of the adversary in the real world.[14]

1. $\mathsf{Sim}_{\mathsf{mpc}}^{\mathsf{sh}}$ begins by corrupting the parties in set $\mathcal{I}$ and obtains the inputs $x_{\mathcal{I}} = \{x_i\}_{i \in \mathcal{I}}$ and database $D_{\mathcal{I}} = \{D_i\}_{i \in \mathcal{I}}$ of the corrupt parties. $\mathsf{Sim}_{\mathsf{mpc}}^{\mathsf{sh}}$ queries trusted functionality for program $P$ on input $(x_{\mathcal{I}}, D_{\mathcal{I}})$ and receives output $y$.
2. $\mathsf{Sim}_{\mathsf{mpc}}^{\mathsf{sh}}$ picks PRF keys $s_i$ and randomness $r_i$ for ORAM for all $i \in \mathcal{I}$. It computes the shares of keys for all the wires of all the circuits $\mathsf{Labels}_i$, wire masks $\lambda_i$ as well as PRG outputs $\mathsf{PRG}_i$ honestly for all the corrupt parties.
3. $\mathsf{Sim}_{\mathsf{mpc}}^{\mathsf{sh}}$ invokes the simulator of the garbled RAM $\mathsf{GramSim}_{\mathsf{bmr}}(1^\kappa, 1^t, 1^{|D|}, (x_{\mathcal{I}}, D_{\mathcal{I}}, \mathsf{Labels}_{\mathcal{I}}), y)$ to get the simulated garbled RAM. Let us denote it by $(\tilde{D}, \tilde{P}, \tilde{x})$.
4. $\mathsf{Sim}_{\mathsf{mpc}}^{\mathsf{sh}}$ now invokes $\mathsf{Sim}_{\mathsf{bmr}}$ for functionality $\mathcal{F}_{\mathsf{GRAM}}$ where the inputs of the corrupt parties are obtained in Step 2 above, that is, $\{(x_i, D_i, r_i, \mathsf{Labels}_i, \lambda_i, \mathsf{PRG}_i)\}_{i \in \mathcal{I}}$ and output is the simulated garbled RAM $(\tilde{D}, \tilde{P}, \tilde{x})$. More precisely, the simulator $\mathsf{Sim}_{\mathsf{mpc}}^{\mathsf{sh}}$ outputs $\mathsf{Sim}_{\mathsf{bmr}}(1^\kappa, \{(x_i, D_i, r_i, \mathsf{Labels}_i, \lambda_i, \mathsf{PRG}_i)\}_{i \in \mathcal{I}}, (\tilde{D}, \tilde{P}, \tilde{x}))$.

Next, we show that the output of the simulator $\mathsf{Sim}_{\mathsf{mpc}}^{\mathsf{sh}}$ is indistinguishable from the real execution via a sequence of hybrids $\mathsf{Hyb}_0, \mathsf{Hyb}_1, \mathsf{Hyb}_2$, where we prove that the output of any pair of consecutive hybrids is indistinguishable. $\mathsf{Hyb}_0$ corresponds to the real execution and $\mathsf{Hyb}_2$ corresponds to the final simulation.

$\mathsf{Hyb}_1$: This is same as the hybrid $\mathsf{Hyb}_0$ except we simulate the protocol for $\Pi_{\mathsf{bmr}}^{\mathcal{F}_{\mathsf{GRAM}}}$ by using $\mathsf{Sim}_{\mathsf{bmr}}$ with real garbled RAM as output. More precisely, this hybrid is as follows:

1. For all $i \in [n]$, let $x_i$ and $D_i$ denote the input and database respectively.
2. Pick PRF keys $s_i$ and randomness $r_i$ for ORAM for all $i \in [n]$.
3. For each $i \in \mathcal{I}$, compute the shares of keys for all the wires of all the circuits, wire masks as well as PRG outputs honestly for the corrupt parties. For the party $Q_i$, denote the collection of wire key shares by $\mathsf{Labels}_i$, collection of wire masks shares by $\lambda_i$ and the collection of PRG outputs by $\mathsf{PRG}_i$.
4. Generate the garbled RAM $(\tilde{D}, \tilde{P}, \tilde{x})$ honestly as $\mathsf{Gram}_{\mathsf{bmr}}(1^\kappa, 1^t, D_{[n]}, P, x_{[n]}, s_{[n]})$.
5. Run $\mathsf{Sim}_{\mathsf{bmr}}$ for functionality $\mathcal{F}_{\mathsf{GRAM}}$ where the inputs of the corrupt parties are obtained in Steps 2 and 3 above, that is, $\{(x_i, D_i, r_i, \mathsf{Labels}_i, \lambda_i, \mathsf{PRG}_i)\}_{i \in \mathcal{I}}$ and output is the garbled RAM $(\tilde{D}, \tilde{P}, \tilde{x})$ obtained above. More precisely, the simulator $\mathsf{Sim}_{\mathsf{mpc}}^{\mathsf{sh}}$ outputs $\mathsf{Sim}_{\mathsf{bmr}}(1^\kappa, \{(x_i, D_i, r_i, \mathsf{Labels}_i, \lambda_i, \mathsf{PRG}_i)\}_{i \in \mathcal{I}}, (\tilde{D}, \tilde{P}, \tilde{x}))$.

---

[14] In the semi-honest setting, outputs of the honest parties are identical in the real world and the ideal world.

Intuitively, $\mathsf{Hyb}_0 \approx_c \mathsf{Hyb}_1$ by the correctness and the semi-honest security of the multi-party protocol from [2] used for computing the garbled RAM (see Theorem 3 for formal security guarantee).

$\mathsf{Hyb}_2$: This is same as simulator $\mathsf{Sim}^{\mathsf{sh}}_{\mathsf{mpc}}$. In other words, this is same as the previous hybrid except instead of using actual circuits of garbled RAM, we use simulated garbled circuits output by $\mathsf{GramSim}_{\mathsf{bmr}}$ on $\mathsf{GramSim}_{\mathsf{bmr}}(1^\kappa, 1^t, 1^{|D|}, (x_\mathcal{I}, D_\mathcal{I}, \mathsf{Labels}_\mathcal{I}), y)$. Note that unlike previous hybrid, this hybrid does not rely on the inputs and the database of honest parties.

Hybrids $\mathsf{Hyb}_1 \approx_c \mathsf{Hyb}_2$ by Lemma 3.

## 5.5  Running More Than One Program on a Persistent Database

In this section, we provide a protocol for executing multiple programs $P^{(1)}, \ldots, P^{(\ell)}$ on a persistent database. For exposition, it suffices to describe the case of $P^{(1)}$ and $P^{(2)}$, and it would be easy to extend to more programs in a natural way.

Recall that the garbled memory consists of a collection of garbled circuits that get consumed when a program is executed. Note that for security each garbled circuit can only be executed on a single input. Hence, the main issue in executing multiple programs on a persistent memory (as already observed by [11]) is that we need a way to replenish the circuits in the memory so that we can allow for more reads/writes by programs. To address this challenge, [11] gave a mechanism for replenishing circuits obliviously where each garbled program running for time $T$ also generates enough garbled circuits for memory to support $T$ more reads. Recall that the garbled memory consists of a tree of garbled circuits where each node consists of a sequence of garbled circuits. In [11], since the execution of the program $P^{(1)}$ as well as which circuits get consumed is hidden from the garbler, they need a more sophisticated oblivious technique for replenishing. This can be simplified in our setting because all the parties execute the garbled RAM for $D, P^{(1)}$ and hence, know which garbled circuits have been consumed and need to be replenished at any node of the tree.

At the time of garbling of the second program $P^{(2)}$, the parties would compute the functionality $\mathcal{F}_{\mathrm{REP}}$ described next.

**The functionality $\mathcal{F}_{\mathsf{Rep}}$.** This functionality guarantees that the parties replenish the garbled memory to support $M$ reads before executing the program $P^{(2)}$, where $M$ is the size of the database. The garbled circuits that need to be replenished is determined by the execution of $P^{(1)}$ and hence, is known to all the parties.

1. **Inputs:** The input of party $Q_i$ consists of $x_i$ (inputs for program $P^{(2)}$), shares of keys, wire masks needed for the garbled circuits of program $P^{(2)}$ as well as share of keys, wire masks for the memory garbled circuits to be added. Note that these keys and wire masks for garbled circuits of the program and memory need to be consistent with previously generated garbled circuits. As before, we denote these by $\mathsf{Labels}_i$ and $\lambda_i$ respectively. Parties also generate the relevant PRG outputs on labels denoted by $\mathsf{PRG}_i$.

2. **Output:** Each party gets the garbled program and garble input $(\tilde{P}^{(2)}, \tilde{x}^{(2)})$ for program $P^{(2)}$, as well as more garbled circuits for the memory which restores the garbled memory to support $M$ more reads.

**Our protocol.** We describe the protocol for running $P^{(1)}$ and $P^{(2)}$ on a persistent database $D$ below. Note that at a high level, the following invariant is maintained. Before executing a program, the memory is replenished to support $M$ reads, where $M$ is the size of the memory (which is w.l.o.g. greater than the running time of the program).

- **Step 1.** The parties $Q_1, \ldots, Q_n$ run the protocol $\Pi_{\mathsf{RAM}}^{\mathsf{sh}}$ to generate the garbled RAM $\tilde{D}^{(1)}, \tilde{P}^{(1)}, \tilde{x}^{(1)}$. Each party $Q_i$ executes the above garbled RAM to obtain the output $y^{(1)}$ and the resulting garbled memory $\tilde{D}'^{(1)}$. (The parties know what all garbled circuits got consumed from the garbled memory and need to be replenished.)

- **Step 2.** Party $Q_i$ computes the inputs for the functionality $\mathcal{F}_{\mathrm{REP}}$ described above. More precisely, party $Q_i$ does the following: It uses its PRF seed $s_i$ to generate its shares for all keys needed for generating the new circuits $\mathsf{Labels}_i$. It picks a random mask for each wire $\lambda_i$, and locally computes the relevant PRG outputs $\mathsf{PRG}_i$ needed for garbled tables. Note that the shares for keys and wire masks need to be consistent with previously generated garbled circuits. Since we are in the semi-honest setting, we can safely assume that the consistency is maintained.

- **Step 3.** The parties run the semi-honest secure protocol $\Pi_{\mathsf{bmr}}^{\mathcal{F}_{\mathrm{REP}}}$ to compute the functionality $\mathcal{F}_{\mathrm{REP}}$ described above. Each party will obtain the garbled program and garbled input $(\tilde{P}^{(2)}, \tilde{x}^{(2)})$. Each party will also get more circuits for the memory which restores the garbled memory to support $M$ more reads, thus obtaining an updated garbled memory $\tilde{D}^{(2)}$.

- **Step 4.** Each party runs $\mathsf{GEval}^{\tilde{D}^{(2)}}(\tilde{P}^{(2)}, \tilde{x}^{(2)})$ to obtain the output $y^{(2)}$ and modified garbled memory $\tilde{D}'^{(2)}$.

The *correctness, constant round complexity,* and *black-box use of one-way functions* of the protocol can be argued similarly as before in Sect. 5.4. A crucial idea in arguing correctness is that the structure of the garbled circuits needed is public and the keys of new circuits should be consistent with previously generated garbled circuits. This is easy to ensure as the parties behave honestly in generating consistent labels using the same PRF keys as before.

*Efficiency.* We argue that the complexity of garbling second program is proportional to the running time of $P^{(2)}$ up to poly-logarithmic factors. The argument follows in a similar manner as Sect. 5.4. First note that the number of garbled circuits in garbled program $\tilde{P}^{(2)}$ is proportional to the running time of $P^{(2)}$. The replenishing mechanism of [11] ensures that the number of new circuits needed is only proportional to the running time of $P^{(1)}$ up to poly-logarithmic factors (since these are the number of circuits consumed from memory while running $P^{(1)}$ program).[15]

It is crucial to note that the cost of replenishing does not grow with the size of the memory. We note that replenishing does not require the database contents or the randomness used by ORAM as input. More precisely, for any node in the tree of the garbled memory, adding more circuits to that node only requires knowledge of the labels used in the previous circuit in that node and being consistent with that.[16]

*Security.* We can argue the semi-honest security of the above protocol as follows: The simulator needs to simulate the view of the adversary for all the program executions using the corresponding simulator of garbled RAM provided by [11] as before. The only non-triviality is that the underlying simulator of garbled RAM would need to know the outputs of all the executions in order to do successful simulation of garbled RAM. Since we are the semi-honest setting, these are easy to compute. This is because since all the parties are semi-honest, the parties choose the inputs for different execution just based on their previous inputs and outputs. This choice does not depend on the garbled RAM given to the parties in the real execution. This will be the major bottleneck for the malicious setting and we will revisit this point later.[17]

At a high level, our simulator for the persistent database would do the following. As before, let $\mathcal{I} \subset [n]$ denote the set of corrupt parties. The simulator will use $D_\mathcal{I}$ and $x_\mathcal{I}^{(1)}$ of corrupt parties to learn the output $y^{(1)}$ from the trusted functionality. Then, it will use the honest party strategy to compute the next round of adversarial inputs $x_\mathcal{I}^{(2)}$ for the second program, and learn the output $y^{(2)}$. This way the simulator learns the outputs $y^{(1)}, \dots, y^{(\ell)}$ for all the executions. Now we can use the simulator of [11] to simulate the garbled RAM consisting of all the garbled circuits. Finally, we use the simulator $\mathsf{Sim}_{\mathsf{bmr}}$ on garbled RAM to simulate the view of adversary for all the program executions. The argument of indistinguishability follows in the same manner as proof of a single program in Sect. 5.4.

### 5.6 Load Balancing

In the protocol we have described above for semi-honest RAM computation, each party gets a garbled RAM, which can be computed locally. This requires each party to store information whose size is at least as large as size of total database. In some settings, this might not be desired. In the semi-honest setting it is easy to guarantee security even when only one party stores the garbled memory and

---

[15] For ease of calculation, we can include of replenishing after running $P^{(1)}$ in the cost of $P^{(1)}$ and cost of replenishing after $P^{(2)}$ in the cost of $P^{(2)}$ and so on.

[16] This is because the keys of the successor circuit are already fed as input to the predecessor and the new circuit just has to be consistent to the previously generated circuit.

[17] Note that a malicious party might choose its inputs for the second program execution based on the garbled memory obtained in the first execution.

computes the program. We can simply modify the protocol such that only one party gets the garbled RAM as output. All the other parties get no output. The party which gets the garbled RAM, computes it, and sends the output of the computation to all the other parties. Since we are in the semi-honest setting, it is easy to see that this protocol is correct and secure given the correctness and security of the original protocol.

# 6    Malicious Setting

In this section, we show how the semi-honest protocol presented in Sect. 5 can be extended using appropriate tools to work for the malicious setting in the case of non-persistent database.[18] We show the following result:

**Theorem 4.** *There exists a constant-round malicious secure multiparty protocol in the OT-hybrid model that makes a black-box use of one-way functions for secure RAM computation for the case of non-persistent database satisfying security and efficiency requirements of Sect. 4.*

**Protocol Overview.** We first recall the semi-honest protocol from Sect. 5.3 at a high level. In the first step, each party picks a seed of a PRF and computes the share of keys and masks for all the circuits in the garbled RAM as output of the PRF as well as the PRG outputs on all the key shares. The parties also pick randomness for statistical ORAM such that final randomness used is the sum of the randomness from all the parties. Next, the parties run the constant round protocol of [2] to generate all the garbled circuits for the garbled RAM of [11] instantiated with distributed garbling scheme (see Lemma 3). A key point was that the functionality $\mathcal{F}_{\text{GRAM}}$ executed via the secure computation protocol is information theoretic that gives us a black-box constant-round protocol. In this section, we would extend these ideas to construct a malicious secure protocol as follows:

To protect against malicious behavior, we need to ensure that the adversary behaves honestly in the above high-level protocol. Towards this, we transform the above protocol as follows: The first step is to replace the execution of $\Pi_{\text{bmr}}^{\mathcal{F}_{\text{GRAM}}}$ with a malicious secure computation protocol. Since the overall goal is to get a constant round protocol for RAM which makes a black-box use of PRG in the OT-hybrid model, we use the protocol from [30] that gives such a protocol for circuits. Denote this protocol by $\Pi_{\text{ips}}^{\mathcal{F}_{\text{GRAM}}}$. More formally, the following theorem was proven by [30].

**Theorem 5 ([30], Theorem 3).** *For any $n \geq 2$ there exists an n-party constant-round secure computation protocol in the OT-hybrid model which makes a black-box use of a pseudorandom generator and achieves computational UC-security against an active adversary which may adaptively corrupt at most $n - 1$ parties.*

---

[18] We address the issue of persistent database at the end of this section.

Let $\mathsf{Sim}_{\mathsf{ips}}$ be the simulator provided by the above theorem. In our case, we only need stand-alone security against a static malicious adversary which is weaker than what is provided by the above theorem.

Next, recall that in a distributed garbling scheme of Sect. 5.2, each party computes shares of labels using a PRF seed as well as PRG outputs on these labels to be used in garbling of individual gates. If we can ensure that a malicious party correctly computes the labels as outputs of a PRF and also the PRG outputs, intuitively the security would follow similarly to the semi-honest scenario by relying on the malicious security of $\Pi_{\mathsf{ips}}^{\mathcal{F}_{\mathrm{GRAM}}}$. Since the goal is to construct a protocol that only makes a black-box of one-way functions, we cannot make the parties prove that they computed the PRF or PRG values correctly. In particular, proving correctness of PRF and PRG outputs would lead to a non-black-box use of cryptography. To solve this issue we make two key observations described below.

First, we observe that in any scheme for circuit garbling as well as garbled RAM, the wire labels are chosen to be outputs of a PRF only for efficiency and not security. That is, even if malicious parties choose these labels as arbitrarily chosen strings, it does not compromise the security of the garbled circuits, or garbled RAM scheme, and hence, our construction. But, the correct computation of PRG values used to encrypt the keys in garbled gate tables, is indeed critical. In fact, if a malicious party feeds the wrong output of PRG in computing of the garbled tables, it can cause the honest parties to abort during evaluation of garbled circuits or garbled RAM.[19] Note that an adversary can choose to cause this abort selectively by computing, let us say, most PRG outputs correctly and only a few incorrectly. This seems highly problematic at first, since this can lead to the problem of selective abort based on the inputs of honest parties and would break security. Our key observation is that the probability of this abort happening is independent of the inputs of the honest parties and is, indeed, simulatable given just the labels and the PRG outputs used by the adversary during the secure computation. This holds because of the following:

1. In the distributed garbling scheme, during evaluation, all parties decrypt the same row during evaluation (See Sects. 5.1 and 5.2 for details). That is, the adversary as well as the honest parties decrypt the same row for all the gate tables. Now, since this scheme is semi-honest secure, which row is decrypted during evaluation is independent of the honest parties' inputs. In fact, it depends on the mask value $\lambda^w$ for the wires which is secret shared among all parties and hence, hidden from all the parties.
2. The protocol $\Pi_{\mathsf{ips}}^{\mathcal{F}_{\mathrm{GRAM}}}$ is a correct and malicious secure protocol. Hence, the simulator of $\Pi_{\mathsf{ips}}^{\mathcal{F}_{\mathrm{GRAM}}}$ would extract an input for the adversary that consists of $x_i, D_i, r_i, \mathsf{Labels}_i, \mathsf{PRG}_i$ for all corrupt parties. Here, $\mathsf{Labels}_i$ and $\mathsf{PRG}_i$ correspond to the label shares and PRG outputs used by $Q_i$. Looking at these,

---

[19] Recall that during evaluation, when a party $Q_i$ decrypts the key for the output of a gate, it aborts if the $i^{th}$ sub-part of the key does not match either the 0 or 1 key of $Q_i$.

the simulator can check which PRG outputs have been computed incorrectly. Moreover, each PRG value is used in exactly one row of one gate table that is fixed. Now, as mentioned before, incorrect PRG values can cause an honest party to abort. But, whether $Q_i$ aborts or not is independent of input of $Q_i$ or any other party because of the following: The adversary feeds a PRG value to mask the keys in this row of garbled table, which acts independently on $k_b^w(i)$ for each $i$. This is because $\mathsf{key}_b^w = k_b^w(1) \circ \ldots \circ k_b^w(n)$. If the PRG value used by the adversary does not match the correct PRG output for masking $k_b^w(i)$, then w.h.p. it would not match the keys of $Q_i$ for both 0 and 1 and the party $Q_i$ would abort. This behavior is completely simulatable just given the input of the adversary.

In short, the adversary can only control whether an honest party $Q_i$ aborts or not on some specific gate. The adversary cannot set up the incorrect PRG values to change the label from 0 to 1 for an honest party because honest labels are chosen to be outputs of a PRF. We can continue the same argument for each gate to conclude that the adversary cannot make an honest party compute a wrong output.

Recall that each garbled gate has 4 rows of encryptions. For any gate, the adversary can behave honestly for $\alpha$ rows of the gate and cheat in $4 - \alpha$ rows. In this case, the honest party would abort with probability $1 - \alpha/4$, again independent of inputs as which row gets decrypted during evaluation is uniform (depends only on mask values). This cheating behavior is simulatable as well.

## 6.1    Our Protocol

We now describe our protocol for constant-round malicious secure RAM computation denoted by $\Pi_{\mathsf{RAM}}^{\mathsf{mal}}$. This protocol is same as the semi-honest secure protocol with one change that we use malicious secure protocol $\Pi_{\mathsf{ips}}^{\mathcal{F}_{\mathrm{GRAM}}}$ from [30] instead of semi-honest secure $\Pi_{\mathsf{bmr}}^{\mathcal{F}_{\mathrm{GRAM}}}$ to compute the garbled RAM in a distributed manner.

Recall the functionality $\mathcal{F}_{\mathrm{GRAM}}$ described in Sect. 5.3 that takes as input the shares of randomness for ORAM, keys for all wires of all the garbled circuits, PRG outputs used in generating garbled tables (see Sect. 5.1), share of wire masks as well as inputs $x_1, \ldots, x_n$ and data base $D_1, \ldots, D_n$ and produces the corresponding garbled RAM for $P$ as output. The randomness used for ORAM is the sum of the shares of randomness from all the parties. Note that as argued in Sect. 5.3, the functionality $\mathcal{F}_{\mathrm{GRAM}}$ is information theoretic and does not use oneway function or any other cryptographic primitives.[20] Now the protocol $\Pi_{\mathsf{RAM}}^{\mathsf{mal}}$ is as follows:

– **Step 1.** Party $Q_i$ computes the inputs for the functionality $\mathcal{F}_{\mathrm{GRAM}}$. More precisely, party $Q_i$ does the following: It picks a seed $s_i$ for a PRF and randomness $r_i$ for ORAM. It generates its shares for keys to all wires of all the circuits $\mathsf{Labels}_i$ by computing the PRF with key $s_i$ on appropriate inputs. It also picks

---

[20] This is because ORAM is statistical and garble table generation is information theoretic once wire labels as well PRG outputs are known.

a random mask for each wire $\lambda_i$. Party $Q_i$ also locally computes the relevant PRG outputs $\mathsf{PRG}_i$ needed for garbled tables (see Sect. 5.1).

- **Step 2.** The parties run the constant round malicious secure protocol $\Pi_{\mathsf{ips}}^{\mathcal{F}_{\text{GRAM}}}$ (provided by [30]) to compute the functionality $\mathcal{F}_{\text{GRAM}}$. Each party will get the garbled RAM $(\tilde{D}, \tilde{P}, \tilde{x})$ as output.
- **Step 3.** Each party runs $\mathsf{GEval}^{\tilde{D}}(\tilde{P}, \tilde{x})$ to obtain the output[21] $y$.

Similar to the semi-honest case, the correctness, the constant round-complexity and the black-box nature of the above protocol follow in a straightforward manner from the corresponding properties of $\Pi_{\mathsf{ips}}^{\mathcal{F}_{\text{GRAM}}}$ as well as garbled RAM of [11].

The formal proof of malicious security of this protocol appears in Sect. 6.2.

**The case of persistent database.** As noted earlier, we do not handle the case of persistent database against malicious adversaries, as remarked below:

*Remark 3.* We leave open the problem of realizing a solution against malicious adversaries for the persistent database setting. Realizing a solution that supports persistent database would involve realizing garbled RAM with *adaptive input security*. Specifically, a garbled RAM solution for which the inputs on which the persistent garbled RAM is invoked can be chosen depending on the provided garbled RAM itself. In order to construct such a scheme, efficient garbled circuit construction satisfying analogous stronger security properties is needed. No construction for such garbled circuits are known based on the standard assumptions (i.e., without random-oracle model).

## 6.2  Proof of Malicious Security

In this section, we prove that the above protocol is secure against a malicious adversary corrupting a set $\mathcal{I} \subseteq [n]$ of parties. We will construct a simulator $\mathsf{Sim}_{\mathsf{mpc}}^{\mathsf{mal}}$ and prove that the joint distribution of the view computed by $\mathsf{Sim}_{\mathsf{mpc}}^{\mathsf{mal}}$ and the outputs of the honest parties is indistinguishable from the real world.

Let us denote the collection of labels used by party $Q_i$ by $\mathsf{Labels}_i$, set of PRG outputs by $\mathsf{PRG}_i$ and the set of wire masks by $\lambda_i$. Note that while generating the garbled RAM, an honest party uses the correct PRG outputs but a malicious party may use arbitrary strings. Let $\widehat{\mathsf{PRG}}_i$ denote the correct PRG outputs corresponding to $\mathsf{Labels}_i$. $\mathsf{Sim}_{\mathsf{mpc}}^{\mathsf{mal}}$ works as follows.

1. $\mathsf{Sim}_{\mathsf{mpc}}^{\mathsf{mal}}$ begins by corrupting the parties in set $\mathcal{I}$.
2. Next, $\mathsf{Sim}_{\mathsf{mpc}}^{\mathsf{mal}}$ runs $\mathsf{Sim}_{\mathsf{ips}}$ for functionality $\mathcal{F}_{\text{GRAM}}$ that would begin by extracting the inputs $x_{\mathcal{I}} = \{x_i\}_{i \in \mathcal{I}}$ and database $D_{\mathcal{I}} = \{D_i\}_{i \in \mathcal{I}}$ of the corrupt parties as well as the collection of labels $\mathsf{Labels}_{\mathcal{I}}$, PRG values $\mathsf{PRG}_{\mathcal{I}}$, wire masks $\lambda_{\mathcal{I}}$ and ORAM randomness share $r_{\mathcal{I}}$.

---

[21] Since we are in the malicious setting, some honest parties may output $\bot$. This would be captured by the ideal world adversary described later.

3. $\mathsf{Sim}_{\mathsf{mpc}}^{\mathsf{mal}}$ queries trusted functionality for program $P$ on input $(x_{\mathcal{I}}, D_{\mathcal{I}})$ and receives output $y$.

4. $\mathsf{Sim}_{\mathsf{mpc}}^{\mathsf{mal}}$ runs a simulator $\mathsf{GramSim}_{\mathsf{bmr}}'$ on $(1^\kappa, 1^t, 1^{|D|}, (x_{\mathcal{I}}, D_{\mathcal{I}}, \mathsf{Labels}_{\mathcal{I}}, \mathsf{PRG}_{\mathcal{I}}), y)$ to obtain the simulated garbled RAM. Let us denote it by $(\tilde{D}, \tilde{P}, \tilde{x})$. Here, $\mathsf{GramSim}_{\mathsf{bmr}}'$ denotes the stronger version[22] of $\mathsf{GramSim}_{\mathsf{bmr}}$ that uses $\mathsf{PRG}_{\mathcal{I}}$ instead of $\widetilde{\mathsf{PRG}}_{\mathcal{I}}$. We describe this simulator formally in full version.

5. $\mathsf{Sim}_{\mathsf{mpc}}^{\mathsf{mal}}$ gives $(\tilde{D}, \tilde{P}, \tilde{x})$ to $\mathsf{Sim}_{\mathsf{ips}}$ as the output of the secure computation protocol of $\Pi_{\mathsf{ips}}^{\mathcal{F}_{\mathsf{GRAM}}}$. $\mathsf{Sim}_{\mathsf{ips}}$ will now simulate the view of the adversary in the protocol $\Pi_{\mathsf{ips}}^{\mathcal{F}_{\mathsf{GRAM}}}$.

6. $\mathsf{Sim}_{\mathsf{mpc}}^{\mathsf{mal}}$ computes the set $\mathcal{J} \subseteq [n] \setminus \mathcal{I}$ of honest parties who receive the output. Details on how to do this are described below. $\mathsf{Sim}_{\mathsf{mpc}}^{\mathsf{mal}}$ will now send $\mathcal{J}$ to the trusted functionality.

**Computing the set $\mathcal{J}$ of honest parties who receive the output in ideal world.** $\mathsf{Sim}_{\mathsf{mpc}}^{\mathsf{mal}}$ runs $\mathsf{GramSim}_{\mathsf{bmr}}((1^\kappa, 1^t, 1^{|D|}, (x_{\mathcal{I}}, D_{\mathcal{I}}, \mathsf{Labels}_{\mathcal{I}}), y)$ to compute the honest garbled RAM $(\hat{D}, \hat{P}, \hat{x})$ and executes it. This defines the set of relevant rows as the ones that need to be decrypted in any gate that is executed for any garbled circuit. Note that an honest party aborts iff decryption of at least one of the relevant row fails. Moreover, decryption fails for party $Q_j$ if the $j^{th}$ part of the decrypted label matches neither the party's label 0 nor label 1.

For simplicity, consider one such relevant row that is used during execution. W.l.o.g. it is enough to consider the xor of PRG values input by the adversary on behalf of all corrupt parties. For this row, let $\hat{\mathsf{prg}} = \hat{a}_1 \circ \ldots \circ \hat{a}_n$ define the xor of correct PRG values from the adversary and $\mathsf{prg} = a_1 \circ \ldots \circ a_n$ denote the xor of PRG values used by the adversary. Here, intuitively, $\hat{a}_j$ or $a_j$ defines the part of PRG value used to mask the part of the label contributed by party $Q_j$. See Sect. 5.1 for details of each gate garbling. The index $j \in [n] \setminus \mathcal{I}$ belongs to set $\mathcal{J}$ iff $\hat{a}_j = a_j$ for all relevant rows. That is, only then the honest party $Q_j$ is able to decrypt all the relevant rows correctly. Note that since honest party chooses its labels as outputs of a PRF, the encryption of label 0 cannot be decrypted as label 1. This happens only when $\hat{a}_j \oplus a_j = \mathsf{k}_0^\mathsf{w}(j) \oplus \mathsf{k}_1^\mathsf{w}(j)$, which happens with negligible probability in $\kappa$.

For a formal proof of indistinguishability of views of real and ideal worlds, refer to the full version of the paper.

# References

1. Beaver, D.: Correlated pseudorandomness and the complexity of private computations. In: 28th ACM STOC, pp. 479–488. ACM Press, May 1996

---

[22] Recall that PRG outputs are only used to mask the keys of the output wire (see gate garbled technique in Sect. 5.1). This simulator uses the PRG values provided by the adversary instead of correct PRG outputs. This change can be described by a simple deterministic transformation on output of $\mathsf{GramSim}_{\mathsf{bmr}}$.

2. Beaver, D., Micali, S., Rogaway, P.: The round complexity of secure protocols (extended abstract). In: 22nd ACM STOC, pp. 503–513. ACM Press, May 1990
3. Bellare, M., Hoang, V.T., Rogaway, P.: Foundations of garbled circuits. In: Yu, T., Danezis, G., Gligor, V.D. (eds.) ACM CCS 2012, pp. 784–796. ACM Press (2012)
4. Bitansky, N., Garg, S., Telang, S.: Succinct randomized encodings and their applications. Cryptology ePrint Archive, Report 2014/771 (2014). http://eprint.iacr.org/2014/771
5. Blum, M., Feldman, P., Micali, S.: Non-interactive zero-knowledge and its applications. In: STOC, pp. 103–112 (1988)
6. Canetti, R., Holmgren, J., Jain, A., Vaikuntanathan, V.: Indistinguishability obfuscation of iterated circuits and RAM programs. Cryptology ePrint Archive, Report 2014/769 (2014). http://eprint.iacr.org/2014/769
7. Cook, S.A., Reckhow, R.A.: Time bounded random access machines. J. Comput. Syst. Sci. **7**(4), 354–375 (1973)
8. Feige, U., Lapidot, D., Shamir, A.: Multiple non-interactive zero knowledge proofs under general assumptions. SIAM J. Comput. **29**(1), 1–28 (1999)
9. Garg, S., Gentry, C., Halevi, S.: Candidate multilinear maps from ideal lattices. In: Johansson, T., Nguyen, P.Q. (eds.) EUROCRYPT 2013. LNCS, vol. 7881, pp. 1–17. Springer, Heidelberg (2013). doi:10.1007/978-3-642-38348-9_1
10. Garg, S., Gentry, C., Halevi, S., Raykova, M., Sahai, A., Waters, B.: Candidate indistinguishability obfuscation and functional encryption for all circuits. In: 54th FOCS, pp. 40–49. IEEE Computer Society Press, October 2013
11. Garg, S., Lu, S., Ostrovsky, R.: Black-box garbled RAM. In: 56th Annual IEEE Symposium on Foundations of Computer Science (2015)
12. Garg, S., Lu, S., Ostrovsky, R., Scafuro, A.: Garbled RAM from one-way functions. In: Servedio, R.A., Rubinfeld, R. (eds.) 47th ACM STOC, pp. 449–458. ACM Press (2015)
13. Gentry, C.: Fully homomorphic encryption using ideal lattices. In: Mitzenmacher, M. (ed.) 41st ACM STOC, pp. 169–178. ACM Press, May/June 2009
14. Gentry, C., Halevi, S., Lu, S., Ostrovsky, R., Raykova, M., Wichs, D.: Garbled RAM revisited. In: Nguyen, P.Q., Oswald, E. (eds.) EUROCRYPT 2014. LNCS, vol. 8441, pp. 405–422. Springer, Heidelberg (2014). doi:10.1007/978-3-642-55220-5_23
15. Gentry, C., Halevi, S., Raykova, M., Wichs, D.: Outsourcing private RAM computation. In: 55th FOCS, pp. 404–413. IEEE Computer Society Press, October 2014
16. Goldreich, O.: Towards a theory of software protection and simulation by oblivious RAMs. In: Aho, A. (ed.) 19th ACM STOC, pp. 182–194. ACM Press (1987)
17. Goldreich, O., Micali, S., Wigderson, A.: How to play any mental game or a completeness theorem for protocols with honest majority. In: Aho, A. (ed.) 19th ACM STOC, pp. 218–229. ACM Press (1987)
18. Goldreich, O., Ostrovsky, R.: Software protection and simulation on oblivious RAMs. J. ACM **43**(3), 431–473 (1996)
19. Goldwasser, S., Kalai, Y.T., Popa, R.A., Vaikuntanathan, V., Zeldovich, N.: How to run turing machines on encrypted data. In: Canetti, R., Garay, J.A. (eds.) CRYPTO 2013. LNCS, vol. 8043, pp. 536–553. Springer, Heidelberg (2013). doi:10.1007/978-3-642-40084-1_30
20. Gordon, S.D., Katz, J., Kolesnikov, V., Krell, F., Malkin, T., Raykova, M., Vahlis, Y.: Secure two-party computation in sublinear (amortized) time. In: CCS (2012)
21. Goyal, V., Lee, C.K., Ostrovsky, R., Visconti, I.: Constructing non-malleable commitments: a black-box approach. In: 53rd FOCS, pp. 51–60. IEEE Computer Society Press, October 2012

22. Goyal, V., Mohassel, P., Smith, A.: Efficient two party and multi party computation against covert adversaries. In: Smart, N. (ed.) EUROCRYPT 2008. LNCS, vol. 4965, pp. 289–306. Springer, Heidelberg (2008). doi:10.1007/978-3-540-78967-3_17

23. Goyal, V., Ostrovsky, R., Scafuro, A., Visconti, I.: Black-box non-black-box zero knowledge. In: Shmoys, D.B. (ed.) 46th ACM STOC. pp. 515–524. ACM Press, May/June 2014

24. Groth, J., Ostrovsky, R., Sahai, A.: Perfect non-interactive zero knowledge for NP. In: Vaudenay, S. (ed.) EUROCRYPT 2006. LNCS, vol. 4004, pp. 339–358. Springer, Heidelberg (2006). doi:10.1007/11761679_21

25. Hazay, C., Yanai, A.: Constant-round maliciously secure two-party computation in the RAM model. In: TCC (2016-B)

26. Impagliazzo, R., Rudich, S.: Limits on the provable consequences of one-way permutations. In: 21st ACM STOC, pp. 44–61. ACM Press, May 1989

27. Impagliazzo, R., Rudich, S.: Limits on the provable consequences of one-way permutations. In: Goldwasser, S. (ed.) CRYPTO 1988. LNCS, vol. 403, pp. 8–26. Springer, Heidelberg (1990). doi:10.1007/0-387-34799-2_2

28. Ishai, Y., Kilian, J., Nissim, K., Petrank, E.: Extending oblivious transfers efficiently. In: Boneh, D. (ed.) CRYPTO 2003. LNCS, vol. 2729, pp. 145–161. Springer, Heidelberg (2003). doi:10.1007/978-3-540-45146-4_9

29. Ishai, Y., Kushilevitz, E., Lindell, Y., Petrank, E.: Black-box constructions for secure computation. In: Kleinberg, J.M. (ed.) 38th ACM STOC, pp. 99–108. ACM Press (2006)

30. Ishai, Y., Prabhakaran, M., Sahai, A.: Founding cryptography on oblivious transfer – efficiently. In: Wagner, D. (ed.) CRYPTO 2008. LNCS, vol. 5157, pp. 572–591. Springer, Heidelberg (2008). doi:10.1007/978-3-540-85174-5_32

31. Lin, H., Pass, R.: Succinct garbling schemes and applications. Cryptology ePrint Archive, Report 2014/766 (2014). http://eprint.iacr.org/2014/766

32. Lu, S., Ostrovsky, R.: How to garble RAM programs? In: Johansson, T., Nguyen, P.Q. (eds.) EUROCRYPT 2013. LNCS, vol. 7881, pp. 719–734. Springer, Heidelberg (2013). doi:10.1007/978-3-642-38348-9_42

33. Miao, P.: Cut-and-choose for garbled RAM. Personal Communication (2016)

34. Nielsen, J.B., Orlandi, C.: LEGO for two-party secure computation. In: Reingold, O. (ed.) TCC 2009. LNCS, vol. 5444, pp. 368–386. Springer, Heidelberg (2009). doi:10.1007/978-3-642-00457-5_22

35. Ostrovsky, R.: Efficient computation on oblivious RAMs. In: 22nd ACM STOC, pp. 514–523. ACM Press, May 1990

36. Ostrovsky, R., Shoup, V.: Private information storage (extended abstract). In: 29th ACM STOC, pp. 294–303. ACM Press, May 1997

37. Pass, R., Wee, H.: Black-box constructions of two-party protocols from one-way functions. In: Reingold, O. (ed.) TCC 2009. LNCS, vol. 5444, pp. 403–418. Springer, Heidelberg (2009). doi:10.1007/978-3-642-00457-5_24

38. Pippenger, N., Fischer, M.J.: Relations among complexity measures. J. ACM **26**(2), 361–381 (1979)

39. Regev, O.: On lattices, learning with errors, random linear codes, and cryptography. In: Gabow, H.N., Fagin, R. (eds.) 37th ACM STOC, pp. 84–93. ACM Press (2005)

40. Wee, H.: Black-box, round-efficient secure computation via non-malleability amplification. In: 51st FOCS, pp. 531–540. IEEE Computer Society Press, October 2010

41. Yao, A.C.C.: Protocols for secure computations (extended abstract). In: 23rd FOCS, pp. 160–164. IEEE Computer Society Press, November 1982

# Constant-Round Maliciously Secure Two-Party Computation in the RAM Model

Carmit Hazay[(✉)] and Avishay Yanai

Bar Ilan University, Ramat Gan, Israel
carmit.hazay@biu.ac.il, ay.yanay@gmail.com

**Abstract.** The *random-access memory (RAM)* model of computation allows program constant-time memory lookup and is more applicable in practice today, covering many important algorithms. This is in contrast to the classic setting of secure 2-party computation (2PC) that mostly follows the approach for which the desired functionality must be represented as a boolean circuit. In this work we design the first *constant round* maliciously secure two-party protocol in the RAM model. Our starting point is the garbled RAM construction of Gentry et al. [16] that readily induces a constant round semi-honest two-party protocol for any RAM program assuming identity-based encryption schemes. We show how to enhance the security of their construction into the malicious setting while facing several challenges that stem due to handling the data memory. Next, we show how to apply our techniques to a more recent garbled RAM construction by Garg et al. [13] that is based on one-way functions.

## 1 Introduction

*Background on Secure Computation.* Secure multi-party computation enables a set of parties to mutually run a protocol that computes some function $f$ on their private inputs, while preserving a number of security properties. Two of the most important properties are privacy and correctness. The former implies data confidentiality, namely, nothing leaks by the protocol execution but the computed output. The latter requirement implies that the protocol enforces the integrity of the computations made by the parties, namely, honest parties learn the correct output. More generally, a rigorous security definition requires that distrusting parties with secret inputs will be able to compute a function of their inputs as if the computation is executed in an ideal setting, where the parties send their inputs to a incorruptible trusted party that performs the computation

Supported by the European Research Council under the ERC consolidators grant agreement no. 615172 (HIPS) and by the BIU Center for Research in Applied Cryptography and Cyber Security in conjunction with the Israel National Cyber Bureau in the Prime Minister's Office. First author's research partially supported by a grant from the Israel Ministry of Science and Technology (grant No. 3-10883). This paper was presented jointly with [11] in proceedings of the 14th IACR Theory of Cryptography Conference (TCC) 2016-B.

© International Association for Cryptologic Research 2016
M. Hirt and A. Smith (Eds.): TCC 2016-B, Part I, LNCS 9985, pp. 521–553, 2016.
DOI: 10.1007/978-3-662-53641-4_20

and returns its result (also known by the ideal/real paradigm). The feasibility of secure computation has been established by a sequence of works [2,7,19,36,48], proving security under this rigorous definition with respect to two adversarial models: the semi-honest model (where the adversary follows the instructions of the protocol but tries to learn more than it should from the protocol transcript), and the malicious model (where the adversary follows an arbitrary polynomial-time strategy).

Following these works much effort was put in order to improve the efficiency of computation with the aim of minimizing the workload of the parties [24–27,30–32,37,39]. These general-purpose protocols are restricted to functions represented by Boolean/arithmetic circuits. Namely, the function is first translated into a (typically Boolean) circuit and then the protocol securely evaluates it gate-by-gate on the parties' private inputs. This approach, however, falls short when the computation involves access to a large memory since in the circuits-based approach, dynamic memory accesses, which depend on the secret inputs, are translated into a linear scan of the memory. This translation is required for every memory access and causes a huge blowup in the description of the circuit.

*The RAM Model of Computation.* We further note that the majority of applications encountered in practice today are more efficiently captured using *random-access memory (RAM)* programs that allow constant-time memory lookup. This covers graph algorithms, such as the known Dijkstra's shortest path algorithm, binary search on sorted data, finding the $k$th-ranked element, the Gale-Shapely stable matching algorithm and many more. This is in contrast to the sequential memory access that is supported by the architecture of Turing machines. Generic transformations from RAM programs that run in time $T$ generate circuits of size $(T^3 \log T)$ which are non-scalable even for cases where the memory size is relatively small [9,40].

To address these limitations, researchers have recently started to design secure protocols directly in the RAM model [1,10,22]. The main underlying idea is to rely on Oblivious RAM (ORAM) [17,20,38], a fundamental tool that supports dynamic memory access with poly-logarithmic cost while preventing any leakage from the memory. To be concrete, ORAM is a technique for hiding all the information about the memory of a RAM program. This includes both the content of the memory as well as the access pattern to it.

In more details, a RAM program $P$ is defined by a function that is executed in the presence of memory $D$ via a sequence of read and write operations, where the memory is viewed as an array of $n$ entries (or blocks) that are initially set to zero. More formally, a RAM program is defined by a "next instruction" function that is executed on an input $x$, a current state state and data element $b^{\text{read}}$ (that will always be equal to the last read element from memory $D$), and outputs the next instruction and an updated state. We use the notation $P^D(x)$ to denote the execution of such a program. To avoid trivial solutions, such as fetching the entire memory, it is required that the space used by the evaluator grows linearly with $\log n$, $|x|$ and the block length. The space complexity of a RAM program on inputs $x, D$ is the maximum number of entries used by $P$ during the course of the execution. The time complexity of a RAM program on the same inputs is the number of instructions issued in the execution as described above.

*Secure Computation for RAM Programs.* An important application of ORAM is in gaining more efficient protocols for secure computation [1,12–16,22,23,28,33, 34,44,45]. This approach is used to securely evaluate RAM programs where the overall input sizes of the parties are large (for instance, when one of the inputs is a database). Amongst these works, only [1] addresses general secure computation for arbitrary RAM programs with security in the presence of malicious adversaries. The advantage of using secure protocols directly for RAM programs is that such protocols imply (amortized) complexity that can be sublinear in the total size of the input. In particular, the overhead of these protocols grows linearly with the time-complexity of the underlying computation on the RAM program (which may be sublinear in the input size). This is in contrast to the overhead induced by evaluating the corresponding Boolean/arithmetic circuit of the underlying computation (for which its size is linear in the input size).

One significant challenge in handling dynamic memory accesses is to hide the actual memory locations being read/written from all parties. The general approach in most of these protocols is of designing protocols that work via a sequence of ORAM instructions using traditional circuit-based secure computation phases. More precisely, these protocols are defined using two phases: (1) initialize and setup the ORAM, a one-time computation with cost depending on the memory size, (2) evaluate the next-instruction circuit which outputs shares of the RAM program's internal state, the next memory operations (read/write), the location to access and the data value in case of a write. This approach leads to protocols with semi-honest security whom their round complexity depends on the ORAM running time. In [22] Gordon et al. designed the first rigorous semi-honest secure protocols based on this approach, that achieves sublinear amortized overhead that is asymptotically close to the running time of the underlying RAM program in an insecure environment.

As observed later by Afshar et al. [1], adapting this approach in the malicious setting is quite challenging. Specifically, the protocol must ensure that the parties use state and memory shares that are consistent with prior iterations, while ensuring that the running time only depends on the ORAM running time rather than on the entire memory. They therefore consider a different approach of garbling the memory first and then propagate the output labels of these garbling within the CPU-step circuits.

The main question left open by their work is the *feasibility of constant round malicious secure computation in the RAM model.* In this work we address this question in the two-party setting.

## 1.1  Our Results

We design the first constant round maliciously secure protocol for arbitrary RAM programs. Our starting point is the garbled RAM construction of Gentry et al. [16], which is the analogue object of garbled circuits [4,47] with respect to

RAM programs. Namely, a user can garble an arbitrary RAM program directly without converting it into a circuit first. A garbled RAM scheme can be used to garble the data, the program and the input in a way that reveals only the evaluation outcome and nothing else. In their work, Gentry et al. proposed two ways to fix a subtle point emerged in an earlier construction by Lu and Ostrovsky [34] that requires a complex "circular" use of Yao garbled circuits and PRFs. For simplicity, we chose to focus on their garbled RAM based on identity based encryption (IBE) schemes. We show how to transform their IBE based protocol into a maliciously secure 2PC protocol at the cost of involving the cut-and-choose technique. Following that, in the full version we show how to achieve the same result using the garbled RAM construction of Garg et al. [13] assuming only the existence of one-way-functions. We state our main theorem below,

**Theorem 1 (Informal).** *Under the standard assumptions for achieving static malicious 2PC security, there exists a constant round protocol securely realizes any RAM program in the presence of malicious adversaries, making only black-box use of an Oblivious RAM construction, where the size of the garbled database is $|D| \cdot \mathsf{poly}(\kappa)^1$, the size of the garbled input is $|x| \cdot O(\kappa) + T \cdot \mathsf{poly}(\kappa)$ and the size of the garbled program and its evaluation time is $|\mathrm{C}_{\mathrm{CPU}}^{P}| \times T \times \mathsf{poly}(\kappa) \times \mathsf{polylog}(|D|) \times s$.*

Where $\mathrm{C}_{\mathrm{CPU}}^{P}$ is a circuit that computes a CPU-step that involves reading/writing to the memory, $T$ is the running time of program $P$ on input $x$, $\kappa$ is the security parameter and $s$ is a statistical cut-and-choose parameter.

*Challenges Faced in the Malicious Setting and RAM Programs.*

1. MEMORY MANAGEMENT. Intuitively speaking, garbled RAM readily induces a two-party protocol with semi-honest security by exchanging the garbled input using oblivious transfer (OT). The natural approach for enhancing the security of a garbled RAM scheme into a maliciously 2PC protocol is by using the cut-and-choose approach [31] where the basic underlying semi-honest protocol is repeated $s$ times (for some statistical parameter $s$), such that a subset of these instances are "opened" in order to demonstrate correct behaviour whereas the rest of "unopened" instances are used to obtaining the final outcome (typically by taking the majority of results). The main challenge in boosting the security of a semi-honest secure protocol into the malicious setting, using this technique in the RAM model, is with handling multiple instances of memory data. That is, since each semi-honest protocol instance is executed independently, the RAM program implemented within this instance is associated with its own instance of memory. Recalling that the size of the memory might be huge compared to the other components in the RAM system, it is undesirable to store multiple copies of the data in the local memory of the parties. Therefore, the first challenge we had to handle

---

[1] The size mentioned is correct when relying on the IBE assumption, while relying on the OWF assumption would incur database size of $|D| \cdot \log |D| \cdot \mathsf{poly}(\kappa)$.

is how to work with multiple copies of the same protocol while having access to a single memory data.

2. HANDLING CHECK/EVALUATION CIRCUITS. The second challenge concerns the cut-and-choose proof technique as well. The original approach to garble the memory is by using encryptions computed based on PRF keys that are embedded inside the garbled circuits. These keys are used to generate a translation mapping which allows the receiver to translate between the secret keys and the labels of the read bit in the next circuit. When employing the cut-and-choose technique, all the secret information embedded within the circuits is exposed during the check process of that procedure which might violate the privacy of the sender. The same difficulty arises when hardwiring the randomness used for the encryption algorithm. A naive solution would be to let the sender choose $s$ sets of keys, such that each set is used within the appropriate copy of the circuit. While this solution works, it prevents the evaluator from determining the majority of the (intermediate) results of all copies.

3. INTEGRITY AND CONSISTENCY OF MEMORY OPERATIONS. During the evaluation of program $P$, the receiver reads and writes back to the memory. In the malicious setting these operations must be backed up with a mechanism that enforces correctness. Moreover, a corrupted evaluator should not be able to roll back the stored memory to an earlier version. This task is very challenging in a scenario where the evaluator locally stores the memory and fully controls its accesses without the sender being able to verify whether the receiver has indeed carried out the required instructions (as that would imply that the round complexity grows linearly with the running time of the RAM program).

*Constant Round 2PC in the RAM Model.* Towards achieving malicious security, we demonstrate how to adapt the garbled RAM construction from [16] into the two-party setting while achieving malicious security. Our protocol is combined of two main components. First, an initialization circuit is evaluated in order to create all the IBE keys (or the PRF keys) that are incorporated in the latter RAM computation, based on the joint randomness of the parties (this phase is not computed locally since we cannot rely on the sender properly creating these keys). Next, the program $P$ is computed via a sequence of small CPU-steps that are implemented using a circuit that takes as input the current CPU state and a bit that was read from the last read memory location, and outputs an updated state, the next location to read, a location to write to and a bit to write into that location. In order to cope with the challenges regarding the cut-and-choose approach, we must ensure that none of the secret keys nor randomness are incorporated into the circuits, but instead given as inputs. Moreover, to avoid memory duplication, all the circuits are given the same sequence of random strings. This ensures that the same set of secret keys/ciphertexts are created within all CPU circuits.

We note that our protocol is applicable to any garbled scheme that supports wire labels and can be optimized using all prior optimizations. Moreover, in a variant of our construction the initialization phase can be treated as a preprocessing phase that does not depend on the input. We further note that our abstraction of garbled circuits takes into account authenticity [4]. Meaning that, a malicious evaluator should not be able to conclude the encoding of a string that is different than the actual output. This requirement is crucial for the security of garbled circuits with reusable labels (namely, where the output labels are used as input labels in another circuit), and must be addressed even in the semi-honest setting (and specifically for garbled RAM protocols). This is because authenticity is not handled by the standard privacy requirement. Yet, all prior garbled RAM constructions do not consider it. We stress that we do not claim that prior proofs are incorrect, rather that the underlying garbled circuits must adhere this security requirement in addition to privacy.

As final remark, we note that our construction employs the underlying ORAM in a black-box manner as the parties invoke it locally. This is in contrast to alternative approaches that compute the ORAM using a two (or multi)-party secure protocol such as in [22].

*Complexity.* The overhead of our protocol is dominated by the complexity induced by the garbled RAM construction of [16] times $s$, where $s$ is the cut-and-choose statistical parameter. The [16] construction guarantees that the size/evaluation time of the garbled program is $|C_{\mathrm{CPU}}^P| \times T \times \mathsf{poly}(\kappa) \times \mathsf{polylog}(n)$. Therefore the multiplicative overhead of our protocol is $\mathsf{poly}(\kappa) \times \mathsf{polylog}(n) \times s$.

*Reusable/Persistent Data.* Reusable/persistent data means that the garbled memory data can be reused across multiple program executions. That is, all memory updates are persist for future program executions and cannot be rolled back by the malicious evaluator. This feature is very important as it allows to execute a sequence of programs without requiring to initialize the data for every execution, implying that the running time is only proportional to the program running time (in a non-secured environment). The [16] garbled RAM allows to garble any sequence of programs (nevertheless, this set must be given to the garbler in advance and cannot be adaptively chosen). We show that our scheme preserves this property in the presence of malicious attacks as well.

*Concurrent Work.* In a concurrent and independent work by Garg, Gupta, Miao and Pandey [11], the authors demonstrate constant-round multi-party computation with the advantage of achieving a construction that is *black-box* in the one-way function. Their work is based on the black-box GRAM construction of [12] and the constant-round MPC construction of [3]. Their semi-honest secure protocol achieves persistent data, whereas their maliciously secure protocol achieves the weaker notion of selectively choosing the inputs in advance, as we do. The core technique of pulling secrets out of the programs and into the inputs is common to both our and their work. Whereas our construction achieves two features which [12] does not. First, we use the ORAM in a black-box way since the parties can locally compute it. Second, only one party locally stores the memory, rather than both parties string shares of the memory. In another paper [35], Miao demonstrates how to achieve persistent data in the two-party setting assuming a random oracle and using techniques from [37] and [4], where the underlying one-way function is used in a black-box manner.

## 2  Preliminaries

### 2.1  The RAM Model of Computation

We follow the notation from [16] verbatim. We consider a program $P$ that has random-access to a memory of size $n$, which is initially empty. In addition, the program gets a "short" input $x$, which we can alternatively think of as the initial state of the program. We use the notation $P^D(x)$ to denote the execution of such program. The program can read/write to various locations in memory throughout the execution. [16] also considered the case where several different programs are executed sequentially and the memory persists between executions. Our protocol follows this extension as well. Specifically, this process is denoted as $(y_1, \ldots, y_c) = (P_1(x_1), \ldots, P_\ell(x_c))^D$ to indicate that first $P_1^D(x_1)$ is executed, resulting in some memory contents $D_1$ and output $y_1$, then $P_2^{D_1}(x_2)$ is executed resulting in some memory contents $D_2$ and output $y_2$ etc.

*CPU-Step Circuit.* We view a RAM program as a sequence of at most $T$ small CPU-steps, such that step $1 \leq t \leq T$ is represented by a circuit that computes the following functionality:

$$C_{\text{CPU}}^P(\text{state}_t, b_t^{\text{read}}) = (\text{state}_{t+1}, i_t^{\text{read}}, i_t^{\text{write}}, b_t^{\text{write}}).$$

Namely, this circuit takes as input the CPU state $\text{state}_t$ and a bit $b_t^{\text{read}}$ that was read from the last read memory location, and outputs an updated state $\text{state}_{t+1}$, the next location to read $i_t^{\text{read}} \in [n]$, a location to write to $i_t^{\text{write}} \in [n] \cap \perp$ (where $\perp$ means "write nothing") and a bit $b_t^{\text{write}}$ to write into that location. The computation $P^D(x)$ starts in the initial state $\text{state}_1 = (x_1, x_2)$, corresponding to the parties "short input" and by convention we will set the initial read bit to $b_1^{\text{read}} := 0$. In each step $t$, the computation proceeds by running $C_{\text{CPU}}^P(\text{state}_t, b_t^{\text{read}}) = (\text{state}_{t+1}, i_t^{\text{read}}, i_t^{\text{write}}, b_t^{\text{write}})$. We first read the requested location $i_t^{\text{read}}$ by stetting $b_{t+1}^{\text{read}} := D[i_t^{\text{read}}]$ and, if $i_t^{\text{write}} \neq \perp$ we write to the location by setting $D[i_t^{\text{write}}] := b_t^{\text{write}}$. The value $y = \text{state}_{T+1}$ output by the last CPU-step serves as the output of the computation.

A program $P$ has a *read-only* memory access, if it never overwrites any values in memory. In particular, using the above notation, the outputs of $C_{\text{CPU}}^P$ always set $i_t^{\text{write}} = \perp$.

**Predictably Time Writes.** Predictably Time Writes (ptWrites) means that whenever we want to read some location $i$ in memory, it is easy to figure out the time (i.e., CPU step) $t'$ in which that location was last written to, given only the current state of the computation and without reading any other values in memory. In [16] the authors describe how to upgrade a solution for ptWrites to one that allows arbitrary writes. More formally,

**Definition 1 (Predictably timed writes [16]).** A program execution $P^D(x_1, x_2)$ has predictably timed writes if there exists a poly-size circuit, denoted WriteTime, such that the following holds for every CPU step $t = 1, \ldots, T$. Let the inputs/outputs of the $t$-th CPU step be $\mathsf{cpu\text{-}step}(\mathsf{state}_t, b_t^{\mathsf{read}}) = (\mathsf{state}_{t+1}, i_t^{\mathsf{read}}, i_t^{\mathsf{write}}, b_t^{\mathsf{write}})$, then $t' = \mathsf{WriteTime}(t, \mathsf{state}_t, i_t^{\mathsf{read}})$ is the largest value of $t' < t$ such that the CPU step $t'$ wrote to memory location $i_t^{\mathsf{read}}$; i.e. $i_{t'}^{\mathsf{write}} = i_t^{\mathsf{read}}$.

As in [16], we also describe a solution for RAM programs that support ptWrites and then show how to extend it to the general case.

## 2.2 Oblivious RAM (ORAM)

ORAM, initially proposed by Goldreich and Ostrovsky [17,20,38], is an approach for making a read/write memory access pattern of a RAM program input-oblivious. More precisely, it allows a client to store private data on an untrusted server and maintain obliviousness while accessing that data, by only storing a short local state. A secure ORAM scheme not only hides the content of the memory from the server, but also the access pattern of which locations in the memory the client is reading or writing in each protocol execution.[2] The work of the client and server in each such access should be small and bounded by a poly-logarithmic factor in the memory size, where the goal is to access the data without downloading it from the server in its entirely. In stronger attack scenarios, the ORAM is also authenticated which means that the server cannot modify the content of the memory. In particular, the server cannot even "roll-back" to an older version of the data. The efficiency of ORAM constructions is evaluated by their bandwidth blowup, client storage and server storage. Bandwidth blowup is the number of data blocks that are needed to be sent between the parties per request. Client storage is the amount of trusted local memory required for the client to manage the ORAM and server storage is the amount of storage needed at the server to store all data blocks. Since the seminal sequence of works by Goldreich and Ostrovsky, ORAM has been extensively studied [21,29,41–43,46], optimizing different metrics and parameters.

Before giving the formal definition let us put down the settings and notations: A Random Access Machine (RAM) with memory size $n$ consists of a CPU with a small number of registers (e.g. $poly(\kappa)$, where $\kappa$ is the security parameter), that each can store a string of length $\kappa$ (called a "word") and external memory of size $n$. A word is either $\perp$ or a $\kappa$ bit string. Given $n$ and $x$, the CPU executes the program $P$ by sequentially evaluating the CPU-step function $C_{\mathsf{CPU}}^P(n, \mathsf{state}_t, b_t^{\mathsf{read}}) = (\mathsf{state}_{t+1}, i_t^{\mathsf{read}}, i_t^{\mathsf{write}}, b_t^{\mathsf{write}})$ where $t = 0, 1, 2, \ldots, T-1$ such that $T$ is the upper bound on the program run time and $\mathsf{state}_0 = x$. The sequence of memory cells and data written in the course of the execution of the program is defined by $\mathsf{MemAccess}(P, n, x) = \{(i_t^{\mathsf{read}}, i_t^{\mathsf{write}}, b_t^{\mathsf{write}})\}_{t \in [T]}$ and the number of memory accesses that were performed during a program execution is denoted by

---

[2] This can always be done by encrypting the memory.

$T(P, n, x)$ (that is, the running time of the program $P$ with memory of size $n$ on input $x$).

In this work we follow a slightly modified version of the standard definition (of [17,20,38]), in which the compiled program $P^*$ is not hardcoded with any secret values, namely, neither secret keys for encryption/authentication algorithms nor the randomness that specifies future memory locations to be accessed by the program, rather, the compiled program obtains these secret values as input. More concretely, $P^*$ is given two inputs: (1) a secret value $k$ that is used to derive the keys for encrypting and authenticating the data, (2) a uniformly random string $r$ which corresponds to the random indices that are accessed during the computation. The formal definition follows:[3]

**Definition 2.** *A polynomial time algorithm $C$ is an Oblivious RAM (ORAM) compiler with computational overhead $c(\cdot)$ and memory overhead $m(\cdot)$, if $C$, when given $n \in \mathbb{N}$ and a deterministic RAM program $P$ with memory size $n$, outputs a program $P^*$ with memory size $m(n) \cdot n$, such that for any input $x \in \{0,1\}^*$, uniformly random key $k \in \{0,1\}^\kappa$ and uniformly random string $r \in \{0,1\}^\kappa$, it follows that $T(P^*(n, x, k, r)) = c(n) \cdot T(P, n, x)$ and there exists a negligible function $\mu$ such that the following properties hold:*

- *Correctness. For any $n \in \mathbb{N}$, any input $x \in \{0,1\}^*$, any key and uniformly random string $k, r \in \{0,1\}^\kappa$, with probability at least $1 - \mu(\kappa)$, $P^*(n, x, k, r) = P(n, x)$.*
- *Obliviousness. For any two programs $P_1, P_2$, any $n \in \mathbb{N}$, any two inputs, uniformly random keys and uniformly random strings: $x_1, x_2 \in \{0,1\}^*, k_1, k_2, r_1, r_2 \in \{0,1\}^\kappa$ respectively, if $T(P_1(n, x_1)) = T(P_2(n, x_2))$ and $P_1^* \leftarrow C(n, P_1, \rho_1)$, $P_2^* \leftarrow C(n, P_2, \rho_2)$ then the access patterns $\mathsf{MemAccess}(P_1^*(n, x_1, k_1, r_1))$ and $\mathsf{MemAccess}(P_2^*(n, x_2, k_2, r_2))$ are computationally indistinguishable, where the random tapes $\rho_1, \rho_2$ that were used by the compiler to generate the compiled programs are given to the distinguisher.[4]*

Note that the above definition (just as the definition of [20]) only requires an oblivious compilation of deterministic programs $P$. This is without loss of generality: We can always view a randomized program as a deterministic one that receives random coins as part of its input.

**Realization of the Modified Definition.** We present here a sketch of an ORAM compiler that meets the above requirements, which is a slightly modified construction of the Simple ORAM that was presented in [8]. The modified compiler is a deterministic algorithm $C$, that is, its random tape $\rho$ is an empty string. When given a program $P$, the compiler outputs a program $P^*$ that takes the inputs $x, k, r$ where $x$ is the input to the original program $P$, $k$ is a uniformly

---

[3] The following definition is derived from the definition given in [8].

[4] The use of $\rho_1, \rho_2$ does not reveal any information about the access pattern nor about the encryption key of the data, these are determined only by the keys $k_1, k_2$ and the random strings $r_1, r_2$.

random string from which the encryption/authentication keys are derived and $r$ is a uniformly random strings of the following form: $r = \{Pos, r_1, r_2, \ldots, r_T\}$ such that $Pos$ is the initial position map of the oblivious program and $r_1, \ldots, r_T$ are the additional random locations that are used for each iteration during the execution of the program $P^*$. The program $P^*$ that $C$ outputs is specified exactly as the oblivious program presented in [8], except that the position map $Pos$ and random paths $r_1, \ldots, r_T$ are not hardcoded within the program, rather, they are given as inputs to the program.

## 2.3  Secure Computation in the RAM Model

We adapt the standard definition for secure two-party computation of [18, Chap. 7] for the RAM model of computation. In this model of computation, the initial input is split between two parties and the parties run a protocol that securely realizes a program $P$ on a pair of "short" inputs $x_1, x_2$, which are viewed as the initial state of the program. In addition, the program $P$ has random-access to a memory of size $n$ which is initially empty. Using the notations from Sect. 2.1, we refer to this (potentially random) process by $P^D(x_1, x_2)$. In this work we prove the security of our protocols in the presence of malicious computationally bounded adversaries.

We next formalize the ideal and real executions, considering $D$ as a common resource.[5] Our formalization induces two flavours of security definitions. In the first (and stronger) definition, the memory accesses to $D$ are hidden, that is, the ideal adversary that corrupts the receiver only obtains (from the trusted party) the running time $T$ of the program $P$ and the output of computation $y$. Given only these inputs, the simulator must be able to produce an indistinguishable memory access pattern. In the weaker, unprotected memory access model described below, the simulator is further given the content of the memory, as well as the memory access pattern produced by the trusted party throughout the computation of $P^D$. We present here both definitions, starting with the definition of full security.

**Full Security**

*Execution in the Ideal Model.* In an ideal execution, the parties submit their inputs to a trusted party that computes the output; see Fig. 1 for the description of the functionality computed by the trusted party in the ideal execution. Let $P$ be a two-party program, let $\mathcal{A}$ be a non-uniform PPT machine and let $i \in \{S, R\}$ be the corrupted party. Then, denote the *ideal execution of $P$* on inputs $(x_1, x_2)$, auxiliary input $z$ to $\mathcal{A}$ and security parameters $s, \kappa$, by the random variable $\mathbf{IDEAL}^{\mathcal{F}_{RAM}}_{\mathcal{A}(z), i}(s, \kappa, x_1, x_2)$, as the output pair of the honest party and the adversary $\mathcal{A}$ in the above ideal execution.

---

[5] Nevertheless, we note that the memory data $D$ will be kept in the receiver's local memory.

---

**Functionality $\mathcal{F}_{RAM}$**

The functionality $\mathcal{F}_{RAM}$ interacts with a sender S and a receiver R. The program $P$ is known and agreed by both parties.

*Input:* Upon receiving input value $(\text{INPUT}_S, x_1)$ from S and input value $(\text{INPUT}_R, x_2)$ from R, store $x_1, x_2$ and initialize the memory data $D$ with $0^n$.

*Output:* If both inputs are recorded execute $y \leftarrow P^D(x_1, x_2)$ and send $(\text{OUTPUT}_R, T, y)$ to R.

---

**Fig. 1.** A 2PC secure evaluation functionality in the RAM model for program $P$.

*Execution in the Real Model.* In the real model there is no trusted third party and the parties interact directly. The adversary $\mathcal{A}$ sends all messages in place of the corrupted party, and may follow an arbitrary PPT strategy. The honest party follows the instructions of the specified protocol $\pi$. Let $P^D$ be as above and let $\pi$ be a two-party protocol for computing $P^D$. Furthermore, let $\mathcal{A}$ be a non-uniform PPT machine and let $i \in \{S, R\}$ be the corrupted party. Then, the *real execution of $\pi$* on inputs $(x_1, x_2)$, auxiliary input $z$ to $\mathcal{A}$ and security parameters $s, \kappa$, denoted by the random variable $\textbf{REAL}^{\pi}_{\mathcal{A}(z), i}(s, \kappa, x_1, x_2)$, is defined as the output pair of the honest party and the adversary $\mathcal{A}$ from the real execution of $\pi$.

*Security as Emulation of a Real Execution in the Ideal Model.* Having defined the ideal and real models, we can now define security of protocols. Loosely speaking, the definition asserts that a secure party protocol (in the real model) emulates the ideal model (in which a trusted party exists). This is formulated by saying that adversaries in the ideal model are able to simulate executions of the real-model protocol.

**Definition 3 (Secure computation).** *Let $\mathcal{F}_{RAM}$ and $\pi$ be as above. Protocol $\pi$ is said to securely compute $P^D$ with abort in the presence of malicious adversary if for every non-uniform PPT adversary $\mathcal{A}$ for the real model, there exists a non-uniform PPT adversary $\mathcal{S}$ for the ideal model, such that for every $i \in \{S, R\}$,*

$$\left\{ \textbf{IDEAL}^{\mathcal{F}_{RAM}}_{\mathcal{S}(z), i}(s, \kappa, x_1, x_2) \right\}_{s, \kappa \in \mathbb{N}, x_1, x_2, z \in \{0,1\}^*}$$
$$\overset{c}{\approx} \left\{ \textbf{REAL}^{\pi}_{\mathcal{A}(z), i}(s, \kappa, x_1, x_2) \right\}_{s, \kappa \in \mathbb{N}, x_1, x_2, z \in \{0,1\}^*}$$

*where $s$ and $\kappa$ are the security parameters.*

We next turn to a weaker definition of secure computation in the unprotected memory access model, and then discuss a general transformation from a protocol that is secure in the UMA model to a protocol that is fully secure.

**The UMA Model.** In [16], Gentry et al. considered a weaker notion of security, denoted by *Unprotected Memory Access* (UMA), in which the receiver may additionally learn the content of the memory $D$, as well as the memory access pattern throughout the computation including the locations being read/written and their contents.[6] In the context of two-party computation, when considering the ideal execution, the trusted party further forwards the adversary the values MemAccess $= \{(i_t^{\text{read}}, i_t^{\text{write}}, b_t^{\text{write}})\}_{t \in [T]}$ where $i_t^{\text{read}}$ is the address to read from, $i_t^{\text{write}}$ is the address to write to and $b_t^{\text{write}}$ is the bit value to be written to location $i_t^{\text{write}}$ in time step $t$. We denote this functionality, described in Fig. 2, by $\mathcal{F}_{\text{UMA}}$. We define security in the UMA model and then discuss our general transformation from UMA to full security.

**Definition 4 (Secure computation in the UMA model).** *Let $\mathcal{F}_{\text{UMA}}$ be as above. Protocol $\pi$ is said to securely compute $P^D$ with UMA and abort in the presence of malicious adversaries if for every non-uniform PPT adversary $\mathcal{A}$ for the real model, there exists a non-uniform PPT adversary $\mathcal{S}$ for the ideal model, such that for every $i \in \{S, R\}$, for every $s \in \mathbb{N}, x_1, x_2, z \in \{0,1\}^*$ and for large enough $\kappa$*

$$\left\{\textbf{IDEAL}_{\mathcal{S}(z),i}^{\mathcal{F}_{\text{UMA}}}(s, \kappa, x_1, x_2)\right\}_{s, \kappa, x_1, x_2, z} \overset{k,s}{\approx} \left\{\textbf{REAL}_{\mathcal{A}(z),i}^{\pi}(s, \kappa, x_1, x_2)\right\}_{s, \kappa, x_1, x_2, z}$$

*where $s$ and $\kappa$ are the security parameters.*

---

**Functionality $\mathcal{F}_{\text{UMA}}$**

The functionality $\mathcal{F}_{\text{UMA}}$ interacts with a sender S and a receiver R. The program $P$ is known and agreed by both parties.

*Input:* Upon receiving input value (INPUT$_S$, $x_1$) from S and input value (INPUT$_R$, $x_2$) from R, set state$_1 = (x_1, x_2)$ and initialize the memory data $D$ with $0^n$.

*Output:* If both inputs are recorded, execute $y \leftarrow P^D(x_1, x_2)$ and send (OUTPUT$_R$, $T$, $y$, MemAccess) to R, where $T$ is the number of memory accesses that were performed during the execution, and MemAccess is the access pattern of the execution.

---

**Fig. 2.** A 2PC secure evaluation functionality in the UMA model for program $P$.

---

[6] Gentry et al. further demonstrated that this weaker notion of security is useful by providing a transformation from this setting into the stronger setting for which the simulator does not receive this extra information. Their proof holds against semi-honest adversaries. A simple observation shows that their proof can be extended for the malicious 2PC setting by considering secure protocols that run the oblivious RAM and the garbling computations; see below our transformation.

**A Transforation from UMA to Full Security.** Below, we present a transformation $\Theta$, that is given (1) a protocol $\pi$ with UMA security for RAM programs that support ptWrites, and (2) a secure ORAM compiler $C$ that satisfies ptWrites,[7] and outputs a two-party protocol for arbitrary RAM programs with full security; see Fig. 3 for the description of $\Theta$. The formal theorem follows:

**Theorem 2.** *Let $\pi$ be a secure two-party protocol that provides UMA security for RAM programs that support ptWrites in the presence of malicious adversaries and $C$ an ORAM compiler that satisfies ptWrites, then $\Theta$ is a two-party protocol that provides full security for arbitrary RAM programs in the presence of malicious adversaries.*

Note that the transformation uses the ORAM compiler $C$ and the UMA-secure protocol $\pi$ in a black-box manner. In addition, the transformation preserves all the properties that are related to the memory management, i.e., the party who handles the memory in $\pi$ is the same one who handles the memory in $\pi' \leftarrow \Theta(P, \pi)$. Note that the efficiency of the resulted protocol $\pi' \leftarrow \Theta(P, \pi)$ is dominated by the efficiency of the UMA-secure protocol $\pi$ and the ORAM compiler $C$. Specifically, the recent ORAM constructions set an additional polylog overhead with respect to all relevant parameters.

---

**Transformation $\Theta$ from UMA to Fully Secure Protocol**

*Inputs:* The program $P$ that the parties wish to compute. The protocol $\pi$ to compute a two-party protocol with a UMA security. A secure ORAM compiler $C$. The sender S has $x_1$ and the receiver R has $x_2$.

*Protocol:*

1. The parties generate the randomness $\rho$ using a coin-tossing protocol.
2. The parties agree on the oblivious program $P^* \leftarrow C(P, n, \rho)$ where $\rho$ is $C$'s random tape.
3. The parties run $\pi(P^*)$ where R's inputs are $x_1, k_1, r_1$, S's inputs are $x_2, k_2, r_2$ and the program $P^*$ is given $n$ (the memory size), the parties' input $x = x_1 \| x_2$, the key and the random tape of the compiled program, i.e. $k_1 \oplus k_2$ and $r_1 \oplus r_2$, respectively. That is, the parties run $\pi(P^*(n, x_1 \| x_2, k_1 \oplus k_2, r_1 \oplus r_2))$.

---

**Fig. 3.** A transformation from UMA to full security.

---

[7] As for RAM programs, ORAM schemes can also support this property. Moreover, the [16] transformation discussed in Sect. 2.1 can be applied to ORAM schemes as well.

*Security.* We next present a proof sketch to the transformation presented in Fig. 3. We consider first a corrupted receiver, which is the more complicated case, and then a corrupted sender.

R *is Corrupted.* Let $\mathcal{S}_{\text{UMA}}$ be the simulator for protocol $\pi$ in the UMA model. The simulator for the general model, $\mathcal{S}_{\text{RAM}}$, works as follows:

1. Let $T$ be the run time of the program $P$, and let $\tilde{P}$ be the program of the form:

   For i=0 To $T$:
     Read(k);

   for some constant $k \in [n]$ (i.e. k is in the range of $D$'s size). Let $\tilde{P}^* \leftarrow C(n, \tilde{P})$ and let $\text{MemAccess}(\tilde{P}^*, n, \varepsilon)$ be the memory access pattern resulted by its execution (where $\varepsilon$ is an empty string, since $\tilde{P}$ gets no input).
2. Given the output of the program $y = P(x, y)$, the simulator $\mathcal{S}_{\text{RAM}}$ outputs the view that is the result of $\mathcal{S}_{\text{UMA}}(y, \text{MemAccess}(\tilde{P}^*, n, \varepsilon))$.

We claim that the view that $\mathcal{S}_{\text{RAM}}$ outputs is indistinguishable from the real view of the receiver in the real execution of the protocol. Assume, by contradiction, that there exist inputs $x_1, x_2$ for which there exists a distinguisher $\mathcal{D}$ who can distinguish between the two views with more than negligible probability. Consider the following hybrid view **Hyb** which is constructed as follows: Given $P, x_1, x_2, n$, compute $P^* \leftarrow C(n, P)$, choose random strings $k$ and $r$ and run $P^*(n, x_1 \| x_2, k, r)$. Denote the access pattern induced by this execution by $\text{MemAccess}(P^*, n, (x_1 \| x_2, k, r))$, then, outputs **Hyb** $= \mathcal{S}_{\text{UMA}}(y, \text{MemAccess}(P^*, n, (x_1 \| x_2, k, r)))$. The indistinguishability between $\mathcal{S}_{\text{RAM}}(y)$ and **Hyb**$(P, x_1, x_2, n)$ is reduced to the obliviousness of the ORAM compiler and the indistinguishability between **Hyb**$(P, x_1, x_2, n)$ and the real view is reduced to the indistinguishability of the simulation of $\mathcal{S}_{\text{UMA}}$ and the real view of the execution of protocol $\pi$.s

S *is Corrupted.* This case is simpler since, by the definitions of functionalities $\mathcal{F}_{\text{UMA}}$ and $\mathcal{F}_{\text{RAM}}$ the sender receives no output from the computation, thus, the same simulator used in the UMA model works in the general RAM model, that is $\mathcal{S}_{\text{RAM}} = \mathcal{S}_{\text{UMA}}$. Specifically, indistinguishability between the output of $\mathcal{S}_{\text{RAM}}$ and the sender's view in the real execution in the RAM model is immediately reduced to the indistinguishability between the output of $\mathcal{S}_{\text{RAM}}$ and the view in the real execution in the UMA model.

Note that our ORAM compiler definition simplifies the transformation to full security as now the result oblivious program $P^*$ gets its randomness $r$ as part of its input, rather than being hardcoded with it. Furthermore, recalling that this randomness is used to determine the future locations in memory for which the oblivious program is going to access, we stress that $r$ is not revealed as part of the "check circuits" when using the cut-and-choose technique.

**On the Capabilities of Semi-honest in a Garbled RAM and ORAM Schemes.** When considering ORAM schemes in the context of two-party computation, it must be ensured that a read operation is carried out correctly.

Namely, that the correct element from the memory is indeed fetched, and that the adversary did not "roll back" to an earlier version of that memory cell. Importantly, this is not just a concern in the presence of malicious adversaries, as a semi-honest adversary may try to execute its (partial) view on inconsistent memory values. Therefore, the scheme must withhold such attacks. Handling the first attack scenario is relatively simply using message authentications codes (MACs), so that a MAC tag is stored in addition to the encrypted data. Handling roll backs is slightly more challenging and is typically done using Merkle trees. In [16] roll backs are prevented by creating a new secret key for each time period. This secret key is used to decrypt a corresponding ciphertext in order to extract the label for the next garbled circuit. By replacing the secret key each time period, the adversary is not able decrypt a ciphertext created in some time period with a secret key that was previously generated.

## 2.4  Timed IBE [16]

TIBE was introduced by Gentry et al. in [16] in order to handle memory data writings in their garbled RAM construction. This primitive allows to create "time-period keys" $\mathsf{TSK}_t$ for arbitrary time periods $t \geq 0$ such that $\mathsf{TSK}_t$ can be used to create identity-secret-keys $\mathsf{SK}_{(t,v)}$ for identities of the form $(t, v)$ for arbitrary $v$, but cannot break the security of any other identities with $t' \neq t$. Gentry et al. demonstrated how to realize this primitive based on IBE [5,6]. Informally speaking, the security of TIBE is as follows: Let $t^*$ be the "current" time period. Given a single secret key $\mathsf{SK}_{(t,v)}$ for every identity $(t, v)$ of the "past" periods $t < t^*$ and a single period key $\mathsf{TSK}_t$ for every "future" periods $t^* < t \leq T$, semantic security should hold for any identity of the form $\mathsf{id}^* = (t^*, v^*)$ (for which neither a period nor secret key were not given). We omit the formal definition due to space limitations.

## 2.5  Garbled RAM Based on IBE [16]

Our starting point is the garbled RAM construction of [16]. Intuitively speaking, garbled RAM [34] is an analogue object of garbled circuits [4,47] with respect to RAM programs. The main difference when switching to RAM programs is the requirement of maintaining a memory data $D$. In this scenario, the data is garbled once, while many different programs are executed sequentially on this data. As pointed out in the modeling of [16], the programs can only be executed in the specified order, where each program obtains a state that depends on prior executions. The [16] garbled RAM proposes a fix to the aforementioned circularity issue raised in [34] by using an Identity Based Encryption (IBE) scheme [5,6] instead of a symmetric-key encryption scheme.

   In more details, the inputs $D, P, x$ to the garbled RAM are garbled into $\tilde{D}, \tilde{P}, \tilde{x}$ such that the evaluator reveals the output $\tilde{P}(\tilde{D}, \tilde{x}) = P(D, x)$ and nothing else. A RAM program $P$ with running time $T$ can be evaluated using $T$ copies of a Boolean circuit $\mathsf{C}_{\mathrm{CPU}}^P$ where $\mathsf{C}_{\mathrm{CPU}}^P$ computes the function $\mathsf{C}_{\mathrm{CPU}}^P(\mathsf{state}_t, b_t^{\mathsf{read}}) = (\mathsf{state}_{t+1}, i_t^{\mathsf{read}}, i_t^{\mathsf{write}}, b_t^{\mathsf{write}})$. Then secure evaluation of $P$ is possible by having the sender S garble the circuits $\{C_{\mathrm{CPU}}^t\}_{t\in[T]}$ (these are called the garbled program $\tilde{P}$), whereas the receiver R sequentially evaluates these circuits. In order for the evaluation to be secure the state of the program should remain secret when moving from one circuit to another. To this end, the garbling is done in a way that assigns the output wires of one circuit with the same labels as the input wires of the next circuit. The main challenge here is to preserve the ability to read and write from the memory while preventing the evaluator from learning anything beyond the program's output, including any intermediate value.

The original idea from [34] employed a clever usage of a PRF for which the secret key is embedded inside all the CPU-steps circuits, where the PRF's role is twofold. For reading from the memory it is used to produce ciphertexts encrypting the labels of the input wire of the input bit of the next circuit, whereas for writing it is used to generate secret keys for particular "identities". As explained in [16], the proof of [34] does not follow without assuming an extra circularity assumption. In order to avoid circularity, Gentry et al. proposed to replace the PRF with a public-key primitive. As it is insufficient to use a standard public-key cryptosystem (since the circuit must still produce secret keys for each memory location $i$, storing the keys $\mathsf{sk}_{i,0}, \mathsf{sk}_{i,1}$), the alternative is to use IBE. Below, we briefly describe their scheme.

*The Read-Only Solution.* The initialized garbled data $\tilde{D}$ contains a secret key $\mathsf{sk}_{i,b}$ in each memory location $i \in [n]$ where $D[i] = b$, such that $i, b$ serves as an identity secret key for the "identity" $(i, b)$. Moreover, each garbled circuit $\mathsf{GC}_{\mathrm{CPU}}^t$ is hardwired with the master public key $\mathsf{MPK}$ of an IBE scheme.[8] This way, the garbled circuit can encrypt the input labels for the next circuit, that are associated with the bit that has just been read from the memory. More specifically, the circuit generates two ciphertexts $\mathsf{ct}_0, \mathsf{ct}_1$ that are viewed as a translation map. Namely, $\mathsf{ct}_b = \mathsf{Enc}_{\mathsf{MPK}}(\mathsf{id} = (i, b); \mathsf{msg} = \mathsf{lbl}_b^{t+1})$ and the correct label is extracted by decrypting the right ciphertext using $\mathsf{sk}_{i,b}$, such that $\mathsf{lbl}_0^{t+1}, \mathsf{lbl}_1^{t+1}$ are the input labels in the next garbled circuit that are associated with the last input bit read from the memory.

*The Read-Write Solution.* A complete solution that allows both reading and writing is slightly more involved. We describe how to garble the data and the program next.

GARBLING THE DATA. The garbled data consists of secret keys $\mathsf{sk}_{(t,i,b)}$ for identities of the form $\mathsf{id} = (t, i, b)$ where $i$ is the location in the memory $D'$, $t$ is the last time step for which that location was written to and $b \in \{0, 1\}$ is the bit that was written to location $i$ at time step $t$. The honest evaluator only needs to keep the most recent secret key for each location $i$.

---

[8] For ease of presentation, Gentry et al. abstract the security properties of the IBE scheme using a new primitive denoted by Timed IBE (TIBE); see Sect. 2.4 for more details.

GARBLING THE PROGRAM. Next, each CPU garbled circuit computes the last time step in which memory location $i$ was written to by computing $t' = \mathsf{WriteTime}(t, \mathsf{state}_t, i_t^{\mathsf{read}})$. Namely, if at time step $t$ the garbled circuit $\mathsf{GC}_{\mathsf{CPU}}^t$ instructs to read from location $i_t^{\mathsf{read}}$, then the circuit further computes the last time step, $u$, in which that $i_t^{\mathsf{read}}$ was written to, it then computes the translation map $\mathsf{translate}_t = (\mathsf{ct}_0, \mathsf{ct}_1)$ by $\mathsf{ct}_b = \mathsf{Enc}_{\mathsf{MPK}}(\mathsf{id} = (u, i_t^{\mathsf{read}}, b); \mathsf{msg} = \mathsf{lbl}_b^{t+1}))$, and outputs it in the clear.

In order to write at time step $t$ to memory location $i = i_t^{\mathsf{write}}$ the value $b = b_t^{\mathsf{write}}$, a naive solution would hardwire $\mathsf{MSK}$ within each garbled circuit and then generate the key $\mathsf{sk}_{(t,i,b)} = \mathsf{KeyGen}_{\mathsf{MSK}}(\mathsf{id} = (t,i,b))$; but this idea re-introduces the circularity problem. Instead, Gentry et al. [16] solve this problem by introducing a new primitive called Timed IBE (TIBE). Informally, this is a two-level IBE scheme in which the first level includes the master public/secret keys $(\mathsf{MPK}, \mathsf{MSK})$ whereas the second level has $T$ timed secret keys $\mathsf{TSK}_1, \ldots, \mathsf{TSK}_T$. The keys $\mathsf{MPK}, \mathsf{MSK}$ are generated by $\mathsf{MasterGen}(1^\kappa)$ and the timed keys are generated by $\mathsf{TSK}_t = \mathsf{TimeGen}(\mathsf{MSK}, t)$.

Then in the garbling phase, the key $\mathsf{TSK}_t$ is hardwired within the $t$th garbled circuit $\mathsf{GC}_{\mathsf{CPU}}^t$ and is used to write the bit $b_t^{\mathsf{write}}$ to memory location $i_t^{\mathsf{write}}$. To do that $\mathsf{GC}_{\mathsf{CPU}}^t$ computes the secret key for identity $(t,i,b)$ by $\mathsf{sk}_{(t,i,b)} \leftarrow \mathsf{KeyGen}(\mathsf{TSK}_t, (t,i,b))$ which is then stored in memory location $i$ by the evaluator. Note that $\mathsf{GC}_{\mathsf{CPU}}^t$ outputs a secret key for only one identity in every time step (for $(t,i,b)$ but not $(t,i,1-b)$). This solution bypasses the circularity problem since the timed secret keys $\mathsf{TSK}_t$ are hardwired only within the garbled circuit computing $\mathsf{C}_{\mathsf{CPU}}^t$, and cannot be determined from either $\mathsf{sk}_{(t,i,b)}$ or the garbled circuit, provided that the TIBE scheme and the garbling schemes are secure.

## 2.6   Garbled Circuits

The idea of garbled circuit is originated in [47]. Here, a sender can encode a Boolean circuit that computes some PPT function $f$, in a way that (computationally) hides from the receiver any information but the function's output. In this work we consider a variant of the definition from [16] that abstracts out the security properties of garbled circuits that are needed via the notion of a *garbled circuit with wire labels*. The definition that we propose below stems from the cut-and-choose technique chosen to deal with a malicious sender. Specifically, the sender uses the algorithm $\mathsf{Garb}$ to generate $s$ garbled versions of the circuit C, namely $\{\tilde{\mathsf{C}}_i\}_{i \in [s]}$ for some statistical parameter $s$. Then, in order to evaluate these circuits the sender sends $\{\tilde{\mathsf{C}}_i\}_{i \in [s]}$ along with the garbled inputs $\{\tilde{x}_i\}_{i \in [s]}$, such that $\tilde{x}_i$ is the garbled input for the garbled circuit $\tilde{\mathsf{C}}_i$. The evaluator then chooses a subset $Z \subset s$ and evaluates the garbled circuits indexed by $z \in Z$ using algorithm $\mathsf{Eval}$. Note that in the notion of garbled circuits with wire labels the garbled inputs $\tilde{x}_i$ are associated with a single label per input wire of the circuit $\tilde{\mathsf{C}}_i$; we denote these labels by $\tilde{x}_i = (\mathsf{lbl}_{\mathsf{in}, x[1]}^{1,i}, \ldots, \mathsf{lbl}_{\mathsf{in}, x[v_{\mathsf{in}}]}^{v_{\mathsf{in}}, i})$ (where $v_{\mathsf{in}}$ is the number of input wires in C and $x = x[1], \ldots, x[v_{\mathsf{in}}]$ is the input to the circuit).

The evaluator learns $s$ sets[9] of output-wire labels $\{\tilde{y}_i\}_{i\in[s]}$ corresponding to the output $y = C(x)$[10], where $\tilde{y}_i = (\mathsf{lbl}^{1,i}_{\mathrm{out},y[1]},\ldots,\mathsf{lbl}^{v_{\mathrm{out}},i}_{\mathrm{in},y[v_{\mathrm{out}}]})$, but nothing else (for example, it does not learn $\mathsf{lbl}^{1,i}_{\mathrm{out},1-y[1]}$). For clarity, in the following exposition the label $\mathsf{lbl}^{j,i}_{\mathrm{in},b}$ is the label that represents the bit-value $b \in \{0,1\}$ for the $j$th input wire ($j \in v_{\mathrm{in}}$) in the $i$th garbled version of the circuit (for $i \in s$), namely $\tilde{C}_i$. Analogously, $\mathsf{lbl}^{j,i}_{\mathrm{out},b}$ represents the same, except that it is associated with an output wire (where $j \in v_{\mathrm{out}}$).

We further abstract two important properties of *authenticity* and *input consistency*. Loosely speaking, the authenticity property ensures that a malicious evaluator will not be able to produce a valid encoding of an incorrect output given the encoding of some input and the garbled circuit. This property is required due to the reusability nature of our construction. Namely, given the output labels of some iteration, the evaluator uses these as the input labels for the next circuit. Therefore, it is important to ensure that it cannot enter an encoding of a different input (obtained as the output from the prior iteration). In the abstraction used in our work, violating authenticity boils down to the ability to generate a set of output labels that correspond to an incorrect output. Next, a natural property that a maliciously secure garbling scheme has to provide is *input consistency*. We formalize this property via a functionality, denoted by $\mathcal{F}_{\mathrm{IC}}$. That is, given a set of garbled circuits $\{\tilde{C}_i\}_i$ and a set of garbled inputs $\{\tilde{x}_i\}_i$ along with the randomness $r$ that was used by Garb; the functionality outputs 1 if the $s$ sets of garbled inputs $\{\tilde{x}_i\}_{i=1}^{s}$ (where $|\tilde{x}_i| = j$) represent the same input value, and 0 otherwise. This functionality is described in Fig. 8 (Appendix A). We now proceed to the formal definition.

**Definition 5 (Garbled circuits).** *A circuit garbling scheme with wire labels consists of the following two polynomial-time algorithms:*

- *The garbling algorithm* Garb:

$$\left(\{\tilde{C}_i\}_i, \{u, b, \mathsf{lbl}^{u,i}_{\mathrm{in},b}\}_{u,i,b}\right) \leftarrow \mathsf{Garb}\left(1^\kappa, s, C, \{v, b, \mathsf{lbl}^{v,i}_{\mathrm{out},b}\}_{v,i,b}\right)$$

*for every* $u \in [v_{\mathrm{in}}], v \in [v_{\mathrm{out}}], i \in [s]$ *and* $b \in \{0,1\}$. *That is, given a circuit* C *with input size* $v_{\mathrm{in}}$, *output size* $v_{\mathrm{out}}$ *and* $s$ *sets of output labels* $\{v, b, \mathsf{lbl}^{v,i}_{\mathrm{out},b}\}_{v,i,b}$, *outputs* $s$ *garbled circuits* $\{\tilde{C}_i\}_{i\in[s]}$ *and* $s$ *sets of input labels* $\{u, b, \mathsf{lbl}^{u,i}_{\mathrm{in},b}\}_{u,i,b}$.
- *The evaluation algorithm* Eval:

$$\left\{\mathsf{lbl}^{1,i}_{\mathrm{out}},\ldots,\mathsf{lbl}^{v_{\mathrm{out}},i}_{\mathrm{out}}\right\}_{i\in[s]} = \mathsf{Eval}\left(\{\tilde{C}_i, (\mathsf{lbl}^{1,i}_{\mathrm{in}},\ldots,\mathsf{lbl}^{v_{\mathrm{in}},i}_{\mathrm{in}})\}_{i\in[s]}\right).$$

*That is, given* $s$ *garbled circuits* $\{\tilde{C}_i\}_i$ *and* $s$ *sets of input labels* $\{\mathsf{lbl}^{1,i}_{\mathrm{in}},\ldots,\mathsf{lbl}^{v_{\mathrm{in}},i}_{\mathrm{in}}\}_i$, *outputs* $s$ *sets of output labels* $\{\mathsf{lbl}^{1,i}_{\mathrm{out}},\ldots,\mathsf{lbl}^{v_{\mathrm{out}},i}_{\mathrm{out}}\}_i$. *Intuitively, if the input labels* $(\mathsf{lbl}^{1,i}_{\mathrm{in}},\ldots,\mathsf{lbl}^{v_{\mathrm{in}},i}_{\mathrm{in}})$ *correspond to some input* $x \in$

---

[9] This $s$ might be different from the $s$ used in the garbling algorithm, still we used the same letter for simplification.

[10] Note that this holds with overwhelming probability since some of the garbled circuits might be malformed.

$\{0,1\}^{v_{in}}$ *then the output labels* $(\mathsf{lbl}_{out}^{1,i}, \ldots, \mathsf{lbl}_{out}^{v_{out},i})$ *should correspond to* $y = C(x)$.

Furthermore, the following properties hold.

*Correctness.* For correctness, we require that for any circuit C and any input $x \in \{0,1\}^{v_{in}}$, $x = (x[1], \ldots, x[v_{in}])$ such that $y = (y[1], \ldots, y[v_{out}]) = C(x)$ and any $s$ sets of output labels $\{v, b, \mathsf{lbl}_{b,out}^{v,i}\}_{v,i,b}$ (for $u \in v_{in}, v \in v_{out}, i \in [s]$ and $b \in \{0,1\}$) we have

$$\Pr\left[\mathsf{Eval}\left(\{\tilde{C}_i, (\mathsf{lbl}_{in,x[1]}^{1,i}, \ldots, \mathsf{lbl}_{in,x[v_{in}]}^{v_{in},i})\}_i\right) = \{\mathsf{lbl}_{out,y[1]}^{1,i}, \ldots, \mathsf{lbl}_{out,y[v_{out}]}^{v_{out},i}\}_i\right] = 1$$

where $(\{\tilde{C}_i\}_i, \{u, b, \mathsf{lbl}_{in,b}^{u,i}\}_{u,i,b}) \leftarrow \mathsf{Garb}(1^\kappa, s, C, \{v, b, \mathsf{lbl}_{out,b}^{v,i}\}_{v,i,b})$ as described above.

*Verifying the correctness of a circuit.* Note that in a cut-and-choose based protocols, the receiver is instructed to check the correctness of a subset of the garbled circuits. This check can be accomplished by the sender sending the receiver the randomness used in $\mathsf{Garb}$. In our protocol this is accomplished by giving the receiver *both* input labels for each input wire of the check circuits, for which it can verify that the circuit computes the agreed functionality. We note that this check is compatible with all prior known garbling schemes.

*Privacy.* For privacy, we require that there is a PPT simulator $\mathsf{SimGC}$ such that for any $C, x, Z$ and $\{\mathsf{lbl}_{out}^{1,z}, \ldots, \mathsf{lbl}_{out}^{v_{out},z}\}_{z\in[Z]}, \{v, b, \mathsf{lbl}_{out,b}^{v,z}\}_{v,z\notin[Z],b}$ (i.e. one output label for wires in circuits indexed by $z \notin Z$ and a pair of output labels for wires in circuits indexed by $z \in Z$), we have

$$\left(\{\tilde{C}_z \, (\mathsf{lbl}_{in,x[1]}^{1,z}, \ldots, \mathsf{lbl}_{in,x[v_{in}]}^{v_{in},z})\}_z\right)$$

$$\stackrel{c}{\approx} \mathsf{SimGC}\left(1^\kappa, \{\mathsf{lbl}_{out}^{1,z}, \ldots, \mathsf{lbl}_{out}^{v_{out},z}\}_{z\in[Z]}, \{v, b, \mathsf{lbl}_{out,b}^{v,z}\}_{v,i\notin[Z],b}\right)$$

where $(\{\tilde{C}_z\}_z, \{u, b, \mathsf{lbl}_{in,b}^{u,z}\}_{u,z,b}) \leftarrow \mathsf{Garb}(1^\kappa, s, C, \{v, b, \mathsf{lbl}_{out,b}^{v,z}\}_{v,z,b})$ and $y = C(x)$.

*Authenticity.* We describe the authenticity game in Fig. 7 (Appendix A) where the adversary is obtained a set of garbled circuits and garbled inputs for which the adversary needs to output a valid garbling of an invalid output. Namely, a garbled scheme is said to have *authenticity* if for every circuit C, for every PPT adversary $\mathcal{A}$, every $s$ and for all large enough $\kappa$ the probability $\Pr[\mathsf{Auth}_{\mathcal{A}}(1^\kappa, s, C) = 1]$ is negligible. Our definition is inspired by the definition from [4] and also adapted for the cut-and-choose approach.

*Input Consistency.* We abstract out the functionality that checks the validity of the sender's input across all garbled circuits. We say that, a garbling scheme has *input consistency* (in the context of cut-and-choose based protocols) if there exists a protocol that realize the $\mathcal{F}_{\text{IC}}$ functionality described in Fig. 8 (Appendix A).

*Realizations of our garbled circuits notion.* We require the existence of a protocol $\Pi_{\text{IC}}$ that securely realizes the functionality $\mathcal{F}_{\text{IC}}$ described in Fig. 8, in the presence of malicious adversaries. We exemplify this realization using [32] in the full version of the paper.

## 3   Building Blocks

In this section we show how to overcome the challenges discussed in the introduction and design the first maliciously secure 2PC protocol that does not require duplication of the data and works for every garbling scheme that supports our definition based on wire labels. Recall first that in [16] Gentry et al. have used a primitive called Timed IBE, where the secret-key for every memory location and stored bit $(i, b)$ is enhanced with another parameter: the last time step $t$ in which it has been written to the memory. The secret-key $\text{sk}_{(t,i,b)}$ for identity $\text{id} = (t, i, b)$ is then generated using the hard-coded time secret-key $\text{TSK}_t$. Now, since algorithm KeyGen is randomized, running this algorithm $s$ times will yield $s$ independent secret timed keys. This results in $s$ different values to be written to memory at the same location, which implies duplication of memory data $D$. In order to avoid this, our solution forces the $s$ duplicated garbled circuits for time step $t$ to use the same random string $r$, yielding that all garbled circuits output the same key for the identity $(t, i, b)$. Importantly, this does not mean that we can hard-code $r$ in all those $s$ circuits, since doing this would reveal $r$ when applying the cut-and-choose technique on these garbled circuits as half of the circuits are opened. Clearly, we cannot reveal the randomness to the evaluator since the security definition of IBE (and Timed IBE) does not follow in such a scenario. Instead, we instruct the sender to input the *same randomness* in all $s$ copies of the circuits and then run an input consistency check to these inputs in order to ensure that this is indeed the case. We continue with describing the components we embed in our protocol. An overview of the circuits involved in our protocol can be found in Fig. 4 and a high-level overview of the protocol can be found in Sect. 4.

### 3.1   Enhanced CPU-Step Function

The enhanced cpustep$^+$ function is essentially the CPU-step functionality specified in Sect. 2.1 enhanced with more additional inputs and output, and defined as follows

$$\text{cpustep}^+(\text{state}_t, b_t^{\text{read}}, \text{MPK}, \text{TSK}_t, r_t) = (\text{state}_{t+1}, i_t^{\text{read}}, i_t^{\text{write}}, b_t^{\text{write}}, \text{translate}_t)$$

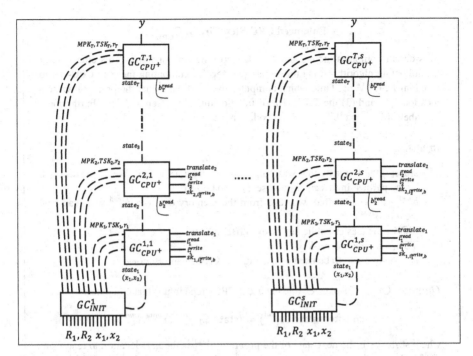

**Fig. 4.** Garbled chains $GC_{INIT}, GC_{CPU+}^{1,i}, \ldots GC_{CPU+}^{T,i}$ for $i \in [s]$. Dashed lines refer to values that are passed privately (as one label per wire) whereas solid lines refer to values that are given in the clear.

where the additional inputs $MSK, TSK_t$ and $r_t$ are the master public-key, a timed secret-key for time $t$ and the randomness $r$ used by the KeyGen algorithm. The output $\text{translate}_t$ is a pair of ciphertexts $ct_1, ct_2$, encrypted under MPK, that allows the evaluator to obtain the appropriate label of the wire that corresponds to the input bit in the next circuit. We denote the circuit that computes that function by $C_{CPU+}^t$. The functionality of $C_{CPU+}^t$ is described in Fig. 5). We later describe how to securely realize this function and, in particular, how these new additional inputs are generated and given to the $T$ CPU-circuits. The enhanced CPU-step circuit wraps the WriteTime algorithm defined in Definition 1.

## 3.2 Initialization Circuit

The initialization circuit generates all required keys and randomness to our solution and securely transfer them to the CPU-step circuits. As explained before, our solution requires the parties to input not only their input to the program but also a share to a randomness that the embedded algorithms would be given (that is, the randomness is not fixed by one of the parties). The circuit is described in Fig. 6.

---

### Enhanced CPU-Step Circuit $C_{CPU+}^t$

This circuit computes the enhanced CPU-step function $\mathsf{cpustep}^+$. This circuit wraps the following algorithms: (1) the usual cpu-step for computing the next CPU-step of program $P$, (2) WriteTime which computes the last time $t'$ that the program wrote to location $i_t^{\mathsf{read}}$ and (3) the TIBE related functionalities KeyGen and Enc. Furthermore, the labels $\mathsf{lbl}_0^{t+1}$ and $\mathsf{lbl}_1^{t+1}$ are hard coded in the circuit.

*Inputs.*

- $\mathsf{state}_t$ - the last state that was output by the previous circuit. We define $\mathsf{state}_1$ to be the parties' inputs $x_1, x_2$ and set $b_0^{\mathsf{read}}$ to be zero.
- $b_t^{\mathsf{read}}$ - the last bit that was read from the memory data (i.e. $b_t^{\mathsf{read}}$ was read from location $i_t^{\mathsf{read}}$).
- MPK - the master public key of the TIBE scheme.
- $\mathsf{TSK}_t$ - a timed secret-key.
- $r_t$ - randomness to be used by algorithms KeyGen and $\mathsf{Enc}_{\mathsf{MPK}}$.

*Outputs.* $C_{CPU+}^t$ invokes $C_{CPU}^t$ (the usual CPU-step circuit) that computes:

$$\mathsf{cpu\text{-}step}(\mathsf{state}_t, b_t^{\mathsf{read}}) = (\mathsf{state}_{t+1}, i_t^{\mathsf{read}}, i_t^{\mathsf{write}}, b_t^{\mathsf{write}})$$

where $\mathsf{state}_{t+1}$ is the next state of the program; $i_{t+1}^{\mathsf{read}}$ is the next location to read from; $i_{t+1}^{\mathsf{write}}$ is the next location to write to and $b_{t+1}^{\mathsf{read}}$ is the bit to write to location $i_{t+1}^{\mathsf{write}}$. The circuit outputs the translation $\mathsf{translate}_t = (\mathsf{ct}_t^0, \mathsf{ct}_t^1)$ defined by:

$$t' = \mathsf{WriteTime}(i_{t+1}^{\mathsf{read}})$$
$$\mathsf{ct}_t^0 = \mathsf{Enc}_{\mathsf{MPK}}(\mathsf{id} = (t, t', 0), \mathsf{msg} = \mathsf{lbl}_{t+1}^0)$$
$$\mathsf{ct}_t^1 = \mathsf{Enc}_{\mathsf{MPK}}(\mathsf{id} = (t, t', 1), \mathsf{msg} = \mathsf{lbl}_{t+1}^1)$$

Finally, the circuit computes $\mathsf{sk}_{(t,i,b)} = \mathsf{KeyGen}(\mathsf{TSK}_t, \mathsf{id} = (t, i_{t+1}^{\mathsf{write}}, b_{t+1}^{\mathsf{write}}))$ and outputs

$$(\mathsf{state}_{t+1}, i_{t+1}^{\mathsf{read}}, i_{t+1}^{\mathsf{write}}, \mathsf{sk}_{(t,i,b)}, \mathsf{translate}_t).$$

---

**Fig. 5.** The CPU-step circuit.

## 3.3 Batch Single-Choice Cut-and-Choose OT

As a natural task in a cut-and-choose based protocol, we need to carry out cut-and-choose oblivious transfers for all wires in the circuit, for which the receiver picks a subset $Z \subset [s]$ and then obtains either both input labels (for circuits indexed with $z \in Z$), or the input label that matches the receiver's input otherwise. It is crucial that the subset of indices for which the receiver obtains both input labels is the same in all transfers. The goal of this functionality is to ensure the input consistency of the receiver and it is named by "batch single-choice cut-and-choose OT" in [32].

---

**Initialization Circuit $C_{\text{INIT}}$**

The circuit generates all keys and randomness for the $T$ CPU step circuits $C^1_{\text{CPU}+}, \ldots, C^T_{\text{CPU}+}$.

*Inputs.*

- The parties input $x_1, x_2$, and
- $(2 \cdot (1 + T + T + 2T)) \cdot m = (8T + 2) \cdot m$ random values where $m$ an upper bound on the length of the randomness required to run the TIBE algorithms: MasterGen, TimeGen, KeyGen and Enc. This particular number of random values is explained below.

*Computation.* Let $R_1$ (resp. $R_2$) be the first (resp. last) $(4t+1) \cdot m$ bits of the inputs for the randomness. The circuit computes $R = R_1 \oplus R_2$ and interprets the result $(4t+1) \cdot m$ bits as follows: (each of the following is a $m$-bit string)

- $r^{\text{MasterGen}}$ used to generate the keys MPK and MSK.
- $r_1^{\text{TimedGen}}, \ldots, r_T^{\text{TimedGen}}$ used to generate the timed secret-keys $\text{TSK}_1, \ldots, \text{TSK}_T$.
- $r_1^{\text{KeyGen}}, \ldots, r_T^{\text{KeyGen}}$ used to generate secret-keys $\{\text{sk}_{t,i,b}\}_{t \in [T], i \in [n], b \in \{0,1\}}$ written to memory.
- $\{r_{t,b}^{\text{Enc}}\}_{t \in [T], b \in \{0,1\}}$ are used by the encryption algorithm within the CPU circuits. (Recall that the $t$th enhanced CPU step circuit $C^t_{\text{CPU}+}$ encrypts the two labels of the input wire that corresponds to the input bit of the next circuit $C^{t+1}_{\text{CPU}+}$.)

Then, the circuit computes:

$$(\text{MPK}, \text{MSK}) = \text{MasterGen}(1^\kappa; r^{\text{MasterGen}})$$

$$\forall_{t \in [T]} : \text{TSK}_t = \text{TimeGen}(\text{MSK}, t; r_t^{\text{TimedGen}})$$

*Outputs.*

$$\left( x_1, x_2, \{\text{MPK}_t\}_{t \in [T]}, \{\text{TSK}_t\}_{t \in [T]}, \{r_t^{\text{KeyGen}}\}_{t \in [T]}, \{r_{t,b}^{\text{Enc}}\}_{t \in [T], b \in \{0,1\}} \right)$$

where $\text{MPK}_1 = \ldots = \text{MPK}_T = \text{MPK}$ (the reason for duplicating MPK will be clearer later).

---

**Fig. 6.** Initialization circuit $C_{\text{INIT}}$.

In addition to the above, our protocol uses the following building blocks: A garbled scheme $\pi_{\text{GC}} = (\text{Garb}, \text{Eval})$ that preserves the security properties from Definition 5; Timed IBE $\pi_{\text{TIBE}} = (\text{MasterGen}, \text{TimeGen}, \text{KeyGen}, \text{Enc}, \text{Dec})$ and a statistically binding commitment scheme Com.

# 4    The Complete Protocol

Given the building blocks detailed in Sect. 3, we are now ready to introduce our complete protocol. Our description incorporates ideas from both [32] and [16]. Specifically, we borrow the cut-and-choose technique and the cut-and-choose OT abstraction from [32] (where the latter tool enables to ensure input consistency for the *receiver*). Moreover, we extend the garbled RAM ideas presented in [16] for a maliciously secure two-party protocol in the sense that we modify their garbled RAM to support the cut-and-choose approach. This allows us to obtain constant round overhead. Before we delve into the details of the protocol, let us present an overview of its main steps:

The parties wish to run the program $P$ on inputs $x_1, x_2$ with the aid of an external random access storage $D$. In addition to their original inputs, the protocol instructs the parties to provide random strings $R_1, R_2$ that suffice for all the randomness needed in the execution of the CPU step circuits.

- **Chains construction.** Considering a sequence of circuits $C_{\text{INIT}}, C^1_{\text{CPU+}}, \ldots, C^T_{\text{CPU+}}$ as a *connected chain of circuits*, the sender S first generates $s$ versions of garbled chains $GC^i_{\text{INIT}}, GC^{1,i}_{\text{CPU+}}, \ldots, GC^{T,i}_{\text{CPU+}}$ for every $i \in [s]$. It does so by iteratively feeding the algorithm Garb with $s$ sets of pairs of output labels, where the first set of output labels $\text{lbl}_{\text{out}}$ are chosen uniformly and are fed, together with the circuit $C^T_{\text{CPU+}}$, to procedure Garb, which in turn, outputs $s$ sets of input labels. This process is being repeated till the first circuit in the chain, i.e. $C_{\text{INIT}}$, the last $s$ sets of input labels are denoted $\text{lbl}_{\text{in}}$.
- **Cut-and-choose.** Then, the parties run the batch Single-Choice Cut-and-choose OT protocol $\Pi_{\text{SCCOT}}$ on the receiver's input labels, which let the receiver obtain a pair of labels for each of its input wires for every *check chain* with an index in $Z \subset [s]$ and a single label for each of its input wires for the *evaluation chains* with an index not in $Z$, where $Z$ is input by the receiver to $\Pi_{\text{SCCOT}}$.
- **Sending chains and commitments.** Then S sends R all garbled chains together with a commitment for every label associated with its input wires in all copies $i \in [s]$.
- **Reveal the cut-and-choose parameter.** The receiver R then notifies S with its choice of $Z$ and proves that indeed that is the subset it used in $\Pi_{\text{SCCOT}}$ (by sending a pair of labels for some of its input wires in every chain with an index in $Z$).
- **Checking correctness of check-chains.** When convinced, S sends R a pair of labels for each input wire associated with the sender's input; this allows R check all the check chains, such that if all found to be built correctly than the majority of the other, evaluation chains, are also built correctly with overwhelming probability.
- **Input consistency.** S then supplies R with a single label for each input wire associated with the sender's input, for all evaluation chains; this step requires checking that those labels are consistent with a *single* input $x_2$ of the sender. To this end, S and R run the input consistency protocol that is provided by the garbling scheme defined in Sect. 2.6.

– **Evaluation.** Upon verifying their consistency, R uses the input labels and evaluates all evaluation chains, such that in every time step $t$ it discards the chains that their outputs ($i_t^{\text{read}}, i_t^{\text{write}}, \text{sk}_t, \text{translate}_t$) do not comply to the majority of the outputs in all evaluation chains. We put a spotlight on the importance of the random strings $R_1, R_2$ that the parties provide to the chains, these allow our protocol to use a *single* block of data $D$ for *all* threads of evaluation, which could not be done in a trivial plugging of the cut-and-choose technique. As explained in Definition 5, verifying the correctness of the check chains can be done given only (both of the) input labels for $C_{\text{INIT}}$ circuits.

## 4.1    2PC in the UMA Model

We proceed with the formal detailed description of our protocol.

*Protocol $\Pi_{\text{UMA}}^P$ executed between sender S and receiver R.* Unless stated differently, in the following parameters $z, i, t, j$ respectively iterate over $[Z], [s], [T], [\ell]$.
*Inputs.* S has input $x_1$ and R has input $x_2$ where $|x_1| = |x_2| = \ell'$. R has a blank

storage device $D$ with a capacity of $n$ bits.

*Auxiliary inputs.*

– Security parameters $\kappa$ and $s$.
– The description of a program $P$ and a set of circuits $C_{\text{INIT}}, C_{\text{CPU+}}^1, \ldots, C_{\text{CPU+}}^T$ (as described above) that computes its CPU-steps, such that the output of the last circuit $\text{state}_{T+1}$ equals $P^D(x_1, x_2)$, given that the read/write instructions output by the circuits are being followed.
– $(\mathbb{G}, g, q)$ where $\mathbb{G}$ is cyclic group with generator $g$ and prime order $q$, where $q$ is of length $\kappa$.
– S and R respectively choose random strings $R_1$ and $R_2$ where $|R_1| = |R_2| = (4t + 1) \cdot m$. We denote the overall input size of the parties by $\ell$, that is, $|x_1| + |R_1| = |x_2| + |R_2| = \ell' + (4t + 1) \cdot m = \ell$. Also, denote the output size by $v_{\text{out}}$.

*The Protocol.*

1. GARBLED CPU-STEP AND INITIALIZATION CIRCUITS.
   (a) Garble the last CPU-step circuit ($t = T$):
       – Choose random labels for the labels corresponding to $\text{state}_{T+1}$.
       – Garble $C_{\text{CPU+}}^t$ by calling

$$\left(\{GC_{\text{CPU}}^{t,i}\}_i, \{\text{lbl}_{\text{in},b}^{u,i,t}\}_{u,i,b}\right) \leftarrow \text{Garb}\left(1^\kappa, s, C_{\text{CPU+}}^t, \{\text{lbl}_{\text{out},b}^{v,i,t}\}_{v,i,b}; r_g^t\right)$$

       for $v \in [v_{\text{out}}], i \in [s], b \in \{0,1\}$ and $r_g^t$ the randomness used within Garb.
       – Interpret the result labels $\{\text{lbl}_{\text{in},b}^{u,i,t}\}_{u,i,b}$ as the following groups of values: $\text{state}_t, b_t^{\text{read}}, \text{MPK}_t, \text{TSK}_t$ and $r_t$, that cover the labels: $\{\text{lbl}_{,,b}^{u,i,t}\}_{u,i,b}$, $\{\text{lbl}_{b_t^{\text{read}},b}^{u,i,t}\}_{u,i,b}, \{\text{lbl}_{\text{MPK}_t,b}^{u,i,t}\}_{u,i,b}, \{\text{lbl}_{\text{TSK}_t,b}^{u,i,t}\}_{u,i,b}, \{\text{lbl}_{r_t,b}^{u,i,t}\}_{u,i,b}$ resp.

(b) Garble the remaining CPU-step circuits. For $t = T - 1, \ldots, 1$:
- Hard-code the labels $\{\mathsf{lbl}_{b_{t+1}^{\text{read}},b}^{u,i}\}_{u,i,b}$ inside $\mathrm{C}_{\text{CPU}+}^t$.
- Choose random labels for the output wires correspond to $i_t^{\text{read}}, i_t^{\text{write}}, \mathsf{sk}_{t,i,b}$ and $\mathsf{translate}_t$ and unite them with the labels $\{\mathsf{lbl}_{,b}^{u,i,t+1}\}_{u,i,b}$ correspond to $\mathsf{state}_{t+1}$ obtained from the previous invocation of Garb; denote the resulting set $\{\mathsf{lbl}_{\text{out},b}^{v,i,t}\}_{v,i,b}$.
- Garble $\mathrm{C}_{\text{CPU}+}^t$ by calling

$$\left(\{\mathrm{GC}_{\text{CPU}}^{t,i}\}_i, \{\mathsf{lbl}_{\text{in},b}^{u,i,t}\}_{u,i,b}\right) \leftarrow \mathsf{Garb}\left(1^\kappa, s, \mathrm{C}_{\text{CPU}+}^t, \{\mathsf{lbl}_{\text{out},b}^{v,i,t}\}_{v,i,b}; r_g^t\right)$$

with $\{\mathsf{lbl}_{\text{out},b}^{v,i,t}\}_{v,i,b}$ the set of labels from above and $r_g^t$ the randomness used within Garb.
- Interpret the result labels $\{\mathsf{lbl}_{\text{in},b}^{u,i,t}\}_{u,i,b}$ as the following groups of values: $\mathsf{state}_t, b_t^{\text{read}}, \mathsf{MPK}_t, \mathsf{TSK}_t$ and $r_t$, that cover the labels: $\{\mathsf{lbl}_{,b}^{u,i,t}\}_{u,i,b}$, $\{\mathsf{lbl}_{b_t^{\text{read}},b}^{u,i,t}\}_{u,i,b}$, $\{\mathsf{lbl}_{\mathsf{MPK}_t,b}^{u,i,t}\}_{u,i,b}$, $\{\mathsf{lbl}_{\mathsf{TSK}_t,b}^{u,i,t}\}_{u,i,b}$ $\{\mathsf{lbl}_{r_t,b}^{u,i,t}\}_{u,i,b}$ resp.

(c) Garble the initialization circuit $\mathrm{C}_{\text{INIT}}$:
- Combine the group of labels $\{\mathsf{lbl}_{,b}^{u,i,1}\}_{i,b}$, that is covered by the value $\mathsf{state}_1$ which resulted from the last invocation of Garb, with the groups of labels $\{\mathsf{lbl}_{\mathsf{MPK}_t,b}^{u,i,t}, \mathsf{lbl}_{\mathsf{TSK}_t,b}^{u,i,t}, \mathsf{lbl}_{r_t,b}^{u,i,t}\}_{u,i,b}$ that are covered by the values $\{\mathsf{MPK}_t, \mathsf{TSK}_t, r_t\}$ for all $t \in [T]$. That is, set $\{\mathsf{lbl}_{\text{out},b}^{v,i}\}_{v,i,b} = \{\mathsf{lbl}_{,b}^{u,i,1} \cup \mathsf{lbl}_{\mathsf{MPK}_t,b}^{u,i,t} \cup \mathsf{lbl}_{\mathsf{TSK}_t,b}^{u,i,t} \cup \mathsf{lbl}_{r_t,b}^{u,i,t}\}_{u,i,b}$ for all $u, i, t, b$.
- Garble the initialization circuit:

$$\left(\{\mathrm{GC}_{\text{INIT}}^i\}_i, \{\mathsf{lbl}_{\text{in},b}^{u,i}\}_{u,i,b}\right) \leftarrow \mathsf{Garb}\left(1^\kappa, s, \mathrm{C}_{\text{INIT}}, \{\mathsf{lbl}_{\text{out},b}^{v,i}\}_{v,i,b}; r_g^0\right).$$

- Interpret the input labels result from that invocation of Garb by $\{\mathsf{lbl}_{\mathsf{S},b}^{u,i}\}_{u,i,b}$ and $\{\mathsf{lbl}_{\mathsf{R},b}^{u,i}\}_{u,i,b}$ which are the input wire labels that are respectively associated with the sender's and receiver's input wires.

2. OBLIVIOUS TRANSFERS.
S and R run the Batch Single-Choice Cut-And-Choose Oblivious Transfer protocol $\Pi_{\text{SCCOT}}$.
(a) S defines vectors $\mathbf{v_1}, \ldots, \mathbf{v_\ell}$ so that $\mathbf{v_j}$ contains the $s$ pairs of random labels associated with R's $j$th input bit $x_2[j]$ in all garbled circuits $\mathrm{GC}_{\text{INIT}}^1, \ldots, \mathrm{GC}_{\text{INIT}}^s$.
(b) R inputs a random subset $Z \subset [s]$ of size exactly $s/2$ and bits $x_2[1], \ldots, x_2[\ell]$.
(c) The result of $\Pi_{\text{SCCOT}}$ is that R receives all the labels associated with its input wires in all circuits $\mathrm{GC}_{\text{INIT}}^z$ for $z \in Z$, and receives a single label for every wire associated with its input $x_2$ in all other circuits $\mathrm{GC}_{\text{INIT}}^z$ for $z \notin Z$.

3. SEND GARBLED CIRCUITS AND COMMITMENTS.

S sends R the garbled circuits chains $GC_{INIT}^i, GC_{CPU+}^{1,i}, \ldots, GC_{CPU+}^{T,i}$ for every $i \in [s]$, and the commitment $,_b^{u,i} = Com(lbl_{S,b}^{u,i}, dec_b^{u,i})$ for every label in $\{lbl_{S,b}^{u,i}\}_{u,i,b}$ where $lbl_{S,b}^{u,i}$ is the $b$th label ($b \in \{0,1\}$) for the sender's $u$th bit ($u \in [\ell]$) for the $i$th garbled circuit $GC_{INIT}$.

4. SEND CUT-AND-CHOOSE CHALLENGE.

R sends S the set $Z$ along with the *pair* of labels associated with its first input bit in every circuit $GC_{INIT}^z$ for every $z \in Z$. If the values received by S are incorrect, it outputs $\bot$ and aborts. Chains $GC_{INIT}^z, GC_{CPU+}^{1,z}, \ldots, GC_{CPU+}^{t,z}$ for $z \in Z$ are called *check-circuits*, and for $z \notin Z$ are called *evaluation-circuits*.

5. SEND ALL INPUT GARBLED VALUES IN CHECK CIRCUITS.

S sends the pair of labels and decommitments that correspond to its input wires for every $z \in Z$, whereas R checks that these are consistent with the commitments received in Step 3. If not R aborts, outputting $\bot$.

6. CORRECTNESS OF CHECK CIRCUITS.

For every $z \in Z$, R has a pair of labels for every input wire for the circuits $GC_{INIT}^z$ (from Steps 2 and 5). This means that it can check the correctness of the chains $GC_{INIT}^z, GC_{CPU+}^{1,z}, \ldots, GC_{CPU+}^{T,z}$ for every $z \in Z$. If the chain was not built correctly for some $z$ then output $\bot$.

7. CHECK GARBLED INPUTS CONSISTENCY FOR THE EVALUATION-CIRCUITS.
   - S sends the labels $\{(lbl_{in,x_1[1]}^{1,z}, \ldots, lbl_{in,x_1[\ell]}^{\ell,z})\}_{z \notin [Z]}$ for its input $x_1$.
   - S and R participate in the input consistency check protocol $\Pi_{IC}$.
     • The common inputs for this protocol are the circuit $C_{INIT}$, its garbled versions $\{GC_{INIT}^i\}_{z \notin Z}$ and the labels $\{(lbl_{in,x_1[1]}^{1,z}, \ldots, lbl_{in,x_1[\ell]}^{\ell,z})\}_{z \notin [Z]}$ that were sent before.
     • S inputs its randomness $r_g^0$ and the set of output labels $\{lbl_{out,b}^{v,i}\}_{v,i,b}$ that were used within Garb on input $GC_{INIT}$, along with the decommitments $\{dec_b^{u,z}\}_{u \in [\ell], z \notin Z, b \in \{0,1\}}$.

8. EVALUATION.

Let $\tilde{Z} = \{z \mid z \notin Z\}$ be the indices of the *evaluation* circuits.

(a) For every $z \in \tilde{Z}$, R evaluate $GC_{INIT}^z$ using Eval and the input wires it obtained in Step 7 and reveal one label for each of its output wires $lbl_{INIT}^{out,z}$.

(b) For $t = 1$ to $T$:

   i. For every $z \in \tilde{Z}$, evaluate $GC_{CPU+}^{t,z}$ using Eval and obtain one output label for each of its output wires, namely, $lbl_{CPU+}^{out,t,z}$. Part of these labels refer to state$_{t+1,z}$. In addition Eval outputs out$_{t,z}$ = $(i_{t,z}^{read}, i_{t,z}^{write}, b_{t,z}^{write}, translate_{t,z})$ in the clear[11]. For $t = T$ Eval outputs state$_{T+1}$ in the clear and we assign out$_{t,z}$ = state$_{T+1,z}$.

   ii. Take the majority out$_t$ = Maj($\{out_{t,z}\}_{z \in \tilde{Z}}$) and remove from $\tilde{Z}$ the indices $\tilde{z}$ for which out$_{t,\tilde{z}} \neq$ out$_t$. Formally set $\tilde{Z} = \tilde{Z} \setminus \{z' \mid out_{t,z'} \neq out_t\}$. This means that R halts the execution thread of the circuit copies that were found flawed during the evaluation.

   iii. Output out$_{T+1}$.

---

[11] Note that if S is honest then out$_{t,z_1}$ = out$_{t,z_2}$ for every $z_1, z_2 \in \tilde{Z}$.

**Theorem 3.** *Assume that* $\pi_{GC}$ *is a garbling scheme (cf. Definition 5), that* $\pi_{TIBE}$ *is TIBE scheme and that* Com *is a statistical binding commitment scheme. Then, protocol* $\Pi_{UMA}^P$ *securely realizes* $\mathcal{F}_{UMA}$ *in the presence of malicious adversaries in the* $\{\mathcal{F}_{SCCOT}, \mathcal{F}_{IC}\}$*-hybrid for all program executions with ptWrites.*

*High-Level Overview of Our Proof.* In this section we present the intuition of why does our protocol secure in the UMA model, while a full proof of Lemma 3 presented in the full version of the paper. With respect to garbled circuits security, we stress that neither the selective-bit-attack nor the incorrect-circuit-construction attack can harm the computation here due to the cut-and-choose technique, which prevents the sender from cheating in more than $\frac{s-|Z|}{2}$ of the circuits without being detected. As explained in [32], the selective-bit attack cannot be carried out successfully since R obtains all the input keys associated with its input in the cut-and-choose oblivious transfer, where the labels associated with both the check and evaluation circuits are obtained together. Thus, if S attempts to run a similar attack for a small number of circuits then it will not effect the majority, whereas if it does so for a large number of circuits then it will be caught with overwhelming probability. In the protocol, R checks that half of the chains and their corresponding input garbled values were correctly generated. It is therefore assured that with high probability the majority of the remaining circuits and their input garbled values are correct as well. Consequently, the result output by the majority of the remaining circuits must be correct.

The proof for the case the receiver is corrupted is based on two secure components: The garbling scheme and the timed IBE scheme. In the proof we reduce the security of our protocol to the security of each one of them. The intuition behind this proof asserts that R receives $|Z|$ opened check circuits and $|\tilde{Z}| = s - |Z|$ evaluation circuits. Such that for each evaluation circuit it only receives a single set of keys for decrypting the circuit. Furthermore, the keys that it receives for each of the $|\tilde{Z}|$ evaluation circuits are associated with the same pair of inputs $x_1, x_2$. This intuitively implies that R can do nothing but correctly decrypt $|\tilde{Z}|$ circuits, obtaining the same value $P^d(x_1, x_2)$. One concern regarding the security proof stems from the use of a TIBE encryption scheme within each of the CPU-step circuits. Consequently, we have to argue the secrecy of the input label that is not decrypted by R. Specifically, we show that this is indeed the case by constructing a simulator that, for each CPU-step, outputs a fake translate table translate that correctly encrypts the active label (namely, the label observed by the adversary), yet encrypts a fake inactive label. We then show, that the real view in which all labels are correctly encrypted, is indistinguishable from the simulated view in which only the active label is encrypted correctly.

# A    Garbled Circuits

The definition of garbled circuits with respect to the cut-and-choose technique is presented in Sect. 2.6. In this section we present the Input Consistency Functionality (Fig. 8) which is realized via a secure 2PC protocol when the underlying garbling scheme is applied using a cut-and-choose based protocol. We next present the authenticity game (Fig. 7) used in the definition of garbled circuits.

---

**The authenticity game** $\text{Auth}_{\mathcal{A}}(1^{\kappa}, s, C)$

**Parameters.** For an arbitrary circuit C, a security parameters $\kappa$ and $s$ the game is as follows.

1. The adversary hands an input $x$ and a subset $Z \in [s]$ to the game.
2. The game chooses $s$ sets of output labels $\{v, b, \mathsf{lbl}^{v,i}_{\text{out},b}\}_{v,i,b}$ for every $v \in [v_{\text{out}}], i \in [s]$ and $b \in \{0, 1\}$ and computes:

$$\left( \{\tilde{C}_i\}_i, \{u, b, \mathsf{lbl}^{u,i}_{\text{in},b}\}_{u,i,b} \right) \leftarrow \mathsf{Garb}\left( 1^{\kappa}, s, C, \{v, b, \mathsf{lbl}^{v,i}_{\text{out},b}\}_{v,i,b} \right)$$

3. The game sends the adversary $s$ sets of garbled circuits $\{\tilde{C}_i\}_{i \in [s]}$ and $s$ sets of garbled inputs $\tilde{x}_i$: For garbled circuits indexed with $z \in Z$ it is given $\tilde{x}_z = (\mathsf{lbl}^{1,z}_{\text{in},b}, \ldots, \mathsf{lbl}^{v_{\text{in}},z}_{\text{in},b})$ for every $b \in \{0, 1\}$; while for garbled circuits indexed with $z \notin Z$ the set is $\tilde{x}_z = (\mathsf{lbl}^{1,z}_{\text{in},x[1]}, \ldots, \mathsf{lbl}^{v_{\text{in}},z}_{\text{in},x[v_{\text{in}}]})$ for some input $x$.
4. The adversary returns a single index $z$ and one set of output labels $\hat{y}_z = (\hat{\mathsf{lbl}}^{1,z}_{\text{out},b}, \ldots, \hat{\mathsf{lbl}}^{v_{\text{out}},z}_{\text{in},b})$.
5. The game concludes as follows:
   (a) If $z \in Z$ return 0. Otherwise continue.
   (b) Compute $(\mathsf{lbl}^{1,z}_{\text{out},y[1]}, \ldots, \mathsf{lbl}^{v_{\text{out}},z}_{\text{in},y[v_{\text{out}}]}) = \mathsf{Eval}(\tilde{C}_z, \tilde{x}_z)$.
   (c) If for some $j \in [v_{\text{out}}]$ it holds that $\hat{\mathsf{lbl}}^{j,z}_{\text{out},b} = \mathsf{lbl}^{1,z}_{\text{out},1-y[1]}$ then output 1. Otherwise, output 0.

---

**Fig. 7.** The authenticity game $\text{Auth}_{\mathcal{A}}(1^{\kappa}, s, C)$.

---

**The Input Consistency Functionality - $\mathcal{F}_{\mathrm{IC}}$**

The functionality checks that the set of garbled inputs $\{\tilde{x}_i\}_i$ that are sent to the receiver represent the same input $x$. Note that this functionality checks the input of the *sender* S, and thus, the variable $x$ in this context actually refers to its input only (and not the receiver's input). Also note that $|x| = v_{\mathrm{in}}$.

*Common inputs.*

- The circuit C and the security parameters $\kappa, s$.
- $s$ garbled versions of C, namely $\{\tilde{C}_i\}_{i\in[s]}$.
- $s$ sets of garbled input $\left\{ (\mathsf{lbl}_{\mathrm{in},x[1]}^{1,i}, \ldots, \mathsf{lbl}_{\mathrm{in},x[v_{\mathrm{in}}]}^{v_{\mathrm{in}},i}) \right\}_{i\in[s]}$.
- $s$ sets of commitments for the sender's input labels, denoted by $\{\mathsf{com}_{1,b}^i, \ldots, \mathsf{com}_{v_{\mathrm{in}},b}^i\}_{b\in\{0,1\},i\in[s]}$.

*Sender's private inputs.* (The receiver has no private input)

- The output labels used in Garb, denoted by $\{\mathsf{lbl}_{\mathrm{out},b}^{v,i}\}_{v,i,b}$.
- The randomness $r$ used in Garb.
- Decommitments $\{\mathsf{dec}_{1,b}^i, \ldots, \mathsf{dec}_{v_{\mathrm{in}},b}^i\}_{b\in\{0,1\},i\in[s]}$ for the above commitments to the input labels.

*Output.* The functionality works as follows:

- Compute

$$
\left( \{\hat{C}_i\}_i, \{u, b, \hat{\mathsf{lbl}}_{\mathrm{in},b}^{u,i}\}_{u,i,b} \right) \leftarrow \mathsf{Garb}\left( 1^\kappa, s, \mathrm{C}, \{v, b, \mathsf{lbl}_{\mathrm{out},b}^{v,i}\}_{v,i,b}; r \right)
$$

- For every $u \in [v_{\mathrm{in}}]$:
  - For every $i \in [s]$ set $b_i$ as
  
  $$
  b_i = \left\{ \begin{array}{ll} 0, & \mathsf{com}(\mathsf{lbl}_{\mathrm{in},x[u]}^{u,i}, \mathsf{dec}_{u,0}^i) = \mathsf{com}_{u,0}^i \\ 1, & \mathsf{com}(\mathsf{lbl}_{\mathrm{in},x[u]}^{u,i}, \mathsf{dec}_{u,1}^i) = \mathsf{com}_{1,0}^i \\ \bot & \text{otherwise} \end{array} \right\}
  $$

  - If $b_i = \bot$ for some $i$ then output 0. Also If $b_1 \neq b_i$ for some $i$ output 0. (This checks that all labels are interpreted as the same input bit in all garbled circuits).
- Given that the above algorithm has not output 0, then output 1.

---

**Fig. 8.** The input consistency functionality $\mathcal{F}_{\mathrm{IC}}$.

# References

1. Afshar, A., Hu, Z., Mohassel, P., Rosulek, M.: How to efficiently evaluate RAM programs with malicious security. In: Oswald, E., Fischlin, M. (eds.) EUROCRYPT 2015. LNCS, vol. 9056, pp. 702–729. Springer, Heidelberg (2015). doi:10.1007/978-3-662-46800-5_27
2. Beaver, D.: Foundations of secure interactive computing. In: Feigenbaum, J. (ed.) CRYPTO 1991. LNCS, vol. 576, pp. 377–391. Springer, Heidelberg (1992). doi:10.1007/3-540-46766-1_31
3. Beaver, D., Micali, S., Rogaway, P.: The round complexity of secure protocols. In: Harriet, O. (ed.) 22nd STOC, pp. 503–513. ACM (1990)
4. Bellare, M., Hoang, V.T., Rogaway, P.: Foundations of garbled circuits. In: CCS, pp. 784–796 (2012)
5. Boneh, D., Boyen, X.: Efficient selective identity-based encryption without random oracles. J. Cryptology 24(4), 659–693 (2011)
6. Boneh, D., Franklin, M.K.: Identity-based encryption from the weil pairing. SIAM J. Comput. 32(3), 586–615 (2003)
7. Canetti, R.: Security and composition of multiparty cryptographic protocols. J. Cryptology 13(1), 143–202 (2000)
8. Chung, K.-M., Pass, R.: A simple oram. Cryptology ePrint Archive, Report 2013/243 (2013). http://eprint.iacr.org/2013/243
9. Cook, S.A., Reckhow, R.A.: Time-bounded random access machines. In: Proceedings of the 4th Annual ACM Symposium on Theory of Computing, Denver, Colorado, USA, 1–3 May 1972, pp. 73–80 (1972)
10. Damgård, I., Meldgaard, S., Nielsen, J.B.: Perfectly secure oblivious RAM without random oracles. In: Ishai, Y. (ed.) TCC 2011. LNCS, vol. 6597, pp. 144–163. Springer, Heidelberg (2011). doi:10.1007/978-3-642-19571-6_10
11. Garg, S., Gupta, D., Miao, P., Pandey, O.: Secure multiparty ram computation in constant rounds. In: TCC (2016, to appear)
12. Garg, S., Lu, S., Ostrovsky, R.: Black-box garbled RAM. In: FOCS 2015, pp. 210–229 (2016)
13. Garg, S., Lu, S., Ostrovsky, R., Scafuro, A.: Garbled RAM from one-way functions. In: STOC, pp. 449–458 (2015)
14. Gentry, C., Goldman, K.A., Halevi, S., Julta, C., Raykova, M., Wichs, D.: Optimizing ORAM and using it efficiently for secure computation. In: De Cristofaro, E., Wright, M. (eds.) PETS 2013. LNCS, vol. 7981, pp. 1–18. Springer, Heidelberg (2013). doi:10.1007/978-3-642-39077-7_1
15. Gentry, C., Halevi, S., Jutla, C.S., Raykova, M.: Private database access with he-over-oram architecture. IACR Cryptology ePrint Archive, 2014:345 (2014)
16. Gentry, C., Halevi, S., Lu, S., Ostrovsky, R., Raykova, M., Wichs, D.: Garbled RAM revisited. In: Nguyen, P.Q., Oswald, E. (eds.) EUROCRYPT 2014. LNCS, vol. 8441, pp. 405–422. Springer, Heidelberg (2014). doi:10.1007/978-3-642-55220-5_23
17. Goldreich, O.: Towards a theory of software protection and simulation by oblivious RAMs. In: STOC, pp. 182–194 (1987)
18. Goldreich, O.: Foundations of Cryptography: Volume 2, Basic Applications. Cambridge University Press, New York (2004)
19. Goldreich, O., Micali, S., Wigderson, A.: How to play any mental game or a completeness theorem for protocols with honest majority. In: STOC, pp. 218–229 (1987)

20. Goldreich, O., Ostrovsky, R.: Software protection and simulation on oblivious rams. J. ACM **43**(3), 431–473 (1996)
21. Goodrich, M.T., Mitzenmacher, M., Ohrimenko, O., Tamassia, R.: Privacy-preserving group data access via stateless oblivious RAM simulation. In: SODA, pp. 157–167 (2012)
22. Gordon, S.D., Katz, J., Kolesnikov, V., Krell, F., Malkin, T., Raykova, M., Vahlis, Y.: Secure two-party computation in sublinear (amortized) time. In: CCS, pp. 513–524 (2012)
23. Hu, Z., Mohassel, P., Rosulek, M.: Efficient zero-knowledge proofs of non-algebraic statements with sublinear amortized cost. In: Gennaro, R., Robshaw, M. (eds.) CRYPTO 2015. LNCS, vol. 9216, pp. 150–169. Springer, Heidelberg (2015). doi:10.1007/978-3-662-48000-7_8
24. Ishai, Y., Kushilevitz, E., Ostrovsky, R., Prabhakaran, M., Sahai, A.: Efficient non-interactive secure computation. In: Paterson, K.G. (ed.) EUROCRYPT 2011. LNCS, vol. 6632, pp. 406–425. Springer, Heidelberg (2011). doi:10.1007/978-3-642-20465-4_23
25. Ishai, Y., Prabhakaran, M., Sahai, A.: Founding cryptography on oblivious transfer – efficiently. In: Wagner, D. (ed.) CRYPTO 2008. LNCS, vol. 5157, pp. 572–591. Springer, Heidelberg (2008). doi:10.1007/978-3-540-85174-5_32
26. Ishai, Y., Prabhakaran, M., Sahai, A.: Secure arithmetic computation with no honest majority. In: Reingold, O. (ed.) TCC 2009. LNCS, vol. 5444, pp. 294–314. Springer, Heidelberg (2009). doi:10.1007/978-3-642-00457-5_18
27. Jarecki, S., Shmatikov, V.: Efficient two-party secure computation on committed inputs. In: Naor, M. (ed.) EUROCRYPT 2007. LNCS, vol. 4515, pp. 97–114. Springer, Heidelberg (2007). doi:10.1007/978-3-540-72540-4_6
28. Keller, M., Scholl, P.: Efficient, oblivious data structures for MPC. In: Sarkar, P., Iwata, T. (eds.) ASIACRYPT 2014, Part II. LNCS, vol. 8874, pp. 506–525. Springer, Heidelberg (2014). doi:10.1007/978-3-662-45608-8_27
29. Kushilevitz, E., Lu, S., Ostrovsky, R.: On the (in)security of hash-based oblivious RAM and a new balancing scheme. In: SODA, pp. 143–156 (2012)
30. Lindell, Y.: Fast cut-and-choose based protocols for malicious and covert adversaries. In: Canetti, R., Garay, J.A. (eds.) CRYPTO 2013, Part II. LNCS, vol. 8043, pp. 1–17. Springer, Heidelberg (2013). doi:10.1007/978-3-642-40084-1_1
31. Lindell, Y., Pinkas, B.: An efficient protocol for secure two-party computation in the presence of malicious adversaries. In: Naor, M. (ed.) EUROCRYPT 2007. LNCS, vol. 4515, pp. 52–78. Springer, Heidelberg (2007). doi:10.1007/978-3-540-72540-4_4
32. Lindell, Y., Pinkas, B.: Secure two-party computation via cut-and-choose oblivious transfer. In: Ishai, Y. (ed.) TCC 2011. LNCS, vol. 6597, pp. 329–346. Springer, Heidelberg (2011). doi:10.1007/978-3-642-19571-6_20
33. Liu, C., Huang, Y., Shi, E., Katz, J., Hicks, M.W.: Automating efficient ram-model secure computation. In: IEEE Symposium on Security and Privacy, pp. 623–638 (2014)
34. Lu, S., Ostrovsky, R.: How to garble RAM programs? In: Johansson, T., Nguyen, P.Q. (eds.) EUROCRYPT 2013. LNCS, vol. 7881, pp. 719–734. Springer, Heidelberg (2013). doi:10.1007/978-3-642-38348-9_42
35. Miao, P.: Cut-and-choose for garbled RAM. Cut-and-Choose for Garbled RAM IACR Cryptology ePrint Archive 2016:907 (2016)
36. Micali, S., Rogaway, P.: Secure computation. In: Feigenbaum, J. (ed.) CRYPTO 1991. LNCS, vol. 576, pp. 392–404. Springer, Heidelberg (1992)

37. Nielsen, J.B., Orlandi, C.: LEGO for two-party secure computation. In: Reingold, O. (ed.) TCC 2009. LNCS, vol. 5444, pp. 368–386. Springer, Heidelberg (2009). doi:10.1007/978-3-642-00457-5_22

38. Ostrovsky, R.: Efficient computation on oblivious RAMs. In: STOC, pp. 514–523 (1990)

39. Pinkas, B., Schneider, T., Smart, N.P., Williams, S.C.: Secure two-party computation is practical. In: Matsui, M. (ed.) ASIACRYPT 2009. LNCS, vol. 5912, pp. 250–267. Springer, Heidelberg (2009). doi:10.1007/978-3-642-10366-7_15

40. Pippenger, N., Fischer, M.J.: Relations among complexity measures. J. ACM **26**(2), 361–381 (1979)

41. Ren, L., Fletcher, C.W., Kwon, A., Stefanov, E., Shi, E., van Dijk, M., Devadas, S.: Constants count: practical improvements to oblivious RAM. In: USENIX, pp. 415–430 (2015)

42. Shi, E., Chan, T.-H.H., Stefanov, E., Li, M.: Oblivious RAM with $O((\log N)^3)$ worst-case cost. In: Lee, D.H., Wang, X. (eds.) ASIACRYPT 2011. LNCS, vol. 7073, pp. 197–214. Springer, Heidelberg (2011). doi:10.1007/978-3-642-25385-0_11

43. Stefanov, E., van Dijk, M., Shi, E., Fletcher, C.W., Ren, L., Xiangyao, Y., Devadas, S.: Path ORAM: an extremely simple oblivious RAM protocol. In: CCS, pp. 299–310 (2013)

44. Wang, X., Hubert Chan, T.-H., Shi, E.: Circuit ORAM: on tightness of the Goldreich-Ostrovsky lower bound. In: CCS, pp. 850–861 (2015)

45. Wang, X.S., Huang, Y., Hubert Chan, T.-H., Shelat, A., Shi, E.: SCORAM: oblivious RAM for secure computation. In: CCS, pp. 191–202 (2014)

46. Williams, P., Sion, R.: Single round access privacy on outsourced storage. In: CCS, pp. 293–304 (2012)

47. Yao, A.C.: Protocols for secure computations (extended abstract). In: FOCS, pp. 160–164 (1982)

48. Yao, A.C.: How to generate and exchange secrets (extended abstract). In: FOCS, pp. 162–167 (1986)

# More Efficient Constant-Round Multi-party Computation from BMR and SHE

Yehuda Lindell[1]([✉]), Nigel P. Smart[2], and Eduardo Soria-Vazquez[2]

[1] Bar-Ilan University, Ramat Gan, Israel
yehuda.lindell@biu.ac.il
[2] University of Bristol, Bristol, UK
nigel@cs.bris.ac.uk, eduardo.soria-vazquez@bristol.ac.uk

**Abstract.** We present a multi-party computation protocol in the case of dishonest majority which has very low round complexity. Our protocol sits philosophically between Gentry's Fully Homomorphic Encryption based protocol and the SPDZ-BMR protocol of Lindell et al. (CRYPTO 2015). Our protocol avoids various inefficiencies of the previous two protocols. Compared to Gentry's protocol we only require Somewhat Homomorphic Encryption (SHE). Whilst in comparison to the SPDZ-BMR protocol we require only a quadratic complexity in the number of players (as opposed to cubic), we have fewer rounds, and we require less proofs of correctness of ciphertexts. Additionally, we present a variant of our protocol which trades the depth of the garbling circuit (computed using SHE) for some more multiplications in the offline and online phases.

## 1 Introduction

*Secure multiparty computation:* In the setting of secure multiparty computation (MPC), a set of mutually distrusting parties wish to compute a joint function of their private inputs. Secure computation has been studied since the 1980s, and it has been shown that any functionality can be securely computed, even in the presence of a dishonest majority [17,35]. Classically, two main types of adversaries have been considered: *passive* (or semi-honest) adversaries follow the protocol specification but try to learn more than allowed from the transcript, and *active* (or malicious) adversaries who run any arbitrary strategy in an attempt to breach security.

*Efficient MPC:* In the last decade, significant effort has been placed on making secure computation efficient, both theoretically (with asymptotic efficiency) and practically. Both in theory and in practice the round complexity of MPC protocols is of interest. The theoretical interest is obvious, but it is in practice that probably the most effect can be felt. It is well known from practical experiments that often round complexity has more of an effect on the performance of MPC systems than communication complexity. This is especially true in networks with high latency (e.g., when the participating parties are on opposite sides of the

M. Hirt and A. Smith (Eds.): TCC 2016-B, Part I, LNCS 9985, pp. 554–581, 2016.
DOI: 10.1007/978-3-662-53641-4_21

world), where protocols with many rounds perform very poorly [33]. In practice, one also finds that constants matter considerably.

Most of the research effort on making secure computation practically efficient has focused on the case of two parties [30,31,35]. The most progress has been with protocols based on Yao's garbled circuits [35]. Extraordinary efficiency has been achieved for both passive adversaries [1,4] and active adversaries [19,23, 25,26,28,29,34]. In contrast, the case of multiple parties is way behind. When considering protocols with many rounds, the protocol of GMW can be used for passive adversaries (see [17] with an implementation in [7]) and the protocols of SPDZ and TinyOT can be used for active adversaries [10,11,22]. However, as mentioned above, these protocols have inherent inefficiency based on the fact that the number of rounds in the protocol is linear in the *depth* of the circuit that the parties compute.

In contrast to the impressive progress made in the garbled circuits area for the case of two parties, very little is known for multiple parties.

*The focus of this paper:* In this paper, we focus on the construction of a concretely efficient actively secure MPC protocol in the case of dishonest majority, which requires a (small) constant number of rounds. From a theoretical standpoint this problem is essentially solved: By combining Gentry's passively secure MPC protocol based on Fully Homomorphic Encryption (FHE) (see [16] and below) with generic non-interactive zero-knowledge proofs (NIZKs), one can obtain a protocol with two rounds of communication. Below we will show that such a protocol using more "practical" interactive zero-knowledge proofs can be realised using five rounds of interaction. In the plain model (with no preprocessing or access to a common reference string) recent work of Garg et al. [14] shows that six rounds are sufficient (using a construction based on iO). This work of Garg et al. builds on earlier work of Katz and Ostrovsky [21] who showed that five rounds are necessary and sufficient for the case of two parties.

From a practical point of view though the use of either general FHE, iO and/or generic NIZKs are clearly not suitable. Our protocol is based on the BMR approach [3]; and will use only *Somewhat* Homomorphic Encryption (SHE) and relatively efficient interactive zero-knowledge proofs. This approach consists of constructing a two phase protocol. In the first phase the parties use a generic MPC protocol to construct a "garbled" version of the function being computed. Then, in a constant round evaluation phase the garbled function is evaluated. The first garbling phase works in a gate-by-gate manner, and so by processing all gates in one go we obtain a constant round protocol for both phases; since this first garbling phase evaluates, via generic MPC, a circuit of constant depth.

In recent work [27] an efficient variant of the BMR protocol is used which utilizes the SPDZ [11] generic MPC protocol in the first garbling phase. In addition the authors introduce other optimizations which make the entire protocol actively secure for very little additional overhead. The SPDZ protocol itself uses a two phase approach, in the first phase, which utilizes Somewhat Homomorphic Encryption, correlated randomness is produced. As opposed to the general *Fully* Homomorphic Encryption of the above theoretical approaches. Then in the second phase this correlated randomness is used to evaluate the desired functionality (which in this case is the BMR garbling). Thus overall this protocol, which we dub SPDZ-BMR, consists of three phases; a phase using SHE, a phase doing generic MPC via the SPDZ online phase, and the final BMR circuit evaluation phase.

As alluded to above, there is another approach to constant round MPC, which utilizes Fully Homomorphic Encryption, namely Gentry's MPC protocol [16]. In this protocol the parties simply input their data using the encryption of the underlying FHE scheme, the parties evaluate the function locally using FHE, and then perform a distributed decryption (which requires $R_{Out} = 2$ rounds of interaction with current FHE schemes). This protocol is essentially optimal in terms of the number of rounds of communication, but it suffers from a number of drawbacks. The major drawback is that it requires FHE, which is itself a prohibitively expensive operation (and currently not practical). In addition, it is not immediately clear how to make the protocol actively secure without incurring significant additional costs. We outline in this paper how to address this latter problem, as a by-product of the analysis of our main protocol.

*Our Contributions:* Returning to the BMR based approach we note that *any* MPC protocol could be used for the BMR garbling phase, as long as it can be made actively secure within the specific context of the BMR protocol. In particular we could utilize Gentry's FHE-based MPC protocol (using only a SHE scheme) to perform the first stage of the BMR protocol; a protocol idea which we shall denote by SHE-BMR. The main observation as to why this is possible is that, as we have mentioned, the depth of the circuit computing the BMR garbled circuit is itself constant (and, in particular, independent of the depth of the circuit computing the function itself). This is due to the fact that in the BMR approach all garbled gates are computed in parallel; thus, the depth of the circuit computing the entire garbled circuit equals the depth of the circuit required to compute a single garble gate. We therefore conclude that *somewhat* homomorphic encryption suffices, with the depth being that sufficient to compute a single garbled gate.

A number of problems arise with this idea, which we address in this paper. First, can we make the resulting protocol actively secure for little additional cost? Second, is the required depth of the SHE scheme sufficiently small to make the scheme somewhat practical? Recall the SPDZ-BMR protocol only requires the underlying SHE scheme to support circuits of multiplicative depth one, and increasing the depth increases the cost of the SHE itself. Third, is the resulting round complexity of the scheme significantly less than that of the SPDZ-BMR protocol? Note that we can only expect a constant factor improvement, but such constants matter in practice. Fourth, can we save on any additional costs of the SPDZ-BMR protocol?

Since we use Gentry's FHE-based protocol (or an SHE version of it), we now outline two key challenges with using Gentry's FHE based protocol, which also apply to our protocol. When entering data we require an actively secure protocol to encrypt the FHE data, in particular we need to guarantee to the receiving parties that each encryption is well formed. The standard technique to do this is to also transmit a zero-knowledge proof of the correctness of encryption. A method to do this is given in [10, Appendix F], or [2, Sect. 3.2]. This is costly, and in practice rather inefficient. We call this protocol ID, and the associated round cost by $R_{\mathsf{ID}}$. In addition if we need to make further *input dependent inputs*, then this round cost will multiply. Thus we also need to introduce a sub-protocol with round cost $R_{\mathsf{Input}+} = 1$, which enables us to place all the zero-knowledge proofs for proving correctness of input into a pre-processing phase.

The second problem with Gentry's protocol is that we need to ensure that the distributed decryption is also actively secure; in the sense that the malicious parties cannot get an honest party to accept an incorrect result. We describe an efficient sub-protocol Out+ for performing this task, which importantly does not require zero-knowledge proofs (after a key generation phase) and has round complexity $R_{\mathsf{Out}+} = 2$.

We present a variant of our protocol which reduces the depth of the required SHE scheme, at the expense of requiring each party to input a larger amount of data. Interestingly, the main aim of the design in the SPDZ-BMR protocol was to reduce the number of multiplications needed (since each multiplication required generating a multiplication tuple for SPDZ, and this was the main cost). In contrast, when using SHE directly, additional multiplications are not expensive as long as they are carried out in parallel. Stated differently, the main concern is the *depth* of the circuit computing the BMR garbled circuit, and not necessarily its size. Of course, for concrete efficiency, one must try to minimize both, as reducing one slightly while greatly increasing the other would not be beneficial. In order to achieve this reduction in the depth of the circuit computing the BMR circuit, we utilize an observation that when computing the garbled circuit it suffices to obtain either the PRF key on the output wire or its additive inverse. This is due to the fact that we can actually take the PRF key to be the *square* of the value obtained in the garbled gate, which is the same whether $k$ or $-k$ is obtained. This allows us to combine the generation of the indicator-bits and the key-vector generation together. The additional flexibility of being able to output either the key or its additive inverse allows us to reduce the required SHE depth by one; in particular, from a depth of *four* to a depth of *three*.

In summary, we actually obtain two distinct protocols $\pi_b$ where $b \in \{0,1\}$; for which $b = 0$ means applying our basic variant protocol and $b = 1$ means applying the modified variant with a reduced depth cost. In some sense we can think of our basic SHE-BMR protocol as the same as the SPDZ-BMR protocol of [27], but it "cuts out the middle man" of producing multiplication triples, and the interaction needed to evaluate the garbling via the online phase of SPDZ. Indeed almost all of our basic protocol is identical to that described in [27]. However, naively applying SHE to the protocol from [27] results in a protocol that is neither efficient nor secure. For example, naively applying Gentry's MPC protocol to the garbling stage would result in needing an SHE scheme which supports a depth logarithmic in the number of parties $n$; whereas we would rather utilize a SHE scheme with constant depth. Thus we need to carefully design the FHE based MPC protocol to realise the BMR garbled circuit.

By utilizing the actively secure input and output routines in Gentry's protocol we also obtain an actively secure variant of Gentry's FHE based protocol which we denote by $G_a$. This is in addition to the original passively secure FHE based protocol of Gentry which we denote by $G_p$.

*Comparison:* By way of comparison we outline in Fig. 1 differences between the variants of our protocol, and those of Gentry and SPDZ-BMR. We let $n$ denote the number of parties, $W$ and $G$ denote the number of wires and gates in the binary circuit respectively, and $W_{in}$ the number of input wires and $W_{out}$ the number of output wires. To ease counting of rounds we consider a secure broadcast to be a single round operation (in the case of a dishonest majority, where parties may abort, a simple two-round echo-broadcast protocol suffices in any case [18]). We will see later that $R_{Out+} = 2$, and $R_{ID} = 3$. In the table the various functions $T_1, T_2, T_3$ describing the number of executions of ID are

$$T_1 = 16 \cdot G \cdot n^3 + (8 \cdot G + 4 \cdot W) \cdot n^2 + 9 \cdot W \cdot n + 156 \cdot G \cdot n,$$
$$T_2 = 4 \cdot G \cdot n^2 + (3 \cdot W + 1) \cdot n,$$
$$T_3 = (4 \cdot G + 2 \cdot W) \cdot n^2 + (W + 1) \cdot n.$$

If we compare the SPDZ-BMR protocol with our protocol variants $\pi_0$ and $\pi_1$ we see that the major difference in computational cost is the number of invocations of the protocol ID. The difference between SPDZ-BMR and $\pi_0$ is equal to $T_1 - T_2 = 16 \cdot G \cdot n^3 + 4 \cdot (G + W) \cdot n^2 + (6 \cdot W - 1) \cdot n + 156 \cdot G \cdot n$ invocations. To be very concrete, for 9 parties, a circuit of size 10,000 gates and wires, the number of ID invocations equals 141,210,000 in SPDZ-BMR versus 3,510,009 in SHE-BMR-$\pi_0$ versus 4,950,009 in SHE-BMR-$\pi_1$. Thus, $\pi_0$ is *one fortieth* of the cost of SPDZ-BMR, and $\pi_1$ is *one twenty-eighth* of the cost of SPDZ-BMR. This gap widens further as the number of parties grows, with the difference for 25 parties being a factor of 100 for $\pi_0$ and 70 for $\pi_1$. We remark, however, that even for just 3 parties, protocols $\pi_0$ and $\pi_1$ are already one twenty-third and one eighteenth of the cost, respectively.

On the downside we require an SHE scheme which will support depth three or four circuits, as opposed to the depth one circuits of the SPDZ-BMR

| Protocol | Security | Rounds of Interaction | Depth of FHE/SHE | Number of ID Execs |
|---|---|---|---|---|
| $G_p$ | passive | $3 = 1 + R_{Out}$ | Depth of $f$ | 0 |
| $G_a$ | active | $5 = R_{ID} + R_{Out+}$ | $1 + $ Depth of $f$ | $n + W_{in}$ |
| SPDZ-BMR | active | $16 = 13 + R_{ID}$ | 1 | $T_1$ |
| $\pi_0$ | active | $9 = R_{ID} + 4 + R_{Out+}$ | 4 | $T_2$ |
| $\pi_1$ | active | $9 = R_{ID} + 4 + R_{Out+}$ | 3 | $T_3$ |

Fig. 1. Comparison of Gentry's, the SPDZ-BMR and our protocol

protocol. The SHE scheme needs to support message spaces of $\mathbb{F}_p$, where $p > 2^\kappa$. We use [9], which gives potential parameter sizes for various SHE schemes supporting depth two and five, and run the experiments there to compare the parameters required for our specific depths here (depth 1 for SPDZ, depth 4 for protocol $\pi_0$ and depth 3 for protocol $\pi_1$). Specifically, assuming ciphertexts live in a ring $R_q$, then the dimension needs to go up by approximately a factor of 1.5 for depth-3 and a factor of 2 for depth-4, and the modulus by a factor of 1.6 for depth-3 and a factor of 2 for depth-4. Assuming standard DCRT representation of $R_q$ elements, this equates to an increase in the ciphertext size by a factor of approximately 2.4 for depth-3, and by approximately a factor of 4 for depth-4. Furthermore, the performance penalty (cost of doing arithmetic) increases by a factor of approximately 3.6 for depth-3, and by a factor of 8 for depth-4. Factoring in this additional cost, we have that when compared to SPDZ-BMR, the relative improvement in the computational cost in the above example becomes a factor of $40/8 = 5$ for $\pi_0$ and $28/3.6 = 7.7$ for $\pi_1$ for 9 parties, and a factor of $100/8 = 12.5$ for $\pi_0$ and $70/3.6 \approx 19.4$ for $\pi_1$ for 25 parties. Thus, both $\pi_0$ and $\pi_1$ significantly outperform BMR-SPDZ, and the depth reduction carried out in $\pi_1$ provides additional speedup (and reduction in bandwidth).

Our work should also be compared to [8] in which a constant round 3PC protocol is given, based on Yao's garbled circuits. However, active security in their case is provided by an expensive cut-and-choose protocol. In addition, their protocol is specifically designed for the three-party case, whereas we consider multiparty computation for any number of parties. Another constant-round multiparty protocol was constructed by [20]. However, although this protocol has good asymptotic complexity, its concrete efficiency is very unclear and no concretely efficient instantiation has been found. In [24] the authors propose a concrete instantiation of [20] for the two-party case, but no analogous proposal exists for the multiparty case.

## 2    Background on MPC and FHE

As a warm up to our main protocol, and to introduce the aspects of the FHE functionality we shall be using in detail we first give an outline of Gentry's FHE based protocol to evaluate the generic MPC functionality.

### 2.1    The Generic MPC Functionality

The goal of all the protocols in this paper is to securely realise the functionality given in Fig. 2. Namely we want protocols which allow $n$ mutually distrusting parties, with a possibly dishonest majority, to evaluate the function $f(x_1, \ldots, x_n)$ on their joint inputs.

### 2.2    A Basic FHE Functionality with Distributed Decryption

We first describe in Fig. 3 a basic FHE functionality which contains a distributed decryption functionality. Two points need to be noted about the functionality:

---

### The General MPC Functionality: $\mathcal{F}_{\mathrm{MPC}}$

The functionality is parametrized by a function $f(x_1, \ldots, x_n)$ which is input as a binary circuit $C_f$. The protocol consists of three externally exposed commands **Initialize**, **InputData**, and **Output** and one internal subroutine **Wait**.

**Initialize:** On input $(init, C_f)$ from all parties, where $C_f$ is a Boolean circuit, the functionality activates and stores $C_f$.

**Wait:** This waits on the adversary to return a $GO/NO\text{-}GO$ decision. If the adversary returns $NO\text{-}GO$ then the functionality aborts.

**InputData:** On input $(input, P_i, \mathsf{varid}, x_i)$ from $P_i$ and $(input, P_i, \mathsf{varid}, ?)$ from all other parties, with $\mathsf{varid}$ a fresh identifier, the functionality stores $(\mathsf{varid}, x_i)$. The functionality then calls **Wait**.

**Output:** On input $(output)$ from all parties, if $(\mathsf{varid}, x_i)$ is stored for each $P_i$, the functionality computes $y = f(x_1, \ldots, x_n)$ and outputs $y$ to the adversary. The functionality then calls **Wait**. If **Wait** does not result in an abort, the functionality outputs $y$ to all parties.

**Fig. 2.** The MPC Functionality: $\mathcal{F}_{\mathrm{MPC}}$

---

Firstly, the distributed decryption operation in **Output** can produce an incorrect result under the control of the adversary, but the "additive error" which is introduced by the adversary is introduced before the adversary learns the correct output. Secondly, the **InputData** routine is actively secure, and so a proof of correctness of its correct decryption is needed for each input ciphertext. The need for such an actively secure input routines is because we need to ensure that parties enter "valid" FHE/SHE encryptions, and that the simulator can "extract" the plaintext values. Within the functionality we denote the depth of a variable $x$ by $D(x)$, and we describe how the depth is altered with each operation which can affect the depth.

A method to perform the required **InputData** operation is given in [10, Appendix F], or [2, Sect. 3.2]. The basic idea is to check a number of executions of **InputData** at the same time. The protocol run in two phases, in the first phase a set of reference ciphertexts are produced and via cut-and-choose one subset is checked for correctness, whilst the other is permuted into buckets; one bucket for each value entered via **InputData**. In the second phase the input ciphertexts are checked for correctness by combining them homomorphically with the reference ciphertexts and opening the result. We denote the round complexity of the protocol implementing **InputData** by $R_{\mathsf{ID}}$. An analysis of the protocol from [10] indicates that it requires $R_{\mathsf{ID}} = 3$ rounds of communication: In the first round of the proof one party broadcasts the reference ciphertexts, in the next round the parties choose which ciphertexts to open, and in the third

---

### The FHE Functionality: $\mathcal{F}_{\text{FHE}}/\mathcal{F}_{\text{SHE}}$

The functionality consists of externally exposed commands **Initialize, InputData, Add, Multiply** and **Output**, and one internal subroutine **Wait**.

**Initialize:** On input $(init,p)$ from all parties, the functionality activates and stores $p$. All additions and multiplications below will be mod $p$.

**Wait:** This waits on the adversary to return a $GO/NO\text{-}GO$ decision. If the adversary returns $NO\text{-}GO$ then the functionality aborts.

**InputData:** On input $(input,P_i,\text{varid},x)$ from $P_i$ and $(input,P_i,\text{varid},?)$ from all other parties, with varid a fresh identifier, the functionality stores $(\text{varid},x)$. The functionality then calls **Wait**.

**Add:** On command $(add,\ \text{varid}_1,\ \text{varid}_2,\ \text{varid}_3)$ from all parties (if $\text{varid}_1$, $\text{varid}_2$ are present in memory and $\text{varid}_3$ is not), the functionality retrieves $(\text{varid}_1,x)$, $(\text{varid}_2,y)$ and stores $(\text{varid}_3,x+y \mod p)$.

**Add-scalar:** On command $(add\text{-}scalar,\ a,\ \text{varid}_1,\ \text{varid}_2)$ from all parties (if $\text{varid}_1$ is present in memory and $\text{varid}_2$ is not), the functionality retrieves $(\text{varid}_1,x)$ and stores $(\text{varid}_2,a+x \mod p)$.

**Multiply:** On command $(multiply,\ \text{varid}_1,\ \text{varid}_2,\ \text{varid}_3)$ *from all parties (if* $\text{varid}_1$, $\text{varid}_2$ are present in memory and $\text{varid}_3$ is not), the functionality retrieves $(\text{varid}_1,x)$, $(\text{varid}_2,y)$ and stores $(\text{varid}_3,x \cdot y \mod p)$.
In the case of the $\mathcal{F}_{\text{SHE}}$ version of this functionality only a limited depth of such commands can be performed; this depth is specified for the functionality. Depth Cost: $D(\text{varid}_3) = \max(D(\text{varid}_1), D(\text{varid}_2)) + 1$.

**Multiply-scalar:** On command $(multiply\text{-}scalar,\ a,\ \text{varid}_1,\ \text{varid}_2)$ from all parties (if $\text{varid}_1$ is present in memory and $\text{varid}_2$ is not), the functionality retrieves $(\text{varid}_1,x)$ and stores $(\text{varid}_2,a \cdot x \mod p)$.

**Output:** On input $(output,\text{varid},i)$ from all honest parties (if varid is present in memory), and a value $e \in \mathbb{F}_p$ from the adversary, the functionality retrieves $(\text{varid},x)$, and if $i = 0$ it outputs $(\text{varid},x)$ to the adversary. The functionality then calls **Wait**. If **Wait** does not result in an abort, then the functionality outputs $x+e$ to all parties if $i=0$, or it outputs $x+e$ only to party $i$ if $i \neq 0$.

**Fig. 3.** The FHE/SHE Functionality: $\mathcal{F}_{\text{FHE}}/\mathcal{F}_{\text{SHE}}$

round the ciphertexts are opened and combined.[1] Thus, overall, three rounds suffice.

In the following, we fix the notation $\langle varid \rangle$ to represent the result stored in the variable varid by the $\mathcal{F}_{\text{FHE}}/\mathcal{F}_{\text{SHE}}$ functionalities. In particular, we will use the arithmetic shorthands $\langle z \rangle = \langle x \rangle + \langle y \rangle$ and $\langle z \rangle = \langle x \rangle \cdot \langle y \rangle$ to represent the result of calling the **Add** and **Multiply** commands in the $\mathcal{F}_{\text{FHE}}/\mathcal{F}_{\text{SHE}}$ functionality,

---

[1] Choosing at random which ciphertexts to open cannot be carried out in a single round. However, it is possible for all parties to commit to the randomness in previous rounds and only decrypt in this round.

and we will slightly abuse those shorthands to denote subsequent additions or multiplications.

The description of **Output** in the case of a passively secure functionality is identical to the behaviour of the standard distributed decryption procedure for FHE schemes such as BGV, again see [10] for how the distributed decryption is performed. The basic protocol is to commit to the distributed decryption shares, and then open the shares. This gives a round complexity for **Output** of $R_{Out} = 2$. We shall provide a simple mechanism to provide active security for the Output command in the next section, which comes at the expense of increasing the required supported depth of the SHE scheme by one.

In the case of a passively secure variant of the FHE functionality, one would always have $e = 0$ in the **Output** routine. Furthermore, we would not need a proof of correctness of the input ciphertexts and so the number of rounds of interaction in the **InputData** routine would be $R_{ID} = 1$.

### 2.3   Gentry's FHE-Based MPC Protcol

In [16] Gentry presents an MPC protocol which has optimal round complexity to implement $\mathcal{F}_{MPC}$. In the $\mathcal{F}_{FHE}$-hybrid model the protocol can be trivially described as follows: The parties enter their data using the **InputData** command of the FHE functionality, the required function is evaluated using the **Add** and **Multiply** commands (i.e. each party locally evaluates the function using the FHE operations). The **Add-scalar** and **Multiply-scalar** commands can be computed by the parties locally encrypting the scalar with a mutually agreed randomness (so that all hold the same ciphertext) and then using the regular FHE **Add** or **Multiply** command, respectively.

Finally, the output is obtained using the **Output** command of the FHE functionality. For passively secure adversaries this gives us an "efficient" MPC protocol, assuming the FHE scheme can actually evaluate the function. For active adversaries we then have to impose complex zero-knowledge proofs to ensure that the **InputData** command is performed correctly, and we need a way of securing the **Output** command (which we will come to later).

## 3   The SPDZ-BMR Protocol

We shall now overview the SPDZ-BMR protocol from [27]. Much of the details we cover here focus on the offline SHE-part of the SPDZ protocol and how it is used in the SPDZ-BMR protocol. Recall the SPDZ protocol makes use of two phases; one an offline phase which uses an SHE scheme (which for our purposes we model via the functionality $\mathcal{F}_{SHE}$ above restricted to functions of multiplicative depth one), and an online phase using (essentially) only information theoretic constructs. These two phases are used to create a shared garbled circuit which is then evaluated in a third phase in the SPDZ-BMR protocol.

*First Phase Cost:* The first phase of the SPDZ-BMR protocol requires an upper bound on the total number of parties $n$, internal wires $W$, gates $G$ and input wires per party $W_{in}$ of the circuit which will be evaluated. The phase then calls the offline phase of the SPDZ engine to produce $M = 13 \cdot G$ multiplication triples, $B = W$ shared random bits, $R = 2 \cdot W \cdot n$ shared random values and $I = 8 \cdot G \cdot n$ shared values for entering data per party.

The main cost of the SPDZ-BMR protocol is actually in computing this initial data; yet the paper [27] does not address this cost in much detail. Delving into the paper [10] we see that each of these operations requires parties to encrypt random data under the SHE scheme and to produce additive sharings of SHE encrypted data. This first operation is identical to our *input* command on the functionality $\mathcal{F}_{SHE}$. We delve into the costs of the operations in more detail:

- **Encrypting (Input) Data ID:** When a party produces an encryption we need to ensure that it is validly formed, so as to protect against active attackers. As remarked above this is done using a zero-knowledge proof of correctness. Whilst the computational costs of this can be amortized due to "packing" in the SHE scheme, it is a non-trivial cost per encryption. We shall denote the computational and round cost in what follows by $C_{ID}$ and $R_{ID}$ respectively, i.e. the computational and round cost of the actively secure *EncCommit* operation from [10].
- **Producing Random ReSharings:** Given a ciphertext encrypting a value $m$ this procedure results in an additive sharing of $m$ amongst the $n$ parties. The computational cost of this procedure is dominated by the invocations of the ID protocol. Since each party needs to encrypt a random value, the computational cost $n \cdot C_{ID}$ and the round complexity is $R_{ID} + 1$. Again, the computational costs can be amortized due to the packing of the SHE scheme.
- **Producing Multiplication Triples:** To produce an unchecked triple this requires (per party) the encryption of two random values (of $a_i$ and $b_i$ in the triple $([a], [b], [c])$), plus four resharings (three of which can be done in parallel, with the fourth only partially in parallel). To produce a checked triple, this needs to be done twice (in parallel), followed by a sacrificing step of one of the triples via a procedure (described in [10]) which requires another two rounds of interaction. Thus the total computational cost is dominated by $12 \cdot n \cdot C_{ID}$; the round complexity is $R_{ID} + 4$.
- **Producing Shared Random Bits:** To produce an unchecked random bit we require (per party) the encryption of one random value, one passively secure distributed decryption (with only one round of interaction), plus two resharings (in parallel). To produce a checked random bit, the above has to be combined with an unchecked multiplication triple in a sacrificing step which requires two rounds of interaction. Thus the total computational cost is dominated by $9 \cdot n \cdot C_{ID}$; and the round complexity is $R_{ID} + 4$.
- **Producing Shared Random Values:** This requires (per party) the encryption of one random value, and two resharings which can be done in parallel. Thus the total computational cost is $2 \cdot n \cdot C_{ID}$, and the round complexity is $R_{ID} + 1$.

- **Producing Input Data:** Per data item which needs to be input for each player this requires the encryption of one random value plus two resharings (which cannot be fully parallelised), as well as one additional round of interaction. Thus the total computational cost is dominated by $C_{\mathsf{ID}} + 2 \cdot n \cdot C_{\mathsf{ID}}$, and the round complexity is $R_{\mathsf{ID}} + 3$.

A major bottleneck in the protocol, for active security, is the cost of encrypting the random data required by the protocol. Combining the costs, using the various formulae above, we see that this cost is given by

$$T_{\mathsf{ID}} \cdot C_{\mathsf{ID}} = 12 \cdot n \cdot C_{\mathsf{ID}} \cdot M + 9 \cdot n \cdot C_{\mathsf{ID}} \cdot B + 2 \cdot n \cdot C_{\mathsf{ID}} \cdot R + (1 + 2 \cdot n) \cdot n \cdot C_{\mathsf{ID}} \cdot I$$
$$= (12 \cdot 13 \cdot G + 9 \cdot W + 4 \cdot W \cdot n + (1 + 2 \cdot n) \cdot n \cdot 8 \cdot G) \cdot n \cdot C_{\mathsf{ID}}$$
$$= (16 \cdot G \cdot n^3 + (8 \cdot G + 4 \cdot W) \cdot n^2 + 9 \cdot W \cdot n + 156 \cdot G \cdot n) \cdot C_{\mathsf{ID}}$$

which is *cubic* in the number of players. In our protocol the same amortization due to SHE packing can be achieved. Thus we do not pay further attention to the constant improvement in performance due to packing, as the same constant can be applied to our protocol.

The total round complexity of the SPDZ offline phase is the maximum round complexity of the various pre-processing operations in the SPDZ offline phase; namely $R_{\mathsf{ID}} + 4$. This holds since the transmission of *all* random encrypted values can occur in one round at the beginning of this phase. We stress that the depth of the SHE needed for SPDZ is just *one*, making it very efficient.

*Second Phase Cost:* A careful analysis of the rest of the garbling phase of SPDZ-BMR implies that it requires six additional rounds of communication.[2]

*Third Phase Cost:* The online phase of the SPDZ-BMR protocol requires three rounds of interaction, one to open the secret shared values and two to verify the associated MACs.

*Summary:* In summary, the round complexity of SPDZ-BMR is $R_{\mathsf{ID}} + 10$ in the offline phase, and 3 in the online phase.

## 4     Extending the $\mathcal{F}_{\mathrm{FHE}}/\mathcal{F}_{\mathrm{SHE}}$ Functionalities

### 4.1     The Extended Functionality Definition

The first step in describing our new offline protocol for constructing the BMR circuit is to extend the functionalities $\mathcal{F}_{\mathrm{FHE}}/\mathcal{F}_{\mathrm{SHE}}$ to new functionalities $\mathcal{F}_{\mathrm{FHE}+}/\mathcal{F}_{\mathrm{SHE}+}$. In Fig. 4 we present the $\mathcal{F}_{\mathrm{FHE}+}$ functionality; the definition of the $\mathcal{F}_{\mathrm{SHE}+}$ functionality is immediate.

---

[2] With reference to [27] this is one round in the preprocessing-I phase and the start of the preprocessing-II phase due to the Output commands, and three to evaluate the required circuits in step 3 of preprocessing-II (since the circuits are of depth three, and hence require three rounds of computation), plus two to verify all the associated MAC values.

---

The Extended Functionality $\mathcal{F}_{\mathrm{FHE}+}$

This functionality runs the same **Initialize, Wait, InputData, Add, Multiply,** and **Output** commands as $\mathcal{F}_{\mathrm{FHE}}$ of Figure 3. It additionally has the four following externally exposed commands:

**Output+:** On input (*output+*,varid, $i$) from all honest parties (if varid is present in memory), the functionality retrieves (varid, $x$), and if $i = 0$ it outputs (varid, $x$) to the adversary. The functionality then calls **Wait**, and only if **Wait** does not abort then outputs $x$ to all parties if $i = 0$, or outputs $x$ only to party $i$ if $i \neq 0$.

**InputData+:** On input (*input+*,$P_i$,varid,$x$) from $P_i$ and (*input+*,$P_i$,varid,?) from all other parties, with varid a fresh identifier, the functionality stores (varid,$x$). The functionality then calls **Wait**.

**RandomElement:** This command is executed on input (*randomelement*, varid) from all parties, with varid a fresh identifier. The functionality then selects uniformly at random $x \in \mathbb{F}_p$ and stores (varid,$x$).

**RandomBit:** This command is executed on input (*randombit*, varid) from all parties, with varid a fresh identifier. The functionality then selects uniformly at random $x \in \{0, 1\}$ and stores (varid,$x$).

---

**Fig. 4.** The Extended Functionality $\mathcal{F}_{\mathrm{FHE}+}$

These new functionalities mimic the output possibilities of the SPDZ offline phase, which were exploited in [27]; by allowing the functionality to produce encryptions of random data and encryptions of random bits. In addition the functionalities provide a version of **Output**, which we call **Output+**, which does not allow the adversary to introduce an error value. There is also a new version of **InputData** called **InputData+** which will enable us to reduce the number of rounds of interaction in our main protocol. Functionally this does nothing different from **InputData** but it will be convenient to introduce a different name for a different implementation within our FHE functionality.

### 4.2 Securely Realising the Extended Functionality

In Fig. 5 we give the protocol $\pi_{\mathrm{FHE}+}$ for realising the $\mathcal{F}_{\mathrm{FHE}+}$ functionality in the $\mathcal{F}_{\mathrm{FHE}}$-hybrid model. Let us start by looking at the **Output+** command in more detail (after first reading Fig. 5). Suppose the adversary tries to make player $P_j$ accept an incorrect value, by introducing errors into the calls to the weakly secure **Output** command from $\mathcal{F}_{\mathrm{FHE}}$. The honest player $P_j$ will receive varid $+ e_1$ instead of varid and authvarid$_j + e_2$ instead of authvarid$_j$, for some adversarially chosen values of $e_1$ and $e_2$. If player $P_j$ is not to abort then these quantities must satisfy authvarid$_j + e_2 = sk_j \cdot (\text{varid} + e_1)$. Now since we know that authvarid$_j = \text{varid} \cdot sk_j$ then this implies that the adversary needs to select $e_1$ and $e_2$ such that $e_2 = sk_j \cdot e_1$, which it needs to do without having any knowledge of $sk_j$. Thus either the adversary needs to select $e_1 = e_2 = 0$, or he needs to guess the correct value of $sk_j$. This will happen with probability at most $1/p$, which is negligible.

We note that in the concurrent independent work of [12] a similar approach to our **Output+** command is taken in order to attain active security. Nevertheless, they use a global MAC key $\langle sk \rangle = \langle sk_1 \rangle + \cdots + \langle sk_n \rangle$ that is revealed to all parties after decryption, which means that $sk$ needs to be renewed after each call to **Output+**. Thus, each call to their similar **Output+** implementation requires $n$ calls to the expensive **InputData** protocol, which does not pay off in terms of concrete efficiency.

The protocol which implements **InputData+** works by first running **Input-Data** with a random value, and then later providing the difference between the random value input and the real input. This enables parallel *preprocessing* of the InputData procedure, thereby reducing the overall number of rounds.

The protocol which implements the **RandomElement** command generates an encrypted random value $\langle x \rangle$, unknown to any party as long as one of the parties honestly chooses his additional share $x_i$ randomly.

The protocol which implements the **RandomBit** command is more elaborate, and borrows much from the equivalent operations in the SPDZ offline phase, see [10]. The basic idea is to generate an encrypted random value $\langle x \rangle$, unknown to any party. This value is then squared to obtain $\langle s \rangle$. The value of $s$ is then publicly revealed and an arbitrary square root $y$ is taken. As long as $s \neq 0$ (which happens with negligible probability due to the size of $p$) we then have that $\langle b \rangle = \langle x \rangle / y$ is an encryption of a value chosen uniformly from $\{-1, 1\}$. Since $p$ is prime, with probability $1/2$ the square root taken will be equal to $x$ and with probability $1/2$ it will be equal to $-x$. This encryption of a value in $\{-1, 1\}$ is turned into an encryption of a value in $\{0, 1\}$ by the final step, by computing $(\langle b \rangle + 1)/2$, which is a linear function and can be thus computed by calling **Add-Scalar** and **Multiply-Scalar**. However, unlike in SPDZ no sacrificing procedure is required as the **Output+** command is actively secure.

**Theorem 1.** *Protocol $\pi_{FHE+}$ securely computes $\mathcal{F}_{FHE+}$ in the $\mathcal{F}_{FHE}$-hybrid model in the UC framework, in the presence of static, active adversaries corrupting any number of parties.*

*Proof (sketch).* By [5], it suffices to prove the security of Protocol $\pi_{FHE+}$ in the SUC (simple UC) framework. We will sketch the proof for each of the processes in the functionality separately. In the $\mathcal{F}_{FHE}$-hybrid model the security follows in a straightforward way utilizing the security of the commands in $\mathcal{F}_{FHE}$.

*Output+:* The security of **Output+** relies on the security of the **InputData** and **Output** commands of $\mathcal{F}_{FHE}$. Namely, by the security of **InputData** we have that all $sk_j$ values are secret, and by the security of **Output** the only change that $\mathcal{A}$ can make to the output is an additive difference $e$ (fixed before the output is given). Thus, $\mathcal{A}$ can only change the output if it chooses additive differences $e_1, e_2$ with $e_1 \neq 0$ such that $(x + e_1) \cdot sk_j = x \cdot sk_j + e_2 \pmod{p}$, where $x$ is the value output. This implies that $e_1 \cdot sk_j = e_2 \pmod{p}$. Since $sk_j$ is secret, the adversary can cause this equality to hold with probability at most $p$.

---

Protocol $\pi_{\text{FHE+}}$

This protocol implements the functionality $\mathcal{F}_{\text{FHE+}}$ in the $\mathcal{F}_{\text{FHE}}$-hybrid model.

**Initialize:** This performs the initialisation routine just as in the $\mathcal{F}_{\text{FHE}}$ functionality. However, in addition, each party executes **InputData** to obtain an encryption $\langle sk_i \rangle$ of a random MAC value $sk_i$ known only to player $P_i$.

**Output+:** On input ($output+$, varid, $i$) from all honest parties, if varid is present in memory, the following steps are executed.

1. If $i \neq 0$, party $P_i$ computes $\text{authvarid}_i = \langle \text{varid} \rangle \cdot \langle sk_i \rangle$, else, each party $P_j$ computes $\text{authvarid}_j = \langle \text{varid} \rangle \cdot \langle sk_j \rangle$.
2. The parties call $\mathcal{F}_{\text{FHE}}$ with the command ($output$, varid, $i$).
3. If $i \neq 0$, they call $\mathcal{F}_{\text{FHE}}$ with the command ($output$, $\text{authvarid}_i$, $i$), else, they use command ($output$, $\text{authvarid}_j$, $j$) for every $j \in [1, \ldots, n]$.
4. Any party $P_j$ aborts if $\text{authvarid}_j \neq \text{varid} \cdot sk_j$.

Depth Needed: $D(\text{varid}) + 1$.
Round Cost: 2 (since steps 2 and 3 can be performed in parallel).

**InputData+:** The first step of this command does not depend on the input, and so can be run in a pre-processing step if the number of values to be input per party are known in advance. Upon input ($input+$, $P_i$, varid, $x$) with $x \in \mathbb{F}_p$ for $P_i$ and ($input+$, $P_i$, varid, ?) for all other parties:

1. Party $P_i$ chooses a random $r_i \in \mathbb{F}_p$ (in the same field as $x$) and sends ($input$, $P_i$, varid-1, $r_i$) to Functionality $\mathcal{F}_{\text{FHE}}$.
2. All parties $P_j$ with $j \neq i$ send ($input$, $P_i$, varid-1, ?) to Functionality $\mathcal{F}_{\text{FHE}}$.
3. Party $P_i$ broadcasts $c_i = x_i - r_i \pmod{p}$ to all parties.
4. All parties send ($add$-$scalar$, $c_i$, varid-1, varid) to Functionality $\mathcal{F}_{\text{FHE}}$.

Depth Needed: $D(x_i) = D(c) = 0$.
Round Cost: $R_{\text{ID}} + 1$. Although all $R_{\text{ID}}$ rounds can be performed in parallel at the start of the protocol.

**RandomElement:**

1. For $i = 1, \ldots, n$, each $P_i$ chooses a random $x_i \in \mathbb{F}_p$, and calls $\mathcal{F}_{\text{FHE}}$ with the command ($input$, $P_i$, $x_i$) from party $P_i$ and ($input$, $P_i$, ?) for the others.
2. Call **Add** as many times as needed to compute $\langle x \rangle = \langle x_1 \rangle + \cdots + \langle x_n \rangle$.

Depth Needed: $D(x_i) = \max\{D(x_i)\} = 0$.
Round Cost: $R_{\text{ID}}$.

**RandomBit:** This command requires a more elaborate implementation

1. For $i = 1, \ldots, n$, call $\mathcal{F}_{\text{FHE}}$ with the command ($input$, $P_i$, $x_i$) from party $P_i$ and ($input$, $P_i$, ?) for the rest of the parties.
2. Call **Add** as many times as needed to compute $\langle x \rangle = \langle x_1 \rangle + \cdots + \langle x_n \rangle$.
3. Call **Multiply** to compute $\langle s \rangle = \langle x \rangle \cdot \langle x \rangle$.
4. Call $\mathcal{F}_{\text{FHE+}}$ on input ($output+$, $s$, 0) so all parties obtain $s$.
5. $y = \sqrt{s} \pmod{p}$, if $s = 0$ then restart the protocol.
6. $\langle b \rangle = \langle x \rangle / y$.
7. Call **Add-scalar** and **Multiply-scalar** to compute $\langle \text{varid} \rangle = (\langle b \rangle + 1)/2$.

Depth Needed: $D(s) + 1 = 2$. Note this is the depth required, but the output encrypted bit has depth zero.
Round Cost: $R_{\text{ID}} + 2$.

---

**Fig. 5.** Protocol $\pi_{\text{FHE+}}$

We remark that the MAC key $sk_j$ is only used for output values given to $P_j$. Thus, it always remains secret (even when used for many outputs).

The simulator for **Output+** works simply by simulating the **Output** interaction with $\mathcal{F}_{\mathrm{FHE}}$ for all honest $P_i$. Regarding a corrupt $P_j$, the simulator receives the value $x$ that is supposed to be output. Furthermore, the simulator receives the value $sk_j$ from the **InputData** instruction, as well as any errors that are introduced in the **Output** calls by corrupted parties. Thus, the simulator can construct the exact value that $\mathcal{A}$ would receive in a real execution.

*InputData+*: The only difference between **InputData+** and **InputData** is that **InputData+** can be run such that the actual input is only known to the party in the last round of the protocol. This is done in a straightforward way by using **InputData** to have a party input a random string, and then using that result to mask the real data (at the end). The simulator for this procedure therefore relies directly on the **InputData** procedure of $\mathcal{F}_{\mathrm{FHE}}$ in a straightforward way. Namely, in the $\mathcal{F}_{\mathrm{FHE}}$-hybrid model when the party $P_i$ is corrupted, the simulator receives the value $r_i$ that party $P_i$ sends to **InputData**. Then, upon receiving $c_i$ as broadcast by $P_i$, the simulator defines $x_i = c_i + r_i \pmod{p}$ and sends $(input+, P_i, \mathsf{varid}, x_i)$ to the ideal functionality as input. In the case that $P_i$ is honest, the simulator chooses a random $c_i \in \mathbb{F}_p$ and simulates $P_i$ broadcasting that value. Furthermore, it simulates the $(input, ...)$ and $(add\text{-}scalar, ...)$ interaction with $\mathcal{F}_{\mathrm{FHE}}$.

The view of the adversary is identical in the simulated and real executions. In addition, since **InputData** is secure and $c_i$ is broadcast and therefore the same for all parties, the protocol fully determines the input value $x_i = c_i + r_i \pmod{p}$, as required.

*RandomElement*: This is a straightforward coin tossing protocol. The security is derived from the fact that $\mathcal{F}_{\mathrm{FHE}}$ provides a secure **InputData** protocol that reveals no information about the input values. Thus, no party knows anything about the $x$-values input by the others. Formally, a simulator just simulates the message interaction with $\mathcal{F}_{\mathrm{FHE}}$ for all of the $(input, P_i, x_i)$ and $(input, P_i, ?)$ messages. As long as at least one party is honest, the distribution over the value $x$ defined is uniform, as required.

*RandomBit*: The first step of this protocol is to essentially run **RandomElement** in order to define a random shared value $x$. Then, the value $s = x^2 \pmod{p}$ is output to all parties, and each takes the same square-root $y$ of $s$. Assume that the square root taken is the one that is between 1 and $(p-1)/2$. Now, if $1 \leq x \leq \frac{p-1}{2}$, then $y = x$ and so $\langle b \rangle = \langle 1 \rangle$, and we have that $\langle \mathsf{varid} \rangle = \langle \frac{1+1}{2} \rangle = \langle 1 \rangle$. Else, if $\frac{p-1}{2} < x \leq p-1$ then $\langle b \rangle = \langle -1 \rangle$ and we have $\langle \mathsf{varid} \rangle = \langle \frac{-1+1}{2} \rangle = \langle 0 \rangle$. The security relies on the fact that the result is *fully determined* from the $(input, ...)$ messages sent in the beginning. Relying on the security of **InputData** and **Add/Multiply** in $\mathcal{F}_{\mathrm{FHE}}$, and on the security of the **Output+** procedure, the value $x$ is uniformly distributed and the value $s$ that is output to all parties equals $x^2$ and no other value. All other steps are deterministic and thus this guarantees that the output is a uniformly distributed bit, as required.

Regarding simulation, the simulator simulates the calls to **InputData**, **Add** and **Multiply** as in the protocol. For the output, the simulator simply chooses a random $s$ as the value received from **Output+**. The view of the parties is clearly identical to in a real execution.

This completes the proof sketch of the theorem.                               $\square$

# 5    The First Variant of the SHE-BMR Protocol: $\pi_0$

In this section we outline our basic protocol, which follows much upon the lines of the SPDZ-BMR protocol. The modifications needed for a variant using only depth three will be left to Sect. 6. We divide our discussion into three subsections. In the first section we outline the offline functionality $\mathcal{F}_{\text{offline}}$ we require. This functionality produces a shared garbled circuit which computes the function amongst the players.

For each wire there are $2 \cdot n$ wire labels, corresponding to two labels for each party. The wire labels are held as encrypted key values $\langle k^i_{w,\beta} \rangle$, where encryption is under the SHE scheme, along with encrypted masking values $\langle \lambda_w \rangle$; where $1 \leq w \leq W$, $\beta \in \{0,1\}$ and $1 \leq i \leq n$. The garbled gates are held as a set of linear combinations of outputs from a suitable Pseudo-Random Function (PRF) which is keyed by the wire labels of all parties. These linear combinations are then used to one-time pad encrypt the output wire label, with the precise linear combination to be used in any given situation determined by the encrypted mask values. The output wire masking values are decrypted towards all parties, and the input wire masking values are decrypted towards the inputting party, but everything else remains held in encrypted form.

We then present the online protocol $\pi_{\text{MPC},0}$ which implements $\mathcal{F}_{\text{MPC}}$ in the $\mathcal{F}_{\text{offline}}$-hybrid model. This first decrypts the input wire labels desired for each party via the distributed decryption functionality, and revealing the associated selector variables $\Lambda_w = \rho_w \oplus \lambda_w$, where $\rho_w$ is the actual intended wire value. The parties are then able, for each gate, to determine which linear combination to apply (using the selector variables), and can then determine the output wire label using the given linear combination. From this they can determine the output selector variable and repeat the process for the next gate, and so on. Once all gates have been processed in this way the players have learnt the selector variables $\Lambda_w$ for the output wires, and so can compute the output wire values from $\Lambda_w \oplus \lambda_w$, where the value of $\lambda_w$, for the output wires, was revealed in the pre-processing phase. At the end of this section we present the offline protocol itself $\pi_{\text{offline},0}$ which implements $\mathcal{F}_{\text{offline}}$ in the $\mathcal{F}_{\text{FHE+}}$-hybrid model.

## 5.1    Functionality $\mathcal{F}_{\text{offline}}$ for the Offline Phase

We first present the offline functionality (see Fig. 6) for our main MPC protocol. This is almost identical to the offline functionality for the SPDZ-BMR protocol of [27]. The main difference is that it is built on top of our $\mathcal{F}_{FHE+}$ functionality from the previous section, as opposed to the SPDZ MPC protocol. In particular this means we have just a single pre-processing step as opposed to the two phases in [27], which are in turn inherited from the two phases of the SPDZ protocol.

## 5.2   The SHE-BMR Protocol Specification $\pi_{MPC,0}$

We can now give our protocol $\pi_{MPC,0}$, described in Fig. 7, which securely computes the functionality $\mathcal{F}_{MPC}$ described in Fig. 2 in the $\mathcal{F}_{offline}$-hybrid model. The computational and communication costs of $\pi_{MPC,0}$ are mainly in the preprocessing step, with the on-line phase adding only a depth of one to the SHE scheme and two more rounds of communication, all of these costs coming from the need for an actively secure **Output+**. The on-line phase is just an adaptation of BMR [3], following the trend of [27], and thus we will not discuss it in details.

## 5.3   The $\pi_{offline,0}$ Protocol

Protocol $\pi_{offline,0}$ in Fig. 8 implements $\mathcal{F}_{offline}$ in the $\mathcal{F}_{FHE+}$-hybrid model.

For completeness, we show how to calculate the output indicators for functions $f_g = AND$ and $f_g = XOR$ in Fig. 9 as shown in [27]. Note that we consume a multiplicative depth of two for both operations.

- For $f_g = AND$, we compute $\langle t \rangle = \langle \lambda_a \rangle \cdot \langle \lambda_2 \rangle$ and then $\langle x_A \rangle = (\langle t \rangle - \langle \lambda_c \rangle)^2$, $\langle x_B \rangle = (\langle \lambda_a \rangle - \langle t \rangle - \langle \lambda_c \rangle)^2$, $\langle x_C \rangle = (\langle \lambda_b \rangle - \langle t \rangle - \langle \lambda_c \rangle)^2$, $\langle x_D \rangle = (1 - \langle \lambda_a \rangle - \langle \lambda_b \rangle + \langle t \rangle - \langle \lambda_c \rangle)^2$.
- For $f_g = XOR$, we first compute $\langle t \rangle = \langle \lambda_a \rangle \oplus \langle \lambda_b \rangle = \langle \lambda_a \rangle + \langle \lambda_b \rangle - 2 \cdot \langle \lambda_a \rangle \cdot \langle \lambda_2 \rangle$, and then $\langle x_A \rangle = (\langle t \rangle - \langle \lambda_c \rangle)^2$, $\langle x_B \rangle = (1 - \langle \lambda_a \rangle - \langle \lambda_b \rangle + 2 \cdot \langle t \rangle - \langle \lambda_c \rangle)^2$, $\langle x_C \rangle = \langle x_B \rangle$, $\langle x_D \rangle = \langle x_A \rangle$.

## 5.4   Security

The security of our protocol follows from the proof of the security of the SPDZ-BMR protocol in [27]. Apart from the use of Gentry's MPC protocol, as opposed to the SPDZ protocol, (which is purely an implementation change) the only difference is that the InputData in SPDZ-BMR is generated in a way that guarantees that it is random. For our basic protocol, this is not the case. However, there is nothing that forces the adversary to input the value it actually gets and security is preserved. In particular, the adversary can ignore the value it obtained and use a different one honestly, and no problem arises. So, it is no different from this case where the adversary can choose the value in InputData.

---

The Offline Functionality - $\mathcal{F}_{\text{offline}}$

This functionality runs the same **Initialize, Wait,** and **Output+** commands as $\mathcal{F}_{\text{FHE+}}$. In addition it has the following command:

**Preprocessing:** On input $(\texttt{preprocessing}, C_f)$, for a circuit $C_f$ with at most $W$ wires and $G$ gates, the functionality performs the following operations.

- For all wires $w \in [1, \ldots, W]$ :
  - The functionality stores a random mask $\langle \lambda_w \rangle$, where $\lambda_w \in \{0, 1\}$.
  - For every value $\beta \in \{0, 1\}$, each party $P_i$ chooses and stores a random key $\langle k^i_{w,\beta} \rangle$, where $k^i_{w,\beta} \in \mathbb{F}_p$.
- For all wires $w$ which are attached to party $P_i$ the functionality decrypts $\langle \lambda_w \rangle$ to party $P_i$ by running **Output+** as in functionality $\mathcal{F}_{\text{FHE+}}$.
- For all output wires $w$ the functionality decrypts $\langle \lambda_w \rangle$ to all parties by running **Output+** as in functionality $\mathcal{F}_{\text{FHE+}}$.
- For every gate $g$ with input wires $1 \le a, b \le W$ and output wire $1 \le c \le W$.
  - Party $P_i$ provides the following values for $x \in \{a, b\}$ on the $4 \cdot G$ values:

$$F_{k^i_{x,0}}(0||1||g), \ldots, F_{k^i_{x,0}}(0||n||g), \qquad F_{k^i_{x,0}}(1||1||g), \ldots, F_{k^i_{x,0}}(1||n||g)$$
$$F_{k^i_{x,1}}(0||1||g), \ldots, F_{k^i_{x,1}}(0||n||g), \qquad F_{k^i_{x,1}}(1||1||g), \ldots, F_{k^i_{x,1}}(1||n||g)$$

  (In our protocols, the parties actually provide sums of pairs of these values; see Figure 9. This reduces the number of values input from 8 per-party per-gate to only 4 per-party per-gate.)
  - Define the selector variables

$$\chi_1 = \begin{cases} 0, & \text{If } f_g(\lambda_a, \lambda_b) = \lambda_c. \\ 1, & \text{Otherwise.} \end{cases} \qquad \chi_2 = \begin{cases} 0, & \text{if } f_g(\lambda_a, \bar{\lambda}_b) = \lambda_c. \\ 1, & \text{Otherwise.} \end{cases}$$

$$\chi_3 = \begin{cases} 0, & \text{If } f_g(\bar{\lambda}_a, \lambda_b) = \lambda_c. \\ 1, & \text{Otherwise.} \end{cases} \qquad \chi_4 = \begin{cases} 0, & \text{If } f_g(\bar{\lambda}_a, \bar{\lambda}_b) = \lambda_c. \\ 1, & \text{Otherwise.} \end{cases}$$

  - Set $\mathbf{A}_g = (A^1_g, \ldots, A^n_g)$, $\mathbf{B}_g = (B^1_g, \ldots, B^n_g)$, $\mathbf{C}_g = (C^1_g, \ldots, C^n_g)$, $\mathbf{D}_g = (D^1_g, \ldots, D^n_g)$ where for $1 \le j \le n$:

$$A^j_g = \left( \sum_{i=1}^n F_{k^i_{a,0}}(0||j||g) + F_{k^i_{b,0}}(0||j||g) \right) + k^j_{c,\chi_1}$$

$$B^j_g = \left( \sum_{i=1}^n F_{k^i_{a,0}}(1||j||g) + F_{k^i_{b,1}}(0||j||g) \right) + k^j_{c,\chi_2}$$

$$C^j_g = \left( \sum_{i=1}^n F_{k^i_{a,1}}(0||j||g) + F_{k^i_{b,0}}(1||j||g) \right) + k^j_{c,\chi_3}$$

$$D^j_g = \left( \sum_{i=1}^n F_{k^i_{a,1}}(1||j||g) + F_{k^i_{b,1}}(1||j||g) \right) + k^j_{c,\chi_4}$$

  - The functionality finally stores the values $\langle \mathbf{A}_g \rangle, \langle \mathbf{B}_g \rangle, \langle \mathbf{C}_g \rangle, \langle \mathbf{D}_g \rangle$.

**Fig. 6.** The Offline Functionality $\mathcal{F}_{\text{offline}}$

The MPC Protocol - $\pi_{\text{MPC},0}$

On input a circuit $C_f$ representing the function $f$, the parties execute the following commands in sequence.

**Preprocessing:** This sub-task is performed as follows.
- Call **Initialize** on $\mathcal{F}_{\text{offline}}$ to initialize the FHE scheme.
- Call **Preprocessing** on $\mathcal{F}_{\text{offline}}$ with input $C_f$.

**Online Computation:** This sub-task is performed as follows.
- For all his input wires $w$, each party computes $\Lambda_w = \rho_w \oplus \lambda_w$, where $\lambda_w$ was obtained in the preprocessing stage, and $\Lambda_w$ is broadcast to all parties.
- Party $i$ calls **Output+** to all parties on $\mathcal{F}_{\text{offline}}$ to decrypt the key $\langle k_w^i \rangle$ associated to $\Lambda_w$, for all his input wires $w$.
- The parties call **Output+** on $\mathcal{F}_{\text{offline}}$ to decrypt $\{\mathbf{A}_g\}$, $\{\mathbf{B}_g\}$, $\{\mathbf{C}_g\}$, and $\{\mathbf{D}_g\}$ for every gate $g$.
- Passing through the circuit topologically, the parties can now locally compute the following operations for each gate $g$. Let the gates input wires be labelled $a$ and $b$, and the output wire be labelled $c$.
  - For $j = 1, \ldots, n$ compute $k_c^j$ according to the following cases:
    $(\Lambda_a, \Lambda_b) = (0,0)$ : Set $k_c^j = A_g^j - \left( \sum_{i=1}^n F_{k_a^i}(0\|j\|g) + F_{k_b^i}(0\|j\|g) \right)$.
    $(\Lambda_a, \Lambda_b) = (0,1)$ : Set $k_c^j = B_g^j - \left( \sum_{i=1}^n F_{k_a^i}(1\|j\|g) + F_{k_b^i}(0\|j\|g) \right)$.
    $(\Lambda_a, \Lambda_b) = (1,0)$ : Set $k_c^j = C_g^j - \left( \sum_{i=1}^n F_{k_a^i}(0\|j\|g) + F_{k_b^i}(1\|j\|g) \right)$.
    $(\Lambda_a, \Lambda_b) = (1,1)$ : Set $k_c^j = D_g^j - \left( \sum_{i=1}^n F_{k_a^i}(1\|j\|g) + F_{k_b^i}(1\|j\|g) \right)$.
  - If $k_c^i \notin \{k_{c,0}^i, k_{c,1}^i\}$, then $P_i$ outputs abort. Otherwise, it proceeds. If $P_i$ aborts it notifies all other parties with that information. If $P_i$ is notified that another party has aborted it aborts as well.
  - If $k_c^i = k_{c,0}^i$ then $P_i$ sets $\Lambda_c = 0$; if $k_c^i = k_{c,1}^i$ then $P_i$ sets $\Lambda_c = 1$.
  - The output of the gate is defined to be $(k_c^1, \ldots, k_c^n)$ and $\Lambda_c$.
- Assuming party $P_i$ does not abort it will obtain $\Lambda_w$ for every circuit-output wire $w$. The party can then recover the actual output value from $\rho_w = \Lambda_w \oplus \lambda_w$, where $\lambda_w$ was obtained in the preprocessing stage.

Depth Needed: $D(Output + (\{\mathbf{A}_g\}, \{\mathbf{B}_g\}, \{\mathbf{C}_g\}, \{\mathbf{D}_g\})) = 3 + 1 = 4$.

Round Cost:The round cost of the online stage is that of the first three steps, which can be done in parallel in two rounds.

**Fig. 7.** The MPC Protocol - $\pi_{\text{MPC},0}$

## 5.5   Analysis of Efficiency

Just as in our analysis of the SPDZ-BMR protocol, we wish to estimate the cost of the most expensive operations; which are the encryptions of input data and random input data.

- Each party calls **InputData** once during the **Initialize** phase of the extended FHE functionality.
- We perform $W$ **RandomBit** operations, each of which consumes a $C_{\text{ID}}$ per party.

---

### The offline Protocol: $\pi_{\text{offline},0}$

The protocol runs the commands **Initialize, Wait,** and **Output+** by calling the equivalent commands on $\mathcal{F}_{\text{FHE}+}$. Thus we only need to describe **Preprocessing** as follows:

1. Call **Initialize** on the functionality $\mathcal{F}_{\text{FHE}+}$ with input a prime $p > 2^k$.
2. **Generate wire masks:** For every circuit wire $w$ we need to generate a random and hidden masking-values $\lambda_w$. Thus for *all* wires $w$ the parties execute **RandomBit** of $\mathcal{F}_{\text{FHE}+}$; the output is denoted by $\langle \lambda_w \rangle$.
   Depth Needed: $D(RandomBit) = 2$
   Round Cost: $R_{\text{ID}} + 2$.
3. **Generate keys:** For every wire $w$, each party $i \in [1, \ldots, n]$ and for $\beta \in \{0,1\}$, the parties execute the command **InputData** of the functionality $\mathcal{F}_{\text{FHE}+}$ to obtain output $\langle k^i_{w,\beta} \rangle$; where player $i$ learns $k^i_{w,\beta}$. For the vector of shares $\left( \langle k^i_{w,\beta} \rangle \right)^n_{i=1}$ we shall abuse the notation and denote it by $\langle \mathbf{k}_{w,\beta} \rangle$.
   Depth Needed: $D(k^i_{w,\beta}) = 0$.
   Round Cost: $R_{\text{ID}}$.
4. **Output masks for circuit-input-wires:** For all wires $w$ which are attached to party $P_i$ we execute the command **Output+** on the functionality $\mathcal{F}_{\text{FHE}+}$ to decrypt $\langle \lambda_w \rangle$ to party $i$.
   Depth Needed: $\max(D(RandomBit), D(Output + \langle \lambda_w \rangle)) = \max(2,1) = 2$.
   Round Cost: 2.
5. **Output masks for circuit-output-wires:** In order to reveal the real values of the circuit-output-wires it is required to reveal their masking values. That is, for every circuit-output-wire $w$, the parties execute the command **Output+** on the functionality $\mathcal{F}_{\text{FHE}+}$ for the stored value $\langle \lambda_w \rangle$.
   Depth Needed: $\max(D(RandomBit), D(Output + \langle \lambda_w \rangle)) = \max(2,1) = 2$.
   Round Cost: 2.
6. **Calculate garbled gates:** See Figure 9 for the details of this step.

We note that steps two and three can be run in parallel, and that steps four and five also can be run in parallel, but need to follow steps two. We also note that the calls to **InputData+** in the last step (detailed in Figure 9) need to be executed after step three. Hence, we have:
Total Depth Needed: 3.
Total Round Cost: $\max(R_{\text{ID}} + 3, R_{\text{ID}} + 4) = R_{\text{ID}} + 4$.

---

**Fig. 8.** The offline Protocol: $\pi_{\text{offline},0}$

- To create the encrypted PRF keys we require an additional $2 \cdot W$ invocations of $C_{\text{ID}}$ per party.
- Finally to enter the garbled labels we require, $4 \cdot n^2 \cdot G$ invocations of the input data routine, which consists of $4 \cdot n \cdot G$ invocations of **InputData** per party.

Thus the cost of encrypting the data for the SHE-BMR protocol is given by the expression $\left( 4 \cdot n^2 \cdot G + (3 \cdot W + 1) \cdot n \right) \cdot C_{\text{ID}}$, which is *quadratic* in $n$ as opposed to the *cubic* complexity of the SPDZ-BMR protocol.

---

Calculate Garbled Gates Step of $\pi_{\text{offline},0}$

This step is operated for each gate $g$ in the circuit in parallel. Specifically, let $g$ be a gate whose input wires are $a, b$ and output wire is $c$. Do as follows:

(a) **Calculate output indicators:** This step calculates four indicators $\langle x_A \rangle, \langle x_B \rangle$, $\langle x_C \rangle, \langle x_D \rangle$ whose values will be $\langle 0 \rangle$ or $\langle 1 \rangle$. Each indicator is determined by some quadratic function $f_g$ on $\langle \lambda_a \rangle, \langle \lambda_b \rangle, \langle \lambda_c \rangle$, depending on the truth table of the gate. See Section 5.3 for details.

$$\langle x_A \rangle = (f_g(\langle \lambda_a \rangle, \langle \lambda_b \rangle) - \langle \lambda_c \rangle)^2 \qquad \langle x_B \rangle = (f_g(\langle \lambda_a \rangle, (1 - \langle \lambda_b \rangle)) - \langle \lambda_c \rangle)^2$$
$$\langle x_C \rangle = (f_g((1 - \langle \lambda_a \rangle), \langle \lambda_b \rangle) - \langle \lambda_c \rangle)^2 \quad \langle x_D \rangle = (f_g((1 - \langle \lambda_a \rangle), (1 - \langle \lambda_b \rangle)) - \langle \lambda_c \rangle)^2$$

Depth Needed: $D(x_*) = D(\lambda_*) + 2 = 2$.

(b) **Assign the correct vector:** The indicators are used to choose, for every garbled label, either $\mathbf{k}_{c,0}$ or $\mathbf{k}_{c,1}$, for $t = A, B, C, D$,

$$\langle \mathbf{v}_{c,x_t} \rangle = (1 - \langle x_t \rangle) \cdot \langle \mathbf{k}_{c,0} \rangle + \langle x_t \rangle \cdot \langle \mathbf{k}_{c,1} \rangle.$$

Depth Needed: $D(\mathbf{v}_{c,x_*}) = \max(D(x_*), D(\mathbf{k}_{c,*})) + 1 = 3$.

(c) **Calculate garbled labels:** Party $i$ can now compute the $2 \cdot n$ PRF values $F_{k^i_{w,\beta}}(0||1||g), \ldots, F_{k^i_{w,\beta}}(0||n||g)$ and $F_{k^i_{w,\beta}}(1||1||g), \ldots, F_{k^i_{w,\beta}}(1||n||g)$, for each input wire $w$ of gate $G$, and $\beta = 0, 1$.

$$F^0_{k^i_{w,\beta}}(g) = \left( F_{k^i_{w,\beta}}(0||1||g), \ldots, F_{k^i_{w,\beta}}(0||n||g) \right)$$
$$F^1_{k^i_{w,\beta}}(g) = \left( F_{k^i_{w,\beta}}(1||1||g), \ldots, F_{k^i_{w,\beta}}(1||n||g) \right).$$

Then, they call $4 \cdot n \cdot G$ times the command **InputData+** on the functionality $\mathcal{F}_{\text{FHE}}$, so all the parties obtain the output:

$$\langle F^0_{k^i_{a,0}} + F^0_{k^i_{b,0}} \rangle, \quad \langle F^1_{k^i_{a,0}} + F^0_{k^i_{b,1}} \rangle, \quad \langle F^0_{k^i_{a,1}} + F^1_{k^i_{b,0}} \rangle, \quad \langle F^1_{k^i_{a,1}} + F^1_{k^i_{b,1}} \rangle.$$

All the parties now compute $\langle \mathbf{A}_g \rangle, \langle \mathbf{B}_g \rangle, \langle \mathbf{C}_g \rangle, \langle \mathbf{D}_g \rangle$ via

$$\langle \mathbf{A}_g \rangle = \langle \mathbf{v}_{c,x_A} \rangle + \sum_{i=1}^{n} \langle F^0_{k^i_{a,0}}(g) + F^0_{k^i_{b,0}}(g) \rangle \quad \langle \mathbf{B}_g \rangle = \langle \mathbf{v}_{c,x_B} \rangle + \sum_{i=1}^{n} \langle F^1_{k^i_{a,0}}(g) + F^0_{k^i_{b,1}}(g) \rangle$$

$$\langle \mathbf{C}_g \rangle = \langle \mathbf{v}_{c,x_C} \rangle + \sum_{i=1}^{n} \langle F^0_{k^i_{a,1}}(g) + F^1_{k^i_{b,0}}(g) \rangle \quad \langle \mathbf{D}_g \rangle = \langle \mathbf{v}_{c,x_D} \rangle + \sum_{i=1}^{n} \langle F^1_{k^i_{a,1}}(g) + F^1_{k^i_{b,1}}(g) \rangle$$

Round Cost: $R_{ID} = R_{ID} + 1$, but the $R_{ID}$ can be done in parallel before. Depth Needed: $D(\mathbf{A}_g) = D(\mathbf{B}_g) = D(\mathbf{C}_g) = D(\mathbf{D}_g) = D(\mathbf{v}_{c,x_*}) = 3$.

---

**Fig. 9.** Calculate garbled gates step of $\pi_{\text{offline},0}$

## 6    A Modified SHE-BMR Protocol of Depth 3: $\pi_1$

In this section we give a description of the protocol $\pi_1$ (see Figs. 10 and 12 for the offline and online modifications respectively) which requires only a multiplicative depth of three rather than four as in $\pi_0$. This reduction on the depth of the SHE scheme comes directly from the reduction of the depth of the circuit used for garbling the actual circuit to evaluate. On the downside, we require additional $2 \cdot W \cdot n \cdot (n-1)$ calls to **InputData** and some more multiplications in the offline and online phases. The new protocol $\pi_1$ is, in fact, just a variant of $\pi_0$, and for which set of parameters one would be preferred in practice over the other remains to be empirically tested.

## 6.1   Protocol $\pi_1$ Description

Our earlier protocol $\pi_0$ securely computes the BMR garbled gates, as follows. For every gate the parties first compute the shares $\langle x_A \rangle, \langle x_B \rangle, \langle x_C \rangle, \langle x_D \rangle$ and then use these shares to compute the shares $\langle \mathbf{v}_{c,x_A} \rangle, \langle \mathbf{v}_{c,x_B} \rangle, \langle \mathbf{v}_{c,x_C} \rangle, \langle \mathbf{v}_{c,x_D} \rangle$ of the keys $\mathbf{k}_{c,0}$ or $\mathbf{k}_{c,1}$ on the output wire of the gate. Finally, these are masked by the pseudorandom values provided by all parties; see Fig. 9. Considering how these equations are computed, we have that the $\langle x_* \rangle$ values require two multiplications and the $\langle \mathbf{v}_{c,x_*} \rangle$ require an additional multiplication. The final multiplication, making it depth-4, is needed for computing **Output+**. Thus, our aim is to compute the $\langle \mathbf{v}_{c,x_*} \rangle$ values directly, with just two multiplications instead of three.

In order to achieve this, we directly considered AND and XOR gates, and provide direct formulae for them. The main idea is that it actually suffices to compute shares of either the key $\mathbf{k}_{c,*}$ on the output wire or its opposite $-\mathbf{k}_{c,*}$ modulo $p$. The reason that this suffices is that the *square* of these values is the same. Thus, we have two versions of each key: the basic-key and the squared-key. The offline protocol works by the parties calling **RandomElement** in order to generate each *basic-key* and then squaring the result and revealing the *squared-key* to the appropriate party. Recall that in BMR, each party has one part of the key, and inputs it in the offline phase to generate the garbled gates. The parties then compute the shares of the *basic-keys* on the output wire of the gate (or their negative) and mask the result with the outputs of the PRF, computed using the revealed *squared-keys*. Observe that in the online phase, the *basic-key* is revealed (since this is what is masked) and the parties then square it in order to compute the PRF values to decrypt the next garbled gate.

Since the *basic-key* is random and was *never revealed*, the parties have no idea if they received the basic-key or it's negative. Otherwise this would leak information about the values on the wires (as we mentioned, we compute either the key or its negative, and this depends on the values on the wires). This adds $2 \cdot W \cdot n \cdot (n-1)$ calls to **InputData** to generate the keys via calls to **RandomElement** to ensure that no party knows them in the offline phase.

*The AND gate.* We now present the equations for computing an AND gate with input wires $a, b$ and output wire $c$. In order to motivate these equations, we build the first equation for computing $\langle \mathbf{v}_{c,x_A} \rangle$, which is the share of the key output from the first ciphertext in the garbled gate, in detail. We denote the basic-keys (before being squared) on the output wire by $\tilde{\mathbf{k}}_{c,0}, \tilde{\mathbf{k}}_{c,1}$. If the indicator-bit $\lambda_c$ on the output wire equals 1 then the roles of the 0-key and 1-key are reversed. Then, if the input-bit on wire $a$ equals 0, then $\langle \lambda_a \rangle$ equals 0 and so the output

---

The Offline Protocol: $\pi_{\text{offline},1}$

This protocol is identical to the $\pi_{\text{offline},0}$ protocol given in Figure 8, except for the following changes:

3 **Generate keys** in Figure 8 is changed as follows:
   (a) For every wire $w$, bit value $\beta \in \{0,1\}$ and party $i \in [1, \ldots, n]$, the parties execute the command **RandomElement** of the functionality $\mathcal{F}_{\text{FHE}+}$ to obtain output $\langle \tilde{k}^i_{w,\beta} \rangle$. We stress that nobody learns $\tilde{k}^i_{w,\beta}$. Let varid be the identifier of $\langle \tilde{k}^i_{w,\beta} \rangle$. In the following, we shall abuse the notation to denote $\langle \tilde{\mathbf{k}}_{w,\beta} \rangle = \left( \langle \tilde{k}^1_{w,\beta} \rangle, \ldots, \langle \tilde{k}^n_{w,\beta} \rangle \right)$.
   (b) The parties call $(multiply, \text{varid}, \text{varid}, \text{varid2})$ where varid2 is a new identifier, in order to share a ciphertext $\langle k^i_{w,\beta} \rangle = \langle \tilde{k}^i_{w,\beta} \rangle^2$.
   (c) The parties call $\mathcal{F}_{\text{FHE}+}$ on input $(output+, \text{varid2}, i)$ for party $P_i$ to obtain $k^i_{w,\beta}$.
   Depth Needed: $D(Output + (k^i_{w,\beta})) = 2$.
   Round Cost: $R_{\text{ID}} + 2$.
4 **Calculate garbled gates** in Figure 9 is changed as follows:
   (a) The **calculate output indicators** and **assign the correct vector** phases are replaced by the following functions, that choose, for every garbled label, either $\tilde{\mathbf{k}}_{c,0}$, $-\tilde{\mathbf{k}}_{c,0}$, $\tilde{\mathbf{k}}_{c,1}$ or $-\tilde{\mathbf{k}}_{c,1}$.
      – For an AND gate, the parties compute shares of the keys on the output wires according to Equations (1)–(4).
      – For a XOR gate, the parties compute shares of the keys on the output wires according to Equations (5)–(7).
   Depth Needed: $D(\mathbf{v}_{c,x_A}) = D(\mathbf{v}_{c,x_B}) = D(\mathbf{v}_{c,x_C}) = D(\mathbf{v}_{c,x_D}) = 2$.

Total Round Cost: $\max(R_{\text{ID}} + 3, R_{\text{ID}} + 4) = R_{\text{ID}} + 4$.
Total Depth Needed: 2.

**Fig. 10.** The modified protocol $\pi_{\text{offline},1}$

is a function of the first row of the equation. Once $a$ equals 0, the output equals 0 irrespective of $b$, since this is an AND gate. Thus, if the output indicator bit equals 0 then the output should be $\langle \tilde{\mathbf{k}}_{c,0} \rangle$; otherwise the output should be $\langle \tilde{\mathbf{k}}_{c,1} \rangle$. In contrast, if the input on wire $a$ equals 1, then the output depends only on the second row of the equation (since $1 - \langle \lambda_a \rangle$ equals 0). The output in this case depends on $b$. If $b = 1$ and $c = 0$ or if $b = 0$ and $c = 1$ then the output should be $\langle \tilde{\mathbf{k}}_{c,1} \rangle$ (since in the first case $a = b = 1$ and the output is the 1-key, and in the second case the output should be the 0-key but $c = 1$ and so the roles are reversed). This is obtained by multiplying $\langle \tilde{\mathbf{k}}_{c,1} \rangle$ by $\langle \lambda_b \rangle - \langle \lambda_c \rangle$ which equals $\pm 1$ in both of these cases (and 0 otherwise). We then multiply $\langle \tilde{\mathbf{k}}_{c,0} \rangle$ by $1 - \langle \lambda_b \rangle - \langle \lambda_c \rangle$, which equals 0 in both of these cases that $b = 0, c = 1$ and $b = 1, c = 0$. In contrast, if $b = c = 0$ or $b = c = 1$ then the output should be $\langle \tilde{\mathbf{k}}_{c,0} \rangle$ (since if $b = c = 0$ then the output is 0, and if $b = c = 1$ then the output is 1 but the 1-key is reversed). Finally leading to the equation

$$\langle \mathbf{v}_{c,x_A} \rangle = (1 - \langle \lambda_a \rangle) \cdot \left( \langle \lambda_c \rangle \cdot \langle \tilde{\mathbf{k}}_{c,1} \rangle + (1 - \langle \lambda_c \rangle) \cdot \langle \tilde{\mathbf{k}}_{c,0} \rangle \right) \tag{1}$$
$$+ \langle \lambda_a \rangle \cdot \left( ((\langle \lambda_b \rangle - \langle \lambda_c \rangle)) \cdot \langle \tilde{\mathbf{k}}_{c,1} \rangle + (1 - \langle \lambda_b \rangle - \langle \lambda_c \rangle) \cdot \langle \tilde{\mathbf{k}}_{c,0} \rangle \right).$$

The remaining three equations are computed similarly, as follows:

$$\langle \mathbf{v}_{c,x_B} \rangle = (1 - \langle \lambda_a \rangle) \cdot \left( \langle \lambda_c \rangle \cdot \langle \tilde{\mathbf{k}}_{c,1} \rangle + (1 - \langle \lambda_c \rangle) \cdot \langle \tilde{\mathbf{k}}_{c,0} \rangle \right)$$
$$+ \langle \lambda_a \rangle \cdot \left( ((\langle \lambda_b \rangle - \langle \lambda_c \rangle)) \cdot \langle \tilde{\mathbf{k}}_{c,0} \rangle + (1 - \langle \lambda_b \rangle - \langle \lambda_c \rangle) \cdot \langle \tilde{\mathbf{k}}_{c,1} \rangle \right) \tag{2}$$

$$\langle \mathbf{v}_{c,x_C} \rangle = \langle \lambda_a \rangle \cdot \left( \langle \lambda_c \rangle \cdot \langle \tilde{\mathbf{k}}_{c,1} \rangle + (1 - \langle \lambda_c \rangle) \cdot \langle \tilde{\mathbf{k}}_{c,0} \rangle \right)$$
$$+ (1 - \langle \lambda_a \rangle) \cdot \left( ((\langle \lambda_b \rangle - \langle \lambda_c \rangle)) \cdot \langle \tilde{\mathbf{k}}_{c,1} \rangle + (1 - \langle \lambda_b \rangle - \langle \lambda_c \rangle) \cdot \langle \tilde{\mathbf{k}}_{c,0} \rangle \right) \tag{3}$$

$$\langle \mathbf{v}_{c,x_D} \rangle = \langle \lambda_a \rangle \cdot \left( \langle \lambda_c \rangle \cdot \langle \tilde{\mathbf{k}}_{c,1} \rangle + (1 - \langle \lambda_c \rangle) \cdot \langle \tilde{\mathbf{k}}_{c,0} \rangle \right)$$
$$+ (1 - \langle \lambda_a \rangle) \cdot \left( ((\langle \lambda_b \rangle - \langle \lambda_c \rangle)) \cdot \langle \tilde{\mathbf{k}}_{c,0} \rangle + (1 - \langle \lambda_b \rangle - \langle \lambda_c \rangle) \cdot \langle \tilde{\mathbf{k}}_{c,1} \rangle \right) \tag{4}$$

In order to prove correctness of these equations, we present the truth table of the outputs in Fig. 11. Observe that all values are correct, but sometimes the negative value of the basic-key is obtained.

| $\lambda_a$ | $\lambda_b$ | $\lambda_c$ | $\langle \mathbf{v}_{c,x_A} \rangle$ | $\langle \mathbf{v}_{c,x_B} \rangle$ | $\langle \mathbf{v}_{c,x_C} \rangle$ | $\langle \mathbf{v}_{c,x_D} \rangle$ |
|---|---|---|---|---|---|---|
| 0 | 0 | 0 | $\langle \tilde{\mathbf{k}}_{c,0} \rangle$ | $\langle \tilde{\mathbf{k}}_{c,0} \rangle$ | $\langle \tilde{\mathbf{k}}_{c,0} \rangle$ | $\langle \tilde{\mathbf{k}}_{c,1} \rangle$ |
| 0 | 0 | 1 | $\langle \tilde{\mathbf{k}}_{c,1} \rangle$ | $\langle \tilde{\mathbf{k}}_{c,1} \rangle$ | $\langle -\tilde{\mathbf{k}}_{c,1} \rangle$ | $\langle -\tilde{\mathbf{k}}_{c,0} \rangle$ |
| 0 | 1 | 0 | $\langle \tilde{\mathbf{k}}_{c,0} \rangle$ | $\langle \tilde{\mathbf{k}}_{c,0} \rangle$ | $\langle \tilde{\mathbf{k}}_{c,1} \rangle$ | $\langle \tilde{\mathbf{k}}_{c,0} \rangle$ |
| 0 | 1 | 1 | $\langle \tilde{\mathbf{k}}_{c,1} \rangle$ | $\langle \tilde{\mathbf{k}}_{c,1} \rangle$ | $\langle -\tilde{\mathbf{k}}_{c,0} \rangle$ | $\langle -\tilde{\mathbf{k}}_{c,1} \rangle$ |
| 1 | 0 | 0 | $\langle \tilde{\mathbf{k}}_{c,0} \rangle$ | $\langle \tilde{\mathbf{k}}_{c,1} \rangle$ | $\langle \tilde{\mathbf{k}}_{c,0} \rangle$ | $\langle \tilde{\mathbf{k}}_{c,0} \rangle$ |
| 1 | 0 | 1 | $\langle -\tilde{\mathbf{k}}_{c,1} \rangle$ | $\langle -\tilde{\mathbf{k}}_{c,0} \rangle$ | $\langle \tilde{\mathbf{k}}_{c,1} \rangle$ | $\langle \tilde{\mathbf{k}}_{c,1} \rangle$ |
| 1 | 1 | 0 | $\langle \tilde{\mathbf{k}}_{c,1} \rangle$ | $\langle \tilde{\mathbf{k}}_{c,0} \rangle$ | $\langle \tilde{\mathbf{k}}_{c,0} \rangle$ | $\langle \tilde{\mathbf{k}}_{c,0} \rangle$ |
| 1 | 1 | 1 | $\langle -\tilde{\mathbf{k}}_{c,0} \rangle$ | $\langle -\tilde{\mathbf{k}}_{c,1} \rangle$ | $\langle \tilde{\mathbf{k}}_{c,1} \rangle$ | $\langle \tilde{\mathbf{k}}_{c,1} \rangle$ |

| $\lambda_a$ | $\lambda_b$ | $\lambda_c$ | $\langle \mathbf{v}_{c,x_A} \rangle$ | $\langle \mathbf{v}_{c,x_B} \rangle$ | $\langle \mathbf{v}_{c,x_C} \rangle$ | $\langle \mathbf{v}_{c,x_D} \rangle$ |
|---|---|---|---|---|---|---|
| 0 | 0 | 0 | $\langle \tilde{\mathbf{k}}_{c,0} \rangle$ | $\langle \tilde{\mathbf{k}}_{c,1} \rangle$ | $\langle \tilde{\mathbf{k}}_{c,1} \rangle$ | $\langle \tilde{\mathbf{k}}_{c,0} \rangle$ |
| 0 | 0 | 1 | $\langle \tilde{\mathbf{k}}_{c,1} \rangle$ | $\langle \tilde{\mathbf{k}}_{c,0} \rangle$ | $\langle \tilde{\mathbf{k}}_{c,0} \rangle$ | $\langle \tilde{\mathbf{k}}_{c,1} \rangle$ |
| 0 | 1 | 0 | $\langle -\tilde{\mathbf{k}}_{c,1} \rangle$ | $\langle -\tilde{\mathbf{k}}_{c,0} \rangle$ | $\langle -\tilde{\mathbf{k}}_{c,0} \rangle$ | $\langle -\tilde{\mathbf{k}}_{c,1} \rangle$ |
| 0 | 1 | 1 | $\langle -\tilde{\mathbf{k}}_{c,0} \rangle$ | $\langle -\tilde{\mathbf{k}}_{c,1} \rangle$ | $\langle -\tilde{\mathbf{k}}_{c,1} \rangle$ | $\langle -\tilde{\mathbf{k}}_{c,0} \rangle$ |
| 1 | 0 | 0 | $\langle \tilde{\mathbf{k}}_{c,1} \rangle$ | $\langle \tilde{\mathbf{k}}_{c,0} \rangle$ | $\langle \tilde{\mathbf{k}}_{c,0} \rangle$ | $\langle \tilde{\mathbf{k}}_{c,1} \rangle$ |
| 1 | 0 | 1 | $\langle \tilde{\mathbf{k}}_{c,0} \rangle$ | $\langle \tilde{\mathbf{k}}_{c,1} \rangle$ | $\langle \tilde{\mathbf{k}}_{c,1} \rangle$ | $\langle \tilde{\mathbf{k}}_{c,0} \rangle$ |
| 1 | 1 | 0 | $\langle -\tilde{\mathbf{k}}_{c,0} \rangle$ | $\langle -\tilde{\mathbf{k}}_{c,1} \rangle$ | $\langle -\tilde{\mathbf{k}}_{c,1} \rangle$ | $\langle -\tilde{\mathbf{k}}_{c,0} \rangle$ |
| 1 | 1 | 1 | $\langle -\tilde{\mathbf{k}}_{c,1} \rangle$ | $\langle -\tilde{\mathbf{k}}_{c,0} \rangle$ | $\langle -\tilde{\mathbf{k}}_{c,0} \rangle$ | $\langle -\tilde{\mathbf{k}}_{c,1} \rangle$ |

**Fig. 11.** The truth table of the vectors for an AND gate (on the left) and for a XOR gate (on the right) computed in Fig. 10.

*The XOR gate.* We use a similar idea as above to compute the XOR gate. Intuitively, in a XOR gate, there are two cases: $\lambda_a = \lambda_b$ and $\lambda_a \neq \lambda_b$. Multiplying by $\lambda_a - \lambda_b$ gives $\pm 1$ if $\lambda_a \neq \lambda_b$ and 0 if $\lambda_a = \lambda_b$. Furthermore, multiplying by $1 - \lambda_a - \lambda_b$ gives the exact reverse case; it equals 0 if $\lambda_a \neq \lambda_b$ and equals $\pm 1$ if $\lambda_a = \lambda_b$. Observe that $\langle \mathbf{v}_{c,x_C} \rangle$ and $\langle \mathbf{v}_{c,x_D} \rangle$ need not be computed at all since $(1 - a) \oplus b = a \oplus (1 - b)$ and $(1 - a) \oplus (1 - b) = a \oplus b$. This yields the following equations, where as above, we prove correctness via the truth table given in Fig. 11.

$$\langle \mathbf{v}_{c,x_A} \rangle = \langle \mathbf{v}_{c,x_D} \rangle = (\langle \lambda_a \rangle - \langle \lambda_b \rangle) \cdot \left( \langle \lambda_c \rangle \cdot \langle \tilde{\mathbf{k}}_{c,0} \rangle + (1 - \langle \lambda_c \rangle) \cdot \langle \tilde{\mathbf{k}}_{c,1} \rangle \right)$$
$$+ (1 - \langle \lambda_a \rangle - \langle \lambda_b \rangle) \cdot \left( \langle \lambda_c \rangle \cdot \langle \tilde{\mathbf{k}}_{c,1} \rangle + (1 - \langle \lambda_c \rangle) \cdot \langle \tilde{\mathbf{k}}_{c,0} \rangle \right) \quad (5)$$

$$\langle \mathbf{v}_{c,x_B} \rangle = \langle \mathbf{v}_{c,x_C} \rangle = (\langle \lambda_a \rangle - \langle \lambda_b \rangle) \cdot \left( \langle \lambda_c \rangle \cdot \langle \tilde{\mathbf{k}}_{c,1} \rangle + (1 - \langle \lambda_c \rangle) \cdot \langle \tilde{\mathbf{k}}_{c,0} \rangle \right)$$
$$+ (1 - \langle \lambda_a \rangle - \langle \lambda_b \rangle) \cdot \left( \langle \lambda_c \rangle \cdot \langle \tilde{\mathbf{k}}_{c,0} \rangle + (1 - \langle \lambda_c \rangle) \cdot \langle \tilde{\mathbf{k}}_{c,1} \rangle \right) \quad (6)$$

**Security of the Modified Protocol:** Observe that in the offline phase, the only difference is that the $\langle \mathbf{v}_{c,x_*} \rangle$ values contain the "tilde" version of the keys; more formally, the $\langle \mathbf{v}_{c,x_*} \rangle$ ciphertexts encrypt the *square root* of the keys, and not the keys themselves. Thus, in the online phase, the parties receive the square roots of the keys and need to square them before proceeding. The only issue that needs to be explained here is that the specific square root provided reveals no information. This needs to be justified because if an adversary could know that $-\tilde{k}$ is computed or $\tilde{k}$, then it would know some information about the masks $\lambda_a, \lambda_b, \lambda_c$. However, since the $\tilde{k}$ values are *uniformly distributed* in $\mathbb{F}_p$, and the keys themselves revealed in the offline phase are $k = \tilde{k}^2$, it follows that each of the two square roots of $k$ are equally probable. Stated differently, given $k$, the distribution over $\tilde{k}$ and $-\tilde{k}$ is identical.

---

**The modified MPC Protocol - $\pi_{\text{MPC},1}$**

This protocol is identical to the $\pi_{\text{MPC},1}$ protocol described in Figure 7, except for the four cases of the **Online Computation** sub-task, in which for $j = 1, \ldots, n$, the values $k_c^j$ are now computed as follows:

Case $(\Lambda_a, \Lambda_b) = (0,0)$: Compute $k_c^j = \left( A_g^j - (\sum_{i=1}^n F_{k_a^i}(0||j||g) + F_{k_b^i}(0||j||g)) \right)^2$.

Case $(\Lambda_a, \Lambda_b) = (0,1)$: Compute $k_c^j = \left( B_g^j - (\sum_{i=1}^n F_{k_a^i}(1||j||g) + F_{k_b^i}(0||j||g)) \right)^2$.

Case $(\Lambda_a, \Lambda_b) = (1,0)$: Compute $k_c^j = \left( C_g^j - (\sum_{i=1}^n F_{k_a^i}(0||j||g) + F_{k_b^i}(1||j||g)) \right)^2$.

Case $(\Lambda_a, \Lambda_b) = (1,1)$: Compute $k_c^j = \left( D_g^j - (\sum_{i=1}^n F_{k_a^i}(1||j||g) + F_{k_b^i}(1||j||g)) \right)^2$.

Depth Needed: $D_{Out+} + D(\{\mathbf{A}_g\}, \{\mathbf{B}_g\}, \{\mathbf{C}_g\}, \{\mathbf{D}_g\}) = 2 + 1 = 3$.

**Fig. 12.** The Modified Protocol $\pi_{\text{MPC},1}$

---

**Analysis of Efficiency of the Modified Protocol:** As we noted in the introduction, the two main sources of overhead that concern our MPC protocol are the number of rounds and the number of calls to the ID protocol. The former is not changed by our $\pi_1$ variant, but the latter does. To generate the keys in $\pi_{\text{offline},1}$, we now perform $2 \cdot W \cdot n^2$ calls to **InputData**, via calls to **RandomElement**. In $\pi_0$ we performed $2 \cdot W \cdot n$ calls to generate the keys, so overall we add $2 \cdot W \cdot n \cdot (n-1)$ calls to **InputData**. To analyse the number of homomorphic multiplications we go through each step of the protocol:

- **Generate keys step:** We perform $4 \cdot W \cdot n$ more multiplications (half of them to square the keys, the other half to **Output+** them).
- **Calculate garbled gates step:**
  1. For every AND gate, we used 13 multiplications in the first variant. Now, by careful rewriting of the equations, we can do this in 20.
  2. For every XOR gate, we used 7 multiplications in the first variant. Now we use 12.
  3. So, on average, we pass from 10 to 16 multiplications per gate.

Thus, overall on average we perform $4 \cdot W \cdot n + 6 \cdot G$ more homomorphic multiplications. However, in practice each homomorphic multiplication will be more efficient since the overall depth of the SHE scheme can now be three rather than four.

**Acknowledgements.** The first author was supported in part by the European Research Council under the European Union's Seventh Framework Programme (FP/2007-2013)/ERC consolidators grant agreement n. 615172 (HIPS), and by the BIU Center for Research in Applied Cryptography and Cyber Security in conjunction with the Israel National Cyber Bureau in the Prime Minster's Office. The second author was supported in part by ERC Advanced Grant ERC-2010-AdG-267188-CRIPTO, DARPA and the US Navy under contract #N66001-15-C-4070, and by EPSRC via grants EP/I03126X and EP/N021940/1. The third author was supported in part by the Marie Sklodowska-Curie ITN ECRYPT-NET (Project Reference 643161). All authors were also supported by an award from EPSRC (grant EP/M012824), from the Ministry of Science, Technology and Space, Israel, and the UK Research Initiative in Cyber Security.

# References

1. Asharov, G., Lindell, Y., Schneider, T., Zohner, M.: More efficient oblivious transfer and extensions for faster secure computation. In: Sadeghi, A.-R., Gligor, V.D., Yung, M. (eds.) 2013 ACM SIGSAC Conference on Computer and Communications Security, CCS 2013, 4–8 November 2013, pp. 535–548. ACM (2013)
2. Baum, C., Damgård, I., Toft, T., Zakarias, R.: Better preprocessing for secure multiparty computation. In: Manulis, M., Sadeghi, A.-R., Schneider, S. (eds.) ACNS 2016. LNCS, vol. 9696, pp. 327–345. Springer, Heidelberg (2016). doi:10.1007/978-3-319-39555-5_18
3. Beaver, D., Micali, S., Rogaway, P.: The round complexity of secure protocols. In: Ortiz, H. (ed.), 22nd STOC, pp. 503–513. ACM (1990)
4. Bellare, M., Hoang, V.T., Keelveedhi, S., Rogaway, P.: Efficient garbling from a fixed-key blockcipher. In: 2013 IEEE Symposium on Security and Privacy, SP 2013, 19–22 May 2013, Berkeley, CA, USA, pp. 478–492. IEEE Computer Society (2013)
5. Canetti, R., Cohen, A., Lindell. Y.: A simpler variant of universally composable security for standard multiparty computation. In: Gennaro and Robshaw [15], pp. 3–22 (2015)
6. Canetti, R., Garay, J.A. (eds.): CRYPTO 2013. LNCS, vol. 8043. Springer, Heidelberg (2013). doi:10.1007/978-3-642-40084-1_2

7. Choi, S.G., Hwang, K.-W., Katz, J., Malkin, T., Rubenstein, D.: Secure multi-party computation of Boolean circuits with applications to privacy in on-line marketplaces. In: Dunkelman, O. (ed.) CT-RSA 2012. LNCS, vol. 7178, pp. 416–432. Springer, Heidelberg (2012). doi:10.1007/978-3-642-27954-6_26
8. Choi, S.G., Katz, J., Malozemoff, A.J., Zikas, V.: Efficient three-party computation from cut-and-choose. In: Garay and Gennaro [13], pp. 513–530 (2014)
9. Costache, A., Smart, N.P.: Which ring based somewhat homomorphic encryption scheme is best? In: Sako, K. (ed.) CT-RSA 2016. LNCS, vol. 9610, pp. 325–340. Springer, Heidelberg (2016). doi:10.1007/978-3-319-29485-8_19
10. Damgård, I., Keller, M., Larraia, E., Pastro, V., Scholl, P., Smart, N.P.: Practical covertly secure MPC for dishonest majority – or: breaking the SPDZ limits. In: Crampton, J., Jajodia, S., Mayes, K. (eds.) ESORICS 2013. LNCS, vol. 8134, pp. 1–18. Springer, Heidelberg (2013). doi:10.1007/978-3-642-40203-6_1
11. Damgård, I., Pastro, V., Smart, N.P., Zakarias, S.: Multiparty computation from somewhat homomorphic encryption. In: Safavi-Naini and Canetti [32], pp. 643–662 (2012)
12. Damgård, I., Polychroniadou, A., Rao, V.: Adaptively secure multi-party computation from LWE (via equivocal FHE). In: Cheng, C.-M., et al. (eds.) PKC 2016. LNCS, vol. 9615, pp. 208–233. Springer, Heidelberg (2016). doi:10.1007/978-3-662-49387-8_9
13. Garay, J.A., Gennaro, R. (eds.): CRYPTO 2014. LNCS, vol. 8617. Springer, Heidelberg (2014). doi:10.1007/978-3-662-44381-1_29
14. Garg, S., Mukherjee, P., Pandey, O., Polychroniadou, A.: The exact round complexity of secure computation. In: Fischlin, M., Coron, J.-S. (eds.) EUROCRYPT 2016. LNCS, vol. 9666, pp. 448–476. Springer, Heidelberg (2016). doi:10.1007/978-3-662-49896-5_16
15. Gennaro, R., Robshaw, M. (eds.): CRYPTO 2015. LNCS, vol. 9216. Springer, Heidelberg (2015). doi:10.1007/978-3-662-48000-7_1
16. Gentry, C.: A fully homomorphic encryption scheme. Ph.D. thesis, Stanford University (2009). http://crypto.stanford.edu/craig
17. Goldreich, O., Micali, S., Wigderson, A.: How to play any mental game or A completeness theorem for protocols with honest majority. In: Aho, A.V. (ed.) Proceedings of the 19th Annual ACM Symposium on Theory of Computing, pp. 218–229. ACM, New York (1987)
18. Goldwasser, S., Lindell, Y.: Secure multi-party computation without agreement. J. Cryptology 18(3), 247–287 (2005)
19. Huang, Y., Katz, J., Evans, D.: Efficient secure two-party computation using symmetric cut-and-choose. In: Canetti and Garay [6], pp. 18–35 (2013)
20. Ishai, Y., Prabhakaran, M., Sahai, A.: Founding cryptography on oblivious transfer – efficiently. In: Wagner, D. (ed.) CRYPTO 2008. LNCS, vol. 5157, pp. 572–591. Springer, Heidelberg (2008). doi:10.1007/978-3-540-85174-5_32
21. Katz, J., Ostrovsky, R.: Round-optimal secure two-party computation. In: Franklin, M. (ed.) CRYPTO 2004. LNCS, vol. 3152, pp. 335–354. Springer, Heidelberg (2004). doi:10.1007/978-3-540-28628-8_21
22. Larraia, E., Orsini, E., Smart. N.P.: Dishonest majority multi-party computation for binary circuits. In: Garay and Gennaro [13], pp. 495–512 (2014)
23. Lindell, Y.: Fast cut-and-choose based protocols for malicious and covert adversaries. In: Canetti and Garay [6], pp. 1–17 (2013)
24. Lindell, Y., Oxman, E., Pinkas, B.: The IPS compiler: optimizations, variants and concrete efficiency. In: Rogaway, P. (ed.) CRYPTO 2011. LNCS, vol. 6841, pp. 259–276. Springer, Heidelberg (2011). doi:10.1007/978-3-642-22792-9_15

25. Lindell, Y., Pinkas, B.: An efficient protocol for secure two-party computation in the presence of malicious adversaries. In: Naor, M. (ed.) EUROCRYPT 2007. LNCS, vol. 4515, pp. 52–78. Springer, Heidelberg (2007). doi:10.1007/978-3-540-72540-4_4

26. Lindell, Y., Pinkas, B.: Secure two-party computation via cut-and-choose oblivious transfer. In: Ishai, Y. (ed.) TCC 2011. LNCS, vol. 6597, pp. 329–346. Springer, Heidelberg (2011). doi:10.1007/978-3-642-19571-6_20

27. Lindell, Y., Pinkas, B., Smart, N.P., Yanai, A.: Efficient constant round multi-party computation combining BMR and SPDZ. In: Gennaro and Robshaw [15], pp. 319–338 (2015)

28. Lindell, Y., Riva, B.: Cut-and-choose Yao-based secure computation in the online/offline and batch settings. In: Garay and Gennaro [13], pp. 476–494 (2014)

29. Lindell, Y., Riva, B.: Blazing fast 2PC in the offline/online setting with security for malicious adversaries. In: Ray, I., Li, N., Kruegel, C. (eds.) Proceedings of the 22nd ACM SIGSAC Conference on Computer and Communications Security, 12–6 October 2015, Denver, CO, USA, pp. 579–590. ACM (2015)

30. Nielsen, J.B., Nordholt, P.S., Orlandi, C., Burra, S.S.: A new approach to practical active-secure two-party computation. In: Safavi-Naini and Canetti [32], pp. 681–700 (2012)

31. Pinkas, B., Schneider, T., Smart, N.P., Williams, S.C.: Secure two-party computation is practical. In: Matsui, M. (ed.) ASIACRYPT 2009. LNCS, vol. 5912, pp. 250–267. Springer, Heidelberg (2009)

32. Safavi-Naini, R., Canetti, R. (eds.): CRYPTO 2012. LNCS, vol. 7417, pp. 643–662. Springer, Heidelberg (2012). doi:10.1007/978-3-642-32009-5_38

33. Schneider, T., Zohner, M.: GMW vs. Yao? efficient secure two-party computation with low depth circuits. In: Sadeghi, A.-R. (ed.) FC 2013. LNCS, vol. 7859, pp. 275–292. Springer, Heidelberg (2013). doi:10.1007/978-3-642-39884-1_23

34. shelat, A., Shen, C.: Two-output secure computation with malicious adversaries. In: Paterson, K.G. (ed.) EUROCRYPT 2011. LNCS, vol. 6632, pp. 386–405. Springer, Heidelberg (2011). doi:10.1007/978-3-642-20465-4_22

35. Yao, A.C.-C.: Protocols for secure computations. In: 23rd Annual Symposium on Foundations of Computer Science, 3–5 November 1982, Chicago, Illinois, USA, pp. 160–164. IEEE Computer Society (1982)

# Cross and Clean: Amortized Garbled Circuits with Constant Overhead

Jesper Buus Nielsen[(⊠)] and Claudio Orlandi

Aarhus University, Aarhus, Denmark
{jbn,orlandi}@cs.au.dk

**Abstract.** Garbled circuits (GC) are one of the main tools for secure two-party computation. One of the most promising techniques for efficiently achieving active-security in the context of GCs is the so called *cut-and-choose* approach, and the main measure of efficiency in cut-and-choose based protocols is the number of garbled circuits which need to be constructed, exchanged and evaluated.

In this paper we investigate the following, natural question: *how many garbled circuits are needed to achieve active security?* and we show that in the amortized setting (for large enough circuits and number of executions), it is possible to achieve active security while using only a constant number of garbled circuits.

## 1 Introduction

Garbled circuits are one of the most widely used and promising tools for secure two-party computation. Garbled circuits were introduced by Yao [Yao82] and they were first implemented in Fairplay by Malkhi et al. [MNPS04].

The basic version of Yao's protocol only guarantees security in the presence of *passive* corruptions (i.e., when the adversary follows the protocol but might try to learn more information from their view). From a very high level point of view, since garbling schemes hide (to some extent) the circuit which is being garbled, a malicious party can garble a different function from the one they are supposed to without the honest party noticing it, therefore breaking the security of the protocol. During the years many approaches have been proposed to construct GC-based protocols with strong security guarantees against adversaries who deviate arbitrarily from the protocol (i.e., *malicious* or *active* corruptions). The main technique for achieving active security in the GC context is the so called *cut-and-choose* approach: in a nutshell, cut-and-choose involves several copies of the same circuit being garbled; afterwards, a random subset of the garbled circuits are checked for correctness, while the rest are evaluated.

There are many different instantiations of the cut-and-choose approach: in 2007, Lindell and Pinkas [LP07] proposed a method which achieves security $2^{-\kappa}$ by garbling approximately $3\kappa$ copies of the circuit. This was improved in 2013 in several works [Lin13, HKE13, Bra13] using the so called *forge-and-lose* technique. In its most efficient instantiation [Lin13] this technique allows to

M. Hirt and A. Smith (Eds.): TCC 2016-B, Part I, LNCS 9985, pp. 582–603, 2016.
DOI: 10.1007/978-3-662-53641-4_22

achieve security $2^{-\kappa}$ using only $\kappa$ garbled circuits (by adding a "small" actively secure computation).

A different approach was taken by Nielsen and Orlandi in 2009 [NO09]. Using "LEGO style" cut-and-choose the overhead decreases logarithmically with the size of the circuit $f$ i.e., it is possible to get security $2^{-\kappa}$ using a replication factor of $O(\kappa/\log|f|)$. While the original LEGO approach required to perform exponentiations for each gate in the circuit, subsequent work [FJN+13,FJNT15] got rid of this limitation and only uses generic assumptions. A variant of the LEGO approach has proven itself particularly useful in the *amortized* setting i.e., when the two parties are evaluating the same circuit $f$ multiple times (say $\ell$) on different inputs. In this case the amortized overhead to get security $2^{-\kappa}$ is $O(\kappa/\log\ell)$ [LR14,HKK+14], and experimental validation achieves "blazing fast" results [LR15].

To summarize, while advanced styles of cut-and-choose techniques have shown that one can achieve practically-efficient actively-secure two-party computation in the amortized setting, in all of the above approaches the number of garbled circuits *grows linearly* with the security parameter. It is natural to ask whether this is an inherent limitation or whether it is possible to achieve actively-secure two-party computation based on garbled circuits *with constant overhead.*

## 1.1   Our Contribution

Before stating our contributions, it is important to clarify the question we are asking as much as possible: it is of course possible to achieve active security using a single garbled circuit and using the GMW compiler [GMW87] (i.e., proving in *zero-knowledge* that the circuit is well-formed). This is not a satisfactory solution since it is not *black-box* in the underlying garbling scheme and, therefore, does not preserve the efficiency of Yao's protocol. Jarecki and Shmatikov [JS07] proposed an instantiation of this paradigm using a specific number theoretic assumption (Pailler's cryptosystem [Pai99]): thanks to the algebraic nature of the underlying cryptosystem, the extra zero-knowledge proofs only add a constant overhead. We do not consider this solution satisfying either, since one of the strengths of Yao's protocol (both in terms of security and efficiency) is that it only requires to perform symmetric key operations per gate in the circuit. Therefore, we are only interested in solutions that can be instantiated using *any projective garbling scheme in a black-box way.* We are now ready to ask our question:

*Can we achieve actively-secure two-party computation protocols in the amortized setting with only a constant overhead over Yao's protocol?*

We answer the question positively: let $p(\kappa)$ be an upper bound on the cost of generating, evaluating or checking a garbled gate[1], and let $A(\kappa)$ be a fixed

---

[1] In all known instantiations of garbled circuits generating a gate is the most expensive operation, requiring at most 4 calls to a PRF. Evaluating typically requires fewer calls while checking is equivalent to garbling.

function (this term describes the cost of performing some "small" fixed actively secure computation and is independent of the circuit size and the number of executions). Then the amortized complexity of our protocol (for large enough circuits and number of executions) is bounded by:

$$O(1) \cdot |f| \cdot p(\kappa) + A(\kappa) \tag{1}$$

i.e., only a constant overhead over Yao's passive protocol and an additive factor independent of $|f|$.

## 1.2   Technical Overview

We give here a high level description of our techniques. We have two parties, Alice and Bob, respectively with inputs $\{x_i^A, x_i^B\}_{i \in [\ell]}$. At the end both parties should learn $y_i = f(x_i^A, x_i^B)$ for all $i \in [\ell]$. (We will assume that both $|f|$ and $\ell$ are large, and that $\ell \geq |f|$).

Our protocol proceeds in five stages: in the first stage we let the parties commit to a key and exchange their inputs in an encrypted format (using a symmetric encryption scheme). In this way the inputs to all computations are well-defined already from this stage. We then let both parties garble $(1 + \epsilon)\ell$ copies of the circuit (for some constant $\epsilon \leq 1$) and, using *cut-and-choose*, verify that $\epsilon\ell$ of them are correct: this guarantees that even if one of the two parties has been actively corrupted, there are at most $O(\kappa)$ incorrect circuits among the unopened ones except with (negligible) probability $2^{-\kappa}$. We then proceed to evaluate these circuits in both directions (as in the *dual-execution* protocol of Franklin and Mohassel [MF06]). Remember that Yao's protocol is "almost" actively secure against corrupt evaluators, which means that at this stage the corrupt party learns the correct output of the function (together with unforgeable output labels) in all positions, while the honest party learns outputs (with corresponding output labels) in at least $\ell - O(\kappa)$ positions.[2] For the remaining $O(\kappa)$ positions the honest party might receive an incorrect output or no output at all. Therefore in the next stage, which we call the *filling-in stage*, we allow each party to ask for at most $O(\kappa)$ re-computations using an actively secure protocol (without disclosing in which positions), and we enforce that the computation is performed on the same inputs (remember that the inputs are provided in encrypted format). This computation also outputs MACs on the new results. Note that the malicious party cannot gain anything in this stage, since the corrupt party has already learned the correct output, and learning it once more does not leak any extra information. After this stage we are ensured that both parties have $\ell$ candidate outputs together with unforgeable certificates of their authenticity (either the output labels from the garbled circuits or the MACs from the do-overs). But still it might be that some of the outputs received by the honest party are incorrect and, therefore, different from the one received

---

[2] As we cannot guarantee fairness nor termination in the two-party setting, the adversary can of course abort the protocol at this stage and prevent the honest party from learning any output at all.

by the corrupt party. In the final stage of the protocol we run a kind of *forge-and-lose* sub-protocol, where both parties input all the received outputs to an actively secure computation. The computation finds the first position in which the outputs differ and recomputes the function for that index. Since the parties cannot lie about their outputs at this point (due to the unforgeable certificates), there is at most one party with the incorrect output, and that party is the honest one. Therefore, the other party must have cheated by garbling an incorrect circuit. To "punish" this party, we release the secret key of the malicious party to the honest party. (Crucially, the malicious party does not learn whether they have been caught or not, since this would open the door for *selective-failure attacks*). This allows the honest party to decrypt all the encrypted inputs and compute all the correct results "in the clear".

To recap, here are the 5 stages of the protocol:

1. *(encrypted input)* Both parties exchange inputs. This is done by using a committed OT using some symmetric keys $(\sigma_A, \sigma_B) \in \{0,1\}^{O(\kappa)}$ as choice bits and then exchanging encrypted inputs

$$X_i^A = E(\sigma_A, x_i^A) \text{ and } X_i^B = E(\sigma_B, x_i^B)$$

2. *(cut-and-choose)* both parties garble $(1 + \epsilon)\ell$ circuits which compute

$$(\sigma_A, X_A, \sigma_B, X_B) \mapsto f(D(\sigma_A, X_A), D(\sigma_B, X_B))$$

and do cut-and-choose by checking $\epsilon\ell$ circuits each.

3. *(dual execution)* The parties evaluate each of the $\ell$ remaining circuits. Thanks to cut-and-choose, at most $O(\kappa)$ circuits are incorrect, which in particular means there are at most $O(\kappa)$ positions in which the honest party did not receive an output.

4. *(filling-in)* The parties run an actively secure protocol which recomputes the function in at most $O(\kappa)$ positions. Using Merkle-trees based commitments we can make sure that a) the functions are recomputed on the same inputs as before and that b) the input to this protocol (and its complexity) does not grow linearly with $\ell$. This protocol also outputs MACs for the recomputed values;

5. *(forge-and-lose)* At this point both parties have $\ell$ outputs, but if one party is dishonest some of the outputs might still be different. So now the parties run an actively secure protocol which (a) finds the first position where the outputs are different and (b) recomputes the function in that position, finds out which party cheated, and reveals the secret key of the corrupt party to the honest party, which can therefore decrypt all inputs and recompute the function in the clear. Since MACs are used, a corrupt party cannot input a wrong output (which would make the honest party to look corrupt).

Stage 1 can be seen as a kind of *committed oblivious transfer* combined with an *oblivious transfer extension*, where we start with a "small" committed OT functionality for $O(\kappa)$ pair of messages which are then used to provide very long inputs to a garbled circuit based computation: if $a(\kappa)$ is the cost of evaluating a gate with an actively secure protocol and $q_1(\kappa)$ is some function describing the complexity of the circuit computing $O(\kappa)$ committed OTs on messages of length $O(\kappa)$, then the complexity of this stage is $q_1(\kappa)a(\kappa)$. If we set e.g., $\epsilon := 1/4$ then the total complexity of Stage 2 and 3 is bounded by $5\ell(|f| + q_2(\kappa))p(\kappa)$, where $q_2$ represents the complexity of the decryption circuit $D$. Then, the complexity of Stage 4 is bounded by $\psi\kappa(|f| + q_3(\kappa, \log \ell))a(\kappa)$ where $q_3$ is the complexity of verifying the Merkle-tree commitment and computing a MAC, and $\psi := \log_{1+\epsilon}(2)$ is a constant picked to guarantee that the probability that there are more than $\psi\kappa$ bad circuits among the unchecked ones is less than $2^{-\kappa}$. Finally the complexity of Stage 5 is $(\ell q_4(\kappa) + |f|)a(\kappa)$ where the $q_4$ factor represents the complexity of verifying the certificates. Since $a(\kappa) > p(\kappa)$ the total cost of the protocol is bounded by:

$$5\ell|f|p(\kappa) + (|f| + \ell)A(\kappa)$$

where $A(\kappa)$ collects all the terms which are independent of the circuit size or the number of computations. Now, when amortizing over $\ell$ executions, and assuming that $\ell$ is at least as large as $|f|$, we achieve the desired amortized complexity stated earlier in (1).

We can actually quantify the constant overhead over Yao's protocol even more precisely, by looking at the actual cost (in PRF calls) for garbling (or checking) vs. evaluating a gate in some of the most common garbling schemes (i.e., instead of upper bounding it with $p$). Let $g$ be the number of calls to a PRF (encryptions) performed during garbling/checking a gate and $e$ be the number of calls to a PRF (decryptions) performed during the evaluation of a gate. Then $(g + e)$ is the exact computational cost (per gate) in the passive version of Yao's protocol. In our protocol the exact cost is $(2 + 4\epsilon)g + 2e$. In Yao's original garbling $(g, e) = (4, 4)$, in *point-and-permute* [BMR90] $(g, e) = (4, 1)$ and finally in the *half-gate* construction [ZRE15] $(g, e) = (4, 2)$ which means that when using $\epsilon = 1/4$ then the concrete overhead over Yao's passive protocol is between 2.5 and 2.8.[3]

We conclude by stressing that this work is of theoretical nature and therefore we have made no attempts in optimizing the concrete efficiency of any of the steps. On the contrary, since our protocol is already quite complex and involves several stages we have chosen at each turn simplicity of presentation over (concrete) efficiency. We leave it as an interesting open direction for future work to investigate whether the approach proposed in this paper might lead to practical efficiency.

## 2     Preliminaries and Notation

We review here the standard tools which are used in our protocol and their syntax.

---

[3] Decreasing $\epsilon$ reduces this overhead but increases the number of potentially unchecked bad circuits, therefore increasing the number of necessary do-overs in Stage 3.

*Commitments.* We use a computationally binding and computationally hiding commitment scheme Com with commitment key $ck \leftarrow \mathsf{CGen}(1^\kappa)$, and we use an informative but slightly abusive notation: we write $\langle x \rangle \leftarrow \mathsf{Com}_{ck}(x, \mathsf{open}(x))$ where $\langle x \rangle$ is a commitment to the value $x$ using randomness $\mathsf{open}(x)$. In the proof we need the commitment to be extractable i.e., we need the simulator to be able to compute $x \leftarrow \mathsf{Ext}(\mathsf{td}, \langle x \rangle)$ using some trapdoor $\mathsf{td}$ associated to the commitment key $ck$.

*Merkle-tree Commitments.* We use Merkle-tree based commitments with the following interface: Given a string of elements from some alphabet $x \in \Sigma^n$ it is possible to compute a short commitment by running $\mathsf{root} \leftarrow \mathsf{MT.C}(1^\kappa, x)$. It is possible to construct a proof for a give position $j \in [\ell]$ by computing $\pi \leftarrow \mathsf{MT.P}(x, j)$ and the proof can be verified running $b \leftarrow \mathsf{MT.V}(\mathsf{root}, j, x', \pi)$ with $b \in \{\top, \bot\}$. We want that $b = \top$ when the prover is honest and $x' = x_j$ (i.e., *correctness*), that the proof is short i.e., $|\pi| = O(\kappa \log \ell)$ (*compactness*) and that no PPT adversary can produce a tuple $(\mathsf{root}, j, x, x', \pi, \pi')$ such that $x \neq x'$ and $\mathsf{MT.V}(\mathsf{root}, j, x, \pi) = \mathsf{MT.V}(\mathsf{root}, j, x', \pi') = \top$ (*computational binding*). (We do not need these commitments to be hiding, since they are only used to reduce the input size of the ideal functionality in the *filling-in* stage of the protocol).

*Symmetric Encryption.* We use an IND-CPA symmetric encryption scheme $(\mathsf{SE.E}, \mathsf{SE.D})$ with key $\sigma \in \{0,1\}^{8\kappa}$. We use lower-case letters for plaintexts and upper-case letters for ciphertexts, so $X \leftarrow \mathsf{SE.E}(\sigma, x)$ and $x \leftarrow \mathsf{SE.D}(\sigma, X)$. We need the encryption scheme to be secure even if $\kappa$ bits of the secret key leak to the adversary (to counteract standard selective-failure attacks during the OT phase). This is done in the following way: we start by generating a uniformly random $\kappa$ bit key $\sigma'$, which is then encoded into a $8\kappa$ bit long key $\sigma$ using the (randomized) encoding scheme $(\mathsf{enc}, \mathsf{dec})$ of Lindell and Pinkas [LP07] i.e., we compute $\sigma \leftarrow \mathsf{enc}(\sigma', r)$ with some randomness $r$. Now given any encryption scheme $E', D'$ which is IND-CPA secure using a $\kappa$-bit long key, we define $\mathsf{SE.E}, \mathsf{SE.D}$ to be $\mathsf{SE.E}(\sigma, m) = E'(\mathsf{dec}(\sigma), m)$ and $\mathsf{SE.D}(\sigma, c) = D'(\mathsf{dec}(\sigma), m)$.

*MAC Scheme.* We use an unforgeable message authentication code (MAC) $(\mathsf{MAC.Tag}, \mathsf{MAC.Ver})$ with key $\tau \in \{0,1\}^\kappa$ and the following interface: one can compute a tag on a message $x$ by computing $t \leftarrow \mathsf{MAC.Tag}(\tau, x)$ and the tag can be verified running $\mathsf{MAC.Ver}(t, \tau, x) \in \{\bot, \top\}$.

*Oblivious Transfer.* We use the following notation for transforming a random OT on short, random messages (of length $\kappa$) into an OT on chosen messages of any lengths (using the same choice bits): we start with the sender knowing $\mathsf{sen} = \{r_0, r_1\}$ (a pair of random strings in $\{0,1\}^\kappa$) and the receiver knowing $\mathsf{rec} = \{\sigma, r_\sigma\}$ (with $\sigma \in \{0,1\}$). Then we write

$$\mathsf{tra}_j \leftarrow \mathsf{OTTransfer}(\mathsf{sen}, j, \{m_0, m_1\})$$

for the process of encrypting the pair of messages $\{m_0, m_1\}$ using keys $r_0, r_1$ respectively (using an IND-CPA symmetric encryption scheme) and

$$m_\sigma \leftarrow \mathsf{OTRetrieve}(\mathsf{rec}, j, \mathsf{tra}_j)$$

for the process of recovering $m_{\sigma,j}$ from $\mathsf{tra}_j$. To ease the notation, we also allow a "vector" version of these OT commands i.e., if $e = \{m_{i,0}, m_{i,1}\}_{i\in[n]}$ is a vector of $n$ pairs of messages and $\sigma \in \{0,1\}^n$ is a vector of $n$ bits then we write $\mathsf{rec} = \{\sigma_i, r_{i,\sigma_i}\}_{i\in[n]}$, $\mathsf{sen} = \{r_{i,0}, r_{i,1}\}_{i\in[n]}$ for the information known to the receiver and sender respectively, $\mathsf{tra}_j \leftarrow \mathsf{OTTransfer}(\mathsf{sen}, j, e)$ for the process of encrypting each pair of messages and finally $M \leftarrow \mathsf{OTRetrieve}(\mathsf{rec}, j, \mathsf{tra}_j)$ with $M = \{m_{i,\sigma_i}\}_{i\in[n]}$. (In the proof of security we also use $e \leftarrow \mathsf{OTRetrieve}(\mathsf{sen}, j, \mathsf{tra}_j)$ to denote the process of recovering all pairs of messages using the keys known to the sender).

*Garbled Circuits.* We use the generalization of the notation introduced by Bellare et al. [BHR12] already used in [JKO13,FNO15]: a garbling scheme is a tuple of algorithms

$$(\mathsf{GC.Gb}, \mathsf{GC.Ev}, \mathsf{GC.En}, \mathsf{GC.De}, \mathsf{GC.Ve})$$

where:

- $(\hat{f}, e, d) \leftarrow \mathsf{GC.Gb}(f; r)$ generates a garbled version $\hat{f}$ of the circuit $f : \{0,1\}^n \rightarrow \{0,1\}^n$ which has $n$ input bits and $n$ output bits. We make explicit the randomness $r$ used to garble since it will be used in the verification process. The $\mathsf{GC.Gb}$ function outputs the garbled version of the function $\hat{f}$, the encoding tables $e$ and the decoding tables for the output wires $d$;
- $\hat{x} \leftarrow \mathsf{GC.En}(e, x)$ outputs an encoding of $x$.
- $\hat{z} \leftarrow \mathsf{GC.Ev}(\hat{f}, \hat{x})$ outputs an encoded version of the output;
- $z' \leftarrow \mathsf{GC.De}(d, \hat{z})$ outputs the plaintext version of an encoded value $\hat{z}$ (or $\perp$ for an invalid encoding);
- $b \leftarrow \mathsf{GC.Ve}(f, r, \hat{f}, e, d)$ allows to verify if a given garbled circuit was garbled correctly and outputs $b \in \{\top, \perp\}$;

As usual we need the garbling scheme to be *projective* – i.e., both $(e, d)$ are vectors of pairs of strings – to be compatible with Yao's protocol. We need the garbling scheme to satisfy *privacy* and *authenticity* as defined in [BHR12]. We need the garbling scheme to be *verifiable* in the standard sense i.e., that an adversary cannot "open" a garbling $\hat{f}$ to any function different than $f$.

**Definition 1 (Correctness).** *We say that a garbling scheme enjoys correctness if for all $n = \mathrm{poly}(\kappa)$, $f : \{0,1\}^n \rightarrow \{0,1\}^n$ and all inputs $x \in \{0,1\}^n$:*

$$\Pr\left( f(x) \neq \mathsf{GC.De}(d, \mathsf{GC.Ev}(\hat{f}, \mathsf{GC.En}(e, x))) : (\hat{f}, e, d) \leftarrow \mathsf{GC.Gb}(1^\kappa, f) \right) = 0$$

*(the probability is taken over the random coins of all algorithms).*

**Definition 2 (Privacy).** *We say that a garbling scheme enjoys privacy if there exists a PPT simulator $S$ such that the two following distributions are computationally indistinguishable:*

$$\{(\hat{f}, \mathsf{GC.En}(e, x), d) : (\hat{f}, e, d) \leftarrow \mathsf{GC.Gb}(1^\kappa, f)\}_x \approx \{S(1^\kappa, f, f(x))\}_x$$

*for all $f, x$.*

**Definition 3 (Authenticity).** *We say that a garbling scheme enjoys authenticity if for all PPT $\mathcal{A}$, for all $f, x \in \{0,1\}^n$*

$$\Pr\left(\mathsf{GC.De}(d, z^*) \neq \bot \wedge z^* \neq \mathsf{GC.Ev}(\hat{f}, \hat{x}): \begin{array}{c} (\hat{f}, e, d) \leftarrow \mathsf{GC.Gb}(1^\kappa, f), \\ \hat{x} \leftarrow \mathsf{GC.En}(e, x), \\ z^* \leftarrow \mathcal{A}(\hat{f}, \hat{x}) \end{array}\right)$$

*is negligible in $\kappa$.*

**Definition 4 (Circuit Verifiability).** *We say a garbling scheme enjoys circuit verifiability if for all PPT $\mathcal{A}$:*

$$\Pr\left(\mathsf{GC.Gb}(f_0; r_0) = \mathsf{GC.Gb}(f_1; r_1) : (f_0, r_0, f_1, r_1) \leftarrow \mathcal{A}(1^\kappa), f_0 \neq f_1\right)$$

*is negligible in $\kappa$.*

*Input Verification.* We also enhance the garbling scheme with two algorithms $(\mathsf{GC.TkG}, \mathsf{GC.TkV})$. The algorithm $\mathsf{GC.TkG}$ allows to generate some "tokens" tk from the input labels $e$. These tokens can be used with the $\mathsf{GC.TkV}$ algorithm to check whether an encoding of an input $\hat{x}$ is correct without leaking any information about the input $x$ itself. In a nutshell, we construct this from any projective garbling scheme in the following way: let $e = (K_0, K_1)$ be the encoding information of the original garbling scheme (for simplicity we assume a single input bit). Then we flip a random bit $r$ and let $\mathsf{tk} = (\langle K_r \rangle, \langle K_{1-r} \rangle)$ and $e' = ((K_0, \mathsf{open}(K_0)), (K_1, \mathsf{open}(K_1)))$ that is, we extend the input labels with some randomness and we compute two commitments, and permute them in a random order. Now given an encoding of an input $\hat{x} \leftarrow \mathsf{GC.En}(e', x)$ (using the extended labels i.e., $\hat{x} = (K_x, \mathsf{open}(K_x)))$ it is possible to verify whether this is a correct encoding by running $\mathsf{GC.TkV}(\mathsf{tk}, \hat{x}) \in \{\top, \bot\}$. The algorithm simply parses $\hat{x} = (K^*, \mathsf{open}(K^*))$, computes $\langle K^* \rangle = \mathsf{Com}_{ck}(K^*, \mathsf{open}(K^*))$ and checks if $\langle K^* \rangle \in \mathsf{tk}$. These tokens satisfy the following properties: (1) adding the tokens does not break the privacy property of the garbling scheme, and (2) if a (possibly malicious) encoding of an input passes the verification against the tokens, then evaluating a (honestly generated) garbled circuit on this input encoding will give an output different than $\bot$.

**Token Generation:** Given any *projective* garbling scheme i.e., one where $e = \{(K_0^i, K_1^i)\}_{i \in [n]}$, we construct a new garbling scheme with *verifiable input* in the following way: First we define the new encoding information $e'$ to be

$$e' = \{(K_0^i, \mathsf{open}(K_0^i)), (K_1^i, \mathsf{open}(K_1^i))\}_{i \in [n]}$$

and then we compute tokens $\mathsf{tk} \leftarrow \mathsf{GC.TkG}(e')$ by sampling random bits $r_1, \ldots, r_n$ and outputting

$$\mathsf{tk} = \{\langle K_{r_i}^i \rangle, \langle K_{1-r_i}^i \rangle\}_{i \in [n]}$$

With $\langle K_b^i \rangle = \mathsf{Com}_{ck}(K_b^i, \mathsf{open}(K_b^i))$ for all $b \in \{0,1\}, i \in [n]$.

**Input Verification:** $b \leftarrow \mathsf{GC.TkV}(\mathsf{tk}, \hat{x})$ is a deterministic algorithm that parses

$$\hat{x} = \{K^i, \mathsf{open}(K^i)\}_{i \in [n]}$$

computes $\langle K^i \rangle = \mathsf{Com}_{ck}(K^i, \mathsf{open}(K^i))$, and outputs $\perp$ if there exists an $i$ such that $\langle K^i \rangle \notin \mathsf{tk}$.

We define the following properties:

**Definition 5 (Token Privacy).** *We say that a garbling scheme enjoys* token privacy *if there exists a PPT simulator $S$ such that the two following distributions are computationally indistinguishable:*

$$\left\{ (\hat{f}, \mathsf{GC.En}(e, x), d, \mathsf{tk}) : \begin{matrix} (\hat{f}, e, d) \leftarrow \mathsf{GC.Gb}(1^\kappa, f), \\ \mathsf{tk} \leftarrow \mathsf{GC.TkG}(e) \end{matrix} \right\}_x \approx \{S(1^\kappa, f, f(x))\}_x$$

**Definition 6 (Input Verifiability).** *We say that a garbling scheme enjoys* input verifiability *if for all PPT $\mathcal{A}$ the following probability*

$$\Pr \left( \mathsf{GC.De}(d, \mathsf{GC.Ev}(\hat{f}, \hat{x}^*)) = \perp \wedge \mathsf{GC.TkV}(\mathsf{tk}, \hat{x}^*) = \top : \begin{matrix} (\hat{f}, e, d) \leftarrow \mathsf{GC.Gb}(1^\kappa, f), \\ \mathsf{tk} \leftarrow \mathsf{GC.TkG}(e), \\ \hat{x}^* \leftarrow \mathcal{A}(1^\kappa, \hat{f}, e), \end{matrix} \right)$$

*is negligible in $\kappa$.*

The proof that our construction satisfies the above requirements is straightforward.

**Lemma 1.** *Any projective garbling scheme can be enhanced to achieve token privacy and input verifiability using computationally hiding and computationally binding commitments as described above.*

*Proof.* For *token privacy*, we simply run the simulator guaranteed by the *privacy* property of the underlying garbling scheme. In addition, our simulator needs to output tk, a vector of pair of commitments. The simulator does so by parsing the encoded input $\hat{x}$ (provided by the privacy simulator) into $\hat{x} = \{K^i\}_{i \in [n]}$, chooses random values $\mathsf{open}(K^i)$, computes commitments $\langle K^i \rangle = \mathsf{Com}_{ck}(K^i, \mathsf{open}(K^i))$. The simulator also constructs $n$ commitments $\langle 0 \rangle$ (using independent randomness), and constructs tk as $n$ pairs of commitments, where each pair is $(\langle K^i \rangle, \langle 0 \rangle)$ if $r_i = 0$ or $(\langle 0 \rangle, \langle K^i \rangle)$ otherwise. Any adversary that can distinguish between the two distributions in the token privacy property can be trivially reduced to an adversary for either the computationally hiding property of the commitment scheme or the privacy property of the underlying garbling scheme.

For the *input verifiability* property, let $\hat{x}^* = \{K^i, \mathsf{open}(K^i)\}_{i \in [n]}$ be the output of the adversary and let $e = \{(K_0^i, \mathsf{open}(K_0^i)), (K_1^i, \mathsf{open}(K_1^i))\}_{i \in [n]}$. Now if $\mathsf{GC.TkV}(\mathsf{tk}, \hat{x}^*) = \top$, it must be the case that $K^i = K_b^i$ for some $b$ or the adversary can be used to break the binding property of the commitment scheme. But then, the property follows from the correctness of the underlying garbling scheme.

*Sub-functionalities.* In some stages of our protocol we let the parties run *any* actively-secure two-party computation protocol to implement a desired functionality. In the protocol description we only describe *what* the functionality should do and not *how* the functionality is implemented (in the proof we will make use of the UC composition theorem [Can01] to replace these subprotocols with hybrid functionalities under the control of the simulator). In particular we describe the private input of both parties, the private output of each party and the computation performed by the functionality. We also describe the *public input* of some functionalities. These are values which are defined previously in the protocol which can be imagined as values which are given as input by both parties (and the functionality aborts if they are different).

# 3   Our Protocol

We are now ready to present our protocol, which we like to call *cross-and-clean*[4]. As our protocol is quite complex, we split its presentation in five stages, which are described in Figs. 1, 2, 3, 4 and 5 respectively.

We have already (in the introduction) argued that the efficiency of the protocol is as desired. So it is now only left to prove that the protocol is secure:

**Theorem 1.** *The protocol described in Figs. 1, 2, 3, 4 and 5 securely evaluates $\ell$ copies of $f$ in the presence of active adversaries.*

*Proof.* Thanks to the UC composition theorem [Can01] it is sufficient to prove security of the protocol where we replace all actively secure subprotocols in the protocol (the *committed OT, coin-flip, filling-in* and *forge-and-lose* subprotocols respectively in Stage 1, 2, 4, 5) with ideal functionalities controlled by the simulator (in order to prove our theorem is enough to know that protocols for these functionalities exist [CLOS02] and we have measured their complexity in terms of the size of the functionalities that they implement). Since the protocol is completely symmetric for A and B, we will assume in the proof that A is corrupt and B is honest. Note that, since we prove the security of the protocol in the standard simulation-based indistinguishability between the real world and the ideal world, we must prove correctness and privacy at the same time – not as separate properties. Note also that, for the sake of presentation, our proof neglects many of the technicalities of the UC-framework [Can01] (such as delayed delivery of messages) but that our simulation strategy is straight-line.

As usual we make a proof by hybrids. We describe the simulator strategy along the way, by making progressive changes from the real protocol towards the simulator strategy, arguing for indistinguishability after every change. We describe the final simulator strategy in Fig. 6.

---

[4] *Cross* since the parties send garbled circuits to each other in the *dual-execution* phase, and *clean* since the subsequent stages "clean-up" the potential discrepancies between the outputs of the two parties.

---

**Stage 1: Providing Inputs:**
1. A, B run an actively UC-secure protocol which implements the following functionality:

**Input:** none.
**Computation:** The functionality runs the following code:
1. Generate commitment keys $ck^A \leftarrow \mathsf{CGen}(1^\kappa)$, $ck^B \leftarrow \mathsf{CGen}(1^\kappa)$;
2. Sample random $\sigma^A, \sigma^B \in \{0,1\}^{8\kappa}$;
3. Compute $\langle \sigma^A \rangle \leftarrow \mathsf{Com}_{ck^A}(\sigma^A, \mathsf{open}(\sigma^A))$;
4. Compute $\langle \sigma^B \rangle \leftarrow \mathsf{Com}_{ck^B}(\sigma^B, \mathsf{open}(\sigma^B))$;
5. Sample random $\mathsf{sen}^A = \{r^A_{0,i}, r^A_{0,i}\}_{i \in [8k]}$ with each $r^A_{b,i} \in \{0,1\}^\kappa$;
6. Sample random $\mathsf{sen}^B = \{r^B_{0,i}, r^B_{0,i}\}_{i \in [8k]}$ with each $r^B_{b,i} \in \{0,1\}^\kappa$;
7. Define $\mathsf{rec}^B = \{\sigma^A[i], r^A_{\sigma^A[i],i}\}_{i \in [8k]}$;
8. Define $\mathsf{rec}^A = \{\sigma^B[i], r^B_{\sigma^B[i],i}\}_{i \in [8k]}$;

**A's output:** The functionality sends as private output to A:
1. $\sigma^A, \mathsf{open}(\sigma^A), \langle \sigma^B \rangle$;
2. $\mathsf{sen}^A, \mathsf{rec}^A$;

**B's output:** The functionality sends as private output to B:
1. $\sigma^B, \mathsf{open}(\sigma^B), \langle \sigma^A \rangle$;
2. $\mathsf{sen}^B, \mathsf{rec}^B$;

2. For all $i \in [\ell]$: A sends $X^A_i = \mathsf{SE.E}(\sigma^A, x^A_i)$ to B;
3. For all $i \in [\ell]$: B sends $X^B_i = \mathsf{SE.E}(\sigma^B, x^B_i)$ to A;

---

Fig. 1. Stage 1: providing inputs

*Hybrid 0.* We start in dream version of the simulation, where we assume the simulator is given the real inputs $x^B_i$ of the honest party $B$. Here the simulator simulates simply by running the real protocol with the adversary controlling $A$ and the simulator running $B$ and the hybrid ideal functionalities. If $B$ aborts in the protocol before receiving encrypted inputs from $A$, instruct the ideal functionality to abort on behalf of the corrupted $A$. Otherwise the simulator extracts $\{x^A_i \leftarrow \mathsf{SE.D}(\sigma^A, X^A_i)\}_{i \in [\ell]}$ and input these values to the ideal functionality. This allows the simulator to learn all the real outputs $\{y_i\}_{i \in [\ell]}$. If $B$ aborts in the protocol after receiving encrypted inputs, instruct the ideal functionality to abort on behalf of the corrupted $A$. Otherwise, let $\{y'_i\}_{i \in [\ell]}$ be the outputs of the protocol. If $\{y_i\}_{i \in [\ell]} \neq \{y'_i\}_{i \in [\ell]}$, the simulator aborts the simulation. Clearly this first hybrid is perfectly indistinguishable from the real protocol execution as long as $\{y_i\}_{i \in [\ell]} = \{y'_i\}_{i \in [\ell]}$. Therefore hybrid 0 is computational indistinguishable from the real protocol execution if the protocol has correctness except with negligible probability. We argue correctness of this at the end of the proof.

*Hybrid 1.* Here we change the inner working of the *committed OT* functionality (Stage 1): from now on the simulator sends A a commitment to 0 instead of $\sigma^B$. Any adversary that can distinguish after this change can be used to break the computationally hiding property of the commitment scheme.

**Stage 2: Cut-n-Choose:**

1. A garbles $(1+\epsilon)\ell$ times (with independent randomness)[a]:

$$(\hat{f}_i^A, e_i^A, d_i^A) \leftarrow \mathsf{GC.Gb}(f_i^A; r_i^A)$$

where $f_i^A$ is the circuit that computes

$$(\sigma^A, X_i^A, \sigma^B, X_i^B) \mapsto f(\mathsf{SE.D}(\sigma^A, X_i^A), \mathsf{SE.D}(\sigma^B, X_i^B))$$

2. A computes $\mathsf{tk}_i^A \leftarrow \mathsf{GC.TkG}(e_i^A)$ and $\langle d_i^A \rangle \leftarrow \mathsf{Com}_{ck^A}(d_i^A, \mathsf{open}(d_i^A))$;
3. A sends $(\hat{f}_i^A, \mathsf{tk}_i^A, \langle d_i^A \rangle)$ to B;
4. A,B perform an actively UC-secure coin flip to determine a random subset

$$CC \subset \{1, \ldots, (1+\epsilon)\ell\}$$

of size $\epsilon\ell$ of circuits to be checked;
5. A sends $(r_i^A, e_i^A, d_i^A, \mathsf{open}(d_i^A))$ to B;
6. B aborts if
$$\mathsf{GC.Ve}(f_i^A, e_i^A, d_i^A, r_i^A, \hat{f}_i^A) = \perp \text{ or}$$
$$\exists i \in CC: \qquad \mathsf{tk}_i^A \neq \mathsf{GC.TkG}(e_i^A) \text{ or}$$
$$\langle d_i^A \rangle \neq \mathsf{Com}_{ck^A}(d_i^A, \mathsf{open}(d_i^A))$$

7. A and B re-index the unopened circuits $1, \ldots, \ell$.[b]
8. A,B repeat with reversed roles;

---

[a] For the sake of notation, we implicitly assume that $\epsilon\ell$ is an integer.
[b] This is simply done to simplify the notation of the upcoming stages.

Fig. 2. Stage 2: cut-n-choose

*Hybrid 2.* Here we replace the commitment sent to A in Stage 4 (*filling in phase*) to be a commitment to 0 instead of the MAC key $\tau^B$. Any adversary that can distinguish after this change can be used to break the hiding property of the commitment.

*Hybrid 3.* Here we replace the abort condition (Step 1.a) in the functionalities for *filling-in* in Stage 4. Let $\sigma^A$ be the values received by A from the *committed OT* functionality in Stage 1 and $\sigma^*$ be the values input by A to the *filling-in* functionality in Stage 4. From now on we always abort if $\sigma^* \neq \sigma^A$ (even if what A inputs is a valid opening of the commitment $\langle \sigma^A \rangle$). Any adversary that can distinguish after this change can be used to break the computationally binding property of the commitment scheme. (Note that from now on we have the guarantee that the output of *filling-in* to the honest party B can only be the real output $y_i$.)

*Hybrid 4.* Here we replace the abort condition (Step 1.a and 1.c in **Computation**) of the functionality for *forge-and-lose* in Stage 5 in a similar way as in the previous hybrid, namely: let $\sigma^*, \tau^*$ be the keys input by A, then in this hybrid we abort if $\sigma^A \neq \sigma^*$ or $\tau^A \neq \tau^*$ (even if A inputs proper commitment

---

**Stage 3: Run Computation $i$:**

1. $A$ computes encoded versions of the inputs $\hat{X}_i^A, \hat{X}_i^B, \hat{\sigma}_i^A$ i.e.,
   (a) $\hat{X}_i^A \leftarrow$ GC.En$(e_i^A, X_i^A)$,
   (b) $\hat{X}_i^B \leftarrow$ GC.En$(e_i^A, X_i^B)$,
   (c) $\hat{\sigma}_i^A \leftarrow$ GC.En$(e_i^A, \sigma^A)$,
   and sends them to B;
2. $A$ runs tra$_i^A \leftarrow$ OTTransfer$(\text{sen}^A, i, e_i^A)$ and sends tra$_i^A$ to B;
3. B runs $\hat{\sigma}_i^B \leftarrow$ OTRetrieve$(\text{rec}^B, i, \text{tra}_i^A)$;
4. B aborts if GC.TkV$(\text{tk}_i^A, (\hat{\sigma}_i^A, \hat{X}_i^A, \hat{\sigma}_i^B, \hat{X}_i^B)) = \perp$;
5. B evaluates $\hat{y}_i^B =$ GC.Ev$(\hat{f}_i^A, \hat{X}_i^A, \hat{X}_i^B, \hat{\sigma}_i^A, \hat{\sigma}_i^B)$;
6. B computes $\langle \hat{y}_i^B \rangle \leftarrow$ Com$_{ckB}(\hat{y}_i^B, \text{open}(\hat{y}_i^B))$
7. B sends $\langle \hat{y}_i^B \rangle$ to A;
8. A sends $d_i^A, \text{open}(d_i^A)$ to B; B aborts if Com$_{ck}(d_i^A, \text{open}(d_i^A)) \neq \langle d_i^A \rangle$
9. Bob computes $y_i^B \leftarrow$ GC.De$(d_i^A, \hat{y}_i^B)$;
10. If $y_i^B = \perp$ add $i$ to $I^B$;
11. If $|I^B| > \psi\kappa$, then abort the protocol.
12. A,B repeat with reversed roles;

---

Fig. 3. Stage 3: run computation $i$

openings). An adversary distinguishing after this change can again be used to break the computationally binding property of the commitment scheme.

*Hybrid 5.* Here we replace the last abort condition (Step 1.c) in the functionality for *filling-in* in Stage 4. For each $i \in I^A$ let $V_i^*$ be the value input by A to this computation. From now on we always abort if $V_i^* \neq V_i$ (even if the MT.V algorithm accepts the proof $\pi_i$). An adversary distinguishing after this change can be used to break the computationally binding property of the Merkle-tree commitment. (Note that at this point, by definition, the output of *filling-in* to A cannot be different from $y_i$).

*Hybrid 6.* Here we change the distribution of tra$_i^B$ (the value sent from B to A in Step 2 of Stage 3): instead of computing tra$_i^B =$ OTTransfer$(\text{sen}^B, i, e_i^B)$, we compute tra$_i^B =$ OTTransfer$(\text{sen}^B, i, e_i^*)$ where $e_i^* = \{K_0^i, K_1^i\}$ is defined as follows: let $\hat{\sigma}_i^A \leftarrow$ GC.En$(e_i^B, \sigma^A)$ and parse $\hat{\sigma}_i^A = \{K^i\}$, then we set

$$K_{\sigma^A[i]}^i = K^i \text{ and } K_{1-\sigma^A[i]}^i = 0.$$

That is, we set all labels not corresponding to the bits of $\sigma^A$ to 0. Since A only has access to the keys rec$^A$ corresponding to the bits of $\sigma^A$, we can use an adversary that distinguishes after this change to break the IND-CPA of underlying symmetric encryption scheme.

*Hybrid 7.* Here we change the distribution of the garbled circuits and garbled inputs sent to A during Stage 2 and 3 for all $i \notin CC$ (since the simulator is controlling the coin flip functionality, this set is known to the simulator from the

---

**Stage 4: Actively Secure Filling-in:**
1. Define $V_i = (X_i^A, X_i^B)$. Both A and B compute $\mathsf{root} = \mathsf{MT.C}(V_1, \ldots, V_\ell)$;
2. Run an actively UC-secure protocol which implements the following functionality:

**Public Input:** The functionality takes public inputs $\mathsf{root}, \langle \sigma^A \rangle, \langle \sigma^B \rangle$.
**A's input** The functionality takes as private input from A:
   1. $\sigma^A, \mathsf{open}(\sigma^A)$,
   2. the set $I^A$ with $|I^A| \leq \psi\kappa$ and
   3. $(V_i, \pi_i \leftarrow \mathsf{MT.P}((V_1, \ldots, V_\ell), i)$ for all $i \in I^A$.
**B's input** The functionality takes as private input from B:
   1. $\sigma^B, \mathsf{open}(\sigma^B)$,
   2. the set $I^B$ with $|I^B| \leq \psi\kappa$ and
   3. $(V_i, \pi_i \leftarrow \mathsf{MT.P}((V_1, \ldots, V_\ell), i)$ for all $i \in I^B$.
**Computation:** The functionality runs the following code:
   1. Abort if any of the following is true:
     (a) $\langle \sigma^A \rangle \neq \mathsf{Com}_{ckA}(\sigma^A, \mathsf{open}(\sigma^A))$,
     (b) $\langle \sigma^B \rangle \neq \mathsf{Com}_{ckB}(\sigma^B, \mathsf{open}(\sigma^B))$,
     (c) $\exists i : I^A \cup I^B : \mathsf{MT.V}(\mathsf{root}, i, V_i, \pi_i) = \bot$;
   2. If all checks pass
     (a) Sample random $\tau^A, \tau^B \in \{0,1\}^\kappa$;
     (b) Compute $\langle \tau^A \rangle \leftarrow \mathsf{Com}_{ckA}(\tau^A, \mathsf{open}(\tau^A))$;
     (c) Compute $\langle \tau^B \rangle \leftarrow \mathsf{Com}_{ckB}(\tau^B, \mathsf{open}(\tau^B))$;
   3. For all $i \in I^A \cup I^B$:
     (a) Parse $V_i = (X_i^A, X_i^B)$ and compute
     (b) $x_i^A \leftarrow \mathsf{Dec}(\sigma^A, X_i^A)$;
     (c) $x_i^B \leftarrow \mathsf{Dec}(\sigma^B, X_i^B)$;
     (d) $y_i = f(x_i^A, x_i^B)$;
     (e) $t_i^A = \mathsf{MAC.Tag}(\tau^B, y_i)$ if $i \in I^A$ or $t_i^B = \mathsf{MAC.Tag}(\tau^A, y_i)$ otherwise;
**A's output:** The functionality sends as private output to A:
   1. $\tau^A, \mathsf{open}(\tau^A), \langle \tau^B \rangle$;
   2. For all $i \in I^A : (y_i^A, t_i^A)$
**B's output:** The functionality sends as private output to B:
   1. $\tau^B, \mathsf{open}(\tau^B), \langle \tau^A \rangle$;
   2. For all $i \in I^B : (y_i^B, t_i^B)$;

---

**Fig. 4.** Stage 4: actively secure filling-in

beginning), by running the simulator (which is guaranteed to exist thanks to the *token privacy* property of the garbling scheme on input the function $f$ and the output $y_i$. (Token privacy is defined in Definition 5.) The simulator provides us with garbled versions of all inputs including $\hat{\sigma}_i^A$, as well as $\mathsf{tk}_i^B$, $d_i^B$ and $\hat{f}_i^B$ which can now replace the values sent to A in Step 3 of Stage 2 and Steps 1, 2 and 8 of Stage 3. Any adversary distinguishing after this step can be used to break the *token privacy* of the underlying garbling scheme. (Note that the simulator can also, running $\mathsf{GC.Ev}$ on the garbled circuit and the garbled inputs, compute the garbled output $\hat{y}_i^A$ which is needed in the next steps.)

---

**Stage 5: Forge-and-Lose:**

1. A defines $t_i^A = \bot$ for all $i \notin I^A$;
2. B defines $t_i^B = \bot$ for all $i \notin I^B$;
3. A and B run an actively UC-secure protocol which implements the following functionality:

**Public Input:**
1. $\langle \sigma^A \rangle, \langle \sigma^B \rangle, \langle \tau^A \rangle, \langle \tau^B \rangle$
2. for all $i \in [\ell] : (d_i^A, d_i^B, X_i^A, X_i^B, \langle \hat{y}_i^A \rangle, \langle \hat{y}_i^B \rangle)$;

**A's input** The functionality takes as private input from A:
1. $\sigma^A$, open($\sigma^A$),
2. $\tau^A$, open($\tau^A$);
3. For all $i \in [\ell] : (y_i^A, \hat{y}_i^A, \mathsf{open}(\hat{y}_i^A), t_i^A)$;

**B's input** The functionality takes as private input from B:
1. $\sigma^B$, open($\sigma^B$),
2. $\tau^B$, open($\tau^B$);
3. For all $i \in [\ell] : (y_i^B, \hat{y}_i^B, \mathsf{open}(\hat{y}_i^B), t_i^B)$;

**Computation:** The functionality:
1. Abort if any of the following is true:
    (a) $\langle \sigma^A \rangle \neq \mathsf{Com}_{ckA}(\sigma^A, \mathsf{open}(\sigma^A))$,
    (b) $\langle \sigma^B \rangle \neq \mathsf{Com}_{ckB}(\sigma^B, \mathsf{open}(\sigma^B))$,
    (c) $\langle \tau^A \rangle \neq \mathsf{Com}_{ckA}(\tau^A, \mathsf{open}(\tau^A))$,
    (d) $\langle \tau^B \rangle \neq \mathsf{Com}_{ckB}(\tau^B, \mathsf{open}(\tau^B))$,
    (e) $\exists i : \langle \hat{y}_i^A \rangle \neq \mathsf{Com}_{ckA}(\hat{y}_i^A, \mathsf{open}(\hat{y}_i^A))$,
    (f) $\exists i : \langle \hat{y}_i^B \rangle \neq \mathsf{Com}_{ckB}(\hat{y}_i^B, \mathsf{open}(\hat{y}_i^B))$,
    (g) $\exists i : y_i^B \neq \mathsf{GC.De}(d_i^A, \hat{y}_i^B) \wedge \mathsf{MAC.Ver}(t_i^B, \tau^A, y_i^B) = \bot$;
    (h) $\exists i : y_i^A \neq \mathsf{GC.De}(d_i^B, \hat{y}_i^A) \wedge \mathsf{MAC.Ver}(t_i^A, \tau^B, y_i^A) = \bot$;
2. If all checks pass
    (a) Find $i : y_i^A \neq y_i^B$;
    (b) If no such $i$ exists conclude that "no-one cheated";
    (c) Else compute $y_i = f(x_i^A, x_i^B)$;
    (d) Conclude that "Alice cheated" if: $y_i^B \neq y_i$;
    (e) Conclude that "Bob cheated" if: $y_i^A \neq y_i$;

**Output:** If
1. "Alice cheated" : A receives $\bot$, B receives $\sigma^A$;
2. "Bob cheated" : A receives $\sigma^B$, B receives $\bot$;
3. else ("No-one cheated"): A receives $\bot$, B receives $\bot$;

4. If A (resp. B) receives an output $\sigma^B \neq \bot$, A computes $x_i^B \leftarrow \mathsf{SE.D}(\sigma^B, X_i^B)$ for all $i \in [\ell]$ and outputs $y_i^A = f(x_i^A, x_i^B)$;

---

**Fig. 5.** Stage 5: forge-and-lose

*Hybrid 8.* Here we replace the abort condition (Step 1.e) in the functionality *forge-and-lose* in Stage 5. For all $i$ the simulator computes $\hat{y}_i^* \leftarrow \mathsf{Ext}(\mathsf{td}^A, \langle \hat{y}_i^* \rangle)$ using the trapdoor $\mathsf{td}^A$ (which the simulator learns as it controls the *committed-OT* sub-protocol) and the commitments $\langle \hat{y}_i^* \rangle$ received during Step 6 of Stage 3.

---

**Simulator Strategy:**

**Stage 1:**   1. Simulate the *committed OT* functionality by sending A a random key $(\sigma^A, \mathsf{open}(\sigma^A))$; sending A a commitment $\langle 0 \rangle$ (instead of $\langle \sigma^B \rangle$); sending A random strings $\mathsf{sen}^A, \mathsf{rec}^A$; learning the trapdoor $\mathsf{td}^A$ corresponding to $ck^A$;

   2. Receive $\{X_i^A\}_{i \in [\ell]}$, decrypt $x_i^A \leftarrow \mathsf{SE.D}(\sigma^A, X_i^A)$ and input these values to the ideal functionality; receive $\{y_i\}_{i \in [\ell]}$ from the ideal functionality;

   3. Send A $\{X_i^B = \mathsf{SE.E}(\sigma^B, 0)\}_{i \in [\ell]}$;

**Stage 2.A:** In Stage 2.A the simulator (when receiving garbled circuits from A):

   1-7. Receive garbled circuits; Sample a random CC and simulate the *coin-flip* functionality; Verify the circuits with $i \in \mathsf{CC}$ and abort as an honest B would;

**Stage 2.B** In Stage 2.B the simulator (when sending garbled circuits to A):

   1-7. Sample a random set CC. Construct honest garbling of the function for all $i \in \mathsf{CC}$; For all $i \notin \mathsf{CC}$, run the simulator of the garbling scheme to receive the garbled circuit $\hat{f}_i^B$, inputs $\hat{\sigma}_i^A, \hat{\sigma}_i^B, \hat{X}_i^A, \hat{X}_i^B$, decoding tables $d_i^A$ and tokens $\mathsf{tk}_i^B$. Force the *coin-flip* sub-protocol to output CC;

**Stage 3.A:** In Stage 3 the simulator (when acting as circuit evaluator):

   1. Receive garbled inputs;

   2-4. Fully decrypt $\mathsf{tra}_i^A$, compute the set $\mathcal{L}$ and abort according to the strategy as described during Hybrid 11;

   5-7. Compute and send A a commitment to $\langle 0 \rangle$;

   8-10. Abort if $\langle d_i^A \rangle$ is not opened correctly;

   11. There is no abort if $|I^B| > \psi \kappa$ as the simulator cannot compute $I^B$.

**Stage 3.B:** In Stage 3 the simulator (when acting as circuit constructor):

   1. Send garbled inputs (from the garbling scheme simulator);

   2-4. Prepare and send $\mathsf{tra}_i^B$ to A as described during Hybrid 6 (i.e., only send labels corresponding to the bits of $\sigma^A$, and 0 in all other positions); Fully decrypt $\mathsf{tra}_i$, compute the set $\mathcal{L}$ and abort according to the strategy as described during Hybrid 11;

   5-7. Receive from A a commitment $\langle \hat{y}_i^B \rangle$ and extract $\hat{y}_i^{**} \leftarrow \mathsf{Ext}(\mathsf{td}^A, \langle \hat{y}_i^A \rangle)$ and abort if $\hat{y}_i^{**}$ breaks *authenticity*;

   8-10. Open $\langle d_i^B \rangle$;

**Stage 4:** In Stage 4 the simulator

   1. Computes root as an honest B would;

   2. Simulate the *filling-in* functionality by aborting if (on top of the original conditions) $\sigma^* \neq \sigma^A$ or $V_i^* \neq V_i$ where $\sigma^*, \{V_i^*\}_{i \in I^A}$ are the value sent by the adversary to the functionality; If the simulator does not abort, it sends A a random MAC key and commitment opening $(\tau^A, \mathsf{open}(\tau^A))$ and a commitment $\langle 0 \rangle$ (instead of $\tau^B$);

**Stage 5:** In Stage 5 the simulator

   1-3. Simulates the *forge-and-lose* functionality by aborting if (on top of the original conditions) $\sigma^* \neq \sigma^A$, $\tau^* \neq \tau^A$, $\hat{y}_i^{**} \neq \hat{y}_i^*$, $y_i^* \neq y_i$ where $y_i^*, \sigma^*, \tau^*, \hat{y}_i^*$ are the value sent by the adversary to the functionality;

---

**Fig. 6.** Simulator strategy

Let $\hat{y}_i^{**}$ be the value input by A to the *forge-and-lose* functionality. The simulator now aborts if $\hat{y}_i^* \neq \hat{y}_i^{**}$ even if A provides a valid commitment opening.

Any adversary distinguishing after this step can be used to break the binding property of the commitment scheme.

*Hybrid 9.* Here we change the last aborting condition (Step 1.h) in the functionality *forge-and-lose* in Stage 5. Let $(y_i^*, \hat{y}_i^*, t_i^*)$ be the value input by the adversary and let $(y_i, \hat{y}_i^A, t_i^A)$ be the values computed by the simulator in the previous hybrids. From now on, instead of aborting if

$$\exists i : y_i^* \neq \mathsf{GC.De}(d_i^B, \hat{y}_i^*) \wedge \mathsf{MAC.Ver}(t_i^*, \tau^B, y_i^*) = \bot$$

the simulator aborts if

$$\exists i : (y_i^* \neq \mathsf{GC.De}(d_i^B, \hat{y}_i^*) \wedge \mathsf{MAC.Ver}(t_i^*, \tau^B, y_i^*) = \bot) \vee (y_i^* \neq y_i).$$

Any adversary that can distinguish after this change can be used to break unforgeability of the MAC scheme (note that the simulator at this point does not need to know $\tau^B$ since it has been replaced by 0 in the commitment that A receives at the end of the *filling-in* stage, and we can therefore successfully run the reduction) or to break the *authenticity* property of the garbling scheme (note that we have already made sure that value $\hat{y}_i^*$ input by A here is the same as the one he commits to in Step 7 of Stage 3 and – since the simulator can extract the value in the commitment using the trapdoor – the reduction can already break the authenticity property *before* having to send $d_i^B$ or the opening of the commitment to A in Step 8 of Stage 3). Note that after this change we are ensured (by definition) that A will never receive $\sigma^B$ as a result of running the *forge-and-lose* sub-protocol.

*Hybrid 10.* Here we change the distribution of the commitment that the simulator sends to A in Step 6 of Stage 3, from being a commitment to $\hat{y}_i^B$ to being a commitment to 0. An adversary that distinguishes after this change can be used to break the hiding property of the commitment scheme.

*Hybrid 11.* Here we let the simulator fully decrypt the transfer message $\mathsf{tra}_i$ from A in Step 3 of Stage 3. That is, instead of running $\hat{\sigma}_i^B \leftarrow \mathsf{OTRetrieve}(\mathsf{rec}^B, i, \mathsf{tra}_i^A)$ the simulator extracts

$$e_i^* \leftarrow \mathsf{OTRetrieve}(\mathsf{sen}^A, i, \mathsf{tra}_i^A)$$

and constructs the set

$$\mathcal{L}_i \subset [8\kappa] \times \{0, 1\}$$

as follows. Parse

$$\mathsf{tk}_i^A = \{\langle A_j \rangle, \langle B_j \rangle\}_{j \in [8\kappa]}$$

and

$$e_i^* = \{(K_{j,0}, \mathsf{open}(K_{j,0})), (K_{j,1}, \mathsf{open}(K_{j,1}))\}_{j \in [8\kappa]}.$$

We add $(j, b)$ to $\mathcal{L}_i$ if

$$\text{Com}_{ck}(K_{j,b}, \text{open}(K_{j,b})) \notin \{\langle A_j \rangle, \langle B_j \rangle\}.$$

We then compute

$$\mathcal{L} = \cup_{i \in [\ell]} \mathcal{L}_i.$$

The set $\mathcal{L}$ represents all the positions in all the OT transfers in which A "cheated" i.e., where A sent some value which is not consistent with the tokens $\text{tk}_i^A$. Since an honest B uses the same input bits $\sigma^B$ in all transfers, we only count each combination of position $j$ and bit $b$ once. In other words, for each index $j$ A has three strategies:

1. Input the right values (i.e., values that make GC.TkV accept) for both $b \in \{0, 1\}$ (for all $i \in [\ell]$): in this case neither $(j, b) \notin \mathcal{L}$ for both $b \in \{0, 1\}$;
2. Input the right value for a single $b \in \{0, 1\}$ (for all $i \in [\ell]$)) and at least a wrong value for $1 - b$ for some $i^* \in [\ell]$: in this case $(j, 1 - b) \in \mathcal{L}$;
3. Input the wrong value for both $b \in \{0, 1\}$ (potentially for different $i^* \in [\ell]$): in this case both $(j, 0) \in \mathcal{L}$ and $(j, 1) \in \mathcal{L}$;

Now we replace the abort condition in Step 4 of Stage 3 to the following: the simulator aborts with probability 1 if $\exists\, j$ such that both $(j, 0) \in \mathcal{L}$ and $(j, 1) \in \mathcal{L}$ (this is consistent with what B would do in the real protocol, since in this case B will detect the wrong labels regardless of the value of $\sigma^B[j]$). Otherwise, the simulator aborts with probability $1 - 2^{|\mathcal{L}|}$ (this is consistent with what B would do in the real protocol, since in this case B detects the wrong labels only if $\sigma^B[j] = b$ for $(j, b) \in \mathcal{L}$ – note on the other hand that if $(j, b) \in \mathcal{L}$ and B does not abort then the corrupt A learns that the value of $\sigma^B[j] \neq b$, and we will take care of this in a moment). At the same time we change the distribution of the encryptions $X_i^B$ sent by B to A in Step 3 of Stage 1 to be all encryptions of 0. Any adversary that can distinguish after this change can be used to break the IND-CPA security of $(\text{SE.E}, \text{SE.D})$. Remember that we required $(\text{SE.E}, \text{SE.D})$ to be secure even against adversaries who learn up to $\kappa$ bits of the secret key. We here use this property to let the reduction ask the IND-CPA challenger for the bits of $\sigma^B[j]$ for all $j : (j, b) \in \mathcal{L}$.

*Hybrid 12.* After encrypting 0s instead of the real input of B, it can easily be seen that there is only one place left where we use the input of B, namely to compute the set $I^B$ for which we need the input of B to evaluate the garbled circuits, as a bad circuit might give an output on some of B's inputs and $\bot$ on some other inputs. We get rid of this last use of the inputs of B by removing the restriction that $|I^B| \leq \psi \kappa$ in the functionalities for *filling-in* and dropping the abort condition in Step 11 in Stage 3. This change is indistinguishable, as $|I^B| \leq \psi \kappa$ except with negligible probability. To see why this is the case, let good be the set of honestly generated circuits among those received by $B$. Thanks to the *input verifiability* property of our garbling scheme we know that for all

$i \in$ good the values received by B as output from the Stage 3 satisfy $y_i^B \neq \bot$. Since we open $\epsilon\ell$ of the circuits in the cut-and-choose the probability that $\psi\kappa$ bad circuits will all survive without any being detected is less than $(1 + \epsilon)^{-\psi\kappa}$, and we have set $\psi$ such that $(1 + \epsilon)^{-\psi\kappa} = 2^{-\kappa}$.

This concludes the description of our simulation strategy. It can be seen that (by construction) at this point the simulator does not use the input of the honest party B and we have argued for indistinguishability after each individual change. The complete description of the simulator after all the hybrids can be found in Fig. 6. The simulator is simply a compilation of all the individual changes done in the above hybrids. Therefore the distribution of Hybrid 12 is identical to the distribution of the simulation.

What remains is therefore only to argue that Hybrid 0 is indistinguishable from the real protocol. In Hybrid $i$ define an event $E^i$ as follows. Let $Y = \bot$ if the ideal functionality aborts and let $Y = \{y_i\}_{i\in[\ell]}$ be the outputs of the ideal functionality otherwise. Let $Y' = \bot$ if the protocol aborts and let $Y' = \{y_i'\}_{i\in[\ell]}$ otherwise. Let $E^i$ be the event that $Y \neq Y'$. To argue that Hybrid 0 is indistinguishable from the real protocol it is clearly enough to argue that $\Pr[E^0]$ is negligible. We have that $\Pr[E^{12}] = 0$ by construction: it has already been argued that since Hybrid 5 the outputs of the *filling-in* for the honest party are correct; and it has been argued that since Hybrid 8 the corrupt party can only input the correct outputs to the *forge-and-lose* functionality, which implies that either the outputs that B received before this sub-protocol are the correct ones or B will receive $\sigma^A$ as a result of *forge-and-lose* and compute the right outputs in the clear.

It then follows from Hybrid 0 and Hybrid 12 being indistinguishable that $\Pr[E^0]$ is indistinguishable from 0, i.e., negligible.

## 4   Dealing with Long Inputs and Outputs

The protocol described and analysed in the previous section allows to compute $f : \{0,1\}^{n_I} \to \{0,1\}^{n_O}$ where $n_I$ is the input size and $n_O$ is the output size. In the previous sections we have, for simplicity, assumed that $n_I = n_O = O(\kappa)$. However in general $n_I$ and $n_O$ can be of size linear in $|f|$. This presents an issue in the forge-and-lose step, since the size of the circuit implementing the functionality is of size:

$$(\ell n_O + \ell n_I + |f|)\kappa = O(\ell|f|\kappa)$$

instead of $O((\ell + |f|)\kappa)$ as desired. We describe here two optimizations which allow to deal with this:

**Dealing with Long Outputs:** It is quite easy to deal with long outputs in the following way: modify the circuit to be garbled so that, in addition to outputting $y_i$, it also outputs $h(y_i)$ with $h$ a collision resistant hash function. Now it is clear that $|h(y_i)| = \kappa < n_O$ and therefore we can modify the protocol in the following way: instead of letting A,B input the values $y_i$ to the forge and lose functionality, they input the hashes instead. The forge and lose finds the first index where the *hashes* differ, recompute the function (and the hash) on that index and determines who cheated.

**Dealing with Long Inputs:** To deal with long inputs, the key is to notice that only a single pair of inputs (in encrypted format) is ever used during the forge and lose functionality, and the only reasons for the parties to input all of the ciphertexts is to guarantee that the adversary cannot learn in which positions (if any) the function is being recomputed. In some sense we are using a very naïve private information retrieval (PIR), which can of course be replaced with a more clever one. We can therefore modify the forge and lose stage in the following way: instead of having A, B input all ciphertexts at the beginning, they only input the outputs (or their hashes as described above). The functionality finds the first $i$ such that $y_i^A \neq y_i^B$ (if any), and then runs a 2-server PIR protocol with A, B. This allows the functionality to learn the ciphertext pair $X_i^A, X_i^B$ necessary to determine the right value of $y_i$ by receiving only $\sqrt{\ell}n_I$ bits from A,B. Since $n_I < |f|$ it is enough to assume that $\ell > |f|^2$ to bound this term with $\ell$.

After these changes, the number of bits which A, B send to the forge and lose functionality (regardless of the input and output size of the function) is bounded by

$$(\ell\kappa + \sqrt{\ell}f + |f|)\kappa = O((\ell + |f|))\operatorname{poly}(\kappa)$$

as desired.

**Acknowledgements.** This project was supported by: the Danish National Research Foundation and The National Science Foundation of China (grant 61361136003) for the Sino-Danish Center for the Theory of Interactive Computation; the European Union Seventh Framework Programme ([FP7/2007-2013]) under grant agreement number ICT-609611 (PRACTICE).

# A    List of symbols

- $\sigma^A, \sigma^B$ global encryptions keys, also used as selection bits used in OTChoose;
- $\tau^A, \tau^B$ global MAC key, generated during *filling in*;
- $x_i^A, x_i^B$ the inputs used in execution $i$;
- $y_i^A, y_i^B$ the output received by A/B in execution $i$;
- $y_i$, the real output of execution $i$ (i.e., $f(x_i^A, x_i^B)$);
- We use $\hat{x}$ to indicate "garbled values/functions" (in the GC context);
- We use capitals for encryptions of the inputs under $\sigma$ ($X_A^i, X_B^i$);
- $t_i^A, t_i^B$ are MACs computed during *filling in*;
- $f : \{0,1\}^n \times \{0,1\}^n \to \{0,1\}^n$ the original function that we are trying to evaluate, with input/output size $n$;
- $\kappa$ the security parameter;
- $\ell$ the number of copies of $f$ we are evaluating i.e., $i = 1, \ldots, \ell$;

- $\epsilon$, the fraction of circuits being checked i.e., the number of garbled circuits which each party generates is $(1 + \epsilon)\ell$, and then $\epsilon\ell$ of those are checked during the cut-and-choose. For simplicity we assume $\epsilon\ell$ to be an integer;
- $\psi = \log_{1+\epsilon}(2)$;
- Given a value $x$, $\langle x \rangle$ is a commitment to $x$ using randomness open($x$).

# References

[BHR12] Bellare, M., Hoang, V.T., Rogaway, P.: Foundations of garbled circuits. In: The ACM Conference on Computer and Communications Security, CCS 2012, Raleigh, NC, USA, 16–18 October 2012, pp. 784–796 (2012)

[BMR90] Beaver, D., Micali, S., Rogaway, P.: The round complexity of secure protocols (extended abstract). In: Proceedings of the 22nd Annual ACM Symposium on Theory of Computing, 13–17 May 1990, Baltimore, Maryland, USA, pp. 503–513 (1990)

[Bra13] Brandão, L.T.A.N.: Secure two-party computation with reusable bit-commitments, via a cut-and-choose with forge-and-lose technique. In: Sako, K., Sarkar, P. (eds.) ASIACRYPT 2013, Part II. LNCS, vol. 8270, pp. 441–463. Springer, Heidelberg (2013). doi:10.1007/978-3-642-42045-0_23

[Can01] Canetti, R.: Universally composable security: a new paradigm for cryptographic protocols. In: 42nd Annual Symposium on Foundations of Computer Science, FOCS 2001, 14–17 October 2001, Las Vegas, Nevada, USA, pp. 136–145. IEEE Computer Society (2001)

[CLOS02] Canetti, R., Lindell, Y., Ostrovsky, R., Sahai, A.: Universally composable two-party and multi-party secure computation. In: Proceedings on 34th Annual ACM Symposium on Theory of Computing, 19–21 May 2002, Montréal, Québec, Canada, pp. 494–503 (2002)

[FJN+13] Frederiksen, T.K., Jakobsen, T.P., Nielsen, J.B., Nordholt, P.S., Orlandi, C.: MiniLEGO: efficient secure two-party computation from general assumptions. In: Johansson, T., Nguyen, P.Q. (eds.) EUROCRYPT 2013. LNCS, vol. 7881, pp. 537–556. Springer, Heidelberg (2013). doi:10.1007/978-3-642-38348-9_32

[FJNT15] Frederiksen, T.K., Jakobsen, T.P., Nielsen, J.B., Trifiletti, R.: Tinylego: an interactive garbling scheme for maliciously. IACR Cryptology ePrint Archive 2015:309 (2015)

[FNO15] Frederiksen, T.K., Nielsen, J.B., Orlandi, C.: Privacy-free garbled circuits with applications to efficient zero-knowledge. In: Oswald, E., Fischlin, M. (eds.) EUROCRYPT 2015. LNCS, vol. 9057, pp. 191–219. Springer, Heidelberg (2015). doi:10.1007/978-3-662-46803-6_7

[GMW87] Goldreich, O., Micali, S., Wigderson, A.: How to play any mental game or a completeness theorem for protocols with honest majority. In: Proceedings of the 19th Annual ACM Symposium on Theory of Computing, 1987, New York, USA, pp. 218–229 (1987)

[HKE13] Huang, Y., Katz, J., Evans, D.: Efficient secure two-party computation using symmetric cut-and-choose. In: Canetti, R., Garay, J.A. (eds.) CRYPTO 2013, Part II. LNCS, vol. 8043, pp. 18–35. Springer, Heidelberg (2013). doi:10.1007/978-3-642-40084-1_2

[HKK+14] Huang, Y., Katz, J., Kolesnikov, V., Kumaresan, R., Malozemoff, A.J.: Amortizing garbled circuits. In: Garay, J.A., Gennaro, R. (eds.) CRYPTO 2014, Part II. LNCS, vol. 8617, pp. 458–475. Springer, Heidelberg (2014). doi:10.1007/978-3-662-44381-1_26

[JKO13] Jawurek, M., Kerschbaum, F., Orlandi, C.: Zero-knowledge using garbled circuits: how to prove non-algebraic statements efficiently. In: 2013 ACM SIGSAC Conference on Computer and Communications Security, CCS 2013, Berlin, Germany, 4–8 November 2013, pp. 955–966 (2013)

[JS07] Jarecki, S.: Efficient two-party secure computation on committed inputs. In: Naor, M. (ed.) EUROCRYPT 2007. LNCS, vol. 4515, pp. 97–114. Springer, Heidelberg (2007). doi:10.1007/978-3-540-72540-4_6

[Lin13] Lindell, Y.: Fast cut-and-choose based protocols for malicious and covert adversaries. In: Canetti, R., Garay, J.A. (eds.) CRYPTO 2013, Part II. LNCS, vol. 8043, pp. 1–17. Springer, Heidelberg (2013). doi:10.1007/978-3-642-40084-1_1

[LP07] Lindell, Y., Pinkas, B.: An efficient protocol for secure two-party computation in the presence of malicious adversaries. In: Naor, M. (ed.) EUROCRYPT 2007. LNCS, vol. 4515, pp. 52–78. Springer, Heidelberg (2007). doi:10.1007/978-3-540-72540-4_4

[LR14] Lindell, Y., Riva, B.: Cut-and-choose yao-based secure computation in the online/offline and batch settings. In: Garay, J.A., Gennaro, R. (eds.) CRYPTO 2014, Part II. LNCS, vol. 8617, pp. 476–494. Springer, Heidelberg (2014). doi:10.1007/978-3-662-44381-1_27

[LR15] Lindell, Y., Riva, B.: Blazing fast 2pc in the offline/online setting with security for malicious adversaries. In: Proceedings of the 22nd ACM SIGSAC Conference on Computer and Communications Security, Denver, CO, USA, 12–16 October 2015, pp. 579–590 (2015)

[MF06] Mohassel, P., Franklin, M.K.: Efficiency tradeoffs for malicious two-party computation. In: Yung, M., Dodis, Y., Kiayias, A., Malkin, T. (eds.) PKC 2006. LNCS, vol. 3958, pp. 458–473. Springer, Heidelberg (2006). doi:10.1007/11745853_30

[MNPS04] Malkhi, D., Nisan, N., Pinkas, B., Sella, Y.: Fairplay - secure two-party computation system. In: Proceedings of the 13th USENIX Security Symposium, 9–13 August 2004, San Diego, CA, USA, pp. 287–302 (2004)

[NO09] Nielsen, J.B., Orlandi, C.: LEGO for two-party secure computation. In: Reingold, O. (ed.) TCC 2009. LNCS, vol. 5444, pp. 368–386. Springer, Heidelberg (2009). doi:10.1007/978-3-642-00457-5_22

[Pai99] Paillier, P.: Public-key cryptosystems based on composite degree residuosity classes. In: Stern, J. (ed.) EUROCRYPT 1999. LNCS, vol. 1592, p. 223. Springer, Heidelberg (1999). doi:10.1007/3-540-48910-X_16

[Yao82] Yao, A.C.-C.: Protocols for secure computations (extended abstract). In: FOCS, pp. 160–164 (1982)

[ZRE15] Zahur, S., Rosulek, M., Evans, D.: Two halves make a whole. In: Oswald, E., Fischlin, M. (eds.) EUROCRYPT 2015. LNCS, vol. 9057, pp. 220–250. Springer, Heidelberg (2015). doi:10.1007/978-3-662-46803-6_8

# Differential Privacy

# Separating Computational and Statistical Differential Privacy in the Client-Server Model

Mark Bun[✉], Yi-Hsiu Chen, and Salil Vadhan

John A. Paulson School of Engineering and Applied Sciences, Center for Research
on Computation and Society, Harvard University, Cambridge, MA, USA
{mbun,yhchen,salil}@seas.harvard.edu

**Abstract.** Differential privacy is a mathematical definition of privacy
for statistical data analysis. It guarantees that any (possibly adversarial)
data analyst is unable to learn too much information that is specific to
an individual. Mironov et al. (CRYPTO 2009) proposed several computational relaxations of differential privacy (CDP), which relax this
guarantee to hold only against computationally bounded adversaries.
Their work and subsequent work showed that CDP can yield substantial
accuracy improvements in various multiparty privacy problems. However, these works left open whether such improvements are possible in
the traditional client-server model of data analysis. In fact, Groce, Katz
and Yerukhimovich (TCC 2011) showed that, in this setting, it is impossible to take advantage of CDP for many natural statistical tasks.

Our main result shows that, assuming the existence of sub-
exponentially secure one-way functions and 2-message witness indistinguishable proofs (zaps) for **NP**, that there is in fact a computational
task in the client-server model that can be efficiently performed with
CDP, but is infeasible to perform with information-theoretic differential
privacy.

## 1 Introduction

*Differential privacy* is a formal mathematical definition of privacy for the analysis of statistical datasets. It promises that a data analyst (treated as an adversary) cannot learn too much individual-level information from the outcome of
an analysis. The traditional definition of differential privacy makes this promise
information-theoretically: Even a computationally unbounded adversary is limited in the amount of information she can learn that is specific to an individual.

©IACR 2016. This article is the final version submitted by the authors to the IACR
and to Springer-Verlag on August 23, 2016.

M. Bun—Supported by an NDSEG Fellowship and NSF grant CNS-1237235. Part
of this work was done while the author was visiting Yale University.

Y.-H. Chen—Supported by NSF grant CCF-1420938.

S. Vadhan—Supported by NSF grant CNS-1237235 and a Simons Investigator
Award. Part of this work was done while the author was visiting the Shing-Tung
Yau Center and the Department of Applied Mathematics at National Chiao-Tung
University in Hsinchu, Taiwan.

© International Association for Cryptologic Research 2016
M. Hirt and A. Smith (Eds.): TCC 2016-B, Part I, LNCS 9985, pp. 607–634, 2016.
DOI: 10.1007/978-3-662-53641-4_23

On one hand, there are now numerous techniques that actually achieve this strong guarantee of privacy for a rich body of computational tasks. On the other hand, the information-theoretic definition of differential privacy does not itself permit the use of basic cryptographic primitives that naturally arise in the practice of differential privacy (such as the use of cryptographically secure pseudorandom generators in place of perfect randomness). More importantly, computationally secure relaxations of differential privacy open the door to designing improved mechanisms: ones that either achieve better *utility* (accuracy) or *computational efficiency* over their information-theoretically secure counterparts.

Motivated by these observations, and building on ideas suggested in [BNO08], Mironov et al. [MPRV09] proposed several definitions of *computational differential privacy* (CDP). All of these definitions formalize what it means for the output of a mechanism to "look" differentially private to a computationally bounded (i.e. probabilistic polynomial-time) adversary. The sequence of works [DKM+06, BNO08, MPRV09] introduced a paradigm that enables *two or more parties* to take advantage of CDP, either to achieve better utility or reduced round complexity, when computing a joint function of their private inputs: The parties use a secure multi-party computation protocol to simulate having a trusted third party perform a differentially private computation on the union of their inputs. Subsequent work [MMP+10] showed that such a CDP protocol for approximating the Hamming distance between two private bit vectors is in fact more accurate than any (information-theoretically secure) differentially private protocol for the same task. A number of works [CSS12, GMPS13, HOZ13, KMS14, GKM+16] have since sought to characterize the extent to which CDP yields accuracy improvements for two-party privacy problems.

Despite the success of CDP in the design of improved algorithms in the multi-party setting, much less is known about what can be achieved in the traditional client-server model, in which a trusted curator holds all of the sensitive data and mediates access to it. Beyond just the absence of any techniques for taking advantage of CDP in this setting, results of Groce, Katz, and Yerukhimovich [GKY11] (discussed in more detail below) show that CDP yields no additional power in the client-server model for many basic statistical tasks. An additional barrier stems from the fact that all known lower bounds against computationally efficient differentially private algorithms [DNR+09, UV11, Ull13, BZ14, BZ16] in the client-server model are proved by exhibiting computationally efficient adversaries. Thus, these lower bounds rule out the existence of CDP mechanisms just as well as they rule out differentially private ones.

In this work, we give the first example of a computational problem in the client-server model which can be solved in polynomial-time with CDP, but (under plausible assumptions) is computationally infeasible to solve with (information-theoretic) differential privacy. Our problem is specified by an efficiently computable *utility function* $u$, which takes as input a dataset $D \in \mathcal{X}^n$ and an answer $r \in \mathcal{R}$, and outputs 1 if the answer $r$ is "good" for the dataset $D$, and 0 otherwise.

**Theorem 1 (Main (Informal)).** *Assuming the existence of sub-exponentially secure one-way functions and "exponentially extractable" 2-message witness indistinguishable proofs (zaps) for* **NP**, *there exists an efficiently computable utility function* $u : \mathcal{X}^n \times \mathcal{R} \to \{0, 1\}$ *such that*

1. *There exists a polynomial time CDP mechanism $M^{\mathrm{CDP}}$ such that for every dataset $D \in \mathcal{X}^n$, we have $\Pr[u(D, M^{\mathrm{CDP}}(D)) = 1] \geq 2/3$.*
2. *There exists a computationally unbounded differentially private mechanism $M^{\mathrm{unb}}$ such that $\Pr[u(D, M^{\mathrm{unb}}(D)) = 1] \geq 2/3$.*
3. *For every polynomial time differentially private $M$, there exists a dataset $D \in \mathcal{X}^n$, such that $\Pr[u(D, M(D)) = 1] \leq 1/3$.*

Note that the theorem provides a task where achieving differential privacy is infeasible – not impossible. This is inherent because the CDP mechanism we exhibit (for item 1) satisfies a simulation-based form of CDP ("SIM-CDP"), which implies the existence of a (possibly inefficient) differentially private mechanism, provided the utility function $u$ is efficiently computable as we require. It remains an intriguing open problem to exhibit a task that can be achieved with a weaker indistinguishably-based notion of CDP ("IND-CDP") but is *impossible* to achieve (even inefficiently) with differential privacy. Such a task would also separate IND-CDP and SIM-CDP, which is an interesting open problem in its own right.

*Circumventing the impossibility results of* [GKY11]. Groce et al. showed that in many natural circumstances, computational differential privacy cannot yield any additional power over differential privacy in the client-server model. In particular, they showed two impossibility results:

1. If a CDP mechanism accesses a one-way function (or more generally, any cryptographic primitive that can be instantiated with a random function) in a black-box way, then it can be simulated just as well (in terms of both utility and computationally efficiency) by a differentially private mechanism.
2. If the output of a CDP mechanism is in $\mathbb{R}^d$ (for some constant $d$) and its utility is measured via an $L_p$-norm, then the mechanism can be simulated by a differentially private one, again without significant loss of utility or efficiency.

(In Sect. 4, we revisit the techniques [GKY11] to strengthen the second result in some circumstances. In general, we show that when error is measured in any metric with doubling dimension $O(\log k)$, CDP cannot improve utility by more than a constant factor. Specifically, respect to $L_p$-error, CDP cannot do much better than DP mechanisms even when $d$ is logarithmic in the security parameter.)

We get around both of these impossibility results by (1) making non-black-box use of one-way functions via the machinery of zap proofs and (2) relying on a utility function that is far from the form in which the second result of [GKY11] applies. Indeed, our utility function is cryptographic and unnatural from a data analysis point view. Roughly speaking, it asks whether the answer $r$ is a valid zap proof of the statement "there exists a row of the dataset $D$ that is a valid message-signature pair" for a secure digital signature scheme. It remains an intriguing problem for future work whether a separation can be obtained from a more natural task (such as answering a polynomial number of counting queries with differential privacy).

*Our Construction and Techniques.* Our construction is based on the existence of two cryptographic primitives: an existentially unforgeable digital signature scheme (Gen, Sign, Ver), and a 2-message witness indistinguishable proof system (zap) $(P, V)$ for **NP**. We make use of complexity leveraging [CGGM00] and thus require a complexity gap between the two primitives: namely, a sub-exponential time algorithm should be able to break the security of the zap proof system, but should not be able to forge a valid message-signature pair for the digital signature scheme.

We now describe (eliding technical complications) the computational task which allows us to separate computational and information-theoretic differential privacy in the client-server model. Inspired by prior differential privacy lower bounds [DNR+09, UV11], we consider a dataset $D$ that consists of many valid message-signature pairs $(m_1, \sigma_1), \ldots, (m_n, \sigma_n)$ for the digital signature scheme. We say that a mechanism $M$ gives a useful answer on $D$, i.e. the utility function $u(D, M(D))$ evaluates to 1, if it produces a proof $\pi$ in the zap proof system that there exists a message-signature pair $(m, \sigma)$ for which $\text{Ver}(m, \sigma) = 1$.

First, let us see how the above task can be performed *inefficiently* with differential privacy. Consider the mechanism $M^{\text{unb}}$ that first confirms (in a standard differentially private way) that its input dataset indeed contains "many" valid message-signature pairs. Then $M^{\text{unb}}$ uses its unbounded computational resources to forge a canonical valid message-signature pair $(m, \sigma)$ and uses the zap prover on witness $(m, \sigma)$ to produce a proof $\pi$. Since the choice of the forged pair does not depend on the input dataset at all, the procedure as a whole is differentially private.

Now let us see how a CDP mechanism can perform the same task efficiently. Our mechanism $M^{\text{CDP}}$ again first checks that it possesses many valid message-signature pairs, but this time it simply outputs a proof $\pi$ using an arbitrary valid pair $(m_i, \sigma_i) \in D$ as its witness. Since the proof system is witness indistinguishable, a computationally bounded observer cannot distinguish $\pi$ from the canonical proof output by the differentially private mechanism $M^{\text{unb}}$. Thus, the mechanism $M^{\text{CDP}}$ is in fact CDP in the strongest (simulation-based) sense.

Despite the existence of the inefficient differentially private mechanism $M^{\text{unb}}$, we show that the existence of an efficient mechanism $M$ for this task would violate the sub-exponential security of the digital signature scheme. Suppose there were such a mechanism $M$. Now consider a sub-exponential time adversary $A$ that completely breaks the security of the zap proof system, in the sense that given a valid proof $\pi$, it is always able to recover a corresponding witness $(m, \sigma)$. Since $M$ is differentially private, the $(m, \sigma)$ extracted by $A$ cannot be in the dataset $D$ given to $M$. Thus, $(m, \sigma)$ constitutes a forgery of a valid message-signature pair, and hence the composed algorithm $A \circ M$ violates the security of the signature scheme.

## 2   Preliminaries

### 2.1   (Computational) Differential Privacy

We first set notations that will be used throughout this paper, and recall the notions of $(\varepsilon, \delta)$-differential privacy and computational differential privacy. The abbreviation "PPT" stands for "probabilistic polynomial-time Turing machine."

*Security Parameter $k$.* Let $k \in \mathbb{N}$ denote a security parameter. In this work, datasets, privacy-preserving mechanisms, and privacy parameters $\varepsilon, \delta$ will all be sequences parameterized in terms of $k$. Adversaries will also have their computational power parameterized by $k$; in particular, efficient adversaries have circuit size polynomial in $k$. A function is said to be *negligible* if it vanishes faster than any inverse polynomial in $k$.

*Dataset $D$.* A dataset $D$ is an ordered tuple of $n$ elements from some data universe $\mathcal{X}$. Two datasets $D, D'$ are said to be *adjacent* (written $D \sim D'$) if they differ in at most one row. We use $\{D_k\}_{k \in \mathbb{N}}$ to denote a sequence of datasets, each over a data universe $\mathcal{X}_k$, with sizes growing with the parameter $k$. The size in bits of a dataset $D_k$, and in particular the number of rows $n$, will always be $\mathrm{poly}(k)$.

*Mechanism $M$.* A mechanism $M : \mathcal{X}^* \rightarrow \mathcal{R}$ is a randomized function taking a dataset $D \in \mathcal{X}^*$ to an output in a range space $\mathcal{R}$. We will be especially interested in ensembles of *efficient* mechanisms $\{M_k\}_{k \in \mathbb{N}}$ where each $M_k : \mathcal{X}_k^* \rightarrow \mathcal{R}_k$, when run on an input dataset $D \in \mathcal{X}_k^n$, runs in time $\mathrm{poly}(k, n)$.

*Adversary $A$.* Given an ensemble of mechanisms $\{M_k\}_{k \in \mathbb{N}}$ with $M_k : X_k^* \rightarrow \mathcal{R}_k$, we model an adversary $\{A_k\}_{k \in \mathbb{N}}$ as a sequence of polynomial-size circuits $A_k : \mathcal{R}_k \rightarrow \{0, 1\}$. Equivalently, $\{A_k\}_{k \in \mathbb{N}}$ can be thought of as a probabilistic polynomial time Turing machine with non-uniform advice.

**Definition 1 (Differential Privacy** [DMNS06, DKM+06]**).** *A mechanism $M$ is $(\varepsilon, \delta)$-differentially private if for all adjacent datasets $D \sim D'$ and every set $S \subseteq \mathrm{Range}(M)$,*

$$\Pr[M(D) \in S] \leq e^\varepsilon \Pr[M(D') \in S] + \delta$$

*Equivalently, for all adjacent datasets $D \sim D'$ and every (computationally unbounded) algorithm $A$, we have*

$$\Pr[A(M(D)) = 1] \leq e^\varepsilon \Pr[A(M(D')) = 1] + \delta \qquad (1)$$

*For consistency with the definition of SIM-CDP, we also make the following definitions for sequences of mechanisms:*

– *An ensemble of mechanisms* $\{M_k\}_{k \in \mathbb{N}}$ *is* $\varepsilon_k$-*DP if for all* $k$, $M_k$ *is* $(\varepsilon_k, \mathrm{negl}(k))$-*differentially private.*
– *An ensemble of mechanisms* $\{M_k\}_{k \in \mathbb{N}}$ *is* $\varepsilon_k$-*PURE-DP if for all* $k$, $M_k$ *is* $(\varepsilon_k, 0)$-*differentially private.*

The above definitions are completely information-theoretic. Several computational relaxations of this definition are proposed by Mironov et al. [MPRV09]. The first "indistinguishability-based" definition, denoted IND-CDP, relaxes Condition (1) to hold against computationally-bounded adversaries:

**Definition 2 (IND-CDP).** *A sequence of mechanisms* $\{M_k\}_{k \in \mathbb{N}}$ *is* $\varepsilon_k$-*IND-CDP if there exists a negligible function* $\mathrm{negl}(\cdot)$ *such that for all sequences of pairs of* $\mathrm{poly}(k)$-*size adjacent datasets* $\{(D_k, D'_k)\}_{k \in \mathbb{N}}$, *and all non-uniform polynomial time adversaries* $A$,

$$\Pr[A(M_k(D_k)) = 1] \leq e^{\varepsilon_k} \Pr[A(M_k(D'_k)) = 1] + \mathrm{negl}(k).$$

Mironov et al. [MPRV09] also proposed a stronger "simulation-based" definition of computational differential privacy. A mechanism is said to be $\varepsilon$-SIM-CDP if its output is computationally indistinguishable from that of an $\varepsilon$-differentially private mechanism:

**Definition 3 (SIM-CDP).** *A sequence of mechanisms* $\{M_k\}_{k \in \mathbb{N}}$ *is* $\varepsilon_k$-*SIM-CDP if there exists a negligible function* $\mathrm{negl}(\cdot)$ *and a family of mechanisms* $\{M'_k\}_{k \in \mathbb{N}}$ *that is* $\varepsilon_k$-*differentially private such that for all* $\mathrm{poly}(k)$-*size datasets* $D$, *and all non-uniform polynomial time adversaries* $A$,

$$|\Pr[A(M_k(D)) = 1] - \Pr[A(M'_k(D)) = 1]| \leq \mathrm{negl}(k).$$

*If* $M'_k$ *is in fact* $\varepsilon_k$-*pure differentially private, then we say that* $\{M_k\}_{k \in \mathbb{N}}$ *is* $\varepsilon_k$-*PURE-SIM-CDP.*

Writing $A \preceq B$ to denote that a mechanism satisfying definition $A$ also satisfies definition $B$ (that is, $A$ is a stricter privacy definition than $B$). We have the following relationships between the various notions of (computational) differential privacy:

$$\mathrm{DP} \preceq \mathrm{SIM\text{-}CDP} \preceq \mathrm{IND\text{-}CDP}.$$

We will state and prove our separation between CDP and differential privacy for the simulation-based definition SIM-CDP. Since SIM-CDP is a stronger privacy notion than IND-CDP, this implies a separation between IND-CDP and differential privacy as well.

## 2.2   Utility

We describe an abstract notion of what it means for a mechanism to "succeed" at performing a computational task. We define a computational task implicitly in terms of an efficiently computable *utility function*, which takes as input a dataset $D \in \mathcal{X}^*$ and an answer $r \in \mathcal{R}$ and outputs a score describing how well $r$ solves a given problem on instance $D$. For our purposes, it suffices to consider binary-valued utility functions $u$, which output 1 iff the answer $r$ is "good" for the dataset $D$.

**Definition 4 (Utility).** *A utility function is an efficiently computable (deterministic) function $u : \mathcal{X}^* \times \mathcal{R} \rightarrow \{0,1\}$. A mechanism $M$ is $\alpha$-useful for a utility function $u : \mathcal{X}^* \times \mathcal{R} \rightarrow \{0,1\}$ if for all datasets $D$,*

$$\Pr_{r \leftarrow M(D)}[u(D,r) = 1] \geq \alpha.$$

Restricting our attention to efficiently computable utility functions is necessary to rule out pathological separations between computational and statistical notions of differential privacy. For instance, let $\{G_k\}_{k \in \mathbb{N}}$ be a pseudorandom generator with $G_k : \{0,1\}^k \rightarrow \{0,1\}^{2k}$, and consider the (hard-to-compute) function $u(0,r) = 1$ iff $r$ is in the image of $G_k$, and $u(1,r) = 1$ iff $r$ is *not* in the image of $G_k$. Then the mechanism $M(b)$ that samples from $G_k$ if $b = 0$ and samples a random string if $b = 1$ is useful with overwhelming probability. Moreover, $M$ is computationally indistinguishable from the mechanism that always outputs a random string, and hence SIM-CDP. On the other hand, the supports of $u(0,\cdot)$ and $u(1,\cdot)$ are disjoint, so no differentially private mechanism can achieve high utility with respect to $u$.

## 2.3   Zaps (2-Message WI Proofs)

The first cryptographic tool we need in our construction is 2-message witness indistinguishable proofs for **NP** ("zaps") [FS90, DN07] in the plain model (with no common reference string). Consider a language $L \in \mathbf{NP}$. A *witness relation* for $L$ is a polynomial-time decidable binary relation $R_L = \{(x,w)\}$ such that $|w| \leq \text{poly}(|x|)$ whenever $(x,w) \in R_L$, and

$$x \in L \iff \exists w \text{ s.t. } (x,w) \in R_L.$$

**Definition 5 (Zap).** *Let $R_L = \{(x,w)\}$ be a witness-relation corresponding to a language $L \in \mathbf{NP}$. A zap proof system for $R_L$ consists of a pair of algorithms $(P,V)$ where:*

- *In the first round, the verifier sends a message $\rho \leftarrow \{0,1\}^{\ell(k,|x|)}$ ("public coins"), where $\ell(\cdot,\cdot)$ is a fixed polynomial.*
- *In the second round, the prover runs a PPT $P$ that takes as input a pair $(x,w)$ and verifier's first message $\rho$ and outputs a proof $\pi$.*
- *The verifier runs an efficient, deterministic algorithm $V$ that takes as input an instance $x$, a first-round message $\rho$, and proof $\pi$, and outputs a bit in $\{0,1\}$.*

*The security requirements of the proof system are:*

1. PERFECT COMPLETENESS. *An honest prover who possesses a valid witness can always convince an honest verifier. Formally, for all $x \in \{0,1\}^{\mathrm{poly}(k)}$, $(x,w) \in R_L$, and $\rho \in \{0,1\}^{\ell(k,|x|)}$,*

$$\Pr_{\pi \leftarrow P(1^k, x, w, \rho)}[V(1^k, x, \rho, \pi) = 1] = 1.$$

2. STATISTICAL SOUNDNESS. *With overwhelming probability over the choice of $\rho$, it is impossible to convince an honest verifier of the validity of a false statement. Formally, there exists a negligible function $\mathrm{negl}(\cdot)$ such that for all sufficiently large $k$ and $t = \mathrm{poly}(k)$, we have*

$$\Pr_{\rho \leftarrow \{0,1\}^{\ell(k,t)}}[\exists x \notin L \cap \{0,1\}^t, \pi \in \{0,1\}^* : V(1^k, x, \rho, \pi) = 1] \leq \mathrm{negl}(k).$$

3. WITNESS INDISTINGUISHABILITY. *For every sequence $\{x_k\}_{k \in \mathbb{N}}$ with $|x_k| = \mathrm{poly}(k)$, every two sequences $\{w_k^1\}_{k \in \mathbb{N}}, \{w_k^2\}_{k \in \mathbb{N}}$ such that $(x_k, w_k^1), (x_k, w_k^2) \in R_L$, and every choice of the verifier's first message $\rho$, we have*

$$\{P(1^k, x_k, w_k^1, \rho)\}_{k \in \mathbb{N}} \overset{c}{\approx} \{P(1^k, x_k, w_k^2, \rho)\}_{k \in \mathbb{N}}.$$

*Namely, for every such pair of sequences, there exists a negligible function $\mathrm{negl}(\cdot)$ such that for all polynomial-time adversaries $A$ and all sufficiently large $k$, we have*

$$|\Pr[A(1^k, P(1^k, x_k, w_k^1, \rho)) = 1] - \Pr[A(1^k, P(1^k, x_k, w_k^2, \rho)) = 1]| \leq \mathrm{negl}(k).$$

In our construction, we will need more fine-grained control over the security of our zap proof system. In particular, we need the proof system to be *extractable* by an adversary running in time $2^{O(k)}$, in that such an adversary can always reverse-engineer a valid proof $\pi$ to find a witness $w$ such that $(x,w) \in R_L$. It is important to note that we require the running time of the adversary to be exponential in the security parameter $k$, but otherwise independent of the statement size $|x|$.

**Definition 6 (Extractable Zap).** *The algorithm triple $(P, V, E)$ is an extractable zap proof system if $(P, V)$ is a zap proof system and there exists an algorithm $E$ running in time $2^{O(k)}$ with the following property:*

4. (EXPONENTIAL STATISTICAL) EXTRACTABILITY. *There exists a negligible function $\mathrm{negl}(\cdot)$ such that for all $x \in \{0,1\}^{\mathrm{poly}(k)}$:*

$$\Pr_{\rho \leftarrow \{0,1\}^{\ell(k,|x|)}}[\exists \pi \in \{0,1\}^*, w \in E(1^k, x, \rho, \pi) :$$
$$(x,w) \notin R_L \land V(1^k, x, \rho, \pi) = 1] \leq \mathrm{negl}(k).$$

While we do not know whether extractability is a generic property of zaps, it is preserved under Dwork and Naor's reduction to NIZKs in the common random string model. Namely, if we plug an extractable NIZK into Dwork and Naor's construction, we obtain an extractable zap.

**Theorem 2.** *Every language in* **NP** *has an extractable zap proof system* $(P, V, E)$, *as defined in Definition 6, if there exists non-interactive zero-knowledge proofs of knowledge for* **NP** [DN07].

For completeness, we sketch Dwork and Naor's construction in Appendix B and argue its extractability.

## 2.4 Digital Signatures

The other ingredient we need in our construction is sub-exponentially strongly unforgeable digital signature schemes. Here "strong unforgeability" [ADR02] means that the adversary in the existential unforgeability game is allowed to forge a signature for a message it has queried before, as long as the signature is different than the one it received.

**Definition 7 (Sub-exponentially Strongly Unforgeable Digital Signature Scheme).** *Let* $c \in (0, 1)$ *be a constant. A* $c$-*strongly unforgeable digital signature is a triple of PPT algorithms* (Gen, Sign, Ver) *where*

- $(sk, vk) \leftarrow \text{Gen}(1^k)$: *The generation algorithm takes as input a security parameter $k$ and generates a secret key and a verification key.*
- $\sigma \leftarrow \text{Sign}(sk, m)$: *The signing algorithm signs a message* $m \in \{0, 1\}^*$ *to produce a signature* $\sigma \in \{0, 1\}^*$.
- $b \leftarrow \text{Ver}(vk, m, \sigma)$: *The (deterministic) verification algorithm outputs a bit to indicate whether the signature* $\sigma$ *is a valid signature of* $m$.

*The algorithms have the following properties:*

1. CORRECTNESS. *For every message* $m \in \{0, 1\}^*$,

$$\Pr_{\substack{(sk,vk) \leftarrow \text{Gen}(1^k) \\ \sigma \leftarrow \text{Sign}(sk,m)}} [\text{Ver}(vk, m, \sigma) = 1] = 1.$$

2. EXISTENTIAL UNFORGEABILITY. *There exists a negligible function* $\text{negl}(\cdot)$ *such that for all adversaries $A$ running in time* $2^{k^c}$,

$$\Pr_{\substack{(sk,vk) \leftarrow \text{Gen}(1^k) \\ (m,\sigma) \leftarrow A^{\text{Sign}(sk,\cdot)}(vk)}} [\text{Ver}(m, \sigma) = 1 \text{ and } (m, \sigma) \notin Q] < \text{negl}(k)$$

*where $Q$ is the set of messages-signature pairs obtained through $A$'s use of the signing oracle.*

**Theorem 3.** *If sub-exponentially secure one-way functions exist, then there is a constant* $c \in (0, 1)$ *such that a $c$-strongly unforgeable digital signature scheme exists.*

The reduction from a one-way function to digital signature [NY89, Rom90, KK05, Gol04] can be applied when both schemes are secure against sub-exponential time adversaries.

# 3   Separating CDP and Differential Privacy

In this section, we define a computational problem in the client-server model that can be efficiently solved with CDP, but not with statistical differential privacy. That is, we define a utility function $u$ for which there exists a CDP mechanism achieving high utility. On the other hand, any efficient differentially private algorithm can only have negligible utility.

**Theorem 4 (Main).** *Assume the existence of sub-exponentially secure one-way functions and extractable zaps for* **NP***. Then there exists a sequence of data universes* $\{\mathcal{X}_k\}_{k\in\mathbb{N}}$*, range spaces* $\{\mathcal{R}_k\}_{k\in\mathbb{N}}$ *and an (efficiently computable) utility function* $u_k : \mathcal{X}_k^* \times \mathcal{R}_k \to \{0,1\}$ *such that*

1.  *There exists a polynomial $p$ such that for any $\varepsilon_k, \beta_k > 0$ there exists a polynomial-time $\varepsilon_k$-PURE-SIM-CDP mechanism $\{M_k^{\mathrm{CDP}}\}_{k\in\mathbb{N}}$ and an (inefficient) $\varepsilon_k$-PURE-DP mechanism $\{M_k^{\mathrm{unb}}\}_{k\in\mathbb{N}}$ such that for every $n \geq p(k, 1/\varepsilon_k, \log(1/\beta_k))$ and dataset $D \in \mathcal{X}_k^n$, we have*

$$\Pr[u_k(D, M^{\mathrm{CDP}}(D)) = 1] \geq 1 - \beta_k \quad \text{and} \quad \Pr[u_k(D, M^{\mathrm{unb}}(D)) = 1] \geq 1 - \beta_k$$

2.  *For every $\varepsilon_k \leq O(\log k)$, $\alpha_k = 1/\operatorname{poly}(k)$, $n = \operatorname{poly}(k)$, and efficient $(\varepsilon_k, \delta = 1/n^2)$-differentially private mechanism $\{M_k'\}_{k\in\mathbb{N}}$, there exists a dataset $D \in \mathcal{X}_k^n$ such that*

$$\Pr[u(D, M'(D)) = 1] \leq \alpha_k \text{ for sufficient large } k.$$

*Remark 1* We can only hope to separate SIM-CDP and differential privacy by designing a task that is *infeasible* with differential privacy but not *impossible*. By the definition of (PURE-)SIM-CDP for a mechanism $\{M_k\}_{k\in\mathbb{N}}$, there exists an $\varepsilon_k$-(PURE-)DP mechanism $\{M_k'\}_{k\in\mathbb{N}}$ that is computationally indistinguishable from $\{M_k\}_{k\in\mathbb{N}}$. But if for every differentially private $\{M_k'\}_{k\in\mathbb{N}}$, there were a dataset $D_k \in \mathcal{X}_k^n$ such that $\Pr[u_k(D_k, M_k'(D_k)) = 1] \leq \Pr[u_k(D_k, M_k(D_k)) = 1] - 1/\operatorname{poly}(k)$, then the utility function $u_k(D_k, \cdot)$ would itself serve as a distinguisher between $\{M_k'\}_{k\in\mathbb{N}}$ and $\{M_k\}_{k\in\mathbb{N}}$.

## 3.1   Construction

Let (Gen, Sign, Ver) be a $c$-strongly unforgeable secure digital signature scheme with parameter $c > 0$ as in Definition 7. After fixing $c$, we define for each $k \in \mathbb{N}$ a reduced security parameter $k_c = k^{c/2}$. We will use $k_c$ as the security parameter for an extractable zap proof system $(P, V, E)$. Since $k$ and $k_c$ are polynomially related, a negligible function in $k$ is negligible in $k_c$ and vice versa.

Given a security parameter $k \in \mathbb{N}$, define the following sets of bit strings:

**Verification Key Space:** $\mathcal{K}_k = \{0,1\}^{\ell_1}$ where $\ell_1 = |vk|$ for $(sk, vk) \leftarrow$ Gen$(1^k)$,
**Message Space:** $\mathcal{M}_k = \{0,1\}^k$,

**Signature Space:** $\mathcal{S}_k = \{0,1\}^{\ell_2}$ where $\ell_2 = |\sigma|$ for $\sigma \leftarrow \mathrm{Sign}(sk, m)$ with $m \in \mathcal{M}_k$,

**Public Coins Space:** $\mathcal{P}_k = \{0,1\}^{\ell_3}$ where $\ell_3 = \mathrm{poly}(\ell_1)$ is the length of first-round zap messages used to prove statements from $\mathcal{K}_k$ under security parameter $k_c$,

**Data Universe:** $\mathcal{X}_k = \mathcal{K}_k \times \mathcal{M}_k \times \mathcal{S}_k \times \mathcal{P}_k$.

That is, similarly to one the hardness results of [DNR+09], we consider datasets $D$ that contain $n$ rows of the form $x_1 = (vk_1, m_1, \sigma_1, \rho_1), \ldots, x_n = (vk_n, m_n, \sigma_n, \rho_n)$ each corresponding to a verification key, message, and signature from the digital signature scheme, and to a zap verifier's public coin tosses.

Let $L \in \mathbf{NP}$ be the language

$$vk \in (L \cap \mathcal{K}_k) \iff \exists (m, \sigma) \in \mathcal{M}_k \times \mathcal{S}_k \text{ s.t. } \mathrm{Ver}(vk, m, \sigma) = 1$$

which has the natural witness relation

$$R_L = \bigcup_k \{(vk, (m, \sigma)) \in \mathcal{K}_k \times (\mathcal{M}_k \times \mathcal{S}_k) \; : \; \mathrm{Ver}(vk, m, \sigma) = 1\}.$$

Define

**Proof Space:** $\Pi_k = \{0,1\}^{\ell_4}$ where $\ell_4 = |\pi|$ for $\pi \leftarrow P(1^{k_c}, vk, (m, \sigma), \rho)$ for $vk \in (L \cap \mathcal{K}_k)$ with witness $(m, \sigma) \in \mathcal{M}_k \times \mathcal{S}_k$ and public coins $\rho \in \mathcal{P}_k$, and

**Output Space:** $\mathcal{R}_k = \mathcal{K}_k \times \mathcal{P}_k \times \Pi_k$.

*Definition of Utility Function u.* We now specify our computational task of interest via a utility function $u : \mathcal{X}_k^n \times \mathcal{R}_k \to \{0,1\}$. For any string $vk \in \mathcal{K}_k$ and $D = ((vk_1, m_1, \sigma_1, \rho_1), \cdots, (vk_n, m_n, \sigma_n, \rho_n)) \in \mathcal{X}_k^n$ define an auxiliary function

$$f_{vk, \rho}(D) = \#\{i \in [n] : vk_i = vk \wedge \rho_i = \rho \wedge \mathrm{Ver}(vk, m_i, \sigma_i) = 1\}.$$

That is, $f_{vk, \rho}$ is the number of elements of the dataset $D$ with verification key equal to $vk$ and public coin string equal to $\rho$ for which $(m_i, \sigma_i)$ is a valid message-signature pair under $vk$. We now define $u(D, (vk, \rho, \pi)) = 1$ iff

$$f_{vk, \rho}(D) \geq 9n/10 \quad \wedge \quad V(1^{k_c}, vk, \rho, \pi) = 1$$

or

$$f_{vk', \rho'}(D) < 9n/10 \quad \text{for all } vk' \in \mathcal{K}_k \text{ and } \rho' \in \mathcal{P}_k.$$

That is, the utility function $u$ is satisfied if either (1) many entries of $D$ contain valid message-signature pairs under the same verification key $vk$ with the same public coin string $\rho$ and $\pi$ is a valid proof for statement $vk$ using $\rho$, or (2) it is not the case that many entries of $D$ contain valid message-signature pairs under the same verification key, with the same public coin string (in which case any response $(vk, \rho, \pi)$ is acceptable).

## 3.2   An Inefficient Differentially Private Algorithm

We begin by showing that there is an inefficient differentially private mechanism that achieves high utility under $u$.

**Proposition 1.** *Let $k \in \mathbb{N}$. For every $\varepsilon > 0$, there exists an $(\varepsilon, 0)$-differentially private algorithm $M_k^{\mathrm{unb}} : \mathcal{X}_k^n \to \mathcal{R}_k$ such that, for every $\beta > 0$, every $n \geq \frac{10}{\varepsilon} \log(2 \cdot |\mathcal{K}_k| \cdot |\mathcal{P}_k|/\beta)) = \mathrm{poly}(1/\varepsilon, \log(1/\beta), k)$ and $D \in (\mathcal{K}_k \times \mathcal{M}_k \times \mathcal{S}_k \times \mathcal{P}_k)^n$,*

$$\Pr_{(vk, \rho, \pi) \leftarrow M_k^{\mathrm{unb}}(D)} [u(D, (vk, \rho, \pi)) = 1] \geq 1 - \beta$$

*Remark 2.* While the mechanism $M^{\mathrm{unb}}$ considered here is only accurate for $n \geq \Omega(\log |\mathcal{P}_k|)$, it is also possible to use "stability techniques" [DL09, TS13] to design an $(\varepsilon, \delta)$-differentially private mechanism that achieves high utility for $n \geq O(\log(1/\delta)/\varepsilon)$ for $\delta > 0$. We choose to provide a "pure" $\varepsilon$-differentially private algorithm here to make our separation more dramatic: Both the inefficient differentially private mechanism and the efficient SIM-CDP mechanism achieve pure $(\varepsilon, 0)$-privacy, whereas no efficient mechanism can even achieve $(\varepsilon, \delta)$-differential privacy with $\delta > 0$.

Our algorithm relies on standard differentially private techniques for identifying frequently occurring elements in a dataset.

*Report Noisy Max.* Consider a data universe $\mathcal{X}$. A predicate $q : \mathcal{X} \to \{0, 1\}$ defines a *counting query* over the set of datasets $\mathcal{X}^n$ as follows: For $D = (x_1, \ldots, x_n) \in \mathcal{X}^n$, we abuse notation by defining $q(D) = \sum_{i=1}^n q(x_i)$. We further say that a collection of counting queries $Q$ is *disjoint* if, whenever $q(x) = 1$ for some $q \in Q$ and $x \in \mathcal{X}$, we have $q'(x) = 0$ for every other $q' \neq q$ in $Q$. (Thus, disjoint counting queries slightly generalize *point functions*, which are each supported on exactly one element of the domain $\mathcal{X}$.)

The "Report Noisy Max" algorithm [DR14], combined with observations of [BV16], can efficiently and privately identify which of a set of disjoint counting queries is (approximately) the largest on a dataset $D$, and release its identity along with the corresponding noisy count. We sketch the proof of the following proposition in Appendix A.

**Proposition 2 (Report Noisy Max).** *Let $Q$ be a set of efficiently computable and sampleable disjoint counting queries over a domain $\mathcal{X}$. Further suppose that for every $x \in \mathcal{X}$, the query $q \in Q$ for which $q(x) = 1$ (if one exists) can be identified efficiently. For every $n \in \mathbb{N}$ and $\varepsilon > 0$ there is an mechanism $F : \mathcal{X}^n \to \mathcal{X} \times \mathbb{R}$ such that*

1. *$F$ runs in time $\mathrm{poly}(n, \log |\mathcal{X}|, \log |Q|, 1/\varepsilon)$.*
2. *$F$ is $\varepsilon$-differentially private.*
3. *For every dataset $D \in \mathcal{X}^n$, let $q_{\mathrm{OPT}} = \mathrm{argmax}_{q \in Q} \, q(D)$ and $\mathrm{OPT} = q_{\mathrm{OPT}}(D)$. Let $\beta > 0$. Then with probability at least $1 - \beta$, the algorithm $F$ outputs a solution $(\hat{q}, a)$ such that $a \geq \hat{q}(D) - \gamma/2$ where $\gamma = \frac{8}{\varepsilon} \cdot (\log |Q| + \log(1/\beta))$. Moreover, if $\mathrm{OPT} - \gamma > \max_{q \neq q_{\mathrm{OPT}}} q(D)$, then $\hat{q} = \mathrm{argmax}_{q \in Q} \, q(D)$.*

We are now ready to describe our unbounded algorithm $M_k^{\mathrm{unb}}$ as Algorithm 1. We prove Proposition 1 via the following two claims, capturing the privacy and utility guarantees of $M_k^{\mathrm{unb}}$, respectively.

---

**Algorithm 1.** $M_k^{\mathrm{unb}}$

---

**Input:** Dataset $D \in (\mathcal{K}_k \times \mathcal{M}_k \times \mathcal{S}_k \times \mathcal{P}_k)^n$
**Output:** Triple $(vk, \rho, \pi) \in \mathcal{K}_k \times \mathcal{P}_k \times \Pi_k$

1. Run the Report Noisy Max algorithm on $D$ with privacy parameter $\varepsilon$ using the set of disjoint counting queries $\{f_{vk,\rho} : vk \in \mathcal{K}_k, \rho \in \mathcal{P}_k\}$, obtaining an answer $((vk, \rho), a)$.
2. If $a < 7n/10$, output $(\perp, \perp, \perp)$ and halt. Otherwise:
3. Choose the lexicographically first $(m^*, \sigma^*) \in \mathcal{M}_k \times \mathcal{S}_k$ such that $\mathrm{Ver}(vk, m^*, \sigma^*) = 1$ (If no such pair exists, output $(\perp, \perp, \perp)$ and halt)
4. Let $\pi = P(1^{k_c}, vk, (m^*, \sigma^*), \rho)$, and output $(vk, \rho, \pi)$.

---

**Lemma 1.** *The algorithm $M_k^{\mathrm{unb}}$ is $\varepsilon$-differentially private.*

*Proof.* The algorithm $M_k^{\mathrm{unb}}$ accesses its input dataset $D$ only through the $\varepsilon$-differentially private Report Noisy Max algorithm (Proposition 2). Hence, by the closure of differential privacy under post-processing, $M_k^{\mathrm{unb}}$ is also $\varepsilon$-differentially private.

**Lemma 2.** *The algorithm $M_k^{\mathrm{unb}}$ is $(1 - \beta)$-useful for any number of rows $n \geq \frac{20}{\varepsilon}(\log(|\mathcal{K}_k| \cdot |\mathcal{P}_k|/\beta))$.*

*Proof.* If $f_{vk,\rho}(D) < 9n/10$ for every $vk$ and $\rho$, then the utility of the mechanism is always 1. Therefore, it suffices to consider the case when there exist $vk, \rho$ for which $f_{vk,\rho}(D) \geq 9n/10$. When such $vk$ and $\rho$ exist, observe that we have $f_{vk',\rho'}(D) \leq n/10$ for every other pair $(vk', \rho') \neq (vk, \rho)$. Thus, as long as

$$\frac{9n}{10} - \frac{n}{10} > \frac{8}{\varepsilon} \cdot (\log(|\mathcal{K}_k| \cdot |\mathcal{P}_k|) + \log(1/\beta)),$$

the Report Noisy Max algorithm successfully identifies the correct $vk, \rho$ in Step 1 with probability all but $\beta$ (Proposition 2). Moreover, the reported value $a$ is at least $7n/10$. By the perfect completeness of the zap proof system, the algorithm produces a useful triple $(vk, \rho, \pi)$ in Step 4. Thus, the mechanism as a whole is $(1 - \beta)$-useful.

### 3.3 A SIM-CDP Algorithm

We define a PPT algorithm $M_k^{\mathrm{CDP}}$ in Algorithm 2, which we argue is an efficient, SIM-CDP algorithm achieving high utility with respect to $u$.

---

**Algorithm 2.** $M_k^{\mathrm{CDP}}$

---

**Input:** Dataset $D \in (\mathcal{K}_k \times \mathcal{M}_k \times \mathcal{S}_k \times \mathcal{P}_k)^n$
**Output:** Triple $(vk, \rho, \pi) \in \mathcal{K}_k \times \mathcal{P}_k \times \Pi_k$

1. Run the Report Noisy Max algorithm on $D$ with privacy parameter $\varepsilon$ using the set of disjoint counting queries $\{f_{vk,\rho} : vk \in \mathcal{K}_k, \rho \in \mathcal{P}_k\}$, obtaining an answer $((vk, \rho), a)$.
2. If $a < 7n/10$, output $(\perp, \perp, \perp)$ and halt. Otherwise:
3. Select the first $(vk_i = vk, m_i, \sigma_i) \in D$ such that $\mathrm{Ver}(vk, m_i, \sigma_i) = 1$ (If there is no such pair in the dataset, output $(\perp, \perp, \perp)$ and halt).
4. Let $\pi = P(1^{k_c}, vk, (m_i, \sigma_i), \rho)$, and output $(vk, \rho, \pi)$.

---

The only difference between $M_k^{\mathrm{CDP}}$ and the inefficient algorithm $M_k^{\mathrm{unb}}$ occurs in Step 3, where we have replaced the inefficient process of finding a canonical message-signature pair $(m^*, \sigma^*)$ with selecting a message-signature pair $(m_i, \sigma_i)$ in the dataset. Since all the other steps (Report Noisy Max and the zap prover's algorithm) are efficient, $M_k^{\mathrm{CDP}}$ runs in polynomial time. However, this change renders $M_k^{\mathrm{CDP}}$ statistically non-differentially private, since a (computationally unbounded) adversary could reverse engineer the proof $\pi$ produced in Step 4 to recover the pair $(m_i, \sigma_i)$ contained in the dataset. On the other hand, the witness indistinguishability of the proof system implies that $M_k^{\mathrm{CDP}}$ is nevertheless computationally differentially private:

**Lemma 3.** *The algorithm $M_k^{\mathrm{CDP}}$ is $\varepsilon$-SIM-CDP provided that $n \geq (20/\varepsilon) \cdot (k + \log |\mathcal{K}_k| + \log |\mathcal{P}_k|) = \mathrm{poly}(k, 1/\varepsilon)$.*

*Proof.* Indeed, we will show that $M_k' = M_k^{\mathrm{unb}}$ is secure as the simulator for $M_k = M_k^{\mathrm{CDP}}$. That is, we will show that for any $\mathrm{poly}(k)$-size adversary $A$, that

$$\Pr[A(M_k^{\mathrm{CDP}}(D)) = 1] - \Pr[A(M_k^{\mathrm{unb}}(D)) = 1] \leq \mathrm{negl}(k).$$

First observe that by definition, the first two steps of the mechanisms are identical. Now define, for either mechanism $M_k^{\mathrm{unb}}$ or $M_k^{\mathrm{CDP}}$, a "bad" event $B$ where the mechanism in Step 1 produces a pair $((vk, \rho), a)$ for which $f_{vk,\rho}(D) = 0$, but does *not* output $(\perp, \perp, \perp)$ in Step 2. For either mechanism, the probability of the bad event $B$ is $\mathrm{negl}(k)$, as long as $n \geq (20/\varepsilon) \cdot (k + \log(|\mathcal{K}_k| \cdot |\mathcal{P}_k|))$. This follows from the utility guarantee of the Report Noisy Max algorithm (Proposition 2), setting $\beta = 2^{-k}$.

Thus, it suffices to show that for any fixing of the coins of both mechanisms in Steps 1 and 2 in which $B$ does not occur, that the mechanisms $M_k^{\mathrm{CDP}}(D)$ and $M_k^{\mathrm{unb}}(D)$ are indistinguishable. There are now two cases to consider based on the coin tosses in Steps 1 and 2:

*Case 1: Both Mechanisms Output $(\perp, \perp, \perp)$ in Step 2.* In this case,

$$\Pr[A(M_k^{\mathrm{CDP}}(D)) = 1] = \Pr[A(\perp, \perp, \perp) = 1] = \Pr[A(M_k^{\mathrm{unb}}(D)) = 1],$$

and the mechanisms are perfectly indistinguishable.

*Case 2: Step 1 Produced a Pair* $((vk, \rho), a)$ *for which* $f_{vk,\rho}(D) > 0$. In this case, we reduce to the indistinguishability of the zap proof system. Let $(vk_i = vk, m_i, \sigma_i)$ be the first entry of $D$ for which $\mathrm{Ver}(vk, m_i, \sigma_i) = 1$, and let $(m^*, \sigma^*)$ be the lexicographically first message-signature pair with $\mathrm{Ver}(vk, m^*, \sigma^*) = 1$. The proofs we are going to distinguish are $\pi_{\mathrm{CDP}} \leftarrow P(1^{k_c}, vk, (m_i, \sigma_i), \rho)$ and $\pi_{\mathrm{unb}} \leftarrow P(1^{k_c}, vk, (m^*, \sigma^*), \rho)$. Let $A^{\mathrm{zap}}(1^{k_c}, \rho, \pi) = A(vk, \rho, \pi)$. Then we have

$$\Pr[A(M_k^{\mathrm{CDP}}(D)) = 1] = \Pr[A^{\mathrm{zap}}(1^{k_c}, \rho, \pi_{\mathrm{CDP}}) = 1]$$

and

$$\Pr[A(M_k^{\mathrm{unb}}(D)) = 1] = \Pr[A^{\mathrm{zap}}(1^{k_c}, \rho, \pi_{\mathrm{unb}}) = 1].$$

Thus, indistinguishability of $M_k^{\mathrm{CDP}}(D)$ and $M_k^{\mathrm{unb}}(D)$ follows from the witness indistinguishability of the zap proof system.

The proof of Lemma 2 also shows that $M_k$ is useful for $u$.

**Lemma 4.** *The algorithm* $M_k^{\mathrm{CDP}}$ *is* $(1 - \beta)$-*useful for any number of rows* $n \geq \frac{20}{\varepsilon}(\log(2 \cdot |\mathcal{K}_k| \cdot |\mathcal{P}_k|/\beta))$.

### 3.4  Infeasibility of Differential Privacy

We now show that any efficient algorithm achieving high utility cannot be differentially private. In fact, like many prior hardness results, we provide an attack $A$ that does more than violate differential privacy. Specifically we exhibit a distribution on datasets such that, given any useful answer produced by an efficient mechanism, $A$ can with high probability recover a row of the input dataset. Following [DNR+09], we work with the following notion of a re-identifiable dataset distribution.

**Definition 8 (Re-identifiable Dataset Distribution).** *Let* $u : \mathcal{X}^n \times \mathcal{R} \to \{0, 1\}$ *be a utility function. Let* $\{\mathcal{D}_k\}_{k \in \mathbb{N}}$ *be an ensemble of distributions over* $(D_0, z) \in \mathcal{X}^{n(k)+1} \times \{0, 1\}^{\mathrm{poly}(k)}$ *for* $n(k) = \mathrm{poly}(k)$. *(Think of* $D_0$ *as a dataset on* $n + 1$ *rows, and* $z$ *as a string of auxiliary information about* $D_0$*). Let* $(D, D', i, z) \leftarrow \tilde{\mathcal{D}}_k$ *denote a sample from the following experiment: Sample* $(D_0 = (x_1, \ldots, x_{n+1}), z) \leftarrow \mathcal{D}_k$ *and* $i \in [n]$ *uniformly at random. Let* $D \in \mathcal{X}^n$ *consist of the first* $n$ *rows of* $D_0$*, and let* $D'$ *be the dataset obtained by replacing* $x_i$ *in* $D$ *with* $x_{n+1}$.

*We say the ensemble* $\{\mathcal{D}_k\}_{k \in \mathbb{N}}$ *is a* re-identifiable dataset distribution *with respect to* $u$ *if there exists a (possibly inefficient) adversary* $A$ *and a negligible function* $\mathrm{negl}(\cdot)$ *such that for all polynomial-time mechanisms* $M_k$,

1. *Whenever* $M_k$ *is useful,* $A$ *recovers a row of* $D$ *from* $M_k(D)$. *That is, for any PPT* $M_k$:

$$\Pr_{\substack{(D,D',i,z) \leftarrow \tilde{\mathcal{D}}_k \\ r \leftarrow M_k(D)}} [u(D, r) = 1 \ \wedge \ A(r, z) \notin D] \leq \mathrm{negl}(k).$$

2. *A cannot recover the row $x_i$ not contained in $D'$ from $M_k(D')$. That is, for any algorithm $M_k$:*

$$\Pr_{\substack{(D,D',i,z)\leftarrow \tilde{\mathcal{D}}_k \\ r\leftarrow M_k(D')}} [A(r,z) = x_i] \leq \text{negl}(k),$$

*where $x_i$ is the $i$-th row of $D$.*

**Proposition 3** ([DNR+09]). *If a distribution ensemble $\{\mathcal{D}_k\}_{k\in\mathbb{N}}$ on datasets of size $n(k)$ is re-identifiable with respect to a utility function $u$, then for every $\gamma > 0$ and $\alpha(k)$ with $\min\{\alpha, (1 - 8\alpha)/8n^{1+\gamma}\} \geq \text{negl}(k)$, there is no polynomial-time $(\varepsilon = \gamma \log(n), \delta = (1 - 8\alpha)/2n^{1+\gamma})$-differentially private mechanism $\{M_k\}_{k\in\mathbb{N}}$ that is $\alpha$-useful for $u$.*

*In particular, for every $\varepsilon = O(\log k), \alpha = 1/\text{poly}(k)$, there is no polynomial-time $(\varepsilon, 1/n^2)$-differentially private and $\alpha$-useful mechanism for $u$.*

*Construction of a Re-identifiable Dataset Distribution.* For $k \in \mathbb{N}$, recall that the digital signature scheme induces a choice of verification key space $\mathcal{K}_k$, message space $\mathcal{M}_k$, and signature space $\mathcal{S}_k$, each on $\text{poly}(k)$-bit strings. Let $n = \text{poly}(k)$. Define a distribution $\{\mathcal{D}_k\}_{k\in\mathbb{N}}$ as follows. To sample $(D_0, z)$ from $\mathcal{D}_k$, first sample a key pair $(sk, vk) \leftarrow \text{Gen}(1^k)$. Sample messages $m_1, \ldots, m_{n+1} \leftarrow \mathcal{M}_k$ uniformly at random. Then let $\sigma_i \leftarrow \text{Sign}(sk, m_i)$ for each $i = 1, \ldots, n+1$. Let the dataset $D_0 = (x_1, \ldots, x_{n+1})$ where $x_i = (vk, m_i, \sigma_i, \rho)$, and set the auxiliary string $z = (vk, \rho)$.

**Proposition 4.** *The distribution $\{\mathcal{D}_k\}_{k\in\mathbb{N}}$ defined above is re-identifiable with respect to the utility function $u$.*

*Proof.* We define an adversary $A : \mathcal{R}_k \times \mathcal{K}_k \rightarrow \mathcal{X}_k$. Consider an input to $A$ of the form $(r, z) = ((vk', \rho', \pi), (vk, \rho))$. If $vk' \neq vk$ or $\rho' \neq \rho$ or $\pi = \bot$, then output $(vk, \bot, \bot, \rho)$. Otherwise, run the zap extraction algorithm $E(1^{k_c}, vk, \rho, \pi)$ to extract a witness $(m, \sigma)$, and output the resulting $(vk, m, \sigma, \rho)$. Note that the running time of $A$ is $2^{O(k_c)}$.

We break the proof of re-identifiability into two lemmas. First, we show that $A$ can successfully recover a row in $D$ from any useful answer:

**Lemma 5.** *Let $M_k : \mathcal{X}_k^n \rightarrow \mathcal{R}_k$ be a PPT algorithm. Then*

$$\Pr_{\substack{(D,D',i,z)\leftarrow \tilde{\mathcal{D}}_k \\ r\leftarrow M_k(D)}} [u(D,r) = 1 \ \wedge \ A(r,z) \notin D] \leq \text{negl}(k).$$

*Proof.* First, if $u(D, r) = u(D, (vk', \rho', \pi)) = 1$, then $vk' = vk$, $\rho' = \rho$, and $V(1^k, vk, \rho, \pi) = 1$. In other words, $\pi$ is a valid proof that $vk \in (L \cup \mathcal{K}_k)$. Hence, by the extractability of the zap proof system, we have that $(m, \sigma) = E(1^{k_c}, vk, \rho, \pi)$ satisfies $(vk, (m, \sigma)) \in R_L$; namely $\text{Ver}(vk, m, \sigma) = 1$ with overwhelming probability over the choice of $\rho$.

---

**Algorithm 3.** Forgery algorithm $A_{\text{forge}}^{\text{Sign}(sk,\cdot)}$

---

**Input:** Verification key $vk$
**Output:** Message-signature pair $(m,\sigma)$

1. Sample public coins $\rho \leftarrow \mathcal{P}_k$.
2. Invoke the signing oracle $n$ times on random messages $m_i \in \mathcal{M}_k$ to get message-signature pairs $(m_1, \sigma_1), \cdots, (m_n, \sigma_n)$, and construct the dataset $D = \{(vk, m_i, \sigma_i, \rho)\}_{i \in [n]}$.
3. Obtain the result $r = (vk, \rho, \pi)$ from $M_k(D)$.
4. Output $(m, \sigma)$ where $(vk, m, \sigma, \rho) \leftarrow A(r, (vk, \rho))$.

---

Next, we use the exponential security of the digital signature scheme to show that the extracted pair $(m, \sigma)$ must indeed appear in the dataset $D$. Consider the following forgery adversary for the digital signature scheme.

The dataset built by the forgery algorithm $A_{\text{forge}}^{\text{Sign}(sk,\cdot)}$ is identically distributed to a sample $D$ from the experiment $(D, D', i, z) \leftarrow \tilde{D}_k$. Since a message-signature pair $(m, \sigma)$ appears in $D$ if and only if the signing oracle was queried on $m$ to produce $\sigma$, we have

$$\Pr_{\substack{(sk,vk) \leftarrow \text{Gen}(1^k) \\ (m,\sigma) \leftarrow A_{\text{forge}}^{\text{Sign}(sk,\cdot)}(vk)}} [\text{Ver}(m,\sigma) = 1 \wedge (m,\sigma) \notin Q]$$

$$= \Pr_{\substack{(D,D',i,z) \leftarrow \tilde{D}_k \\ r \leftarrow M_k(D)}} [u(D,r) = 1 \wedge (vk, m, \sigma, \rho) = A(r,z) \notin D].$$

The running time of the algorithm $A$, and hence the algorithm $A_{\text{forge}}^{\text{Sign}(sk,\cdot)}$, is $2^{O(k_c)} = 2^{o(k^c)}$. Thus, by the existential unforgeability of the digital signature scheme against $2^{k^c}$-time adversaries, this probability is negligible in $k$.

We next argue that $A$ cannot recover row $x_i = (vk, m_i, \sigma_i, \rho)$ from $M_k(D')$, where we recall that $D'$ is the dataset obtained by replacing row $x_i$ in $D$ with row $x_{n+1}$.

**Lemma 6.** *For every algorithm $M_k$:*

$$\Pr_{\substack{(D,D',i,z) \leftarrow \tilde{D}_k \\ r \leftarrow M_k(D')}} [A(r,z) = x_i] \leq \text{negl}(k),$$

*where $x_i$ is the $i$-th row of $D$.*

*Proof.* Since in $D_0 = ((vk, m_1, \text{Sign}_{vk}(m_1), \rho) \cdots, (vk, m_{n+1}, \text{Sign}_{vk}(m_{n+1}), \rho))$, the messages $m_1, \cdots, m_{n+1}$ are drawn independently, the dataset $D' = (D_0 - \{(vk, m_i, \sigma_i, \rho)\}) \cup \{(vk, m_{n+1}, \sigma_{n+1}, \rho)\}$ contains no information about message $m_i$. Since $m_i$ is drawn uniformly at random from the space $\mathcal{M}_k = \{0,1\}^k$, the probability that $A(r,z) = A(M_k(D'), (vk, \rho))$ outputs row $x_i$ is at most $2^{-k} = \text{negl}(k)$.

Re-identifiability of the distribution $\tilde{\mathcal{D}}_k$ follows by combining Lemmas 5 and 6.

# 4   Limits of CDP in the Client-Server Model

We revisit the techniques of [GKY11] to exhibit a setting in which efficient CDP mechanisms cannot do much better than information-theoretically differentially private mechanisms. In particular, we consider computational tasks with output in some discrete space (or which can be reduced to some discrete space) $\mathcal{R}_k$, and with utility measured via functions of the form $g : \mathcal{R}_k \times \mathcal{R}_k \to \mathbb{R}$. We show that if $(\mathcal{R}_k, g)$ forms a metric space with $O(\log k)$-doubling dimension (and other properties described in detail later), then CDP mechanisms can be efficiently transformed into differentially private ones. In particular, when $\mathcal{R}_k = \mathbb{R}^d$ for $d = O(\log k)$ and utility is measured by an $L_p$-norm, we can transform a CDP mechanism into a differentially private one.

The result in this section is incomparable to that of [GKY11]. We incur a constant-factor blowup in error, rather than a negligible additive increase as in [GKY11]. However, in the case that utility is measured by an $L_p$ norm, our result applies to output spaces of dimension that grow logarithmically in the security parameter $k$, whereas the result of [GKY11] only applies to outputs of constant dimension. In addition, we handle IND-CDP directly, while [GKY11] prove their results for SIM-CDP, and then extend them to IND-CDP by applying a reduction of [MPRV09].

## 4.1   Task and Utility

Consider a computational task with discrete output space $\mathcal{R}_k$. Let $g : \mathcal{R}_k \times \mathcal{R}_k \to \mathbb{R}$ be a metric on $\mathcal{R}_k$. We impose the following additional technical conditions on the metric space $(\mathcal{R}_k, g)$:

**Definition 9 (Property $\mathcal{L}$).** *A metric space formed by a discrete set $\mathcal{R}_k$ and a metric $g$ has property $\mathcal{L}$ if*

1. *The doubling dimension of $(\mathcal{R}_k, g)$ is $O(\log k)$. That is, for every $a \in \mathcal{R}_k$ and radius $r > 0$, the ball $B(a, r)$ centered at $a$ with radius $r$ is contained in a union of $\mathrm{poly}(k)$ balls of radius $r/2$.*
2. *The metric space is uniform. Namely, for any fixed radius $r$, the size of a ball of radius $r$ is independent of its center.*
3. *Given a center $a \in \mathcal{R}_k$ and a radius $r > 0$, the membership in the ball $B(a, r)$ can be checked in time $\mathrm{poly}(k)$.*
4. *Given a center $a \in \mathcal{R}_k$ and a radius $r > 0$, a uniformly random point in $B(a, r)$ can be sampled in time $\mathrm{poly}(k)$.*

Given a metric $g$, we can define a utility function measuring the accuracy of a mechanism with respect to $g$:

**Definition 10 ($\alpha$-accuracy).** *Consider a dataset space $\mathcal{X}_k$. Let $q_k : \mathcal{X}_k^n \to \mathcal{R}_k$ be any function on datasets of size $n$. Let $M_k : \mathcal{X}_k^n \to \mathbb{N}_k^d$ be a mechanism for approximating $q_k$. We say that $M_k$ is $\alpha_k$-accurate for $q_k$ with respect to $g$ if with overwhelming probability, the error of $M_k$ as measured by $g$ is at most $\alpha_k$. Namely, there exists a negligible function $\mathrm{negl}(\cdot)$ such that*

$$\Pr[g(q_k(D), M_k(D)) \leq \alpha_k] \geq 1 - \mathrm{negl}(k).$$

We take the failure probability here to be negligible primarily for aesthetic reasons. In general, taking the failure probability to be $\beta_k$ will yield in our result below a mechanism that is $(\varepsilon_k, \beta_k + \mathrm{negl}(k))$-differentially private.

Moreover, for reasonable queries $q_k$, taking the failure probability to be negligible is essentially without loss of generality. We can reduce the failure probability of a mechanism $M_k$ from constant to negligible by repeating the mechanism $O(\log^2 k)$ times and taking a median. By composition theorems for differential privacy, this incurs a cost of at most $O(\log^2 k)$ in the privacy parameters. But we can compensate for this loss in privacy by first increasing the sample size $n$ by a factor of $O(\log^2 k)$, and then applying a "secrecy-of-the-sample" argument [KLN+11] – running the original mechanism on a random subsample of the larger dataset. This step maintains accuracy as long as the query $q_k$ generalizes from random subsamples.

## 4.2 Result and Proof

**Theorem 5.** *Let $(\mathcal{R}_k, g)$ be a metric space with property $\mathcal{L}$. Suppose $M_k : \mathcal{X}_k^n \to \mathcal{R}_k$ is an efficient $\varepsilon_k$-IND-CDP mechanism that is $\alpha_k$-accurate for some function $q_k$ with respect to $g$. Then there exists an efficient $(\varepsilon, \mathrm{negl}(k))$-differentially private mechanism $\hat{M}_k$ that is $O(\alpha_k)$-accurate for $q_k$ with respect to $g$.*

*Proof.* We denote a ball centered at $a$ with radius $r$ in the metric space $(\mathcal{R}_k, g)$ by

$$B(a, r) = \{x \in \mathcal{R}_k : g(a, x) \leq r\}.$$

We also let $V(r) \stackrel{\text{def}}{=} |B(a, r)|$ for any $a \in \mathcal{R}_k$, which is well-defined due to the uniformity of the metric space. Now we define a mechanism $\hat{M}_k$ which outputs a uniformly random point from $B(M_k(x), c_k)$, where $c_k > 0$ is a parameter be determined later. Note that $\hat{M}_k$ can be implemented efficiently due to the efficient sampling condition of property $\mathcal{L}$. Since $g$ satisfies the triangle inequality, $\hat{M}_k$ is $(\alpha_k + c_k)$-accurate. Thus it remains to prove that $\hat{M}_k$ is $(\varepsilon, \mathrm{negl}(k))$-DP.

The key observation is that, for every $D \in \mathcal{X}_k^n$ and $s \in \mathcal{R}_k$,

$$\Pr[\hat{M}_k(D) = s] = \frac{1}{V(c_k)} \Pr[M_k(D) \in B(s, c_k)]$$

For all sets $S \subseteq \mathcal{R}_k$, we thus have

$$\Pr[\hat{M}_k(D) \in S]$$

$$\leq \left( \sum_{s \in S \cap B(q_k(D), \alpha_k + c_k)} \Pr[\hat{M}_k(D) = s] \right) + \Pr[\hat{M}_k(D) \notin B(q_k(D), \alpha_k + c_k)]$$

$$\leq \left( \sum_{s \in S \cap B(q_k(D), \alpha_k + c_k)} \frac{1}{V(c_k)} \Pr[M_k(D) \in B(s, c_k)] \right) + \mathrm{negl}(k)$$

(by the above observation and $\alpha_k$-accuracy of $M_k$)

$$\leq \left( \sum_{s \in S \cap B(q_k(D), \alpha_k + c_k)} \frac{1}{V(c_k)} \left( e^\varepsilon \Pr[M_k(D') \in B(s, c_k)] + \mathrm{negl}'(k) \right) \right) + \mathrm{negl}(k)$$

(since $M_k$ is IND-CDP, and testing containment in $B(s, c_k)$ is efficient)

$$\leq \sum_{s \in S \cap B(q_k(D), \alpha_k + c_k)} \left[ e^{\varepsilon_k} \Pr[\hat{M}_k(D') = s] + \frac{1}{V(c_k)} \mathrm{negl}'(k) \right] + \mathrm{negl}(k)$$

$$\leq e^{\varepsilon_k} \Pr[M_k(D') \in S] + \frac{V(\alpha_k + c_k)}{V(c_k)} \cdot \mathrm{negl}'(k) + \mathrm{negl}(k).$$

By the bounded doubling dimension of $(\mathcal{R}_k, g)$, we can set $c_k = O(\alpha_k)$ to make $V(\alpha_k + c_k)/V(c_k) = \mathrm{poly}(k)$. Hence $\hat{M}_k$ is a $(\varepsilon_k, \mathrm{negl}(k))$-differentially private algorithm.

$L_p$-*norm Case.* Many natural tasks can be captured by outputs in $\mathbb{R}^d$ with utility measured by an $L_p$ norm (e.g. counting queries). Since we work with efficient mechanisms, we may assume that our mechanisms always have outputs represented by $\mathrm{poly}(k)$ bits of precision. The level of precision is unimportant, so we may assume an output space represented by $k$ bits of precision for simplicity. By rescaling, we may assume all responses are integers and take values in $\mathbb{N}_k \overset{\mathrm{def}}{=} \mathbb{N} \cap [0, 2^k]$. When $d = O(\log k)$, the doubling dimension of the new discrete metric space induced by the $L_p$-norm on integral points is $O(\log k)$ ([GKL03] shows that the subspace of $\mathbb{R}^d$ equipped with $L_p$ norm has doubling dimension $O(d)$). Now the metric space almost satisfies property $\mathcal{L}$, with the exception of the uniformity condition. This is because the sizes of balls close the boundary of $\mathbb{N}_k$ are smaller than those in the interior. However, we can apply Theorem 5 to first construct a statistically DP mechanism with outputs in the larger uniform metric space $\mathbb{N}^d$. Then we may construct the final statistical mechanism $\hat{M}_k$, by projecting answers that are not in $\mathbb{N}_k^d$ to the closest point in $\mathbb{N}_k^d$. By post-processing, the modified mechanism $\hat{M}_k$ is still differentially private. Moreover, its utility is only improved since $\hat{M}_k$ can only get closer to the true query answer in every coordinate. Therefore, we have the following corollary.

**Corollary 1.** *Let $M_k : \mathcal{X}_k^n \to \mathbb{R}^d$ with $d = O(\log k)$ be an efficient $\varepsilon_k$-IND-CDP mechanism that is $\alpha_k$-accurate for some function $q_k$ when error is measured* *by an $L_p$-norm. Then there exists an efficient $(\varepsilon, \mathrm{negl}(k))$-differentially private mechanism $\hat{M}_k$ that is $O(\alpha_k)$-accurate for $q_k$.*

**Acknowledgements.** We are grateful to an anonymous reviewer for pointing out that our original construction based on non-interactive witness indistinguishable proofs could be modified to accommodate 2-message proofs (zaps).

# A    Missing Proofs

## A.1    Proof of Proposition 2

**Proposition 2 (Report Noisy Max).** *Let $Q$ be a set of efficiently computable and sampleable disjoint counting queries over a domain $\mathcal{X}$. Further suppose that for every $x \in \mathcal{X}$, the query $q \in Q$ for which $q(x) = 1$ (if one exists) can be identified efficiently. For every $n \in \mathbb{N}$ and $\varepsilon > 0$ there is an mechanism $F$ : $\mathcal{X}^n \to \mathcal{X} \times \mathbb{R}$ such that*

1. *$F$ runs in time $\mathrm{poly}(n, \log |\mathcal{X}|, \log |Q|, 1/\varepsilon)$.*
2. *$F$ is $\varepsilon$-differentially private.*
3. *For every dataset $D \in \mathcal{X}^n$, let $q_{\mathrm{OPT}} = \mathrm{argmax}_{q \in Q} q(D)$ and $\mathrm{OPT} = q_{\mathrm{OPT}}(D)$. Let $\beta > 0$. Then with probability at least $1 - \beta$, the algorithm $F$ outputs a solution $(\hat{q}, a)$ such that $a \geq \hat{q}(D) - \gamma/2$ where $\gamma = \frac{8}{\varepsilon} \cdot (\log |Q| + \log(1/\beta))$. Moreover, if $\mathrm{OPT} - \gamma > \max_{q \neq q_{\mathrm{OPT}}} q(D)$, then $\hat{q} = \mathrm{argmax}_{q \in Q} q(D)$.*

The proof of Proposition 2 relies on the existence of an efficient sanitizer for the disjoint query class $Q$. Such a sanitizer appears in [Vad16], and is based on ideas of [BV16]. (There, it is stated for the specific class of point functions, but immediately extends to disjoint counting queries).

**Proposition 3** ([Vad16, Theorem 7.1]). *Let $Q$ be a set of efficiently computable and sampleable disjoint counting queries over a domain $\mathcal{X}$. Suppose that for every element $x \in \mathcal{X}$, the query $q \in Q$ for which $q(x) = 1$ (if one exists) can be identified in time $\mathrm{polylog}(|X|)$. Let $\beta > 0$. Then there exists an algorithm San running in time $\mathrm{poly}(n, \log |X|, 1/\varepsilon)$ for which the following holds. For any database $D \in \mathcal{X}^n$, with probability at least $1 - \beta$, the algorithm San produces a "synthetic database" $\hat{D} \in \mathcal{X}^m$ such that*

$$\left| q(D) - \frac{n}{m} q(\hat{D}) \right| < \frac{4(\log |Q| + \log(1/\beta))}{\varepsilon}$$

*for every $q \in Q$.*

*Proof (Proof of Proposition 2).* Consider the algorithm $F$ which first runs the algorithm San on its input dataset to obtain a synthetic dataset $\hat{D}$, and then outputs the pair $(\hat{q}, \frac{n}{m} \hat{q}(\hat{D}))$ where $\hat{q} = \mathrm{argmax}_{q \in Q} q(\hat{D})$. The algorithm $F$ inherits its efficiency and differential privacy from San. To see that it useful, suppose San indeed produces a database $\hat{D} \in \mathcal{X}^m$ for which

$$\left| q(D) - \frac{n}{m} q(\hat{D}) \right| < \frac{4(\log |Q| + \log(1/\beta))}{\varepsilon}$$

for every $q \in Q$. Let $q_{\text{OPT}} = \text{argmax}_{q \in Q} \, q(D)$, and $\gamma = 8(\log|Q| + \log(1/\beta))/\varepsilon$. Then $\frac{n}{m}\hat{q}(\hat{D}) \geq \frac{n}{m}q_{\text{OPT}}(\hat{D}) \geq q_{\text{OPT}}(D) - \gamma/2$. Moreover, suppose $q_{\text{OPT}}(D) - \gamma > \max_{q \neq q_{\text{OPT}}} q(D)$. Then for any $q' \neq q_{\text{OPT}}$, we have

$$\frac{n}{m}q'(\hat{D}) < q'(D) + \gamma/2 < q_{\text{OPT}}(D) - \gamma/2 \leq \frac{n}{m}\hat{q}(\hat{D}).$$

Hence $q'(\hat{D}) < \hat{q}(\hat{D})$ for every $q' \neq q_{\text{OPT}}$, and hence $\hat{q} = q_{\text{OPT}}$.

# B    Extractability for Zap Proof Systems

## B.1    Non-interactive Zero Knowledge Proofs

Most known constructions of zaps, as defined in Definition 5, are based on constructions of non-interactive zero knowledge proofs or arguments in the common reference string model. We review the requirements of such proof systems below.

**Definition 11 (NIZK Proofs and Arguments).** *Let* $R_L = \{(x, w)\}$ *be a witness-relation corresponding to a language* $L \in \mathbf{NP}$. *A non-interactive zero-knowledge proof (or argument) system for* $R_L$ *consists of a triple of algorithms* $(\text{Gen}, P, V)$ *where:*

- *The generator* Gen *is a PPT that takes as input a security parameter* $k$ *and statement length* $t = \text{poly}(k)$, *and produces a common reference string* crs. *An important special case is where* $\text{Gen}(1^k, 1^t)$ *outputs a uniformly random string, in which case we say the proof (or argument) system operates in the* common random *string model.*
- *The prover* $P$ *is a PPT that takes as input a* crs *and a pair* $(x, w)$ *and outputs a proof* $\pi$.
- *The verifier* $V$ *is an efficient, deterministic algorithm that takes as input a* crs, *an instance* $x$ *and proof* $\pi$, *and outputs a bit in* $\{0, 1\}$.

*Various security requirements we can impose on the proof system are:*

PERFECT COMPLETENESS. *An honest prover who possesses a valid witness can always convince an honest verifier. Formally, for all* $(x, w) \in R_L$,

$$\Pr_{\substack{\text{crs} \leftarrow \text{Gen}(1^k, 1^{|x|}) \\ \pi \leftarrow P(\text{crs}, x, w)}}[V(\text{crs}, x, \pi) = 1] = 1.$$

STATISTICAL SOUNDNESS. *It is statistically impossible to convince an honest verifier of the validity of a false statement. There exists a negligible function* $\text{negl}(\cdot)$ *such that for every sequence* $\{x_k\}_{k \in \mathbb{N}}$ *of* $\text{poly}(k)$-*size statements* $x_k \notin L$,

$$\Pr_{\text{crs} \leftarrow \text{Gen}(1^k, 1^{|x_k|})}[\exists \pi \in \{0, 1\}^* \text{ s.t. } V(\text{crs}, x_k, \pi) = 1] \leq \text{negl}(k).$$

COMPUTATIONAL ZERO-KNOWLEDGE. *Proofs do not reveal anything to the verifier beyond their validity. Formally, a proof system is* computational zero-knowledge *if there exists a PPT simulator* $(S_1, S_2)$ *where* $S_1$ *produces a simulated common reference string* crs *with associated trapdoor* $\tau$. *The pair* $(\text{crs}, \tau)$ *allows* $S_2$ *to simulate accepting proofs without knowledge of a witness* $w$. *That is, there exists a negligible function* negl *such that for all (possibly cheating) PPT verifiers* $V^*$ *and sequences* $\{(x_k, w_k)\}_{k \in \mathbb{N}}$ *of* poly($k$)-*size statement-witness pairs* $(x_k, w_k) \in R_L$,

$$\left| \Pr_{\substack{\text{crs} \leftarrow \text{Gen}(1^k, 1^{|x_k|}) \\ \pi \leftarrow P(\text{crs}, x_k, w_k)}} [V^*(\text{crs}, x_k, \pi) = 1] \right.$$

$$\left. - \Pr_{\substack{(\text{crs}, \tau) \leftarrow S_1(1^k, 1^{|x_k|}) \\ \pi \leftarrow S_2(\text{crs}, \tau, x_k)}} [V^*(\text{crs}, x_k, \pi) = 1] \right| \leq \text{negl}(k).$$

STATISTICAL KNOWLEDGE EXTRACTION. *A proof system is additionally a* proof of knowledge *if a witness can be extracted from a valid proof. That is, there exists a polynomial-time knowledge extractor* $E = (E_1, E_2)$ *such that* $E_1$ *produces a simulated common reference string* crs *with associated extraction key* $\xi$, *which we assume to have length* $O(k)$.[1] *The pair* $(\text{crs}, \xi)$ *allows the deterministic algorithm* $E_2$ *to extract a witness from a proof. Formally, the first component of* $(\text{crs}, \xi) \leftarrow E_1(1^k, 1^{|x|})$ *is identically distributed to* crs $\leftarrow$ Gen($1^k, 1^{|x|}$). *Moreover, there exists a negligible function* negl *such that for every* $x \in \{0, 1\}^{\text{poly}(k)}$,

$$\Pr_{\text{crs} \leftarrow \text{Gen}(1^k, 1^{|x|})} \left[ \exists \xi \in \{0, 1\}^*, \pi \in \{0, 1\}^*, w \in E_2(\text{crs}, \xi, x, \pi) : \right.$$

$$\left. (\text{crs}, \xi) \in E_1(1^k, 1^{|x|}) \ \wedge \ (x, w) \notin R_L \ \wedge \ V(1^k, x, \pi) = 1 \right] \leq \text{negl}(k).$$

*For technical reasons, we also require that the relation* $\{(\text{crs}, \xi) \in E_1(1^k, 1^{|x|})\}$ *be recognizable in polynomial-time, which will always be the case for our constructions.*

## B.2    Extractability of Zaps Based on Exponentially Extractable NIZKs

We next describe Dwork and Naor's original construction of zaps [DN07]. Here, we show that extractable zaps can be based on the existence of NIZK proofs of knowledge in the common *random* string model, which can in turn be built from various number theoretic assumptions [DP92, DDP00, GOS12]. (Recall that in

---

[1] Such a constraint which depends only on the security parameter $k$ will be important for meeting our definition of exponentially extractable zaps.

the common random string model for NIZK proofs, the crs generation algorithm simply outputs a uniformly random string.) The discussion in this section can be summarized by the following theorem.

**Theorem 6.** *Let $R_L$ be a witness relation for a language $L \in$ **NP**. Then $R_L$ has an extractable zap proof system if:*

*There exists a non-interactive zero-knowledge proof of knowledge for $R_L$ (in the common random string model) with perfect completeness, statistical soundness, computational zero-knowledge, and statistical extractability.*

The existence of such proofs of knowledge for **NP** can be based on any of the following assumptions:

1. The existence of NIZK proofs of membership for **NP** and "dense secure public-key encryption schemes" [DP92]. NIZK proofs of membership can in turn be constructed from trapdoor permutations [FLS99] or indistinguishability obfuscation and one-way functions [BP15]. Dense secure public-key encryption schemes can be constructed under the hardness of factoring Blum integers [DDP00] or the Diffie-Hellman assumption [DP92].
2. The decisional linear assumption for groups equipped with a bilinear map [GOS12].

The remainder of this section is devoted to the proof of Theorem 6. Let $R_L$ be a witness relation for a language $L \in$ **NP**. Let $(P_{\text{NIZK}}, V_{\text{NIZK}})$ be a NIZK proof system in the common random string model. We now describe Dwork and Naor's [DN07] zap proof system for $R_L$ based on $(P_{\text{NIZK}}, V_{\text{NIZK}})$.

For simplicity, assume we are interested in proving statements $x$ having length which is a fixed polynomial in $k$. Let $\ell = \ell(k)$ be a fixed polynomial. (This depends on the length of $x$ and on the soundness error of the NIZK proof system. We defer discussion of its value to the proof of Proposition 6, where it will also depend on the knowledge error of the NIZK knowledge extractor $E_2$.) The verifier's first message is a string $\rho \in \{0, 1\}^{\ell \cdot m}$, which should be interpreted as a sequence of random strings $\rho_1, \ldots, \rho_\ell$ each in $\{0, 1\}^m$. Here, $m = \text{poly}(k)$ is the length of the crs used in the proof system $(P_{\text{NIZK}}, V_{\text{NIZK}})$. The prover and verifier algorithms appear as Algorithms 4 and 5 respectively.

---

**Algorithm 4.** Zap Prover $P(1^k, x, w, \rho)$

**Input:** Security parameter $k$, instance $x$, witness $w$ such that $(x, w) \in R_L$, first message $\rho$

**Output:** Proof $\pi$

1. Choose a random $m$-bit string $b \in \{0, 1\}^m$. For each $j = 1, \ldots, \ell$, let $\text{crs}_j = b \oplus \rho_j$ be the bitwise exclusive-OR of $b$ with $\rho_j$
2. For each $j = 1, \ldots, \ell$, let $\pi_j \leftarrow P_{\text{NIZK}}(\text{crs}_j, x, w)$
3. Send the verifier $\pi = (b, \pi_1, \ldots, \pi_\ell)$

---

**Algorithm 5.** Zap Verifier $V(1^k, x, \rho, \pi)$

**Input:** Security parameter $k$, instance $x$, first message $\rho$, proof $\pi = (b, \pi_1, \ldots, \pi_\ell)$
**Output:** Accept or reject decision

1. Let $\mathrm{crs}_j = b \oplus \rho_j$ for each $j = 1, \ldots, \ell$
2. Accept iff $V_{\mathrm{NIZK}}(\mathrm{crs}_j, x, \pi) = 1$ for all $j = 1, \ldots, \ell$

**Theorem 7** ([DN07]). *Suppose* $(P_{\mathrm{NIZK}}, V_{\mathrm{NIZK}})$ *is a perfectly complete and statistically sound NIZK proof system for* $R_L$ *in the common random string model. Then* $(P, V)$ *is a perfectly complete, statistically sound zap proof system for* $R_L$.

Our goal now is to show that if $(P_{\mathrm{NIZK}}, V_{\mathrm{NIZK}})$ is also a statistically sound *proof of knowledge*, then the zap proof system $(P, V)$ is extractable in the sense of Definition 6.

**Proposition 6.** *If, in addition,* $(P_{\mathrm{NIZK}}, V_{\mathrm{NIZK}})$ *is statistically knowledge extractable, then* $(P, V)$ *is also an extractable zap for* $R_L$.

*Proof (Proof).* Consider the extraction Algorithm 6.

**Algorithm 6.** Zap Extractor $E(1^k, x, \rho, \pi)$

**Input:** Security parameter $k$, instance $x$, first message $\rho$, proof $\pi = (b, \pi_1, \ldots, \pi_\ell)$
**Output:** Witness $w$

For each $j = 1, \ldots, \ell$:

1. Via brute force, identify (and verify) an extraction key $\xi_j$ corresponding to a common random string $\mathrm{crs}_j = b \oplus \rho_j$
2. Run the NIZK knowledge extractor $E_2(\mathrm{crs}_j, \xi_j, x, \pi_j)$ to obtain a witness $w$
3. If $(x, w) \in R_L$, halt and output $w$

Let $x \in \{0, 1\}^*$. We say a common random string $\mathrm{crs} \in \{0, 1\}^k$ is *knowledge-sound* for $x$ if there does *not* exist a pair $(\pi, \xi)$ such that

1. $V_{\mathrm{NIZK}}(\mathrm{crs}, x, \pi) = 1$,
2. $(\mathrm{crs}, \xi)$ is in the support of $E_1(1^k, 1^{|x|})$, and
3. $(x, w) \notin R_L$ for $w \leftarrow E_2(\mathrm{crs}, \xi, x, \pi)$.

**Lemma 7.** *There exists a polynomial* $\ell(k)$ *for which the following holds. Let* $x \in \{0, 1\}^{\mathrm{poly}(k)}$ *and let* $\rho_1, \ldots, \rho_\ell$ *be random $m$-bit strings. Then with overwhelming probability over the choice of $\rho$, for every $b \in \{0, 1\}^m$, there exists an index $j$ for which $\mathrm{crs}_j = b \oplus \rho_j$ is knowledge-sound for $x$.*

*Proof.* Let $q(k)$ denote the knowledge error of the NIZK proof system, i.e.

$$q(k) = \Pr_{\text{crs}\leftarrow\text{Gen}(1^k,1^{|x|})} [\exists \xi, \pi : (\text{crs}, \xi) \in E_1(1^k, 1^{|x|})$$

$$\wedge (x, E_2(\text{crs}, \xi, x, \pi)) \notin R_L \wedge V_{\text{NIZK}}(\text{crs}, x, \pi) = 1].$$

Statistical extractability of the NIZK proof system requires that $q(k) = \text{negl}(k)$ for any $|x| = \text{poly}(k)$. For any fixed $b$, the strings $\text{crs}_j = b \oplus \rho_j$ are independent and uniformly random. Therefore, the probability that all $\ell$ copies fail to be knowledge-sound for $x$ is at most $q^\ell$. The number of possible assignments to $b \in \{0, 1\}^m$ is $2^m$. Therefore, it suffices to take $\ell = 2m$ to make $2^m q^\ell < \text{negl}(k)$.

We may now complete the proof of Proposition 6.

By Lemma 7, with overwhelming probability over the choice of $\rho$, there exists an index $j$ for which $\text{crs}_j = b \oplus \rho_j$ is knowledge-sound for $x$. If the zap verifier $V$ accepts, then in particular, $V_{\text{NIZK}}(\text{crs}_j, x, \pi) = 1$. Thus, the zap knowledge extractor $E_2(\text{crs}_j, \xi_j, x, \pi_j)$ recovers a valid witness $w$ for $x$. Since the number of strings $\text{crs}_j$ that need to be checked is polynomial in $k$, and each extraction key has length $O(k)$, the extractor runs in time $2^{O(k)}$.

# References

[ADR02] An, J.H., Dodis, Y., Rabin, T.: On the security of joint signature and encryption. In: Knudsen, L.R. (ed.) EUROCRYPT 2002. LNCS, vol. 2332, pp. 83–107. Springer, Heidelberg (2002). doi:10.1007/3-540-46035-7_6

[BNO08] Beimel, A., Nissim, K., Omri, E.: Distributed private data analysis: simultaneously solving how and what. In: Wagner, D. (ed.) CRYPTO 2008. LNCS, vol. 5157, pp. 451–468. Springer, Heidelberg (2008). doi:10.1007/978-3-540-85174-5_25

[BP15] Bitansky, N., Paneth, O.: ZAPs and non-interactive witness indistinguishability from indistinguishability obfuscation. In: Dodis, Y., Nielsen, J.B. (eds.) TCC 2015, Part II. LNCS, vol. 9015, pp. 401–427. Springer, Heidelberg (2015). doi:10.1007/978-3-662-46497-7_16

[BV16] Balcer, V., Vadhan, S.: Efficient algorithms for differentially private histograms with worst-case accuracy over large domains (2016). Manuscript

[BZ14] Boneh, D., Zhandry, M.: Multiparty key exchange, efficient traitor tracing, and more from indistinguishability obfuscation. In: Garay, J.A., Gennaro, R. (eds.) CRYPTO 2014, Part I. LNCS, vol. 8616, pp. 480–499. Springer, Heidelberg (2014). doi:10.1007/978-3-662-44371-2_27

[BZ16] Bun, M., Zhandry, M.: Order-revealing encryption and the hardness of private learning. In: Kushilevitz, E., Malkin, T. (eds.) TCC 2016-A. LNCS, vol. 9562, pp. 176–206. Springer, Heidelberg (2016). doi:10.1007/978-3-662-49096-9_8

[CGGM00] Canetti, R., Goldreich, O., Goldwasser, S., Micali, S.: Resettable zero-knowledge. In: Proceedings of the Thirty-Second Annual ACM Symposium on Theory of Computing, pp. 235–244. ACM (2000)

[CSS12] Chan, T.-H.H., Shi, E., Song, D.: Privacy-preserving stream aggregation with fault tolerance. In: Keromytis, A.D. (ed.) FC 2012. LNCS, vol. 7397, pp. 200–214. Springer, Heidelberg (2012). doi:10.1007/978-3-642-32946-3_15

[DDP00] Santis, A., Crescenzo, G., Persiano, G.: Necessary and sufficient assumptions for non-interactive zero-knowledge proofs of knowledge for all NP relations. In: Montanari, U., Rolim, J.D.P., Welzl, E. (eds.) ICALP 2000. LNCS, vol. 1853, pp. 451–462. Springer, Heidelberg (2000). doi:10.1007/3-540-45022-X_38

[DKM+06] Dwork, C., Kenthapadi, K., McSherry, F., Mironov, I., Naor, M.: Our data, ourselves: privacy via distributed noise generation. In: Vaudenay, S. (ed.) EUROCRYPT 2006. LNCS, vol. 4004, pp. 486–503. Springer, Heidelberg (2006). doi:10.1007/11761679_29

[DL09] Dwork, C., Lei, J.: Differential privacy and robust statistics. In: Proceedings of the 41st Annual ACM Symposium on Theory of Computing, STOC 2009, Bethesda, 31 May–2 June 2009, pp. 371–380 (2009)

[DMNS06] Dwork, C., McSherry, F., Nissim, K., Smith, A.: Calibrating noise to sensitivity in private data analysis. In: Halevi, S., Rabin, T. (eds.) TCC 2006. LNCS, vol. 3876, pp. 265–284. Springer, Heidelberg (2006). doi:10.1007/11681878_14

[DN07] Dwork, C., Naor, M.: Zaps, their applications. SIAM J. Comput. 36(6), 1513–1543 (2007). Preliminary version in FOCS 2000

[DNR+09] Dwork, C., Naor, M., Reingold, O., Rothblum, G.N., Vadhan, S.P.: On the complexity of differentially private data release: efficient algorithms and hardness results. In: STOC, pp. 381–390 (2009)

[DP92] De Santis, A., Persiano, G.: Zero-knowledge proofs of knowledge without interaction (extended abstract). In: 33rd Annual Symposium on Foundations of Computer Science, Pittsburgh, 24–27 October 1992, pp. 427–436 (1992)

[DR14] Dwork, C., Roth, A.: The algorithmic foundations of differential privacy. Found. Trends Theor. Comput. Sci. 9(3–4), 211–407 (2014)

[FLS99] Feige, U., Lapidot, D., Shamir, A.: Multiple noninteractive zero knowledge proofs under general assumptions. SIAM J. Comput. 29(1), 1–28 (1999)

[FS90] Feige, U., Shamir, A.: Witness indistinguishable and witness hiding protocols. In: Proceedings of the Twenty-Second Annual ACM Symposium on Theory of Computing, STOC 1990, pp. 416–426. ACM, New York (1990)

[GKL03] Gupta, A., Krauthgamer, R., Lee, J.R.: Bounded geometries, fractals, and low-distortion embeddings. In: Proceedings of 44th Symposium on Foundations of Computer Science (FOCS 2003), 11–14 October 2003, Cambridge, pp. 534–543 (2003)

[GKM+16] Goyal, V., Khurana, D., Mironov, I., Pandey, O., Sahai, A.: Do distributed differentially-private protocols require oblivious transfer? In: 43rd International Colloquium Automata, Languages, and Programming, ICALp 2016, Rome, 12–15 July 2016, Proceedings, Part I (2016, to appear)

[GKY11] Groce, A., Katz, J., Yerukhimovich, A.: Limits of computational differential privacy in the client/server setting. In: Ishai, Y. (ed.) TCC 2011. LNCS, vol. 6597, pp. 417–431. Springer, Heidelberg (2011). doi:10.1007/978-3-642-19571-6_25

[GMPS13] Goyal, V., Mironov, I., Pandey, O., Sahai, A.: Accuracy-privacy tradeoffs for two-party differentially private protocols. In: Canetti, R., Garay, J.A. (eds.) CRYPTO 2013, Part I. LNCS, vol. 8042, pp. 298–315. Springer, Heidelberg (2013). doi:10.1007/978-3-642-40041-4_17

[Gol04] Goldreich, O.: Foundations of Cryptography: Basic Applications. Cambridge University Press, Cambridge (2004)

[GOS12] Groth, J., Ostrovsky, R., Sahai, A.: New techniques for noninteractive zero-knowledge. J. ACM (JACM) **59**(3), 11 (2012)

[HOZ13] Haitner, I., Omri, E., Zarosim, H.: Limits on the usefulness of random oracles. In: Sahai, A. (ed.) TCC 2013. LNCS, vol. 7785, pp. 437–456. Springer, Heidelberg (2013). doi:10.1007/978-3-642-36594-2_25

[KK05] Katz, J., Koo, C.-Y.: On constructing universal one-way hash functions from arbitrary one-way functions. IACR Cryptology ePrint Archive 2005:328 (2005)

[KLN+11] Kasiviswanathan, S.P., Lee, H.K., Nissim, K., Raskhodnikova, S., Smith, A.D.: What can we learn privately? SIAM J. Comput. **40**(3), 793–826 (2011)

[KMS14] Khurana, D., Maji, H.K., Sahai, A.: Black-box separations for differentially private protocols. In: Sarkar, P., Iwata, T. (eds.) ASIACRYPT 2014, Part II. LNCS, vol. 8874, pp. 386–405. Springer, Heidelberg (2014). doi:10.1007/978-3-662-45608-8_21

[MMP+10] McGregor, A., Mironov, I., Pitassi, T., Reingold, O., Talwar, K., Vadhan, S.: The limits of two-party differential privacy. In: 2010 51st Annual IEEE Symposium on Foundations of Computer Science (FOCS), pp. 81–90. IEEE (2010)

[MPRV09] Mironov, I., Pandey, O., Reingold, O., Vadhan, S.: Computational differential privacy. In: Halevi, S. (ed.) CRYPTO 2009. LNCS, vol. 5677, pp. 126–142. Springer, Heidelberg (2009). doi:10.1007/978-3-642-03356-8_8

[NY89] Naor, M., Yung, M.: Universal one-way hash functions and their cryptographic applications. In: Proceedings of the Twenty-First Annual ACM Symposium on Theory of Computing, STOC 1989, pp. 33–43. ACM, New York (1989)

[Rom90] Rompel, J.: One-way functions are necessary and sufficient for secure signatures. In: Proceedings of the Twenty-Second Annual ACM Symposium on Theory of Computing, STOC 1990, pp. 387–394. ACM, New York (1990)

[TS13] Thakurta, A., Smith, A.D.: Differentially private feature selection via stability arguments, and the robustness of the Lasso. In: The 26th Annual Conference on Learning Theory. COLT 2013, 12–14 June 2013, Princeton University, pp. 819–850 (2013)

[Ull13] Ullman, J.: Answering $n^{2+o(1)}$ counting queries with differential privacy is hard. In: Proceedings of the Forty-Fifth Annual ACM Symposium on Theory of Computing, pp. 361–370. ACM (2013)

[UV11] Ullman, J., Vadhan, S.: PCPs and the hardness of generating private synthetic data. In: Ishai, Y. (ed.) TCC 2011. LNCS, vol. 6597, pp. 400–416. Springer, Heidelberg (2011). doi:10.1007/978-3-642-19571-6_24

[Vad16] Vadhan, S.: The complexity of differential privacy (2016). http://privacytools.seas.harvard.edu/publications/complexity-differential-privacy

# Concentrated Differential Privacy: Simplifications, Extensions, and Lower Bounds

Mark Bun[1]([✉]) and Thomas Steinke[2]([✉])

[1] John A. Paulson School of Engineering and Applied Sciences,
Harvard University, Cambridge, MA, USA
mbun@seas.harvard.edu
[2] IBM, Almaden Research Center, San Jose, CA, USA
Thomas.Steinke@ibm.com

**Abstract.** "Concentrated differential privacy" was recently introduced by Dwork and Rothblum as a relaxation of differential privacy, which permits sharper analyses of many privacy-preserving computations. We present an alternative formulation of the concept of concentrated differential privacy in terms of the Rényi divergence between the distributions obtained by running an algorithm on neighboring inputs. With this reformulation in hand, we prove sharper quantitative results, establish lower bounds, and raise a few new questions. We also unify this approach with approximate differential privacy by giving an appropriate definition of "approximate concentrated differential privacy".

## 1 Introduction

Differential privacy [DMNS06] is a formal mathematical standard for protecting individual-level privacy in statistical data analysis. In its simplest form, (pure) differential privacy is parameterized by a real number $\varepsilon > 0$, which controls how much "privacy loss"[1] an individual can suffer when a computation (i.e., a statistical data analysis task) is performed involving his or her data.

One particular hallmark of differential privacy is that it degrades smoothly and predictably under the *composition* of multiple computations. In particular, if one performs $k$ computational tasks that are each $\varepsilon$-differentially private and combines the results of those tasks, then the computation as a whole is $k\varepsilon$-differentially private. This property makes differential privacy amenable to the

---

The full version of this work appears at https://arXiv.org/abs/1605.02065

M. Bun—Supported by an NDSEG Fellowship and NSF grant CNS-1237235. Part of this work was done while the author was visiting Yale University.

T. Steinke—Part of this work was done while the author was a Harvard University, supported by NSF grants CCF-1116616, CCF-1420938, and CNS-1237235.

[1] The privacy loss is a random variable which quantifies how much information is revealed about an individual by a computation involving their data; it depends on the outcome of the computation, the way the computation was performed, and the information that the individual wants to hide. We discuss it informally in this introduction and define it precisely in Definition 2 on p. 637.

M. Hirt and A. Smith (Eds.): TCC 2016-B, Part I, LNCS 9985, pp. 635–658, 2016.
DOI: 10.1007/978-3-662-53641-4_24

type of modular reasoning used in the design and analysis of algorithms: When a sophisticated algorithm is comprised of a sequence of differentially private steps, one can establish that the algorithm as a whole remains differentially private.

A widely-used relaxation of pure differential privacy is *approximate* or $(\varepsilon, \delta)$-differential privacy [DKM+06], which essentially guarantees that the probability that any individual suffers privacy loss exceeding $\varepsilon$ is bounded by $\delta$. For sufficiently small $\delta$, approximate $(\varepsilon, \delta)$-differential privacy provides a comparable standard of privacy protection as pure $\varepsilon$-differential privacy, while often permitting substantially more useful analyses to be performed.

Unfortunately, there are situations where, unlike pure differential privacy, approximate differential privacy is not a very elegant abstraction for mathematical analysis, particularly the analysis of composition. The "advanced composition theorem" of Dwork, Rothblum, and Vadhan [DRV10] (subsequently improved by [KOV15, MV16]) shows that the composition of $k$ tasks that are each $(\varepsilon, \delta)$-differentially private is $(\approx\sqrt{k}\varepsilon, \approx k\delta)$-differentially private. However, these bounds can be unwieldy; computing the tightest possible privacy guarantee for the composition of $k$ arbitrary mechanisms with differing $(\varepsilon_i, \delta_i)$-differential privacy guarantees is #P-hard [MV16]! Moreover, these bounds are not tight even for simple privacy-preserving computations. For instance, consider the mechanism that approximately answers $k$ statistical queries on a given database by adding independent Gaussian noise to each answer. Even for this basic computation, the advanced composition theorem does not yield a tight analysis.[2]

Dwork and Rothblum [DR16] recently put forth a different relaxation of differential privacy called *concentrated differential privacy*. Roughly, a randomized mechanism satisfies concentrated differentially privacy if the privacy loss has small mean and is subgaussian. Concentrated differential privacy behaves in a qualitatively similar way as approximate $(\varepsilon, \delta)$-differential privacy under composition. However, it permits sharper analyses of basic computational tasks, including a tight analysis of the aforementioned Gaussian mechanism.

Using the work of Dwork and Rothblum [DR16] as a starting point, we introduce an alternative formulation of the concept of concentrated differential privacy that we call "zero-concentrated differential privacy" (zCDP for short). To distinguish our definition from that of Dwork and Rothblum, we refer to their definition as "mean-concentrated differential privacy" (mCDP for short). Our definition uses the Rényi divergence between probability distributions as a different method of capturing the requirement that the privacy loss random variable is subgaussian.

---

[2] In particular, consider answering $k$ statistical queries on a dataset of $n$ individuals by adding noise drawn from $\mathcal{N}(0, (\sigma/n)^2)$ independently for each query. Each individual query satisfies $(O(\sqrt{\log(1/\delta)}/\sigma), \delta)$-differential privacy for any $\delta > 0$. Applying the advanced composition theorem shows that the composition of all $k$ queries satisfies $(O(\sqrt{k}\log(1/\delta)/\sigma), (k+1)\delta)$-differential privacy for any $\delta > 0$. However, it is well-known that this bound can be improved to $(O(\sqrt{k\log(1/\delta)}/\sigma), \delta)$-differential privacy.

## 1.1    Our Reformulation: Zero-Concentrated Differential Privacy

As is typical in the literature, we model a dataset as a multiset or tuple of $n$ elements (or "rows") in $\mathcal{X}^n$, for some "data universe" $\mathcal{X}$, where each element represents one individual's information. A (privacy-preserving) computation is a randomized algorithm $M : \mathcal{X}^n \rightarrow \mathcal{Y}$, where $\mathcal{Y}$ represents the space of all possible outcomes of the computation.

**Definition 1 (Zero-Concentrated Differential Privacy (zCDP)).** *A randomised mechanism $M : \mathcal{X}^n \rightarrow \mathcal{Y}$ is $(\xi, \rho)$-zero-concentrated differentially private (henceforth $(\xi, \rho)$-zCDP) if, for all $x, x' \in \mathcal{X}^n$ differing on a single entry and all $\alpha \in (1, \infty)$,*

$$D_\alpha \left( M(x) \| M(x') \right) \leq \xi + \rho\alpha, \tag{1}$$

*where $D_\alpha \left( M(x) \| M(x') \right)$ is the $\alpha$-Rényi divergence[3] between the distribution of $M(x)$ and the distribution of $M(x')$.*

*We define $\rho$-zCDP to be $(0, \rho)$-zCDP.[4]*

Equivalently, we can replace (1) with

$$\mathbb{E} \left[ e^{(\alpha-1)Z} \right] \leq e^{(\alpha-1)(\xi+\rho\alpha)}, \tag{2}$$

where $Z = \mathsf{PrivLoss} \left( M(x) \| M(x') \right)$ is the privacy loss random variable:

**Definition 2 (Privacy Loss Random Variable).** *Let $Y$ and $Y'$ be random variables on $\Omega$. We define the privacy loss random variable between $Y$ and $Y'$ – denoted $Z = \mathsf{PrivLoss}(Y \| Y')$ – as follows. Define a function $f : \Omega \rightarrow \mathbb{R}$ by $f(y) = \log(\mathbb{P}[Y = y] / \mathbb{P}[Y' = y])$. Then $Z$ is distributed according to $f(Y)$.*

Intuitively, the value of the privacy loss $Z = \mathsf{PrivLoss} \left( M(x) \| M(x') \right)$ represents how well we can distinguish $x$ from $x'$ given only the output $M(x)$ or $M(x')$. If $Z > 0$, then the observed output of $M$ is more likely to have occurred if the input was $x$ than if $x'$ was the input. Moreover, the larger $Z$ is, the bigger this likelihood ratio is. Likewise, $Z < 0$ indicates that the output is more likely if $x'$ is the input. If $Z = 0$, both $x$ and $x'$ "explain" the output of $M$ equally well.

A mechanism $M : \mathcal{X}^n \rightarrow \mathcal{Y}$ is $\varepsilon$-differentially private if and only if $\mathbb{P}[Z > \varepsilon] = 0$, where $Z = \mathsf{PrivLoss} \left( M(x) \| M(x') \right)$ is the privacy loss of $M$ on arbitrary inputs $x, x' \in \mathcal{X}^n$ differing in one entry. On the other hand, $M$ being $(\varepsilon, \delta)$-differentially

---

[3] Rényi divergence has a parameter $\alpha \in (1, \infty)$ which allows it to interpolate between KL-divergence ($\alpha \rightarrow 1$) and max-divergence ($\alpha \rightarrow \infty$). It should be thought of as a measure of dissimilarity between distributions. We define it formally in Sect. 2. Throughout, we assume that all logarithms are natural unless specified otherwise — that is, base $e \approx 2.718$.

[4] For clarity of exposition, we consider only $\rho$-zCDP in the introduction and give more general statements for $(\xi, \rho)$-zCDP later. We also believe that having a one-parameter definition is desirable.

private is equivalent, up to a small loss in parameters, to the requirement that $\mathbb{P}[Z > \varepsilon] \leq \delta$.

In contrast, zCDP entails a bound on the *moment generating function* of the privacy loss $Z$ — that is, $\mathbb{E}\left[e^{(\alpha-1)Z}\right]$ as a function of $\alpha - 1$. The bound (2) implies that $Z$ is a *subgaussian* random variable with small mean. Intuitively, this means that $Z$ resembles a Gaussian distribution with mean $\xi + \rho$ and variance $2\rho$. In particular, we obtain strong tail bounds on $Z$. Namely (2) implies that

$$\mathbb{P}[Z > \lambda + \xi + \rho] \leq e^{-\lambda^2/4\rho}$$

for all $\lambda > 0$.[5]

Thus zCDP requires that the privacy loss random variable is concentrated around zero (hence the name). That is, $Z$ is "small" with high probability, with larger deviations from zero becoming increasingly unlikely. Hence we are unlikely to be able to distinguish $x$ from $x'$ given the output of $M(x)$ or $M(x')$. Note that the randomness of the privacy loss random variable is taken only over the randomnesss of the mechanism $M$.

**Comparison to the Definition of Dwork and Rothblum.** For comparison, Dwork and Rothblum [DR16] define $(\mu, \tau)$-*concentrated differential privacy* for a randomized mechanism $M : \mathcal{X}^n \to \mathcal{Y}$ as the requirement that, if $Z = \mathsf{PrivLoss}\,(M(x)\|M(x'))$ is the privacy loss for $x, x' \in \mathcal{X}^n$ differing on one entry, then

$$\mathbb{E}[Z] \leq \mu \quad \text{and} \quad \mathbb{E}\left[e^{(\alpha-1)(Z - \mathbb{E}[Z])}\right] \leq e^{(\alpha-1)^2 \frac{1}{2}\tau^2}$$

for all $\alpha \in \mathbb{R}$. That is, they require both a bound on the mean of the privacy loss and that the privacy loss is tightly concentrated around its mean. To distinguish our definitions, we refer to their definition as *mean-concentrated differential privacy* (or mCDP).

Our definition, zCDP, is a *relaxation* of mCDP. In particular, a $(\mu, \tau)$-mCDP mechanism is also $(\mu - \tau^2/2, \tau^2/2)$-zCDP (which is tight for the Gaussian mechanism example), whereas the converse is not true. (However, a partial converse holds; see Lemma 24.)

## 1.2  Results

**Relationship Between zCDP and Differential Privacy.** Like Dwork and Rothblum's formulation of concentrated differential privacy, zCDP can be thought of as providing guarantees of $(\varepsilon, \delta)$-differential privacy *for all* values of $\delta > 0$:

---

[5] We only discuss bounds on the upper tail of $Z$. We can obtain similar bounds on the lower tail of $Z = \mathsf{PrivLoss}\,(M(x)\|M(x'))$ by considering $Z' = \mathsf{PrivLoss}\,(M(x')\|M(x))$.

**Proposition 3.** *If $M$ provides $\rho$-zCDP, then $M$ is $(\rho + 2\sqrt{\rho \log(1/\delta)}, \delta)$-differentially private for any $\delta > 0$.*

There is also a partial converse, which shows that, up to a loss in parameters, zCDP is equivalent to differential privacy with this $\forall \delta > 0$ quantification (see Lemma 22).

There is also a direct link from pure differential privacy to zCDP:

**Proposition 4.** *If $M$ satisfies $\varepsilon$-differential privacy, then $M$ satisfies $(\frac{1}{2}\varepsilon^2)$-zCDP.*

Dwork and Rothblum [DR16, Theorem 3.5] give a slightly weaker version of Proposition 4, which implies that $\varepsilon$-differential privacy yields $(\frac{1}{2}\varepsilon(e^\varepsilon - 1))$-zCDP; this improves on an earlier bound [DRV10] by the factor $\frac{1}{2}$.

Propositions 3 and 4 show that zCDP is an intermediate notion between pure differential privacy and approximate differential privacy. Indeed, many algorithms satisfying approximate differential privacy do in fact also satisfy zCDP.

**Gaussian Mechanism.** Just as with mCDP, the prototypical example of a mechanism satisfying zCDP is the *Gaussian mechanism*, which answers a real-valued query on a database by perturbing the true answer with Gaussian noise.

**Definition 5 (Sensitivity).** *A function $q : \mathcal{X}^n \to \mathbb{R}$ has sensitivity $\Delta$ if for all $x, x' \in \mathcal{X}^n$ differing in a single entry, we have $|q(x) - q(x')| \leq \Delta$.*

**Proposition 6 (Gaussian Mechanism).** *Let $q : \mathcal{X}^n \to \mathbb{R}$ be a sensitivity-$\Delta$ query. Consider the mechanism $M : \mathcal{X}^n \to \mathbb{R}$ that on input $x$, releases a sample from $\mathcal{N}(q(x), \sigma^2)$. Then $M$ satisfies $(\Delta^2/2\sigma^2)$-zCDP.*

We remark that either inequality defining zCDP — (1) or (2) — is exactly tight for the Gaussian mechanism for all values of $\alpha$. Thus the definition of zCDP seems tailored to the Gaussian mechanism.

**Basic Properties of zCDP.** Our definition of zCDP satisfies the key basic properties of differential privacy. Foremost, these properties include smooth degradation under composition, and invariance under postprocessing:

**Lemma 7 (Composition).** *Let $M : \mathcal{X}^n \to \mathcal{Y}$ and $M' : \mathcal{X}^n \to \mathcal{Z}$ be randomized algorithms. Suppose $M$ satisfies $\rho$-zCDP and $M'$ satisfies $\rho'$-zCDP. Define $M'' : \mathcal{X}^n \to \mathcal{Y} \times \mathcal{Z}$ by $M''(x) = (M(x), M'(x))$. Then $M''$ satisfies $(\rho + \rho')$-zCDP.*

**Lemma 8 (Postprocessing).** *Let $M : \mathcal{X}^n \to \mathcal{Y}$ and $f : \mathcal{Y} \to \mathcal{Z}$ be randomized algorithms. Suppose $M$ satisfies $\rho$-zCDP. Define $M' : \mathcal{X}^n \to \mathcal{Z}$ by $M'(x) = f(M(x))$. Then $M'$ satisfies $\rho$-zCDP.*

These properties follow immediately from corresponding properties of the Rényi divergence outlined in Lemma 15.

We remark that Dwork and Rothblum's definition of mCDP is not closed under postprocessing; we provide a counterexample in the full version of this work. (However, an arbitrary amount of postprocessing can worsen the guarantees of mCDP by at most constant factors.)

**Group Privacy.** A mechanism $M$ guarantees *group privacy* if no small group of individuals has a significant effect on the outcome of a computation (whereas the definition of zCDP only refers to individuals, which are groups of size 1). That is, group privacy for groups of size $k$ guarantees that, if $x$ and $x'$ are inputs differing on $k$ entries (rather than a single entry), then the outputs $M(x)$ and $M(x')$ are close.

Dwork and Rothblum [DR16, Theorem 4.1] gave nearly tight bounds on the group privacy guarantees of concentrated differential privacy, showing that a $(\mu = \tau^2/2, \tau)$-concentrated differentially private mechanism affords $(k^2\mu \cdot (1 + o(1)), k\tau \cdot (1 + o(1)))$-concentrated differential privacy for groups of size $k = o(1/\tau)$. We are able to show a group privacy guarantee for zCDP that is exactly tight and works for a wider range of parameters:

**Proposition 9.** *Let $M : \mathcal{X}^n \to \mathcal{Y}$ satisfy $\rho$-zCDP. Then $M$ guarantees $(k^2\rho)$-zCDP for groups of size $k$ — i.e. for every $x, x' \in \mathcal{X}^n$ differing in up to $k$ entries and every $\alpha \in (1, \infty)$, we have*

$$\mathrm{D}_\alpha\left(M(x)\|M(x')\right) \leq (k^2\rho) \cdot \alpha.$$

In particular, this bound is achieved (simultaneously for all values $\alpha$) by the Gaussian mechanism. Our proof is also simpler than that of Dwork and Rothblum; see Sect. 5.

**Lower Bounds.** The strong group privacy guarantees of zCDP yield, as an unfortunate consequence, strong lower bounds as well. We show that, as with pure differential privacy, zCDP is susceptible to information-based lower bounds, as well as to so-called packing arguments [HT10, MMP+10, De12]:

**Theorem 10.** *Let $M : \mathcal{X}^n \to \mathcal{Y}$ satisfy $\rho$-zCDP. Let $X$ be a random variable on $\mathcal{X}^n$. Then*

$$I\left(X; M(X)\right) \leq \rho \cdot n^2,$$

*where $I(\cdot; \cdot)$ denotes the mutual information between the random variables (in nats, rather than bits). Furthermore, if the entries of $X$ are independent, then $I(X; M(X)) \leq \rho \cdot n$.*

Theorem 10 yields strong lower bounds for zCDP mechanisms, as we can construct distributions $X$ such that $M(X)$ reveals a lot of information about $X$ (i.e. $I(X; M(X))$ is large) for any accurate $M$.

In particular, we obtain a strong separation between approximate differential privacy and zCDP. For example, we can show that releasing an accurate approximate histogram (or, equivalently, accurately answering all point queries) on a data domain of size $k$ requires an input with at least $n = \Theta(\sqrt{\log k})$ entries to satisfy zCDP. In contrast, under approximate differential privacy, $n$ can be *independent* of the domain size $k$ [BNS13]! In particular, our lower bounds show that "stability-based" techniques (such as those in the propose-test-release framework [DL09]) are not compatible with zCDP.

Our lower bound exploits the strong group privacy guarantee afforded by zCDP. Group privacy has been used to prove tight lower bounds for pure differential privacy [HT10, De12] and approximate differential privacy [SU15a]. These results highlight the fact that group privacy is often the limiting factor for private data analysis. For $(\varepsilon, \delta)$-differential privacy, group privacy becomes vacuous for groups of size $k = \Theta(\log(1/\delta)/\varepsilon)$. Indeed, stability-based techniques exploit precisely this breakdown in group privacy.

As a result of this strong lower bound, we show that any mechanism for answering statistical queries that satisfies zCDP can be converted into a mechanism satisfying pure differential privacy with only a quadratic blowup in its sample complexity. More precisely, the following theorem illustrates a more general result we prove in Sect. 7.

**Theorem 11.** *Let $n \in \mathbb{N}$ and $\alpha \geq 1/n$ be arbitrary. Set $\varepsilon = \alpha$ and $\rho = \alpha^2$. Let $q : \mathcal{X} \to [0,1]^k$ be an arbitrary family of statistical queries. Suppose $M : \mathcal{X}^n \to [0,1]^k$ satisfies $\rho$-zCDP and*

$$\mathop{\mathbb{E}}_{M} \left[\|M(x) - q(x)\|_\infty\right] \leq \alpha$$

*for all $x \in \mathcal{X}^n$. Then there exists $M' : \mathcal{X}^{n'} \to [0,1]^k$ for $n' = 5n^2$ satisfying $\varepsilon$-differential privacy and*

$$\mathop{\mathbb{E}}_{M'} \left[\|M'(x) - q(x)\|_\infty\right] \leq 10\alpha$$

*for all $x \in \mathcal{X}^{n'}$.*

For some classes of queries, this reduction is essentially tight. For example, for $k$ one-way marginals, the Gaussian mechanism achieves sample complexity $n = \Theta(\sqrt{k})$ subject to zCDP, whereas the Laplace mechanism achieves sample complexity $n = \Theta(k)$ subject to pure differential privacy, which is known to be optimal. For more details, see Sects. 6 and 7.

**Approximate zCDP.** To circumvent these strong lower bounds for zCDP, we consider a relaxation of zCDP in the spirit of approximate differential privacy that permits a small probability $\delta$ of (catastrophic) failure:

**Definition 12 (Approximate zCDP).** *A randomized mechanism $M : \mathcal{X}^n \to \mathcal{Y}$ is $\delta$-approximately $(\xi, \rho)$-zCDP if, for all $x, x' \in \mathcal{X}^n$ differing on a single entry, there exist events $E$ (depending on $M(x)$) and $E'$ (depending on $M(x')$) such that $\mathbb{P}[E] \geq 1 - \delta$, $\mathbb{P}[E'] \geq 1 - \delta$, and*

$$\forall \alpha \in (1, \infty) \quad D_\alpha \left(M(x)|_E \| M(x')|_{E'}\right) \leq \xi + \rho \cdot \alpha$$
$$\wedge \quad D_\alpha \left(M(x')|_{E'} \| M(x)|_E\right) \leq \xi + \rho \cdot \alpha,$$

*where $M(x)|_E$ denotes the distribution of $M(x)$ conditioned on the event $E$. We further define $\delta$-approximate $\rho$-zCDP to be $\delta$-approximate $(0, \rho)$-zCDP.*

In particular, setting $\delta = 0$ gives the original definition of zCDP. However, this definition unifies zCDP with approximate differential privacy:

**Proposition 13.** *If $M$ satisfies $(\varepsilon, \delta)$-differential privacy, then $M$ satisfies $\delta$-approximate $\frac{1}{2}\varepsilon^2$-zCDP.*

Approximate zCDP retains most of the desirable properties of zCDP, but allows us to incorporate stability-based techniques and bypass the above lower bounds. This also presents a unified tool to analyse a composition of zCDP with approximate differential privacy; see Sect. 8.

**Related Work.** Our work builds on the aforementioned prior work of Dwork and Rothblum [DR16].[6] We view our definition of concentrated differential privacy as being "morally equivalent" to their definition of concentrated differential privacy, in the sense that both definitions formalize the same concept.[7] (The formal relationship between the two definitions is discussed in Sect. 4.) However, the definition of zCDP generally seems easier to work with than mCDP. In particular, our formulation in terms of Rényi divergence simplifies many analyses.

Dwork and Rothblum prove several results about concentrated differential privacy that are similar to ours. Namely, they prove analogous properties of mCDP as we prove for zCDP. However, as noted, some of their bounds are weaker than ours; also, they do not explore lower bounds.

Several of the ideas underlying concentrated differential privacy are implicit in earlier works. In particular, the proof of the advanced composition theorem of Dwork, Rothblum, and Vadhan [DRV10] essentially uses the ideas of concentrated differential privacy.

We also remark that Tardos [Tar08] used Rényi divergence to prove lower bounds for cryptographic objects called *fingerprinting codes*. Fingerprinting codes turn out to be closely related to differential privacy [Ull13, BUV14, SU15b], and Tardos' lower bound can be (loosely) viewed as a kind of privacy-preserving algorithm.

**Further Work.** We believe that concentrated differential privacy is a useful tool for analysing private computations, as it provides both simpler and tighter bounds. We hope that CDP will be prove useful in both the theory and practice of differential privacy.

Furthermore, our lower bounds show that CDP can really be a much more stringent condition than approximate differential privacy. Thus CDP defines a "subclass" of all $(\varepsilon, \delta)$-differentially private algorithms. This subclass includes most differentially private algorithms in the literature, but not all — the most

---

[6] Although Dwork and Rothblum's work only appeared publicly in March 2016, they shared a preliminary draft of their paper with us before we commenced this work. As such, our ideas are heavily inspired by theirs.

[7] We use "concentrated differential privacy" (CDP) to refer to the underlying *concept* formalized by both definitions.

notable exceptions being algorithms that use the propose-test-release approach [DL09] to exploit low local sensitivity.

This "CDP subclass" warrants further exploration. In particular, is there a "complete" mechanism for this class of algorithms, in the same sense that the exponential mechanism [MT07, BLR13] is complete for pure differential privacy? Can we obtain a simple characterization of the sample complexity needed to satisfy CDP? The ability to prove stronger and simpler lower bounds for CDP than for approximate DP may be useful for showing the limitations of certain algorithmic paradigms. For example, any differentially private algorithm that only uses the Laplace mechanism, the exponential mechanism, the Gaussian mechanism, and the "sparse vector" technique, along with composition and postprocessing will be subject to the lower bounds for CDP.

There is also room to examine how to interpret the zCDP privacy guarantee. In particular, we leave it as an open question to understand the extent to which $\rho$-zCDP provides a stronger privacy guarantee than the implied $(\varepsilon, \delta)$-DP guarantees (cf. Proposition 3).

In general, much of the literature on differential privacy can be re-examined through the lens of CDP, which may yield new insights and results.

## 2 Rényi Divergence

Recall the definition of Rényi divergence:

**Definition 14 (Rényi Divergence [Ren61, Eq. (3.3)]).** *Let $P$ and $Q$ be probability distributions on $\Omega$. For $\alpha \in (1, \infty)$, we define the* Rényi divergence *of order $\alpha$ between $P$ and $Q$ as*

$$
\begin{aligned}
\mathrm{D}_\alpha \left( P \| Q \right) &= \frac{1}{\alpha - 1} \log \left( \int_\Omega P(x)^\alpha Q(x)^{1-\alpha} \mathrm{d}x \right) \\
&= \frac{1}{\alpha - 1} \log \left( \underset{x \sim P}{\mathbb{E}} \left[ \left( \frac{P(x)}{Q(x)} \right)^{\alpha - 1} \right] \right),
\end{aligned}
$$

*where $P(\cdot)$ and $Q(\cdot)$ are the probability mass/density functions of $P$ and $Q$ respectively or, more generally, $P(\cdot)/Q(\cdot)$ is the Radon-Nikodym derivative of $P$ with respect to $Q$.*

*We also define the* KL-divergence

$$
\mathrm{D}_1 \left( P \| Q \right) = \lim_{\alpha \to 1} \mathrm{D}_\alpha \left( P \| Q \right) = \int_\Omega P(x) \log \left( \frac{P(x)}{Q(x)} \right) \mathrm{d}x
$$

*and the* max-divergence

$$
\mathrm{D}_\infty \left( P \| Q \right) = \lim_{\alpha \to \infty} \mathrm{D}_\alpha \left( P \| Q \right) = \sup_{x \in \Omega} \log \left( \frac{P(x)}{Q(x)} \right).
$$

Alternatively, Rényi divergence can be defined in terms of the privacy loss (Definition 2) between $P$ and $Q$:

$$e^{(\alpha-1)D_\alpha(P\|Q)} = \mathop{\mathbb{E}}_{Z\sim\mathsf{PrivLoss}(P\|Q)} \left[e^{(\alpha-1)Z}\right]$$

for all $\alpha \in (1,\infty)$. Moreover, $D_1(P\|Q) = \mathop{\mathbb{E}}_{Z\sim\mathsf{PrivLoss}(P\|Q)} [Z]$.

We record several useful and well-known properties of Rényi divergence. We refer the reader to [vEH14] for proofs and discussion of these (and many other) properties.

**Lemma 15.** *Let $P$ and $Q$ be probability distributions and $\alpha \in [1,\infty]$.*

- Non-negativity: $D_\alpha(P\|Q) \geq 0$ *with equality if and only if* $P = Q$.
- Composition: *Suppose $P$ and $Q$ are distributions on $\Omega \times \Theta$. Let $P'$ and $Q'$ denote the marginal distributions on $\Omega$ induced by $P$ and $Q$ respectively. For $x \in \Omega$, let $P'_x$ and $Q'_x$ denote the conditional distributions on $\Theta$ induced by $P$ and $Q$ respectively, where $x$ specifies the first coordinate. Then*

$$D_\alpha\left(P'\|Q'\right) + \min_{x\in\Omega} D_\alpha\left(P'_x\|Q'_x\right) \leq D_\alpha\left(P\|Q\right) \leq D_\alpha\left(P'\|Q'\right) + \max_{x\in\Omega} D_\alpha\left(P'_x\|Q'_x\right).$$

*In particular if $P$ and $Q$ are product distributions, then the Rényi divergence between $P$ and $Q$ is just the sum of the Rényi divergences of the marginals.*

- Quasi-Convexity: *Let $P_0, P_1$ and $Q_0, Q_1$ be distributions on $\Omega$, and let $P = tP_0 + (1-t)P_1$ and $Q = tQ_0 + (1-t)Q_1$ for $t \in [0,1]$. Then $D_\alpha(P\|Q) \leq \max\{D_\alpha(P_0\|Q_0), D_\alpha(P_1\|Q_1)\}$. Moreover, KL divergence is convex:*

$$D_1(P\|Q) \leq tD_1(P_0\|Q_0) + (1-t)D_1(P_1\|Q_1).$$

- Postprocessing: *Let $P$ and $Q$ be distributions on $\Omega$ and let $f : \Omega \to \Theta$ be a function. Let $f(P)$ and $f(Q)$ denote the distributions on $\Theta$ induced by applying $f$ to $P$ or $Q$ respectively. Then $D_\alpha(f(P)\|f(Q)) \leq D_\alpha(P\|Q)$.*
  *Note that quasi-convexity allows us to extend this guarantee to the case where $f$ is a randomized mapping.*
- Monotonicity: *For $1 \leq \alpha \leq \alpha' \leq \infty$, $D_\alpha(P\|Q) \leq D_{\alpha'}(P\|Q)$.*

## 2.1   Gaussian Mechanism

The following lemma gives the Rényi divergence between two Gaussian distributions with the same variance.

**Lemma 16.** *Let $\mu, \nu, \sigma \in \mathbb{R}$ and $\alpha \in [1,\infty)$. Then*

$$D_\alpha\left(\mathcal{N}(\mu,\sigma^2)\|\mathcal{N}(\nu,\sigma^2)\right) = \frac{\alpha(\mu-\nu)^2}{2\sigma^2}$$

Consequently, the Gaussian mechanism, which answers a sensitivity-$\Delta$ query by adding noise drawn from $\mathcal{N}(0,\sigma^2)$, satisfies $\left(\frac{\Delta^2}{2\sigma^2}\right)$-zCDP (Proposition 6).

For the multivariate Gaussian mechanism, Lemma 16 generalises to the following.

**Lemma 17.** *Let* $\mu, \nu \in \mathbb{R}^d$, $\sigma \in \mathbb{R}$, *and* $\alpha \in [1, \infty)$. *Then*

$$D_\alpha \left( \mathcal{N}(\mu, \sigma^2 I_d) \| \mathcal{N}(\nu, \sigma^2 I_d) \right) = \frac{\alpha \| \mu - \nu \|_2^2}{2\sigma^2}$$

Thus, if $M : \mathcal{X}^n \to \mathbb{R}^d$ is the mechanism that, on input $x$, releases a sample from $\mathcal{N}(q(x), \sigma^2 I_d)$ for some function $q : \mathcal{X}^n \to \mathbb{R}^d$, then $M$ satisfies $\rho$-zCDP for

$$\rho = \frac{1}{2\sigma^2} \sup_{\substack{x, x' \in \mathcal{X}^n \\ \text{differing in one entry}}} \| q(x) - q(x') \|_2^2. \tag{3}$$

# 3  Relation to Differential Privacy

We now discuss the relationship between zCDP and the traditional definitions of pure and approximate differential privacy. There is a close relationship between the notions, but not an exact characterization.

**Definition 18 (Differential Privacy (DP) [DMNS06, DKM+06]).** *A randomized mechanism* $M : \mathcal{X}^n \to \mathcal{Y}$ *satisfies* $(\varepsilon, \delta)$-*differential privacy if, for all* $x, x' \in \mathcal{X}$ *differing in a single entry, we have*

$$\mathbb{P}\left[ M(x) \in S \right] \le e^\varepsilon \mathbb{P}\left[ M(x') \in S \right] + \delta$$

*for all (measurable)* $S \subset \mathcal{Y}$. *Further define* $\varepsilon$-*differential privacy to be* $(\varepsilon, 0)$-*differential privacy.*

## 3.1  Pure DP versus zCDP

We now show that $\varepsilon$-differential privacy implies $(\frac{1}{2}\varepsilon^2)$-zCDP (Proposition 4).

**Proposition 19.** *Let* $P$ *and* $Q$ *be probability distributions on* $\Omega$ *satisfying* $D_\infty(P\|Q) \le \varepsilon$ *and* $D_\infty(Q\|P) \le \varepsilon$. *Then* $D_\alpha(P\|Q) \le \frac{1}{2}\varepsilon^2\alpha$ *for all* $\alpha > 1$.

**Remark 20.** *In particular, Proposition 19 shows that the KL-divergence* $D_1(P\|Q) \le \frac{1}{2}\varepsilon^2$. *A bound on the KL-divergence between random variables in terms of their max-divergence is an important ingredient in the analysis of the advanced composition theorem [DRV10]. Our bound sharpens (up to lower order terms) and, in our opinion, simplifies the previous bound of* $D_1(P\|Q) \le \frac{1}{2}\varepsilon(e^\varepsilon - 1)$ *proved by Dwork and Rothblum [DR16].*

*Proof (Proof of Proposition 19).* We may assume $\frac{1}{2}\varepsilon\alpha \le 1$, as otherwise $\frac{1}{2}\varepsilon^2\alpha > \varepsilon$, whence the result follows from monotonicity. We must show that

$$e^{(\alpha-1)D_\alpha(P\|Q)} = \mathop{\mathbb{E}}_{x \sim Q} \left[ \left( \frac{P(x)}{Q(x)} \right)^\alpha \right] \le e^{\frac{1}{2}\alpha(\alpha-1)\varepsilon^2}.$$

We know that $e^{-\varepsilon} \le \frac{P(x)}{Q(x)} \le e^{\varepsilon}$ for all $x$. Define a random function $A : \Omega \to \{e^{-\varepsilon}, e^{\varepsilon}\}$ by $\mathbb{E}_A [A(x)] = \frac{P(x)}{Q(x)}$ for all $x$. By Jensen's inequality,

$$\mathbb{E}_{x \sim Q} \left[ \left( \frac{P(x)}{Q(x)} \right)^{\alpha} \right] = \mathbb{E}_{x \sim Q} \left[ \left( \mathbb{E}_A [A(x)] \right)^{\alpha} \right] \le \mathbb{E}_{x \sim Q} \left[ \mathbb{E}_A [A(x)^{\alpha}] \right] = \mathbb{E}_A [A^{\alpha}],$$

where $A$ denotes $A(x)$ for a random $x \sim Q$. We also have $\mathbb{E}_A [A] = \mathbb{E}_{x \sim Q} \left[ \frac{P(x)}{Q(x)} \right] = 1$. From this equation, we can conclude that

$$\mathbb{P}_A \left[ A = e^{-\varepsilon} \right] = \frac{e^{\varepsilon} - 1}{e^{\varepsilon} - e^{-\varepsilon}} \quad \text{and} \quad \mathbb{P}_A \left[ A = e^{\varepsilon} \right] = \frac{1 - e^{-\varepsilon}}{e^{\varepsilon} - e^{-\varepsilon}}.$$

Thus

$$e^{(\alpha-1) D_{\alpha}(P \| Q)} \le \mathbb{E}_A [A^{\alpha}]$$

$$= \frac{e^{\varepsilon} - 1}{e^{\varepsilon} - e^{-\varepsilon}} \cdot e^{-\alpha \varepsilon} + \frac{1 - e^{-\varepsilon}}{e^{\varepsilon} - e^{-\varepsilon}} \cdot e^{\alpha \varepsilon}$$

$$= \frac{\sinh(\alpha \varepsilon) - \sinh((\alpha - 1)\varepsilon)}{\sinh(\varepsilon)}.$$

The result now follows from the following inequality, which is proved in the full version of this work.

$$0 \le y < x \le 2 \implies \frac{\sinh(x) - \sinh(y)}{\sinh(x - y)} \le e^{\frac{1}{2} xy}.$$

## 3.2  Approximate DP versus zCDP

The statements in this section show that, up to some loss in parameters, zCDP is equivalent to a family of $(\varepsilon, \delta)$-DP guarantees for all $\delta > 0$.

**Lemma 21.** *Let $M : \mathcal{X}^n \to \mathcal{Y}$ satisfy $(\xi, \rho)$-zCDP. Then $M$ satisfies $(\varepsilon, \delta)$-DP for all $\delta > 0$ and*

$$\varepsilon = \xi + \rho + \sqrt{4\rho \log(1/\delta)}.$$

Thus to achieve a given $(\varepsilon, \delta)$-DP guarantee it suffices to satisfy $(\xi, \rho)$-zCDP with

$$\rho = \left( \sqrt{\varepsilon - \xi + \log(1/\delta)} - \sqrt{\log(1/\delta)} \right)^2 \approx \frac{(\varepsilon - \xi)^2}{4 \log(1/\delta)}.$$

*Proof.* Let $x, x' \in \mathcal{X}^n$ be neighbouring. Define $f(y) = \log(\mathbb{P}[M(x) = y] / \mathbb{P}[M(x') = y])$. Let $Y \sim M(x)$ and $Z = f(Y)$. That is, $Z = \mathsf{PrivLoss}(M(x) \| M(x'))$ is the privacy loss random variable. Fix $\alpha \in (1, \infty)$ to be chosen later. Then

$$\mathbb{E}\left[e^{(\alpha-1)Z}\right] = \mathop{\mathbb{E}}_{Y \sim M(x)}\left[\left(\frac{\mathbb{P}\left[M(x) = Y\right]}{\mathbb{P}\left[M(x') = Y\right]}\right)^{\alpha-1}\right] = e^{(\alpha-1)\mathrm{D}_\alpha\left(M(x)\|M(x')\right)} \leq e^{(\alpha-1)(\xi+\rho\alpha)}.$$

By Markov's inequality

$$\mathbb{P}\left[Z > \varepsilon\right] = \mathbb{P}\left[e^{(\alpha-1)Z} > e^{(\alpha-1)\varepsilon}\right] \leq \frac{\mathbb{E}\left[e^{(\alpha-1)Z}\right]}{e^{(\alpha-1)\varepsilon}} \leq e^{(\alpha-1)(\xi+\rho\alpha-\varepsilon)}.$$

Choosing $\alpha = (\varepsilon - \xi + \rho)/2\rho > 1$ gives

$$\mathbb{P}\left[Z > \varepsilon\right] \leq e^{-(\varepsilon-\xi-\rho)^2/4\rho} \leq \delta.$$

This implies that for any measurable $S \subset \mathcal{Y}$,

$$\mathbb{P}\left[M(x) \in S\right] \leq e^\varepsilon \mathbb{P}\left[M(x') \in S\right] + \delta.$$

Lemma 21 is not tight, and we give a quantitative refinement in Lemma 38 (setting $\delta = 0$ there). There, we also show a partial converse to Lemma 21:

**Lemma 22.** *Let $M : \mathcal{X}^n \to \mathcal{Y}$ satisfy $(\varepsilon, \delta)$-DP for all $\delta > 0$ and*

$$\varepsilon = \hat{\xi} + \sqrt{\hat{\rho}\log(1/\delta)} \tag{4}$$

*for some constants $\hat{\xi}, \hat{\rho} \in [0, 1]$. Then $M$ is $\left(\hat{\xi} - \frac{1}{4}\hat{\rho} + 5\sqrt[4]{\hat{\rho}}, \frac{1}{4}\hat{\rho}\right)$-zCDP.*

Thus zCDP and DP are equivalent up to a (potentially substantial) loss in parameters and the quantification over all $\delta$.

## 4    Zero- versus Mean-Concentrated Differential Privacy

We begin by recalling the definition of mean-concentrated differential privacy:

**Definition 23 (Mean-Concentrated Differential Privacy [DR16]).** *A randomized mechanism $M : \mathcal{X}^n \to \mathcal{Y}$ satisfies $(\mu, \tau)$-mCDP if, for all $x, x' \in \mathcal{X}^n$ differing in one entry, and letting $Z = \mathsf{PrivLoss}\left(M(x)\|M(x')\right)$, we have*

$$\mathbb{E}\left[Z\right] \leq \mu \quad and \quad \mathbb{E}\left[e^{\lambda\left(Z - \mathbb{E}[Z]\right)}\right] \leq e^{\lambda^2 \cdot \tau^2/2}$$

*for all $\lambda \in \mathbb{R}$.*

In contrast $(\xi, \rho)$-zCDP requires that, for all $\alpha \in (1, \infty)$, $\mathbb{E}\left[e^{(\alpha-1)Z}\right] \leq e^{(\alpha-1)(\xi+\rho\alpha)}$, where $Z \sim \mathsf{PrivLoss}\left(M(x)\|M(x')\right)$ is the privacy loss random variable. In the full version of this work, we show that these definitions are equivalent up to a (potentially significant) loss in parameters.

**Lemma 24.** *If $M : \mathcal{X}^n \to \mathcal{Y}$ satisfies $(\mu, \tau)$-mCDP, then $M$ satisfies $(\mu - \tau^2/2, \tau^2/2)$-zCDP. Conversely, if $M : \mathcal{X}^n \to \mathcal{Y}$ satisfies $(\xi, \rho)$-zCDP, then $M$ satisfies $(\xi + \rho, O(\sqrt{\xi + 2\rho}))$-mCDP.*

Thus we can convert $(\mu, \tau)$-mCDP into $(\mu - \tau^2/2, \tau^2/2)$-zCDP and then back to $(\mu, O(\sqrt{\mu + \tau^2/2}))$-mCDP. This may result in a large loss in parameters, which is why, for example, pure DP can be characterised in terms of zCDP, but not in terms of mCDP.

We view zCDP as a relaxation of mCDP; mCDP requires the privacy loss to be "tightly" concentrated about its mean and that the mean is close to the origin. The triangle inequality then implies that the privacy loss is "weakly" concentrated about the origin. (The difference between "tightly" and "weakly" accounts for the use of the triangle inequality.) On the other hand, zCDP direcly requires that the privacy loss is weakly concentrated about the origin. That is, zCDP gives a subgaussian bound on the privacy loss that is centered at zero, whereas mCDP gives a subgaussian bound that is centered at the mean and separately bounds the mean.

There may be some advantage to the stronger requirement of mCDP, either in terms of what kind of privacy guarantee it affords, or how it can be used as an analytic tool. However, it seems that for most applications, we only need what zCDP provides.

## 5    Group Privacy

In this section we show that zCDP provides privacy protections to small groups of individuals.

**Definition 25 (zCDP for Groups).** *We say that a mechanism $M : \mathcal{X}^n \to \mathcal{Y}$ provides $(\xi, \rho)$-zCDP for groups of size $k$ if, for every $x, x' \in \mathcal{X}^n$ differing in at most $k$ entries, we have*

$$\forall \alpha \in (1, \infty) \qquad D_\alpha \left( M(x) \| M(x') \right) \le \xi + \rho \cdot \alpha.$$

The usual definition of zCDP only applies to groups of size 1. Here we show that it implies bounds for all group sizes. We begin with a technical lemma.

**Lemma 26 (Triangle-like Inequality for Rényi Divergence).** *Let $P$, $Q$, and $R$ be probability distributions. Then*

$$D_\alpha \left( P \| Q \right) \le \frac{k\alpha}{k\alpha - 1} D_{\frac{k\alpha - 1}{k - 1}} \left( P \| R \right) + D_{k\alpha} \left( R \| Q \right) \tag{5}$$

*for all $k, \alpha \in (1, \infty)$.*

*Proof.* Let $p = \frac{k\alpha-1}{\alpha(k-1)}$ and $q = \frac{k\alpha-1}{\alpha-1}$. Then $\frac{1}{p} + \frac{1}{q} = \frac{\alpha(k-1)+(\alpha-1)}{k\alpha-1} = 1$. By Hölder's inequality,

$$
\begin{aligned}
e^{(\alpha-1)\mathrm{D}_\alpha(P\|Q)} &= \int_\Omega P(x)^\alpha Q(x)^{1-\alpha}\mathrm{d}x \\
&= \int_\Omega P(x)^\alpha R(x)^{-\alpha} \cdot R(x)^{\alpha-1} Q(x)^{1-\alpha} \cdot R(x)\mathrm{d}x \\
&= \mathbb{E}_{x\sim R}\left[\left(\frac{P(x)}{R(x)}\right)^\alpha \cdot \left(\frac{R(x)}{Q(x)}\right)^{\alpha-1}\right] \\
&\leq \mathbb{E}_{x\sim R}\left[\left(\frac{P(x)}{R(x)}\right)^{p\alpha}\right]^{1/p} \cdot \mathbb{E}_{x\sim R}\left[\left(\frac{R(x)}{Q(x)}\right)^{q(\alpha-1)}\right]^{1/q} \\
&= e^{(p\alpha-1)\mathrm{D}_{p\alpha}(P\|R)/p} \cdot e^{q(\alpha-1)\mathrm{D}_{q(\alpha-1)+1}(R\|Q)/q}.
\end{aligned}
$$

Taking logarithms and rearranging gives

$$
\mathrm{D}_\alpha(P\|Q) \leq \frac{p\alpha-1}{p(\alpha-1)}\mathrm{D}_{p\alpha}(P\|R) + \mathrm{D}_{q(\alpha-1)+1}(R\|Q).
$$

Now $p\alpha = \frac{k\alpha-1}{k-1}$, $q(\alpha-1)+1 = k\alpha$, and $\frac{p\alpha-1}{p(\alpha-1)} = \frac{k\alpha}{k\alpha-1}$.

**Proposition 27.** *If $M : \mathcal{X}^n \to \mathcal{Y}$ satisfies $(\xi, \rho)$-zCDP, then $M$ gives $(\xi \cdot k\sum_{i=1}^k \frac{1}{i}, \rho \cdot k^2)$-zCDP for groups of size $k$.*

In particular, $(\xi, \rho)$-zCDP implies $(\xi \cdot O(k \log k), \rho \cdot k^2)$-zCDP for groups of size $k$. The Gaussian mechanism shows that $k^2\rho$ is the optimal dependence on $\rho$. However, $O(k \log k)\xi$ is not the optimal dependence on $\xi$: $(\xi, 0)$-zCDP implies $(k\xi, 0)$-zCDP for groups of size $k$.

*Proof.* We show this by induction on $k$. The statement is clearly true for groups of size 1. We now assume the statement holds for groups of size $k - 1$ and will verify it for groups of size $k$.

Let $x, x' \in \mathcal{X}^n$ differ in $k$ entries. Let $\hat{x} \in \mathcal{X}^n$ be such that $x$ and $\hat{x}$ differ in $k - 1$ entries and $x'$ and $\hat{x}$ differ in one entry.

Then, by the induction hypothesis,

$$
\mathrm{D}_\alpha(M(x)\|M(\hat{x})) \leq \xi \cdot (k-1)\sum_{i=1}^{k-1}\frac{1}{i} + \rho \cdot (k-1)^2 \cdot \alpha
$$

and, by zCDP,

$$
\mathrm{D}_\alpha(M(\hat{x})\|M(x')) \leq \xi + \rho \cdot \alpha.
$$

for all $\alpha \in (1, \infty)$. By (5), for any $\alpha \in (1, \infty)$,

$$D_\alpha \left( M(x) \| M(x') \right)$$

$$\leq \frac{k\alpha}{k\alpha - 1} D_{\frac{k\alpha - 1}{k-1}} \left( M(x) \| M(\hat{x}) \right) + D_{k\alpha} \left( M(\hat{x}) \| M(x') \right)$$

$$\leq \frac{k\alpha}{k\alpha - 1} \left( \xi \cdot (k-1) \sum_{i=1}^{k-1} \frac{1}{i} + \rho \cdot (k-1)^2 \cdot \frac{k\alpha - 1}{k-1} \right) + \xi + \rho \cdot k\alpha$$

$$= \xi \cdot \left( 1 + \frac{k\alpha}{k\alpha - 1}(k-1) \sum_{i=1}^{k-1} \frac{1}{i} \right) + \rho \cdot \left( \frac{k\alpha}{k\alpha - 1}(k-1)^2 \frac{k\alpha - 1}{k-1} + k\alpha \right)$$

$$\leq \xi \cdot k \sum_{i=1}^{k} \frac{1}{i} + \rho \cdot k^2 \cdot \alpha,$$

where the last inequality follows from the fact that $\frac{k\alpha}{k\alpha - 1}$ is a decreasing function of $\alpha$ for $\alpha > 1$. $\qquad \blacksquare$

## 6   Lower Bounds

In this section we develop tools to prove lower bounds for zCDP. We will use group privacy to bound the mutual information between the input and the output of a mechanism satisfying zCDP. Thus, if we are able to construct a distribution on inputs such that any accurate mechanism must reveal a high amount of information about its input, we obtain a lower bound showing that no accurate mechanism satisfying zCDP can be accurate for this data distribution.

We begin with the simplest form of our mutual information bound, which is an analogue of the bound of [MMP+10] for pure differential privacy:

**Proposition 28.** *Let $M : \mathcal{X}^n \to \mathcal{Y}$ satisfy $(\xi, \rho)$-zCDP. Let $X$ be a random variable in $\mathcal{X}^n$. Then*

$$I(X; M(X)) \leq \xi \cdot n(1 + \log n) + \rho \cdot n^2,$$

*where $I$ denotes mutual information (measured in nats, rather than bits).*

*Proof.* By Proposition 27, $M$ is $(\xi \cdot n \sum_{i=1}^{n} \frac{1}{i}, \rho \cdot n^2)$-zCDP for groups of size $n$. Thus

$$D_1 \left( M(x) \| M(x') \right) \leq \xi \cdot n \sum_{i=1}^{n} \frac{1}{i} + \rho \cdot n^2 \leq \xi \cdot n(1 + \log n) + \rho \cdot n^2$$

for all $x, x' \in \mathcal{X}^n$. Since KL-divergence is convex,

$$I(X; M(X)) = \mathop{\mathbb{E}}_{x \leftarrow X} [D_1 (M(x) \| M(X))]$$

$$\leq \mathop{\mathbb{E}}_{x \leftarrow X} \left[ \mathop{\mathbb{E}}_{x' \leftarrow X} [D_1 (M(x) \| M(x'))] \right]$$

$$\leq \mathop{\mathbb{E}}_{x \leftarrow X} \left[ \mathop{\mathbb{E}}_{x' \leftarrow X} [\xi \cdot n(1 + \log n) + \rho \cdot n^2] \right]$$

$$= \xi \cdot n(1 + \log n) + \rho \cdot n^2.$$

The reason this lower bound works is the strong group privacy guarantee — even for groups of size $n$, we obtain nontrivial privacy guarantees. While this is good for privacy it is bad for usefulness, as it implies that even information that is "global" (rather than specific to a individual or a small group) is protected. These lower bounds reinforce the connection between group privacy and lower bounds [HT10, De12, SU15a].

In contrast, $(\varepsilon, \delta)$-DP is not susceptible to such a lower bound because it gives a vacuous privacy guarantee for groups of size $k = O(\log(1/\delta)/\varepsilon)$. This helps explain the power of the propose-test-release paradigm.

Furthermore, we obtain even stronger mutual information bounds when the entries of the distribution are independent:

**Lemma 29.** *Let $M : \mathcal{X}^m \to \mathcal{Y}$ satisfy $(\xi, \rho)$-zCDP. Let $X$ be a random variable in $\mathcal{X}^m$ with independent entries. Then*

$$I(X; M(X)) \leq (\xi + \rho) \cdot m,$$

*where $I$ denotes mutual information (measured in nats, rather than bits).*

*Proof.* First, by the chain rule for mutual information,

$$I(X; M(X)) = \sum_{i \in [m]} I(X_i; M(X) | X_{1 \cdots i-1}),$$

where

$$I(X_i; M(X) | X_{1 \cdots i-1}) = \mathop{\mathbb{E}}_{x \leftarrow X_{1 \cdots i-1}} [I(X_i | X_{1 \cdots i-1} = x; M(X) | X_{1 \cdots i-1} = x)]$$

$$= \mathop{\mathbb{E}}_{x \leftarrow X_{1 \cdots i-1}} [I(X_i; M(x, X_{i \cdots m}))] ,$$

by independence of the $X_i$s.
We can define mutual information in terms of KL-divergence:

$$I(X_i; M(x, X_{i \cdots m})) = \mathop{\mathbb{E}}_{y \leftarrow X_i} [D_1 (M(x, X_{i \cdots m}) | X_i = y \| M(x, X_{i \cdots m}))]$$

$$= \mathop{\mathbb{E}}_{y \leftarrow X_i} [D_1 (M(x, y, X_{i+1 \cdots m}) \| M(x, X_{i \cdots m}))] .$$

By zCDP, we know that for all $x \in \mathcal{X}^{i-1}$, $y, y' \in \mathcal{X}$, and $z \in \mathcal{X}^{m-i}$, we have

$$D_1\left(M(x, y, z) \| M(x, y', z)\right) \le \xi + \rho.$$

Thus, by the convexity of KL-divergence,

$$D_1\left(M(x, y, X_{i+1 \cdots m}) \| M(x, X_{i \cdots m})\right) \le \xi + \rho$$

for all $x$ and $y$. The result follows.

More generally, we can combine dependent and independent rows as follows.

**Theorem 30.** *Let* $M : \mathcal{X}^n \to \mathcal{Y}$ *satisfy* $(\xi, \rho)$-*zCDP. Take* $n = m \cdot \ell$. *Let* $X^1, \cdots, X^m$ *be independent random variables on* $\mathcal{X}^\ell$. *Denote* $X = (X^1, \cdots, X^m) \in \mathcal{X}^n$. *Then*

$$I\left(X; M(X)\right) \le m \cdot \left(\xi \cdot \ell(1 + \log \ell) + \rho \cdot \ell^2\right),$$

*where* $I$ *denotes the mutual information (measured in nats, rather than bits).*

### 6.1    Example Applications of the Lower Bound

We informally discuss a few applications of our information-based lower bounds to some simple and well-studied problems in differential privacy.

*One-Way Marginals.* Consider $M : \mathcal{X}^n \to \mathcal{Y}$ where $\mathcal{X} = \{0, 1\}^d$ and $\mathcal{Y} = [0, 1]^d$. The goal of $M$ is to estimate the attribute means, or one-way marginals, of its input database $x$:

$$M(x) \approx \overline{x} = \frac{1}{n} \sum_{i \in [n]} x_i.$$

It is known that this is possible subject to $\varepsilon$-DP if and only if $n = \Theta(d/\varepsilon)$ [HT10, SU15a]. This is possible subject to $(\varepsilon, \delta)$-DP if and only if $n = \tilde{\Theta}(\sqrt{d \log(1/\delta)}/\varepsilon)$, assuming $\delta \ll 1/n$ [BUV14, SU15a].

We now analyze what can be accomplished with zCDP. Adding independent noise drawn from $\mathcal{N}(0, d/2n^2\rho)$ to each of the $d$ coordinates of $\overline{x}$ satisfies $\rho$-zCDP. This gives accurate answers as long as $n \gg \sqrt{d/\rho}$.

For a lower bound, consider sampling $X_1 \in \{0, 1\}^d$ uniformly at random. Set $X_i = X_1$ for all $i \in [n]$. By Proposition 28,

$$I(X; M(X)) \le n^2 \rho$$

for any $\rho$-zCDP $M : (\{0, 1\}^d)^n \to [0, 1]^d$. However, if $M$ is accurate, we can recover (most of) $X_1$ from $M(X)$, whence $I(X; M(X)) \ge \Omega(d)$. This yields a lower bound of $n \ge \Omega(\sqrt{d/\rho})$, which is tight up to constant factors.

*Histograms (a.k.a. Point Queries).* Consider $M : \mathcal{X}^n \to \mathcal{Y}$, where $\mathcal{X} = [T]$ and $\mathcal{Y} = \mathbb{R}^T$. The goal of $M$ is to estimate the histogram of its input:

$$M(x)_t \approx h_t(x) = |\{i \in [n] : x_i = t\}|$$

For $\varepsilon$-DP it is possible to do this if and only if $n = \Theta(\log(T)/\varepsilon)$; the optimal algorithm is to independently sample

$$M(x)_t \sim h_t(x) + \mathsf{Laplace}(2/\varepsilon).$$

However, for $(\varepsilon, \delta)$-DP, it is possible to attain sample complexity $n = O(\log(1/\delta)/\varepsilon)$ [BNS16, Theorem 3.13]. Interestingly, for zCDP we can show that $n = \Theta(\sqrt{\log(T)/\rho})$ is sufficient and necessary:

Sampling

$$M(x)_t \sim h_t(x) + \mathcal{N}(0, 1/\rho)$$

independently for $t \in [T]$ satisfies $\rho$-zCDP. Moreover,

$$\mathbb{P}\left[\max_{t \in [T]} |M(x)_t - h_t(x)| \geq \lambda\right] \leq T \cdot \mathbb{P}\left[|\mathcal{N}(0, 1/\rho)| > \lambda\right] \leq T \cdot e^{-\lambda^2 \rho/2}.$$

In particular $\mathbb{P}\left[\max_{t \in [T]} |M(x)_t - h_t(x)| \geq \sqrt{\log(T/\beta)/\rho}\right] \leq \beta$ for all $\beta > 0$. Thus this algorithm is accurate if $n \gg \sqrt{\log(T)/\rho}$.

On the other hand, if we sample $X_1 \in [T]$ uniformly at random and set $X_i = X_1$ for all $i \in [n]$, then $I(X; M(X)) \geq \Omega(\log T)$ for any accurate $M$, as we can recover $X_1$ from $M(X)$ if $M$ is accurate. Proposition 28 thus implies that $n \geq \Omega(\sqrt{\log(T)/\rho})$ is necessary to obtain accuracy. This gives a strong separation between approximate DP and zCDP.

*Lower Bounds with Accuracy.* The above examples can be easily discussed in terms of a more formal and quantitative definition of accuracy. For instance, in the full version of this work, we revisit the histogram example:

**Proposition 31.** *If $M : [T]^n \to \mathbb{R}^T$ satisfies $\rho$-zCDP and*

$$\forall x \in [T]^n \qquad \underset{M}{\mathbb{E}}\left[\max_{t \in [T]} |M(x)_t - h_t(x)|\right] \leq \alpha n,$$

*then $n \geq \Omega(\sqrt{\log(\alpha^2 T)/\rho\alpha^2})$.*

We remark that our lower bounds for zCDP can be converted to lower bounds for mCDP using Lemma 24.

# 7  Obtaining Pure DP Mechanisms from zCDP

We now establish limits on what more can be achieved with zCDP over pure differential privacy. In particular, we prove that any mechanism satisfying zCDP can be converted into a mechanism satisfying pure DP with at most a quadratic blowup in sample complexity. Formally, we show the following theorem.

**Theorem 32.** *Fix $n \in \mathbb{N}$, $n' \in \mathbb{N}$, $k \in \mathbb{N}$ $\alpha > 0$, and $\varepsilon > 0$. Let $q : \mathcal{X} \to \mathbb{R}^k$ and let $\| \cdot \|$ be a norm on $\mathbb{R}^k$. Assume $\max_{x \in \mathcal{X}} \|q(x)\| \leq 1$. Suppose there exists a $(\xi, \rho)$-zCDP mechanism $M : \mathcal{X}^n \to \mathbb{R}^k$ such that for all $x \in \mathcal{X}^n$,*

$$\mathop{\mathbb{E}}_{M} \left[ \|M(x) - q(x)\| \right] \leq \alpha.$$

*Assume $\xi \leq \alpha^2$, $\rho \leq \alpha^2$, and*

$$n' \geq \frac{4}{\varepsilon \alpha} \left( \rho \cdot n^2 + \xi \cdot n \cdot (1 + \log n) + 1 \right).$$

*Then there exists a $(\varepsilon, 0)$-differentially private $M' : \mathcal{X}^{n'} \to \mathbb{R}^k$ satisfying*

$$\mathop{\mathbb{E}}_{M'} \left[ \|M'(x) - q(x)\| \right] \leq 10\alpha$$

*and*

$$\mathop{\mathbb{P}}_{M'} \left[ \|M'(x) - q(x)\| > 10\alpha + \frac{4}{\varepsilon n'} \log \left( \frac{1}{\beta} \right) \right] \leq \beta$$

*for all $x \in \mathcal{X}^{n'}$ and $\beta > 0$.*

Before discussing the proof of Theorem 32, we make some remarks about its statement:

- Unfortunately, the theorem only works for families of statistical queries $q : \mathcal{X} \to \mathbb{R}^k$. However, it works equally well for $\| \cdot \|_\infty$ and $\| \cdot \|_1$ error bounds.
- If $\xi = 0$, we have $n' = O(n^2 \rho / \varepsilon \alpha)$. So, if $\rho$, $\varepsilon$, and $\alpha$ are all constants, we have $n' = O(n^2)$. This justifies our informal statement that we can convert any mechanism satisfying zCDP into one satisfying pure DP with a quadratic blowup in sample complexity.
- The requirement that $\xi, \rho \leq \alpha^2$ is only used to show that

$$\max_{x \in \mathcal{X}^{n'}} \min_{\hat{x} \in \mathcal{X}^n} \|q(x) - q(\hat{x})\| \leq 2\alpha. \tag{6}$$

However, in many situations (6) holds even when $\xi, \rho \gg \alpha^2$. For example, if $n \geq O(\log(k)/\alpha^2)$ or even $n \geq O(VC(q)/\alpha^2)$ then (6) is automatically satisfied. The technical condition (6) is needed to relate the part of the proof with inputs of size $n$ to the part with inputs of size $n'$.

The proof of Theorem 32 is not constructive. Rather than directly constructing a mechanism satisfying pure DP from any mechanism satisfying zCDP, we show the contrapositive statement: any lower bound for pure DP can be converted into a lower bound for zCDP. Pure DP is characterized by so-called packing lower bounds and the exponential mechanism.

In the full version of this work, we use a greedy argument to show that for any output space and any desired accuracy, there is a set $T$ that is simultaneously a "packing" and a "net:"

**Lemma 33.** *Let $(\mathcal{Y}, d)$ be a metric space. Fix $\alpha > 0$. Then there exists a countable $T \subset \mathcal{Y}$ such that both of the following hold.*

- *(Net:) Either $T$ is infinite or for all $y' \in \mathcal{Y}$ there exists $y \in T$ with $d(y, y') \le \alpha$.*
- *(Packing:) For all $y, y' \in T$, if $y \ne y'$, then $d(y, y') > \alpha$.*

It is well-known that a net yields a pure DP algorithm:

**Lemma 34 (Exponential Mechanism [MT07, BLR13]).** *Let $\ell : \mathcal{X}^n \times T \to \mathbb{R}$ satisfy $|\ell(x, y) - \ell(x', y)| \le \Delta$ for all $x, x' \in \mathcal{X}^n$ differing in one entry and all $y \in T$. Then, for all $\varepsilon > 0$, there exists an $\varepsilon$-differentially private $M : \mathcal{X}^n \to T$ such that*

$$\mathbb{P}_M \left[ \ell(x, M(x)) \le \min_{y \in T} \ell(x, y) + \frac{2\Delta}{\varepsilon} \log \left( \frac{|T|}{\beta} \right) \right] \ge 1 - \beta$$

*and*

$$\mathbb{E}_M \left[ \ell(x, M(x)) \right] \le \min_{y \in T} \ell(x, y) + \frac{2\Delta}{\varepsilon} \log |T|$$

*for all $x \in \mathcal{X}^n$ and $\beta > 0$.*

On the other hand, in the full version of this work we use Proposition 28 to show that a packing yields a lower bound for zCDP:

**Lemma 35.** *Let $(\mathcal{Y}, d)$ be a metric space and $q : \mathcal{X}^n \to \mathcal{Y}$ a function. Let $M : \mathcal{X}^n \to \mathcal{Y}$ be a $(\xi, \rho)$-zCDP mechanism satisfying*

$$\mathbb{P}_M \left[ d(M(x), q(x)) > \alpha/2 \right] \le \beta$$

*for all $x \in \mathcal{X}^n$. Let $T \subset \mathcal{Y}$ be such that $d(y, y') > \alpha$, for all $y, y' \in T$ with $y \ne y'$. Assume that for all $y \in T$ there exists $x \in \mathcal{X}^n$ with $q(x) = y$. Then*

$$(1 - \beta) \log |T| - \log 2 \le \xi \cdot n(1 + \log n) + \rho \cdot n^2.$$

*In particular, if $\xi = 0$, we have*

$$n \ge \sqrt{\frac{(1 - \beta) \log |T| - \log 2}{\rho}} = \Omega(\sqrt{\log |T|/\rho}).$$

In the full version of this work, we combine these lemmas to prove Theorem 32.

# 8   Approximate zCDP

Recall our definition of approximate zCDP:

**Definition 36 (Approximate zCDP).** *A randomised mechanism* $M : \mathcal{X}^n \to \mathcal{Y}$ *is* $\delta$-*approximately* $(\xi, \rho)$-*zCDP if, for all* $x, x' \in \mathcal{X}^n$ *differing on a single entry, there exist events* $E = E(M(x))$ *and* $E' = E'(M(x'))$ *such that, for all* $\alpha \in (1, \infty)$,

$$D_\alpha \left( M(x)|_E \| M(x')|_{E'} \right) \leq \xi + \rho \cdot \alpha \quad and \quad D_\alpha \left( M(x')|_{E'} \| M(x)|_E \right) \leq \xi + \rho \cdot \alpha$$

*and* $\underset{M(x)}{\mathbb{P}} [E] \geq 1 - \delta$ *and* $\underset{M(x')}{\mathbb{P}} [E'] \geq 1 - \delta$.

    Clearly 0-approximate zCDP is simply zCDP. Hence we have a generalization of zCDP. As we will show later in this section, $\delta$-approximate $(\varepsilon, 0)$-zCDP is equivalent to $(\varepsilon, \delta)$-DP. Thus we have also generalized approximate DP. Hence, this definition unifies both relaxations of pure DP.

    Approximate zCDP is a three-parameter definition which allows us to capture many different aspects of differential privacy. However, three parameters is quite overwhelming. We believe that use of the one-parameter $\rho$-zCDP (or the two-parameter $\delta$-approximate $\rho$-zCDP if necessary) is sufficient for most purposes.

    It is easy to verify that the definition of approximate zCDP satisfies the usual composition and post-processing properties. However, the strong group privacy guarantees of Sect. 5 no longer apply to approximate zCDP and, hence, the strong lower bounds of Sect. 6 also no longer hold. Circumventing these lower bounds is part of the motivation for considering approximate zCDP.

    In the full version of this work, we use techniques developed in [KOV15, MV16] to show that approximate DP can be converted to approximate zCDP.

**Lemma 37.** *If* $M : \mathcal{X}^n \to \mathcal{Y}$ *satisfies* $(\varepsilon, \delta)$-DP, *then* $M$ *satisfies* $\delta$-*approximate* $(\varepsilon, 0)$-*zCDP, which, in turn, implies* $\delta$-*approximate* $(0, \frac{1}{2}\varepsilon^2)$-*zCDP.*

    Conversely, approximate zCDP also implies approximate DP. The following result sharpens Lemma 21.

**Lemma 38.** *Suppose* $M : \mathcal{X}^n \to \mathcal{Y}$ *satisfies* $\delta$-*approximate* $(\xi, \rho)$-*zCDP. If* $\rho = 0$, *then* $M$ *satisfies* $(\xi, \delta)$-*DP. In general,* $M$ *satisfies* $(\varepsilon, \delta + (1 - \delta)\delta')$-*DP for all* $\varepsilon \geq \xi + \rho$, *where*

$$\delta' = e^{-(\varepsilon - \xi - \rho)^2/4\rho} \cdot \min \begin{cases} 1 \\ \sqrt{\pi \cdot \rho} \\ \frac{1}{1 + (\varepsilon - \xi - \rho)/2\rho} \\ \frac{2}{1 + \frac{\varepsilon - \xi - \rho}{2\rho} + \sqrt{\left(1 + \frac{\varepsilon - \xi - \rho}{2\rho}\right)^2 + \frac{4}{\pi\rho}}} \end{cases} .$$

    A result is that we can give a sharper version of the so-called advanced composition theorem [DRV10]. Note that the following results are subsumed by the bounds of Kairouz, Oh, and Viswanath [KOV15] and Murtagh and Vadhan

[MV16]. However, these bounds may be extended to analyse the composition of mechanisms satisfying CDP with mechanisms satisfying approximate DP. We believe that such a "unified" analysis of composition will be useful.

Applying Lemmas 37 and 38 yields the following result.

**Corollary 39.** *Let* $M_1, \cdots, M_k : \mathcal{X}^n \to \mathcal{Y}$ *and let* $M : \mathcal{X}^n \to \mathcal{Y}^k$ *be their composition. Suppose each* $M_i$ *satisfies* $(\varepsilon_i, \delta_i)$*-DP. Then* $M$ *satisfies*

$$\left( \frac{1}{2} \|\varepsilon\|_2^2 + \sqrt{2}\lambda \|\varepsilon\|_2, \sqrt{\frac{\pi}{2}} \cdot \|\varepsilon\|_2 \cdot e^{-\lambda^2} + \|\delta\|_1 \right) \text{-}DP$$

*for all* $\lambda \geq 0$. *Alternatively* $M$ *satisfies*

$$\left( \frac{1}{2} \|\varepsilon\|_2^2 + \sqrt{2\log(\sqrt{\pi/2} \cdot \|\varepsilon\|_2/\delta')} \cdot \|\varepsilon\|_2, \delta' + \|\delta\|_1 \right) \text{-}DP$$

*for all* $\delta' \geq 0$.

In comparison to the composition theorem of [DRV10], we save modestly by a constant factor in the first term and, in most cases $\sqrt{\pi/2}\|\varepsilon\|_2 < 1$, whence the logarithmic term is an improvement over the usual advanced composition theorem.

**Acknowledgements.** We thank Cynthia Dwork and Guy Rothblum for sharing a preliminary draft of their work with us. We also thank Ilya Mironov, Kobbi Nissim, Adam Smith, Salil Vadhan, and the Harvard Differential Privacy Research Group for helpful discussions and suggestions.

# References

[BLR13]  Blum, A., Ligett, K., Roth, A.: A learning theory approach to noninteractive database privacy. J. ACM **60**(2), 12 (2013)

[BNS13]  Beimel, A., Nissim, K., Stemmer, U.: Private learning and sanitization: pure vs. approximate differential privacy. In: Raghavendra, P., Raskhodnikova, S., Jansen, K., Rolim, J.D.P. (eds.) APPROX/RANDOM -2013. LNCS, vol. 8096, pp. 363–378. Springer, Heidelberg (2013). doi:10.1007/978-3-642-40328-6_26

[BNS16]  Bun, M., Nissim, K., Stemmer, U.: Simultaneous private learning of multiple concepts. In: Proceedings of the 2016 ACM Conference on Innovations in Theoretical Computer Science, ITCS 2016, pp. 369–380. ACM, New York (2016)

[BUV14]  Bun, M., Ullman, J., Vadhan, S.P.: Fingerprinting codes and the price of approximate differential privacy. In: Symposium on Theory of Computing, STOC 2014, New York, NY, USA, 31 May - 03 June 2014, pp. 1–10 (2014)

[De12]  De, A.: Lower bounds in differential privacy. In: Cramer, R. (ed.) TCC 2012. LNCS, vol. 7194, pp. 321–338. Springer, Heidelberg (2012). doi:10.1007/978-3-642-28914-9_18

[DKM+06] Dwork, C., Kenthapadi, K., McSherry, F., Mironov, I., Naor, M.: Our data, ourselves: privacy via distributed noise generation. In: Vaudenay, S. (ed.) EUROCRYPT 2006. LNCS, vol. 4004, pp. 486–503. Springer, Heidelberg (2006). doi:10.1007/11761679_29

[DL09] Dwork, C., Lei, J.: Differential privacy and robust statistics. In: Proceedings of the 41st Annual ACM Symposium on Theory of Computing, STOC 2009, Bethesda, MD, USA, 31 May - 2 June 2009, pp. 371–380 (2009)

[DMNS06] Dwork, C., McSherry, F., Nissim, K., Smith, A.: Calibrating noise to sensitivity in private data analysis. In: Halevi, S., Rabin, T. (eds.) TCC 2006. LNCS, vol. 3876, pp. 265–284. Springer, Heidelberg (2006). doi:10.1007/11681878_14

[DR16] Dwork, C., Rothblum, C.: Concentrated differential privacy. CoRR, abs/1603.01887 (2016)

[DRV10] Dwork, C., Rothblum, G.N., Vadhan, S.P.: Boosting and differential privacy. In: IEEE Symposium on Foundations of Computer Science (FOCS 2010), pp. 51–60. IEEE, 23–26 October 2010

[HT10] Hardt, M., Talwar, K.: On the geometry of differential privacy. In: Proceedings of the Forty-Second ACM Symposium on Theory of Computing, STOC 2010, pp. 705–714, New York, NY, USA. ACM (2010)

[KOV15] Kairouz, P., Oh, S., Viswanath, P.: The composition theorem for differential privacy. In: Proceedings of the 32nd International Conference on Machine Learning, ICML 2015, Lille, France, pp. 1376–1385, 6–11 July 2015

[MMP+10] McGregor, A., Mironov, I., Pitassi, T., Reingold, O., Talwar, K., Vadhan, S.P.: The limits of two-party differential privacy. In: 51th Annual IEEE Symposium on Foundations of Computer Science, FOCS 23–26, 2010, Las Vegas, Nevada, USA, pp. 81–90, October 2010

[MT07] McSherry, F., Talwar, K.: Mechanism design via differential privacy. In: 48th Annual IEEE Symposium on Foundations of Computer Science, FOCS 2007, pp. 94–103, October 2007

[MV16] Murtagh, J., Vadhan, S.P.: The complexity of computing the optimal composition of differential privacy. In: Proceedings of Theory of Cryptography - 13th International Conference, TCC2016-A, Tel Aviv, Israel, 10-13 January 2016, Part I, pp. 157–175 (2016)

[Ren61] Rényi, A.: On measures of entropy, information. In: Proceedings of the Fourth Berkeley Symposium on Mathematical Statistics, Probability: Contributions to the Theory of Statistics, vol. 1, pp. 547–561. University of California Press (1961)

[SU15a] Steinke, T., Ullman, J.: Between pure and approximate differential privacy. CoRR, abs/1501.06095 (2015)

[SU15b] Steinke, T., Ullman, J.: Interactive fingerprinting codes, the hardness of preventing false discovery. In: COLT (2015). http://arXiv.org/abs/1410.1228

[Tar08] Tardos, G.: Optimal probabilistic fingerprint codes. J. ACM **55**(2), 10 (2008)

[Ull13] Ullman, J.: Answering n {2+ o (1)} counting queries with differential privacy is hard. In: Proceedings of the Forty-Fifth Annual ACM Symposium on Theory of Computing, pp. 361–370. ACM (2013)

[vEH14] van Erven, T., Harremos, P.: Rényi divergence and kullback-leibler divergence. IEEE Trans. Inf. Theory **60**(7), 3797–3820 (2014)

# Strong Hardness of Privacy from Weak Traitor Tracing

Lucas Kowalczyk[1]($\boxtimes$), Tal Malkin[1], Jonathan Ullman[2], and Mark Zhandry[3,4]

[1] Columbia University, New York, USA
luke@cs.columbia.edu
[2] Northeastern University, Boston, USA
[3] MIT, Cambridge, USA
[4] Princeton University, Princeton, USA

**Abstract.** A central problem in differential privacy is to accurately answer a large family $Q$ of *statistical queries* over a *data universe* $X$. A statistical query on a dataset $D \in X^n$ asks "what fraction of the elements of $D$ satisfy a given predicate $p$ on $X$?" Ignoring computational constraints, it is possible to accurately answer exponentially many queries on an exponential size universe while satisfying differential privacy (Blum et al., STOC'08). Dwork et al. (STOC'09) and Boneh and Zhandry (CRYPTO'14) showed that if both $Q$ and $X$ are of polynomial size, then there is an efficient differentially private algorithm that accurately answers all the queries. They also proved that if $Q$ and $X$ are *both* exponentially large, then under a plausible assumption, no efficient algorithm exists.

We show that, under the same assumption, if *either* the number of queries *or* the data universe is of exponential size, then there is no differentially private algorithm that answers all the queries. Specifically, we prove that if one-way functions and indistinguishability obfuscation exist, then:

1. For every $n$, there is a family $Q$ of $\tilde{O}(n^7)$ queries on a data universe $X$ of size $2^d$ such that no poly$(n, d)$ time differentially private algorithm takes a dataset $D \in X^n$ and outputs accurate answers to every query in $Q$.

2. For every $n$, there is a family $Q$ of $2^d$ queries on a data universe $X$ of size $\tilde{O}(n^7)$ such that no poly$(n, d)$ time differentially private algorithm takes a dataset $D \in X^n$ and outputs accurate answers to every query in $Q$.

In both cases, the result is nearly quantitatively tight, since there is an efficient differentially private algorithm that answers $\tilde{\Omega}(n^2)$ queries on an exponential size data universe, and one that answers exponentially many queries on a data universe of size $\tilde{\Omega}(n^2)$.

Our proofs build on the connection between hardness of differential privacy and traitor-tracing schemes (Dwork et al., STOC'09; Ullman, STOC'13). We prove our hardness result for a polynomial size query set (resp., data universe) by showing that they follow from the existence of a special type of traitor-tracing scheme with very short ciphertexts (resp., secret keys), but very weak security guarantees, and then constructing such a scheme.

The full version of this work appears on the IACR Crypto ePrint [26].

© International Association for Cryptologic Research 2016
M. Hirt and A. Smith (Eds.): TCC 2016-B, Part I, LNCS 9985, pp. 659–689, 2016.
DOI: 10.1007/978-3-662-53641-4_25

# 1    Introduction

The goal of privacy-preserving data analysis is to release rich statistical information about a sensitive dataset while respecting the privacy of the individuals represented in that dataset. The past decade has seen tremendous progress towards understanding when and how these two competing goals can be reconciled, including surprisingly powerful differentially private algorithms as well as computational and information-theoretic limitations. In this work, we further this agenda by showing a strong new computational bottleneck in differential privacy.

Consider a dataset $D \in X^n$ where each of the $n$ elements is one individual's data, and each individual's data comes from some *data universe* $X$. We would like to be able to answer sets of *statistical queries* on $D$, which are queries of the form "What fraction of the individuals in $D$ satisfy some property $p$?" However, *differential privacy* [14] requires that we do so in such a way that no individual's data has significant influence on the answers.

If we are content answering a relatively small set of queries $Q$, then it suffices to perturb the answer to each query with independent noise from an appropriate distribution. This algorithm is simple, very efficient, differentially private, and ensures good accuracy—say, within $\pm.01$ of the true answer—as long as $|Q| \lesssim n^2$ queries [5,13,14,16].

Remarkably, the work of Blum et al. [6] showed that it is possible to output a summary that allows accurate answers to an *exponential* number of queries—nearly $2^n$—while ensuring differential privacy. However, neither their algorithm nor the subsequent improvements [15,17,22,23,29,30,35] are computationally efficient. Specifically, they all require time at least $\text{poly}(n, |X|, |Q|)$ to privately and accurately answer a family of statistical queries $Q$ on a dataset $D \in X^n$. Note that the size of the input is $n \log |X|$ bits, so a computationally efficient algorithm runs in time $\text{poly}(n, \log |X|)$.[1] For example, in the common setting where each individual's data consists of $d$ binary attributes, so $X = \{0,1\}^d$, the size of the input is $nd$ but $|X| = 2^d$. As a result, all known private algorithms for answering arbitrary sets of statistical queries are inefficient if either the number of queries or the size of the data universe is superpolynomial.

This accuracy vs. computation tradeoff has been the subject of extensive study. Dwork et al. [15] showed that the existence of cryptographic *traitor-tracing schemes* [11] yields a family of statistical queries that cannot be answered accurately and efficiently with differential privacy. Applying recent traitor-tracing schemes [8], we conclude that, under plausible cryptographic assumptions (discussed below), if both the number of queries and the data universe can be superpolynomial, then there is no efficient differentially private algorithm. [34] used variants of traitor-tracing schemes to show that in the interactive setting, where

---

[1] It may require exponential time just to describe and evaluate an arbitrary counting query, which would rule out efficiency for reasons that have nothing to do with privacy. In this work, we restrict attention to queries that are efficiently computable in time $\text{poly}(n, \log |X|)$, so they are not the bottleneck in the computation.

the queries are not fixed but are instead given as input to the algorithm, assuming one-way functions exist, there is no private and efficient algorithm that accurately answers more than $\tilde{O}(n^2)$ statistical queries. All of the algorithms mentioned above work in this interactive setting, but for many applications we only need to answer a fixed family of statistical queries.

Despite the substantial progress, there is still a basic gap in our understanding. The hardness results for Dwork et al. apply if *both* the number of queries and the universe are large. But the known algorithms require exponential time if *either* of these sets is large. Is this necessary? Are there algorithms that run in time $\text{poly}(n, \log|X|, |Q|)$ or $\text{poly}(n, |X|, \log|Q|)$?

Our main result shows that under the same plausible cryptographic assumptions, the answer is no—if either the data universe or the set of queries can be superpolynomially large, then there is some family of statistical queries that cannot be accurately and efficiently answered while ensuring differential privacy.

## 1.1   Our Results

Our first result shows that if the data universe can be of superpolynomial size then there is some fixed family of polynomially many queries that cannot be efficiently answered under differential privacy. This result shows that the efficient algorithm for answering an arbitrary family of $|Q| \lesssim n^2$ queries by adding independent noise is optimal up to the specific constant in the exponent.

**Theorem 1 (Hardness for small query sets).** *Assume the existence of indistinguishability obfuscation and one-way functions. Let $\lambda \in \mathbb{N}$ be a computation parameter. For any polynomial $n = n(\lambda)$, there is a sequence of pairs $\{(X_\lambda, Q_\lambda)\}$ with $|X_\lambda| = 2^\lambda$ and $|Q_\lambda| = \tilde{O}(n^7)$ such that there is no polynomial time differentially private algorithm that takes a dataset $D \in X_\lambda^n$ and outputs an accurate answer to every query in $Q_\lambda$ up to an additive error of $\pm 1/3$.*

Our second result shows that, even if the data universe is required to be of polynomial size, there is a fixed set of superpolynomially many queries that cannot be answered efficiently under differential privacy. When we say that an algorithm efficiently answers a set of superpolynomially many queries, we mean that it efficiently outputs a summary such that there is an efficient algorithm for obtaining an accurate answer to any query in the set. For comparison, if $|X| \lesssim n^2$, then there is a simple $\text{poly}(n, |X|)$ time differentially private algorithm that accurately answers superpolynomially many queries. Our result shows that this efficient algorithm is optimal up to the specific constant in the exponent.

**Theorem 2 (Hardness for small query sets).** *Assume the existence of indistinguishability obfuscation and one-way functions. Let $\lambda \in \mathbb{N}$ be a computation parameter. For any polynomial $n = n(\lambda)$, there is a sequence of pairs $\{(X_\lambda, Q_\lambda)\}$ with $|X_\lambda| = \tilde{O}(n^7)$ and $|Q_\lambda| = 2^\lambda$ such that there is no polynomial time differentially private algorithm that takes a dataset $D \in X_\lambda^n$ and outputs an accurate answer to every query in $Q_\lambda$ up to an additive error of $\pm 1/3$.*

Before we proceed to describe our techniques, we make a few remarks about these results. In both of these results, the constant 1/3 in our result is arbitrary, and can be replaced with any constant smaller than 1/2. We also remark that, when we informally say that an algorithm is differentially private, we mean that it satisfies $(\varepsilon, \delta)$-differential privacy for some $\varepsilon = O(1)$ and $\delta = o(1/n)$. These are effectively the largest parameters for which differential privacy is a meaningful notion of privacy. That our hardness results apply to these parameters only makes our results stronger. Finally, we remark that it is possible to show that our results also rule out the weaker notion of *computational differential privacy* [28].

*On Indistinguishability Obfuscation.* Indistinguishability obfuscation (iO) has recently become a central cryptographic primitive. The first candidate construction, proposed just a couple years ago [19], was followed by a flurry of results demonstrating the extreme power and wide applicability of iO (cf., [4,8,19,24,31]). However, the assumption that iO exists is currently poorly understood, and the debate over the plausibility of iO is far from settled. While some specific proposed iO schemes have been attacked [12,27], other schemes seem to resist all *currently known* attacks [1,20]. We also do not know how to base iO on a solid, simple, natural computational assumption (some attempts based on multilinear maps have been made [21], but they were broken with respect to all current multilinear map constructions).

Nevertheless, our results are meaningful whether or not iO exists. If iO exists, our results show that certain tasks in differential privacy are intractable. Interestingly, unlike many previous results relying on iO, these conclusions were not previously known to follow from even the much stronger (and in fact, false) assumption of virtual black-box obfuscation. If, on the other hand, iO does not exist, then our results still demonstrate a barrier to progress in differential privacy—such progress would need to *prove* that iO does not exist. Alternatively, our results highlight a possible path toward proving that iO does not exist. We note that other "incompatibility" results are known for iO; for example, iO and certain types of hash functions cannot simultaneously exist [3,9].

## 1.2 Techniques

**(Weak) PLBE Schemes and the Hardness of Privacy.** We prove our results by building on the connection between differentially private algorithms for answering statistical queries and traitor-tracing schemes discovered by Dwork et al. [15]. Traitor-tracing schemes were introduced by Chor et al. [11] for the purpose of identifying pirates who violate copyright restrictions.

Although previous results are described in the language of traitor-tracing, our results are simpler to describe in the language of *private linear broadcast encryption (PLBE)*, which is a simpler primitive that implies traitor-tracing in a very direct way (e.g. [7]). We will thus refer to PLBE rather than traitor-tracing in all technical discussions going forward. A PLBE scheme allows a sender to generate keys for $n$ users so that (1) the sender can broadcast an encrypted message that can be decrypted by any subset of users $[1, i]$ for $0 \le i \le n$,[2] so that any user outside of $[1, i]$ will decrypt 0, and (2) the index $i$ describing the set of users is *hidden* in the sense that any coalition of users that excludes user $i$ cannot distinguish messages sent to the set $[1, i]$ from messages sent to the set $[1, i - 1]$.

Dwork et al. show that the existence of traitor-tracing schemes implies hardness results for differential privacy. In the language of PLBE, the reduction is as follows: Suppose a coalition of users takes their keys and builds a dataset $D \in X^n$ where each element of the dataset contains one of their user keys. The family $Q$ will contain a query $q_c$ for each possible ciphertext $c$. The query $q_c$ asks "What fraction of the elements (user keys) in $D$ would decrypt the ciphertext $c$ to the message 1?"

Suppose there were an efficient algorithm that accurately answers every query $q_c$ in $Q$. Then the coalition could run it on the dataset $D$ to produce a summary that can efficiently decrypt the ciphertexts. That means if $c$ encrypts the message 1 to all users $[1, n]$, the summary outputs an answer close to 1, and if $c$ encrypts a message 1 to the empty set of users, the summary outputs an answer close to 0. Thus, there exists a user $i$ such that the summary is distinguishing encryptions to the group $[1, i]$ from encryptions to $[1, i - 1]$. Differential privacy requires that the summary's behavior is essentially the same even if run it on the dataset $D'$ that excludes the secret key of user $i$. However, that means there is an efficient algorithm that takes the keys of all users excluding $i$ and distinguishes encryptions to the group $[1, i]$ from encryptions to the group $[1, i - 1]$, which violates the second property of the PLBE scheme.

To instantiate this result, we need a PLBE. Observe that the data universe contains one element for every possible user key, and the set of queries contains one query for every ciphertext, and we want to minimize the size of these sets. Boneh and Zhandry constructed a traitor-tracing scheme where both the keys and the ciphertexts have length equal to the security parameter $\lambda$, which under the Dwork et al. reduction yields hardness for a data universe and query set each of size $2^\lambda$. The main contribution of this work is to show that we can reduce either the number of possible ciphertexts or the number of possible keys to $\text{poly}(n)$ while the other remains of size $2^\lambda$.

But how is it possible to have a secure PLBE scheme with $\text{poly}(n)$ ciphertexts (resp., keys)? Even a semantically secure private key encryption scheme requires superpolynomially many ciphertexts (resp., keys)! Here we rely on observations from [34] showing that in order to show hardness for differential privacy, it suffices to have a PLBE scheme with very weak functionality and security. First, in the reduction, we only encrypt the message 1, so only the group $[1, i]$ is actually hidden. Second, in the reduction, the differentially private algorithm only has access to the user's keys, and there does not need to be a public encryption key or access to an encryption oracle. Thus, the adversary does not have the ability to generate encryptions to arbitrary groups $[1, i]$. Finally, the quantitative

---

[2] We use $[1, i]$ to denote the discrete interval $\{1, 2, \ldots, i\}$, with the convention that $[1, 0] = \emptyset$.

version of the reduction only requires that the coalition has advantage $o(1/n)$ in distinguishing encryptions to different groups, rather than negligible. All three of these relaxations are necessary for making the number of ciphertexts (resp., keys) poly($n$), and, as we show, are sufficient as well.

**Weak PLBE Schemes from Obfuscation.** In order to provide intuition for how we can achieve PLBE with a ciphertext or key space of size poly($n$), we will assume the existence of virtual black-box obfuscation (VBB). While our actual results use iO, we emphasize that a PLBE scheme with the right properties to establish our results was previously not even known to follow from VBB.

**Polynomially Many Ciphertexts.** Consider the following simple scheme: Let the set of ciphertexts be $[m]$ for an appropriate $m = \text{poly}(n)$. Choose a pseudorandom function $f : [m] \rightarrow \{0, 1, \ldots, n\}$ and associate each ciphertext $c \in [m]$ with the group of users $[1, f(c)]$. Pseudorandomness is only used to keep the description of $f$ short, and for intuition it's fine to think of $f$ as truly random. To encrypt to a set $[1, i]$, choose a random ciphertext $c \in f^{-1}(i)$ and send it. Each user $i$ will get a secret key containing an obfuscation of the program $P_i(c)$ that computes $j = f(c)$ and, if outputs 1 if $j \geq i$ and otherwise outputs 0.

Consider a coalition with keys for every user except some user $i$. Since there are only poly($n$) ciphertexts, we may as well assume that these users evaluate each of their obfuscated programs on every ciphertext $c$, and VBB security of the obfuscation ensures that they "cannot learn anything else" from their keys. By evaluating their programs on every ciphertext, they can determine the value of $f(c)$ exactly on every ciphertext $c$ such that $f(c) < i - 1$ or $f(c) > i$. Since $f$ is pseudorandom, for ciphertexts such that $f(c) \in \{i - 1, i\}$, they have at most a negligible advantage in guessing whether $f(c) = i$ or $f(c) = i - 1$, for any ciphertext $c$. Thus, if the coalition guesses the value of $f(c)$ on all ciphertexts $c \in f^{-1}(\{i - 1, i\})$, a simple Chernoff bound shows that they will guess at most $1/2 + O(\sqrt{\log(n)/T})$ of them correctly, where $T$ is the size of $f^{-1}(\{i - 1, i\})$. For a PRF, the size of this set will be at least $m/2n$ with overwhelming probability. Thus, in order to ensure that his overall advantage is $o(1/n)$, it suffices to choose $m = \tilde{O}(n^3)$.

In this straw-man scheme, the length of the ciphertext will clearly be $O(\log(n))$. The user keys contain an obfuscation of a poly($\lambda + \log(n)$) time program, so the user keys are poly($\lambda + \log(n)$), so this scheme satisfies our efficiency requirements.

While the scheme is very simple to describe using VBB, replacing VBB with iO introduces some additional technicalities, and requires a new notion of puncturable PRF (Sect. 5.2). These technicalities are also the reason we use $m = \tilde{O}(n^7)$ ciphertexts.

**Polynomially Many Keys.** Our scheme with polynomially many keys is roughly "dual" to the scheme with polynomially many ciphertexts. Let the set of user keys be $[n] \times [m]$ for an appropriate choice of $m = \text{poly}(n)$. Each user $i = 1, \ldots, n$ will receive a secret key $(i, sk_i)$ for a random $s \leftarrow_R [m]$. To encrypt a message to a group $[1, i]$ produce a VBB obfuscation $\mathcal{O}$ of the following program: The input is a pair $(j, s) \in [n] \times [m]$. If $s = sk_j$, then output 1 if $j \in [1, i]$ and otherwise output 0. Otherwise, output the value $r(j, s)$ for a pseudorandom function $r$. Again, pseudorandomness is only used to keep the description of the $r$ short, and for intuition it's fine to think of $r$ as an independent random value for each input $(j, s)$.

Suppose the coalition has the keys for every user except $i$ and a ciphertext encrypted to the group $[1, i - b]$ for $b \in \{0, 1\}$. We want to claim that the coalition has advantage at most $o(1/n)$ in trying to determine $b$. Since there are only polynomially many pairs $(j, s)$, the coalition might as well evaluated the obfuscated program on every one of the inputs, and VBB security of the obfuscation ensures that they "cannot learn anything else" from their keys and the ciphertext. Observe that by evaluating the ciphertext on all inputs $(j, s)$, they will actually evaluate the ciphertext on the input $(i, sk_i)$ belonging to user $i$, but they do not actually know which input of the form $(i, s)$ was the correct one.

By (pseudo)randomness of $r$, the only values that contain any information about the bit $b$ are $o = (\mathcal{O}(i, s))_{s \in [m]}$. In the case that $b = 0$, meaning the ciphertext was for group $[1, i - 1]$, $o$ is distributed as a (pseudo)random vector in $\{0, 1\}^m$ except that one random entry corresponding to the pair $(i, sk_i)$ is set to 1. Similarly, if $b = 1$, then $o$ is distributed as a (pseudo)random vector in $\{0, 1\}^m$ with one random entry set to 0. A simple argument based on Renyi-divergence shows that these two distributions are $O(1/\sqrt{m})$-close in statistical distance, so the coalition's advantage in determining $b$ is at most $O(1/\sqrt{m})$. Thus, it suffices to take $m = \tilde{O}(n^2)$ to obtain the level of security we need, corresponding to $nm = \tilde{O}(n^3)$ keys.

As before, moving from VBB to iO introduces additional technicalities, leading to $\tilde{O}(n^7)$ keys. We remark that for both the short-ciphertext and short-key schemes, obtaining the optimal $\tilde{O}(n^2)$ ciphertexts or keys seems to require both coming up with a more efficient VBB scheme and avoiding the loss in efficiency from moving to iO, or using another approach entirely.

## 1.3   Related Work

Theorem 1 should be contrasted with the line of work showing that differentially private algorithms can efficiently answer many more than $n^2$ *simple* queries. These results include algorithms for highly structured queries like point queries, threshold queries, and conjunctions (see e.g. [2,33] and the references therein).

Ullman and Vadhan [36] (building on Dwork et al. [15]) show that, assuming one-way functions, no differentially private and computationally efficient algorithm that outputs a *synthetic dataset* can accurately answer even the very simple family of 2-way marginals. This result is incomparable to ours, since it applies to a very small and simple family of statistical queries, but necessarily only applies to algorithms that output synthetic data.

There is also a line of work using *fingerprinting codes* to prove *information-theoretic* lower bounds on differentially private mechanisms [10,18,32]. Namely, that if the data universe is of size $\exp(n^2)$, then there is no differentially private algorithm, even a computationally unbounded one, that can answer more than $n^2$ statistical queries. Fingerprinting codes are essentially the information-theoretic analogue of traitor-tracing schemes, and thus these results are technically related, although the models are incomparable.

### 1.4  Paper Outline

In Sect. 2 we will give the necessary background on differential privacy. In Sect. 3 we will give our definition of weak PLBE schemes, and in Sect. 4 we will connect them to differential privacy. In Sect. 5 we will define some cryptographic tools that we use to construct PLBE schemes. In Sect. 6 we will construct the short-ciphertext scheme we use to prove Theorem 1 and in Sect. 7 we will construct the short-key scheme we use to prove Theorem 2.

## 2   Differential Privacy Preliminaries

### 2.1  Differentially Private Algorithms

A *dataset* $D \in X^n$ is an ordered set of $n$ rows, where each row corresponds to an individual, and each row is an element of some the *data universe* $X$. We write $D = (D_1, \ldots, D_n)$ where $D_i$ is the $i$-th row of $D$. We will refer to $n$ as the *size* of the dataset. We say that two datasets $D, D' \in X^*$ are *adjacent* if $D'$ can be obtained from $D$ by the addition, removal, or substitution of a single row, and we denote this relation by $D \sim D'$. In particular, if we remove the $i$-th row of $D$ then we obtain a new dataset $D_{-i} \sim D$. Informally, an algorithm $A$ is differentially private if it is randomized and for any two adjacent datasets $D \sim D'$, the distributions of $A(D)$ and $A(D')$ are similar.

**Definition 3 (Differential Privacy [14]).** *Let* $A : X^n \to S$ *be a randomized algorithm. We say that* $A$ *is* $(\varepsilon, \delta)$*-differentially private if for every two adjacent datasets* $D \sim D'$ *and every subset* $T \subseteq S$, $\mathbb{P}[A(D) \in T] \leq e^\varepsilon \cdot \mathbb{P}[A(D') \in T] + \delta$. *In this definition,* $\varepsilon, \delta$ *may be a function of* $n$.

### 2.2  Algorithms for Answering Statistical Queries

In this work we study algorithms that answer *statistical queries* (which are also sometimes called *counting queries, predicate queries*, or *linear queries* in the literature). For a data universe $X$, a statistical query on $X$ is defined by a predicate $q : X \to \{0, 1\}$. Abusing notation, we define the evaluation of a query $q$ on a dataset $D = (D_1, \ldots, D_n) \in X^n$ to be $\frac{1}{n} \sum_{i=1}^n q(D_i)$.

A single statistical query does not provide much useful information about the dataset. However, a sufficiently large and rich set of statistical queries is sufficient to implement many natural machine learning and data mining algorithms [25], thus we are interesting in differentially private algorithms to answer such sets. To this end, let $Q = \{q : X \to \{0,1\}\}$ be a set of statistical queries on a data universe $X$.

Informally, we say that a mechanism is accurate for a set $Q$ of statistical queries if it answers every query in the family to within error $\pm\alpha$ for some suitable choice of $\alpha > 0$. Note that $0 \le q(D) \le 1$, so this definition of accuracy is meaningful when $\alpha < 1/2$.

Before we define accuracy, we note that the mechanism may represent its answer in any form. That is, the mechanism outputs may output a *summary* $S \in \mathcal{S}$ that somehow represents the answers to every query in $Q$. We then require that there is an *evaluator* $Eval : \mathcal{S} \times Q \to [0,1]$ that takes the summary and a query and outputs an approximate answer to that query. That is, we think of $Eval(S, q)$ as the mechanism's answer to the query $q$. We will abuse notation and simply write $q(S)$ to mean $Eval(S, q)$.[3]

**Definition 4 (Accuracy).** *For a family $Q$ of statistical queries on $X$, a dataset $D \in X^n$ and a summary $s \in S$, we say that $s$ is $\alpha$-accurate for $Q$ on $D$ if $\forall q \in Q$   $|q(D) - q(s)| \le \alpha$. For a family of statistical queries $Q$ on $X$, we say that an algorithm $A : X^n \to S$ is $(\alpha, \beta)$-accurate for $Q$ given a dataset of size $n$ if for every $D \in X^n$, $\mathbb{P}[A(D)$ is $\alpha$-accurate for $Q$ on $X] \ge 1 - \beta$.*

In this work we are typically interested in mechanisms that satisfy the very weak notion of $(1/3, o(1/n))$-accuracy, where the constant $1/3$ could be replaced with any constant $< 1/2$. Most differentially private mechanisms satisfy quantitatively much stronger accuracy guarantees. Since we are proving hardness results, this choice of parameters makes our results stronger.

## 2.3  Computational Efficiency

Since we are interested in asymptotic efficiency, we introduce a computation parameter $\lambda \in \mathbb{N}$. We then consider a sequence of pairs $\{(X_\lambda, Q_\lambda)\}_{\lambda \in \mathbb{N}}$ where $Q_\lambda$ is a set of statistical queries on $X_\lambda$. We consider databases of size $n$ where $n = n(\lambda)$ is a polynomial. We then consider algorithms $A$ that take as input a dataset $X_\lambda^n$ and output a summary in $S_\lambda$ where $\{S_\lambda\}_{\lambda \in \mathbb{N}}$ is a sequence of output ranges. There is an associated evaluator $Eval$ that takes a query $q \in Q_\lambda$ and a summary $s \in S_\lambda$ and outputs a real-valued answer. The definitions of differential privacy and accuracy extend straightforwardly to such sequences.

---

[3] If we do not restrict the running time of the algorithm, then it is without loss of generality for the algorithm to simply output a list of real-valued answers to each queries by computing $Eval(S, q)$ for every $q \in Q$. However, this transformation makes the running time of the algorithm at least $|Q|$. The additional generality of this framework allows the algorithm to run in time sublinear in $|Q|$. Using this framework is crucial, since some of our results concern settings where the number of queries is exponential in the size of the dataset.

We say that such an algorithm is *computationally efficient* if the running time of the algorithm and the associated evaluator run in time polynomial in the computation parameter $\lambda$. We remark that in principle, it could require at many as $|X|$ bits even to specify a statistical query, in which case we cannot hope to answer the query efficiently, even ignoring privacy constraints. In this work we restrict attention exclusively to statistical queries that are specified by a circuit of size $\mathrm{poly}(\log|X|)$, and thus can be evaluated in time $\mathrm{poly}(\log|X|)$, and so are not the bottleneck in computation. To remind the reader of this fact, we will often say that $\mathcal{Q}$ is a family of *efficiently computable statistical queries*.

# 3  Weakly Secure Private Linear Broadcast Schemes

We now describe a very relaxed notion of private linear broadcast schemes whose existence will imply the hardness of differentially private data release.

## 3.1  Syntax and Correctness

For a function $n : \mathbb{N} \to \mathbb{N}$ and a sequence $\{K_\lambda, C_\lambda\}_{\lambda \in \mathbb{N}}$, a $(n, \{K_\lambda, C_\lambda\})$-*private linear broadcast scheme* is a tuple of efficient algorithms $\Pi = (\mathsf{Setup}, \mathsf{Enc}, \mathsf{Dec})$ with the following syntax.

- $\mathsf{Setup}$ takes as input a security parameter $\lambda$, runs in time $\mathrm{poly}(\lambda)$, and outputs $n = n(\lambda)$ secret *user keys* $sk_1, \ldots, sk_n \in K_\lambda$ and a secret *master key* $mk$. We will write $k = (sk_1, \ldots, sk_n, mk)$ to denote the set of keys.
- $\mathsf{Enc}$ takes as input a master key $mk$ and an *index* $i \in \{0, 1, \ldots, n\}$, and outputs a ciphertext $c \in C_\lambda$. If $c \leftarrow_{\mathrm{R}} \mathsf{Enc}(j, mk)$ then we say that $c$ is *encrypted to index $j$*.
- $\mathsf{Dec}$ takes as input a ciphertext $c$ and a user key $sk_i$ and outputs a single bit $b \in \{0, 1\}$. We assume for simplicity that $\mathsf{Dec}$ is deterministic.

Correctness of the scheme asserts that if $k$ are generated by $\mathsf{Setup}$, then for any pair $i, j$, $\mathsf{Dec}(sk_i, \mathsf{Enc}(mk, j)) = \mathbb{I}\{i \leq j\}$. For simplicity, we require that this property holds with probability 1 over the coins of $\mathsf{Setup}$ and $\mathsf{Enc}$, although it would not affect our results substantively if we required only correctness with high probability.

**Definition 5 (Perfect Correctness).** *An $(n, \{K_\lambda, C_\lambda\})$-private linear broadcast scheme is perfectly correct if for every $\lambda \in \mathbb{N}$, and every $i, j \in \{0, 1, \ldots, n\}$*

$$\mathop{\mathbb{P}}_{k=\mathsf{Setup}(\lambda),\, c=\mathsf{Enc}(mk,j)} [\mathsf{Dec}(sk_i, c) = \mathbb{I}\{i \leq j\}] = 1.$$

## 3.2  Weak Index-Hiding Security

Intuitively, the security property we want is that any computationally efficient adversary who is missing one of the user keys $sk_{i^*}$ cannot distinguish ciphertexts encrypted with index $i^*$ from index $i^* - 1$, even if that adversary holds all $n - 1$ other keys $sk_{-i^*}$. In other words, an efficient adversary cannot infer anything about the encrypted index beyond what is implied by the correctness of decryption and the set of keys he holds.

More precisely, consider the following two-phase experiment. First the adversary is given every key except for $sk_{i^*}$, and outputs a decryption program $S$. Then, a challenge ciphertext is encrypted to either $i^*$ or to $i^* - 1$. We say that the private linear broadcast scheme is secure if for every polynomial time adversary, with high probability over the setup and the decryption program chosen by the adversary, the decryption program has small advantage in distinguishing the two possible indices.

**Definition 6 (Weak Index Hiding).** *A private linear broadcast scheme $\Pi$ satisfies weak index-hiding security if for every sufficiently large $\lambda \in \mathbb{N}$, every $i^* \in [n(\lambda)]$, and every adversary $A$ with running time* $\mathrm{poly}(\lambda)$,

$$\mathop{\mathbb{P}}_{\substack{k=\mathrm{Setup}(\lambda) \\ S=A(sk_{-i^*})}} \left[ \mathbb{P}\left[S(\mathrm{Enc}(mk, i^*)) = 1\right] - \mathbb{P}\left[S(\mathrm{Enc}(mk, i^* - 1)) = 1\right] > \frac{1}{2en} \right] \leq \frac{1}{2en} \quad (1)$$

*In the above, the inner probabilities are taken over the coins of* Enc *and* $S$.

Note that in the above definition we have fixed the success probability of the adversary for simplicity. Moreover, we have fixed these probabilities to relatively large ones. Requiring only a polynomially small advantage is crucial to achieving the key and ciphertext lengths we need to obtain our results, while still being sufficient to establish the hardness of differential privacy.

**The Index-Hiding and Two-Index-Hiding Games.** While Definition 6 is the most natural, in this section we consider some related ways of defining security that will be easier to work with when we construct and analyze our schemes. Consider the following **IndexHiding** game (Fig. 1).

---

The challenger generates keys $k = (sk_1, \ldots, sk_n, mk) \leftarrow_R \mathrm{Setup}(\lambda)$.
The adversary $A$ is given keys $sk_{-i^*}$ and outputs a decryption program $S$.
The challenger chooses a bit $b \leftarrow_R \{0, 1\}$
The challenger generates an encryption to index $i^* - b$, $c \leftarrow_R \mathrm{Enc}(mk, i^* - b)$
The adversary makes a guess $b' = S(c)$

---

**Fig. 1. IndexHiding$[i^*]$**

Let **IndexHiding$[i^*, k, S]$** be the game **IndexHiding$[i^*]$** where we fix the choices of $k$ and $S$. Also, define

$$\mathrm{Adv}[i^*, k, S] = \mathop{\mathbb{P}}_{\mathsf{IndexHiding}[i^*, k, S]} [b' = b] - \frac{1}{2}.$$

so that

$$
\mathop{\mathbb{P}}_{\substack{\text{IndexHiding}[i^*]}} [b' = b] - \frac{1}{2} = \mathop{\mathbb{E}}_{\substack{k=\text{Setup}(\lambda) \\ S=A(sk_{-i^*})}} [\text{Adv}[i^*, k, S]]
$$

Then the following is equivalent to (1) in Definition 6 as

$$
\mathop{\mathbb{P}}_{k=\text{Setup}(\lambda),\, S=A(sk_{-i^*})} \left[ \text{Adv}[i^*, k, S] > \frac{1}{4en} \right] \le \frac{1}{2en} \tag{2}
$$

In order to prove that our schemes satisfy weak index-hiding security, we will go through an intermediate notion that we call two-index-hiding security. To see why this is useful, In our constructions it will be fairly easy to prove that $\text{Adv}[i^*]$ is small, but because $\text{Adv}[i^*, k, S]$ can be positive or negative, that alone is not enough to establish (2). Thus, in order to establish (2) we will analyze the following variant of the index-hiding game (Fig. 2).

---

The challenger generates keys $k = (sk_1, \ldots, sk_n, mk) \leftarrow_\text{R} \text{Setup}$.
The adversary $A$ is given keys $sk_{-i^*}$ and outputs a decryption program $S$.
Choose $b_0 \leftarrow_\text{R} \{0,1\}$ and $b_1 \leftarrow_\text{R} \{0,1\}$ independently.
Let $c_0 \leftarrow_\text{R} \text{Enc}(i^* - b_0; mk)$ and $c_1 \leftarrow_\text{R} \text{Enc}(i^* - b_1; mk)$.
Let $b' = S(c_0, c_1)$.

---

**Fig. 2.** TwoIndexHiding$[i^*]$

Analogous to what we did with **IndexHiding**, we can define the quantity **TwoIndexHiding**$[i^*, k, S]$ to be the game **TwoIndexHiding**$[i^*]$ where we fix the choices of $k$ and $S$, and define

$$
\text{TwoAdv}[i^*] = \mathop{\mathbb{P}}_{\textbf{TwoIndexHiding}[i^*]} [b' = b_0 \oplus b_1] - \frac{1}{2}
$$

$$
\text{TwoAdv}[i^*, k, S] = \mathop{\mathbb{P}}_{\textbf{TwoIndexHiding}[i^*, k, S]} [b' = b_0 \oplus b_1] - \frac{1}{2}
$$

so that

$$
\mathop{\mathbb{P}}_{\textbf{TwoIndexHiding}[i^*]} [b' = b_0 \oplus b_1] - \frac{1}{2} = \mathop{\mathbb{E}}_{k=\text{Setup}(\lambda),\, S=A(sk_{-i^*})} [\text{TwoAdv}[i^*, k, S]]
$$

The crucial feature is that if we can bound the expectation of TwoAdv then we get a bound on the expectation of $\text{Adv}^2$. Since $\text{Adv}^2$ is always positive, we can apply Markov's inequality to establish (2). Formally, we have the following claim.

**Claim 1.** Suppose that for every efficient adversary $A$, $\lambda \in \mathbb{N}$, and index $i^* \in [n(\lambda)]$, $\mathrm{TwoAdv}[i^*] \leq \varepsilon$. Then for every efficient adversary $A$, $\lambda \in \mathbb{N}$, and index $i^* \in [n(\lambda)]$,

$$\mathop{\mathbb{E}}_{\substack{k=\mathsf{Setup}(\lambda),\\ S \leftarrow A(sk_{-i^*})}} \left[\mathrm{Adv}[i^*, k, S]^2\right] \leq \frac{\varepsilon}{2}. \tag{3}$$

*Proof.* Given any adversary $A$ in the **IndexHiding** game, consider the following adversary $A_2$ in the **TwoIndexHiding** game, which, when given a set of keys, runs $A$ with the same keys to get program $S_A$, then creates and outputs the program $S_{A_2}$, which on input $c_0, c_1$, runs $S$ on $c_0$ to get output $b'_0$, runs $S$ on $c_1$ to get output $b'_1$, then outputs $b' = b'_0 \oplus b'_1$. Then, for this $A_2$,

$$\mathrm{TwoAdv}[i^*] = \mathop{\mathbb{E}}_{\substack{k=\mathsf{Setup}(\lambda),\\ S_{A_2} \leftarrow A_2(sk_{-i^*})}} \left[\mathrm{TwoAdv}[i^*, k, S_{A_2}]\right]$$

$$= \mathop{\mathbb{E}}_{\substack{k=\mathsf{Setup}(\lambda),\\ S_{A_2} \leftarrow A_2(sk_{-i^*})}} \left[ \mathop{\Pr}_{\substack{b_i \leftarrow_{\mathrm{R}} \{0,1\},\\ c_i \leftarrow \mathsf{Enc}(i^* - b_i)}} [b' = b_0 \oplus b_1 : b' = S_{A_2}(c_0, c_1)] - \frac{1}{2} \right]$$

$$= \mathop{\mathbb{E}}_{\substack{k=\mathsf{Setup}(\lambda),\\ S_A \leftarrow A(sk_{-i^*})}} \left[ \mathop{\Pr}_{\substack{b_i \leftarrow_{\mathrm{R}} \{0,1\},\\ c_i \leftarrow \mathsf{Enc}(i^* - b_i)}} [b'_0 \oplus b'_1 = b_0 \oplus b_1 : b'_i = S_A(c_i)] - \frac{1}{2} \right]$$

$$= \mathop{\mathbb{E}}_{\substack{k=\mathsf{Setup}(\lambda),\\ S_A \leftarrow A(sk_{-i^*})}} \left[2 \cdot \mathrm{Adv}[i^*, k, S_A]^2\right]$$

So if every efficient adversary $A'$, $\lambda \in \mathbb{N}$, and index $i^* \in [n(\lambda)]$ satisfies $\mathrm{TwoAdv}[i^*] \leq \varepsilon$, then this condition also holds for $A_2$'s $\mathrm{TwoAdv}[i^*] = \mathbb{E}_{\substack{k=\mathsf{Setup}(\lambda),\\ S_A \leftarrow A(sk_{-i^*})}} \left[2 \cdot \mathrm{Adv}[i^*, k, S_A]^2\right]$, which implies $\mathbb{E}_{\substack{k=\mathsf{Setup}(\lambda),\\ S_A \leftarrow A(sk_{-i^*})}} \left[\mathrm{Adv}[i^*, k, S_A]^2\right] \leq \frac{\varepsilon}{2}$.

Using this claim we can prove the following lemma.

**Lemma 7.** *Let $\Pi$ be a private linear broadcast scheme such that for every efficient adversary $A$, $\lambda \in \mathbb{N}$, and index $i^* \in [n(\lambda)]$, $\mathrm{TwoAdv}[i^*] \leq \frac{1}{200n^3}$. Then $\Pi$ satisfies weak index-hiding security.*

*Proof.* By applying Claim 1 to the assumption of the lemma, we have that for every efficient adversary $A$,

$$\mathop{\mathbb{E}}_{k=\mathsf{Setup}(\lambda), S=A(sk_{-i^*})} \left[\mathrm{Adv}[i^*, k, S]^2\right] \leq \frac{1}{400n^3}$$

Now we have

$$\underset{k=\mathsf{Setup}(\lambda),S=A(sk_{-i*})}{\mathbb{E}}\left[\mathrm{Adv}[i^*,\boldsymbol{k},S]^2\right] \le \frac{1}{400n^3}$$

$$\implies \underset{k=\mathsf{Setup}(\lambda),S=A(sk_{-i*})}{\mathbb{P}}\left[\mathrm{Adv}[i^*,\boldsymbol{k},S]^2 > \frac{1}{(4en)^2}\right] \le \frac{(4en)^2}{400n^3} \le \frac{1}{2en}$$

$$\implies \underset{k=\mathsf{Setup}(\lambda),S=A(sk_{-i*})}{\mathbb{P}}\left[\mathrm{Adv}[i^*,\boldsymbol{k},S] > \frac{1}{4en}\right] \le \frac{1}{2en}$$

To complete the proof, observe that this final condition is equivalent to the definition of weak index-hiding security (Definition 6).

In light of this lemma, we will focus on proving that the schemes we construct in the following sections satisfying the condition $\mathrm{TwoAdv}[i^*] \le \frac{1}{200n^3}$, which will be easier than directly establishing Definition 6.

# 4   Hardness of Differential Privacy from PLBE

In this section we prove that a private linear broadcast scheme satisfying perfect correctness and index-hiding security yields a family of statistical queries that cannot be answered accurately by an efficient differentially private algorithm. The proof is a fairly straightforward adaptation of the proofs in Dwork et al. [15] and Ullman [34] that various sorts of traitor-tracing schemes imply hardness results for differential privacy. We include the result for completeness, and to verify that our very weak definition of private linear broadcast is sufficient to prove hardness of differential privacy.

**Theorem 8.** *Suppose there is an* $(n, \{K_\lambda, C_\lambda\})$*-private linear broadcast scheme that satisfies perfect correctness (Definition 5) and weak index-hiding security (Definition 6). Then there is a sequence of pairs* $\{X_\lambda, Q_\lambda\}_{\lambda \in \mathbb{N}}$ *where* $Q_\lambda$ *is a set of statistical queries on* $X_\lambda$, $|Q_\lambda| = |C_\lambda|$, *and* $|X_\lambda| = |K_\lambda|$ *such that there is no algorithm A that is simultaneously,*

1. $(1, 1/2n)$*-differentially private,*
2. $(1/3, 1/2n)$*-accurate for* $Q_\lambda$ *on datasets* $D \in X_\lambda^{n(\lambda)}$, *and*
3. *computationally efficient.*

Theorems 1 and 2 in the introduction follow by combining Theorem 8 above with the constructions of private linear broadcast schemes in Sect. 6. The proof of Theorem 8 closely follows the proofs in Dwork et al. [15] and Ullman [34]. We give the proof both for completeness and to verify that our definition of private linear broadcast suffices to establish the hardness of differential privacy.

*Proof.* Let $\Pi = (\mathsf{Setup}, \mathsf{Enc}, \mathsf{Dec})$ be the promised $(n, \{K_\lambda, C_\lambda\})$ private linear broadcast scheme. For every $\lambda \in \mathbb{N}$, we can define a distribution on datasets $D \in X_\lambda^{n(\lambda)}$ as follows. Run $\mathsf{Setup}(\lambda)$ to obtain $n = n(\lambda)$ secret user keys

$sk_1, \ldots, sk_n \in K_\lambda$ and a master secret key $mk$. Let the dataset be $D = (sk_1, \ldots, sk_n) \in X_\lambda^n$ where we define the data universe $X_\lambda = K_\lambda$. Abusing notation, we'll write $(D, mk) \leftarrow_R \mathsf{Setup}(\lambda)$.

Now we define the family of queries $Q_\lambda$ on $X_\lambda$ as follows. For every ciphertext $c \in C_\lambda$, we define the predicate $q_c \in Q_\lambda$ to take as input a user key $sk_i \in K_\lambda$ and output $\mathsf{Dec}(sk_i, c)$. That is, $Q_\lambda = \{q_c(sk) = \mathsf{Dec}(sk, c) \mid c \in C_\lambda\}$. Recall that, by the definition of a statistical query, for a dataset $D = (sk_1, \ldots, sk_n)$, we have

$$q_c(D) = (1/n) \sum_{i=1}^{n} \mathsf{Dec}(sk_i, c).$$

Suppose there is an algorithm $A$ that is computationally efficient and is $(1/3, 1/2n)$-accurate for $Q_\lambda$ given a dataset $D \in X_\lambda^n$. We will show that $A$ cannot satisfy $(1, 1/2n)$-differential privacy. By accuracy, for every $\lambda \in \mathbb{N}$ and every fixed dataset $D \in X_\lambda^n$, with probability at least $1 - 1/2n$, $A(D)$ outputs a summary $S \in \mathcal{S}_\lambda$ that is $1/3$-accurate for $Q_\lambda$ on $D$. That is, for every $D \in X_\lambda^n$, with probability at least $1 - 1/2n$,

$$\forall q_c \in Q_\lambda \quad |q_c(D) - q_c(S)| \leq 1/3. \tag{4}$$

Suppose that $S$ is indeed $1/3$-accurate. By perfect correctness of the private linear broadcast scheme (Definition 5), and the definition of $Q$, we have that since $(D, mk) = \mathsf{Setup}(\lambda)$,

$$(c = \mathsf{Enc}(mk, 0)) \implies (q_c(D) = 0) \qquad (c = \mathsf{Enc}(mk, n)) \implies (q_c(D) = 1). \tag{5}$$

Combining Eqs. (4) and (5), we have that if $(D, mk) = \mathsf{Setup}(\lambda)$, $S \leftarrow_R A(D)$, and $S$ is $1/3$-accurate, then we have both $\underset{c \leftarrow_R \mathsf{Enc}(mk,0)}{\mathbb{P}} [q_c(S) \leq 1/3] = 1$ and $\underset{c \leftarrow_R \mathsf{Enc}(mk,n)}{\mathbb{P}} [q_c(S) \leq 1/3] = 0$ Thus, for every $(D, mk)$ and $S$ that is $1/3$-accurate, there exists an index $i \in \{1, \ldots, n\}$ such that

$$\left| \underset{c \leftarrow_R \mathsf{Enc}(mk,i)}{\mathbb{P}} [q_c(S) \leq 1/3] - \underset{c \leftarrow_R \mathsf{Enc}(mk,i-1)}{\mathbb{P}} [q_c(S) \leq 1/3] \right| > \frac{1}{n} \tag{6}$$

By averaging, using the fact that $S$ is $1/3$-accurate with probability at least $1 - 1/2n$, there must exist an index $i^* \in \{1, \ldots, n\}$ such that

$$\underset{\substack{(D,mk)=\mathsf{Setup}(\lambda) \\ S \leftarrow_R A(D)}}{\mathbb{P}} \left[ \left| \underset{c=\mathsf{Enc}(mk,i^*)}{\mathbb{P}} \left[ q_c(S) \leq \frac{1}{3} \right] - \underset{c=\mathsf{Enc}(mk,i^*-1)}{\mathbb{P}} \left[ q_c(S) \leq \frac{1}{3} \right] \right| > \frac{1}{n} \right] \geq \frac{1}{n} \tag{7}$$

Assume, for the sake of contradiction that $A$ is $(1, 1/2n)$-differentially private. For a given $i, mk$, let $\mathcal{S}_{i,mk} \subseteq \mathcal{S}_\lambda$ be the set of summaries $S$ such that (6) holds. Then, by (7), we have $\underset{(D,mk) \leftarrow_R \mathsf{Setup}(\lambda)}{\mathbb{P}} [A(D) \in \mathcal{S}_{i^*,mk}] \geq \frac{1}{n}$. By differential privacy of $A$, we have

$$\underset{(D,mk) \leftarrow_R \mathsf{Setup}}{\mathbb{P}} [A(D_{-i^*}) \in \mathcal{S}_{i^*,mk}] \geq \frac{1}{e} \left( \frac{1}{n} - \frac{1}{2n} \right) = \frac{1}{2en}$$

Thus, by our definition of $\mathcal{S}_{i^*, mk}$, and by averaging over $(D, mk) \leftarrow_R \mathsf{Setup}(\lambda)$, we have

$$\mathop{\mathbb{P}}_{\substack{(D, mk)=\mathsf{Setup} \\ S=A(D_{-i^*})}} \left[ \left| \mathop{\mathbb{P}}_{c=\mathsf{Enc}(mk, i^*)} \left[ q_c(S) \le \frac{1}{3} \right] - \mathop{\mathbb{P}}_{c=\mathsf{Enc}(mk, i^*-1)} \left[ q_c(S) \le \frac{1}{3} \right] \right| > \frac{1}{n} \right] \ge \frac{1}{2en} \quad (8)$$

But this violates the weak index hiding property of the private linear broadcast scheme. Specifically, if we consider an adversary for the private linear broadcast scheme that runs $A$ on the keys $sk_{-i^*}$ to obtain a summary $S$, then decrypts a ciphertext $c$ by computing $q_c(S)$ and rounding the answer to $\{0, 1\}$, then by (8) this adversary violates weak index-hiding security (Definition 6).

Thus we have obtained a contradiction showing that $A$ is not $(1, 1/2n)$-differentially private. This completes the proof.

# 5    Cryptographic Primitives

We will make use of several cryptographic tools and information-theoretic primitives. Due to space, we will omit a formal definition of standard concepts like almost-pairwise-independent hash functions, pseudorandom generators, and pseudorandom functions and defer these to the full version.

## 5.1    Puncturable Pseudorandom Functions

A pseudorandom function family $\mathcal{F}_\lambda = \{\mathsf{PRF} : [m] \to [n]\}$ is *puncturable* if there is a deterministic procedure $\mathsf{Puncture}$ that takes as input $\mathsf{PRF} \in \mathcal{F}_\lambda$ and $x^* \in [m]$ and outputs a new function $\mathsf{PRF}^{\{x^*\}} : [m] \to [n]$ such that $\mathsf{PRF}^{\{x^*\}}(x) = \mathsf{PRF}(x)$ if $x \ne x^*$ and $\mathsf{PRF}^{\{x^*\}}(x) = \perp$ if $x = x^*$.

The definition of security for a punctured pseudorandom function states that for any $x^*$, given the punctured function $\mathsf{PRF}^{\{x^*\}}$, the missing value $\mathsf{PRF}(x^*)$ is computationally unpredictable. Specifically, we define the game **Puncture** to capture the desired security property (Fig. 3).

---

The challenger chooses $\mathsf{PRF} \leftarrow_R \mathcal{F}_\lambda$
The challenger chooses uniform random bit $b \in \{0, 1\}$, and samples

$$y_0 \leftarrow_R \mathsf{PRF}(x^*), \quad y_1 \leftarrow_R [n].$$

The challenger punctures $\mathsf{PRF}$ at $x^*$, obtaining $\mathsf{PRF}^{\{x^*\}}$.
The adversary is given $(y_b, \mathsf{PRF}^{\{x^*\}})$ and outputs a bit $b'$.

---

**Fig. 3. Puncture$[x^*]$**

**Definition 9 (Puncturing Secure PRF).** *A pseudorandom function family* $\mathcal{F}_\lambda = \{\text{PRF} : [m] \to [n]\}$ *is* $\varepsilon$-puncturing secure *if for every* $x^* \in [m]$,

$$\mathbb{P}_{\text{Puncture}[x^*]} [b' = b] \leq \frac{1}{2} + \varepsilon.$$

## 5.2 Twice Puncturable PRFs

A *twice puncturable PRF* is a pair of algorithms (*PRFSetup*, Puncture).

- *PRFSetup* is a randomized algorithm that takes a security parameter $\lambda$ and outputs a function PRF : $[m] \to [n]$ where $m = m(\lambda)$ and $n = n(\lambda)$ are parameters of the construction. Technically, the function is parameterized by a seed of length $\lambda$, however for notational simplicity we will ignore the seed and simply use PRF to denote this function. Formally PRF $\leftarrow_R$ *PRFSetup*($\lambda$).
- Puncture is a deterministic algorithm that takes a PRF and a pair of inputs $x_0, x_1 \in [m]$ and outputs a new function $\text{PRF}^{\{x_0,x_1\}} : [m] \to [n]$ such that

$$\text{PRF}^{\{x_0,x_1\}} = \begin{cases} \text{PRF}(x) & \text{if } x \notin \{x_0, x_1\} \\ \bot & \text{if } x \in \{x_0, x_1\} \end{cases}$$

Formally, $\text{PRF}^{\{x_0,x_1\}} = \text{Puncture}(\text{PRF}, x_0, x_1)$.

In what follows we will always assume that $m$ and $n$ are polynomial in the security parameter and that $m = \omega(n \log(n))$.

In addition to requiring that this family of functions satisfies the standard notion of cryptographic pseudorandomness, we will now define a new security property for twice puncturable PRFs, called *input matching indistinguishability*. For any two distinct outputs $y_0, y_1 \in [n], y_0 \neq y_1$, consider the following game (Fig. 4).

---

The challenger chooses PRF such that $\forall y \in [n]$, $\text{PRF}^{-1}(y) \neq \emptyset$.
The challenger chooses independent random bits $b_0, b_1 \in \{0, 1\}$, and samples

$$x_0 \leftarrow_R \text{PRF}^{-1}(y_{b_0}), \quad x_1 \leftarrow_R \text{PRF}^{-1}(y_{b_1}).$$

The challenger punctures PRF at $x_0, x_1$, obtaining $\text{PRF}^{\{x_0,x_1\}}$.
The adversary is given $(x_0, x_1, \text{PRF}^{\{x_0,x_1\}})$ and outputs a bit $b'$.

---

**Fig. 4.** InputMatching$[y_0, y_1]$

Notice that in this game, we have assured that every $y \in [n]$ has a preimage under PRF. We need this condition to make the next step of sampling random preimages well defined. Technically, it would suffice to have a preimage only for

$y_{b_0}$ and $y_{b_1}$, but for simplicity we will assume that every possible output has a preimage. When $f : [m] \to [n]$ is a random function, the probability that some output has no preimage is at most $n \cdot \exp(-\Omega(m/n))$ which is negligible when $m = \omega(n \log(n))$. Since $m, n$ are assumed to be a polynomial in the security parameter, we can efficiently check if every output has a preimage, thus if PRF is pseudorandom it must also be the case that every output has a preimage with high probability. Since we can efficiently check whether or not every output has a preimage under PRF, and this event occurs with all but negligible probability, we can efficiently sample the pseudorandom function in the first step of InputMatching$[y_0, y_1]$.

**Definition 10 (Input-Matching Secure PRF).** *A function family* $\{\text{PRF} : [m] \to [n]\}$ *is* $\varepsilon$-*input-matching secure if the function family is a secure pseudorandom function and additionally for every* $y_0, y_1 \in [n]$ *with* $y_0 \neq y_1$,

$$\Pr_{\text{InputMatching}[y_0, y_1]} [b' = b_0 \oplus b_1] \leq \frac{1}{2} + \varepsilon.$$

In the full version of this work we show that input-matching secure twice puncturable pseudorandom functions with suitable parameters exist.

**Theorem 11.** *Assuming the existence of one-way functions, if* $m, n$ *are polynomials such that* $m = \omega(n \log(n))$, *then there exists a pseudorandom function family* $\mathcal{F}_\lambda = \{\text{PRF} : [m(\lambda)] \to [n(\lambda)]\}$ *that is twice puncturable and is* $\tilde{O}(\sqrt{n/m})$-*input-matching secure.*

### 5.3 Indistinguishability Obfuscation

We use the following formulation of Garg et al. [19] for indistinguishability obfuscation:

**Definition 12 (Indistinguishability Obfuscation).** *A indistinguishability obfuscator* O *for a circuit class* $\{C_\lambda\}$ *is a probabilistic polynomial-time uniform algorithm satisfying the following conditions:*

1. $O(\lambda, C)$ *preserves the functionality of* $C$. *That is, for any* $C \in C_\lambda$, *if we compute* $C' = O(\lambda, C)$, *then* $C'(x) = C(x)$ *for all inputs* $x$.
2. *For any* $\lambda$ *and any two circuits* $C_0, C_1$ *with the same functionality, the circuits* $O(\lambda, C_0)$ *and* $O(\lambda, C_1)$ *are indistinguishable. More precisely, for all pairs of probabilistic polynomial-time adversaries* $(Samp, D)$, *if*

$$\Pr_{(C_0, C_1, \sigma) \leftarrow Samp(\lambda)} [(\forall x),\ C_0(x) = C_1(x)] > 1 - \text{negl}(\lambda)$$

*then*

$$|\Pr[D(\sigma, O(\lambda, C_0)) = 1] - \Pr[D(\sigma, O(\lambda, C_1)) = 1]| < \text{negl}(\lambda)$$

*The circuit classes we are interested in are polynomial-size circuits - that is, when* $C_\lambda$ *is the collection of all circuits of size at most* $\lambda$. *When clear from context, we will often drop* $\lambda$ *as an input to* O *and as a subscript for* $C$.

# 6   A PLBE Scheme with Very Short Ciphertexts

In this section we construct a private linear broadcast scheme for $n$ users where the key length is polynomial in the security parameter $\lambda$ and the ciphertext length is only $O(\log(n))$. This scheme will be used to establish our hardness result for differential privacy when the data universe can be exponentially large but the family of queries has only polynomial size. The construction of a weak private linear broadcast scheme with user keys of length $O(\log(n))$ is in Sect. 7.

## 6.1   Construction

Let $n = \text{poly}(\lambda)$ denote the number of users for the scheme. Let $m = \tilde{O}(n^7)$ be a parameter. Our construction will rely on the following primitives:

- A pseudorandom generator $\text{PRG} : \{0,1\}^{\lambda/2} \to \{0,1\}^\lambda$.
- A puncturable PRF family $\mathcal{F}_{\lambda,sk} = \{\text{PRF}_{sk} : [n] \to \{0,1\}^\lambda\}$.
- A twice-puncturable PRF family $\mathcal{F}_{\lambda,\text{Enc}} = \{\text{PRF}_{\text{Enc}} : [m] \to [n]\}$.
- An iO scheme Obfuscate.

---

$\text{Setup}(\lambda)$ :
  Choose $\text{PRF}_{sk} \leftarrow_R \mathcal{F}_{\lambda,sk}$
  Choose $\text{PRF}_{\text{Enc}} \leftarrow_R \mathcal{F}_{\lambda,\text{Enc}}$ such that for every $i \in [n]$, $\text{PRF}_{\text{Enc}}^{-1}(i) \neq \emptyset$
  For $i = 1, \ldots, n$, let $s_i = \text{PRF}_{sk}(i)$.
  Let $O \leftarrow_R \text{Obfuscate}(P_{\text{PRF}_{sk},\text{PRF}_{\text{Enc}}})$.
  Let each user's secret key be $sk_i = (i, s_i, O)$
  Let the master key be $mk = \text{PRF}_{\text{Enc}}$.

$\text{Enc}(j, mk = \text{PRF}_{\text{Enc}})$ :
  Let $c$ be chosen uniformly from $\text{PRF}_{\text{Enc}}^{-1}(j)$.
  Output $c$.

$\text{Dec}(sk_i = (i, s_i, O), c)$:
  Output $O(c, i, s_i)$.

$P_{\text{PRF}_{sk},\text{PRF}_{\text{Enc}}}(c, i, s)$ :
  If $\text{PRG}(s) \neq \text{PRG}(\text{PRF}_{sk}(i))$, halt and output $\bot$.
  Output $\mathbb{I}\{i \leq \text{PRF}_{\text{Enc}}(c)\}$.

---

**Fig. 5.** Our scheme $\Pi_{\text{short}-\text{ctext}}$.

**Theorem 13.** *Assuming the existence of one-way functions and indistinguishability obfuscation. For every polynomial $n$, the scheme $\Pi_{\text{short}-\text{ctext}}$ (Fig. 5) is an $(n, d, \ell)$-private linear broadcast scheme for $d = \text{poly}(\lambda)$ and $2^\ell = \tilde{O}(n^7)$ and satisfies:* $\text{TwoAdv}[i^*] \leq \frac{1}{200n^3}$.

Combining this theorem with Lemma 7 and Theorem 8 establishes Theorem 1 in the introduction.

**Parameters.** First we verify that $\Pi_{\text{short-ctext}}$ is an $(n, d, \ell)$-private linear broadcast scheme for the desired parameters. Observe that the length of the secret keys is $\log(n) + \lambda + |O|$. By the efficiency of the pseudorandom functions and the specification of P, the running time of P is $\text{poly}(\lambda + \log(n))$. Thus, by the efficiency of Obfuscate, $|O| = \text{poly}(\lambda + \log(n))$. Therefore the total key length is $\text{poly}(\lambda + \log(n))$. Since $n$ is assumed to be a polynomial in $\lambda$, we have that the secret keys have length $d = \text{poly}(\lambda)$ as desired. By construction, the ciphertext is an element of $[m]$. Thus, since $m = \tilde{O}(n^7)$ the ciphertexts length $\ell$ satisfies $2^\ell = \tilde{O}(n^7)$ as desired.

## 6.2   Proof of Weak Index-Hiding Security

In light of Lemma 7, in order to prove that the scheme satisfies weak index-hiding security, it suffices to show that for every sufficiently large $\lambda \in \mathbb{N}$, and every $i^* \in [n(\lambda)]$, $\mathbb{P}_{\textbf{TwoIndexHiding}[i^*]}[b' = b_0 \oplus b_1] - \frac{1}{2} = o(1/n^3)$. We will demonstrate this using a series of hybrids to reduce security of the scheme in the **TwoIndexHiding** game to input-matching security of the pseudorandom function family $\text{PRF}_{\lambda,\text{Enc}}$.

Before we proceed with the argument, we remark a bit on how we will present the hybrids. Note that the view of the adversary consists of the keys $sk_{-i^*}$. Each of these keys is of the form $(i, s_i, O)$ where $O$ is an obfuscation of the same program P. Thus, for brevity, we will discuss only how we modify the construction of the program P and it will be understood that each user's key will consist of an obfuscation of this modified program. We will also rely crucially on the fact that, because the challenge ciphertexts depend only on the master key $mk$, we can generate the challenge ciphertexts $c_0$ and $c_1$ can be generated before the users' secret keys $sk_1, \ldots, sk_n$. Thus, we will be justified when we modify P in a manner that depends on the challenge ciphertexts and include an obfuscation of this program in the users' secret keys. We also remark that we highlight the changes in the hybrids in green.

**Breaking the Decryption Program for Challenge Index.** We use a series of hybrids to ensure that the obfuscated program reveals no information about the secret $s_{i^*}$ for the specified user $i^*$. First, we modify the program by hard-coding the secret $s_{i^*}$ into the program. The obfuscated versions of P and P1 (Fig. 6) are indistinguishable because the input-output behavior of the programs are identical, thus the indistinguishability obfuscation guarantees that the obfuscations of these programs are computationally indistinguishable.

Next we modify the setup procedure to give a uniformly random value for $s_{i^*}$. The new setup procedure is indistinguishable from the original setup procedure by the pseudorandomness of $s_{i^*} = \text{PRF}_{sk}(i^*)$. Finally, we modify the decryption program to use a truly random value $x^*$ instead of $x^* = \text{PRG}(\text{PRF}_{sk}(i^*))$. The new decryption program is indistinguishable from the original by pseudorandomness of PRG and $\text{PRF}_{sk}$.

$P^1_{PRF^{\{i^*\}}_{sk},PRF_{Enc},i^*,x^*}(c,i,s):$
  If $i = i^*$ and $PRG(s) \neq x^*$, halt and output $\perp$.
  If $i \neq i^*$ and $PRG(s) \neq PRG(PRF^{\{i^*\}}_{sk}(i))$, halt and output $\perp$.
  Output $\mathbb{I}\{i \leq PRF_{Enc}(c)\}$.

**Fig. 6.** Modified program $P^1$. $i^*$ and $x^* = PRG(PRF_{sk}(i^*))$ are hardcoded values.

$P^2_{PRF^{\{i^*\}}_{sk},PRF_{Enc},i^*}(c,i,s):$
  If $i = i^*$, halt and output $\perp$.
  If $i \neq i^*$ and $PRG(s) \neq PRG(PRF^{\{i^*\}}_{sk}(i))$, halt and output $\perp$.
  Output $\mathbb{I}\{i \leq PRF_{Enc}(c)\}$.

**Fig. 7.** Modified program $P^2$.

$P^3_{PRF^{\{i^*\}}_{sk},PRF^{\{c_0,c_1\}}_{Enc},i^*,c_0,b_0,c_1,b_1}(c,i,s):$
  If $i = i^*$, halt and output $\perp$.
  If $i \neq i^*$ and $PRG(s) \neq PRG(PRF^{\{i^*\}}_{sk}(i))$, halt and output $\perp$.
  If $c = c_0$, output $\mathbb{I}\{i \leq i^* - b_0\}$
  If $c = c_1$, output $\mathbb{I}\{i \leq i^* - b_1\}$
  Output $\mathbb{I}\{i \leq PRF^{\{c_0,c_1\}}_{Enc}(c)\}$.

**Fig. 8.** Modified program $P^3$. $c_0, b_0, c_1, b_1$ are hardcoded values.

After making these modifications, with probability at least $1 - 2^{-\lambda/2}$, the random value $x^*$ is not in the image of PRG. Thus, with probability at least $1 - 2^{-\lambda/2}$, the condition $PRG(sk) = x^*$ will be unsatisfiable. Therefore, we can simply remove this test without changing the program on any inputs. Thus, the obfuscation of $P^1$ will be indistinguishable from the obfuscation of the following program $P^2$ (Fig. 7).

**Breaking the Decryption Program for the Challenge Ciphertexts.** First we modify the program so that the behavior on the challenge ciphertexts is hardcoded and $PRF_{Enc}$ is punctured on the challenge ciphertexts. The new decryption program is as follows. Note that the final line of the program is never reached when the input satisfies $c = c_0$ or $c = c_1$, so puncturing $PRF_{Enc}$ at these points does not affect the output of the program on any input. Thus, $P^3$ (Fig. 8) is indistinguishable from $P^2$ by the security of indistinguishability obfuscation.

Next, since, $b_0, b_1 \in \{0,1\}$, and the decryption program halts immediately if $i = i^*$, the values of $b_0, b_1$ do not affect the output of the program. Thus, we

$P^4_{PRF_{sk}^{\{i^*\}}, PRF_{Enc}^{\{c_0, c_1\}}, i^*, c_0, c_1}(c, i, s)$ :

    If $i = i^*$ , halt and output $\perp$.

    If $i \neq i^*$ and $PRG(s) \neq PRG(PRF_{sk}^{\{i^*\}}(i))$, halt and output $\perp$.

    If $c = c_0$, output $\mathbb{I}\{i \leq i^*\}$

    If $c = c_1$, output $\mathbb{I}\{i \leq i^*\}$

    Output $\mathbb{I}\{i \leq PRF_{Enc}^{\{c_0, c_1\}}(c)\}$.

**Fig. 9.** Modified program $P^4$. $c_0, c_1$ are hardcoded values.

can simply drop them from the description of the program without changing the program on any input. So, by security of the indistinguishability obfuscation, $P^3$ is indistinguishable from the following program $P^4$ (Fig. 9).

**Reducing to Input-Matching Security.** Finally, we claim that if the adversary is able to win at **TwoIndexHiding** then he can also win the game **InputMatching**$[i^* - 1, i^*]$, which violates input-matching security of $\mathcal{F}_{\lambda, Enc}$.

Recall that the challenge in the game **InputMatching**$[i^* - 1, i^*]$ consists of a tuple $(c_0, c_1, PRF^{\{c_0, c_1\}})$ where $PRF_{Enc}$ is sampled subject to 1) $PRF_{Enc}(c_0) = i^* - b_0$ for a random $b_0 \in \{0, 1\}$, 2) $PRF_{Enc}(c_1) = i^* - b_1$ for a random $b_1 \in \{0, 1\}$, and 3) $PRF_{Enc}^{-1}(i) \neq \emptyset$ for every $i \in [n]$. Given this input, we can precisely simulate the view of the adversary in **TwoIndexHiding**$[i^*]$. To do so, we can choose $PRF_{sk}$ and give the keys $sk_{-i^*}$ and obfuscations of $P^4_{PRF_{sk}^{\{i^*\}}, PRF_{Enc}^{\{c_0, c_1\}}, i^*, c_0, c_1}$ to the adversary. Then we can user $c_0, c_1$ as the challenge ciphertexts and obtain a bit $b'$ from the adversary. By input-matching security, we have that $\mathbb{P}[b' = b_0 \oplus b_1] - \frac{1}{2} = o(1/n^3)$. Since, as we argued above, the view of the adversary in this game is indistinguishable from the view of the adversary in **TwoIndexHiding**$[i^*]$, we conclude that $\mathbb{P}_{\textbf{TwoIndexHiding}[i^*]}[b' = b_0 \oplus b_1] - \frac{1}{2} = o(1/n^3)$, as desired. This completes the proof.

# 7 A Private Linear Broadcast Scheme with Very Short Keys

In this section we construct a different private linear broadcast scheme for $n$ users where the parameters are essentially reversed—the length of the secret user keys is $O(\log(n))$ and the length of the ciphertexts is $poly(\lambda)$. This scheme will be used to establish our hardness result for differential privacy when the number of queries is exponentially large but the data universe has only polynomial size.

## 7.1 Construction

Let $n = poly(\lambda)$ denote the number of users for the scheme. Let $m = \tilde{O}(n^6)$ be a parameter. Our construction will rely on the following primitives:

---

Setup($\lambda$) :
   Choose a pseudorandom function $\mathsf{PRF}_{sk} \leftarrow_\mathrm{R} \mathcal{F}_{\lambda,sk}$.
   For $i = 1, \ldots, n$, let $s_i = \mathsf{PRF}_{sk}(i)$, and let each user's secret key be $sk_i = (i, s_i) \in [n] \times [m]$.
   Let the master key be $mk = \mathsf{PRF}_{sk}$.

Enc($j, mk = \mathsf{PRF}_{sk}$) :
   Choose a pseudorandom function $\mathsf{PRF}_{\mathsf{Enc}} \leftarrow_\mathrm{R} \mathcal{F}_{\lambda,\mathsf{Enc}}$.
   Let $\mathsf{O} = \mathsf{Obfuscate}(\mathsf{P}_{j,\mathsf{PRF}_{sk},\mathsf{PRF}_{\mathsf{Enc}}})$
   Output $c = \mathsf{O}$.

Dec($sk_i = (i, s_i), c = \mathsf{O}$):
   Output $\mathsf{O}(i, sk_i)$.

$\mathsf{P}_{j,\mathsf{PRF}_{sk},\mathsf{PRF}_{\mathsf{Enc}}}(i, s)$:
   If $s \neq \mathsf{PRF}_{sk}(i)$, output $\mathsf{PRF}_{\mathsf{Enc}}(i, s)$.
   Else, output $\mathbb{I}\{i \leq j\}$.

**Fig. 10.** Our scheme $\Pi_{\mathrm{short-key}}$

- A puncturable PRF family $\mathcal{F}_{\lambda,sk} = \{\mathsf{PRF}_{sk} : [n] \to [m]\}$.
- A puncturable PRF family $\mathcal{F}_{\lambda,\mathsf{Enc}} = \{\mathsf{PRF}_{\mathsf{Enc}} : [n] \times [m] \to \{0,1\}\}$.
- An iO scheme $\mathsf{Obfuscate}$.

**Theorem 14.** *Assuming the existence of one-way functions and indistinguishability obfuscation, for every polynomial $n$, the scheme $\Pi_{\mathrm{short-key}}$ (Fig. 10) is an $(n, d, \ell)$-private linear broadcast scheme for $2^d = \tilde{O}(n^7)$ and $\ell = \mathrm{poly}(\lambda)$, and is weakly index-hiding secure.*

Combining this theorem with Lemma 7 and Theorem 8 establishes Theorem 2 in the introduction.

**Parameters.** First we verify that $\Pi_{\mathrm{short-key}}$ is an $(n, d, \ell)$-private linear broadcast scheme for the desired parameters. Observe that the length of the secret keys is $d$ such that $2^d = nm$. By construction, since $m = \tilde{O}(n^6)$, $2^d = \tilde{O}(n^7)$. The length of the ciphertext is $|\mathsf{O}|$, which is $\mathrm{poly}(|\mathsf{P}|)$ by the efficiency of the obfuscation scheme. By the efficiency of the pseudorandom function family and the pairwise independent hash family, the running time of $\mathsf{P}$ is at most $\mathrm{poly}(\lambda + \log(n))$. Since $n$ is assumed to be a polynomial in $\lambda$, the ciphertexts have length $\mathrm{poly}(\lambda)$.

### 7.2 Proof of Weak Index-Hiding Security

Just as in Sect. 6, we will rely on Lemma 7 so that we only need to show that for every $\lambda \in \mathbb{N}$, and every $i^* \in [n(\lambda)]$,

$$\Pr_{\mathsf{TwoIndexHiding}[i^*]}[b' = b_0 \oplus b_1] - \frac{1}{2} = o(1/n^3).$$

$\mathsf{Enc}^1(i^*, b_0, mk = \mathsf{PRF}_{sk})$ :

Choose a pseudorandom function $\mathsf{PRF}_{\mathsf{Enc}} \leftarrow_{\mathrm{R}} \mathcal{F}_{\lambda, \mathsf{Enc}}$.

Let $s^* = \mathsf{PRF}_{sk}(i^*)$, $\mathsf{PRF}_{sk}^{\{i^*\}} = \mathsf{Puncture}(\mathsf{PRF}_{sk}, i^*)$.

$$\text{Let } \mathsf{O} = \mathsf{Obfuscate}\left(\mathsf{P}^1_{i^*, b_0, s^*, \mathsf{PRF}_{sk}^{\{i^*\}}, \mathsf{PRF}_{\mathsf{Enc}}}\right).$$

Output $c_0 = \mathsf{O}$.

$\mathsf{P}^1_{i^*, b_0, s^*, \mathsf{PRF}_{sk}^{\{i^*\}}, \mathsf{PRF}_{\mathsf{Enc}}}(i, s)$:

  If $i = i^*$

    If $s \neq s^*$, output $\mathsf{PRF}_{\mathsf{Enc}}(i^*, s)$

    If $s = s^*$, output $1 - b_0$

  Else If $i \neq i^*$

    If $s \neq \mathsf{PRF}_{sk}^{\{i^*\}}(i)$, halt and output $\mathsf{PRF}_{\mathsf{Enc}}(i, s)$.

    Output $\mathbb{I}\{i \leq i^* - 1\}$.

**Fig. 11.** Hybrid $(\mathsf{Enc}^1, \mathsf{P}^1)$.

We will demonstrate this using a series of hybrids to reduce security of the scheme in the **TwoIndexHiding** game to the security of the pseudorandom function families.

In our argument, recall that the adversary's view consists of the keys $sk_{-i^*}$ and the challenge ciphertexts $c_0, c_1$. In our proof, we will not modify how the keys are generated, so we will present the hybrids only by how the challenge ciphertexts are generated. Also, for simplicity, we will focus only on how $c_0$ is generated as a function of $i^*, b_0$ and $mk$. The ciphertext $c_1$ will be generated in exactly the same way but as a function of $i^*, b_1$ and $mk$. We also remark that we highlight the changes in the hybrids in green.

**Hiding the Missing User Key.** First we modify the encryption procedure to one where $\mathsf{PRF}_{sk}$ is punctured on $i^*$ and the value $s^* = \mathsf{PRF}_{sk}(i^*)$ is hardcoded into the program (Fig. 11).

We claim that, by the security of the iO scheme, the distribution of $c_0, c_1$ under $\mathsf{Enc}^1$ is computationally indistinguishable from the distribution of $c_0, c_1$ under $\mathsf{Enc}$. The reason is that the obfuscation $\mathsf{P}$ and $\mathsf{P}^1$ compute the same function. Consider two cases, depending on whether $i = i^*$ or $i \neq i^*$. If $i \neq i^*$, since $b_0 \in \{0, 1\}$, and $i \neq i^*$, replacing $\mathbb{I}\{i \leq i^* - b_0\}$ with $\mathbb{I}\{i \leq i^* - 1\}$ does not change the output. Moreover, since we only reach the branch involving $\mathsf{PRF}_{sk}^{\{i^*\}}$ when $i \neq i^*$, the puncturing does not affect the output of the program. If $i = i^*$, then the program either outputs $\mathsf{PRF}_{\mathsf{Enc}}(i^*, s)$ as it did before when $s \neq s^*$ or it outputs $1 - b_0$: equivalent to $\mathbb{I}\{i \leq i^* - b_0\}$. Thus, by iO, the obfuscated programs are indistinguishable.

---

$\mathsf{Enc}^2(i^*, b_0, mk = \mathsf{PRF}_{sk})$:

   Choose a pseudorandom function $\mathsf{PRF}_{\mathsf{Enc}} \leftarrow_{\mathrm{R}} \mathcal{F}_{\lambda, \mathsf{Enc}}$.

   $\mathsf{PRF}_{sk}^{\{i^*\}} = \mathsf{Puncture}(\mathsf{PRF}_{sk}, i^*)$, Let $\tilde{s} \leftarrow_{\mathrm{R}} [m]$.

$$\text{Let } \mathsf{O} = \mathsf{Obfuscate}\left(\mathsf{P}^2_{i^*, b_0, \tilde{s}, \mathsf{PRF}_{sk}^{\{i^*\}}, \mathsf{PRF}_{\mathsf{Enc}}}\right).$$

   Output $c_0 = \mathsf{O}$.

$\mathsf{P}^2_{i^*, b_0, \tilde{s}, \mathsf{PRF}_{sk}^{\{i^*\}}, \mathsf{PRF}_{\mathsf{Enc}}}(i, s)$:

   If $i = i^*$

      If $s \neq \tilde{s}$, output $\mathsf{PRF}_{\mathsf{Enc}}(i^*, s)$

      If $s = \tilde{s}$, output $1 - b_0$

   Else If $i \neq i^*$

      If $s \neq \mathsf{PRF}_{sk}^{\{i^*\}}(i)$, halt and output $\mathsf{PRF}_{\mathsf{Enc}}(i, s)$.

      Output $\mathbb{I}\{i \leq i^* - 1\}$.

**Fig. 12.** Hybrid $(\mathsf{Enc}^2, \mathsf{P}^2)$.

Next, we argue that, since $\mathsf{PRF}_{sk}^{\{i^*\}}$ is sampled from a puncturable pseudorandom function family, and the adversary's view consists of $s_{-i^*} = \{\mathsf{PRF}_{sk}(i)\}_{i \neq i^*}$ but not $\mathsf{PRF}_{sk}(i^*)$, the value of $\mathsf{PRF}_{sk}(i^*)$ is computationally indistinguishable to the adversary from a random value. Thus, we can move to another hybrid $(\mathsf{Enc}^2, \mathsf{P}^2)$ where the value $s^*$ is replaced with a uniformly random value $\tilde{s}$ (Fig. 12).

*Hiding the Challenge Index.* Now we want to remove any explicit use of $b_0$ from $\mathsf{P}^2$. The natural way to try to do this is to remove the line where the program outputs $1 - b_0$ when the input is $(i^*, \tilde{s})$, and instead have the program output $\mathsf{PRF}_{\mathsf{Enc}}(i^*, \tilde{s})$. However, this would involve changing the program's output on one input, and indistinguishability obfuscation does not guarantee any security in this case. We get around this problem in two steps. First, we note that the value of $\mathsf{PRF}_{\mathsf{Enc}}$ on the point $(i^*, \tilde{s})$ is never needed in $\mathsf{P}^2$, so we can move to a new procedure $\mathsf{P}^3$ where we puncture at that point without changing the program functionality. Indistinguishability obfuscation guarantees that $\mathsf{P}^2$ and $\mathsf{P}^3$ are computationally indistinguishable (Fig. 13).

Next, we define another hybrid $\mathsf{P}^4$ where change how we sample $\mathsf{PRF}_{\mathsf{Enc}}$ and sample it so that $\mathsf{PRF}_{\mathsf{Enc}}(i^*, \tilde{s}) = 1 - b_0$. Observe that the hybrid only depends on $\mathsf{PRF}_{\mathsf{Enc}}^{\{(i^*, \tilde{s})\}}$. We claim the distributions of $\mathsf{PRF}_{\mathsf{Enc}}^{\{(i^*, \tilde{s})\}}$ when $\mathsf{PRF}_{\mathsf{Enc}}$ is sampled correctly versus sampled conditioned on $\mathsf{PRF}_{\mathsf{Enc}}(i^*, \tilde{s}) = 1 - b_0$ are computationally indistinguishable. This follows readily from punctured PRF security. Suppose to the contrary that the two distributions were distinguishable with non-negligible advantage $\delta$ by adversary $A$. Then consider a punctured PRF adversary $B$ that is given $\mathsf{PRF}_{\mathsf{Enc}}^{\{(i^*, \tilde{s})\}}, b$ where $b$ is chosen at random, or $b = \mathsf{PRF}_{\mathsf{Enc}}(i^*, \tilde{s})$. $B$ distinguishes the two cases as follows. If $b \neq 1 - b_0$, then $B$

$\mathsf{Enc}^3(i^*, b_0, mk = \mathsf{PRF}_{sk})$ :

Let $\tilde{s} \leftarrow_R [m]$.

Choose a pseudorandom function $\mathsf{PRF}_{\mathsf{Enc}} \leftarrow_R \mathcal{F}_{\lambda, \mathsf{Enc}}$

$\mathsf{PRF}_{\mathsf{Enc}}^{\{(i^*, \tilde{s})\}} = \mathsf{PuncturePRF}_{\mathsf{Enc}}, (i^*, \tilde{s})$.

$\mathsf{PRF}_{sk}^{\{i^*\}} = \mathsf{Puncture}(\mathsf{PRF}_{sk}, i^*)$.

$$\text{Let } \mathsf{O} = \mathsf{Obfuscate}\left(\mathsf{P}^3_{i^*, b_0, \tilde{s}, \mathsf{PRF}_{sk}^{\{i^*\}}, \mathsf{PRF}_{\mathsf{Enc}}^{\{(i^*, \tilde{s})\}}}\right).$$

Output $c_0 = \mathsf{O}$.

$\mathsf{P}^3_{i^*, b_0, \tilde{s}, \mathsf{PRF}_{sk}^{\{i^*\}}, \mathsf{PRF}_{\mathsf{Enc}}^{\{(i^*, \tilde{s})\}}}(i, s)$:

If $i = i^*$

 If $s \neq \tilde{s}$, output $\mathsf{PRF}_{\mathsf{Enc}}^{\{(i^*, \tilde{s})\}}(i^*, s)$

 If $s = \tilde{s}$, output $1 - b_0$

Else If $i \neq i^*$

 If $s \neq \mathsf{PRF}_{sk}^{\{i^*\}}(i)$, halt and output $\mathsf{PRF}_{\mathsf{Enc}}^{\{(i^*, \tilde{s})\}}(i, s)$.

 Output $\mathbb{I}\{i \leq i^* - 1\}$.

**Fig. 13.** Hybrid $(\mathsf{Enc}^3, \mathsf{P}^3)$.

outputs a random bit and stops. Otherwise, it runs $A$ on $\mathsf{PRF}_{\mathsf{Enc}}^{\{(i^*, \tilde{s})\}}$, and outputs whatever $A$ outputs. If $b$ is truly random and independent of $\mathsf{PRF}_{\mathsf{Enc}}$, then conditioned on $b = 1 - b_0$, $\mathsf{PRF}_{\mathsf{Enc}}$ is sampled randomly. However, if $b = \mathsf{PRF}_{\mathsf{Enc}}(i^*, \tilde{s})$, then conditioned on $b = 1 - b_0$, $\mathsf{PRF}_{\mathsf{Enc}}$ is sampled such that $\mathsf{PRF}_{\mathsf{Enc}}(i^*, \tilde{s}) = 1 - b_0$. These are exactly the two cases that $A$ distinguishes. Hence, conditioned on $b = 1 - b_0$, $B$ guesses correctly with probability $\frac{1}{2} + \delta$. Moreover, by PRF security, $b = 1 - b_0$ with probability $\geq \frac{1}{2} - \varepsilon$ for some negligible quantity $\varepsilon$, and in the case $b \neq 1 - b_0$, $B$ guess correctly with probability $\frac{1}{2}$. Hence, overall $B$ guesses correctly with probability $\geq \frac{1}{2}(\frac{1}{2} + \varepsilon) + (\frac{1}{2} + \delta)(\frac{1}{2} - \varepsilon) = \frac{1}{2} + \frac{\delta}{2} - \varepsilon\delta$. Hence, $B$ has non-negligible advantage $\frac{\delta}{2} - \varepsilon\delta$. Thus, changing how $\mathsf{PRF}_{\mathsf{Enc}}$ is sampled is computationally undetectable, and $\mathsf{P}$ is otherwise unchanged. Therefore $\mathsf{P}^3$ and $\mathsf{P}^4$ are computationally indistinguishable (Fig. 14).

Next, since $\mathsf{PRF}_{\mathsf{Enc}}(i^*, \tilde{s}) = 1 - b_0$, we can move to another hybrid $\mathsf{P}^5$ where we delete the line "If $s = \tilde{s}$, output $1 - b_0$" without changing the functionality. Thus, by indistinguishability obfuscation, $\mathsf{P}^4$ and $\mathsf{P}^5$ are computationally indistinguishable (Fig. 15).

Now notice that $\mathsf{P}^5$ is independent of $b_0$. However, $\mathsf{Enc}^5$ still depends on $b_0$. We now move to the final hybrid $\mathsf{P}^6$ where we remove the condition that $\mathsf{PRF}_{\mathsf{Enc}}(i^*, \tilde{s}) = 1 - b_0$, which will completely remove the dependence on $b_0$ (Fig. 16).

To prove that $\mathsf{Enc}^6$ is indistinguishable from $\mathsf{Enc}^5$, notice that they are independent of $\tilde{s}$, except through the sampling of $\mathsf{PRF}_{\mathsf{Enc}}$. Using this, and the following lemma, we argue that we can remove the condition that $\mathsf{PRF}_{\mathsf{Enc}}(i^*, \tilde{s}) = 1 - b_0$.

$\mathsf{Enc}^4(i^*, b_0, mk = \mathsf{PRF}_{sk})$ :

Let $\tilde{s} \leftarrow_R [m]$.

Choose a pseudorandom function $\mathsf{PRF}_{\mathsf{Enc}} \leftarrow_R \mathcal{F}_{\lambda, \mathsf{Enc}}$ conditioned on $\mathsf{PRF}_{\mathsf{Enc}}(i^*, \tilde{s}) = 1 - b_0$.

$\mathsf{PRF}_{\mathsf{Enc}}^{\{(i^*, \tilde{s})\}} = \mathsf{PuncturePRF}_{\mathsf{Enc}}, (i^*, \tilde{s})$.

$\mathsf{PRF}_{sk}^{\{i^*\}} = \mathsf{Puncture}(\mathsf{PRF}_{sk}, i^*)$.

$$\text{Let } \mathsf{O} = \mathsf{Obfuscate}\left(\mathsf{P}^4_{i^*, b_0, \tilde{s}, \mathsf{PRF}_{sk}^{\{i^*\}}, \mathsf{PRF}_{\mathsf{Enc}}^{\{(i^*, \tilde{s})\}}}\right).$$

Output $c_0 = \mathsf{O}$.

$\mathsf{P}^4_{i^*, b_0, \tilde{s}, \mathsf{PRF}_{sk}^{\{i^*\}}, \mathsf{PRF}_{\mathsf{Enc}}^{\{(i^*, \tilde{s})\}}}(i, s)$:

If $i = i^*$

   If $s \neq \tilde{s}$, output $\mathsf{PRF}_{\mathsf{Enc}}^{\{(i^*, \tilde{s})\}}(i^*, s)$

   If $s = \tilde{s}$, output $1 - b_0$

Else If $i \neq i^*$

   If $s \neq \mathsf{PRF}_{sk}^{\{i^*\}}(i)$, halt and output $\mathsf{PRF}_{\mathsf{Enc}}^{\{(i^*, \tilde{s})\}}(i, s)$.

   Output $\mathbb{I}\{i \leq i^* - 1\}$.

**Fig. 14.** Hybrid $(\mathsf{Enc}^4, \mathsf{P}^4)$.

$\mathsf{Enc}^5(i^*, b_0, mk = \mathsf{PRF}_{sk})$ :

Let $\tilde{s} \leftarrow_R [m]$.

Choose a pseudorandom function $\mathsf{PRF}_{\mathsf{Enc}} \leftarrow_R \mathcal{F}_{\lambda, \mathsf{Enc}}$ such that $\mathsf{PRF}_{\mathsf{Enc}}(i^*, \tilde{s}) = 1 - b_0$

$\mathsf{PRF}_{sk}^{\{i^*\}} = \mathsf{Puncture}(\mathsf{PRF}_{sk}, i^*)$.

$$\text{Let } \mathsf{O} = \mathsf{Obfuscate}\left(\mathsf{P}^5_{i^*, \mathsf{PRF}_{sk}^{\{i^*\}}, \mathsf{PRF}_{\mathsf{Enc}}}\right).$$

Output $c_0 = \mathsf{O}$.

$\mathsf{P}^5_{i^*, \mathsf{PRF}_{sk}^{\{i^*\}}, \mathsf{PRF}_{\mathsf{Enc}}}(i, s)$:

If $i = i^*$

   Output $\mathsf{PRF}_{\mathsf{Enc}}(i^*, s)$

Else If $i \neq i^*$

   If $s \neq \mathsf{PRF}_{sk}^{\{i^*\}}(i)$, halt and output $\mathsf{PRF}_{\mathsf{Enc}}(i, s)$.

   Output $\mathbb{I}\{i \leq i^* - 1\}$.

**Fig. 15.** Hybrid $(\mathsf{Enc}^5, \mathsf{P}^5)$.

---

$\mathsf{Enc}^6(i^*, mk = \mathsf{PRF}_{sk})$ :

  Choose a pseudorandom function $\mathsf{PRF}_{\mathsf{Enc}} \leftarrow_R \mathcal{F}_{\lambda, \mathsf{Enc}}$

  $\mathsf{PRF}_{sk}^{\{i^*\}} = \mathsf{Puncture}(\mathsf{PRF}_{sk}, i^*)$.

  Let $\mathsf{O} = \mathsf{Obfuscate}\left(\mathsf{P}^6_{i^*, \mathsf{PRF}_{sk}^{\{i^*\}}, \mathsf{PRF}_{\mathsf{Enc}}}\right)$.

  Output $c_0 = \mathsf{O}$.

$\mathsf{P}^6_{i^*, \mathsf{PRF}_{sk}^{\{i^*\}}, \mathsf{PRF}_{\mathsf{Enc}}}(i, s)$:

  If $i = i^*$

    Output $\mathsf{PRF}_{\mathsf{Enc}}(i^*, s)$

  Else If $i \neq i^*$

    If $s \neq \mathsf{PRF}_{sk}^{\{i^*\}}(i)$, halt and output $\mathsf{PRF}_{\mathsf{Enc}}(i, s)$.

    Output $\mathbb{I}\{i \leq i^* - 1\}$.

---

**Fig. 16.** Hybrid $(\mathsf{Enc}^6, \mathsf{P}^6)$.

**Lemma 15.** *Let $\mathcal{H} = \{h : [T] \to [K]\}$ be a $\delta$-almost pairwise independent hash family. Let $y \in [K]$ and $M \subseteq [T]$ of size $m$ be arbitrary. Define the following two distributions.*

- *$D_1$: Choose $h \leftarrow_R \mathcal{H}$.*
- *$D_2$: Choose a random $x \in M$, and then choose $h \leftarrow_R (\mathcal{H} \mid h(x) = y)$.*

*Then $D_1$ and $D_2$ are $(\frac{1}{2}\sqrt{K/m + 7K^2\delta})$-close in statistical distance.*

We defer the proof to the full version. The natural way to try to show that $(\mathsf{Enc}^6, \mathsf{P}^6)$ is $o(1/n^3)$ statistically close to $(\mathsf{Enc}^5, \mathsf{P}^5)$ is to apply this lemma to the hash family $\mathcal{H} = \mathcal{F}_{\lambda, \mathsf{Enc}}$. Recall that a pseudorandom function family is also $\mathsf{negl}(\lambda)$-pairwise independent. Here, the parameters would be $[T] = [n] \times [m]$, $M = \{(i^*, s) \mid s \in [m]\}$ and $b = 1 - b_0$, and the random choice $x \in M$ is the pair $(i^*, \tilde{s})$.

However, recall that the adversary not only sees $c_0 = \mathsf{Enc}^5(i^*, b_0, mk)$, but also sees $c_1 = \mathsf{Enc}^5(i^*, b_1, mk)$, and these share the same $\tilde{s}$. Hence, we cannot directly invoke Lemma 15 on the $\mathsf{PRF}_{\mathsf{Enc}, 0}$ sampled in $c_0$, since $\tilde{s}$ is also used to sample $\mathsf{PRF}_{\mathsf{Enc}, 1}$ when sampling $c_1$, and is therefore not guaranteed to be random given $c_1$.

Instead, we actually consider the function family $\mathcal{H} = \mathcal{F}^2_{\lambda, \mathsf{Enc}}$, where we define

$$h(i, s) = (\mathsf{PRF}_{\mathsf{Enc}, 0}, \mathsf{PRF}_{\mathsf{Enc}, 1})(i, s) = (\mathsf{PRF}_{\mathsf{Enc}, 0}(i, s), \mathsf{PRF}_{\mathsf{Enc}, 1}(i, s)).$$

In $\mathsf{Enc}^5$, $h$ is drawn at random conditioned on $h(i^*, \tilde{s}) = (1 - b_0, 1 - b_1)$, whereas in $\mathsf{Enc}^6$, it is drawn at random.

$\mathcal{H}$ is still a pseudorandom function family, so it must be $\mathsf{negl}(\lambda)$-almost pairwise independent with $\delta$ negligible. In particular, $\delta = o(1/m)$. Hence, the conditions of Lemma 15 are satisfied with $K = 4$. Since the description of $\mathsf{P}^5, \mathsf{P}^6$ is the tuple $(i^*, \tilde{s}, \mathsf{PRF}_{sk}^{\{i^*\}}, \mathsf{PRF}_{\mathsf{Enc},0}, \mathsf{PRF}_{\mathsf{Enc},1})$, and by Lemma 15 the distribution on these tuples differs by at most $O(\sqrt{1/m})$ in statistical distance, we also have that the distribution on obfuscations of $\mathsf{P}^5, \mathsf{P}^6$ differs by at most $O(\sqrt{1/m})$. Finally, we can choose a value of $m = \tilde{O}(n^6)$ so that $O(\sqrt{1/m}) = o(1/n^3)$.

Observe that when we generate user keys $sk_{-i^*}$ and the challenge ciphertexts according to $(\mathsf{Enc}^6, \mathsf{P}^6)$, the distribution of the adversary's view is completely independent of the random values $b_0, b_1$. Thus no adversary can output $b' = b_0 \oplus b_1$ with probability greater than $1/2$. Since the distribution of these challenge ciphertexts is $o(1/n^3)$-computationally indistinguishable from the original distribution on challenge ciphertexts, we have that for every efficient adversary,

$$\underset{\mathsf{TwoIndexHiding}[i^*]}{\mathbb{P}} [b' = b_0 \oplus b_1] - \frac{1}{2} = o(1/n^3),$$

as desired. This completes the proof.

**Acknowledgments.** We thank Dan Boneh for helpful discussions in the early stages of this work. The first author is supported by an NSF Graduate Research Fellowship #DGE-11-44155. The first and second authors are supported in part by the Defense Advanced Research Project Agency (DARPA) and Army Research Office (ARO) under Contract #W911NF-15-C-0236, and NSF grants #CNS-1445424 and #CCF-1423306. Part of this work was done while the third author was a postdoctoral fellow in the Columbia University Department of Computer Science, supported by a junior fellowship from the Simons Society of Fellows. Any opinions, findings and conclusions or recommendations expressed are those of the authors and do not necessarily reflect the views of the the Defense Advanced Research Projects Agency, Army Research Office, the National Science Foundation, or the U.S. Government.

# References

1. Badrinarayanan, S., Miles, E., Sahai, A., Zhandry, M.: Post-zeroizing obfuscation: new mathematical tools, and the case of evasive circuits. In: Fischlin, M., Coron, J.-S. (eds.) EUROCRYPT 2016. LNCS, vol. 9666, pp. 764–791. Springer, Heidelberg (2016). doi:10.1007/978-3-662-49896-5_27
2. Beimel, A., Nissim, K., Stemmer, U.: Private learning and sanitization: pure vs. approximate differential privacy. In: Raghavendra, P., Raskhodnikova, S., Jansen, K., Rolim, J.D.P. (eds.) APPROX/RANDOM -2013. LNCS, vol. 8096, pp. 363–378. Springer, Heidelberg (2013). doi:10.1007/978-3-642-40328-6_26
3. Bellare, M., Stepanovs, I., Tessaro, S.: Contention in cryptoland: obfuscation, leakage and UCE. In: Kushilevitz, E., Malkin, T. (eds.) TCC 2016. LNCS, vol. 9563, pp. 542–564. Springer, Heidelberg (2016). doi:10.1007/978-3-662-49099-0_20
4. Bitansky, N., Paneth, O., Wichs, D.: Perfect structure on the edge of chaos. In: Kushilevitz, E., Malkin, T. (eds.) TCC 2016. LNCS, vol. 9562, pp. 474–502. Springer, Heidelberg (2016). doi:10.1007/978-3-662-49096-9_20
5. Blum, A., Dwork, C., McSherry, F., Nissim, K.: Practical privacy: the SuLQ framework. In: PODS (2005)
6. Blum, A., Ligett, K., Roth, A.: A learning theory approach to noninteractive database privacy. J. ACM 60(2), 12 (2013)

7. Boneh, D., Sahai, A., Waters, B.: Fully collusion resistant traitor tracing with short ciphertexts and private keys. In: Vaudenay, S. (ed.) EUROCRYPT 2006. LNCS, vol. 4004, pp. 573–592. Springer, Heidelberg (2006). doi:10.1007/11761679_34

8. Boneh, D., Zhandry, M.: Multiparty key exchange, efficient traitor tracing, and more from indistinguishability obfuscation. In: Garay, J.A., Gennaro, R. (eds.) CRYPTO 2014. LNCS, vol. 8616, pp. 480–499. Springer, Heidelberg (2014). doi:10.1007/978-3-662-44371-2_27

9. Brzuska, C., Farshim, P., Mittelbach, A.: Indistinguishability obfuscation and UCEs: the case of computationally unpredictable sources. In: Garay, J.A., Gennaro, R. (eds.) CRYPTO 2014. LNCS, vol. 8616, pp. 188–205. Springer, Heidelberg (2014). doi:10.1007/978-3-662-44371-2_11

10. Bun, M., Ullman, J., Vadhan, S.P.: Fingerprinting codes and the price of approximate differential privacy. In: STOC (2014)

11. Chor, B., Fiat, A., Naor, M.: Tracing traitors. In: Desmedt, Y.G. (ed.) CRYPTO 1994. LNCS, vol. 839, pp. 257–270. Springer, Heidelberg (1994). doi:10.1007/3-540-48658-5_25

12. Coron, J.-S., Gentry, C., Halevi, S., Lepoint, T., Maji, H.K., Miles, E., Raykova, M., Sahai, A., Tibouchi, M.: Zeroizing without low-level zeroes: new MMAP attacks and their limitations. In: Gennaro, R., Robshaw, M. (eds.) CRYPTO 2015. LNCS, vol. 9215, pp. 247–266. Springer, Heidelberg (2015). doi:10.1007/978-3-662-47989-6_12

13. Dinur, I., Nissim, K.: Revealing information while preserving privacy. In: PODS (2003)

14. Dwork, C., McSherry, F., Nissim, K., Smith, A.: Calibrating noise to sensitivity in private data analysis. In: Halevi, S., Rabin, T. (eds.) TCC 2006. LNCS, vol. 3876, pp. 265–284. Springer, Heidelberg (2006). doi:10.1007/11681878_14

15. Dwork, C., Naor, M., Reingold, O., Rothblum, G.N., Vadhan, S.P.: On the complexity of differentially private data release: efficient algorithms and hardness results. In: STOC (2009)

16. Dwork, C., Nissim, K.: Privacy-preserving datamining on vertically partitioned databases. In: Franklin, M. (ed.) CRYPTO 2004. LNCS, vol. 3152, pp. 528–544. Springer, Heidelberg (2004). doi:10.1007/978-3-540-28628-8_32

17. Dwork, C., Rothblum, G.N., Vadhan, S.P.: Boosting and differential privacy. In: FOCS. IEEE (2010)

18. Dwork, C., Smith, A.D., Steinke, T., Ullman, J., Vadhan, S.P.: Robust traceability from trace amounts. In: FOCS (2015)

19. Garg, S., Gentry, C., Halevi, S., Raykova, M., Sahai, A., Waters, B.: Candidate indistinguishability obfuscation and functional encryption for all circuits. In: FOCS, pp. 40–49 (2013)

20. Garg, S., Mukherjee, P., Srinivasan, A.: Obfuscation without the vulnerabilities of multilinear maps. Cryptology ePrint Archive, Report 2016/390 (2016). http://eprint.iacr.org/

21. Gentry, C., Lewko, A.B., Sahai, A., Waters, B.: Indistinguishability obfuscation from the multilinear subgroup elimination assumption. In: FOCS (2015)

22. Gupta, A., Roth, A., Ullman, J.: Iterative constructions and private data release. In: Cramer, R. (ed.) TCC 2012. LNCS, vol. 7194, pp. 339–356. Springer, Heidelberg (2012). doi:10.1007/978-3-642-28914-9_19

23. Hardt, M., Rothblum, G.N.: A multiplicative weights mechanism for privacy-preserving data analysis. In: FOCS (2010)

24. Hohenberger, S., Sahai, A., Waters, B.: Replacing a random oracle: full domain hash from indistinguishability obfuscation. In: Nguyen, P.Q., Oswald, E. (eds.) EUROCRYPT 2014. LNCS, vol. 8441, pp. 201–220. Springer, Heidelberg (2014). doi:10.1007/978-3-642-55220-5_12
25. Kearns, M.J.: Efficient noise-tolerant learning from statistical queries. J. ACM 45(6), 983–1006 (1998)
26. Kowalczyk, L., Malkin, T., Ullman, J., Zhandry, M.: Strong hardness of privacy from weak traitor tracing. IACR Cryptology ePrint Archive 2016/721 (2016)
27. Miles, E., Sahai, A., Zhandry, M.: Annihilation attacks for multilinear maps: cryptanalysis of indistinguishability obfuscation over GGH13. In: Robshaw, M., Katz, J. (eds.) CRYPTO 2016. LNCS, vol. 9815, pp. 629–658. Springer, Heidelberg (2016). doi:10.1007/978-3-662-53008-5_22
28. Mironov, Ilya, Pandey, Omkant, Reingold, Omer, Vadhan, Salil: Computational Differential Privacy. In: Halevi, Shai (ed.) CRYPTO 2009. LNCS, vol. 5677, pp. 126–142. Springer, Heidelberg (2009). doi:10.1007/978-3-642-03356-8_8
29. Nikolov, A., Talwar, K., Zhang, L.: The geometry of differential privacy: the sparse and approximate cases. In: STOC (2013)
30. Roth, A., Roughgarden, T.: Interactive privacy via the median mechanism. In: STOC, pp. 765–774. ACM, 5–8 June 2010
31. Sahai, A., Waters, B.: How to use indistinguishability obfuscation: deniable encryption, and more. In: STOC (2014)
32. Steinke, T., Ullman, J.: Between pure and approximate differential privacy. CoRR abs/1501.06095 (2015). http://arxiv.org/abs/org/abs/1501.06095
33. Thaler, J., Ullman, J., Vadhan, S.: Faster algorithms for privately releasing marginals. In: Czumaj, A., Mehlhorn, K., Pitts, A., Wattenhofer, R. (eds.) ICALP 2012. LNCS, vol. 7391, pp. 810–821. Springer, Heidelberg (2012). doi:10.1007/978-3-642-31594-7_68
34. Ullman, J.: Answering $n^{2+o(1)}$ counting queries with differential privacy is hard. In: STOC (2013)
35. Ullman, J.: Private multiplicative weights beyond linear queries. In: PODS (2015)
36. Ullman, J., Vadhan, S.: PCPs and the hardness of generating private synthetic data. In: Ishai, Y. (ed.) TCC 2011. LNCS, vol. 6597, pp. 400–416. Springer, Heidelberg (2011). doi:10.1007/978-3-642-19571-6_24

# Author Index

Alon, Bar   I-307
Agrawal, Shashank   II-269
Ananth, Prabhanjan   II-3
Apon, Daniel   II-299
Applebaum, Benny   I-27

Baum, Carsten   I-461
Ben-Sasson, Eli   II-31
Bitansky, Nir   I-57, II-391
Blocki, Jeremiah   II-517
Bogdanov, Andrej   II-471
Brakerski, Zvika   I-57, II-330
Bun, Mark   I-607, I-635

Canetti, Ran   II-61
Cascudo, Ignacio   I-204
Cash, David   II-330
Chen, Yi-Hsiu   I-607
Chen, Yilei   II-61
Chen, Yu-Chi   II-3
Chiesa, Alessandro   II-31
Chung, Kai-Min   II-3
Cohen, Aloni   I-84

Dachman-Soled, Dana   II-169
Damgård, Ivan   I-204, II-547
Devadas, Srinivas   I-262

Fan, Xiong   II-299
Fiore, Dario   I-108

Garg, Sanjam   I-491, II-241, II-419
Genkin, Daniel   I-336
Goyal, Rishab   II-361
Guo, Siyao   II-471
Gupta, Divya   I-491

Haagh, Helene   II-547
Hazay, Carmit   I-367, I-400, I-521
Hofheinz, Dennis   II-121, II-146
Holmgren, Justin   II-61

Impagliazzo, Russell   I-235
Ishai, Yuval   I-336

Jafargholi, Zahra   I-433
Jager, Tibor   II-146
Jaiswal, Ragesh   I-235

Kabanets, Valentine   I-235
Kalai, Yael   I-57, II-91
Kapron, Bruce M.   I-235
King, Valerie   I-235
Klein, Saleet   I-84
Komargodski, Ilan   I-139, II-471, II-485
Koppula, Venkata   II-361
Kowalczyk, Lucas   I-659

Lacerda, Felipe   I-204
Li, Baiyu   II-443
Lin, Huijia   II-3
Lin, Wei-Kai   II-3
Lindell, Yehuda   I-554
Liu, Feng-Hao   II-299

Malkin, Tal   I-659
Maurer, Ueli   I-3
Miao, Peihan   I-491
Micciancio, Daniele   II-443
Miles, Eric   II-241
Mukherjee, Pratyay   II-241

Naor, Moni   II-485
Nielsen, Jesper Buus   I-582
Nishimaki, Ryo   II-391
Nitulescu, Anca   I-108

Omri, Eran   I-307
Orlandi, Claudio   I-582, II-547
Orsini, Emmanuela   I-461

Pandey, Omkant   I-491
Paneth, Omer   I-57, II-91
Passelègue, Alain   II-391

Peikert, Chris   II-217
Pietrzak, Krzysztof   I-183
Polychroniadou, Antigoni   I-367
Prabhakaran, Manoj   II-269

Ranellucci, Samuel   I-204
Rao, Vanishree   II-121
Raykov, Pavel   I-27
Raykova, Mariana   II-61
Ren, Ling   I-262
Renner, Renato   I-3
Rupp, Andy   II-146

Sahai, Amit   II-241
Scholl, Peter   I-461
Shiehian, Sina   II-217
Skórski, Maciej   I-159, I-183
Smart, Nigel P.   I-554
Soria-Vazquez, Eduardo   I-554
Spini, Gabriele   I-286
Spooner, Nicholas   II-31
Srinivasan, Akshayaram   II-241, II-419
Steinke, Thomas   I-635

Targhi, Ehsan Ebrahimi   II-192
Tessaro, Stefano   I-235
Tsabary, Rotem   II-330

Ullman, Jonathan   I-659
Unruh, Dominique   II-192

Vadhan, Salil   I-607
Vaikuntanathan, Vinod   I-57
Venkitasubramaniam, Muthuramakrishnan
    I-367, I-400

Waters, Brent   II-361
Wee, Hoeteck   II-330
Weiss, Mor   I-336
Wichs, Daniel   I-433, II-121, II-391

Yanai, Avishay   I-521
Yogev, Eylon   II-485
Yu, Ching-Hua   II-269

Zémor, Gilles   I-286
Zhandry, Mark   I-659, II-241
Zhou, Hong-Sheng   II-517

Pr
By e United States
ers